AF580446

Neuro-Control Systems

IEEE PRESS
445 Hoes Lane, PO Box 1331
Piscataway, NJ 08855-1331

IEEE Neural Networks Council, *Sponsor*

R. C. Eberhart
President
Research Triangle Institute

NNC Liaison to IEEE PRESS
Stamatios Kartalopoulos
AT&T Bell Laboratories

Technical Reviewers for NNC
Susamma Barua
California State University

Djuro Koruga
The University of Arizona

Richard D. Lilley
Harris Corp.

Mohan J. Malkani
Tennessee State University

Major Peter G. Raeth
Wright-Patterson AFB, Ohio

Manjula A. B. Waldron
Columbus, Ohio

Neuro-Control Systems

Theory and Applications

Edited by
Madan M. Gupta
Dandina H. Rao
Intelligent Systems Research Laboratory
College of Engineering
University of Saskatchewan
Saskatoon, Canada

A Selected Reprint Volume
IEEE Neural Networks Council, *Sponsor*

The Institute of Electrical and Electronics Engineers, Inc., New York

This book may be purchased at a discount from the publisher when ordered in bulk quantities. For more information contact:

IEEE PRESS Marketing
Attn: Special Sales
PO Box 1331
445 Hoes Lane
Piscataway, NJ 08855-1331
Fax: (908) 981-8062

Printed in the United States of America
10 9 8 7 6 5 4 3 2 1

ISBN 0-7803-1041-1

IEEE Order Number: PC0392-1

Library of Congress Cataloging-in-Publication Data

Neuro-control systems : theory and applications / Madan M. Gupta and Dandina H. Rao (editors).
p. cm.
Includes bibliographical references and index.
ISBN 0-7803-1041-1
1. Neural networks (Computer science) I. Gupta, Madan M. II. Rao, Dandina H.
QA76.87.N493 1994
629.8′9—dc20 93-29435
CIP

To:

The synergy of many of our graduate students and research colleagues around the globe, who have inspired our thinking in this emerging field of neuro-control systems;

AND

To:

Dr. Peter N. Nikiforuk, who has provided continuous inspiration and challenges to our research group through his untiring, but controlled, positive feedback.

स्पर्शान्कृत्वा बहिर्बाह्यांश्चक्षुश्चैवान्तरे भ्रुवोः ।
प्राणापानौ समौ कृत्वा नासाभ्यन्तरचारिणौ ॥
यतेन्द्रियमनोबुद्धिर्मुनिर्मोक्षपरायणः ।
विगतेच्छाभयक्रोधो यः सदा मुक्त एव सः ॥

A transcendentalist (a contemplative individual) achieves the supreme state of mind (free from desire, fear, and anger) by switching off all the natural sensors from the external environment, concentrating the vision between the eyebrows, and by suspending (equalizing) the inward and outward flow of breath within the nostrils.

Such a supreme (meditative) state of mind brings synergy to the faculty of cognition, thinking, and memory.

[*Bhagwat Gita* V-26,27]

Contents

FOREWORD **xi**

PREFACE **xiii**

MAJOR CURRENT BIBLIOGRAPHICAL SOURCES ON COMPUTATIONAL NEURAL NETWORKS AND APPLICATIONS **xv**

NEURO-CONTROL SYSTEMS: A TUTORIAL **1**
M. M. Gupta and D. H. Rao

PART 1 NEURO-CONTROL SYSTEMS: SOME PERSPECTIVES **45**

1.0 Introduction 45
1.1 Neurocomputing: Picking the Human Brain 47
R. Hecht-Nielsen (*IEEE Spectrum,* March 1988)
1.2 Learning 53
I. Kupfermann (*Principles of Neural Science,* 1985)
1.3 A Hierarchical Neural-Network Model for Control and Learning of Voluntary Movement 64
M. Kawato, K. Furukawa, and R. Suzuki (*Biological Cybernetics,* 1989)
1.4 An Introduction to Autonomous Control Systems 81
P. J. Antsaklis, K. M. Passino, and S. J. Wang (*IEEE Control Systems Magazine,* June 1991)
1.5 Intelligent Control Using Neural Networks 90
K. S. Narendra and S. Mukhopadhyay (*IEEE Control Systems Magazine,* April 1992)

PART 2 NEURONAL MORPHOLOGY OF BIOLOGICAL PROCESSES **99**

2.0 Introduction 99
2.1 The Neuron 101
C. F. Stevens (*Scientific American,* 1979)
2.2 Cognitive and Psychological Computation with Neural Models 112
J. A. Anderson (*IEEE Transactions on Systems, Man, and Cybernetics,* September/October 1983)

2.3 Linear Analysis of the Dynamics of Neural Masses 129
W. J. Freeman (*Biophysical Journal*, 1972)
2.4 Excitatory and Inhibitory Interactions in Localized Populations of Model Neurons 153
H. R. Wilson and J. D. Cowan (*Biophysical Journal*, 1972)

PART 3 COMPUTATIONAL NEURAL SYSTEMS: FOUNDATIONS 169

3.0 Introduction 169
3.1 Neural Networks for Control Systems—A Survey 171
K. J. Hunt, D. Sbarbaro, R. Zbikowski, and P. J. Gawthrop (*Automatica*, 1992)
3.2 The Self-Organizing Map 201
T. Kohonen (*Proceedings of the IEEE*, September 1990)
3.3 Cognitron: A Self-Organizing Multilayered Neural Network 218
K. Fukushima (*Biological Cybernetics*, 1975)
3.4 Neurons with Graded Response Have Collective Computational Properties Like Those of Two-State Neurons 234
J. J. Hopfield (*Proceedings of the National Academy of Sciences*, May 1984)
3.5 Neurons with Hysteresis Form a Network That Can Learn without Any Changes in Synaptic Connection Strengths 239
G. W. Hoffman and M. W. Benson (*American Institute of Physics*, 1986)

PART 4 THEORY OF NEURONAL APPROXIMATIONS 245

4.0 Introduction 245
4.1 On the Approximate Realization of Continuous Mappings by Neural Networks 247
K-I. Funahashi (*Neural Networks*, 1989)
4.2 Networks and the Best Approximation Property 257
F. Girosi and T. Poggio (*Biological Cybernetics*, 1990)
4.3 Multilayer Feedforward Networks Are Universal Approximators 265
K. Hornik, M. Stinchcombe, and H. White (*Neural Networks*, 1989)
4.4 Approximation by Superpositions of a Sigmoidal Function 273
G. Cybenko (*Mathematics of Control Signals and System*, 1989)
4.5 The Stone-Weierstrass Theorem and Its Application to Neural Networks 283
N. E. Cotter (*IEEE Transactions on Neural Networks*, December 1990)
4.6 Dynamic Neural Units and Function Approximation 289
D. H. Rao and M. M. Gupta (*IEEE Conference on Neural Networks*, March 28–April 1, 1993)

PART 5 NEURONAL MORPHOLOGY FOR CONTROL SYSTEMS: STATIC AND DYNAMIC STRUCTURES 295

5.0 Introduction 295
5.1 Backpropagation through Time: What It Does and How to Do It 297
P. J. Werbos (*Proceedings of the IEEE*, October 1990)
5.2 Neural Networks for Self-Learning Control Systems 308
D. H. Nguyen and B. Widrow (*IEEE Control Systems Magazine*, April 1990)
5.3 Direct Adaptive Output Tracking Control Using Multilayered Neural Networks 314
L. Jin, P. N. Nikiforuk, and M. M. Gupta (*IEE Proceedings-D*, November 1993)
5.4 Neural Networks for Nonlinear Internal Model Control 321
K. J. Hunt and D. Sbarbaro (*IEE Proceedings-D*, September 1991)
5.5 Identification and Control of Dynamical Systems Using Neural Networks 329
K. S. Narendra and K. Parthasarathy (*IEEE Transactions on Neural Networks*, March 1990)
5.6 Dynamic Neural Units with Applications to the Control of Unknown Nonlinear Systems 352
M. M. Gupta and D. H. Rao (*Journal of Intelligent and Fuzzy Systems*, 1993)

PART 6 FUZZY-NEURAL CONTROL SYSTEMS 373

6.0 Introduction 373
6.1 Fuzzy Sets 375
L. A. Zadeh (*Information and Control,* 1965)
6.2 Outline of a New Approach to the Analysis of Complex Systems and Decision Process 386
L. A. Zadeh (*IEEE Transactions on Systems, Man and Cybernetics,* January 1973)
6.3 Fuzzy Logic and Neural Networks 403
M. M. Gupta (*Proceedings of the Tenth International Conference on Multiple Criteria Decision Making (TAIPEI '92*), July 19–24 1992)
6.4 Self-Learning Fuzzy Controllers Based on Temporal Back Propagation 417
J.-S. R. Jang (*IEEE Transactions on Neural Networks,* September 1992)
6.5 A Neo Fuzzy Neuron and Its Application to System Identification and Prediction of the System Behavior 427
T. Yamakawa, E. Uchino, T. Miki, and H. Kusanagi (*Proceedings of the 2nd Int. Conference on Fuzzy Logic and Neural Networks (IIZUKA-92),* July 17–22, 1992)

PART 7 NEURO-CONTROL SYSTEMS: APPLICATIONS 435

7.0 Introduction 435
7.1 General Learning Scheme for Robot Coordinate Transformations Using Dynamic Neural Network 439
M. M. Gupta and D. H. Rao (*SPIE's Conference on Intelligent Robots and Computer Vision XI,* Boston 1993)
7.2 Intelligent Coordination of Multiple Systems with Neural Networks 451
X. Cui and K. G. Shin (*IEEE Transactions on Systems, Man, and Cybernetics,* November/December 1991)
7.3 A Distributed Adaptive Control System for a Quadruped Mobile Robot 460
B. L. Digney and M. M. Gupta (*IEEE Conference on Neural Networks,* March 1993)
7.4 Integrating Neural Networks and Knowledge-Based Systems for Intelligent Robotic Control 466
D. A. Handelman, S. H. Lane, and J. J. Gelfand (*IEEE Control Systems Magazine,* April 1990)
7.5 A Network Model for the Control of the Movement of a Redundant Manipulator 476
M. Brüwer and H. Cruse (*Biological Cybernetics,* 1990)
7.6 A Neural Network Regulator for Turbogenerators 483
Q. H. Wu, B. W. Hogg, and G. W. Irwin (*IEEE Transactions on Neural Networks,* January 1992)
7.7 A Neural Network Approach to MVDR Beamforming Problem 489
P.-R. Chang, W.-H. Yang, and K.-K. Chan (*IEEE Transactions on Antennas and Propagation,* March 1992)
7.8 Artificial Neural Networks in Process Engineering 499
M. J. Willis, C. Di Massimo, G. A. Montague, M. T. Tham, and A. J. Morris (*IEE Proceedings-D,* May 1990)
7.9 Modeling Chemical Process Systems via Neural Computation 510
N. V. Bhat, P. A. Minderman, Jr., T. McAvoy, and N. Sun Wang (*IEEE Control Systems Magazine,* April 1990)
7.10 CMAC: An Associative Neural Network Alternative to Backpropagation 516
W. T. Miller III, F. H. Glanz, and L. G. Kraft III (*Proceedings of the IEEE,* October 1990)

PART 8 NEURAL HARDWARE: ARCHITECTURES AND IMPLEMENTATIONS 523

8.0 Introduction 523
8.1 Neuronmorphic Electronic Systems 525
C. Mead (*Proceedings of the IEEE,* October 1990)
8.2 Computing with Neural Circuits: A Model 533
J. J. Hopfield and D. W. Tank (*Science,* August 1986)
8.3 The Design, Fabrication, and Test of a New VLSI Hybrid Analog-Digital Neural Processing Element 542
M. R. DeYong, R. L. Findley, and C. Fields (*IEEE Transactions on Neural Networks,* May 1992)
8.4 Neural Network Architectures for Robotic Applications 554
S-Y. Kung and J-N. Hwang (*IEEE Transactions on Robotics and Automation,* October 1989)

8.5 A Fuzzy Inference Engine in Nonlinear Analog Mode and Its Applications to a Fuzzy Logic Control 571
T. Yamakawa (*IEEE Transactions on Neural Networks,* May 1993)

AUTHOR INDEX **597**

SUBJECT INDEX **599**

EDITORS' BIOGRAPHIES **607**

Foreword

This book brings together a number of crucial historical papers important to the practical control engineer, such as applying neural networks to confront the class of control problems which have been addressed by control engineers today. These are problems in which we seek stable controllers designed to make a plant, such as a robot arm, follow a desired reference model or trajectory, which is typically defined by a human expert or high-level controller. Some of these methods can also be used to find an optimal trajectory by using recursive learning involving backpropagation through time. These methods are discussed in the papers by myself, by Nguyen and Widrow, and by Kawato et al., among others. There are many practical applications where engineers think that they need real-time learning when, in fact, they would be better off using backpropagation through time, in an off-line mode, to learn recurrent connections which adapt quickly to normal changes in the parameters of a plant.

As noted in the preface, the field of neuro-control systems has grown enormously in the past few years. No one book could possibly encompass the entire field. For example, the function approximation and stability in tracking systems are very large topics in themselves, and this book covers it more completely than any other book I have seen.

The reader should also be aware that the definition of ''control'' is very broad in principle. Above and beyond the popular and rigorous domain of tracking theory, neural nets have also been used for pattern recognition and diagnostics within control systems, for the ''cloning'' of pre-existing experts, and for real-time optimization (without requiring backpropagation through time or truncated calculations which simply ignore cross-time effects on derivatives). The editors have wisely chosen to focus on one topic at a time, and place emphasis on the tracking area. The book on intelligent control systems by Gupta and Sinha to be published by IEEE PRESS emphasizes more ''intelligent'' or biologically relevant architectures. These books combined complement each other very nicely and provide a very broad introduction to the field.

During the past few months, the real-world applications of neuro-control, in the broadest sense, have expanded tremendously. Thanks to new mass-production of chips and circuits boards from Motorola and Adaptive Solutions, as well as others, there is good reason to expect these techniques to be of pervasive, practical value across a very wide range of big industries—with the aerospace, chemical, automotive, and financial industries leading the way. (There are specific products heading for the mass market in the near future, but the details are still proprietary.) In order for the universities to remain relevant to the needs of industry, it is essential that a wider range of students be trained in these concepts, especially as part of the basic core courses in control engineering. I hope that this book will contribute in a major way to that critical endeavor.

Paul J. Werbos
Past President, International Neural Network Society
Program Director
Neuroengineering and Emerging Technology Initiation
National Science Foundation
Washington, D.C., 20550

Preface

The recent emergence of the new and complex control systems in space, manufacturing, and health sciences has created a demand for better control techniques. These systems should be able to perceive and adapt their behavior—perhaps as we humans do. This has led to some appreciation of the neuronal morphology of biological control mechanisms. Scientists from various disciplines, such as systems science, computer science, and mathematics, are formulating the theories of neuronal control motivated by the incredible flexibility and adaptability of biological processes. Thus, these recent studies have led to a better appreciation and understanding of the neuronal morphology of biological control, as well as better computational neural structures for the construction of industrial control systems. Some of these studies have evolved over the last four decades, but recent studies, especially during the last decade, have provided some interesting neural paradigms, models, neural architectures, and hardware.

This IEEE Press volume is about the principles, architectures, applications, and hardware aspects of neuro-control systems. There are many research publications in the field but they are scattered throughout many conference proceedings and scientific journals. In our own research studies and classroom teaching, we have not found a manageable source to which we can refer students. We are sure that many other researchers in the field must have faced similar frustration in their own work. These frustrations have motivated us to develop a book in the field, and here is the birth of this volume on neuro-control systems.

In designing the present book, our goal has been to present a pedagogically sound reprint volume on neuro-control systems that would be useful as a supplementary or even as a main text for graduate students. Additionally, this collection of literature should have conceptual, theoretical, and practical information, with a comprehensive view of the general field of neuronal morphology of biological processes and artificial neuro-control systems. Hopefully, our efforts in this volume will stimulate the research interest of the readers. This work will also provide a comprehensive view of the field for practical engineers, researchers, and students.

In order to meet these objectives, 46 of the most significant articles from hundreds of articles published since 1965 (but mostly in the last decade) were chosen for inclusion, and we hope that these articles will provide a wide perspective of the field. There are several sources for the articles included in the book, such as *IEEE Transactions on Neural Networks, Neural Networks, Biological Cybernetics,* and more, mostly scientific journals that publish only refereed articles. This collection contains a wide breadth of classical papers dealing with the philosophical aspects of biological control, survey papers on computational neural networks, and a variety of current papers dealing specifically with neural network architectures, algorithms, hardware implementations, and a variety of applications.

This collection thus provides an ''instant library'' from which students, researchers, and practicing engineers may obtain an overall picture of the early and recent developments of this important field. The authors of these research articles are from some very well known schools distributed over a wide geographical area around the globe.*

The field of neuro-control made slow progress initially, but is advancing very rapidly during recent years as evidenced by the increasing number of publications. It was, therefore, an extremely difficult task for the editors to select a relatively few articles from a large amount of potential candidates. We are aware of the many excellent papers that had to be excluded in order to produce a balanced collection of manageable size.

*The countrywide distribution of the authors for these papers is as follows:

Canada	(7)	Japan	(5)
China	(1)	Taiwan	(1)
Czechoslovakia	(1)	United Kingdom	(4)
Finland	(1)	United States	(28)
Germany	(1)		

However, our editorial efforts, in consultation with some researchers working in the field and feedback received from many reviewers, have produced the volume in this present form. This set of articles, however, does provide a balanced mathematical formulation along with different perspectives of neuro-control systems during the evolutionary phases of the field.

The set of 46 articles in this volume are arranged in a pedagogical style, dividing the volume into eight parts. Part 1 presents some perspectives in the field of neuro-control systems. Part 2 provides biological motivation of neuronal morphology, and Part 3, some foundation for the field of computational neural systems.

The strength of neural computation lies in its ability to approximate various static and dynamic functions. Part 4 of this volume presents some introduction to the theory of neuronal approximations. In Part 5, we provide several static and dynamic structures for system identification and control applications. Part 6 contains some introductory and some advanced articles in the emerging field of fuzzy–neural control systems.

The last two parts of this volume are devoted to applied topics. In Part 7, we present ten articles on selected applications of neuro-control systems. In the last part, Part 8, we present all the various neural hardware architectures that can possibly be used for neural-network-based applications.

We have written an extensive introduction to neuro-control systems, focusing especially on the neuronal morphology of biological control using simple schematic diagrams and a systems-type of explanation, and the neural paradigms using block diagrams and uncluttered mathematics.

This work is self-contained, and it is our hope that this volume will provide the reader not only with valuable conceptual and technical information, but also with a comprehensive view of the general field of neuro-control systems and its problems, accomplishments, and future potential.*

*Entropy is a measure of uncertainty. The process of true learning is always driven by the process of curiosity. This curiosity that arises during the process of learning creates some uncertainty in the minds of careful readers. Therefore, an *effective process of learning always raises the level of entropy in intelligent minds.* We will consider our efforts in this collective work as successful only if its reading helps to increase the entropy level of a curious reader.

Acknowledgments

The ideas for this volume on neuro-control systems were conceived during classroom and research discussions; these ideas were nurtured in the warm and fertile atmosphere that exists at the Intelligent Systems Research Laboratory, College of Engineering, University of Saskatchewan. And, after many learning and adaptive iterations, we are able to produce a set of collective knowledge in the form of this book.

During the editorial phases, we discussed the matter with many of our research colleagues. We started with a long list of potential articles (over 300), and, during the $(n-1)$th iteration, we ended up with an appropriate size of the volume, arranged and divided into a pedagogical style. Since the size of the book (about 90 articles) was unmanageable in a single volume, during the nth iteration we cut down the size of the volume to half (it was a painful process), but we preserved the basic theme and structure of the book. We have appended each part of the volume with an extensive reading list; these reading lists contain papers that we were forced to remove during the various editorial phases.

We are grateful to many research colleagues around the globe who have inspired our own thinking in this emerging field of neuro-control systems. Also, we are grateful to Mr. Dudley Kay, Director of Book Publishing, who continuously provided useful feedback, and Ms. Karen Miller, Production Editor, IEEE Press, who is responsible for the final physical production of the book.

Indeed, we are very much indebted to our families and wives (M. M. Gupta: Suman; D. H. Rao: Alaka) who have generously supported this project at each step by letting us use family time during evenings, weekends, and holidays.

Madan M. Gupta
Dandina H. Rao

Major Current Bibliographical Sources on Computational Neural Networks and Applications

Societies (Neural Networks)

- Neural Information Processing Systems (Natural and Synthetic) (NIPS)
- World Congress on Neural Networks (WCNN)
- IEEE International Conference on Neural Networks (IEEE ICNN)
- International Joint Conference on Neural Networks (IJCNN)
- Japanese Neural Network Society (JNNS)
- International Neural Network Society (INNS)
- European Neural Network Society (ENNS)

Major Journals (Neural Networks)

- *IEEE Transactions on Neural Networks*
- *Neural Networks* (ICNN)
- *Neural Computation*
- *Neurocomputing and Networks*
- *International Journal on Neural and Mass-Parallel Computing and Information Systems* (Neural Network World)
- *IEEE Transactions on Fuzzy Systems*
- *Biophysical Journal*
- *Biological Cybernetics*
- *International Journal of Neural Systems*
- *Neural, Parallel, and Scientific Computations*

Neuro-Control Systems: A Tutorial

Madan M. Gupta
Dandina H. Rao
Intelligent Systems Research Laboratory, College of Engineering
University of Saskatchewan, Saskatoon, Canada, S7N 0W0

***Abstract*—Neuro-control systems, the subject of this volume, has great potential in providing some new challenges in both teaching and research endeavors. In this introductory chapter, we attempt to give a basic motivation to the subject and provide readers with an overview of related topics, such as biological neuronal morphology, learning algorithms, static and dynamic neural networks, fuzzy neurons, and hardware implementations of neural networks. This tutorial is prepared with the intention of providing readers with a basic and unified view of the concepts of neural networks. This chapter will help readers understand and appreciate the advanced material presented in the main part of this volume. An extensive list of references and bibliography is appended to each part.**

1. Introduction [**1.1–1.5**]*

1.1 Biological Neuronal-Control: A Motivation

Since the evolution of machines, it has been a desire of system scientists to create a machine that can operate with increasing independence from human control in an unstructured and uncertain environment; such a machine may be called an *autonomous, intelligent,* or *cognitive* machine. The successful operation of an autonomous machine depends on its ability to cope with a variety of unexpected events in its operating environment. Such an autonomous machine would need only to be presented with a goal; it would achieve its objective through continuous interaction with its environment and with continuous feedback about its response. By having machines possess such a level of autonomy (intelligence), they would be able to learn higher-level cognitive tasks not easily handled by existing machines. Also, they would continue to adapt and perform the tasks with increasing efficiency even under changing and unpredictable environmental conditions. The autonomous machines would prove useful where direct human interaction would be hazardous, tedious, or impossible. A hazardous task would be one where a human operator is in physical danger, such as in nuclear reactors, fire fighting, mining, and military operations. A tedious task would be one where the level of concentration required to perform tasks, such as mowing grass or assembly operations, is so low that a human operator would become bored. A task impossible or difficult for humans would be the unmanned exploration of space, where a spacecraft is beyond the remote control range of an earth-based operator due to the time lag in sending information. It should be clarified that the name "machine," as used here, is a generic term for any artificial system and includes automatic navigation and guidance systems, robots, manufacturing systems, and process controllers.

Biological systems may be considered as a plausible source of motivation and framework for the design of such an autonomous machine. Biology may provide not only a motivation but also several clues for the development of robust learning and adaptation algorithms in machines. By doing so, it is hoped that some of the robust learning and adaptive capabilities of the biological system can be realized in a machine. In the present technology, the lack of these robust and adaptive abilities is due to the fact that the biological methods of processing information is fundamentally different from those used in conventional control techniques. Also, the design procedures of conventional control techiques, such as proportional, integral, and derivative (PID) controller and model reference adaptive controller, are *model based* in the sense that the design methods involve the construction of an explicit mathematical model of the dynamic system to be controlled. The premises of conventional design techniques are briefly discussed in the next subsection.

Biological neuronal control mechanisms, on the other hand, are *nonmodel based* and are quite successful at dealing with uncertainty and complexity, and can smoothly coordinate many degrees of freedom during the execution of manipulative tasks in an unstructured environment. Neuronal control mechanisms are usually very complex and defy exact mathematical formulation of their operations. They carry out complex tasks without having to develop a mathematical model of the task and the environment, and without solving any integral, differ-

*References appearing in bold, such as [**1.1**], refer to papers in this volume, and references in normal print refer to references and bibliography appearing at the end of this tutorial.

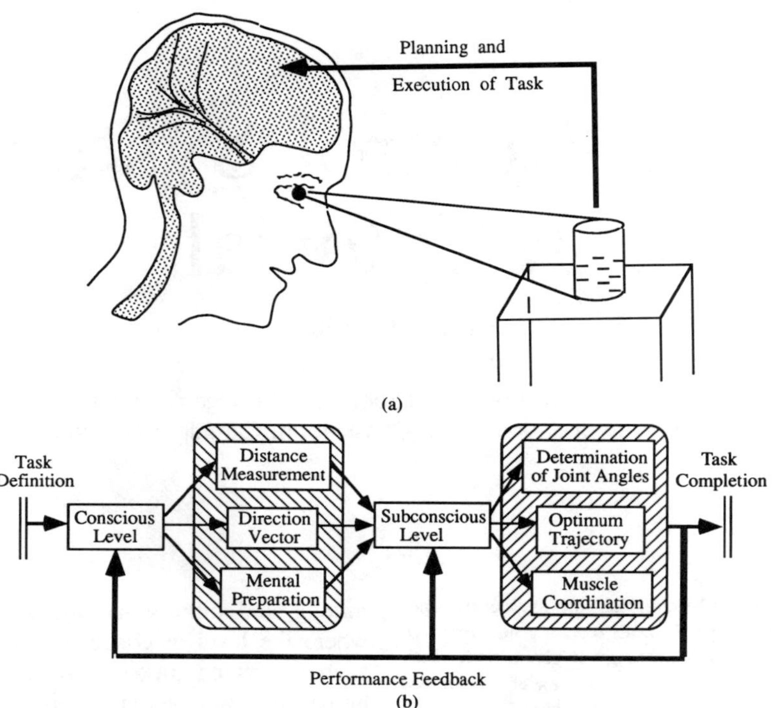

Fig. 1.1(a) Schematic representation of a biological control action. (b) A simplified block schematic of biological control actions involved in the task execution.

ential, or complex mathematical equations. In executing a particular control task, for example, "pick-up-a-glass-of-water," the plan to execute the task is carried out at the conscious level. To pick up a glass of water, it is necessary to measure the position of the hand relative to the glass and compute the direction vector to move the hand toward the glass, Fig. 1.1. The biological system executes this task at the conscious level. Most of the subsequent computations are entirely subconscious. We do not view this control task in terms of joint angles of elbow or shoulder with respect to wrist, or how hard each individual muscle is pulling. The detailed computations of joint angles and muscle coordination are carried out by the lower level—the subconscious computing centers in the central nervous system (CNS). Indeed the biological control system can learn to perform a new task, and can adapt to the changing environment with ease.

On the other hand, if we want to make a robot arm perform the same task, "pick-up-a-glass-of-water," we need a great amount of computations and *a priori* knowledge of the environment and that of the system. The angles required to coordinate different robot joints to produce a desired trajectory may be computed by solving trigonometric relationships between different structural members of the robot itself. The control methodology developed in this traditional way may completely fail should the desired task or the environment or length of arms change.

It is our hypothesis that if the fundamental principles of neural computation used by biological control systems are understood, it seems likely that an entirely new generation of control methodologies can be developed that are more robust and intelligent, far beyond the capabilities of the present techniques based on explicit mathematical modeling.

If we want a machine to emulate the capabilities found in biological controls, then we must learn from the structural, functional, and behavioral aspects of biological neural systems. Biological neuronal morphology (structure), indeed, provides a clue as well as a challenge in the design of intelligent machines that can emulate capabilities for dealing with uncertainty and execution of complex tasks in an unstructured environment.

Although many biologists and psychologists share the view that the brain has a modular neuronal morphology, there is no general agreement on the number of neuronal modules, or the manner in which the modules are structured. One reason for this diversity of opinion is the modular nature of the brain with a large number of interacting components. Even building a system with a few interacting components, a very small fraction compared to that of the brain's complexity, presents formidable computational and analytical difficulties. In many cases, implicit mathematical and computer models provide essential tools for understanding the various aspects of these systems. One class of models that has the potential for helping to resolve the difficulties of complex modular systems is the class of *connectionist models,* also known as *artificial neural network* (ANN), or *computational neural network* (CNN) models **[1.1]**.

Nature has developed a very complex neural structure in biological species. The biological neurons, over one hundred billion in the CNS of humans, play a very important role in various complex sensory, control, and cognitive aspects of information processing and decision making. In neuronal information processing there are a variety of complex operations and mapping functions involved that synergically act in a parallel-cascade structure, forming a complex pattern of neuronal layers evolving into a sort of pyramidal pattern. The information flows from one neuronal layer to another in the forward direction with continuous feedback evolving into a dynamic pyramidal structure. The pyramidal structure is in the sense of extraction and convergence of information at each point in the forward direction.

Modern neural science has evolved from the research studies of many individuals working in a variety of disciplines over a relatively large span of time. These studies have extended from philosophical dialogue on the cognitive (thought processes) and the affective (emotional processes) aspects of the brain to the basic understanding of the *neuron*, the elementary building block of our CNS. Indeed, the neural morphology of the human central nervous system is too complex to analyze using the current techniques of neurophysiology, though we always present certain speculative analogies for engineering applications. Based on these superficial understandings, researchers have developed different computational neural networks.

It is a challenge to capture complex functions of biological neural systems and use this knowledge to generalize and emulate some of the biological functions for the benefit of scientific and engineering problems in designing intelligent systems.

1.2 Premises of Conventional Control System Design

The conventional design methods of control systems involve the construction of a mathematical model describing the dynamic behavior of the plant to be controlled and the application of analytical techniques to this model to derive a control law. Usually, such a mathematical model consists of a set of linear or nonlinear differential/difference equations, most of which are derived under some forms of approximation and simplification. These conventional techniques break down, however, when a representative model is difficult to obtain due to uncertainty or sheer complexity, or when the model thus produced violates the underlying assumptions of the control law design techniques. Also, modeling of a physical system for feedback control involves a trade-off between the simplicity of the model and its accuracy in matching the behavior of the physical system. On the other hand, human operators do not always handle the system control problem with a detailed mathematical model but with a qualitative feeling of the process and approximate reasoning and knowledge of the control process.

Two approaches are usually described in the literature to achieve satisfactory performance from a vaguely known dynamic plant. One approach is robust stabilizers or robust controllers. In this approach, if the actual physical system is a member of a class of systems that are close to the nominal plant, a robust controller guarantees to stabilize it. Since one

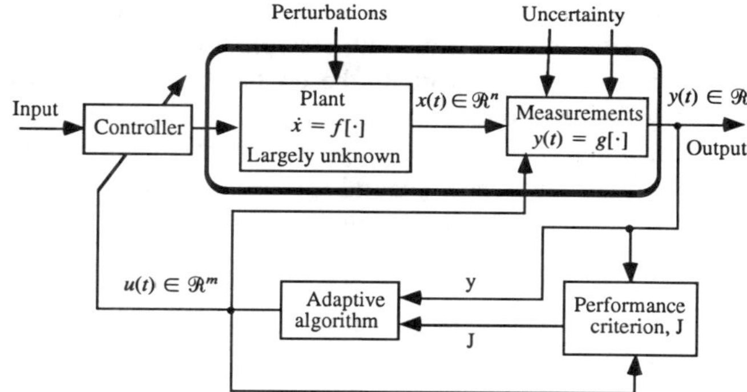

Fig. 1.2 A basic configuration of an adaptive system.

fixed controller is expected to stabilize the whole set, the price might be that the controller thus designed is highly complex compared to the complexity required to stabilize any single plant in this class.

Adaptive control is another approach to the solution of the control problem for complex plants. The parameters of the adaptive controller are made to adapt in accordance with some adaptive algorithm in order to keep the system performance at the desired level. In general, the adaptive approach is applicable to a wider range of uncertainties, but robust controllers are simpler to implement, and no time is required to tune the controller parameters to the plant variations. Detailed descriptions of adaptive and robust control techniques may be found in [20–27].

A schematic representation of a general adaptive control system is shown in Fig. 1.2. An adaptive control system measures certain index of performance using the inputs, the states, and the outputs of the dynamic system under control (plant). From the comparison of the measured index of performance values with those of the desired ones, the adaptation mechanism (adaptive rule) modifies the parameters of the controller in order to maintain the plant response close to the desired one.

Although adaptive control techniques, such as model reference adaptive control (MRAC) and automatic tuning regulators, have been widely used, their applications to realistic problems have been slow. One reason for this may be attributed to the fact that prior knowledge of the plant under control is necessary to guarantee the stability of the adaptive system. The complexity of implementation of the traditional adaptive methods to large and complex systems may also discourage the practitioner [21, 22]. Furthermore, these methods assume knowledge of the upper bound on the plant order. This prior knowledge is necessary to prove stability and also for the implementation of reference models, identifiers, or observer-based controllers of about the same orders as the plant. Since order of the complex plants in the real world may be very large or unknown, implementation of the conventional adaptive methods may be difficult or sometimes impossible [24].

The need to control complex systems under significant uncertainties has led to a reevaluation of the existing control methodologies. Evolution in the control paradigm has been fueled by two major concerns: the need to deal with increasingly complex systems, and the need to accomplish increasingly demanding design requirements with less precise knowledge of the plant and its environment. In these situations, it is almost mandatory for the control schemes to enforce learning and adaptive features. The trend of adaptive control, therefore, can be directed to develop more general approaches and be utilized in as many applications as possible.

1.3 Neural Controller with Learning Algorithm

To cope with uncertainties regarding plant dynamics and its environment, the controller has to estimate the unknown information during its operation. If this estimated information gradually approaches the true information as time proceeds, then the controller thus designed will approach an optimal controller. Because of the gradual improvement of performance due to improvement of the estimated information, this controller may be viewed as a learning adaptive controller. The controller learns the unknown information during operation, and this information, in turn, is used as an experience for future decisions and controls [**1.2,** 28].

Figure 1.3 represents one type of neural learning and control scheme. A control system is called a *learning control system* if the information pertaining to the unknown features of the plant or its environment is acquired during operation, and the obtained information is used for future estimation, recognition, classification, control, or decision such that the overall system performance is improved. For example, a control system may change the type of controller used, or vary the parameters of the controller after learning that the present controller does not perform satisfactorily.

By enhancing the controller with learning, one can effectively expand the operating region of the controller, and can create more robust controllers that may ultimately lead to autonomous controllers. The control laws need not be explicitly stated as learning can be through examples. The control system can then compensate for a larger number of changes in the plant and its environmental conditions. These operating conditions may cause the nominal adaptive system to exceed the tolerances of its design, for no general analytical solution can be determined for such inherently complex and uncertain systems. The learning system, on the other hand, determines neural controller parameter values for optimal performance for given operating conditions. Because of the learning feature, a model of the plant under control is not mandatory to make the plant follow a desired trajectory. A learning system has the ability to improve its performance in the future, based on information it has gained in the past. Both adaptive and learning control systems can be implemented using parameter adjustment algorithms, and both make use of performance feedback information [19]. The difference between adaptive and learning systems lies in the fact that the former treats every distinct operating situation as novel, whereas the latter correlates the past experience with the present situations and accordingly adapts its behavior. Since a learning system is capable of adjusting its actions, it is also an adaptive system.

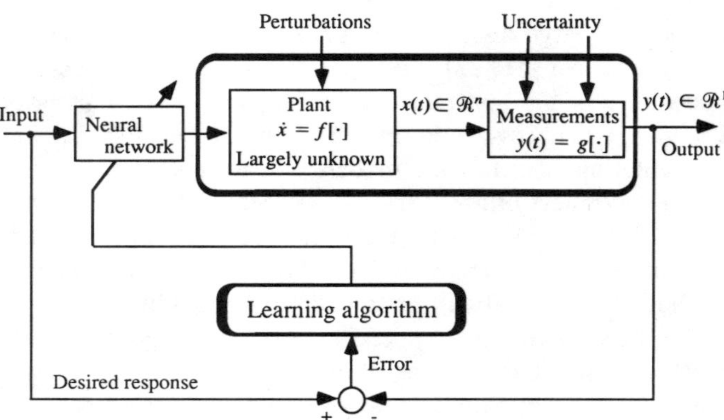

Fig. 1.3 A typical neural learning and control scheme.

The need for learning capability in the control of complex systems operating in the presence of significant uncertainties has made avenues for new control techniques quite apparent. The use of neural networks in control systems can be viewed as a natural step in the evolution of control methodologies [**1.3–1.5, 5.2–5.6**]. Neural networks, with their massive parallelism and their ability to learn, offer exciting possibilities for the introduction of much superior control techniques for complex situations. However, it should be noted that neural-network-based control systems are not panacea, and do not provide solutions to all problems.

The field of artificial neural networks is not new, but it has only recently become an active area of research. Some of the pioneering work in this field is due to McCulloch and Pitts [2] in 1943, when they published an abstract model of a simple neuron. The neuron had a finite number of inputs and a single output. The inputs were characterized by excitatory $(+1)$ and inhibitory (-1) states, the neuron had an internal threshold, and the threshold function was binary. It was thought that by connecting many of these simple devices it would be possible to model the human brain. Although their model proved too simple to achieve human-like abilities, it did influence others to use neural networks of logic elements to build what are now known as digital computers.

The next major development occurred in 1949 when Hebb [4] conjectured a learning mechanism in the brain. He postulated that as the brain learns, it changes its connectivity patterns. More specifically, during the learning process, as a neuron repeatedly activates another neuron, the conductance of the synapse between the two neurons increases. This idea of a learning mechanism was first incorporated in an artificial neural network by Rosenblatt [5] in 1959. He combined the simple McCulloch and Pitts neuron with the adjustable synaptic weights based on Hebbian learning scheme to form the first artificial neural network with the capability to learn.

In 1960, by introducing the least-mean squares (LMS) learning algorithm, Widrow and Hoff [6] developed a model of a neuron that learned quickly and accurately. This model was called ADALINE, for ADAptive LInear NEuron. This learning algorithm first introduced the concept of supervised learning using a "teacher" that guides the learning process. The LMS algorithm attempts to minimize the error at each neuron, subject to the constraint that the weights are disturbed the least. It is the recent generalization of this learning rule into backpropagation that has led to the resurgence in biologically-based neural-network research today.

In 1969, research in the field of neural networks suffered a serious setback. Minsky and Papert [7] published a book entitled *Perceptrons,* in which they proved that single-layer neural networks were limited in their abilities to process data and argued that the study of multilayer neural networks would be unproductive. As a result of this influential book, little progress was made in this area until the early 1980s.

Many of the early applications of neural networks have been in computationally intensive areas of signal processing, such as adaptive pattern recognition, real-time speech recognition, and image interpretation. In control systems, there are also computationally intensive applications, such as real-time system identification and control. Traditional control systems that operate with large uncertainty typically depend on human intervention to function properly. However, human intervention is unacceptable in many real-time autonomous applications, and automatic techniques for handling uncertainty need to be developed. Indeed, artificial intelligence techniques for handling uncertainty, such as expert systems, have been employed to alleviate this problem. Neural networks with their parallel distributed processing may be a better alternative for high-speed, real-time control because they can avoid time-consuming calculations and can adapt to system changes.

The general merits of neural networks are described in the next chapter. With specific reference to neural networks in control systems, neural networks have shown great potential in the realm of nonlinear control problems. While major advances have been made in the design of adaptive controllers for linear systems with unknown parameters, such controllers cannot be used for the global control of nonlinear systems. A range of ''conventional'' methods exist, such as phase plane, describing functions, and feedback linearization, for the analysis and synthesis of nonlinear controllers for specific classes of nonlinear systems. The existing control methods are system specific; in other words, a control methodology suitable for a class of nonlinear systems may be completely unacceptable for some other class of nonlinear systems.

The most significant characteristic of neural networks is their ability to approximate arbitrary nonlinear functions. This ability of neural networks has made them useful to model nonlinear systems, which is of primary importance in the synthesis of nonlinear controllers [**3.1**]. A neuro-controller (neural network-based control system), in general, performs a specific form of adaptive control, with the controller taking the form of a multilayer network and the adaptable parameters being defined as the adjustable weights. In general, neural networks represent parallel-distributed processing structures, which make them prime candidates for use in multivariable control systems. The neural-network approach defines the problem of control as the mapping of measured signals for ''change'' into calculated controls for ''actions'', as shown in Fig. 1.4.

Computational (artificial) neural networks represent massively parallel distributed processing capability with the potential for ever-improving performance through dynamic learning. The development of very-large-scale-integrated (VLSI) network architectures or opto-electronic networks makes implementation in real-time conceivable.

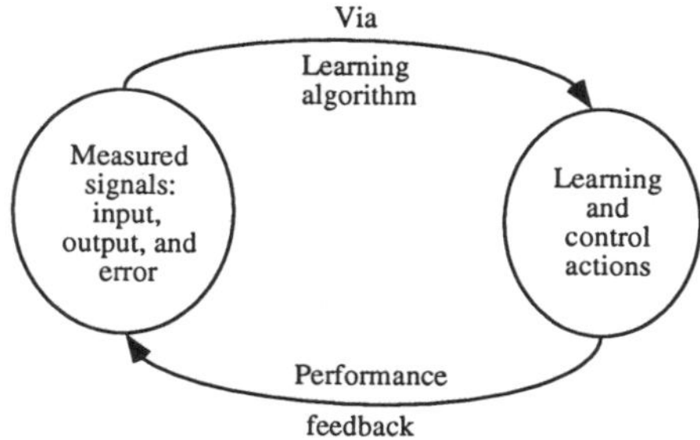

Fig. 1.4 Representation of learning and control actions in a neural network approach: Mapping of the measured signals onto the learning and control space.

1.4 Summary

The incredible flexibility and adaptability of biological neuro-control mechanisms may provide the inspiration for developing robust control mechanisms. Biological systems do provide the framework from which comparable autonomous systems can be developed. From the perspective of system and computer scientists, it is not necessary to emulate the precise neurophysiological behavior of biological control processes. Rather, it is desirable to replicate some of the computational operations involved in biological learning and adaptation. In this way, the scientific principles derived from the understanding of biological processes may be employed for the design of more effective intelligent machines.

The outline of this tutorial chapter is as follows. A brief description of biological neuronal morphology is described in the next section, followed by an overview of neuronal taxonomy in Section 3. Static neural networks are described in Section 4. Mathematical development of a single neuron, followed by the description of multilayer neural networks are also elucidated in this section. Functional approximation theory occupies a predominant place in the neural network paradigm and, hence, a brief overview of approximation theory is presented in Section 5. A natural extension of static neural networks is dynamic neural networks, which are described in Section 6. The morphology, learning algorithms, and, in general, the theory of dynamic neural networks is not well developed as compared to its static counterpart. Most of the developments with regard to theory and applications in neural paradigm have been concentrated around static neural networks. The dynamic neural networks offer a great potential for research endeavors. One can also find some totally different neural architectures, what we call *unconventional neural structures,* such as neurons with hysteresis and Cerebellar Model Articulation Controller (CMAC). A brief description of these neural structures is also given in this section. Another field that has developed in parallel with neural networks is the *fuzzy logic* to capture human cognition. Integration of neural networks and fuzzy logic is another branch of neural science that is developing quite rapidly. An introduction to fuzzy logic and fuzzy neural networks is presented in Section 7. Hardware implementations of neural networks is another branch of research developing very quickly. An overview of different implementation techniques are discussed in Section 8. Finally, this tutorial chapter concludes with perspectives focusing on the problems and future challenges in the neural-network paradigm. An extensive bibliography and list of references are appended to this introductory chapter.

2. Biological Neuronal Morphology [2.1–2.4]

In the previous section, we presented some aspects of biological neural control that provide the basic motivation and framework to neuro-computing systems. We also discussed the premises of conventional control methodologies that lead to the necessity of using computational neural networks in control systems. In this section, we briefly describe the biological neuronal morphology (structure) that forms the basis for the study of computational (artificial) neural networks.

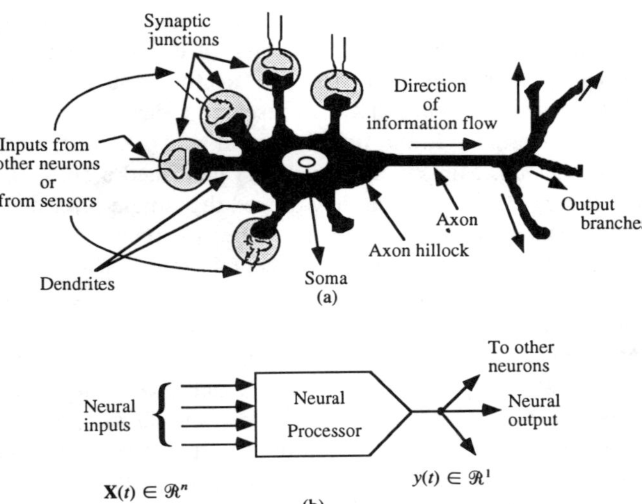

Fig. 2.1(a) A schematic view of the biological neuron. The soma of each neuron receives parallel inputs through its synapses and dendrites, and transmits a common output via the axon to other neurons. (b) Model representation of a biological neuron with multiple inputs $\mathbf{X}(t) \in \Re^n$, and a single output, $y(t) \in \Re^1$.

2.1 The Neuron: The Basic Unit of the Central Nervous System (CNS)

The human brain, the most amazing carbon-based computer in existence, weighing a little over three pounds, consists of approximately 10 billion individual nerve cells called *neurons.* All human activities and behavior can ultimately be traced to the activity of these tiny cells. Each neuron is interconnected to many other neurons, forming a densely connected network called a *neural network.* These massive interconnections provide an exceptionally large computing power and memory.

The basic building block of the CNS is the neuron, the cell that processes and communicates information to and from various parts of the body. From an information-processing point of view, an individual neuron consists of the following three parts, each associated with a particular mathematical function:

1. the dendrites are a receiving area for information from other neurons;
2. the cell body, called a *soma,* collects and combines incoming information received from other neurons; and
3. the neuron transmits information to other neurons through a single fiber called an *axon.* The range of lengths of axons is 50 μm to several meters. The axon is a tubular structure bounded by a typical cell membrane.

The junction point of an axon with a dendrite of another neuron is called a *synapse.* Synapses provide memory to the past accumulated experience or knowledge. A single axon may be involved with hundreds of other synaptic connections. A schematic diagram of the biological neuron is shown in Fig. 2.1(a). From a systems theoretic point of view, the neuron can be considered as a multiple-input–single-output (MISO) system as depicted in Fig. 2.1(b)

2.2 Action Potential

Each neuron can be thought of as a tiny biological battery full of ions and ready to be discharged. Neurons are filled with, and surrounded by, fluids containing dissolved chemicals. Both inside and around the soma are sodium (Na^+), calcium (Ca^{++}), potassium (K^+), and chloride (CL^-) ions. Na^+ and K^+ ions are largely responsible for generating the active neural response called an *action potential,* also called the *nerve impulse.* K^+ ions are concentrated inside the cell of the neuron, whereas the Na^+ ions are concentrated outside the cell membrane.

The process of generating an action potential either in the neuron (where the processing of information takes place) or in the axons (through which the transmission of information takes place) is caused by an exchange of ions (K^+ and Na^+), due to a change in the permeability of the cell membrane.

If the soma is electrically stimulated by a voltage greater than a certain threshold, there is an exchange of ions. This movement of ions causes the soma to change its internal state. More specifically, the flow of Na^+ ions into the nerve membrane and K^+ ions out of the membrane generates an action potential in the neuron. The flow of ionic current is triggered by changes in the permeability of the nerve membrane. Before the action potential occurs—an unexcited state—the membrane permeability to both Na^+ and K^+ ions is low such that there is only a minimal flow of these ionic currents across the membrane. In an unexcited (rest) state, the voltage inside with respect to the outside is at a constant of about −70 mV. The action potential is generated when, due to a traveling stimulus, the axon membrane suddenly becomes permeable to Na^+, enabling this positive ion to rush into the membrane and increasing the inside potential to about +30 mV. After about 0.5 ms, the membrane's permeability to Na^+ decreases and its permeability to K^+ increases, which causes the K^+ ions to

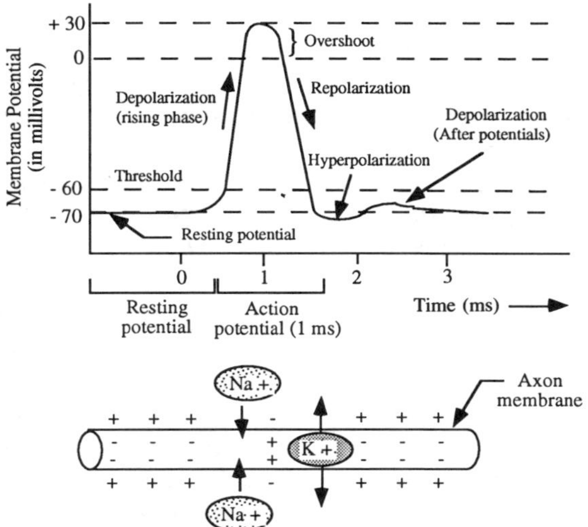

Fig. 2.2 Changes in membrane potential during action potential because of the inflow of Na^+ ions and outflow of K^+ ions. The sudden rise of the membrane potential to a positive state and then its return to its normal negative state causes the action potential. After the occurrence of action potential, an outward flow of potassium (K^+) ions restores the resting potential.

flow out of the membrane fiber. Since K^+ also has a positive charge, the electrode records a decrease in the positive charge inside the axon membrane until the charge returns to its original level. The overall process of ion exchange occurs within a couple of milliseconds.

In summary, the action potentials (nerve impulses) are traveling positive charges generated by the flow of charged ions (Na^+, K^+) across the axon membrane as shown in Fig. 2.2. The speed of the nerve impulse depends on a number of factors, including the size of the axon. Very fine nerve fibers may carry a nerve impulse at a rate of about 8 feet per second (or less). Longer and larger fibers that connect the brain to the body average about 330 feet per second (about 225 miles per hour). If someone steps on our toes, our brain will receive this information in about one-fiftieth of a second.

2.3 Neuron Firing

The neuronal morphology in the CNS is a very complex structure for the processing and transmission of information. Both processing and transmission of information takes place through the flow of ions across the axon and neuron membrane.

In the transmission of information from one neuron to another neuron, axons are involved that transmit the information at about 225 miles per hour. In the processing of information, synapses (the junction points between axons and dendrites) and the main body of the neuron—the soma—are involved. In the following paragraphs, we provide a brief description of information processing and information transmission activities of the neuron.

The neuron is basically a computing node that receives information (signals) and does some processing. The axon of one neuron is connected to the dendrite of another neuron

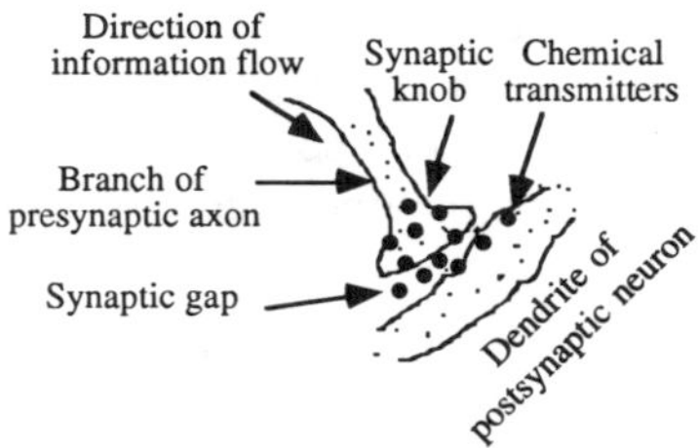

Fig. 2.3 A magnified view of a synapse. Transmitter chemicals that are either excitatory or inhibitory cross the synaptic gap.

through a synaptic junction called a synapse, as depicted in Fig. 2.1(a). This synapse employs a chemical transmitter substance to convey a signal across the boundary of the junction. At the synaptic junction, the action potentials conducted along the axon are converted to a voltage potential called the *postsynaptic potential* (PSP). The PSP is proportional to the amount of transmitter substance released, which, in turn, is proportional to the frequency of the axonic action potential. The PSP becomes saturated for large amounts of transmitter substance. A temporal summation occurs because the synaptic junction has a large time constant that is far greater than the time between the pulses being transmitted along the axon. Any new action potentials are simply added to the partially decayed remains of the past PSPs. This accumulating effect at the dendrite side of the synapse results in a slow amplitude potential. The magnitude of this dendritic depolarization is proportional to the average frequency at which the pulses arrive at the synaptic junction. In terms of information processing, the synapse performs a crude pulse-frequency-to-voltage conversion. These synaptic junctions usually occur between axons and dendrites, but they may also occur between two axons, two dendrites, and even between axons and cell bodies.

When the action potentials (nerve impulses) reach the synaptic junction at the end of the axon, the nerve impulse causes a release of neurotransmitters, Fig. 2.3. These chemicals cross the synapse between two neurons and alter the activity of the next neuron. Neurons transmit information through a combination of events occurring within and between cells. The nerve impulse is basically an electrical signal, whereas communication between neurons is primarily a chemical process. Neurotransmitters may *excite* the neuron or *inhibit* it, depending on the type of transmitter released and the nature of the dendrite's membrane. The dendritic inputs originating from the excitatory synapses will increase the overall rate of neuron firing, whereas inputs from the inhibitory synapses will decrease this firing rate.

The dendrites transmit numerous slow potentials to the soma of the neuron. Each soma will receive, on the average, about 10,000 excitatory and inhibitory inputs. The role of the soma is to perform a spatio-temporal summation of all excitatory and inhibitory slow potentials by means of a weighted average. If this weighted average exceeds a threshold, it is then converted at the *axon hillock* into action potentials with an appropriate output frequency, and the action potentials are transmitted along the axon to the other nerve cells in order to repeat the process. The generation of these action potentials depends, therefore, on the interplay of the excitatory (+) and inhibitory (−) input signals received by the soma.

In brief, there are three components involved in the transmission of information from one neuron to another:

- Action potentials (train of impulses, the information being transmitted is modulated into the frequency of action potentials;
- Synaptic operation (chemical transmitters or post synaptic potential)

  postsynaptic potential (PSP) $\propto$ frequency of action potentials $(f) = w \cdot f$

  where w is a constant of proportionality, the weight that represents the synaptic strength. It may be positive for excitatory signals or negative for inhibitory actions. The synaptic weight w, as a matter of fact, represents the past accumulated experience, thus serving as a memory;
- Somatic operation:
  - aggregation of positive (excitatory) and negative (inhibitory) potentials,
  - thresholding at the axon hillock,
  - firing of the neuron.

In the transmission of an action potential from a neuron, a certain amount of time is consumed in the process of (1) discharge of the transmitter substance by the presynaptic neuron, (2) diffusion of the transmitter to the neuronal membrane, (3) action of the transmitter on the membrane, and (4) inward diffusion of sodium to raise the potential to a high enough value to elicit an action potential. The minimum period of time required for all these events to take place is approximately 0.5 ms. This is called the *synaptic delay.*

In summary, each neuron is activated by the flow of biochemicals across the synapses. The transmission of these biochemicals across the synaptic junction causes a change in the ionic concentration within the neuron, which, in turn, produces a change in its electrochemical potential. These inputs may be excitatory (positive) and increase the electrochemical potential of the postsynaptic neuron, or conversely, they may be inhibitory (negative) and reduce the electrochemical potential. If the net potential at the axon hillock is above a certain threshold level then the neuron will "fire" a sequence of pulses, called the action potentials, along an axon leading to the synaptic junction of another neuron. The electrochemical activities at these synaptic junctions exhibit complex behavior because each neuron makes several hundred interconnections with outer neurons. Each neuron acts as a parallel processor because it receives pulses in parallel from neighboring neurons and then transmits pulses in parallel to all neighboring synapses. Readers will find a detailed description of the biological neuron in **[2.1]**.

From this description, it may be concluded that the processing of information within the biological neuron involves two distinct operations:

1. synaptic operation: This provides a weight to the neural inputs. Thus, the synaptic operation assigns a relative weight (significance) to each incoming signal according to the past experience (knowledge or memory) stored in the synapse.
2. somatic operation: This provides aggregation, thresholding, and nonlinear activation to the dendritic inputs. If the weighted aggregation of the neural inputs exceeds a certain threshold, the soma will produce an output signal.

More details of synaptic and somatic mathematical operations are given in Section 4.

Finally, from experimental studies in neurophysiology, it has been shown that the action potential response of a neuron is largely a random variable, and only by averaging many observations of action potential behavior is it possible to obtain a predictable result **[2.4]**. This observed variability in the response, or firing, of a neuron is a function of both the uncontrolled extraneous electrical signals that are being received from the activated neurons in other parts of the nervous system and intrinsic fluctuations of the electrical membrane potential within the individual neuron. The only way to achieve fast and efficient neural information processing by using slow (i.e., each neuron requires 2 ms in order to fire; in other words, the neural bandwidth is of the order of 500 Hz) and unpredictable processing elements is to employ large numbers of such processors in order to perform some parallel computation. This is how the central nervous system as a whole operates. The exact function of each constituent neuron cannot, therefore, be known precisely. The purpose of the study of biological neuronal morphology is, however, to receive some motivation from nature in the design of robust computing systems—the "computational neural networks."

2.4 Keyword Definitions

Action potential: The pulse of electric potential generated across the membrane of a neuron (or an axon) following the application of a stimulus greater than the threshold value.

Axon: The output fiber of a neuron that carries the information in the form of action potentials to other neurons in the network.

Dendrite: The input line of the neuron that carries a temporal summation of action potentials to the soma.

Excitatory neuron: A neuron that transmits an action potential that has excitatory (positive) influence on the recipient nerve cells.

Inhibitory neuron: A neuron that transmits an action potential that has inhibitory (negative) influence on the recipient nerve cells.

Lateral inhibition: The local spatial interaction where the neural activity generated by one neuron is suppressed by the activity of its neighbors.

Latency: The time between the application of the stimulus and the peak of the resulting action potential output.

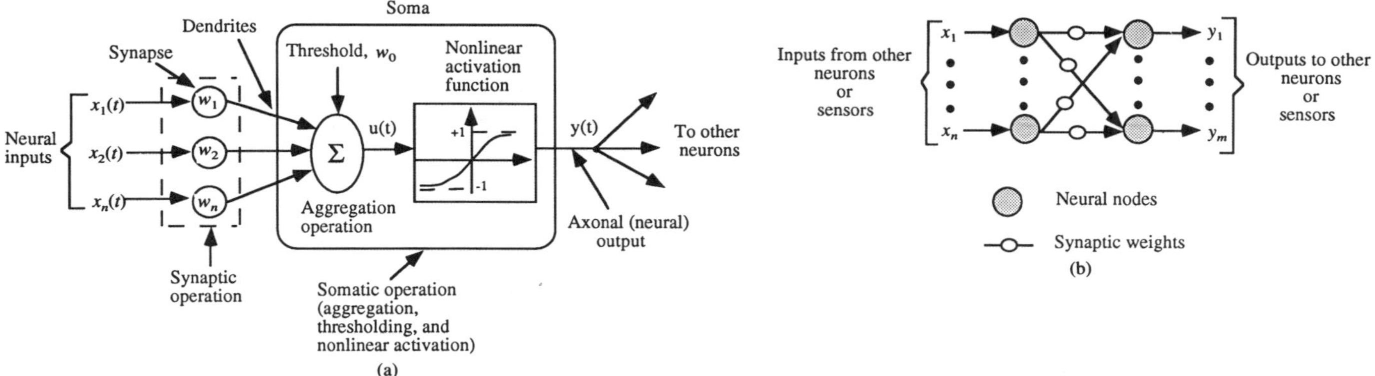

Fig. 3.1(a) A static neural node. $[x_1(t), \ldots, x_n(t)$ represent neural inputs, $[w_1, \ldots, w_n]$ are the synaptic weights, w_0 is the threshold, and $y(t)$ is the axonal (neural) output. (b) A static (feedforward) neural network with n-inputs and m-outputs.

Refractory period: The minimum time required for the axon to generate two consecutive action potentials.

Neural state: A neuron is active if it is firing a sequence of action potentials.

Neuron: The basic nerve cell for processing biological information.

Soma: The body of a neuron that provides aggregation, thresholding, and nonlinear activation to dendritic inputs.

Synapse: The junction point between the axon (of a presynaptic neuron) and the dendrite (of a postsynaptic neuron). This acts as a memory (storage) to the past accumulated experience (knowledge).

3. An Overview of Neuronal Morphologies [1.1, 3.1]

In the previous section, we presented a brief description of the biological neuron and its functional complexity. The purpose of this and the following sections is to provide an overview of the morphologies of computational neural networks.

3.1 Introduction

Due to the complexity and diversity of the properties of biological neurons, the task of compressing their complicated characteristics into a model is extremely difficult. Toward this goal, a model of the biological neuron—also called a neural unit, or simply a neuron—has been developed in the neural-network paradigm. The neuron receives inputs from a number of other neurons or from the external world. A weighted sum of these inputs constitutes the argument of a nonlinear "activation" function, as depicted in Fig. 3.1(a). The neuron is said to have been fired if the weighted sum of its inputs exceeds a certain threshold, w_0. Mathematically, the function of a neuron can be modeled as

$$y(t) = \Psi\left[\sum_{i=1}^{n} w_i x_i - w_0\right] \tag{3.1}$$

where $[x_1, \ldots, x_n]$ represent neural inputs, $[w_1, \ldots, w_n]$ are the synaptic weights, $y(t)$ is the neural output, and $\Psi[\cdot]$ is some nonlinear activation function with threshold w_0.

This is a simple, but useful, first approximation of the biological neuron **[2.2]**. Using this model, many neural morphologies, usually referred to as feedforward neural networks, have been reported in the literature. These feedforward networks respond instantaneously to inputs because they possess no dynamic elements in their structure. Therefore, these neural structures are also called static neural networks. A schematic representation of a static neural network is shown in Fig. 3.1(b). A detailed description of static neural structure is given in the next section.

A natural extension of static (feedforward) networks is the dynamic (feedback) neural networks that incorporate feedback and dynamical elements in their structure. There are several dynamic neural structures based on different neural paradigms. With the parallel growth in the field of fuzzy logic, many neural models encompassing the principles of neural networks and fuzzy logic are also being developed. Although the static, dynamic, and fuzzy–neural networks are being used in many control and machine vision applications, the basic neural models remain a feeble imitation of their biological counterparts.

Over the last few years, a host of static, dynamic, and fuzzy–neural morphologies have been proposed. A detailed description of these morphologies is beyond the scope of this work. However, we will make an attempt to provide the basic concepts of different neural morphologies that are commonly used in the control paradigms.

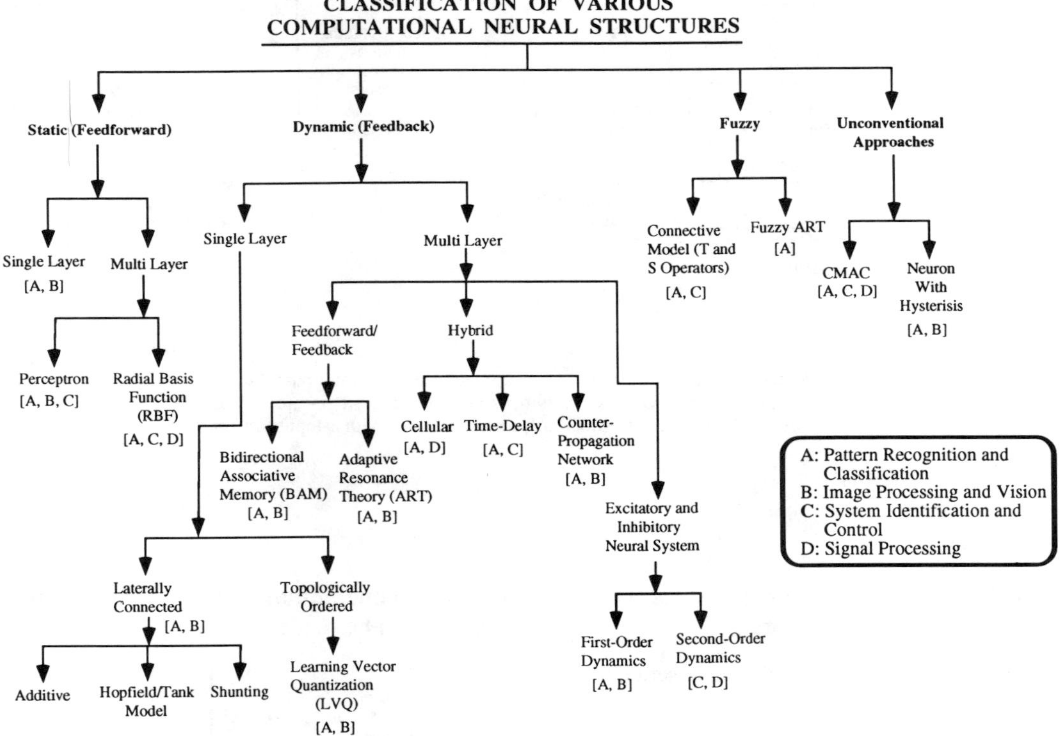

Fig. 3.2 The taxonomy of neural structures illustrating the interrelationships between different neural structures. Letters in brackets under each category indicate the applications a particular neural structure has been used for.

3.2 Computational Neural Networks (CNN)

The goal of computational neural network research is to develop mathematical neural morphologies, not necessarily observed in biology, that can perform various computational functions. Artificial neural networks (ANNs), computational neural networks (CNNs), or simply neural networks go by many descriptions, such as connectionist models, parallel distributed processing networks, and neuromorphic systems. In general, neural models are composed of many nonlinear computational elements operating in parallel and arranged in patterns reminiscent of biological neural nets. They have been shown to perform, on a small scale, such higher cognitive functions as learning, memory and recall, and pattern recognition.

The computational process envisioned for neural networks is as follows. It starts with the development of a "computational" or "artificial" neuron based on the understanding of biological neuronal structures, followed by learning mechanisms for a given set of applications. This leads to the following three steps in a neural computational process:

1. development of neural models motivated by the biological neurons,
2. models of synaptic connections and structures (i.e., network topology and weights), and
3. the learning rules (i.e., the method of adjusting the weights or internodal connection strengths).

Due to differences in either or all of these entities, different structures of neural networks [2, 5, 16, 44, 65–78, **3.2–3.5, 5.6, 6.5, 7.1**] are being explored by researchers. For example, from a structural point of view, neural network architectures may be classified as either static, dynamic, or fuzzy, and single-layer or multilayer. Furthermore, computational differences in the neural networks arise from the different types of synaptic connections that are assumed to exist among the neurons. These synaptic connections can be strictly feedforward, laterally connected, topologically ordered, feedforward/feedback, and hybrid. The taxonomy of neural structures depicting the interrelationship of the various computational neural structures is shown in Fig. 3.2. As illustrated in this figure, the applications of neural networks are categorized into four classes: A: Pattern Recognition and Classification, B: Image Processing and Vision, C: System Identification and Control, and D: Signal Processing. The letters in brackets under each neural structure indicate some of the applications of that particular neural structure. It should be noted, however, that the use of neural structures need not be restricted to the specified applications.

In spite of numerous neural structures and learning algorithms proposed by different researchers, neural networks share many features that are unique to biological systems. These features are in contrast to the traditional computing methods. Computational neural networks can accommodate many inputs in parallel and encode the information in a distributed fashion. Typically, the information that is stored in a neural net is shared by many of its processing units. This type of coding is in sharp contrast to traditional memory schemes, where a particular piece of information is stored in one memory location. The recall process is time consuming and generalization is usually absent. The distributed storage scheme in neural networks provides many advantages, the most important being that the information representation can be redundant. Thus, a neural network can undergo partial destruction of its structure and still be able to function well. Although redundancy can be built into other types of systems, the neural network has a natural way of implementing this redundancy. The result is naturally a fault-tolerant system that is very similar to biological systems.

The neural network attributes, such as learning from examples, generalization redundancy, and fault-tolerance, provide strong incentives for choosing neural networks as an appropriate approach to modeling biological systems. The potential benefits of such a network can be summarized as follows.

1. The neural network models have many neurons (the computational units) linked via the adaptive (synaptic) weights arranged in a massive parallel structure. Because of its high parallelism, failures of a few neurons do not cause significant effects on the overall system performance. This characteristic is known as fault-tolerance.
2. The main strength of neural network structures lies in their learning and adaptive abilities. The ability to adapt and learn from the environment means that the neural network models can deal with imprecise data and ill-defined situations. A suitably trained network has the ability to generalize when presented with inputs not appearing in the trained data.
3. The most significant characteristic of neural networks is in their ability to approximate any nonlinear continuous function to the desired degree of accuracy. This ability of neural networks has made them useful to model nonlinear systems in the synthesis of nonlinear controllers **[5.3–5.6]**.
4. Neural networks can have many inputs and many outputs; they are easily applicable to multivariable systems.
5. With advances in hardware technology, many vendors have recently introduced dedicated VLSI hardware implementations of neural networks. This brings additional speed in neural computing.

Although a large number of computational neural network architectures and learning algorithms are reported in the literature, most of these neural networks have certain features in common with biological neural systems. The primary structural characteristics of a CNN are an organized morphology containing numerous parallel distributed neurons, a method of encoding (i.e., learning) information within the synaptic connections found between the individual neurons, and a method of recalling information when presented with a stimulus input pattern. The neural networks can learn the association or similarities between different patterns. That is, they extract empirically the regularities in the data. Developing a neural-network-based scheme is unlike developing computer software, because the neural network is made to learn by adapting its synaptic connections, not by programming. Incidentally, it is these very features that are absent in the traditional sequential computer paradigm.

With this brief general introduction to computational neural structures, we will describe the three types of neural models, namely, static, dynamic, and fuzzy–neural structures, in the following sections.

4. Static (Feedforward) Neural Networks [3.1, 5.1–5.5]

A brief description of the taxonomy and functional roles of neural networks was presented in Section 3. A detailed discussion of the various neural network architectures described in the literature is far beyond the scope of this introductory chapter. However, to unify this diverse and abundant material, we will now present a generalized interpretation of computational neural mechanisms from a systems-science perspective. It is hoped that these paragraphs will simplify the overwhelming volume of research on neural computing for the novice reader. In this context, we will consider only a single neuron and study its intrinsic properties. As a sequel to this development, we will describe multilayer neural networks.

4.1 Mathematical Model of a Neuron: Synaptic and Somatic Operations

A biological neuron consists of synapses (junction points) and a soma—the main body of the neuron. The numerous synapses that adjoin a neuron receive neural inputs from other neurons and transmit modified (weighted) versions of these signals to the soma via the dendrites. Each soma receives, on the average, 10^4 dendritic inputs. The role of the soma is to perform a spatio-temporal weighted aggregation (often a summation) of all these inputs. If this weighted aggregation is greater than an intrinsic threshold, then the weighted signal is converted into an action potential yielding a neural output. These action potentials are transmitted along the axon to the other neurons for further processing.

From a signal-processing point of view, the biological neuron has two key elements—the synapse and soma—that are responsible for performing computational tasks, such as learning, acquiring knowledge (storage or memory of past experience), and recognizing patterns. Each synapse is a storage element that contains some attribute of the past experience. The synapse learns by continuously adapting its strength (or weight) to the new neuronal inputs. The soma combines the weighted inputs such that if it exceeds a certain threshold then the neuron will fire. This axonal (output) signal undergoes a nonlinear transformation prior to leaving the axonic hillock in the soma. Mathematically, the synapses and early stage of the soma provide a confluence operation between the fresh neuronal inputs and stored knowledge (past experience). The latter part of the soma, the nonlinear activation operation, provides a nonlinear bounded mapping to the aggregated signal.

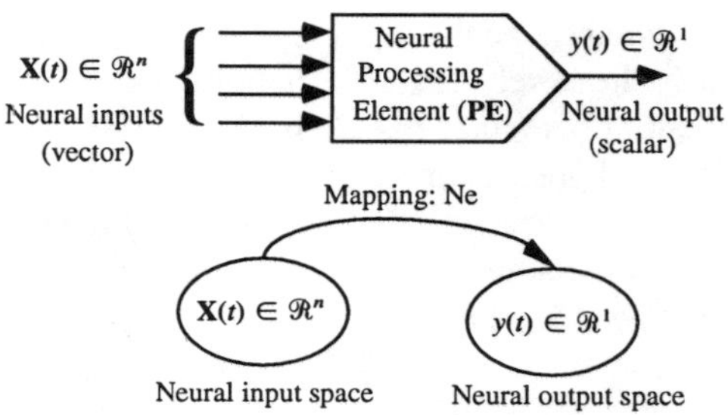

Fig. 4.1 The information-processing ability of a neuron (or processing element), as represented by the nonlinear mapping function Ne: $\mathbf{X}(t) \in \Re^n \rightarrow y(t) \in \Re^1$.

In simple terms, a neuron can be depicted as an information *processing element* (PE), which receives an n-dimensional neural input vector,

$$\mathbf{X}(t) = [x_1(t), x_2(t), \ldots, x_i(t), \ldots, x_n(t)]^T \in \Re^n \tag{4.1}$$

and yields a scalar neural output $y(t) \in \Re^1$. The input vector $\mathbf{X}(t) \in \Re^n$ represents the signals being transmitted from the n-neighboring neurons (including self-feedback signal and/or the outputs (measurements)) from the sensory neurons.

Mathematically, the information-processing ability of a neuron can be represented as a nonlinear mapping operation, Ne, from the input vector $\mathbf{X}(t) \in \Re^n$ to the scalar output $y(t) \in \Re^1$; that is,

$$\text{Ne: } \mathbf{X}(t) \in \Re^n \rightarrow y(t) \in \Re^1 \tag{4.2}$$

The nonlinear mapping operator (Ne), from the n-dimensional input space to the one-dimensional output space, is shown in Fig. 4.1. Alternatively, we can rewrite (4.2) as

$$y(t) = \text{Ne}\,[\mathbf{X}(t) \in \Re^n] \in \Re^1 \tag{4.3}$$

The mathematical operations given by (4.1) to (4.3) for a typical neuron are shown in Fig. 4.2. In the neural network literature, these computational neurons are interchangeably called neuronal PEs, neural populations, nodes, or threshold logic units.

Mathematically, the neuronal nonlinear mapping function Ne can be divided into two parts: (1) *confluence* and (2) *nonlinear activation* operations. The confluence operation provides the weighting, aggregating, and thresholding operations to the neural inputs. In order to account for the thresholding operation, we will define the augmented vectors of neural inputs and synaptic weights as follows:

$$\mathbf{X}_a(t) = [x_0(t), x_1(t), \ldots, x_i(t), \ldots, x_n(t)]^T \in \Re^{n+1}, \quad x_0(t) = 1 \tag{4.4a}$$

and

$$\mathbf{W}_a(t) = [w_0(t), w_1(t), \ldots, w_i(t), \ldots, w_n(t)]^T \in \Re^{n+1} \tag{4.4b}$$

where $w_0(t)$ introduces a thresholding (bias) term in the confluence operation. The confluence operation © essentially provides a measure of similarity between the augmented neural input vector $\mathbf{X}_a(t)$(new information) and the augmented synaptic weight vector $\mathbf{W}_a(t)$ (accumulated knowledge base). The nonlinear activation operation then performs a nonlinear mapping on the similarity measure. These two basic mathematical operations of a computational neuron, Fig. 4.3, will now be described in greater detail.

4.1.1 Confluence Operation: Measure of Similarity

From a biological perspective, the confluence operation represents the weighting of the input signals, $\mathbf{X}_a(t) \in \Re^{n+1}$, with the accumulated knowledge stored at the synapses, $\mathbf{W}_a(t)$, and the spatio-temporal aggregation of these weighted inputs, as performed by the soma. The synaptic weighting assigns a relative weight to each incoming signal component $x_i(t)$ according to an attribute of the past experience (knowledge or memory) stored in synaptic weight $w_i(t)$.

One can mathematically view this confluence operation as a linear weighted mapping from the $(n + 1)$-dimensional neural input space $\mathbf{X}_a(t) \in \Re^{n+1}$ to the one-dimensional space $u(t) \in \Re^1$. The synaptic (weighting) and somatic (aggregation and thresholding) linear mapping can be modeled as

$$u(t) = \mathbf{W}_a(t) \copyright \mathbf{X}_a(t) \tag{4.5}$$

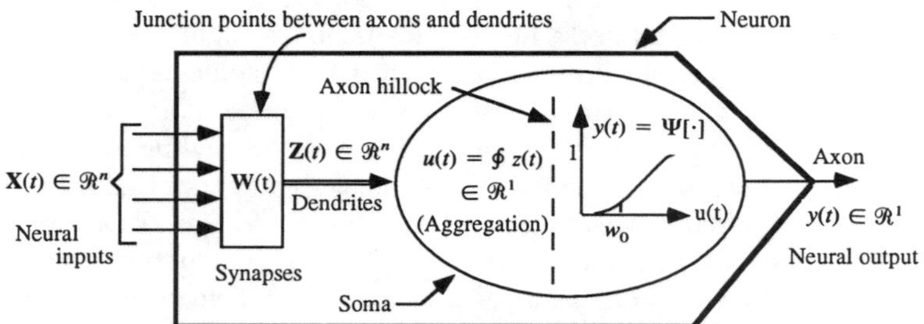

Fig. 4.2 Summary of the computational structure of a neuron: $\mathbf{X}(t) \in \Re^n$, neuronal input vector (from sensors or other neurons); $\mathbf{W}(t) \in \Re^n$, synaptic weight vector (storage of the past experience); $\mathbf{Z}(t) \in \Re^n$, dendritic input vector, $\oint$, somatic aggregation operator; $\Psi[u(t)]$, nonlinear activation operator; and w_0, somatic thresholding.

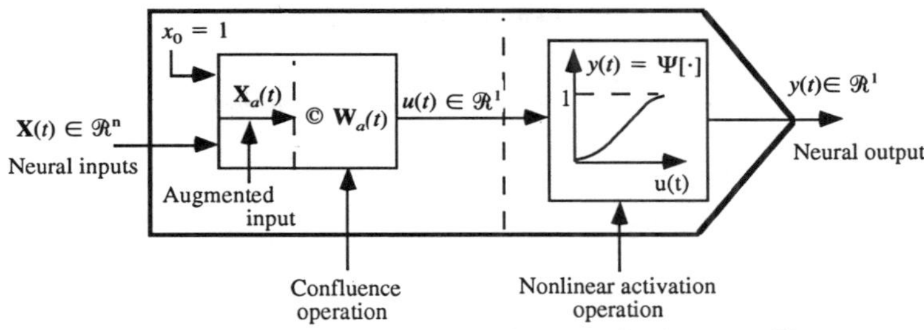

Fig. 4.3 Mathematical representation of a generalized neuron. The confluence operation © compares new neural information $\mathbf{X}_a(t)$ with the past experience stored in the synaptic weights $\mathbf{W}_a(t)$, and the nonlinear activation operation $\Psi[\cdot]$ provides a bounded neural output $y(t)$.

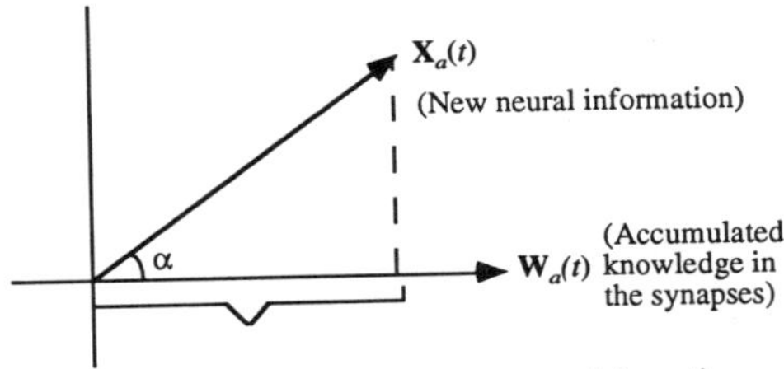

Fig. 4.4 A measure of similarity based on the projection (inner product) of the augmented neural vector $\mathbf{X}_a(t)$ onto the augmented synaptic weight vector $\mathbf{W}_a(t)$. Note that if angle $\alpha = 0°$ in the vector space, then $u(t)$ becomes a maximum value. Alternatively, if $\alpha = 90°$, then the two vectors are orthogonal and the similarity measure is $u(t) = 0$.

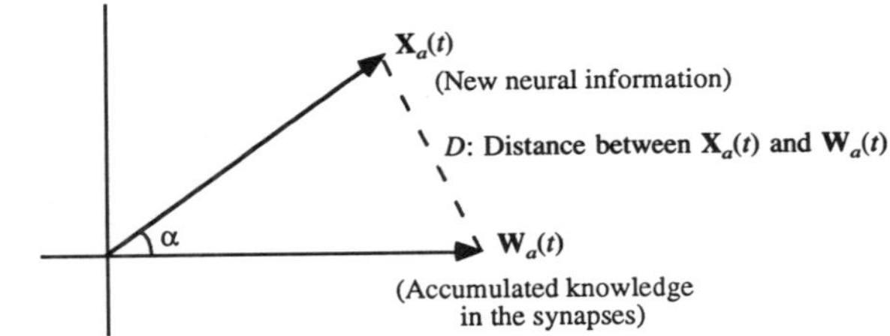

Fig. 4.5 Euclidean distance measure of similarity between the new neural information and the previously accumulated synaptic knowledge. Note that if $D = 0$, then $\mathbf{X}_a(t)$ has a lot in common with $\mathbf{W}_a(t)$, and $u(t) = 1.0$. Conversely, if $D = 1$, then $\mathbf{X}_a(t)$ and $\mathbf{W}_a(t)$ have zero commonality, yielding $u(t) = 0$.

where © is a confluence operation.* Equation (4.5) represents a measure of the similarity between $\mathbf{X}_a(t)$ (input vector) and $W_a(t)$ (synaptic weight vector). We will present two types of similarity measures: (1) the scalar (inner) product of the vectors $\mathbf{X}_a(t)$ and $\mathbf{W}_a(t)$, and (2) the Euclidean distance between vectors $\mathbf{X}_a(t)$ and $\mathbf{W}_a(t)$. The computational neurons of most neural networks described in the literature assume a confluence operation to be a scalar product. A popular exception to this is the radial basis function (RBF) network, which employs the distance measure for describing the confluence between the inputs and weights [9, 43, 48]. These two models of the confluence operation will now be described in detail.

Inner Product of $\mathbf{X}_a(t)$ *and* $\mathbf{W}_a(t)$. The inner product of $\mathbf{X}_a(t)$ and $\mathbf{W}_a(t)$ is defined geometrically as the projection of the neural inputs $\mathbf{X}_a(t)$ (new information) onto the synaptic weights $\mathbf{W}_a(t)$ (the accumulated knowledge) in the vector space, as graphically illustrated in Fig. 4.4; that is,

$$u(t) = \mathbf{W}_a(t)^T\mathbf{X}_a(t) = \sum_{i=0}^{n} w_i x_i \tag{4.6}$$

where $\mathbf{X}_a(t)$ and $\mathbf{W}_a(t)$ are defined in (4.4).

*The confluence operation defined in (4.5) is a combination of the synaptic weighting, somatic aggregating, and somatic thresholding operations. This linear weighted mapping yields a scalar output $u(t)$, which is a measure of the similarity between the augmented neural input vector $\mathbf{X}_a(t)$ and the augmented knowledge stored in the augmented synaptic weight vector $\mathbf{W}_a(t)$.

Euclidean Distance Measure Between $\mathbf{X}_a(t)$ *and* $\mathbf{W}_a(t)$. An alternative approach for measuring the similarity between the vectors $\mathbf{X}_a(t)$ and $\mathbf{W}_a(t)$ is to use the distance measure as shown in Fig. 4.5. The Euclidean distance between the new neural information $\mathbf{X}_a(t)$ and the accumulated knowledge $\mathbf{W}_a(t)$ is given by

$$D = \beta \sqrt{[\mathbf{W}_a(t) - \mathbf{X}_a(t)]^T [\mathbf{W}_a(t) - \mathbf{X}_a(t)]} \in \Re^1 \tag{4.7}$$

where β is a normalization constant such that $0 \leq D \leq 1$. The measure of similarity between $\mathbf{X}_a(t)$ and $\mathbf{W}_a(t)$ may then be defined as

$$u(t) = [1 - D] \tag{4.8}$$

4.1.2 Somatic Nonlinear Activation Function

The somatic nonlinear activation function $\Psi[\cdot]$ maps the confluence value $u(t) \in [-\infty, \infty]$ to a bounded neural output. In general, the neural output is in the range of [0, 1] for unipolar signals and [−1, 1] for bipolar signals. The nonlinear activation operator transforms the aggregate $u(t)$ into a bounded neural output $y(t)$; that is,

$$y(t) = \Psi\,[u(t)] \tag{4.9a}$$

$$= \Psi\,[\mathbf{W}_a(t) © \mathbf{X}_a(t)] \in \Re^1 \tag{4.9b}$$

TABLE 4.1 Examples of Typical Nonlinear Activation Operators $\Psi[\cdot]$

Type	Equation	Functional Form
(1) Linear	$\Psi[u(t)] = g\,u$ $g > 0$, activation gain	Ψ[u(t)], g, 0, u(t)
(2) Piecewise linear	$\Psi[u(t)] = \begin{cases} +1 \text{ if } g\,u > 1 \\ g\,u \text{ if } \lvert g\,u\rvert < 1 \\ -1 \text{ if } g\,u > -1 \end{cases}$ $g > 0$, activation gain	Ψ[u(t)], +1, 0, -1, u(t)
(3) Hard limiter	$\Psi[u(t)] = \text{sgn}[u]$	Ψ[u(t)], +1, 0, -1, u(t)
(4) Unipolar sigmoidal	$\Psi[u(t)] = \dfrac{1}{1 + \exp(-g\,u)}$ $g > 0$, activation gain	Ψ[u(t)], 1, 0.5, 0, u(t)

Many different forms of mathematical functions can be used to model the nonlinear activation function. Some of the possible geometrical shapes are shown in Table 4.1.

The sigmoidal form is a widely used activation function. However, the selection of a nonlinear function in neuronal models needs more careful study than what is presently given in the neural-network paradigms. A brief discussion of this is presented in the following subsection.

4.1.3 Comments on Nonlinear Activation Function

The proportion of neurons in a neural network that receive inputs greater than threshold can be modeled by a nonlinear transformation function $\Psi[u(k)]$, which is related to the distribution of the neural thresholds $\sigma[u(k)]$ within the neural unit **[2.4]**. If the probability distribution of these neural thresholds

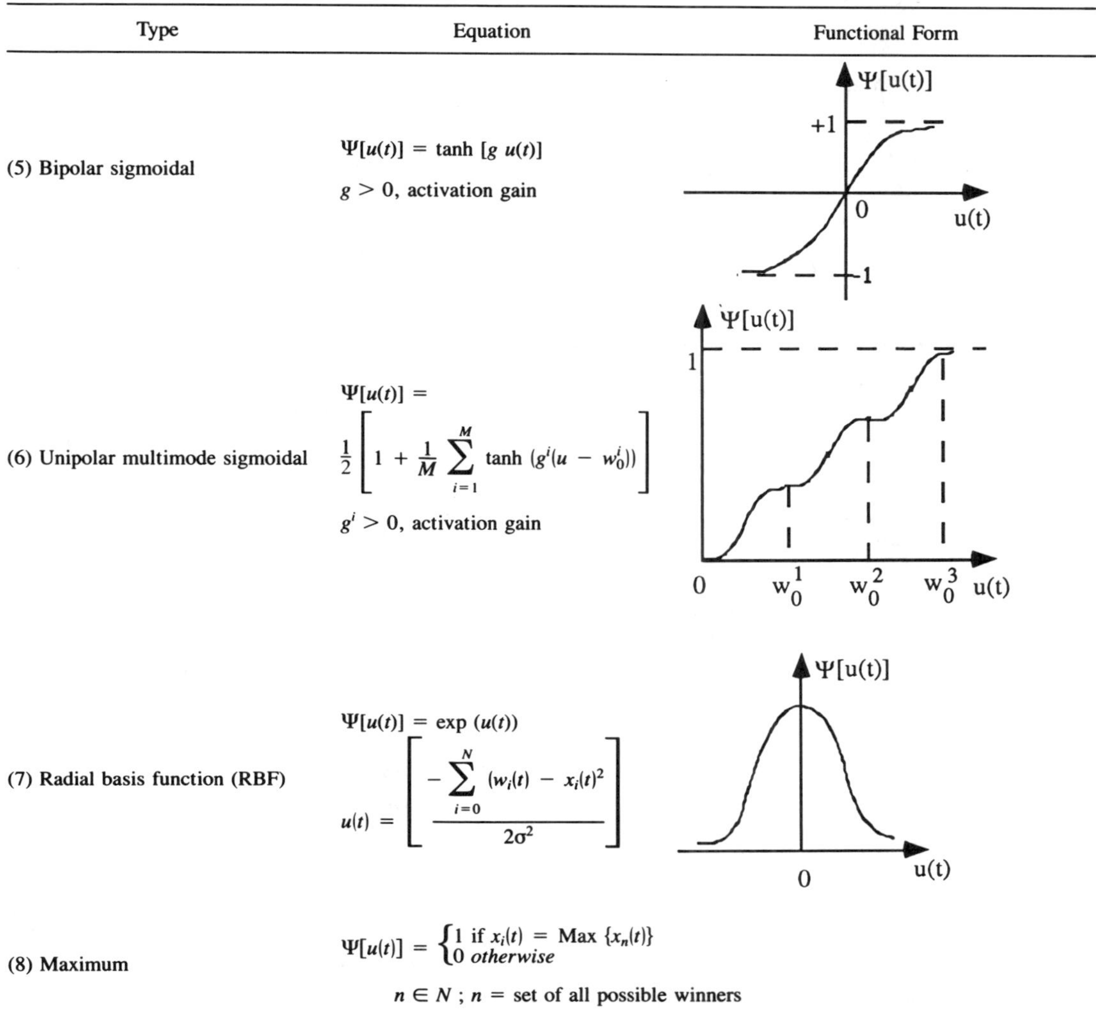

TABLE 4.1 (continued)

Type	Equation	Functional Form
(5) Bipolar sigmoidal	$\Psi[u(t)] = \tanh\,[g\, u(t)]$ $g > 0$, activation gain	
(6) Unipolar multimode sigmoidal	$\Psi[u(t)] =$ $\frac{1}{2}\left[1 + \frac{1}{M}\sum_{i=1}^{M} \tanh\,(g^i(u - w_0^i))\right]$ $g^i > 0$, activation gain	
(7) Radial basis function (RBF)	$\Psi[u(t)] = \exp\,(u(t))$ $u(t) = \left[\frac{-\sum_{i=0}^{N} (w_i(t) - x_i(t))^2}{2\sigma^2}\right]$	
(8) Maximum	$\Psi[u(t)] = \begin{cases} 1 \text{ if } x_i(t) = \text{Max } \{x_n(t)\} \\ 0 \textit{ otherwise} \end{cases}$ $n \in N$; n = set of all possible winners	

about an aggregate value w_0 is given by a *unimodal* distribution, then the nonlinear input transformation may be represented by a *sigmoidal* function. Thus, the proportion of neurons in a neural network receiving inputs greater than the intrinsic threshold may be modeled by the expression [2.4, 76]

$$y(t) = \Psi[u, g, w_0] = \int_{-\infty}^{u(t)} \sigma\,[u(t)]\,du(t) \tag{4.10}$$

where the pair $[g, w_0]$ determines the transformational properties of the function $\Psi[\cdot]$.The activation gain g is defined as the maximum slope of the sigmoidal relationship at the point of inflection w_0. In other words, for a particular distribution of neural thresholds, it is possible to determine the proportion of neurons receiving inputs exceeding the threshold by integrating the neural threshold distribution over the total applied input, eq. (4.10).

An important assumption in deriving this sigmoidal function is that each neural unit in a densely connected neural network is comprised of only one "type" of neuron. This enables the distribution of neural thresholds to be defined as a unimodal function. However, if the network is assumed to be comprised of m different types of neurons, then the distribution of thresholds must be redefined as an m-sigmoidal function that produces a nonlinear activation function with m inflection points, as depicted in Table 4.1, item (6), for $m = 3$. In general, an m-modal probability distribution for the neural thresholds is expressed as

$$\sigma[u(t)] = \frac{1}{2m}\sum_{i=1}^{m} g^i\,\text{sech}^2\,[g^i\,[u(t) - w_0^i)] \tag{4.11a}$$

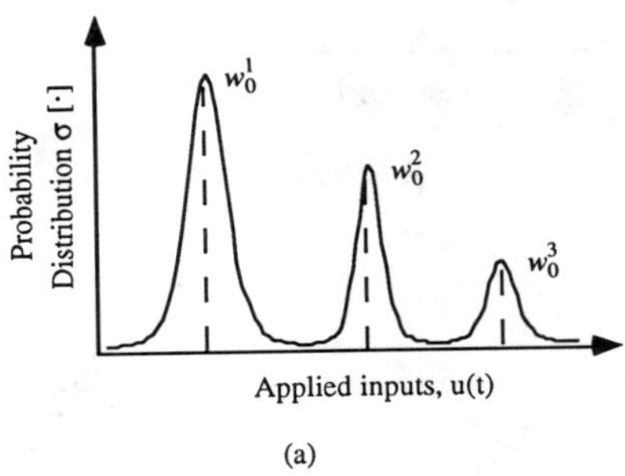

(a)

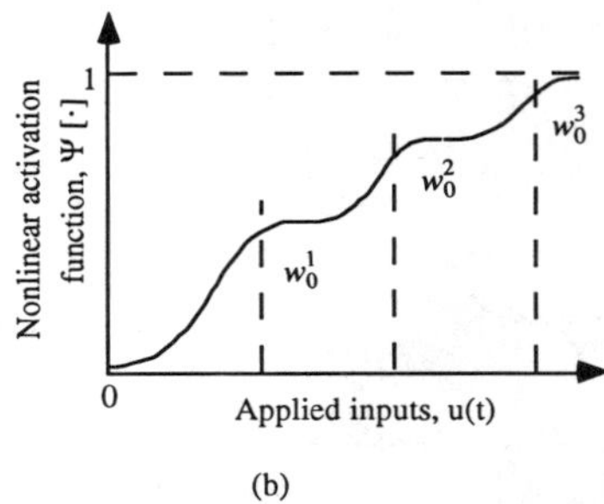

(b)

Fig. 4.6(a) Multimodal distribution of neural threshold for $m = 3$. (b) The corresponding activation function with three inflection points w_0^i, $i = 1, 2, 3$.

and the corresponding monotonically increasing input transformation function is given by

$$\Psi[u(t)] = \frac{1}{2}\left[1 + \frac{1}{m}\sum_{i=1}^{m} \tanh\left(g^i\left[u(t) - w_0^i\right]\right)\right] \tag{4.11b}$$

where for each mode there is a slope parameter g^i and a corresponding inflection point w_0^i, as depicted in Fig. 4.6.

Physiologically, a multimodal distribution would be expected to correspond to the presence of a number of distinct neural-cell types in a neural population [2.3, 2.4]. In a neural-network paradigm, it is normally assumed (though explicitly not mentioned) that the neural network consists of one type of neuron, and the corresponding distribution of thresholds is defined as a unimodal function.

Any function $\Psi[\cdot]$ is said to belong to the class of sigmoidal functions if (a) $\Psi[u(k)]$ is a monotonically increasing function of $u(t)$ in the interval $(-\infty, \infty)$; (b) $\Psi[u(t)]$ approaches or attains the asymptotic values, say, -1 and 1, as $u(k)$ approaches $-\infty$ and ∞, respectively, and; (c) $\Psi[u(k)]$ has one and only one inflection point.

Usually, to extend the mathematical operations to both the positive and negative neural outputs, (i.e., for both the positive (excitatory) and negative (inhibitory) input values of $u(t)$) the nonlinear operator is defined by the bipolar sigmoidal function as

$$\Psi[gu(t)] = \frac{e^{gu(t)} - e^{-gu(t)}}{e^{gu(t)} + e^{-gu(t)}} = \tanh\left[g\, u(t)\right] \tag{4.12}$$

where g is the activation gain. In the limit $g \rightarrow \infty$, the sigmoid function $\Psi[\cdot] = \tanh[g\, u(t)]$ tends to become a sign function. The hyperbolic tangent function provides the following properties: (a) $\Psi[\cdot]$ is a monotonically increasing function; that is, for $u_1 < u_2$

$$\Psi[u_1] \leq \Psi[u_2] \tag{4.13a}$$

(b) $\Psi[u(k)]$ is uniformly Lipschiz; that is, there exists a constant $C > 0$ such that

$$\left|\Psi[u_1] - \Psi[u_2]\right| \leq C\left|u_1 - u_2\right|, \forall\, u_1, u_2 \in \Re \tag{4.13b}$$

Observe that (a) and (b) hold iff (if and only if)

$$0 < \frac{\Psi[u_1] - \Psi[u_2]}{u_1 - u_2} \leq C, \quad \forall\, u_1,\, u_2 \in \Re, \text{ and } u_1 \neq u_2 \tag{4.13c}$$

A generalized static neural model and its mathematical operations are shown in Fig. 4.7. As shown in this figure, the first operation provides a linear mapping from $\mathbf{X}_a(t) \in \Re^n$ to $u(t) \in \Re^1$ through the weighting vector $\mathbf{W}_a(t) \in \Re^n$. The second operation provides a nonlinear mapping from $u(t) \in \Re^1$ to $y(t) \in \Re^1$ through a nonlinear activation function $\Psi[\cdot]$.

4.1.4 Learning: Adapting the Knowledge Base

The weighting and spatio-temporal aggregation operations performed by the synapses and soma, respectively, provide a similarity measure between the input vector $\mathbf{X}_a(t)$ (new neural information) and the synaptic weight vector $\mathbf{W}_a(t)$ (accumulated knowledge base). When a new input pattern that is significantly different from the previously learned patterns is presented to the neural network, the similarity between this input and the existing knowledge base is small. As the neural network learns this new pattern (by changing the strength of the synaptic weights) the distance between the new information and accumulated knowledge decreases. In other words, the purpose of learning is to make $\mathbf{W}_a(t)$ very similar to a given pattern $\mathbf{X}_a(t)$.

Most of the neural network structures (described in Fig. 3.2) undergo a "learning" procedure during which the synaptic weights (connection strengths) are adapted. Algorithms for varying these connection strengths such that learning ensues are called *learning rules.* The objective of learning rules depends on the applications. For example, the objective in pattern classification from sample data is to classify and predict successfully on new data, while the objective in control applications is to approximate nonlinear functions, and/or to make unknown systems follow the desired response. In classification and functional approximation problems, each cycle of presentation of all cases is usually referred to as a *learning epoch*. However, there has been no generalization as to how a neural network can be trained. A flow diagram illustrating the different learning algorithms [36, 44, 57] normally employed for the adaptation of synaptic weights is shown in

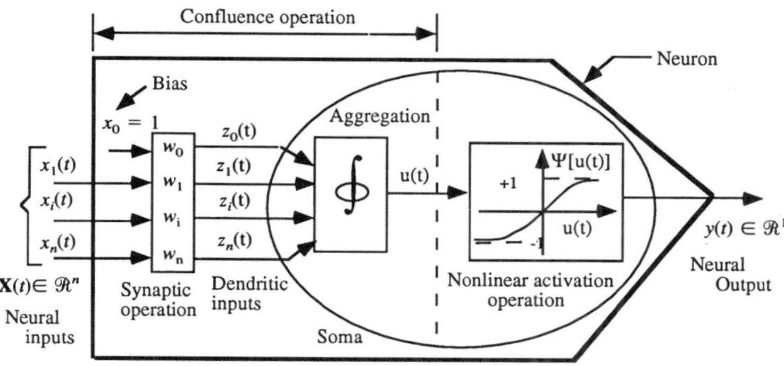

Fig. 4.7 **Mathematical representation of a generalized neuron. The confluence operation, ©, compares new neural information $\mathbf{X}_a(t)$ with the past experience stored in the synaptic weights $\mathbf{W}_a(t)$, and the nonlinear activation operation, $\Psi[\cdot]$, provides a bounded neural output $y(t)$.**

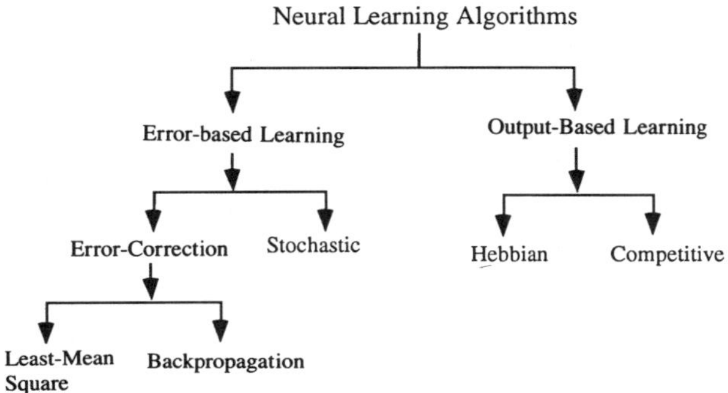

Fig. 4.8 **A flow diagram of learning algorithms employed in different neural structures to adapt the synaptic connections between neurons.**

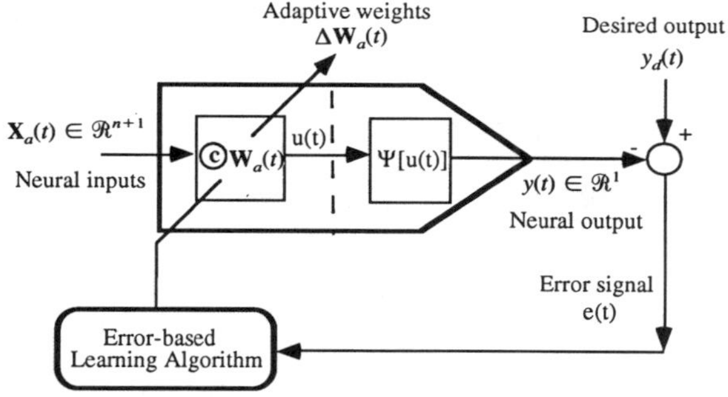

Fig. 4.9 **An error-based (supervised) learning scheme where the learning process is guided by the error signal $e(t)$.**

Fig. 4.8. As shown in this figure, learning algorithms may be broadly categorized as "error-based (supervised)" or "output-based (unsupervised)."

Error-based (also known as supervised) learning algorithms employ an external reference signal (teacher) and generate an error signal by comparing the reference with the obtained response. Based on the error signal, a neural network modifies its synaptic connections to improve the system performance. In this scheme, it is assumed that the desired answer is known *a priori*. The error-based learning procedure is schematically shown in Fig. 4.9.

Examples of error-based learning algorithms include stochastic learning and error-correction learning [44]. Stochastic learning algorithms make random changes to the synaptic weights and then determine the resultant "energy" created by this change. The changes in the weights are retained if the energy relationship for the neural layer is lowered. However, to escape local energy minima it is often necessary to accept some random changes that do not result in lower energy. An additional criterion that employs a predefined probability distribution may be used to accept the weight change, even if energy is not lowered.

A general equation for the error-based learning algorithm is

$$w_i(t+1) = w_i(t) + \Delta w_i(t) \tag{4.14a}$$

where

$$\Delta w_i(t) = \mu\, x_i(t)\, [y_d(t) - y(t)] \tag{4.14b}$$

and $w_i(t)$ is the synaptic weight corresponding to the input $x_i(t)$. The parameter $\Delta w_i(t)$ is the change in synaptic connection $w_i(t)$ over an instant in time, μ is the learning rate, $y_d(t)$ is the desired neural output, and $y(t)$ is the actual neural response. The proper selection of μ is of critical importance in these learning rules. A very small value of μ will result in extremely slow learning. On the other hand, a large value of μ will make learning faster, but it may also result in oscillations or make the system unstable.

In contrast, output-based learning algorithms do not incorporate a reference signal, and generally involve self-organization principles that rely only upon local information and internal control mechanisms in order to discover emergent collective properties. The two most important forms of output-based learning are ***Hebbian learning*** and ***competitive learning***. Hebbian learning [4, 79, 83], Fig. 4.10, involves the adjustment of a synaptic weight according to the correlation of the response of the two neurons that adjoin it. A simple Hebbian learning rule used to describe the correlation of the input $x_i(t)$ with the neuron output $y(t)$ is

$$\Delta w_i(t) = \mu\, x_i(t)\, y(t) \tag{4.15}$$

where $\Delta w_i(t)$ represents the temporal change of the synaptic weight $w_i(t)$ and μ is the learning rate. A large number of vari-

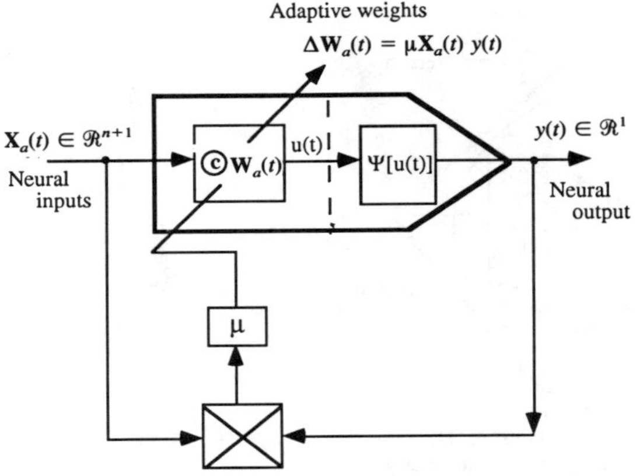

Fig. 4.10 An output-based (unsupervised) learning scheme, often called *Hebbian learning,* is guided by the neural output rather than output error as in an error-based (supervised) learning scheme.

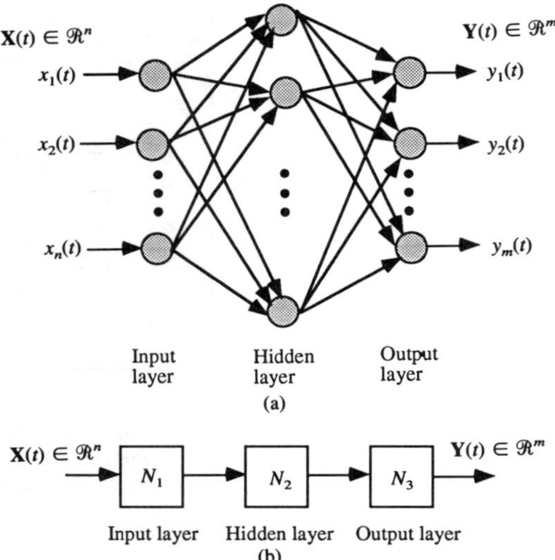

Fig. 4.11(a) A densely interconnected three-layer static neural network. Each shaded circle, or node, represents the neuron shown in Fig. 4.7. This neural network consists of an input layer (stage) with input vector, $\mathbf{X}(t) = [x_1(t), \ldots, x_i(t), \ldots, x_n(t)]^T \in \Re^n$, and the output layer with output vector $\mathbf{Y}(t) = [y_1(t), \ldots, y_i(t), \ldots, y_m(t)]^T \in \Re^m$. Layers between the input and output layers are normally referred to as hidden (intermediate) layers. (b) A block-diagram representation of a three-layer static neural network (MNN), with the input vector $\mathbf{X}(t) \in \Re^n$ and output vector $\mathbf{Y}(t) \in \Re^m$.

ations of this simple learning law are described in the neural-network literature [**3.1**, 44].

On the other hand, competitive learning algorithms are applied to neural layers that contain extensive interneuron connections. In a very simple sense, it is based on the notion of "winner take all." An input pattern is presented to the neural layer. Each neuron competes with all others by transmitting a positive signal to itself using self-excitatory recurrent connections, and sending negative signals to all its neighboring neurons via lateral inhibitory (competitive) connections. After a period of time, the neuron with the greatest activation state will remain active, called the *winner,* and all others will be nullified.

In summary, there are basically two learning rules, namely, error-based (supervised) and output-based (unsupervised) algorithms. Error-based learning algorithms need desired responses or data labeled with target results. If desired (target) results are unknown, then error-based learning algorithms are useless. This means that output-based learning algorithms become useful. Nearly all the neural networks incorporate either of these two rules or variations thereof. Some neural networks, however, have fixed weights; these networks operate by changing the activity levels of their neurons without changing the weights.

4.2 Multilayer Static Neural Networks

In the preceding subsection, the mathematical details of a single neuron were described. Although a single neuron can perform certain simple pattern-detection functions, the power of neural computation comes from the number of neurons connected in a network structure. Larger networks generally offer greater computational capabilities. Arranging neurons in layers or stages is supposed to mimic the layered structure of a certain portion of the brain. These multilayer networks have been proven to have capabilities beyond those of a single layer. The most commonly used neural-network architecture in applications such as pattern recognition, system identification, and control is the multilayer neural network (MNN) with an error backpropagation (BP) algorithm.

A typical MNN consists of an input layer, an output layer, and one hidden layer of neurons, as shown in Fig. 4.11(a). A simplified block-diagram representation of the MNN is given in Fig. 4.11(b).

The input–output mapping of the MNN shown in Fig. 4.11 can be mathematically represented by

$$\mathbf{Y}(t) = N_3\,[N_2[N_1[\mathbf{X}(t) \in \Re^n]]] \in \Re^m \tag{4.16}$$

In terms of the confluence and nonlinear activation operators, (4.16) can be rewritten as

$$\mathbf{Y}(t) = \Psi^3[\mathbf{W}_a^3(t) © \Psi^2[\mathbf{W}_a^2(t) © \Psi^1[\mathbf{W}_a^1(t) © \mathbf{X}_a(t)]]] \tag{4.17}$$

where $\Psi^i[\cdot]$ is the nonlinear activation operator, © is the confluence operator (scalar product or distance measure), and $\mathbf{W}_a^1(t)$, $\mathbf{W}_a^2(t)$ and $\mathbf{W}_a^3(t)$ are the augmented synaptic weight vectors for the input, hidden, and output layers, respectively.

All information is stored in the synaptic weights of the feedforward neural network. During the learning process, the elements of the synaptic matrices $\mathbf{W}_a^1(t)$, $\mathbf{W}_a^2(t)$ and $\mathbf{W}_a^3(t)$ are continuously updated to the new information. Supervised learning algorithms based on an error-correction procedure are often used to determine $\Delta\mathbf{W}_a^1(t)$, $\Delta\mathbf{W}_a^2(t)$ and $\Delta\mathbf{W}_a^3(t)$. One method to adapt the feedforward weights of the neurons in a static neural network is to minimize the least-mean square (LMS) error [37, 80] between the computed and desired outputs for each neuron in the network. The feedforward connections of static networks can also be updated using a gradient descent error-correction algorithm that is commonly called *backpropagation* [**5.1**, 40, 44, 81, 83]. The backpropagation procedure is employed by propagating the error backwards from the output nodes through the hidden layers to adjust the weights by a gradient descent approach. Another unsupervised learning technique can be employed by a static neural network if the feedforward weights are adapted using the Hebbian learning rule.

For multilayer static neural networks, the most popular learning rule in use is the backpropagation algorithm. Backpropagation is a generalization of the least-squares rule for a multilayer neural network. It attempts to reduce the error at each neural node in such a way that it minimizes the disturbance of the weights and improves the information content previously encoded in the weights. The error at each output node is easily determined knowing the target output for each neural node. However, for the hidden layers where outputs are internal to the network and have no explicit target output, the error calculation is much more difficult. The error at a hidden node is literally defined as the amount that its own output is responsible for the error in each neuron in the adjacent layer. It is these hidden layers that serve as abstract domains, into which inputs are mapped. These hidden layers emphasize the differences and de-emphasize similarities between inputs to allow the network to differentiate between trajectories with only subtle differences. Backpropagation can be applied to networks with any number of hidden layers by first calculating the output, determining the output error, then recursively propagating the error backwards to each layer and adapting the weights to minimize the error. The principle of the backpropagation learning algorithm may be summarized as follows [82, 83].

1. A typical backpropagation neural-network structure consists of input, hidden, and output layers. Hidden layers may have more than one layer. It is not very clear in the neural-network paradigm how many hidden layers are necessary for a particular application. There is not much computation taking place at the input layer. The number of neurons in the input layer equals the number of input vector components (measured data values). During learning (training), the input layer sends input information to all hidden nodes, as shown in Fig. 4.12(a).

2. The hidden neurons broadcast their results to all output neurons. Each output neuron calculates a weighted sum and subtracts its actual results (actual response) from its desired results (targeted output) to produce the error vector (output error). The connections active at this stage of learning are shown with thick arrows in Fig. 4.12(b).

3. The output nodes calculate the partial derivatives of error-vector components with respect to the weights, and pass these derivatives back to the hidden layer. This computation during learning gives the algorithm its name: the *backpropagation*. Each hidden neuron calculates the sum of the error derivatives to find its contribution to the output error. This backpropagation of error to the hidden layer is shown in Fig. 4.12(c). Each neuron in the hidden and output layers changes its weight according to a predetermined rule as described by (4.14).

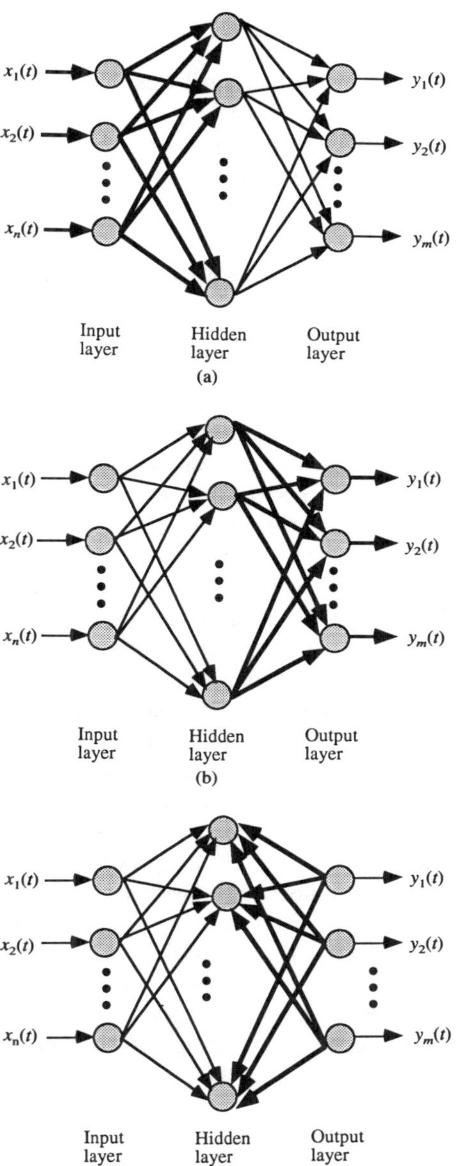

Fig. 4.12(a) During learning, the input layer sends input information to all hidden nodes, as shown with thick arrows. (b)With the input information received from the input layer, neurons in the hidden layer compute and broadcast the results to the output layer. At the output stage, the error signal is computed by subtracting the actual outputs (signals obtained from the output neurons) from the desired signals. The connections active at this stage of computation are shown with thick arrows. (c) Backpropagation of error signal to the hidden layer, which gives the algorithm its name.

As just explained, the backpropagation algorithm requires a desired response during learning to compute the error signal, and then adjusts the neural-network weights. After this initial learning, the neural network should be presented with a new, but known, set of data that was not used during the learning process. The network's accuracy with data outside the learning set give the generalization ability to the neural network, and this indicates reliability of the network. After this phase of learning and testing, the neural network can be utilized for pattern classification and to model unknown nonlinear functions and complex processes.

One of the most important attributes of neural networks is their ability to approximate nonlinear functions. The study of functional approximation occupies an important place in the neural-network paradigm. A brief introduction to the theory of functional approximation is presented in the next section.

5. FUNCTIONAL APPROXIMATION [4.1–4.6]

5.1 Introduction

Galileo said "the book of nature is written in the language of mathematics." The study of nature, for whose phenomena, constant variation, and interdependence are characterized, reduces to the concepts of variables and functions—the most important concepts of modern mathematics. But, how do we approximate such a function?

We say that a function $f(x)$ is defined by real numbers x on a set **X** if a real number y is associated to the very value of x of the set **X**, $y = f(x)$. In other words, the function is defined if (a) some set **X** of real numbers x is given, and (b) a law is prescribed according to which a real number y, $y = f(x)$, is associated with every number x of the set **X**. The aim of function approximation theory is to find a sequence of either algebraic polynomials that converge uniformly to functions continuous in the interval $[a, b]$, or trigonometric polynomials that converge uniformly to a periodic (with period 2π) continuous function. The terms *functional* and *function* are used interchangeably in the literature. The distinction is only in the difference of domain of existence of these quantities: the set of points is the domain of existence of the function, and the set of functions is the domain of existence of the functional.

The study of function approximation is of primary importance in the neural-network paradigm. The intent of this section is to provide readers with the basic concepts of the theory of functional approximation, and to briefly explain the approximation theorems that are normally used for the theoretical development of neural approximation theory.

5.2 Definition of Functional Approximation Problem

Given a point g and a set M in a normal linear space S, a point of M of minimum distance from g is called a *best approximation,* and the problem of determining such a point is called a *best approximation problem.*

To measure the quality of the approximation, one introduces a distance function $d(\Psi, f)$ to determine the distance of an approximation function $\Psi[\mathbf{W}_a, x]$ from the difference function $f(x)$, where $\mathbf{W}_a(t)$ is the augmented vector of synaptic weights. The approximation problem can then be stated as follows.

> If $f(x)$ is a continuous function defined on set S, and $\Psi[\mathbf{W}_a, x]$ is an approximating function that depends continuously on x, the approximation problem is to determine the parameter $\mathbf{W}_a^*$ such that

$$d[\Psi[\mathbf{W}_a^*, x], f(x)] \leq d[\Psi[\mathbf{W}_a, x], f(x)] \tag{5.1}$$

A solution to this problem, if it exists, is said to be a best approximation. The following points are of primary importance in a functional approximation problem [85].

- The existence of the best approximation
- The uniqueness of the best approximation
- The characterization of the best approximation
- The construction of methods for determining the best approximations

5.3 Basic Theorems of Functional Approximation

Recently, many researchers have proved that multilayer static (feedforward) neural networks can approximate arbitrary continuous functions [**4.1–4.6**] to the desired degree of accuracy. Either the Stone–Weierstrass theorem or the Kolmogorov theorem has been employed for the theoretical development of functional approximation capabilities of neural networks. Here, we briefly state these two theorems.

5.3.1 The Stone–Weierstrass Theorem

Theorem: Let domain Ω be a compact space of N-dimensions, and let $\Re$ ([a, b]) be a set of continuous real-valued functions on Ω, defined on the interval $[a, b]$ with the norm of $f \in \Omega$ ([a, b]) defined by

$$\|f\| = \sup_t \left\{ |f| : t \in [a, b] \right\} \tag{5.2}$$

(The notation "sup" denotes *supremum.*)

The approximation theorem of Stone–Weierstrass [85–87, **4.5**] states that:

If a function $f(x)$ is continuous in the interval $a \leq x \leq b$ and $\epsilon > 0$, then we can find a function $g(x)$ such that the inequality

$$|f(x) - g(x)| < \epsilon \tag{5.3}$$

would hold for values of x in this interval, satisfying the following criteria.

1. Identity function: The constant function $f(x) = 1$ is in $\Re$.
2. Separability: For any two points $x_1 \neq x_2$ in Ω, there is an f in $\Re$ such that $f(x_1) \neq f(x_2)$.
3. Algebraic closure: If f and g are any two functions in $\Re$, then $f \cdot g$ and $af + bg$ are in $\Re$ for any two real numbers a and b.

Then $\Re$ is dense in $C(\Omega)$, the set of continuous real-valued functions on Ω. In other words, for any $\epsilon > 0$ and any function in $C(\Omega)$, there is a function f in $\Re$ such that

$$|g[x] - f[x]| < \epsilon, \quad \forall [x] \in \Omega \tag{5.4}$$

5.3.2 The Kolmogorov Theorem

This theorem states that one can express continuous functions defined on an n-dimensional cube by sums and super-

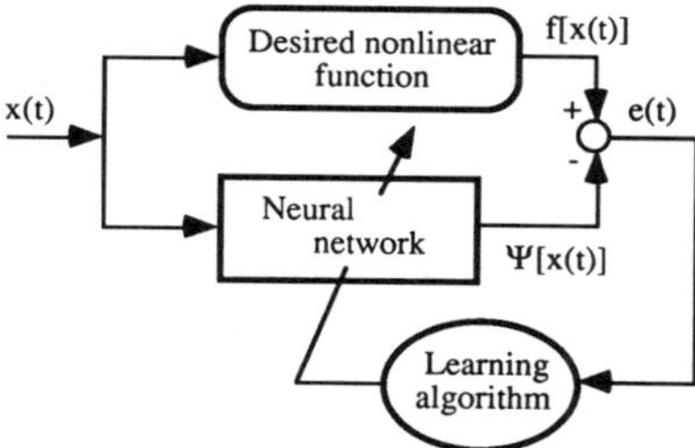

Fig. 5.1 A general learning scheme for function approximation using neural networks.

positions of continuous functions of single variable functions [91–94]. Furthermore, the number of single-variable functions required is finite.

Theorem: There exist fixed continuous increasing functions $\phi_{pq}(x)$ on $I = [0, 1]$ such that each continuous function f on I^m can be expressed in the form

$$f(x_1, \ldots, x_n) = \sum_{q=1}^{2n+1} g_q\left(\sum_{p=1}^{n} \phi_{pq}(x_p)\right) \tag{5.5}$$

where the g_q are properly chosen continuous functions of one variable.

Sprecher showed that ϕ_{pq} could be replaced by $\lambda_q\phi_p$. This yields a modified Kolmogorov's Theorem, which is stated as follows.

There exist constants λ_p and fixed continuous increasing functions $\phi_{pq}(x)$ on $I = [0, 1]$ such that each continuous function f on I^m can be written in the form

$$f(x_1, \ldots, x_n) = \sum_{q=1}^{2n+1} g_q\left(\sum_{p=1}^{n} \lambda_q \phi_p(x_p)\right) \tag{5.6}$$

where g is a properly chosen continuous function of one variable.

The most significant characteristic of neural networks is their ability to approximate arbitrary nonlinear functions. This ability of neural networks has made them useful to model nonlinear systems, which is of primary importance in the synthesis of nonlinear controllers. Neural networks potentially offer a general framework for modeling and control of nonlinear systems. The problem of learning a mapping using neural networks between an input and an output space is equivalent to the problem of estimating the system that transforms inputs and outputs given a set of examples of input–output pairs. Training a neural network using input–output data from a nonlinear dynamic system can be considered as a *nonlinear functional approximation problem* [**4.2,** 84], as depicted in Fig. 5.1.

It is proved by Funahashi [**4.1**], Hornik et al. [**4.3**], Cybenko [**4.4**], Cotter [**4.5**], and Blum and Li [88], using the Weierstrass theorem as a basis, that a continuous function can be well approximated by a static neural network with one hidden layer, where each neuron in the hidden layer has a continuous sigmoidal nonlinearity. Gallant and White [89] showed that a static neural network with a single hidden layer using the monotone "cosine activation function" is capable of embedding a Fourier network, which yields a Fourier series approximation to a given function at its output. Such networks thus possess all the approximation properties of a Fourier series representation. In particular, these networks are capable of approximation, to any degree of accuracy, of any square integrable function on a compact set using an infinite number of hidden units [**4.3**]. Cardaliaguet and Euvrard [90] developed a noise-resistant approximation formula for a function and its derivative. They also addressed the limitations of neural-network architecture on the accuracy of function approximation. Hecht-Nielsen [91], Cotter and Guillerm [92], and Kurkova [93] employed the Kolmogorov theorem to demonstrate the function approximation capabilities of static networks. However, it has been recently pointed out by Hornik et al. [**4.3**] and Girosi and Poggio [94] that Kolmogorov's theorem requires a different nonlinear processing function for each unit in the network, and that functions in the second hidden layer depend on the function being approximated.

Although many significant results are published in the literature demonstrating that multilayer feedforward (static) neural networks can approximate any arbitrary nonlinear functions, many questions remain unanswered. For example, what is the relationship between the accuracy of the function being approximated with the number of hidden layers? Chester [95] has pointed out that neural nets with two hidden layers appear to provide higher accuracy and better generalization than a network with a single hidden layer. More work needs to be done in this area to generalize this characteristic of static neural networks. Another important question is, how many neurons are required in hidden layers to achieve a desired degree of accuracy?

The theory of functional approximation using static neural networks has been extensively studied. As a result, the static networks have been used to represent dynamic systems. Rao and Gupta [**4.6**] used linear and trigonometric polynomials to develop the functional approximation theory of the dynamic neural units. The use of dynamic networks to represent dynamic systems is of more significance and practical importance than using static neural networks. However, this involves the development of dynamic neural-network structures, approximation theory, and the necessary learning algorithms. In the next section, we attempt to address some of these problems.

6. Dynamic (Feedback) Neural Networks [5.5, 5.6, 7.1]

6.1 Introduction

In the preceding sections, we presented a brief description of single- and multilayer feedforward neural-network structures, the structures without any feedback. This class of neural networks are called static, feedforward, or nonrecurrent neural networks. Such networks have no dynamic memory as the response of the network depends on its current inputs and the

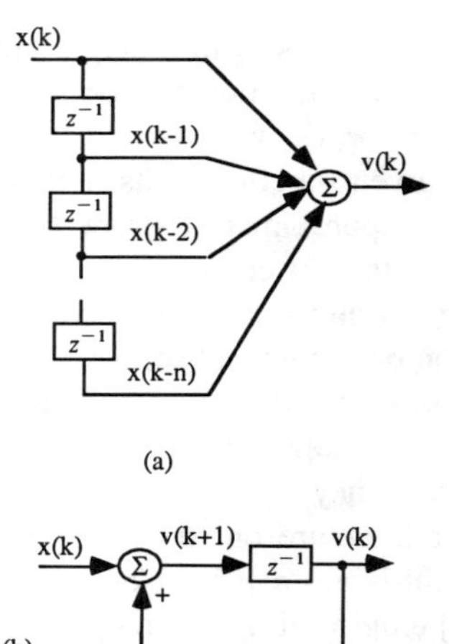

Fig. 6.1 (a) An infinite-order feedforward (FIR) structure (k denotes discrete time index). (b) An equivalent feedback structure with a single pole.

values of synaptic weights. Such networks, since they do not have any feedback, are inherently stable.

The feedback interactions do occur in feedfoward networks during the learning phase. The adjustment of snyaptic weights during learning reduces the output error as the learning trials increase. This feedback interaction is externally imposed rather than occurring within the neural structure.

It is well established that feedforward neural networks can approximate nonlinear functions to a desired degree of accuracy. This attribute of feedforward neural networks has made many researchers use them to model dynamic systems. However, these networks suffer from many limitations [79].

In this section, we describe neural structures with feedback. Such neural networks are known as dynamic, feedback, or recurrent neural networks. These networks not only provide some robust computing characteristics, but also bring about greater insights into biological neural structures. The dynamics in neural networks, or dynamic neural computing, does provide some functional basis of the cerebellum and its associated circuitry. Also, the dynamic neural networks can offer great computational advantages over purely static neural networks. For example, it is well known that an infinite-order FIR (finite impulse response) filter, which is only a feedforward network, is equivalent to a single-pole IIR (infinite impulse response) filter, shown in Fig. 6.1.

From Fig. 6.1(a), the response of an FIR filter with n delay lines may be written as

$$v(k) = x(k) + x(k-1) + x(k-2) + \dots + x(k-n) = \sum_{i=0}^{n} x(k-i),\ n \to \infty \tag{6.1}$$

Alternatively, in transfer function form we may write

$$\frac{V(z)}{X(z)} = 1 + z^{-1} + z^{-2} + \dots + z^{-n}, \quad n \to \infty, \quad |z| < 1 \tag{6.2}$$

The difference, or recursive, equation that describes the behavior of a first-order IIR structure, shown in Fig. 6.1(b), may be written as

$$v(k+1) = x(k) + v(k). \tag{6.3}$$

Alternatively, in transfer function form we may write

$$\frac{V(z)}{X(z)} = \frac{1}{1 - z^{-1}} = 1 + z^{-1} + z^{-2} + \dots + z^{-n}, \quad n \to \infty, \quad |z| < 1 \tag{6.4}$$

From (6.2) and (6.4), it is clear that the two structures are functionally equivalent. From a computational viewpoint, a network with feedback is equivalent to a large, or possibly an infinite, feedforward structure.

Unlike the static neural networks, a dynamic neural network employs extensive feedback between the neurons. This feedback implies that the network has local memory characteristics. The node equations in dynamic networks are described by differential or difference equations. Typically, a dynamic neural network employs feedforward inputs and output feedback, as shown in Fig. 6.2. Mathematically, this neural layer can be described in discrete time as

$$\mathbf{Y}(k) = \Psi\,[\,\mathbf{W}_a^1(k) © \mathbf{X}_a(k) + \mathbf{W}_a^2(k) © \mathbf{Y}_a(k-1)\,] \tag{6.5}$$

where © is the confluence operator (dot product or distance measure), k is the discrete instant in time, $\Psi[\cdot]$ is the nonlinear activation operator that may have one of the functional forms given in Table 4.1, $\mathbf{X}_a(k) \in \Re^{n+1}$ is the augmented neural input vector at time k, $\mathbf{Y}_a(k-1) \in \Re^{m+1}$ is the augmented output vector at time k, $\mathbf{W}_a^1(k)$ is the augmented feedforward synaptic weight matrix, and $\mathbf{W}_a^2(k)$ is the feedback augmented synaptic weight matrix. The generalized topology of a dynamic neural network is illustrated in Fig. 6.3.

Because feedback neural networks have feedback paths from their outputs to the inputs, the response of such networks is *dynamic* or *recursive;* that is, after applying a new input, the output is calculated and fed back to modify the input. The output is then recalculated, and the process is repeated. For a stable network, successive iterations produce smaller and smaller output changes until eventually the outputs become constant. Under some situations, the process may never end, and such networks are said to be unstable. Unstable networks have interesting properties, and one example of such a network is the *chaotic system.*

Neural architectures with feedback are particularly appropriate for system modeling (identification), control, and filtering applications. These networks are important because many of the systems that we system scientists wish to model in the real world are nonlinear dynamic systems. Examples of control systems that we wish to model are the forward or inverse dynamics of systems such as airplanes, rockets, spacecraft, and robots.

In this section, we describe two types of dynamic neural architectures. The first type is developed as an extension of

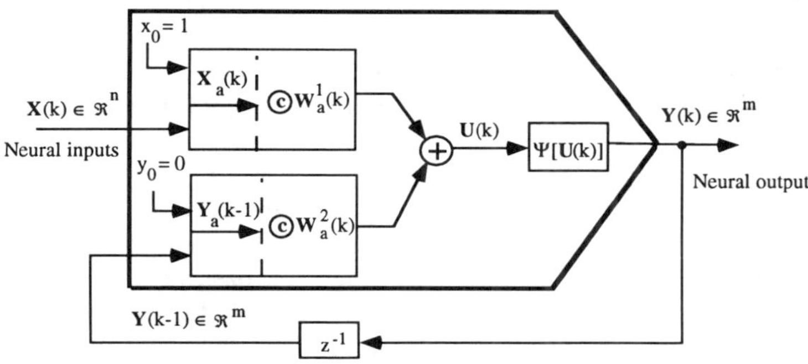

Fig. 6.2 An example of a dynamic neural network with feedforward inputs $\mathbf{W}_a^1(k) \copyright \mathbf{X}_a(k)$ and output feedback $\mathbf{W}_a^2(k) \copyright \mathbf{Y}_a(k-1)$. z^{-1} is the unit time-delay operator.

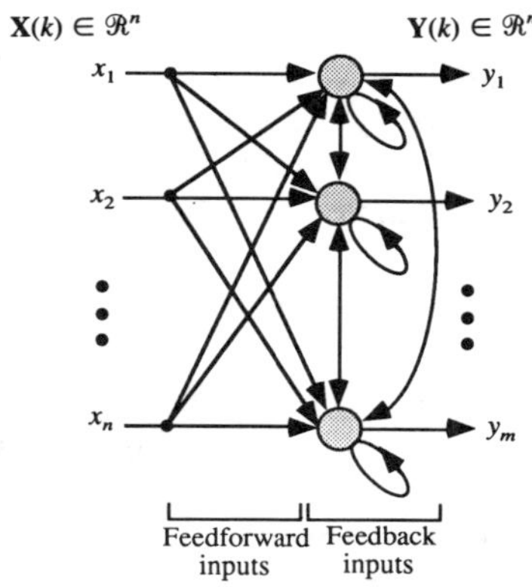

Fig. 6.3 A generalized topology of a dynamic neural network with extensive feedforward and feedback inputs. The feedforward inputs arise from a source outside the neural layer, whereas the feedback inputs are a result of dense lateral, self-excitatory, and self-inhibitory connections between the neurons in the layer. This representation of the dynamic neural network is equivalent to Fig. 6.2.

static networks described in Section 6.2. In this category, we briefly describe two types, namely, recurrent and time-delay neural structures. In Section 6.3, the second type is developed based on the physiological evidence that neural activities in the CNS are dependent on the interaction of excitatory and inhibitory neural subpopulations [**2.3, 2.4,** 10]. Finally, in Section 6.4, we briefly describe two neural structures (what we call unconventional neural structures), namely, a neuron with hysteresis phenomenon and the Cerebellar Model Articulation Controller (CMAC).

6.2 Extension of Static Neural Networks

6.2.1 Recurrent Neural Networks

The recurrent neural structure introduced by Hopfield [155] provided an alternative model to a static neural network. This structure consists of a single layer network included in a feedback configuration with a time delay, as shown in Fig. 6.4(a). This feedback network represents a discrete-time dynamical system and can be described by the following equation

$$y(k + 1) = \Psi [w(k) \cdot y(k)],\ x(0) = x_0 \tag{6.6a}$$

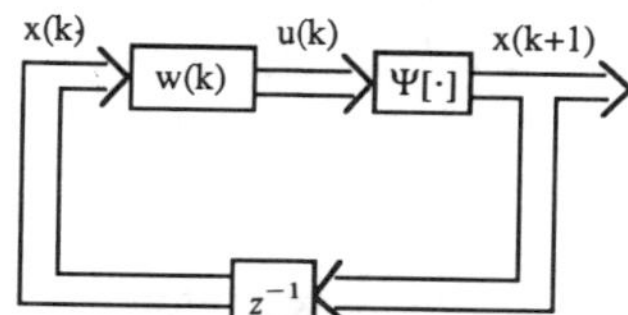

Fig. 6.4(a) The state-space model of the Hopfield neural structure. $y(k)$ and $y(k+1)$ represent the states of the neural network at k and $k+1$ instants of time, x_0 represents the initial value, $w(k)$ denotes the vector of neural weights, and z^{-1} represents the unit delay operator.

where $y(k)$ and $y(k+1)$ represent the states of the neural network at k and $k+1$ instants of time, x_0 represents the initial value, $w(k)$ denotes the vector of neural weights, and $\Psi[\cdot]$ is the nonlinear activation function.

Given an initial value x_0 the dynamic system evolves to an equilibrium state if $\Psi[\cdot]$ is suitably chosen. The set of initial conditions in the neighborhood of x_0 that converge to the same equilibrium state is then identified with that state. The term *associative memory* is used to describe such systems. The feedback networks with or without constant inputs are merely nonlinear dynamical systems, and the asymptotic behavior of such systems depends on the initial conditions, specific inputs as well as on the nonlinear function. Readers will find the computational and stability aspects of this network as applied to system identification and control of nonlinear dynamic systems in [**5.5**].

The single-layer recurrent neural network consists of n computing elements (neurons) having thresholds w_{0i} as depicted in Fig. 6.4(b).

The feedback input to the i-th neuron is equal to the weighted sum of neural outputs y_j, where $j = 1, 2, \ldots, n$. Denoting w_{ij} as the weight value connecting the output of the j-th neuron with the input of the i-th neuron, we can express the total input u_i of the i-th neuron as

$$u_i = \sum_{\substack{j=1 \\ j\neq 1}}^{n} w_{ij} y_{j+x_i} - w_{0i}, \quad i = 1, 2 \ldots, n. \tag{6.6b}$$

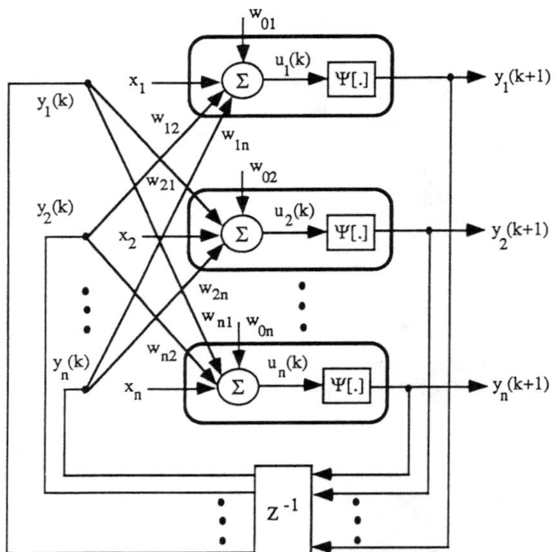

Fig. 6.4(b) Single-layer feedback neural network.

In vector form, (6.6b) can be rewritten as

$$u_i = \mathbf{w}_i^T y + x_i - w_{0i},\ i = 1,2, \ldots, n \tag{6.6c}$$

$$\text{where } \mathbf{w}_i \stackrel{\Delta}{=} \begin{bmatrix} w_{i1} \\ w_{i2} \\ \cdots \\ w_{in} \end{bmatrix} \text{ and } \mathbf{y} \stackrel{\Delta}{=} \begin{bmatrix} y_1 \\ y_2 \\ \cdots \\ y_n \end{bmatrix}.$$

The linear portion of the recurrent neural network can be described in matrix form as

$$\mathbf{U} = \mathbf{W}\,\mathbf{y} + \mathbf{X} - \mathbf{w}_0 \tag{6.6d}$$

$$\text{where } \mathbf{U} \stackrel{\Delta}{=} \begin{bmatrix} u_1 \\ u_2 \\ \cdots \\ u_n \end{bmatrix},\ \mathbf{X} \stackrel{\Delta}{=} \begin{bmatrix} x_1 \\ x_2 \\ \cdots \\ x_n \end{bmatrix} \text{ and}$$

$$\mathbf{w}_0 \stackrel{\Delta}{=} \begin{bmatrix} w_{01} \\ w_{02} \\ \cdots \\ w_{0n} \end{bmatrix}.$$

The matrix **W** in Eqn. (6.6d), also called *connectivity matrix*, is an $(n \times n)$ matrix and may be written as

$$\mathbf{W} = \begin{bmatrix} 0 & w_{12} & w_{13} & \cdots & w_{1n} \\ w_{21} & 0 & w_{23} & \cdots & w_{2n} \\ w_{31} & w_{32} & 0 & \cdots & w_{3n} \\ \cdot & \cdot & \cdot & \cdots & \cdot \\ \cdot & \cdot & \cdot & \cdots & \cdot \\ w_{n1} & w_{n2} & w_{n3} & \cdots & 0 \end{bmatrix}.$$

This matrix is symmetrical, i.e., $w_{ij} = w_{ji}$, and with diagonal entries equal to zero, $w_{ii} = 0$ indicating that no connection exists from any neuron back to itself. This condition is equivalent to the lack of the self-feedback in the neural structure as shown in Fig. 6.4(b).

6.2.2 *Time-Delay Neural Networks (TDNN)*

It is possible to use a static network to process time series data by simply converting the temporal sequence into a static pattern by unfolding the sequence over time. That is, time is treated as another dimension in the problem. From a practical point of view, it is possible to unfold the sequence over a finite period of time. This can be accomplished by feeding the input sequence into a tapped delay line and then into a static neural-network architecture. An architecture like this is often referred to as a time-delay neural network (TDNN) [154]. This neural structure is basically a feedforward network with dynamic (delay) elements. Because of the delay operators, the TDNN is categorized under dynamic neural networks. This is equivalent to a FIR filter whose output forms an argument to a nonlinear function (Fig. 6.5).

The basic time-delay neural unit shown in Fig. 6.5 can function as an adaptive filter by computing the scalar product of the augmented input vector $\mathbf{X}_a(k)$ and augmented synaptic weight vector $\mathbf{W}_a(k)$. The elements of the synaptic weight vector $\mathbf{W}_a(k)$ are modified by the least-mean square (LMS) learning algorithm [37]. This neural unit is adapted (trained) using noise-contaminated samples for which the correct uncontaminated signal values $y_d\,(k)$ are known. In other words, the desired neural output is known for each input sample.

This neural structure can be used to either amplify or attenuate each of the frequency components of the incoming time series $\mathbf{X}_a(k)$. This approach will do a nearly perfect job if the desired signal and noise do not overlap in frequency. However, if the signal and noise do overlap in frequency, then the process will also tend to filter out the signal. In other words, this structure will significantly degrade the system performance.

The dynamics of a time-delay neural unit are described by the following equations:

$$v(k) = \sum_{i=0}^{n} w_i\, x(k - i) \tag{6.7a}$$

$$y(k) = \Psi\,[v(k)] \tag{6.7b}$$

This neural structure has been used in many applications. For example, Sejnowski and Rosenberg [160] and TDNN for text-to-speech conversion, while Waibel et al. [154] employed this structure for phoneme recognition.

6.2.3 *Dynamic Neural Unit (DNU)*

The dynamic neural unit (DNU), proposed by Gupta and Rao [**5.6**], is a dynamic model of the biological neuron. The neural structure was motivated by the fact that biological neuronal systems always function with continuous feedback. One example of such a system is the reverberating circuit in the neuronal pool of the CNS [58], which functions as follows: An incoming signal stimulates the first neuron, which then stimulates the second, third, and so forth. However, branches return to the first neuron, providing feedback and restimulate it, as depicted in Fig. 6.6.

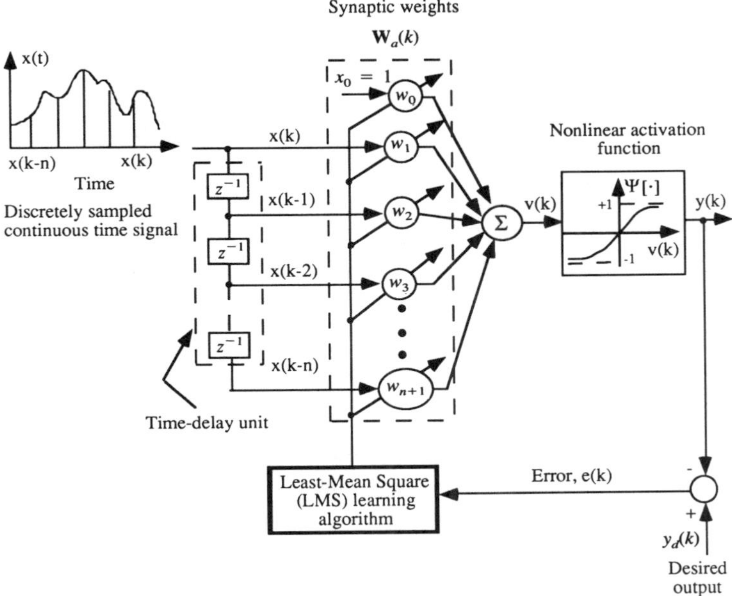

Fig. 6.5 A time-delay unit based on a static neural architecture. The feed-forward weights $\mathbf{W}_a(k)$ are adjusted based on the least-mean square (LMS) learning algorithm.

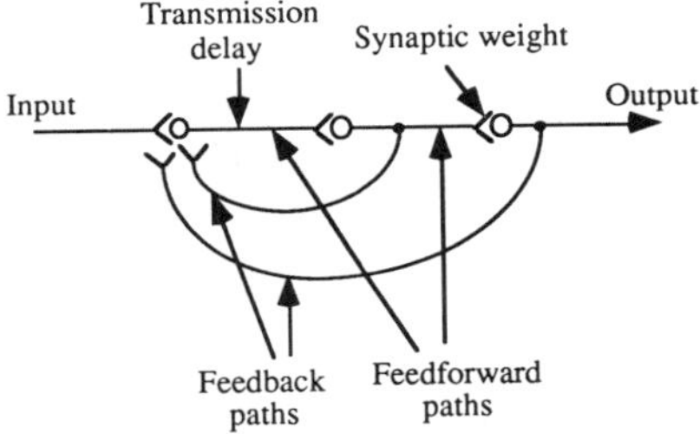

Fig. 6.6 A reverberating circuit in a neuronal pool for the central nervous system (CNS).

The reverberating circuit is the basis of innumerable CNS activities, for it allows a single input signal to elicit a response lasting a few seconds, minutes, or even hours. Almost all rhythmic muscular activities, including the rhythmic movements of walking, are mainly controlled by reverberating circuits.

The dynamic structure of the DNU (Fig. 6.7) is only analogous to that of the reverberating circuit and does not represent any specific anatomical region within the biological nervous system. The output of this dynamic structure constitutes the argument to a time-varying nonlinear activation function. Thus, the DNU performs two distinct operations: (a) the synaptic operation and (b) the somatic operation. The first operation corresponds to adaptation of feedforward and feedback synaptic weights, while the second corresponds to adaptation of gain (slope) of the nonlinear activation function. The DNU consists of delay elements, and feedforward, and feedback paths weighted by the synaptic weights $\mathbf{a}_{ff}$ and $\mathbf{b}_{fb}$, respectively—representing a second-order structure followed by a nonlinear activation function.

The linear dynamics of the DNU can be expressed in the form of a transfer relation

$$w(k, \mathbf{a}_{ff}, \mathbf{b}_{fb}) = \frac{v_1(k)}{s(k)} = \frac{[a_0 + a_1 z^{-1} + a_2 z^{-2}]}{[1 + b_1 z^{-1} + b_2 z^{-2}]} \tag{6.8}$$

where $s(k) = [\sum_{i=1}^{n} w_i s_i - \theta]$ is the neural input to DNU, $s_i \in \Re^n$ are the inputs from other neurons or from sensors, $w_i \in \Re^n$ are the corresponding input weights, θ is the bias term, $v_1(k) \in \Re^1$ is the output of the dynamic structure, $u(k) \in \Re^1$ is the neural output, and $\mathbf{a}_{ff} = [a_0, a_1, a_2]^T$ and $\mathbf{b}_{fb} = [b_1, b_2]^T$ are the vectors of adaptable feedforward and feedback weights, respectively. Alternatively, (6.8) may be described by the following difference equation

$$v_1(k) = -b_1 v_1(k-1) - b_2 v_1(k-2) + a_0 s(k) + a_1 s(k-1) + a_2 s(k-2) \tag{6.9}$$

Let us define the vectors of signals and adaptable weights of the DNU as

$$\Gamma(k, v_1, s) = [v_1(k-1) \quad v_1(k-2) \quad s(k) \quad s(k-1) \quad s(k-2)]^T \tag{6.10}$$

and

$$\zeta(\mathbf{a}_{ff}, \mathbf{b}_{fb}) = [-b_1 \; -b_2 \; a_0 \; a_1 \; a_2]^T \tag{6.11}$$

where the superscript T in these equations denotes transpose. Using (6.10) and (6.11), (6.9) is rewritten as

$$v_1(k) = \zeta^T(\mathbf{a}_{ff}, \mathbf{b}_{fb})\, \Gamma(k, v_1, s) \tag{6.12}$$

The nonlinear mapping operation on $v_1(k)$ yields a neural output $u(k)$ given by

$$u(k) = \Psi[g_s v_1(k)] \tag{6.13}$$

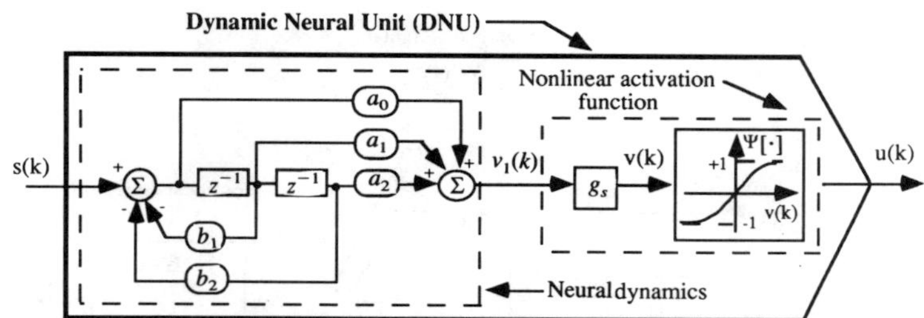

Fig. 6.7 The basic structure of DNU consists of synaptic and somatic components. z^{-1} is the unit delay operator, k is the discrete-time index, and $\mathbf{a}_{ff} = [a_0, a_1, a_2]^T$ and $\mathbf{b}_{fb} = [b_1, b_2]^T$ are the adjustable feedforward and feedback weights, respectively.

where $\Psi[\cdot]$ is a nonlinear activation function (usually the sigmoidal function) and g_s is the somatic gain that controls the slope of the activation function. Many different forms of mathematical functions can be used to model the nonlinear behavior of the biological neuron. We use a bounded, monotonically increasing and differentiable sigmoidal activation function. To extend the mathematical operations to both the excitatory and inhibitory inputs, the bipolar activation (sigmoidal) function is defined over $[-1,1]$ as

$$\Psi[v(k)] = \frac{[e^{(g_s v_1)} - e^{-(g_s v_1)}]}{[e^{(g_s v_1)} + e^{-(g_s v_1)}]} = \tanh[g_s v_1(k)] \qquad (6.14)$$
$$= \tanh[v(k)]$$

where $v(k) = g_s v_1(k)$. Details of this neural structure, the development of the learning algorithm, and its application to the control of unknown nonlinear dynamic systems may be found in [**5.6**].

6.3 Dynamic Neural Structures Based on Neural Subpopulations

The computational neural-network structures described in the literature are often developed based on the concept of a single static neuron. These neural morphologies are almost a parody of their biological counterparts, for they do not reflect many of the functional properties of biological neurons observed in the neurophysiological experiments. For example, the static neural model does not take into account time delays that affect the dynamics of the system; inputs produce an instantaneous output with no memory involved. Furthermore, it does not include the effects of synchronism or the frequency modulation function of biological neurons. Biological neurons continually aggregate, on the average, up to 10,000 synaptic inputs, which do not add up in a simple linear manner. Each neuron is a complex computing element, and they perform much more computation than just summation.

The neural-network structures described in the earlier sections consider the behavior of a single neuron as the basic computing unit for describing neural information-processing operations. Each computing unit in the network is based on an idealized neuron. An ideal neuron is assumed to respond optimally to the applied units. However, experimental studies in neurophysiology show that the response of a biological neuron appears random [**2.3, 2.4**], and only by ensemble-averaging many observations is it possible to obtain deterministic results. In general, a biological neuron is a stochastic mechanism for processing information. However, mathematical analysis has shown that these random cells can transmit reliable information if they are sufficiently redundant in numbers. It is postulated [**2.4**], therefore, that the collective activity generated by large numbers of locally redundant neurons is more significant in a computational context than the activity generated by a single neuron.

Furthermore, the study of neurons at the individual cell level may be appropriate to emulate some functions of the biological neural network. However, it is not necessarily suited for investigations of those parts that are associated with higher cortical functions, such as sensory information processing and the attendant complexities of learning, memory storage, and vision. This shift in emphasis is warranted since the sensory information is introduced into the nervous system in the form of large-scale spatio-temporal activity in sheets of cells (tissue layers); the number of cells involved is simply too large for any approach starting at the single cell to be tractable [**2.4**].

The total neural activity generated within a tissue layer is a result of spatially localized assemblies of densely interconnected nerve cells called *neural populations,* or *neural mass.* The neural population is composed of neurons, and its properties have a generic resemblance to those of individual neurons. But it is not identical to them and its properties cannot be predicted from measurements on single neurons. This is due to the fact that the properties of a neural population depend on various parameters of individual neurons and also on the interconnections between neurons. The study of neural networks based on single-neuron analysis precludes these two facets of biological neural structures. The conceptual gap between the functions of single neurons and those of numbers of neurons—a neural mass—is still very wide.

Each neural population may be further divided into several coexisting *subpopulations.* A subpopulation contains a large class of similar neurons that lie in close spatial proximity. The neurons in each subpopulation are assumed to receive a common set of inputs and provide corresponding outputs. The individual synaptic connections within any subpopulation are random, but dense enough to ensure that at least one mutual connection exists between any two neurons. The most common neural mass is the mixture of excitatory (positive) and in-

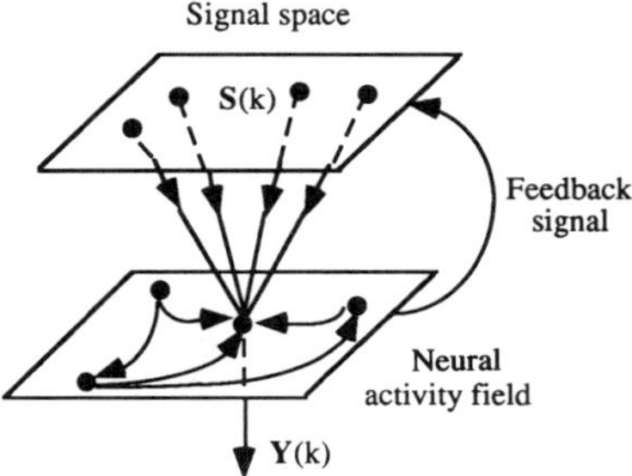

Fig. 6.8 Schematic diagram of a neural activity field **Y**(k) in response to a signal space **S**(k); this structure represents the functional dynamics of a nervous-tissue layer.

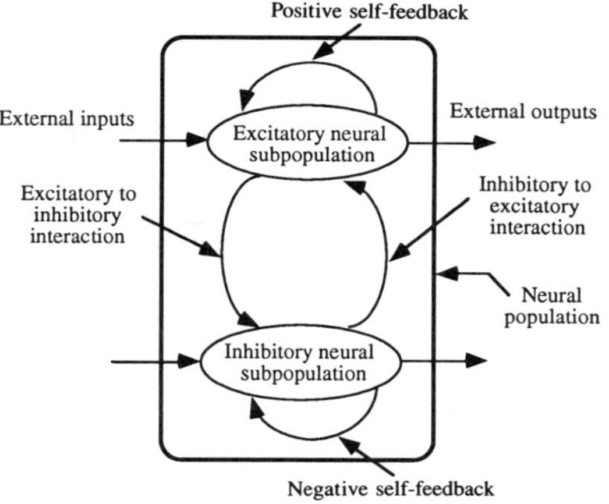

Fig. 6.9 A schematic diagram of the interactions between excitatory and inhibitory neural subpopulations within a neural population.

hibitory (negative) subpopulations of neurons. The excitatory neural subpopulation increases the electrochemical potential of the postsynaptic neuron, while the inhibitory subpopulation reduces the electrochemical potential. The individual neurons within the subpopulations that generate **Y**(k) receive stimuli from other neurons in the nervous-tissue layer, self-feedback signal, and from a common signal space **S**(k) external to the tissue layer, as depicted in Fig. 6.8.

The minimum topology of a neural mass contains excitatory (positive), inhibitory (negative), excitatory–inhibitory (synaptic connection from excitatory to inhibitory), and inhibitory–excitatory (synaptic connection from inhibitory to excitatory) feedback loops. The morphology of a neural mass is shown in Fig. 6.9. One of the important attributes of a neural mass is that some of its fundamental characteristics may be described in terms of linear systems theory [**2.3**]. This property implies that the operations in a neural mass, within appropriate limits of amplitude, conform to the principle of superposition that makes the responses to two or more inputs additive.

Strumillo and Durani have used this neural model to study the cardiac arrhythmia [156]. Based on this hypothesis, Gupta and Knopf proposed a neural model named *PN* (positive–negative) *neural processor* [157–159] for machine vision applications, and Rao and Gupta proposed the *dynamic neural processor* (DNP) for robotics and control applications [**7.1**, 161]. A brief description of these two dynamic neural structures will now be given.

6.3.1 PN Neural Processor

The computational role performed by the PN neural processor emulates the spatio-temporal information-processing capabilities of certain neural activity fields found along the human visual pathway. The state-space model of this visual information processor corresponds to a bilayered two-dimensional array of densely interconnected nonlinear PEs. An individual PE represents the neural activity exhibited by a spatially localized subpopulation of excitatory (positive) or inhibitory (negative) nerve cells. The feedforward connections to each PE transforms the external signal information into informative features. Correspondingly, the feedback connections between the various PEs within the array generate dynamic characteristics that are useful for performing a variety of computational roles such as spatio-temporal filtering (STF), motion detection, spatio-temporal stabilization (STS), short-term visual memory (STVM), content-addressable memory (CAM), and pulse-frequency modulation (PFM). Computationally efficient vision systems can be developed if numerous such neuro-vision processors are incorporated within a pyramidal computing architecture that utilizes both parallel and serial information processing techniques. Potential applications for such active neuro-vision systems include intelligent robotics, automated surveillance, medical robots, and manufacturing systems. A diagram for the state-space model of the PN neural processor is shown in Fig. 6.10.

The state evolution of the PEs within a PN neural processor is given by the first-order difference equation

$$\mathbf{X}(k+1) = \alpha\mathbf{X}(k) + (1-\alpha)\mathbf{Y}(k) \tag{6.15}$$

where α is the rate of decay in the present state, $\mathbf{X}(k) \in [0,1]$, and $(1-\alpha)$ is the rate of growth in new state activity due to the inputs, $\mathbf{Y}(k) \in [0, 1]$. If $\alpha \rightarrow 0$, then the new state $\mathbf{X}(k+1)$ is directly proportional to the present input state $\mathbf{Y}(k)$. The state input $\mathbf{Y}(k)$ is given by

$$\mathbf{Y}(k) = \Phi[\mathbf{U}(k)] \tag{6.16}$$

where $\Phi[\cdot]$ is a nonlinear mapping operator that transforms the unbounded inputs received by the PEs, $\mathbf{U}(k) \in [-\infty, \infty]$, to a bounded input, $\mathbf{Y}(k) \in [0, 1]$. The total applied input to the various PEs of the neuro-vision processor is given by

$$\mathbf{U}(k) = \mathbf{AS}(k) + \mathbf{WX}(k) \tag{6.17}$$

Two types of inputs are received by the constituent PEs. One type arises from the external inputs originating in the signal space $\mathbf{S}(k)$, and the second type originates from the lateral and recurrent connections between the PEs within a common network $\mathbf{X}(k)$. The spatial transformation of the external input signal is given by the matrix $\boldsymbol{A}$ of the feedforward subnet, and the strength of lateral and recurrent connections between individual PEs is given by the matrix $\boldsymbol{W}$ of the feedback subnet.

A variety of information-processing operations associated with the early stages of machine vision can be realized by the basic PN neural processor architecture. These diverse operations are achieved by selectively programming the system matrices $\boldsymbol{A}$ and $\boldsymbol{W}$, and the parameters of the nonlinear mapping

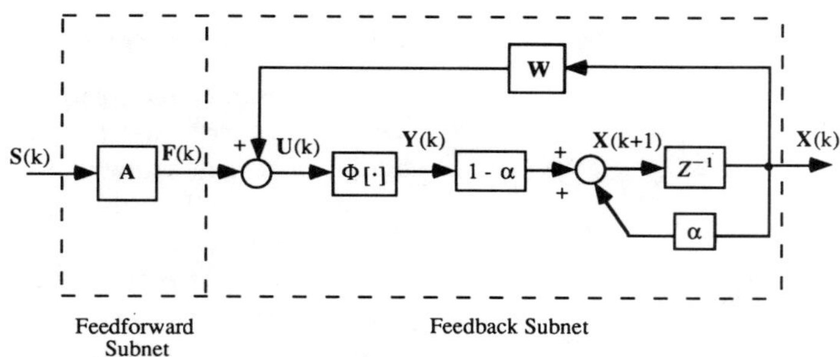

Fig. 6.10 The state-space model for describing the information-processing capabilities of the multitask PN neural processor.

operator $\Phi[\cdot]$. The coefficients of $\boldsymbol{A}$ in the feedforward subnet are made to act as a linear spatial filter that either smooth or enhance features embedded within the external input $\mathbf{S}(k)$. The coefficients selected for $\boldsymbol{W}$ in the feedback subnet, in conjunction with properly chosen parameters for $\Phi[\cdot]$, will determine the computational role of the processor in terms of STF, motion detection, STVM, STS, PFM.

Control applications of the PN neural processor have not yet been reported.

6.3.2 Dynamic Neural Processor (DNP)

The DNP is developed based on physiological evidence that nervous activity of any complexity depends on the interaction of the excitatory and inhibitory neural subpopulations. The DNP thus functionally mimics the aggregate dynamic properties of a neural population or neural mass. The DNP seeks to replicate, on a small scale, the natural and powerful information-processing attributes of biological systems, such as coordinate transformations and learning and adaptation. The DNP consists of two basic nodes called the dynamic neural units (briefly described in Section 6.2.3) and are coupled in excitatory and inhibitory modes. The architecture of the DNU embodies delay elements—feedforward and feedback weights followed by a time-varying nonlinear activation function; thus, it is different from the conventionally assumed structure of neurons. The morphology of a DNP is shown in Fig. 6.11.

In this structure, $s_\lambda(k)$ and $u_\lambda(k)$ represent respectively the neural stimulus (input) and neural output of the computing unit, where the subscript λ indicates either an excitatory (E) or inhibitory (I) state. $s_{t\lambda}(k)$ denotes the total input to the neural unit, $w_{\lambda\lambda}$ is the strength of self-synaptic connections (w_{EE}, w_{II} in Fig. 6.11), and $w_{\lambda\lambda}'$ is the strength of cross synaptic connections from one neural unit to another (w_{IE}, w_{EI} in Fig. 6.11). The z^{-1} elements denote communication delays in the self- and the intersynaptic paths.

The functional dynamics of the neural computing unit, the DNU, is defined by a second-order difference equation as represented by (6.9). The state variables $u_{\mathrm{E}}(k+1)$ and $u_{\mathrm{I}}(k+1)$ generated at time $(k+1)$ by the excitatory and inhibitory neural units are modeled by the nonlinear functional relationships

$$u_{\mathrm{E}}(k+1) = \mathbf{E}\,[u_{\mathrm{E}}(k), v_{\mathrm{E}}(k)] \tag{6.18a}$$

and

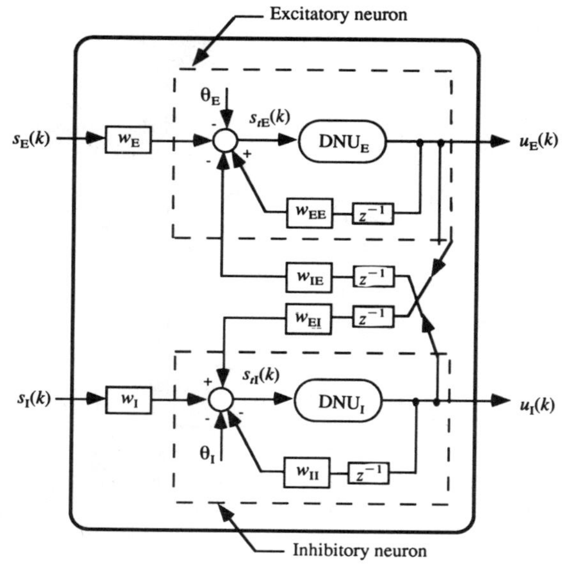

Fig. 6.11 The dynamic neural processor with two dynamic neural units coupled as excitatory and inhibitory neurons represented as $\mathrm{DNU_E}$ and $\mathrm{DNU_I}$, respectively. w_{EE} and w_{II} represent the strength of self-synaptic connections, while w_{IE} and w_{EI} are the interneuron connections. w_{E} and w_{I} are the input weights for excitatory and inhibitory neural inputs $s_{\mathrm{E}}(k)$ and $s_{\mathrm{I}}(k)$, respectively. $u_{\mathrm{E}}(k)$ and $u_{\mathrm{I}}(k)$ represent the responses of the excitatory and inhibitory neural subpopulations. θ_{E} and θ_{I} are the thresholds of excitatory and inhibitory neurons, respectively.

$$u_{\mathrm{I}}(k+1) = [u_{\mathrm{I}}(k), v_{\mathrm{I}}(k)] \tag{6.18b}$$

where $v_{\mathrm{E}}(k)$ and $v_{\mathrm{I}}(k)$ represent the proportion of neurons in the neural unit that receive inputs greater than an intrinsic threshold, and **E** and **I** represent the nonlinear excitatory and inhibitory actions of the neurons. The neurons that receive inputs greater than a certain threshold value are given by a nonlinear function of $v_\lambda(k)$, $\Psi\,[v_\lambda(k)]$. The total inputs incident on the excitatory and inhibitory neural units are, respectively,

$$s_{t\mathrm{E}}(k) = w_{\mathrm{E}}s_{\mathrm{E}}(k) + w_{\mathrm{EE}}u_{\mathrm{E}}(k-1) - w_{\mathrm{IE}}u_{\mathrm{I}}(k-1) - \theta_{\mathrm{E}} \tag{6.19a}$$

and

$$s_{t\mathrm{I}}(k) = w_{\mathrm{I}}s_{\mathrm{I}}(k) - w_{\mathrm{II}}u_{\mathrm{I}}(k-1) + w_{\mathrm{EI}}u_{\mathrm{E}}(k-1) - \theta_{\mathrm{I}} \tag{6.19b}$$

where w_{E} and w_{I} are the weights associated with the excitatory and inhibitory neural inputs, respectively, w_{EE} and w_{II} repre-

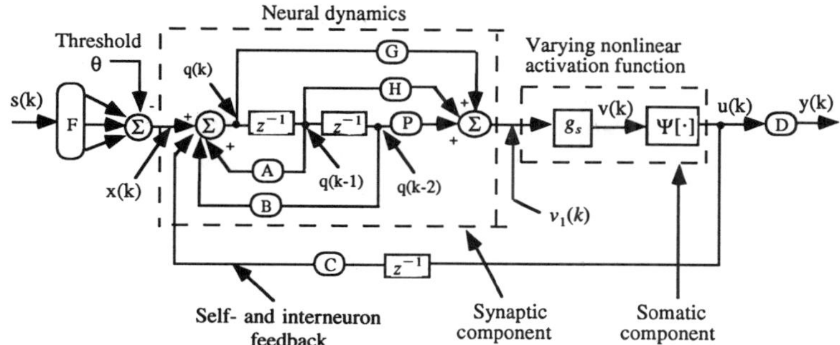

Fig. 6.12 A generalized neural model based on excitatory–inhibitory neuronal morphology. The feedforward synaptic matrices *G, H, P,* and the feedback synaptic matrices *A, B* form the neural dynamics with an internal threshold θ. The output of this dynamic structure forms an argument to a nonlinear activation function with a varying slope g_s. The matrix *C* denotes the self- and interneuron feedback strength. The matrices *F* and *D* represent the scaling matrices of the input and output signals, respectively.

sent the self-synaptic connection strengths, w_{IE} and w_{EI} are the interneuron synaptic strengths, and Φ_E and Φ_I are the thresholds of excitatory and inhibitory neurons, respectively. Equations (6.19a) and (6.19b) may also be written in a matrix form as

$$\begin{bmatrix} s_{tE}(k) \\ s_{tI}(k) \end{bmatrix} = \begin{bmatrix} w_E & 0 \\ 0 & w_I \end{bmatrix} \begin{bmatrix} s_E(k) \\ s_I(k) \end{bmatrix} + \begin{bmatrix} w_{EE} & -w_{IE} \\ w_{EI} & -w_{II} \end{bmatrix} \begin{bmatrix} u_E(k-1) \\ u_I(k-1) \end{bmatrix} - \begin{bmatrix} \theta_E \\ \theta_I \end{bmatrix} \quad (6.20)$$

A direct analytical solution for determining the steady-state and temporal behavior exhibited by the DNP is not possible because of the inherent nonlinearities in (6.19a) and (6.19b). However, these nonlinear equations can be analyzed qualitatively by obtaining the phase trajectories in the $u_E - u_I$ phase plane. These trajectories enable the system characteristics to be observed without solving the nonlinear equations. The locus of points where the phase trajectories have a given slope is called an *isocline curve*. The steady-state activity exhibited by DNUs of the neural processor can be investigated by determining the isocline curves corresponding to $u_E(k+1) = u_E(k)$ and $u_I(k+1) = u_I(k)$. This study needs further investigations.

The DNP has been used for dynamic computational purposes, such as functional approximation, computation of inverse kinematic transformations of robots, and control of unknown nonlinear dynamic systems [**7.1**, 142, 163]. The learning algorithm to modify the self- and inter-feedback synaptic weights is discussed in [**7.1**, 142].

6.3.3 A Generalized Dynamic Neural Model

It is necessary in the neural-network paradigm to develop general computational neural morphologies that can better represent the characteristics and emulate the functional capabilities of biological neural structures. In this section, we develop a generalized neural model based on the concept of neural mass described earlier. It is demonstrated in this section that the existing neural models, such as feedforward (static) neural networks, feedback (recurrent) neural networks, TDNN, and DNU, are a subclass of this generalized model.

A generalized dynamic model is schematically shown in Fig. 6.12. The model consists of second-order dynamics whose output forms an argument to a time-varying nonlinear activation function. This part of the generalized model is nothing but the DNU discussed earlier. The generalized neural model is an extension of the DNU, and incorporates delayed feedback that represents the soft (adaptable) connectivity between subpopulations of neurons.

As shown in Fig. 6.12, the feedforward synaptic matrices are denoted as *G, H, P,* while the feedback synaptic matrices *A, B* form the neural dynamics with an internal threshold θ. The matrix *C* denotes the self- and inter-neuron feedback strength. The matrices *F* and *D* represent the scaling matrices of the input and output signals, respectively. In the conventional static neural networks, the matrix *F* would represent the synaptic weights. The output of the neural dynamics forms an argument to a nonlinear activation function, usually sigmoidal, with a varying slope. This adaptation in sigmoidal slope, what we call somatic adaptation, provides a self-tuning feature to the neural model. The effect of somatic adaptation on neural-network behavior has been discussed in [**5.6**, 122, 162].

The functional dynamics describing neural architecture are represented by the following difference equations:

$$x(k) = \mathrm{F}\, s(k) - \theta \quad (6.21a)$$

$$q(k) = x(k) + \mathrm{A}\, q(k-1) + \mathrm{B}\, q(k-2) + \mathrm{C}\, u(k-1) \quad (6.21b)$$

$$v_1(k) = \mathrm{G}[q(k)] + \mathrm{H}\,[q(k-1)] + \mathrm{P}[q(k-2)] \quad (6.21c)$$

$$v(k) = g_s v_1(k) \quad (6.21d)$$

$$u(k) = \Psi[v(k)] \quad (6.21e)$$

$$y(k) = \mathrm{D}\, u(k) \quad (6.21f)$$

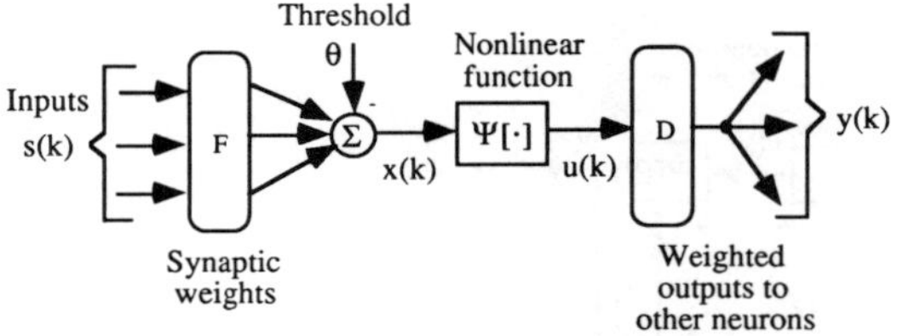

Fig. 6.13 The structure of a static neuron as a special case of the dynamic structure shown in Fig. 6.12.

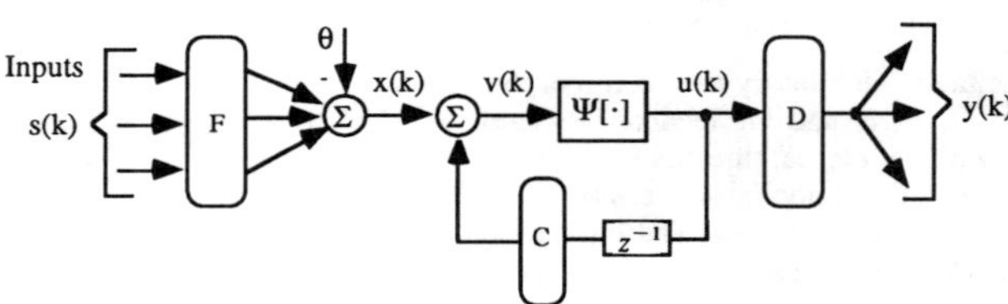

Fig. 6.14 A feedback (recurrent) neural network as a special case of the dynamic structure shown in Fig. 6.12.

We will now show in the following paragraphs that the existing neural morphologies can be derived from this generalized neural model.

Feedforward (Static) Neural Network. As has been described earlier, the static structure of an artificial neuron receives its inputs from a number of other neurons or from the sensors. A weighted sum of these inputs constitutes the argument of an ''activation'' function. The resulting value of the activation function, if it exceeds an internal threshold θ, is the neural output. This output is distributed along weighted connections to other processing units. The static neural network is a subset of the generalized structure with $A = B = C = H = P = C = 0$, $G = 1$, and $g_s = 1$ (nonlinear function with constant slope), as shown in Fig. 6.13.

Feedback (Recurrent) Neural Network. The conventional dynamic neural structure, shown in Fig. 6.14, can be obtained as a special case of the generalized structure shown in Fig. 6.12, with $A = B = H = P = 0$, $G = 1$, and $g_s = 1$.

Time-Delay Neural Network (TDNN). The dynamics of a time-delay neural network can be described as a special case of generalized dynamic structure, shown in Fig. 6.12, with $A = B = C = 0$, and $g_s = 1$. These equations lead to a TDNN structure as shown in Fig. 6.15.

Dynamic Neural Unit (DNU). The structure of the DNU is identical to that shown in Fig. 6.12 except that there is no feedback path from the neural output, $u(k)$, to the input; that is, $C = 0$. The DNU structure thus obtained is shown in Fig. 6.16.

It is envisaged that the neural models developed based on the concept of neural population may present a better representation of biological neural systems.

6.4 Some Unconventional Neural Structures

6.4.1 Neuron with Hysteresis Phenomenon

Based on an analogy with the hysteresis effect observed in ferromagnetic materials, Cragg and Temperley [178] suggested that an assembly of localized neurons situated in the cerebral cortex could produce multistable neural activity in the form of hysteresis behavior. Functionally, hysteresis was suggested as a physiological basis for short-term memory. In addition, some physiological evidence [**2.4**] suggests that the activity generated by certain neural populations exhibits spatially inhomogeneous stable states. Such a possibility is evident, for any input of sufficient intensity and duration will cause the activity in the neural population to jump from the lowest (resting) state into one of the stable excited states. The neural activity will reverberate, or circulate, within the closed-loop circuitry of the population such that the overall response will remain relatively constant, even if the stimulus is altered or removed. This ongoing neural activity is maintained by the dense excitatory feedback among the nerve cells in the population. In this way, any stimuli can be recalled while this reverberation continues, and it will continue until a strong inhibitory influence changes the reverberating activity.

The hysteresis phenomenon is of significant importance in biological vision system. A classic problem in psychology is to investigate how visual information obtained from a temporally discontinuous series of eye fixations [179] is integrated into a coherent view of a scene. One biological plausibility is that certain neural structures along the visual pathway generate a memory image that temporarily preserves the information that is consistently present within the successive fixations. In other words, a form of brief neural memory is required to temporally store and spatially reconcile separate glimpses of a common scene [180]. A plausible neural mechanism for this type of a stabilizing memory is based on the hypothesis that neurons in certain tissue layers within the brain exhibit localized hysteresis phenomenon [**2.4,** 180, 182, 183]. Fender and Julesz [184] demonstrated that hysteresis operative in the fusion of binocularly presented patterns produces single vision.

At a cellular level, Hoffman [182, 183] postulated that the firing rate of individual neurons may exhibit hysteresis char-

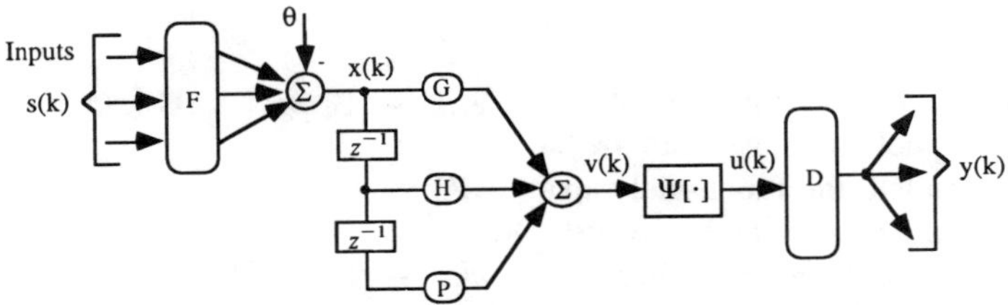

Fig. 6.15 A time-delay neural network (TDNN) as a special case of Fig. 6.12.

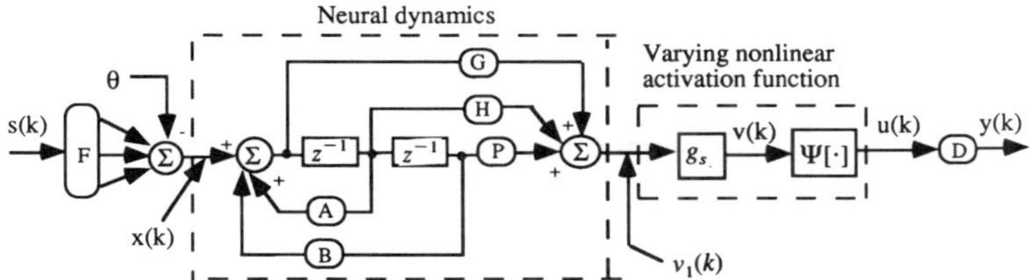

Fig. 6.16 The structure of DNU as a special case of Fig. 6.12.

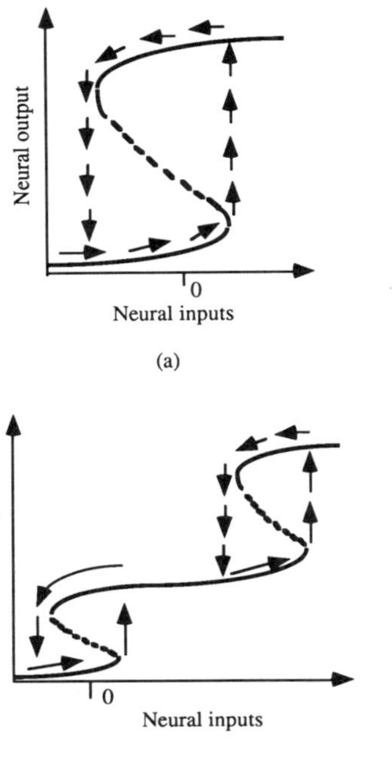

Fig. 6.17 (a) Hysteresis characteristic of a neuron with a unimodal nonlinear activation function. (b) Hysteresis characteristic of a neuron with a multimodal nonlinear activation function [Table 4.1, 158]. Solid lines indicate stable states, while the dashed lines indicate unstable states.

acteristics as the input level to the neuron is varied, Fig. 6.17. Based on this, a neural-network concept derived from an analogy between the immune system and the CNS was proposed by Hoffmann and Benson [185]. The theory is based on the hypothesis that a neuron exhibits hysteresis at the single-cell level. This neural-network architecture is different from the conventional approach in that a network of such neurons (that exhibit hysteresis) can learn without any changes in synaptic strengths. The learning occurs as a natural consequence of interactions between the network and its environment until a point in phase-plane is reached such that the network's responses are appropriate for dealing with the stimuli. Due to the hysteresis association with each neuron, the network tends to stay in the region of phase space where it is located.

The neural network with each neuron exhibiting hysteresis characteristics has been implemented for a machine-vision task by Gupta and Knopf [76]. No control applications have yet been reported.

6.4.2 Cerebellar Model Articulation Controller (CMAC)

One part of the brain that seems to be intimately involved in motor control processes is the cerebellum. The cerebellum has long been called a silent area of the brain [58], principally because electrical excitation of the cerebellum does not cause any sensation and rarely any motor movement. However, removal of cerebellum does make motor movements highly abnormal.

The cerebellum collects continual information from different pathways about the instantaneous physical status of all parts of the body even though the cerebellum is operating at a subconscious level. The summated signal of these inputs constitutes an address, the contents of which are the appropriate muscle actuator signals required to carry out the desired movement. The cerebellum automatically assesses the rate of movement and calculates the length of time that will be required to reach the point of intention. Then appropriate inhibitory signals are transmitted to the motor cortex to inhibit the agonist muscle and to excite the antagonist muscle. In this way, appropriate "brakes" are applied to stop the movement at the precise point of intention. The result is a smooth trajectory of the limb or hand through space.

Motivated by the functional capabilities of the cerebellum, Albus [65] proposed a novel architecture called the *Cerebellar Model Articulation Controller* (CMAC). The CMAC uses the encoding system to approximate nonlinear functions, and can perform these approximations much faster than networks using sigmoids. Only a small subset of the parameters are adjusted at each point in the input space, which makes the CMAC robust to the order in which training data are presented. The CMAC principle is depicted in Fig. 6.18.

Identical to the functioning of the cerebellum, when the input vector $\mathbf{X} = [x_1, x_2, \ldots, x_n]^T$ is presented to the sensory cells, it is mapped into an association cell vector **A**. The response cell adds up values of all the weights attached to active association cells, denoted as **A***, to produce the output vector **Y**. Only the components of the vector of active cells, **A***, affect this sum. The input vector **X** can be treated as an address, and the response vector **Y** as the contents of that address. By adjusting the weights attached to association cells in **A***, it is possible to change the output vector **Y** for any given input vector **X**. Association cells constitute the hidden layer of the CMAC.

The CMAC computes control functions by referring to a table rather than by solution of analytic equations. Functional

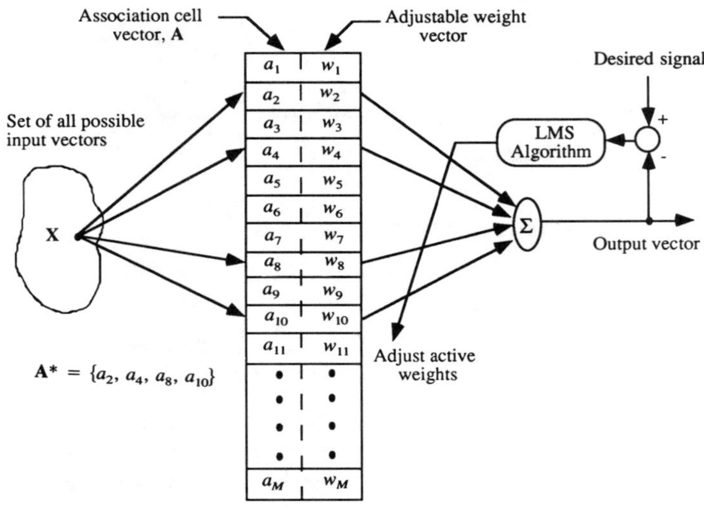

Fig. 6.18 A block diagram of the CMAC system in which the input vector **X** is mapped onto an association cell vector **A**. **A*** is the set of all active (nonzero) elements in **A**. The sum of all weights attached to active cells produces the response. The active weights can be adjusted using the least-mean square (LMS) algorithm.

values are stored in a distributed fashion such that the value of the function at any point in input-space is derived by summing the contents over a number of memory locations. The CMAC is thus a memory-management technique that causes similar inputs to tend to generalize so as to produce similar outputs; yet dissimilar inputs result in outputs that are independent [65]. The CMAC reduces computational calculations per output even when there is a large number of weights. Because of this property, the CMAC can be trained in considerably less time, which is of significant importance for real-time control. Miller et al. **[7.10]**, who have referred to the CMAC as the Cerebellar Model Arithmetic Computer, have shown that the convergence time of the CMAC is very low compared to that of a multilayer neural network with the backpropagation learning algorithm. Furthermore, since the CMAC uses a linear output stage, the principle of superposition is valid in the weight space. The price, however, is a greater number of signal paths between neurons [186]. The salient features of the CMAC are described in **[7.10]**.

Ersu and Tolle have applied the CMAC architecture to nonlinear control problems. Miller et al. **[7.10]** have used the CMAC for real-time robotic control, pattern recognition, and signal processing applications. They have also discussed advantages and disadvantages of the CMAC approach.

It should be noted, however, that due to the complexity and diversity of dynamic neural structures, no general theory similar to that for static neural networks, has yet been developed. Development of the theory of dynamic neural structures based on the understanding of biological neural morphologies poses a challenging task to system and computer scientists. The major drawback of dynamic neural networks is their very limited explanation capability [181]. The solutions offered by these networks are hard to track back, unlike in feedforward neural networks. System-type procedures have to be developed to explain the internal functioning of dynamic networks. Dynamic neural networks are in an early development stage, and more work needs to be done in this area.

7. Fuzzy–Neural Structure **[6.1–6.5]**

7.1 Introduction

Over the last decade or so, several parallel advances have been made in two distinct disciplines: fuzzy logic and neural networks. Fuzzy logic provides an inference morphology that enables approximate human reasoning capabilities to be applied to knowledge-based systems, such as perceptual and linguistic attributes. Also, this theory provides a mathematical strength to capture the uncertainties associated with human cognitive processes like thinking and reasoning.

While fuzzy theory provides an inference mechanism under cognitive uncertainty, computational neural networks offer exciting advantages such as learning, adaptation, fault-tolerance, parallelism and generalization. The computational neural networks, comprising of neuron like processing elements, are capable of coping with computational complexity, nonlinearity and uncertainty. In view of this versatility of neural networks, it is believed that they hold great potential as building blocks for a variety of behaviors associated with biological neural networks.

A brief comparative study between fuzzy systems and neural networks in their operations in the context of knowledge acquisition, uncertainty, reasoning and adaptation is presented in Table 7.1.

Neural-network structures can deal with imprecise data and ill-defined activities. However, the subjective phenomena such as reasoning and perceptions are often regarded beyond the domain of conventional neural network theory. It is interesting to note that fuzzy logic is another powerful tool for modeling uncertainties associated with human thinking and perception.

TABLE 7.1 A Comparative Study Between Fuzzy Systems and Neural Networks.

Skills		Fuzzy Systems	Neural Networks
Knowledge acquisitions	Inputs Tools	Human experts Interaction	Sample sets Algorithms
Uncertainty	Information	Quantitative and qualitative	Quantitative
	Cognition	Decision making	Perception
Reasoning	Mechanism	Heuristic search	Parallel computations
	Speed	Low	High
Adaptation	Fault-tolerance Learning	Low Induction	Very high Adjusting synaptic weights
Natural language	Implementation Flexibility	Explicit High	Implicit Low

In fact, the neural network approach fuses well with fuzzy logic [187, 188] and some research endeavors have given birth to the so called *fuzzy–neural networks* or *fuzzy–neural systems*. These are believed to have considerable potential in the areas of expert systems, medical diagnosis, control systems, pattern recognition, and system modeling.

To enable a system deal with cognitive uncertainties in a manner more like humans, one may incorporate the concept of fuzzy sets into the neural network. Although fuzzy logic is a natural mechanism for modeling cognitive uncertainty, it may involve an increase in the amount of computation required (compared with a system using classical binary logic). This can be readily offset by using fuzzy–neural-network approaches having the potential for parallel computation with high flexibility.

A fuzzy neuron is designed to function in much the same way as a nonfuzzy neuron (static and dynamic neurons discussed in earlier sections), except that it reflects the fuzzy nature of a neuron and has the ability to cope with fuzzy information. Inputs to the fuzzy neuron are fuzzy sets $(x_1, x_2, \ldots, x_N)$ in the universe of discourse $(\mathbf{X}_1, \mathbf{X}_2, \ldots, \mathbf{X}_N)$ respectively. These fuzzy sets may be labeled by such linguistic terms as "high," "large," "warm," "medium," etc. The inputs are then "weighted" in much different ways from those used in a nonfuzzy case. The weighted inputs are then aggregated not by the summation but by the fuzzy aggregation operations (fuzzy union, weighted mean, or intersection).

An important difference between the computational aspects of a nonfuzzy neuron and that of a fuzzy neuron is the definition of synaptic connections. The synaptic connections of a nonfuzzy neuron are formed by having confluence operation (usually a dot product) of adjustable weights $\mathbf{W}_a(t)$ and neural inputs $\mathbf{X}_a(t)$. Any adaptation or learning occurring within the neuron involves modifying these weights. For a fuzzy neuron, the synaptic connection is represented by a two-dimensional fuzzy relation between the synaptic weights and neural inputs $\mathbf{X}_a(t)$. In this situation, learning involves changing the two-dimensional relation surface at each synapse.

The term *fuzzy–neural network* (FNN) has existed for more than a decade. However, the recent resurgence of interest in this area is motivated by the increasing recognition of the potential of fuzzy logic and neural networks as two of the most promising approaches for exploring the functioning of the human brain. Many researchers are currently investigating ways and means of building fuzzy neural networks by incorporating the notion of fuzziness into a neural framework.

Yamakawa et al. **[6.5]** described a fuzzy–neural-network model and applied it successfully to a pattern recognition problem. Kuncicky and Kandel [189] proposed a fuzzy–neuron model in which the output of one neuron is represented by a fuzzy level of confidence and the firing process is regarded as an attempt to find a typical value among inputs. Kiszka and Gupta [191] studied a fuzzy–neuron model described by the logic equations. However, no specific learning algorithms were developed in these three cases. Gupta and Knopf [188] proposed a fuzzy–neuron model which is similar to the first two cases except that a specific modification scheme was proposed for weight adaptation during learning. Nakanishi et al. [192] and Hayashi et al. [190] used the nonfuzzy–neural approach for the design of fuzzy logic controllers with adaptive and learning features. Similarly, Cohen and Hudson [187] used nonfuzzy–neural-network learning techniques to determine the weights of antecedents for use in fuzzy expert systems. However, no fuzzy–neural models were used.

One way to incorporate fuzziness into the neural network is by arranging the integrator/transfer functions at each node to perform some sort of fuzzy aggregation on the numerical information arriving at each node [193]. Another way to introduce fuzziness into the neural network is through the input data itself, which may be "fuzzified" in one of several ways. Gupta **[6.3]** and Qi [194] proposed three different fuzzy neural models. Jang **[6.4]** developed a self-organizing fuzzy controller based on temporal backpropagation. Recently, Carpenter et al. [195] proposed Fuzzy ARTMAP as an extension of their well-known adaptive resonance theory (ART)-based neural network. The Fuzzy ARTMAP is a synthesis of fuzzy logic and ART network. The neural-network structure realizes a new min–max learning rule that minimizes predictive error and improves generalization. Furthermore, it learns each input as it is received online, rather than performing an offline optimization of a certain function. The applications of Fuzzy ARTMAP have concentrated on pattern classification and image recognition. No control applications have yet been reported.

In this section, we present a brief introduction to fuzzy logic, fuzzy set definition, and a model of fuzzy neuron.

7.2 Fuzzy Logic: Basic Introduction

The concept of fuzzy sets was introduced by Zadeh **[6.1, 6.2]** in order to represent and manipulate data that were not precise, but were, instead, "fuzzy." The fuzzy set theory provides a mechanism for representing linguistic constructs such as "many," "low," "medium," "often," and "few." In general, fuzzy logic provides an inference structure that enables appropriate human reasoning capabilities [196]. On the contrary, the traditional binary set theory describes crisp events, events that either do or do not occur. It uses probability theory

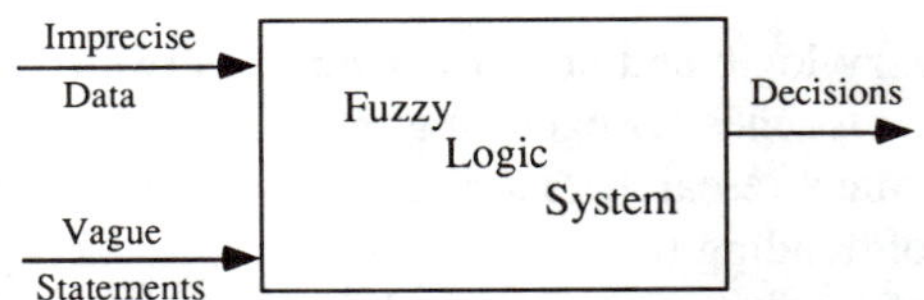

Fig. 7.1 A fuzzy logic system that accepts imprecise data and vague statements, such as "many," "low," "medium," "often," "few," and provides decisions.

to explain if an event will occur, measuring the chance with which a given event is expected to occur. The theory of fuzzy logic is based on the notion of relative graded membership and so are the functions of mentation and cognitive processes [197]. The utility of fuzzy sets lies in their ability to model uncertain or ambiguous data, as depicted in Fig. 7.1, so often encountered in real life.

Fuzzy Set Definition [**6.1**, 198]

Let X be a space of points (or objects) with a generic element of X denoted by x. X is often referred to as the universe of discourse. A fuzzy set (class) A in X characterized by a membership (characteristic) function $\mu_A(x)$, which associates with each point in X a real number in the interval [0, 1], with the value of $\mu_a(x)$ to unity, the higher the grade of membership of x in A.

A fuzzy set A is a subset of the universe of discourse X that admits partial membership. The fuzzy set A is defined as an ordered pair

$$A = \{(x, \mu_A(x)\} \tag{7.1}$$

where $x \in X$ and $0 \leq \mu_A(x) \leq 1$. The membership function $\mu_A(x)$ describes the degree to which the object x belongs to the set A. $\mu_A(x)$ is also referred to as the "characteristic function" or "graded membership" of x in A. If $\mu_A(x) = 0$ then it is certain that x is not in A, and $\mu_A(x) = 1$ then it is certain that x is in A. For x over $0 < \mu_A(x) < 1$, there is an uncertainty associated with x, that is, x belongs to A with the possibility $\mu_A(x)$.

7.3 Fuzzy–Neural Architectures

As mentioned earlier, concepts of neural networks and fuzzy logic integrate very well to give birth to an emerging area of research—"fuzzy neural networks" or "fuzzy neural systems." Paradigms based upon this integration are believed to have considerable potential in the areas of expert systems, medical diagnosis, control systems, pattern recognition, and system modeling. Two possible models of fuzzy neural systems are schematically shown in Figs. 7.2(a) and 7.2(b).

The computational process envisioned for fuzzy–neural systems is as follows. It starts with the development of a "fuzzy neuron" based on the understanding of biological neuronal morphologies, followed by learning mechanisms. This leads to the following three steps in a fuzzy-neural computational process:

(i) development of fuzzy neural models motivated by biological neurons,

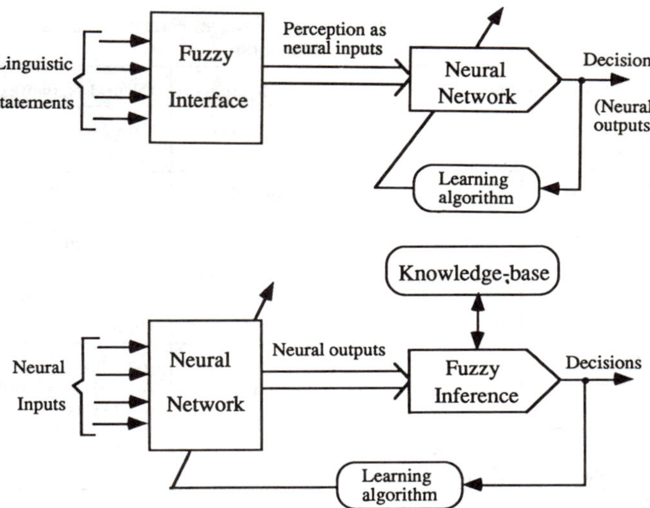

Fig. 7.2 Two models of fuzzy–neural systems. (a) In response to linguistic statements, the "fuzzy interface" block provides an input vector to a multilayered neural network. The neural network can be adapted (trained) to yield desired command outputs or decisions. (b) In this scheme, a multilayered neural network drives the fuzzy inference mechanism.

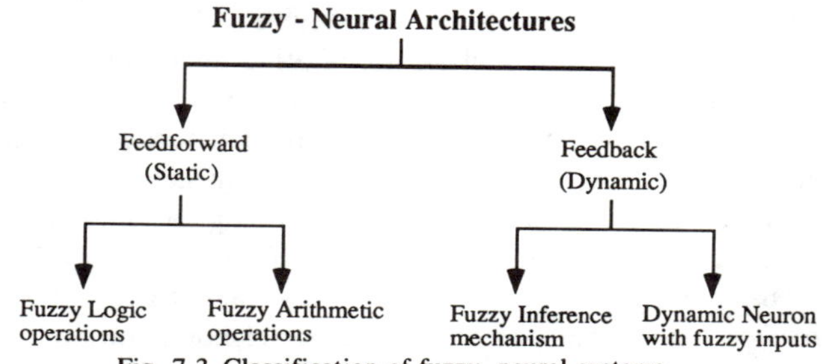

Fig. 7.3 Classification of fuzzy–neural systems.

(ii) models of synaptic connections that incorporates "fuzziness" into neural network, and

(iii) development of learning algorithms (that is, the method of adjusting the synaptic weights).

Based upon the computational process involved in a fuzzy–neural system, one may broadly classify the fuzzy neural structures as feedforward (static) and feedback (dynamic), as shown in Fig. 7.3.

In a feedforward (static) architecture, the neuron responds instantaneously to the fuzzy inputs because of the absence of dynamic elements in the structure. The neural mathematical operations in a feedforward network can be performed either by fuzzy arithmetic or fuzzy logic operations. As described in the earlier sections, the function of a non-fuzzy neuron can be modeled as

$$y(t) = \Psi\left[\sum_{i=0}^{n} w_i\, x_i\right] \tag{7.2}$$

where $[x_1, \ldots, x_n]$ represent neural inputs, $[w_1, \ldots, w_n]$ the synaptic weights, $y(t)$ the neural output and $\Psi[\cdot]$ is some nonlinear activation function. From (7.2) it may be observed that the mathematical operations involved in a computational neuron are: (i) the scalar product between the neural inputs and the synaptic weights, and (ii) the summation of these products.

The scalar product in (7.2) can be replaced by fuzzy multiplication and the summation operation by fuzzy addition. Detailed descriptions of fuzzy arithmetic operations may be found in [198]. These modifications lead to a fuzzy–neural architecture based on fuzzy arithmetic operations. The function of such a fuzzy neuron can be modeled by the following equation

$$y(t) = \Psi[\overset{n}{\underset{i=0}{(+)}} w_i \,(\cdot)\, x_i] \tag{7.3}$$

where $(+)$ and $(\cdot)$ are fuzzy addition and fuzzy multiplication operators respectively.

Alternatively, fuzzy logic operations, such as OR, AND, NOT, or their generalized versions can be introduced in Eqn. (7.2). In this section, we confine our discussion to a neural architecture based upon fuzzy logic operations.

The other classification, as shown in Fig. 7.3, is the feedback (dynamic) architecture. The dynamic networks not only provide some robust computing characteristics but also bring about greater insights into biological neural structures. The dynamics in fuzzy neural computing do provide some functional basis of the cerebellum and its associated circuitry, and can offer great computational advantages over purely feedforward architectures.

7.4 The Fuzzy–Neural Model [6.3]

If we express the neural input signals in terms of their membership functions each over the interval [0, 1], rather than in their absolute amplitudes, then we can write the augmented vector of neural inputs as

$$X_a(t) = [x_0(t), x_1(t) \ldots x_i(t) \ldots x_n(t)]^T \in [0,1]^{n+1} \tag{7.4}$$

where these neural signals (including the bias term, x_0) are bounded by the $(n + 1)$ dimensional hypercube $[0, 1]^{n+1}$. Similarly, the augmented synaptic weighting vector $\mathbf{W}_a(t)$ can be expressed over the unit hypercube $[0, 1]^{n+1}$.

We perform mathematical operations on these signals using logical operations (connectives) such as OR, AND (or their generalized form based on triangular norm or T-operators) and NEGATION.

Let us express the inputs x_1 and x_2 over [0, 1], then we define the generalized AND (T-operation) as a **T** mapping function [197, 198],

$$\mathbf{T}: [0, 1] \times [0, 1] \rightarrow [0, 1] \text{ given by} \tag{7.5}$$

$$y_1 = [x_1 \text{ AND } x_2] \overset{\Delta}{=} [x_1 \mathbf{T} x_2] = \mathbf{T}[x_1, x_2]$$

Similarly, we define the generalized OR, (T-conorm) as an **S** mapping function **S:** [0, 1] × [0, 1] → [0, 1] given by

$$y_2 = [x_1 \text{ OR } x_2] \overset{\Delta}{=} [x_1 \mathbf{S} x_2] = \mathbf{S}[x_1, x_2] \tag{7.6}$$

Negation **N** on $x_1 \in [0, 1]$ is defined as a mapping **N**: [0, 1] → [0, 1] with the following properties:

$$y_3 = \mathbf{N}[x_1] = 1 - x_1 \tag{7.7}$$

Thus, $N(0) = 1$, $N(0) = 1$, and $N(N(x)) = x$.

Now, we describe some important properties of the **T** and **S** operators.

$$\mathbf{T}(0, 0) = 0, \quad \mathbf{T}(1, 1) = 1 \tag{7.8}$$
$$\mathbf{T}(1, x) = x, \quad \mathbf{T}(x, y) = \mathbf{T}(y, x)$$
$$\mathbf{S}(0, 0) = 0, \quad \mathbf{S}(1, 1) = 1$$

$$\mathbf{S}(0, x) = x, \quad \mathbf{S}(x, y) = \mathbf{S}(y, x). \tag{7.9}$$

Also, De'Morgan's Theorems are stated as follows:

$$\mathbf{T}(x_1, x_2) = 1 - \mathbf{S}(1 - x_1, 1 - x_2)$$

and

$$\mathbf{S}(x_1, x_2) = 1 - \mathbf{T}(1 - x_1, 1 - x_2) \tag{7.10}$$

In the development of fuzzy-logic-based neural morphology, we will use the following combined synaptic and somatic operations.

Let the augmented vector of neural inputs and synaptic weights be represented by

$$\mathbf{X}_a(t) \in [0, 1]^{n+1}$$

and

$$\mathbf{W}_a(t) \in [0, 1]^{n+1}$$

respectively. Then by replacing the © (confluence)-operation by **T**-operation, and the Σ-operation by **S**-operation in (4.6), we get

$$u(t) = \overset{n}{\underset{i=0}{\mathbf{S}}} [w_i(t) \,\mathbf{T}\, x_i(t)] \in [0, 1] \tag{7.11a}$$

and

$$y(t) = \Psi[u(t)] \in [0, 1] \tag{7.11b}$$

where the nonlinear mapping $\Psi[\cdot]$ may be one of the functions defined in Table 4.1.

7.5 Unipolar to Bipolar Transformation

The logical operations defined in the preceding section are unipolar signals over the positive unit interval [0, 1]. Such logical operations provided only the neural state corresponding to the excitatory (positive) interactions. In order to account for both the excitatory (positive) and the inhibitory (negative) interactions of the neural input vector, we must consider both $\mathbf{X}_a(t)$ and its negated values $N[\mathbf{X}_a(t)]$, thus making the neural inputs of dimensions $(2n + 2)$.

Alternatively, we may express the neural inputs and synaptic weights as bipolar signals and weights over the interval [−1, 1] and redefine the logical operations over this interval. We will provide a brief description of this transformation for

TABLE 7.2 Summary of Logical Operations (T-Operations) on Unipolar and Bipolar Signals

Unipolar Signals	Bipolar Signals
$x(t) \in [0, 1]$	$z(t) \in [-1, 1]$
Bipolar to unipolar transformation	Unipolar to bipolar transformation
$x(t) = \frac{z(t) + 1}{2}$	$z(t) = 2x(t) - 1$
Negation	
$N[x(t)] = \overline{x(t)} = 1 - x(t)$	$N[z(t)] = \overline{z(t)} = -z(t)$
Boundary Conditions	
T-operator (generalized AND*)*	
$\mathbf{T}(0, 0) = 0$; $\mathbf{T}(1, 1) = 1$ $\mathbf{T}(1, x) = x$; $\mathbf{T}(x_1, x_2) = \mathbf{T}(x_2, x_1)$	$\mathbf{T}(-1, -1) = -1$; $\mathbf{T}(1, 1) = 1$ $\mathbf{T}(1, z) = z$; $\mathbf{T}(z_1, z_2) = \mathbf{T}(z_2, z_1)$
S-operator (generalized OR*)*	
$\mathbf{S}(0, 0) = 0$; $\mathbf{S}(1, 1) = 1$ $\mathbf{S}(0, x) = x$; $\mathbf{S}(x_1, x_2) = \mathbf{S}(x_2, x_1)$	$\mathbf{S}(-1, -1) = -1$; $\mathbf{S}(1, 1) = 1$ $\mathbf{S}(-1, z) = z$; $\mathbf{S}(z_1, z_2) = \mathbf{S}(z_2, z_1)$
Generalized De'Morgan's Theorem	
$\mathbf{T}(x_1, x_2) = 1 - \mathbf{S}(1 - x_1, 1 - x_2)$ $\mathbf{S}(x_1, x_2) = 1 - \mathbf{T}(1 - x_1, 1 - x_2)$	$\mathbf{T}(z_1, z_2) = -\mathbf{S}(-z_1, -z_2)$ $\mathbf{S}(z_1, z_2) = -\mathbf{T}(-z_1, -z_2)$

TABLE 7.3 Summary of Some Important Logical Functions and Operations*

*These logical functions and operations can be defined on both the unipolar [0, 1] and bipolar [−1, 1] signals, but here we will consider only the bipolar signals.

Godel's implication Ⓖ: Godel's implication $[z_1 Ⓖ z_2]$ (read as, z_1 implies z_2) is defined as

$$[z_1 Ⓖ z_2] = [z_1 \rightarrow z_2] = 1, \quad z_1 \leq z_2$$
$$= z_2, \quad z_1 > z_2$$

Degree of equality (using Godel's implication) $\eta(z_1, z_2)$

Given z_1 and z_2 over [−1, 1], *to what degree they are equal* is defined as

$$\eta(z_1, z_2) = \frac{1}{2}[\{z_1 Ⓖ z_2\} \mathbf{T} \{z_2 Ⓖ z_1\} + \{\bar{z}_1 Ⓖ \bar{z}_2 \mathbf{T} \{\bar{z}_2 Ⓖ \bar{z}_1\}] \in [-1, 1]$$

where $\bar{z} = -z$.

Degree of equality (using Lukasiewicz's conjunction): $\eta(z_1, z_2)$

Again, given z_1 and z_2 over [−1, 1], *to what degree they are equal* is defined as

$$\eta(z_1, z_2) = [1 - | z_1 - z_2 |] \in [-1, 1]$$

Degree of inequality (degree of error): $E(z_1, z_2) \in [-1, 1]$

Given z_1 and z_2 over [−1, 1], in order to find the *degree of difference* or *degree of inequality,* we define $E[z_1, z_2]$ as the negation on the degree of equality; that is,

$$E[z_1, z_2] = N[\eta(z_1, z_2)] = -\eta(z_1, z_2)$$

Thus, the degree of error, using Lukasiewicz's conjunction can be defined as

$$E(z_1, z_2) = | z_1 - z_2 | - 1$$

unipolar [0, 1] to bipolar [−1, 1], and of the definition of logical operations over the interval [−1, 1]. Let $x(t) \in [0, 1]$ be a unipolar signal. The corresponding bipolar signal $z(t)$ is defined as

$$z(t) = 2\,x(t) - 1 \tag{7.12}$$

The negation is defined as

$N[x] = 1 - x$, for unipolar signals,

and

$N[z] = -z$, for bipolar signals.

The **T** and **S** functions defined in the interval [0, 1] can be transformed to the interval [−1, 1] using (7.12).

In Table 7.2, we give a summary of logical **T**- and **S**-operations for both unipolar and bipolar signals. Also, in Table 7.3, we define some important logical functions and operations such as Godel's implication, degree of equality using Godel's implication, degree of equality using Lukasiewicz conjunction, and degree of error (inequality) for two bipolar signals z_1 and $z_2 \in [-1, 1]$.

7.6 Learning and Adaptation in Fuzzy Neuron

Figure 7.4 shows a fuzzy neuron for bipolar signals and bipolar synaptic weights. The augmented neural input signals $\mathbf{X}_a(t)$ are defined over the unit hypercube $[0, 1]^{n+1}$. Using the transformation given in (7.12), we transform the unipolar neural inputs $\mathbf{X}_a(t)$ into bipolar signals $\mathbf{Z}_a(t) \in [-1, 1]^{n+1}$.

The logical operation of this neuron is summarized as follows

$$u(t) = \mathop{\mathbf{S}}_{i=0}^{n} [w_i(t)\ \mathbf{T}\ z_i(t)] \tag{7.13a}$$

which is equivalent to

$$u(t) = \mathbf{W}_a^T(t) \text{ AND } \mathbf{Z}_a(t) \in [-1, 1] \tag{7.13b}$$

(a logical scalar product operation), where $w_o(t)$ and $z_o(t)$ correspond to the bias terms and $z_o = 1$. The neural output is defined as

$$y(t) = \Psi[u(t)] \in [-1, 1] \tag{7.14a}$$

where $\Psi[v]$ is defined as

$$\Psi[u(t)] = |u(t)|\ g_s \cdot sgn\,[u(t)], \quad g_s > 0 \tag{7.14b}$$

(g_s *is the activation gain*)

Let us define an error signal with respect to the desired neural output, $y_d(t) \in [-1,1]$, as $e(t) = y_d(t) - y(t) \in [-1, 1]$. The objective of learning and adaptation in neural networks is to adapt the parameters of the neural structure, in the case $\mathbf{W}_a(t)$ and g_s in (7.14a) and (7.14b), in order to minimize an error function. The adaptive rules to modify $\mathbf{W}_a(t)$ and $g_s(t)$ are as follows:

$$\mathbf{W}_a(t+1) = \mathbf{W}_a(t) \text{ OR } \Delta\ \mathbf{W}_a(t) \tag{7.15a}$$

and

$$g_s(t+1) = g_s(t) \text{ OR } \Delta\ g_s(t) \tag{7.15b}$$

where

$$\Delta \mathbf{W}_a(t) = \mathbf{Z}_a(t) \text{ AND } e(t) \tag{7.16a}$$

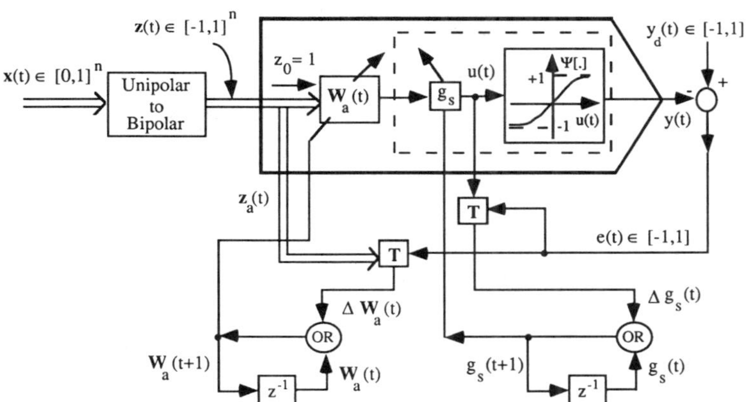

Fig. 7.4 Implementation of the learning scheme, Eqns. (7.15) and (7.16), to modify the synaptic weights, $\mathbf{W}_a$, and the somatic gain, g, of a fuzzy neuron.

and

$$\Delta\ g_s(t) = u(t) \text{ AND } e(t). \qquad (7.16b)$$

This description provides a learning scheme to update the neural weights of a fuzzy neuron. It represents one particular model of a fuzzy–neural architecture. It is postulated that the fuzzy–neural networks can learn by experience if their synaptic connections are interpreted as fuzzy relations between the external inputs and the dendritic inputs. Complex decisions may be derived from a shallow hierarchy of fuzzy neurons and their related network architectures with electronic circuitry. With the development of fuzzy–neural networks, it is envisaged that learning schemes for complex control systems will have the following features:

(i) easy-to-implement natural language features so that the structure of knowledge is very clear and efficient,

(ii) any changes in the task and environment can be easily taken care of by adapting the neural weights, and,

(iii) since a fuzzy system is one kind of interpolation, drastic reduction of data and software/hardware overheads can be achieved.

However, it should be noted that more research endeavors are necessary to develop a general topology of fuzzy–neural models, learning algorithms, and approximation theory so that these models are made applicable in system modeling and control of complex systems. The area of fuzzy–neural networks is still in its infancy, and is a very fertile area of theoretical and applied research.

8. Hardware Implementations [8.1–8.5]

A large number of software- and hardware-based neural-network implementations have been developed in recent years that attempt to streamline the number of computations and take advantage of the inherent parallelism found within neural-network architectures. In general, these neural-network implementations have three key characteristics. First, they are computationally intensive. That is, the output of each neuron (processing element) in a layer is the summation of several products. Every synaptic connection in the neural layer requires a separate multiplication operation. Second, these neural layers employ massive parallelism. Each neuron of the layer can be treated as a local processor acting in parallel with all other neurons. As a result, computational speed of the network is dependent on the number of mathematical operations performed by each neuron and not the overall number of neurons. Third, the neural layers require immense memory. Individual neurons in the layer have many synaptic connections, where each synaptic connection has an associated weight that must be stored. This situation becomes a serious problem as the size of the network increases. For example, doubling the number of neurons will result in a factor-of-four increase in the number of synaptic connections. Many control applications require large numbers of neurons. In order for neural networks to be useful for engineering applications, it is therefore necessary that they have an adequate amount of storage.

The three board areas of neural-network implementation are computer-based software, electronics hardware, and optical/opto-electronics. The various classes and subclasses of neural-network implementations are shown in Fig. 8.1. Computer implementations are largely software-based algorithms that utilize existing computing machines that were not designed explicitly for neural-network processing. Supercomputers, massively parallel computers, and conventional digital computers are examples of hardware computing systems that can be programmed by adequate software to simulate neural-network functions for various applications. Electronic neural-network implementations involve bus-oriented processors, coprocessors, CCD [212, 213] and VLSI circuit designs [**8.1**, 214–221]. In general, this includes any electronic hardware that is designed specifically for neural-network implementations. Finally, optical/opto-electronic implementations are neural networks that involve either optical [222, 223] or a mixture of optical and electronic components in their hardware in order to achieve real-time parallel processing [224].

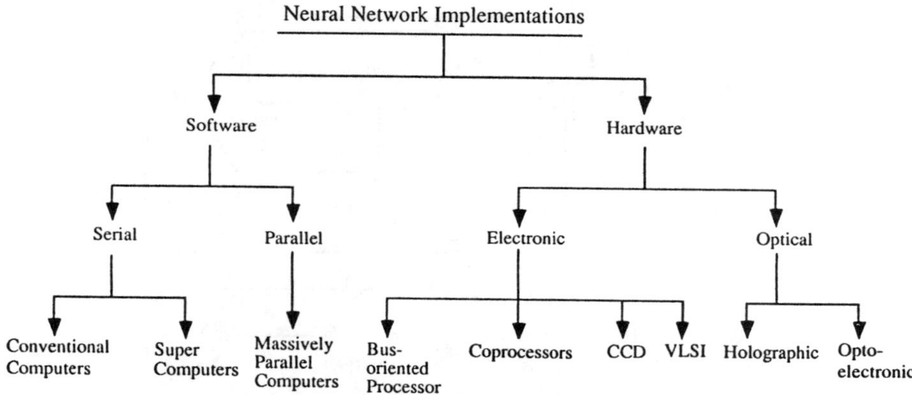

Fig. 8.1 Flow chart of various classes and subclasses for neural-network implementations.

Nearly all applications of neural networks are still running as software applications on conventional digital computers. They run rather slowly, compared to the speed that will be achieved when neural networks are fully implemented in hardware. However, different device technologies have different limiting features down at the level of their physics. One of the limiting features of electronics is that the connections between neurons have to be carried out on wires, which will fill up the space with wires. Optical technology provides a solution in the sense that one light beam can pass through another freely. A single aperture can have multiple light beams going through it, so one can get many more connections in the same amount of space. A number of vendors have recently introduced dedicated VLSI hardware implementations. This brings additional speed, and increases the scale of neural networks that can be implemented.

9. Perspectives

In conventional structure of a computational neural network, the neuron receives its inputs either from other neurons or from the neural sensors. A weighted sum of these inputs constitutes the argument of a fixed nonlinear "activation" function. The weights correspond to the synapses in a biological neuron, while the activation function is analogous to its intercellular current conduction mechanism. The resulting value of the activation function forms the neural output. This output is distributed with weighted connections to other processing units. This is an oversimplified but useful first approximation of the biological neuron. As has been pointed out by Hopfield, the present models of neural networks are almost a parody of biological neural structures. They have abstracted a few properties from what we know of the brain's functioning. It is essential that we gain more insight into how single biological neurons function, how masses of neurons are structured, and how they coordinate to achieve the results we experience with our brains every day. Then we have to abstract the essential features of biological neurons into neural models.

Although there are a number of neural-network models, the two dominating models used at present are the feedforward and feedback (recurrent) networks. The type of architecture most commonly used for engineering applications is the multilayer feedforward neural network where no information is fed back during operation. Although this neural-network structure has been used successfully for system identification and the control of dynamic systems, the major disadvantage is the very slow learning rate. Attempts to accelerate the learning process by increasing the values of the gains (constants) in learning algorithms have resulted in unstable systems. Furthermore, the feedforward networks have no memory; that is, their output is solely determined by the current inputs and the values of the weights. This is quite a contrast to the biological neural systems that always have feedback in their functioning. The field of neural networks suffers from the lack of good feedback models. Therefore, feedback neural architectures will not only result in significant development in neural networks but also bring about greater insights into biological control mechanisms. The concept of feedback is strongly related to the cerebellum and its associated circuitry. Although a number of dynamic neural architectures have been proposed by different researchers, general theory with regard to architecture, learning algorithms, and functional approximation has yet to be developed.

The general focus of this introduction was to transfer the basic knowledge of biological neuro-control mechanisms toward the development of computational neural-network-based control techniques, algorithms, and hardware implementations. The biological foundation, computational architectures, and software/hardware implementation of neural networks were briefly summarized in this introductory chapter. It is impossible to address all related theoretical issues, mathematical models, and computational paradigms in such a short presentation. Rather, it was the objective of the authors to present a *holistic* view of neural-network research in an effort to stress the interdisciplinary relationship between biological and neuro-control paradigms.

The remaining sections of this book are collected papers that follow more or less the basic outline presented in this introduction. The first set of articles provides a variety of perspectives on employing computational models derived from theoretical neurophysiology and psychophysics to control-system design. The second set of papers presents a basic understanding of neural morphology and corresponding mathematical models. The objective of the third part is to provide an overview of important articles that deal with the computational neural-network architectures and algorithms. The development of a functional approximation theory occupies a significant place in the neural-network paradigm. Several journal papers that describe the theory of functional approximation are presented in the fourth section. The fifth set of papers describes different static and dynamic neural morphologies. The field of fuzzy–neural networks is a developing branch of the neural-network paradigm. The intent of the sixth set of papers is to provide an overview of this emerging technology. The seventh set of papers elucidates the wide range of neural-network applications, from system identification, control of nonlinear dynamical systems, robotics, and antennas to process engineering. Finally, the eighth set of papers provides an overview of the different hardware implementations of neural networks.

The key features of these articles and how they relate to neuro-control research is described in an introduction to each part of this book. A bibliography of related books and journal and conference papers is included for each section.

We know intuitively what encompasses a biological neural system and its salient features. We, the system scientists, desire to develop intelligent systems based on the understanding of biological systems. In this volume, a small but representative sample from a variety and a large amount of literature in the area of neural networks is reprinted. The editors hope that both the novice and the experienced researcher will find this volume useful.

10. References and Bibliography

General Background

[1] M. Black, "Vagueness: An exercise in logical analysis," *Phil. Sci.*, vol. 4, pp. 427–455, 1937.

[2] W. S. McCulloch and W. Pitts, "A logical calculus of the ideas immanent in nervous activity," *Bull. Math. Biophys.*, vol. 5, pp. 115–133, 1943.

[3] N. Wiener, *Cybernetics.* New York: John Wiley & Sons, 1948.

[4] D. O. Hebb, *The Organization of Behavior.* New York: John Wiley & Sons, 1949.

[5] F. Rosenblatt, "The Perceptron: A probabilistic model for information storage and organization in the brain," *Psychol. Rev.*, vol. 65, pp. 386–408, 1959.

[6] B. Widrow and M. E. Hoff, "Adaptive switching circuits," in *IREWESCON Convention Record*, vol. 4, pp. 96–104, New York: IRS, 1960.

[7] M. L. Minsky and S. A. Papert, *Perceptrons.* Cambridge, MA: MIT Press, 1969.

[8] M. A. Arbib, *The Metaphorical Brain.* New York: John Wiley & Sons, 1972.

[9] R. C. Conant, "Law of information which govern systems," *IEEE Trans. Syst., Man, Cybern.*, vol. 6, pp. 334–338, 1976.

[10] S. -I. Amari, "A mathematical approach to neural systems," in J. Metzler, Ed., *Systems Neuroscience,* New York: Academic Press, pp. 67–117, 1977.

[11] A. C. Scott, *Neurophysics.* New York: John Wiley & Sons, 1977.

[12] C. O. Lovejoy, "The origin of man," *Science,* vol. 211, pp. 341–350, 1981.

[13] E. R. Kandel and J. H. Schwartz, *Principles of Neural Science,* New York: North-Holland, 1985.

[14] V. Brooks, *The Neural Basis for Motor Control.* New York: Oxford University Press, 1986.

[15] M. A. Arbib, *Brains, Machines and Mathematics.* New York: Springer-Verlag, 1987.

[16] S. Grossberg, *Neural Networks and Natural Intelligence.* Cambridge, MA: MIT Press, 1987.

[17] M. M. Gupta, "On the cognitive computing: Perspectives," in *Fuzzy Computing: Theory, Hardware and Applications,* New York: North-Holland, 1988.

[18] J. S. Albus, "Outlines for a theory of intelligence," *IEEE Trans. Syst., Man, Cybern.*, vol. 21, no. 3, pp. 473–509, May/June 1991.

[19] J. Farrell, T. Berger, and B. Appleby, "Using learning techniques to accommodate unanticipated faults," *IEEE Contr. Syst. Mag.*, pp. 40–49, June 1993.

Adaptive and Robust Control

[20] M. Vidyasagar, *Control Systems Synthesis: A Factorization Approach.* Cambridge, MA: MIT Press, 1985.

[21] M. M. Gupta, Ed., *Adaptive Methods for Control System Design.* New York: IEEE Press, 1986.

[22] K. J. Astrom and B. Wittenmark, *Adaptive Feedback Control.* New York: Addison-Wesley, 1989.

[23] K. S. Narendra and A. M. Annaswamy, *Stable Adaptive Systems.* Englewood Cliffs, NJ: Prentice Hall, 1989.

[24] R. Ortega and Y. Tang, "Robustness of adaptive controllers—A survey," *Automatica,* vol. 25, no. 5, pp. 651–677, 1989.

[25] D. C. McFarlance and K. Glover, "Robust controller design using normalized coprime factor plant descriptions," in *Lecture Notes in Control and Information Sciences,* M. Thoma and A. Wyner, Eds., Berlin: Springer-Verlag, no. 138, 1989.

[26] P. Dorato and R. K. Yedavalli, Eds., *Recent Advances in Robust Control.* New York: IEEE Press, 1990.

[27] C. Abdallah, D. Dawson, and M. Jamshidi, "Survey of robust control for rigid machines," *IEEE Contr. Syst. Mag.*, pp. 24–30, Feb. 1991.

Intelligent and Neuro-Control

[28] K. S. Fu, "Learning control systems—Review and outlook," *IEEE Trans. Automat. Contr.*, pp. 210–221, April 1970.

[29] E. Ersu and H. Tolle, "A new concept for learning control inspired by brain theory," in *Proc. Ninth World Congress IFAC,* pp. 245–250, 1984.

[30] K. M. Passino, "Bridging the gap between conventional and intelligent control," *IEEE Contr. Syst. Mag.*, pp. 12–18, June 1993.

[31] M. D. Peek and P. J. Antsaklis, "Parameter learning for performance adaptation," *IEEE Contr. Syst. Mag.*, pp. 3–11, Dec. 1990.

[32] G. E. Hinton, "How neural networks learn from experience," *Sci. Amer.*, pp. 145–151, Sept. 1992.

[33] A. G. Barto, "Connectionist learning for control," in *Neural Networks for Control,* T. Miller, R. S. Sutton and P. J. Werbos, Eds., Cambridge, MA: MIT Press, pp. 6–58, 1991.

[34] A. Guez, J. L. Eilbert, and M. Kam, "Neural network architecture for control," *IEEE Contr. Syst. Mag.*, pp. 22–25, April 1988.

[35] W. P. Jones and J. Hoskins, "Backpropagation: A generalized delta learning rule," *Byte,* pp. 155–62, Oct. 1987.

[36] D. R. Hush and B. G. Horne, "Progress in supervised neural networks," *IEEE Signal Process. Mag.*, Jan., vol. 10, no. 1, pp. 8–39, 1993.

[37] B. Widrow and M. A. Lehr, "30 Years of adaptive neural networks: Perceptron, Madaline, and backpropagation," *Proc. IEEE,* vol. 78, no. 9, pp. 1415–1442, Sept. 1990.

[38] M. M. Gupta, "Virtual cognitive systems (VCS)," *Neural Network World,* vol. 2, no. 6, pp. 621–628, 1992.

[39] Y. H. Pao, *Adaptive Pattern Recognition and Neural Networks.* Redwood City, CA: Addison-Wesley, 1989.

[40] P. D. Wasserman, *Neural Computing: Theory and Practice*. New York: Van Nostrand Reinhold, 1989.

[41] W. T. Miller III, R. S. Sutton, and P. J. Werbos, Eds., *Neural Networks for Control*. Cambridge, MA: MIT Press, 1990.

[42] J. S. Judd, *Neural Network Design and the Complexity of Learning*. Cambridge, MA: MIT Press, 1990.

[43] T. Khanna, *Foundations of Neural Networks*. Redwood City, CA: Addison-Wesley, 1990.

[44] P. K. Simpson, *Artificial Neural Systems: Foundations, Paradigms, Applications, and Implementations*, New York: Pergamon Press, 1990.

[45] S. Y. Kung, *Digital Neural Computing: From Theory to Application*. Englewood Cliffs, NJ: Prentice Hall, 1991.

[46] R. J. Mammone and Y. Y. Zeevi, *Neural Networks: Theory and Applications*. San Diego, CA: Academic Press, 1991.

[47] B. Kosko, *Neural Networks for Signal Processing*. Englewood Cliffs, NJ: Prentice Hall, 1992.

[48] J. M. Zurada, *Introduction to Artificial Neural Systems*. St. Paul, MN: West Publishing Company, 1992.

[49] Special Issues on Neural Networks, *IEEE Contr. Syst. Mag.*, April 1988, April 1989, April 1990.

Biological Motivation

[50] B. Widrow, "Generalization and information storage in networks of Adaline neurons," in *Self-Organizing Systems*, M. Yovitz, G. Jacobi, and C. Goldstein, Eds., Washington, DC: Spartan Books, pp. 435–461, 1962.

[51] C. F. Stevens, "Synaptic physiology," *Proc. IEEE*, vol. 79, no. 9, pp. 916–930, June 1968.

[52] M. Ito, "Neurophysiological aspects of the cerebellar motor control system," *Int. J. Neurology*, vol. 7, no. 2.3–4, pp. 162–176, 1970.

[53] P. A. Anninos, B. Beek, T. J. Csermel, E. E. Harth, and G. Pertile, "Dynamics of neural structures," *J. Theoret. Biol.*, vol. 26, pp. 121–148, 1970.

[54] S. -I. Amari, "Neural theory of association and concept-formation," *Biol. Cybern.*, 26, pp. 175–185, 1977.

[55] M. Fujita, "Adaptive filter model of the cerebellum," *Biol. Cybern.*, 45, pp. 195–206, 1982.

[56] D. E. Rumelhart and J. L. McClelland, Eds., *Parallel Distributed Processing: Explorations in the Microstructures of Cognition*. Cambridge, MA: MIT Press, vol. 1, 1986.

[57] G. A. Carpenter and S. Grossberg, "A massively parallel architecture for a self-organizing neural pattern recognition machine," *Comput. Vision, Graphics, Image Process.* vol. 37, pp. 54–115, 1987.

[58] A. C. Guyton, *Text Book of Medical Physiology*. Philadelphia: W.B. Saunders Company, 1987.

[59] R. P. Lipmann, "An introduction to computing with neural nets," *ASSP Mag.*, vol. 4, no. 2, pp. 4–22, 1987.

[60] D. S. Melkonian, "Mathematical theory of chemical synaptic transmission," *Biol. Cybern.*, 62, pp. 539–548, 1990.

[61] J. A. Anderson and E. Rosenfield, *Neurocomputing: Foundations of Research*. Cambridge, MA: MIT Press, 1991.

[62] D. S. Touretzky, Ed., *Advances in Neural Information Processing Systems*. San Mateo, CA: Morgan Kaufmann, 1992.

[63] B. Pansky, D. J. Allen, and G. C. Budd, *Review of Neuroscience*, 2nd Ed., New York: Macmillan, 1992.

[64] *Sci. Amer.*, Sept. 1992.

Neuronal Morphologies

[65] J. S. Albus, "A new approach to manipulator control: The cerebellar model articulation controller (CMAC)," *J. Dynamic Syst., Meas., Contr.*, pp. 220–227, Sept. 1975.

[66] K. Fukushima, S. Miyake, and T. Ito, "Neocognitron: A neural network model for a mechanism of visual pattern recognition," *IEEE Trans. Syst., Man, Cybern.*, vol. 13, no. 5, pp. 826–834, Sept./Oct. 1983.

[67] J. S. Denker, "Neural network models of learning and adaptation," *Physica-22D*, pp. 216–232, 1986.

[68] F. J. Pinda, "Dynamics and architecture for neural computation," *J. Complexity*, vol. 4, pp. 216–245, 1988.

[69] L. O. Chua and L. Yang, "Cellular neural networks: Theory," *IEEE Trans. Circuits Syst.*, vol. 35, pp. 1257–1272, 1988.

[70] F. Crick, "The recent excitement about neural networks," *Nature*, vol. 337, pp. 129–132, 1989.

[71] K. J. Lang, A. H. Waibel, and G. E. Hinton, "A time-delay neural network architecture from isolated word recognition," *Neural Networks*, vol. 3, no. 1, pp. 23–44, 1990.

[72] S. D. Wang and M. S. Yeh, "Self-adaptive neural architectures for control applications," in *Proc. Int. Joint Conf. Neural Networks, (IJCNN)*, pp. 309–314, June 1990.

[73] L. Tarassenko, J. N. Tombs, and J. H. Reynolds, "Neural network architectures for content-addressable memory," *IEEE Proc.-F*, vol. 138, no. 1, pp. 33–39, Feb. 1991.

[74] G. Nagy, "Neural networks—Then and now," *IEEE Trans. Neural Networks*, vol. 2, no. 2, pp. 316–318, March 1991.

[75] M. M. Gupta, "Fuzzy logic and neural networks," presented at Int. Conf. Fuzzy Logic and Neural Networks, IIZUKA, Japan, vol. A-6a, pp. 157–160, July 17–21, 1992.

[76] M. M. Gupta and G. K. Knopf, "A multitask visual information processor with a biologically motivated design," *J. Visual Commun. Image Representation*, 3, no. 3, pp. 230–246, Sept. 1992.

[77] Q. Zhang and A. Benveniste, "Wavelet networks," *IEEE Trans. Neural Networks*, vol. 3, no. 6, pp. 889–898, Nov. 1992.

[78] R. Hecht-Nielsen, "Counterpropagation networks," *Appl. Optics*, vol. 26, pp. 4979–4984, 1987.

[79] J. J. Hopfield, "Artificial neural networks are coming," *IEEE Expert*, An Interview by W. Myers, pp. 3–6, April 1990.

[80] B. Widrow and R. G. Winter, "Neural nets for adaptive filtering and adaptive pattern recognition," *IEEE Computer*, pp. 25–39, March 1988.

[81] P. J. Werbos, *Beyond Regression: New Tools for Prediction and Analysis in the Behavior Sciences*. Ph.D. Thesis, Harvard Univ., 1974.

[82] D. Hammerstrom, "Working with neural networks," *IEEE Spectrum*, vol. 30, no. 6 pp. 26–32, June 1993.

[83] D. Hammerstrom, "Working with neural networks," IEEE Spectrum, vol. 30, no. 7 pp. 46–53, July 1993.

Functional Approximation

[84] T. Poggio and F. Girosi, "Networks for approximation and learning," *Proc. IEEE*, vol. 78, no. 9, pp. 1481–1497, Sept. 1990.

[85] G. A. Watson, *Approximation Theory and Numerical Methods*. New York: John Wiley & Sons, 1980.

[86] L. V. Kantorovich and G. P. Akilov, *Functional Analysis*. Translated by H. L. Silcock, Elmsford, NY: Pergamon Press, 1982.

[87] A. N. Kolmogorov and S. V. Fomin, *Introductory Real Analysis*. Translated by R. A. Silverman, Englewood Cliffs, NJ: Prentice-Hall, 1970.

[88] E. K. Blum and L. K. Li, "Approximation theory and feedforward networks," *Neural Networks*, vol. 4, pp. 511–515, 1991.

[89] A. R. Gallant and H. White, "There exists a neural network that does not make avoidable mistakes," in *Proc. IEEE Conf. Neural Networks*, vol. I, pp. 657–664, San Diego, 1988.

[90] P. Cardaliaguet and G. Euvrard, "Approximation of a function and its derivative with a neural network," *Neural Networks*, vol. 5, pp. 207–220, 1992.

[91] R. Hecht-Nielsen, "Kolmogorov's mapping neural network existence theorem," in *Proc. IEEE Conf. Neural Networks*, vol. II, pp. 11–14, San Diego, 1987.

[92] N. E. Cotter and T. J. Gullerm, "The CMAC and a theorem of Kolmogorov," *Neural Networks*, vol. 5, pp. 221–228, 1992.

[93] V. Kurkova, "Kolmogorov's theorem and multilayer neural networks," *Neural Networks*, vol. 5, pp. 501–506, 1992.

[94] F. Girosi and T. Poggio, "Representation of properties of networks: Kolmogorov's theorem is irrelevant," *Neural Computat.* vol. 63, pp. 169–176, 1990.

[95] D. Chester, "Why two hidden layers are better than one," in *Proc. IEEE Int. Joint Conf. Neural Networks (IJCNN)*, pp. 265–268, 1990.

[96] D. S. Broomhead and D. Lowe, ''Multivariable functional interpolation and adaptive networks,'' *Complex Syst.*, vol. 2, pp. 321–355, 1988.

[97] W. J. Daunicht, ''Defanet—A deterministic approach to function approximation by neural networks,'' in *Proc. IEEE Int. Joint Conf. Neural Networks (IJCNN)*, pp. 161–164, 1990.

[98] P. J. Gawthrop and D. G. Sbarbo, ''Stochastic approximation and multilayer Perceptrons: The gain back-propagation algorithm, *Complex Syst.*, vol. 4, pp. 51–74, 1990.

[99] C. L. Giles and T. Maxwell, ''Learning, invariance, and generalization in higher-order neural networks,'' *Appl. Optics*, vol. 26, pp. 4972–4978, 1987.

[100] G. Josin, ''Neural-space generalization of a topological transformation,'' *Biol. Cybern.*, vol. 59, pp. 238–290, 1988.

[101] E. J. Hartman, J. D. Keeler, and J. M. Kowalski, ''Layered neural networks with Gaussian hidden units as universal approximators,'' *Neural Computat.*, vol. 2, no. 2, pp. 210–215, 1990.

[102] E. D. Sontag, ''Sigmoids distinguish more efficiently than Heavisides,'' *Neural Computat.*, vol. 1, pp. 470–472, 1989.

[103] J. A. Leonard, M. A. Kramer, and L. H. Ungar, ''Using radial basis functions to approximate a function and its error bounds,'' *IEEE Trans. Neural Networks*, vol. 3, no. 4, pp. 624–626, July 1992.

[104] K. -Y. Siu and J. Bruck, ''Neural computation of arithmetic functions,'' *Proc. IEEE*, vol. 78, no. 10, Oct. 1990.

Neural Networks and their Applications

[105] S. -I. Amari, ''Mathematical foundations of neurocomputing,'' *Proc. IEEE*, vol. 78, no. 9, pp. 1443–1462, Sept. 1990.

[106] W. P. Jones and J. Hoskins, ''Backpropagation: A generalized delta learning rule,'' *Byte*, pp. 155–62, Oct. 1987.

[107] A. Guez, J. L. Eilbert, and M. Kam, ''Neural network architecture for control,'' *IEEE Contr. Syst. Mag.*, pp. 22–25, April 1988.

[108] S. C. Huang and Y. F. Hyang, ''Bounds on the number of hidden neurons in multilayer Perceptrons,'' *IEEE Trans. Neural Networks*, vol. 2, no. 1, pp. 47–55, 1991.

[109] W. Y. Huang and R. P. Lippmann, ''Neural net and traditional classifiers,'' in *Neural Information Processing Systems*, San Mateo CA: Morgan Kaufman Publishers, pp. 387–396, 1988.

[110] A. D. Kulkarni, ''Solving ill-posed problems with artificial neural networks,'' *Neural Networks*, vol. 4, pp. 477–484, 1991.

[111] D. F. Specht, ''Probabilistic neural networks,'' *Neural Networks*, vol. 3, pp. 109–118, 1990.

[112] H. White, ''Learning in artificial neural networks: A statistical perspective,'' *Neural Computat.*, vol. 1, no. 4, pp. 425–464, 1989.

[113] S. Lee and R. M. Kil, ''A Gaussian potential function network with hierarchically self-organizing learning,'' *Neural Networks*, vol. 4, pp. 207–224, 1991.

[114] M. T. Musavi, W. Ahmed, K. H. Chan, K. B. Faris, and D. M. Hummels, ''On the training of radial basis function classifiers,'' *Neural Networks*, vol. 5, pp. 595–603, 1992.

[115] S. Chen, C. F. N. Cowan, and P. M. Grant, ''Orthogonal least squares learning algorithm for radial basis function networks,'' *IEEE Trans. Neural Networks*, vol. 2, no. 2, pp. 302–309, March 1991.

System Identification and Control

[116] A. Guez and J. Selinsky, ''A trainable neuromorphic controller,'' *J. Robotic Syst.*, vol. 5, no. 4, pp. 363–388, 1988.

[117] S. R. Chi, R. Shoureshi, and M. Tenorio, ''Neural networks for system identification,'' *IEEE Contr. Syst. Mag.*, vol. 10, pp. 31–34, 1990.

[118] S. Chen, S. A. Billings, and P. M. Grant, ''Nonlinear system identification using neural networks,'' *Int. J. Contr.*, vol. 51, no. 6, pp. 1191–1214, 1990.

[119] S. A. Billings, H. B. Jamaluddin, and S. Chen, ''Properties of neural networks with applications to modeling nonlinear dynamical systems,'' *Int. J. Contr.*, vol. 55, no. 1, pp. 193–224, 1992.

[120] S. Mukhopadhyay and K. S. Narendra, ''Disturbance rejection in nonlinear systems using neural networks,'' *IEEE Trans. Neural Networks*, vol. 4, no. 1, pp. 63–72, Jan. 1993.

[121] A. U. Levin and K. S. Narendra, ''Control of nonlinear dynamical systems using neural networks: Controllability and stabilization,'' *IEEE Trans. Neural Networks*, vol. 4, no. 2, pp. 192–206, March 1993.

[122] T. Yabuta and T. Yamada, ''Neural network controller characteristics with regard to adaptive control,'' *IEEE Trans. Syst., Man, Cybern.*, vol. 22, no. 1, pp. 170–176, Jan./Feb. 1991.

[123] R. M. Sanner and J.-J. E. Slotine, ''Gaussian networks for direct adaptive control,'' *IEEE Trans. Neural Networks*, vol. 3, no. 6, pp. 837–863, Nov. 1992.

[124] D. H. Rao, M. M. Gupta, and H. C. Wood, ''Neural networks in control systems,'' in *Proc. IEEE Conf. Communicat. Computers, Power in Modern Environment*, Saskatoon, pp. 313–319, May 17–18, 1993.

[125] Y. Ichikawa and T. Sawa, ''Neural network applications for direct feedback controllers,'' *IEEE Trans. Neural Networks*, vol. 3, no. 2, pp. 224–231, March 1992.

[126] D. A. Hoskins, J. N. Hwang, and J. Vagners, ''Iterative inversion of neural networks and its application to adaptive control,'' *IEEE Trans. Neural Networks*, vol. 3, no. 2, pp. 292–301, March 1992.

[127] J. G. Kuschewski, S. Hui, and S. H. Zak, ''Application of feedforward neural networks to dynamical system identification and control,'' *IEEE Trans. Contr. Syst. Technol.*, vol. 1, no. 1, pp. 37–49, March 1993.

[128] M. M. Gupta and D. H. Rao, ''Adaptive control of unknown nonlinear systems using multi-stage dynamic neural networks,'' in *Proc. SPIE's Conf. Intelligent Robots and Computer Vision XI*, Boston, pp. 130–142, Nov. 18–22, 1992.

[129] M. M. Gupta, D. H. Rao, and P. N. Nikiforuk, ''Neuro-controller with dynamic learning and adaptation,'' *Int. J. Intell. Robotic Syst.*, vol. 7, no. 2, pp. 151–173, April 1993.

[130] D. H. Rao and M. M. Gupta, ''Dynamic neural adaptive control schemes,'' in *Proc. Amer. Contr. Conf.*, San Francisco, pp. 1450–1454, June 2–4, 1993.

[131] D. Sbarbaro-Hofer, D. Neumerkel, and K. Hunt, ''Neural control of a steel rolling mill,'' *IEEE Contr. Syst. Mag.*, pp. 69–75, June 1993.

Robotics

[132] H. Miyamota, M. Kawato, T. Setoyama, and R. Suzuki, ''Feedback-error-learning neural network for trajectory control of a robotic manipulator,'' *Neural Networks*, vol. 1, pp. 251–265, 1988.

[133] M. Kawato, Y. Uno, M. Isobe, and R. Suzuki, ''Hierarchical neural network model for voluntary movement with application to robotics,'' *IEEE Contr. Syst. Mag.*, pp. 8–15, April 1988.

[134] D. F. Bassi and G. A. Beckey, ''Decomposition of neural network model of robot dynamics: A feasibility study,'' *Simulation and AI*, vol. 220, pp. 8–13, 1989.

[135] H. Wang, T. T. Lee, and W. A. Gruver, ''A neuromorphic controller for a three-link biped robot,'' *IEEE Trans. Syst., Man, Cybern.*, vol. 22, no. 1, pp. 164–169, Jan./Feb. 1991.

[136] A. Guez and Z. Ahmad, ''Solution to the inverse kinematics problem in robotics by neural networks,'' in *Proc. IEEE Int. Conf. Neural Networks*, San Diego, CA, pp. 617–624, March 1988.

[137] J. Barhen, S. Gulati, and M. Zak, ''Neural learning of constrained nonlinear transformations,'' *IEEE Computer*, pp. 67–76, June 1989.

[138] M. M. Gupta, D. H. Rao, and P. N. Nikiforuk, ''Dynamic neural network based inverse-kinematics transformation of two- and three-linked robots,'' presented at IFAC Conf., Sydney, Australia, vol. 3, pp. 289–296, July 19–23, 1993.

[139] M. M. Gupta and D. H. Rao, ''Neural learning of robot inverse kinematics transformations,'' in *Neural and Fuzzy Systems: The Emerging Science of Intelligent Computing*. SPIE Press Series, S. Mitra, W. Kraske, and M. M. Gupta, Eds. (in press, 1994).

[140] W. J. Daunicht, ''Control of manipulators by neural networks,'' *IEE Proc.*, vol 136, pt. E, no. 5, pp. 395–399, Sept. 1989.

[141] L. C. Rabelo and X. J. R. Avula, ''Hierarchical neurocontroller architecture for robotic manipulation,'' *IEEE Contr. Syst. Mag.*, pp. 37–41, April 1992.

[142] M. M. Gupta and D. H. Rao, "Dynamic neural processor and its applications to robotics and control," in *Intelligent Control,* IEEE Press, M. M. Gupta and N. K. Sinha, Eds., (in press).

Miscellaneous

[143] B. P. Yuhas, M. H. Goldstein, Jr., T. J. Sejnowski, and R. E. Jenkins, "Neural network models of sensory integration for improved vowel recognition," *Proc. IEEE,* vol. 78, no. 10, pp. 1658–1668, Oct. 1990.

[144] D. J. Burr, "Experiments on neural net recognition of spoken and written text," *IEEE Trans. Acoust., Speech, Signal Process.,* vol. 36, no. 7, pp. 1162–1168, July 1988.

[145] R. P. Gorman and T. J. Sejnowski, "Learned classification of sonar targets using a massively parallel network," *IEEE Trans. Acoust., Speech, Signal Process.,* vol. 36, no. 7, pp. 1135–1140, July 1988.

[146] M. J. Willis, G. A. Montague, C. D. Massimo, M. T. Tham, and A. J. Morris, "Artificial neural networks in process estimation and control," *Automatica,* vol. 28, no. 6, pp. 1181–1187, 1992.

[147] D. H. Rao, P. N. Nikiforuk, M. M. Gupta, and H. C. Wood, "Neural equalization of communication channels," in *Proc. IEEE Conf. Communicat., Computers, Power in Modern Environment,* Saskatoon, pp. 282–290, May 17–18, 1993.

[148] D. H. Rao, P. N. Nikiforuk, and M. M. Gupta, "A central pattern generator model using dynamic neural processor," presented at World Congress on Neural Networks, vol. IV, pp. 533–536, Portland, OR, July 11–15, 1993.

[149] C. Moallemi, "Classifying cells for cancer diagnosis using neural networks," *IEEE Expert,* vol. 6, no. 6, pp. 8–12, Dec. 1991.

[150] L. Udapa and S. S. Udapa, "Neural networks for the classification of nondestructive evaluation signals," *IEE Proc.-F,* vol. 138, no. 1, pp. 41–45, Feb. 1991.

[151] J. Graf, "Long-term stock market forecasting using neural networks," *Neural Network World,* vol. 2, no. 6, pp. 615–620, 1992.

[152] M. Sabourin and A. Mitiche, "Optical character recognition by a neural network," *Neural Networks,* vol. 5, pp. 843–852, 1992.

[153] B. Kosko, "Bidirectional associative memories," *IEEE Trans. Syst., Man, Cybern.,* vol. 18, no. 1, pp. 49–60, Jan./Feb. 1990.

Dynamic Neural Networks

[154] A. Waibel, T. Hanazawa, G. Hinton, K. Shikano, and K. J. Lang, "Phoneme recognition using time-delay neural networks," *IEEE Trans. Acoust., Speech, Signal Process.,* vol. 37, no. 3, pp. 328–339, March 1989.

[155] J. J. Hopfield, "Neurons with graded response have collective computational properties like those of two-state neurons," *Proc. Nat. Acad. Sci.,* vol. 81, pp. 3088–3092, 1984.

[156] P. Strumillo and T. S. Durani, "Simulations of cardiac arrhythmia based on dynamical interactions between neural models of cardiac pacemakers," *IEE Publication No. 349, Second Int. Conf. Artificial Neural Networks,* pp. 195–199, Nov. 18–20, 1991.

[157] M. M. Gupta and G. K. Knopf, "A neural network with multiple hysteresis capabilities for short-term visual memory," in *Proc. Int. Joint Conf. Neural Networks,* vol. 1, Seattle, WA, July 8–12, 1991, pp. 671–676.

[158] M. M. Gupta and G. K. Knopf, "A multitask visual information processor with a biologically motivated design," *J. Visual Communicat., Image Representation,* vol. 3, no. 3, pp. 230–246, Sept. 1992.

[159] G. K. Knopf and M. M. Gupta, "A multi-purpose neural processor for machine vision systems," *IEEE Trans. Neural Networks,* (in press).

[160] T. Sejnowski and C. R. Rosenberg, "NETtalk: A neural network that learns to read aloud," Tech. Rep. JHU/EECS-86/01, Johns Hopkins Univ., 1986.

[161] D. H. Rao and M. M. Gupta, "A neural processor for coordinating multiple systems with dynamic uncertainties," in *Proc. Int. Symp. Uncertainty and Management* (ISUMA), Maryland, pp. 633–640, April 25–28, 1993.

[162] D. H. Rao and M. M. Gupta, "Dynamic neural network with somatic adaptation," in *Proc. IEEE Conf. Neural Networks,* San Francisco, pp. 558–563, March 28–April 1, 1993.

[163] D. H. Rao and M. M. Gupta, "A multi-functional dynamic neural processor for control applications," in *Proc. Amer. Contr. Conf.,* San Francisco, pp. 2902–2906, June 2–4, 1993.

[164] C. L. Giles, C. B. Miller, D. Chen, G. Z. Sun, and Y. C. Lee, "Learning and extracting finite state automata with second-order recurrent neural networks," *Neural Computat.,* vol. 4, no. 3, pp. 393–405, 1992.

[165] K. J. Lang, A. H. Waibel, and G. E. Hinton, "A time-delay neural network architecture from isolated word recognition," *Neural Networks,* vol. 3, no. 1, pp. 23–44, 1990.

[166] E. A. Wan, "Temporal backpropagation for FIR neural networks," in *Proc. Inter. Joint. Conf. Neural Networks (IJCNN),* pp. 575–580, June 1990.

[167] G. Tsutsumidani, N. Ohnishi, and N. Sugie, "Properties and learning algorithm of discrete neural network with time delay," in *Proc. Int. Joint Conf. Neural Networks (IJCNN),* pp. 529–534, Nov. 1991.

[168] S. I. Sudharsanan and M. K. Sundareshan, "Training of a three-layer dynamical recurrent neural network for nonlinear input-output mapping," in *Proc. Int. Joint Conf. Neural Networks (IJCNN),* pp. 111–115, Nov. 1991.

[169] R. J. Williams and D. Zipser, "A learning algorithm for continually running fully recurrent neural networks," *Neural Computat.,* vol. 1, no. 2, pp. 270–280, 1989.

[170] S. Gardellam, T. Kumagai, R. Hashimoto, and M. Wada, "On the dynamics and potentialities of a discrete-time binary neural network with time delay," in *Proc. Second Int. Conf. Fuzzy Logic and Neural Networks,* Fukuoka, Japan, pp. 493–499, July 17–22, 1992.

[171] F. J. Pinda, "Recurrent backpropagation and the dynamical approach to adaptive neural computation," *Neural Computat.,* vol. 1, pp. 161–172, 1989.

[172] H. Li and B. Xu, "A learning algorithm for MLN with dynamic neurons," in *Proc. Int. Joint Conf. Neural Networks (IJCNN),* pp. 523–528, Nov. 1991.

[173] A. Parlos, A. Atiya, and K. Chong, "Recurrent multilayer perceptron for nonlinear system identification," in *Proc. Int. Joint Conf. Neural Networks (IJCNN),* pp. 537–540, Nov. 1991.

[174] P. A. Anninos, B. Beek, T. J. Csermel, E. E. Harth, and G. Pertile, "Dynamics of neural structures," *J. Theoret. Biol.,* vol. 26, pp. 121–148, 1970.

[175] D. Zipser, "A subgrouping strategy that reduces complexity and speeds up learning in recurrent neural networks," *Neural Computat.,* vol. 1, pp. 552–558, 1989.

[176] Y. Fang and T. Sejnowski, "Faster learning for dynamic recurrent backpropagation," *Neural Computat.,* vol. 2, pp. 270–274, 1990.

[177] R. Krisnapuram and L.-F. Chen, "Implementation of parallel thinning algorithms using recurrent neural networks," *IEEE Trans. Neural Networks,* vol. 4, no. 1, pp. 142–147, Jan. 1993.

[178] B. C. Cragg and H. N. V. Temperley, "Memory: The analogy with ferro-magnetic hysteresis," *Brain,* vol. 78, pp. 304–316, 1955.

[179] J. Jondies, D. E. Irwin, and S. Yantis, "Integrating visual information from successive fixations," *Science,* vol. 25, pp. 192–194, 1982.

[180] N. H. Farhat, "Microwave diversity imaging and automated target identification based on models of neural networks," *Proc. IEEE,* vol. 77, no. 5, pp. 670–681, 1989.

[181] J. M. Zurada, *Introduction to Artificial Neural Systems,* St. Paul, MN: West Publishing Company, 1992.

[182] G. W. Hoffmann, "Neuron with hysteresis?" in *Computer Simulation in Brain Science.* R. Cotterill, Ed., Cambridge: Cambridge University Press, pp. 74–87, 1988.

[183] G. W. Hoffmann, "A neural network based on the analogy with the immune system," *J. Theoret. Biol.,* vol. 122, pp. 33–67, 1986.

[184] D. Fender and B. Julesz, "Extension of Panum's fusional area in binocularly stabilized vision," *J. Opt. Soc. Amer.,* vol. 57, no. 6, pp. 819–830, 1967.

[185] G. W. Hoffmann and M. W. Benson, "Neurons with hysteresis from a network that can learn without any changes in synaptic connection strengths," *Amer. Inst. Phys.*, pp. 219–225, 1986.

[186] N. E. Cotter and T. J. Guillerm, "The CMAC and a theorem of Kolmogorov," *Neural Networks*, vol. 5, pp. 221–228, 1992.

Fuzzy Logic and Fuzzy Neural Networks

[187] M. E. Cohen and D. L. Hudson, "An expert system on neural network techniques," in *The Proc. NAFIP*, I. B. Turksen, Ed., pp. 117–112, Toronto, June 1990.

[188] M. M. Gupta and G. K. Knopf, "Fuzzy neural network approach to control systems," in *Proc. First Int. Symp. Uncertainty Modeling and Analysis*, Maryland, pp. 483–488, Dec. 3–5, 1990.

[189] D. C. Kuncicky and A. Kandel, "A fuzzy interpretation of neural networks," in *The Proc. Third IFSA Congress*, J. C. Bezdek, Ed., pp. 113–116, Seattle, WA, 1989.

[190] I. Hayashi, H. Nomura, and N. Wakami, "Artificial neural network driven fuzzy control and its application to learning of inverted pendulum system," in *The Proc. Third IFSA Congress*, J. C. Bezdek, Ed., pp. 610–613, Seattle, WA, 1989.

[191] J. B. Kiszka and M. M. Gupta, "Fuzzy logic neural network," *BUSEFAL*, no. 4, pp. 104–109, 1990.

[192] S. Nakanishi, T. Takagi, K. Uehara, and Y. Gotoh, "Self-organizing fuzzy controllers by neural networks," in *Proc. Int. Conf. Fuzzy Logic and Neural Networks*, IIZUKA '90, pp. 187–192, Japan 1990.

[193] J. C. Bezdek, *Pattern Recognition with Fuzzy Objective Function Algorithms.* New York: Plenum Press, 1991.

[194] M. M. Gupta and J. Qi, "On fuzzy neuron models," in *Proc. Int. Joint Conf. Neural Networks (IJCNN)*, Seattle, pp. 431–456, July 1991.

[195] G. A. Carpenter, S. Grossberg, N. Markuzon, J. H. Reynolds, and D. B. Rosen, "Fuzzy ARTMAP: A neural network architecture for incremental supervised learning of analog multidimensional maps," *IEEE Trans. Neural Networks*, vol. 3, no. 5, pp. 698–713, Sept. 1992.

[196] P. K. Simpson, "Fuzzy min–max neural networks—Part I: Classification," *IEEE Trans. Neural Networks*, vol. 3, no. 5, pp. 776–786, Sept. 1992.

[197] M. M. Gupta, "Uncertainty and information: The emerging paradigms," *Int. J. Neuro and Mass-Parallel Compu., Informat. Syst.*, vol. 2, pp. 65–70, 1991.

[198] A. Kaufmann and M. M. Gupta, *Introduction to Fuzzy Arithmetic: Theory and Applications*, 2nd Ed. New York: Van Nostrand Reinhold, 1991.

[199] S. K. Paul and S. Mitra, "Multilayer Perceptron, fuzzy sets, and classification," *IEEE Trans. Neural Networks*, vol. 3, no. 5, pp. 683–697, Sept. 1992.

[200] M. M. Gupta and D. H. Rao, "Virtual cognitive systems (VCS): Neural–fuzzy logic approach," in *Proc. IFAC Conf.*, Sydney, Australia, vol. 8, pp. 323–330, July 19–23, 1993.

[201] Special Issue on Fuzzy Logic and Neural Networks, *IEEE Trans. Neural Networks*, vol. 3, no. 5, Sept. 1992.

[202] M. M. Gupta and E. Sanchez, Eds., *Fuzzy Information and Decision Processes.* New York, Amsterdam, Oxford: North-Holland, 1982.

[203] M. M. Gupta and E. Sanchez, Eds., *Approximate Reasoning in Decision Analysis.* New York, Amsterdam, Oxford: North-Holland, 1982.

[204] M. M. Gupta, A. Kandel, and W. Bandler, Eds., *Approximate Reasoning in Expert Systems.* New York, Amsterdam, Oxford: North-Holland, 1985.

[205] A. Kaufmann and M. M. Gupta, *Introduction to Fuzzy Arithmetic: Theory and Applications.* 2nd Ed., New York: Van Nostrand Reinhold, 1991, [Japanese Translation 1992, Tokyo: Ohmsha Ltd.

[206] M. M. Gupta and T. Yamakawa, Eds., *Fuzzy Computing: Theory, Hardware and Applications.* New York: Amsterdam, Oxford: North-Holland, 1988.

[207] M. M. Gupta, and T. Yamakawa, Eds., *Fuzzy Logic in Knowledge-Based Systems, Decision and Control.* New York, Amsterdam, Oxford: North-Holland, 1988.

[208] A. Kaufmann and M. M. Gupta, *Fuzzy Mathematical Models in Engineering and Management Science.* New York, Amsterdam, Oxford: North-Holland Japanese Translation by M. Matsuka, Tokyo: Ohmsha Ltd., 1992.

[209] B. M. Ayyub, M. M. Gupta, and L. N. Kanal, Eds., *Analysis and Management of Uncertainty: Theory and Application.* The Netherlands: Kluver Academic Publishers, 1992.

[210] M. M. Gupta, "Fuzzy neural computing systems," *Neural Network World*, vol. 2, no. 6, pp. 629–648, 1992.

[211] B. Kosko, *Neural Networks and Fuzzy Systems.* Englewood Cliffs, NJ: Prentice Hall, 1992.

Hardware Implementations

[212] A. M. Chiang and M. L. Chuang, "A CCD programmable image processor and its neural network applications," *IEEE J. Solid-State Circuits*, vol. 26, pp. 1894–1901, 1991.

[213] S. Kemeny, H. Torbey, H. Meadows, R. Bredthauer, M. La Shell, and E. Fossum, "CCD focal-plane image reorganization processors for lossless image compression," *IEEE J. Solid-State Circuits*, vol. 27, pp. 398–405, 1992.

[214] B. E. Boser, E. Sackinger, J. Bromley, Y. Le Cun, and L. D. Jackel, "An analog neural network processor with programmable topology," *IEEE J. Solid-State Circuits*, vol. 26, no. 12, pp. 2017–2025, 1991.

[215] H. P. Graf, L. D. Jackel, and W. E. Hubbard, "VLSI implementation of a neural network model," *IEEE Computer*, vol. 21, no. 3, pp. 41–49, 1988.

[216] J. Hutchinson, C. Koch, J. Luo, and C. Mead, "Computing motion using analog and binary resistive networks," *IEEE Computer*, vol. 21, pp. 52–64, 1981.

[217] H. Kobayashi, J. L. White, and A. A. Abidi, "An active resistor network for Gaussian filter of images," *IEEE J. Solid-State Circuits*, vol. 26, pp. 738–748, 1991.

[218] H. Li and C. H. Chen, "Simulating a function of visual peripheral processes with an analog VLSI network," *IEEE Micro*, vol. 11, pp. 8–15, 1991.

[219] M. A. Maher, S. P. Deweerth, M. A. Mahowald, and C. A. Mead, "Implementing neural architectures using analog VLSI circuits," *IEEE Trans. Circuits Syst.*, vol. 36, no. 5, pp. 643–653, 1989.

[220] M. A. Mahowald and C. Mead, "The silicon retina," *Sci. Amer.*, pp. 76–82, 1991.

[221] B. A. White and M. I. Elmasry, "The digi-Neocognitron: A digital Neocognitron neural network model for VLSI," *IEEE Trans. Neural Networks*, vol. 3, no. 1, pp. 73–81, 1992.

[222] J. Caulfield, J. Kinser, and N. K. Rogers, "Optical neural networks," *Proc. IEEE*, vol. 77, no. 10, pp. 1573–1583, 1989.

[223] K. Y. Hsu, H. Y. Li, and D. Psaltis, "Holographic implementation of a fully connected neural network," *Proc. IEEE*, vol. 78, no. 10, pp. 1637–1645, 1990.

[224] D. Casasent, "Multifunctional hybrid neural net," *Neural Networks*, vol. 5, pp. 361–370, 1992.

[225] A. F. Murray, "Silicon implementations of neural networks," *IEE Proc.-F*, vol. 138, no. 1, pp. 3–12, Feb. 1991.

[226] H. C. Card, C. R. Schneider, and W. R. Moore, "Hebbian plasticity in MOS synapses," *IEE Proc.-F.*, vol. 138, no. 1, pp. 13–16, Feb. 1991.

[227] C. Mead and M. Ismail, Eds., *Analog VLSI Implementation of Neural Systems.* Boston: Kluwer Academic Publishers, 1989.

[228] B. E. Boser, E. Sackinger, J. Bromely, Y. L. Cun, and L. D. Jackel, "Hardware requirements for neural network pattern classifiers: A case study and implementation," *IEEE Micro*, pp. 32–40, Feb. 1992.

[229] D. W. Tank and J. J. Hopfield, "Simple neural optimization networks: An A/D converter, signal decision circuit, and a linear programming circuit," *IEEE Trans. Circuits Syst.*, vol. 33, no. 5, pp. 533–541, 1986.

[230] S. W. Tsay and R. W. Newcomb, "VLSI implementation of ART1 memories," *IEEE Trans. Neural Networks*, vol. 2, no. 2, pp. 214–221, 1991.

[231] Special Issues on Neural Networks Hardware, *IEEE Trans. Neural Networks*, May 1992, May 1993.

Part 1
Neuro-Control Systems: Some Perspectives

Motivated by the biological neuronal processes and neural control mechanisms, researchers are exploring the field of computational neural systems—a new nonalgorithmic approach to information processing.

MAN has always dreamed of creating a portrait of himself, a machine that can walk, see, and think intelligently. The *neuron,* the basic information processing element in the central nervous system (CNS), plays an important and diverse role in human sensory processing, locomotion, control, and cognition (thinking, learning, adaptation, perception, etc.).

The field of *neuro-control,* the subject of this volume, has evolved over the last decade, more specifically over the last few years, and the intent of the researchers working in this field is to create an intelligent machine with several levels of control, just as nature does in the control of various biological functions.

In this first part of the volume, we present articles that deal with subjects such as the brain and biological learning, and control and adaptation. These articles, authored by some leading authorities in the field, present a biological perspective as well as the biological motivation to readers (students, teachers, and researchers). It is shown that biological neurons, each with a bandwidth of the order of about 400 Hz or so, possess some tremendous capacities and capabilities that are unrealizable even by the nano- and pico-silicon-based technologies. These capabilities, for almost real-time and on-line processing, are due to the layered nature of the network of neurons, with a high degree of parallelism of the biological computing processes.

The selection of the articles for this first part was motivated by the information processing, learning, and control aspects of biological processes. These articles also provide a wide perspective on the neuro-control research that has evolved during recent years. These articles lead the reader into some philosophical, scientific, and mathematical issues pertaining to the theory, design, and applications of neuro-control systems.

Just imagine a machine that can learn and recognize human speech with natural accents or handwriting with a fuzzy flow of characters and translate it into typed text. Think also about a computerized slave-robotic system that has learned the living habits of its master, and does all the household tasks (cooking, vacuuming, cleaning, gardening etc.) according to its master's wishes. It would be wonderful to have a robotic gardener that can water the flowers and vegetables, but that also can prune and weed the garden without damaging the useful plants. Questions arise whether algorithmic-based computer can do all the wonderful things that we all can do so easily. The human brain follows a nonalgorithmic approach with some wonderful attributes such as "genetics" and "learning."

Motivated by the organic machine—the *brain* and its *attributes*—Robert Hecht-Nielsen, in article (1.1), surveys the various theoretical and applied developments in the field of neurocomputing around the world. This article, which appeared in *IEEE Spectrum* in 1988, is responsible, to a large extent, for creating our interest in the field of neurocomputing, and that is the very reason why this article occupies the leading position in this volume.

Learning and storage of knowledge (memory) are two of the main attributes of the biological neuronal processes. In article (1.2), I. Kupfermann provides a vivid and stimulating description of the process of learning in biological processes. The author gives the biological plausibility for learning and memory, and operant conditioning for treating severe behavioral problems. Memory has two stages: long-term memory (LTM) and short-term memory (STM). LTM may be represented by plastic changes in the synapses, and STM is encoded into the reverberating circuits. It is reported that all regions of the CNS appear to contain neurons with properties of plasticity needed for memory. Although the physical changes representing learning are likely to be localized to specific neurons, the complex nature of parallel processing and learning ensures that these neurons are widely distributed in the nervous system. Therefore, even after extensive lesions, some traces of memory can remain. Furthermore, the brain has the capacity to take the limited remaining information, work it over, and reconstruct a good reproduction of the original events/scenes. In spite of several advances in the field of neurobiology, a major challenge confronting scientists is to determine the mechanisms of *plasticity* during the learning process. In the next article, we present the use of plasticity in neuro-control problems.

The neuronal mechanism of control and voluntary movement is very complex and not well understood. However, in article (1.3), M. Kawato, K. Furukawa, and R. Suzuki discuss how the CNS acquires the ability to control movement by making use of synaptic plasticity, not only from the standpoint of neuroscience, but also from that of robotics. In this article, the authors also propose a computational model of voluntary movement. The control and learning performance of the proposed model is investigated using computer simulations. The proposed model accounts for the control and the learning ability of the CNS, and also provides a promising parallel-distributed control scheme for large-scale complex robotic systems in space and manufacturing.

Next, we provide a basic introduction to the subject of intelligent control. In article (1.4), P. J. Antsaklis, K. M. Passino, and S. J. Wang present a discussion on the fundamental issues in autonomous system modeling and analysis. In article (1.5), K. S. Narendra and S. Mukhopadhyay present a two-level control mechanism. The higher level detects a failure using pattern recognition techniques, which activates a stabilizing control action at the lower level.

This part of the volume contains a small set of articles and should provide the reader with biological motivation and artificial neural network perspectives in the design of intelligent control systems. In the following parts, we will provide some details of the overview presented in this part.

Further Reading

[1] W. S. McCulloch and W. Pitts, "A logical calculus of the ideas imminent in nervous activity," *Bull. Math. Biophys.*, vol. 5, pp. 115–133, 1943.

[2] D. O. Hebb, *The Organization of Behavior.* New York: John Wiley & Sons, 1949.

[3] M. Black, "Vagueness: An exercise in logical analysis," *Phil. Sci.*, vol. 4, pp. 427–455, 1937.

[4] L. Brillouin, *Science and Information Theory.* New York: Academic Press, 1956.

[5] S. Cajal and Y. Ramon, "Les preuves objectives de l' unite anatomique des cellules nerveuses," *Trob. Lab. Inest. Biol. Univ. Madrid,* vol. 29, pp. 1–37, 1934, (Translation: M.V. Purkiss and C.A. Fox, Madrid: Instituto "Ramon y Cajal").

[6] N. Wiener, *Cybernetics.* New York: John Wiley & Sons, 1948.

[7] F. M. Reza, *An Introduction to Information Theory.* New York: McGraw-Hill, 1961.

[8] N. Wiener, *The Human Use of Human Beings.* New York: Avon Books, 1954.

[9] R. Penrose, *The Emperor's New Mind: Concerning Computers, Minds and the Laws of Physics.* Oxford: Oxford University Press, 1989.

[10] M. A. Arbib, *The Metaphorical Brain.* New York: John Wiley & Sons, 1972.

[11] M. A. Arbib, *Brains, Machines and Mathematics.* New York: Springer-Verlag, 1987.

[12] E. R. Kandel and J. H. Schwartz, *Principles of Neural Science.* New York: North-Holland, 1985.

[13] A. C. Scott, *Neurophysics.* New York: John Wiley & Sons, 1977.

[14] A. C. Guyton, *Text Book of Medical Physiology.* Philadelphia: W. B. Saunders Company, 1987.

[15] V. Brooks, *The Neural Basis for Motor Control.* New York: Oxford University Press, 1986.

[16] R. A. Schmidt, *Motor Control and Learning.* Champaign, IL: Human Kinetics Publishers, 1982.

[17] C. O. Lovejoy, "The origin of man," *Science,* vol. 211, pp. 341–350, 1981.

[18] W. S. McCulloch, *Embodiments of Minds.* Cambridge, MA: MIT Press, p. 20 of Introduction by S. Papert, 1965.

[19] K. Kornwachs and W. Von Lucadou, "Pragmatic information as a non-classical concept to describe cognitive processes," *Cognitive Syst.*, vol. 1, pp. 79–84, 1985.

[20] R. Hecht-Nielsen, *Neurocomputing.* Reading, MA: Addison-Wesley, 1990.

[21] S. I. Amari, "A Mathematical Approach to Neural Systems," in J. Metzler, Ed., *Systems Neuroscience.* New York: Academic Press, pp. 67-117, 1977.

[22] S. I. Amari and M. A. Arbib, Eds., "Competition and cooperation in neural nets," in *Lecture Notes in Biomathematics,* vol. 45. New York: Springer-Verlag, 1982.

[23] R. C. Conant, "Law of information which govern systems," *IEEE Trans. Syst., Man, Cybern.*, vol. 6, pp. 334–338, 1976.

[24] D. J. Amit, H. Gutfreund, and Y. Sompolinsky, "Spin-glass models of neural networks," *Phys Rev. A*, vol. 32, pp. 1007–1018, 1985.

[25] J. A. Anderson, "Cognitive and psychological computation with neural models," *IEEE Trans. Syst., Man, Cybern.*, vol. 13, pp. 799–815, 1983.

[26] S. Grossberg, *Neural Networks and Natural Intelligence.* Cambridge, MA: MIT Press, 1987.

[27] I. R. Goodman and H. T. Nguyen, *Uncertainty Models for Knowledge-Based Systems.* New York: North-Holland, 1985.

[28] L. A. Zadeh, "Fuzzy sets," *Inform. Contr.*, vol. 8, pp. 338–353, 1965.

[29] M. M. Gupta, "Fuzzy automata and decision processes: The first decade," presented at Sixth Triennial World IFAC Congress, Boston, Cambridge, 1975.

[30] M. M. Gupta, "On the cognitive computing: Perspectives," in *Fuzzy Computing: Theory, Hardware and Applications,* Amsterdam, NY: North Holland, 1988.

[31] J. S. Albus, "Outlines for a theory of intelligence," *IEEE Trans. Syst., Man, Cybern.*, vol. 21, no. 3, pp. 473–509, May/June 1991.

Neurocomputing: picking the human brain

Borrowing from biology, researchers are exploring neural networks—a new, nonalgorithmic approach to information processing

Imagine a computer that learns. Information is fed into it, along with examples of the conclusions it should be reaching or feedback on how it is doing—or the machine may even be left to its own devices. The computer simply runs through the material again and again, making myriads of mistakes but learning from them, until finally it gets itself into proper shape to carry out the task successfully. Such behavior is quite human, and naturally so; for the design of the machine's information-processing system, a neural network, was inspired by the structure of the human brain—its nerve cells, their interconnections, and their interactions—and by envy of what the brain can do.

As an alternative form of information processing, neurocomputing is fast becoming an established discipline, and some neural networks are already on the market. Neural networks are good at some things that conventional computers are bad at. They do well, for instance, at solving complex pattern-recognition problems implicit in understanding continuous speech, identifying handwritten characters, and determining that a target seen from different angles is in fact one and the same object.

Neural networks parallel-process immense quantities of information. Yet for a long time the only way to implement them was by simulating them laboriously, inefficiently, and at great expense on standard, serial computers. That situation is changing. Neurocomputers—hardware on which neural networks can be implemented efficiently—have reached the prototype stage at several companies, and some are already commercially available. All are coprocessor boards that plug into conventional machines. Developers include Hecht-Nielsen Neurocomputer Corp. (HNC), IBM Corp., Science Applications International Corp. (SAIC), Texas Instruments Corp., and TRW Inc.

Meanwhile, researchers at Boston University, the California Institute of Technology, the Helsinki University of Technology, Johns Hopkins University, the University of California at San Diego, and other universities have been investigating the theory behind neural networks and exploring their potential to solve problems that have stumped algorithmic computing for decades.

What is neurocomputing?

Nearly all automated information processing is at present based upon John von Neumann's "glorified adding machine" concept. But before such a computer can be programmed to carry out an information-processing function, some person has both to understand that function and to devise an algorithm for implementing it. For complex functions, such as computed axial tomography, development waits on the birth of geniuses capable of propounding the needed algorithm—in this case, Johan Radon and Alan Cormack.

Even worse, there may be tasks for which algorithms do not yet exist, or for which it is virtually impossible to write down a series of logical or arithmetic steps that will arrive at the answer. Yet in some of these cases it is possible to specify the tasks exactly and even develop an endless set of examples of the function being carried out. Many such tasks exist. There is no algorithmic software as yet for an automobile autopilot, a handwritten-character reader, a spoken-language translator, a system that can identify enemy airplanes or ships, or a system capable of recognizing continuous speech, regardless of who is speaking.

These tasks do have three important characteristics in common, however: humans know how to do them; large sets of examples of the tasks being carried out can be generated; and each task involves associating objects in one set with objects in another set. Such associations are known as mappings or transformations. For example, a computer that can read aloud must somehow associate groups of written letters, spaces, and punctuation with specific sounds, pauses, and intonations.

Robert Hecht-Nielsen
Hecht-Nielsen Neurocomputer Corp.

Defining terms

Adaptive coefficients: values of the previous computations (weights) of a processing element stored in its local memory, which modify subsequent computations.
Connection: a signal transmission pathway between processing elements, corresponding to the axons and synapses of neurons in a human brain, that connects the processing elements into a network.
Learning law: an equation that modifies all or some of the adaptive coefficients (weights) in a processing element's local memory in response to input signals and the values supplied by the transfer function. The equation enables the network to adapt itself to examples of what it should be doing and to organize information within itself, and thereby learn.
Processing element: an artificial neuron in a neural network, consisting of a small amount of local memory and processing power. The output from a processing element is fanned out and becomes the input to many other elements.
Scheduling function: a function that determines if and how often a processing element is to apply its transfer function.
Transfer function: a mathematical formula that, among other things, determines a processing element's output signal as a function of the most recent input signals and the adaptive coefficients (weights) in local memory. The transfer function includes the learning law of the processing element.
Transformations: mappings or associations of objects or representations in one set (such as written words) with objects or representations in another set (such as spoken sounds) according to some rule, which is typically not known to a human programmer, but is implicit in the training data.
Weight: within a processing element, an adaptive coefficient associated with a single input connection. The weight determines the intensity of the connection, depending on the network's design and the information it has learned.

Reprinted from *IEEE Spectrum*, vol. 25, no. 3, pp. 36–41, March 1988.

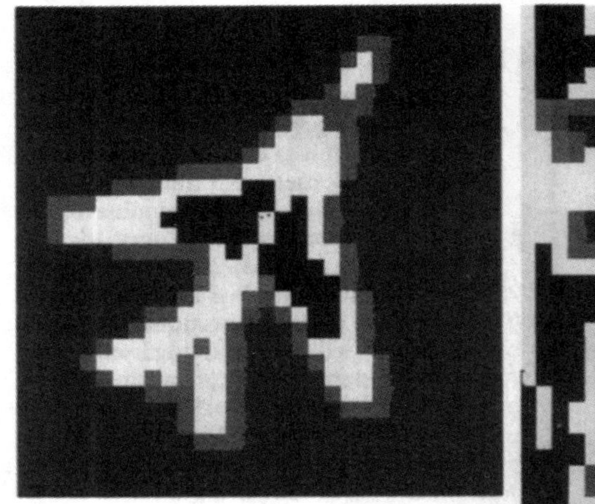

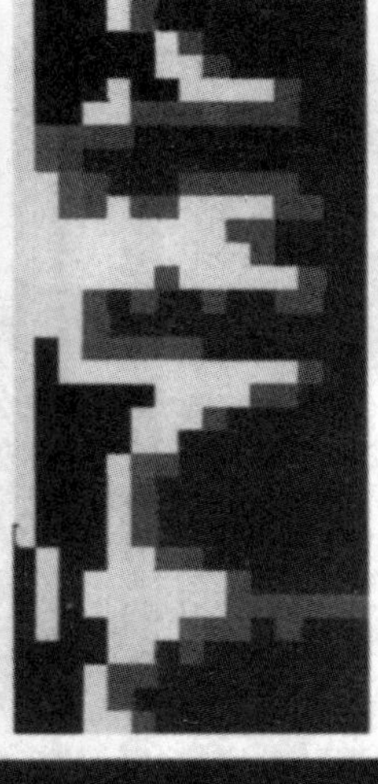

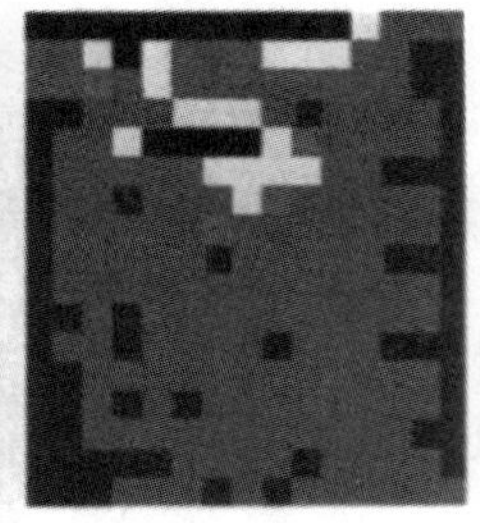

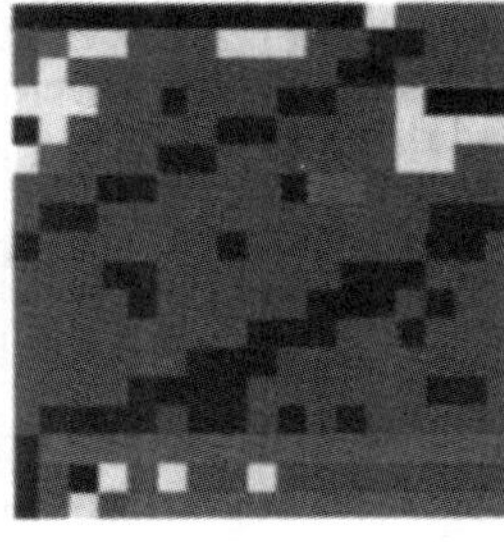

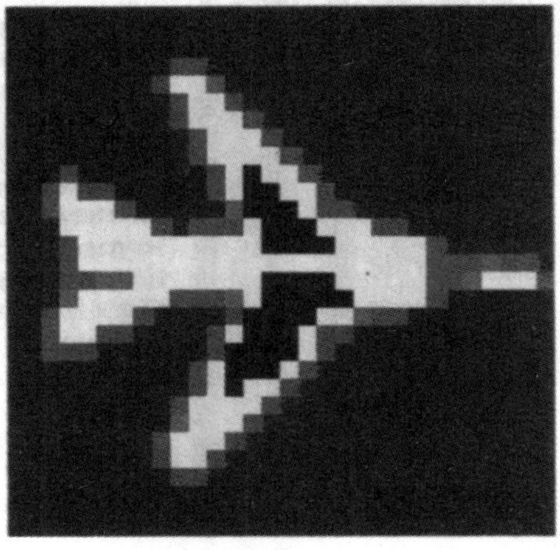

Michael Myers, TRW Inc.

Types of aircraft can be identified with 95-percent accuracy by the TRW Mark IV neurocomputer. The input image of an airplane taken from directly overhead with the nose at any angle (top left), *is digitized into picture elements (pixels), and their brightness is color coded. A neural network takes each pixel from the image, calculates its radius from the center and its angular coordinate on a polar scale, and enters this information on a graph* (top middle). *Next, the network does a Fourier transform on each vertical column, giving a processed image* (top right) *that is independent of the rotation of the plane in the input image. The processed image is fed into a second neural network, one trained to recognize images and classify them; this network mulls the problem* (bottom left), *picks a category for the new image, and produces one of the training images of a plane in that category* (bottom right). *(All the training images had the aircraft noses at the 3 o'clock position.)*

In formal terminology, neurocomputing is the engineering discipline concerned with nonprogrammed adaptive information-processing systems—neural networks—that develop associations (transformations or mappings) between objects in response to their environment. Instead of being given a step-by-step procedure for carrying out the desired transformation, the neural network itself generates its own internal rules governing the association, and refines those rules by comparing its results to the examples. Through trial and error, the network literally teaches itself how to do the task.

Neurocomputing is a fundamentally new and different information-processing paradigm—the first alternative to algorithmic programming. Wherever it is applicable, totally new information-processing capabilities can be developed, and development costs and time often shrink by an order of magnitude.

Neurocomputing does not, however, replace algorithmic programming. For one thing, neurocomputing is still in its infancy and is currently applicable to only certain classes of problems. More important, on a philosophical level, it is now suspected that neurocomputing and algorithmic programming may be conceptually incompatible. Transformations often prove impossible to describe satisfactorily in terms of an algorithm, and vice versa. This fact often unsettles people who think solely in procedural terms, since it means that neurocomputing may solve important information-processing problems without disclosing the rules used in the solution (at least in terms of present-day information-processing concepts). Neuroscience itself may run into this same problem in the future: an accurate understanding of how individual nerve cells interact in the brain may reveal virtually nothing about how brains process information.

How a neural network works

A neural network is modeled on the gross structure of the brain: a collection of nerve cells, or neurons, each of which is connected to as many as 10 000 others, from which it receives stimuli—inputs and feedback—and to which it sends stimuli. Some of those connections are strong; others are weak. The brain accepts inputs and generates responses to them, partly in accordance with its genetically programmed structure, but mainly through learning, organizing itself in reaction to input rather than by doing only by rote what it is told.

The neural networks used by engineers are only loosely based upon biology. At best, the only fair comparison is that they behave in a vaguely similar way. Since we have almost no idea how brains work, it will be a long time before we can re-create in a machine all the capabilities of the brain. Even so, neurocomputing is already offering some valuable, specialized, brain-like capabilities that in all likelihood lie beyond the reach of algorithmic programming.

A neural network consists of a collection of processing elements. Each processing element has many input signals, but only

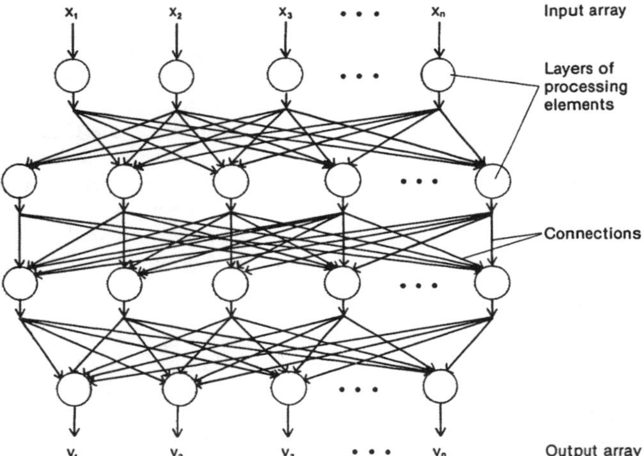

Conceptually, a neural network consists of many processing elements (circles), *each connected to many others. An input array, or sequence of numbers, is entered into the network. Each processing element in the first layer takes a component of the input array, operates on it in parallel with the other processing elements in the layer according to the transfer function, and delivers a single output to processing elements in a layer below. The result is an output array representing some characteristic associated with the input. Since inputs and adaptive coefficients (weights) can change over time, the network adapts and learns.*

a single output signal. The output signal fans out along many pathways to provide input signals to other processing elements. These pathways connect the processing elements into a network [see figure, lower right, p. 37].

Each processing element typically has its own small local memory, which stores the values of some previous computations along with the adaptive coefficients basic to neural-network learning. The processing that each element does is determined by a transfer function—a mathematical formula that defines the element's output signal as a function of whatever input signals have just arrived and the adaptive coefficients present in the local memory. Often a neural network is divided into layers—groups of processing elements all having the same transfer function.

Depending on the design of the neural network, the processing elements either operate continuously or are updated episodically. A scheduling function determines in which way and how often each processing element is to apply its transfer function.

Each processing element is completely self-sufficient and works away in total disregard of the processing going on inside its neighbors. In any neural network a great deal of independent parallel computation is usually under way.

At the same time, all the processing elements intimately affect the behavior of the entire network, since each element's output becomes the input to many others. The topology of the connections among processing elements influences what information-processing functions a neural network can carry out, as it determines what data each processing element receives and therefore the information on which it can act.

By and large, every connection entering a processing element has an adaptive coefficient called a weight assigned to it. This weight, which is stored in the local memory of the processing element, is generally used to amplify, attenuate, and possibly change the sign of the signal in the incoming connection. Often, the transfer function sums this and other weighted input signals to determine the value of the processing element's next output signal. Thus the weights determine the strength of the connections from neighboring processing elements.

Learning without programming

The weights, moreover, are not fixed but may change. Most transfer functions include a learning law—an equation that modifies all or some of the weights in the local memory in response to the input signals and the values supplied by the transfer function. In effect, the learning law allows the processing element's response to change with time, depending on the nature of the input signals. It is the means by which the network adapts itself to the answers desired and so organizes information within itself—in short, learns.

A neural network learns how to process information usually by being given either supervised training or graded training. In both, it runs through a series of trials. In supervised training, the network is supplied with both input data and desired output data (correct answers as examples). After each trial, the network compares its own output with the right answers, corrects any differences, and tries again, iterating until the output error reaches an acceptable level. In graded training, the network is given input data but no desired output data; instead, after each trial or series of trials it is given a grade or performance score that tells it how well it is doing.

Bernard Widrow, Stanford University

Broomstick balancing has become a classic test of a neural network's performance in adaptive control. The original experiment, conducted in 1962 by Bernard Widrow at Stanford University (now professor of electrical engineering), used his Madaline (Multiple ADAptive LINear Elements) neurocomputer (left), *with sensors on the broomstick and cart indicating position, angle, velocity, and acceleration.*

In either case, after training, the network is ready to process genuine inputs. At this point, depending upon the task to be done, a human operator may disable the learning law and "freeze the weights" of the connections, so that the network will stop adapting itself to new data and speed up its processing. For example, if the network has been well trained to read aloud from written text, it need not continue to learn on real data. On the other hand, if the network is to control the attitude of an orbiting spacecraft whose mass will decrease as fuel is spent, it should continue to adapt itself to changing conditions.

From concept to hardware

Neural networks are varied. At least 50 different types are being explored in research or being developed for applications. Of these, 13 are in common use [Table I]. Although all consist of processing elements joined by a multiplicity of connections, they differ in the learning laws incorporated into their transfer functions, the topology of their connections, and the weights assigned to their connections. In fact, some neural networks that learn may not be trained (self-organizing map) and some do not even learn at all (Hopfield network). The result: a host of networks, each suited to different types of tasks.

Any neural-network architecture can take different physical forms: electronic (in which everything consists of electronic devices and circuitry), electrooptical (in which optical signals link electronic processing elements), or entirely optical (in which light signals link optical processing elements made out of some nonlinear optical material). Experimenters at various institutions are working on all of these approaches. But problems with materials make the physical hardware of a network less straightforward than the conceptual architecture ["Neural networks: the physical reality," p. 40].

A number of neurocomputers—specialized machines able to efficiently and cost-effectively implement neural networks—have been built, and a few are now becoming commercially available [Table II]. All of them are configured as coprocessors to a standard serial computer, which acts as the host. The neurocomputer coprocessor, usually a board that looks much like any other circuit board, is connected to the host through a shared data bus or through a standard peripheral interconnection—PC-bus, Ethernet, or DRV-11.

As coprocessors, neurocomputers can be thought of as just another type of peripheral, like a printer or external disk drive. Data are shuffled into and out of the neurocomputer by the host computer through software routines supplied by the neurocomputer's manufacturer, even though the neurocomputer does its actual work on the data nonalgorithmically. No one has proposed a stand-alone neurocomputer—mainly because present networks do not handle input-output processing, and host computers already do that very well.

Most of the neurocomputer coprocessors built so far have been hard-wired designs optimized for implementing one type of neural network or a small selection of types. A few neurocomputers—notably those developed at IBM, Texas Instruments, and TRW—can implement several classes of networks, and a very few—such as those developed at HNC and SAIC—can implement essentially any neural network.

Such flexibility is valuable because the ability to quickly modify the neural network being used is critically important in research, application studies, and early applications. As with von Neumann machines, though, the more general-purpose a neurocomputer is, the slower it is. After the applications of neural networks are understood more fully, it may be possible to substitute specialized neurocomputers in those applications where only a particular network, or small range of networks, is needed.

For those who wish to get deeply involved in altering the structure and thus the behavior of a network, there are languages designed expressly for describing neural networks in a high-level, machine-independent way. The field is very new, and so far only four neurosoftware languages have been introduced: P3, Panspec, AnSpec, and Axon.

Neural networks applied

Several organizations are trying to apply neural networks to information-processing problems in commerce and industry that have proved intractable or far too expensive with algorithmic computers. Preliminary results have been encouraging.

Behavioristics Inc., Silver Spring, Md., has demonstrated a neural network for scheduling airline flights. Airlines sell seats at

I. Thirteen best-known neural networks

Name of network	Inventors and developers	Years introduced	Primary applications	Limitations	Comments
Adaptive resonance theory	Gail Carpenter, Northeastern U.; Stephen Grossberg, Boston U.	1978–86	Pattern recognition, especially when pattern is complicated or unfamiliar to humans (radar or sonar readouts, voiceprints)	Sensitive to translation, distortion, changes in scale	Very sophisticated; not yet applied to many problems
Avalanche	Stephen Grossberg, Boston U.	1967	Continuous-speech recognition; teaching motor commands to robotic arms	Literal playback of motor sequences—no simple way to alter speed or interpolate movements	Class of networks—no single network can do all these tasks
Back propagation	Paul Werbos, Harvard U.; David Parker, Stanford U.; David Rumelhart, Stanford U.	1974–85	Speech synthesis from text; adaptive control of robotic arms; scoring of bank loan applications	Supervised training only—correct input-output examples must be abundant	The most popular network today—works well, simple to learn
Bidirectional associative memory	Bart Kosko, U. of Southern California	1985	Content-addressable associative memory	Low storage density; data must be properly coded	Easiest network to learn—good educational tool; associates fragmented pairs of objects with complete pairs
Boltzmann and Cauchy machines	Jeffrey Hinton, U. of Toronto; Terry Sejnowsky, Johns Hopkins U.; Harold Szu, Naval Research Lab	1985–6	Pattern recognition for images, sonar, radar	Boltzmann machine: long training time. Cauchy machine: generating noise in proper statistical distribution	Simple networks in which noise function is used to find a global minimum
Brain state in a box	James Anderson, Brown U.	1977	Extraction of knowledge from data bases	One-shot decision making—no iterative reasoning	Similar to bidirectional associative memory in completing fragmented inputs
Cerebellatron	David Mar, MIT; James Albus, NBS; Andres Pellionez, NYU	1969–82	Controlling motor action of robotic arms	Requires complicated control input	Similar to avalanche network; can blend several command sequences with different weights to interpolate motions smoothly as needed
Counterpropagation	Robert Hecht-Nielsen, Hecht-Nielsen Neurocomputer Corp.	1986	Image compression; statistical analysis; loan application scoring	Large number of processing elements and connections required for high accuracy for any size of problem	Functions as a self-programming look-up table; similar to back propagation only simpler, although also less powerful
Hopfield	John Hopfield, California Inst. of Technology and AT&T Bell Labs	1982	Retrieval of complete data or images from fragments	Does not learn—weights must be set in advance	Can be implemented on a large scale
Madaline	Bernard Widrow, Stanford U.	1960–62	Adaptive nulling of radar jammers; adaptive modems; adaptive equalizers (echo cancellers) in telephone lines	Assumes a linear relationship between input and output	Acronym stands for multiple adaptive linear elements; powerful learning law; in commercial use for more than 20 years
Neocognitron	Kunihiko Fukushima, NHK Labs	1978–84	Handprinted-character recognition	Requires unusually large number of processing elements and connections	Most complicated network ever developed; insensitive to differences in scale, translation, rotation; able to identify complex characters (such as Chinese)
Perceptron	Frank Rosenblatt, Cornell U.	1957	Typed-character recognition	Cannot recognize complex characters (such as Chinese); sensitive to difference in scale, translation, distortion	The oldest neural network known; was built in hardware; rarely used today
Self-organizing map	Teuvo Kohonen, Helsinki U. of Technology	1980	Maps one geometrical region (such as a rectangular grid) onto another (such as an aircraft)	Requires extensive training	More effective than many algorithmic techniques for numerical aerodynamic flow calculations

Self-organizing map, a neural network devised by Teuvo Kohonen of the Helsinki University of Technology, can map a rectangular grid onto a nonrectangular shape. Here, the map is approximating a triangle, having arranged itself so that the squares remain equal in area so far as is possible. In modeling three-dimensional shapes (such as airframes) for numerical analysis, the network performs the mappings more accurately than most algorithmic techniques.

different fares depending how far in advance a reservation is made. The system, called the Airline Marketing Tactician, optimizes over time the allocation of seats between discount and standard fare classes to maximize the airline's profits. At present, several major airlines are considering the system.

Murray Smith, president of Adaptive Decision Systems Inc., Andover, Mass., has shown how accurately neural networks can score applications for bank loans. The network is given relevant data from a loan application form, and judges the applicant as a good or bad credit risk in terms of the examples on which it has been trained. The scoring system based on neurocomputing performed much more accurately than an existing operational point-scoring system based upon a combination of an expert system and a statistical model. A neurocomputer loan-scoring application has been developed for a major finance company, which plans to install it in the field around midyear.

A great challenge has been to get a computer to recognize human speech and translate it into written text—especially when a person speaks naturally, running words together in a continuous stream rather than articulating every syllable and pausing between individual words. Jeffrey Elman, associate professor of linguistics of the University of California at San Diego, has shown that neural networks can pick individual words out of connected streams and devise representations for them. The neural-network speech-recognition system with the highest accuracy and largest vocabulary was developed by Teuvo Kohonen, research professor in technical physics at the Helsinki University of Technology, under contract to Asahi Chemical Co. of Tokyo, Japan. The system has a front-end network that recognizes short, phoneme-like fragments of speech and a back-end network that recognizes strings of the fragments as words. A post-processing system uses context to distinguish between words that sound alike.

Neural networks have also broken ground in image recognition. Kohonen has shown that an associative memory network can take the image of a partially obscured face, complete it, and identify it with an image in a memory of 500 different people's faces. Kunihiko Fukushima, senior research scientist of NHK Laboratories in Tokyo, and Sei Miyake, a research director at the Automated Telecommunications Research Center in Osaka, have demonstrated a network, the neocognitron, that can identify hand-printed characters with 95-percent accuracy, regardless of shifts in position, changes in scale, and even small distortions.

Adaptive-control problems have been solved by neural networks for more than 25 years [see photograph, p. 38]. The early demonstrations have prompted efforts to apply neural networks to the practical "eye-hand coordination" of robot arms moving in response to feedback from camera images. Remarkable progress has been made by Andres Pellionez, research associate professor of physiology and biophysics of New York University, New York City, who has demonstrated the ability to move a finger of a robot arm in a beautifully coordinated perfect straight line at any angle.

Complementary to algorithmic computing

Algorithmic computing and neurocomputing complement each other nicely. The first is ideal for accounting, aerodynamic and hydrodynamic modeling, and the like. The second is ideal for pattern recognition, fuzzy knowledge processing, and adaptive control. Neurocomputers should not be used to balance checkbooks. Algorithmic computers should not be used to recognize speech. The two can also be integrated easily in hardware.

A technology succeeds, it seems, if it fits smoothly into an existing infrastructure and important applications are developed quickly. Neurocomputing appears to fill the bill. And if the his-

Neural networks: the physical reality

A neural network should in principle be simple to build. In the most basic networks, a processing element needs only to take all its incoming signals, multiply them by the weights of the connections over which they entered, add up the intermediate answers, and multiply the total by a nonlinear function to give its single output. If the incoming signals are voltages, then the weights can be represented by resistors, and by Ohm's law the intermediate answers are currents; by Kirchoff's law, all these currents can be summed by connecting the currents together at one terminal to give the output. Seemingly, then, that processing elements in the simplest neural networks should be resistors at the intersections of connecting wires.

A number of investigators—notably John Hopfield and others at AT&T Bell Laboratories—have indeed built neural networks of wires and resistors. But translating them into chips has proven difficult because it is virtually impossible to build accurate resistors on silicon wafers.

As a result, some neural-network engineers have resorted to "faking" accurate resistances on silicon. For example, Hans Peter Graf, member of the technical staff, and Larry Jackel, head of the device structure research department, at AT&T Bell Laboratories at Holmdel, N.J., have designed a combination analog-and-digital chip in which a digital gate at the crosspoint of the conductors acts in effect like a resistor with a value appropriate to the processing element. Carver Mead, the Gordon and Eddy Moore professor of computer science at the California Institute of Technology in Pasadena, has designed a neurocomputing chip on which each accurate linear "resistor" is simulated by a group of seven digital transistors. Jay Sage, staff member of the analog device technology group at MIT Lincoln Laboratory, Lexington, Mass., has designed a chip in which the same kind of field-effect transistor used in electrically erasable programmable ROMs acts as a resistor whose value is determined by charge buried under the gate.

Other neural network researchers have gotten around the problem by abandoning traditional VLSI processing materials. For example, Satish Khanna, technical group supervisor of advanced materials and devices section, and his colleagues at Jet Propulsion Laboratory, Pasadena, Calif., have devised resistors out of silicon hydride compounds, which are deposited on the spots where the rows and columns of connections meet. For any given network, current pulses sent along the pairs of vertical and horizontal conductors heat the spots of silicon hydride and drive off hydrogen, tailoring the resistance by up to three orders of magnitude.

Still other researchers have circumvented the problem by creative computation techniques. For example, in the Anza Plus neurocomputer manufactured by the Hecht-Nielsen Neurocomputer Corp., a processing element is not a physical entity at all. Instead, it is a slice of time on a VLSI digital arithmetic processing chip: its values are computed in a few microseconds, and then those values are stored in memory until needed again. Thus many processing elements share the same physical hardware, creating a "virtual network"—an approach that has proven to be economic, since VLSI is so fast and powerful. —*R.H.-N.*

II. Neurocomputers built to date*

Neurocomputer	Year introduced	Technology	Capacity: Number of processing elements	Capacity: Number of connections	Capacity: Number of networks†	Speed: Connections per second‡	Developers	Status§
Perceptron	1957	Electromechanical and electronic	8	512	1	10^3	Frank Rosenblatt, Charles Wightman, Cornell Aeronautical Laboratory	Experimental
Adaline/Madaline	1960/62	Electrochemical (now electronic)‖	1/8	16/128	1	10^4	Bernard Widrow, Stanford U.	Commercial
Electro-optic crossbar	1984	Electro-optic	32	10^3	1	10^5	Demitri Psaltis, California Inst. of Technology	Experimental
Mark III	1985	Electronic	8×10^3	4×10^5	1	3×10^5	Robert Hecht-Nielsen, Todd Gutschow, Michael Myers, Robert Kuczewski, TRW	Commercial
Neural emulation processor	1985	Electronic	4×10^3	1.6×10^4	1	4.9×10^5	Claude Cruz, IBM	Experimental
Optical resonator	1985	Optical	6.4×10^3	1.6×10^7	1	1.6×10^5	Bernard Soffer, Yuri Owechko, Gilbert Dunning, Hughes Malibu Research Labs	Experimental
Mark IV	1986	Electronic	2.5×10^5	5×10^6	1	5×10^6	Robert Hecht-Nielsen, Todd Gutschow, Michael Myers, Robert Kuczewski, TRW	Experimental
Odyssey	1986	Electronic	8×10^3	2.5×10^5	1	2×10^6	Andrew Penz, Richard Wiggins, Texas Instruments Central Research Labs	Commercial
Crossbar chip	1986	Electronic	256	6.4×10^4	1	6×10^9	Larry Jackel, John Denker and others, AT&T Bell Labs	Experimental
Optical novelty filter	1986	Optical	1.6×10^4	2×10^6	1	2×10^7	Dana Anderson, U. of Colorado	Experimental
Anza	1987	Electronic	3×10^4	5×10^5	No limit	2.5×10^4 (1.4×10^5)	Robert Hecht-Nielsen, Todd Gutschow, Hecht-Nielsen Neurocomputer Corp.	Commercial
Parallon 2	1987	Electronic	10^4	5.2×10^4	No limit	1.5×10^4 (3×10^4)	Sam Bogoch, Oren Clark, Iain Bason, Human Devices	Commercial
Parallon 2x	1987	Electronic	9.1×10^4	3×10^5	No limit	1.5×10^4 (3×10^4)		Commercial
Delta floating-point processor	1987	Electronic	10^6	10^6	No limit	2×10^6 (10^7)	George A. Works, William L. Hicks, Stephen Deiss, Richard Kasbo, Science Applications Int'l Corp.	Commercial
Anza plus	1988	Electronic	10^6	1.5×10^6	No limit	1.5×10^6 (6×10^6)	Robert Hecht-Nielsen, Todd Gutschow, Hecht-Nielsen Neurocomputer Corp.	Commercial

*Numbers given pertain to individual boards or chips. More than one board may be used to build an individual machine.
†Number of networks that can be simultaneously resident on the board, without going to an outside memory peripheral.
‡Speed outside parentheses is with learning; speed inside parentheses is without learning.
§"Experimental" describes a one-of-a-kind device or machine built to explore an idea or prove a point; "commercial" describes a device or machine that has been offered for sale.
‖Early versions required continuous electroplating lasting about a minute for full-scale change.

tory of computing is any guide, the capabilities of neural networks are likely to grow with every successive generation.

To probe further

An excellent review of neurocomputing in the 1950s and 1960s can be found in *Learning Machines* by Nils Nilsson, McGraw-Hill, N.Y., 1965. James Anderson and Edward Rosenfeld have collected classic papers about neural networks in their book *Neurocomputing*, MIT Press, Cambridge, Mass., 1988.

David Tank and John Hopfield review their work in neurocomputing in "Collective Computation in Neuronlike Circuits," *Scientific American*, December 1987, pp. 104–114.

The IEEE 1988 International Conference on Neural Networks, cosponsored by half a dozen IEEE societies, will be held July 24–27 in San Diego. Contact Nomi Feldman, IEEE ICNN-88 Conference Secretariat, 3770 Tansy St., San Diego, Calif. 92121.

A new quarterly journal, *Neural Networks*, began publication in January 1988. A subscription to it is included in the annual membership fee ($45 regular, $35 student) of the International Neural Network Society, founded last year; for information, write to the society's secretary-treasurer Harold Szu, Naval Research Laboratory, Code 5756, Washington, D.C. 20375. The society will hold its 1988 annual meeting in Boston, Sept. 6–10.

The state of the art in neurocomputing is detailed in the four-volume *Proceedings of the IEEE First International Conference on Neural Networks*, held in San Diego, Calif., June 21–24, 1987. The set, IEEE Catalog No. 87TH0191-7, is available from the IEEE Service Center, 445 Hoes Lane, Piscataway, N.J. 08854.

Article 1.2

Irving Kupfermann

Learning

Certain Elementary Forms of Learning Are Nonassociative

Classical Conditioning Involves Associating a Conditioned and an Unconditioned Stimulus

Conditioning Involves the Learning of Predictive Relationships

Operant Conditioning Involves Associating an Organism's Own Behavior with a Subsequent Reinforcing Environmental Event

Food-Aversion Conditioning Illustrates How Biological Constraints Influence the Efficacy of Reinforcers

Conditioning Is Used as a Therapeutic Technique

Classical Conditioning Has Been Applied in Systematic Desensitization

Operant Conditioning Has Been Used to Treat Severe Behavioral Problems

Learning and Memory Can Be Classified as Reflexive or Declarative on the Basis of How Information Is Stored and Recalled

The Neural Basis of Memory Can Be Summarized in Four Principles

Memory Has Stages

Long-term Memory May Be Represented by Plastic Changes in the Brain

Memory Traces Are Often Localized in Different Places Throughout the Nervous System

Reflexive and Declarative Memories May Involve Different Neuronal Circuits

Reflexive Memory
Declarative Memory
Reflexive Versus Declarative Memory in Amnesic Patients

Selected Readings

References

In Chapter 60 we considered how inborn and environmental factors interact to produce behavior. The environmental factor most important in altering behavior in humans is learning. *Learning* is the acquisition of knowledge about the world. *Memory* is the retention or storage of that knowledge. The study of learning has taught us about the logical capabilities of the brain and is therefore an objective and powerful approach to evaluating mental processing. In the study of learning we can ask several related questions: What types of environmental relationships are learned most easily? What conditions optimize learning? How many different forms of learning are there? What are the stages of memory formation?

Learning is a process that can occur in the absence of overt behavior but its occurrence can only be inferred by seeing changes in behavior. The overt behavioral changes, and the other changes that cannot be detected simply by the organism's overt behavior, all reflect alterations in the brain produced by learning. For this reason, although learning has been very fruitfully studied by purely behavioral techniques, many of the fundamental questions about learning will require direct examination of the brain.

The study of learning is central to the understanding of both normal and abnormal behavior. Learning is thought to contribute to the genesis of certain mental and somatic diseases, and the principles governing learning that have emerged from laboratory studies are used in the treatment of patients with these diseases. Moreover, behavioral techniques based on learning are now used widely in neurobiological and clinical research to assess the effects of brain lesions and drugs.

Psychologists study learning by exposing organisms to information about the world, usually specific types of controlled sensory experiences. By

Reprinted with permission from *Principles of Neural Science*, II Edition, pp. 805–815, 1985.

this means, two major procedures (or paradigms) have been discovered, which give rise to two major classes of learning: nonassociative learning and associative learning. In *nonassociative learning* the organism is exposed once or repeatedly to a single type of stimulus. This procedure provides an opportunity for the organism to learn about the properties of that stimulus. In *associative learning* the organism learns about the relationship of one stimulus to another (classical conditioning) or about the relationship of a stimulus to the organism's behavior (operant conditioning).

Certain Elementary Forms of Learning Are Nonassociative

The most common forms of learning are nonassociative and include habituation and sensitization. *Habituation*, first systematically studied by the Russian biologist Ivan Pavlov, is a decrease in a behavioral reflex response to a repeated, nonnoxious stimulus. An example of habituation is the failure of a person to show a startle response to a loud noise that has been regularly presented. In *sensitization* (or *pseudoconditioning*) there is an increased reflex response to a wide variety of stimuli for a period of time after an intense or noxious stimulus has been delivered. A sensitized animal responds more vigorously to a mild tactile stimulus after it has received a painful pinch. Moreover, a sensitizing stimulus can override the effects of habituation. For example, after a startle response to a noise has become habituated, it can be restored by delivering a strong pinch. This process is called *dishabituation*. Sensitization and dishabituation occur whether or not the intense stimulus is presented soon after the weaker stimulus; no close association between the two stimuli is needed.

Not all examples of nonassociative learning are simple. There are many types of more complex learning in which there is no obvious associational element, although hidden forms of association may be present. These types of learning include *sensory learning*, in which a continuous record of sensory experience is formed, and *imitation learning*, which includes aspects of the acquisition of language.

Classical Conditioning Involves Associating a Conditioned and an Unconditioned Stimulus

There are many types of associative learning. One useful way to classify them is on the basis of the experimental procedures used to establish the learning. Two experimental paradigms have been extensively studied and used clinically: classical and operant conditioning. Associative learning also often occurs in forms that do not readily fit into an operant–classical schema.

Classical conditioning was introduced into behavioral science by Ivan Pavlov at the turn of the century, when he recognized that learning frequently consists of the acquisition of responsiveness to a stimulus that originally was ineffective. Aristotle had earlier suggested the theory that learning involves the association of ideas, a proposal developed further by John Locke and the British empiricist philosophers, the forerunners of modern psychologists. Pavlov's brilliant insight was to combine the philosophers' concept that *learning involves the association of ideas* with Charles Sherrington's concept of the *reflex act*. With this framework Pavlov was able to deal with unobserved mental phenomena—ideas—and to study them objectively by examining behavioral acts, which are external and observable phenomena. Pavlov's theories marked a permanent shift in the study of learning from introspective inferences about unobservable ideas to the objective analysis of stimulus and response. According to Pavlov, what animals and humans learn is not the association of ideas but the association of stimuli.

The essence of classical conditioning is the pairing of two stimuli, an unconditioned stimulus, or US, and a conditioned stimulus, or CS. The *conditioned stimulus*, such as a light or tone, is chosen because it produces either no overt responses or weak responses unrelated to the response that eventually will be learned. On the other hand, the *unconditioned stimulus*, such as food or a shock to the leg, is chosen because it always produces an overt response, the *unconditioned response*, such as salivation or leg withdrawal. Indeed, the reason the response is called unconditioned is because it is innate; it is produced by the eliciting (unconditioned) stimulus without learning. When the conditioned stimulus is repeatedly followed by the unconditioned stimulus in a precise temporal sequence, the conditioned stimulus begins to elicit responses, called *conditioned responses*, that resemble the unconditioned responses. The conditioned stimulus ultimately becomes an anticipatory signal for the occurrence of the unconditioned stimulus, and the animal responds to the conditioned stimulus as if it were preparing for the unconditioned stimulus.

Thus, classical conditioning is a means by which animals learn to predict relationships between events in the environment. For example, if a light is followed repeatedly by the presentation of meat, after several learning trials the animal will respond to the light in the same way as it responds to meat: the light itself will produce salivation. The conditioned response is not precisely identical to the unconditioned response, but the two responses are so similar that it is useful to think of conditioning as a process by which the animal learns to react to the conditioned stimulus as a substitute for the unconditioned stimulus. Classical conditioning is further subdivided into appetitive conditioning and defensive conditioning: if the unconditioned stimulus is rewarding (food, water), the conditioning is considered *appetitive*; if the stimulus is noxious (shock), it is considered *defensive*.

As previously mentioned, Pavlov regarded classical conditioning not only as a way to study learning but also as a way to approach the mind—the inner workings of the brain. He was keenly aware that if he could train animals to respond selectively to stimuli, he could discover which aspects of a stimulus an animal is capable of recognizing and processing. For example, psychologists have explored whether an animal can recognize and distinguish colors by determining whether lights of different colors can serve as discriminative stimuli for classical conditioning. During *discriminative training*, one conditioned stimulus (CS^+) is presented in association with reinforcement on some trials. On other trials, another conditioned stimulus (CS^-) is presented but is never followed by reinforcement. If the CS^+ and CS^- are similar in certain respects, the animal will initially exhibit generalization; that is, it will show conditioned responses to both the reinforced and nonreinforced stimuli. If the animal can discriminate between the stimuli, then after continued training it will show conditioned responses primarily or exclusively to the CS^+ and not to the CS^-. By appropriately manipulating the hue and intensity of visual stimuli, psychologists can determine whether the animal is responding to color rather than to differences in brightness. By this means they can determine the perceptual capacities of any animal capable of being conditioned.

An important principle of conditioning is that an established conditioned response decreases in intensity or probability of occurrence if the conditioned stimulus is repeatedly presented without the unconditioned stimulus. This process is known as *extinction*. Thus, a light that has been paired with an unconditioned stimulus of food will gradually cease to evoke salivation if the light is repeatedly shown in the absence of food. Extinction is just as important an adaptive mechanism as conditioning, because a continued response to cues that are no longer significant is not in the animal's best interest. The available evidence indicates that extinction does not simply involve the fading of previous learning; rather, during the extinction process the animal learns something new—the conditioned stimulus no longer predicts that the unconditioned stimulus will occur; instead, it comes to predict that the unconditioned stimulus will *not* occur.

Conditioning Involves the Learning of Predictive Relationships

Until quite recently, many animal psychologists thought that classical conditioning depends only on temporal contiguity. According to this view, each time a conditioned stimulus is followed by a reinforcing or unconditioned stimulus, an internal stimulus–response or stimulus–stimulus bond is strengthened, until eventually the bond becomes strong enough to produce conditioning. The only relevant variable determining the strength of conditioning was thought to be the number of contiguous CS–US events. This theory is inadequate for two reasons: first, it does not make sense from an adaptive point of view. If animals learned to derive predictive information simply from the occurrence of two events in close temporal contiguity, they might develop erroneous notions about the true causal relationship between signals in the environment. Second, a substantial body of empirical evidence indicates that learning cannot be adequately explained by such simple contiguity.

A striking example of the inadequacy of simple contiguity to produce conditioning is the so-called blocking phenomenon, discovered by Leon Kamin at Princeton University. Kamin discovered blocking by carrying out a three-part experiment. First, he conditioned a stimulus, a light, by pairing it repeatedly with an aversive unconditioned stimulus, a strong electric shock. He then assessed conditioning by determining the ability of the light to suppress ongoing behavior (a reflection of its ability to evoke a strong conditioned fear response in the animal similar to that initially evoked by the electrical shock). In the second part of the blocking experiment, Kamin presented the conditioned stimulus simultaneously with a new stimulus, a tone, and the light-tone compound stimulus was then repeatedly paired with the shock. When now, in the third part of the experiment, Kamin presented the tone alone, he found that little or no conditioning had occurred to the tone. Despite repeated pairings of the light-tone compound stimulus with shock, the tone, when presented alone, failed to suppress behavior and did not evoke a fear response. These findings are consistent with Robert Rescorla and Alan Wagner's theory of classical conditioning, according to which the amount of conditioning on a trial is dependent on the degree to which the unconditioned stimulus is unexpected. If the unconditioned stimulus is completely unexpected because it has not been previously paired with a conditioned stimulus, maximal learning can occur. But when the unconditioned stimulus is fully expected because it is already well predicted by the conditioned stimulus, learning reaches an asymptote and no further learning can occur. Thus, the tone component of the light-tone stimulus is an ineffective conditioning stimulus because the other element of the compound conditioned stimulus (the light) already successfully and fully predicts the occurrence of the unconditioned stimulus. This notion has been formalized in simple mathematical terms by Rescorla and Wagner, and predicts a surprising number of the properties of classical conditioning.

Another line of experiments also demonstrates the inadequacy of simple contiguity in producing conditioning and has revealed that classical conditioning develops best when, in addition to *contiguity*, there is also a *contingency*—a truly predictive relationship—between the conditioned stimulus and the unconditioned stimulus. If an animal is presented with a long sequence of conditioned and unconditioned stimuli each occurring randomly and completely independently, some contiguous CS–US sequences will occur just by chance. Nevertheless, a conditioned response to the CS does not develop. Clearly, the animal is not just counting the number of CS–US pairings, but rather is determining the *overall correlation* or *predictive relationship* between the CS and US. In fact, if a stimulus is presented repeatedly so that it specifically does *not* occur in association with a US, that stimulus comes to predict the absence of the US. When that stimulus is later paired with a US, conditioning occurs only very slowly, presumably because the animal must first unlearn the previous predictive property of the stimulus. In some instances stimuli that have been associated with the absence of the US actually acquire *inhibitory* properties, and their presence can suppress the occurrence of behavioral responses. Thus, in addition to being paired in time, the CS and reinforcer (the US) need to be positively correlated; the CS must indicate an increased probability that the US will occur.

These considerations suggest why animals and humans acquire classical conditioning so readily. It appears likely that animals exhibit classical conditioning, and perhaps all forms of associative learning, because *the brain has evolved to enable animals to distinguish events that reliably and predictably occur together from those that are unrelated.* In other words, the brain has evolved as a detector of causal relationships in the environment.

All animals that exhibit associative conditioning, from snails to humans, seem to learn by detecting environmental contingencies rather than detecting the simple contiguity of a CS and US. Why is the recognition of contingent relationships similar in humans and in simpler animals? One good reason lies in the conservation that is characteristic of evolution and in the consequences of evolutionary pressure on adaptation. All animals, regardless of habitat and heritage, face common problems of adaptation and survival, problems for which learning and flexible decision-making are useful. When different species face common environmental pressures, they often manifest similar patterns of adaptation. These patterns are likely to involve homologous mechanisms because a successful biological solution to an environmental challenge, once evolved in a common primitive ancestor, continues to be inherited as long as it remains useful.

What environmental conditions might have shaped or maintained a common learning mechanism in a wide variety of species? To function effectively, animals need to recognize certain key relationships between external events. They must be able to recognize and distinguish prey from predators; they must search out food that is nutritious and avoid food that is poisonous. There are two ways in which an animal arrives at such knowledge. As discussed in Chapter 60, the correct information can be preprogrammed into the animal's nervous system. The ability to choose correctly among alternatives can also be acquired through learning. Genetic and developmental programming may suffice for all of the behavior of very simple organisms, such as parasites, but more complex animals must be capable of extensive learning to cope efficiently with varied or novel situations. Complex animals need to recognize order in the world. An effective way to do this is to be able to detect causal or predictive relationships between stimulus events, or between behavior and subsequent stimuli.

Operant Conditioning Involves Associating an Organism's Own Behavior with a Subsequent Reinforcing Environmental Event

A second major paradigm of associative learning, introduced by Edward Thorndike of Columbia University, is *operant conditioning* (sometimes called *instrumental conditioning* or *trial-and-error learning*). In a typical laboratory example of operant conditioning, an investigator begins by placing a hungry rat in a test chamber that has a lever protruding from one wall. Because of previous learning as well as innate response tendencies and random activity, the rat will occasionally press the lever. If the rat promptly receives food when it presses the lever, its subsequent rate of lever pressing will increase above the spontaneous rate. The animal can be described as having learned that a certain response (lever pressing) among the many it has made (for example, grooming, rearing, and walking) is rewarded with food. With this information, whenever the rat is hungry and finds itself in the same chamber, it is likely to make the appropriate response.

If we think of classical conditioning as the formation of a predictive relationship between two stimuli (the conditioned stimulus and the unconditioned stimulus), operant conditioning can be considered to consist of the formation of a predictive relationship between a response and a stimulus. Unlike classical conditioning, which is restricted to specific reflex responses that are evoked by specific identifiable stimuli, operant conditioning involves behaviors (called *operants*) that apparently occur spontaneously or with no recognizable eliciting stimuli. Thus, operant behaviors are said to be emitted rather than elicited, and when the behaviors produce favorable changes in the environment (that is, when they either are rewarded or lead to the removal of noxious stimuli), the animal tends to repeat them. This process is known as *reinforcement;* it encompasses the more general observation that behaviors that are rewarded tend to be repeated at the expense of behaviors that are not, while behaviors followed by aversive, though not necessarily painful, consequences *(punishment)* are generally not repeated. Experimental psychologists agree that this simple idea, called the *law of effect,* probably reflects an important principle that governs much voluntary behavior.

Superficially, operant and classical conditioning appear to be dissimilar, involving completely different stimulus and response relationships. However, the laws that govern operant conditioning and those that govern classical conditioning are quite similar, suggesting that the two forms of learning are manifestations of a common underlying neural mechanism. For example, in both forms of conditioning, timing is critical: typically, the reinforcer must closely follow the operant response. If the reinforcer in operant conditioning is delayed, only weak conditioning occurs. Similarly, in classical conditioning, the learning is generally poor when there is a long delay between the conditioned and the unconditioned stimulus. Finally, predictive relationships are equally important in both types of learning. In classical conditioning the animal learns that a certain stimulus predicts a subsequent event. In operant conditioning the animal learns to predict the consequences of its own behavior.

Food-Aversion Conditioning Illustrates How Biological Constraints Help Determine the Efficacy of Reinforcers

The two forms of associative learning discovered by Pavlov and Thorndike—classical conditioning and operant conditioning—are so general and so prominent that for many years it was thought that classical conditioning could occur simply by arbitrarily associating any two stimuli or, in the case of operant conditioning, any response and any reinforcer. Recent studies have indicated, however, that there are important biological (evolutionary) constraints on learning. As we have seen, animals generally learn to associate stimuli that are relevant to their survival; they will not learn to associate events that are biologically meaningless. These findings illustrate nicely a principle we have encountered in the study of the development of behavior. The brain is not a *tabula rasa*, but has inherent predispositions toward the detection and manipulation of certain environmental contingencies. For example, not all reinforcers are equally effective with all stimuli. This principle is dramatically illustrated in studies of *food aversion* (also called *bait shyness*, as it seems to be the means by which animal pests such as rats and mice learn to avoid poisoned bait foods). If a distinctive taste stimulus, such as vanilla, is followed by nausea produced by a poison, an animal will quickly develop a strong aversion to the taste of vanilla. Unlike most other forms of conditioning, food aversion develops even when the unconditioned stimulus (poison-induced nausea) occurs with a delay of up to hours after the conditioned stimulus (specific taste). This makes biological sense, since the ill effects of naturally occurring toxins usually follow ingestion only after some delay.

The food-aversion paradigm has been applied in the treatment of chronic alcoholism. The patient is first given alcoholic beverages to smell and taste, and then a powerful emetic such as apomorphine. The pairing of alcohol and nausea rapidly results in aversion to alcohol. Food-aversion learning has several other important implications in medicine. First, it may be a means by which people unintentionally learn to regulate their diets to avoid the unpleasant consequences of inappropriate or nonnutritious food. Second, the malaise associated with certain forms of cancer may induce aversion conditioning to foods in the ordinary diet of the patient. This, in part, might account for depressed appetites in cancer patients. Furthermore, the nausea that follows chemotherapy for cancer can produce aversion to foods that were tasted shortly before the treatment.

For most species, including humans, food-aversion conditioning is restricted to *taste stimuli* associated with subsequent *illness*. Food aversion develops poorly, or not at all, if the salient taste is followed by a painful stimulus. Conversely, if a visual or auditory stimulus, instead of a taste stimulus, is paired with nausea, an animal does not develop an aversion to that stimulus. Thus, the choice of an appropriate reinforcer depends on the nature of the response to be learned. Evolutionary pressures have predisposed the brains of different species of animals to learn an association between certain stimuli, or between a certain stimulus and a response, much more readily than between others. Within a given species, genetic and experiential factors also can modify the effectiveness of a reinforcer. The results obtained with a particular class of reinforcer vary enormously from species to species and from individual to individual within a species, particularly in humans.

Conditioning Is Used as a Therapeutic Technique

Various psychotherapeutic procedures involve reeducation of the patient in the context of a trusting relationship with the therapist. Aspects of therapeutic change are likely to involve components of classical and operant conditioning, but the specific contribution that each of these procedures makes to therapy has been delineated in only a few relatively simple instances.

Classical Conditioning Has Been Applied in Systematic Desensitization

The process of extinction, characteristic of classical conditioning, may underlie the therapeutic changes resulting from a clinical technique known as *systematic desensitization* (although other interpretations of this method have been offered). Systematic desensitization was introduced into psychiatry by Joseph Wolpe, a South African physician who used it to decrease neurotic anxiety or phobias evoked by certain definable environmental situations, such as heights, crowds, or public speaking. The patient is first taught a technique of muscular relaxation. Then, over a period of days, the patient is told to imagine a series of progressively more severe anxiety-provoking situations while using relaxation to inhibit any anxiety that might be elicited. At the end of the series, the strongest potentially anxiety-provoking situations can be brought to mind without anxiety. This desensitization, induced in the therapeutic situation, often generalizes to real-life situations that the patient encounters.

Operant Conditioning Has Been Used to Treat Severe Behavioral Problems

Principles of operant conditioning also have been applied to the management of psychiatric disorders. One important therapeutic application is in the management of severely disturbed institutionalized patients with behavioral problems, such as shouting obscenities, messiness, or poor hygienic habits. The goal of conditioning these patients is to increase the frequency of positive, constructive behaviors. These behaviors are first defined precisely, and an effective reinforcement is found (compliments, privileges, money, or food). Nurses and orderlies are then trained to provide reinforcements when the patients behave in the desired way.

Biofeedback is another form of operant conditioning that has proved useful clinically. Biofeedback is used to enhance (or suppress) responses of which the patient is unaware. The behavior of interest, such as very slight muscle contractions in a stroke patient, is recorded by an electronic device that provides the patient with immediate reinforcement in the form of an auditory or visual cue signaling that the response has occurred.

Learning and Memory Can Be Classified as Reflexive or Declarative on the Basis of How Information Is Stored and Recalled

The classification of associative learning into either operant or classical conditioning is based on the experimental procedures used to establish the conditioning. This distinction is therefore useful to clinicians or experimentalists who wish to apply a well-defined, reproducible methodology to their work. Alternative classification schemes of learning are based not on what the experimenter does, but rather on the type of knowledge acquired by the subject. These classifications cut across the operant–classical distinction and take into account that a single procedure may produce different forms of learning depending on how the experimental subject codes and recalls the information that is learned. Endel Tulving, at the University of Toronto, was one of the first to appreciate that the memory for many different types of learning can be divided into two categories. Different authors have used various terms to reflect this dichotomy or closely related dichotomies. For our purposes, we shall refer to the two categories as *reflexive* and *declarative* memory. Later in this chapter, we shall consider evidence indicating that these two types of memory can be differentially affected by brain damage and that they may involve different neuronal systems of the brain.

Reflexive memory has an automatic or reflexive quality, and its formation or readout is not dependent on awareness, consciousness, or cognitive processes such as comparison and evaluation. Reflexive memory accumulates slowly through repetition over many trials. This type of memory is expressed primarily by improved performance on certain tasks and is poorly expressed by declarative sentences. Examples of reflexive memory include perceptual and motor skills and the learning of procedures and rules, such as those of grammar. Reflexive memory, however, is not limited to learning of procedures and skills. Certain verbal learning tasks, if repeated often enough, assume the characteristics of reflexive learning. These tasks can then be performed automatically without the participation of other cognitive devices.

Declarative memory depends on conscious reflection for its acquisition and recall, and it relies on cognitive processes such as evaluation, comparison, and inference. Declarative memory encodes information about specific autobiographical events as well as the temporal and personal associations for those events. It often is established in a single trial or experience, and it can be concisely expressed in verbal declarative statements, such as "I saw a yellow canary yesterday." Declarative memory involves the processing of bits and pieces of information that the brain can then use to reconstruct past events or episodes. As we noted above, in certain instances declarative memory may be transformed into the reflexive type by constant repetition. For example, learning to drive a car at first involves conscious cognitive processes, but eventually driving becomes automatic and nonconscious.

How do elementary forms of learning such as classical conditioning fit into this reflexive and declarative scheme? Although classical conditioning often results in reflexive memory, even this ostensibly simple form of conditioning may, under some circumstances, lead to declarative memory and involve mediation by cognitive processes. Consider the following experiment. A subject lays his hand, palm down, on an electrical grill; a light (conditioned stimulus) is turned on and he is immediately shocked on a finger. His finger withdraws (unconditioned response), and, after several light–shock conditioning trials, he withdraws his finger when the light alone is presented. The subject has been conditioned; but what exactly has been conditioned? It appears as though the light is triggering a specific pattern of muscle activity that results in withdrawal. However, what if the subject now places his hand on the grill upside down, and the light is presented? If a specific pattern of muscle activity has been conditioned, the light should produce a response that moves the finger *into* the grill. On the other hand, if the subject has acquired the information that the light means grill shock, he may make a different response appropriate to that information. In fact, when this experiment is done, the subject moves his finger *away* from the grill; that is, he makes an adaptive response, even though it involves motor movements antagonistic to the original ones. Therefore, the subject did not originally learn a fixed response to a fixed stimulus, but, rather, acquired information that the brain could use to solve specific problems.

In another study of the nature of declarative memory, researchers analyzed remembered versions of previously learned stories. The versions that the subjects recalled were shorter and more coherent, and they contained reconstructions not present in the original. The subjects were unaware of where they were substituting, and they often felt most certain about the reconstructed part. The subjects were not confabulating; they were merely recalling in a way that interpreted the original material so it made sense.

Observations such as these lead us to believe that the accumulation of knowledge about past events is an *active* process involving cognitive events. Initially, what goes into the memory store is a representation of information that has been changed as a result of processing by our perceptual apparatus. Optical illusions illustrate that we do not perceive the world precisely as it is but, rather, as a modified version that is altered on the basis of past experience as well as on principles and limits of perceptual analysis. Moreover, once the information is stored, what is recalled from the declarative memory store is not a faithful reproduction of the internal store. Recall of declarative memory involves a process in which past experiences are used in the present as clues to help the brain reconstruct a significant past event. During this reconstruction, the brain uses a variety of cognitive processes—comparison, inferences, shrewd guesses, and suppositions—to generate a consistent and coherent picture.

The Neural Basis of Memory Can Be Summarized in Four Principles

Although the literature on the neurobiology of memory is extensive, much of what is known can be summarized in just four principles: (1) memory has stages and is continually changing; (2) long-term memory may be represented by physical changes in the brain; (3) the traces for memories are localized in multiple regions throughout the nervous system; and (4) reflexive and declarative memories may involve different neuronal circuits. Here we shall consider information obtained by gross techniques, such as brain lesions, electrical stimulation, and drugs. Studies of the cellular mechanisms of learning will be considered in Chapter 62.

Memory Has Stages

It is an old clinical observation that a person who has been knocked unconscious can have selective memory loss for events that occurred shortly before the blow *(retrograde amnesia)* and shortly after regaining consciousness *(anterograde amnesia).* This phenomenon has been documented thoroughly in animal studies using such traumatic agents as electroconvulsive shock, physical trauma to the brain, and drugs that depress neuronal activity or inhibit protein synthesis in the brain. Clinical studies also indicate that brain trauma can produce amnesia that is particularly prominent for recent events, typically within a few days of the trauma. Thus, recently acquired memories are readily disrupted, whereas older memories remain quite undisturbed. Once something has been learned, the extent of potential retrograde amnesia—the span of time during which memory is labile—varies from several seconds to several years, depending on the nature and strength of the learning and on the species of animal.

Studies of memory disruption have contributed to a commonly used model of the memory storage system (Figure 61–1). Input to the brain is processed into a short-term memory store. This information is later transformed by some process into a more permanent long-term store. To complete the model, a system has been added that functions to search the memory store and to read out the information as demanded by specific tasks. In this model, interference with the retention of previous experience can occur either by partial destruction of the contents of a memory store or by disruption of the search and read-out mechanism. In traumatic amnesia, at least part of the interference must be due to a disturbance of the search and read-out mechanism. This conclusion stems from the observation that, after trauma, some memory for once-forgotten events gradually returns. If the stored memory had been completely destroyed, it obviously could not have been recovered.

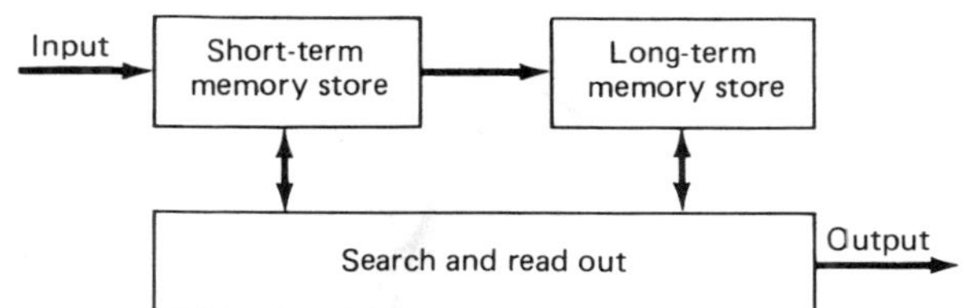

61–1 Model of memory storage system.

Observations of patients undergoing a series of electroconvulsive shocks for the treatment of depression have confirmed and extended the findings of experiments made on animals. Larry Squire and his associates at the University of California Veterans Hospital studied patients who had been given shock therapy. They used a memory test that could reliably quantify the degree of memory for relatively recent events (1–2 years old), old events (3–9 years old), and very old events (9–16 years old). Patients were asked to identify various television programs that were broadcast during a single year between 1957 and 1972. The patients were initially tested and then tested again (with a different set of television programs) after the electroconvulsive shock therapy. The results of this experiment are shown in Figure 61–2. Both before and after shock therapy, correct memory for the programs steadily decreased with time after the memory was first formed. This is a reflection of the all-too-familiar process of forgetting. After the shock therapy, however, the patients showed a significant but transitory memory loss for programs that had gone off the air 1 or 2 years previously, but their memory for the older programs was the same as it was before the shock therapy.

One interpretation of these observations is that the read-out of recent memories is easily disrupted until the memories have been converted into a long-term memory form. Once converted, they are relatively stable, but with time, even without external trauma, there is a gradual loss of the stored information or a diminished capacity to retrieve it. Thus the memory process, at least as assessed by susceptibility to disruption, is *always undergoing continual change with time.*

Several experiments on the effects of drugs on learning support the idea that the memory process is time dependent and is subject to modification when the memory is first formed. James McGaugh and associates at the

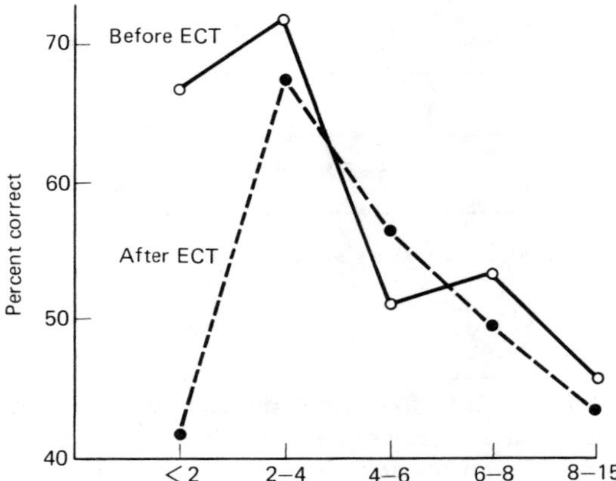

61–2 Recent memories are more susceptible to disruption by electroconvulsive shock therapy **(ECT)** than older memories. Patients were tested on their ability to recognize correctly the name of television programs that were on the air during 1 year between 1957 and 1972. Testing was done before and after the patients received ECT for treatment of depression. After the ECT, the patients showed a significant (but transitory) loss of memory for recent programs (1–2 years old) but not for old programs. (Adapted from Squire, Slater, and Chace, 1975.)

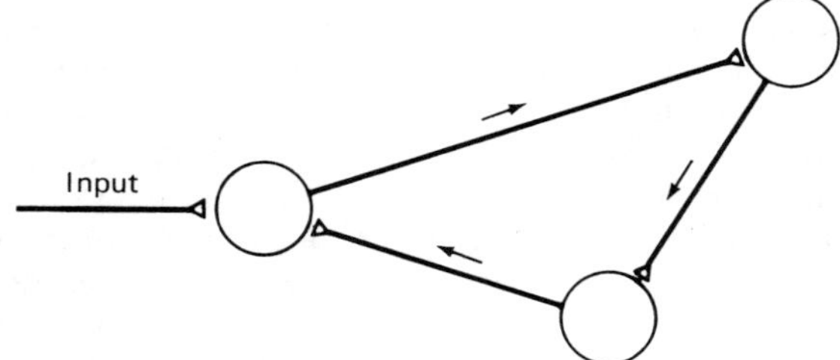

61–3 A reverberating circuit might be used to encode short-term memory. Brief excitatory input can produce long-lasting neural activity through the circulation of spikes among neurons that excite one another.

University of California at Irvine have shown in animals that subconvulsant doses of excitant drugs such as strychnine can improve the retention of learning even when the drug is administered after the training trials. If the drug is given to the animal *soon* after training, retention tested the next day is facilitated. If, however, the drug is given several hours after training, it has no effect.

Long-term Memory May Be Represented by Plastic Changes in the Brain

How is information stored? One type of very brief short-term memory for visual events, called *iconic memory,* is probably due to brief retinal afterimages that follow exposure to visual stimuli. If a person is briefly allowed to view a matrix of many letters and numbers, he can accurately recall specific elements of the matrix; but, unlike most forms of learning, accuracy of recall diminishes extremely rapidly, typically in less than 1 sec. The time before accuracy diminishes can be extended by increasing the brightness of the visual stimulus, and the time course for the decline of accuracy parallels the decay of the visual afterimages. Photochemical processes in the retina can account for visual afterimages. Thus, one very simple form of short-term memory appears to be encoded by a transient physical change in the sensory receptor.

Slightly longer lasting short-term memory that persists for minutes to hours could be mediated by a variety of short-term neural plastic events that we considered in Chapter 11, such as posttetanic potentiation and presynaptic inhibition.

Another possible mechanism for encoding short-term memory is the storage of information in the form of ongoing neural activity that is maintained by excitatory feedback connections between neurons (Figure 61–3). This type of activity could reverberate within a closed loop of neurons and might be sustained for some period of time. The idea of reverberatory circuits is interesting because it does not involve any enduring physical changes in nerve cells; the memory for the event is maintained simply by ongoing neuronal activity. There are few experiments to support this notion, and it seems unlikely that this will prove to be a common mechanism for memory.

But how is long-term memory stored? How are changes maintained for years? Two possibilities exist. First, but not likely, is the possibility that reverberating circuits like those postulated to underlie short-term memory may persist and represent long-term memory as well. The second possibility is that long-term memory is related to some *plastic* rather than *dynamic* change—that is, to a permanent functional or structural change in the brain. A simple experiment can distinguish between these alternatives. If all neuronal activity is temporarily stopped, memories represented by the dynamic mechanism of reverberating circuits should be permanently abolished. Neuronal activity can be silenced by the use of deep anesthesia, by anoxia, or by cooling the brain. When this is done, short-term or recent memories are disrupted, but older memories are not. Thus, it is safe to conclude that at least older memories are not mediated by reverberatory activity but, more likely, involve physical changes in the brain. Because of the enduring nature of memory, it seems reasonable to postulate that in some way the changes must be reflected in long-term alterations of the connections between neurons. We shall consider this question again in Chapter 62.

Memory Traces Are Often Localized in Different Places Throughout the Nervous System

Whatever their nature, it seems clear that the memory traces for many different types of learning are not localized to any one brain structure. Pavlov, the father of physiological studies of learning, believed that all learning processes are limited to the neocortex. The psychologist Karl Lashley, at Harvard University, spent most of his scientific career making lesions in the cortex to define precisely where the representation of learning (called the *engram*) was located. He never succeeded. Lashley and others found that although cortical lesions can seriously disrupt learning, animals can relearn certain tasks even when they are completely decorticated. Classical conditioning of certain simple reflexes can be mediated by the spinal cord even after it has been isolated surgically from the brain, as was first shown by P. S. Shurrager and Elmer Culler in 1940. Thus, many and perhaps all regions of the nervous system appear to contain neurons with the properties of plasticity needed for memory storage.

Even for a simple learning task, several parallel channels of information are used. There is ample opportunity, therefore, for information to be stored in different regions of the brain. For example, David Cohen at the State University of New York at Stony Brook, in studies of conditioned heart rate response in pigeons, found that any one of three visual pathways can sustain the learned response.

Parallel processing may explain in part why a given limited lesion does not eliminate specific learning, but a more important factor may reside in the very nature of the learning process. As we have seen from behavioral studies, learning involves neither the simple formation of stimulus–response bonds nor a faithful reproduction of sensory experience. Although the physical changes representing learning are likely to be localized to specific neurons, the complex nature of learning ensures that these neurons are widely distributed in the nervous system. Therefore, even after extensive lesions, some trace can remain. Furthermore, the brain has the capacity to take even the limited information remaining, work it over, and reconstruct a good reputation of the original.

Reflexive and Declarative Memories May Involve Different Neuronal Circuits

Although many memory traces are typically widely distributed, some learning tasks are profoundly affected by circumscribed lesions of the brain. The most striking evidence indicating that specific brain structures can exert specific effects on learning comes from studies of the cerebellum and of the temporal lobes. These studies suggest the intriguing hypothesis that the brain may possess two classes of neural circuits, one of which is concerned with reflexive memory, the other with declarative memory.

Reflexive Memory. Lesions to several regions of the brain have been found to affect simple classically conditioned responses, and these regions probably represent loci for reflexive types of learning. For example, lesions of the amygdala interfere with conditioned heart rate responses, apparently by interrupting pathways close to the motor end of the reflex arc. Another example of a specific lesion affecting a classically conditioned response comes from the research of Richard Thompson, David McCormick, and their associates at Stanford University. They have been studying the eyeblink (or nictitating membrane) protective reflex in rabbits. By pairing an auditory stimulus with a puff of air to the eye, a conditioned eyeblink reflex can be established to the auditory stimulus. The conditioned response is totally abolished by a lesion to the medial dentate and lateral interpositus nuclei of the cerebellum. After this region is lesioned, the previously effective conditioned auditory stimulus is no longer capable of producing an eyeblink, although the unconditioned eyeblink response that follows the unconditioned stimulus (air puff) remains intact. These results have been confirmed and extended by Mitchell Glickstein in London and John Moore and his colleagues at the University of Massachusetts. Furthermore, the dentate-interpositus nuclei also show learning-dependent increases in neuronal activity that closely parallel the development of the conditioned behavioral response. The results of these experiments, taken as a whole, indicate that the cerebellum plays an essential role in mediating conditioned eyeblink and perhaps other simple forms of classical conditioning.

Declarative Memory. In humans, lesions of the temporal lobe and closely associated structures of the limbic system or of the diencephalon dramatically affect learning. These lesions have weak effects on specific prior memories; they primarily interfere with the retention of new ones. Thus, these structures are not themselves registers or banks for memory storage, but are somehow involved in the process by which memories are placed into storage or are retrieved and read out from storage.

A significant clue that the temporal lobes are important for memory came from the observations by the neurosurgeon Wilder Penfield at the Montreal Neurological Institute. In the course of temporal lobe surgery for the control of epilepsy, Penfield electrically stimulated the exposed temporal lobes in fully conscious patients. The patients reported vividly experiencing past events. For example, stimulation of one point on the temporal lobe caused a patient to hear a specific melody that she believed she had heard in the past. Repeated stimulation of the same point evoked successive experiences of hearing the same melody.

Additional evidence of a role for the temporal lobes in memory has come from the study of a few epileptic patients who underwent bilateral removal of the hippocampus and associated structures in the temporal lobes. Brenda Milner found that these patients exhibit a profound and irreversible deficit of recent memory. They lose the capacity to form new long-term memories, but previously acquired long-term memories remain relatively intact; for example, they remember their names and how to talk. Short-term memory is also unaffected; but the transition from short-term to long-term memory is virtually absent for most types of learning. For example, if the patient was told to remember the number 7, he could repeat the number immediately. If, however, the patient was distracted, even briefly, he had no recollection of the number. The extent of the deficit is indicated by the observation that one patient failed to recognize individuals whom he had known closely for years. In addition to anterograde amnesia, these patients often show some retrograde amnesia; that is, they have a loss of memories that were stored and available for recall before surgery. Often, memories lost retrogradely because of lesions or trauma gradually return, but anterograde amnesia is permanent.

An amnesic syndrome superficially similar to that seen in patients who have undergone bilateral removal of temporal lobe structures occurs in patients suffering from *Korsakoff's psychosis*. This disease, which results from chronic alcoholism and its nutritional deficiency, is characterized by confusion and severe memory deficits. Patients with Korsakoff's psychosis exhibit pathological changes in diencephalic structures that are part of the limbic system. Typically, they have damage to the mammillary bodies of the hypothalamus as well as to the medial dorsal nucleus of the thalamus. Careful study of the precise nature of the deficit in patients with Korsakoff's psychosis supports the idea that the memory deficit is due, at least in part, to defective encoding at the time of original learning rather than exclusively to a defect in the retrieval mechanism. Patients with Korsakoff's psychosis learn slowly, but once the material is learned, they appear to forget at a normal rate.

Elizabeth Warrington and Lawrence Weiskrantz in England have found that when patients with Korsakoff's psychosis are given a list of words to remember, they do poorly on a simple recall task, but their performance is greatly improved when retention is tested by the use of prompts or partial cues. For example, their performance is normal if, following the original learning, they are tested for retention by a completion task rather than being asked simply to recall the words. In the completion task the patients are given a list of letter sequences, each of which has the first few letters of a word in the original list. Then, on the basis of their memory for the words in the original list, the patients must complete the words.

Peter Graf, George Mandler, and Patricia Haden, at the University of California, found that this disjunction between performance on simple recall and completion tests can be demonstrated in normal subjects with no memory defects if the subjects are required to learn a list of words in a task that minimizes the opportunity to understand the meaning of the word. In this experiment, a list of 20 words was presented with the instruction to detect certain vowels in the words. The subjects were not asked to memorize the words or to understand their meaning. After the vowel detection task, they were unable to recall the words in the list. However, like the patients with Korsakoff's psychosis, the subjects were able to recall many of the words if they were given the initial letters of the words and were asked to complete them. Subjects in a second group were presented with a list of words and were instructed to determine if they liked each word, a task that requires an understanding of its semantic content. These subjects subsequently remembered the words just as well on simple recall as on a completion task. These findings support the suggestion that patients with Korsakoff's psychosis fail to encode the semantic component of material properly on initial learning.

Reflexive Versus Declarative Memory in Amnesic Patients. A remarkable finding in studies of amnesic patients, either with temporal lobe or diencephalic damage, is that they can learn certain tasks perfectly well. Although these patients cannot master tasks involving declarative memory, they perform well on tasks involving reflexive memory. A given learning task often involves aspects of both types of learning, and in these instances amnesic patients remember some aspects of the problem, but not others. Thus, if the patient is given a highly complex mechanical puzzle to solve, the patient may learn it as quickly as a normal person but on questioning will not remember seeing the puzzle or having worked on it previously. In other words, amnesic patients can learn a complex skill and yet be unable to recall the specific events that allowed them to learn the rules and procedures that make up the skill. Furthermore, even in instances in which they remember some experience of the past, the experience lacks the feeling of familiarity that accompanies recall in normal individuals.

Warrington and Weiskrantz suggested that the fundamental deficit of amnesic patients is due to some type of disconnection between memory storage systems and a cognitive mediational system in the brain that aids in the retrieval and storage of memory. It is possible that in amnesic patients the cognitive system functions normally, but it lacks access to the declarative learning system. Therefore, amnesic patients often show totally unimpaired intelligence and yet are virtually incapable of new declarative learning. This idea helps explain why amnesic patients, when they perform a particular task, are often not aware that they actually had learned it a few days or weeks earlier.

A major challenge confronting the neurobiology of learning is to determine how reported alterations in the brain are causally related to behavioral changes. A second task is to determine the mechanisms underlying the relevant plastic changes. To this end, a number of simplified vertebrate and invertebrate animal preparations are being investigated, and some of the information deriving from these studies is reviewed in Chapter 62.

Selected Readings

Dickinson, A. 1980. Contemporary Animal Learning Theory. Cambridge, England: Cambridge University Press.

Domjan, M., and Burkhard, B. 1982. The Principles of Learning and Behavior. Monterey, Calif.: Brooks/Cole.

Hilgard, E. R., and Bower, G. H. 1975. Theories of Learning, 4th ed. Englewood Cliffs, N. J.: Prentice-Hall.

Kandel, E. R. 1983. From metapsychology to molecular biology: Explorations into the nature of anxiety. Am. J. Psychiatry 140:1277–1293.

Kanfer, F. H., and Phillips, J. S. 1970. Learning Foundations of Behavior Therapy. New York: Wiley.

Klatzky, R. L. 1980. Human Memory. Structures and Processes. 2nd ed. San Francisco: W. H. Freeman.

Lashley, K. S. 1950. In search of the engram. Symp. Soc. Exp. Biol. 4:454–482.

Mackintosh, N. J. 1983. Conditioning and Associative Learning. Oxford: Clarendon Press.

Rescorla, R. A. 1978. Some implications of a cognitive perspective on Pavlovian conditioning. In S. H. Hulse, H. Fowler, and W. K. Honig (eds.), Cognitive Processes in Animal Behavior. Hillsdale, N. J.: Erlbaum, pp. 15–50.

Squire, L. R., Cohen, N. J., and Nadel, L. 1984. The medial temporal region and memory consolidation: A new hypothesis. In H. Weingartner and E. S. Parker (eds.), Memory Consolidation: Psychobiology of Cognition. Hillsdale, N.J.: Erlbaum, pp. 185–210.

Thompson, R. F., Berger, T. W., and Madden, J., IV. 1983. Cellular processes of learning and memory in the mammalian CNS. Annu. Rev. Neurosci. 6:447–491.

Woody, C. D. (ed.). 1982. Conditioning: Representation of Involved Neural Functions. New York: Plenum Press.

Yates, A. J. 1970. Behavior Therapy. New York: Wiley.

References

Cohen, D. H. 1982. Central processing time for a conditioned response in a vertebrate model system. In C. D. Woody (ed.), Conditioning: Representation of Involved Neural Functions. New York: Plenum Press, pp. 517–534.

Glickstein, M., Hardiman, M. J., and Yeo, C. H. 1983. The effects of cerebellar lesions on the conditioned nictitating membrane response of the rabbit. J. Physiol. (Lond.) 341:30P–31P.

Graf, P., Mandler, G., and Haden, P. E. 1982. Simulating amnesic symptoms in normal subjects. Science 218:1243–1244.

McGaugh, J. L., and Herz, M. J. 1972. Memory Consolidation. San Francisco: Albion.

Milner, B. 1966. Amnesia following operation on the temporal lobes. In C. W. M. Whitty and O. L. Zangwill (eds.), Amnesia. London: Butterworths, pp. 109–133.

Moore, J. W., Desmond, J. E., and Berthier, N. E. 1982. The metencephalic basis of the conditioned nictitating membrane response. In C. D. Woody (ed.), Conditioning: Representation of Involved Neural Functions. New York: Plenum Press, pp. 459–482.

Pavlov, I. P. 1927. Conditioned Reflexes: An Investigation of the Physiological Activity of the Cerebral Cortex. G. V. Anrep (trans.). London: Oxford University Press.

Penfield, W. 1958. Functional localization in temporal and deep Sylvian areas. Res. Publ. Assoc. Res. Nerv. Ment. Dis. 36:210–226.

Rescorla, R. A., and Wagner, A. R. 1972. A theory of Pavlovian conditioning: Variations in the effectiveness of reinforcement and nonreinforcement. In A. H. Black and W. F. Prokasy (eds.), Classical Conditioning II: Current Research and Theory. New York: Appleton-Century-Crofts, pp. 64–99.

Shurrager, P. S., and Culler, E. 1940. Conditioning in the spinal dog. J. Exp. Psychol. 26:133–159.

Squire, L. R., Slater, P. C., and Chace, P. M. 1975. Retrograde amnesia: Temporal gradient in very long term memory following electroconvulsive therapy. Science 187:77–79.

Thompson, R. F., McCormick, D. A., Lavond, D. G., Clark, G. A., Kettner, R. E., and Mauk, M. D. 1983. The engram found? Initial localization of the memory trace for a basic form of associative learning. Prog. Psychobiol. Physiol. Psychol. 10:167–196.

Thorndike, E. L. 1911. Animal Intelligence: Experimental Studies. New York: Macmillan.

Tulving, E. 1984. Précis of elements of episodic memory. Behav. Brain Sci. 7:223–268.

Warrington, E. K., and Weiskrantz, L. 1982. Amnesia: A disconnection syndrome? Neuropsychologia 20:233–248.

Wolpe, J. 1958. Psychotherapy by Reciprocal Inhibition. Stanford, Calif.: Stanford University Press.

A Hierarchical Neural-Network Model for Control and Learning of Voluntary Movement

M. KAWATO*, KAZUNORI FURUKAWA**, AND R. SUZUKI

DEPARTMENT OF BIOPHYSICAL ENGINEERING, FACULTY OF ENGINEERING SCIENCE, OSAKA UNIVERSITY, TOYONAKA, OSAKA, 560 JAPAN

Abstract.—In order to control voluntary movements, the central nervous system (CNS) must solve the following three computational problems at different levels: the determination of a desired trajectory in the visual coordinates, the transformation of its coordinates to the body coordinates and the generation of motor command. Based on physiological knowledge and previous models, we propose a hierarchical neural network model which accounts for the generation of motor command. In our model the association cortex provides the motor cortex with the desired trajectory in the body coordinates, where the motor command is then calculated by means of long-loop sensory feedback. Within the spinocerebellum-magnocellular red nucleus system, an internal neural model of the dynamics of the musculoskeletal system is acquired with practice, because of the heterosynaptic plasticity, while monitoring the motor command and the results of movement. Internal feedback control with this dynamical model updates the motor command by predicting a possible error of movement. Within the cerebrocerebellum—parvocellular red nucleus system, an internal neural model of the inverse-dynamics of the musculo-skeletal system is acquired while monitoring the desired trajectory and the motor command. The inverse-dynamics model substitutes for other brain regions in the complex computation of the motor command. The dynamics and the inverse-dynamics models are realized by a parallel distributed neural network, which comprises many sub-systems computing various nonlinear transformations of input signals and a neuron with heterosynaptic plasticity (that is, changes of synaptic weights are assumed proportional to a product of two kinds of synaptic inputs). Control and learning performance of the model was investigated by computer simulation, in which a robotic manipulator was used as a controlled system, with the following results (*1*) Both the dynamics and the inverse-dynamics models were acquired during control of movements. (*2*) As motor learning proceeded, the inverse-dynamics model gradually took the place of external feedback as the main controller. Concomitantly, overall control performance became much better. (*3*) Once the neural network model learned to control some movement, it could control quite different and faster movements. (*4*) The neural network model worked well even when only very limited information about the fundamental dynamical structure of the controlled system was available. Consequently, the model not only accounts for the learning and control capability of the CNS, but also provides a promising parallel-distributed control scheme for a large-scale complex object whose dynamics are only partially known.

* To whom correspondence should be addressed.

** Present address: Fujitsu Limited, Kawasaki, Japan.

1. INTRODUCTION

Although the neural connections in the brain are basically stable and rigid after their formation at an early developmental stage, some of them have plasticity. This "plasticity" is to be thought the neural basis for adaptive behavior. Most voluntary movements, which are executed at will, are learned movements. It is very important and interesting to investigate how the central nervous system (CNS) acquires the ability to control movements by making use of synaptic plasticity, not only from the standpoint of neuroscience but also from that of robotics.

Based on detailed knowledge on the neural circuits in the cerebellum, Marr (1969) and Albus (1971) proposed learning network models of the cerebellum. In these "perceptron" models, the efficacy of a parallel fiber-Purkinje cell synapse was assumed to change when conjunction of the parallel-fiber input and the climbing-fiber input occurs. Ito et al. (1982) demonstrated the presence of the putative heterosynaptic plasticity of Purkinje cells in the flocculus of the cerebellum, which plays an essential role in the adaptive control of the vestibulo-ocular reflex. Although the Marr-Albus model of the cerebellum as a spatial pattern discriminator does not give a sufficient account of the processing of temporal patterns, which is essential for the control of movement, Fujita (1982a) expanded the Marr-Albus model and proposed an adaptive filter model of the cerebellar cortex. His model reproduced several experimental features in an adaptive modification of the vestibulo-ocular reflex (Fujita 1982b). Consequently, a splendidly comprehensive understanding of the adaptive control of the vestibulo-ocular reflex was provided by these works (see Ito 1984 for a review), which accounts for all of the following three levels for understanding complex information-processing systems proposed by Marr (1982): (*1*) computational theory, (*2*) representation and algorithm, and (*3*) hardware implementation.

The investigation of neural mechanisms involved in the control and learning of voluntary movement seems much more

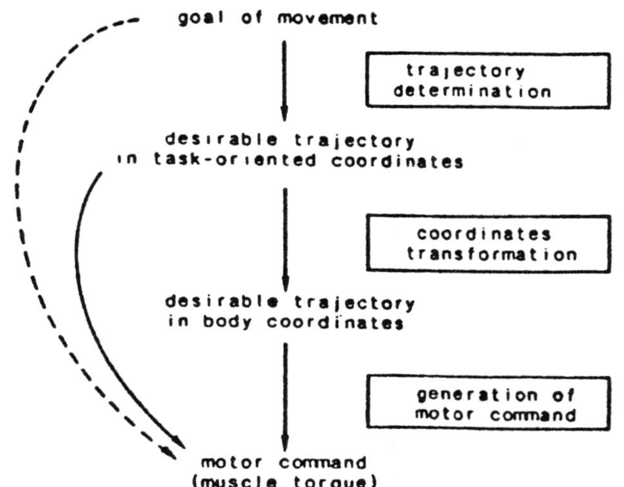

Fig. 1 A computational model for control of voluntary movement

difficult than that of the vestibulo-ocular reflex for the following reasons. First, the control object of voluntary movement (e.g., hand, leg, or trunk) has highly nonlinear dynamics with multiple degrees of freedom. Second, many neural networks and pathways are hierarchically involved (Allen and Tsukahara (1974). Third, volition participates in the highest level.

We propose a computational model of voluntary movement in Fig. 1, which accounts for Marr's first level (1982). Consider a thirsty person reaching for a glass of water on a table. The goal of the movement is moving the arm toward the glass to reduce thirst. First, one desirable trajectory in the task-oriented coordinates must be selected from out of an infinite number of possible trajectories which lead to the glass, whose spatial coordinates are provided by the visual system (determination of trajectory). Second, the spatial coordinates of the desired trajectory must be reinterpreted in terms of a corresponding set of body coordinates, such as joint angles or muscle lengths (transformation of coordinates). Finally, motor commands (e.g., torque) must be generated to coordinate the activity of many muscles so that the desired trajectory is realized (generation of motor command). We do not adhere to the hypothesis of the step-by-step information processing shown by the three straight arrows in Fig. 1. Rather, Uno et al. (1987) proposed a learning algorithm which calculates the motor command directly from the goal of the movement represented by some performance index (broken and curved arrow in Fig. 1). Further, as shown by a curved arrow in Fig. 1, motor command can be obtained directly from the desired trajectory represented in the task-oriented coordinates by an iterative learning algorithm (Kawato et al. 1987). In this respect, our model differs from the three-level hierarchical movement plan proposed by Hollerbach (1982). However, several lines of experimental evidence suggest that the information in Fig. 1 is internally represented in the brain. First, Flash and Hogan (1985) provided strong evidence to indicate that movement is planned at the task-oriented coordinates (visual coordinates) rather than at the joint or muscle level. Second, the presence of the transcortical loop (i.e., the negative feedback loop via the cerebral cortex; Evarts 1981) indicates that the desired trajectory must be represented also in the body coordinates, since signals from the proprioceptors are expressed in the body coordinates. Finally, Cheney and Fetz (1980) showed that discharge frequencies of primate corticomotoneuronal cells in the motor cortex were fairly proportional to active forces (torque). Consequently the CNS must adopt, at least partly, the step-by-step strategy for the control of voluntary movement.

The problem of the determination of the trajectory was investigated by Uno et al. (1987), and the problem of the transformation of the coordinates will be dealt with in our next paper (Kawato et al., in preparation). In this paper, we concentrate on the problem of the generation of motor command. First, a hierarchical neural network model with heterosynaptic plasticity is proposed for the control and learning of voluntary movement. Second, the capability of learning control of a robotic manipulator is demonstrated by computer simulation.

2. Hierarchical Neural Network for the Control of and Learning of Voluntary Movement

In learning a movement, we first execute the movement very slowly because it cannot be adequately preprogramed. Instead, it is performed largely by cerebral intervention with use of long-loop sensory feedback. With practice, a greater amount of the movement can be preprogramed and the movement can be executed more rapidly. Ito (1970) proposed the hypothesis that the cerebrocerebellar communication loop is used as a reference model for the open-loop control of voluntary movement. Allen and Tsukahara (1974) proposed a comprehensive model, which accounts for the functional roles of several brain regions (association cortex, motor cortex, lateral cerebellum, intermediate cerebellum, basal ganglia) in the control of voluntary movement. Tsukahara and Kawato (1982) proposed a theoretical model of the cerebro-cerebello-rubral learning system based on recent experimental findings of the synaptic plasticity in the red nucleus, especially on the sprouting phenomena (see Tsukahara 1981, for a review). Expanding on these previous models, we propose a hierarchical neural network model for the control of and learning of voluntary movement, shown in Fig. 2. This model provides concrete algorithms and neural networks for the problem of the generation of motor command, raised in Fig. 1. In this section we explain the global structure of the model and the information flow in it.

In our model, the association cortex sends the desired motor pattern, that is trajectory x_d expressed in the body coordinates, to the motor cortex, where the motor command, that is torque u to be generated by muscles, is then somehow computed. Here, for simplicity, we identify the motor command with the active torque based on the experimental data of Cheney and Fetz (1980) that discharge frequencies of primate corticomotoneuronal cells in the motor cortex were fairly proportional to

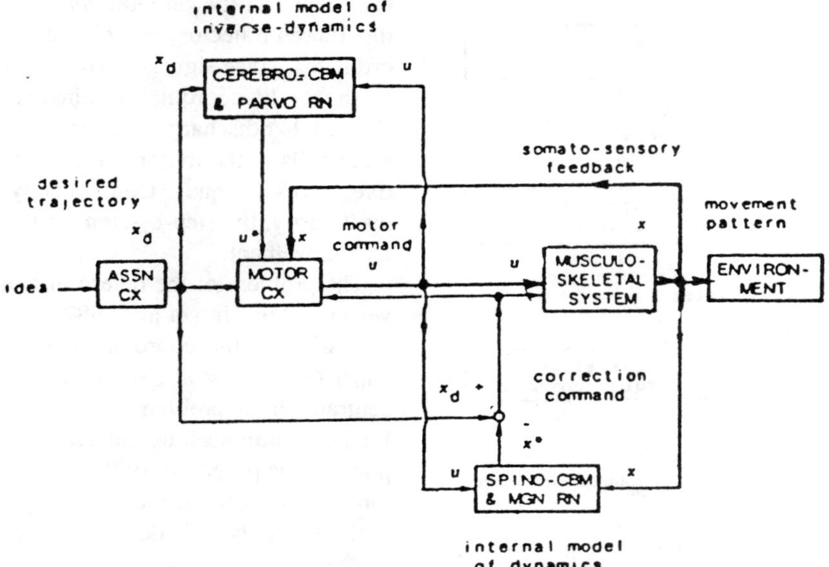

Fig. 2 A hierarchical neural network model for control and learning of voluntary movement. The model is composed of the following three parts: (*1*) The main descending pathway and the transcortical loop designated by heavy lines, (*2*) The spinocerebellum-magnocellular red nucleus system as an internal neural model of dynamics of the musculoskeletal system, (*3*) The cerebrocerebellum-parvocellular red nucleus system as an internal neural model of inverse-dynamics of the musculoskeletal system. See text for detail

active forces. The motor command is transmitted to muscles via spinal motoneurons. The musculoskeletal system interacts with its environment and realizes some kind of motor pattern, x. In general, x does not coincide with x_d. The actual motor pattern (x) and its time derivative (dx/dt) (e.g., muscle length and its derivative) are measured by proprioceptors and sent back to the motor cortex, for example via the transcortical loop. Then, feedback control can be performed utilizing error in the movement trajectory ($x_d - x$). However, severe limitations are imposed on the biological feedback system. There are substantial delays in the feedback loop; for example, the transcortical loop requires 40–60 ms (Evarts 1981). Further, experimental results indicate that the contribution of the supraspinal loop to load compensation is insubstantial (Evarts 1981). Feedback delays and small gains both limit controllable speeds of motions. Consequently, in learning a movement, we first must execute the movement very slowly, because otherwise the control system becomes unstable.

The spinocerebellum (vermis and intermediate part of the hemisphere)—magnocellular part of the red nucleus system receives information about the results of the movement (x) as afferent input from the proprioceptors, as well as an efference copy of the motor command (u). Within this spinocerebellum-magnocellular red nucleus system, an internal neural model of the musculoskeletal system is acquired. Once the internal model is formed by motor learning, it can provide an approximated prediction x^* (see Appendix) of the actual movement x when it receives the motor command u. A predicted possible movement error $x_d - x^*$ is transmitted to the motor cortex and to muscles via the rubrospinal tract. Since the loop time of the cerebro-cerebellar communication loop is 10–20 ms (Eccles 1979) and is shorter than that of the supraspinal loop, the performance of the feedforward control with the internal model and the internal feedback loop is better than that of long-loop sensory feedback. In summary, the spinocerebellum-magnocellular red nucleus system updates the motor command by predicting a possible error of movement.

The cerebrocerebellum (lateral part of the hemisphere)-parvocellular part of the red nucleus system, which develops extensively in primates, especially in man, receives its input from wide areas of the cerebral cortex and does not receive peripheral sensory input. That is, it monitors both the desired trajectory x_d and the motor command u but it does not receive information about the actual movement x. Within the cerebro-cerebellum-parvocellular red nucleus system, an internal neural model of the inverse-dynamics of the musculoskeletal system is acquired, with practice, by making use of the synaptic plasticity. The inverse-dynamics of the musculoskeletal system is defined as the dynamical system whose input and output are inverted (trajectory x is the input and motor command (i.e. torque) u is the output). Note that the spinocerebellum-magnocellular red nucleus system provides a model of the dynamics of the musculoskeletal system (motor command (i.e. torque) u is the input and trajectory x is the output). The inverse-dynamics model is not a model of the external world; rather it is a model of the information processing done in other brain regions such as the motor cortex and the spinocerebellum which computes the motor command from the desired trajectory. A germ of this idea can already be seen in Ito (1970). Once the inverse-dynamics model is acquired by motor learning, it can compute a good motor command u^* directly from the desired trajectory x_d. This motor command is transmitted to the motor cortex via the ventrolateral nucleus of the thalamus. In summary, the cerebrocerebellum-parvocellular red nucleus system substitutes for other brain regions in the complex computation of motor commands.

The neural network model shown in Fig. 2 is based on various physiological and morphological information (Ito 1984; Ghez and Fahn 1985), especially on the importance of synaptic plasticity (Gilbert and Thach 1977) and of the cerebrocerebellar communication loop (Sasaki et al. 1982; Sasaki and Gemba 1982) in the motor learning of voluntary limb movements. The model predicts that if the rubro-olivo-cerebellar loop is destroyed, the internal model would be destroyed also, and motor performance would be severely disturbed. This prediction is in accord with the frequently reported symptoms of "tremor" after lesions anywhere in the rubro-olivary pathway (Poirier et al 1969). This may be interpreted as being the oscillation due to the delay of feedback. Another prediction of the model is that after lesions of the rubro-olivo-cerebellar loop, motor learning does not take place. This, of course, is the deficit of motor learning as reported by Ito and his collaborators (Ito et al. 1974, 1982) and also by Llinás (Llinás et al. 1975), although the interpretation of this deficit has been related to Marr's hypothesis in the case of Ito's paradigm (Ito 1984).

3. Internal Neural Model with Heterosynaptic Plasticity

The internal dynamics model and the internal inverse-dynamics model proposed in the previous section can be realized by a parallel-distributed-processing neural network with heterosynaptic plasticity. They can be regarded, from an engineering point of view, as identifiers of unknown dynamics and inverse-dynamics of the musculoskeletal system. Arbib (1981) pointed out that such a neural identifier is essential for motor learning. Let us consider an identifier which approximates the output $z(t)$ of an unknown nonlinear system by monitoring both the input $u(t)$ and the output $z(t)$ of this system (Fig. 3). This type of identifier can be realized by a neural network, as shown in Fig. 3 (Tsukahara and Kawato 1982), which comprises many subsystems computing various nonlinear transformations of the input $u(t)$, and a neuron with heterosynaptic plasticity.

The input $u(t)$ to the unknown nonlinear system is also fed to n subsystems and is nonlinearly transformed into n different inputs $x_l(t)$ $(l = 1, \ldots, n)$ to the neuron with plasticity. That is, instantaneous firing frequencies of n input fibers to the neuron are designated by $x_1(t), \ldots, x_n(t)$. Let w_l denote a synaptic weight of the l-th input. Membrane potential $y(t)$ of the neuron is the sum of n postsynaptic potential. For simplicity, we assume that the output signal of the neuron is equal to its membrane potential $y(t)$. In vector notation, we have the following equations.

$$\mathbf{x}(t) = [x_1(t), x_2(t), \ldots, x_n(t)]^T,$$
$$\mathbf{w} = [w_1, w_2, \ldots, w_n]^T,$$
$$y(t) = \mathbf{w}^T\mathbf{x}(t) = \mathbf{x}(t)^T\mathbf{w}. \qquad (1)$$

Here, T denotes transpose. The second synaptic input to the neuron is an error signal (e.g., climbing fiber input for the Purkinje cell), and is given as an error between the output of the neuron and the output from the unknown system, $s(t) = z(t) - y(t)$. Based on physiological knowledge of the heterosynaptic plasticity of the red nucleus neurons (Tsukahara et al. 1981) and the Purkinje cells (Ito 1984), we assume that the l-th synaptic weight w_l changes when the conjunction of the l-th input $x_l(t)$ and the teaching signal $s(t)$ occurs. Further, the rate of change of the synaptic weight is assumed proportional to the product of the two inputs:

$$\tau d\mathbf{w}(t)/dt = \mathbf{x}(t)s(t) = \mathbf{x}(t)[z(t) - \mathbf{x}(t)^T\mathbf{w}(t)]. \qquad (2)$$

Here, τ is a time constant of change of the synaptic weight. This learning rule is closely related to Amari's (1977) orthogonal learning, although it was used for the association of static spatial patterns.

If $u(t)$ is a stochastic process, then $\mathbf{x}(t)$ and $z(t)$ are also stochastic processes and Eq. (2) must be rewritten as a stochastic differential equation:

$$\tau d\mathbf{w}(t,\omega)/dt = \mathbf{x}(t,\omega)[z(t,\omega) - \mathbf{x}(t,\omega)^T\mathbf{w}(t,\omega)]. \qquad (3)$$

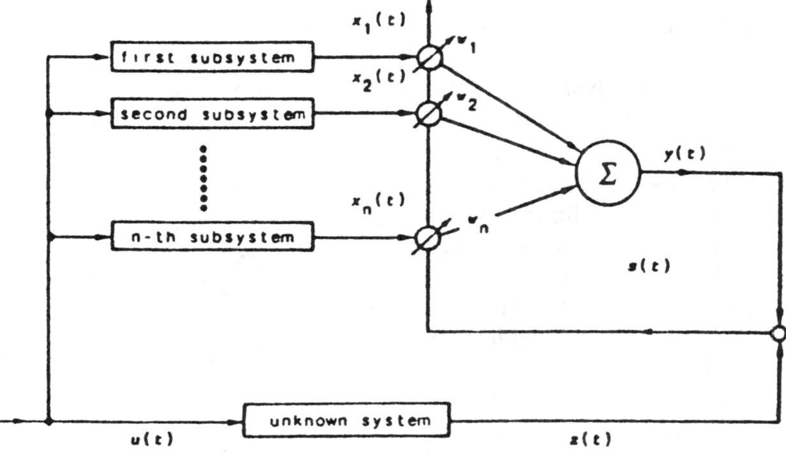

Fig. 3 A neural identifier of an unknown system compromised of n subsystems and a neuron with heterosynaptic plasticity.

Here, ω is a sample point in a stochastic space. Let us consider the following averaged equation of (3), which is obtained by taking an expected value.

$$\tau d\mathbf{m}(t)/dt = E[\mathbf{x}(t,\omega)z(t,\omega)] - E[\mathbf{x}(t,\omega)\mathbf{x}(t,\omega)^{\mathrm{T}}]\mathbf{m}(t). \qquad (4)$$

Here, $\mathbf{m}$ is an averaged vector of $\mathbf{w}$. We can prove the following theorem about the convergence of the synaptic weight $\mathbf{w}(t)$ using Geman's (1979) result, if $\mathbf{x}$ and $\mathbf{z}$ are mixing random processes (see Appendix for proof).

Theorem 1. If the time constant τ of change of the synaptic weight is sufficiently long compared with the "rate" of mixing of $\mathbf{x}$ and z, *then the synaptic weight* $\mathbf{w}$ *converges in mean to the value for which a mean square error of the output* $E[(z - y)^2]$ *is minimum.*

Because the time constants of physiologically known synaptic plasticities are sufficiently long (from a few hours to a few weeks) compared with temporal patterns of movement (several hundred ms), the assumption of the theorem is satisfied. It is worthwhile to note that the averaged Eq. (4) gives the steepest descent method and the convergence is global (i.e. there is no local minimum: see Appendix). The basic organization of the neural identifier shown in Fig. 3 is the same as the LMS (Least Mean Square) adaptive filter of Widrow (Widrow et al. 1976) and the adaptive filter model of Fujita (1982a), although they dealt with a linear system and the proof of convergence was different.

Although the best synaptic weights can always be obtained by the learning rule (2) within a given set of subsystems, Theorem 1 does not necessarily guarantee that the output error tends to become zero as learning proceeds. Asymptotic performance of the neural identifier critically depends on the selection of the set of subsystems (see Appendix).

4. Control of Robotic Manipulator by Model Neural Network

We examined whether the proposed neural-network model is efficient in learning control of an object with highly nonlinear dynamics and multiple degrees of freedom. In computer simulation, a usual industrial robotic manipulator was chosen as a controlled system. Although it is much simpler than musculo-skeletal systems such as the human arm, they both have several essential features (nonlinear dynamics, multiple degrees of freedom, and interactions between different freedoms) in common.

By computer simulation, Furukawa (1984) and Miyamoto (1985) showed that the internal neural model for the dynamics of a robotic manipulator with two degrees of freedom was actually acquired during a 2000–5000 s learning period while monitoring both the input torque [$u(t)$ in Fig. 3] and the output joint angles [$z(t)$ in Fig. 3] of the manipulator. In both studies, the learning rule in (2) was used. However, as subsystems of the neural identifier, Furukawa used Wilson-Cowan's neuronpool models (1972) with different synaptic-connection parameters, while Miyamoto chose different models of the manipulator dynamics with different viscosity-coefficient parameters at the joints. Once the dynamics model was obtained by the learning, the internal feedback loop with this internal model was found to control the manipulator much better than the long-loop feedback via the external world (see the last paragraph of Appendix).

4.1 Hierarchical Control of Manipulator by Inverse-Dynamics Model

To examine learning and control performance, extensive computer simulations were made of the hierarchical neural network model, excluding the spinocerebellum-magnocellular red nucleus system (dynamics model). We omitted the internal dynamics model because of the limitation of computer resources. This part of the model is the most time consuming because we need to numerically integrate many differential equations which describe the subsystems of the dynamics model.

Figure 4a shows a block diagram of a simulated neural network model and a manipulator. Let $T(t)$, $T_i(t)$, and $T_f(t)$ denote a torque fed to the manipulator, a torque calculated by the inverse-dynamics model and a feedback torque, respectively. The total torque was a sum of the feedforward and the feedback torques:

$$T(t) = T_i(t) + T_f(t). \qquad (5)$$

The inverse-dynamics model receives the desired trajectory $q_d(t)$ represented as joint angles as an input [$u(t)$ in Fig. 3] and monitors the total torque $T(t)$ as the output of the unknown dynamical system [$z(t)$ in Fig. 3]. If the simulated neural network would be related to the percepton, the total torque might correspond to the teaching signal. From (5), the error signal $s(t)$ in the previous section equals the feedback torque $T_f(t)$, and it is expected that $T_f(t)$ tends to zero as learning proceeds. But this is by no means guaranteed because we cannot simply take out an unknown dynamical system from the block diagram of Fig. 4a. In other words, the inverse-dynamics model does not receive the real trajectory $q(t)$ represented as joint angles; instead, it receives only the desired trajectory $q_d(t)$. Consequently, it is very important to examine the learning and control performance of the hierarchical control system shown in Fig. 4a.

The configuration of the three-link manipulator shown in Fig. 4b was chosen so that it resembles a human arm (see Table 1 for physical parameters). Let $q_k (k = 1, 2, 3)$ denote the k-th joint angle and $T_k (k = 1, 2, 3)$ denote the torque fed to the k-th joint. Using the Lagrangian, we can derive the following dynamics equation of the manipulator.

$$R(\mathbf{q})\ddot{\mathbf{q}} - \left(\sum_k \dot{q}_k \partial R/\partial q_k\right)\dot{\mathbf{q}} - (1/2)\dot{\mathbf{q}}^{\mathrm{T}}(\partial R/\partial \mathbf{q})\dot{\mathbf{q}} + B\dot{\mathbf{q}} + \mathbf{G}(\mathbf{q}) = \mathbf{T}(t), \qquad (6)$$

where,

$$\mathbf{q} = (q_1, q_2, q_3)^{\mathrm{T}} \text{ and } \mathbf{T} = (T_1, T_2, T_3)^{\mathrm{T}}.$$

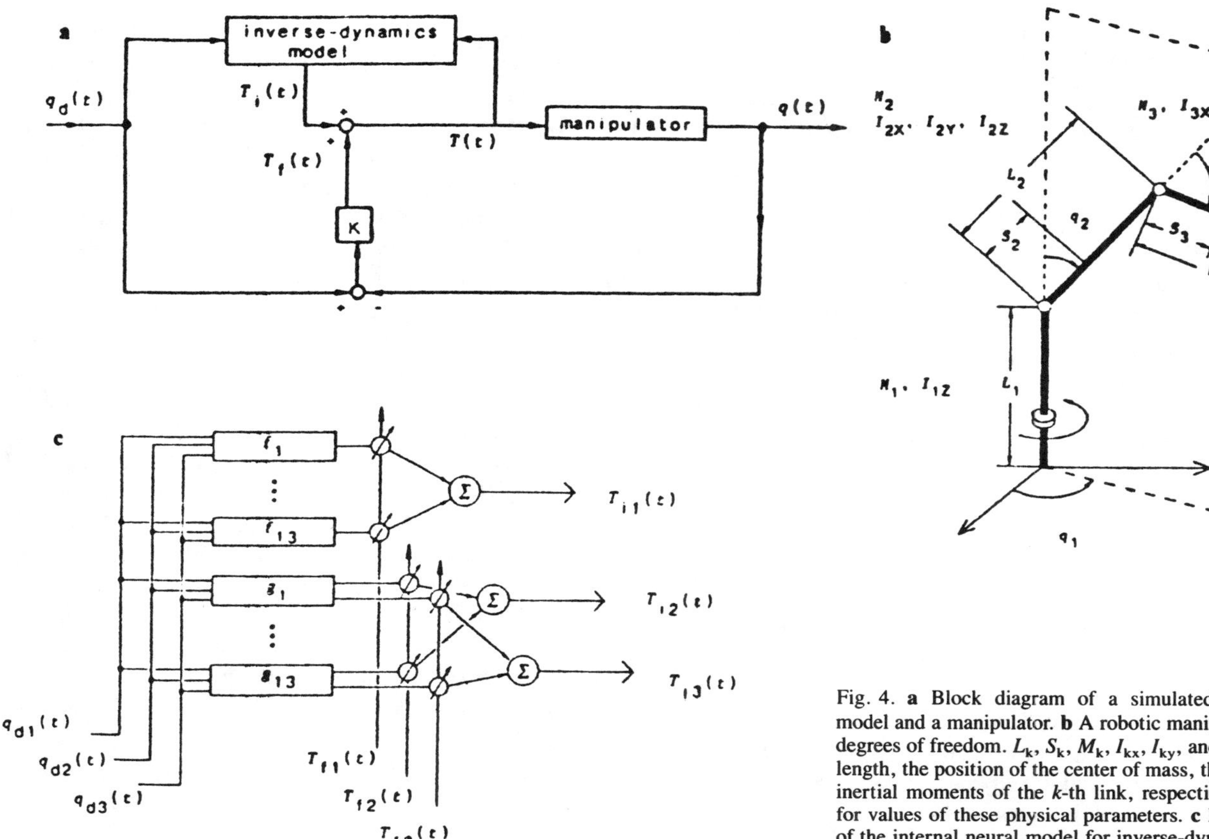

Fig. 4. **a** Block diagram of a simulated neural network model and a manipulator. **b** A robotic manipulator with three degrees of freedom. L_k, S_k, M_k, I_{kx}, I_{ky}, and I_{kz} represent the length, the position of the center of mass, the mass, the three inertial moments of the k-th link, respectively. See Table 1 for values of these physical parameters. **c** Detailed structure of the internal neural model for inverse-dynamics of the manipulator with three degrees of freedom.

$R(\mathbf{q})$ is a 3×3 inertia matrix, which is not diagonal. B is a 3×3 diagonal matrix representing viscosity coefficients, $B = \text{diag}(b_k)$. $\mathbf{G}(\mathbf{q})$ is a 3-dimensional nonlinear vector function and represents gravitational forces. The first term represents the inertia force, the second and third terms represent the centripetal and Coriolis forces, the fourth term is the frictional force and the fifth term is the gravitational force. It can be seen that the manipulator dynamics is nonlinear and there are interactions between differential freedoms.

Figure 4c shows the detailed structure of the internal model for the inverse-dynamics of the manipulator. It receives the three inputs $[q_{d1}(t), q_{d2}(t), q_{d3}(t)]$ and the three error signals $[T_{f1}(t), T_{f2}(t), T_{f3}(t)]$, and the outputs of three torques $[T_{i1}(t), T_{i2}(t), T_{i3}(t)]$. $q_{dk}(t)$, $T_{fk}(t)$, and $T_{ik}(t)$ $(k = 1, 2, 3)$ denote the desired joint angle, the feedback torque and the torque generated by the inverse-dynamics model, respectively, regarding the k-th joint. This is an expansion of the neural identifier of Fig. 4a to the multiple inputs-outputs case. Each subsystem receives the three inputs, q_{d1}, q_{d2}, and q_{d3}, and nonlinearly transforms them into its output $f_l(q_{d1}, q_{d2}, q_{d3})$ for the first joint, or $g_l(q_{d1}, q_{d2}, q_{d3})$ for the second and the third joints. The subsystems f_l and g_l were conveniently chosen as shown in Table 2 from manipulator dynamics Eq. (6). That is, each subsystem corresponds to an expanded term of the left-hand side of (6). Although, in general, different sets of subsystems need to be prepared for different outputs $T_{ik}(k = 1, 2, 3)$, the same set of subsystems $g_l(l = 1, \ldots, 13)$ were used in common by the second and third outputs since their subsystems almost overlap with each other. In summary, the torques generated by the inverse-dynamics model are expressed as follows.

$$T_{i1}(t) = \sum_{l=1}^{13} w_{1l} f_l(q_{d1}(t), q_{d2}(t), q_{d3}(t))$$

$$T_{ik}(t) = \sum_{l=1}^{13} w_{kl} g_l(q_{d1}(t), q_{d2}(t), q_{d3}(t)) \quad (k = 2, 3). \tag{7}$$

Here w_{kl} is the weight of a synapse from the l-th subsystem to the k-th output neuron. If a perfect inverse-dynamics model is formed, Eq. (7) must coincide with Eq. (6). Let the left-hand

TABLE 1. Values of physical parameters of the robotic manipulator shown in Fig. 4b. b_k is a viscosity coefficient at the k-th joint. See explanation of Fig. 4 for definitions of other symbols

Parameter	First link	Second link	Third link
L_k(m)	0.4	0.4	0.4
S_k(m)	—	0.15	0.15
M_k(kg)	15.0	7.0	3.0
I_{kx}(kg·m^2)	—	0.589	0.251
I_{ky}(kg·m^2)	—	0.584	0.253
I_{kz}(kg·m^2)	0.0170	0.00673	0.00340
b_k(kg·m/s)	20.0	15.0	5.0
Mass of payload (kg)	1.0		

Table 2. Nonlinear transformations of 26 subsystems used in the inverse-dynamics model shown in Fig. 4c. q_{dk} is simply denoted as q_k here

l	$f_l(q_1, q_2, q_3)$	$g_l(q_1, q_2, q_3)$
1	$\bar{q}_1$	$\bar{q}_2$
2	$\sin^2\bar{q}_2 \cdot \bar{q}_1$	$\bar{q}_3$
3	$\cos^2 q_2 \cdot \bar{q}_1$	$\cos q_3 \cdot \bar{q}_2$
4	$\sin^2(q_2 + q_3) \cdot \bar{q}_1$	$\cos q_3 \cdot \bar{q}_3$
5	$\cos^2(q_2 + q_3) \cdot \bar{q}_1$	$\sin q_2 \cdot \cos q_2 \cdot \dot{q}_1^2$
6	$\sin q_2 \sin(q_2 + q_3) \cdot \bar{q}_1$	$\sin(q_2 + q_3)\cos(q_2 + q_3) \cdot \dot{q}_1^2$
7	$\sin q_2 \cos q_2 \cdot \dot{q}_1\dot{q}_2$	$\sin q_2 \cos(q_2 + q_3) \cdot \dot{q}_1^2$
8	$\sin(q_2 + q_3)\cos(q_2 + q_3) \cdot \dot{q}_1\dot{q}_2$	$\cos q_2 \sin(q_2 + q_3) \cdot \dot{q}_1^2$
9	$\sin q_2 \cos(q_2 + q_3) \cdot \dot{q}_1\dot{q}_2$	$\sin q_3 \cdot \dot{q}_2^2$
10	$\cos q_2 \sin(q_2 + q_3) \cdot \dot{q}_1\dot{q}_2$	$\sin q_3 \cdot \dot{q}_3^2$
11	$\sin(q_2 + q_3)\cos(q_2 + q_3) \cdot \dot{q}_1\dot{q}_3$	$\sin q_3 \cdot \dot{q}_2\dot{q}_3$
12	$\sin q_2 \cos(q_2 + q_3) \cdot \dot{q}_1\dot{q}_3$	$\dot{q}_2$
13	$\dot{q}_1$	$\dot{q}_3$

side of (6) be expanded by the same functions f_l and g_l as in (7), and w^*_{kl} denote a coefficient of the l-th function f_l or g_l in this expansion of the k-th torque. If uniqueness of convergence of **w** holds (see Appendix), we expect that each w_{kl} converges to w^*_{kl} as learning proceeds.

The fourth order Runge-Kutta-Gill method with a time step of 2 ms was used to numerically integrate the learning Eq. (2) and the manipulator dynamics Eq. (6). The first and the second time derivatives of joint angles, which were used in nonlinear transformations of subsystems, were calculated by central difference methods. Initial values of the synaptic weights w_{kl} at the beginning of learning were all set at 0; that is, the inverse-dynamics model did not output any torque at the beginning of learning. The program was written in Fortran 77 for a ACOS 1000 computer at Osaka University.

4.2 Learning with Repeated Movement Pattern

We first studied control performance of the neural network model when one movement pattern was repeatedly learned. A desired trajectory, shown in Fig. 5, with a duration of 30 s was given for 20 min (i.e. for 40 times). In this pattern, the three joints moved cooperatively at a maximum speed of about 500 deg/s, which was quite fast for industrial robotic manipulators. As can be seen in Fig. 5, the time course of each joint angle $q_{dk}(t)$ was smooth, since it was composed of trigonometric functions smoothly jointed with constant parts, while the manipulator held a fixed posture. The time constant τ of learning was chosen as 1000 s. For simplicity, the gravitational force $\mathbf{G}(\mathbf{q})$ in (6) was compensated beforehand.

The feedback torque was made of a proportional component and a local derivative component:

$$T_{fk}(t) = K_{pk}[q_{dk}(t) - q_k(t)] + K_{vk}dq_k(t)/dt, \qquad k = 1, 2, 3,$$
$$K_{vk} = 0 \text{ unless } |q_k(t) - q_{dk} \text{ (objective point)}| < \epsilon.$$

Proportional and derivative feedback gains K_{pk} and K_{vk} were selected as (517.2, 746.0, 191.4) and (16.2, 37.2, 8.4) so that the natural angular frequency $\omega_n = 20$ and the damping ratio $\xi = 0.7$ were attained. These values were calculated based on a linearization of the manipulator dynamics Eq. (6). The velocity feedback was applied only around the stopping points (i.e. only after the joint angle got into some small bounds of the objective point). Organisms hold a posture by the cocontraction of flexor and extensor muscles around the same joint, which induces an increase of viscous friction of the muscles. The local velocity feedback simulates this effect and reduces overshoots of movements.

Figure 6 compares the total torques fed to the third joint (T_3, top), feedback torques (T_{f3}, middle) and the torques generated by the inverse-dynamics model (T_{i3}, lower), during the first 30 s at the beginning of learning (left) and during the final 30 s at the end of 20 min learning (right). At the beginning of learning T_3 was composed mainly of T_{f3} and was considerably spiky. As the learning proceeded, T_{f3} decreased while T_{i3}, gradually increased. After 20 min of training, T_{f3} was very small and was composed only of the local velocity feedback; hence T_3 was almost identical to T_{i3}. The time course of T_3 was smoother at the end of learning than at the beginning.

Changes of mean square errors of the outputs (i.e. mean square of feedback torques) during 20 min of learning are semilogarithmically plotted in Fig. 7, top; and mean square errors of the joint angles are plotted in Fig. 7, bottom. They were averaged values over 30 s of one training session. Note that scales of the ordinates for the three different joints were different. The mean square errors of both the torques and trajectories decreased gradually, but those of the first joint decreased more rapidly than the second and the third joints. At the end of 20 min of learning, the errors of trajectory were considerably small. Figure 5 actually plots not only the desired trajectory but also a realized trajectory during the last 30 s of the learning. But desired and real trajectories almost overlapped and could not be seen separately at this resolution.

Figure 8 shows a change of the synaptic weight w_{14} during the first 5 min of learning. The synaptic weight gradually approached an asymptomatic value while showing a damped oscillation with a 30 s period of one training session. Table 3 compares the values of the synaptic weights w_{1l} at the end of 20 min training with the corresponding expansion coefficients w^*_{1l} of (6), which were calculated from physical parameters in Table 1. Some synaptic weights (e.g. w_{14}, w_{16}, w_{19}, w_{113}) have already converged quite closely to the corresponding values of w^*_{kl}; but some other synaptic weights (e.g. w_{11}, w_{13}, w_{17}) were still considerably different from them. These discrepancies could, of course, be ascribed to insufficient learning time; but they were partially due to the linear dependency of outputs of different subsystems (e.g. $f_1 = f_2 + f_3$). This linear dependency is irrelevant, since our concern is not whether w_{kl} converges to w^*_{kl}, but whether the inverse-dynamics model generates good torque (see Appendix).

We then examined whether the neural network model after 20 min of learning a single pattern could control a quite different movement which was about twice as fast as the training pattern. Figure 9 shows the desirable trajectory of this test movement (the third joint angle, broken curve), and trajectories (solid curves) realized by the neural network model

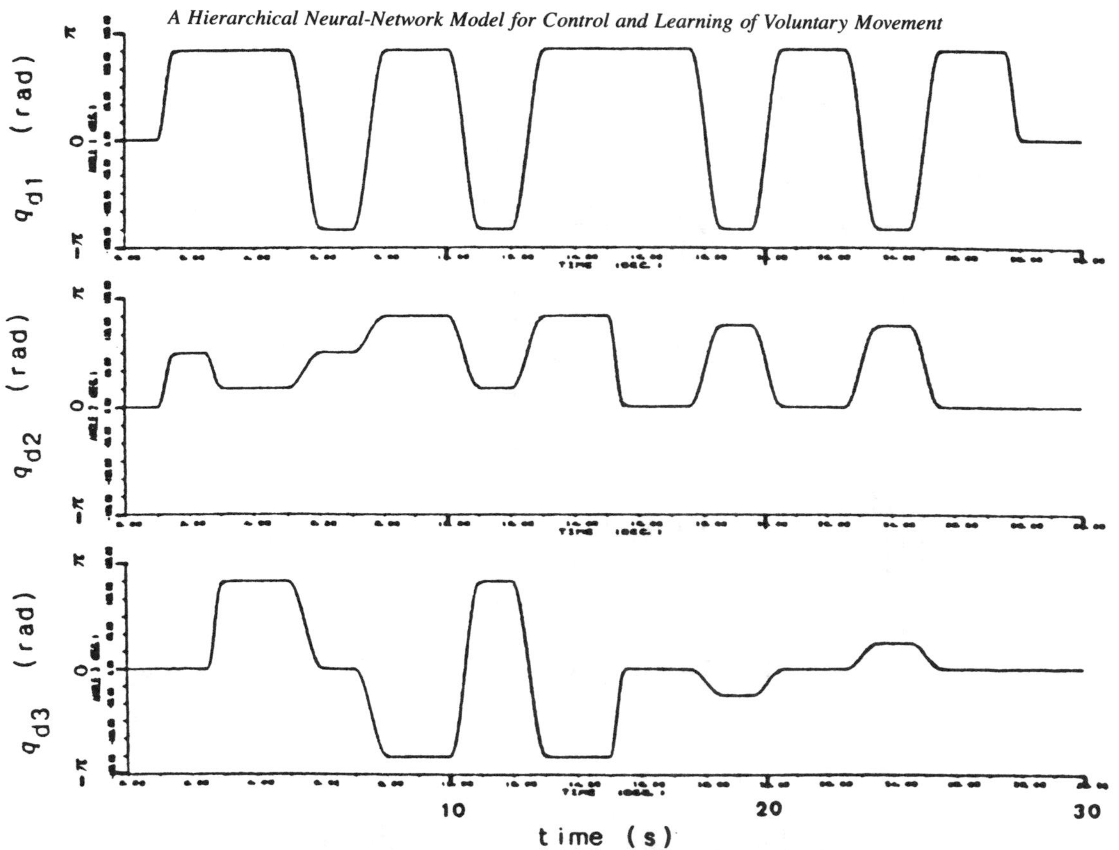

Fig. 5 Time courses of the three joint angles during the desirable movement pattern. The realized trajectory after 20 min learning is also plotted but can not be separately seen.

without learning (top), and after learning (bottom). Before learning, delays and overshoots were often observed, and the realized trajectory deviated considerably from the desired trajectory. However, after learning, the actual trajectory almost coincided with the desired trajectory. This control capability for quite different and faster movements than the training pattern is one of the most outstanding features of our neural network model (see Discussion).

In the above experiments, we incorporated the local velocity feedback into the feedback torque, and hence we could set comparatively high feedback gains. So, the feedback control before learning was reasonably good (see Fig. 9 top). We examined whether the neural network model could learn movements even when the feedback was poorer. The feedback torque was generated only by the proportional term, and the feedback gains K_{p1}, K_{p2}, and K_{p3} were set (309.4, 120.6, 52.3) so that the damping ratio ξ was 0.5. The resulting natural frequencies ω_n were between 8 and 15. With this poorer feedback, the neural network model could not learn the desirable movement shown in Fig. 5; that is, the synaptic weights diverged. However, when a training movement pattern the same as that shown in Fig. 5 but with a duration of 1 min was chosen (i.e the speed of the movement was reduced by half), the learning went well. After 40 min of learning, the network controlled the test pattern shown in Fig. 9 quite well. Consequently, even if the feedback was poor, the neural network model could learn slow movements, and after learning, it could also control different and faster movements.

Although the gravitational force was compensated for beforehand in almost all computer simulations, this is not indispensable. When additional subsystems which correspond to expanded terms of the gravitational force $\mathbf{G}(\mathbf{q})$ were prepared in the inverse-dynamics model, the neural network model without gravity compensation could learn the movement shown in Fig. 5.

4.3 Learning with Quasi-Periodic Movement Pattern

The movement pattern in Fig. 5 contained various elements of coordinated multi-joint movements. But, in general, it is not easy to determine such a pattern; and further, it is very unnatural for organisms to learn movement by repetitions of a single pattern. We examined a learning performance when a simple and quasi-periodic movement pattern was given as a desirable trajectory. In this pattern, the three joint angles changed as $\sin(\omega_k t)$ and the ratio of angular frequencies $\omega_1:\omega_2:\omega_3$ were set as $1:\sqrt{2}:\sqrt{3}$, so that various coordinated movements were experienced during learning. The feedback gains, the time constant of learning and the learning duration were the same as the first simulation experiment (Figs. 6–8). However, since the manipulator did not hold a constant posture in this training pattern, the local velocity feedback did not work during learning.

Figure 10 shows changes of mean square errors of the output (i.e. the mean square of the feedback torques) during 20 min

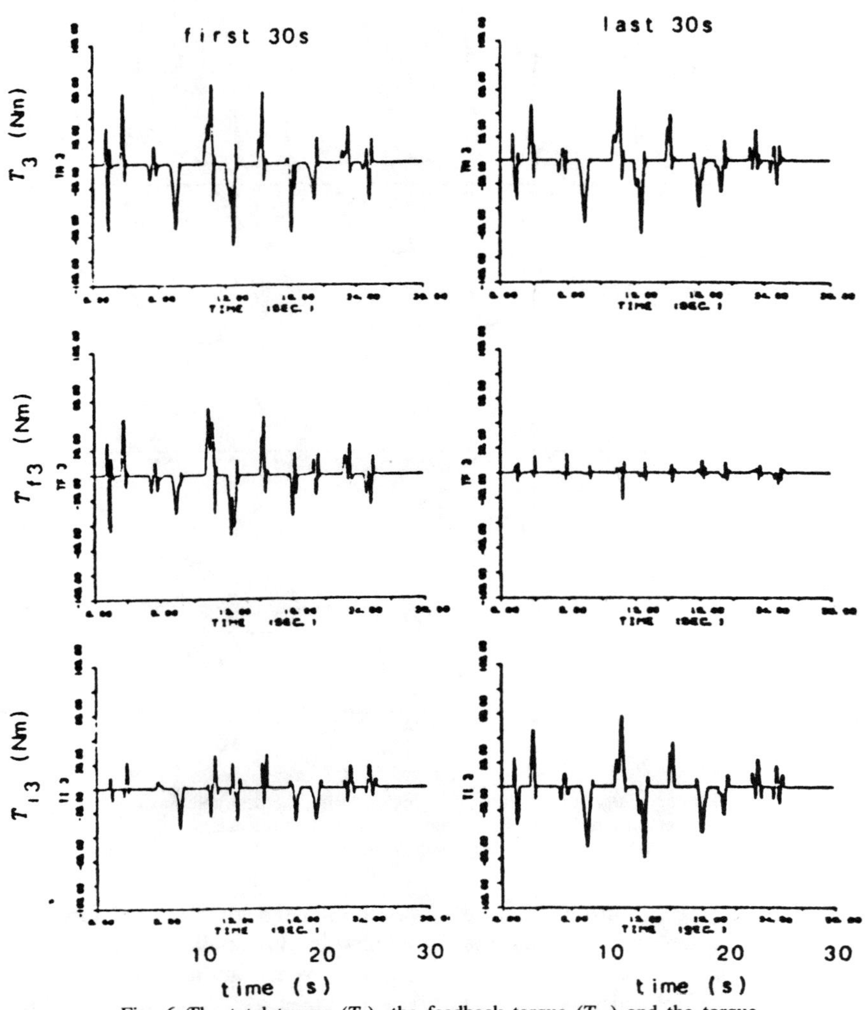

Fig. 6 The total torque (T_3), the feedback torque (T_{f3}) and the torque computed by the inverse-dynamics model (T_{i3}), which were fed to the third joint during the first 30 s of learning (left) and the last 30 s (right).

of learning. The mean square errors about the three joints all decreased monotonically and the learning went well. Since the final feedback torques were smaller than those in Fig. 7 by order of 10^{2-3}, it first appears that the learning with the quasi-periodic pattern was much more efficient. However, we can not compare control performances using the two different training patterns. Thus, we examined control performance using the test pattern of Fig. 9 after 20 min of quasi-periodic learning. The trajectory realized was almost identical to the desired trajectory, and we could not simply determine whether this result was better than the result shown in Fig. 9 (repetitive learning). Regarding the first and the second joints, the quasi-periodic learning seemed better; while for the third joint, the repetitive learning seemed better.

4.4 Learning with Redundant Subsystems

We conveniently chose necessary and sufficient subsystems, as shown in Table 2, from an expansion of the manipulator dynamics Eq. (6) (although some of them were linearly dependent; see Sect. 4.2 and Appendix). From an engineering point of view, we sometimes need to control a very complex system, even when its fundamental dynamical structure is unknown. Furthermore, as a neural network model, it is too idealistic to assume that necessary and sufficient subsystems are inherently prepared in the CNS. Thus, we examined learning performance when 20 extra subsystems, shown in Table 4 (f_l and g_l were the same) were added to those of Table 2. The idea is that the CNS can inherently prepare a very large number of subsystems which are highly redundant but which include a relatively small number of the essential subsystems.

The feedback torques, the feedback gains, the time constant of learning and the desirable trajectory were the same as the first simulation experiment (Figs. 5–7). Figure 11 shows changes of the mean-square feedback torque fed to the third joint (left), and the mean square error of the third joint angle (right) during 20 min of learning. Both of them decreased monotonically and were not significantly larger than the errors shown in Fig. 7. In Table 5, 23 synaptic weights w_{1l} at the end

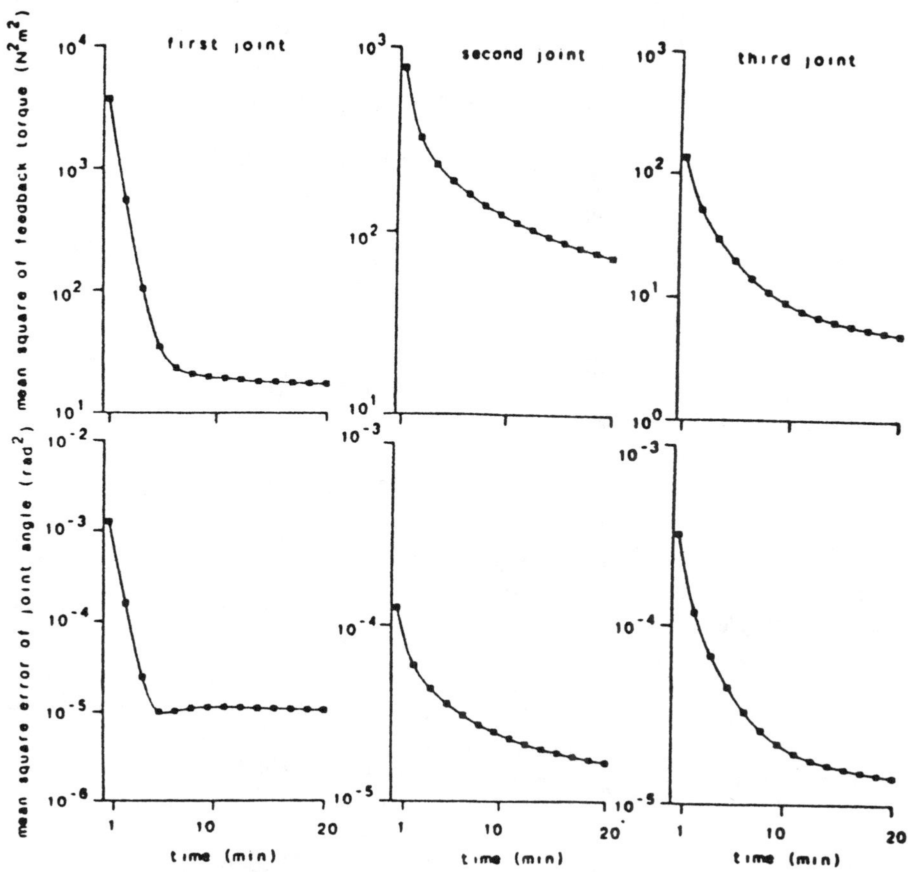

Fig. 7 Mean squares of feedback torques (top) and mean square errors of joint angles (below) during 20 min learning.

of learning are shown. The corresponding coefficients w^*_{1l} are the same as in Table 3 for $l = 1 - 13$ and are 0 for $l = 14 - 23$. Most of the unnecessary synaptic weights were close to zero; but, for example, w_{114} was considerably large. Probably, this did not severely interfere with the overall performance of the inverse-dynamics model as shown in Fig. 11, since the following approximation might hold within a working range of the manipulator: $20q_1 = 16.47q_1 + 6.42q_1^3$ (compare $w^*_{113} = 20$ and $w^*_{114} = 0$ with w^*_{113} and w^*_{114} of Table 5). In summary, if a sufficient number of subsystems are prepared, which can be done even from a very incomplete knowledge of the dynamics of a controlled system, the neural network model can efficiently learn and control movements.

4.5 Change of Manipulator Dynamics during Learning

We studied the control performance of the neural network model when a physical parameter of the manipulator suddenly changed. The feedback torques, the feedback gains, the time constant of learning and the desirable trajectory were the same as the first simulation experiment (Figs. 5–7). After 20 min

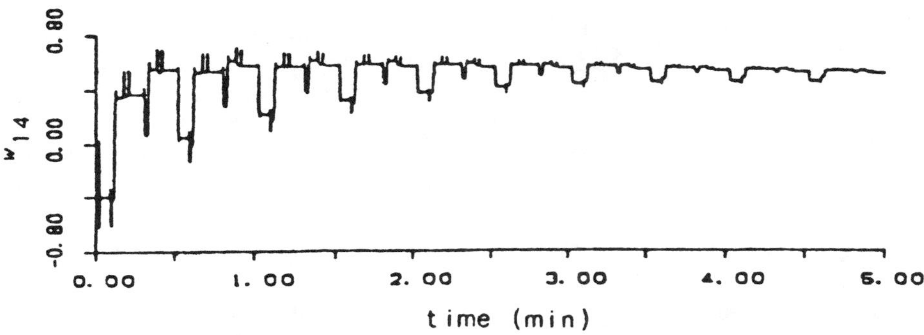

Fig. 8 Change of the synaptic weight w_{14} during the first 5 min of learning

TABLE 3. Comparison of synaptic weights w_{1l} at the end of 20 min training with the corresponding expansion coefficients w^*_{1l} of (6)

l	w_{1l}	w^*_{1l}
1	0.526	0.017
2	0.976	0.382
3	−0.451	0.007
4	0.478	0.480
5	0.048	0.003
6	0.655	0.680
7	0.655	2.75
8	1.955	0.953
9	0.718	0.680
10	0.817	0.680
11	0.466	0.953
12	1.040	0.680
13	19.917	20.0

of learning, the mass of the payload changed from 1 kg to 3 kg, and the simulation was continued for another 20 min. Figure 12 shows a change of the mean square of the feedback torque fed to the first joint (rectangles), and that of a naive neural network model without pre-learning (triangles). As can be seen, with pre-learning, the error quickly returned to a previous level within 5 min, and it was significantly smaller than the error without pre-learning. It can be said that the model has good adaptability to changes of the physical parameters of the manipulator.

5. Discussion

A hierarchical neural network model was proposed, based on physiological information and previous models. It contains internal neural models of dynamics and inverse-dynamics of the musculoskeletal system as essential learning parts. The potential of this model to control a complex nonlinear object was demonstrated by computer simulation with the following results.

(*1*) The dynamics model (spinocerebellum-magnocellular red nucleus) was acquired by learning while monitoring the motor command and the resulting movement.

(*2*) The inverse-dynamics model (cerebrocerebellum-parvocellular red nucleus) was acquired by repetitively experiencing a single motor pattern, while receiving the desired trajectory and monitoring the feedback torque as an "internal" error signal.

(*3*) As motor learning proceeded, the inverse-dynamics model gradually took the place of the external feedback as a main controller.

(*4*) Once the neural network model learned some movement, it could control quite different and faster movements. So, the present model is totally different from previous "table look-up" learning proposed by Albus (1975) or Raibert (1978), because of its capability of generalizing learned movements. The reason is because the present model learns the dynamics and inverse-dynamics of a control object instead of a specific motor command for a specific movement pattern.

(*5*) The total torque fed to the manipulator was by no means a good teacher at the beginning of learning; hence, quick movements could not be realized by it. However once the inverse-dynamics model was acquired while being supervised by this teacher, it could control fast movements smoothly. The student eventually surpassed the teacher.

(*6*) The model had adaptability to a sudden change in the dynamics of the controlled system.

(*7*) Even when redundant subsystems were added to the inverse-dynamics model, learning performance remained essentially unchanged.

Although we did not simulate the learning performance of the whole neural network model shown in Fig. 2, as learning proceeds, the internal feedback loop is first expected to take the role of the external feedback loop as the main controller; then the inverse-dynamics model is expected to take the part of the internal feedback loop. This upward shift of a dominant controller in the hierarchical neural network might reflect the phylogenesis of the motor nervous system in vertebrates. If a perfect inverse-dynamics model is formed, neither the internal nor the external feedback controls function, since movement error is absent. On the other hand, even when a perfect dynamics model is formed, the internal feedback loop suffers from limitations inherent in feedback control. So, at first sight, the dynamics model seems to do things by halves. However, it plays an essential role in providing the inverse-dynamics model with a good teaching signal. In the second simulation experiment described in Sect. 4.2, the synaptic weights diverged because the feedback torque was too poor. In order to avoid the failure of learning, either the training movement must be slowed down, the time constant of learning must be lengthened or the teaching signal must be improved. Consequently, the dynamics model makes it possible for the inverse-dynamics model to learn a relatively quick movement during a relatively short time.

In this paper, we have mainly studied free movements. It is worthwhile considering constrained movements also, such as carrying a book or exerting a force against a door, since most of our voluntary movements are related to the manipulation of external objects. In Fig. 13 we propose an expansion of the inverse-dynamics model of Fig. 4c. Figure 13 shows the single input-output-case; but it is easily expanded to the multiple inputs-outputs case. The basic idea is that several sets of subsystems are prepared for different situations of movement. The selection of the sets used in control and learning of a certain movement is specified by higher motor centers (e.g. association cortex). During free movement, only a fundamental set, which is equivalent to Fig. 4c, is used both for control and learning. On the other hand, during movement when holding a light-weight object, the second set of subsystems is also recruited, and only the synaptic weights of this set are subjected to modification, while the synaptic weights of the fundamental

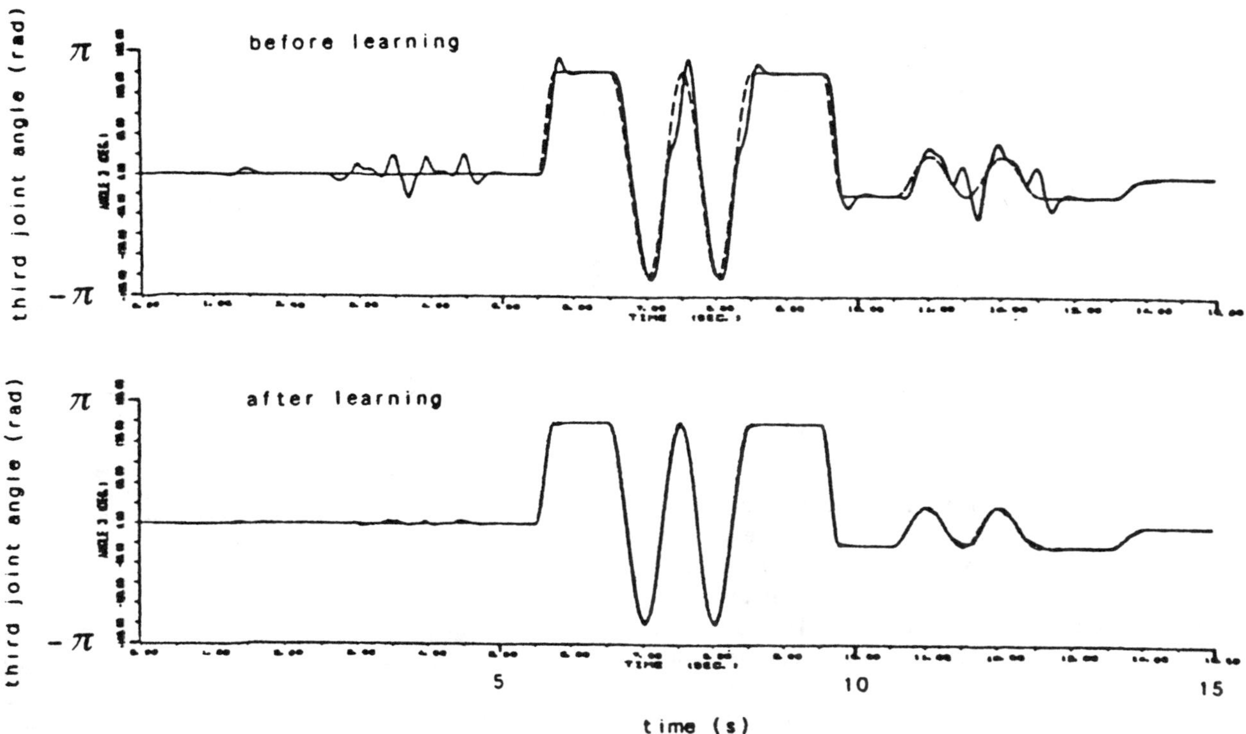

Fig. 9 Results of control for a faster and different movement from the training pattern, before (top) and after (bottom) the 20 min of learning. The desired trajectory of the third joint angle is plotted by a broken curve and the realized time courses are plotted by solid curves. They almost overlapped after learning

set remain unchanged. The second set compensates for torques necessary to carry a light-weight object. When the hand releases the object, the second set of subsystems is instantaneously suppressed (or not excited); then, the inverse-dynamics model can control free arm-movement as well as before. Similarly, during movement while gripping a heavy object, the third set is recruited (it might be even more efficient if higher motor centers transmit the estimated mass of a gripped object to the inverse-dynamics model). Further, if a hand needs to exert a force on an object, a quite different set of subsystems is recruited, which receives informations about the desirable forces as well as the desirable trajectory, and generates torques necessary to exert the required force at the hand. The merit of possessing several sets of subsystems is twofold. First, the internal model can preserve various dynamical information about the musculoskeletal system combined with its different environments (e.g. grasped objects or the knob of a door) as synaptic weights (on the contrary, dynamical information about free movement was lost in Fig. 12 after the payload changed). Second, a good motor performance is instantaneously realized as soon as an experienced behavioral situation is once again given.

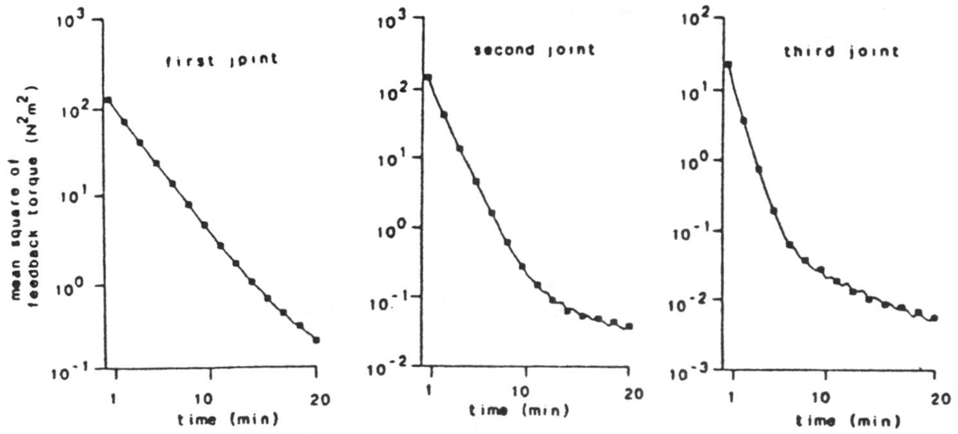

Fig. 10 Changes of mean squares of the feedback torques during 20 min of learning with the quasi-periodic training pattern

TABLE 4. Nonlinear transformations of 20 extra subsystems which were added to the original subsystems of Table 2

l	$f_l(q_1, q_2, q_3)$	$g_l(q_1, q_2, q_3)$
14	$0.01 \cdot \dot{q}_1^3$	
15	$0.01 \cdot \dot{q}_2^3$	
16	$0.01 \cdot \dot{q}_3^3$	
17	q_1^2	
18	q_2^2	
19	q_3^2	
20	$\sin q_2 \cos(q_2 + q_3)$	
21	$\cos q_2 \sin(q_2 + q_3)$	
22	$\sin q_2 \cos q_3$	
22	$\cos q_2 \sin q_3$	

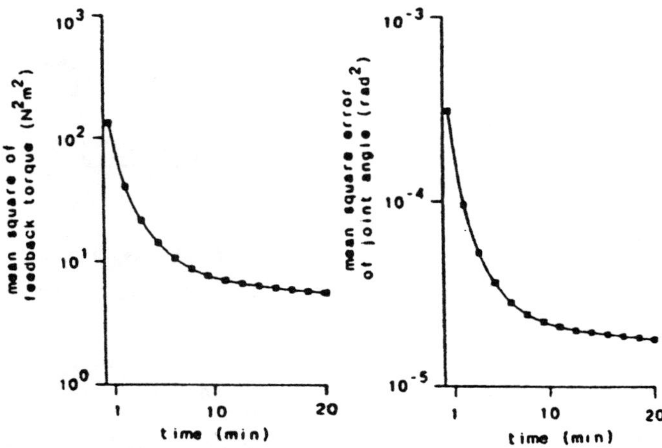

Fig. 11 Changes of the mean square of the feedback torque fed to the third joint (left) and the mean square error of the third joint angle (right) when the extra subsystems were added to the original subsystems

For control of the 3 degrees of freedom (d.g.f) manipulator, 3 output neurons and 26 subsystems were used in the inverse-dynamics model. In general, n output neurons are required to control an object with n d.g.f. For a 6 d.g.f. manipulator, about 900 subsystems are required (Setoyama 1987). The number of necessary subsystems is expected to increase in an order of n^4. A human arm has 7 d.g.f. but the number of muscles is much higher. So, the number of necessary subsystems for the inverse-dynamics model of a human arm may reach astronomical figures. Based on these considerations we can predict possible divergence and convergence of neural projections within the internal neural models. Each subsystem needs to receive the desirable time courses of lengths of several muscles whose contractions interact with other in movements. Hence, input signals to the subsystems (desirable trajectory) extensively diverge with moderate convergence. Then, the outputs of the subsystems intensively converge onto a relatively small number of output neurons. There are two possibilities about how the CNS computes nonlinear transformations of the subsystems. One is that they are realized by neural circuits. The other is that they are computed by nonlinear information processing within the dendrites of neurons (Poggio and Torre 1981; Kawato et al. 1984). Although it is possible that the subsystems are generally prepared, the other possibility, that the subsystems themselves are acquired by the synaptic plasticity of the Purkinje cells is appealing, since within the internal neural models both the Purkinje cells and red nucleus neurons have heterosynaptic plasticity. In this case, we need to investigate the performance of learning in the two-layered adaptive neural network. This is one of our problems for the future.

TABLE 5. Values of 23 synaptic weights w_{1l} regrading the first joint at the end of 20 min of learning when the extra subsystems were added to the original subsystems

l	w_{1l}	l	w_{1l}
1	0.440	13	16.472
2	0.892	14	6.417
3	−0.452	15	−0.350
4	0.365	16	−0.575
5	0.075	17	−0.001
6	0.933	18	0.099
7	2.065	19	−0.065
8	0.516	20	0.240
9	1.064	21	−0.034
10	1.506	22	0.142
11	0.891	23	0.011
12	0.824		

Finally, we assess the present model from an engineering point of view. Feedback controls have been nearly adequate for the slow control of usual industrial robotic manipulators with high reduction ratios between joints and actuators, which dramatically reduces the nonlinearity of manipulator dynamics and interactions between different freedoms. However, for a directdrive manipulator such as used in the present simulation study, new control methods have been demanded. Although several learning and adaptive control schemes were proposed (model reference adaptive control: Dubowsky and DesForges 1979; betterment process: Arimoto et al. 1984a and 1984b; table look-up method: Albus 1975; Raibert 1978) they were at most perturbation-learning schemes. That is, experiences obtained during learning can not be used for the execution of a quite different movement. The method of computed torque (Luh et al. 1980; Hollerbach 1980) requires both strict modelling of the manipulator dynamics and the precise estimation of physical parameters, which are difficult in practice. In contrast, the present method requires neither an accurate model (see Sect. 4.4) nor parameter estimation. Further, it possesses a great ability to generalize learning. We also note that it can be easily implemented in a parallel distributed processing machine, since both the nonlinear transformations in subsystems and the synaptic modifications are essentially in parallel. Consequently, the computation time in a parallel machine using the present control scheme is expected to be much shorter (shorter than 1/100) than the serial methods such as the recursive scheme for computed torque of Luh et al. (1980). Moreover, it does not require the enormous memory size of

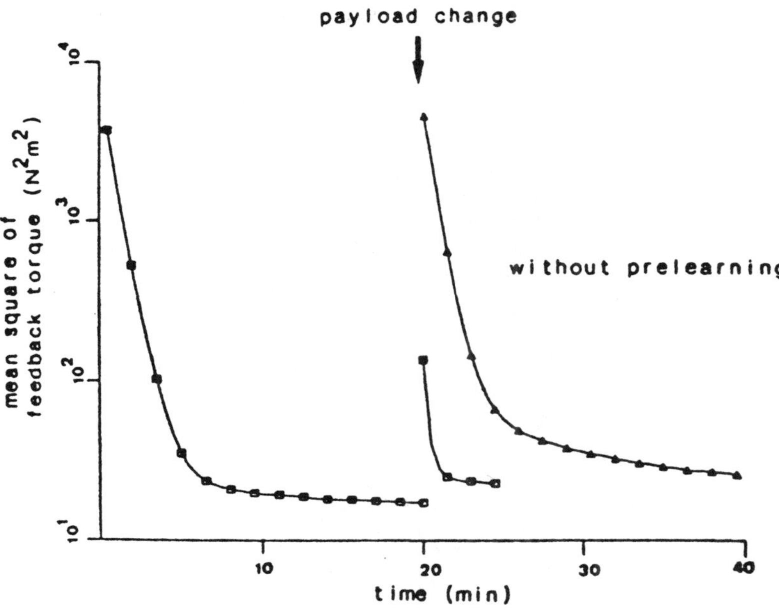

Fig. 12 Changes of the mean square of the feedback torque fed to the first joint during 40 min of learning. The mass of the payload suddenly changed from 1 kg to 3 kg 20 min after the beginning of learning. Triangles show results without the 20 min of pre-learning

the table look-up method (only 925 synaptic weights are necessary for a 6 d.g.f. manipulator; Setoyama 1987). Miyamoto, Kawato and Suzuki (1987) has successfully applied the present method to control an industrial robotic manipulator (Kawasaki-Unimate PUMA 260) with the neural network model in a microcomputer (Hewlett Packard 9000-300-320). In summary, the present method is one of the promising schemes for the future control of not only a direct drive manipulator, but also of a large-scale complex system, whose dynamics is only partially known.

Acknowledgment. We would like to thank Drs. F. Murakami and Y. Oda for discussing this work on several occasions and reading earlier versions of the manuscript.

Appendix

In this appendix we prove the convergence of the synaptic weights to optimal values, for which the mean square error of the output of the neural identifier shown in Fig. 3b is minimum. We assume that $u(t,\omega)$, and hence $x(t,\omega)$ and $z(t,\omega)$ of Fig. 3b, are strongly mixing stochastic processes (stochastic processes for which the "past" and the "future" are asymptomatically independent). This assumption seems sufficiently reasonable when we consider the "randomness" of our voluntary movements. Geman (1979) showed that when the "rate" of mixing is rapid relative to the rate of change of the solution processes, the averaged deterministic Eq. (4) is a good approximation of (3). Geman proved the following theorem in a general manner.

Theorem For all τ sufficiently large, $\sup_{t\geqq 0} E[W(t,\omega)^2 < \infty$, and $\lim_{\tau\to\infty} \sup_{t\geqq 0} E[\{M(t) - W(t,\omega)\}^2] = 0.$

Consequently, we only need to study the averaged Eq. (4) if the time constant τ of synaptic modification is sufficiently long. Let P denote the cross-correlation vector between the input signals $X(t,\omega)$ and the desired output $z(t,\omega)$:

$$P = E[x_i(t, \omega)\, z(t, \omega)]^T.$$

Q denotes the symmetric and positive definite correlations matrix of the input signals $X(t,\omega)$.

$$Q \equiv \{q_{ij}\} = \{E[x_i(t, \omega)x_j(t, \omega)]\}.$$

Then, Eq. (4) can be rewritten as follows.

$$\tau dM(t)/dt = P - QM(t). \tag{i}$$

For simplicity, P and Q are assumed to be constant in time. This is equivalent to assuming the stationary of the stochastic processes $X(t,\omega)$ and $z(t,\omega)$. Further, if Q is invertible, the averaged equation has the following solution

$$M(t) = \{1 - \exp(- Qt/\tau)\}Q^{-1}P.$$

Therefore, M(t) asymptomatically converges to $Q^{-1}P$ since Q is positive definite. From the above theorem, the synaptic weights $W(t,\omega)$ also converge to $Q^{-1}P$ in mean.

$$\text{m.s.e.} = E[s(t)^2] = E[\{z(t,\omega) - X(t,\omega)^T W(t,\omega)\}^2]$$
$$= E[z^2] - 2P^T W^+ W^T QW.$$

Substituting the solution $M(t)$ into this equation we obtain:

$$\text{m.s.e.}_{\text{mass}} = E[z^2] - P^T\{1 - \exp(-2Qt/\tau)\}Q^{-1}P.$$

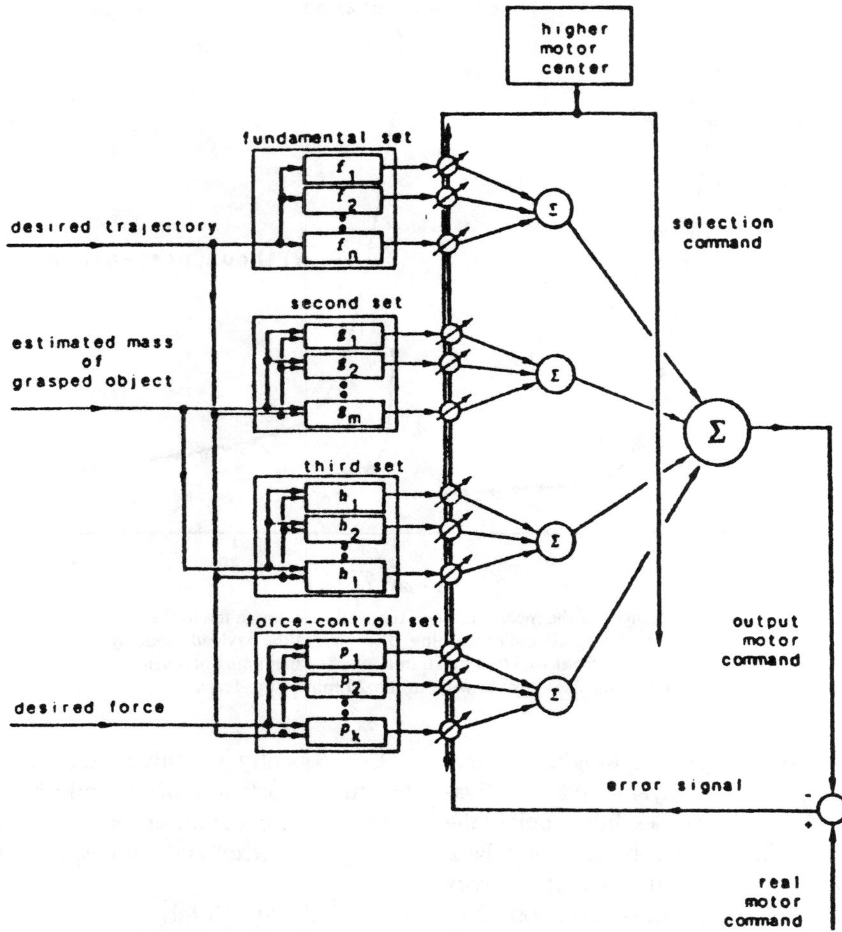

Fig. 13 Internal structure of an expanded inverse-dynamics model, which can cope with various behavioral situations such as carrying an object or exerting a force against an object, as well as with free movement

It may be observed that the m.s.e. performance is a quadratic function of the synaptic weights, that is, a "bowl-shaped" surface. The gradient at any point of the performance surface is obtained by differentiating the m.s.e. with respect to the synaptic weights as follows.

$$\nabla = -2\mathrm{P} + 2Q\mathrm{W}.$$

From this equation, it is easy to see that the averaged Eq. (4) gives the steepest descent method. The optimal synaptic weights W* and the minimum m.s.e. are obtained by setting $\nabla = 0$.

$$\mathrm{W}^* = Q^{-1}\mathrm{P},$$
$$\text{m.s.e.}_{\min} = \mathrm{E}[z^2] - \mathrm{P}Q^{-1}\mathrm{P}.$$

Consequently, we can conclude that the synaptic weights converge in mean to the optimal values for which the mean square output error is minimum.

It is worthwhile to note that the m.s.e.$_{\min}$ critically depends on the choice of subsystems. For example if $\mathrm{X}(t) = (z(t), 0, \ldots, 0)^{\mathrm{T}}$ then the m.s.e.$_{\min}$ is zero. On the other hand, if there are no correlations between X(t) and $x(t)$, then P equals zero and the m.s.e.$_{\min}$ equals $\mathrm{E}[x^2]$.

If we feed a delayed input $u(t - \Delta t)$ to the subsystems and the desired output $z(t)$ to the modifiable synapses, then the learning rule (3) attains optimal synaptic weights $\mathrm{W}^*_{\Delta t}$ with which $y(t) = \mathrm{W}^{\mathrm{T}}\mathrm{X}(t - \Delta t)$ best approximates $z(t)$. If we feed input $u(t)$ to the subsystems after $\mathrm{W}^*_{\Delta t}$ is learned and fixed, $y(t)$ clearly approximates $(z(t + \Delta t)$. Consequently, in this case, the internal neural model of Fig. 3b acts as a predictor. These two separate steps (identifier mode and prediction mode) can be simultaneously realized by the following modified learning equation of (2)

$$\tau d\mathrm{W}(t)/dt = \mathrm{X}(t - \Delta t)\,[z(t) - y(t - \Delta t)]$$
$$= \mathrm{X}(t - \Delta t)\,[z(t) - \mathrm{X}(t - \Delta t)^T\mathrm{W}(t - \Delta t)].$$

In this modified learning rule, the rate of change of the synaptic weights is proportional to the product of delayed input signal $\mathrm{X}(t - \Delta t)$ and the error signal between desired output $z(t)$ and delayed output signal $y(t - \Delta t)$. It is worthwhile to note that the heterosynaptic modification in the red nucleus is most prominent when the conditioned stimulus ($u(t)$) preceeds the unconditioned stimulus ($z(t)$) by about 100 ms (Tsukahara et al. 1981).

References

Albus JS (1971) A theory of cerebellar functions. Math Biosci 10:25–61

Albus JS (1975) A new approach to manipulator control: the cerebellar model articulation controller (CMAC). J Dyn Syst Meas Control 97:270–277

Allen GI, Tsukahara N (1974) Cerebrocellular communication systems. Physiol Rev 54:957–1006

Amari S (1977) Neural theory of association and concept-formation. Biol Cybern 26:175–185

Arbib MA (1981) Perceptual structures and distributed motor control. In: Brooks VB (ed) Handbook of physiology, sect 1: vol 11, part 2. American Physiol Soc, Bethesda, pp 1449–1480

Arimoto S, Kawamura S, Miyazaki F (1984a) Bettering operation of dynamic systems by learning: a new control theory for sevomechanism or mechatronics systems. 23 IEEE Conf Des Control 2:1064–1069

Arimoto S, Kawamura S, Miyazaki F (1984a) Can mechanical robots learn by themselves; Proceedings of the 2nd International Symposium on Robotics Research, Kyoto, Japan

Cheney PD, Fetz EE (1980) Functional classes of primate corticomotoneuronal cells and their relation to active force. J Neurophysiol 44:773–791

Dubowsky S, DesForges DT (1979) The application of model reference adaptive control to robotic manipulators. J Dyn Syst Meas Control 101:193–200

Eccles JC: Introductory remarks. In: Massion J, Sasaki K (eds) Cerebro-cerebellar interactions, pp 1–18. North-Holland Elsevier, Amsterdam Oxford New York, pp 1–18

Evarts EV (1981) Role of motor cortex in voluntary movements in primates. In: Brooks VB (ed) Handbook of physiology, sect 1: vol 11, part 2. American Physiol Soc, Bethesda, pp 1083–1120

Geman S (1979) Some averaging and stability results for random differential equations. SIAM J Appl Math 36:86–105

Ghez C, Fahn S (1985) The cerebellum. In: Kandel ER, Schwartz JH (eds) Principles of neural science. Elsevier, New York, pp 502–522

Gilbert PFC, Thach WT (1977) Purkinje cell activity during motor learning. Brain Res 128:309–328

Flash T, Hogan N (1985) The coordination of arm movements; an experimentally confirmed mathematical model. J Neurosci 5:1688–1703

Fujita M (1982a) Adaptive filter model of the cerebellum. Biol Cybern 45:195–206

Fujita M (1982b) Simulation of adaptive modification of the vestibulo-ocular reflex with an adaptive filter model of the cerebellum. Biol Cybern 45:207–214

Furukawa K (1984) Identification of a robotic manipulator by a neural model. Osaka Univ, Bachelor's Thesis

Hollerbach JM (1980) A recursive Lagrangian formulation of manipulator dynamics and a comparative study of dynamics formulation complexity. IEEE Trans SMC-10:730–736

Hollerbach JM (1982) Computers, brains and the control of movement. Trends Neuro Sci 5:189–192

Ito M (1970) Neurophysiological aspects of the cerebellar motor control system. Int J Neurol 7:162–176

Ito M (1984) The cerebellum and neural control. Raven Press, New York

Ito M, Shiida T, Yagi N, Yamamoto M (1974) The cerebellar modification of a rabbit's horizontal vestibulo-ocular reflex induced by sustained head rotation combined with visual stimulation. Proc Jpn Acad 50:85–89

Ito M, Jastreboff PJ, Miyashita Y (1982) Specific effects of unilateral lesions in the flocculus upon eye movements in albino rabbits. Exp Brain Res 45:233–242

Ito M, Sakurai M, Tongroach P (1982) Climbing fibre induced depression of both mossy fibre responsiveness and gultamate sensitivity of cerebellar Purkinje cells. J Physiol 324:113–134

Kawato M, Hamaguchi T, Murakami F, Tsukahara N (1984) Quantitative analysis of electrical properties of dendritic spines. Biol Cybern 50:447–454

Llinás R, Walton K, Hillman D, Sotelo C (1975) Inferior olive: its role in motor learning. Science 190:1230–1231

Luh JYS, Walker MW, Paul RPC (1980) On-line computational scheme for mechanical manipulations. J Dyn Syst Meas Control 102:69–76

Marr D (1969) A theory of cerebellar cortex. J Physiol 202:437–470

Marr D (1982) Vision. Freeman, New York

Miyamoto H (1985) A motor learning model based on synaptic plasticity. Osaka University, Bachelor's Thesis

Miyamoto H, Kawato M, Suzuki R (1987) Hierarchical learning control of an industrial manipulator using a model of the central nervous system, Japan IEICE Technical Report, MBE-86-81:25–32

Poggio T, Torre V (1981) A theory of synaptic interactions. In: Reichardt WE, Poggio T (eds) Theoretical approaches in neurobiology. MIT Press, Cambridge, pp. 28–46

Poirier LJ, Bouvier G, Bédard P, Bouchard R, Larochelle L, Olivier A, Singh P (1969) Essai sur les cirvuits neuronaux mipliqués dans le tremblement postural et l'hypokinesie. Rev Neurol 120:15–40

Raibert MH (1978) A model for sensorimotor control and learning. Biol Cybern 29:29–36

Sasaki K, Gemba H (1982) Development and change of cortical field potentials during learning processes of visually initiated hand movements in the monkey. Exp Brain Res 48:429–437

Sasaki K, Gemba H, Mizuno N (1982) Cortical field potentials preceding visually initiated hand movements and cerebellar actions in the monkey. Exp Brain Res 46:29–36

Setoyama T (1987) Symbolic calculation of subsystems of the neural inverse-dynamics model for a 6 degrees of freedom manipulator using REDUCE. Osaka University, Bachelor's Thesis

Tsukahara N (1981) Synaptic plasticity in the mammalian central nervous system. Annu Rev Neurosci 4:351–379

Tsukahara N, Kawato M (1982) Dynamic and plastic properties of the brain stem neuronal networks as the possible neuronal basis of learning and memory. In: Amari S, Arbib MA (eds) Competition and cooperation in neural nets. Springer, Berlin Heidelberg New York, pp. 430–441

Tsukahara N, Oda Y, Notsu T (1981) Classical conditioning mediated by the red nucleus in the cat. J Neurosci 1:72–79

Uno Y, Kawato M, Suzuki R (1987) Formation of optimum trajectory in control of arm movement—minimum torque-change model—Japan IEICE Technical Report MBE 86–79:9–16

Widrow B, McCool JM, Larimore MG, Johnson CR (1976) Stationary and nonstationary learning characteristics of the LMS adaptive filter. Proc IEEE 64:1151–1162

Wilson HR, Cowan JD (1972) Excitatory and inhibitory interactions in localized populations of model neurons. Biophys J 12:1–24

Received: April 3, 1987

Dr. M. Kawato
Department of Biophysical Engineering
Faculty of Engineering Science
Osaka University, Toyonaka
Osaka
560 Japan

Article 1.4

An Introduction to Autonomous Control Systems

PANOS J. ANTSAKLIS, KEVIN M. PASSINO, AND S. J. WANG

Autonomous control systems are designed to perform well under significant uncertainties in the system and environment for extended periods of time, and they must be able to compensate for significant system failures without external intervention. Intelligent autonomous control systems use techniques from the field of artificial intelligence (AI) to achieve this autonomy. Such control systems evolve from conventional control systems by adding intelligent components, and their development requires interdisciplinary research. Here, we provide an introduction to the area of intelligent autonomous control. The fundamental issues in autonomous control system modeling and analysis are discussed, with emphasis on mathematical modeling. Some recent results in relevant research areas are summarized.

Introduction

Autonomous means having the power for self government. *Autonomous controllers* have the power and ability for self governance in the performance of control functions. They are composed of a collection of hardware and software, which can perform the necessary control functions, without external intervention, over extended time periods. There are several *degrees of autonomy.* A fully autonomous controller should perhaps have the ability to even perform hardware repair, if one of its components fails. Note that conventional fixed controllers can be considered to have a *low degree* of autonomy since they can only tolerate a restricted class of plant parameter variations and disturbances. To achieve a *high degree* of autonomy, the controller must be able to perform a number of functions in addition to the conventional control functions such as tracking and regulation. These additional functions, which include the ability to accommodate for drastic system failures, are discussed in this article. This article is based on the developments in [1]-[3].

Autonomous controllers can of course be used in a variety of systems from manufacturing to unmanned space, atmospheric, ground, and underwater exploratory vehicles (for a description of several applications see [4]). This introduction to autonomous control will be developed around a space vehicle application so that a) concrete examples for the various control functions, and fundamental characteristics of autonomous control can be given, and b) so that the development addresses relatively well defined control needs rather than abstract requirements. Furthermore, the autonomous control of space vehicles is highly demanding; consequently the developed architecture is general enough to encompass all related autonomy issues. It should be stressed that all the results presented here apply to any autonomous control system. In other classes of applications, the architecture, or parts of it, can be used directly and the same fundamental concepts and characteristics identified here are valid.

We begin by describing the architecture of the autonomous controller necessary for the operation of future advanced space vehicles that was developed in [2],[3]. The concepts and methods needed to successfully design such an autonomous controller are introduced and discussed. A hierarchical functional autonomous controller architecture is described; it is designed to ensure the autonomous operation of the control system and it allows interaction with the pilot/ground station and the systems on board the autonomous vehicle. A command by the pilot or the ground station is executed by dividing it into appropriate subtasks which are then performed by the controller. The controller can deal with unexpected situations, new control tasks, and failures within limits. To achieve this, high level decision making techniques for reasoning under uncertainty and taking actions must be utilized. These techniques, if used by humans, are attributed to *intelligent* behavior. Hence, one way to achieve autonomy, for some applications, is to utilize high level decision making techniques, "intelligent" methods, in the autonomous controller. *Autonomy is the objective, and "intelligent" controllers are one way to achieve it.* The fields of artificial intelligence (AI) [5],[6] and operations research offer some of the tools to add the higher level decision making abilities.

Autonomous Control Functions

Autonomous control systems must perform well under significant uncertainties in the plant and the environment for extended periods of time and they must be able to com-

Presented at the Fifth IEEE International Symposium on Intelligent Control, Philadelphia, PA, Sept. 5-7, 1990. Panos J. Antsaklis is with the Dept. of Electrical Engineering, University of Notre Dame, Notre Dame, IN 46556. Kevin M. Passino is with the Dept. of Electrical Engineering, The Ohio State University, Columbus, OH 43210. S.J. Wang is with the Jet Propulsion Laboratory, California Inst. of Tech., Pasadena, CA 91109. This work was supported in part by the Jet Propulsion Laboratory under Contract No.957856.

Reprinted from *IEEE Contr. Syst. Mag.*, vol. 11, no. 4, pp. 5–13, June 1991.

pensate for system failures without external intervention. Such autonomous behavior is a very desirable characteristic of advanced systems. An autonomous controller provides high level *adaptation* to changes in the plant and environment. To achieve autonomy the methods used for control system design should utilize both a) algorithmic-numeric methods, based on the state-of-the-art conventional control, identification, estimation, and communication theory, and b) decision making-symbolic methods, such as the ones developed in computer science (e.g., automata theory), and specifically in the field of AI. In addition to supervising and tuning the control algorithms, the autonomous controller must also provide a high degree of tolerance to failures. To ensure system reliability, failures must first be detected, isolated, and identified (and if possible *contained*), and subsequently a new control law must be designed if it is deemed necessary. The autonomous controller must be capable of planning the necessary sequence of control actions to be taken to accomplish a complicated task. It must be able to interface to other systems as well as with the operator, and it may need learning capabilities to enhance its performance while in operation. It is for these reasons that advanced planning, learning, and expert systems, among others, must work together with conventional control systems in order to achieve autonomy.

The need for quantitative methods to model and analyze the dynamical behavior of such autonomous systems presents significant challenges well beyond current capabilities. It is clear that the development of autonomous controllers requires significant interdisciplinary research effort as it integrates concepts and methods from areas such as control, identification, estimation, and communication theory, computer science, artificial intelligence, and operations research. It is also important to note that autonomous controllers are *evolutionary* and not *revolutionary*. They evolve from existing controllers in a natural way fueled by actual needs, as is now discussed.

Design Methodology - History

Conventional control systems are designed using mathematical models of physical systems. A mathematical model which captures the dynamical behavior of interest is chosen and then control design techniques are applied, aided by CAD packages, to design the mathematical model of an appropriate controller. The controller is then realized via hardware or software and it is used to control the physical system. The procedure may take several iterations. The mathematical model of the system must be "simple enough" so that it can be analyzed with available mathematical techniques, and "accurate enough" to describe the important aspects of the relevant dynamical behavior. It approximates the behavior of a plant in the neighborhood of an operating point.

The first mathematical model to describe plant behavior for control purposes is attributed to J.C. Maxwell who in 1868 used differential equations to explain instability problems encountered with James Watt's flyball governor; the governor was introduced in 1769 to regulate the speed of steam engine vehicles. Control theory made significant strides in the past 120 years, with the use of frequency domain methods and Laplace transforms in the 1930s and 1940s and the introduction of the state space analysis in the 1960s. Optimal control in the 1950s and 1960s, stochastic, robust and adaptive control methods in the 1960s to today, have made it possible to control more accurately significantly more complex dynamical systems than the original flyball governor.

The control methods and the underlying mathematical theory were developed to meet the ever increasing control needs of our technology. The evolution in the control area was fueled by three major needs:

a) The need to deal with increasingly complex dynamical systems.

b) The need to accomplish increasingly demanding design requirements.

c) The need to attain these design requirements with less precise advanced knowledge of the plant and its environment, that is, the need to control under increased uncertainty.

The need to achieve the demanding control specifications for increasingly complex dynamical systems has been addressed by using more complex mathematical models such as nonlinear and stochastic ones, and by developing more sophisticated design algorithms for, say, optimal control. The use of highly complex mathematical models however, can seriously inhibit our ability to develop control algorithms. Fortunately, simpler plant models, for example linear models, can be used in the control design; this is possible because of the feedback used in control which can tolerate significant model uncertainties. Controllers can then be designed to meet the specifications around an operating point, where the linear model is valid and then via a scheduler a controller emerges which can accomplish the control objectives over the whole operating range. This is, for example, the method typically used for aircraft flight control. In autonomous control systems we *need to significantly increase the operating range*. We must be able to deal effectively with significant uncertainties in models of increasingly complex dynamical systems in addition to increasing the validity range of our control methods. This will involve the use of intelligent decision making processes to generate control actions so that a performance level is maintained even though there are drastic changes in the operating conditions.

There are needs today that cannot be successfully addressed with the existing conventional control theory. They mainly pertain to the area of uncertainty. Heuristic methods may be needed to tune the parameters of an adaptive control law. New control laws to perform novel control functions should be designed while the system is in operation. Learning from past experience and planning control actions may be necessary. Failure detection and identification is needed. These functions have been performed in the past by human operators. To increase the speed of response, to relieve the pilot from mundane tasks, to protect operators from hazards, autonomy is desired. It should be pointed out that several functions proposed in later sections, to be part of the autonomous controller, have been performed in the past by separate systems; examples include fault trees in chemical process control for failure diagnosis and hazard analysis, and control system design via expert systems.

Summary

In the next section the functions, characteristics, and benefits of autonomous control are outlined. Next it is explained that plant complexity and design requirements dictate how sophisticated a controller must be. From this it can be seen that often it is appropriate to use methods from operations research or computer science to achieve autonomy. Such methods are studied in *intelligent control theory*. An overview of some relevant research literature in the field of intelligent and autonomous control is given together with references that outline research directions. An autonomous control functional architecture for future space vehicles is then presented, which incorporates the concepts and characteristics described earlier. The controller is hierarchical, with three levels, the execution level (lowest level), the coordination level (middle level), and the management and organization level (highest level). The general characteristics of the overall architecture, including those of the three levels are explained, and an example to illustrate their functions is given.

In the following section the fundamental issues and attributes of intelligent autonomous systems are described. Then we discuss mathematical models for autonomous systems including "logical" discrete event system models. An approach to the quantitative, systematic modeling, analysis, and design of autonomous controllers is also discussed. It is a "hybrid" approach since it is proposed to use both conventional analysis techniques based on difference and differential equations, together with new techniques for the analysis of systems described with a symbolic formalism such as finite automata. The more global, macroscopic, view of dynamical systems taken in the development of autonomous controllers, suggests the use of a model with a hybrid or nonuniform structure, which in turn requires the use of a hybrid analysis. Finally, several major relevant research areas are indicated. In particular, some interesting recent results from the areas of planning and expert systems, machine learning, artificial neural networks and the area of restructurable controls are briefly outlined. The last section provides some concluding remarks.

Functional Architecture of an Autonomous Controller

Intelligent Autonomous Control

Motivation: Sophistication and Complexity in Control: The complexity of a dynamical system model and the increasingly demanding closed loop system performance requirements, necessitate the use of more complex and sophisticated controllers. For example, highly nonlinear systems normally require the use of more complex controllers than low order linear ones when goals beyond stability are to be met. The increase in uncertainty, which corresponds to the decrease in how well the problem is structured or how well the control problem is formulated, and the necessity to allow human intervention in control, also necessitate the use of increasingly sophisticated controllers. Controller complexity and sophistication is then directly proportional to both the complexities of the plant model and of the control design requirements.

Based on these ideas, the authors in [7] and [8] suggest a hierarchical ranking of increasing controller sophistication on the path to *intelligent* controls. At the lowest level, deterministic feedback control based on conventional control theory is utilized for simple linear plants. As plant complexity increases, such controllers will need for instance, state estimators. When process noise is significant, Kalman or other filters may be needed. Also, if it is required to complete a control task in minimum time or with minimum energy, optimal control techniques are utilized. When there are many quantifiable, stochastic characteristics in the plant, stochastic control theory is used. If there are significant variations of plant parameters, to the extent that linear robust control theory is inappropriate, adaptive control techniques are employed. For still more complex plants, self-organizing or learning control may be necessary. At the highest level in their hierarchical ranking, plant complexity is so high, and performance specifications so demanding, that intelligent control techniques are used.

In the hierarchical ranking of increasingly sophisticated controllers described above, the decision to choose more sophisticated control techniques is made by studying the control problem using a controller of a certain complexity belonging to a certain class. When it is determined that the class of controllers being studied (e.g., adaptive controllers) is inadequate to meet the required objectives, a more sophisticated class of controllers (e.g., intelligent controllers) is chosen. That is, if it is found that certain higher level decision making processes are needed for the adaptive controller to meet the performance requirements, then these processes can be incorporated via the study of intelligent control theory. These intelligent autonomous controllers are the next level up in sophistication. They are *enhanced adaptive controllers,* in the sense that they can adapt to more significant global changes in the plant and its environment than conventional adaptive controllers, while meeting more stringent performance requirements.

One turns to more sophisticated controllers only if simpler ones cannot meet the required objectives. The need to use intelligent autonomous control stems from the need for an increased level of autonomous decision making abilities in achieving complex control tasks. In the next section a number of intelligent and autonomous control research results which have appeared in the literature are outlined.

A Literature Overview: In [2],[3] the authors provided a relatively complete list of references for the field of autonomous control. Here we provide references which we feel will provide the reader with an introduction to autonomous control. First, there are several relevant books: Hierarchical systems are treated in [9],[10]. In [11] the authors explain how a wide variety of AI techniques will be useful in enhancing space station autonomy, capability, safety, etc. Aerospace applications are also discussed in [12]. For a book on AI and autonomous systems see [13], and for one on cybernetics and intelligent systems see [14]. For a book on intelligent manufacturing systems see [15].

Journals with papers relevant to the area of intelligent autonomous control are *The Journal of Intelligent and Robotic Systems*, *IEEE Transactions on Systems, Man, and Cybernetics*, *IEEE Transactions on Pattern Analysis and Machine Intelligence*, *Journal of Applied Artificial Intelligence*, and the standard AI and control theoretic journals. The reader should also consult some of the recent conference proceedings: *Proceedings of the 1985 IEEE Workshop on Intelligent Control*, *Proceedings of the 1986 Intelligent Autonomous Systems Conference*, *Proceedings of the Space Telerobotics Workshop*, and the *Proceedings of the IEEE International Symposium on Intelligent Control* in 1987, 1988, 1989, and 1990.

In [2],[3] the authors introduce an intelligent autonomous controller and discuss in detail the fundamental characteristics of autonomous control. In [16] the author offers a decentralized control-theoretic view on intelligent control. Functional and structural hierarchies are studied in [17]. Fundamentals of intelligent systems such as the principle of increasing intelligence with decreasing precision, are discussed in [18],[19], and [20]. The work in [18],[19] and [21]-[26] probably represents the most complete mathematical approach to the analysis of intelligent machines. In [27] and the references therein the authors study distributed intelligent systems. In [28] the author introduces a theory of intelligent control that has received considerable attention since then. There have been numerous studies on the use of expert systems to control various processes; in [29] expert systems have been used in chemical process control. There are interesting relationships between the type of problems examined in intelligent autonomous control, "fuzzy control" [30], and "automated reasoning" [31]. Simulation of autonomous systems and related issues has been studied extensively in [32],[33] and the references therein.

An Intelligent Autonomous Control Architecture For Future Space Vehicles

Here, a functional architecture of an autonomous controller for future space vehicles is introduced and discussed. This hierarchical architecture has three levels, the execution level, the coordination level, and the management and organization level. The architecture exhibits certain characteristics, as discussed below, which have been shown in the literature to be necessary and desirable in autonomous systems. Based on this architecture we identify the important fundamental issues and concepts that are needed for an autonomous control theory.

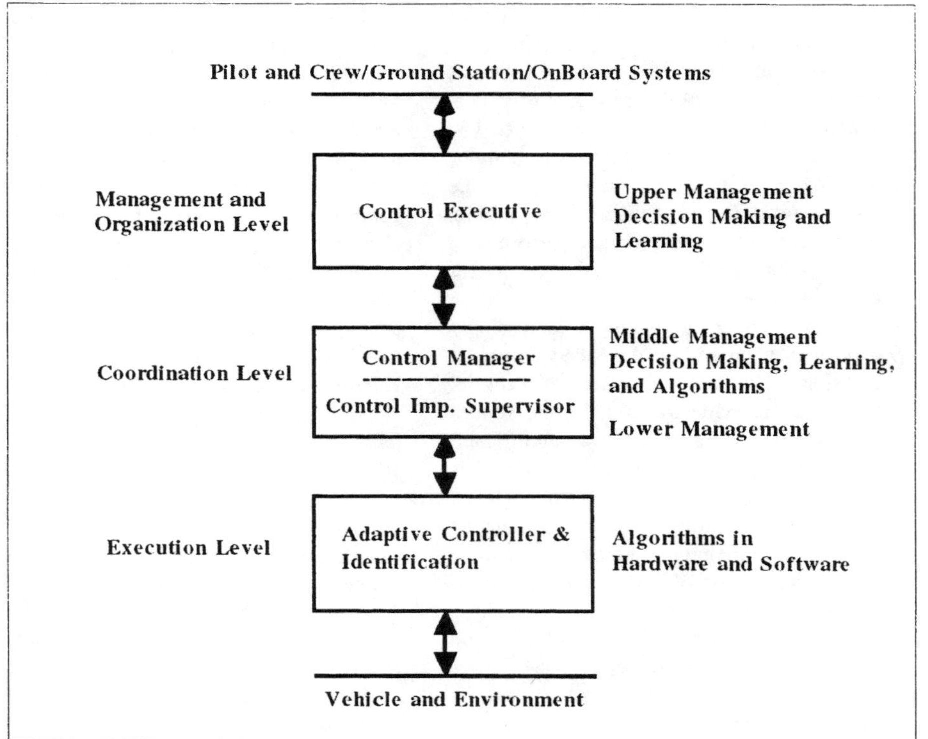

Fig. 1. Autonomous controller functional architecture.

Architecture Overview: Structure and Characteristics: The overall functional architecture for an autonomous controller is given by the architectural schematic of Fig. 1; for more detailed description see [2],[3]. This is a functional architecture rather than a hardware processing one, therefore it does not specify the arrangement and duties of the hardware used to implement the functions described. Note that the processing architecture also depends on the characteristics of the current processing technology; centralized or distributed processing may be chosen for function implementation depending on available computer technology.

The architecture in Fig. 1 has three levels. At the lowest level, the execution level, there is the interface to the vehicle and its environment via the sensors and actuators. At the highest level, the management and organization level, there is the interface to the pilot and crew, ground station, or onboard systems. The middle level, called the coordination level, provides the link between the execution level and the management level. Note that we follow the somewhat standard viewpoint that there are three major levels in the hierarchy. *It must be stressed that the system may have more or fewer than three levels.* For instance, see the architecture developed in [34]. Some characteristics of the system which dictate the number of levels are the extent to which the operator can intervene in the system's operations, the degree of autonomy or level of intelligence in the various subsystems, the dexterity of the subsystems, the hierarchical characteristics of the plant. Note however that the three levels shown here in Fig. 1 are applicable to most architectures of autonomous controllers, by grouping together sublevels of the architecture if necessary. Notice that as it is indicated in the figure, the lowest, execution level involves conventional control algorithms, while the highest, management and organization level involves only higher level, intelligent, decision making methods. The middle, coordination level is the level which provides the interface between the actions of the other two levels and it uses a combination of conventional and intelligent decision making methods.

The sensors and actuators are implemented mainly with hardware. They are the connection between the physical system and the controller. Software and perhaps hardware are used to implement the execution level. Mainly software is used for both the coordination and management levels. There are *multiple copies* of the control functions at each level, more at the lower and fewer at the higher levels. For example, there may be one control manager which directs a number of different adaptive control algorithms to control the flexible modes of the vehicle via appropriate sensors and actuators. Another control manager is responsible for the control functions of a robot arm for satellite repair. The control executive issues commands to the managers and coordinates their actions.

Note that the autonomous controller is only one of the autonomous systems on the vehicle. It is responsible for all the functions related to the control of the physical system and allows for continuous online development of the autonomous controller and to provide for various phases of mission operations. The tier structure of the architecture allows us to build on existing advanced control theory. Development progresses, creating each time, higher level adaptation and a new system which can be operated and tested independently. The autonomous controller performs many of the functions currently performed by the pilot, crew, or ground station. The pilot and crew are thus relieved from mundane tasks and some of the ground station functions are brought aboard the vehicle. In this way the degree of autonomy of the vehicle is increased.

Functional Operation: Commands are issued by higher levels to lower levels and response data flows from lower levels upwards. Parameters of subsystems can be altered by systems one level above them in the hierarchy. There is a delegation and distribution of tasks from higher to lower levels and a layered distribution of decision making authority. At each level, some preprocessing occurs before information is sent to higher levels. If requested, data can be passed from the lowest subsystem to the highest, e.g., for display. All subsystems provide status and health information to higher levels. Human intervention is allowed even at the control implementation supervisor level, with the commands however passed down from the upper levels of the hierarchy.

The specific functions at each level are described in detail in [2],[3]. Here we present a simple illustrative example to clarify the overall operation of the autonomous controller. Suppose that the pilot desires to repair a satellite. After dialogue with the control executive, the task is refined to "repair satellite using robot A". This is arrived at using the capability assessing, performance monitoring, and planning functions of the control executive. The control executive decides if the repair is possible under the current performance level of the system, and in view of near term planned functions. The control executive, using its planning capabilities, sends a sequence of subtasks sufficient to achieve the repair to the control manager. This sequence could be to order robot A to: "go to satellite at coordinates xyz", "open repair hatch", "repair". The control manager, using its planner, divides say the first subtask, "go to satellite at coordinates xyz", into smaller subtasks: "go from start to $x_1y_1z_1$," then "maneuver around obstacle," "move to $x_2y_2z_2$,"..., "arrive at the repair site and wait." The other subtasks are divided in a similar manner. This information is passed to the control implementation supervisor, which recognizes the task, and uses stored control laws to accomplish the objective. The subtask "go from start to $x_1y_1z_1$," can for example, be implemented using stored control algorithms to first, proceed forward 10 m, to the right 15°, etc. These control algorithms are executed in the controller at the execution level utilizing sensor information; the control actions are implemented via the actuators.

Some Design Guidelines for Autonomous Controllers

There are certain functions, characteristics, and behaviors that autonomous systems should possess [10],[34]. These are outlined below. Some of the important characteristics of autonomous controllers are that they relieve humans from time consuming mundane tasks thus increasing efficiency, enhance reliability since they monitor health of the system, enhance performance, protect the system from internally induced faults, and they have consistent performance in accomplishing complex tasks.

There are autonomy guidelines and goals that should be followed and sought after in the development of an autonomous system. Autonomy should reduce the work load requirements of the operator or, in the space vehicle case discussed here, of the pilot/crew-/ground station, for the performance of routine functions, since the gains due to autonomy would be superficial if the maintenance and operation of the autonomous controller taxed the operators. Autonomy should enhance the functional capability of the system. Since the autonomous controller will be performing the simpler routine tasks, persons will be able to dedicate themselves to even more complex tasks.

There are certain autonomous system architectural characteristics that should be sought after in the design process. The autonomous control architecture should be amenable to evolving future needs and updates in the state of the art. The autonomous control architecture should be functionally hierarchical; for lower level subsystems to take some actions, they have to clear it with a higher level authority. The system must, however, be able to have lower level subsystems, that are monitoring and reconfiguring for failures, act autonomously to certain extent to enhance system safety. There are also certain operational characteristics of autonomous controllers. Persons should have ultimate supervisory override control of autonomy functions. Autonomous activities should be highly visible, "transparent", to the operator the maximum extent possible.

Finally, there must be certain features inherent in the autonomous system design. Autonomous design features should prevent failures that would jeopardize the overall system mission goals or safety. These features should enhance safety, and avoid false alarms and unnecessary hardware reconfiguration. This implies that the controller should have self-test capability. Autonomous design features should also be tolerant of transient errors, they should not degrade the reliability or operational lifetime of functional elements, they should include adjustable fault detection thresholds, avoid irreversible state changes, and provide protection from erroneous or invalid external commands.

Characteristics of Autonomous Control Systems

Based on the architecture described above we identify the important fundamental concepts and characteristics that are needed for an autonomous control theory. Note that several of these have been discussed in the literature as outlined above. Here, these characteristics are brought together for completeness. Furthermore, the fundamental issues which must be addressed for a quantitative theory of intelligent autonomous control are introduced and discussed.

There is a *successive delegation of duties* from the higher to lower levels; consequently the *number of distinct tasks* increases as we go down the hierarchy. Higher levels are concerned with slower aspects of the system's behavior and with its larger portions, or broader aspects. There is then a *smaller contextual horizon at lower levels*, i.e. the control decisions are made by considering less information. Also notice that higher levels are concerned with *longer time horizons* than lower levels. Due to the fact that there is the need for high level decision making abilities at the higher levels in the hierarchy, there is *increasing intelligence* as one moves from the lower to the higher levels. This is reflected in the use of fewer conventional numeric-algorithmic methods at higher levels as well as the use of more symbolic-decision making methods. This is the "principle of increasing intelligence with decreasing precision" described in [23]. The decreasing precision is reflected by a decrease in *time scale density,* decrease in *bandwidth* or *system rate,* and a decrease in the *decision (control action) rate.* (These properties have been studied for a class of hierarchical systems in [35],[36].) All these characteristics lead to a decrease in *granularity of models* used, or equivalently, to an *increase in model abstractness.* Model granularity also depends on the *dexterity* of the autonomous controller as discussed in [2],[3]. The execution level of a highly dexterous controller is very sophisticated and it can accomplish complex control tasks. The control implementation supervisor can issue high level commands to a dexterous controller, or it can completely dictate each command in a less dexterous one. The simplicity, and level of abstractness of macro commands in an autonomous controller depends on its dexterity. The more sophisticated the execution level is, the simpler are the commands that the control implementation supervisor needs to issue. Notice that a very dexterous robot arm may itself have a number of autonomous functions. If two such dexterous arms were used to complete a task which required the coordination of their actions then the arms would be considered to be two dexterous actuators and a new supervisory autonomous controller would be placed on top for the supervision and coordination task. In general, this can happen recursively, adding more intelligent autonomous controllers as the lower level tasks, accomplished by autonomous systems, need to be supervised.

There is an ongoing *evolution* of the intelligent functions of an autonomous controller and this is now discussed. It was pointed out above that complex control problems required a controller sophistication that involved the use of AI methodologies. It is interesting to observe the following [37]: Although there are characteristics which separate intelligent from non-intelligent systems, as intelligent systems evolve, the distinction becomes less clear. Systems which were originally considered intelligent evolve to gain more character of what are considered to be non-intelligent, numeric-algorithmic systems. An example is a route planner. Although there are AI route planning systems, as problems like route planning become better understood, more conventional numeric-algorithmic solutions are developed. The AI methods which are used in intelligent systems, help us to understand complex problems so we can organize and synthesize new approaches to problem solving, in addition to being problem solving techniques themselves. AI techniques can be viewed as research vehicles for solving very complex problems. As the problem solution develops, purely algorithmic approaches, which have desirable implementation characteristics, substitute AI techniques and play a greater role in the solution of the problem. It is for this reason that we concentrate on achieving autonomy and not on whether the underlying system can be considered "intelligent".

Mathematical Models for Autonomous Systems

For autonomous control problems, normally the plant is so complex that it is either impossible or inappropriate to describe it with conventional system models such as differential or difference equations. Even though it might be possible to accurately describe some system with highly complex nonlinear differential equations, it may be inappropriate if this description makes subsequent analysis too difficult to be useful. The complexity of the plant model needed in design depends on both the complexity of the physical system and on how demanding the design specifications are. There is a tradeoff between model complexity and our ability to perform analysis on the system via the model. However, if the control performance specifications are not too demanding, a more abstract, higher level, model can be utilized, which will make subsequent analysis simpler. This model intentionally ignores some of the system characteristics, specifically those that need not be considered in attempting to meet the particular performance specifications. For example, a simple temperature controller could ignore almost all dynamics of the house or the office and consider only a temperature threshold model of the system to switch the furnace off or on.

Logical discrete event system (DES) models such as those used in the Ramadge-Wonham framework (e.g., [38]) or such as Petri nets [39] are quite useful for modeling the higher level decision making processes in the intelligent autonomous controller. It was shown in [40],[41] that DES-theoretic models can be used to represent AI planning systems which are an important component of the intelligent autonomous controller. Also, it was shown in [42] that Petri nets can be used as knowledge representation tools in AI. In particular the authors showed that knowledge that can be represented with *semantic networks*, *scripts*, and *production rules* in an expert system can also be clearly represented with Petri net models. The "timed" or "performance" models from DES-theoretic research will also prove useful in modeling components of the higher levels in the intelligent autonomous controller. For instance, queuing network models, Markov chains, etc. will be useful. The choice of whether to use such models will, of course, depend on what properties of the autonomous system need to be studied.

The quantitative, systematic techniques for modeling, analysis, and design of control systems are of central and utmost practical importance in conventional control theory. Similar techniques for intelligent autonomous controllers do not exist. This is of course because of their novelty, but for the most part, it is due to the *"hybrid"* structure (nonuniform, nonhomogeneous nature) of the dynamical systems under consideration. The systems are hybrid since in order to examine autonomy issues, a more global, macroscopic view of a dynamical system must be taken than in conventional control theory. Modeling techniques for intelligent autonomous systems must be able to support this macroscopic view of the dynamical system, hence it is necessary to represent both numeric and symbolic information. We need modeling methods that can gather all information necessary for analysis and design. For example, we need to model the dynamical system to be controlled (e.g., a space platform), we need models of the failures that might occur in the system, of the conventional adaptive controller, and of the high level decision making processes at the management and organization level of the intelligent autonomous controller (e.g., an AI planning system performing actions that were once the responsibility of the ground station). The nonuniform components of the intelligent controller all take part in the generation of the low level control inputs to the dynamical system, therefore they all must be considered in a complete analysis. For an extended discussion on the modeling of hybrid systems consult [43].

It is our viewpoint that research should begin by using different models for different components of the intelligent autonomous controller. Full hybrid models that can represent large portions or even the whole autonomous system should be examined but much can be attained by using the best available models for the various components of the architecture and joining them via some appropriate interconnecting structure. For instance, research in the area of systems that are modeled with a logical DES model at the higher levels and a difference equation at the lower level should be examined. In any case, our modeling philosophy requires the examination of *hierarchical* models. Much work needs to be done on hierarchical DES modeling, analysis, and design, let alone the full study of hybrid hierarchical dynamical systems. Some research has begun to address hierarchical DES [38].

A practical but very important issue is the simulation of hybrid systems. This requires simulation of both conventional differential equations and symbolic decision making processes or DES. Normally, numeric-algorithmic processing is done with languages like FORTRAN and symbolic decision making can be implemented with LISP or PROLOG while DES are often simulated with SLAM. Sometimes several types of processing are done on computers with quite different architectures. There is then the problem of combining symbolic and numeric processing on one computer. If the computing is done on separate computers, the communication link normally presents a serious bottleneck. Combining AI, DES, and conventional numeric processing is currently being addressed by many researchers and some promising results have been reported. Some very promising results have been reported in [32],[33] and the references therein.

Planning and Expert Systems, Learning and Neural Networks, Restructurable Control

In this section we will discuss results obtained on the analysis and design of several components of the intelligent autonomous controller architecture. One can roughly categorize research in the area of intelligent autonomous control into two areas: conventional control theoretic research, addressing the control functions at the execution and coordination levels, and the modeling, analysis, and design of higher level decision making systems found in the management and organization level, and the coordination level. Below we provide only a sampling of the results to introduce the reader to these research areas.

To determine how to utilize AI techniques it is productive to study the relationships between AI and conventional control methods. In this way one can determine what AI techniques have to offer over conventional control methods. For instance, the authors in [40] have provided a systems and control theoretic perspective on AI planning (and expert) systems. In this work, the authors explain how AI planning systems are in fact control systems where the input and output variables are symbols rather than numbers. It is shown that the techniques used in the implementation of AI planning systems are actually generalized open and closed loop control, state estimation, system identification, and adaptive control.

It is also important to study how to use conventional control techniques in conjunction with AI techniques to perform autonomous control functions. For instance, in [44],[45] the authors introduce a fault detection and identification (FDI) system that is composed of AI decision making mechanisms and conventional FDI algorithms. The "hybrid" algorithmic-decision making FDI system detects and identifies failures for an intelligent restructurable controller on board an advanced aircraft.

Some control theoretic techniques offer modeling, analysis, and design techniques for the higher level decision making mechanisms in the intelligent autonomous controller. For instance, in [41],[46],[47] the authors show that AI planning problems can be studied in a discrete event system (DES) theoretic framework by utilizing the A^* algorithm. Moreover, there are many recent results developed in a DES-theoretic framework that can be used for the study of components of the intelligent autonomous controller (e.g., results from the Ramadge-Wonham formulation for the study of "logical" DES models).

It is important to note that in order to obtain a high degree of autonomy it is absolutely necessary to, in some way, adapt or learn [48]. Although the literature on higher level learning performed in conjunction with low level adaptation is limited, in [49]-[51] the authors show how an expert learning system can be used to tune the parameters of an adaptive controller for a large flexible space antenna so as to optimize its performance and then also enhance the operating range of the system by storing this information for future use. Neural networks also appear to offer methodologies to perform learning functions in the intelligent autonomous controller (see for instance, the April issues of the *IEEE Control Systems Magazine* in 1987, 1988 and the Special Issue of April 1989 [52] ; also the new *IEEE Trans. on Neural Networks).* Neural networks can also be used to implement certain components of the intelligent autonomous controller. For instance, the authors in [53],[54] investigate how to implement the *match phase* of expert systems with a "multi-layer perceptron".

We stress that in autonomous control we seek only to significantly widen the operating range of the system so that significant failures and environmental changes can occur and performance will still be maintained. All of the conventional control techniques are useful in the development of autonomous controllers and they are relevant to the study of autonomous control. It is the case however, that certain techniques are more suitable for interfacing to the autonomous controller and for compensating for significant system failures. For instance the area of "restructurable" or "reconfigurable" control systems [45],[55] studies techniques to reconfigure controllers when significant failures occur. Recently there have been advances in the theory of restructurable controls [56],[57] where the authors develop stability bounds on the allowable parameter variations, induced by system failures.

It is our viewpoint that conventional modeling, analysis, and design methods should be used whenever they are applicable for the components of the intelligent autonomous controller. For instance, they should be used at the execution level of many autonomous controllers. We propose to augment and enhance existing theories rather than develop a completely new theory for the hybrid systems described above; we wish to build upon existing, well understood and proven conventional methods. The symbolic/-numeric interface is a very important issue; consequently it should be included in any analysis. There is a need for systematically generating less detailed, more abstract models from differential/difference equation models to be used in higher levels of the autonomous controller (coordination level). There is also a need for systematically extracting the necessary information from lower level symbolic models to generate higher level symbolic models to be used in the hierarchy where appropriate. Tools for the implementation of this *information extraction* also need to be developed (see for instance [58]). In this way conventional analysis can be used in conjunction with the developed analysis methods to obtain an overall quantitative, systematic analysis paradigm for intelligent autonomous control systems. In short, we propose to use hybrid modeling, analysis, and design techniques for nonuniform systems. This approach is not unlike the approaches used in the study of any complex phenomena by the scientific and engineering communities.

Concluding Remarks

The fundamental issues in autonomous control system modeling and analysis were identified and briefly discussed, thus providing an introduction to the research problems in the area. A hierarchical functional autonomous controller architecture was also presented. It was proposed to utilize a hybrid approach to modeling and analysis of autonomous systems. This will incorporate conventional control methods based on differential equations and new techniques for the analysis of systems described with a symbolic formalism. In this way, the well developed theory of conventional control can be fully utilized. It should be stressed that autonomy is the design requirement and intelligent control methods appear, at present, to offer some of the necessary tools to achieve autonomy for some classes of applications. A conventional approach may evolve and replace some or all of the "intelligent" functions. Note that this paper is based on the development in [2],[3].

References

[1] P.J. Antsaklis, K.M. Passino, and S.J. Wang, "Autonomous control systems: Architecture and fundamental issues", in *Proc. 1988 Amer. Control Conf.*, Atlanta, GA, June 15-17, 1988, pp. 602-607; see also, P.J. Antsaklis and K.M. Passino, "Autonomous control systems: Architecture and concepts for future space vehicles," Final Rep., Jet Propulsion Laboratory Contract 957856, Oct. 1987.

[2] P.J. Antsaklis, K.M. Passino, and S.J. Wang, "Towards intelligent autonomous control systems: Architecture and fundamental issues," *J. Intelligent Robotic Syst.*, Vol.1, pp.315-342, 1989.

[3] P.J. Antsaklis, K.M. Passino, and S.J. Wang, "An introduction to autonomous control systems," in *Proc. IEEE Int. Symp. Intelligent Control*, Philadelphia, PA, Sept. 1990, pp. 21-26.

[4] Special Issue on Autonomous Intelligent Machines, *IEEE Computer*, Vol. 22, June 1989.

[5] E. Charniak and D. McDermott, *Introduction to Artificial Intelligence*. Reading, MA: Addison Wesley, 1985.

[6] S.C. Shapiro, Ed., *Encyclopedia of Artificial Intelligence*. New York, NY: Wiley, 1987.

[7] G.N. Saridis, "Toward the realization of intelligent controls," *Proc. IEEE*, Vol. 67, pp. 1115-1133, Aug. 1979.

[8] W.B. Gevarter, *Artificial Intelligence.* Park Ridge, NJ: Noyes, 1984.

[9] M. Mesarovic, D. Macko, and Y. Takahara, *Theory of Hierarchical, Multilevel, Systems.* Orlando, FL: Academic, 1970.

[10] W. Findeisen *et al.*, *Control and Coordination in Hierarchical Systems.* New York, NY: Wiley, 1980.

[11] O. Firschein *et al.*, Artificial Intelligence for Space Station Automation., Park Ridge, NJ: Noyes, 1986.

[12] E. Heer and H. Lum, Eds., *Machine Intelligence and Autonomy for Aerospace Systems.* Washington, DC: AIAA, 1988.

[13] E.R. Dougherty and C.R. Giardina, *Mathematical Methods for Artificial Intelligence and Autonomous Systems.* Englewood Cliffs, NJ: Prentice Hall, 1988.

[14] R.M. Glorioso and F.C. Colon Osorio, *Engineering Intelligent Systems.* Bedford, MA: Digital, 1980.

[15] A. Kusiak, *Intelligent Manufacturing Systems.* Englewood Cliffs, NJ: Prentice Hall, 1990.

[16] U. Ozguner,"Decentralized and distributed control approaches and algorithms," in *Proc. 28th IEEE Conf. Decision and Control*, Tampa, FL, Dec. 1989, pp. 1289-1294.

[17] L. Acar and U. Ozguner, "Design of knowledge-rich hierarchical controllers for large functional systems," *IEEE Trans. Syst., Man, Cybern.*, Vol. 20, pp. 791-803, July/Aug. 1990.

[18] G.N. Saridis,"Foundations of the theory of intelligent controls," in *Proc. IEEE Workshop on Intelligent Control*, pp 23-28, 1985.

[19] G.N. Saridis, "Knowledge implementation: Structures of intelligent control systems," in *Proc. IEEE Int. Symp. Intelligent Control*, pp. 9-17, 1987.

[20] A. Meystel, "Intelligent control: Issues and perspectives," in *Proc. IEEE Workshop on Intelligent Control*, pp. 1-15, 1985.

[21] G.N. Saridis,"Intelligent controls for advanced automated processes," in *Proc. Automated Decision Making and Problem Solving Conf.*, NASA CP-2180, May 1980.

[22] G.N. Saridis, "Intelligent robot control," *IEEE Trans. Auto. Control*, Vol. AC-28, pp. 547-556, May 1983.

[23] G.N. Saridis, "Analytic formulation of the principle of increasing precision with decreasing intelligence for intelligent machines," *Automatica*, Vol.25, pp. 461-467, 1989.

[24] K.P. Valavanis, "A mathematical formulation for the analytical design of intelligent machines," Ph.D. Diss., Elec. & Comp. Eng. Dept., Rensselaer Polytechnic Institute, Troy, NY, Nov. 1986.

[25] K.P. Valavanis and G.N. Saridis, "Information theoretic modelling of intelligent robotic systems, Part I: The organization level," in *Proc. 26th Conf. Decision and Control*, Los Angeles, CA, pp. 619-626, Dec. 1987.

[26] K.P. Valavanis and G.N. Saridis, "Information theoretic modelling of intelligent robotic systems, Part II: The coordination and execution levels," in *Proc. 26th Conf. Decision and Control*, Los Angeles, CA, pp. 627-633, Dec. 1987.

[27] V.Y. Jin and A.H. Levis, "Compensatory behavior in team decision making," in *Proc. IEEE Int. Symp. on Intelligent Control*, pp. 107-112, Philadelphia, PA, Sept. 1990.

[28] J. Albus *et al.*, "Theory and practice of intelligent control," in *Proc. 23rd IEEE COMPCON*, pp. 19-39, 1981.

[29] K.J. Astrom *et al.*, "Expert control," *Automatica*, Vol. 22, pp. 277-286, 1986.

[30] L.A. Zadeh, "Fuzzy logic," *Computer*, pp. 83-93, Apr. 1988.

[31] L. Wos, *Automated Reasoning: 33 Basic Research Problems.* Englewood Cliffs, NJ: Prentice Hall, 1988.

[32] B.P. Zeigler, "DEVS representation of dynamical systems: Event based intelligent control," *Proc. IEEE*, Vol. 77, pp. 72-80, 1989.

[33] B.P. Zeigler and S.D. Chi, "Model-based concepts for autonomous systems," *Proc. IEEE Int. Symp. on Intelligent Control*, Philadelphia, PA, Sept. 1990, pp. 27-32.

[34] P.R. Turner *et al.*, "Autonomous systems: Architecture and implementation," Jet Propulsion Laboratories, Rep. JPL D-1656, Aug. 1984.

[35] K.M. Passino and P.J. Antsaklis, "Relationships Between Event Rates and Aggregation in Hierarchical Discrete Event Systems," Proc. Allerton Conf. on Comm., Control, and Computing, Univ. of Illinois at Champaign-Urbana, pp. 475-484, Oct. 1990.

[36] K.M. Passino and P.J. Antsaklis, "Timing characteristics of hierarchical discrete event systems," to appear in *Proc. Amer. Control Conf.*, Boston, MA, 1991.

[37] J. Mendel and J. Zapalac,"The application of techniques of artificial intelligence to control system design," in *Advances in Control Systems*, C.T. Leondes, Ed. New York, NY: Academic, 1968.

[38] H. Zhong and W.M. Wonham, "On the consistency of hierarchical supervision in discrete-event systems," *IEEE Trans. Auto. Control*, Vol. 35, pp. 1125-1134, 1990.

[39] J.L. Peterson, *Petri Net Theory and the Modelling of Systems.* Englewood Cliffs, NJ: Prentice-Hall, 1981.

[40] K.M. Passino and P.J. Antsaklis, "A system and control theoretic perspective on artificial intelligence planning systems," *J. Appl. Artific. Intell.*, Vol. 3, pp. 1-32, 1989; see also, P.J. Antsaklis and K.M. Passino, "Artificial intelligence planning and control theory relationships," Final Rep. for McDonnell Douglas Corp., Contract Z71145, Oct. 1987.

[41] K.M. Passino and P.J. Antsaklis, "On the optimal control of discrete event systems," in *Proc. 28th IEEE Conf. Decision and Control*, Tampa, FL, Dec. 13-15, 1989, pp. 2713-2718.

[42] Antsaklis P.J, K.M. Passino, and M.A. Sartori, "Modelling and analysis of artificial intelligence planning systems," Final Rep. for McDonnell Douglas, Contract Z81014, Oct. 1988.

[43] B.P. Zeigler, "Knowledge representation from Newton to Minsky and beyond," *J. Appl. Artific. Intell.*, Vol. 1, pp. 87-107, 1987.

[44] K.M. Passino and P.J. Antsaklis, "Fault detection and identification in an intelligent restructurable controller," *J. Intell. Robotic Syst.*, Vol. 1, pp. 145-161, June 1988; see also, K.M. Passino and P.J. Antsaklis, "Restructurable controls study: An artificial intelligence approach to the fault detection and identification problem," Final Rep. for McDonnell Douglas, Contract Z60110, Oct. 1986.

[45] K.M. Passino, "Restructurable controls and artificial intelligence," McDonnell Aircraft Internal Rep. IR-0392, Apr. 1986.

[46] K.M. Passino and P.J. Antsaklis, "Artificial intelligence planning problems in a Petri net framework," in *Proc. Amer. Control Conf.*, pp. 626-631, June 1988.

[47] K.M. Passino and P.J. Antsaklis, "Planning via heuristic search in a Petri net framework," *Proc. 3rd IEEE Int. Symp. on Intelligent Control*, Arlington, VA, Aug. 24-26, 1988, pp. 350-355.

[48] P.J. Antsaklis, "Learning in control," *in Proc. 3rd IEEE Int. Symp. Intelligent Control*, pp. 500-507, Arlington, VA, Aug. 24-26, 1988.

[49] Z. Gao, M.D. Peek, and P.J. Antsaklis, "Learning for the adaptive control of a large flexible structure," in Proc. 3rd IEEE Int. Symp. Intelligent Control, Arlington, VA, Aug. 24-26, 1988, pp. 508-512; see also, P.J. Antsaklis, Z. Gao, K.M. Passino, M.D. Peek, and M. Sartori, "Learning and decision making models for higher level adaptation," Final Rep., Jet Propulsion Laboratory Contract 957856, Nov. 1988.

[50] M.D. Peek and P.J. Antsaklis, "Parameter learning for performance adaptation in large space structures," in *Proc. 4th IEEE Int. Symp. Intelligent Control*, Albany, NY, Sept. 25-27, 1989; see also, P.J. Antsaklis, M.D. Peek, Z. Gao, and K.M. Passino, "Learning control for higher level adaptation," Final Rep., Jet Propulsion Laboratory Contract 957856, Mod. 3, Nov. 1989.

[51] M.D. Peek and P.J. Antsaklis, "Parameter learning for performance adaptation," *IEEE Control Syst. Mag.*, Vol. 10, pp. 3-11, Dec. 1990.

[52] P.J. Antsaklis, "Neural networks in control systems," *IEEE Control Syst. Mag.*, Vol.10, pp.3-5, Apr. 1990; also Special Issue on Neural Networks in Control Systems, *IEEE Control Syst. Mag.*, Vol.10, pp.3-87, Apr. 1990.

[53] M.A. Sartori, K.M. Passino, and P.J. Antsaklis, "Artificial neural networks in the match phase of rule based expert systems," in

Proc. 27th Annu. Allerton Conf. Commun., Control and Comput., Urbana, IL, Sept. 27-29, 1989, pp. 1037-1046.

[54] M.A. Sartori, K.M. Passino, and P.J. Antsaklis, "An artificial neural network solution to the match phase problem in rule based artificial intelligence systems," *IEEE Trans. Knowledge and Data Eng.*, to be published, 1991.

[55] R.F. Stengel, "AI theory and reconfigurable flight control systems," Princeton Univ., Princeton, NJ, Rep. 1664-MAE, June 1984.

[56] Z. Gao and P.J. Antsaklis, "On the stability of the pseudo-inverse method for reconfigurable control systems," in *Proc. Nat. Aerospace and Electron. Conf.*, Dayton, OH, May 22-26, 1989, pp. 333-337; also *Int. J. Control,* to be published.

[57] Z. Gao and P.J. Antsaklis, "Pseudo-inverse methods for reconfigurable control with guaranteed stability," in *Proc. 1990 IFAC 11th World Cong.*, Tallinn, U.S.S.R., Aug. 13-17, 1990.

[58] K.M. Passino, M.A. Sartori, and P.J. Antsaklis, "Neural computing for numeric to symbolic conversion in control systems," *IEEE Control Syst. Mag.*, pp. 44-52, Apr. 1989.

Article 1.5

Intelligent Control Using Neural Networks

Kumpati S. Narendra and Snehasis Mukhopadhyay

Intelligent control, using neural networks, can be applied to complex dynamical systems in the presence of structural failures. After a failure occurs, the dynamical system is assumed to be in one of a finite number of configurations, corresponding to each of which exists a stabilizing controller. A two level controller structure is used, in which the higher level detects a failure using pattern recognition techniques and activates a stabilizing controller. At the second level, an adaptive controller is used to optimize the stabilized system. Multilayer neural networks are used at the lower level for the identification and control of the plant and as a pattern recognizer in the higher level for the detection of a failure in the system. Simulation results presented reveal that efficient multilevel controllers using neural networks can be designed for nonlinear dynamical systems.

Intelligent Control

As control methods have found their way into standard practice, they have opened the door to a wide spectrum of complex applications. Such complex systems are characterized by poor models, high dimensionality of the decision space, distributed sensors and decision makers, high noise levels, multiple subsystems, levels, time-scales and/or performance criteria, complex information patterns, overwhelming amount of data and stringent performance requirements. Since the ability of living systems to cope with similar situations is well known, systems theorists attempted to incorporate such characteristics in artificial systems. As a result, over the past three decades numerous terms borrowed from psychology and biology, such as pattern recognition, adaptation, learning and self organization, have been introduced into the systems literature.

A preliminary version of the paper was presented at the American Control Conference, Boston, 1991. The authors are with the Center for Systems Science, Department of Electrical Engineering, Yale University, CT 06520. This work was supported by the National Science Foundation under Grant EET-8814747 and by Sandia National Laboratories under Contract 84-1791.

Intelligent control is merely the latest in this series of terms. Qualitatively, a system which includes the ability to sense its environment, process the information to reduce uncertainty, plan, generate and execute control action in the situations mentioned earlier, constitutes an intelligent control system. The greater the ability to deal with the above difficulties, the more intelligent is the control system.

The difficulties that arise in the control of complex systems can be broadly classified into three categories. The first is complexity, the second is the presence of nonlinearities and the third is uncertainty. In recent years it has been demonstrated that artificial neural networks are ideally suited to cope with all three categories of difficulties. Further, it has been shown that neural networks are extremely efficient as pattern recognizers. In 1990 [1], one of the authors introduced several dynamical models containing artificial neural networks which have proved to be very efficient for the identification and control of nonlinear dynamical systems. This versatility of neural networks is used in this paper for the design of a two-level controller for dynamical plants.

Changes in environments and performance criteria, unmeasurable disturbances and component failures are some of the charateristics which necessitate intelligent control. Adaptive control is a natural choice when only parametric uncertainty exists. However, adaptive controllers do not have long term memory and hence do not "remember" the optimal control parameters corresponding to different configurations of the plant. In such cases, pattern recognizers have been used to classify the plant and choose the corresponding control parameters. Learning controllers can be used to choose between different adaptive algorithms, reference models or performance criteria. As more intelligent control systems are designed, it becomes necessary to combine adaptation, learning and pattern recognition in novel ways to make decisions at various levels of abstraction. This paper describes such a multilevel controller for use in situations where a dynamical system can fail in a finite number of ways. The plant in each failure mode, said to belong to a configuration, is described by a discrete-time nonlinear difference equation and the state variables are assumed accessible. The objective is to detect a fault when it occurs and use a suitable controller so that the system responds in a satisfactory way. In the present approach, after stability is assured, an adaptive controller is used to improve the performance of the overall system. The novelty of the approach suggested is that identification, pattern recognition and control are carried out using multilayer neural networks when the plant is nonlinear. Hence a common framework can be used to deal with the different concepts encountered. The problem of controlling a nonlinear dynamical plant, in general, is a formidable one. To make the problem tractable, numerous assumptions have to be made and these are stated explicitly later in the paper. Some of the theoretical questions that arise in this context are also indicated. The primary contribution of the paper is the approach used, which is justified by simulation studies. The theoretical questions raised are currently being investigated.

Statement of the Problem

A general fault detection and control problem may be qualitatively stated as follows. A dynamical system together with a controller performs satisfactorily in an environment (or configuration) E_1. At some unspecified time t_0, a failure occurs and the configuration of the system is altered. It is known *a priori* that the system can be in only a finite number of configurations (E_i, $i = 2,3,\ldots, N$) after failure, where each configuration corresponds to a compact set in the space of system parameters. For the satisfactory control of the plant, the exact values of the new system parameters have to be estimated. However, since a continuous adaptive scheme is generally a slow process, a two stage approach is used. In the first stage only the configuration E_i to which the plant belongs is determined. It is assumed that all elements of a configuration E_i can be stabilized using a controller C_i. Once the stability of the overall system is guaranteed, an adaptive controller is initiated to identify the values of the system parameters after failure and update the controller parameters on-line to achieve the desired response.

The control problem, as stated above, consequently consists of three parts : i) the

Reprinted from *IEEE Contr. Syst. Mag.*, vol. 12, no. 2, pp. 11–18, April 1992.

monitoring of the system response to detect a failure when it occurs and estimating the configuration E_i to which the modified plant belongs ii) activating a fixed controller to stabilize the system in its new configuration iii) activating an adaptive controller to improve the response of the system. From a theoretical viewpoint, the fixed controller can be combined with the adaptive controller and viewed as merely providing the initial values of the adaptive controller parameters after the fault is detected. The nature of the system (i.e. linear or nonlinear), and the nature of the configurations (i.e. whether $E_i (i = 2, 3, \ldots, N)$ contain a finite or an infinite number of elements), determine the complexity of the control problem.

In this paper, we consider the above problem in three stages. In the first stage (Problem 1), the plant is assumed to be linear and time-invariant in its various configurations so that the overall system can be considered as piecewise (over time) linear and time-invariant. In view of this, the various questions that arise in the design of the controller can be stated precisely. In the following two stages the plant is assumed to be nonlinear in the different configurations and hence is substantially more complex. Hence the problem is attempted in two stages (Problem 2 and Problem 3) of increasing complexity,

In Problem 1, each configuration corresponds to a compact set of points in the parameter space. It is assumed that linear controllers which stabilize the different configurations are known. Hence, the first part of the problem lies in determining the specific configuration to which the system belongs at any instant. Since the system is linear, the detection problem can be solved using well established linear techniques. However, we propose a pattern recognition method based on neural networks to motivate the approach used for the nonlinear control problems discussed later. Once the configuration is identified and the overall system is stabilized, a standard linear adaptive controller is used to improve the response of the system.

In the second stage (Problem 2), the plant dynamics in different configurations are nonlinear. However, each configuration is assumed to consist of only one such plant. In this case, it is assumed that the nonlinear plant can be identified off-line in the different configurations using neural networks. In each case a stabilizing controller is also designed off-line. Hence, only the detection problem is carried out on-line using a neural network.

Problem 3 deals with the most complex problem, when the nonlinear plant is suitably parametrized and each of the different configurations contains an infinite number of elements. Here, in addition to the fault detection and stabilization problems discussed in Problem 2, an adaptive controller has to be used to compensate for the uncertainty in the parameters.

The precise statements of the problems follow.

Statement of Problem 1

The discretized linearized dynamics of a helicopter (henceforth referred to as the plant) in the vertical plane around an operating condition representing typical loading and flight conditions at an airspeed of 135 knots configuration E_1) are described by the set of difference equations (1), where $x(k) \in \mathrm{IR}^4$, $A \in \mathrm{IR}^{4\times4}$, $B \in \mathrm{IR}^{4\times2}$ and $u(k) \in \mathrm{IR}^2$ [2].

$$x(k+1) = (I + A\Delta t)x(k) + (B_1 \Delta t)u(k) \quad (1)$$

The sampling time Δt is set to be 0.01 and I is the 4 x 4 identity matrix. We assume that after failure the system can belong to one of the two configurations E_2 and E_3. Each of the configurations E_2 and E_3 corresponds to a set of input matrices. These sets (denoted by $\boldsymbol{B}_2$ and $\boldsymbol{B}_3$) are chosen to include nominal plant values given, respectively, by the two difference equations

$$x(k+1) = (I + A\Delta t)x(k) + (B_2^* \Delta t)u(k) \quad (2)$$
$$x(k+1) = (I + A\Delta t)x(k) + (B_3^* \Delta t)u(k) \quad (3)$$

where

$$A = \begin{bmatrix} -0.0366 & 0.0271 & 0.0188 & -0.4558 \\ 0.0482 & -1.01 & 0.0024 & -4.0208 \\ 0.1002 & 0.3681 & -0.707 & 1.420 \\ 0 & 0 & 1 & 0 \end{bmatrix}$$

$$B_1 = \begin{bmatrix} 0.4422 & 0.1761 \\ 3.5446 & -7.5922 \\ -5.52 & 4.49 \\ 0 & 0 \end{bmatrix}$$

$$B_2^* = \begin{bmatrix} 0.4422 & 0.1761 \\ 0 & 0 \\ -5.52 & 4.49 \\ 0 & 0 \end{bmatrix}$$

$$B_3^* = \begin{bmatrix} 0.4422 & 0.1761 \\ 3.5446 & -7.5922 \\ 0 & 0 \\ 0 & 0 \end{bmatrix}$$

The set $\boldsymbol{B}_2$ is assumed to be such that the elements b_{ij} of a matrix B belonging to $\boldsymbol{B}_2$ satisfy $-2 \le b_{21} \le 2$, $-2 \le b_{22} \le 2$, $-7.52 \le b_{31} \le -3.52$, $2.29 \le b_{32} \le 6.49$, where the remaining elements of B are the same as in B_2^*.

The elements b_{ij} of a matrix B belonging to $\boldsymbol{B}_3$ satisfy $3.3446 \le b_{21} \le 3.7446$, $-7.7922 \le b_{22} \le -7.3922$, $-0.2 \le b_{31} \le 0.2$, $-0.2 \le b_{32} \le 0.2$ with the remaining elements being the same as those of B_3^*. Throughout this paper, the reference input is assumed to be a constant with $r_i(k) = 1$. The objective is to control the system in the presence of such failures so as to achieve the desired response to the reference trajectory.

Statement of Problem 2

In this case the dynamical system is nonlinear and has a configuration E_1 described by the following difference equation, where $x(k)$, I, A, Δt, B_1 and $u(k)$ are as defined earlier and $f_1 : \mathrm{IR}^4 \rightarrow \mathrm{IR}^4$.

$$x(k+1) = (I + A\Delta t)x(k) + f_1(x(k)) + (B_1 \Delta t)u(k) \quad (4)$$

We assume that after failure the system can be described by one of the two difference equations

$$x(k+1) = (I + A\Delta t)x(k) + f_1(x(k)) + (B_2^* \Delta t)u(k) \quad (5)$$

$$x(k+1) = (I + A\Delta t)x(k) + f_2(x(k)) + (B_3^* \Delta t)u(k) \quad (6)$$

The matrices A, B_1, B_2^* and B_3^* are chosen to be the same as those in Problem 1. The nonlinearities are given by

$$f_1^{\,T(x)} = [0, 0, (0.88 \sin x_1 - 0.74)x_2, 0]$$

and

$$f_2^{\,T(x)} = [0, (0.88 \sin x_1 - 0.74)x_2, 0, 0]$$

and are assumed to be unknown. Hence, when the configuration changes from E_1, to E_2, only the matrix B_1 changes to B_2^*, while in the second case where the new configuration is E_3, the nonlinear part changes from $f_1(.)$ to $f_2(.)$ and B_1 changes to B_3^*. Once again the objective is to assure satisfactory performance in the presence of failures.

Statement of Problem 3

In this case, the dynamical system in E_1 is given by the same nonlinear difference equation as in Problem 2. However, unlike Problem

2, the configurations E_2 and E_3 are asssumed to correspond to sets of parameterized nonlinear plants. In particular, configurations E_2 and E_3 correspond to the following sets of difference equations

Configuration E_2:

$$x(k+1) = (I + A\Delta t)x(k) + \theta_1 f_1(x(k)) + (B_2^* \Delta t)u(k) \quad (7)$$
$$S_1 : \{0.95 \leq \theta_1 \leq 5\}$$

Configuration E_3:

$$x(k+1) = (I + A\Delta t)x(k) + \theta_2 f_2(x(k)) + (B_3^* \Delta t)u(k) \quad (8)$$
$$S_2 : \{0.65 \leq \theta_2 \leq 1.5\}$$

The simple linear parametrization used here permits us to illustrate the main features of the design of a two-level controller for a complex nonlinear control problem. The objective, once again, is to assure satisfactory plant behavior in the presence of failures. The uncertainty in the plant parameter values (i.e., θ_i, $i = 1, 2$) once again necessitates the use of an adaptive controller.

Problem 1

Given the configurations E_1, E_2 and E_3 defined earlier, matrices K_1, K_2 and K_3 with

$$K_1 = \begin{bmatrix} -0.8142 & -1.2212 & 0.2662 & 0.8263 \\ -0.2582 & 1.1777 & 0.0623 & -0.2123 \end{bmatrix}$$

$$K_2 = \begin{bmatrix} -0.5679 & -1.2749 & 0.9317 & 3.8383 \\ -0.4148 & 0.3868 & -0.2995 & -1.0962 \end{bmatrix}$$

$$K_2 = \begin{bmatrix} -1.3598 & -1.6167 & -9.2455 & -14.1461 \\ 0.0126 & 1.2201 & 4.8390 & 7.4463 \end{bmatrix}$$

exist such that

i) the undamaged plant is stable with a desired response when a matrix K_1 is used in the feedback path,

ii) the configurations E_2 and E_3 are stable respectively with feedback gain matrices K_2 and K_3 and

iii) configurations E_2 and E_3 are unstable when the matrix K_1 is in the feedback path.

In addition, the response of any element in E_2 with a feedback gain matrix K_1 is distinctly different from that of the response of a typical element of E_3 with the same feedback gain. The above facts are made use of, both to detect a failure using pattern recognition techniques and to classify the resulting configuration.

Pattern Recognition

For the problem under consideration, it was found convenient to use two neural networks in the detection process. The first distinguishes between configurations E_1 and E_2, while the second decides between configurations E_1 and E_3. The first network monitors whether the configuration has changed from E_1 to E_2 every 20 time units (0.2 s) on the basis of the state vectors at the beginning and the end of the interval. The second network does the same for the transition of configuration E_1 to E_3 every 200 time units (2 s). The length of the intervals were chosen after observing the corresponding responses. Empirical studies also revealed that during these intervals, state variables x_1 and x_4 do not change appreciably and that $x_2(k)$ and $x_3(k)$ are adequate to detect and classify the configuration of the plant. The fact that these outputs are significantly different for the two configurations with the same initial state and refernce input, is used in the pattern recognition process. This implies that in the training stage, if the values of x_2 and x_3 are known at the beginning and end of a suitably chosen interval, they can be used to determine the configuration of the system. Hence these four quantities are made the inputs to a neural network. The networks were trained off-line with randomiy generated initial states within a suitably defined domain of interest in the state space. These domains include the steady-state values in the normal configuration. For each network, two models, corresponding to the nominal plant descriptions for the two configurations required to be distinguished, are intialized with the randomly generated initial state. The final states are observed for both the models at the end of a test period. If the states of the two models differ in norm from each other by a predefined threshold, the corresponding output values are considered as samples of the corresponding configurations, and used in the training process. If the threshold condition is not satisfied, state vectors of both the models are used as samples for the undamaged class (configuration E_1). The process is continued until adequate number of training samples of the two configurations are obtained. Each neural network has 4 inputs and an additional bias input, two hidden layers with 8 and 4 nodes, respectively, and a single output. Each network is trained with 10 000 sample points belonging to each of the two classes to be distinguished.

Comment 1: A question that arises naturally at this stage is whether pattern recognition is needed to detect a failure when it occurs. In fact, depending upon the application, many other methods may be available for detection and classification of a failure. In [3], a method based on dynamic models of the plant is described in detail. However, such models may become computationally less efficient when the dimensionality of the system is high. Pattern recognition becomes attractive in such cases, particularly when the fault can be detected using relatively few values of critical output variables.

Adaptive Control

While the feedback matrix K_2 (K_3) is adequate to stabilize any system belonging to E_2 (E_3), it does not yield the desired response for all the systems belonging to a configuration. Hence, to improve the system performance, some form of adaptive control is needed. If indirect control is used, the elements of the B matrix have to be estimated on-line and a corresponding feedback control matrix chosen. The estimation of B as $\hat{B}(k)$ is carried out using standard identification techniques with an identification model and the updating law shown below, where $e(k) = x(k) - \hat{x}(k)$ is the identification error. Λ is a matrix with $0 < \lambda_{ij} < 2$ and $\otimes$ denotes the Kronecker product:

$$\hat{x}(k+1) = (I + A\Delta t)x(k) + \hat{B}(k)u(k) \quad (9)$$

$$\hat{B}(k+1) = \hat{B}(k) - \Lambda \otimes (e(k) \frac{u^t(k)}{1 + \|u(k)\|^2}) \quad (10)$$

The desired response is assumed to be achieved by minimizing a quadratic performance index of the following form, and Q and R are chosen to be the following diagonal matrices:

$$j = \sum_{k=t_0}^{\infty} (X^T(k)Qx(k) + u^T(k)Ru(k))$$
$$Q \geq 0, R > 0 \quad (11)$$

$$Q = \begin{bmatrix} 0.04 & 0 & 0 & 0 \\ 0 & 0.25 & 0 & 0 \\ 0 & 0 & 0 & 0 \\ 0 & 0 & 0 & 0 \end{bmatrix}$$

$$R = \begin{bmatrix} 0.04 & 0 \\ 0 & 0.11 \end{bmatrix}$$

At the end of an interval of 500 time units, an iterative algorithm is used to determine the

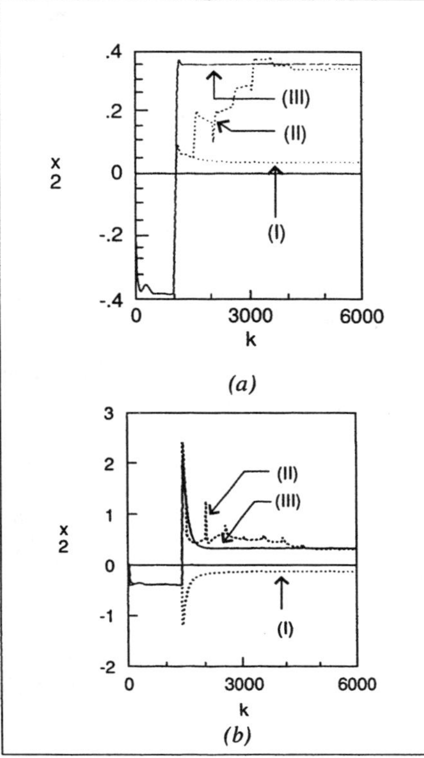

Fig. 1. The operation of the two level controller with and without adaptation at the lower level (a) x_2 when fault 2 occurs : (I) nonoptimal fixed controller (II) adaptive controller (III) optimal controller (b) x_2 when fault 3 occurs : (I) nonoptimal fixed controller (II) adaptive controller (III) optimal controller.

solution of an algebraic Riccati equation to compute the feedback matrix F^i as

$$G^{i^T} K^{i+1} + K^{i+1} G^i + Q + F^{i^T} R F^i = 0$$
$$F^{i+1} = -R^{-1} \hat{B}^T K^{i+1}$$
$$G^{i+1} = G^i + \hat{B} F^{i+1}$$

where $\hat{B}$ is the estimate of the matrix B and the intial value F^0 is chosen to make the matrix $A + \hat{B} F^0$ asymptotically stable [4].

Simulation Results

After training, the higher level pattern recognizer, consisting of two neural networks as described earlier, was tested with a test set composed of 1000 patterns. The patterns were chosen randomly with a uniform distribution. The first network yielded an accuracy of 76% at the end of the first interval and 92% at the end of the subsequent interval. The corresponding performances for the second network were 71% and 73% respectively. In contrast to this off-line testing, the classifications given by both networks on samples encountered on-line were 100% accurate. This can be accounted for by the fact that on-line failures are assumed to take place only when almost stationary conditions prevail.

As mentioned earlier, transitions from E_1 to E_2 were checked every 20 time units. When a fault occurred within the first 16 units of such an interval, it was detected at the end of that interval. However, faults occurring later in the interval were detected only at the end of the following interval. Detection of fault 3 (transition from E_1 to E_3) was more involved than the detection of fault 2. Even though the network was trained using data over intervals of length 200 units, detection was possible over two or more intervals of time (i.e., 4 to 6 s). In particular, if a failure occurred within the first 21 units of an interval (of length 200), it was detected at the end of the following interval. Failures occurring from 21 to 200 units were detected two intervals later. The maximum time interval for detection was less than 6 s and was found to be satisfactory for the overall performance of the system. The first network never detected fault 2 when fault 3 occurred and vice versa, so that the resolution of conflicting decisions was not called for.

Fig. 1 shows the on-line operation of the two-level system. Throughout the paper, in the simulation results, only one of the four state variables is shown due to space limitations. The state variable, which is most significant in terms of illustrating the point made in a particular experiment, is chosen for this purpose. In this experiment, the state variable x_2 is shown for both Fig. 1(a) and (b). The system is in normal mode (configuration E_1) with the matrix K_1 in the feedback path for the first 1000 time units. The higher level pattern recognizer monitors the system responses starting from 900 time units. In Fig. 1(a), a fault 2 occurs at time step 1000 and the system configuration changes to E_2, with the B matrix changing to

$$B_2 = \begin{bmatrix} 0.4422 & 0.1761 \\ -0.4872 & 1.0520 \\ -6.9366 & 2.9211 \\ 0 & 0 \end{bmatrix}$$

Without corrective action, the overall system is unstable and the state variables diverge. However, the higher level decision maker detects the fault correctly at time instant 1020. As a consequence of this decision, the feedback matrix K_1 is changed to K_2 at time instant 1020. This is a stabilizing controller for the new configuration. However, this is not the optimal controller for the given performance index and for the given B matrix. The trajectory marked (I) corresponds to the case when only the stabilizing controller K_2 is used after detection and classification of the fault. To improve the system response an adaptive controller is started at time instant 1020 and is updated every 500 time units. Trajectory (II) corresponds to the response obtained with the adaptive controller. Trajectory (III) is the response that would have resulted if the B matrix were exactly known and the corresponding optimal feedback matrix were used. It is clear from the results that the fault detection and classification are achieved within a very short time before the system performance degrades appreciably. Also, once the overall system is stabilized, the adaptive controller results in an asymptotic performance very close to the optimal value. However, the trajectories in the two cases are seen to be distinctly different.

Fig. 1(b) shows similar results when a fault 3 occurs. At time instant 1000, the B matrix is changed from B_1 to B_3 where

$$B_3 = \begin{bmatrix} 0.4422 & 0.1761 \\ 3.6060 & -7.6255 \\ -0.0466 & 0.0630 \\ 0 & 0 \end{bmatrix}$$

This fault is detected at time instant 1400 by the pattern recognizer. This is followed by stabilization using feedback matrix K_3 and the initialization of the adaptive controller. Once again, the rapid detection of the failure and stabilization prevents the performance from degrading, and the use of the adaptive controller results in near optimal response in the new configuration.

Comment 2: It should be noted that pattern classification is not carried out at every instant of time but only at the end of an interval. If a failure were to occur at any point in an interval, (depending upon its position) the fault may not be detected at the end of that interval. In such cases, one may have to wait till the end of a second interval to detect and successfully classify the failure so that suitable control action can be taken.

Problem 2

In this case, the plant dynamics before and after the fault are described by the nonlinear difference equations given earlier for Problem 2. The various steps to be carried out in this case are i) identification of the nonlinear systems in the three configurations ii) design of suboptimal nonlinear stabilizing controllers for all the three configurations iii) training of the higher level pattern recognizer to detect and classify failures. A schematic diagram of the structure of the overall system is shown in Fig. 2. As discussed before, the different components of the systems (i.e., identifiers, controllers, pattern recognizers) are realized using multilayer neural networks. The procedure adopted is discussed in the following.

Identification

The difference equations (4)-(6) describing the plant in the three configurations are unstable when no control is used. Before identifying the three nonlinear models, it is therefore necessary to stabilize them in some fashion. Since the problem of determining a stabilizing controller even for a known nonlinear plant is a formidable one, it becomes significantly more complex when the nonlinearities $f_i(x)$ $(i = 1, 2)$ in (4)-(6) are unknown. Hence, we assume considerable prior knowledge concerning the nonlinearities to design stabilizing controllers. Following this, the three stabilized configurations of the plant are identified off-line using random inputs and an additional nonlinear controller is designed to achieve satisfactory performance in each case.

Let K_i be a feedback matrix of gains which optimizes a quadratic criterion for configuration E_i when only the linear part of the configuration is present. We assume that K_i also stabilizes the true nonlinear plant in configuration E_i. This implies that the nonlinear terms are not significant as far as stabilization is concerned but are significant for performance. The identification of the stabilized system in the first configuration is described below. A similar procedure is also used for the other two cases.

For configuration E_1, the stabilized system is described by the nonlinear differential equation

$$x(k+1) = (I + A\Delta t)x(k) + f_1(x(k)) + (B_1\,\Delta t)[K_1\,x(k) + r(k)] \quad (12)$$

We assume that the system (12) has bounded outputs for a bounded input r, when $r(t)$ belongs to a compact set. Since A_1, B_1, and K_1 are known, the nonlinear function $f_1(x)$ can be estimated as $\hat{F}_1(x)$ off-line to any desired degree of accuracy using a multilayer neural network.

Using a linear model of the form $z(k+1) = (I + A\Delta t)x(k) + (B_1\Delta t)[K_1x(k) + r(k)]$ of the unknown plant, the nonlinear function $f_1(x)$ is estimated using the error $e(k) = x(k) - z(k)$ between the system and model outputs. The values of $f_1(x)$ were stored for 10 000 equally spaced points of x in the state space and the input to the multilayer neural network was chosen randomly with a uniform distribution in this set. A similar procedure was adopted to identify the nonlinearity $f_2(x)$.

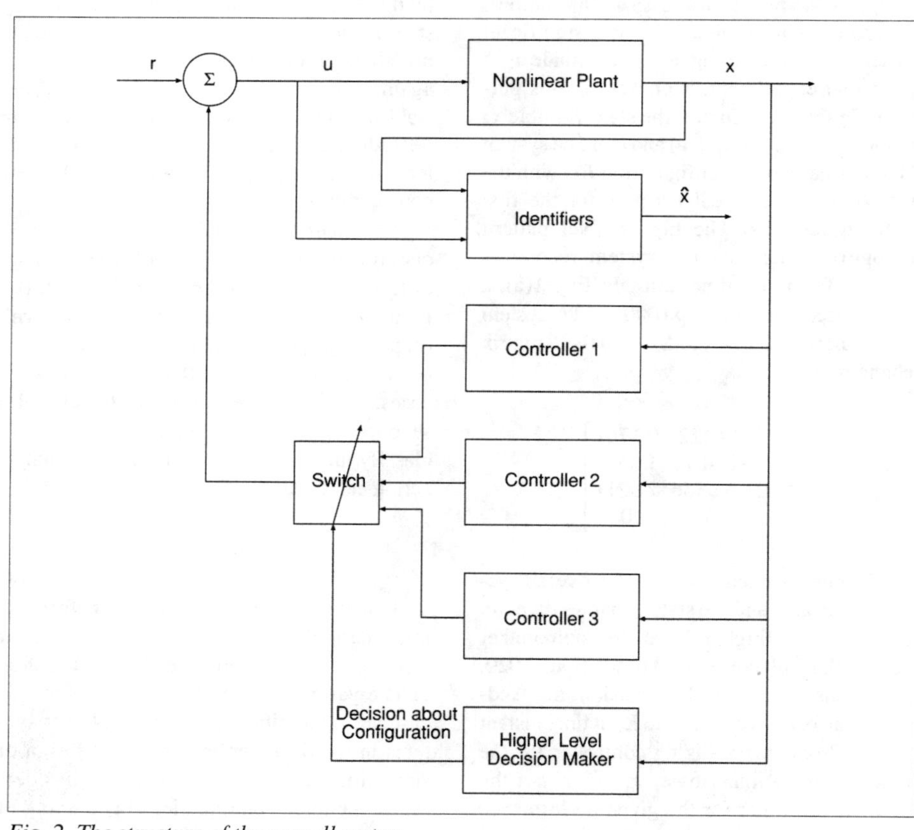

Fig. 2. The structure of the overall system.

Design of Controllers

The linear feedback control matrix K_i merely stabilizes the nonlinear plant in configuration E_i. To obtain satisfactory performance an additional nonlinear controller $N_i(x)$ $(i = 1, 2, 3)$ is used so that the control input to the plant has the form $u_i(k) = K_ix(k) + N_i(x(k)) + r(k)$. The three nonlinear controllers were realized off-line using multilayer neural networks. In each case, the parameters of the neural network N_i were adjusted so as to mininuze the effect of the nonlinearities. This involved minimizing $||\hat{F}_i(x) + B_iN_i(x)||$ over the range of interest of the state x. In configuration E_1, since perfect cancellation of the nonlinearity is not possible, the error was minimized by using back propagation through the matrix B_1 to adjust the weights of N_1. In configurations E_2 and E_3, $N_i(x)$ $(i = 2, 3)$ could be determined algebraically to make $\hat{F}_i(x) + B_iN_i(x) = 0$. The resulting controllers are given by

$$N_2(x) = \frac{1}{\Delta t}\begin{bmatrix} 0 & 0 & -0.0594 & 0 \\ 0 & 0 & 0.1495 & 0 \end{bmatrix}\hat{F}_2(x)$$

and

$$N_3(x) = \frac{1}{\Delta t}\begin{bmatrix} 0 & 0.0442 & 0 & 0 \\ 0 & -0.1111 & 0 & 0 \end{bmatrix}\hat{F}_3(x).$$

Fig. 3 shows the improvement resulting from the use of the nonlinear controller in configuration E_2. As before, only a typical state variable (*i.e.* x_1) is shown. Since exact cancellation of the nonlinear term is possible,

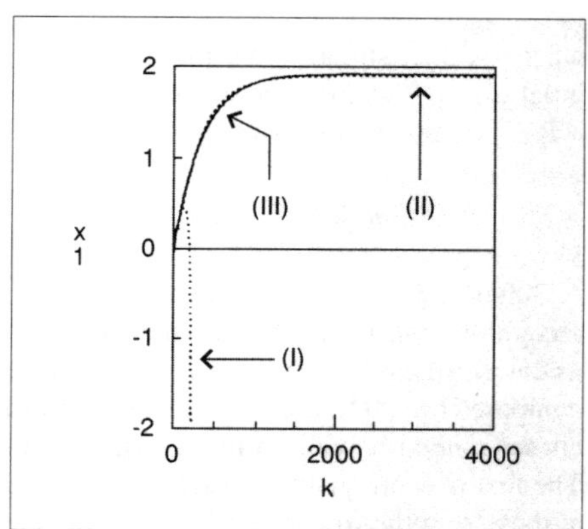

Fig. 3. Effect of the nonlinear controller in configuration 2 : (I) nonlinear system with linear controller, (II) nonlinear system with nonlinear controller, (III) linear system with linear controller.

the desired response (*i.e.* the response of the linear part of the plant with the optimal linear controller K_2) is indistinguishable from the response of the actual nonlinear system with the nonlinear controller. However, the response of the actual system with only K_2 in the feedback path, in spite of being bounded, shows large oscillations and is very different from the desired response. Similar improvements are observed by the use of the nonlinear controller in the other configurations.

Pattern Recognition

In earlier discussions, the identification of the plant and the design of stabilizing controllers in its various configurations were described. In the following, the design of a pattern recognizer which detects and classifies a fault when it occurs, is described.

The recognition of the true configuration is carried out by distinguishing between the responses of the plant in the three possible configurations with the same controller $K_1x + N_1(x) + r$ in the feedback path. These are described by the three difference equations

$$x(k+1) = (I + A\,\Delta t)x(k) + f_1(x(k)) + (B_1\,\Delta t)[K_1\,x(k) + N_1(x(k)) + r(k)]$$

$$x(k+1) = (I + A\,\Delta t)x(k) + f_1(x(k)) + (B_2\,\Delta t)[K_1\,x(k) + N_1(x(k)) + r(k)]$$

$$x(k+1) = (I + A\Delta\,t)x(k) + f_2\,(x(k)) + (B_3\,\Delta t)[K_1\,x(k) + N_1(x(k)) + r(k)]$$

If a fault occurs at some time t_1, and is detected and classified as belonging to one of configurations E_2 or E_3 at time t_2 ($t_2 > t_1$), the appropriate controller is chosen at time t_2.

The detection and classification of a fault when it occurs depends upon the nature of the evolution of the trajectories. As in Problem 1, neural networks are used in this case also as pattern recognizers. Extensive simulations revealed that the state variables in configuration E_2 grow at a much faster rate than those in configuration E_3. Hence, once again, two neural networks are used in the detection process.

Both the networks used eight inputs which were the values of the state before and after an interval of 5 time units (0.05 s). The two networks were identical in structure, having two intermediate layers with 40 and 20 nodes respectively and a single node at the output layer. Denoting the input vector as z, the outputs of the networks can be represented as $N_1(z)$ and $N_2(z)$, respectively. The first network was used to distinguish between "no fault" (configuration E_1) and "fault 2" (configuration E_2) situations, while the second network distinguished between the "no fault" case and "fault 3" (configuration E_3). In both networks $N_i(z) < 0$ corresponded to a "no fault" case. Each network was trained with a total of 2000 samples (1000 samples belonging to each of the two classes) drawn randomly with a uniform distribution. The samples used in the training of the networks, which critically affect the performance of the recognition procedure, were collected exactly in the same fashion as was used in Problem 1.

When no fault occurred, it was properly identified by both neural networks. Fault 2 occurring at any point of time was detected by the first network in approximately 30 time instants. Similarly, fault 3 occurring at any point of time was detected by the second network in approximately 130 time units. None of the networks made wrong decisions regarding failures.

Fig. 4 shows the entire system in operation when the state variables are observed over 3000 time units. As in Fig. 3, only one state variable (x_1) is shown. The system is in configuration E_1, for the first 1000 units of time with the control $K_1x + N_1(x) + r$ in operation when fault 2 occurs at time 1001. The fault is detected at the time instant 1030. It is classified as belonging to configuration E_2, following which the control $K_2x + N_2(x) + r$ is activated. In Fig. 4, the behavior of the system with and without such corrective action is shown. With correct detection and activation of the controller, the system is seen to perform satisfactorily, stabilizing around a new operating point. However, without such control, all the state variables are seen to grow without bound.

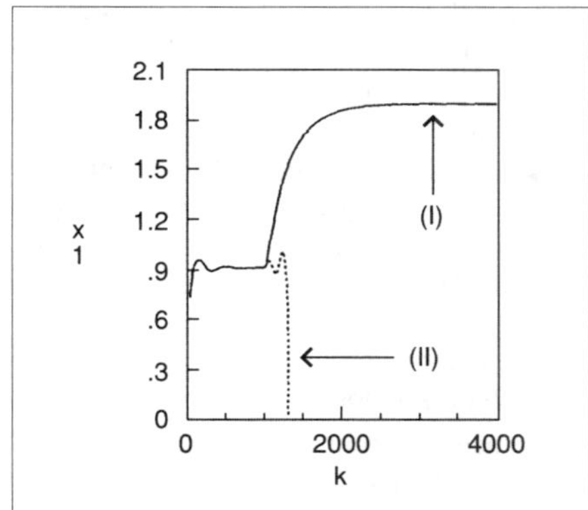

Fig. 4. Comparison of the system responses with fault 2 : (I) with fault detection, (II) without fault detection.

Problem 3 : Problem 2 with Adaptive Control

In Problem 2, when failure occurred, the resulting system could be described by one of two nonlinear difference equations (i.e., each configuration contained a single element). In Problem 3, the nonlinear systems are parametrized and each configuration contains an infinite number of elements. As in problems 1 and 2, pattern recognition is used to detect a new configuration which, in turn, is used for stabilization. However, since the specific value of the parameter θ_1, (or θ_2) is not known, adaptation is needed to achieve the desired response.

The two equations representing the plant in configurations E_2 and E_3 in Problem 2 now correspond to the nominal values of the plant in the same configurations. As in Problem 2, these are used for the design of a pattern recognizer. The stabilizing feedback controllers used in Problem 2 are found to stabilize the entire configurations E_2 and E_3. This implies that the parametric uncertainty in the nonlinear term does not significantly affect either fault detection or system stabilization. However, as seen from the simulation studies described later, the performance of the overall system is significantly affected by them. Improvement in performance is achieved by a nonlinear adaptive controller which attempts to compensate for the nonlinearity. The design of such a controller is described below.

In each of the configurations E_2 and E_3 described earlier in the statement of Problem 3, only a single parameter θ is seen to be unknown. This is estimated (for example in configuration E_2) by using the identification model given below, where $\hat{F}_1$ (obtained off-line using a neural network) is assumed to be a good enough approximation of f_1 :

$$\hat{x}(k+1) = Ax(k) + \hat{\theta}_1\hat{F}_{1(x(k))} + B_2^*\,u(k) \tag{13}$$

The estimate $\hat{\theta}$ is updated using the following adaptive law, where $e_{i3}(k) = x_3(k) - \hat{x}_3(k)$is the identification error in the third state variable, $\hat{F}_{13}$ is the third element of the vector $\hat{F}$ and λ is a positive gain.

$$\hat{\theta}(k+1) = \hat{\theta}(k) - \lambda e_{i3}\;(k)\hat{F}_{13}\;(x(k)) \tag{14}$$

The stabilizing control input for configuration E_i, as described earlier for Problem 2, has the form $K_ix(k) + N_i(x(k)) + r(k)$. In the present case, using equation (14), it is updated every

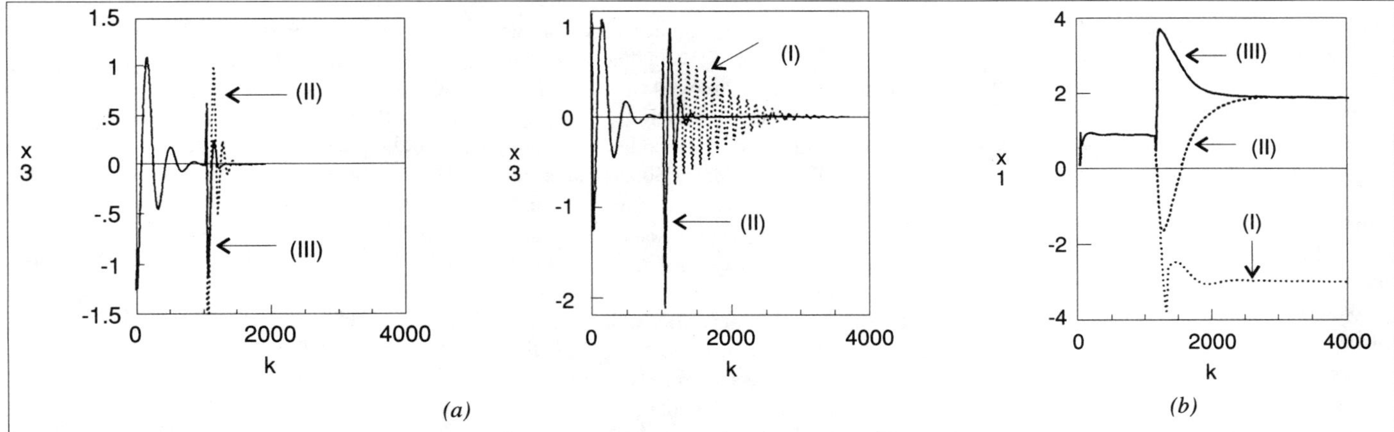

Fig. 5. Response with and without the adaptive component of the lower level controller in Problem 3 : (a) Fault 2 : (I) nonoptimal fixed controller (II) adaptive controller (III) optimal controller (b) Fault 3 : (I) nonoptimal fixed controller (II) adaptive controller (III) optimal controller.

100 time instants as $K_i x(k) + \hat{\theta}(k) N_i(x(k)) + r(k)$.

The simulation results for Problem 3 are presented in Fig. 5. The mechanism of the operation of the two-level controller as well as the explanations of the different trajectories are the same as those for Fig. 1 (Problem 1). In Fig. 5(a) the response $x_3(t)$ is shown when fault 2 occurs. The response resulting from a fixed non-optimal controller (denoted by (I)) is seen to be very oscillatory. The response (III) when an optimal controller is used, is seen to have a large overshoot (i.e., low damping) but settles down to a constant value in a relatively short time. The use of an adaptive controller results in a response (denoted by (II)) which is much closer to the optimal response than that obtained by the non-adaptive stabilizing controller. Responses (II) and (III) and responses (I) and (II) are shown separately in Fig. 5(a) for the sake of clarity.

When fault 3 occurs, the response $x_1(t)$ obtained using the three different controllers is shown in the single Fig. 5(b). While (II) and (III) approach the same constant value asymptotically with time, the trajectories are seen to be distinctly different. The stabilizing controller yields a response which approaches a constant value different from that of responses (II) and (III).

Assumptions and Theoretical Questions

Numerous assumptions were made throughout the paper in the design of the different components of the multilevel controllers for Problem 1 and problems 2 and 3. In the following, we briefly review these assumptions and discuss both their theoretical and practical implications. Since Problem 1 deals with the case where the system is linear in each of the configurations, it is substantially simpler to indicate the difficulties involved in this case. Hence we shall discuss Problem 1 first before considering the more realistic cases of problems 2 and 3.

In Problem 1, methods for the determination of a state feedback gain matrix to stabilize a known plant are well known and standard design tools can be used in the choice of K_i ($i = 1, 2, 3$) to obtain satisfactory response. However, the existence of a single gain matrix to stabilize an entire configuration is not assured. Even if it were to exist, its determination can be quite involved and may depend upon recent results in robust control theory. In Problem 1, the sets in parameter space are made to be sufficiently small so that the problem of determining the stabilizing controllers is simplified. In practical problems, it may be necessary to divide a given configuration into subconfigurations to assure this property.

Pattern recognition techniques, for the determination of the configuration to which the plant belongs, are simple when the responses of plants in different configurations for the same reference input and initial conditions are significantly different. Configurations E_2 and E_3 were chosen to assure that this condition was satisfied. In practical fault detection problems, it is not unreasonable to expect this to be true. However, as the distance between the two sets of parameters decreases (e.g., when subconfigurations are used), this assumption may no longer be valid.

The adaptive controller used for Problem 1 is based on the solution of an algebraic Riccatti equation on-line using the estimated value of the B matrix. There is no guarantee that the overall system will be stable. However, if the feedback gain matrix is updated over sufficiently large intervals of time, such stability can be assured.

Problems 2 and 3 are substantially more complex than Problem 1. But such problems arise quite often in practice where nonlinearities in the system dynamics become significant when failures occur. Since determination of a stabilizing controller for a known nonlinear system is itself a difficult task, the determination of a fixed controller to stabilize a family of plants is truly a formidable one and in general analytically intractable. The same also applies to the design of adaptive controllers. Hence, we assume that substantial prior information concerning the plant in its various configurations is available and that the responses of the different elements in any configuration are not significantly different.

In all the problems considered, it is assumed that the pattern recognizer recognizes the correct configuration with 100% accuracy. This was indeed the case in all the simulation studies. However, in practice, this may not be the case and both pattern recognizer and controller may have to be updated on-line. While such solutions may not be practically acceptable, they give rise to interesting theoretical questions in switching adaptive systems.

Comments and Conclusions

Fault detection and control as applied to both linear and nonlinear systems are considered in this paper. The problem is to detect a failure when it occurs and use an appropriate controller to realize satisfactory performance. In Problem 1, the plant is linear in all its configurations. When a failure occurs, this is detected by a pattern recognizer and used to stabilize the system. Following this, an adaptive controller is used to improve its performance. In Problem 2, the plant is nonlinear in all its configurations. Unlike Problem 1, only a special case where all the configurations contain only one element is considered in this problem. Finally, in Problem 3, the on-line identification and control of the nonlinear plant, when each configuration contains an infinite number of nonlinear elements, is considered.

In problems 2 and 3, in view of the nonlinear nature of the plant and the controllers, very little theory currently exists to guide us in the design of a two level controller. The emphasis of the paper is on the use of multilayer neural networks in the identification of the plant in its different configurations, determination of suboptimal controllers for each configuration and pattern recognition of failures when they occur.

In problems 1 and 3, it is shown that improved performance can be obtained by using adaptive controllers in conjunction with fixed controllers designed off-line to stabilize the various configurations. If the nonlinearities are parametrized in more general ways (e.g., the parameters appearing nonlinearly), the problem becomes substantially more complex. Such problems are closer in spirit to those which arise in practice in reconfigurable control of aircraft systems. In light of the results obtained thus far in the control of nonlinear plants, the authors feel that the latter class of problems are also tractable in restricted cases. Work in this direction is currently in progress using real data from aircraft systems.

Acknowledgment

The authors thank K. Parthasarathy for many useful comments and discussions.

References

[1] Kumpati S. Narendra and Kannan Parthasarathy, "Identification and control of dynamical systems using neural networks", *IEEE Trans. Neural Networks,* vol. 1, pp. 4–27, Mar. 1990.

[2] Kumpati S. Narendra and Shiva S. Tripathy, "Identification and optimization of aircraft dynamics", *J. Aircraft,* vol. 10, no. 4, pp. 193–199, Apr. 1973.

[3] Kumpati S. Narendra and Snehasis Mukhopadhyay, "Multilevel control of dynamical systems using neural networks", Tech. Rep. 9102, Feb. 1991, Center for Systems Science, Electrical Engineering, Yale University, New Haven, CT 06520.

[4] D. L. Kleinman, "On an iterative technique for Riccatti equation computations", *IEEE Trans. Automatic Control,* vol. AC-13, pp. 114–115, Feb. 1968.

Part 2
Neuronal Morphology of Biological Processes

Understanding the morphology of biological neural systems is a challenge, and scientists from many disciplines are thriving on this challenge.

THE carbon-based cognitive faculty—the brain—is a mysterious machine with a very complex neuronal morphology. All our actions and emotions are controlled by this mysterious organ. We perceive, think, see, and learn. We compose and recite poems and play musical instruments. We devise mechanisms for solving complex problems, we think about what we know, and we investigate new things. We enjoy the beauty of snow peaks and that of the blue sky. Some events make us happy and we laugh, others make us unhappy. Intuition tells us that the neuronal morphology of organs doing all these wonderful things must be very complex. Indeed, this brain is too complex to understand. It is wrong to call it a computer because, unlike a computer, it does things beyond simple numerical computations, such as cognition and perception. Nature has endowed the brain with a marvelous and a complex neuronal morphology that is beyond human comprehension. Yet we know that it is composed of a large number of nerve (neural) cells with a high degree of interconnectivity. There are over 10^{11} (one hundred billion) neural cells, and each neuron, on the average, receives information from about 10,000 neighboring neurons. Thus, there are typically over 10^{15} connections (synapses) in the brain. The anatomical morphology of these neurons and their connections are what make the brain so complex, yet it is very precise in conducting the various cognitive tasks.

The purpose of Part Two of this volume is to provide a broad view of the biological neuronal morphology that forms the basis for our neuro-control processes. Let us look at the neural mechanism in our own vision and control mechanisms. When we write and read these lines, the photonic energy emitting from these characters strikes the photoreceptors—125 million rods and 5 million cones—in each retina. Complex biochemical reactions in the photoreceptors change the photonic energy into equivalent electrical impulses. The task of the retina and the rest of the brain is not only to coordinate the function of our hands (in writing) and eyes (in reading), but also to think and extract useful cognitive information from these lines. It would be wonderful if we could explain this neuronal computing phenomenon in our retina and brain. In spite of tremendous progress in neurophysiology, our knowledge about biological neuronal computing is shrouded by ignorance. However, over the last decade or so, scientists and engineers have embarked on creating a computational neural machine.

In this part, we present a set of four articles. The first article, by C. F. Stevens, provides a tutorial overview of the biological neurons and their information-processing capabilities in our sensory and control processes. Because synapses in biological neural processes are the sites of knowledge storage and information transfer from one neuron to another, the synaptic mechanism has been the center of interest for neurophysiologists for a long time. With modern experimental techniques, our knowledge of synaptic functions has increased rapidly and many aspects of neural information transfer are well understood. Readers will find a brief description of synaptic neuronal morphology in this article.

The desire to build artificial neural machines that can perform cognitive tasks as we humans do has long existed. The cognitive faculty, which provides us with the ability to speak, to perceive, to reason, to speculate, to learn, and to memorize, is highly developed in humans as opposed to other animals. Cognition has hardware (neuroscience) and software (neuropsychology) aspects, and neuro scientists and psychologists have expended great amounts of effort in the understanding of neural functions. In article (2.2), J. A. Anderson discusses some of the hardware aspects of our central nervous systems (CNS). He also presents in this article neural models for cognition with inherent natural constraints. The approach to cognitive computation involves a methodology to use these distributed parallel associative models for information processing. Models presented in this article have more similarities with natural logic found in human cognition than the formal logic found in mathematical books: "Socrates is a man, men are mortal, therefore Socrates is mortal."

The neuron is the basic hardware in our central nervous system. One of the aims of neurophysiology is the understanding of brain functions in terms of the properties of a single neuron. The conceptual gap between the functions of single neurons and those of masses of neurons is still very wide. In article (2.3), W. J. Freeman reviews the problems of how to identify and characterize empirically masses of neurons as dynamic entities that have priorities related to, but that are distinct from, those of the component neurons. The author emphasizes how little we really know about neural masses, and how rich the opportunities are for studying them.

It is probably true that studies of primitive nervous systems should focus on individual nerve cells (neurons) and their precise, genetically determined interactions with other neural cells. It is difficult to study higher brain functions such as cognition at an individual-cell level. Pattern recognition is, in some sense, a global process. It is most unlikely that the neural approaches that emphasize only local properties will provide much insight. Neural processes at a local level may give rise to random interactions, but viewed macroscopically, the same processes may provide a deterministic behavior. The neural activity of any complexity is dependent upon the interaction of excitatory and inhibitory neural cells. In the last article of this part, (2.4), H. R. Wilson and J. D. Cowan provide a dynamic neural model that represents the interactions between two distinct classes of localized neural populations: *excitatory* and *inhibitory.* This model exhibits some interesting neural phenomena such as multineural hysteresis and limit cycle oscillations.

This set of four articles represents a mere sample of articles taken from the vast literature that exists in the area of morphology of biological neural systems. Hopefully, this small sample of articles along with the accompanying reading list will provide a direction to readers for future work in the field. In the next part, we present a small sample of articles related to artificial neural systems.

Further Reading

[1] B. Pansky, D. J. Allen, and G. C. Budd, *Review of Neuroscience,* II Edition. New York: Macmillan, 1992.

[2] I. Kuperfmann, "Learning," in E. Kandel and J. Schwartz, Eds., *Principles of Neural Science.* New York: Elsevier, 1985.

[3] R. Hecht-Nielsen, "Neurocomputing: Picking the human brain," *IEEE Spectrum,* vol. 25, pp. 36–41, 1988.

[4] F. Rosenblatt, "The Perceptron: A probabilistic model for information storage and organization in the brain," *Psychol. Rev.,* vol. 65, pp. 386–408, 1959.

[5] M. L. Minsky and S. A. Papert, *Perceptrons.* Cambridge, MA: MIT Press, 1969.

[6] M. Fujita, "Simulation of adaptive modification of the vestibule-ocular reflex with an adaptive filter model of the cerebellum," *Biol. Cybern.,* vol. 45, pp. 207–214, 1982.

[7] M. Ito, "Neurophysiological aspects of the cerebellar motor control system," *Int. J. Neurol.,* vol. 7, no. 2.3-4, pp. 162–176, 1970.

[8] M. Fujita, "Adaptive filter model of the cerebellum," *Biol. Cybern.,* vol. 45, pp. 195–206, 1982.

[9] D. S. Melkonian, "Mathematical theory of chemical synaptic transmission," *Biol. Cybern.,* vol. 62, pp. 539–548, 1990.

[10] S. I. Amari, "Neural theory of association and concept-formation," *Biol. Cybern.,* vol. 26, pp. 175–185, 1977.

[11] R. P. Lipmann, "An introduction to computing with neural nets," *ASSP Mag.,* vol. 4, no. 2, pp. 4–22, 1987.

[12] B. Widrow and M. E. Hoff, "Adaptive switching circuits," in IREWESCON *Convention Rec.,* IRS, New York, 1960.

[13] J. J. Hopfield, "Artificial neural networks are coming," *IEEE Expert,* An Interview by W. Myers, pp. 3–6, April 1990.

[14] R. C. Eberhart and R. W. Dobbins, "Early neural network development history: The age of Camelot," *IEEE Eng. Med. Biol. Mag.,* vol. 9, no. 3, pp. 15–18, Sept. 1990.

[15] D. E. Rumelhart, G. E. Hinton, and R. J. Williams, "Learning representations by error propagation," in *Parallel Distributed Processing: Explorations in the Microstructure of Cognition,* Vol. 1. D. E. Rumelhart and J. L. McClelland, Eds., Cambridge, MA: MIT Press, 1986.

[16] D. S Touretzky, Ed., *Advances in Neural Information Processing Systems,* San Mateo, CA: Morgan Kaufmann Publishers, 1992.

[17] J. Diederich, Ed., *Artificial Neural Networks: Concept Learning,* Los Alamitos, CA: IEEE Computer Society Press, 1990.

[18] D. E. Rumelhart and J. L. McClelland, Eds., *Paràllel Distributed Processing: Explorations in the Microstructures of Cognition.* Cambridge, MA: MIT Press, Vol. 1, 1986.

[19] G. A. Carpenter and S. Grossberg, "A Massively Parallel Architecture for a Self-Organizing Neural Pattern Recognition Machine," *Comput. Vision, Graphics, Image Process.,* vol. 37, pp. 54–115, 1987.

[20] T. Kohonen, "A New Model for Randomly Organized Associative Memory," *Int. J. Neurosci.,* vol. 5, pp. 27–29, 1973.

[21] K. Fukushima, S. Miyake, and T. Ito, "Neocognitron: A neural network model for a mechanism of visual pattern recognition," *IEEE Trans. Syst., Man, Cybern.,* vol. 13, no. 5, pp. 826–834, Sept/Oct. 1983.

[22] E. Harth, "Order and chaos in neural systems: An approach to the dynamics of higher brain functions," *IEEE Trans. Syst., Man, Cybern.,* vol. 13, no. 5, pp. 49–60, Sept/Oct 1983.

[23] B. C. Cragg and H. N. V. Temperley, "Memory: The analogy with ferro-magnetic hysteresis," *Brain,* vol. 78, pp. 304–316, 1955.

[24] J. Jondies, D. E. Irwin, and S. Yantis, "Integrating visual information from successive fixations," *Science,* vol. 25, pp. 192–194, 1982.

[25] P. A. Anninos, B. Beek, T. J. Csermel, E. E. Harth, and G. Pertile, "Dynamics of Neural Structures," *J. Theoret. Biol.,* vol. 26, pp. 121–148, 1970.

[26] G. W. Hoffmann, "Neuron with hysteresis?" in *Computer Simulation in Brain Science.* R. Cotterill, Ed., Cambridge: Cambridge University Press, pp. 74–87, 1988.

Article 2.1

The Neuron

It is the individual nerve cell, the building block of the brain. It transmits nerve impulses over a single long fiber (the axon) and receives them over numerous short fibers (the dendrites)

by Charles F. Stevens

Neurons, or nerve cells, are the building blocks of the brain. Although they have the same genes, the same general organization and the same biochemical apparatus as other cells, they also have unique features that make the brain function in a very different way from, say, the liver. The important specializations of the neuron include a distinctive cell shape, an outer membrane capable of generating nerve impulses, and a unique structure, the synapse, for transferring information from one neuron to the next.

The human brain is thought to consist of 10^{11} neurons, about the same number as the stars in our galaxy. No two neurons are identical in form. Nevertheless, their forms generally fall into only a few broad categories, and most neurons share certain structural features that make it possible to distinguish three regions of the cell: the cell body, the dendrites and the axon. The cell body contains the nucleus of the neuron and the biochemical machinery for synthesizing enzymes and other molecules essential to the life of the cell. Usually the cell body is roughly spherical or pyramid-shaped. The dendrites are delicate tubelike extensions that tend to branch repeatedly and form a bushy tree around the cell body. They provide the main physical surface on which the neuron receives incoming signals. The axon extends away from the cell body and provides the pathway over which signals can travel from the cell body for long distances to other parts of the brain and the nervous system. The axon differs from the dendrites both in structure and in the properties of its outer membrane. Most axons are longer and thinner than dendrites and exhibit a different branching pattern: whereas the branches of dendrites tend to cluster near the cell body, the branches of axons tend to arise at the end of the fiber where the axon communicates with other neurons.

The functioning of the brain depends on the flow of information through elaborate circuits consisting of networks of neurons. Information is transferred from one cell to another at specialized points of contact: the synapses. A typical neuron may have anywhere from 1,000 to 10,000 synapses and may receive information from something like 1,000 other neurons. Although synapses are most often made between the axon of one cell and the dendrite of another, there are other kinds of synaptic junction: between axon and axon, between dendrite and dendrite and between axon and cell body.

At a synapse the axon usually enlarges to form a terminal button, which is the information-delivering part of the junction. The terminal button contains tiny spherical structures called synaptic vesicles, each of which can hold several thousand molecules of chemical transmitter. On the arrival of a nerve impulse at the terminal button, some of the vesicles discharge their contents into the narrow cleft that separates the button from the membrane of another cell's dendrite, which is designed to receive the chemical message. Hence information is relayed from one neuron to another by means of a transmitter. The "firing" of a neuron—the generation of nerve impulses—reflects the activation of hundreds of synapses by impinging neurons. Some synapses are excitatory in that they tend to promote firing, whereas others are inhibitory and so are capable of canceling signals that otherwise would excite a neuron to fire.

Although neurons are the building blocks of the brain, they are not the only kind of cell in it. For example, oxygen and nutrients are supplied by a dense network of blood vessels. There is also a need for connective tissue, particularly at the surface of the brain. A major class of cells in the central nervous system is the glial cells, or glia. The glia occupy essentially all the space in the nervous system not taken up by the neurons themselves. Although the function of the glia is not fully understood, they provide structural and metabolic support for the delicate meshwork of the neurons.

One other kind of cell, the Schwann cell, is ubiquitous in the nervous system. All axons appear to be jacketed by Schwann cells. In some cases the Schwann cells simply enclose the axon in a thin layer. In many cases, however, the Schwann cell wraps itself around the axon in the course of embryonic development, giving rise to the multiple dense layers of insulation known as myelin. The myelin sheath is interrupted every millimeter or so along the axon by narrow gaps called the nodes of Ranvier. In axons that are sheathed in this way the nerve impulse travels by jumping from node to node, where the extracellular fluid can make direct contact with the cell membrane. The myelin sheath seems to have evolved as a means of conserving the neuron's metabolic energy. In general myelinated nerve fibers conduct nerve impulses faster than unmyelinated fibers.

Neurons can work as they do because their outer membranes have special

NEURON FROM A CAT'S VISUAL CORTEX has been labeled in the photomicrograph on the opposite page by injection with the enzyme horseradish peroxidase. The cell bodies in the background are counterstained with a magenta dye. All the fibers extending from the cell body are dendrites, which receive information from other neurons. The fiber that transmits information, the axon, is much finer and not readily visible at this magnification. The thickest fiber, extending vertically upward, is known as the apical dendrite, only a small portion of which falls within this section. At this magnification (about 500 diameters) the complete apical dendrite would be about 75 centimeters long. (It can be traced through adjacent sections.) The activity of this particular cell was recorded in the living animal and was found to respond optimally to a light-dark border rotated about 60 degrees from the vertical. The neuron is classified as a pyramidal cell because of its form. It is one of two major types in cortex of mammals. Micrograph was made by Charles Gilbert and Torsten N. Wiesel of Harvard Medical School.

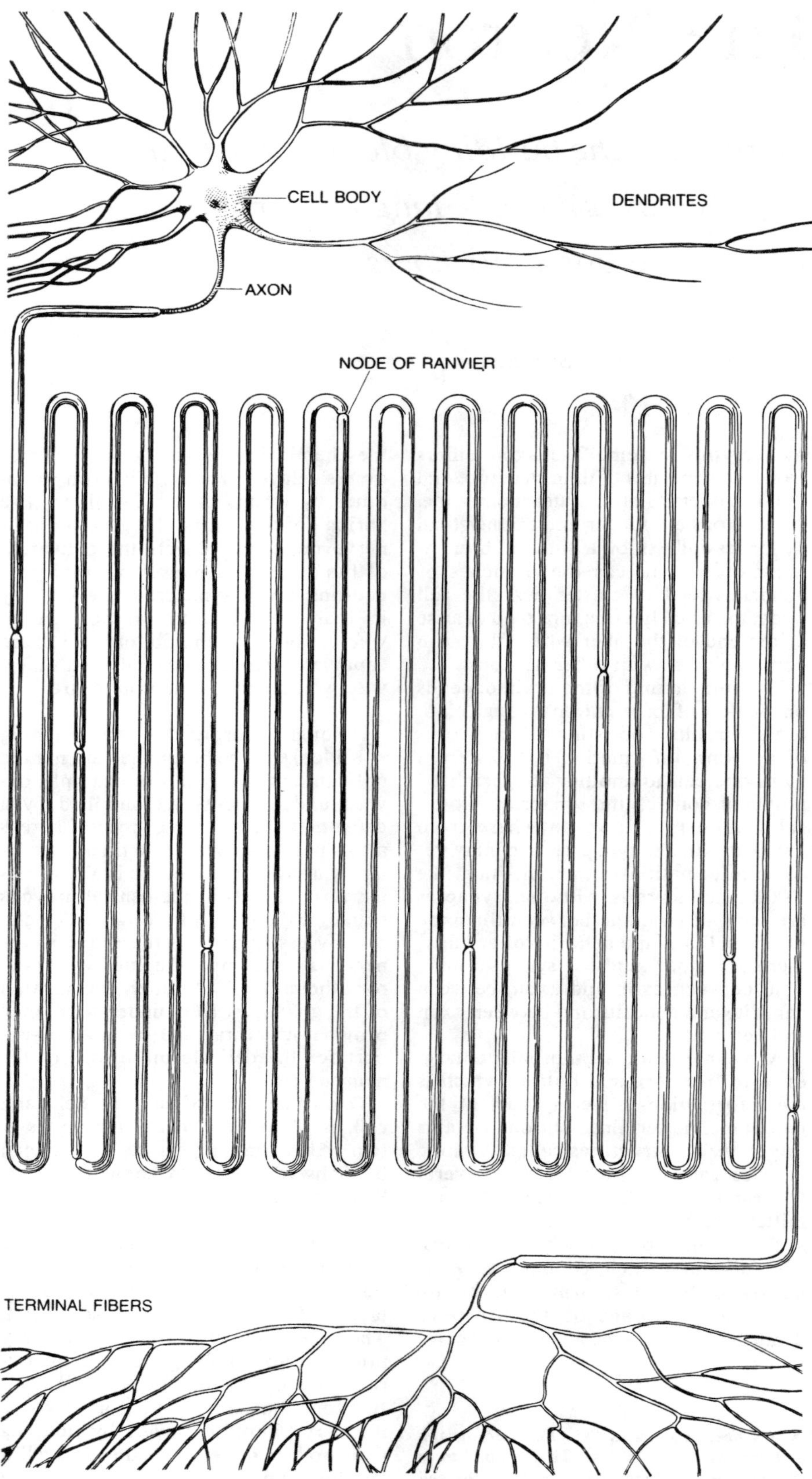

TYPICAL NEURON of a vertebrate animal can carry nerve impulses for a considerable distance. The neuron depicted here, with its various parts drawn to scale, is enlarged 250 times. The nerve impulses originate in the cell body and are propagated along the axon, which may have one or more branches. This axon, which is folded for diagrammatic purposes, would be a centimeter long at actual size. Some axons are more than a meter long. The axon's terminal branches form synapses with as many as 1,000 other neurons. Most synapses join the axon terminals of one neuron with the dendrites forming a "tree" around the cell body of another neuron. Thus the dendrites surrounding the neuron in the diagram might receive incoming signals from tens, hundreds or even thousands of other neurons. Many axons, such as this one, are insulated by a myelin sheath interrupted at intervals by the regions known as nodes of Ranvier.

properties. Along the axon the membrane is specialized to propagate an electrical impulse. At the terminal of the axon the membrane releases transmitters, and on the dendrites it reponds to transmitters. In addition the membrane mediates the recognition of other cells in embryonic development, so that each cell finds its proper place in the network of 10^{11} cells. Much recent investigation therefore focuses on the membrane properties responsible for the nerve impulse, for synaptic transmission, for cell-cell recognition and for structural contacts between cells.

The neuron membrane, like the outer membrane of all cells, is about five nanometers thick and consists of two layers of lipid molecules arranged with their hydrophilic ends pointing toward the water on the inside and outside of the cell and with their hydrophobic ends pointing away from the water to form the interior of the membrane. The lipid parts of the membrane are about the same for all kinds of cells. What makes one cell membrane different from another are various specific proteins that are associated with the membrane in one way or another. Proteins that are actually embedded in the lipid bilayer are termed intrinsic proteins. Other proteins, the peripheral membrane proteins, are attached to the membrane surface but do not form an integral part of its structure. Because the membrane lipid is fluid even the intrinsic proteins are often free to move by diffusion from place to place. In some instances, however, the proteins are firmly fastened down by a substructure.

The membrane proteins of all cells fall into five classes: pumps, channels, receptors, enzymes and structural proteins. Pumps expend metabolic energy to move ions and other molecules against concentration gradients in order to maintain appropriate concentrations of these molecules within the cell. Because charged molecules do not pass through the lipid bilayer itself cells have evolved channel proteins that provide selective pathways through which specific ions can diffuse. Cell membranes must recognize and attach many types of molecules. Receptor proteins fulfill these functions by providing binding sites with great specificity and high affinity. Enzymes are placed in or on the membrane to facilitate chemical reactions at the membrane surface. Finally, structural proteins both interconnect cells to form organs and help to maintain subcellular structure. These five classes of membrane proteins are not necessarily mutually exclusive. For example, a particular protein might simultaneously be a receptor, an enzyme and a pump.

Membrane proteins are the key to understanding neuron function and therefore brain function. Because they play such a central role in modern views of the neuron, I shall organize my discussion around a description of an ion pump, various types of channel and some other proteins that taken together endow neurons with their unique properties. The general idea will be to summarize the important characteristics of the membrane proteins and to explain how these characteristics account for the nerve impulse and other complex features of neuron function.

Like all cells the neuron is able to maintain within itself a fluid whose composition differs markedly from that of the fluid outside it. The difference is particularly striking with regard to the concentration of the ions of sodium and potassium. The external medium is about 10 times richer in sodium than the internal one, and the internal medium is about 10 times richer in potassium than the external one. Both sodium and potassium leak through pores in the cell membrane, so that a pump must operate continuously to exchange sodium ions that have entered the cell for potassium ions outside it. The pumping is accomplished by an intrinsic membrane protein called the sodium-potassium adenosine triphosphatase pump, or more often simply the sodium pump.

The protein molecule (or complex of protein subunits) of the sodium pump has a molecular weight of about 275,000 daltons and measures roughly six by eight nanometers, or slightly more than the thickness of the cell membrane. Each sodium pump can harness the energy stored in the phosphate bond of adenosine triphosphate (ATP) to exchange three sodium ions on the inside of the cell for two potassium ions on the outside. Operating at the maximum rate, each pump can transport across the membrane some 200 sodium ions and 130 potassium ions per second. The actual rate, however, is adjusted to meet the needs of the cell. Most neurons have between 100 and 200 sodium pumps per square micrometer of membrane surface, but in some parts of their surface the density is as much as 10 times higher. A typical small neuron has perhaps a million sodium pumps with a capacity to move about 200 million sodium ions per second. It is the transmembrane gradients of sodium and potassium ions that enable the neuron to propagate nerve impulses.

Membrane proteins that serve as channels are essential for many aspects of neuron function, particularly for the nerve impulse and synaptic transmission. As an introduction to the role played by channels in the electrical activity of the brain I shall briefly describe the mechanism of the nerve impulse and then return to a more systematic survey of channel properties.

Since the concentration of sodium and potassium ions on one side of the cell membrane differs from that on the

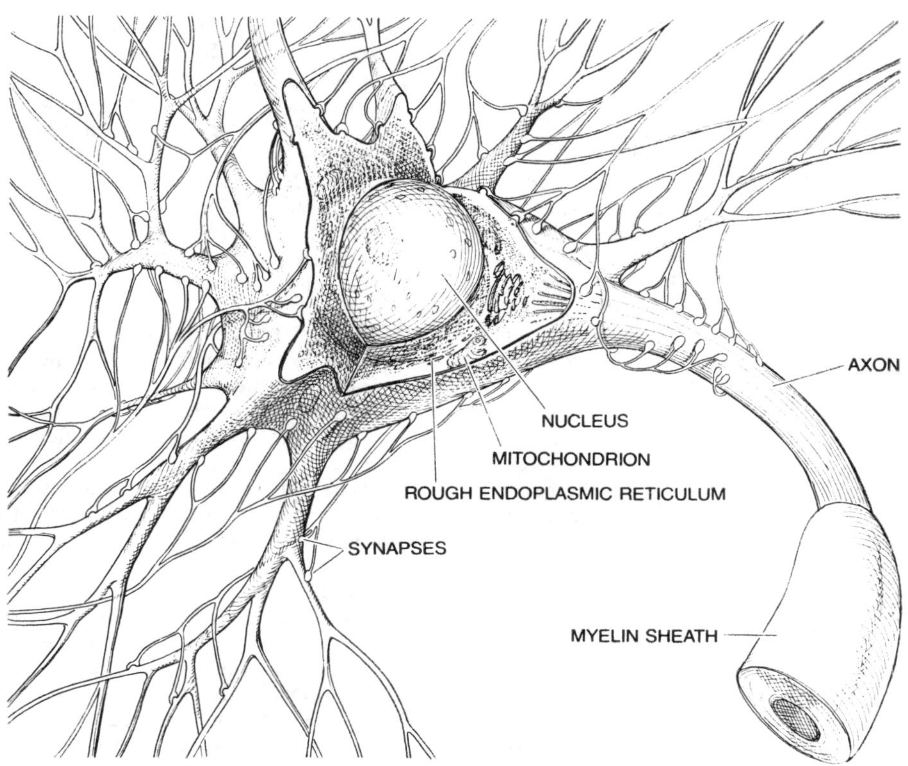

CELL BODY OF A NEURON incorporates the genetic material and complex metabolic apparatus common to all cells. Unlike most other cells, however, neurons do not divide after embryonic development; an organism's original supply must serve a lifetime. Projecting from the cell body are several dendrites and a single axon. The cell body and dendrites are covered by synapses, knoblike structures where information is received from other neurons. Mitochondria provide the cell with energy. Proteins are synthesized on the endoplasmic reticulum. A transport system moves proteins and other substances from cell body to sites where they are needed.

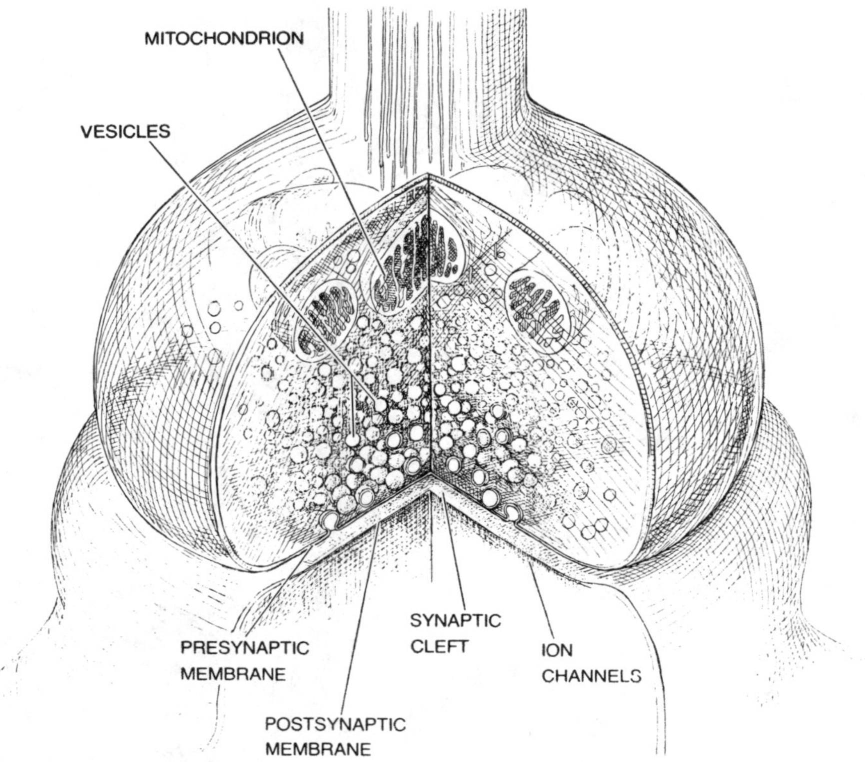

SYNAPSE is the relay point where information is conveyed by chemical transmitters from neuron to neuron. A synapse consists of two parts: the knoblike tip of an axon terminal and the receptor region on the surface of another neuron. The membranes are separated by a synaptic cleft some 200 nanometers across. Molecules of chemical transmitter, stored in vesicles in the axon terminal, are released into the cleft by arriving nerve impulses. Transmitter changes electrical state of the receiving neuron, making it either more likely or less likely to fire an impulse.

other side, the interior of the axon is about 70 millivolts negative with respect to the exterior. In their classic studies of nerve-impulse transmission in the giant axon of the squid a quarter of a century ago, A. L. Hodgkin, A. F. Huxley and Bernhard Katz of Britain demonstrated that the propagation of the nerve impulse coincides with sudden changes in the permeability of the axon membrane to sodium and potassium ions. When a nerve impulse starts at the origin of the axon, having been triggered in most cases by the cell body in response to dendritic synapses, the voltage difference across the axon membrane is locally lowered. Immediately ahead of the electrically altered region (in the direction in which the nerve impulse is propagated) channels in the membrane open and let sodium ions pour into the axon.

The process is self-reinforcing: the flow of sodium ions through the membrane opens more channels and makes it easier for other ions to follow. The sodium ions that enter change the internal potential of the membrane from negative to positive. Soon after the sodium channels open they close, and another group of channels open that let potassium ions flow out. This outflow restores the voltage inside the axon to its resting value of −70 millivolts. The sharp positive and then negative charge, which shows up as a "spike" on an oscilloscope, is known as the action potential and is the electrical manifestation of the nerve impulse. The wave of voltage sweeps along until it reaches the end of the axon much as a flame travels along the fuse of a firecracker.

This brief description of the nerve impulse illustrates the importance of channels for the electrical activity of neurons and underscores two fundamental properties of channels: selectivity and gating. I shall discuss these two properties in turn. Channels are selectively permeable and selectivities vary widely. For example, one type of channel lets sodium ions pass through and largely excludes potassium ions, whereas another type of channel does the reverse. The selectivity, however, is seldom absolute. One type of channel that is fairly nonselective allows the passage of about 85 sodium ions for every 100 potassium ions; another more selective type passes only about seven sodium ions for every 100 potassium ions. The first type, known as the acetylcholine-activated channel, has a pore about .8 nanometer in diameter that is filled with water. The second type, known as the potassium channel, has a much smaller opening and contains less water.

The sodium ion is about 30 percent smaller than the potassium ion. The exact molecular structure that enables the larger ion to pass through the cell membrane more readily than the smaller one is not known. The general principles that underlie the discrimination, however, are understood. They involve interactions between ions and parts of the channel structure in conjunction with a particular ordering of water molecules within the pore.

The gating mechanism that regulates the opening and closing of membrane channels takes two main forms. One type of channel, mentioned above in the description of the nerve impulse, opens and closes in response to voltage differences across the cell membrane; it is therefore said to be voltage-gated. A second type of channel is chemically gated. Such channels respond only slightly if at all to voltage changes but open when a particular molecule—a transmitter—binds to a receptor region

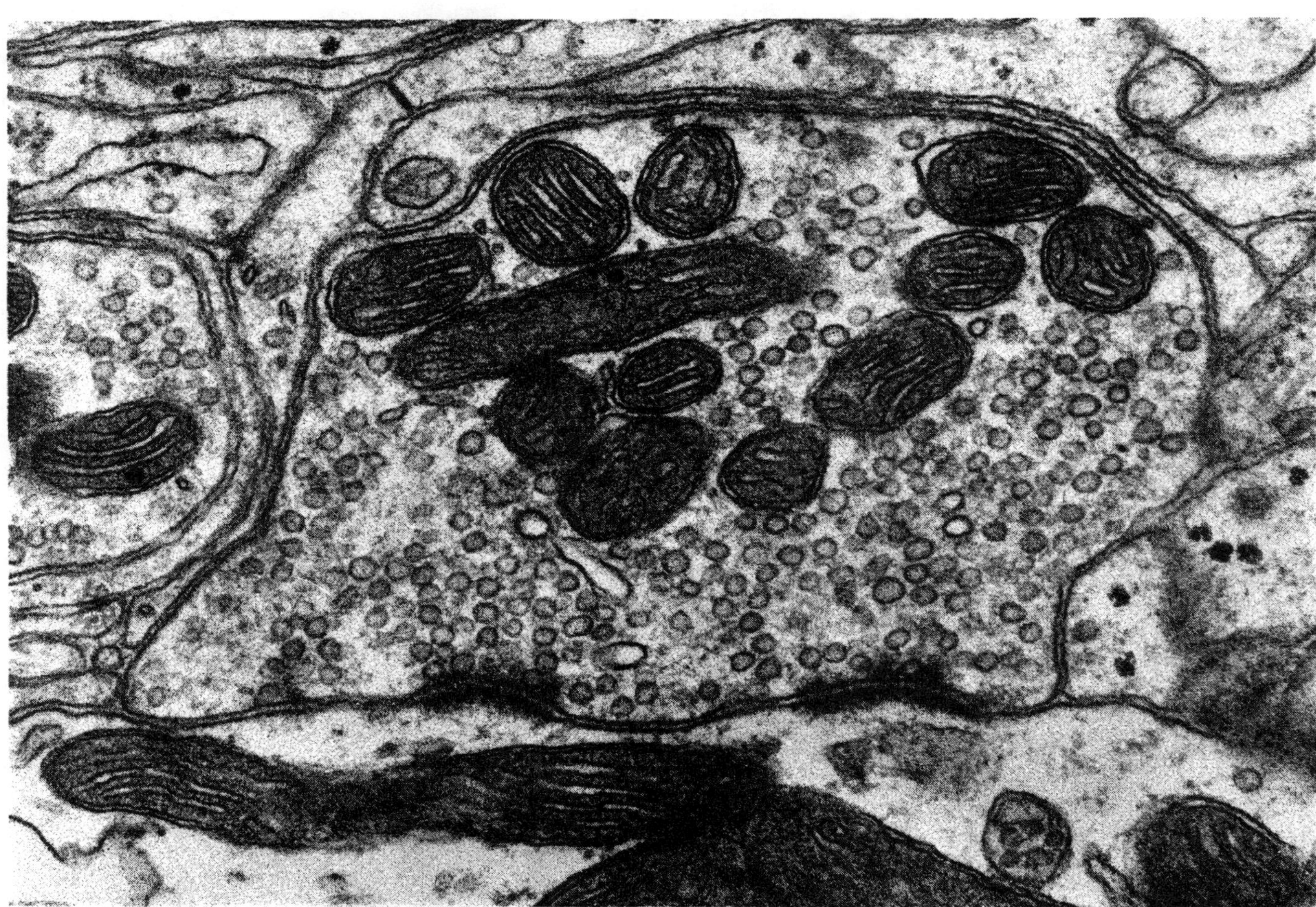

SYNAPTIC TERMINAL occupies most of this electron micrograph made by John E. Heuser of the University of California School of Medicine in San Francisco and Thomas S. Reese of the National Institutes of Health. The cleft separating the presynaptic membrane from the postsynaptic one undulates across the lower part of the picture. The large dark structures are mitochondria. The many round bodies are vesicles that hold transmitter. The fuzzy dark thickenings along the cleft are thought to be principal sites of transmitter release.

on the channel protein. Chemically gated channels are found in the receptive membranes of synapses and are responsible for translating the chemical signals produced by axon terminals into ion permeability changes during synaptic transmission. It is customary to name chemically gated channels according to their normal transmitter. Hence one speaks of acetylcholine-activated channels or GABA-activated channels. (GABA is gamma-aminobutyric acid.) Voltage-gated channels are generally named for the ion that passes through the channel most readily.

Proteins commonly change their shape as they function. Such alterations in shape, known as conformational changes, are dramatic for the contractile proteins responsible for cell motion, but they are no less important in many enzymes and other proteins. Conformational changes in channel proteins form the basis for gating as they serve to open and close the channel by slight movements of critically placed portions of the molecule that unblock and block the pore.

When either voltage-gated or chemically gated channels open and allow ions to pass, one can measure the resulting electric current. Quite recently it has become possible in a few instances to record the current flowing through a single channel, so that the opening and closing can be directly detected. One finds that the length of time a channel stays open varies randomly because the opening and closing of the channel represents a change in the conformation of the protein molecule embedded in the membrane. The random nature of the gating process arises from the haphazard collision of water molecules and other molecules with the structural elements of the channel.

In addition to ion pumps and channels neurons depend on other classes of membrane proteins for carrying out essential nervous-system functions. One of the important proteins is the enzyme adenylate cyclase, which helps to regulate the intracellular substance cyclic adenosine monophosphate (cyclic AMP). Cyclic nucleotides such as cyclic AMP take part in cell functions whose mechanisms are not yet understood in detail. The membrane enzyme adenylate cyclase appears to have two chief subunits, one catalytic and the other regulatory. The catalytic subunit promotes the formation of cyclic AMP. Various regulatory subunits, which are thought to be physically distinct from the catalytic one, can bind specific molecules (including transmitters that open and close channels) in order to control intracellular levels of cyclic AMP. The various types of regulatory subunit are named according to the molecule that normally binds to them; one, for example, is called serotonin-activated ade-

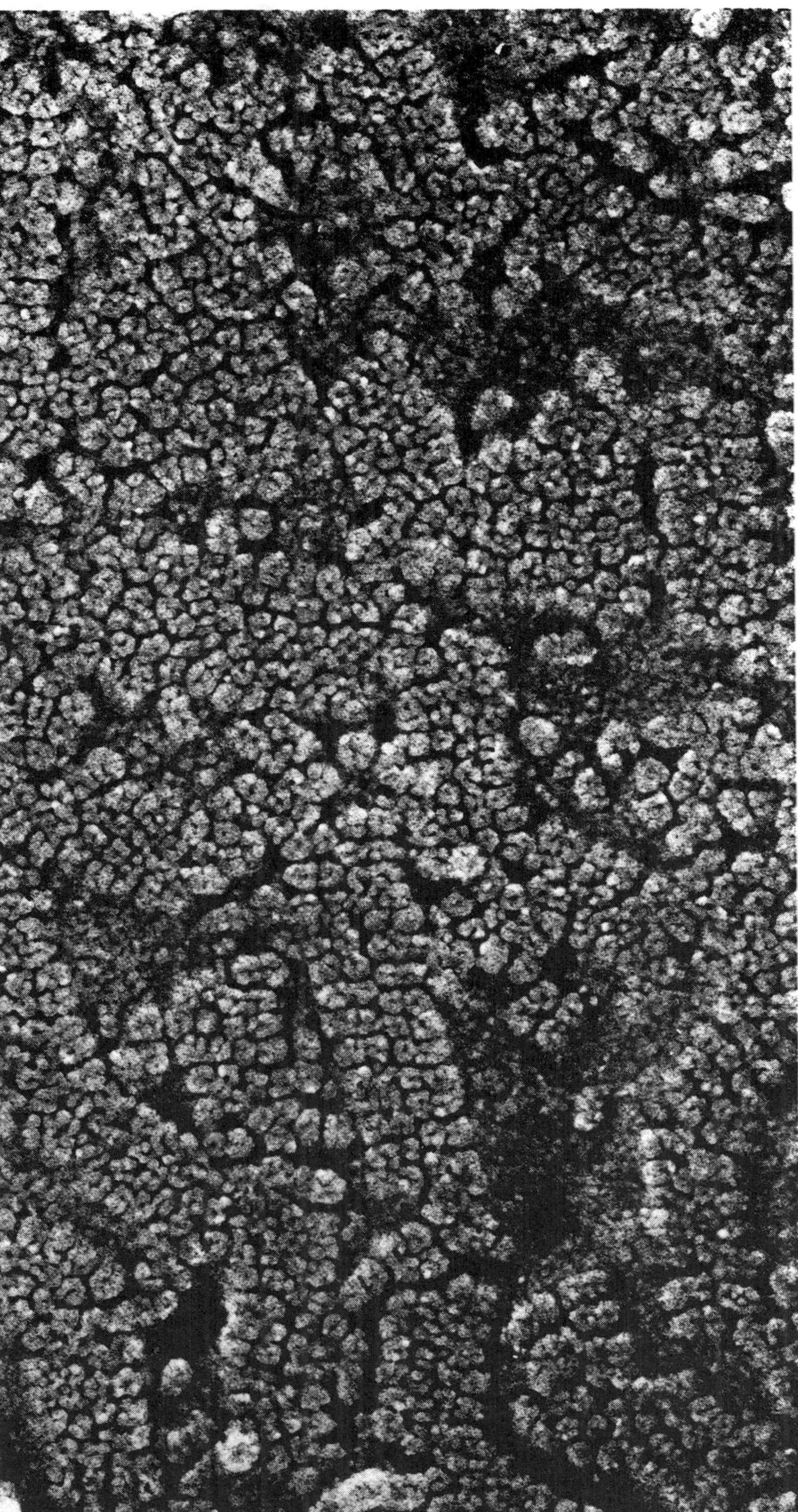

ACETYLCHOLINE-ACTIVATED CHANNELS are densely packed in the postsynaptic membrane of a cell in the electric organ of a torpedo, a fish that can administer an electric shock. This electron micrograph shows the platinum-plated replica of a membrane that had been frozen and etched. The size of the platinum particles limits the resolution to features larger than about two nanometers. According to recent evidence the channel protein molecule, which measures 8.5 nanometers across, consists of five subunits surrounding a channel whose narrowest dimension is .8 nanometer. The micrograph was made by Heuser and S. R. Salpeter.

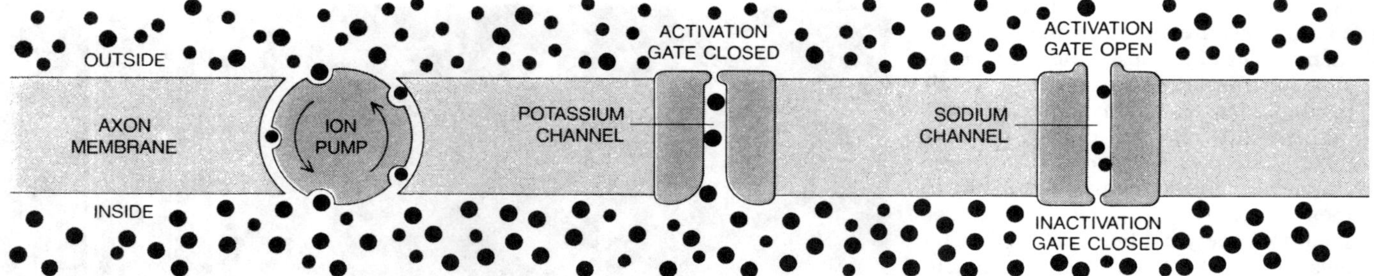

AXON MEMBRANE separates fluids that differ greatly in their content of sodium ions (*lighter dots*) and potassium ions (*black dots*). The exterior fluid is about 10 times richer in sodium ions than in potassium ions; in the interior fluid the ratio is the reverse. The membrane is penetrated by proteins that act as selective channels for preferentially passing either sodium or potassium ions. In the resting state, when no nerve impulse is being transmitted, the two types of channel are closed and an ion pump maintains the ionic disequilibrium by pumping out sodium ions in exchange for potassium ions. The interior of the axon is normally about 70 millivolts negative with respect to the exterior. If this voltage difference is reduced by the arrival of a nerve impulse, the sodium channel opens, allowing sodium ions to flow into the axon. An instant later the sodium channel closes and the potassium channel opens, allowing an outflow of potassium ions. The sequential opening and closing of the two kinds of channel effects the propagation of the nerve impulse, which is illustrated below.

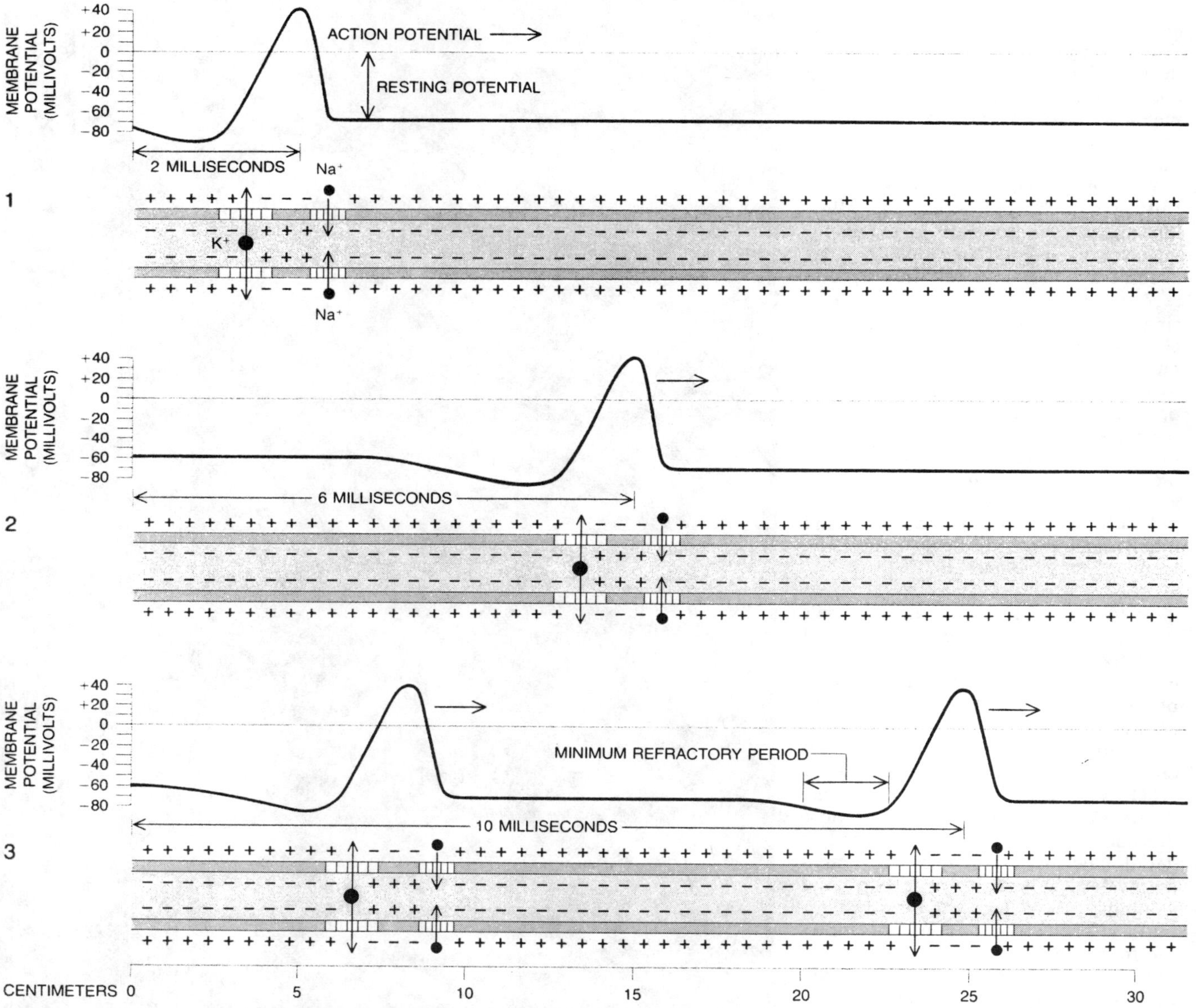

PROPAGATION OF NERVE IMPULSE along the axon coincides with a localized inflow of sodium ions (Na^+) followed by an outflow of potassium ions (K^+) through channels that are "gated," or controlled, by voltage changes across the axon membrane. The electrical event that sends a nerve impulse traveling down the axon normally originates in the cell body. The impulse begins with a slight depolarization, or reduction in the negative potential, across the membrane of the axon where it leaves the cell body. The slight voltage shift opens some of the sodium channels, shifting the voltage still further. The inflow of sodium ions accelerates until the inner surface of the membrane is locally positive. The voltage reversal closes the sodium channel and opens the potassium channel. The outflow of potassium ions quickly restores the negative potential. The voltage reversal, known as the action potential, propagates itself down the axon (*1, 2*). After a brief refractory period a second impulse can follow (*3*). The impulse-propagation speed is that measured in the giant axon of the squid.

nylate cyclase. Adenylate cyclase and related membrane enzymes are known to serve a number of regulatory functions in neurons, and the precise mechanisms of these actions are now under active investigation.

In the course of the embryonic development of the nervous system a cell must be able to recognize other cells so that the growth of each cell will proceed in the right direction and give rise to the right connections. The process of cell-cell recognition and the maintenance of the structure arrived at by such recognition depend on special classes of membrane proteins that are associated with unusual carbohydrates. The study of the protein-carbohydrate complexes associated with cell recognition is still at an early stage.

The intrinsic membrane proteins I have been describing are neither distributed uniformly over the cell surface nor all present in equal amounts in each neuron. The density and the type of protein are governed by the needs of the cell and differ among types of neuron and from one region of a neuron to another. Thus the density of channels of a particular type ranges from zero up to about 10,000 per square micrometer. Axons generally have no chemically gated channels, whereas in postsynaptic membranes the density of such channels is limited only by the packing of the channel molecules. Similarly, dendritic membranes typically have few voltage-gated channels, whereas in axon membranes the density can reach 1,000 channels per square micrometer in certain locations.

The intrinsic membrane proteins are synthesized primarily in the body of the neuron and are stored in the membrane in small vesicles. Neurons have a special transport system for moving such vesicles from their site of synthesis to their site of function. The transport system seems to move the vesicles along in small jumps with the aid of contractile proteins. On reaching their destination the proteins are inserted into the surface membrane, where they function until they are removed and degraded within the cell. Precisely how the cell decides where to put which membrane protein is not known. Equally unknown is the mechanism that regulates the synthesis, insertion and destruction of the membrane proteins. The metabolism of membrane proteins constitutes one of cell biology's central problems.

How do the properties of the various membrane proteins I have been discussing relate to neuron function? To approach this question let us now return to the nerve impulse and examine more closely the molecular properties that underlie its triggering and propagation. As we have seen, the interior of the neuron is about 70 millivolts negative with respect to the exterior. This "resting po-

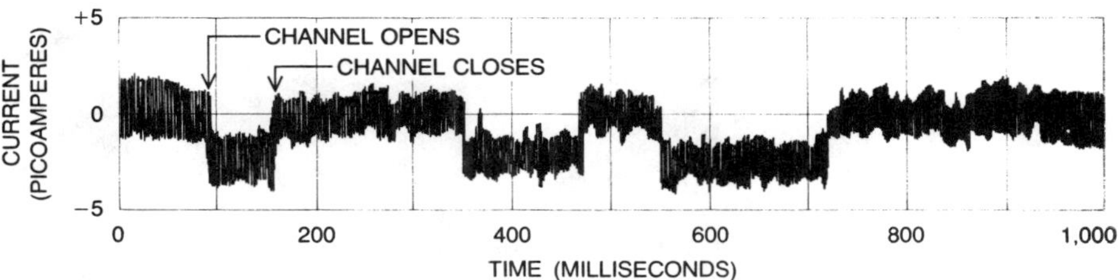

RESPONSE OF A SINGLE MEMBRANE CHANNEL to the transmitter compound acetylcholine is revealed by a recently developed technique that has been applied by Erwin Neher and Joseph H. Steinbach of the Yale University School of Medicine. Acetylcholine-activated channels, which are present in postsynaptic membranes, allow the passage of roughly equal numbers of sodium and potassium ions. The record shows the flow of current through a single channel in the postsynaptic membrane of a frog muscle activated by the compound suberyldicholine, which mimics action of acetylcholine but keeps channels open longer. Experiment shows that channels open on an all-or-none basis and stay open for random lengths of time.

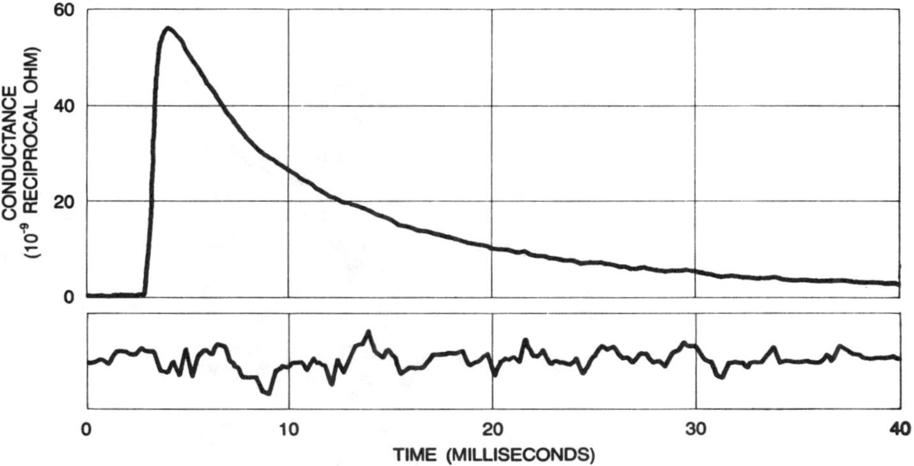

SODIUM CHANNELS IN AN AXON also operate in simple open-or-shut manner as well as independently of one another, according to investigations conducted by Frederick J. Sigworth of the Yale University School of Medicine. During the propagation of a nerve impulse about 10,000 channels normally open in a myelin-free region of the axon membrane, namely a node of Ranvier. The upper trace depicts the sodium permeability at such a node as a function of time. The lower trace, recorded at a 12-fold amplification of the upper one, shows fluctuations in permeability around the average due to the random opening and closing of channels.

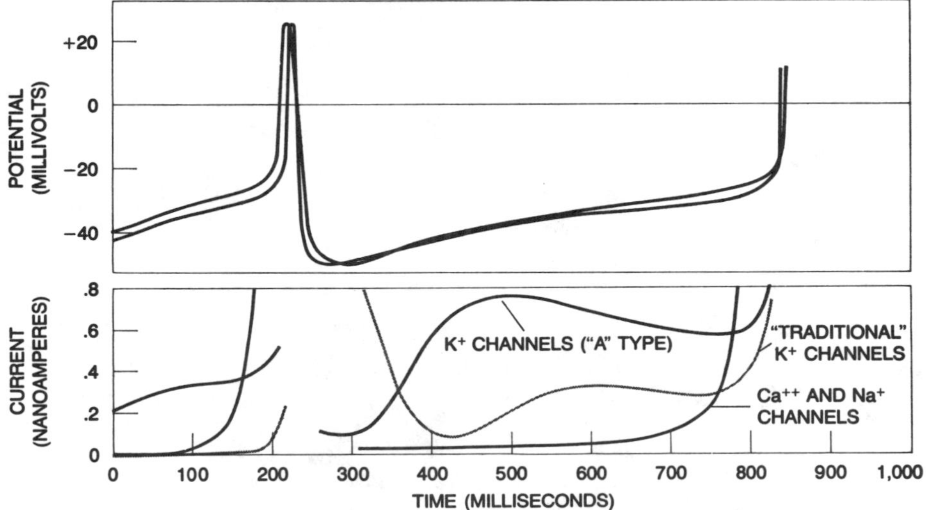

NERVE IMPULSES IN BODIES OF NEURONS require the coordinated opening and closing of five types of channel permeable to various kinds of ion (sodium, potassium or calcium). The contribution of the different channels to the nerve impulse can be represented by simultaneous nonlinear differential equations. The upper pair of curves represent an actual recording of voltage changes as a function of time in the body of a neuron *(dark)* and changes computed from equations *(light)*. The lower curves depict the current carried by the principal types of channel as a function of time. A complicated interaction of channel types is required to achieve a train of nerve impulses. The study on which curves are based was carried out by John A. Connor at the University of Illinois and by the author at the Yale University School of Medicine.

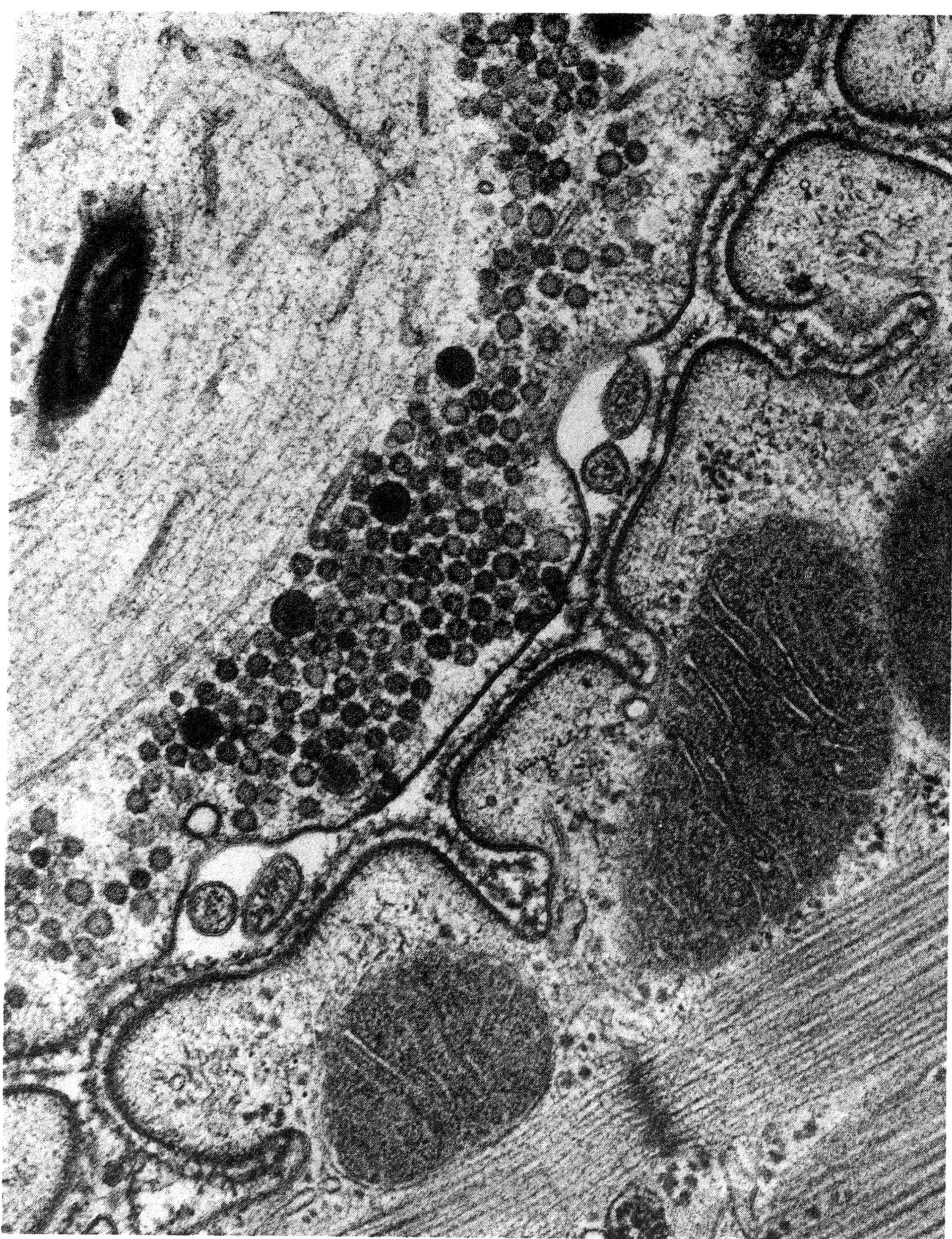

FROG NEUROMUSCULAR JUNCTION appears in this electron micrograph made by Heuser. The synaptic cleft separates the axon at the upper left from the muscle cell at the lower right. Synaptic vesicles cluster along the presynaptic membrane, with two synaptic contacts visible near the center. Postsynaptic membrane of the muscle cell exhibits a feature that is not seen at other synapses: the membrane forms postjunctional folds opposite each contact. Freeze-fracture replicas of presynaptic membrane are shown on opposite page.

tential" is a consequence of the ionic disequilibrium brought about by the sodium pump and by the presence in the cell membrane of a class of permanently open channels selectively permeable to potassium ions. The pump ejects sodium ions in exchange for potassium ions, making the inside of the cell about 10 times richer in potassium ions than the outside. The potassium channels in the membrane allow the potassium ions immediately adjacent to the membrane to flow outward quite freely. The permeability of the membrane to sodium ions is low in the resting condition, so that there is almost no counterflow of sodium ions from the exterior to the interior even though the external medium is tenfold richer in sodium ions than the internal medium. The potassium flow therefore gives rise to a net deficit of positive charges on the inner surface of the cell membrane and an excess of positive charges on the outer surface. The result is the voltage difference of 70 millivolts, with the interior being negative.

The propagation of the nerve impulse depends on the presence in the neuron membrane of voltage-gated sodium channels whose opening and closing is responsible for the action potential. What are the characteristics of these important channel molecules? Although the sodium channel has not yet been well characterized chemically, it is a protein with a molecular weight probably in the range of 250,000 to 300,000 daltons. The pore of the channel measures about .4 by .6 nanometer, a space through which sodium ions can pass in association with a water molecule. The channel has many charged groups critically placed on its surface. These charges give the channel a large electric dipole moment that varies in direction and magnitude when the molecular conformation of the channel changes as the channel goes from a closed state to an open one.

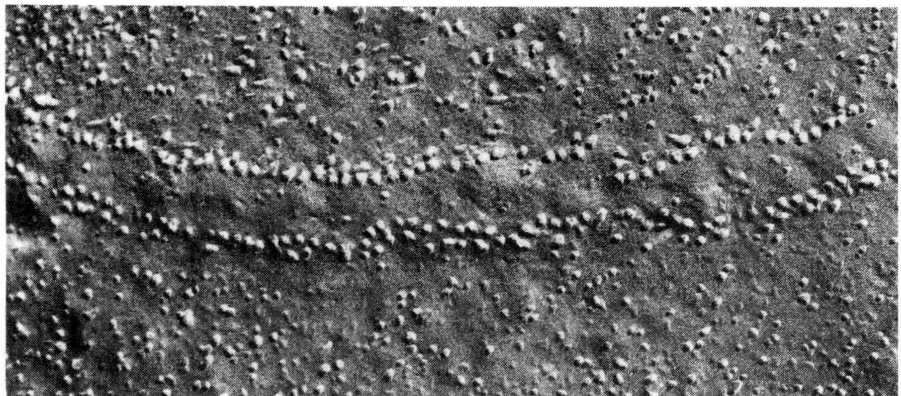

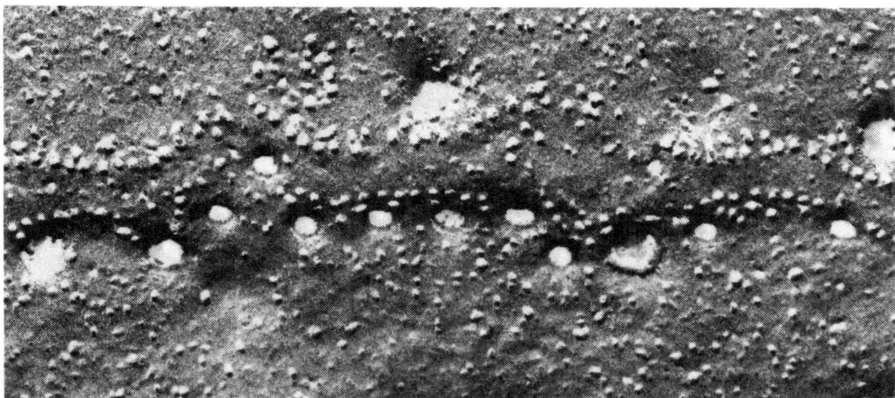

FREEZE-FRACTURE REPLICAS of the presynaptic membrane of the frog neuromuscular junction were made by Heuser. The upper micrograph shows the membrane three milliseconds after the muscle had been stimulated. Running across the axon membrane is a double row of particles: membrane proteins that may be calcium channels or structural proteins to which vesicles attach. The lower micrograph shows the membrane five milliseconds after stimulation. The stimulation has caused synaptic vesicles to fuse with presynaptic membrane and form pits.

Because the surface membrane of the cell is so thin the difference of 70 millivolts across the resting membrane gives rise to a large electric field, on the order of 100 kilovolts per centimeter. In the same way that magnetic dipoles tend to align themselves with the lines of force in a magnetic field, the electric dipoles in the sodium-channel protein tend to align themselves with the membrane electric field. Changes in the strength of the membrane field can therefore drive the channel from the closed conformation to the open one. As the inner surface of the membrane is made more positive by the entering vanguard of sodium ions the sodium channels tend to spend an increasing fraction of their time in the open conformation. The process in which the channels are opened by a change in the membrane voltage is known as sodium-channel activation.

The process is terminated by a phenomenon called sodium inactivation. Voltage differences across the membrane that cause sodium channels to open also drive them into a special closed conformation different from the conformation characteristic of the channel's resting state. The second closed conformation, called the inactivated state, develops more slowly than the activation process, so that channels remain open briefly before they are closed by inactivation. The channels remain in the inactivated state for some milliseconds and then return to the normal resting state.

The complete cycle of activation and inactivation normally involves the opening and closing of thousands of sodium channels. How can one tell whether the increase in overall membrane permeability reflects the opening and closing of a number of channels in an all-or-none manner or whether it reflects the operation of channels that have individually graded permeabilities? The question has been partly answered by a new technique that relates fluctuations in membrane permeability to the inherently probabilistic nature of conformational changes in the channel proteins. One can trigger repeated episodes of channel opening and calculate the average permeability at a particular time and also the exact permeability on a given trial. The exact permeability fluctuates 10 percent or so around a mean value. Analysis of the fluctuations shows that the sodium channels open in an all-or-none manner and that each channel opening increases the conductance of the membrane by 8×10^{-12} reciprocal ohms. One of the principal challenges in understanding the neuron is the development of a complete theory that will describe the behavior of the sodium channels and relate it to the molecular structure of the channel protein.

As I noted briefly above, axons also have voltage-gated potassium channels that help to terminate the nerve impulse by letting potassium ions flow out of the axon, thereby counteracting the inward flow of sodium ions. In the cell body of the neuron the situation is still more complex, because there the membrane is traversed by five types of channel. The different channels open at different rates, stay open for various intervals and are preferentially permeable to different species of ions (sodium, potassium or calcium).

The presence of the five types of channel in the cell body of the neuron, compared with only two in the axon, gives rise to a more complex mode of nerve-impulse generation. If an axon is presented with a maintained stimulus, it generates only a single impulse at the onset of the stimulus. Cell bodies, however, generate a train of impulses with a frequency that reflects the intensity of the stimulus.

Neurons are able to generate nerve impulses over a wide range of frequencies, from one or fewer per second to several hundred per second. All nerve impulses have the same amplitude, so that the information they carry is represented by the number of impulses generated per unit of time, a system known as frequency coding. The larger the magnitude of the stimulus to be conveyed, the faster the rate of firing.

When a nerve impulse has traveled the length of the axon and has arrived at a terminal button, one of a variety of transmitters is released from the presynaptic membrane. The transmitter diffuses to the postsynaptic membrane, where it induces the opening of chemically gated channels. Ions flowing through the open channels bring about the voltage changes known as postsynaptic potentials.

Most of what is known about synaptic mechanisms comes from experiments on a particular synapse: the neuromuscular junction that controls the contraction of muscles in the frog. The axon of the frog neuron runs for several hundred micrometers along the surface of the muscle cell, making several hundred synaptic contacts spaced about a micrometer apart. At each presynaptic region the characteristic synaptic vesicles can be recognized readily.

Each of the synaptic vesicles contains some 10,000 molecules of the transmitter acetylcholine. When a nerve impulse reaches the synapse, a train of events is set in motion that culminates in the fusion of a vesicle with the presynaptic membrane and the resulting release of acetylcholine into the cleft between the presynaptic and the postsynaptic membranes, a process termed exocytosis. The fused vesicle is subsequently reclaimed from the presynaptic membrane and is quickly refilled with acetylcholine for future release.

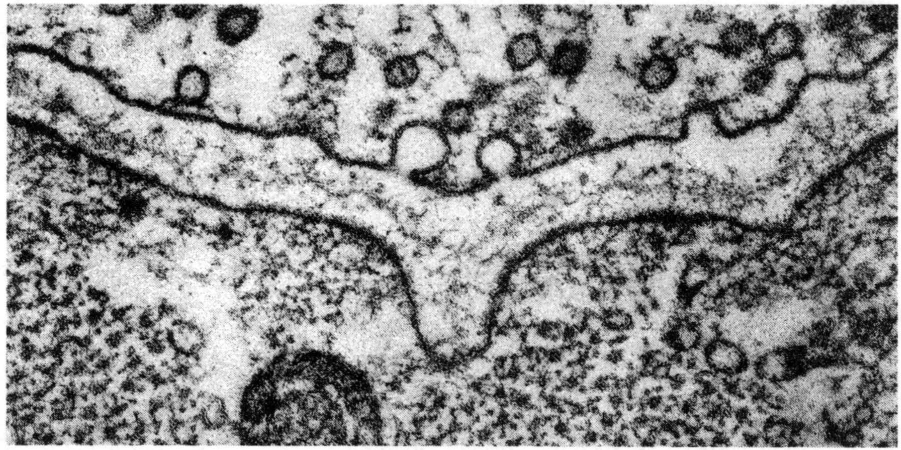

TRANSMITTER IS DISCHARGED into the synaptic cleft at the synaptic junctions between neurons by vesicles that open up after they fuse with the axon's presynaptic membrane, a process called exocytosis. This electron micrograph made by Heuser has caught the vesicles in the terminal of an axon in the act of discharging acetylcholine into the neuromuscular junction of a frog. The structures that appear in the micrograph are enlarged some 115,000 diameters.

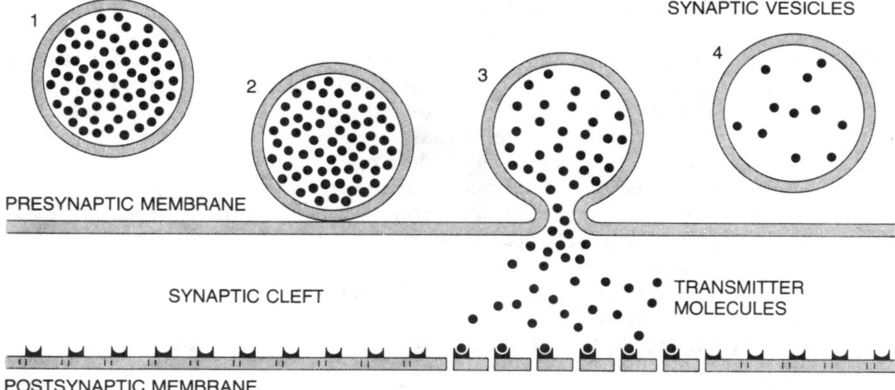

SYNAPTIC VESICLES are clustered near the presynaptic membrane. The diagram shows the probable steps in exocytosis. Filled vesicles move up to synaptic cleft, fuse with the membrane, discharge their contents and are reclaimed, re-formed and refilled with transmitter.

Many details of the events leading to exocytosis have recently been elucidated. The fusion of vesicles to the presynaptic membrane is evidently triggered by a rapid but transient increase in the concentration of calcium in the terminal button of the axon. The arrival of a nerve impulse at the terminal opens calcium channels that are voltage-gated and allows calcium to flow into the terminal. The subsequent rise in calcium concentration is brief, however, because the terminal contains a special apparatus that rapidly sequesters free calcium and returns its concentration to the normal very low level. The brief spike in the free-calcium level leads to the fusion of transmitter-filled vesicles with the presynaptic membrane, but the precise mechanism of this important process is not yet known.

Interesting details of the structure of the terminal membrane have been revealed by the freeze-fracture technique, a method that splits the layers of the bilayer membrane and exposes the intrinsic membrane proteins for examination by electron microscopy. In the frog neuromuscular junction a double row of large membrane proteins runs the width of each synapse. Synaptic vesicles become attached on or near the proteins. Only these vesicles then fuse to the membrane and release their transmitter; other vesicles seem to be held in reserve some distance away. The fusion of vesicles is a random process and occurs independently for each vesicle.

In less than 100 microseconds acetylcholine released from fused vesicles diffuses across the synaptic cleft and binds to the acetylcholine receptor: an intrinsic membrane protein embedded in the postsynaptic membrane. The receptor is also a channel protein that is chemically gated by the presence of acetylcholine. When two acetylcholine molecules attach themselves to the channel, they lower the energy state of the open conformation of the protein and thereby increase the probability that the channel will open. The open state of the channel is a random event with an average lifetime of about a millisecond. Each packet of 10,000 acetylcholine molecules effects the opening of some 2,000 channels.

During the brief period that a channel is open about 20,000 sodium ions and a roughly equal number of potassium ions pass through it. As a result of this ionic flow the voltage difference between the two sides of the membrane tends to approach zero. How close it approaches to zero depends on how many channels open and how long they stay open. The acetylcholine released by a typical nerve impulse produces a postsynaptic potential, or voltage change, that lasts for only about five milliseconds. Because postsynaptic potentials are produced by chemically gated channels rather than by voltage-gated ones they have properties quite different from those of the nerve impulse. They are usually smaller in amplitude, longer in duration and graded in size depending on the quantity of transmitter released and hence on the number of channels that open.

Different types of chemically gated channels exhibit different selectivities. Some resemble the acetylcholine channel, which passes sodium and potassium ions with little selectivity. Others are highly selective. The voltage change that results at a particular synapse depends on the selectivity of the channels that are opened. If positive ions move into the cell, the voltage change is in the positive direction. Such positive-going voltage channels tend to open voltage-gated channels and to generate nerve impulses, and so they are known as excitatory postsynaptic potentials. If positive ions (usually potassium) move out of the cell, the voltage change is in the negative direction, which tends to close voltage-gated channels. Such postsynaptic potentials oppose the production of nerve impulses, and so they are termed inhibitory. Excitatory and inhibitory postsynaptic potentials are both common in the brain.

Brain synapses differ from neuromuscular-junction synapses in several ways. Whereas at the neuromuscular junction the action of acetylcholine is always excitatory, in the brain the action of the same substance is excitatory at some synapses and inhibitory at others. And whereas acetylcholine is the usual transmitter at neuromuscular junctions, the brain synapses have channels gated by a large variety of transmitters. A particular synaptic ending, however, releases only one type of transmitter, and channels gated by that transmitter are present in the corresponding postsynaptic membrane. In contrast with neuromuscular channels activated by acetylcholine, which stay open for about a millisecond, some types of brain synapses have channels that stay open for less than a millisecond and others have channels that remain open for hundreds of milliseconds. A final major difference is that whereas the axon makes hundreds of synaptic contacts with the muscle cell at the frog's neuromuscular junction, axons in the brain usually make only one or two synaptic contacts on a given neuron. As might be expected, such different functional properties are correlated with significant differences in structure.

As we have seen, the intensity of a stimulus is coded in the frequency of nerve impulses. Decoding at the synapse is accomplished by two processes: temporal summation and spatial summation. In temporal summation each postsynaptic potential adds to the cumulative total of its predecessors to yield a voltage change whose average amplitude reflects the frequency of incoming nerve impulses. In other words, a neuron that is firing rapidly releases more transmitter molecules at its terminal junctions than a neuron that is firing less rapidly. The more transmitter molecules that are released in a given time, the more channels that are opened in the postsynaptic membrane and therefore the larger the postsynaptic potential is. Spatial summation is an equivalent process except that it reflects the integration of nerve impulses arriving from all the neurons that may be in synaptic contact with a given neuron. The grand voltage change derived by temporal and spatial summation is encoded as nerve-impulse frequency for transmission to other cells "downstream" in the nerve network.

I have described what is usually regarded as the normal flow of information in neural circuits, in which postsynaptic voltage changes are encoded as nerve-impulse frequency and transmitted over the axon to other neurons. In recent years, however, a number of instances have been discovered where a postsynaptic potential is not converted into a nerve impulse. For example, the voltage change due to a postsynaptic potential can directly cause the release of transmitter from a neighboring site that lacks a nerve impulse. Such direct influences are thought to come into play in synapses between dendrites and also in certain reciprocal circuits where one dendrite makes a synaptic contact on a second dendrite, which in turn makes a synaptic contact back on the first dendrite. Such direct feedback seems to be quite common in the brain, but its implications for information processing remain to be worked out.

Much current investigation of the neuron focuses on the membrane proteins that endow the cell's bilayer membrane, which is otherwise featureless, with the special properties brain function depends on. With regard to channel proteins there are many unanswered questions about the mechanisms of gating, selectivity and regulation. Within the next five or 10 years it should be possible to relate the physical processes of gating and selectivity to the molecular structure of the channels. The basis of channel regulation is less well understood but is now coming under intensive investigation. It seems that hormones and other substances play a role in channel regulation that is now becoming appreciated. The central problems at synaptic junctions involve exocytosis and other activities related to the metabolism and release of transmitters. One can expect increasing attention to be focused on the role of the surface membrane in the growth and development of neurons and their synaptic connections, the remarkable process that establishes the integration of the nervous system.

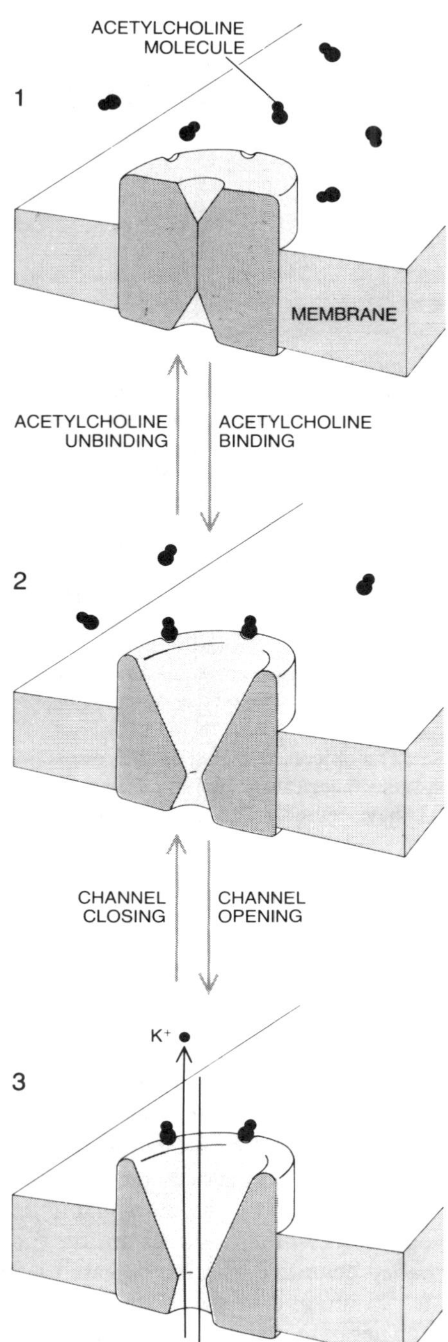

ACETYLCHOLINE CHANNEL in a postsynaptic membrane is opened by acetylcholine molecules' discharging into the synaptic cleft. The drawing shows the acetylcholine receptor at the frog neuromuscular junction. Two acetylcholine molecules bind rapidly to the resting closed channel to form a receptor-acetylcholine complex (*1, 2*). The complex undergoes a change in its conformation that opens the channel to the passage of sodium and potassium ions (*3*). The time required for conformational change in the complex limits the speed of the reaction. The channel remains open for about a millisecond on the average and then reverts to the receptor-acetylcholine complex. While it is open the channel passes about 20,000 sodium ions and an equal number of potassium ions. The acetylcholine rapidly dissociates and is destroyed by the enzyme acetylcholine esterase. Acetylcholine receptor appears in micrograph on page 105.

Article 2.2

Cognitive and Psychological Computation with Neural Models

JAMES A. ANDERSON

Abstract—**Biological support exists for the idea that large-scale models of the brain should be parallel, distributed, and associative. Some of this neurobiology is reviewed. It is then assumed that state vectors, large patterns of activity of groups of individual somewhat selective neurons, are the appropriate elementary entities to use for cognitive computation. Simple neural models using this approach are presented that will associate and will respond to prototypes of sets of related inputs. Some experimental evidence supporting the latter model is discussed. A model for categorization is then discussed. Educating the resulting systems and the use of error correcting techniques are discussed, and an example is presented of the behavior of the system when diffuse damage occurs to the memory, with and without compensatory learning. Finally, a simulation is presented which can learn partial information, integrate it with other material, and use that information to reconstruct missing information.**

> The object of science is the connection of phenomena; but the theories are like dry leaves which fall away when they have ceased to be the lungs of the tree of science.
>
> Ernst Mach (1872)

I. Introduction

THE DESIRE to build artificial systems that do the kinds of interesting things that we do has long existed. From mechanical automata in past centuries to electronic devices now, we have tried to make hardware and software that acts like us, or at least some significant part of us. Much of the current work in this tradition now tries to model with computers various aspects of human cognition. The things that make us most interesting to each other and which seem to be the most highly developed in humans as opposed to other animals are the faculties that are usually called cognitive, that is, our abilities to speak, to perceive, to reason, and to speculate.

There are many ways to understand cognition, in particular, to understand it well enough to mimic it with models or gadgets. We have access to a number of examples of a cognizing organism (i.e., us). We can study us in detail, both in terms of our system performance (psychology) and in terms of our hardware (neuroscience). We can also study us in the abstract, asking essentially, how we would (preferably from first principles) build a system that performs the cognitive functions that we can. The result of such a design process may bear little or no relationship to the system that nature has evolved, though it has been claimed that there are powerful constraints on intelligence, so that all systems that can do the same intelligent things are somehow related since they have solved the same problems.

My own bias, however, based more on faith than concrete accomplishment, is that the best approach to understanding and constructing intelligent devices is to study carefully the one that we know works. The limitations of this approach are obvious: birds fly. Airplanes are neither feathered nor flap their wings. Studying flying from first principles might have given rise to hot air balloons and rockets, but it is unlikely that studying birds in order to fly would have done so.

This paper will discuss some of the hardware of real nervous systems. We will then develop some simple neural models for cognition that try to work within the constraints that nature has had to work with. At the end of this paper we show the beginnings of an approach to cognitive computation: that is, how it is possible to use these distributed parallel associative models to compute and what they can be used for.

II. Biological Assumptions

State Vectors

Our claim is that biology places severe restrictions on the kinds of computations done by our brains. A great deal is currently known about neuroscience that bears on this point. A particularly good introduction to neuroscience for nonbiologists is an issue of *Scientific American* now available as a book [16]. An excellent textbook has also recently appeared [55].

Two key conclusions must be mentioned. First, neurons are analog devices. That is, they take their synaptic inputs, perform a computation on these inputs, and generate an output which is almost always a continuous valued firing frequency, represented as the time between discrete pulses called action potentials. A weighted integration of the synaptic inputs over a brief period of time is an oversimplified but useful first approximation of a neuron model. The neuron typically does not act like a digital device such as a McCulloch–Pitts neural logic element, but as a pulse-code modulation system.

Manuscript received August 1, 1982; revised April 4, 1983. This work was supported in part by the National Science Foundation under Grants BNS-79-23900 and BNS-82-14728, administered by the Memory and Cognitive Processes section, in part by the Alfred P. Sloan Foundation, in part by the Digital Equipment Corporation, and in part by Contract N-00014-81-K-0136 from the U.S. Office of Naval Research.

The author is with the Department of Psychology and Center for Neural Science, Brown University, Providence, RI 02912.

Reprinted from *IEEE Trans. Syst., Man., Cybern.*, vol. 13, no. 5, pp. 799–815, Sept./Oct. 1983.

Second, ten billion or more individual neurons exist in the mammalian nervous system. This means that the computational strategies used by the nervous system can take advantage of the presence of very large numbers of elements. However, since neurons are slow devices, operating with integration times in the millisecond or tens of milliseconds range, no time exists for the long strings of elementary computations that characterize digital computers. A highly parallel strategy is employed. The brain's "machine operations" must be of a very powerful kind since not many of them will have time to execute during a single "program."

When a stimulus of any complexity is presented, many neurons respond. (Not all of them, but not a single one either.) When a motor action of any significant kind is made, many motor neurons respond. Therefore, a pattern of activity of many neurons represents response to the input and many neurons respond as the output of the system. Internal communication between brain areas has the same many-to-many architecture. Therefore, we become interested in elementary operations involving the simultaneous activities of many individual neurons which give rise to the activities of many neurons. In the models to be presented, we represent these activities as state vectors of simultaneous neuron activities, and we claim that elementary operations involving transformations of state vectors form a useful approach to nervous system models.

This approximation is one way of avoiding the "homunculus" problem. No internal CPU (a high-tech homunculus) abstractly processes information. Activity pattern may follow activity pattern in lawful sequence, but information is not represented in a form other than as neuron activities or as connection strengths between neurons.

Neurons

Large state vectors are a biologically justifiable way to represent information in the human nervous system. The elements of these vectors correspond with something of the size and properties of single neurons.

At the lowest level, single neurons devote great care to analyzing what is important to the organism. In primary visual cortex, many cells analyze orientation, binocular interactions, movement, color, spatial frequency, and spatial location. Less or no analysis is made of absolute light intensity, large areas with no change in intensity, and stationary stimuli in general. The implication of this is that interesting things potentially affect a number of cells strongly though only a small number are actually excited or inhibited by a stimulus. The cells not affected are also contributing information of a kind. Interesting aspects of the stimulus are "richly coded" in that they may make profound effects on potentially very many elements of the state vectors.

As one example, higher mammals are born with what are apparently inbuilt orientation detectors in their visual system whose properties can be modified (usually for the worse) by environmental manipulation. However, these orientation selective units are also affected by other physical aspects of the stimulus such as binocularity, spatial frequency, wavelength, or movement. The biological approach taken seems to be to have many cells responding somewhat selectively to important aspects of the environment. An alternative design would be to have a few high-quality feature detectors, but this design does not seem to be used by mammals, though it is by invertebrates and perhaps by some nonmammalian vertebrates. The equivalents of the exquisitely selective neural responses to pheromones, say, or to particular patterns and frequencies of sound found in invertebrate species may also exist in mammals, but they seem to be outnumbered by less selective cells, where the emphasis has shifted to developing processing selectivity at the group level.

The question of specificity and distribution in the nervous system is important for neuroscientists and for theoreticians as well. As a recent example, Feldman and Ballard [18] have suggested that information is represented in the nervous system by a very small number of active neurons. Each potential value of a stimulus parameter (say size, brightness, color) is represented by a single neuron. Combinations of parameters may be represented by single cells also, though Feldman and Ballard devote some time to discussing ways of avoiding the obvious combinatorial explosion of the required number of units. They develop the nice idea of "winner take all" networks where only one of a number of contending values is excited and the rest are inhibited. The final representation is that stable coalitions are formed: they give as an example of a stable coalition one containing three active units.

Interestingly, Barlow [8], a neurophysiologist, suggested a similar idea: very selective cells ("on the order of selectivity of a word") are present in the nervous system, and most cells are quiet most of the time. However, Barlow came to the same conclusion as Feldman and Ballard that, to represent information of any complexity, more than one active cell was necessary. Barlow concluded there were no "pontifical" cells, but there was a distributed "college of cardinals."

The physiology supports a degree of selectivity in neural coding. Suppose one percent of cells were active in a complex concept or perception. This would correspond to many millions of cells, yet a microelectrode would reveal very little electrical activity in such a brain. The conclusions that representation of information in the nervous system is contained in simultaneous discharge of a number of neurons and that information is distributed in this sense are difficult to avoid.

Cerebral Cortex

The cerebral cortex is a flat thin two-dimensional structure on the order of a fifth of a square meter in area [12]. It is extensively folded in higher mammals to fit inside a skull of reasonable size. The neocortex is relatively homogeneous; the similarities of cell type and circuitry between different areas are more striking than the differences. (See the collection of essays on cortical organization edited by

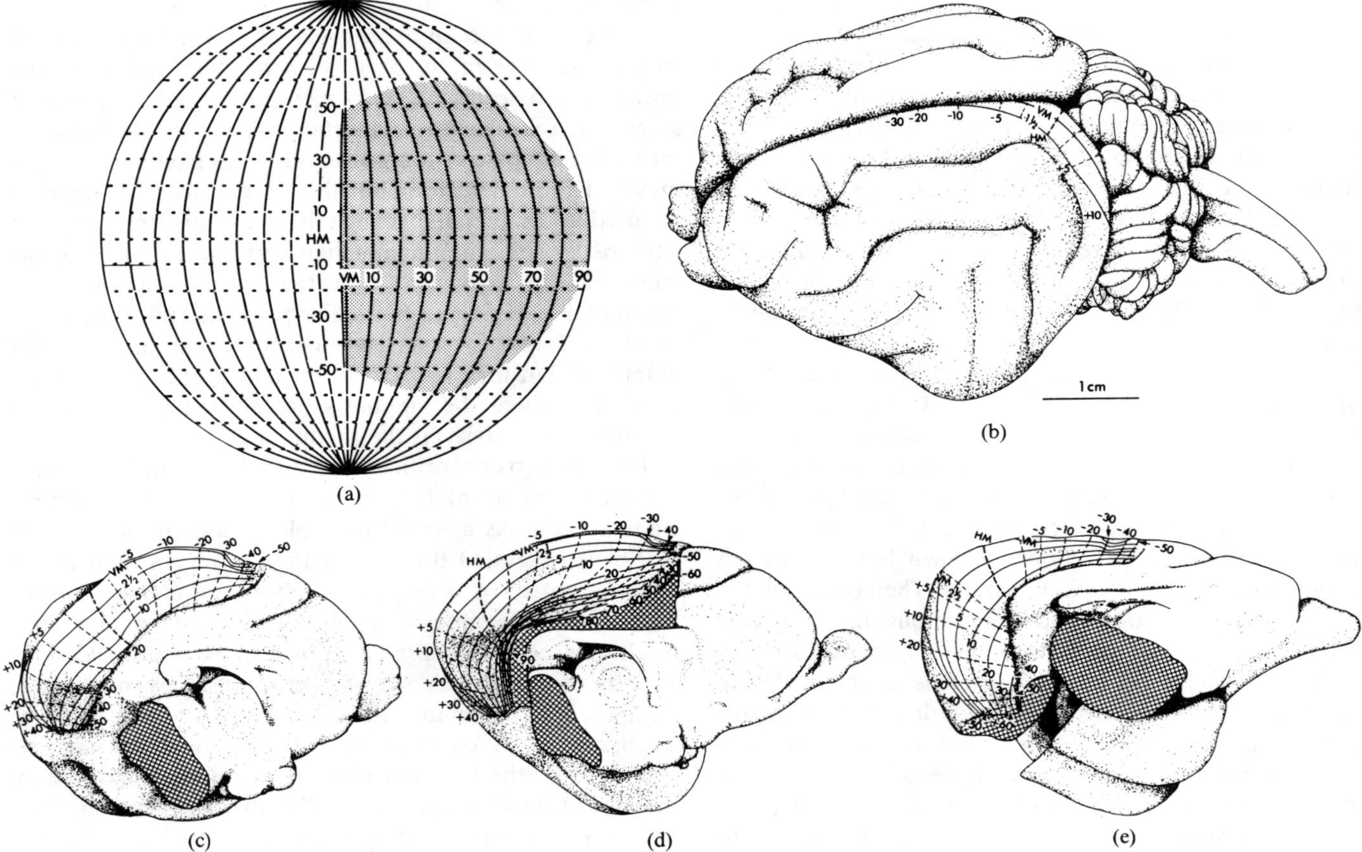

Fig. 1. Diagram of representation of visual field in area 17 (primary visual cortex) of cat. (a) Perimeter chart showing extent of visual field represented in area 17. (b)–(e) Location of visual field in area 17 of cat brain. From Tusa *et al.* [60], reprinted by permission.

Schmitt *et al.* [52]). The already two-dimensional cortex is strongly layered. There are numerous subareas (perhaps 50 or so) of cerebral cortex, which seem to be functional areas, though details of organization and function are often quite obscure. Thus Area 17 (located in humans at the back of the head) is the primary cortical receiving area for visual inputs. Area 3 receives somatic sensory inputs (skin senses). The cortex is exquisitely structured. Areas associated with a sensory system have topographic organization, that is, the visual field in area 17 is represented as a distorted map of visual space. Fig. 1 shows a diagram of the topography of the map of visual space onto the surface of cortex. If the location of a cell is known, its general area of maximum visual responsiveness can be inferred. The body surface is represented in the somatosensory areas, frequency is loosely plotted on the surface of the auditory cortex, and so on. Although the overall outlines of the map are lawful in the large, individual cells may show local variability.

The maps show striking distortions. The human retina has an area of greatest optical quality called the fovea. The fovea has a higher receptor density than other regions of the retina and is correspondingly overrepresented in the cortex. It is possible to get a good first impression of the relative importance of a structure in the life of a mammal by looking at its cortical map: in our own somatosensory system we have a very large cortical area devoted to our fingers and a small area devoted to the toes. A rhesus monkey has more equal representation of toes and fingers. Any model of the cortex must be consistent with this "more is better" philosophy because it is not immediately obvious why it should be so. The brain could have simply paid more attention to input signals from the fovea, for example and not had more of them. Nervous tissue is very costly in terms of its biological overhead: it consumes enormous amounts of energy, it is very sensitive mechanically and biochemically and is generally more of a burden than other tissue types. Therefore, if it physically expands to the extent it has in us, it must earn its keep in enhanced processing power. Whatever organizational scheme is used in the cortex must be such as to add on power by expansion in a simple way, without requiring too much in the way of detailed interconnection specifications.

Connections between sensory receptors and cortex, and between one cortical area and another are physically parallel. One sheet of cells projects to another, with very many fibers and considerable convergence and divergence in the projections. This striking parallelism in the anatomy has led to interest in parallel models for brain function over the past decades, from the Perceptron onward.

These points are worth briefly mentioning here, because I feel that our minds are much less of a general purpose cognitive device than we might like to think. What we seem to be is a somewhat flexible analog processor with enormous memory capacity, which is good at performing a class of tasks that interest us as a species and which are important for our success in our particular world. Our attempts at general purpose computation (logic, say, or even language) are often inconsistent. They are unnatural. Far more complex tasks that are biologically relevant (throwing a ball, recognizing a face, understanding speech) are so effortless that we do not realize how hard they are until we try to make a machine do them. On the other hand, the pitiful mess most humans make of formal logical reasoning or arithmetic would embarrass a $10 pocket calculator. Yet we can recognize a face with speed and accuracy no computer can match.

William James made this point 90 years ago.

> In the main, if a phenomenon is important for our welfare, it interests and excites us the first time we come into its presence. Dangerous things fill us with involuntary fear; poisonous things with distaste; indispensible things with appetite. Mind and world in short have been evolved together, and in consequence are something of a mutual fit [27, p. 17].

III. Associative Models

Several sections of this paper will contain reviews of previously published material. Several general references have been published for this area. Kohonen's book [30] is essential. A recent collection of papers [26] contains some related and alternate approaches. The Perceptron of Rosenblatt and related models in the late 1950's and early 1960's pioneered the use of models for cognition inspired by parallel nervous system architecture. Nilsson [42] summarizes this literature. Minsky and Papert [40] pointed out the considerable limitations in processing ability of simple Perceptrons. (We argue later that limitations in ability are to be expected from brainlike models and allow the strongest experimental tests of such models.) McCulloch and Pitts, after their immensely influential paper on neurons as discrete logical devices [38], published a paper [44] proposing a parallel model for eye movements using continuous mathematics and based on the topographic organization of the superior colliculus.

A recent paper by Sutton and Barto [57] contains a fine review of work in the area along with an application of a learning model to psychological classical conditioning. A specifically parallel model using state vectors and tensors to model the cerebellum has been described by Pellionisz and Llinas [43]. Papers by Bienenstock *et al.* [9] and by Cooper [13] deal with application of the learning rule used in this paper and extensions of it to plasticity in the visual cortex, with careful fitting of neurophysiological data to theoretical predictions.

A somewhat different approach to cognitive questions is taken by Grossberg [21] but with significant similarities in direction. Arbib's book [6] and his work with Szentagothai [59] contain many valuable insights and interesting material. These sources will provide more detailed references to the journal literature, as will many of the papers in this journal. The visual system lends itself in a very obvious way to parallel analysis. The well-known work of Marr [36] discusses in detail the kind of parallel computation, tied closely to physiology, that may be used in the early stages of visual information processing. Marr's early work on cerebellum [34] and neocortex [35] assumes highly parallel architecture combined with simple conjunctional learning rules.

Minsky [39] has proposed a distributed model with centralized elements where information is represented in states of many low-level agents whose activities constitute a "mental state." Memory is the reconstruction of a past state. Some powerful and selective elements (K-lines) control states of many agents and can reconstruct past states. The agents are not specifically neurons, but they communicate by means of excitation and inhibition, and the mental state notion is similar to the state vectors used in this paper. The model is a hybrid of localized and distributed computation.

These models as a group involve massive parallelism of many simple elements and often have simple rules for modifying strengths of connections between elements. Variations between them come in specifying the assumed rules and operations. The details of wiring can be specifically brainlike or much more abstract.

In this paper, we will adhere less closely to the details of the neuroscience than some would like. We will focus our attention on the implications for cognition of models which seem to us to capture the appropriate parallel, distributed essence of most of the models proposed to date, yet which are simple enough to analyze and simulate in some detail. This means we will start with a linear model which demonstrates how associative learning can arise naturally in parallel neural models. We will show that even this simple, rather unrealistic model is capable of some striking psychological predictions. Then we will introduce simple nonlinearities as we need them to make a first step at curing some of the obvious defects of the linear model, always hoping that each increase in complexity pays for itself in explaining a new psychological phenomenon or giving us more cognitive computing power. Such successive refinement seems to us to be one valid way of approaching a system with the complexity of the brain.

Synaptic Connectivity

Neurons talk to one another. The connections between neurons are called synapses. We have argued that many neurons talk to many neurons. The connectivity of cortical neurons is extensive; a single large cortical pyramidal cell is estimated to have thousands of synapses. The exact value is a function of the type and location of the cell. Although competing hypotheses have been seriously considered, almost every neuroscientist believes that changes in synaptic strength are the location of memory. In some cases, it has been possible to demonstrate convincingly that synaptic changes occur in learning-related contexts: the best studied example of this is the marine mollusk *Aplysia* which has been studied by Kandel and coworkers for a number of years [28].

One might first think that learning is simply a matter of strengthening synapses by use: the more a synapse is used (or not used) the stronger (or weaker) it gets. Indeed, the *Aplysia* has an inverted version of this in the habituation paradigm, where recurrent stimulation causes a diminution (habituation) of the resulting response. However, this kind of learning, though interesting, present, and important, seems to be inadequate for most complex and cognitively interesting kinds of learning. For millenia, since Aristotle, those interested in memory recognized its associative aspect. That is, events tended to become linked together because, "... one (event) is of a nature to occur after another." [7, p. 54]. Association of a sufficiently flexible kind seems not to be possible with a simple stimulus directed change-by-use rule such as habituation. A close connection with the response is required.

The rule that seems to be the starting point for virtually every recent model of associative memory seems to have first been formulated by Hebb [22]. Hebb's proposal for cellular learning was

> When an axon of cell A is near enough to excite a cell B and repeatedly or persistently takes part in firing it, some growth process or metabolic change takes place in one or both cells such that A's efficiency as one of the cells firing B, is increased [22, p. 62].

This rule suggests that a correlation between pre- and postsynaptic cell will develop, and such a synapse is called a correlational synapse. Such a rule is indeed adequate to build an associative memory that does a number of quite interesting things. The rest of this paper will be devoted to exploring some of the specifically psychological and cognitive implications of networks using correlational synapses.

Simple Association

The basic system that we shall discuss in one variant or another throughout this review is shown in Fig. 2. We assume one set of simple model neurons projects to another set (or to the same set, a special case). This architecture is specifically inspired by projection systems in the brain, where one set of elements projects to another over highly parallel pathways.

Suppose we have two sets of N neurons, called alpha and beta, where every neuron in beta projects to every neuron in alpha. A neuron j in alpha is connected to neuron i in beta by way of a modifiable synapses with strength $A(i, j)$, forming an $N \times N$ connectivity matrix A. We are interested in the set of simultaneous individual neuron activities in a group of neurons. We represent these large patterns as state vectors. We assume these components can have positive or negative values. This could occur if we build inhibition as well as excitation into the system and if we assume that the nervous system is concerned with deviations from spontaneous level, positive as well as negative.

If pattern f occurs in alpha and pattern g occurs in beta, we can associate these two patterns using a simple learning

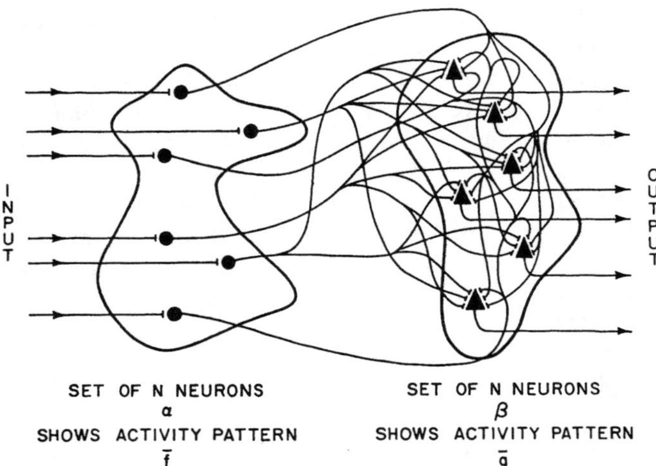

Fig. 2. Models assume two sets of N neurons, alpha projecting to beta. Every neuron in alpha projects to every neuron in beta. This drawing has $N = 6$. From Anderson *et al.* [4], reprinted by permission.

rule, a generalization of a Hebb synapse. We need to change the connectivity matrix according to the rule,

$$\Delta A(i, j) = \eta f(j) g(i).$$

We have introduced a learning parameter η. Note that this is information locally available to the junction: it is proportional to the product of pre- and postsynaptic activity. This defines the matrix ΔA to be of the form

$$\Delta A = \eta g f^T.$$

This matrix now acts like an associator. Consider the simplest case: Initially, $A = 0$, $\eta = 1$, and f and g are normalized. If now

$$A = g f^T,$$

we have established connections between the first and second set of neurons. Now, if an input pattern of activation is impressed on f, a pattern will appear on g. If we assume as an initial approximation a simple linear integrator model for the way neurons respond to their inputs, we can calculate the output pattern as the product of the connectivity matrix A, and the input pattern. Suppose the pattern is f. Then the output pattern will be g, since

$$g = g f^T f.$$

The classic neural system showing simple linearity of this kind is the Limulus eye, where the approximation is quite accurate [10], [11]. Other systems show various nonlinearities, but often (referring to communication from one neuron to another) a simple linear model is quite good as a first approximation. Sensory transduction of the physical stimulus can be quite nonlinear, however, masking what may be a simpler relationship at the neuron level. Linearity is an adequate approximation only up to a point. The relationship of linearity and the nervous system is a complex one; see Anderson and Silverstein [5] for a few examples and caveats. For a fuller discussion of this issue in the visual system, see Ratliff [47].

In general, we want to couple more than one set of patterns. Suppose we have a set of associations that we wish to teach the system $(f_1, g_1), (f_2, g_2), \cdots, (f_k, g_k)$. Suppose we teach our matrix these pairs of patterns with each pair having associated with it an incremental matrix of the form

$$\Delta A_i = g_i f_i^T.$$

Let us then assume that the overall synaptic connectivity matrix is given by the sum of all the incremental matrices so that

$$A = \sum_i g_i f_i^T.$$

Single matrix elements (synaptic contacts) can do multiple duty in that they may participate in storing information about associations between any pairs of statea vectors. This means information may not be localized or localizable, and the joint operation of many synapses is required for function. Information is distributed in the state vectors (since the simultaneous pattern of many cells is required for meaning) and also in the actual locus of memory. This is a holographic property, though these models are not Fourier transform holograms.

Suppose that the input vectors are orthonormal. Then, if one of the stored items is impressed on alpha, we have

$$\begin{aligned}
(\text{pattern on beta}) &= Af_i \\
&= gf_i^T f_i + g\left(\sum_{i \neq j} f_j^T f_i\right) \\
&= g.
\end{aligned}$$

This means that vector g_i can be generated at the output if vector f_i is presented at the input. Note that the state vectors are large, and if components are statistically independent, then the resulting vectors will be close to orthogonal. We assume as a fundamental coding assumption that stimuli very different from each other have uncorrelated state vectors, that is, orthogonal to each other on the average.

Many intriguing properties emerge from the interaction between learned vectors. Two recent psychological papers [41], [17] use a vector approach similar to the foregoing to explain a good many psychological phenomena. The model they both use stores associations by a convolution operation and retrieves them by correlation, extensions of a model of Liepa [31]. The memory vectors are "superimposed in a composite memory trace" [17, p. 627]. Murdock and Eich discuss, simulate, and suggest explanations for some classic list learning experiments, some short-term memory phenomena, the qualitative effects found during the learning of lists of associations, and prototype formation. Many of the most interesting effects in their model arise because "... the events stored in such a memory combine and interfere with one another ..." [17, p. 657]. As Eich comments, "... it is precisely because CHARM [Eich's model] transforms and combines events that the model is psychologically interesting" [17, p. 654].

General Properties

Matrix and vector associative models have some pronounced strengths and limitations. Their strengths are, first, that they are intrinsically parallel. Second, they are very tolerant of noise and partial connectivity. Since they contain correlational and averaging elements, the resulting systems are often optimal or close to optimal in a signal processing sense. (See the section on error correction.) They tend to be computationally robust. Third, they work better in the sense of better discrimination and signal-to-noise ratio as the state vectors increase in size. Since the models are parallel, this can be done with no increase in processing time if the hardware is also parallel.

Their primary limitation is their limited storage capacity relative to the size of the system. Clearly, only N orthogonal vectors can exist. Second, they can generate noise and inappropriate behavior of an unpredictable nature since things mix together in storage. (This can be a virtue or a problem, depending on context.) Third, they are poorly suited to rapid computation using traditional digital computers. Fourth, they tend to be rather ponderous and inflexible. Fifth, as linear models, they are subject to a host of essential limitations. As Sejnowski comments, "The matrix model resembles memory in the way a toy glider resembles a bird. It does fly, in a rigid sort of way, but it lacks dynamics and grace" [53, p. 203]. However, if it flies even crudely, let us see if it takes us anywhere interesting.

III. Categorization

Concepts and Prototypes

The nervous system is faced with the problem of using information from many moderately selective analyzers. Another problem is intrinsic to the functions of a cognitive system: analysis of the world cannot be too precise. It is necessary to form equivalence classes of events and things of a convenient size. For example, many real things are described by the words "dog" or "bird." Not only are particular individuals different from each other, but the same individual at different times is quite different in its exact physical description. A simple change in lighting can cause great changes in the physical properties of the reflections from the object. We are constructed to ignore these differences, a photocell is not. This process is so basic that we are often not aware of its operation.

A malfunction of this mechanism described in the psychological literature is the subject burdened with an exceptional memory described by Luria [32]. Luria's mnemonist had difficulty recognizing people because, as Luria commented:

> S. often complained that he had a poor memory for faces: "They're so changeable," he had said. "A person's expression depends on his mood and on the circumstances under which you happen to meet him. People's faces are constantly changing; it's the different shades of expression that confuse me and make it so hard to remember faces." [32, p. 64]

We are faced with the problem of forming equivalence classes in a natural way. A tremendous amount of biology is involved in this. We discriminate what we are built to be good at discriminating.

In cognitive science and cognitive psychology, the equivalence classes that result are usually called concepts. The study of concepts is difficult because different kinds exists and because concepts, by their nature, are not consistent or totally stable entities. Good introductions to the modern study of concepts are contained in books by Rosch and Lloyd [50] and by Smith and Medin [56]. One idea supported by a good deal of evidence is that humans will normally operate at a "natural" level of concept complexity, often related to sensory aspects of the stimulus. We will say in normal speech, "Look at that bird on the lawn," as opposed to, "Look at that organism on the flat area of Kentucky bluegrass, clover, and creeping red fescue." We can say the last, but a default level exists which seems to correspond to a natural concept level. The second sentence is both too general and too specific.

We will be concerned in the next sections with two particular aspects of simple concept formation: prototypes and categories. We make the fundamental assumption that items that belong in the same natural categories will have similar neural codings. Their state vectors will be correlated in terms of our models. The study of natural psychological categories does suggest this: many familiar birds look and behave similarly to each other; birds that do not (penguins or ostriches) are often handled as concepts by themselves. One will say, "There is a bird on the lawn," if the bird is a sparrow, pigeon, or robin but, "There is a penguin on the lawn." This observation is related to the model of concepts usually associated with the work of Rosch, which holds that many natural concepts (dogs, birds, vegetables, etc.) are represented by prototypical members, i.e., best examples of the class. People agree on how close objects are to the prototype. There are "good" birds (robins, sparrows, etc.) and birds which are not good examples, such as turkeys, ostriches, and buzzards. To use one of Rosch's more picturesque examples, one would not be bothered in the least by the occurrence in a novel of a sentence such as, "Twenty or so birds often perch on the telephone wires outside my window and twitter in the morning" [49, p. 39] until one replaces the word "birds" with "turkeys."

Prototype Formation in a Neural Model

Suppose we have a number of examples of a category. Suppose we call the category name the state vector g and the different example vectors $f_1, f_2, \cdots, f_k$. Each incremental matrix is generated as before, and we arrive at an overall connectivity matrix given by

$$A = g\sum_i f_i^T.$$

This expression contains the sum of the f's. This term acts like an average response computer. The central tendency of the f's will emerge. The amplitude of response to a new state vector will give a measure of distance from the central tendency. (See [2,], [17], [29] for further discussions of this effect and its implications from a psychological perspective.)

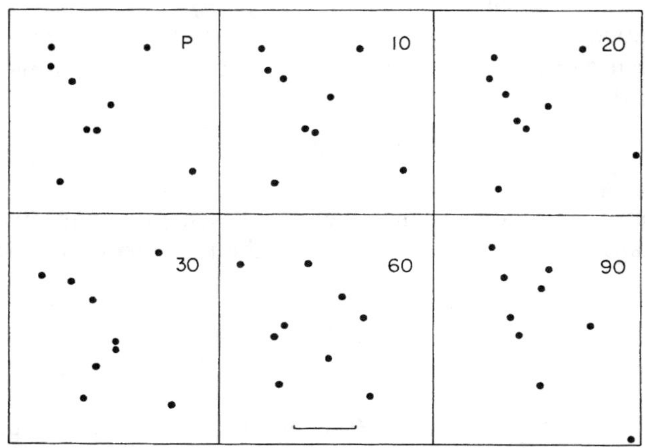

Fig. 3. Prototype dot pattern (P), followed by five examples at various degrees of distortion. Dots were generated on 512 × 512 array and presented to subjects on CRT screen. Number refers to average number of locations moved on the array. Distance of 100 array locations is indicated. In experiments distortions of about 24 units were used. From Knapp and Anderson [29], reprinted by permission.

Posner–Keele Experiments

Psychologists Posner and Keele have demonstrated what might be a simple example of this process [45], [46]. These experiments, extensions of them, and a theoretical discussion, of which the following is a summary, can be found in Knapp and Anderson [29].

A pattern of nine random dots is generated on an oscilloscope screen. These initial patterns are denoted "prototypes." Then examples of the prototypes are made by moving the dots random directions and distances. Fig. 3 shows a prototype and different examples of the prototype, as the average distance a dot moved is increased. Subjects classify examples of a prototype together in the learning phase of the experiment by pressing one of several buttons, each button associated arbitrarily with distortions of a particular prototype by the experimenter. After the response, they are told whether their classification was correct. They do not see the prototype. They are then given a test where they are asked to classify a set of patterns. Classification is correct if a new example is associated with the same response as old distortions of the same prototype. In the testing phase, they can be given 1) the examples they saw, 2) the prototypes, and 3) new examples of the prototypes. Depending on experimental conditions, the prototype (which subjects never saw) may be the best classified, both in terms of reaction time, percent correct classification, and confidence the pattern had been seen before.

In our experiments, patterns of nine random dots were used. Distortions were generated in such a way as to ensure the prototype was extremely unlikely to occur as a stimulus, yet the prototype was often categorized most accurately. This is a common result in the concept literature.

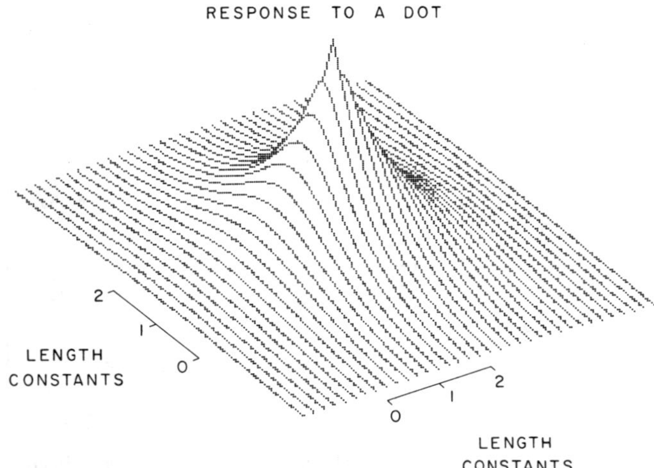

Fig. 4. Activity pattern on hypothetical cortex from single dot in the real world. This is exponential falloff of activity from a single location. From Knapp and Anderson [29], reprinted by permission.

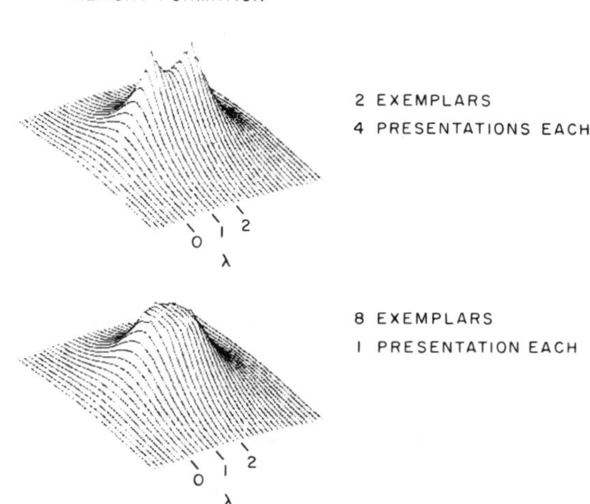

Fig. 5. Sum of eight activity patterns. In one case, two exemplars (presented four times each) form memory; in other case, eight different exemplars are used. In both cases, significant representation is seen at prototype location, but in two exemplar case is also stronger representation of patterns actually seen ("old" patterns). From Knapp and Anderson [29], reprinted by permission.

Application of our concept abstraction model is straightforward. Suppose the neural representation of a dot in the visual system is somewhat localized. (The spectacular map of the visual field shown in Fig. 1 would certainly suggest this.) Suppose that the amount of activity in a two-dimensional representation of visual space falls off exponentially from a central location on the topographic map. Fig. 4 shows the activity pattern due to a single dot on a hypothetical flat cortex. The model predicts that we should sum activities due to different examples of the same prototype, since they are associated with the same response. Fig. 5 shows activity patterns of a single dot in eight patterns when added together. In one case, there were eight different dot locations, in the other only two different dot locations. Note that the first case has "averaged out," few irregularities appear on the resulting summed activity; the second case has strong representations at the locations of the two presented examples. Both have strong representations at the location of the prototype, even though the prototype had never actually been seen.

It is convenient to simplify our calculations. Since we do not know and are not really interested in the details of the output of the system, we want to compute interactions between the stimuli (the vector f's). These will be the coefficients multiplying the g's in the vector model. We assume the output pattern with the largest coefficient will be the categorization made. This is a nonlinear decision rule.

Ultimately, we could use a nonlinear classification model to generate a model response directly, but for initial studies this seemed premature. We want to know the strength of response to different patterns given a memory constructed from the activities. These will be given by dot products of the stored vectors with the input vectors, that is, if similarity of f_1 and f_2 are to be judged.

$$(\text{output strength}) = f_1^T f_2.$$

These are interactions at the level of the rows of the connectivity matrix. A memory composed of K items, $f_1 \cdots f_K$ associated with the same response would have, in response to input f,

$$(\text{output strength}) = \sum_i f_i^T f.$$

A series of experiments was done to collect similarities between dot patterns generated with different degrees of distortion. The prediction is that subjective similarity should be proportional to the inner product between one pattern of activity and the other pattern. The fit using the model, assuming an exponential falloff of activity as shown in Fig. 4, is quite good (0.97 correlation between theory and prediction) and allows determination of the spatial falloff length constant of the activity.

We can see that a number of different examples learned and average displacement of examples from prototype should be powerful determiners of the classification ability of the system. One experimental manipulation was to change the number of different examples learned and study the resulting behavior.

The experimental results are shown in Fig. 6 for an experiment where prototypes were formed from nine random dots presented on a CRT screen. One, six, and 24 different examples were learned. In any given experiment three different prototypes were used. Results show a pronounced change in behavior. If 24 examples are used, pronounced prototype enhancement occurs. With one example, the example presented gives the largest correct classification as would be expected. Of course, a one-example category does not, strictly speaking, show prototype averaging but similarity between one pattern and a different one, since the resulting memory representation will be a multiple of activity at a single dot location.

A direct computer simulation of this experiment using the model presented is possible using this length constant determined from the similarity experiment. The resulting

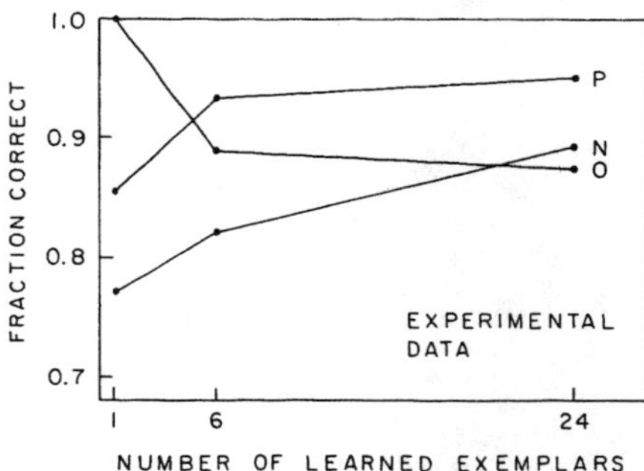

Fig. 6. Experimental results from categorization experiment using dot patterns like those in Fig. 3. Subjects received during learning part same total number of patterns, but generated from one, six, or 24 different exemplars. Three prototypes were used during each experiment. Subjects were taught to classify distortions of given prototype together. During test phase, percent correct categorization was measured for old, new, and prototype patterns. From Knapp and Anderson [29], reprinted by permission.

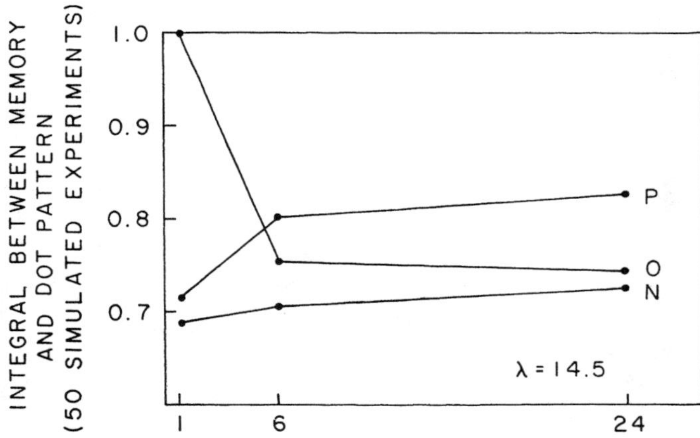

Fig. 7. Simulation of experimental results for prototype experiments. Dot patterns were generated exactly as in experiment, length constant of exponential falloff was measured in similarity judgment experiment, and model computed integral between resulting memory and input pattern (old, new, or prototype, as in experiments). Simulation was repeated 50 times; averages are plotted. Percent correct does not map directly into this integral, but qualitative agreement is quite good. From Knapp and Anderson [29], reprinted by permission.

activity patterns for all conditions of the experiment are generated, summed, and theoretical similarities between memory and the prototype, new examples, and old examples, are computed. These computations are shown in Fig. 7. A very strong qualitative similarity exists between the two figures. There is not necessarily a simple relationship between computed similarity and percent correct classification, but we should expect relative relationships to be maintained. Since we have accounted for the similarity data with the same constants and model, we find that a reasonably powerful model for actual data can be evolved essentially from the first principles of the associative model.

Teaching Methods

Cognitively oriented neural models become problems in teaching. We assume that when an input state vector is presented, an output state vector appears. The problem is how to arrange synaptic interactions to ensure that the output state vector is the one we want. Many of the difficult problems in distribution systems arise in a context that can only be called educational. Putting information into the system is tricky. Once the system works correctly, however, information retrieval is easy, since only simple operations need be performed independent of what other or how much information has been put into the system. Since the system is parallel, everything is searched at once. The combinatorially growing search trees, common in some information retrieval methods, do not occur.

This leads to a difficult problem for distributed systems: how to debug an incorrectly functioning system. In a complex computer program that malfunctions there is faith that the cause is reasonable, discoverable, and correctable. In a distributed system, the error is also distributed and is no more localizable than the correct information. There are some ways around this problem; the error correction procedures discussed next are one example. In general, this is a major difficulty for distributed systems because stored information interacts in so many unpredictable and subtle ways. This problem will not yield to a quick fix—first, since it is fundamental, and second, since some of the more desirable features of distributed systems (such as inference, prototype extraction, and concept formation) are due to the same interactions in a constructive role.

Error Correction

The state vector models presented up to now can learn arbitrary associations in a few favorable cases. Correct operation of the system consists of generating the proper output paired with the input during learning. Suppose due to the nonorthogonality of the learned inputs, or noise, the correct output does not appear. That is, if g is the correct association, what can be done if Af does not equal g?

We have described a classical statistical problem. Because of the correlational nature of the synaptic modification scheme we have assumed, it has been pointed out [30], [19], [20] that the associative model discussed and simple variants are often optimal in the least mean squares sense: the output is often the best linear estimator of what the completely correct response pattern would be, and many useful neural models realize known statistical techniques.

How can we modify our system to converge to the correct, or nearly correct, association? As Sutton and Barto [57] have pointed out, a slight modification of the association model implements the well-known Widrow–Hoff procedure (see, for example, Duda and Hart [15]). Suppose that, instead of incrementing the connectivity matrix with the outer product gf^T, we use the error signal

$$\Delta A = \eta(g - Af)f^T.$$

Here, we learn the difference between what the output of the system ought to be (g) and what it actually is (Af). We assume the learning system has access to both patterns: trying to match a pronunciation during foreign language learning could be one example of such a situation.

This algorithm is computationally robust and converges well for many situations. It can be shown in many cases to converge to the best mean square error approximation possible for a linear system. We can see immediately that the system can be stable if there is no error since there is then no learning.

Dynamic error correction procedures have an additional interesting property: they contain a kind of short-term memory. This approach has been applied with success by Heath [23], who has shown that some psychological reaction time data can be quantitatively modeled by assuming a rapid adaptive process in memory that acts like an error correction procedure.

The short-term memory aspect arises in the following way. If we consider the foregoing formula when a sequence of associated inputs and outputs that are not orthogonal are learned, the immediately last pair learned will (with appropriate parameters) have small or no error. As more pairs are learned, pairs further in the past will develop an error because of the later material stored in the system. In situations where perfect association does not occur, the system gives its most accurate response to items presented in the immediate past: a recency effect. Fig. 8 shows a computational example of this as a demonstration. Interestingly, the amount of recency is powerfully affected by the nature of the learned vectors. The top trace is the correlation between desired and actual output vectors for random vectors as the last presentation of the pair recedes into the past. The bottom trace plots the same thing for a highly structured set of vectors generated for a cognitive example, where a number of the input vectors were highly correlated in parts, causing interference between associations. The qualitative prediction is that a correlated group of inputs should be more difficult to keep straight in short-term memory and should show a relatively stronger recency effect than uncorrelated inputs. This is consistent with much psychological data.

A Nonlinear Algorithm for Categorization

Throughout this review, the emphasis has been on state vectors; nothing is analysed or represented, but one state vector is transformed into another, eventually to become a pattern of motor neuron discharges. When psychologists or linguists describe the component parts of a complex stimulus, the word most frequently used is "feature." To claim the existence of "feature detectors" in the nervous system (single elements) that respond to psychological or linguistic features is then only a small step.

Neurons, though selective, do not show this kind of extreme selectivity. Features, a very useful concept, must correspond to more complex and abstract aspects of the stimulus. We have argued elsewhere [4] that it is possible to discriminate two different kinds of featurelike entities in a way that seems consistent with single neuron properties. We differentiate microfeatures (selective single units) and macrofeatures (vector valued activity patterns that act in a featurelike manner).

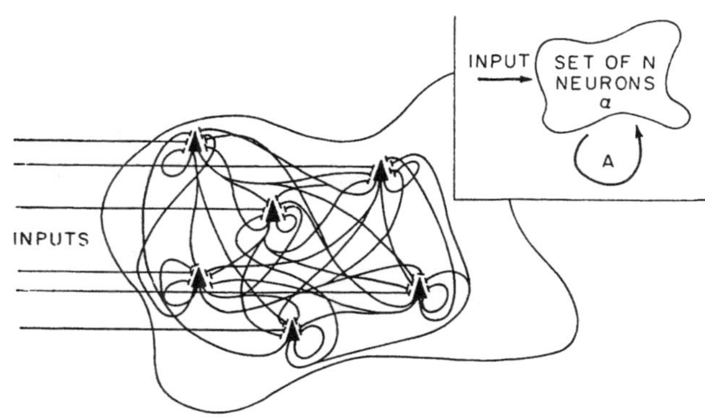

1. SET OF N NEURONS, α
2. EVERY NEURON IN α IS CONNECTED TO EVERY OTHER NEURON IN α THROUGH LEARNING MATRIX OF SYNAPTIC CONNECTIVITIES A

Fig. 8. Set of neurons projects to itself through synapses. Used as basis of nonlinear categorization model. From Anderson *et al.* [4], reprinted by permission.

The engineering literature on features sometimes defines features as vector valued quantities: the chapter in [63] on feature extraction was influential in developing the categorization model described next. Having features to be more than single elements in the vectors (i.e., selective single neurons) is neither the terminology or the tradition in psychology or neuroscience, hence the attempt on our part to emphasize this distinction with the microfeature versus macrofeature dichotomy.

We can develop a vector feature model quite easily if we assume a kind of connection known to exist in the cerebral cortex. The cortex contains an extensive set of collateral connections so that one pyramidal cell can influence another over a distance of millimeters. These collateral connections are not inhibitory (i.e., they are not "lateral inhibition," though this exists also) but probably excitatory, as indicated by the shape of the neurotransmitter containing vesicles in the collateral synapses. (For more details of the relevant anatomy see Szentagothai [58] or Shepherd [54].) Note that the system of extensive lateral interconnections forces the system to become closely related to many relaxation type models, models which give a generally good account of themselves on the types of problems that concern us here [48], [24].

Feedback Models

Let us assume that we have a single set of neurons and that this set of neurons projects to itself over a set of modifiable synapses. This anatomy is shown in Fig. 9. Let us assume that this set of lateral interconnections shows the same kind of Hebbian modification discussed earlier. When a pattern of activity learns itself, the synaptic incremental matrix is given by

$$A(i, j) = f(i)f(j).$$

The form of the resulting matrix is essentially that of the sample covariance matrix. This means that the kinds of

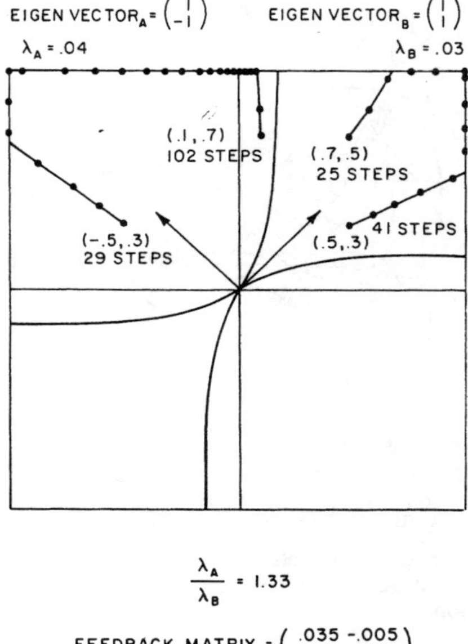

Fig. 9. Simple example of two-dimensional brain-state-in-a-box model. x and y axes correspond to activities in two-neuron system. Feedback is applied through feedback matrix, which has eigenvectors pointing toward corners and eigenvalues as shown. Curved lines passing through origin are boundaries of equivalence regions corresponding to one or another corner. Dots are placed on trajectories every five iterations, and total number of steps required to reach corner is placed next to starting point. From Anderson and Silverstein [5], reprinted by permission.

results obtained in principal component analysis (factor analysis) hold, and the eigenvectors, with the largest positive eigenvectors of the resulting matrix, contain the largest amount of the variance of the system. This is another example of the correlation synapse giving us a statistically useful result. The cognitive importance of these eigenvectors is that if we wish to discriminate members of a stimulus set, these particular eigenvectors are the ones to use.

However, this is also a feedback system. The coefficient of feedback of a pattern corresponding to an eigenvector is a function of the eigenvalue of that eigenvector. This positive feedback will cause a relative enhancement of eigenvectors with large positive eigenvalue over the others. We have created a system with differential weighting of a useful set of eigenvectors. This looks something like feature analysis (representation is a better word), but at no point was the stimulus actually analyzed into its component parts.

Nonlinearities

We should discriminate two aspects of this model: learning and performance. Feedback, eigenvector enhancement, takes place in real time. Learning takes place on a different time scale and may or may not affect feedback, depending on the learning time constant.

Let us consider a particular example of such a system. Suppose we have a feedback system which has generated a feedback matrix A. An input $x(0)$ is presented to the system. Suppose decay of activity is quite long (i.e., the membrane time constants are long, say). Left to itself, the activity pattern would be unchanged. When feedback through the matrix enters the system, then we have, assuming linear addition of activities again, after t time periods,

$$x(t+1) = x(t) + Ax(t)$$
$$= (I + A)x(t)$$

as a convenient discrete representation of the underlying continuous system.

This is positive feedback system, and any nonzero eigenvector with a positive eigenvalue will increase its activity without bound. Such instability is undesirable and untypical of the nervous system which is normally exceedingly stable under almost all conditions. The simplest (traditional) way of stabilizing the system is simply to put limits on element activity. This is quite consistent with physiology where cells can fire no slower than zero or faster than a limiting rate. This has the effect of putting the state vector in a hypercube, leading to the nickname for this model of the "brain-state-in-a-box."

Qualitative Dynamics

Once the system has learned, the qualitative dynamics of this nonlinear system are quite intuitive (and easy to simulate on a computer). If we start with an activity pattern inside the box, it receives positive feedback on certain components which have the effect of forcing it outward. When its elements start to limit (i.e., when it hits the walls of the box), it moves into a corner of the box where it remains for eternity. The corners then become particularly salient aspects of the system, and the elements take on a hybrid aspect, being partially continuous (when within the box) and partly discrete (when limited).

Fig. 10 shows a simple two-dimensional system of this kind. The connectivity matrix has two eigenvectors pointing toward corners. The box becomes divided into regions, where the final state of every point in the region is the same corner. This system forms a categorizer in a strict sense. All information about the starting point is lost, and the final state only contains the category information.

This is a classification algorithm. Its novelty is that it actually constructs the classification rather than represents the classification. We have suggested that such a system might perform a useful function as a preprocessor of noisy data, since moderate noise in the starting point is supressed in the final state.

Simulations

We have done a large number of simulations of this system over the past few years, experimenting with parameters, learning assumptions, etc. As might be expected, because of powerful feedback coupled with a limiting nonlinearity, the system is robust, and most variants work similarly. (A detailed analysis of a simulation is given in [3] and another in [2].)

A typical simulation would start with a set of vectors to be discriminated, say codings of letters, or whatever. This

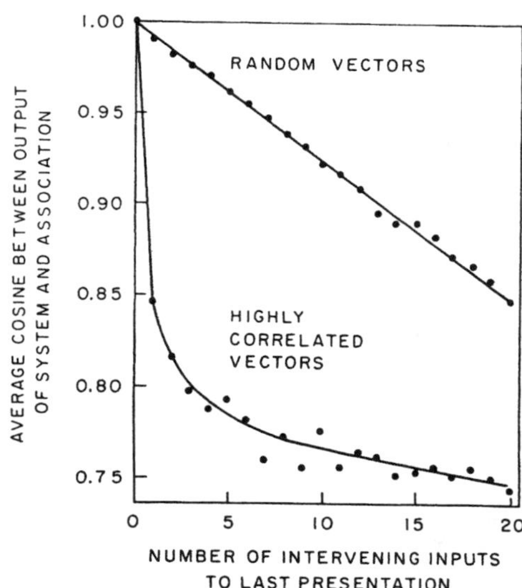

Fig. 10. Short-term changes in response can be seen when error correction techniques is used. Here input is presented, and as additional inputs from set are learned in error correction procedure, response to first input becomes less and less accurate as last presentation recedes in time. Loss of accuracy with time is function of nature of input set. Random, nearly orthogonal vectors gradually lose accuracy; highly correlated input set shows more rapid dropoff.

starting state vector $x(t)$ would change under the influence of feedback according to the rule given earlier, so that

$$x(t+1) = (I + A)x(t).$$

After seven iterations (seven is chosen arbitrarily as being large enough to approximate a continuous system and small enough to compute quickly), the process is stopped. The final state is then learned according to the rule

$$\Delta A = \eta x x^T.$$

The main technical problem is making learning "turn off" when the synapses have learned enough. The simplest assumption is that no synapse associated with a cell learns when the cell limits. (This means the corresponding row and column of the incremental matrix are set to zero.) Limitation rules related to error correction models work somewhat better. The matrix is no longer the sample covariance matrix, but often is closely related to it, sufficiently enough to have many of the same desirable properties.

In the first few presentations very few elements saturate, and the matrix learns rapidly. Typically, all members of the input set finish in the same corner if the matrix is allowed to operate until all elements are limited. As learning goes on, members of the input set start to separate, so different final states emerge. When learning has ceased (i.e., all elements of the input vector saturate in seven iterations) all or almost all of the input set have separate corners.

This model has several psychologically interesting aspects. The first is its categorization behavior. Anderson *et al.* [4] pointed out that this model duplicates some of the qualitative properties of categorical perception in speech perception. Second, the model tends to work both faster and better as it learns. The most common aspects of the stimulus set are learned first. Third, the macrofeatures that the model develops are indeed satisfactory in representing the input set (see Anderson and Mozer [3]). Fourth, the model has been used to give a quantitative account of a classical effect in statistical learning theory called "probability matching." Fifth, it can be used to generate concepts and compute with conceptlike elements in cognitive applications, as we discuss next.

Brain Damage

One of the properties claimed for distributed models is that they are damage resistant. If function is spread over many elements, loss of a few will not do much harm. Wood [62] has done simulations on the neural model presented here studying the effect of selective "ablations" of matrix elements. The effects are more complex than at first might be thought. Although the statistical predictions are clear-cut, removing some elements may be harmful for particular associations, giving a mixed picture of distribution of function (sometimes) and localization (sometimes). Also, real brains have topographic representations of sensory and motor areas, and localized lesions there will often give rise to very specific deficits, no matter how distributed is the rest of the system.

We have also performed ablation studies on a brain-state-in-a-box simulation, confirming Wood's observations [1]. We used a 50-dimensional system, presented with an input set of vectors representing 26 letters, using oriented line segments to represent the letters. (Details are available in Anderson and Mozer [3].) After a few thousand learning trials, a very stable set of final categories was formed, so that all input vectors representing different letters were either in corners by themselves or with a small number of other letters. (The codings used by design contained some very difficult discriminations with highly correlated input vectors.) Initially, the connectivity matrix was 90-percent connected; that is, ten percent of the matrix elements were permanently zero. After 45 000 more learning trials, no significant changes occurred in the final corners. The matrix at 5000 learning trials was used for the ablation. Small numbers of elements were randomly set to zero, and categorization behavior of the system was observed. Physiologically, this corresponds to loss of synapses rather than neurons. The connectivity was changed in four-percent decrements to a final value of ten percent. Fig. 11 shows the average value of correlation between corners at 90 percent and corners at intermediate stages. Essentially, no change in categorization occurred until 20 percent of the matrix elements had been removed. It was possible to see an increase in the number of iterations required to reach a corner before the change in categorization.

Ablation studies in animals usually show that slow damage is much less harmful than rapid damage. The brain, like most organs, is powerfully homeostatic and will resist change. Suppose we allow learning between ablations. When the matrix was allowed to learn for 1000 trials

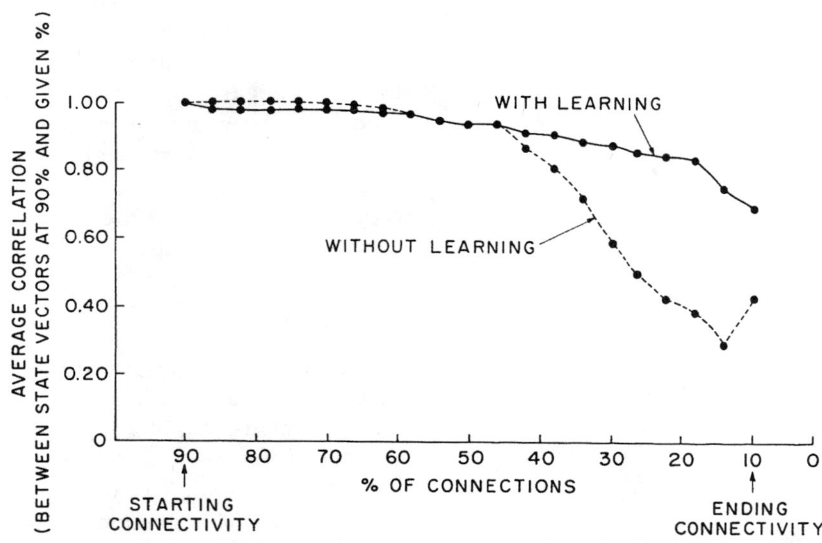

Fig. 11. Computer ablation study. Set of codings designed to represent letters was categorized by brain-state-in-a-box model with 90-percent connectivity. Connections were removed, four percent at time, and resulting categorizations were observed. With learning, system was allowed to learn for 1000 presentations of letters between each ablation. From Anderson [1], reprinted with permission.

between each four-percent ablation, we saw similar behavior in the simulation. As can be seen from Fig. 11, learning maintained the category structure of the system in the face of significantly larger amounts of diffuse damage.

IV. An Example of a Cognitive Computation

Let us give an example of a simple cognitive computation. We have discussed simple models for association, abstraction, and feature analysis, as well as a few of the biological constraints that our models must fulfill. One way to proceed is to use these models as tools to build complex structures to perform the kinds of operations that we perform. This allows us to see if our models are computationally adequate.

An important cognitive function is the following: can the models make use of partial information to generalize or infer facts not specifically presented? Humans are extremely good at this. A successful model for distributed inference using state vectors is given in [25]. Hinton's elements were somewhat more intelligent than the model neurons we used, but the distributed representation of information is similar. See also [14] for an example of a parallel system which can perform mathematical conjecture, studying many examples of numerical behavior and inferring regularities from its experience. McClelland and Rumelhart [37], [51] have described a distributed word and letter recognition simulation which uses a network of lateral and hierarchical connections between elements. Their model does an impressive job of accounting for many effects found in the psychological literature on the perception of words and letters, and its reconstructive behavior acts similarly to the model to be described

Suppose we have some information we want to represent in a small nervous system: specifically, we have several words. These words represent entities in the real world. Because of computer limitations and for ease in presentation, we will arrange our state vectors in a convenient way. We use a 50-dimensional system. Individual words are sets of 16 nonzero elements. We use our general assumption about neural coding: things that are different in the world are different in their neural codes. The coding is sufficiently rich so that different things will have orthogonal codings for their state vectors. We assume the state vectors are partitioned so that elements 1–16 contain names (Socrates, Alcibiades, Plato, Zeus, Apollo, Diana), elements 17–32 contain supernatural status (man or god), and elements 33–48 contain life span (immortal or mortal). Elements 49–50 were not used in this simulation and were initially set to zero. We assumed that all state vectors were orthogonal and values could only be $+1$ or -1, represented by $+$ and $-$ in the figures. Other elements were set to zero, with "." representing zero.

Orthogonality in this system can be achieved in two ways. Members of groups (i.e., Zeus and Socrates) use the same neurons but have orthogonal codings (in this case, Walsh functions). Members of different groups (i.e., Zeus and immortal) use different neurons entirely. This structure conforms to the psychological truth that antonyms (black–white or mortal–immortal) are actually very similar, since they make use of the same set of features in different ways, often differing from one another along a single dimension. Things dissimilar (shoes–sealing wax or cabbages–kings) share no or very few common conceptual elements. The convenient partition of the state vector is artificial. A more realistic simulation would take the active elements, mix them together arbitrarily, and add in several hundred zero elements to represent nonresponding cells. The ten different word vectors used are presented in Fig. 12.

```
                    Words -- Inference Program

Name              Species           Life span            Title

+++++++--------- ................ ................ ..  Socrates
----++++----++++ ................ ................ ..  Alcibiades
+-+-+-+-+-+-+-+- ................ ................ ..  Plato
--++--++--++--++ ................ ................ ..  Zeus
++++--------++++ ................ ................ ..  Apollo
+--++--++--++--+ ................ ................ ..  Diana

................ +-+-+-+-+-+-+-+- ................ ..  Man
................ --++--++--++--++ ................ ..  God

................ ................ +--++--++--++--+ ..  Mortal
................ ................ --------++++++++ ..  Immortal
```

Fig. 12. Fifty-dimensional state vectors learned by system for interference simulation. Plus is value of +1, minus is −1, and "." corresponds to zero. These small 16-dimensional vectors were words that matrix learned first.

Teaching the Matrix

We wish our system to be able to infer missing bits of information when appropriate. We will use the brain-state-in-a-box model to do the learning. The essence of this model is that it is an automatic classification algorithm, that generates an output state vector from an input one, driven by the synaptic connectivity matrix. It is a categorizer with only a discrete number of final states and thus can perfectly reconstruct missing vector information in appropriate situations. Again, we emphasize the importance of this behavior to the nervous system: the nervous system is not trying to analyze anything, it is trying to generate appropriate behavior. It is action oriented, as Arbib has said, and its output is a suitable set of neuron discharges.

We use a partially connected matrix (from 50 to 75 percent in this simulation). Zeros appear along the main diagonal, for aesthetic reasons. Feedback of a neuron on itself exists in the nervous system but is almost always "special" in some way, often serving an inhibitory gain control or gating function. Renshaw cells which feedback on and gate by inhibition spinal motor neurons provides one example. A small amount of Gaussian noise is present, for realism.

First the system is taught the words. The developing matrix was presented the words randomly for 500 learning trials. Second, since we wish to test ability to regenerate missing information, we will teach the matrix pairs of words. We present "Socrates mortal," "Socrates man," "Zeus immortal," and "Zeus god," for example, randomly for 1000 trials. The complete set of assertions learned is give in Fig. 13.

The entire teaching process given here takes about 5 min of CPU time on a VAX 11/780. A number of parameters exist, but exact values are not critical. After a few false starts getting the parameters in the right range, the matrix learned satisfactorily three times in a row using different initial conditions. The critical variable was to ensure that there was sufficient synaptic flexibility to learn the new information being presented. This was achieved most directly by learning the words with 50-percent connectivity and then learning the pairs of words with 75-percent connectivity, ensuring 25 percent uninstructed synapses.

```
              Learned Patterns -- Inference Program

Name              Species           Life span            Title

+++++++--------- ................ +--++--++--++--+ ..  Socrates mortal
----++++----++++ ................ +--++--++--++--+ ..  Alcibiades mortal
+-+-+-+-+-+-+-+- ................ +--++--++--++--+ ..  Plato mortal

+++++++--------- +-+-+-+-+-+-+-+- ................ ..  Socrates man
----++++----++++ +-+-+-+-+-+-+-+- ................ ..  Alcibiades man
+-+-+-+-+-+-+-+- +-+-+-+-+-+-+-+- ................ ..  Plato man

--++--++--++--++ ................ --------++++++++ ..  Zeus immortal
++++--------++++ ................ --------++++++++ ..  Apollo immortal
+--++--++--++--+ ................ --------++++++++ ..  Diana immortal

--++--++--++--++ --++--++--++--++ ................ ..  Zeus god
++++--------++++ --++--++--++--++ ................ ..  Apollo god
+--++--++--++--+ --++--++--++--++ ................ ..  Diana god

................ +-+-+-+-+-+-+-+- +--++--++--++--+ ..  Man mortal
................ --++--++--++--++ --------++++++++ ..  God immortal
```

Fig. 13. Fifty-dimensional state vectors representing associations between pairs of words. After learning words, system next learned pairs of words.

```
          Response to Learned Patterns -- Inference Program

Name              Species           Life span            Title

-+++++++----+--- +-+-+-+-+-+-+-+- +--++--++--++--+ --  Socrates mortal
----++++----++++ +-+-+-+-+-+-+-+- +--++--++--++--+ --  Alcibiades mortal
+-+-+-+-+-+-+-+- +-+-+-+-+-+-+-+- +--++--++--++--+ --  Plato mortal

-+++++++----+--- +-+-+-+-+-+-+-+- +--++--++--++--+ --  Socrates man
----++++----++++ +-+-+-+-+-+-+-+- +--++--++--++--+ --  Alcibiades man
+-+-+-+-+-+-+-+- +-+-+-+-+-+-+-+- +--++--++--++--+ --  Plato man

---+---+--++--++ --++--++--++--++ --------++++++++ --  Zeus immortal
++++---+----+++- --++--++--++--++ --------++++++++ --  Apollo immortal
+--++--++--++--+ --++--++--++--++ --------++++++++ --  Diana immortal

---+---+--++--++ --++--++--++--++ --------++++++++ --  Zeus god
++++---+----+++- --++--++--++--++ --------++++++++ --  Apollo god
+--++--++--++--+ --++--++--++--++ --------++++++++ --  Diana god

+-+-+-+++---+-++ +-+-+-+-+-+-+-+- +--++--++--++--+ --  Man mortal
---+---+--++--++ --++--++--++--++ --------++++++++ --  God immortal
```

Fig. 14. Fifty-dimensional state vectors corresponding to final states of system after learning. Limits of box were set at 1.5, so plus now corresponds to 1.5, and minus to −1.5. Note that system has filled in missing parts of input state vectors (Fig. 13) in way consistent with what it has learned.

Operation of the System

Having learned that Socrates is a man and that men are mortal, we would like to generate the triple, "Socrates is a man and mortal." This would be a satisfactory output from the program. Given a new person, say, Herb, who is also a man, if we present "Herb is a man," we would like to retrieve at the output "Herb is a man and mortal." The matrix we have generated will do this. Fig. 14 shows responses to the pairs of words the system originally learned.

The system supplies the missing information in all the cases the system learned. The system was never actually presented with a triple, but with pairs of words. In every pair containing a name, the missing information is filled in correctly. For names, the output pattern is not exactly the same as the input for the name but may have changed in one or another sign; the pattern of signs for the second and third word locations were identical to those actually learned in the case presented here, but this is not always the case. If the correct triples are presented, the final states are exactly those presented here as the final states for incomplete information. It would be possible to force the system to a desired final state using a correction procedure as described earlier if this was felt to be necessary.

Two pairs do not contain a name: "man mortal" and "god immortal." In the first case, the final state contains nonsense in the name location (a pattern which does not correspond to any name the system saw). In the second case, the final state reads "Zeus god immortal." If we present only the pattern "god," we obtain this triple. Zeus has become the prototypical god, whereas the prototypical man was not one of those actually learned. Further analysis seemed unpropitious.

If we present a new example of a man, say Herb, with a state vector orthogonal to those learned, then the final state corresponds to "Herb man mortal," but the name location contains several unsaturated elements, corresponding to long reaction time (time to reach a corner) and what might be interpreted as uncertainty. Vector Frank, not orthogonal to the other names, is transmuted in the memory process to be close to Plato. Odin, also not orthogonal, ends near to Zeus, suggesting Odin and Zeus are different names for the same thing. Statements the system saw or which conform to the information presented reach corners rapidly. If we deliberately present a state vector containing an error, the system can respond in several ways. Fig. 15 give examples of system responses to different test vectors which contain partial or erroneous information.

If we present "Socrates immortal," then the final state contains "god immortal" but a nonsense pattern containing an unsaturated element in the name location. If we present "Plato god immortal," then the system ends up containing "god immortal," but the name location contains nonsense with three unsaturated elements. If we present "Zeus man immortal," then we get nonsense in both the name and "man" location. "Zeus man mortal" generates several errors in the name location.

```
Test Stimuli -- Inference Program

Name            Species          Life span            Title

Learned Patterns - Mortals:

-++++++----+--- +-+-+-+-+-+-+-+- +--++--++--++--+ --  Socrates
---+++----++++ +-+-+-+-+-+-+-+- +--++--++--++--+ --  Alcibiades
+-+-+-+-+-+-+-+- +-+-+-+-+-+-+-+- +--++--++--++--+ --  Plato

New mortals:

--.+++++--..---+ +-+-+-+-+-+-+-+- +--++--++--++--+ --  Herb
-..+++++--.-.--+ +-+-+-+-+-+-+-+- +--++--++--++--+ --  Herb man
-..+++++--.-.--+ +-+-+-+-+-+-+-+- +--++--++--++--+ --  Herb mortal
+-+-+-+-+-+-+-+- +-+-+-+-+-+-+-+- +--++--++--++--+ --  Frank man

Learned patterns - Gods:

---+---+--++--++ --++--++--++--++ -------+++++++ --  Zeus
++++---+----+++- --++--++--++--++ -------+++++++ --  Apollo
+--++--++--++--+ --++--++--++--++ -------+++++++ --  Diana

New god:

--.+---+--++--++ --++--++--++--++ -------+++++++ --  Odin immortal

Learned patterns:

+-+-+-++---+-++ +-+-+-+-+-+-+-+- +--++--++--++--+ --  Man
---+---+-++--++ --++--++--++--++ -------+++++++ --  God
+-+-+-++---+-++ +-+-+-+-+-+-+-+- +--++--++--++--+ --  Mortal
---+---+-++--++ --++--++--++--++ -------+++++++ --  Immortal

Deliberate errors:

+-+----++-++--++ +-+-+-+-+-+-+-++ -------+++++++ --  Man immortal
+.+++-.++---+--. --++--++--++--+- +--++--++--++--+ --  God mortal

-+-+++-+---.++-- --++--++--++--++ -------+++++++ --  Socrates immortal
-+-+++-+---.++-- --++--++--++--++ -------+++++++ --  Socrates god

+--..---+-+++.+- --++--++--++--++ -------+++++++ --  Plato god immortal
+---+-+-+-+-+-+- +-+-+-+-+-+-+-+. -------+++++++ --  Plato man immortal

--++---+--++--++ +-+-+-+-+-+-+-++ -------+++++++ --  Zeus man immortal
--+-.-++.-+---++ +-+-+-+-+-+-+-+- +--++--++--++--+ --  Zeus man mortal
```

Fig. 15. Fifty-dimensional state vectors used to test stimuli after learning. These special, partial, or erroneous stimuli were used to check different aspects of resulting inference system (see text). Limits of box were set at 1.5 as in Fig. 14. "." now means that element did not reach limit in 50 interations.

Conclusion

We have produced a system which can generate missing information. This could easily be considered to be inference, under one interpretation, since presentation of partial information generated a consistent whole. It is also similar to what is called "property inheritance," another interpretation. This is because partial information automatically brings in and presents associated information.

Relations to Formal Logic

The reason this particular set of names was used was obviously that it conforms in structure to the most famous of the classic syllogisms: "Socrates is a man, men are mortal, therefore Socrates is mortal." Clearly, our system is not doing formal logic. It is not clear that humans do formal logic, either. Logic is extremely difficult for most humans, and we need machines, diagrams, and extensive use of memory to work with formal logic. Human reasoning and logic is highly memory-oriented, using analogy, probabalistic inference, and a very effective but nonlogical set of strategies to infer information about a real, not an abstract, world. Our model has more similarities to human logic than formal logic.

V. Summary

We have used vectorlike quantities as elementary entities in all these models. We have showed that these entities can be used to perform operations reminscent of simplified version of a few psychological abilities of humans. Using state vectors in this way may not be the most effective or efficient way to do the computation. The resulting systems have serious trouble with logic and accuracy and have problems with unpredictable errors. However, in return, they provide properties like abstraction, concepts, and possibly inference that looks somewhat like ours. Use of systems like these is a constraint forced on us, I feel, by nervous system organization. Perhaps the virtues of the approach for certain kinds of computations make their drawbacks worthwhile.

Acknowledgment

Some of this work has appeared in more detail elsewhere. My collaborators Jack Silverstein, Stephen Ritz, Randall Jones, Andrew Knapp, and Michael Mozer have had a major part in the development of these ideas. I would like to acknowledge my debt to Geoffrey Hinton who suggested to me that it was indeed feasible to do computations with parallel systems. I would also like to express special thanks to Teuvo Kohonen and his collaborators whose pioneering work in the field of parallel, associative models has benefited us all. At Brown I have learned a great deal from discussions with Stuart Geman, Barry Davis, Gregory Murphy, and Richard Heath. A summer at U.S.C.D. allowed me to talk with Geoffrey Hinton. Donald Norman, David Rumelhart, and Jay McClelland and suggested a host of new problems to think about. Leon Cooper and the Center for Neural Science have provided immense help over the past few years. Computer simulations described here were performed on the VAX 11/780 of the Center for Cognitive Science, Richard Millward, Director.

References

[1] J. A. Anderson, "Neural models and a little about language," in *Biological Bases of Language*, D. Caplan, A. Smith, and A. Roche-Lecour, Eds. Cambridge, MA: MIT Press (to appear).

[2] ____, "Neural models with cognitive implications," in *Basic Processes in Reading*, D. LaBerge and S. J. Samuels, Eds. Hillsdale, NJ: Erlbaum, 1977.

[3] J. A. Anderson and M. Mozer, "Categorization and selective neurons," in *Parallel Models of Associative Memory*, G. Hinton and J. A. Anderson, Eds. Hillsdale, NJ: Erlbaum, 1981.

[4] J. A. Anderson, J. W. Silverstein, S. A. Ritz, and R. S. Jones, "Distinctive features, categorical perception, and probability learning: Some applications of a neural model," *Psychol. Rev.*, vol. 84, pp. 413–451, 1977.

[5] J. A. Anderson and J. W. Silverstein, "Reply to Grossberg," *Psychol. Rev.*, vol. 85, pp. 597–603, 1978.

[6] M. A. Arbib, *The Metaphorical Brain*. New York: Wiley, 1972.

[7] R. Sorabji, tr., *Aristotle on Memory*. Providence, RI: Brown Univ. Press, 1972.

[8] H. B. Barlow, "Single units and sensation," *Perception*, vol. 1, pp. 371–394, 1972.

[9] E. L. Bienenstock, L. N. Cooper, and P. W. Monro, "A theory for the development of neuron selectivity: Orientation selectivity and binocular interactions in visual cortex," *J. Neurosci.*, vol. 2, pp. 32–48, 1982.

[10] S. E. Brodie, B. W. Knight, and F. Ratliff, "The responses of the Limulus retina to moving stimuli: Prediction by Fourier synthesis," *J. General Physiol.*, vol. 72, pp. 129–166, 1978.

[11] ____, "The spatio-temporal transfer function of the Limulus lateral eye," *J. Gen. Physiol.*, vol. 72, pp. 167–202, 1978.

[12] M. Colonnier, "The electron-microscopic analysis of the neuronal organization of the cerebral cortex," in *The Organization of the Cerebral Cortex*, F. O. Schmitt, F. G. Worden, G. Adelman, and S. G. Dennis, Eds. Cambridge, MA: MIT Press, 1981.

[13] L. N. Cooper, "Distributed memory in the central nervous system: Possible test of assumptions in visual cortex," in *The Organization of the Cerebral Cortex*, F. O. Schmitt, F. G. Worden, G. Adelman, and S. G. Dennis, Eds. Cambridge, MA: MIT Press, 1981.

[14] B. Davis, "A neurobiological approach to machine intelligence," Ph.D. thesis, Div. of Appl. Math., Brown Univ., Providence, RI, June 1982.

[15] R. O. Duda and P. E. Hart, *Pattern Classification and Scene Analysis*. New York: Wiley, 1973.

[16] Editors of *Scientific American*, *The Brain*. San Francisco, CA: Freeman, 1979.

[17] J. M. Eich, "A composite holographic associative recall model," *Psychol. Rev.*, vol. 89, pp. 627–661, 1982.

[18] J. A. Feldman and D. H. Ballard, "Connectionist models and their properties," *Cognitive Sci.*, vol. 6, pp. 205–254, 1982.

[19] S. Geman, "Application of stochastic averaging to learning systems," *Brain Theory Newsletter*, vol. 3, pp. 69–71, 1978.

[20] S. Geman, "The law of large numbers in neural modelling," *SIAM-AMS Proc.*, vol. 13, pp. 91–105, 1981.

[21] S. Grossberg, "How does the brain build a cognitive code?," *Psychol. Rev.*, vol. 87, pp. 1–51, 1980.

[22] D. O. Hebb, *The Organization of Behavior*. New York: Wiley, 1949.

[23] R. A. Heath, "A model for signal detection based on an adaptive filter," *Biol. Cybern.*, in press.

[24] G. E. Hinton, "Relaxation and its role in vision," Ph.D. dissertation, Univ. of Edinburgh, 1976.

[25] ____, "Implementing semantic networks in parallel hardware," in *Parallel Models of Associative Memory*, G. E. Hinton and J. A. Anderson, Ed. Hillsdale, NJ: Erlbaum, 1981.

[26] G. Hinton and J. A. Anderson, Eds., *Parallel Models of Associative Memory*. Hillsdale, NJ: Erlbaum, 1981.

[27] W. James, *Psychology (Briefer Course)*. New York: Collier, 1962 (originally published 1890).

[28] E. Kandel, "Small systems of neurons," in *The Brain*, Editors of *Scientific American*, Eds. San Francisco, CA: Freeman, 1979.

[29] A. Knapp and J. A. Anderson, "A signal averaging model for concept formation," *J. Exp. Psychol., Learning, Memory, and Cognition*, submitted.

[30] T. Kohonen, *Associative Memory: A System Theoretic Approach*. Berlin, Germany: Springer-Verlag, 1977.

[31] P. Liepa, "Models of content addressable distributed associative memory (CADAM)," unpublished manuscript, Univ. of Toronto, Toronto, ON, Canada, 1977.

[32] A. R. Luria, *The Mind of a Mnemonist*. New York: Basic Books, 1968.

[33] E. Mach, "Ernst Mach," in *Dictionary of Scientific Biography*, vol. VIII, C. C. Gillespie, Ed. New York: Scribners, 1973.

[34] D. Marr, "A theory of cerebellar cortex," *J. Physiol.*, vol. 202, pp. 437–470 1969.

[35] ____, "A theory for cerebral neocortex," *Proc. Roy. Soc.*, Ser. B, vol. 176, pp. 161–234, 1970.

[36] ____, *Vision*. San Francisco, CA: Freeman, 1982.

[37] J. L. McClelland, and D. E. Rumelhart, "An interactive activation model of context effects in letter perception: Part 1. An account of basic findings," *Psychol. Rev.*, vol. 88, pp. 375–497, 1981.

[38] W. S. McCulloch and W. Pitts, "A logical calculus of the ideas immanent in nervous activity," *Bull. Math. Biophys.*, vol. 5, pp. 115–133, 1943.

[39] M. Minsky, "K-lines: A theory of memory," *Cognitive Sci.*, vol. 4, pp. 117–133, 1980.

[40] M. Minsky and S. Papert, *Perceptrons*. Cambridge, MA: MIT Press, 1969.

[41] B. B. Murdock, Jr., "A theory for the storage and retrieval of item and associative information," *Psychol. Rev.*, pp. 609–626, 1982.

[42] N. J. Nilsson, *Learning Machines*. New York: McGraw-Hill, 1965.

[43] A. Pellionisz and R. Llinas, "Brain modelling by tensor network theory and computer simulation. The Cerebellum: Distributed processor for predictive coordination," *Neurosci.*, vol. 4, pp. 323–348, 1979.

[44] W. Pitts and W. S. McCulloch, "How we know universals: The perception of auditory and visual forms," *Bull. Math. Biophys.*, vol. 9, pp. 127–147, 1947.

[45] M. I. Posner and S. W. Keele, "On the genesis of abstract ideas," *J. Exp. Psychol.*, vol. 77, pp. 353–363, 1968.

[46] ____, "Retention of abstract ideas," *J. Exp. Psychol.*, vol. 83, pp. 304–308, 1970.

[47] F. Ratliff, "Form and function: Linear and nonlinear analyses of neural networks in the visual system," in *Neural Mechanisms in Behavior*, D. McFadden, Ed. New York: Springer, 1980.

[48] A. Rosenfeld, "Iterative methods in image analysis," *Pattern Recognition*, vol. 10, pp. 181–187, 1978.

[49] E. Rosch, "Principles of categorization," in *Cognition and Categorization*, E. Rosch and B. B. Lloyd, Eds. Hillsdale, NJ: Erlbaum, 1978.

[50] E. Rosch and B. B. Lloyd, Eds., *Cognition and Categorization*. Hillsdale, NJ: Erlbaum, 1978.

[51] D. E. Rumelhart and J. L. McClelland, "An interactive activation model of context effects in letter perception: Part 2. The contextual enhancement effect and some tests and extensions of the model," *Psychol. Rev.*, vol. 89, pp. 60–94, 1982.

[52] F. O. Schmitt, F. G. Worden, G. Adelman, and S. G. Dennis, *The Organization of the Cerebral Cortex*. Cambridge, MA: MIT Press, 1981.

[53] T. J. Sejnowski, "Skeleton filters in the brain," in *Parallel Models of*

Associative Memory, G. E. Hinton and J. A. Anderson, Eds. Hillsdale, NJ: Erlbaum, 1981.

[54] G. Shepherd, *The Synaptic Organization of the Brain*, 2nd ed. New York: Oxford Univ. Press, 1979.

[55] _____, *Neurobiology*. New York: Oxford Univ. Press, 1983.

[56] E. E. Smith and D. L. Medin, *Categories and Concepts*. Cambridge, MA: Harvard Univ. Press, 1981.

[57] R. S. Sutton and A. G. Barto, "Toward a modern theory of adaptive networks: expectation and prediction," *Psychol. Rev.*, vol. 88, pp. 135–170, 1981.

[58] J. Szentagothai, "Specificity versus (quasi) randomness in cortical connectivity," in *Architectonics of the Cerebral Cortex*, M. A. B. Brazier and H. Petsche, Eds. New York: Raven Press, 1978.

[59] J. Szentagothai and M. Arbib, "Conceptual models of neural organization," *Neurosci. Res. Program Bull.*, vol. 12, pp. 307–510, 1974.

[60] R. J. Tusa, L. A. Palmer, and A. C. Rosenquist, "The retinotopic organization of area 17 (Striate Cortex) in the cat," *J. Comp. Neurol.*, vol. 177, pp. 213–235, 1978.

[61] C. Wood, "Variations on a theme by Lashley: Lesion experiments on the neural models of Anderson, Silverstein, Ritz, and Jones," *Psychol. Rev.*, vol. 85, 582–591, 1978.

[62] _____, "Implications of simulated lesion experiments for the interpretation of lesions in real nervous systems," in *Neural Models of Language Processes*, M. A. Arbib, D. Caplan, and J. C. Marshall, Eds. New York: Academic, 1983.

[63] T. Y. Young and T. W. Calvert, *Classification, Estimation, and Pattern Recognition*. New York: American Elsevier, 1974.

Article 2.3

LINEAR ANALYSIS OF THE DYNAMICS OF NEURAL MASSES[1]

WALTER J. FREEMAN

University of California, Berkeley, California

This review is concerned with the problem of how to identify and characterize masses of neurons empirically as dynamic entities that have properties related to but distinct from those of the component neurons. This can only be done by observing and measuring experimentally the responses of neural masses to known stimuli. But measurement presupposes some theoretical framework to provide the necessary basis functions and the units. For example, a value for frequency is predicated on the presence of a periodic wave form; a rate constant presumes an exponential decay; a numerical estimate of dispersion requires some distribution function; and so forth. The purpose of this review is to suggest that linear systems analysis can provide a very useful structure for measuring compound neural responses and for comprehending some of the elementary underlying dynamics. It does not define the entities nor explain what they do or how they do it. It merely provides the basis for measurement, which is the first step beyond uncorrelated observations and the prerequisite for testing theories of neural masses.

WHAT IS A NEURAL MASS?

Granted that nervous systems are composed of neurons, the most compelling fact about even the simpler brains is that the numbers of neurons are extremely large. Because the neuron is conceived as the elementary unit of neural function, and because it is accessible to measurement by an array of microtechniques, the proper aim of neurophysiology is the analysis of brain function in terms of the properties of single neurons. Yet most of what is known about the operation of the brain in relation to behavior has come from studies based on stimulation (electrical or chemical), field potential recordings, or ablation (by selective surgery or disease) of masses of brain tissue containing tens or hundreds of millions of neurons. The conceptual gap between the functions of single neurons and those numbers of neurons is still very wide.

The parts of nervous systems accessible to experimental observation may be arranged in a logarithmic scale, as in Table 1, which contains examples of estimated numbers of neurons in some well-known preparations (1–3). Listed in the lower third of the table, extending roughly from 10^1 to 10^4 neurons, are some isolatable parts of invertebrate nervous systems, including the cardiac ganglion of the lobster (2), the eye of the horseshoe crab (5), and the visceral ganglion of the sea-snail (4). To those might be added the monosynaptic sensorimotor relays of the vertebrate spinal cord (6). These are systems for which the relatively small numbers of neurons have offered hope of analysis and understanding by models based on discrete networks of single simulated neurons.

The upper third of the scale, 10^8 to 10^{10} neurons, is represented by structures from the vertebrate brain, which on surgical removal or destruction by disease leave recognizable and reproducible deficits in behavior, such as anosmia, cortical blindness, hemiplegia, etc (7).

Between these levels is a range of numbers, 10^4 to 10^7 neurons, which are too few to leave notable behavioral deficits on ablation, unless they comprise a projection pathway such as the optic nerve. On the other hand the numbers are too great to conceive of modeling in terms of networks of finite numbers of cells.

[1] This work was supported by NIMH Grant MH 06686.

This is the region over which a conceptual span is needed, to account for the behaviorally related properties of brains in terms of single neurons. It is the domain of neural masses.

WHY USE LINEAR SYSTEMS ANALYSIS?

The neural masses in the central numerical region can be conceived empirically as occupying a few mm² of cortical surface, or a few mm³ of nuclear volume in the brain stem or spinal cord. They have three properties of particular interest in the present context.

First, the output of a neural mass is often accessible to measurement as a holistic event, such as a compound action potential of a nerve trunk, a field of potential in the volume of the mass owing to the extracellular spread and summation of dendritic current, or some derivative event such as the strength of a muscle contraction. For neural masses in the brain the electrical field potentials may be the single most valuable source of information about their dynamic properties, because they manifest the weighted instantaneous sum of extracellular potentials generated by large numbers of neurons. The problems of determining the locations, distributions, and active states of neurons in the masses generating such fields have been discussed in numerous articles and reviews (8–13), and the subject is still grossly underdeveloped. Yet it seems undeniable that such events, when properly analyzed, provide an essential key to the understanding of neural masses.

TABLE 1. Numbers of neurons in logarithmic scale[a]

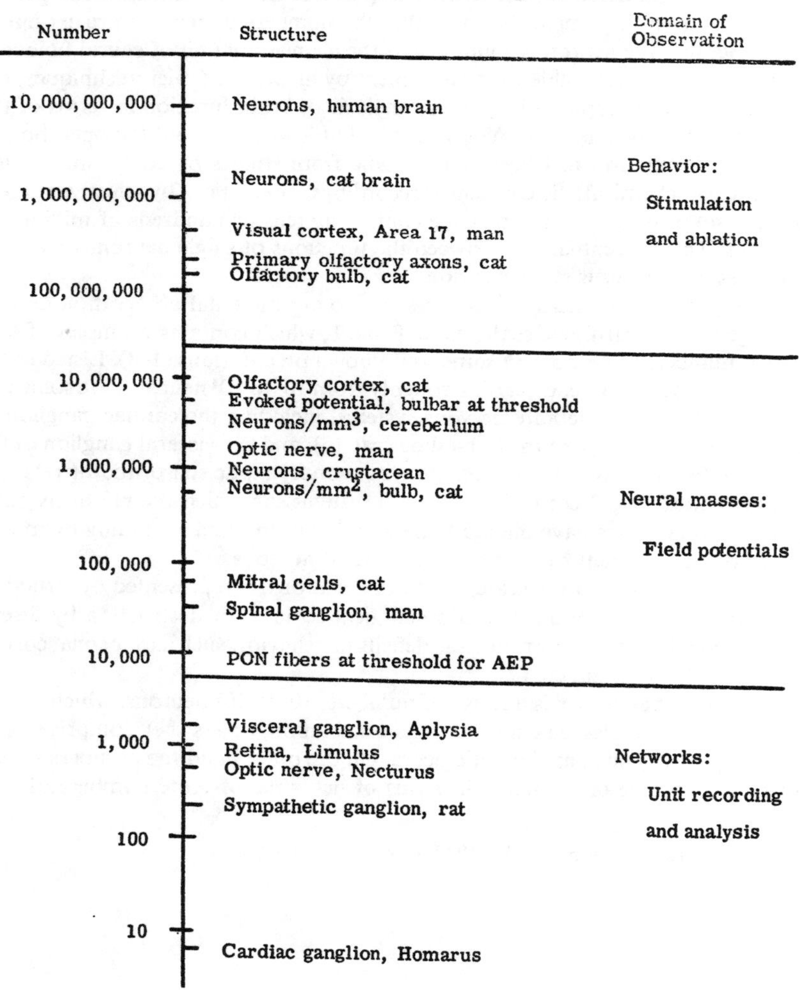

Number	Structure	Domain of Observation
10,000,000,000	Neurons, human brain	
	Neurons, cat brain	Behavior:
1,000,000,000		Stimulation
	Visual cortex, Area 17, man	and ablation
	Primary olfactory axons, cat	
	Olfactory bulb, cat	
100,000,000		
10,000,000	Olfactory cortex, cat	
	Evoked potential, bulbar at threshold	
	Neurons/mm^3, cerebellum	
	Optic nerve, man	
1,000,000	Neurons, crustacean	
	Neurons/mm^2, bulb, cat	Neural masses:
100,000		Field potentials
	Mitral cells, cat	
	Spinal ganglion, man	
10,000	PON fibers at threshold for AEP	
	Visceral ganglion, Aplysia	
1,000	Retina, Limulus	Networks:
	Optic nerve, Necturus	
	Sympathetic ganglion, rat	Unit recording
100		and analysis
10		
	Cardiac ganglion, Homarus	

[a] Some examples are listed on a logarithmic scale of estimated numbers of neurons in some common preparations. From references 1–5 and the author's unpublished data.

Second, the output of neural masses characteristically is graded over the physiological range of function. Output is proportional to input, within limits, and the responses to two or more inputs are additive. This set of properties was explored and documented in quantitative detail by Sherrington (14–16) in his studies of reflex mechanisms in the brain stem and spinal cord, which were the immediate precursor for the analysis of the synaptic potentials of neurons by use of the intracellular microelectrode (6, 17, 18). Sherrington described his principle results as demonstrating the algebraic summation of excitatory and inhibitory influences in the neurons of the spinal cord.

These properties imply that within appropriate limits of amplitude, the operations of neural masses conform to the principle of superposition, and to that extent their dynamics can be described by means of linear differential equations. The principle has been directly verified for spinal motorneurons by Granit (17) and is the basis for the quantitative analysis of the dynamics of the array of receptor neurons comprising the eye of *Limulus* (19, 20).

The relevance of these properties to some aspects of transmission in the higher nervous systems has been documented among others by Stark & Sherman (21) in the analysis of the pupillary reflex as a servomechanism; by Lopes da Silva (22), Cleland & Enroth-Cugell (23), Maffei (24), and Regan (25) in the analysis of visual cortical potentials evoked by sine wave modulated light; and by Tielen et al (26) in the observation of electrical responses of auditory cortex to sinusoidally modulated sound.

Third, neural masses are readily accessible to electrical stimulation with extracellular electrodes on peripheral afferent nerves or in the tracts, nuclei, and cortical surfaces of the brain. Characteristically such stimulation activates large numbers of axons in the vicinity of an electrode, i.e., it is a multicellular input leading to the activation of masses of neurons. Mathematically it can be approximated by a delta function in the time dimension although not in the spatial dimensions of the stimulus. This form of activation has the additional advantage of bypassing the receptors through which sinusoidal stimuli are delivered (21–26) and for which transfer functions are difficult to obtain.

These three properties form the basis for the expectation that some of the fundamental characteristics of neural masses might be described in terms of linear systems analysis. Very simply, the neural mass under study is subjected to an electrical pulse, which is superimposed on the background activity normally present in the input pathways to all neural masses. The output field potential, after suitable averaging to remove the background activity generated by the mass (comprising the EEG waves of the mass), is treated as the impulse response of the system.

By use of paired shocks or repetitive trains of stimuli (27, 28) a linear range of function is defined. In this range the impulse response can be treated as the sum of a set of terms, which are the solution to a set of linear differential equations having the boundary conditions corresponding to an impulse input (29). These terms serve to generate exponential curves and damped sine waves, which become the basis functions (30) or elementary curves required for measurement. An appropriate set is chosen by trial and error, and the sum is fitted by use of nonlinear regression to the digitized averaged evoked potential (AEP) or impulse response (31). The Laplace transform of the evaluated curve yields the linear differential equation best describing the dynamics of the neural mass in the designated experimental state.

The combination of electrical stimulation at some reproducible site with field potential recording near the center of some evoked activity provides a useful empirical description of a neural mass. Again, it does not provide a definition; it provides a platform for observation and measurement. The most effective use of the approach requires detailed specification of the anatomy of the stimulated tract and the target mass; of the spatial distribution of the field potential and its relation to the gross and microscopic structures of the generating neurons; identification of the neuron types responsible for unit and field potentials; and so forth. One of the seeming disadvantages of the use of evoked potentials and linear analysis is the heavy requirement for detailed correlative field mapping and histological measurement. On the contrary, even with initially poor specification of stimulus and recording conditions, which is usually the case in the beginning, it is easy to get immediate results, and to feed these back into the experiment as a basis for improving the specification. By successive modifications the initially hazy conception of the neural mass under study can and should become progressively more sharply defined and richly detailed.

How Might Linear Analysis Be Applied?

The derivation of transfer functions in this approach (thus far described) is of very limited value, owing to the fact that the waveforms of AEPs vary with changes in stimulus intensity. The amplitudes of responses change in linear proportion to the input intensity, but the coefficients describing the frequencies and exponential decay rates of the responses are often exquisitely sensitive to changes in the input magnitude.

The limitation with respect to amplitude is further reflected in the fact that the modulation amplitude for sinusoidal stimulation must in most cases be restricted to some fraction of a sustained "dc" input bias, e.g. on the order of 20% for visual cortex in dogs in response to sine wave modulated light (22). The limits are similar though less stringent for retinal ganglion cells and peripheral receptors, e.g. up to 60% for *Limulus* (20). When the limits are exceeded, harmonics become prominent, and the frequency-response curves tend to change with amplitude (32).

The principal reason for this pervasive nonlinearity in the function of neural masses lies in the limitations on output of the neurons in the masses over a range of input amplitudes. Sherrington described these limits (15) with the terms "facilitation" and "occlusion." The former was attributed to the threshold of each neuron and implied that two stimuli together might cause a neuron to discharge, whereas either alone might not. The latter was attributed to the refractory properties of neurons, such that if one stimulus caused a neuron to discharge, the addition of another stimulus too soon after the first would not augment its output.

More recently the nonlinearity has been described as bilateral saturation (22, 32–38). This is a static nonlinearity (35), which can be described as an amplitude-dependent property of neural masses. Conceptually the transference of the neural mass can be separated into a linear frequency-dependent part $C(s)$ and a nonlinear amplitude-dependent part $P(V)$. The latter function serves to specify the values for the coefficients of the former over a range of input magnitudes. In this manner, using linear differential equations having state-dependent coefficients, the linear systems approach can be extended to cover a broad range of dynamic function of neural masses, in so far as AEPs and their functional correlates are concerned (28, 39).

Additionally, owing to the fact that neurons in masses communicate within themselves by dendritic currents (often but not always giving rise to EEG waves and to AEPs) but between each other by propagated action potentials (spikes or pulses), it is essential to observe and measure the pulse trains of representative neurons. These are either random pulse trains of background activity or induced responses to electrical stimulation, which are best observed in the form of post-stimulus time (PST) histograms. In either case, because observation is restricted to one or at most a few neurons at a time, it is necessary to assume that the performance of the neuron is representative of others in the mass, provided the period of observation is long enough. This quasi-ergodic hypothesis seems to work well enough for purposes of linear analysis, though how far it can be pushed is unclear. Simultaneously recorded EEG waves and background pulse trains on the one hand and of AEPs and PST histograms on the other constitute the essential raw materials for the analysis of the dynamics of neural masses.

In What Way Does the Mass Differ from the Neuron?

The neural mass is composed of neurons, and its properties have a generic resemblance to those of single neurons. But it is not identical to them, and its properties cannot be predicted or evaluated wholly from measurements on single neurons. This is partly because the properties of the mass depend to some extent on distributions of various parameters of single neurons and partly because the properties of the mass depend on the massive connectivity of large numbers of neurons. Both of these facets are largely inaccessible to single-unit analysis.

In order to illustrate this proposition examples will now be given of the calculation of the amplitude-dependent transference $P(V)$, where V is wave amplitude, and the frequency-dependent transference $C(s)$, where s is the Laplacian operator, for the neural masses in the olfactory bulb and cortex of the cat and rabbit. These structures have been thoroughly studied both anatomically (40, 41) and electrophysiologically (28, 36–39, 42–48) and have been shown to be well-suited to description using linear differential equations (34, 36, 48).

What is the Meaning of Forward Gain?

Communication within and between neural masses is assumed here to be solely on the basis of synaptic transmission. This implies that in general a neural mass receives impulses of some space-time density and transmits impulses of some differing space-time density. Forward gain is defined as the ratio of the instantaneous magnitudes of output and input. Because both take the form of pulse density functions (pulses/unit time/unit area or volume) the gain is a dimensionless factor. However, each neural mass converts its pulse density input to a dendritic current or wave function and reconverts some spatiotemporally transformed wave function into another pulse density function. The conversion for each of the two stages must be described, the first in terms of the magnitude of pulse-to-wave conversion (P-V) and the second in terms of the magnitude of wave-to-pulse conversion (V-P). The product of the two magnitudes is the forward gain of the mass.

In general there is a central amplitude range of function for both stages of a neural mass, in which the amplitude of output is proportional to that of input, and is additive as well, so that the conversions are linear. If the pulse density functions are estimated from the pulse frequencies of single-neuron pulse trains in pulses/second (pps), and the wave function is estimated from the amplitude of the extracellular field potential in microvolts (μV), then (P-V) and (V-P) conversions can be described using coefficients in units of μV/pps and pps/μV respectively. The product of the coefficients is the dimensionless gain.

In fact neither stage is linear over the achievable physiological range. Each conversion undergoes progressively stronger saturation with increasing departure from the central part of the range. The effect of saturation is to reduce the gain, so that forward gain is an amplitude-dependent nonlinearity (36, 38). The dependency of gain for the impulse response is on the input pulse magnitude or on the initial response amplitude, and the initial value for the gain holds throughout the duration of the impulse response. For this reason the nonlinearity is static and is readily susceptible to piece-wise linear approximation. It is then feasible to separate the frequency-dependent from the amplitude-dependent properties,and to describe the former with linear differential equations in time and the latter with linear differential equations in amplitude. Owing to the remarkable degree to which superposition holds in neural masses, these considerations are also valid for "spontaneous" or background activity. They do not hold for the full description of seizures or convulsive neural discharges, although the conditions for instability leading to seizures can be described (36, 48).

For (P-V) conversion the limits are imposed by the ionic mechanisms of dendritic current (49). The greater the membrane depolarization in response to an excitatory input, the less the difference between membrane potential and the equilibrium potential for the excitatory postsynaptic potential (EPSP). This reduces the effective electromotive force for the dendritic current operating into a high resistance current path, so the current increment is reduced proportionately for equal increases in excitatory input. At some level of input, were it achievable, the current would asymptotically approach a limiting value.[2]

[2] The occasional failure of superposition of intracellular postsynaptic potentials and the demonstration that conductance changes in dendritic membrane are the basis for synaptic potentials (49) has lead to the conclusion that variable membrane conductance can operate as an amplitude-dependent voltage divider for some neuronal geometries. This nonlinearity would give rise not only to saturation but also to variable passive membrane rate constants. Reasons have been given elsewhere (36) for believing that over the "physiological" range of function these rate constants are invariant. Therefore, it is concluded that even if the voltage divider effects can be demonstrated to hold for single neurons in some conditions, they do not account for the saturation effects observed in neural masses.

The same properties hold for inhibition, except that the limit is closer to the resting state, because the difference between resting membrane potential and the equilibrium potential for the inhibitory postsynaptic potential (IPSP) is (from the illustrations cited in reference 49) roughly one-sixth that for the EPSP. The input-output curve for (P-V) conversion is therefore sigmoidal with sharper curvature on the inhibitory side (Figure 1, upper right: this graph must be viewed after rotation 90° counterclockwise, for the reason given below).

The stage for (V-P) conversion is similarly bounded, on the inhibitory side by the thresholds for the trigger zones and on the excitatory side by the maximum mean sustained firing rate for the mass, which in turn is determined by the relative refractory periods and the hyperpolarizing and depolarizing after-potentials of the single neurons. For single neurons the relation between pulse output rate and imposed transmembrane current is linear or nearly so (17, 19, 32, 50) over a much wider range of pulse rates than is expected for a neural mass. There are two reasons for this. On the inhibitory side the thresholds or firing times for the mass are distributed (51–53). On the excitatory side neurons can be driven to very high pulse rates for brief periods, provided they are not challenged to fire during a subsequent rest period (37). For a neural mass the maximum firing rate must be computed over both active and rest periods. On these grounds the input-output curve for (V-P) conversion must also be sigmoidal, with sharper curvature on the inhibitory side (Figure 1, upper left).

How Can Pulse-to-Wave Conversion by Dendrites Be Quantified?

The relationships between pulse input P and wave output V can be expressed in the form of two first-order differential equations, which state that the magnitude of output decreases in proportion to the rate of change in output with respect to input between two limiting values, V_i and V_e,

$$\frac{dV}{dP} = -\zeta(V - V_i) \qquad V < 0 \tag{1}$$

$$\frac{dV}{dP} = \frac{\zeta}{r_d}(V - V_e) \qquad V > 0 \tag{2}$$

where $V_i(<0)$ is the level of extracellular wave potential corresponding to the inhibitory equilibrium potential for the dendritic membranes of the neurons generating the wave, and $V_e(>0)$ is the level of extracellular wave potential corresponding to the excitatory equilibrium potential. Both are expressed as the difference from extracellular rest potential, which is taken as zero for background activity. The empirical rate constants, ζ on the inhibitory side ($V<0$) and ζ/r_d on the excitatory side, are in units of 1/pps, and r_d is dimensionless. The derivatives are set equal to each other at $V=0$, so that

$$V_e = -r_d V_i \tag{3}$$

The solutions to the differential equations are

$$V = V_i(1 - \exp[-\zeta(P - \overline{P}_0)]) \qquad P > \overline{P}_0,\ V < 0 \tag{4}$$

$$V = r_d V_i(1 - \exp[-\zeta/r_d(P - \overline{P}_0)]) \qquad P < \overline{P}_0,\ V > 0 \tag{5}$$

where $\overline{P}_0$ is the overall mean pulse rate of the neural mass at $V=0$.

How Can Wave-to-Pulse Conversion by Axons Be Quantified?

For (V-P) conversion the rate of change in pulse output P with respect to wave input V is likewise characterized as proportional to the output:

$$\frac{dP}{dV} = r_a \gamma P \qquad V < 0 \tag{6}$$

$$\frac{dP}{dV} = -\gamma P \qquad V > 0 \tag{7}$$

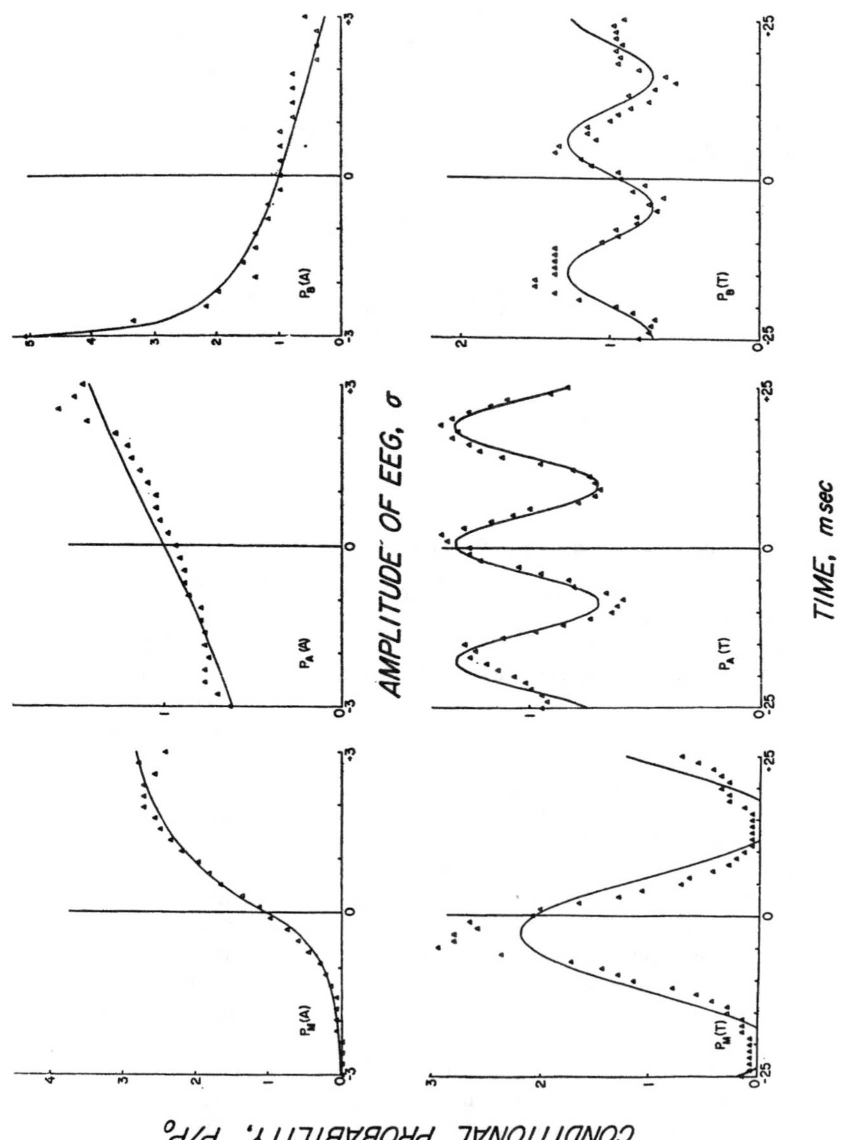

CONDITIONAL PROBABILITY, P/P₀
AMPLITUDE OF EEG, σ
TIME, msec
P_M(A)
P_A(A)
P_B(A)
P_M(T)
P_A(T)
P_B(T)

where γ and $r_a\gamma$ are respectively excitatory and inhibitory rate constants in units of $1/\mu V$.

The solutions to the differential equations are

$$P = \overline{P}_0 e^{r_a \gamma V} \qquad V < 0 \qquad 8.$$

$$P = P_{max} - (P_{max} - \overline{P}_0)e^{-\gamma V} \qquad V > 0 \qquad 9.$$

$$P = \overline{P}_0 \qquad V = 0 \qquad 10.$$

The value for P_{max} is found by setting the derivatives equal to each other at $V=0$, and solving for P_{max}:

$$\frac{dP}{dV} = r_a \gamma \overline{P}_0 e^{r_a \gamma V} \qquad V < 0 \qquad 11.$$

$$\frac{dP}{dV} = \gamma(P_{max} - \overline{P}_0)e^{-\gamma V} \qquad V > 0 \qquad 12.$$

$$P_{max} = \overline{P}_0(r_a + 1) \qquad V = 0 \qquad 13.$$

Equation 9 then becomes

$$P = \overline{P}_0 + r_a\overline{P}_0(1 - e^{-\gamma V}) \qquad V > 0 \qquad 14.$$

What Is Pulse Probability Conditional on Wave Amplitude?

The experimental evaluation of these two sets of equations 4–5 and 8–14 (54) is based on fitting curves generated from them to the pulse probability of single neurons in the olfactory bulb or cortex $\hat{P}(T, A)$, conditional (55) on the amplitude and time of the EEG recorded from a closely neighboring point in the bulb or cortex.

The conditional probability tables are constructed from long records of EEG amplitudes and single-neuron pulse trains measured simultaneously at 1.0 msec intervals. Three correlations have been made. The first is between the pulse probability of the mitral and tufted cells in the olfactory bulb, $\hat{P}_M(T, A)$, and the wave generated by the bulbar granule cells, $V_G(T)$ (40). The second is between the pulse probability of superficial pyramidal cells in the olfactory cortex, $\hat{P}_A(T, A)$, and the wave generated by the same cells, $V_A(T)$ (28, 39). The third is between the pulse probability of cortical granule cells $\hat{P}_B(T, A)$ and $V_A(T)$ (37).

Owing to delays between the actions of the mitral-tufted and granule cells, and between types B and A neurons, it is essential to establish the optimum time lag T for each correlation. This is done by asking, for each measured pulse value (0 or 1), what is the value for wave amplitude in each of the 25 msec preceding the present value, and for each wave amplitude, what is the pulse value in each of the 25 msec preceding? For each time lag and amplitude value the total number of times a pulse occurred is divided by the total number of times the wave amplitude occurred. The resulting conditional probability $\hat{P}(T, A)$ is multiplied by 1000 to express it in pps. For graphic display it is divided by P_0 (mean pulse rate).

←

Figure 1 (instructions for reading these graphs should be followed upon orienting the time and amplitude abscissas in the horizontal plane). *Above.* The calculated curves in the three upper frames are the predicted input-output curves $P(A)$ of the olfactory neural masses. The derivatives of these curves suffice to describe the amplitude-dependent forward gains. The triangles are the pulse probability of single neurons conditional on the amplitude of EEG, $\hat{P}(A)$. P_0 is the mean pulse rate. The curves are: upper left—(V-P) conversion for type M cells (Equations 17 and 18); upper middle—(V-P) conversion for type A cells (Equations 17 and 18); upper left—(P-V) conversion for type B cells (rotate graph 90° counterclockwise, Equations 15 and 16).

Below. Experimental [triangles, $\hat{P}(T)$] and theoretical [curves, $P(T)$] pulse probabilities conditional on time lag from EEG are shown. The EEG crest occurs at $T=0$. $P_M(T)$ leads the EEG; $P_A(T)$ shows a slight lag from the EEG, though on the average it is in phase; $P_B(T)$ lags the EEG.

The EEG amplitude probability histogram for bulb and cortex almost always conforms to a normal density function with standard deviation σ. The limits of the table, $\hat{P}(T, A)$, are placed at $\pm 3\sigma$ and at ± 25 msec. The pulse probability conditional on time, $\hat{P}(T)$, is determined by averaging across values for $\hat{P}(T, A)$ between $+1$ to 3σ for each value of T. The oscillatory time courses for $\hat{P}(T)$ have the same frequency as the dominant peak of the power spectrum of the EEG. The crests of $\hat{P}_M(T)$ lead the crest of the EEG by about one quarter cycle (at the center line in Figure 1, lower set); those for $\hat{P}_A(T)$ are in phase, and those for $\hat{P}_B(T)$ lag by about one quarter cycle.

The experimental pulse probability conditional on amplitude, $\hat{P}(A)$, is taken at time of the crest of $\hat{P}_M(T)$ preceding the EEG crest, at the crest of $\hat{P}_A(T)$ at the EEG crest, and at the trough of $\hat{P}_B(T)$ preceding the EEG crest (Figure 1, upper set).

The choice of equations for the curves $\hat{P}(A)$ to fit these data is based on the premise that in all cases the independent variable is V, owing to the fact that the instantaneous pulse probability is indeterminate. Therefore, Equations 4 and are solved for P as a function of V.

$$P = \overline{P}_0 - \frac{1}{\zeta} \ln \frac{1 - V}{V_i} \qquad P > \overline{P}_0,\ V < 0 \qquad 15.$$

$$P = \overline{P}_0 - \frac{r_d}{\zeta} \ln \frac{1 - V}{r_d V_i} \qquad P < \overline{P}_0,\ V > 0 \qquad 16.$$

From unpublished theoretical results, which lie beyond the scope of this limited review, the value for r_a is 2.0, so Equations 8 and 14 are modified accordingly:

$$P = P_0 e^{2\gamma V} \qquad V \leq 0 \qquad 17.$$

$$P = P_0(3 - 2e^{-\gamma V}) \qquad V \geq 0 \qquad 18.$$

The value for P_0 (as an estimator for $\overline{P}_0$) conforms to the mean pulse rate for each neuron (the total number of pulses divided by the number of thousands of observations), so that only a single unspecified variable, γ, suffices to fit the theoretical curves to experimental data (Figure 1, upper left).

For any neuron, one of three conditions is assumed to hold. If the limits on (V-P) conversion are dominant, then the asymptotes for $P(A)$ must be horizontal, because at some high positive or negative values of amplitude the pulse probability does not change. If the limits on (P-V) conversion dominate the neuron, the asymptotes are vertical, because excessive values for pulse probability are required to yield wave amplitudes nearing the asymptotic limit. If the limits on the input function are well within the limits for both conversions, then the relationship between P and V is linear or nearly so. Equations 17 and 18 apply to the first case (V-P), Equations 15 and 16 to the second (P-V), and either pair to the third.

Examples of each type of curve are shown in Figure 1, upper row. The pattern for $\hat{P}_M(A)$ reflects (V-P) conversion. That for $\hat{P}_B(A)$ reflects (P-V) conversion. The linear curves for $\hat{P}_A(A)$ are fitted with a curve $P(A)$ from (V-P) conversion, as the converse of $P_B(A)$ for the same neural mass.

How Is Forward Gain Calculated?

The conversion rate for each stage is given by the slope of the input-output curve for each stage, respectively, for curves from Equations 15–16 and 17–18. For each neuron population having two stages it is the product of the two derivatives. The forward gain is denoted $K_i{}^{.5}$ for inhibition by an inhibitory neural mass and disexcitation by an excitatory neural mass. The forward gain is $K_e{}^{.5}$ for excitation by an excitatory neural mass and disinhibition by an inhibitory neural mass.

$$K_i{}^{.5} = -\frac{dP}{dV}\frac{dV}{dP} \qquad V < 0 \qquad 19.$$

$$K_e{}^{.5} = -\frac{dP}{dV}\frac{dV}{dP} \qquad V > 0 \qquad 20.$$

The derivatives of Equations 17 and 18 are

$$\frac{dP}{dV} = 2\gamma \bar{P}_0 e^{2\gamma V} \qquad V < 0 \tag{21.}$$

$$\frac{dP}{dV} = 2\gamma \bar{P}_0 e^{-\gamma V} \qquad V > 0 \tag{22.}$$

From the derivations of Equations 15 and 16, we have

$$\frac{dV}{dP} = \zeta(V_i - V) \qquad V < 0 \tag{23.}$$

$$\frac{dV}{dP} = \zeta(V_i + V/r_d) \qquad V > 0 \tag{24.}$$

Therefore

$$K_i = -2\gamma \bar{P}_0 \zeta (V_i - V) e^{2\gamma V} \qquad V < 0 \tag{25.}$$

$$K_e = -2\gamma \bar{P}_0 \zeta (V_i + V/r_d) e^{-\gamma V} \qquad V > 0 \tag{26.}$$

The "reference gain" is K_0 at $V=0$.

$$K_0{}^{.5} = -2\gamma \bar{P}_0 \zeta V_i \qquad V = 0 \tag{27.}$$

Some representative values of the coefficients are as follows (56). The mean value of P_0 for 22 bulbar mitral and tufted cells is 10.1 pps and the mean value for γ_M is .00512/μV. For 10 type A neurons the mean for P_0 is 13.8 pps and that for γ_A is .00169/μV (57). In addition to P_0 (9.6 pps for 12 type B units), three coefficients are evaluated from fitting $P_B(A)$ to $\hat{P}_B(A)$. The mean for r_d is 5.7. The mean for V_i is -138 μV or $-3.09\ \sigma$, slightly lower than three standard deviations of EEG amplitude. The mean for the rate constant ζ_B is 0.41/pps.

The estimated gain factor for (V-P) conversion of mitral-tufted cells is $2\gamma_M P_0 = .10$ pps/μV. That for (V-P) conversion type A neurons is $2\gamma_A P_0 = .046$ pps/μV. That for (P-V) conversion from type B pulses to type A waves is $-\zeta_B V_i = 56$ pps/μV. The estimated value for $K_0{}^{.5}$ for the product of (P-V) and (V-P) conversion factors for superficial pyramidal cells (A) is 2.6. The factor for (P-V) conversion in the bulbar relay can not be estimated because bulbar-granule cells do not generate detectable action potentials.

These numerical estimates are without confidence intervals, particularly in regard to the use of the mean for observed values of P_0 as an estimator of the mean $\bar{P}_0$. There is some bias in the experimenter toward selecting relatively fast-firing neurons for statistical analysis, because they give smoother pictures at lower cost, so that the estimate of the type A forward gain based on $\bar{P}_0$ is probably about two-fold too high. The point is that conversion factors and forward gains can be defined by theory and measured by experiment for neural masses. Confidence limits can be established only by extensive experimental follow-up and further moulding of the theoretical infrastructure.

Equations 25–27 imply that there is a central quasi-linear amplitude range, in which forward gain is maximal. With increasing positive (excitatory) amplitudes both stages undergo progressive saturation, the (V-P) stage exponentially and the (P-V) stage linearly. With decreasing negative (inhibitory) amplitudes, the (V-P) stage saturates twice as rapidly in the exponential mode and the (P-V) stage six times as rapidly in the linear mode on the inhibitory side as on the excitatory side. The forward gain appears to depend on four parameters or system variables; the population mean pulse rate $\bar{P}_0$; the ambient maintained degree of depolarization V_i; and the two rate constants γ and ζ for which an interpretation at the cellular level has not been attempted. Whether these four system variables are or are not independent of each other has not been determined.

What Is the Open-Loop Response?

Neurons in masses are densely interconnected by countless numbers of synapses. Normally these are capable of transmitting output depending on the magnitude of input, so that electrical stimulation of a nerve or tract leading to a neural mass leads to multiple sequential synaptically transmitted events in the mass. Repetitive excitation and inhibition of the initially excited or inhibited neurons is the rule. However, by pharmacological means it is feasible to reduce the transmission effectiveness of synapses in the mass, to the extent that background EEG and pulse trains are totally suppressed. In this state the afferent volley, electrically induced, activates the dendrites of neurons at the first synapse, and perhaps the second, but no further. Feedback interaction is reduced to zero, and only forward transmission to the first one or two subsets of neurons in the mass is present. This is referred to as the *open-loop state* (36, 37).

The experimental proof of this state depends on demonstrating the absence of background unit activity and all but one brief volley of induced firing, if any (37). (This is another example of the manner in which the conjoint recording of pulse and wave activity is essential to the analysis of neural masses.) The averaged dendritic response manifested as the AEP is the open-loop impulse response of a neural mass.

An example of the open-loop AEP of the olfactory bulb is shown in Figure 2 (top) as the sets of triangles. The response is induced by stimulation of the axons of the mitral cells in the lateral olfactory tract (LOT) antidromically. The volley is delivered by the mitral cells to the granule cells, and the dendrites on synaptic activation generate the field potential yielding the AEP. There is no further transmitted event. A virtually identical open-loop response occurs in the olfactory cortex, generated by the type A neurons in response to an orthodromic LOT volley (34, 38, 57).

In essence[3] these experimental data have been fitted by the sum of three exponentials:

$$a(t) = \sum_{i=1}^{3} A_i e^{-a_i t} \qquad 28.$$

where a_1 is the rate constant of the decay of the response, a_2 is that of the rising phase, and a_3 is that determining the curvature of the foot of the response. The Laplace transform of Equation 29 gives the differential equation for the dendritic open-loop response:

$$A(s) = A_0 \prod_{i=1}^{3} \left(\frac{a_i}{s + a_i} \right) \qquad 29.$$

Mean values for the rate constants are $a_1 = 220$/sec (equivalent to a time constant of 4.55 msec), $a_2 = 720$/sec (1.38 msec), and $a_3 = 2300$/sec (0.43 msec).

The similarity of these rate constants to those computed from intracellular measurements on single cells (49, 58–62) strongly suggests that a_1 can be taken to represent mainly the passive membrane RC decay rate, a_2 the lumped synaptic and cable delays (both equivalent to one-dimensional diffusion processes), and a_3 the effect of axonal delays within the mass, which are very short (37). However, the rate constants of the mass cannot be uniquely identified with these known sources of delay in single neurons on a one-to-one basis.

This procedure to identify and measure the rate constants (and to interpret them in terms of underlying processes) is formally identical to that used to define

[3] The experimental determination of the open-loop rate constants is complicated by the presence of dispersion in afferent axonal pathways, which is not part of the delays found within the loops of the neural mass. For LOT input to the bulb and cortex, this dispersion $A_x(s)$ is negligible, but for PON input it is rather strong (47). Therefore, the total transference for PON input is $A_x(s)A_m(s)A(s)$. The forward transference within the bulb is $A_m(s)A(s)$. The separation of $A_x(s)$ from $A_m(s)$ requires measurements in both open- and closed-loop states, using both orthodromic (forward limb) and antidromic (feedback limb) inputs.

A simple experimental test to determine whether $A_x(s)$ can be ignored is to search for the afferent axonal or cell body compound action potential in the mass preceding the dendritic response. If it cannot be detected (other than in the form of intracellular or extracellular unit potentials), then it has been degraded by temporal dispersion acting as a low pass filter (47), and $A_x(s)$ must be explicitly evaluated in order to evaluate $A(s)$ with adequate precision.

"passive" membrane resistance and capacitance (59). For an axon or a group of axons a range of response amplitudes is defined over which additivity and proportionality hold. Within this range the response in membrane potential $v(t)$ to a current step $I \cdot U(t)$ is recorded. It is fitted with a rising exponential curve having the equation $v(t)=k \cdot I \cdot U(t) \cdot (1-e^{t/\tau})$ where $\tau = 1/a$ is the time constant in sec and k is in units of ohms and depends on the value for the potential as $t \to \infty$. The Laplace transform is $V(s)=I \cdot k/s(s\tau+1)$. The transfer function is the ratio of the transforms of the input and output functions, $V(s)/(I/s)=k/(s\tau+1)$. A differential equation is then written to describe the discharge of a capacitor C through a resistor R after C has been charged by a current pulse $I \cdot \delta(t)$. This can be written $Cdv(t)/dt=I \cdot \delta(t)-v(t)/R$. The Laplace transform is $V(s)/I=R/(sRC+1)$. It is next inferred that $R=k$ and $C=\tau/k$. Numerical estimates for R and C are then

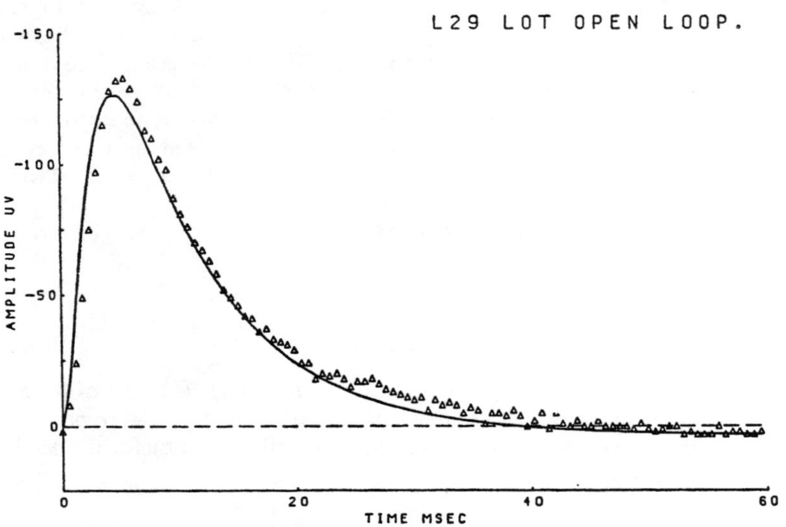

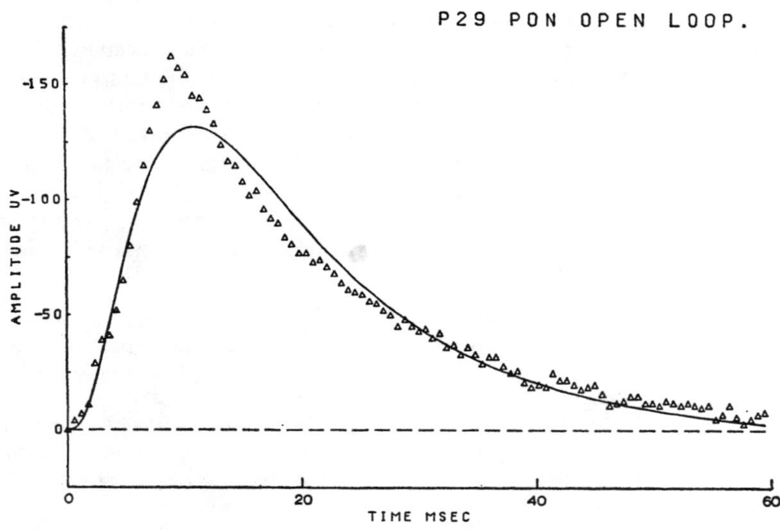

FIGURE 2. These are open-loop responses of bulbar neurons on antidromic LOT (above) or orthodromic PON (below) single-shock electrical stimulation. AEPs (triangles) $N=100$. Curves are from the inverse transforms of Equation 28 (above) and of Equation 30 (below).

obtained in a variety of experimental conditions to confirm the dynamic range of linearity for the preparation, which is the range of validity for the differential equation and its evaluated coefficients.

It is well known that both R and C of biological membranes vary with frequency (60), but the approximation of their behavior within the usual experimental range of function to ohmic resistance and coulombic capacitance is close enough for most purposes. Moreover, each is an average measurement over a large ensemble of membrane structures, such as the sodium or potassium channels, which on more intensive analysis outside the linear range are identified and measured by elaborate equations, e.g., the Hodgkin–Huxley equations (61). Or the linear model is retained but the lumped circuit approximation is dropped, and the spatial dimensions are introduced, e.g., in the form of a cylinder (8, 62) or a branching dendritic tree (8, 9, 63).

In all these cases the response waveform for a severely limited input guides the selection of an appropriate set of linear basis functions and the construction of a linear differential equation. The equation is modified and elaborated to extend prediction and observation into a broader functional range. The essential difference in the present usage is that linear equations are used to describe and measure the properties of the neural mass, and the interpretation of the results is directed toward known components at the next lower level of complexity, that is, the properties of neurons rather than the properties of membranes.

On the other hand the same results are used as the basis for interpretation of the next highest level of complexity. An immediate example of neural masses is the AEP response of the bulbar granule cells $A(s)$ to excitation of the primary olfactory nerve (PON), shown in Figure 2 (bottom). The afferent volley is transmitted orthodromically through the mitral-tufted cell pool having the transference $A_m(s)$. It is found experimentally (W. J. Freeman, unpublished results) that $A_m(s)$ is almost identical to $A(s)$, so that the overall transfer function for the two neural masses in series, $A_g(s)$, is

$$A_g(s) = A^2(s) \qquad 30.$$

The inverse transform of Equation 30, which contains double poles after substitution of Equation 29, is too cumbersome to reproduce here. The predicted waveform is shown as the curve in Figure 2 (bottom). The interpretation at the first sublevel is that the delays introduced by the two neural masses in series are about equal to each other, and at the second sublevel that the passive membrane decay rates for the mitral-tufted and granule cells are equal, despite the gross differences between virtually all other properties of the two cell types. These same rate constants and interpretations hold for types A and B neurons in the cortex as well (34, 36, 37, 48).

What Are the Characteristics of Negative Feedback?

The mitral and tufted cells in the bulb (type M) are excitatory, whereas the granule cells (type G) are inhibitory. Excitation of type M cells by single-shock stimulation normally leads to excitation of type G cells and to feedback inhibition of type M cells (46). The inhibited type M cells disexcite type G cells which disinhibit or reexcite type M cells, and so on, such that the impulse response in the closed-loop state is predictably oscillatory. The same prediction holds for the interaction of types A and B neutrons in the cortex (34, 36, 57).

In a narrowly limited range of function (to be described in a later section), the transfer function for these negative feedback loops for orthodromic input to the excitatory neural mass and output (Figure 3, upper right) from the same neural mass (the forward limb) is

$$C_{ee}(s) = \frac{A(s)}{1 + K_n A^2(s)} \qquad 31.$$

For input to the forward limb and output (Figure 3, lower right) from the feedback limb (the inhibitory bulbar or cortical granule cells), the transference is

$$C_{ei}(s) = \frac{A^2(s)}{1 + K_n A^2(s)} \qquad 32.$$

For antidromic input to the feedback limb and output from the forward limb (Figure 3, upper left) or output from the feedback limb (Figure 3, lower left) the transfer functions are respectively

$$C_{ie}(s) = \frac{C_{ee}(s)}{A(s)} \tag{33.}$$

$$C_{ii}(s) = \frac{C_{ei}(s)}{A(s)} \tag{34.}$$

In each of Equations (31–34) the feedback gain is $K_n = (K_e K_i)^{.5}$, where $K_e^{.5}$ and $K_i^{.5}$ are defined by Equations 19 and 20.

Following substitution of Equation 29 into any of Equations 31–34, factoring of the denominator, and partial fraction expansion, the inverse Lapace transform yields the equation for a damped sine wave with a sigmoidal inflection at the foot of the first upward peak (34, 36, 48):

$$c(t) = \sum_{i=1}^{2} V_i \sin(\omega_i t + \phi_i) e^{-\alpha_i t} \tag{35.}$$

The same equation holds for predictions of the state variables of both forward and feedback limbs. The frequencies and decay rates are identical, but the phase of the feedback limb transient characteristically displays a quarter cycle phase lag from that of the forward limb transient.

These predicted transients are shown as curves in Figure 3 fitted to the PST histogram (triangles, above) of a mitral cell (excitatory, forward limb) and to the AEP (triangles, below) of the granule cells (inhibitory, feedback limb). The common frequency and decay rate are apparent, as well as the phase lag of each AEP from the corresponding PST histogram (48).

The bulbar responses to PON stimulation (Figure 3, right), which is orthodromic and to the forward limb, show approximately one quarter cycle phase lag over the bulbar responses to LOT stimulation (Figure 3, left), which is antidromic and to the feedback limb, as predicted by Equations 31–34.

The comparison of AEPs generated by type A neurons in the cortex (37) on orthodromic (LOT) stimulation with PST histograms shows that the AEP is in phase with the oscillation in PST histograms from type A neurons (forward limb) and leads the oscillation in PST histograms for type B neurons (feedback limb) by about one quarter cycle. These same phase relationships are shown in Figure 1 (lower row) for background pulse and wave activity.

The mean value for K_n, obtained from the Laplace transform of Equation 35 after fitting curves to the AEPs and PST histograms for both bulb and cortex, is between 1.75 and 2.25 (36). The value for K_n, representing effective connection density within the mass, is a property only of the mass and not of the single neurons in the mass. The measured values for the rate constants are the same in the open-loop and closed-loop states. These and related results (36, 37) imply that the rate constants can be treated as invariants, and that the amplitude-dependent nonlinearity $P(V)$ can be introduced into the linear equations $C(s)$ as a variable gain coefficient.

What Are the Characteristics of Positive Feedback?

The example given for negative feedback is based on the assumption that a neural mass contains large numbers of excitatory and inhibitory neurons having reciprocal connections. If the mass contains excitatory neurons maintaining significant feedback connections with each other, a different kind of feedback loop must be considered. This is a positive feedback loop, in which excitatory neurons excite and reexcite each other in the mass upon initial excitation. The pattern is familiar among physiologists as "avalanche conduction" or the "reverberating circuit."

Such loops usually appear in neural masses having negative feedback as well, and seldom occur in isolation. An example of the latter is to be found in the periglomerular neurons of the olfactory bulb (40, 41, 45, 48). An illustration of the impulse response of one of these neurons is shown in Figure 4, left. The three

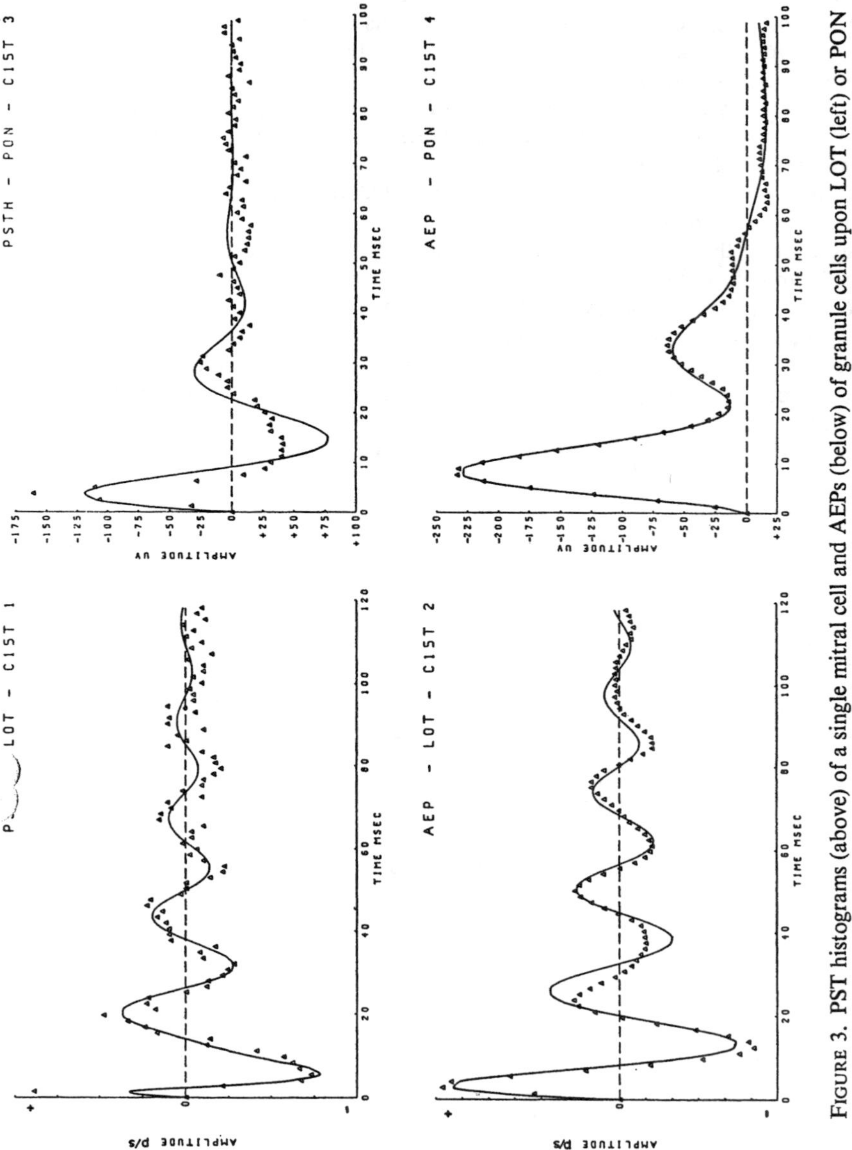

FIGURE 3. PST histograms (above) of a single mitral cell and AEPs (below) of granule cells upon LOT (left) or PON (right) stimulation are fitted with curves for predicted responses from Equation 35 (closed loop, negative feedback). There is phase lag of the AEP from the PST histogram of about one quarter cycle (see Figure 1, lower left). Transfer functions: upper right, Equation 31; lower right, Equation 32; upper left, Equation 33; lower left, Equation 34.

sets of triangles illustrate three PST histograms at low, medium, and high PON stimulus intensity. (The background pulse rate of the neuron, which is constant, is indicated by the height of the baseline above the abscissa. This gives the change in scale from which to measure the increase in pulse rate at the crest of the response with increasing stimulus intensity.)

In qualitative terms the response of the neural mass to an impulse input by way of the PON is a rapid increase in mean pulse rate above the baseline of background activity, which then decays with a slow rate constant. When the input intensity is increased, the induced pulse rate is augmented, but so also is the decay rate of the response. There is no terminal overshoot.

The response of this neural mass cannot be directly detected in the form of an extracellular field potential. The concomitantly recorded AEPs shown in Figure 4 (right) display a sinusoidal oscillation generated by the bulbar negative-feedback loop, to which both the PON and the periglomerular neurons project. The oscillation is superimposed on a monotonic shift in baseline, which is the granule cell response to periglomerular input. It is not owing to a field potential of periglomerular cells.

The dynamics of this neural mass can be described in terms of a positive-

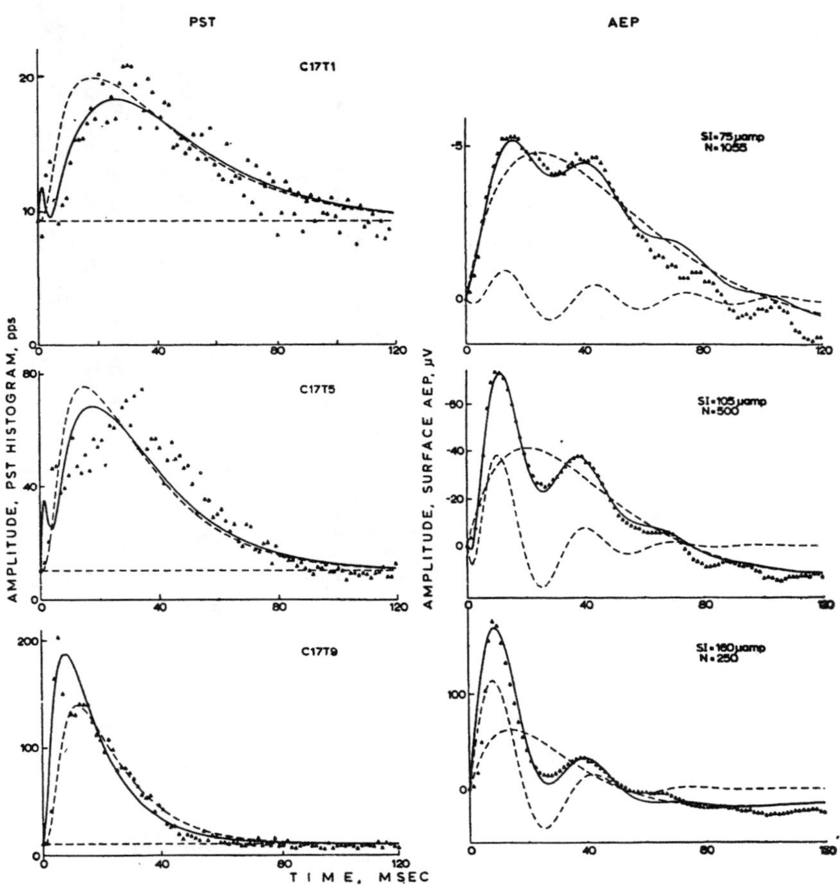

FIGURE 4 (from Freeman 48). *Left.* The PST histograms (triangles) of pulses from a single periglomerular neuron represent the output of a neural mass having internal positive feedback. The lowest rate constant increases with increasing response amplitude. The change in scale for display of the histograms is reflected in the decreased scale for the constant background activity (Equation 37).

Right. AEPs from the bulbar granule cells (concomitantly recorded field potential) show the granule cell response to the PON impulse input (the oscillatory component) and to the periglomerular cell input (the monotonic component). Initial response amplitudes: 22 μV, 324 μV, 986 μV (Equations 35 and 37).

feedback loop, having input to a subset (B_1) of the periglomerular cells constituting the forward limb, interacting with another subset (B_2) of the same mass not receiving the initial impulse input and constituting the feedback limb. The transfer function for output from the forward limb is

$$B_1(s) = \frac{A_0 \cdot A(s)}{1 - K_p A^2(s)} \qquad 36.$$

where $A(s)$ is defined in Equation 29 and $K_p = K_e$ is the square of excitatory forward gain defined by Equation 20; A_0 is a forward gain constant.

After substituting Equation 29 into Equation 36, reducing fractions, and expanding the denominator in partial fractions, the inverse Laplace transform yields

$$b_1(t) = \sum_{j=1}^{4} B_j e^{-b_j t} + B_5 \cos(\omega t + \phi) e^{-b_5 t} \qquad 37.$$

where the amplitude coefficients B_j and B_5 depend on the open loop rate constants a_i, the feedback gain K_e, and a forward gain constant A_0.

The rate constants[4] are invariant with stimulus intensity at the values $a_1 = 230$/sec, $a_2 = 550$/sec, and $a_3 = 2300$/sec. The value for A_0 increases in proportion to stimulus magnitude. The change in the closed-loop rate constants, most notably that for the decay rate of the response b_1, is owing solely to the change in K_e predicted by Equation 26, where either A_0 or B_1 or the crest amplitude of the PST histogram is used to estimate V.

These and related results (36, 48) show that positive feedback among neurons in masses leads to monotonic responses with decay rates that are not the same as those of the component neurons. The rate constants of the neural mass with internal positive feedback have much lower values than those usually assigned to passive membrane. Furthermore, whereas by inference the rate constants of the component neurons in the mass are invariant with respect to response amplitude over the designated range of observation, the rate constants of the neural mass are amplitude-dependent, primarily owing to the presence of saturation in the feedback path.

This type of long-lasting neural response has been observed to follow electrical stimulation in many parts of the nervous system, most prominently in the spinal cord (64, 65), where the prolonged impulse responses have been associated with presynaptic inhibition. However, the dependencies of the rate constants on response amplitudes have not been measured with adequate precision, and the associations with concomitantly recorded PST histograms have not been well enough established, to permit strong inference that these responses manifest positive excitatory feedback, although it seems likely that many of them do.

Mutual inhibition can be modeled using Equations 36 and 37. An example is the lateral eye of *Limulus* (19), which consists of an array of about 10^3 densely interconnected receptor neurons having a common input (light) and a common sign of output (inhibition). They form a positive inhibitory-feedback loop in which the gain characteristic is linear over a certain range but is bounded by saturation (threshold) on the inhibitory side. Above that level the function is readily approximated by linear analysis (20).

A similarly isolated example of an inhibitory neural mass has not been identified in the mammalian nervous system. Inhibitory neurons seem always to be

[4] Owing to the inaccessibility of a dendritic field potential from those neurons, the open-loop rate constants were determined by trial and error. A sum of exponential curves and a damped sine wave (Equation 37) was fitted to the data (Figure 4, left, solid curves, and Figure 5, open triangles). The transferences for $A_x(s)$ and $A(s)$ were evaluated by initial guesses, and root locus plots were calculated. When optimal values for the open-loop poles had been found, the results were checked by generating new curves using these invariants (Figure 4, left, dashed curves). The neural mass was found to have the same rate constants as the other masses in the bulb and cortex, within the limits of experimental error (34).

This case illustrates a general principle in the analysis of neural dynamics, that sets of AEPs and PST histograms, taken in conjunction with systematic variation of an antecedent variable, can be used to compensate for the limitations imposed by the inaccessibility of some of the state variables of neural masses to direct measurement.

densely connected with excitatory neurons as well as with each other. Therefore, information about them is indirect. Their predicted patterns of behavior correspond to those for excitatory interactions in most aspects. But whereas the outputs for the forward and feedback limbs of an excitatory neural mass during the impulse response both increase and decrease together, in the inhibitory neural mass they change in opposite directions. On initial excitation, for example, the activity of the forward limb increases, which inhibits the activity of the feedback limb. The latter disinhibits or further excites the forward limb, which further inhibits the feedback limb. The time courses of the two parts are parallel and monotonic, but have opposite polarity.

Why Use Root Locus Display?

The dynamic relation between the rate constants of the closed-loop response and the underlying physiological variable, the closed-loop gain, is best displayed by a root locus representation in the *s* plane (29, 35). Such a diagram is shown in Figure 5 for Equation 36, which on substitution of Equation 29 and reduction of fractions becomes

$$B_1(s) = \frac{\prod_{i=1}^{3} (s + a_i)}{\prod_{i=1}^{3} (s + a_i)^2 - K_e a_1^2 a_2^2 a_3^2} \qquad 38.$$

On expansion and factoring of the polynomial in the denominator, this becomes

$$B_1(s) = \frac{\prod_{i=1}^{3} (s + a_i)}{\prod_{j=1}^{6} (s + b_j)} \qquad 39.$$

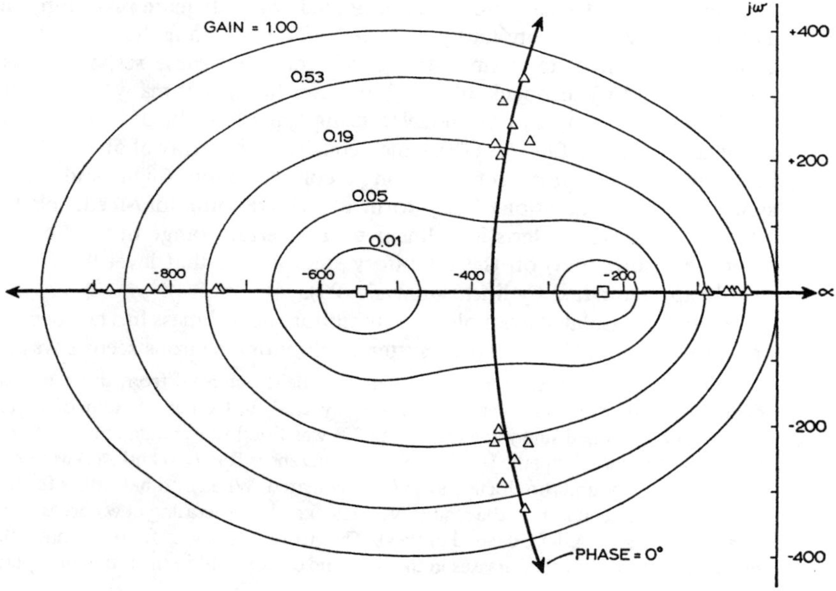

FIGURE 5. This is a root locus diagram in the *s* plane for neural positive feedback. The triangles show the rate constants of the solid curves in Figure 4 (left). The open squares show the locations of the open-loop poles and closed-loop zeroes. The solid curves are the root loci. The light curves are representative gain contours.

where the closed-loop rate constants b_j are determined by the open-loop rate constants a_i and K_e. The closed-loop zeroes at $s=-a_1$ and $s=-a_2$ appear as open squares on the negative real axis in Figure 5, at the locations of the open-loop poles. The poles and zeroes at $s=-a_3$ are far to the left. Experimental closed-loop roots appear as the four sets of triangles where the four loci intersect representative gain contours between $K_e=0.53$ and 0.19. For each value of K_e there are four real roots and two complex roots. The latter predict the over-damped cosine component in the rise of the output. The rate constant specified by the pole b_1 nearest the $j\omega$ axis determines the decay rate of the impulse response.

The method displays in a single graph the open-loop rate constants (which are invariant), the closed-loop rate constants, the ambient level of gain (which is amplitude-dependent), and the predicted change in the impulse response with changes in amplitude or other determinant of gain. Also shown are the locations of the closed-loop zeroes, which play an important role in precision analysis, and the stability characteristics of both real and model systems.

The technique is especially well adapted for use in conjunction with impulse stimulation. Owing to the fact that direct access to most neural masses in the brain is by nerves and tracts, which are accessible only to electrical pulses, the impulse response (the AEP and the PST histogram) is the most common source of information on their active states.

The curve fitted to each impulse response yields coefficients, which form a constellation of poles and zeroes in the s plane. Many physiological variables, but particularly response amplitude variation, generate successive values for the coefficients, which define sets of "physiological" root loci as indicated by the triangles in Figure 5. These loci are matched by theoretical root loci, which are generated from differential equations based on the topology of connections and the open-loop rate constants of the neurons. They serve to evaluate the closed-loop gain.

The ultimate justification for reliance on the root locus diagram (29) as a basic analytic tool in neural dynamics is the feasibility of separation of the transference into a linear frequency-dependent part and a nonlinear amplitude-dependent part, the former evaluated by fixed rate constants and the latter by variable gain coefficients. No other method of system representation seems so well adapted to this feature of neural masses.

What Are the Characteristics of Multiple-Loop Feedback?

Doubtless the most common neural mass by far is the mixture of excitatory and inhibitory neurons, which are densely interconnected with each other without restriction as to cell type. The minimum topology of such a mass contains multiple loops of three kinds: negative, positive-excitatory, and positive-inhibitory feedback (36). Three feedback gain coefficients are required: K_e, K_i, and K_n.

A twelfth-order differential equation is required to represent the connections of such a neural mass, in which the open-loop transference of the subsets composing it, $A(s)$, is specified and evaluated by Equation 29 and in the text following. The techniques for formulating and solving the equation have been given elsewhere (36, 57). The solution for the impulse response predicts a damped sine wave in the output, which is superimposed on a monotonic transient (similar to that in Figure 4, right). The oscillation is due to the negative feedback loop, and the monotonic transient is the output of whichever of the two positive feedback loops has the higher gain.

Changing the intensity of the stimulus delivered to a mixed neural mass characteristically alters the frequency ω and decay rate α of its oscillatory response (36–39). Measurement of the series of AEPs such as that shown in Figure 4, right, yields sets of values for ω and α that in the s plane define a physiological root locus for the neural mass with changing input intensity (Figure 5, circles). The frequency is characteristically reduced and the decay rate is augmented with increased input magnitude. The changes imply, in accordance with Equations 25 and 26, that the feedback gains are reduced with increasing amplitude owing to saturation.

Calculations to fit these physiological root loci are complicated by two factors. First, all three gain coefficients must be expressed as dependents on one variable (input or response amplitude, which has both positive and negative extremes). Second, the input to the mass may consist not merely of an impulse, but of a prolonged monotonic input function, such as that from periglomerular neurons to mitral and bulbar granule cells (Figure 4). The degrees of saturation on the excitatory and inhibitory sides and their ratio may vary widely, depending on the magnitude of this baseline shift.

An interim technique to achieve the calculation is based on the use of an empirical dimensionless factor δ, which serves to define the dependence of K_e and K_i on K_n:

$$K_e = K_n \left(\frac{K_n}{K_0}\right)^{\delta} \qquad 40.$$

$$K_i = K_n \left(\frac{K_n}{K_0}\right)^{-\delta} \qquad 41.$$

where K_0 is a reference gain (see Equation 27) at which $K_n = K_e = K_i$. When this condition holds (even for $\delta \neq 0$), the mixed neural mass is reduced to a single negative-feedback loop, so that K_0 has a value between 1.75 and 2.25.

From the product of the left-hand terms of Equations 40 and 41 set equal to the product of the right-hand terms, $K_n = (K_e K_i)^{.5}$.

It is found empirically (W. J. Freeman, unpublished data) that the locus for a neural mass with an impulse input is replicated by a value for $\delta = -0.5$. For a mass with an excitatory monotonic function or bias in the input, δ approaches zero. This serves to describe the bulbar physiological root locus for PON input. For an inhibitory bias, δ approaches -1. The latter value serves to describe the physiological root locus of the olfactory cortex for LOT input (36, 39).

These characteristic curves for neural masses are shown as a family in Figure 6 for a value of $K_0 = 2.24$. The curvilinear segments from upper left to lower right designate values for gain expressed as the $\log_{10} (K_n/K_0)$. The stability limits to the left of the $j\omega$ axis (at the high-frequency, low-amplitude, high-gain ends of the curves) are determined by a root locus (not shown) on the real axis of the s plane, which crosses the $j\omega$ axis to the right with decreasing amplitude. The stability characteristics at the low-frequency, high-amplitude ends of the curves have been discussed elsewhere (36).

The characteristic curves are used as follows. The open-loop rate constants are determined for a neural mass. Then in some normal physiological state a set of AEPs over a range of stimulus intensities is obtained and measured. The frequencies and decay rates define the physiological root locus. The position and orientation of the locus serve to evaluate K_0 and δ. The values for α and ω serve to evaluate K_n. Equations 40 and 41 evaluate K_e and K_i. By this means the closed-loop responses suffice to specify the three feedback gains (functional connection densities) in the mass. The logarithms of the gains (with reference gain at K_0) are plotted as a function of the amplitude of the oscillatory component of the impulse response to determine conformance of the dynamics of the mass to the type of saturation predicted by Equations 25–27. The value for δ serves to predict the sign and magnitude of the monotonic component of the impulse response (Figure 4, right).

An interesting alternative approach to the description of this multiple-loop neural configuration has been developed by Wilson & Cowan (66). They derived coupled nonlinear differential equations to predict the responses of a neural mass having negative feedback and the two kinds of positive feedback. A single rate constant was used for the delay [$A(s)$, $i=1$], and the logistic curve was used to represent the input–output curve (Figure 1). Using phase-plane methods and numerical integration, they found multiple stable states as well as limit-cycle oscillation, corresponding to the real and complex roots of Equations 35 and 37. The frequency of oscillation was also found to be a monotonic increasing function of stimulus intensity.

The intensity referred to in their model is the sustained input to either or both subpopulations, which is equivalent here to a background bias maintaining the mean activity level V_0 at some level other than zero. As the bias is increased their limit-cycle frequency goes monotonically from some minimum to some maximum value, above which the oscillation is suppressed. Experimentally (38), in bulb or

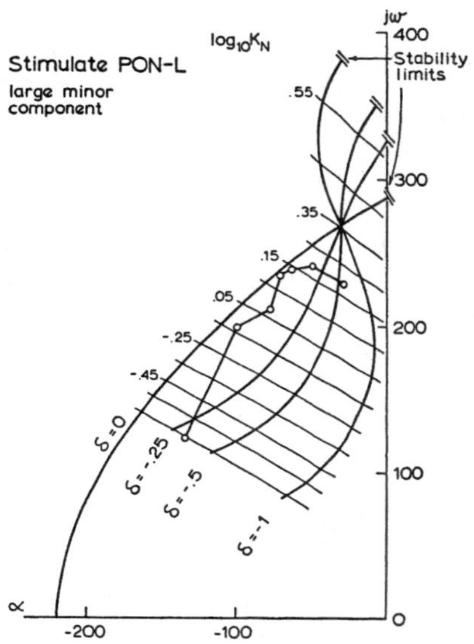

FIGURE 6. This is a composite diagram of four root loci in the upper left quadrant of the s plane. Each heavy curve is a locus for a single complex pole. Open circles are from an experimental set of AEPs, some of which are reproduced in Figure 4, right.

With increased stimulus intensity the AEP frequency and the gain K_N both decrease. Corresponding values for K_N are connected by light arcs. For a single negative-feedback loop, $\delta=0$. For both negative and positive feedback with an impulse input, $\delta=-.5$. For an impulse input superimposed on an excitatory bias to a mass having both positive and negative feedback (Figure 4, right), $\delta>-.5$. For inhibitory bias, $\delta<-.5$.

The normal midrange operating condition for the olfactory neural masses (Figure 3) appears to be δ near $-.5$ and K_N near 2.24($\log_{10} K_0=0.35$). The open loop double pole (Figure 2) is at $s=-220$/sec. Root loci on the negative real axis (as in Figure 5) are not shown here.

cortex the bias can be increased by use of tetanizing electrical stimulus pulse trains or decreased by administration of pentobarbital or other anesthetic. The frequency changes in the predicted manner (cf. Figure 12a in 66 and Figure 2 in 38). The characteristic curves shown in Figure 6 ($\delta \leq .5$) hold equally for increasing bias or for decreasing test pulse intensity (39). That is, frequency is determined by the ratio of test pulse to bias amplitude, not by either alone.

The Wilson-Cowan model displays some remarkable hysteresis properties not apparent from linear analysis. On the other hand the replication of the observed "physiological" root loci with calculated root loci (36, 39) could not be achieved unless the approximation for the open-loop response contained at least three poles [$A(s)$, $i=3$]. It is unlikely that phase-plane methods can readily be adapted to twelfth-order systems, to provide the means for identification and measurement of experimental observations. Each approach has its advantages and the interaction between linear and nonlinear analysis is bound to be fruitful.

How Widely Might these Techniques Be Applied?

AEPs often seem complex in appearance, particularly when comparisons are being made between those from different parts of the brain. This is deceptive. When the proper basis functions are used in conjunction with signal detection theory (30, 35), the typical AEP can be measured, stored, and reconstructed using a rather small set of numbers (67). The real complexity of AEPs resides in the sensitivity of their waveforms to a host of experimental factors, including the sites of stimulation and recording, the parameters of the input, the conditions of the animal, and the local state of the neural mass (14, 39, 42–46). When any one of these antecedent variables is changed, several or all of the numerical coefficients of the basis functions change in a coordinated way (39). The patterns of variation reveal more information about the neural mass than do the mean values of the coefficients. This is why root locus techniques combined with factor analysis (39, 67) and analysis of variance (68) of AEP sets are so useful (see footnote 4).

The characteristic curves shown in Figures 5 and 6 and related curves (36) hold for neural masses in the olfactory system over a response amplitude range from several microvolts to several millivolts, and a response frequency range from 0 to 60 Hz and above. It is suggested that any neural mass having characteristic frequencies in this range is likely to have dynamics closely related to those of the olfactory system. These include such structures as the hippocampus (69), the superior colliculus (70), the thalamus (71), and most if not all areas of the neocortex (13, 22, 25, 26, 72, 73). The impulse responses of the hippocampus (69) and superior colliculus (70) have been shown to conform to damped sine waves, for which the frequency decreases with increasing input intensity. The neocortical areas generate EEG waves in the β range (15 to 30 Hz and above). The presence of a β frequency selective system in the visual analyzer has been thoroughly documented by Lopes da Silva and his colleagues (13, 22) from its responses to sine wave input.

The outstanding difficulty in determining whether the characteristic curves in Figures 5 and 6 apply also to these other structures is experimental. The open-loop rate constants must be measured to determine whether the values conform to those used to evaluate $A(s)$. This in turn requires that the transference of the input pathway $A_x(s)$ be specified (see footnote 3). This is not straightforward for neocortical neural masses in terms of electrical stimulation, because the input and output axons are not physically separated as they are in the olfactory system, so that an afferent volley is likely to be mixed orthodromic and antidromic. Even so, it seems feasible and should be attempted. Otherwise the interaction densities (feedback gains) cannot be defined and evaluated.

What Is the Significance of Linear Analysis?

The application of linear analysis does not of itself yield a theory of neural masses. It is an empirical tool for observation, description, measurement, and prediction. It provides the basis functions for fitting the AEPs and PST histograms; it helps to define appropriate ranges for the input parameters; it forces consideration of the details of topologies of interconnection; it provides the means for using the values for frequencies and decay rates of responses in differing physiological states to estimate the numerical magnitudes of the interconnections; and it gives an effective framework in which to study the relations between single neural pulse trains and dendritic currents from large masses of cells, both evoked AEPs and "spontaneous" or background EEG.

These features in themselves amply justify further applications of the technique to larger and more complexly organized neural masses. But to what end? Granted that neurons in large numbers undergo correlated changes in activity following natural or artificial stimulation of the nervous system, does the mere fact of correlation justify the conception of a neural mass? Does the covariance of large-scale activity in itself have any significance for behavior, or is it epiphenomenal? Is sensory or perceptual information conveyed in broadly distributed spatio-temporal patterns of covariance among neurons (74–76), or in the pulse trains of single neurons (77–79)? Might the apparent properties of neural masses be significant merely in terms of the large-scale normalizing, scaling, and smoothing functions required for the operations of single neurons, thus providing the "ground" against which the "figure" is inscribed? Are the results of neuroelectrical measurements in the conditions prescribed by linear analysis relevant to any of these possible functions of neural masses? Is it feasible to construct valid theories of neural information processing, storage, and retrieval without prior working knowledge of the properties of neural masses?

These questions cannot be answered yet, primarily because so little is known about the dynamics of neural masses. The results briefly summarized here imply that neural masses do exist in at least an operational sense, that they have properties distinct from those of single neurons, that these reside in poorly understood distributions of firing rates, thresholds, interconnections, etc, and that these properties must be defined and measured in terms of statistical averages of neural activity.

Linear analysis can best serve now to open a door to the experimental study of neural masses, as it did 40 years ago to the study of membranes, axons, and dendrites of single neurons (59). Beyond this there is the expectation that when the essential nonlinearities of neural masses have been clarified, linear and quasilinear equations will be supplanted by the "real" equations, as the nonlinear Hodgkin–Huxley equations superseded older "two-factor" theories (61) for axon function. However, the axolemma and the neural mass are complex in different ways, primarily in the relative looseness of the coupling among the component parts for the latter. The alternative possibility must be considered seriously that matrices of linear equations may become the basic working tools for the articulation of our knowledge of how brains work. This point serves to emphasize how little we really know about neural masses, and how rich the opportunities are for studies of them in both theoretical and experimental neurophysiology.

LITERATURE CITED

1. Blinkov, S. M., Glezer, I. I. 1968. *The Human Brain in Figures and Tables.* New York: Plenum
2. Bullock, T. H., Horridge, G. 1965. *Structure and Function in the Nervous Systems of Invertebrates.* San Francisco: W. H. Freeman
3. Sholl, D. A. 1956. *The Organization of the Cerebral Cortex.* London: Methuen
4. Coggeshall, R. E. 1967. *J. Neurophysiol.* 30:1288
5. Barlow, R. B., Jr. 1969. *J. Gen. Physiol.* 54:383
6. Eccles, J. C. 1964. *The Physiology of Synapses.* New York: Academic
7. Gardner, E. 1964. *Fundamentals of Neurology.* Philadelphia: Saunders
8. Lorente de Nó, R. 1947. *J. Cell Comp. Physiol.* 29:207
9. Rall, W. 1962. *Ann. NY Acad. Sci.* 96:1071
10. Horowitz, J. M., Freeman, W. J. 1968. *Bull. Math. Biophys.* 28:519.
11. Freeman, W. J., Patel, H. H. 1968. *Electroencephalogr. Clin. Neurophysiol.* 24:444
12. Plonsey, R. 1969. *Bioelectric Phenomena,* Ch. 5. New York: McGraw-Hill
13. MacKay, D. M., Ed. 1969. *Necrosci. Res. Progr. Bull.,* Vol. 7, No. 3, Ch. 1, 4. Brookline, Mass.: Neurosci. Res. Progr.
14. Sherrington, C. S. 1906. *The Integrative Action of the Nervous System.* New Haven: Yale Univ. Press
15. Sherrington, C. S. 1929. *Proc. Roy. Soc. London* 105B:332
16. Denny-Brown, D. 1940. *Selected Writings of Sir Charles Sherrington.* New York: Hoeber
17. Granit, R. 1963. *Progr. Brain Res.* 1:23
18. Brookhart, J. M., Kubota, K. 1963. *Progr. Brain Res.* 1:38
19. Hartline, H. K., Ratliff, F. 1958. *J. Gen. Physiol.* 41:1049
20. Knight, B. W., Toyoda, J., Dodge, F. A., Jr. 1970. *J. Gen. Physiol.* 56:421
21. Stark, L., Sherman, P. M. 1957. *J. Neurophysiol.* 20:17
22. Lopes da Silva, F. H., van Rotterdam, A., Storm van Leeuwen, W., Tielen, A. M. 1970. *Electroencephalogr. Clin. Neurophysiol.* 29:260
23. Cleland, B., Enroth-Cugell, C. 1968. *Acta Physiol. Scand.* 68:365
24. Maffei, L. 1968. *J. Neurophysiol.* 31:283
25. Regan, D. 1968. *Electroencephalogr. Clin. Neurophysiol.* 25:231
26. Tielen, A. M., Kamp, A., Lopes da Silva, F. H., Reneau, J. P., Storm van Leeuwen, W. 1969. *Electroencephalogr. Clin. Neurophysiol.* 26:381
27. Freeman, W. J. 1962. *Exp. Neurol.* 5:477
28. Freeman, W. J. 1963. *Int. Rev. Neurobiol.* 5:53
29. Harris, L. D. 1961. *Introduction to Feedback Systems.* New York: Wiley
30. Huggins, W. H. 1960. *Johns Hopkins Univ. Report No. AFCRC-TN-60-360*
31. Freeman, W. J. 1964. *Exp. Neurol.* 10:475
32. Hermann, H. T., Stark, L. 1963. *J. Neurophysiol.* 26:215
33. Houk, J., Simon, W. 1967. *J. Neurophysiol.* 30:1466
34. Biedenbach, M. A., Freeman, W. J. 1965. *Exp. Neurol.* 11:400
35. Smith, O. J. M. 1958. *Feedback Control Systems.* New York: McGraw-Hill
36. Freeman, W. J. 1967. *Logistics Rev.* 3:5
37. Freeman, W. J. 1968. *J. Neurophysiol.* 31:337
38. Ibid. 349
39. Ibid. 1
40. Ramon y Cajal, S. 1955. *Studies on the Cerebral Cortex (Limbic Structures),* trans. L. M. Kraft. Chicago: Year Book
41. Valverde, F. 1965. *Studies of the Piriform Lobe.* Cambridge, Mass.: Harvard Univ. Press
42. Green, J. D., Mancia, M., von Baumgarten, R. 1962. *J. Neurophysiol.* 25:367
43. Yamamoto, C., Yamamoto, T., Iwama, K. 1963. *J. Neurophysiol.* 26:403
44. Phillips, C. G., Powell, T. P. S., Shepherd, G. M. 1963. *J. Physiol.* 168:65
45. Shepherd, G. M. 1963. *J. Physiol.* 168:101
46. Rall, W., Shepherd, G. M. 1968. *J. Neurophysiol.* 31:884
47. Freeman, W. J. 1969. *Physiologist* 12:229
48. Freeman, W. J. 1970. In *Approaches to Neural Modeling,* ed. M. A. B.

Brazier, D. Walter. Los Angeles: Brain Information Service, UCLA. In press
49. Eccles, J. C. 1957. *The Physiology of Nerve Cells,* Chaps. 2 (Fig. 21), 3 (Figs. 39, 45). Baltimore: Johns-Hopkins
50. Granit, R., Kernell, D. Shortess, G. K. 1963. *J. Physiol.* 168:911
51. Rall, W. 1955. *J. Cell. Comp. Physiol.* 46:373
52. Ten Hoopen, M., Verveen, A. A. 1963. *Progr. Brain Res.* 2:8
53. Calvin, W. H., Stevens, C. F. 1968. *J. Neurophysiol.* 31:524
54. Freeman, W. J. 1967. *Physiologist* 10: 172
55. Parzen, E. 1960. *Modern Probability Theory and Its Applications,* p. 60. New York: Wiley
56. Freeman, W. J. 1971. Unpublished data
57. Freeman, W. J. 1968. *Math. Biosci.* 2: 181
58. Rall, W. 1960. *Exp. Neurol.* 2:503
59. Katz, B. 1939. *Electric Excitation of Nerve.* London: Oxford Univ. Press
60. Schwan, H. P. 1957. In *Advances in Biological and Medical Physics,* ed. J. H. Lawrence, C. A. Tobias, pp. 148–209. New York: Academic
61. Katz, B. 1966. *Nerve, Muscle, and Synapse.* New York: McGraw-Hill
62. Hodgkin, A. L., Rushton, W. A. H. 1946. *Proc. Roy. Soc. London* 133B:444
63. Rall, W. 1959. *Exp. Neurol.* 1:491
64. Eccles, J. C. 1964. *The Physiology of Synapses.* New York: Academic
65. Wall, P. D. 1962. *J. Physiol.* 164:508
66. Wilson, H. R., Cowan, J. D. 1972. *Biophys. J.* 12:1
67. Freeman, W. J. 1964. *Recent Advan. Biol. Psychiat.* 7:235
68. Emery, J., Freeman, W. J. 1969. *Physiol. Behav.* 4:69
69. Horowitz, J. M. 1972. *Electroencephalogr. Clin. Neurophysiol.* 32:227
70. Pickering, S., Freeman, W. J. 1968. *Exp. Neurol.* 19:127
71. Poggio, G. P., Viernstein, L. J. 1964. *J. Neurophysiol.* 27:517
72 Mimura, K., Sato, K. 1970. *Int. J. Neurosci.* 1:75
73. Brazier, M. A. B. 1958. *The Electrical Activity of the Nervous System.* New York: MacMillan
74. John, E. R. 1967. *Mechanisms of Memory.* New York: Academic
75. Anderson, J. A. 1968. *Kybernetik* 5: 113
76. Longuet-Higgins, H. C. 1968. *Proc. Roy. Soc. London.* 171B:327
77. Hubel, D. H., Wiesel, T. N. 1959. *J. Physiol.* 148:574
78. Mountcastle, V. B. 1961. In *Sensory Communication,* ed. W. A. Rosenblith, Chap. 22. Cambridge: MIT Press
79. Barlow, H. B. 1969. In *Information Processing in the Nervous System,* ed. K. N. Liebovic, Chap. 11. New York: Springer-Verlag

Article 2.4

EXCITATORY AND INHIBITORY INTERACTIONS IN LOCALIZED POPULATIONS OF MODEL NEURONS

HUGH R. WILSON *and* JACK D. COWAN

From the Department of Theoretical Biology, The University of Chicago, Chicago, Illinois 60637

ABSTRACT Coupled nonlinear differential equations are derived for the dynamics of spatially localized populations containing both excitatory and inhibitory model neurons. Phase plane methods and numerical solutions are then used to investigate population responses to various types of stimuli. The results obtained show simple and multiple hysteresis phenomena and limit cycle activity. The latter is particularly interesting since the frequency of the limit cycle oscillation is found to be a monotonic function of stimulus intensity. Finally, it is proved that the existence of limit cycle dynamics in response to one class of stimuli implies the existence of multiple stable states and hysteresis in response to a different class of stimuli. The relation between these findings and a number of experiments is discussed.

INTRODUCTION

It is probably true that studies of primitive nervous systems should be focused on individual nerve cells and their precise, genetically determined interactions with other cells. Although such an approach may also be appropriate for many parts of the mammalian nervous system, it is not necessarily suited to an investigation of those parts which are associated with higher functions, such as sensory information processing and the attendant complexities of learning, memory storage, and pattern recognition. There are several reasons why a shift in emphasis is warranted in the investigation of such problems. There is first of all the pragmatic point that since sensory information is introduced into the nervous system in the form of large-scale spatiotemporal activity in sheets of cells, the number of cells involved is simply too vast for any approach starting at the single cell level to be tractable. Closely related to this is the observation that since pattern recognition is in some sense a global process, it is unlikely that approaches which emphasize only local properties will provide much insight. Finally, it is at least a reasonable hypothesis that local interactions between nerve cells are largely random, but that this local randomness gives rise to quite precise long-range interactions. Here an example from physics suggests itself. If a fluid is observed at the molecular level, what is seen is brownian motion, whereas the same fluid, viewed macroscopically, may be undergoing very orderly streamlined flow. Following up this analogy, we shall develop a deterministic model for the dynamics of neural populations. This may be interpreted as a treatment of the mean values of the underlying statistical processes.

In view of these remarks, we introduce a model which emphasizes not the individual cell but rather the properties of populations. The cells comprising such populations are assumed to be in close spatial proximity, and their interconnections are assumed to be random, yet dense enough so that it is very probable that there will be at least one path (either direct or via interneurons) connecting any two cells within the population. Under these conditions we may neglect spatial interactions and deal simply with the temporal dynamics of the aggregate.[1] Consistent with this approach, we have chosen as the relevant variable (following Beurle, 1956) the proportion of cells in the population which become active per unit time. This implies that the relevant aspect of single cell activity is not the single spike but rather spike frequency. Furthermore, time will be treated as a continuous variable so as to

[1] The neglect of spatial interactions is only temporary; a paper dealing with the extension of the present model to spatially distributed neural populations is in preparation (Wilson and Cowan).

Reproduced from the *Biophysical Journal*, 1972, vol. 12, pp. 1–24, by copyright permission of the Biophysical Society.

avoid the introduction of the spurious oscillations often found when a differential dynamical system is treated by finite difference equations.

Physiological evidence for the existence of spatially localized neural populations is provided by the work of Mountcastle (1957) and Hubel and Wiesel (1963, 1965). Their findings indicate that even within relatively small volumes of cortical tissue there exist many cells with very nearly identical responses to identical stimuli: there is a high degree of *local redundancy*.[2] It is just such local redundancy which must be invoked to justify characterizing spatially localized neural populations by a single variable. Local redundancy in the cerebral cortex has also been inferred from anatomical evidence (Szentágothai, 1967; Colonnier, 1965).

There is one final and crucial assumption upon which this study rests: *all nervous processes of any complexity are dependent upon the interaction of excitatory and inhibitory cells.* This assertion is supported by the work of Hartline and Ratliff (1958), Hubel and Wiesel (1963, 1965), Freeman (1967, 1968 *a*, *b*), Szentágothai (1967), and many others. In fact, this assumption is virtually a truism at this point, yet many neural modelers have dealt with nets composed entirely of excitatory cells (Beurle, 1956; Farley and Clark, 1961; ten Hoopen, 1965; Allanson, 1956). It was just this failure to consider inhibition that led Ashby et al. (1962) to conclude that the dynamical stability of the brain was paradoxical, and it was the introduction of inhibition by Griffith (1963) which dissolved the paradox. Consequently, we take it to be essential that there be both excitatory and inhibitory cells within any local neural population. We shall therefore speak of a localized neural population as being composed of an excitatory subpopulation and an inhibitory subpopulation. This will require a two-variable description of the population.[3]

THE MODEL

In accordance with the preceding remarks, we define as the variables characterizing the dynamics of a spatially localized neural population:

$E(t)$ = proportion of excitatory cells firing per unit time at the instant t;

$I(t)$ = proportion of inhibitory cells firing per unit time at the instant t.

The state $E(t) = 0, I(t) = 0$, the resting state, will be taken to be a state of low-level background activity, since such activity seems ubiquitous in neural tissue. Therefore, small negative values of E and I will have physiological significance, representing depression of resting activity. $E(t)$ and $I(t)$ will be referred to as the activities in the respective subpopulations.

We now derive the equations satisfied by $E(t)$ and $I(t)$. By assumption the value of these functions at time $(t + \tau)$ will be equal to the proportion of cells which are sensitive (i.e., not refractory) *and* which also receive at least threshold excitation at time t. We shall first obtain independent expressions for the proportion of sensitive cells and for the proportion of cells receiving at least threshold excitation.

If the absolute refractory period has a duration of r msec, then the proportion of excitatory cells which are refractory will evidently be given by[4]

$$\int_{t-r}^{t} E(t')\,dt'.$$

Consequently, the proportion of excitatory cells which are sensitive is just

$$1 - \int_{t-r}^{t} E(t')\,dt'.$$

Similar expressions are obtained for the inhibitory subpopulation.

The functions giving the expected proportions of the subpopulations receiving at

---

[2] Local redundancy has been used before by Von Neumann (1956) and Winograd and Cowan (1963) to account for the reliability of information processing in neural nets. In the latter work the neural nets had properties analogous to the ones discussed in this paper: excitatory and inhibitory cells, densely interconnected in a redundant fashion.

[3] Cowan (1970) has previously developed a two-variable treatment of neural activity in which, however, the fundamental variables are the mean rates of firing of *individual* excitatory and inhibitory cells rather than of subpopulations.

[4] No account is taken of relative refractoriness. An extended model which includes a refractory period after any desired time course is given in the Appendix; however, the complexity of the extended model is such that any detailed examination of the effects of relative refractoriness must await a thorough investigation of the present simpler model.

least threshold excitation per unit time as a function of the average levels of excitation within the subpopulations will be called subpopulation response functions and designated by $\mathcal{S}_e(x)$ and $\mathcal{S}_i(x)$. We call $\mathcal{S}_e(x)$ and $\mathcal{S}_i(x)$ response functions because they give the expected proportion of cells in a subpopulation which would respond to a given level of excitation if none of them were initially in the absolute refractory state. The general form of these functions can be derived in several ways.

Assume first that there is a distribution of individual neural thresholds within a subpopulation characterized by the distribution function $D(\theta)$. If it is further assumed that all cells receive the same numbers of excitatory and inhibitory afferents, then on the average all cells will be subjected to the same average excitation $x(t)$, and the subpopulation response function $\mathcal{S}(x)$ will take the form:

$$\mathcal{S}(x) = \int_0^{x(t)} D(\theta)\, d\theta. \tag{1}$$

Alternatively, assume that all cells within a subpopulation have the same threshold θ, but let there be a distribution of the number of afferent synapses per cell. If $C(w)$ is the synaptic distribution function and $x(t)$ the average excitation per synapse, then all cells with at least $\theta/x(t)$ synapses will be expected to receive sufficient excitation. Thus, the subpopulation response function takes the form:

$$\mathcal{S}(x) = \int_{\theta/x(t)}^{\infty} C(w)\, dw. \tag{1 a}$$

The validity of both of these formulas of course rests on the assumption that the total number of afferents reaching a cell is sufficiently large, for it is only in this case that all cells will be subjected to approximately the same $x(t)$.

$\mathcal{S}(x)$ as defined in either equations 1 or 1 *a* is readily seen to be a monotonically increasing function of $x(t)$ with a lower asymptote of 0 and an upper asymptote of 1. If in addition $D(\theta)$ or $C(w)$ is a unimodal distribution, the response function will assume a sigmoid form such as that shown in Fig. 1. It is this sigmoid form which will be taken to be characteristic of any acceptable subpopulation response function. Any function $f(x)$ will be said to belong to the class of sigmoid functions if:

(*a*) $f(x)$ is a monotonically increasing function of x on the interval $(-\infty, \infty)$,

(*b*) $f(x)$ approaches or attains the asymptotic values 0 and 1 as x approaches $-\infty$ and ∞ respectively, or

(*c*) $f(x)$ has one and only one inflection point. This inflection point will be termed the subpopulation threshold, although it is related to the single cell thresholds only through equations 1 or 1 *a*.

There are several points to be made concerning the sigmoid shape of the subpopulation response function. First, the phenomenological significance of the sigmoid shape is intuitively clear: in a population of threshold elements too low a level of excitation will fail to excite any elements, while very strong excitation can do more than excite all of the elements in the population. Second, a number of experimental studies have shown that single cell response curves are sigmoid functions of excitation (Kernell, 1965 *a*, *b*) as well as population response curves (Rall, 1955 *a*, *b*, *c*). Finally, it may be noted that the response function is essentially the event density of renewal theory (Cox, 1962). The event density is known to be related to a sum of convolutions of first-passage time densities for the single units of the population. The relationship of the subpopulation response function to the first-passage density has been explored more fully by Cowan (1971).

Before proceeding it should be mentioned that if $D(\theta)$ or $C(w)$ is multimodal, $\mathcal{S}(x)$ will still be monotonic but will not be sigmoid as defined above. Rather than a unique inflection point, there will be one inflection point for each mode of the distribution, as is shown in Fig. 2 for the bimodal case. For an n-modal distribution, however, $\mathcal{S}(x)$ can always be written as a weighted sum of n sigmoid functions having different inflection points. Physiologically, a multimodal distribution would be expected to correspond to the presence of a number of distinct cell types within the subpopulation. For the present we shall take $\mathcal{S}(x)$ to be a single sigmoid function, but we will return briefly to the more complex case later.

An expression for the average level of excitation generated in a cell of each subpopulation must now be obtained. If it is assumed that individual cells sum their inputs and that the effect of stimulation decays with a time course $\alpha(t)$, then the average level of excitation generated in an excitatory cell at time t will be:

$$\int_{-\infty}^{t} \alpha(t - t')[c_1 E(t') - c_2 I(t') + P(t')] \, dt'. \qquad (2)$$

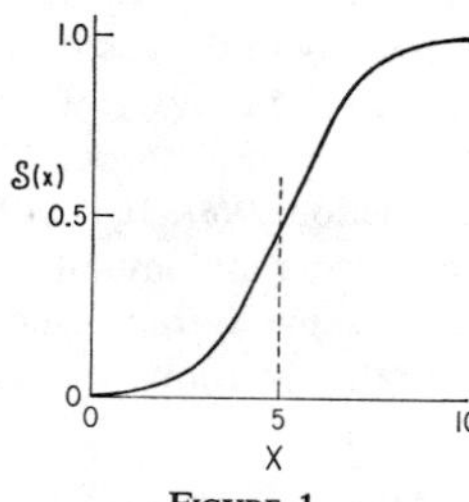

FIGURE 1

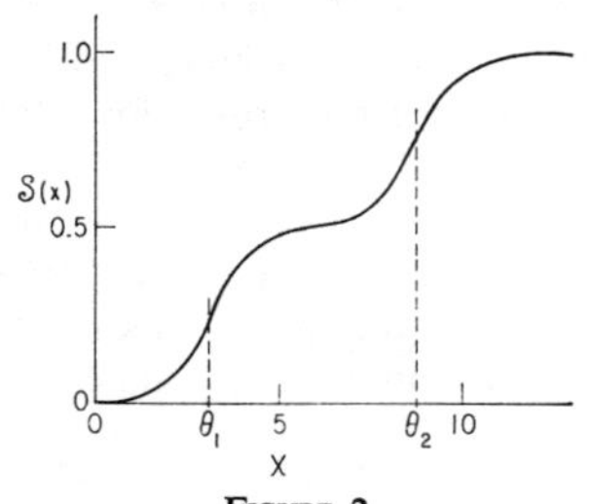

FIGURE 2

FIGURE 1 Plot of typical sigmoid subpopulation response function. X is average level of excitation in threshold units. The particular function shown here is the logistic curve: $\mathcal{S}(x) = 1/[1 + e^{-a(x-\theta)}]$ with $\theta = 5$, $a = 1$.

FIGURE 2 Subpopulation response function resulting from bimodal distribution of thresholds or afferent synapses. X is excitation in threshold units, while θ_1 and θ_2 are the two local maxima of the underlying distribution. Note that this curve may be decomposed into a weighted sum of two sigmoid functions.

The connectivity coefficients c_1 and c_2 (both positive) represent the average number of excitatory and inhibitory synapses per cell, while $P(t)$ is the external input to the excitatory subpopulation. A similar expression but with different coefficients and a different external input will apply to the inhibitory subpopulation. The differing coefficients reflect differences in axonal and dendritic geometry between the excitatory and inhibitory cell types, while the difference in external inputs assumes the existence of cell-type specific afferents to the population.

Given these expressions for the subpopulation response functions, the average excitation, and the proportion of sensitive cells in each subpopulation, we may now obtain equations for the activities $E(t)$ and $I(t)$. As we have noted, the activity in a subpopulation at time $(t + \tau)$ will be equal to the proportion of cells which are both sensitive and above threshold at time t. If the probability that a cell is sensitive is independent of the probability that it is currently excited above its threshold, then the desired expression for the excitatory subpopulation is just:

$$\left[1 - \int_{t-\tau}^{t} E(t') \, dt'\right] \mathcal{S}_e(x)\delta t.$$

In general, however, there will be some correlation between the level of excitation of a cell and the probability that it is sensitive. Furthermore, this correlation will tend to reduce the value of the expression just obtained. This is so because cells which are currently highly excited are more likely to have been highly excited in the recent past and thus are more likely to have already fired and be refractory. Designating this correlation between excitation and sensitivity by

$$\gamma\left[\int_{t-\tau}^{t} E(t') \, dt', \mathcal{S}_e(x)\right],$$

the previous expression becomes:

$$\left[1 - \int_{t-\tau}^{t} E(t') \, dt'\right] \mathcal{S}_e(x) \left\{1 - \gamma\left[\int_{t-\tau}^{t} E(t') \, dt', \mathcal{S}_e(x)\right]\right\} \delta t.$$

Although the particular functional form of γ will depend on the details of the connectivity or threshold distribution within the population, it will always have the following properties:

$$(a) \quad \lim_{\mathcal{S}_e(x) \to 0,1} \gamma = 0;$$

$$(b) \quad \lim_{\int_{t-\tau}^{t} E(t')dt' \to 0,1} \gamma = 0;$$

$$(c) \quad 0 \leq \max(\gamma) \leq 1.$$

The first two conditions follow from the observation that the uncorrelated and correlated expressions will coincide when *all* of the cells are below threshold, above threshold, sensitive, or refractory.

In the case we are considering, that of a richly interconnected population of cells, max (γ) will generally be very small. There are two reasons for this. The first is the presence of spatial and temporal fluctuations in the average level of excitation within the population caused both by the presence of fluctuations in the inputs and by the activity due to firing of cells within the population. The second is the existence of fluctuations in the thresholds of the individual cells themselves (Frishkopf and Rosenblith, 1958; Verveen and Derksen, 1969; Rall and Hunt, 1956). In the present study, therefore, we shall take γ to be zero, thus dealing with the case in which sensitivity is not correlated with level of excitation. It follows that the equations governing the dynamics of a localized population of neurons are:

$$E(t + \tau) = \left[1 - \int_{t-r}^{t} E(t')\,dt'\right] \cdot \mathcal{S}_e\left\{\int_{-\infty}^{t} \alpha(t - t')[c_1 E(t') - c_2 I(t') + P(t')]\,dt'\right\}, \quad (3)$$

and

$$I(t + \tau') = \left[1 - \int_{t-r'}^{t} I(t')\,dt'\right] \cdot \mathcal{S}_i\left\{\int_{-\infty}^{t} \alpha(t - t')[c_3 E(t') - c_4 I(t') + Q(t')]\,dt'\right\}, \quad (4)$$

for the excitatory and inhibitory subpopulations.

TIME COARSE GRAINING

Equations 3 and 4 are intuitively simple in that each term has been shown to have a clear physiological interpretation. Mathematically, however, they are extremely complex both because of their strongly nonlinear character and because they involve temporal integrals. The nonlinearity is a fundamental characteristic of this as well as most other biological control systems. The presence of temporal integrals, however, is an aspect of lesser significance biologically, as will now be shown. The mathematical advantage to be gained from the removal of the time integrals will be the applicability of phase plane analysis to extract significant qualitative features of the solutions of equations 3 and 4 for various parameter ranges and initial conditions.

The technique we will use to simplify equations 3 and 4 is a form of temporal coarse graining, which was first applied by Kirkwood (1946) to some problems in statistical physics. Although we shall not follow the original arguments precisely, the basis of the method is the replacement of the dependent variable, e.g. $f(t)$, by the moving time average of this quantity over some appropriately chosen interval s. The coarse-grained variable, $\bar{f}(t)$, is thus given by:

$$\bar{f}(t) = \frac{1}{s}\int_{t-s}^{t} f(t')\,dt'. \quad (5)$$

Obviously, the effect of this change of variable is to average out rapid temporal variations taking place on a time scale shorter than s. To justify the use of the temporal coarse-graining approximation, therefore, it is necessary to show that the behavior which is lost through averaging is not of significance for the problem at hand.

To obtain the appropriate coarse-grained forms for equations 3 and 4, notice first that $E(t)$ and $I(t)$ appear on the right side of these equations only in the form of time-averaged quantities. If $\alpha(t)$ is close to unity for $0 \leq t \leq r$ and drops fairly rapidly to zero for $t > r$, then it is a reasonable approximation to replace both these integrals by the same coarse-grained variables. That is,

$$\int_{t-r}^{t} E(t')\,dt' \rightarrow r\bar{E}(t),$$

$$\int_{-\infty}^{t} \alpha(t - t')E(t')\,dt' \rightarrow k\bar{E}(t), \quad (6)$$

with k and r constant. Similar replacements apply to $I(t)$.

As time coarse graining has a marked smoothing effect on temporal variation, it is appropriate to replace $E(t + \tau)$ and $I(t + \tau')$ in equations 3 and 4 by Taylor expansions in the coarse-grained variable about the value $\tau = 0$. Thus, we arrive at the time coarse-grained form of equations 3 and 4:

$$\tau \frac{d\bar{E}}{dt} = -\bar{E} + (1 - r\bar{E})\mathcal{S}_e[kc_1\bar{E} - c_2k\bar{I} + kP(t)], \tag{7}$$

$$\tau' \frac{d\bar{I}}{dt} = -\bar{I} + (1 - r\bar{I})\mathcal{S}_i[k'c_3\bar{E} - c_4k'\bar{I} + k'Q(t)]. \tag{8}$$

In order to assess the appropriateness of the coarse-graining approximation, we have compared computer solutions to equation 3 with those obtained from equation 7. For this purpose equation 3 was expanded to lowest order in τ, and $\alpha(t - t')$ was taken to be an exponential decay. Interaction with the inhibitory subpopulation was excluded from both equations 3 and 7 to simplify the comparison. Thus we are concerned with the equations:

$$\tau \frac{dE}{dt} = -E + \left[1 - \int_{t-r}^{t} E(t')\, dt'\right] \mathcal{S}_e\left\{\int_{-\infty}^{t} e^{-\alpha(t-t')}[c_1E(t') + P(t')]\, dt'\right\}, \tag{9}$$

$$\tau \frac{d\bar{E}}{dt} = -\bar{E} + (1 - r\bar{E})\mathcal{S}_e[kc_1\bar{E}(t) + kP(t)]. \tag{10}$$

The major difference that is observed between the two cases is that the solution to equation 9 generally involves a damped oscillation with period equal to twice the refractory period, whereas the solution to equation 10 approaches the same asymptotic value monotonically. A typical example is shown in Fig. 3.

We suggest that this damped oscillation is not of great significance, for the following reasons. First, as the period of the oscillation is dependent almost entirely on the length of the absolute refractory period, it cannot transmit information concerning the nature of a stimulus. Thus, the damped oscillation is not likely to have any functional significance. Second, although damped oscillations are often observed in evoked potential studies (Freeman, 1967, 1968 *a*, *b*; Andersen and Eccles, 1962; MacKay, 1970), such oscillations typically have periods of 40 msec or longer, whereas an oscillation produced by absolute refractoriness alone would be unlikely to have a period of more than about 6 msec. In addition, Freeman's work makes it reasonably certain that the oscillations observed in evoked potential studies result from interactions among excitatory and inhibitory neurons. Finally, if it is the long-time behavior of a neural population (on the order of 100 msec) that is functionally significant, then the coarse-grained equation 10 provides correct results.

There is, however, an apparent exception to the last point: for certain parameter ranges the solution to equation 9 can be shown to give sustained oscillations. A necessary condition for this is that the summation constant α be much smaller than the refractory period. This is usually not the case, for physiological studies show α to be around 4 msec and r around 1–2 msec (Eccles, 1964). Furthermore, we would roughly expect that for equation 6 to be valid, α would have to be somewhat greater than r, for otherwise the exponentially weighted integral would approach zero too fast for significant temporal averaging to take place. Thus, we conclude that when α and r are given physiologically reasonable values, the temporally coarse-grained equations are valid.

PHASE PLANE ANALYSIS

Before proceeding to analyze equations 7 and 8, one minor adjustment will be made for conceptual and mathematical convenience. As previously mentioned, the state $E = 0$, $I = 0$ will be chosen to be the state of low-level background activity so ubiquitous in the nervous system. The source of this activity (be it spontaneous, reverberatory, or driven) is of no consequence to the present investigation. The mathematical consequence of this choice of resting state is that $E = 0$, $I = 0$ must be a steady-state solution to equations 7 and 8 for $P(t) = Q(t) = 0$, i.e., in the absence of external inputs. Furthermore, the resting state must be stable to be of physiological significance.

The first of these requirements is readily fulfilled by transforming $\mathcal{S}_e$ and $\mathcal{S}_i$ so that $\mathcal{S}_e(0) = \mathcal{S}_i(0) = 0$. Given any sigmoid function, this may be done by subtracting $\mathcal{S}(0)$ from the original function. Now, however, the maximum values of the response functions will in general be less than unity. Designating these values by k_e and k_i, the refractory terms must be modified, giving the final result:

$$\tau_e \frac{dE}{dt} = -E + (k_e - r_e E)\mathcal{S}_e(c_1 E - c_2 I + P), \tag{11}$$

$$\tau_i \frac{dI}{dt} = -I + (k_i - r_i I)\mathcal{S}_i(c_3 E - c_4 I + Q). \tag{12}$$

(The bars denoting coarse graining have been dropped for convenience.)

The equations may be analyzed qualitatively using the E, I phase plane. From the mathematical properties of sigmoid functions, it is evident that $\mathcal{S}_e$ and $\mathcal{S}_i$ have unique inverses. Denoting these inverses by $\mathcal{S}_e^{-1}$ and $\mathcal{S}_i^{-1}$ it is possible to write the equations for the isoclines corresponding to $dE/dt = 0$ and $dI/dt = 0$ as:

$$c_2 I = c_1 E - \mathcal{S}_e^{-1}\left(\frac{E}{k_e - r_e E}\right) + P \quad \text{for} \quad \frac{dE}{dt} = 0, \tag{13}$$

$$c_3 E = c_4 I + \mathcal{S}_i^{-1}\left(\frac{I}{k_i - r_i I}\right) - Q \quad \text{for} \quad \frac{dI}{dt} = 0. \tag{14}$$

Notice that c_2 and c_3 must always be nonvanishing for the isoclines to be nontrivial, thus making negative feedback between the subpopulations an essential feature of the model. A typical plot of these two equations for $P = 0$, $Q = 0$ is shown in Fig. 4. In this case there are three steady-state solutions corresponding to the three intersections of the two curves. Depending on the parameter values chosen there may be either one or five steady states instead of three, a point to which we shall return.

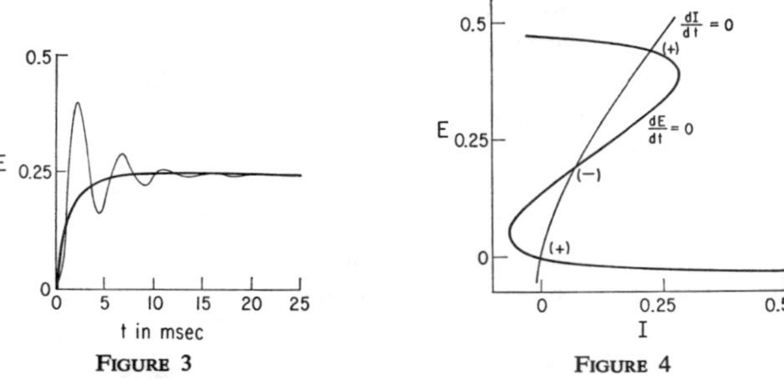

FIGURE 3 Comparison of solution to equation 9 (lighter line) with solution with the temporal coarse-grained equation 10 (heavier line). Duration of refractory period: $r = 3$ msec.

FIGURE 4 Phase plane and isoclines (equations 13 and 14). (+) denotes stability and (−), instability of steady state. Parameters: $c_1 = 12$, $c_2 = 4$, $c_3 = 13$, $c_4 = 11$, $a_e = 1.2$, $\theta_e = 2.8$, $a_i = 1$, $\theta_i = 4$, $r_e = 1$, $r_i = 1$, $P = 0$, $Q = 0$.

Before going further, let us choose a particular form of sigmoid function to make matters more definite. The form we shall choose is the logistic curve [shifted downward by a constant amount so that $\mathcal{S}(0) = 0$]:

$$\mathcal{S}(x) = \frac{1}{1 + \exp\left[-a(x - \theta)\right]} - \frac{1}{1 + \exp(a\theta)}. \tag{15}$$

Here a and θ are parameters, the latter giving the position of maximum slope and the former determining the value of the maximum slope through the relationship:

$$\max\left[\mathcal{S}'(x)\right] = \mathcal{S}'(\theta) = \frac{a}{4}. \tag{16}$$

No particular significance is to be attached to the choice of the logistic curve; any other function with the defining sigmoid properties would be equally suitable. A different function would, of course, lead to different detailed dynamics, but qualitative properties of the solutions such as number and stability of steady states, hysteresis effects, presence of limit cycles, etc., may be obtained from equations 11 and 12 for any particular function chosen.[5]

Returning now to our discussion of the isoclines defined by equations 13 and 14, we first observe that the inverse of a sigmoid function is a monotonically increasing function of its argument ranging from $-\infty$ to $+\infty$. Therefore, E as defined by equation 14 will always be a monotonically increasing function of I. On the other hand, because of the negative sign before $\mathcal{S}_e^{-1}$ in equation 13, I will be a generally decreasing function of E except over a short range where it may temporarily increase. This is observed in the curve $dE/dt = 0$ in Fig. 4. This qualitative difference between the two isoclines is a direct manifestation of the antisymmetry between excitation and inhibition.

As it is the "kink" in the isocline for $dE/dt = 0$ which gives rise to the possibility of multiple steady states, hysteresis phenomena, and maintained oscillations, it is important to know for what values of the parameters this temporary reversal in the slope of equation 13 can occur. A necessary and sufficient condition for this is that the maximum slope of this curve of I as a function of E be greater than zero. The maximum slope is not easy to calculate, but a sufficient condition may be simply obtained by requiring that the slope of equation 13 at the inflection point of $\mathcal{S}_e^{-1}$ be greater than zero. The slope of the isocline at this point is

$$\left(\frac{c_1}{c_2} - \frac{9}{a_e c_2}\right),$$

thus leading to the condition:

$$c_1 > 9/a_e, \tag{17}$$

where a_e is the slope parameter for the excitatory response function. In obtaining condition 17 r_e and r_i have been set equal to unity in order to simplify the result. As a matter of convenience we shall adopt this value for r_e and r_i from now on, as nothing essential is lost thereby.

A physiological interpretation of condition 17 is possible once it is realized that $1/a_e$ is directly related to the variance of the distribution of thresholds or synaptic connections from which the excitatory subpopulation response function was derived (see equations 1 and 1 *a*). That is, for the maximum slope of the response function (see equation 16) to increase, it is necessary that the variance of the underlying distribution decrease. Thus, condition 17 implies that a sufficient condition for the existence of multiple steady states is that the average number of synapses between excitatory neurons must exceed a function of the variance in the distribution of these connections (or alternatively, the variance in the distribution of thresholds).

Assuming that condition 17 is satisfied, under what conditions will there exist multiple steady states? If P and Q are restricted to the value zero, this is a difficult question to answer, for the conditions will depend in complex ways on all of the parameters of the population. If P and Q are not so restricted, however, then we may state the following theorem.

Theorem 1. If $c_1 > 9/a_e$, then there is a class of stimulus configurations such that the isoclines defined by equations 13 and 14 will have at least three intersections. That is, equations 11 and 12 will have at least three steady-state solutions.

A stimulus configuration is defined to be any particular choice of *constant* values for P and Q.

Proof: The condition $c_1 > 9/a_e$ is sufficient to insure that there will be a region in which the isocline for $dE/dt = 0$ can be intersected at three points by a line parallel to the E-axis in the phase plane. As the isocline for $dI/dt = 0$ approaches asymptotes parallel to the E-axis, and as the effect of changing P and Q is to translate their respective isoclines parallel to the I- and E-axes respectively, one can always choose values of P and Q for which there are at least three intersections.

Once the number and locations of steady-state solutions to equations 11 and 12 have been determined, the stability of each steady state can readily be determined by linearization around each state and solution of the resulting characteristic equation. The procedure is simple but tedious, and no real insight is to be gained from displaying the equations. Accordingly, we shall simply indicate stability characteristics where appropriate.

---

[5] This assertion may be proved through considerations of the general shape of inverse sigmoid functions and the resulting shapes of the isoclines, equations 13 and 14.

HYSTERESIS

In the example illustrated in Fig. 4 two of the three steady states can be shown to be stable and are separated by an unstable state. This fact, plus the observation that the effect of a change in the value of P or Q is to translate the appropriate isocline parallel to one of the phase plane axes, suggests the existence of hysteresis phenomena. (It will be recalled that P and Q represent external inputs to the excitatory and inhibitory subpopulations.) This is indeed the case, and a graph of the hysteresis loop obtained from Fig. 4 as P is varied and Q held constant is shown in Fig. 5. Only excitatory activity has been plotted, although a corresponding plot could be made for the accompanying inhibitory activity. Had P been held constant and Q varied the resulting hysteresis loop would have been reversed: excitement of inhibitory cells leads to a decrease in excitatory cell activity.

The hysteresis phenomenon illustrated in Fig. 5 is a simple one, as only two stable states are involved. Since stability of two of the three states is easy to prove, a sufficient condition for the existence of such a loop is given by condition 17. Simple hysteresis loops have been demonstrated and discussed by Harth and coworkers in model neural populations containing mainly excitatory cells (Harth, et al., 1970; Anninos, et al., 1970). Consequently, it is not surprising that condition 17 contains only parameters of the excitatory subpopulation of the present model.

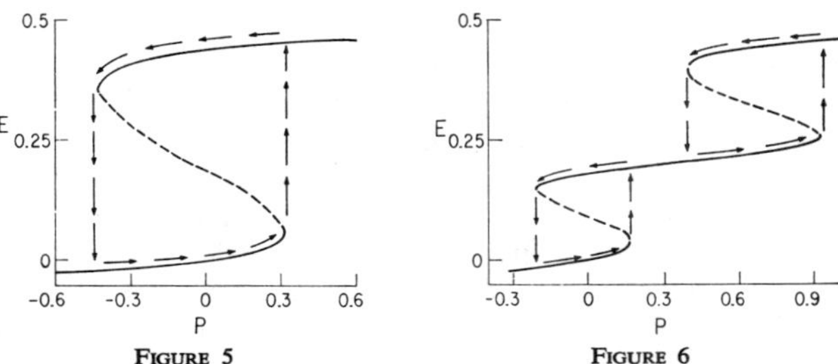

FIGURE 5 FIGURE 6

FIGURE 5 Steady-state values of E as a function of P ($Q = 0$). Solid lines indicate stable states, while the dashed line indicates an unstable state. Hysteresis loop indicated by arrows is generated if P is varied slowly back and forth through the range shown on the graph. Parameters are those given in Fig. 4.

FIGURE 6 Steady-state values of E as a function of P ($Q = 0$). Solid lines indicate stability, dashed line, instability. Here two simple hysteresis loops (arrows) are separated by a region with a single stable state. Parameters: $c_1 = 13$, $c_2 = 4$, $c_3 = 20$, $c_4 = 2$, $a_e = 1.2$, $\theta_e = 2.7$, $a_i = 5$, $\theta_i = 3.7$, $r_e = 1$, $r_i = 1$.

The presence of inhibitory cells can lead to more complex hysteresis phenomena. Two examples of this are shown in Figs. 6 and 7. In the former case two separated loops occur, while in the latter three simultaneous stable steady states are observed. Parameters may be chosen so that the points at which the intermediate stable state in Fig. 7 appears and vanishes bear any desired relation to the bifurcation points for the upper and lower stable states. Thus, as P increases the intermediate stable state may vanish before the lowest state, etc.

A sufficient condition for the existence of five steady states may be derived by examining the phase plane and isoclines in Fig. 8. Parameters here are those used to obtain the double hysteresis loop in Fig. 7. It will be seen that such a configuration of isoclines can only be obtained if the minimum slope of the isocline for $dE/dt = 0$ is less than the reciprocal of the maximum slope of the kink in the isocline for $dE/dt = 0$. (The reciprocal slope must be taken in the latter case because equation 13 defines I as a function of E.) A sufficient condition for this is that:

$$\frac{a_e c_2}{a_e c_1 - 9} > \frac{a_i c_4 + 9}{a_i c_3}. \qquad (18)$$

It is obvious that condition 18 can only be satisfied if $a_e c_1$ is greater than 9, so condition 17 must also be satisfied. We state this as a second theorem.

Theorem 2. Let the parameters of a neural population satisfy equation 18. Then five steady states will exist, though not necessarily concurrently (see Fig. 6), for some class of stimulus configurations.

This is not a sufficient condition for multiple hysteresis phenomena, since the intermediate state may be unstable in some cases.

A physiological interpretation of condition 18 is more apparent if it is rewritten as

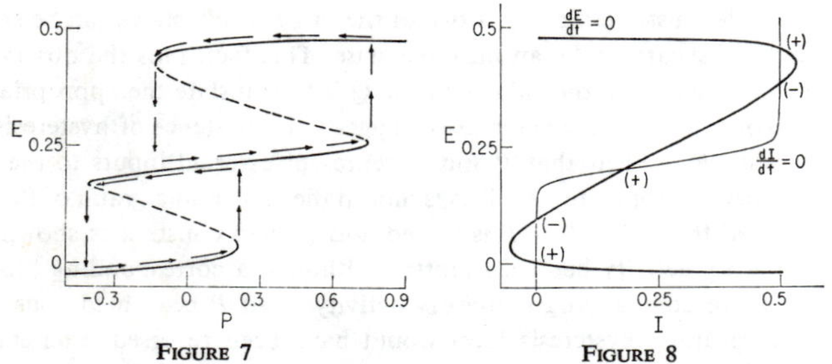

FIGURE 7 FIGURE 8

FIGURE 7 Steady-state values of E as a function of P ($Q = 0$). Solid lines indicate stability and dotted lines, instability. Here two overlapping hysteresis loops (arrows) are present: note the existence of three stable states in an interval around $P = 0$. Parameters: $c_1 = 13$, $c_2 = 4$, $c_3 = 22$, $c_4 = 2$, $a_e = 1.5$, $\theta_e = 2.5$, $a_i = 6$, $\theta_i = 4.3$, $r_e = 1$, $r_i = 1$.
FIGURE 8 Phase plane and isoclines with parameters chosen to give three stable (+) and two unstable (−) steady states. Parameters are the same as those in Fig. 7 with $P = 0$.

$$a_e a_i c_2 c_3 > (a_e c_1 - 9)(a_i c_4 + 9). \tag{19}$$

Neglecting a_e and a_i, which appear on both sides of the expression, it will be noted that $c_2 c_3$ is a measure of the strength of the negative feedback loop in the population. Similarly, the right side of condition 19 is a product of factors measuring the strengths of interactions *within* the excitatory and inhibitory subpopulations respectively. Thus, it may be said that condition 19 requires that there be a relatively strong negative feedback loop within the neural population.

In contrast to the requirements for simple hysteresis, it is evident from the foregoing that multiple hysteresis phenomena are dependent upon the inclusion of inhibition as an essential part of the present model. Although Smith and Davidson (1962) and Griffith (1963) did exhibit special cases in which an intermediate state of activity was stabilized by inhibition, we are not aware of any previous discussion of multiple hysteresis phenomena in model neural populations.

Functionally, hysteresis was first suggested as a physiological basis for short-term memory by Cragg and Temperley (1955). Such a possibility is evident, for any input of sufficient intensity and duration will cause the activity in the neural population to jump from the lowest (resting) state into one of the stable excited states, and the activity will remain in this state even after the input ceases. It may also be noted in this context that stable high-level activity of this type may be interpreted as resulting from reverberation: activity may circulate in the population in such a manner that the total activity is constant. Hysteresis as a form of short-term memory is therefore consistent with the work of Hebb (1949).

In addition to these conjectures linking hysteresis to short-term memory, there is at least one experimental verification of the existence of hysteresis within the central nervous system. This is the work of Fender and Julesz (1967), in which it is demonstrated that hysteresis is operative in the fusion of binocularly presented patterns to produce single vision. Since the earliest interactions between patterns presented to the two eyes occur in area 17, it is clear that Fender and Julesz have demonstrated the existence of hysteresis phenomena in cortical tissue. Our model provides a neural interpretation of these results.

It is important to note that hysteresis has two important forms of noise insensitivity. Observe first that in a loop such as that in Fig. 5 a large change in P is necessary to excite the population to the higher stable state: there is a population threshold. Secondly, because of the response time of the population even suprathreshold inputs will fail to alter the state of the population if they are of insufficient duration. To measure this time-intensity relationship a stimulus of intensity P was applied to an excitatory subpopulation which was initially in the resting state or passed to a state of maintained self-excitation. The plot in Fig. 9 represents the time-intensity threshold for the initiation of maintained activity. This curve is of the Block type which is commonly observed in the visual system (Le Grand, 1957). For a noisy system such as the brain subjected to a noisy environment these features are of obvious significance.

Finally, let us consider briefly the case in which the excitatory subpopulation has a bimodal distribution of thresholds or connections and consequently a response function such as that shown in Fig. 2. Clearly, this response function may give rise to an isocline for $dE/dt = 0$ in which there are two kinks, i.e., two regions of positive slope of the isocline separated by a region of negative slope (see equation 13). This additional kink will increase the number of possible intersections of the two isoclines by two, one of which will be stable. Therefore, one additional loop will be added to the hysteresis phenomenon. In general, an n-modal excitatory subpopulation response function will give rise to complex hysteresis phenomena composed of n simple hysteresis loops. The existence of a multimodal distribution of thresholds or synapses within the excitatory subpopulation will, of course, yield similar results.

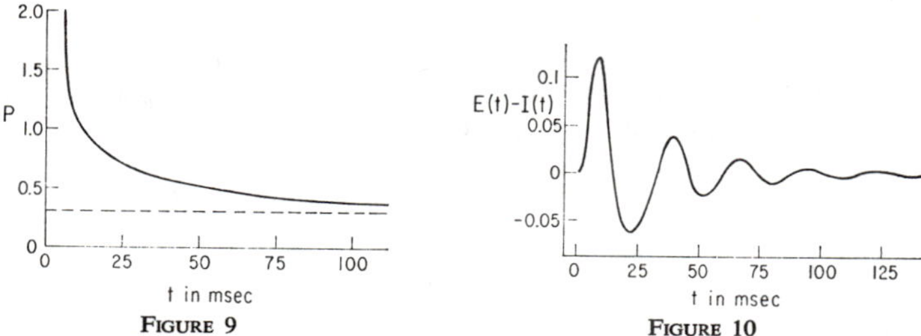

FIGURE 9

FIGURE 10

FIGURE 9 Strength-duration curve for excitation of population from lower to upper stable steady state in Fig. 5. Curve indicates intensity (P) and duration (t) of rectangular impulse which is just sufficient for population to become self-exciting and pass to upper excited state. Parameters are those for Fig. 5, with $\tau = 8$ msec. Dashed line indicates asymptotic value of P below which population cannot become self-exciting regardless of the duration of stimulation.

FIGURE 10 Damped oscillatory behavior of $[E(t) - I(t)]$ in response to brief stimulating impulse. It is suggested that this function is related to the average evoked potential (see text). Parameters: $c_1 = 15$, $c_2 = 15$, $c_3 = 15$, $c_4 = 3$, $a_e = 1$, $\theta_e = 2$, $a_i = 2$, $\theta_i = 2.5$, $\tau = 10$ msec.

TEMPORAL PHENOMENA: LIMIT CYCLES

So far discussion has been limited to steady states, and nothing has been said of the transient behavior of the neural population. This is because the approach to a stable steady state has been found to be monotonic and uneventful in most cases. There are, however, two types of temporal behavior exhibited by our model which are of considerable physiological interest.

There are a number of physiological systems which, in response to impulse stimulation, produce an average evoked potential in the form of a damped oscillation. Among such systems are the thalamus (Andersen and Eccles, 1962) and the olfactory bulb and cortex (Freeman, 1967, 1968 *a*, *b*). Further examples are given in MacKay (1970). Such oscillations typically show periods of 25–40 msec or more.

The usual interpretation is that the potential seen by the recording electrode represents the net difference between excitatory and inhibitory postsynaptic potentials in the neighborhood of the electrode. This suggests that the function in our model most closely related to the average evoked potential would be proportional to $[\bar{E}(t) - \bar{I}(t)]$. For an appropriate choice of parameters, this function will respond to impulse stimulation in a damped oscillatory manner as shown in Fig. 10. To obtain a period similar to that obtained experimentally, it was necessary to choose the time constants τ_e and τ_i to be about 10 msec. This value is in the range for the delays associated with the propagation of postsynaptic potentials from the dendrites of a neuron to the axon hillock (Oshima, 1969).

The ability of our model to reproduce the general form of the average evoked potential should not be taken too seriously, as systems may be readily designed to give damped oscillations in response to brief stimulation, and as no attempt has been made to reproduce details of the experimental curves. Rather, we regard the reproduction of a damped oscillatory average evoked potential as a constraint to be satisfied by any neural model claiming physiological plausibility.

There is a second form of temporal behavior exhibited by our model which is potentially of greater functional significance: the limit cycle. Limit cycles will arise whenever there is only one steady state determined by the intersection of the isoclines, and when this steady state is unstable. As all trajectories must remain within the unit square in the phase plane, these conditions are necessary and sufficient for the existence of a limit cycle. Linear stability analysis can be used to show that a sufficient (but not necessary) condition for the instability of such a steady state is that:

$$c_1 a_e > c_4 a_i + 18. \qquad (20)$$

This expression follows from the linear analysis plus the observation that the requirement of a single unstable steady state can only be realized when the isoclines intersect at a point in the vicinity of the inflection points of the sigmoid response functions. Expression 20 may be interpreted to mean that the existence of limit cycles in a neural population requires that the interactions *within* the excitatory subpopulation be significantly stronger than those *within* the inhibitory subpopulation. This is reasonable, since strong interactions within the inhibitory subpopulation will tend to damp out the negative feedback which is responsible for the oscillation.

The requirement that there exist a single stready state for some choice of P and Q and that it occur for values of E and I near the inflection points of the sigmoid response functions leads to the conditions:

$$\frac{a_e c_2}{a_e c_1 - 9} > \frac{a_i c_4 + 9}{a_i c_3}, \qquad (21)$$

$$\frac{a_e c_1 - 9}{a_e c_2} < 1. \qquad (22)$$

Requirement 21 is identical with condition 18 and is derived in exactly the same way. Requirement 22 insures that there is one steady state rather than five. We may therefore state a theorem encompassing both limit cycle phenomena and multiple hysteresis.

Theorem 3. Let parameters be chosen so that requirement 21 is satisfied. Then if expression 20 is *not* satisfied, multiple hysteresis phenomena will occur for some class of stimulus configurations. If, on the other hand, requirements 20 and 22 are satisfied, then for some class of stimulus configurations limit cycle dynamics will be obtained.

The proof of this theorem follows directly from a consideration of the shapes of the isoclines defined in equations 13 and 14 plus an enumeration of the possible ways in which they can intersect. It is straightforward but tedious and will not be reproduced.

Typical of the limit cycle activity found is that shown in Figs. 11 *a* and 11 *b*. As we have required the resting state $E = 0, I = 0$ to be stable in the absence of a driving force, the neural population will only exhibit limit cycle activity in response to constant stimulation. We therefore felt it appropriate to investigate the manner in which the limit cycle depends on the value of P (Q being set equal to zero). Typical results are shown in Figs. 12 *a* and 12 *b*. The important observations are:

(*a*) There is a threshold value of P below which limit cycle activity cannot occur.

(*b*) There is a higher value of P above which the system saturates and limit cycle activity is extinguished.

(*c*) Between these two values both the frequency of the limit cycle and the average value of $E(t)$ increase monotonically with increasing P.

Although limit cycle activity as a result of the constant stimulation of neural populations has not been looked for experimentally to our knowledge, our results do seem to be directly related to microelectrode studies. In particular, we cite the

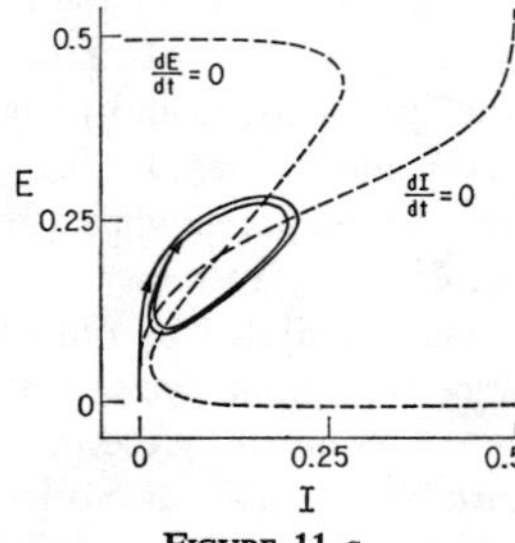

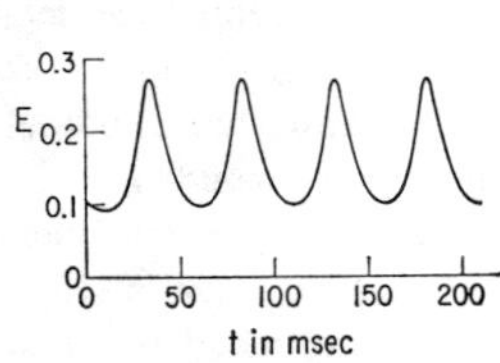

FIGURE 11 *a* FIGURE 11 *b*

FIGURE 11 *a* Phase plane showing limit cycle trajectory in response to constant stimulation $P = 1.25$. Dashed lines are isoclines. Parameters: $c_1 = 16, c_2 = 12, c_3 = 15, c_4 = 3, a_e = 1.3, \theta_e = 4, a_i = 2, \theta_i = 3.7, r_e = 1, r_i = 1$.

FIGURE 11 *b* $E(t)$ for limit cycle shown in Fig. 11 *a*. $\tau = 8$ msec.

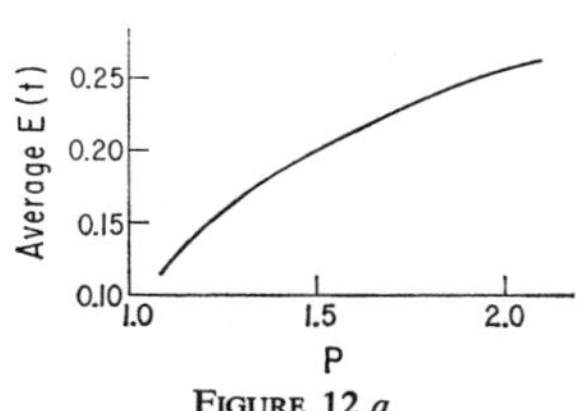

FIGURE 12 *a*

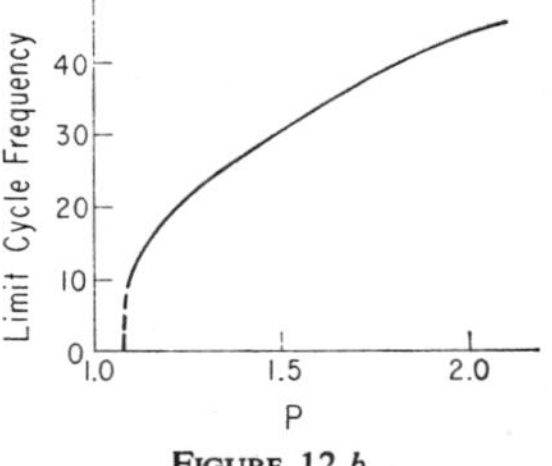

FIGURE 12 *b*

FIGURE 12 *a* $E(t)$ averaged over one period of limit cycle as a function of stimulation at constant intensity P.

FIGURE 12 *b* Frequency of limit cycle (in Hz) for different levels of constant stimulation P. For very low values of P no cycle is obtained, i.e., frequency drops to zero. For very high values of P the oscillation is extinguished and only high-level, constant activity is observed. Parameters are those given in Fig. 11 *a*.

work of Poggio and Viernstein (1964) on thalamic somatosensory neurons. The constant stimulation in their study was provided by a constant angle of flection of the wrist joint of a monkey. When the expectation density function of neurons driven by joint angle receptors was plotted, it was found to be an undamped periodic function of time. Both the average firing rate and the frequency of the oscillation in the expectation density function were found to increase monotonically with in-increasing (constant) angle of flection (see Fig. 9, Poggio and Viernstein, 1964).

Thus, it is seen that $E(t)$ in our model reproduces qualitatively the characteristics of averaged single unit firing patterns of certain thalamic neurons. Whether localized groups of neurons in the thalamus are set into a collective limit cycle oscillation in response to a constant stimulus is unknown but certainly worthy of experimental investigation.

The implication of both our model study and the work of Poggio and Viernstein is clear: stimulus intensity may be coded into both average spike frequency and the frequency of periodic variations in average spike frequency. How such redundancy in coding may be used in the nervous system is at present unknown, but it is hoped that extensions of the present model to include spatial interactions between neural populations may lead to some insight into the matter. Certainly both the existence of a stimulus threshold for initiation of limit cycle activity and the stability of the limit cycle itself are important forms of noise insensitivity.

Limit cycles have also been used as a model for some of the characteristics of electroencephalogram (EEG) rhythms (Dewan, 1964). In this work the existence of limit cycle oscillations within the central nervous system was assumed without independent evidence. Our present results, therefore, provide a more concrete physiological basis for this approach to the study of EEG rhythms.

Before leaving the subject of limit cycles it may be asked whether a neural population which is capable of limit cycle activity for a certain class of stimulus configurations will exhibit hysteresis under different stimulus conditions. The answer to this may be obtained by comparing requirements 20 and 21 with condition 17. As the minimum value of the right-hand side of equation 20 is 18, and as the left-hand side of equation 21 must be greater than zero, it follows that whenever requirements 20 and 21 are satisfied condition 17 will also be satisfied. This proves the following theorem.

Theorem 4. Any neural population which exhibits limit cycle activity for some class of stimulus configurations will also display simple hysteresis phenomena for some other class of stimulus configurations.[6]

The converse of this theorem is, of course, false. Also, the coexistence of limit cycle phenomena and multiple hysteresis is precluded by Theorem 3.

Theorem 4 is very strong in light of the suggested functional significance of both limit cycles and hysteresis. For example, the theorem shows that nonspecific biasing inputs to a neural population from other parts of the central nervous system may completely change the character of the response of that population to specific sensory (or experimental) stimulation. Furthermore, the theorem is in principle testable, although this might be difficult in practice. One probem is that the experi-

[6] It must be remembered that a stimulus configuration involves inputs to *both* subpopulations. To pass from limit cycle activity to hysteresis it will generally be necessary to change both P and Q.

menter would need independent control over the inputs to both the excitatory and the inhibitory subpopulations.

CONCLUSIONS

There have been a number of previous studies and simulations of spatially localized neural populations (Allanson, 1956; Smith and Davidson, 1962; Griffith, 1963; ten Hoopen, 1965; Anninos et al., 1970). These treatments have, of course, differed from each other in various ways: some use discrete time, others continuous time, etc. In common to all these studies, however, has been the description of the state of the population at time t by a single variable: e.g., the fraction of cells becoming active per unit time. This has been true even of those studies in which a number of connections have been designated as inhibitory. The most fundamental difference between this study and previous work, therefore, is in the treatment of inhibition as arising from exclusively inhibitory neurons. It thus becomes necessary to deal with interactions between two distinct subpopulations explicitly, and this requires the use of the two variables $E(t)$ and $I(t)$ to characterize the state of the population.[7]

The assumption that the influence of one neuron upon all others is either exclusively excitatory or exclusively inhibitory is known as Dale's law (Eccles, 1964). Although this law is probably not universally true, it is certainly true in most instances. If one accepts the fact that there are no exclusively excitatory subsystems within the central nervous system, then a two-variable approach such as ours is required even in the study of spatially localized populations.

Results of our study which follow directly from the explicit treatment of excitatory-inhibitory interactions are the existence of multiple hysteresis loops and limit cycles. (It will be remembered that simple hysteresis is dependent only upon characteristics of the excitatory subpopulation.) To these may be added the extremely important result in Theorem 4 stating that any neural population exhibiting limit cycle behavior for one class of stimuli will show simple hysteresis for some other class of stimuli. All of these results have been shown to be of potential functional significance. A paper extending the current model to deal with spatial interactions within sheets of neural tissue is in preparation and will deal with some of the information processing capabilities resulting from the phenomena cited above (Wilson and Cowan, in preparation).

Finally, it is to be emphasized that the qualitative results obtained, i.e. simple and multiple hysteresis, limit cycles, and Theorem 4, are independent of the particular choice of the logistic curve for the subpopulation response functions. The arguments leading to these results depend essentially only on the general shapes of the isoclines as defined in equations 13 and 14. The particular constraints on the parameters given in relationships 17–22 will, of course, differ for differing sigmoid functions, but completely general relations may be obtained by relating the connectivity constants to the maximum slopes of the response functions. This independence of our model from the particular choice of sigmoid response function is extremely important, both because of the difficulty in obtaining experimental determinations of the distributions in equations 1 and 1 *a* and because of the likelihood that these distributions will differ in different parts of the nervous system.

APPENDIX

In this appendix we extend our basic model to include relative refractoriness. We will deal only with excitatory cells, since an identical equation is obtained for inhibitory cells. Furthermore, we will assume that the relative refractory period is much longer than either the absolute refractory period or the effective summation time. This will permit us to assume the results of the temporal coarse-graining argument for absolute refractoriness and temporal summation. This latter assumption is for convenience only and does not play any essential part in the derivation. Finally, we assume that the resting threshold θ_0 is the same for all cells in the population. The sigmoid response function is therefore assumed to relate to a distribution of synapses as shown by equation 1 *a*.

We assume that after firing a cell is initially absolutely refractory and then relatively refractory for a time r, at the end of which it has returned to its totally sensitive state. During the relative refractory period the cell may of course be fired by supernormal stimulation, after which it will again become absolutely refractory, etc. We assume, therefore, that the relative refractory period can be completely characterized by specifying the time course of the return of its threshold from an initially very high value ultimately to its resting value (Adrian, 1928; Fuortes and Mantegazzini, 1962). Let the function which describes this

[7] Anninos et al. (1970) actually specify that some neurons are inhibitory, but they still describe the state of their population using only one variable.

return of the threshold to its resting value be called $\theta(t - t')$, where t' is the time at which the cell last fired. The only restrictions on $\theta(t - t')$ are that it be continuous and that it reach the resting value in a finite time r. In particular, $\theta(t - t')$ need not be monotonic, thus allowing for rebound depolarization, etc.

Let $R(t, t')\delta t$ be the proportion of those cells which fired during the interval $(t', t' + \delta t)$ which are still in the relative refractory state. Clearly, these cells will have a threshold of $\theta(t - t')$. If the net excitation is designated by $x(t)$ and the sigmoid response function for cells with this threshold by $\mathcal{S}[x(t), \theta(t - t')]$ then the equation which governs the evolution of $R(t, t')$ will be:

$$\frac{\mathrm{d}R}{\mathrm{d}t} = -\mathcal{S}[x(t), \theta(t - t')]R. \qquad \text{(A 1)}$$

This equation accounts for the loss of cells from the relative refractory state through refiring. Equation A 1 may be solved formally for $R(t, t')$. As the initial condition is that $R(t', t') = E(t')$, the result is

$$R(t, t') = E(t') \exp\left\{-\int_{t'}^{t} \mathcal{S}[x(t''), \theta(t'' - t')]\, \mathrm{d}t''\right\}. \qquad \text{(A 2)}$$

Using this result, it is evident that the contribution of the firing of refractory cells to the total activity in the population at time t is just:

$$\int_{t-r}^{t} \mathcal{S}[x(t), \theta(t - t')]R(t, t')\, \mathrm{d}t'.$$

The fraction of cells which have completely recovered from firing is:

$$1 - r_e E(t) - \int_{t-r}^{t} R(t, t')\, \mathrm{d}t',$$

where the term $r_e E(t)$ gives the fraction of cells which have just fired and are therefore absolutely refractory.

Putting all these results together we arrive at the equation:

$$\tau \frac{\mathrm{d}E}{\mathrm{d}t} = -E + \int_{t-r}^{t} \mathcal{S}[x(t), \theta(t - t')]R(t, t')\, \mathrm{d}t' + \left[1 - r_e E - \int_{t-r}^{t} R(t, t')\, \mathrm{d}t'\right] \mathcal{S}[x(t), \theta_0], \qquad \text{(A 3)}$$

where $R(t, t')$ is given by equation A 2. This is the equation for a population of excitatory cells having both absolute and relative refractory periods. The second term on the right gives the contribution to $E(t)$ of relatively refractory cells, while the third term gives the contribution of cells which are totally sensitive. Note that for $r = 0$, i.e. for a relative refractory period of zero duration, equation A 3 reduces to the same form as equation 7 in the text This shows that equation A 3 is indeed the proper extension of our model to account for relative refractoriness.

This research was supported in part by the Alfred P. Sloan Foundation and the Otho S. A. Sprague Memorial Institute.

Received for publication 4 June 1971.

REFERENCES

ADRIAN, E. D. 1928. The Basis of Sensation. Christophers Ltd., London.

ALLANSON, J. T. 1956. *In* Information Theory (Third London Symposium). Butterworth and Co. Ltd., London. 303.

ANDERSEN, P., and J. ECCLES. 1962. *Nature (London)*. **196**:645.

ANNINOS, P. A., B. BEEK, T. J. CSERMELY, E. M. HARTH, and G. PERTILE. 1970. *J. Theor. Biol.* **26**:121.

ASHBY, W. R., H. VON FOERSTER, and C. C. WALKER. 1962. *Nature (London)*. **196**:561.

BEURLE, R. L. 1956. *Phil. Trans. Roy. Soc. London Ser. B. Biol. Sci.* **240**:55.

COLONNIER, M. L. 1965. *In* Brain and Conscious Experience. J. C. Eccles, editor. Springer-Verlag New York Inc., New York.

COWAN, J. D. 1970. *In* Lectures on Mathematics in the Life Sciences. M. Gerstenhaber, editor. American Mathematical Society, Providence, R.I. **2**:1.

COWAN, J. D. 1971. *In* Proceedings of the International Union of Pure and Applied Physics Conference on Statistical Mechanics, 1971. S. Rice, J. Light, and K. Freed, editors. University of Chicago Press, Chicago. In press.

COX, D. R. 1962. Renewal Theory. Methuen and Co. Ltd., London.

CRAGG, B. G., and H. N. V. TEMPERLEY. 1955. *Brain*. **78**(Pt. II):304.

DEWAN, E. M. 1964. *J. Theor. Biol.* **7**:141.

ECCLES, J. 1964. The Physiology of Synapses. Academic Press, Inc., New York.

FARLEY, B. G., and W. A. CLARK. 1961. *In* Information Theory (Fourth London Symposium). C. Cherry, editor. Butterworth and Co. Ltd., London. 242.

FENDER, D., and B. JULESZ. 1967. *J. Opt. Soc. Amer.* **57**:819.
FREEMAN, W. J. 1967. *Logistics Rev.* **3**:5.
FREEMAN, W. J. 1968 *a. Math. Biosci.* **2**:181.
FREEMAN, W. J. 1968 *b. J. Neurophysiol.* **31**:337.
FRISHKOPF, L. S., and W. A. ROSENBLITH. 1958. *In* Symposium on Information Theory in Biology. H. P. Yockey, R. L. Platzman, and H. Quastler, editors. Pergamon Press, Ltd., Oxford, England.
FUORTES, M. G. F., and F. MANTEGAZZINI. 1962. *J. Gen. Physiol.* **45**:1163.
GRIFFITH, J. S. 1963. *Biophys. J.* **3**:299.
HARTH, E. M., T. J. CSERMELY, B. BEEK, and R. D. LINDSAY. 1970. *J. Theor. Biol.* **26**:93.
HARTLINE, H. K., and F. RATLIFF. 1958. *J. Gen. Physiol.* **41**:1049.
HEBB, D. O. 1949. The Organization of Behavior. John Wiley & Sons, Inc., New York.
HUBEL, D. H., and T. N. WIESEL. 1963. *J. Neurophysiol.* **26**:994.
HUBEL, D. H., and T. N. WIESEL. 1965. *J. Neurophysiol.* **28a**:229.
KERNELL, D. 1965 *a. Acta Physiol. Scand.* **65**:65.
KERNELL, D. 1965 *b. Acta Physiol. Scand.* **65**:74.
KIRKWOOD, J. G. 1946. *J. Chem. Phys.* **14**:180.
LE GRAND, Y. 1957. Light, Color, and Vision. John Wiley & Sons, Inc., New York.
MACKAY, D. M. 1970. *In* Neurosciences Research Symposium Summaries. F. O. Schmitt, T. Melnechuk, G. C. Quarton, and G. Adelman, editors. The M.I.T. Press, Cambridge, Mass. **4**:397.
MOUNTCASTLE, V. B. 1957. *J. Neurophysiol.* **20**:408. 253.
OSHIMA, T. 1969. *In* Basic Mechanisms of the Epilepsies. H. Jasper, A. Ward, and A. Pope, editors. Little, Brown and Company, Boston. 253.
POGGIO, G. F., and L. J. VIERNSTEIN. 1964. *J. Neurophysiol.* **27**:517.
RALL, W. 1955 *a. J. Cell. Comp. Physiol.* **46**:3.
RALL, W. 1955 *b. J. Cell. Comp. Physiol.* **46**:373.
RALL, W. 1955 *c. J. Cell. Comp. Physiol.* **46**:413.
RALL, W., and C. C. HUNT. 1956. *J. Gen. Physiol.* **39**:397.
SMITH, D. R., and C. H. DAVIDSON. 1962. *J. Soc. Comput. Mach.* **9**:268.
SZENTÁGOTHAI, J. 1967. *In* Recent Development of Neurobiology in Hungary. K. Lissák, editor. Akadémiai Kiadó, Budapest. **1**:9.
TEN HOOPEN, M. 1965. Cybernetics of Neural Processes. E. R. Caianiello, editor. C.N.R., Rome.
VERVEEN, A. A., and H. E. DERKSEN. 1969. *Acta Physiol. Pharmacol. Neer.* **15**:353.
VON NEUMANN, J. 1956. *In* Automata Studies. C. E. Shannon and J. McCarthy, editors. Princeton University Press, Princeton, N. J. 43.
WINOGRAD, S., and J. D. COWAN. 1963. Reliable Computation in the Presence of Noise. The M.I.T. Press, Cambridge, Mass.

Part 3
Computational Neural Systems: Foundations

Early research in the field of neurophysiology and more recent studies on the neuronal morphology of biological vision and control mechanisms have inspired mathematicians, computer scientists, and engineers to replicate some of the neuronal functions for the development of a new class of computing systems—the neural computing systems.

ALTHOUGH the subject of neurophysiology is relatively old, it was only about a decade ago that we started thinking in terms of computational neural systems—a science of developing neural models for computational applications. Since 1985, the field has enjoyed an exponential growth, giving rise to a vast volume of literature in the form of books, scientific journals, and international scientific conferences and symposia sponsored by some major scientific societies. At the same time, the field of neural computing has generated enormous commercial interests, which in turn have resulted in many major computer-oriented companies becoming involved in the development of neural hardwares and softwares. Interestingly, at the same time, they gave birth to many new commercial outfits around the globe, and caused keen competition. Some of them have survived and are doing well, while others experienced an untimely demise.

In Part Three of the volume, we present an introduction to computational neural systems. We have assembled a set of five representative articles, most of them written during relatively recent years by experts in the field.

In the first article, (3.1), we present an extensive survey, authored by K. J. Hunt et al., which focuses on the promise of artificial neural networks in the realm of modeling, identification, and control of nonlinear dynamical systems. In this article, the authors present the basic ideas and techniques of artificial neural systems, and discuss various static and dynamic neural structures in language and notations familiar to system scientists.

Among the architecture and algorithms suggested for artificial neural networks (ANNs), the self-organizing maps have the special properties of effectively creating spatially organized "internal representations" of various features of input patterns and their abstractions. One novel property is that the self-organizing process can also discover semantic relationships in sentences. In this respect, the resulting maps very closely resemble the topographical organized maps found in the cortices of the more developed animal brains. In the second article, T. Kohonen, another pioneer in this field, describes the role of self-organizing maps (SOMs) among neural network models. He shows that SOMs have been particularly successful in various pattern recognition tasks involving noisy signals. These maps have been used in practical speech recognition, and several works are in progress leading to applications in robotics, process control, telecommunications, and so on. This article contains a survey of several basic facts and results. Although the author does not say so, this work should motivate readers to extend this methodology to fuzzy processes using the genetic–neural approach. Such extensions seem to have potential applications in robotics, medical diagnosis, and pattern recognition, especially when the patterns do not have well-defined boundaries.

It is thought that the synaptic connections between neurons in the brain are not completely inherited, but are modified through the process of learning after birth. For example, it has been reported that the visual cortex adjusts itself during maturation to the nature of the visual experience. Perhaps the nervous systems are modified to match the features in the retinal stimulus. Such a self-organizing neural network would be more prominent in the higher center of the brain. The mechanism of self-organization during our sensory and control actions is not well understood. Although several hypotheses exist, none of them are physiologically substantiated. In article (3.3), K. Fukushima discusses some of his early work on Cognitron. Cognitron is a multilayered neural network structure with self-organizing capability. In this article, a new hypothesis for the self-organization of the synapses is given. Since the Cognitron is a multilayered structure, it seems to have a large capacity for information processing. The author conjectures that the hypothesis on the synapse organization proposed in his article holds not only in the higher center of the brain but also in the distal parts of the sensory systems.

The next two articles present the collective behavior of neurons. In fact, a neural system with a high degree of connectivity has some wonderful computational capabilities in addition to the fault tolerance capability. Associate memories are one of the collective properties of the neuron.

In article (3.4), J. J. Hopfield presents a study of a model for a large network of neurons with a graded response using a sigmoidal function. The content-addressable memory and other emergent collective properties are also based upon graded response properties. The author presents a simple electronic model for such a neural network.

Sometimes analogies help in the development of new theories. For example, the theory of electromagnetism in the last century involved the use of complicated mechanical analogies. Similarities between the CNS and the immune system led to a new way of viewing the immune system. An exciting possibility is that we can now reverse the process, and apply insights gained in the study of the immune system to CNS problems. In article (3.5), G. W. Hoffmann and M. W. Benson present a neural-network concept derived from an analogy between the immune system and the CNS. The theory is based on a "neuron" that exhibits hysteresis at the single-cell level. Such a neuron can learn without any changes in the synaptic weights. The process of learning is purely natural: it is the interaction between the neural network and its environment. Due to the hysteresis associated with the neuron, the network tends to stay in the region of the phase space. This new, but speculative, theory has potential in developing a new class of computational networks, and that is the very reason for including this article here.

The foundations of computational neural systems discussed in these articles have profound applications in the development of neuro-control mechanisms. The importance of neural networks lies in their ability to approximate continuous nonlinear functions to any degree of accuracy. This ability has made neural networks very useful in the design of controllers for nonlinear dynamical systems. The functional approximation capability of neural networks is discussed in detail in the following part.

Further Reading

[1] P. K. Simpson, *Artificial Neural Systems: Foundations, Paradigms, Applications, and Implementations.* Elmsford, NY: Pergamon Press, 1990.

[2] P. D. Wasserman, *Neural Computing: Theory and Practice.* New York: Van Nostrand, 1989.

[3] D. S Touretzky, Ed., *Advances in Neural Information Processing Systems.* San Mateo, CA: Morgan Kaufmann Publishers, 1992.

[4] J. Diederich, Ed., *Artificial Neural Networks: Concept Learning.* Los Alamitos, CA: IEEE Computer Society Press, 1990.

[5] E. R. Kandel and J. H. Schwartz, *Principles of Neural Science.* New York: North-Holland, 1985.

[6] J. Jondies, D. E. Irwin and S. Yantis, "Integrating visual information from successive fixations," *Science,* vol. 25, pp. 192–194, 1982.

[7] M. M. Gupta and G. K. Knopf, "A multitask visual information processor with a biologically motivated design," *J. Visual Commun. Image Representation,* vol. 3, no. 3, pp. 230–246, Sept. 1992.

[8] P. A. Anninos, B. Beek, T. J. Csermel, E. E. Harth, and G. Pertile, "Dynamics of neural structures," *J. Theoret. Biol.,* vol. 26, pp. 121–148, 1970.

[9] P. Stolorz and G. W. Hoffmann, "Learning by selection using energy functions" in *Systems with Learning and Memory Abilities.* J. Delacour and J. C. S. Levy, Eds., New York: North-Holland, pp. 437–452, 1989.

[10] S. P. Grossman, *The Motor System and Mechanics of Basic Sensory-Motor Integration.* New York: John Wiley & Sons, 1967.

[11] J. S. Albus, "A new approach to manipulator control: The cerebellar model articulation controller," *J. Dynamic Syst., Meas., Cont.,* pp. 220–227, Sept. 1975.

[12] W. T. Miller, F. H. Glanz, and L. G. Kraft, "CMAC: An associative neural network alternative to backpropagation," *Proc. IEEE,* vol. 78, no. 10, pp. 1561–1567, Oct. 1990.

[13] E. Ersu and H. Tolle, "A new concept for learning control inspired by the brain theory," in *Proce. 9th World Congress on IFAC,* vol. 7, pp. 245–250, 1984.

[14] G. W. Hoffmann, M. W. Benson, G. M. Bree and P. E. Kinahan, "A teachable neural network based on an unorthodox neuron," *Physica,* 22D, pp. 233–246, 1986.

[15] W. J. Freeman, *Mass Action in the Nervous System.* New York: Academic Press, 1975.

[16] M. M. Gupta, "Introduction to neuro-computing systems with applications to control and vision," *Tutorial Notes.* Intelligent Systems Research Laboratory, University of Saskatchewan, Canada, 1992.

Article 3.1

Neural Networks for Control Systems—A Survey*

K. J. HUNT,† D. SBARBARO,† R. ŻBIKOWSKI† and P. J. GAWTHROP†

Neural networks potentially provide a general framework for modelling and control of nonlinear systems. A survey of neural networks from a control systems perspective illustrates that deep insight into this potential can be achieved.

Key Words—Neural networks; nonlinear control systems; nonlinear systems modelling; systems identification.

Abstract—This paper focuses on the promise of artificial neural networks in the realm of modelling, identification and control of nonlinear systems. The basic ideas and techniques of artificial neural networks are presented in language and notation familiar to control engineers. Applications of a variety of neural network architectures in control are surveyed. We explore the links between the fields of control science and neural networks in a unified presentation and identify key areas for future research.

1. INTRODUCTION

No-one active in the field of control systems can be unaware of the growth of papers, journals, conferences and conference sessions devoted to the topic of Artificial Neural Networks. This is clearly indicative of a wide intellectual interest in the concepts associated with Artificial Neural Networks together with a desire to use the corresponding algorithms within a range of application areas. Control systems form one such application area as witnessed in part by recent special issues on the subject (*IEEE Control Systems*, 1988, 1989, 1990).

As with other growth areas in the past (for example, adaptive control and expert systems to name but two) it is possible to take two extreme positions either of which equally obstructs progress. The first such position is to embrace the new subject in an undiscerning way and use it unscientifically. This leads to papers of the form: "x is the hot topic of the decade and therefore supersedes all other approaches. We embedded it in a controller to give us an x controller which is, by definition, the best controller. We simulated it with impressive results". Here, x is self-tuning, expert system, Artificial Neural Network or any other hot topic. Such papers tend to advance anthropomorphic arguments to explain apparent successes. The second extreme is to dismiss such growth areas as consisting of mere hyperbole and not attempt to use them.

Both these extreme positions must be avoided if progress is to be made. Instead, a rational evaluation of new ideas in the context of established techniques is needed. This should avoid the two extremes and rather steer a middle course allowing new ideas and established principles to fuse thus allowing real (as opposed to illusory) advances to be made.

Those who have read Wiener's seminal book "Cybernetics" (Wiener (1948)) will know that control, information and neural science were once regarded as a common subject under the banner of Cybernetics. Since that time, the disciplines of control, computing science (including Artificial Intelligence) and neurobiology have tended to go their own separate ways. This has lead to an unfortunate breakdown in communication between the disciplines; in particular the differing jargon and notation now poses a barrier to effective interchange of ideas.

This survey is a contribution to removing this barrier for researchers currently working in the area of control systems in that we attempt to present the basic ideas and techniques of Artificial Neural Networks in language and notation familiar to control engineers. This has the twofold advantage of making such techniques more readily accessible to control engineers and, conversely, allowing control systems insights to be applied to Artificial Neural Network techniques.

We note that an early comprehensive survey of adaptive and learning systems and their application in control and pattern analysis can be found in the classic book by Tsypkin (1971).

1.1. *Origins of connectionist research*

In recent years there has been an increasing interest in studying the mechanisms and structure of the brain. This has led to the development of new computational models, based on this biological background, for

* Received 25 June 1991; revised 24 February 1992; revised 2 April 1992; received in final form 14 May 1992. The original version of this paper was not presented at any IFAC meeting. This paper was recommended for publication in revised form by Editor K. J. Åström.

† Control Group, Department of Mechanical Engineering, University of Glasgow, Glasgow G12 8QQ, U.K.

Reprinted with permission from *Automatica*, vol. 28, no. 6, K. J. Hunt, D. Sbarbaro, R. Zbikowski, and P. J. Gawthrop, "Neural Networks for Control Systems—A Survey," pp. 1083–1112, 1992, Pergamon Press Ltd., Oxford, England.

solving complex problems like pattern recognition, fast information processing and adaptation.

In the early 1940s the pioneers of the field (McCulloch and Pitts (1943)) studied the potential and capabilities of the interconnection of several basic components based on the model of a neuron. Others, like Hebb (1949), were concerned with the adaptation laws involved in neural systems. Rosenblatt (1958) coined the name Perceptron and devised an architecture which has subsequently received much attention. Minsky and Papert (1969) introduced a rigorous analysis of the Perceptron; they proved many properties and pointed out limitations of several models. In the 1970s the work of Grossberg came to prominence (Grossberg (1976)). This work, based on biological and psychological evidence, proposed several architectures of nonlinear dynamic systems with novel characteristics. Hopfield applied a particular nonlinear dynamic structure to solve technical problems like optimization (Hopfield (1982)). In 1986 the parallel distributed processing (PDP) group published a series of results and algorithms (Rumelhart *et al.* (1986)). This work gave a strong impulse to the area and provided the catalyst for much of the subsequent research in this field. For a fine collection of key papers in the development of models of neural networks see *Neurocomputing: Foundations of Research* (Anderson and Rosenfeld (1988)). Since the work of the PDP group much research has been done and today there are several well-defined architectures to tackle a variety of problems. Many examples of real-world applications ranging from finance to aerospace are currently being explored (Hecht-Nielsen (1988)).

1.2. *Neural networks and control*

To provide a rational assessment of new methods it is important to compare and contrast the emerging technologies with established and traditional techniques. One way to do this is to list the key characteristics of competing methods in the context of the field under study.

With specific reference to neural networks in control the following characteristics and properties of neural networks are important:

- Nonlinear systems. Neural networks have greatest promise in the realm of nonlinear control problems. This stems from their theoretical ability to approximate arbitrary nonlinear mappings. Networks may also achieve more parsimonious modelling than alternative approximation schemes; this question requires further study.
- Parallel distributed processing. Neural networks have a highly parallel structure which lends itself immediately to parallel implementation. Such an implementation can be expected to achieve a higher degree of fault tolerance than conventional schemes. The basic processing element in a neural network has a very simple structure. This, in conjunction with parallel implementation, results in very fast overall processing.
- Hardware implementation. This is closely related to the preceding point. Not only can networks be implemented in parallel, a number of vendors have recently introduced dedicated VLSI hardware implementations. This brings additional speed and increases the scale of networks which can be implemented.
- Learning and adaptation. Networks are trained using past data records from the system under study. A suitably trained network then has the ability to generalize when presented with inputs not appearing in the training data. Networks can also be adapted on-line.
- Data fusion. Neural networks can operate simultaneously on both quantitative and qualitative data. In this respect networks stand somewhere in the middle ground between traditional engineering systems (quantitative data) and processing techniques from the artificial intelligence field (symbolic data).
- Multivariable systems. Neural networks naturally process many inputs and have many outputs; they are readily applicable to multivariable systems.

It is clear that a modelling paradigm which has all of the above features has great promise.

From the control theory viewpoint the ability of neural networks to deal with nonlinear systems is perhaps most significant. The great diversity of nonlinear systems is the primary reason why no systematic and generally applicable theory for nonlinear control design has yet evolved. A range of "traditional" methods for the analysis and synthesis of nonlinear controllers for specific classes of nonlinear systems exist: phase plane methods, linearization techniques and describing functions are three examples.

However, it is the ability of neural networks to represent nonlinear mappings, and hence to model nonlinear systems, which is the feature to be most readily exploited in the synthesis of nonlinear controllers.

In this respect more recent methods of nonlinear control design including nonlinear operator–theoretic methods and optimization techniques have an important role to play. One promising possibility is to use neural networks to provide the nonlinear system models required by these techniques. This possibility is explored in depth in Section 5.3.

A further established area in control science with great relevance here is adaptive systems theory. This area has produced many fruitful theoretical results in the past 15 years. Although the established adaptive systems theory is founded upon the assumption of linear time-invariant systems, the concepts and theoretical challenges addressed are likely to find strong parallels in the neural networks field. As emphasised in Section 7, however, the nonlinear nature of the subsystems involved in the neural networks case means that the problems encountered are likely to be more complex and many fundamental theoretical questions will need to be addressed.

The compilation book (Miller *et al.* (1990)) provides a broad overview of the field of neural networks in control. That book also contains a number of benchmark problems.

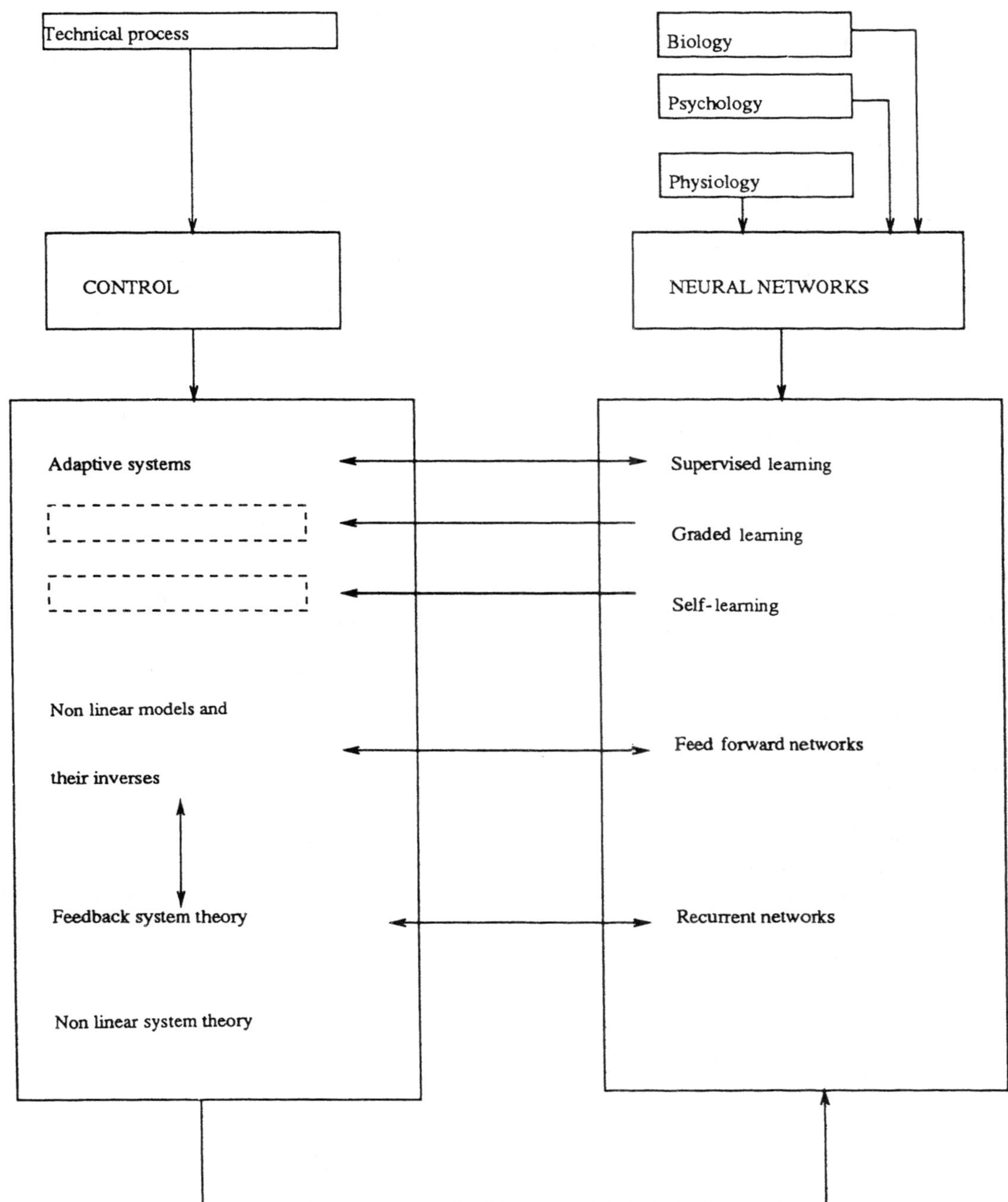

FIG. 1. Relation between control and neural networks.

Our view of the general relationship between the fields of control science and neural networks is shown in Fig. 1. Blank boxes are used when no obvious parallel exists. The technical terms used in the diagram are defined and discussed in later sections of the paper. In this paper we attempt to expand upon the links between the two fields in a unified presentation, and to identify key areas for future research. Key abbreviations used in the paper are summarised in Table 1.

TABLE 1. ABBREVIATIONS USED IN THE PAPER

Abbreviation	Meaning
ANN	Artificial neural network
ART	Adaptive resonance theory
BPTT	Backpropagation through time
CAM	Content-addressable memory
CMAC	Cerebellar model articulation controller
CMN	Cellular neural network
CPN	Counterpropagation network
IMC	Internal model control
LVQ	Learning vector quantization

1.3. *Organization of the paper*

The paper has two main parts:

- An exposition of Artificial Neural Network (ANN) techniques in the notation and language of control systems. Section 2 discusses the architecture of ANNs. Section 3 (static networks) and Section 4 (dynamic networks) discuss how these can be made adaptive and trained using learning algorithms.

- A comparative survey of Artificial Neural Network applications in control systems set in the context of the first part. Section 5 discusses system representation and modelling by ANNs and applies these representations to system identification and hence to nonlinear and adaptive control. Section 6 discusses a variety of further applications of ANNs.

In Section 7 we discuss a range of future theoretical research directions for ANNs in the context of control.

2. NETWORK ARCHITECTURES

The network architecture is defined by the basic processing elements and the way in which they are interconnected. These two aspects are discussed in turn.

2.1. *Basic elements*

The basic processing element of the connectionist architecture is often called a neuron by (loose) analogy with neurophysiology, but other names such as perceptron (Rosenblatt (1958)) or adaline (Widrow and Hoff (1960)) are also used. Below we describe a standard and unifying model for neural networks and later in this section show that many well-known structures fall within the standard model. Many of the basic processing elements may be considered to have three components (see Fig. 2):

(1) A weighted summer.
(2) A linear dynamic SISO system.
(3) A non-dynamic nonlinear function.

These elements are considered in turn in the following sections.

2.1.1. *Weighted summer.* The *weighted summer* is described by

$$v_i(t) = \sum_{j=1}^{N} a_{ij} y_j(t) + \sum_{k=1}^{M} b_{ik} u_k(t) + w_i, \qquad (1)$$

giving a weighted sum v_i in terms of the outputs of all elements y_j, external inputs u_k and corresponding weights a_{ij} and b_{ik} together with constants w_i. N of these weighted summer elements can be conveniently expressed in vector–matrix notation.

Stacking N weighted sums v_i into a column vector v, the N outputs y_j into a vector y and M inputs u_k into a vector u and the N constants w_i into a vector w, equation (1) may be rewritten in vector matrix form as:

$$v(t) = Ay(t) + Bu(t) + w, \qquad (2)$$

where the ijth element of the $N \times N$ matrix A is a_{ij} and the ikth element of the $N \times M$ matrix B is b_{ik} (the constants w_i could be incorporated with the inputs u_k, but it is useful to represent them explicitly).

2.1.2. *Linear dynamic system.* The linear dynamic SISO system has input v_i and output x_i. In transfer function form it is described by

$$\bar{x}_i(s) = H(s)\bar{v}_i(s), \qquad (3)$$

where the bar denotes Laplace transformation. In the time domain, equation (3) becomes:

$$x_i(t) = \int_{-\infty}^{t} h(t - t')v_i(t')\, dt', \qquad (4)$$

where $H(s)$ and $h(t)$ form a Laplace transform pair.

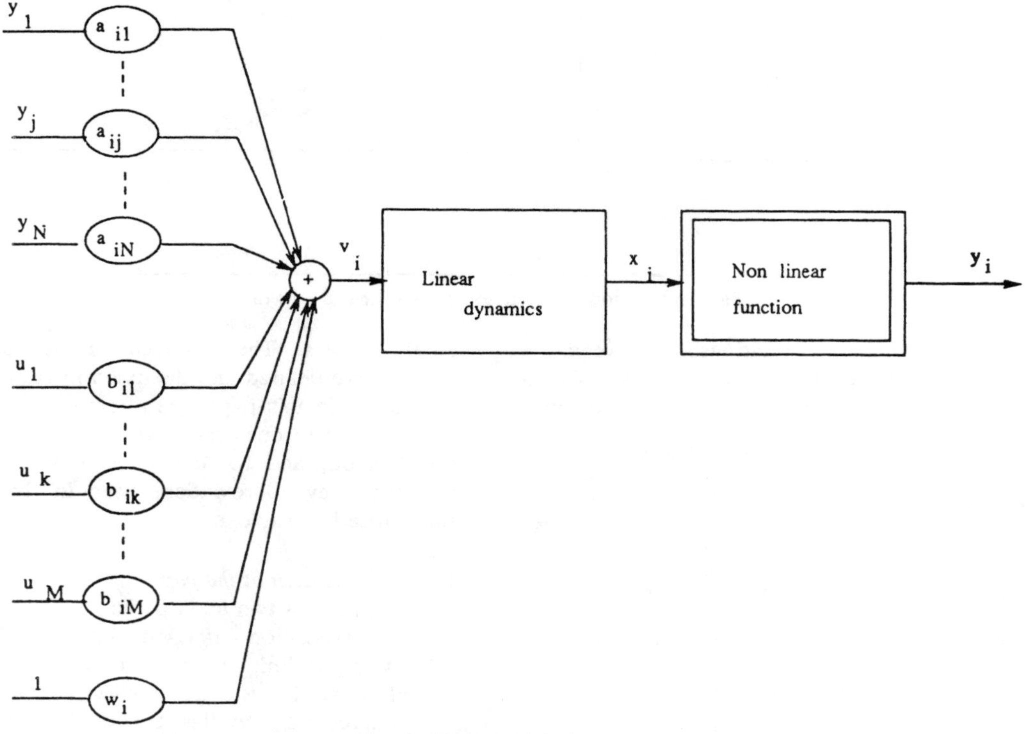

FIG. 2. Basic model of a neuron.

Five common choices of $H(s)$ are:

$$H(s) = 1,$$
$$H(s) = \frac{1}{s},$$
$$H(s) = \frac{1}{1 + sT}, \tag{5}$$
$$H(s) = \frac{1}{\alpha_0 s + \alpha_1},$$
$$H(s) = e^{-sT},$$

corresponding to

$$h(t) = \delta(t),$$
$$h(t) = \begin{cases} 0 & t < 0 \\ 1 & t \geq 0 \end{cases},$$
$$h(t) = \frac{1}{T} e^{-t/T}, \tag{6}$$
$$h(t) = \frac{1}{\alpha_0} e^{-(\alpha_1/\alpha_0)t},$$
$$h(t) = \delta(t - T),$$

where δ is the Dirac delta function. In the time domain the corresponding input–output relations are:

$$x_i(t) = v_i(t),$$
$$\dot{x}_i(t) = v_i(t),$$
$$T\dot{x}_i(t) + x_i(t) = v_i(t), \tag{7}$$
$$\alpha_0 \dot{x}_i(t) + \alpha_1 x_i(t) = v_i(t),$$
$$x_i(t) = v_i(t - T).$$

The first, second, and third versions are clearly just special cases of the fourth.

Discrete-time dynamic systems are also used. For example

$$\alpha_0 x_i(t + 1) + \alpha_1 x_i(t) = v_i(t), \tag{8}$$

where t is now an integer time index.

2.1.3. *Non-dynamic nonlinear function.* The non-dynamic nonlinear function $g(\cdot)$ gives the element output y_i in terms of the transfer function output x_i:

$$y_i = g(x_i). \tag{9}$$

There are a number of twofold classifications of these functions:

(1) Differentiable/non-differentiable.
(2) Pulse-like/step-like.
(3) Positive/zero-mean.

Classification 1 distinguishes smooth from sharp functions. Smooth functions are needed for some adaptation algorithms such as backpropagation (Rumelhart *et al.* (1986)) (Section 3.2), whereas discontinuous (e.g. signum) functions are needed to give a true binary output.

Classification 2 distinguishes functions which only have a significant output value for inputs near to zero from functions which only change significantly around zero.

Classification 3 refers to step like functions. Positive functions change from 0 at $-\infty$ to 1 at ∞; zero-mean changes from -1 at $-\infty$ to 1 at ∞.

Some standard functions are given in Table 2. Note that in the table there are strong relations between the given functions. The sigmoid and tanh functions are similar: sigmoid ranges from 0 to 1 while tanh ranges from -1 to 1. Secondly, the threshold functions correspond to the high gain limits of the sigmoid and tanh functions.

It is of course possible to create pulse-like functions from differentiable step-like functions by differentiation, and vice-versa, although this does not often seem to be done. Alternatively, pulse-like functions may be created by differentiating shifted step-like functions. This is sometimes achieved by interconnecting two neurons appropriately.

It is known that neurons located in different parts in the nervous system have different characteristics. For example, the neurons of the ocular motor system have a sigmoid characteristic while those located in the visual area have a Gaussian characteristic (Ballard (1988)). The former is called variable-encoding and exhibits monotonic behaviour. The second exhibits a response maximum that spans the measurement space. This is called value-encoded (Ballard (1988)). There are other types of function such as logarithmic and exponential which are useful (Daunicht (1990)), although their biological basis has not been established.

Each representation of the basic unit has advantages and some disadvantages. For example, the neuron with Gaussian encoding seems to have the ability to represent multidimensional variables and functions more easily than the sigmoid. This is further discussed in Section 5.

2.2. *Connections*

The neurons by themselves are not very powerful in terms of computation or representation but their interconnection allows us to encode relations between the variables giving different powerful processing capabilities. The three components of the neuron discussed in Section 2.1 can be combined in various ways. For example, if the neurons are all non-dynamic

TABLE 2. NONLINEAR FUNCTIONS $g(x)$

Name	Formula	Characteristics
Threshold	$+1$ if $x > 0$ else 0	Non-differentiable, step-like, positive
Threshold	$+1$ if $x > 0$ else -1	Non-differentiable, step-like, zero-mean
Sigmoid	$1/1 + e^{-x}$	Differential, step-like, positive
Hyperbolic tangent	$\tanh(x)$	Differential, step-like, zero-mean
Gaussian	$e^{(-x^2/\sigma^2)}$	Differential, pulse-like

$(H(s) = 1)$ then an assembly of neurons can be written as the set of *algebraic* equations obtained by combining equations (2), (3), and (9):

$$\begin{aligned} x(t) &= Ay(t) + Bu(t) + w, \\ y(t) &= g(x(t)), \end{aligned} \tag{10}$$

where x is a vector of $N x_i$ elements and $g(x)$ is a vector whose components are $g(x_i)$. If, on the other hand, each neuron has first order low-pass dynamics $H(s) = \frac{1}{Ts+1}$ then an assembly of neurons can be written as the set of *differential* equations:

$$\begin{aligned} T\dot{x}(t) + x(t) &= Ay(t) + Bu(t) + w, \\ y(t) &= g(x(t)). \end{aligned} \tag{11}$$

Clearly, the solutions of equation (10) form possible steady-state solutions of equation (11).

A discrete-time version of equation (11) is

$$\begin{aligned} Tx(t+1) + (1-T)x(t) &= Ay(t) + Bu(t) + w, \\ y(t) &= g(x(t)), \end{aligned} \tag{12}$$

where t is the integer time index.

The behaviour of such a network clearly depends on the interconnection matrix A and on the form of $H(s)$. Some important forms of the interconnection matrix are discussed in the following sections.

2.3. *Static multi-layer feedforward networks*

The connection of several layers gives the possibility of more complex nonlinear mapping between the inputs and the outputs.

This capability can be used to implement classifiers or to represent complex nonlinear relations among the variables (Section 5.1).

Such networks are typically non-dynamic; that is $H(s) = 1$ in equation (3). The connection matrix A is such that the outputs are partitioned into layers so that a neuron in one layer receives inputs only from neurons in the previous layer (or, in the case of the first layer, from the network input). There is no feedback in such networks. For example, in a three layer network, each layer containing N neurons, we may partition the network x, y, u and w vectors from equation (10) as:

$$\begin{bmatrix} x^1(t) \\ x^2(t) \\ x^3(t) \end{bmatrix} = A \begin{bmatrix} y^1(t) \\ y^2(t) \\ y^3(t) \end{bmatrix} + B \begin{bmatrix} u^1(t) \\ u^2(t) \\ u^3(t) \end{bmatrix} + \begin{bmatrix} w^1 \\ w^2 \\ w^3 \end{bmatrix}, \tag{13}$$

where the superscripts denote the corresponding layer in the network. The structure of the A and B matrices for this network are as follows:

$$A = \begin{bmatrix} 0_{NN} & 0_{NN} & 0_{NN} \\ A^2 & 0_{NN} & 0_{NN} \\ 0_{NN} & A^3 & 0_{NN} \end{bmatrix}; \quad B = \begin{bmatrix} B^1 & 0_{NM} & 0_{NM} \\ 0_{NM} & 0_{NM} & 0_{NM} \\ 0_{NM} & 0_{NM} & 0_{NM} \end{bmatrix}, \tag{14}$$

where 0_{NN} is the $N \times N$ zero matrix and 0_{NM} the $N \times M$ zero matrix. A^2 and A^3 are $N \times N$ matrices of weights while B^1 is a $N \times M$ matrix of weights. For the first layer we have

$$\begin{aligned} x^1(t) &= B^1 u^1(t) + w^1, \\ y^1(t) &= g(x^1(t)), \end{aligned} \tag{15}$$

and for the second and third layers

$$\begin{aligned} x^l(t) &= A^l y^{l-1}(t) + w^l, \\ y^l(t) &= g(x^l(t)), \end{aligned} \tag{16}$$

where $l = 2, 3$.

Different characteristics are obtained using different nonlinearities $g(\cdot)$ from Table 2. Use of the sigmoid function is traditional Rumelhart *et al.* (1986).

2.4. *Dynamical networks*

The introduction of feedback produces a dynamic network with several stable points. The general equation can be expressed as

$$\begin{aligned} \dot{x}(t) &= F(x(t), u(t), \theta), \\ y(t) &= G(x(t), \theta). \end{aligned} \tag{17}$$

Here, x represents the state, u the external inputs, and θ the parameters of the network. F is a function that represents the structure of the network and G is function which represents the relation between the state variables and the outputs.

Originally, feedback (recurrent) networks were introduced in the context of associative or content-addressable memory (CAM) problems (Kohonen (1987); Hopfield (1984)) for pattern recognition. The uncorrupted pattern is used as a stable equilibrium point and its noisy versions should lie in its basin of attraction. In this way, a dynamical system associated with a set of patterns is created. If the whole working space is corrrectly partitioned by such a CAM, then any initial condition (corresponding to a sample pattern) should have a steady-state solution corresponding to the uncorrupted pattern. The dynamics of such a classifier serve as a filter.

2.4.1. *Hopfield nets.* The best-known example of a CAM is the Hopfield net. Hopfield published two fundamental papers (Hopfield (1982, 1984)). However, the discrete model in Hopfield (1982) is different from the continuous one elaborated in Hopfield (1984).

The discrete model assumes the step-like nonlinearity

$$g(x_i(t)) \rightarrow \begin{cases} 1, & \text{if } x_i(t) > 0; \\ 0, & \text{if } x_i(t) < 0 \end{cases} \quad \text{with}$$

$$x_i(t) = \sum_{j=1}^{N} a_{ij} y_j(t) - w_i, \tag{18}$$

and works in the asynchronous mode, i.e. only one neuron output is calculated at a time, leaving the others unchanged. The active neuron, p, is chosen randomly. The systems evolves with weights a_{ij} established earlier (with the Hebbian rule, see below) and held fixed during output calculation. The update rule is as follows:

$$y_i(t+1) = \begin{cases} g(x_i(t)), & \text{if } i = p,\ x_i(t) \neq 0, \\ y_i(t), & \text{otherwise.} \end{cases} \tag{19}$$

If $a_{ii} = 0$ and $a_{ij} = a_{ji}$ then the function

$$E(y) = -\tfrac{1}{2}\sum_{i=1}^{N}\sum_{j=1}^{N} a_{ij}y_iy_j + \sum_{i=1}^{N} w_iy_i \quad \text{or}$$

$$E(y) = -\tfrac{1}{2}y^TAy + w^Ty, \quad (20)$$

will decrease with every asynchronous change of y_p according to

$$\Delta E = -\Delta y_p\left[\sum_{j=1}^{N} a_{pj}y_j - w_p\right], \quad (21)$$

where $\Delta y_p(t) = y_p(t+1) - y_p(t)$. The expression (20) is *not* a Lyapunov function, since it is a quadratic form with an indefinite matrix (the zero diagonal). Moreover, Lyapunov theory for difference equations cannot be applied here, because the system of equations (18) is solved one equation at a time. However, the network will always reach an equilibrium because (20) is bounded and (21) is non-positive, and the system does not change when $\Delta E = 0$. It will settle after a finite time, since the domain of E is finite.

The Hebbian rule is an attempt to encode P patterns, y^k, $k = 1, \ldots, P$, as equilibrium points of the system represented by equations (18)–(19) by choosing

$$a_{ij} = \begin{cases} \sum_{k=1}^{P} (2y_i^k - 1)(2y_j^k - 1), & \text{if } i \neq j, \\ 0, & \text{otherwise}, \end{cases}$$

$$\text{and} \quad w_i = \tfrac{1}{2}\sum_{j=1}^{N} a_{ij}, \quad (22)$$

since the E is, within a multiple and a constant,

$$E \sim -(2y - \vec{1})^T\left[\sum_{k=1}^{P} (2y^k - \vec{1})^T(2y^k - \vec{1})\right](2y^k - \vec{1}), \quad (23)$$

where $\vec{1}$ is a vector of 1s. If the $(2y^k - \vec{1})$ are all orthogonal, E will have a minimum at each y^k and, hopefully, the dynamics of the system will have a region of attraction about each y^k that associates initial values of y that are near y^k and y^k. Detailed discussion of these issues can be found in Bruck (1989). It should be remarked here that the Hebbian rule encoding the patterns generates automatically the equilibrium points of (18)–(19), placing them in the locations resulting from y^k and (22) and unknown *a priori* to the designer.

The name Hopfield net is widely used for the continuous model described by

$$T_i\dot{x}_i = -x_i + \sum_{j=1}^{N} a_{ij}y_j + u_i, \qquad y_i = g_i(x_i), \quad i = 1, \ldots, N, \quad (24)$$

where $x_i = x_i(t)$, $y_i = y_i(t)$ and $g_i(\cdot)$ are sigmoid functions. The system (24) is, unlike (18), just a system of ordinary differential equations. Hopfield suggested that Lyapunov function

$$E = -\tfrac{1}{2}\sum_{i=1}^{N}\sum_{j=1}^{N} a_{ij}y_iy_j + \sum_{i=1}^{N} \rho_i\int_0^{y_i} g_i^{-1}(\xi)\, d\xi - \sum_{i=1}^{N} u_iy_i, \quad (25)$$

where $\rho_i > 0$ are constants, $g_i(\cdot)$ are monotone increasing functions and $a_{ij} = a_{ji}$ $\forall i, j$. It is straightforward to show that $\dot{E} \leq 0$.

According to Hopfield a_{ij} can be of both signs (or 0), which means A can be indefinite in the general case. Neglected discussion of positivity conditions for (25) in Hopfield (1984) was addressed in Michel *et al.* (1989) and Li *et al.* (1988) where alternative means of analysis were described.

2.4.2. *Cohen–Grossberg Theorem*. The central issue of dynamic networks stability did not receive as much attention as did simulations and experiments. Hopfield nets stability, even after the work of Michel *et al.*, Li *et al.* (1988) and Michel *et al.* (1989) does not seem to be perfectly clear. The most general result in the area, the Cohen–Grossberg Theorem (Cohen and Grossberg (1983)), was obtained under conservative assumptions (see below), often violated by practitioners without much harm (Pineda (1987)). Other special cases of stability are also considered in the literature Sudharsanan and Sundareshan (1990), Kelly (1990), Tan *et al.* (1990) and Johnson (1991) and Poteryaiko (1991) to name a few. Below we present the Cohen–Grossberg Theorem and relate it to Hopfield nets stability.

The Cohen–Grossberg result applies to the general model

$$\dot{x}_i = r_i(x_i)\left[s_i(x_i) - \sum_{j=1}^{N} a_{ij}g_j(x_j)\right], \quad i = 1, \ldots, N. \quad (26)$$

The following assumptions are made for $i, j = 1, \ldots, N$:

1° $a_{ij} \geq 0$ and $a_{ij} = a_{ji}$;
2° $r_i(\xi)$ is continuous for $\xi \geq 0$ and positive for $\xi > 0$;
3° $s_i(\xi)$ is continuous;
4° $g_i(\xi) \geq 0$ for $\xi \in (-\infty, +\infty)$, differentiable and monotone non-decreasing for $\xi \geq 0$;

5° $\limsup_{\xi\to\infty} [s_i(\xi) - a_{ii}g(\xi)] < 0$;

6° either

a) $\lim_{\xi\to 0^+} s_i(\xi) = \infty$,

or

b) $\lim_{\xi\to 0^+} s_i(\xi) < \infty$ and $\int_0^{\varepsilon} \frac{d\xi}{r_i(\xi)} = \infty$ for some $\varepsilon > 0$.

The assumption of weights symmetry, 1°, is popular in dynamic network models, but here it is strengthened by the non-negativity of a_{ij}. The continuity assumptions 2°–3° and positivity $r_i(\cdot)$ for positive argument are needed together to prove that if the system (26) starts from positive initial conditions then its trajectories remain positive. Thus the state space is contracted to the positive orthant $\mathscr{R}_+^N$, which together with the constraint $a_{ij} \geq 0$ is a strong condition. The assumptions about network nonlinearity, 4°, are mild in practice, unless the step-like function is used.

The Lyapunov function (recall 1°, 4° and $\mathscr{R}_+^N$

restriction) for the system (26) is defined as

$$V(x) = -\sum_{i=1}^{N} \int_0^{x_i} s_i(\xi) g_i'(\xi)\, d\xi + \tfrac{1}{2} \sum_{i=1}^{N} \sum_{j=1}^{N} a_{ij} g_i(x_i) g_j(x_j), \qquad (27)$$

and its time derivative has the property $\dot{V}(x) \le 0$, since it has the form

$$\dot{V}(x) = -\sum_{i=1}^{N} r_i(x_i) g_i'(x_i) \left[s_i(x_i) - \sum_{j=1}^{N} a_{ij} g_j(x_j) \right]^2. \qquad (28)$$

In (28) Assumptions 2^0 and 4^0 were applied.

The price for the generality of the theorem is the introduction of technical (5° and 6°) and strong (1°) conditions, often violated in practice. The violations do not always result in instability, because the theorem is a sufficient condition.

2.5. *Other architectures*

There exist numerous neural network architectures and learning algorithms. Many of them are refinement/extensions of the ones described in this paper. In this section we include some of the original schemes with certain potential for control/optimization applications.

2.5.1. *Simulated annealing and Boltzmann machines.* Artificial neural networks learning is an optimization process (see Section 4.3.3), aimed at minimization of the error function E with respect to weights a_{ij}. Due to the nonlinearity of networks, $E(A)$ possesses many local minima (which is desirable for CAM, but not for control), resulting in suboptimal solutions when gradient methods are applied. An alternative approach is to allow occasional search directions increasing E, in this way trying to escape local minima.

The method is possible in the stochastic context (Metropolis *et al.* (1953)), inspiration coming from mathematical models describing melted metals annealing (Aarts and Korst (1989)). Originally, the physical interpretation says that in a high temperature (T) the probability that metal is in a given energy state is uniform. With the temperature dropping, the average chaotic movement of atoms becomes less intense, so that minima surrounded by "high" energy barriers are more likely for metal to assume. Provided the cooling process is slow, initial big perturbations (with T high) should allow the system to escape "shallow" local minima and, after lowering the temperature T, small perturbations should prevent it from jumping out of a "deep" (possibly global) minimum. Thus, controlling one parameter, T, it is possible to change "energy" probability distribution to steer the system toward the global solution, exciting it randomly to overcome local ones.

The Boltzmann machine (Ackley *et al.* (1985)) is an application of these ideas to the asynchronous Hopfield net (see Section 2.4.1). Interpreting E in (20) as an "energy" function, the key element is the calculation of ΔE of (21). The algorithm is as follows. Excite the network (18) (weights fixed) with a random initial condition and set the parameter T ("temperature") to a high value. Pick a random node, p. Calculate ΔE from (21) by setting $\Delta y_p = 1$. Generate a random number, $\gamma \in [0, 1]$ (uniform distribution). If $\gamma \le (1 + \exp(\Delta E/T))^{-1}$ set $y_p(t+1) = 1$. After a prescribed number of changes of E and/or x, T is lowered and the process is started over from the state it reached in the previous iteration. In the equilibrium the relative probability of two global states y^{k_1}, y^{k_2}, will follow the Boltzmann distribution $P(y = y^{k_1})/P(y = y^{k_2}) = \exp([E(y^{k_2}) - E(y^{k_1})]/T)$.

Since E is non-increasing (see Section 2.4.1) the network will head to the nearest local minimum. However, if T is high then $\gamma \ge \exp(-\Delta E/T)$ will hold more often than not. The system will be perturbed on the way to a local solution and pushed randomly (with the probability $\exp(-\Delta E/T)$) towards another local minimum. This "dither signal" will fade away with dropping temperature T, possibly allowing the network to settle close to the global extremum.

The method is statistically convergent but this requires an infinite number of trials and continuous change of T. In practice this is violated, so *ad hoc* methods of termination must be employed (e.g. $|\Delta E| < \epsilon$). This, however, prevents the network from reaching the true global solution. Because of its trial-and-error character the method is very slow.

2.5.2. *Lateral connections.* Lateral connections are connections among units in the same layer, as shown in Fig. 3. This type of connection is vital in many different applications, such as maximum selection, competitive learning and contrast enhancement. There are two kinds of lateral connections: non-recurrent and recurrent. The first class can be described using

$$A = 0,$$

$$B = B_1 - B_2,$$

where $A = 0$ implies no feedback, B_1 is the identity matrix, and B_2 (having diagonal entries equal to zero) represents explicitly the inhibition of other inputs. This kind of lateral connection has been shown to be useful in estimating acoustic spectra (Shamma (1989)). The recurrent architecture uses the interconnection matrix A to provide inhibition. In this case we

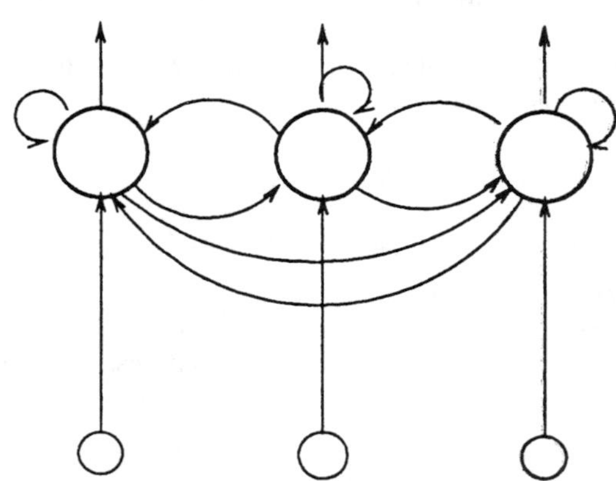

FIG. 3. Network with lateral connections.

$$A = \begin{cases} a_{ij} & i = j, \\ a_{ij} \leq 0 & i \neq j \end{cases}$$
$$B = B_1.$$

In addition, all elements of B here are positive. The feedback now brings the nonlinear function $g(\cdot)$ into play and so the properties of the recurrent network are richer than those of the non-recurrent network. This kind of network is useful in producing quantization of the inputs, or can serve as a local or global peak detector. The combination of several types of connections can create complex architectures like analogue Adaptive Resonance Theory (ART, and its later derivates) (Grossberg (1988)). This system is a two layer architecture with feedback and lateral interaction capable of creating different classes to classify signals according to their similarities.

The idea of encoding inputs has been used to generate nonlinear mappings. A network which uses this characteristic is the counterpropagation network (CPN) (Hecht-Nielsen (1987a)). This network has a first layer of competitive units, known as the Kohonen or LVQ layer (Kohonen (1987)), and a second layer of output units. In this case the output of the network can be regarded as a linear combination of functions that represent the characteristic of the encoder, i.e.

$$y_i = \sum_{j=1}^{N} a_{ij}\phi_j(u_j), \tag{29}$$

where ϕ_j can have different forms, one of which uses lateral inhibition.

2.5.3. *Cerebellar model articulation controller*. Another network which uses the encoding system to approximate nonlinear functions is the Cerebellar Model Articulation Controller (CMAC) (Albus (1975a, b)), even though CMAC does not use a competitive network to produce an encoding of the input to reach the same general objective (radial basis function networks and splines also fall within this framework). These networks use the concept of locally tuned overlapping receptive fields. They have the advantage that they can approximate complex nonlinear functions much faster than networks using sigmoids. Only a small subset of the parameters are adjusted at each point in the input space. This reduces sensitivity to the order in which the training data are presented. A complete and detailed account of the relationship between splines and CMAC is given in Lane *et al.* (1991).

The CMAC architecture has been applied to nonlinear control problems (Ersü and Tolle (1984)). The use of the system as an encoder not only gives the possibility of nonlinear mapping but also a fault tolerant characteristic.

2.5.4. *Cellular neural networks*. (Chua and Yang (1988a, b)) proposed an architecture derived from cellular automata (Wolfram (1986)), called the Cellular Neural Network (CNN). Each layer of a CNN is a two-dimensional structure with sparse interconnections and local properties based on spatial-invariant templates. The underlying idea was to propose a two-dimensional fast parallel image processing filter (Dzieliński *et al.* (1990)), although optimization applications can be found (*Proc. CNNA*'90 (1990)). No control applications have yet been reported.

3. LEARNING IN STATIC NETWORKS

The application of artificial neural networks normally proceeds in two phases; for the architectures described in Section 2 we assumed that the networks had already been trained using suitable data. A trained network with constant coefficients then provides a fixed behaviour (in the context of associative memory this is called the recall phase). In this section (static networks) and in the following section (dynamic networks) we discuss the training phase and present algorithms for automatic adjustment of network coefficients—a process called learning in the ANN literature and adaptation in the control literature.

A learning algorithm is associated with any change in the memory as represented by the weights; learning does not in this sense imply a change in the structure of the memory (structural learning is a separate issue which is receiving some attention). From this point of view learning can be regarded as a parametric adaptation algorithm.

Learning algorithms can be classified into two main groups: supervised learning, which incorporates an external reference signal (teacher) and/or global information about the system, and unsupervised learning, which incorporates no external reference signal and relies only upon local information and internal signals.

In the following we present, in a standard and unifying fashion, a learning algorithm based on systems identification theory (Ljung and Söderström (1983)). Within this framework we show that backpropagation is a special case of the general algorithm.

3.1. *Non-dynamic single-layer architecture*

From equation (15) the single-layer non-dynamic architecture is described by:

$$\begin{aligned} x(t) &= Bu(t), \\ y(t) &= g(x(t)), \end{aligned} \tag{30}$$

where we assume $w = 0$. The standard ANN learning problem can be posed as follows:

- y and u contain signals available for measurement.
- The function g is prespecified, typically taken from Table 2.
- A desired output value, or reference value, r (corresponding to each u) is known.
- Find a parameter estimate $\hat{B}$ such that the square of the error $e(t)$ between r and y (i.e. $e^2(t) = (r(t) - y(t))^2$) is minimized over all pairs y, u.

Backpropagation, a learning algorithm common in the ANN literature, is a special case of this general problem. In the linear case when $g(x) = x$ (i.e. $y = x$) this corresponds to the standard linear in the

TABLE 3. SUMMARY OF ARCHITECTURES

Architecture	Type of connection	Main characteristic	References
Perceptron	1 layer feedforward	Linear system with on–off output	Fu (1976)
Associative Reward–Penalty	3 layers feedforward	Use of low quality information as feedback	(Barto (1988); Barto *et al.* (1983); Helferty *et al.* (1989); Anderson (1989); Porcino and Collins (1990); Guhua and Mathur (1990); Chen (1990); Franklin (1989); Lee and Berenji (1989))
Backpropagation	several layers feedforward	Nonlinear system decision regions	(Goldberg and Pearlmutter (1988); Yeung and Beckey (1989); Bassi and Beckey (1989); Josin (1988); Sekiguchi *et al.* (1989); Chen and Pao (1989); Psaltis *et al.* (1988); Guez and Selinsky (1990); Josin (1990); Grant and Zhang (1989); Sanner and Akin (1990); Chen (1989); Karsai *et al.* (1989); Troudet and Merril (1989); Ciliz and Isik (1989); Bhart and McAvoy (1990); Ydstie (1990); Ungar *et al.* (1990); Savic and Tan (1989); Bhat *et al.* (1990))
Cohen–Grossberg	1 layer feedback	Dynamic nonlinear system	(Guez *et al.* (1988); Bavarian (1988); Zhao and Mendel (1988); Pearlmutter (1988); Żak (1990); Chi *et al.* (1990))
LVQ	lateral connections	Quantization	(Graf and Lalonde (1988); Martinez *et al.* (1988); Ritter and Schulten (1986, 1988))
CPN	lateral connections feedforward	Nolinear system (look-up table)	(Miller *et al.* (1987); Atkenson and Reinkensmeyer (1989); Ersü and Tolle (1984); Cheok and Huang (1989); Kraft and Campagna (1990))
ART II	2 layers feedback lateral connections	Adaptive classifer (look-up table)	(Kumar and Guez (1990); Bullock (1988))
CMAC	2 feedfoward layers, overlapping receptive fields	Look-up table	(Albus (1975b); Ersü and Tolle (1984); Kraft and Campagna (1990); Miller *et al.* (1987))

parameters parameter identification problem. It would then typically be solved in continuous time by the equation

$$\dot{\hat{B}}(t) = P(t)u(t)[r(t) - y(t)], \tag{31}$$

and in discrete time by

$$\hat{B}(t+1) = \hat{B}(t) + P(t)u(t)[r(t) - y(t)], \tag{32}$$

where t in the last equation is the integer time index. Various choices of P could be used. For example:

- $P = \alpha I$, α is a scalar constant—(gradient algorithm).
- $P = \frac{\alpha}{\|u\|} I$, α is a scalar constant—(normalized gradient algorithm (Widrow and Lehr (1990)).
- $\dot{P}(t) = P(t)u(t)u(t)^T P(t)$, if P^{-1} is non-singular—(least-squares).

These methods increase in sophistication at the expense of increased computation. In particular the least-squares approach may be unfeasible for a large number of units N.

In the ANN literature it is very common to use the linear in the parameters approach as exemplified by equations (31) and (32) even though nonlinear functions are involved. In addition, the simple gradient algorithm is normally used. Note, however, that the neuron nonlinearity precludes simple application of linear least-squares algorithms for adaptation of neuron weights.

3.2. *Non-dynamic multiple layers*

The methods of Section 3.1 cannot be applied to the multilayer ANN of Section 2.3 as neither the input to, nor the reference signals for, the hidden layers are known. For many years this hindered this branch of ANNs until the backpropagation (BP) algorithm, developed by Werbos (1974), was rediscovered and popularized by Rumelhart *et al.* (1986). An interesting perspective on this is given by Minsky and Papert in the epilogue of the second edition of their book "Perceptrons" (Minsky and Papert (1988)).

The BP algorithm can be seen as a gradient algorithm applied to a nonlinear optimization problem. Briefly, the BP algorithm solves the missing information problem as follows:

- Inputs to the hidden layers are taken as the inputs to the first layers propagated forward through the network.
- The effective reference signals for the hidden layers are obtained by backpropagation of the error through the network. This is achieved by taking the partial derivative of the squared error against the parameters.

3.2.1. *Derivation of the backpropagation algorithm.* Backpropagation (BP) has brought to the attention of neural network researchers by the PDP Group Rumelhart and McClelland (1986). Rumelhart *et al.* (1986) rediscovered, however, the algorithm derived in an entirely different context by Werbos (1974). As outlined above, BP solves the problem of hidden layers learning, which is why the PDP contribution is widely recognised. Below we give Werbos' derviation of the algorithm, which is more general and mathematically more rigorous than the one given in Rumelhart and McClelland (1986).

Let an ordered system of equations (Werbos (1989)) be given as

$$x_i = f_i(x_{i-1}, \ldots, x_r), \quad i = 1, \ldots, N+1. \tag{33}$$

The formally proper method of calculating the partial derivatives of x_{N+1} is given by:

$$\frac{\partial^+ x_{N+1}}{\partial x_i} = \sum_{j>i}^{N+1} \frac{\partial^+ x_{N+1}}{\partial x_j} \frac{\partial f_j}{\partial x_i}. \tag{34}$$

Formula (34) is a recursive definition of the ordered derivative $\partial^+ x_{N+1}/\partial x_i$ for the systems (33).

A feedforward network is given a reference signal d only for the output and it is required to minimize

$$E = \tfrac{1}{2} \sum_{k=1}^{P} (d^k - x)^T (d^k - x), \tag{35}$$

which is the nonlinear least-squares fitting problem (Eykhoff (1974)) for P patterns. The gradient algorithm for adjusting weights yields

$$a_{ij}^{\text{new}} = a_{ij}^{\text{old}} - \alpha \frac{\partial E}{\partial a_{ij}}, \quad \alpha > 0, \tag{36}$$

where ($P = 1$ assumed for simplicity)

$$\frac{\partial E}{\partial a_{ij}} = \frac{\partial E}{\partial x_i} \frac{\mathrm{d}x_i}{\mathrm{d}s_i} \frac{\partial s_i}{\partial a_{ij}} = \frac{\partial E}{\partial x_i} g'(s_i) x_j. \tag{37}$$

In (37) s_i is the total input to ith neuron, i.e. a weighted sum of the outputs of neurons of the previous layer. For the output layer $\partial E/\partial x_i = x_i - d_i$, but it is not obvious how to find the error derivative for the hidden layers. The backpropagation algorithm assumes linear propagation of the error derivative

$$\frac{\partial E}{\partial x_i} = \sum_j \frac{\partial E}{\partial x_j} \frac{\mathrm{d}x_j}{\mathrm{d}s_j} \frac{\partial s_j}{\partial x_i} = \sum_j \frac{\partial E}{\partial x_j} g'(s_j) a_{ji}, \tag{38}$$

where x_i belongs to layer l, and x_j to $l + 1$. Beginning with the output layer this can be recursively solved. This formulation (Rumelhart and McClelland (1986)) requires care with neuron indexing and is imprecise as far as the definition of the partial derivative is concerned.

On the other hand, the network is an ordered system. Thus, (38) can be expressed as

$$\frac{\partial^+ E}{\partial x_i} = \frac{\partial E}{\partial x_i} + \sum_{j>i}^{N} \frac{\partial^+ E}{\partial x_j} \frac{\partial x_j}{\partial x_i}, \tag{39}$$

with E substituted for x_{N+1} in (34), and is no longer an heuristic assumption. Then the BP algorithm for feedforward networks may be written as

$$\begin{aligned} z_i &= \sum_{j>i} a_{ji} g'(s_j) z_j + \epsilon_i, \\ \frac{\partial E}{\partial a_{ij}} &= g'(s_i) z_i x_j, \\ a_{ij}^{\text{new}} &= a_{ij}^{\text{old}} - \alpha \frac{\partial E}{\partial a_{ij}}, \end{aligned} \tag{40}$$

where $z_i = \partial^+ E/\partial x_i$ and $\epsilon_i = \partial E/\partial x_i$, or in vector–matrix form:

$$\begin{aligned} \Gamma &= \operatorname{diag}[g'(s_1), \ldots, g'(s_N)], \\ z &= A^T \Gamma z + \epsilon, \\ A^{\text{new}} &= A^{\text{old}} - \alpha \Gamma z x^T. \end{aligned} \tag{41}$$

Here, neurons are labelled with consecutive numbers, irrespective of the layer structure, i.e. $x_1, x_2, \ldots, x_i, \ldots, x_N$.

The procedure can be summarized as follows:

(1) Assign random values to all a_{ij} (initialize A).

(2) Input P patterns and calculate all intermediate and output signals of the network (x_js and s_js).

(3) Apply (40) to update a_{ij}.

The forward pass encompasses steps (1)–(2) and the backward pass is given by (3).

Backpropagation is an off-line technique as the forward pass must be computed for all P before (41) is applied (backward pass).

3.2.2. *Control interpretation of backpropagation.* Consider a general network represented by a nonlinear function f and a set of parameters θ, containing the coefficients of (A, B) in equation (14), described by the following equation:

$$y(t) = f(u(t), \theta), \tag{42}$$

where $y(t)$ and $u(t)$ are the output and the input of the system. Here we consider discrete time systems having the integer time index t. The problem is, given a set of points $[y(t), u(t)]$, try to find a set of weights $\hat{\theta}$ such that the mean square error E is minimized, where

$$E = \sum_{t=1}^{T} \lambda^{T-1} \|(y(t) - f(u(t), \hat{\theta}))\|^2. \tag{43}$$

Here, λ is an exponential forgetting factor which gives different weights to different observations and T is the number of presentations of different training sets.

Applying a nonlinear estimation algorithm (Gawthrop and Sbarbaro (1990) and Albert and Gardner (1967)), the estimates $\hat{\theta}(t)$ can be calculated as

$$\hat{\theta}(t+1) = \hat{\theta}(t) + P^+(t)\bar{x}(t)[y(t) - f(u(t), \hat{\theta})], \tag{44}$$

where $\bar{x}(t)$ is the partial derivative of the output against the parameter, that is $\frac{\partial F}{\partial \hat{\theta}}$, and $P^+(t)$ is the pseudo-inverse of a matrix P given by

$$P(t) = \sum_{\tau=1}^{T} \lambda^{T-\tau} \bar{x}(\tau)\bar{x}(\tau)^T. \tag{45}$$

The backpropagation algorithm is a special case of this learning algorithm obtained by setting $P^+(t)$ to a constant diagonal matrix.

There are many variations of the algorithm for off-line as well as on-line use, and some decoupled versions to avoid inverting the information matrix $P(t)$, and to utilize the parallel structure and local information (Chen *et al.* (1990b); Scalero and Tepedelenlioglu (1990); Kollias (1989)). These algorithms are like BP with varying gain.

Some problems which must be faced are the definition of the structure of the network, the possibility of local minima due to data dependencies, and selecting the appropriate input signal such that the parameters converge to values which produce a good approximation of the system.

3.2.3. *Static networks representing temporal processing.* There are two approaches to processing temporal signals. The first and the most used is to window the input and then treat the time domain like another spatial domain. Here, tapped delay input is fed to a standard multilayer network, producing in this way a nonlinear ARMAX (NARMAX) model

(Chen *et al.* (1990b)). The second possibility is to use internal storage to maintain the current state having in this way a dynamic network. Dynamic networks are discussed in the following section.

The tapped delay input approach has two main disadvantages. Firstly, there is a hard limit as to the amount of temporal context. Secondly, these models have no inbuilt mechanism for dealing with variations in the rate of input (Robinson (1989)).

4. LEARNING IN DYNAMIC NETWORKS

Dynamic or recurrent networks are qualitatively different from feedforward ones, because their structure incorporates feedback. In general, the output of every neuron is fed back with varying gains (weights) to the inputs of all neurons (fully connected network). The architecture is inherently dynamic and usually one-layered, since its complex dynamics give it powerful representation capabilities. Other characteristics of dynamic networks will be highlighted below.

There are two basic dynamic network descriptions. The first (Hopfield (1984)) assumes separate state and output

$$\begin{aligned} T &= \operatorname{diag}[T_1, \ldots, T_N], \\ T\dot{x} &= -x + Ay + u, \\ y &= g(x), \end{aligned} \tag{46}$$

while the other gives linear output equal to state

$$\begin{aligned} s &= Ay, \\ T\dot{x} &= -x + g(s) + u, \\ y &= x, \end{aligned} \tag{47}$$

or, in discrete time,

$$\begin{aligned} s(t) &= Ay(t), \\ Tx(t+1) &= -x(t) + g(s(t)) + u(t), \\ y(t) &= x(t). \end{aligned} \tag{48}$$

In this section we shall use equations (47)–(48). For the sake of simplicity we assume T to be the identity matrix ($T = I$), unless otherwise stated.

4.1. *Dynamic behaviour learning*

Recurrent networks possessing the same structure can exhibit different dynamic behaviour, due to the use of distinct learning algorithms. The network is defined when its architecture and learning rule are given. In other words, it is a composition of two dynamic systems: transmission and adjusting systems. The overall input–output behaviour is thus a result of the interaction of both.

There are two general concepts of recurrent structures training. Fixed point learning is aimed at making the network reach the prescribed equilibria or perform steady-state matching. The only requirement on the transients is that they die out. Trajectory learning trains the network to follow the desired trajectories in time. In particular, when $t \to \infty$, it will also reach the prescribed steady-state, so it can be viewed as a generalization of fixed point algorithms.

4.2. *Fixed point learning*

For the network (47) the error is defined as

$$E = \tfrac{1}{2} \sum_{k=1}^{P} (y_\infty^k - y)^T (y_\infty^k - y), \tag{49}$$

where y_∞^k are vectors of the desired equilibria, being the solutions of (47) with the left-hand side set to 0:

$$0 = -y_\infty^k + g(s) + u, \quad k = 1, \ldots, P. \tag{50}$$

Comparison of (49) with (35) shows that similar error functions are minimized. However, d^k in (35) do not affect network's dynamics (because it has none) and y_∞^k in (49) define with (50) the steady state of a dynamic network.

During learning the network does not receive any external inputs. It is excited by initial conditions corresponding to the expected workspace and evolves with y_∞ as a constant reference signal, which is called relaxation (Pineda (1989)). Practically, this is done using recurrent backpropagation (Pineda (1987); Almeida (1988)) (formulae given assume $P = 1$ in (49) for simplicity):

$$\begin{aligned} e_i &= \begin{cases} y_{\infty_i} - y_i, & \text{if } y_i \text{ is a network output;} \\ 0, & \text{otherwise,} \end{cases} \\ \dot{z}_i &= -z_i + \sum_j a_{ji} g'(s_j) z_j + e_i, \\ \dot{a}_{ij} &= \alpha g'(s_i(\infty)) z_i(\infty) y_i(\infty), \end{aligned} \tag{51}$$

or, in vector–matrix form

$$\begin{aligned} e &= [0 \,\vdots\, y_\infty - y]^T, \\ \Gamma &= \operatorname{diag}[g'(s_1), \ldots, g'(s_n)], \\ \dot{z} &= -z + A^T \Gamma z + e, \\ \dot{A} &= \alpha \Gamma(\infty) z(\infty) y^T(\infty). \end{aligned} \tag{52}$$

Since only $\lim_{t \to \infty} z(t) = z(\infty)$ is of interest, the initial condition for the differential equation for z in (52) can theoretically be arbitrary. Practically, numerical soundness should be borne in mind.

The idea exploited in (50) is used to calculate z, which corresponds to the ordered derivative (34); also e_i from (51) is equal to $-\epsilon_i$ of (40) (see Section 3.2.1). The equation for A can be solved after the network has settled, since $y(\infty)$ and $z(\infty)$ (steady-state values of $y(t)$ and $z(t)$, respectively) are needed.

This algorithm differs from Hopfield nets (Hopfield (1982, 1984)) (compare Section 2.4.1), since there the learning was performed by the Hebbian rule (cf. (22)) and the relaxation is done only during the recall phase (the Hopfield net is a content-addressable memory). Recurrent backpropagation creates basins of attraction around the given equilibria by adjusting the weights, learning them through relaxation (52). The Hopfield net generates its attractors and basins in applying the Hebbian rule and thus produces equilibria on its own, representing input patterns.

If the forward pass in recurrent backpropagation is stable, so is the backward one (Almeida (1988)). On the other hand, the Hopfield net should be stable, because defining it requires guessing a Lyapunov function appropriate to an application, (e.g. Park and Lee (1990)).

4.3. *Trajectory learning*

In this case, the error for the network (47) is given by

$$E = \frac{1}{2}\int_{t_0}^{t_1} [(d(\tau) - y(\tau))^T(d(\tau) - y(\tau))]\,d\tau, \quad (53)$$

where $d(\tau)$ is a vector of the desired trajectory and t_1 may be a constant (off-line techniques) or a variable (on-line algorithms). The discrete-time version (see (48)) yields:

$$E = \frac{1}{2}\sum_{\tau=t_0}^{t_1} [(d(\tau) - y(\tau))^T(d(\tau) - y(\tau))], \quad (54)$$

with the previous remarks valid. Both (53) and (54) may include extra summation to cover multiple trajectories $d^k(\tau)$. For simplicity, this is omitted in the following discussion.

4.3.1. *Backpropagation through time.* The simplest method, already mentioned in Rumelhart and McClelland (1986), is to unfold the network through time, i.e. replace a one-layer recurrent network with a feedforward one with t_1 layers. The equivalent static network has as many layers as time instants, which is easily understood in the discrete time context and therefore we start with this. Standard backpropagation will be applied (see Section 3.2.1), and since each layer contains the same weights "seen" in different moments, their final value is the sum of intermediate values (Werbos (1990)). Thus, we have

$$E = \frac{1}{2}\sum_{\tau=t_0}^{t_1}\sum_i (d_i(\tau) - y_i(\tau))^2, \quad t_1 = \text{const},$$

$$e_i(\tau) = \begin{cases} d_i(\tau) - y_i(\tau), & \text{if } y_i(\tau) \text{ is a network output;} \\ 0, & \text{otherwise,} \end{cases}$$

$$z_i(\tau) = \sum_j a_{ji} g'(s_j(\tau+1)) z_j(\tau+1) - e_i(\tau), \quad z_i(t_1) = 0,$$

$$\frac{\partial E}{\partial a_{ij}} = \sum_{\tau=t_0}^{t_1} g'(s_i(\tau)) z_i(\tau) y_j(\tau),$$

$$a_{ij}^{t_1} = a_{ij}^{t_0} - \alpha \frac{\partial E}{\partial a_{ij}}, \quad (55)$$

or, in vector–matrix form,

$$\begin{aligned} &E = \frac{1}{2}\sum_{\tau=t_0}^{t_1} [(d(\tau) - y(\tau))^T(d(\tau) - y(\tau))], \\ &t_1 = \text{const}, \\ &e(\tau) = [0 \vdots d(\tau) - y(\tau)]^T, \\ &\Gamma(\tau) = \text{diag}\,[g'(s_1(\tau)), \ldots, g'(s_n(\tau))], \\ &z(\tau) = A^T(t_0)\Gamma(\tau+1)z(\tau+1) - e(\tau), \quad z(t_1) = 0, \\ &\nabla_A E = \sum_{\tau=t_0}^{t_1} \Gamma(\tau) z(\tau) y^T(\tau), \\ &A^{t_1} = A^{t_0} - \alpha \nabla_A E. \end{aligned} \quad (56)$$

The definition of $e_i = e_i(\tau)$ in (55) and/or $e = e(\tau)$ in (56) will be further used without additional explanation.

Theoretically, there exists the possibility (Williams and Zipser (1990)) of an on-line algorithm, but the above calculations would have to be performed at every step, requiring unlimited memory and computational power.

The continuous time version was introduced heuristically by Pearlmutter (1989), and then rediscovered by Sato (1990), whose derivation, based on the calculus of variations, is mathematically rigorous and will be briefly presented here. Introduce the Lagrangian

$$L = \int_{t_0}^{t_1} \Big\{ \frac{1}{2}\sum_i (d_i - y_i)^2 - \sum_i z_i [T_i \dot{y}_i + y_i - g(s_i) - u_i] \Big\}\,d\tau, \quad (57)$$

where z_i are Lagrange multipliers (note that we show a derivation for $T \neq I$; see remarks after (47)–(48)). The first variation yields

$$\begin{aligned} \delta L = \int_{t_0}^{t_1} \Big\{ &\sum_i [(y_i - d_i) - z_i]\,\delta y_i + \sum_i z_i g'(s_i) \sum_j a_{ij}\,\delta y_j - \sum_i z_i T_i\,\delta \dot{y}_i \\ &+ \sum_i z_i g'(s_i) \sum_j y_j\,\delta a_{ij} \Big\}\,d\tau, \end{aligned} \quad (58)$$

which can be reduced by defining the auxiliary equation

$$T_i \dot{z}_i = z_i - \sum_j a_{ji} g'(s_j) z_j - e_i. \quad (59)$$

Multiplying both sides of (59) by δy_i and summing with respect to i gives

$$\delta L = \sum_i [T_i z_i(t_0)\,\delta y_i(t_0) - T_i z_i(t_1)\,\delta y_i(t_1)] + \int_{t_0}^{t_1} \Big[\sum_i z_i g'(s_i) \sum_j y_j\,\delta a_{ij} \Big]\,d\tau. \quad (60)$$

Since the initial conditions $y_i(t_0)$ do not depend on weights, $\delta y_i(t_0) = 0$. If additionally the boundary values

$$z_i(t_1) = 0, \quad (61)$$

are imposed then

$$\delta L = \int_{t_0}^{t_1} \Big[\sum_i z_i g'(s_i) \sum_j y_j\,\delta a_{ij} \Big]\,d\tau, \quad (62)$$

and hence (compare (55))

$$\frac{\partial E}{\partial a_{ij}} = \int_{t_0}^{t_1} g'(s_i) z_i y_j\,d\tau, \quad (63)$$

$$\dot{a}_{ij} = -\alpha \frac{\partial E}{\partial a_{ij}}, \quad (64)$$

which together with (59) completes the algorithm. The vector–matrix form looks as follows:

$$\begin{aligned} &E = \frac{1}{2}\int_{t_0}^{t_1} [(d(\tau) - y(\tau))^T(d(\tau) - y(\tau))]\,d\tau, \\ &t_1 = \text{const}, \\ &T\dot{z} = z - A^T\Gamma z - e, \\ &z(t_1) = 0, \\ &\nabla_A E = \int_{t_0}^{t_1} \Gamma(\tau) z(\tau) y^T(\tau)\,d\tau, \\ &\dot{A} = -\alpha \nabla_A E. \end{aligned} \quad (65)$$

Notice that the auxiliary equation (59) is defined on $[t_0, t_1]$ with $z_i(t_1) = 0$ (see (61)), which means that it must be integrated backwards in time. The method for adjusting time constants and delays can be found in Pearlmutter (1990). A speeding up modification of the algorithm was proposed in Fang and Sejnowski (1990).

The method is essentially non-recurrent, because it uses a feedforward equivalent for learning, so there is no explicit use of the network's feedback. It requires recording of the whole trajectories and playing them back to calculate the weights. However, once the weights are established the network runs as a recurrent one.

4.3.2. *Forward propagation.* Backpropagation through time was inherently an off-line technique. The following algorithm (Robinson and Fallside (1987); Williams and Zipser (1989b)) overcomes the difficulty. For the recurrent network (T_k suppressed for clarity; compare (47))

$$\dot{y}_k = -y_k + g(s_k) + u_k, \tag{66}$$

with

$$\frac{\partial \dot{y}_k}{\partial a_{ij}} = -\frac{\partial y_k}{\partial a_{ij}} + g'(s_k)\left[\delta_{ik} y_i + \sum_l a_{kl} \frac{\partial y_l}{\partial a_{ij}}\right], \tag{67}$$

where δ_{ik} is the Kronecker symbol, define error

$$E = \int_{t_0}^{t} \left[\tfrac{1}{2} \sum_k (d_k(\tau) - y_k(\tau))^2\right] \mathrm{d}\tau, \quad t \to \infty, \tag{68}$$

i.e. with free termination time (t is a variable). The rationale for introducing the subscript k in (66) is that each neuron is fully connected, so it is influenced by all weights a_{ij} (cf. (67)).

Consider the first variation of the error

$$\delta E = \sum_k \left[\int_{t_0}^{t} [e_k(\tau)\, \delta y_k]\, \mathrm{d}\tau\right], \tag{69}$$

where $e_k(\tau)$ is defined as in (55). Since a_{ij}s are independent variables, the following holds:

$$\delta y_k = \sum_i \sum_j \frac{\partial y_k}{\partial a_{ij}} \delta a_{ij}. \tag{70}$$

Taking into account constraints (66) gives, with $p_{ij}^k = \partial y_k / \partial a_{ij}$ substituted to (67),

$$\frac{\partial E}{\partial a_{ij}} = \sum_k \left[\int_{t_0}^{t} [e_k(\tau) p_{ij}^k(\tau)]\, \mathrm{d}\tau\right],$$

$$\dot{p}_{ij}^k = -p_{ij}^k + g'(s_k)\left[\delta_{ik} y_i + \sum_l a_{kl} p_{ij}^l\right], \tag{71}$$

$$\dot{a}_{ij} = -\alpha \frac{\partial E}{\partial a_{ij}},$$

and

$$p_{ij}^k(t_0) = 0, \tag{72}$$

because initial conditions, $y_k(t_0)$, do not depend on weights. The replacement of the boundary values (61) with the initial conditions (72) allows on-line calculations in (71). It is not straightforward to write down (71) and (72) in the vector–matrix form, because variables p_{ij}^k require a three-dimensional array. One possibility is:

$$E = \tfrac{1}{2} \int_{t_0}^{t} [(d(\tau) - y(\tau))^T (d(\tau) - y(\tau))]\, \mathrm{d}\tau,$$

$$t \to \infty,$$

$$\nabla_A E = \sum_k \left[\int_{t_0}^{t} [e_k(\tau) P_k(\tau)]\, \mathrm{d}\tau\right], \tag{73}$$

$$\dot{P}_k = -P_k + g'(s_k)[\lambda^k y^T + \Pi_k], \quad k = 1, \ldots, N,$$

$$P_k(t_0) = 0, \quad k = 1, \ldots, N,$$

$$\dot{A} = -\alpha \nabla_A E,$$

where $P_k = [p_{ij}^k]$ and $\Pi_k = [\pi_{ij}]$, $\pi_{ij} = \sum_l a_{kl} p_{ij}^l$ are square matrices of dimension N and $\lambda^k = [0, \ldots, \underbrace{1}_{\lambda_k}, \ldots, 0]^T$ is a vector.

A modification of the method can be found in Zipser (1989).

Forward propagation, called also real-time recurrent learning (RTRL) (Williams and Zipser (1989a)), requires non-local computations, since to calculate $\partial E / \partial a_{ij}$ knowledge of all p_{ij}^k is needed (see (71)). Real-time solving of the auxiliary equation on p_{ij}^k contributes to computing complexity, but the algorithm is truly on-line.

It is worth noting that recurrent networks allow an elegant solution of the key problem of multilayer feedforward networks, viz. the reference signal for hidden units. The answer is straightforward (see (51)) by setting the error to 0. It does not make impossible calculation of weights for hidden units (compare (71)) since through feedback they get information about the system performance.

4.3.3. *Trajectory learning: identification and optimal control.* Most of the problems of neural networks learning can be expressed in terms of an index minimization, which is usually a squared error function. A typical solution is an application of the gradient algorithm whose convergence properties are quite poor (Fletcher (1987)). However, the problem involved is not a classical optimization task, because the function to be minimized, $E(A)$, is not given explicitly. Thus, an N^2-component gradient (A is an $N \times N$ matrix) must be estimated which explains why algorithms making use of the second-order information (Newton methods exploiting Hessian-related data) are not developed despite their quadratic rate of convergence.

Dynamic neural networks are parametric models, which raises the question of their relation to identification. It should be borne in mind that the systems (47)–(48) are nonlinear state-space models. In the theory of nonlinear identification there exist several methods for certain special cases (Haber and Unbehauen (1990)). However, they are limited to particular input–output models. Hence dynamic networks can be seen as an alternative approach of capacity that remain to be investigated.

5. REPRESENTATION, IDENTIFICATION AND CONTROL STRUCTURES

In this section we review the possibilities for using neural networks directly in nonlinear control strategies. In this context neural networks are viewed as a process modelling formalism, or even a knowledge representation framework; our knowledge about the plant dynamics and mapping characteristics is implicitly stored within the network.

The ability of networks to approximate nonlinear mappings is thus of central importance in this task. For this reason we first review a body of theoretical work which has characterized, from an approximation theory viewpoint, the possibilities of nonlinear functional approximation using neural networks.

Second, we discuss learning structures from training networks to represent forward and inverse dynamics of nonlinear systems. Finally, a number of control structures in which such models play a central role are reviewed. These established control structures provide a basis for nonlinear control using neural networks.

In the same way that transfer functions provide a generic representation for linear black-box models, ANNs potentially provide a generic representation for nonlinear black-box models.

5.1. *Networks, approximation and modelling*

The nonlinear functional mapping properties of neural networks are central to their use in control. Training a neural network using input–output data from a nonlinear plant can be considered as a nonlinear functional approximation problem. Approximation theory is a classical field of mathematics; from the famous Weierstrass Theorem (Burkill and Burkill (1970); Rudin (1976)) it is known that polynomials, and many other approximation schemes, can approximate arbitrarily well a continuous function. Recently, considerable effort has gone into the application of a similar mathematical machinery in the investigation of the approximation capabilities of networks.

A number of results have been published showing that a feedforward network of the multilayer perceptron type can approximate arbitrarily well a continuous function (Cybenko (1988, 1989); Funahashi (1989); Hornik *et al.* (1989); Carrol and Dickinson (1989)). To be specific, these papers prove that a continuous function can be arbitrarily well approximated by a feedforward network with only a single internal hidden layer, where each unit in the hidden layer has a continuous sigmoidal nonlinearity. The use of nonlinearities other than sigmoids is also discussed.

These results provide no special motivation for the use of networks in preference to, say, polynomial methods for approximation; both approaches share the "Weierstrass Property". Such comparitive judgements must be made on the basis of issues such as parsimony. Namely, do networks or polynomials require fewer parameters? For network approximators, key questions are how many layers of hidden units should be used, and how many units are required in each layer? Of course, the implementation characteristics of approximation schemes also provide a further basis for comparison (for example, the parallel, distributed nature of networks is important). For the moment, however, we concentrate purely on approximation properties.

Although the results referred to above at first sight appear attractive they do not provide much insight into these practical questions. In fact, the degree of arbitrariness of the approximation achieved using a sigmoidal network with one hidden layer is reflected in a corresponding arbitrariness in the number of units required in the hidden layer; the results were achieved by placing no restriction on the number of units used. Cybenko himself says (Cybenko (1989)) "we suspect quite strongly that the overwhelming majority of approximation problems will require astronomical numbers of terms".

What is needed now is an indication of the numbers of layers/units required to achieve a specific degree of accuracy for the function being approximated. Some work along these lines is given in Chester (1990). That paper gives theoretical support to the empirical observation that networks with two hidden layers appear to provide higher accuracy and better generalization than a single hidden layer network, and at a lower cost (i.e. fewer total processing units). Guidelines derived from a mix of theoretical and heuristic considerations are given in the introductory paper by Lippmann (1987).

From the theoretical point of view the work of Kolmogorov (1957) (see also Lorentz (1976))) did appear to throw some light on the problem of exact approximation (Poggio (1982); Hecht-Nielsen (1987b); Lippmann (1987)). Kolmogorov's theorem (a negative resolution of Hilbert's 13th problem) states that any continuous function of N variables can be computed using only linear summations and nonlinear but continuously increasing functions of only one variable. In the network context the theorem can be interpreted as explicitly stating that to approximate any continuous function of N variables requires a network having $N(2N+1)$ units in a first hidden layer and $(2N+1)$ units in a second hidden layer. However, it has recently been pointed out (Hornik *et al.* (1989); Girosi and Poggio (1989)) that the practical value of this result is tenuous for a number of reasons:

(1) Kolmogorov's Theorem requires a different nonlinear processing function for each unit in the network.
(2) The functions in the first hidden layer are required to be highly non-smooth. In practice this would lead to problems with generalization and noise robustness.
(3) The functions in the second hidden layer depend upon the function being approximated.

It is apparent that these conditions may be violated in the practical situations of interest here. However, a well motivated dissenting viewpoint to that presented in Girosi and Poggio (1989) has been put forward by Kurkova (1991). In that work it is argued that the above points are nullified if one is only trying to approximate a function. We leave it to the reader to explore these conflicting viewpoints further in this rapidly developing area.

From the foregoing discussion it is clear that the property of approximating functions arbitrarily well is not sufficient for characterizing good approximation schemes (since many schemes have this property), nor does this property help in justifying the use of one particular approximation scheme in preference to another. This observation has been deeply considered by Girosi and Poggio (1990) (see also the expository paper (Poggio and Girosi (1990)). These authors propose that the key property is not that of arbitrary approximation, but the property of *best approximation*. An approximation scheme is said to have this property if in the set of approximating functions there is one which has the minimum distance from the given function (a precise mathematical formulation is given in the paper). The first main result of their paper (Girosi and Poggio (1990)) is that multilayer perceptron networks do not have the best approximation property. Secondly, they prove that Radial Basin Function networks (for example, Gaussian networks) do have the best approximation property. Thus, although one must bear in mind the precise mathematical formulation of "best", there is theoretical support for favouring RBF networks. Moreover, these networks may always be structured with only a single hidden layer and trained using linear optimization techniques with a guaranteed global solution (Broomhead and Lowe (1988)).

As with sigmoidal feedforward networks, however, there remain open questions regarding the network complexity required (i.e. the number of units). For RBF networks this question is directly related to the size of the training data. A related issue is that of choosing the centres of the basis functions (for results on automatically selecting the centres (see Moody and Darken (1989); Chen *et al.* (1990a)). Girosi and Poggio (1990) and Broomhead and Lowe (1988) discuss methods for achieving almost best approximation using restricted complexity RBF networks. This is discussed further in Sbarbaro and Gawthrop (1991).

Thus, although RBF networks have the best approximation property, it is necessary to consider the cost of utilizing this approach. For high-dimensional input spaces the number of nodes needed grows with the dimension of the input space (Hartman and Keeler (1992); Weigand *et al.* (1990)). This is in contrast to sigmoids which slice up the space into feature regions and hence are more economical with the number of nodes. This belief is rigorously supported by recent results of Barron (1991) which show that members of a certain class of "smooth" functions can be approximated by a two-layer network and that the error between the target function and the trained network can be bounded in terms of the network's size. It is shown in Barron (1991) that the error grows linearly with the dimension of the input space.

In summary, a large body of theoretical results relating to approximation using neural networks exist. These results provide theoretically important possibility theorems and deep insight into the ultimate performance of networks. They are not constructive results which define the type and structure of a suitable network for a given problem. Restraint must therefore be exercised when citing these results in support of the practical application of neural networks.

5.2. *Identification*

Although a number of key theoretical problems remain, the results discussed above do demonstrate that neural networks have great promise in the modelling of nonlinear systems. Without reference to any particular network structure we now discuss architectures for training networks to represent nonlinear dynamical systems and their inverses.

An important question in system identification is that of system identifiability (see Ljung and Söderström (1983), Ljung (1987), and Söderström and Stoica (1989)) i.e. given a particular model structure, can the system under study be adequately represented within that structure? In the absence of such concrete theoretical results for neural networks we proceed under the assumption that all systems we are likely to study belong to the class of systems that the chosen network is able to represent.

5.2.1. *Forward modelling*. The procedure of training a neural network to represent the forward dynamics of a system will be referred to as forward modelling. A structure for achieving this is shown schematically in Fig. 4. The neural network model is placed in parallel with the system and the error between the system and network outputs (the prediction error) is used as the network training signal. As pointed out by Jordan and Rumelhart (1991) this learning structure is a classical supervised learning problem where the teacher (i.e. the system) provides target values (i.e. its outputs) directly in the output coordinate system of the learner (i.e. the network model). In the particular case of a multilayer perceptron type network straightforward back-propagation of the prediction error through the network would provide a possible training algorithm.

An issue in the context of control is the dynamic nature of the systems under study. One possibility is to introduce dynamics into the network itself. This can be done either using recurrent networks (as discussed in a previous section) or by introducing dynamic behaviour into the neurons (see Willis *et al.* (1991)). A straightforward approach, and the one which for purposes of exposition will be followed here, is to augment the network input with signals corresponding to past inputs and outputs (this is also discussed in Section 3.2.3).

We assume that the system is governed by the following nonlinear discrete-time difference equation:

$$y^p(t+1) = f[y^p(t), \ldots, y^p(t-n+1); \quad u(t), \ldots, u(t-m+1)]. \quad (74)$$

Thus, the system output y^p at time $t+1$ depends (in the sense defined by the nonlinear map f) on the past n output values and on the past m values of the input u. We concentrate here on the dynamical part of the system response; the model does not explicitly represent plant disturbances (for a method of including disturbances (see Chen *et al.* (1990a)). Special cases of the model (74) have been considered

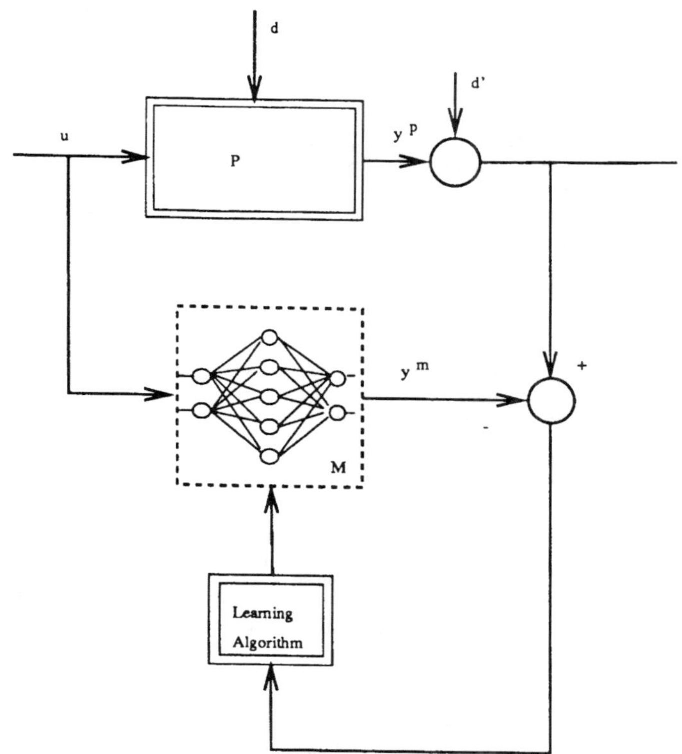

FIG. 4. Identification.

by Narendra and Parthasarathy (1990) (see also Narendra (1990)). These authors consider particular cases where the system output is linear in either the past values of y^p or u.

An obvious approach for system modelling is to choose the input–output structure of the neural network to be the same as that of the system. Denoting the output of the network as y^m we then have

$$y^m(t+1) = \hat{f}[y^p(t), \ldots, y^p(t-n+1); u(t), \ldots, u(t-m+1)]. \quad (75)$$

Here, $\hat{f}$ represents the nonlinear input–output map of the network (i.e. the approximation of f). Notice that the input to the network includes the past values of the real system output (the network has no feedback). This dependence on the system output is not included in the schematic of Fig. 4 for simplicity. If we assume that after a suitable training period the network gives a good representation of the plant (i.e. $y^m \approx y^p$) then for subsequent post-training purposes the network output itself (and its delayed values) can be fed-back and used as part of the network input. In this way the network can be used independently of the plant. Such a network model is described by

$$y^m(t+1) = \hat{f}[y^m(t), \ldots, y^m(t-n+1); u(t), \ldots, u(t-m+1)]. \quad (76)$$

The structure in (76) may also be used for training the network. This possibility has been discussed by Narendra (Narendra and Parthasarathy (1990); Narendra (1990)).

In the context of the identification of linear time invariant systems the two possibilities have been extensively considered by Narendra and Annaswamy (1989). The two structures have also been discussed in the signal processing literature (see Widrow and Stearns (1985)). The structure of equation (75) (referred to as the series–parallel model by Narendra) is supported in the identification context by stability results. On the other hand, (76) (referred to by Narendra as the parallel model) may be preferred when dealing with noisy systems since it avoids problems of bias caused by noise on the real system output (Widrow and Stearns (1985); Widrow (1986)).

5.2.2. *Inverse modelling.* Inverse models of dynamical systems play a crucial role in a range of control structures. This will become apparent in Section 5.3. However, obtaining inverse models raises several important issues which will be discussed.

Conceptually the simplest approach is direct inverse modelling as shown schematically in part (a) of Fig. 5 (this structure has also been referred to as generalized inverse learning (Psaltis *et al.* (1988)). Here, a synthetic training signal is introduced to the system. The system output is then used as input to the network. The network output is compared with the training signal (the system input) and this error is used to train the network. This structure will clearly tend to force the network to represent the inverse of the plant. However, there are drawbacks to this approach:

- The learning procedure is not "goal directed" (Jordan and Rumelhart (1991)); the training signal must be chosen to sample over a wide range of system inputs, and the actual operational inputs may be hard to define *a priori*. The actual

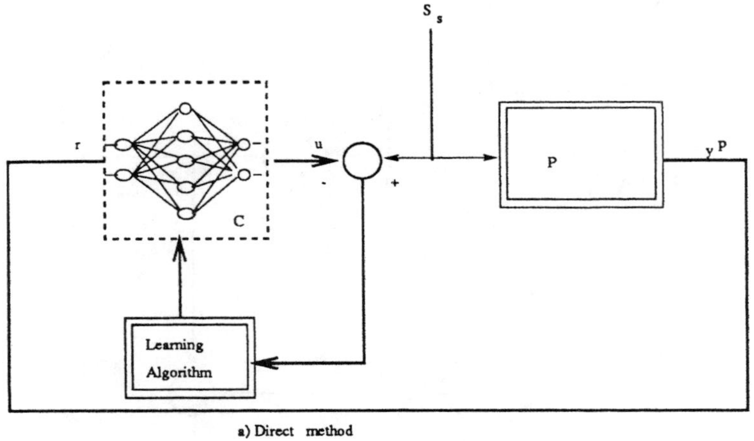

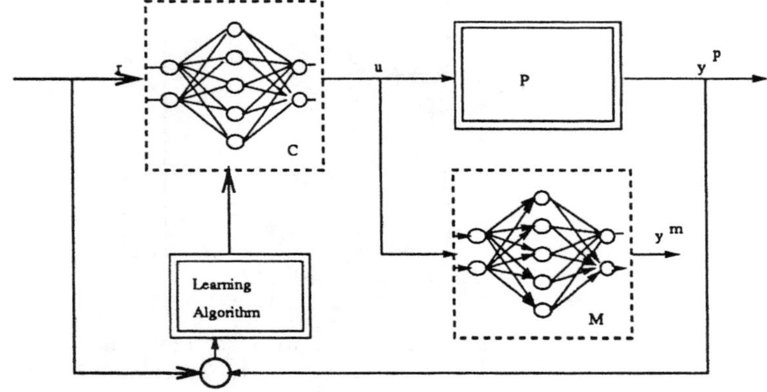

FIG. 5. Structures for inverse identification.

goal in the control context is to make the system output behave in a desired way, and thus the training signal in direct inverse modelling does not correspond to the explicit goal.

- Second, if the nonlinear system mapping is not one–one then an incorrect inverse can be obtained.

The first point above is strongly related with the general concept of persistent excitation; the importance of the inputs used to train learning systems is widely appreciated. In the adaptive control literature conditions for ensuring persistent excitation persistent excitation which will result in parameter convergence are well established (see, for example, Åström and Wittenmark (1989) and the references therein). For neural networks, methods of characterizing persistent excitation are highly desirable. A preliminary discussion on this question can be found in Narendra (1990).

A second approach to inverse modelling which aims to overcome these problems is known as specialized inverse learning (Psaltis *et al.* (1988)) (somewhat confusingly, Jordan and Rumelhart (1991) refer to this structure as forward modelling). The specialized inverse learning structure is shown in part (b) of Fig. 5. In this approach the network inverse model precedes the system and receives as input a training signal which spans the desired operational output space of the controlled system (i.e. it corresponds to the system reference or command signal). This learning structure also contains a trained forward model of the system (for example, a network trained as described in Section 5.2.1) placed in parallel with the plant. The error signal for the training algorithm in this case is the difference between the training signal and the system output (it may also be the difference between the training signal and the forward model output in the case of noisy systems; this obviates the need for the real system in the training procedure which is important in situations where using the real system is not viable). Jordan and Rumelhart (1991) show that using the real system output can produce an exact inverse even when the forward model is inexact; this is not the case when the forward model output is used. The error may then be propagated back through the forward model and then the inverse model; only the inverse network model weights are adjusted during this procedure. Thus, the procedure is effectively directed at learning an identity mapping across the inverse model and the forward model; the inverse model is learned as a side effect (Jordan and Rumelhart (1991)).

In comparison with direct inverse modelling the specialized inverse learning approach has the following features:

- The procedure is goal directed since it is based on the error between desired system outputs and actual outputs. In other words, the system receives inputs during training which correspond to the actual operational inputs it will subsequently receive.
- In cases in which the system forward mapping is not one–one a particular inverse will be found (Jordan and Rumelhart (1991) discuss ways in which learning can be biased to find particular inverse models with desired properties).

We now consider the input–output structure of the network modelling the system inverse. From equation (74) the inverse function f^{-1} leading to the generation of $u(t)$ would require knowledge of the future value $y^p(t+1)$. To overcome this problem we replace this future value with the value $r(t+1)$ which we assume is available at time t. This is a reasonable assumption since r is typically related to the reference signal which is normally known one step ahead. Thus, the nonlinear input–output relation of the network modelling the plant inverse is

$$u(t) = \widehat{f^{-1}}[y^p(t), \ldots, y^p(t-n+1), r(t+1); u(t-1), \ldots, u(t-m+1)], \quad (77)$$

i.e. the inverse model network receives as inputs the current and past system outputs, the training (reference) signal, and the past values of the system input. In case where it is desirable to train the inverse without the real system (as discussed above) the values of y^p in the above relation are simply replaced by the forward model outputs y^m.

5.3. *Control structures*

Models of dynamic systems and their inverses have immediate utility for control. In the control literature a number of well-established and deeply analysed structures for the control of nonlinear systems exist; we focus on those structures having a direct reliance on system forward and inverse models. We assume that such models are available in the form of neural networks which have been trained using the techniques outlined above.

In the literature on neural network architectures for control a large number of control structures have been proposed and used; it is beyond the scope of this work to provide a full survey of all architectures used. In the sequel we give particular emphasis to those structures which, from the mainstream control theory viewpoint, are well-established and whose properties have been deeply analysed. First, we briefly discuss two direct approaches to control: supervised control and direct inverse control.

5.3.1. *Supervised control.* There are many control situations where a human provides the feedback control actions for a particular task and where it has proven difficult to design an automatic controller using standard control techniques (e.g. it may be impossible to obtain an analytical model of the controlled system). In some situations it may be desirable to design an automatic controller which mimics the action of the human (this has been called supervised control (Werbos (1990)).

A neural network provides one possibility for this (as an alternative approach expert systems can be used to provide the knowledge representation and control formalisms). Training the network is similar in principle to learning a system forward model as described above. In this case, however, the network input corresponds to the sensory input information received by the human. The network target outputs used for training correspond to the human control input to the system. This approach has been used in the standard pole-cart control problem (Grant and Zhang (1989)), among others.

5.3.2. *Direct inverse control.* Direct inverse control utilizes an inverse system model. The inverse model is simply cascaded with the controlled system in order that the composed system results in an identity mapping between desired response (i.e. the network inputs) and the controlled system output. Thus, the network acts directly as the controller in such a configuration. Direct inverse control is common in robotics applications; the compilation book (Miller *et al.* (1990)) provides a number of examples.

Clearly, this approach relies heavily on the fidelity of the inverse model used as the controller. For general purpose use serious questions arise regarding the robustness of direct inverse control. This lack of robustness can be attributed primarily to the absence of feedback. This problem can overcome to some extent by using on-line learning: the parameters of the inverse model can be adjusted on-line.

5.3.3. *Model reference control.* Here, the desired performance of the closed-loop system is specified through a stable reference model M, which is defined by its input–output pair $\{r(t), y^r(t)\}$. The control system attempts to make the plant output $y^p(t)$ match the reference model output asymptotically, i.e.

$$\lim_{t\to\infty} \|y^r(t) - y^p(t)\| \le \epsilon,$$

for some specified constant $\epsilon \ge 0$. The model reference control structure for nonlinear systems utilizing connectionist models is shown in Fig. 6 (Narendra and Parthasarathy (1990)). In this structure the error defined above is used to train the network acting as the controller. Clearly, this approach is related to the training of inverse plant models as outlined above. In the case when the reference model is the identity mapping the two approaches coincide. In general, the training procedure will force the controller to be a "detuned" inverse, in a sense defined by the reference model. Previous and similar work in this area has been considering linear in the control structures (Kosikov and Kurdyukov (1987)).

5.3.4. *Internal Model Control.* In Internal Model Control (IMC) the role of system forward and inverse models is emphasised (Garcia and Morari (1982)). In this structure a system forward and inverse model are used directly as elements within the feedback loop. IMC has been thoroughly examined and shown to yield transparently to robustness and stability analysis (Morari and Zafiriou (1989)). Moreover, IMC extends

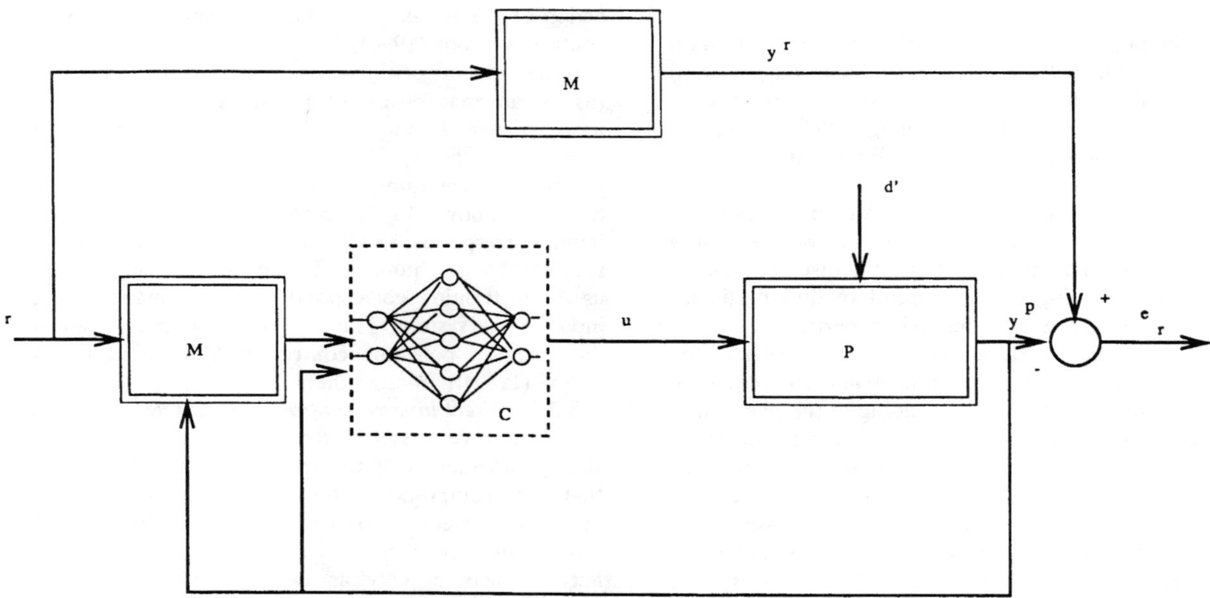

FIG. 6. Model reference structure.

readily to nonlinear systems control (Economou *et al.* (1986)).

In internal model control a system model is placed in parallel with the real system. The difference between the system and model outputs is used for feedback purposes. This feedback signal is then processed by a controller subsystem in the forward path; the properties of IMC dictate that this part of the controller should be related to the system inverse (the neural network realization of IMC is illustrated in Fig. 7).

Given network models for the system forward and inverse dynamics the realization of IMC using neural networks is straightforward (Hunt and Sbarbaro (1991)); the system model M and the controller C (the inverse model) are realized using the neural network models as shown in Fig. 7. The subsystem F is usually a linear filter which can be designed to introduce desirable robustness and tracking response to the closed-loop system.

It should be noted that the implementation structure of IMC is limited to open-loop stable systems. However, the technique has been widely applied in process control. From the control theoretic viewpoint there is strong support for the IMC approach. Examples of the use of neural networks for

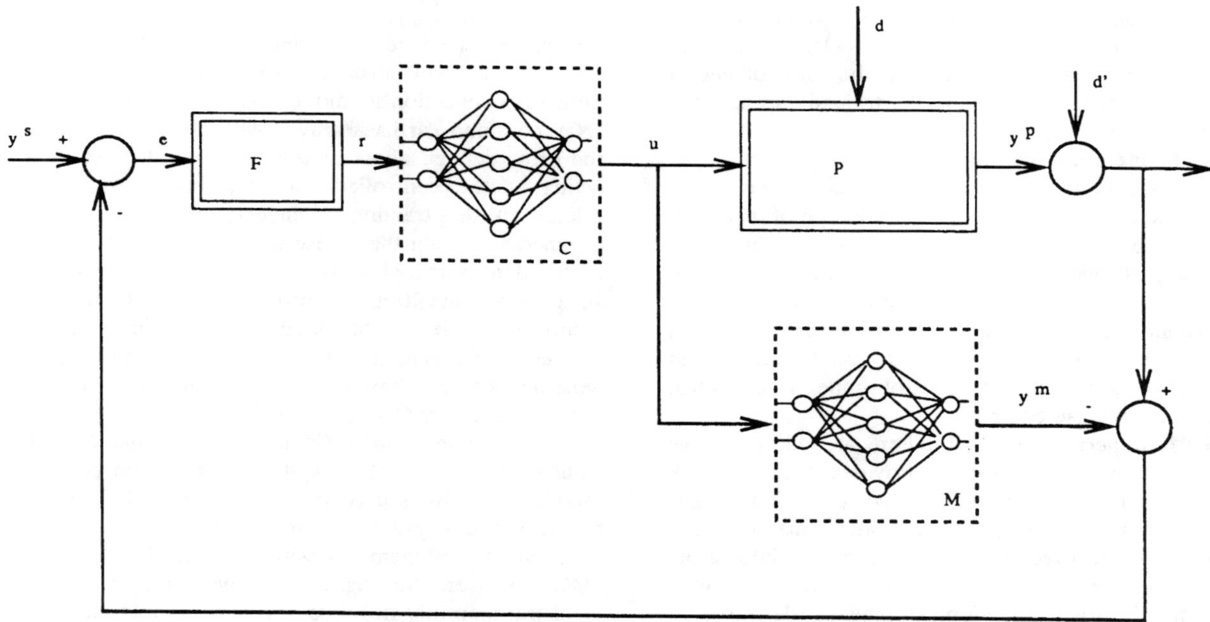

FIG. 7. Structure for internal model control.

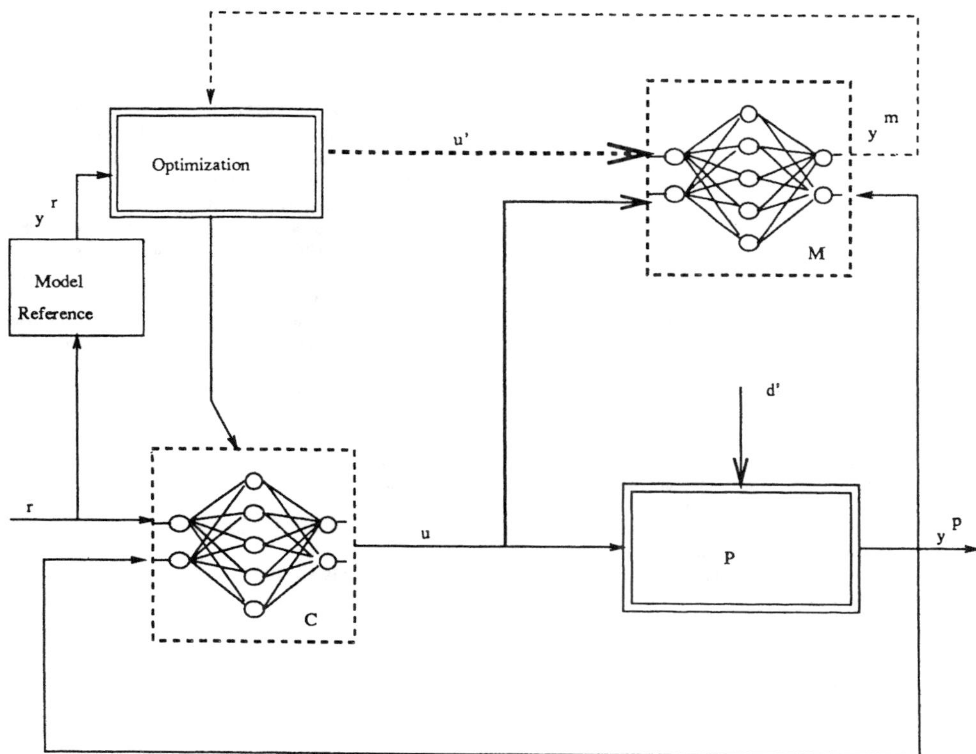

FIG. 8. Structure for predictive control.

nonlinear IMC can be found in Hunt and Sbarbaro (1991).

5.3.5. *Predictive control.* In the realm of optimal and predictive control methods the receding horizon technique has been introduced as a natural, computationally feasible feedback law. It has been proven that the method has desirable stability properties for nonlinear systems (Keerthi and Gilbert (1986); Mayne and Michalska (1990)).

In this approach a neural network model provides prediction of the future plant response over the specified horizon (see Fig. 8). The predictions supplied by the network are passed to a numerical optimization routine which attempts to minimize a specified performance criterion in the calculation of a suitable control signal.

The control signal u' is chosen to minimize the quadratic performance criterion

$$J = \sum_{j=N_1}^{N_2} (y^r(t+j) - y^m(t+j))^2 + \sum_{j=1}^{N_2} \lambda_j (u'(t+j-1) - u'(t+j-2))^2, \quad (78)$$

subject to the constraint of the dynamical model. Here, the constants N_1 and N_2 define the horizons over which the tracking error and control increments are considered. The values of λ are the control weights.

One further possibility, as illustrated in Fig. 8, is to train a further network to mimic the action of the optimization routine. This controller network C is trained to produce the same control output u, for a given plant output, as the optimization routine (u'). An advantage of this approach is that the outer loop consisting of plant model and optimization routine is no longer needed when training is complete.

6. OTHER APPLICATIONS

In the preceding section we described the use of neural networks in nonlinear control structures which are well established in the control theory literature. In this section we review alternative approaches to nonlinear control from the connectionist standpoint. We also take a more general look at system identification using networks and briefly discuss prediction and filtering.

A resume of the control applications of neural networks can be found in Table 4. Table 5 gives an idea of the actual architectures and their use in control. Below, we give a more detailed explanation of some of the possibilities.

6.1. *System identification*

The identification of plant forward and inverse models was discussed in the previous section. These models were developed with a view to utilizing them in a model based control strategy. In this section we take a more general look at identification using neural networks.

The first step in solving a system identification problem is to select an appropriate class of model structures. The connectionist paradigm provides a set of structures to represent nonlinear systems. In general, however, it is possible that a complete description of the physical system is not feasible, or is

TABLE 4. SUMMARY OF APPLICATIONS

Application	Architecture	Learning	References
Adaptive control using generic index	BP	Reinforcement learning	(Barto (1988); Barto *et al.* (1983); Helferty *et al.* (1989); Anderson (1989); Porcino and Collins (1990); Guhua and Mathur (1990); Chen (1990); Franklin (1989); Lee and Berenji (1989))
Adaptive filtering and prediction	BP	BP learning alg.	(Lapedes and Farber (1987); Casdagli (1989); Chen *et al.* (1990b); Ydstie (1990); Weigand *et al.* (1990))
Adaptive nonlinear control	BP	BP learning alg.	(Goldberg and Pearlmutter (1988); Yeung and Beckey (1989); Bassi and Beckey (1989); Josin (1988); Sekiguchi *et al.* (1989); Chen and Pao (1989); Psaltis *et al.* (1988); Guez and Selinsky (1990); Josin (1990); Grant and Zhang (1989); Sanner and Akin (1990); Chen (1989); Karsai *et al.* (1989); Troudet and Merril (1989); Ciliz and Isik (1989); Bhat and McAvoy (1990); Ydstie (1990); Ungar *et al.* (1990); Savic and Tan (1989); Bhat *et al.* (1990))
	CMAC	Delta rule	(Atkenson and Reinkensmeyer (1989); Ersü and Tolle (1984); Kraft and Campagna (1990); Miller *et al.* (1987))
	Kohonen and linear system	Kohonen alg. and delta rule	(Graf and Lalonde (1988); Martinez *et al.* (1988); Ritter and Schulten (1986, 1988))
	CPN	Kohonen alg. and delta rule	(Cheok and Huang (1989))
Optimal control	Kohonen and linear system	Kohonen alg. and delta rule	(Fu (1970))
Nolinear modelling	BP	BP learning alg.	(Bhat and McAvoy (1990); Bhat *et al.* (1990))
Adaptive linear control	Hopfield		(Chi *et al.* (1990))
	ART II		(Kumar and Guez (1990))
	Hopfield		(Żak (1990))
Gain scheduling	Hopfield		(Guez *et al.* (1988))

not practical, and a compromise has to be made between the model complexity and its adequacy for a given application (Goodwin and Sin (1984)). The questions of overparametrization and identifiability are discussed in Section 7.1.2.

The connectionist approach can be located in a corner of a triangle (Bassi (1990)) where the different approaches to the modelling of nonlinear systems are shown (Fig. 9). Each architecture oriented to nonlinear representation can be located in some part of this triangle. It should be noted, however, that the connectionist approach is not confined to the corner of the triangle shown; in fact, neural networks can be used as the implementation mechanism for the other approaches.

The lookup table (tabular) approach is based on a position of a table that contains the value of the function. To address this position an algorithm to detect features in the inputs signals can be used (i.e. LVQ). Some authors (Ritter and Schulten (1986, 1988)) have refined this mechanism allowing each entry be a linear model. A more general approach is the use of overlapping regions instead of disjoint regions. This approach gives the possibility of generalization around a neighbourhood. CMAC and CPN provide an alternative to this approach. Other methods that have been used widely include Potential Functions (Aizerman *et al.* (1964)) and, related with this, Gaussian networks. In the Gaussian case the functions depend upon the distance from a determined point in the space. In this way an interpolation capability is achieved.

In the algebraic representation, on the other hand, there is a set of basic nonlinear functions that forms a base for the model. In some cases it is necessary to use powers of the inputs and their combinations (Billings (1985); Giles and Maxwell (1987)). In other cases it is possible to include some knowledge of the

TABLE 5. CROSS-REFERENCE TABLE

Applications	BP	CMAC	Kohonen	ART II	Hopfield
	Architectures				
Adaptive control using generic index	X				
Adaptive filtering and prediction	X				
Adaptive nonlinear control	X	X	X		
Optimal control			X		
Nonlinear modelling	X				
Estimation for time varying systems					X
Adaptive linear control				X	X
Gain scheduling	X				

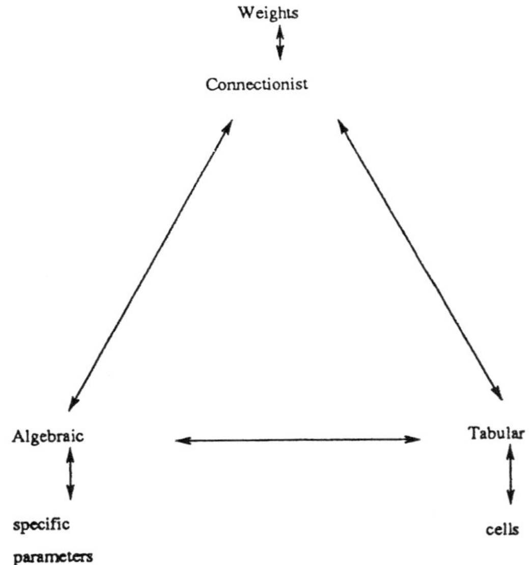

FIG. 9. Connectionist approach in the representation framework.

plant (Kawato *et al.* (1988); Johanson (1990)). In all these cases the nonlinear system is transformed to a linear in the parameters model.

Finally, looking at the third corner of our triangle, the connectionist approach has the advantage that it uses a type of generic nonlinearity and allows all the parameters to be adjusted. In this way it can deal with a very wide range of nonlinearities. Techniques for estimation of the network parameters vary according to the particular architecture used. However, the general structure, which is shown in Fig. 4, can be regarded as invariant.

All these architectures are used in temporal processing provided that the input is a temporal window with data. As has been mentioned in Section 4, however, there are some drawbacks with this approach. The extension to temporal learning in control is an open field to be addressed.

6.2. *Optimal decision control*

In this application, the state space is partitioned into regions (feature space) corresponding to various control situations (pattern classes). The realization of the control surface is accomplished through a training procedure. Since the time-optimal surface in general is nonlinear it is necessary to use an architecture capable to approximating a nonlinear surface. One possibility is to first quantize the state space into elementary hypercubes in which the control action is assumed constant. This process can be carried out with a LVQ architecture. It is then necessary to have another network acting as a classifier. If continuous signals are required a standard back-propagation architecture can be used (see Fig. 10).

The switching surface is not known *a priori* but is defined implicitly by a training set of points in state space whose optimal control action is known. During the training process the learning algorithm makes changes in its weights based only on the training pattern vector presently being shown to it, together with the desired control situation. The training pattern vectors are presented to the controller sequentially several times until all the pattern vectors in the training set are being correctly classified, or until the number of classification errors has reached some steady-state value (Fu (1970)).

6.3. *Adaptive linear control*

The connectionist approach can be used in nonlinear control but also as a part of a controller for linear plants.

The Hopfield network may be utilized as a dynamical controller (Żak (1990)) for a linear system. In this case, elements of variable structure theory were utilized to construct the controller, and as a result the proposed controller is characterized by its robustness. It is possible to include adaptation using some standard method (Demircioglu and Gawthrop (1988)). The general structure is shown in Fig. 11.

The Hopfield network can also be used as a part of the adaptation mechanism (Chi *et al.* (1990)). In this case the network is included in the adaptation path and is used to minimize simultaneously square-error rates of all the states. Here, the output of the network represents the parameters of a linear model for the plant. This approach has been shown to be useful for implementation of a least-squares estimation for time-varying and time-invariant systems.

6.4. *Adaptive control using a generic index (reinforcement learning control)*

This kind of network makes use of a low quality feedback signal. While the performance measure for a

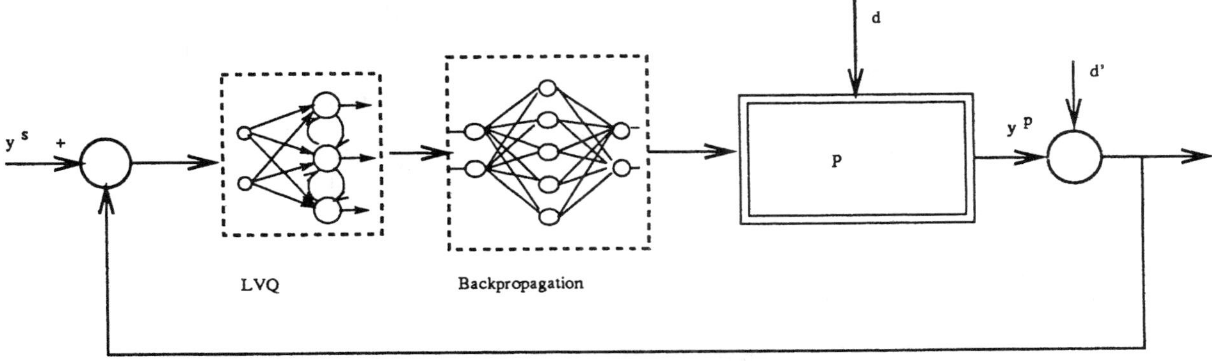

FIG. 10. Structure for optimal control.

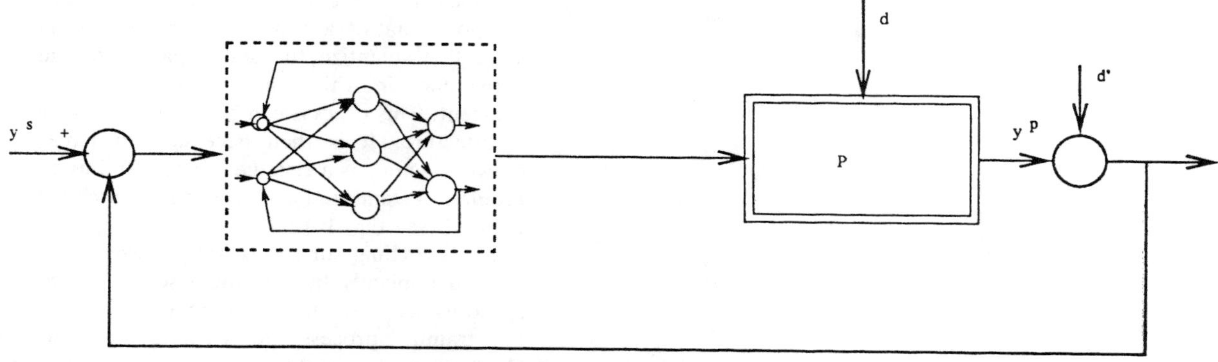

FIG. 11. Structure for feedback networks as controllers.

supervised system is defined in terms of a set of targets by means of a known error, reinforcement learning addresses the problem using any measure whose value can be supplied to the learning algorithm.

Instead of trying to determine target controller outputs from target plant responses, one tries to determine target controller outputs that would lead to an increase in a measure of plant performance (a measure not necessarily defined in terms of target plant responses). The "critic" block is capable of evaluating the plant performance and generates an "evaluation signal" which is used by the reinforcement learning algorithm. This approach is appropriate when there is a genuine lack of knowledge required for applying more specialized learning methods. An example is the pole-cart problem where in a reinforcement scheme the evaluation is zero throughout a trial and becomes -1 when a failure occurs. In the more knowledge intensive controller a reference trajectory is necessary to control the plant.

A comprehensive discussion of reinforcement learning control and a list of references can be found in Barto (1990).

6.5. *Gain scheduling*

This is an old control engineering technique which uses process variable related with its dynamics to compensate the effect of working in different operating regions. Here, the neural network is used as an associative memory that relates the parameters of the controller with the state of the plant. The Hopfield network has been used in this case (Guez *et al.* (1988)) where its output corresponds to the parameters of the controller. Thus, the stable-state topology can be designed so that the local minima correspond to optimal control laws for the parameters of the controller. Other connectionist architectures, such as those based on receptive fields (for example, CMAC—see Section 2.5) can be used to obtain the same results.

6.6. *Filtering and prediction*

Filtering is concerned with the extraction of signals from noise. In this case the connectionist approach is used to produce a nonlinear filter such that the effect of the noise is cancelled. In the past the general least-mean-square approach has been extended on the basis of a truncated Wiener–Volterra series representation. This series is a general representation for a large class of nonlinear systems, and has the advantage that the output is linear with respect to the filter weights. This approach suffers from the drawback that the computation and implementation cost increases sharply with the nonlinearity. Thus, it has been used only in weak-nonlinear cases (Lin and Unbehauen (1990)). In this case the connectionist approach has the advantage of saving computation and implementation costs, especially in strongly-nonlinear control. In the special case that the model of the noise is a chaotic system in a linear process it is possible to find an expression for the equivalent linear minimum variance filter (Ydstie (1990)). For a presentation of some non-ANN approaches to the control of chaotic systems the reader is directed to Ditto *et al.* (1990) and Ott *et al.* (1990).

Prediction is defined as the problem of extrapolating a given time series into the future. As for filtering, if the underlying model of the time series is known it is possible in principle to design an optimal predictor for future values. The design is based in this case on connectionist models. Prediction in the context of receding horizon control was discussed in the preceding section.

As mentioned in Lapedes and Farber (1987) a competing method for processing nonlinear signals is the algebraic approach using polynomials (see also Farmer and Sidorowich (1987)). This has the advantage that polynomial nonlinearities can be modelled exactly. On the other hand, non-polynomial nonlinearities must also be modelled by polynomials. This can lead to a rapid oscillation of the polynomials, and as the system size increases there is an explosion in the number of polynomials. The reason that connectionist networks seem to work well can be related to the fact that the network has the capability to perform a kind of generalized mode decomposition of underlying maps. That is, the parameters of the basis function are adjusted to change completely their input/output relations.

A discussion of filtering and prediction using neural networks can be found in the paper by Widrow and Winter (1988).

7. OPEN RESEARCH TOPICS

The field of neural networks in control is proceeding, as are most other emergent scientific disciplines, through a mix of empirical and theoretical studies. Not surprisingly, thus far more effort has been concentrated on the former aspect. On the theoretical side progress has been limited; a large number of crucial and highly challenging questions remain open. In this section we make an initial assessment of the major research topics.

Much insight can be gained from adaptive control theory (Åström and Wittenmark (1989)). Twenty years ago this area was characterized by a large body of experimental work which showed great promise. This initial flurry has been followed by a consolidation period where some of the major theoretical questions were formulated and solved; the theory of adaptive systems is now relatively mature. A strong interplay between theory and practice has also been evident in the development of adaptive control.

Adaptive control theory is generally based on the assumption of linear time-invariant plant. Even though the adaptation process results in the overall system being nonlinear the cornerstone of this work lies in linear systems theory (Narenda and Annaswamy (1989)). Many results additionally require a significant level of prior knowledge about the plant.

It is evident, on the other hand, that the primary focus of neural networks will be on nonlinear control problems. The strong analogy between traditional adaptive control and neural network based control indicates that this new area will develop along similar lines taken by adaptive control theory and that similar theoretical difficulties will emerge. Thus, the established results and techniques of linear adaptive control may well provide insight for the case where system components are nonlinear. It is also reasonable to conjecture that the problems encountered will be more complex due to the nonlinearity of the subsystems involved. In order to have a well-developed theory of control using neural networks many fundamental questions will have to be addressed.

7.1. *Systems theory*

A rigorous characterization of theoretical properties such as stability, controllability and observability should be developed.

7.1.1. *Stability*. The crucial question of stability of adaptive systems has generated a large literature. Lyapunov methods and hyperstability theory have been applied to prove the stability of adaptive systems where the underlying plant is assumed linear and time-invariant. Linearization techniques also play an important role.

For adaptive systems composed of nonlinear plants and neural networks new concepts and techniques for stability theory will need to be explored. The results obtained in the context of dynamic networks are far from satisfactory (see Sections 2.4.1 and 2.4.2). Stability issues should also be elaborated for control-orientated neural network structures.

7.1.2. *Persistent excitation and convergence*. The importance of the signals used to train a network has already been emphasised. The training data must sufficiently cover the range of operation signals. This will help to ensure that the network will operate in the desired way when presented with inputs not included in the training set. The issue of how well the model output approximates that of the plant after training is known as generalization. A related concept in the adaptive control and system identification literature is persistent excitation. Having persistently exciting training data ensures convergence of the model to the "true" plant.

A related concept is that of system identifiability: can the system under study be adequately represented within the chosen model structure? For example, it is well known in adaptive control that using an overparametrized model can lead to convergence problems. This is indicative of a trade-off between identifiability and generalization: using more parameters to get a better approximation makes it more difficult to actually estimate (from a given training set) the values of parameters that give good generalization.

A rigorous theory covering these issues needs to be developed in the context of neural network models.

7.1.3. *Robustness*. As already mentioned, we believe that results from the field of adaptive control have a role to play in the analysis and design of Neural Networks. Perhaps the most promising category of such results relates to robustness: that is, the study of when properties such as stability and convergence are retained in the presence of system dynamics which cannot be completely modelled by the adaptive controller (Rohrs *et al.* (1985); Kosut and Friedlander (1985); Kosut and Johnson (1984); Anderson *et al.* (1986); Gawthrop (1987); Cluett *et al.* (1988); Sastry and Bodson (1989); Åström (1989); Gawthrop (1990)).

7.2. *Network theory*

In a previous section a body of existence theorems relating to the approximation capabilities of neural networks was described. What is now needed is the development of constructive procedures for design and development of neural networks. The ultimate aim is to achieve a systematic engineering procedure which will guide the selection of a network architecture for a specific problem.

7.2.1. *Suitable applications*. It is desirable to develop a characterization of the kinds of real problems for which neural networks are suitable, and for which they can be expected to improve on the performance of "traditional" control techniques in a meaningful way. Neural networks will be aimed primarily at nonlinear systems. It is important to classify the nonlinear systems for which current nonlinear control techniques have been developed, and hence to identify those classes of nonlinear systems at which neural networks should be aimed.

7.2.2. *Network architecture*. A relationship needs to be established between the system under study (i.e. the nonlinear function to be approximated) and the structure of a suitable network. By structure here we are referring to characteristics such as number of layers, number of nodes, and type of nonlinear activation function. A link also needs to be established between these network parameters and the required approximation accuracy.

The implications of using finite numbers of layers/nodes for function approximation also need examined. This raises questions of robustness of the underlying control strategy to modelling error.

7.2.3. *Structural learning.* Structural learning algorithms aim to empirically determine the network structure in terms of numbers of nodes and layers. Such techniques are complementary to the theoretical (or analytical) approach described above.

7.2.4. *Dynamic systems.* In preceding sections we presented one possibility for using static networks to represent dynamic systems. The use of dynamic recurrent networks to represent dynamic systems will we believe become increasingly important. This raises questions of approximation theory and learning algorithms for dynamical networks.

8. CONCLUSIONS

This paper has provided a survey of artificial neural networks in the realm of modelling, identification and control of nonlinear systems. The basic ideas and techniques of artificial neural networks have been presented in language and notation familiar to control engineers. Applications of a variety of neural network architectures in control were surveyed. We explored the links between the fields of control science and neural networks in a unified presentation and identified key areas for future research.

This survey is aimed at researchers currently working in the area of control systems, by presenting the basic ideas and techniques of artificial neural networks in language and notation familiar to control engineers we hope to have made these techniques more readily accessible to control engineers. Control systems insights can be applied to artificial neural network techniques.

Acknowledgements—Kenneth Hunt currently holds a Royal Society of Edinburgh Personal Research Fellowship. Daniel Sbarbaro is supported by a Chilean government research scholarship. Rafał Żbikowski is supported under SERC research grant GR/G/53569, *Neural Networks for Modelling and Control of Nonlinear Dynamic Systems.* He previously worked on this paper while on leave from Warsaw University of Technology; support for his visit was furnished by the TEMPUS programme of the European Community (grant number IMG-PLS-0630-90). Peter Gawthrop is Wylie Professor of Mechanical Engineering.

The authors would finally like to express their gratitude to the reviewers of this paper who put considerable effort into their task and made many deep and constructive comments.

REFERENCES

Aarts, E. and J. Korst (1989). *Simulated Annealing and Boltzmann Machines.* Wiley, New York.

Åström, K. J. (1989). Application of the robust and adaptive pole placement design technique *Int. J. of Adaptive Control and Signal Processing,* **3,** 167–189.

Åström, K. J. and B. Wittenmark (1989). *Adaptive Control.* Addison-Wesley, New York.

Ackley, D. H., G. H. Hinton and T. J. Sejnowski (1985). A learning algorithm for Boltzmann machines. *Cognitive Science,* **9,** 147–169.

Aizerman, M. A., E. M. Braverman and L. I. Rozonoer (1964). Theoretical foundation of the potential function method in pattern recognition learning. *Automatika i Telemekhanika,* **25,** 1917–1936.

Albert, A. E. and L. A. Gardner (1967). *Stochastic Approximation and Nonlinear Regression.* MIT Press, Cambridge, MA.

Albus, J. S. (1975a). Data storage in the cerebellar model articulation controller (CMAC). *Trans. of the ASME, J. of Dynamic Systems, Measurement, and Control,* **97,** 228–233.

Albus, J. S. (1975b). A new approach to manipulator control: The cerebellar model controller (CMAC). *Trans. of the ASME, J. of Dynamic Systems, Measurement, and Control,* **97,** 220–227.

Almeida, L. B. (1988). Backpropagation in perceptrons with feedback. In R. Eckmiller and Ch. v. d. Malsburg (Eds), *Neural Computers.* Springer-Verlag, Berlin.

Anderson, B. D. O., R. R. Bitmead, C. R. Johnson, P. V. Kokotovic, R. L. Kosut, I. M. Y. Mareels, L. Praly and B. D. Reidle (1986). *Stability of Adapative Systems: Passivity and Averaging Analysis.* MIT Press, Cambridge, MA.

Anderson, C. W. (1989). Learning to control an inverted pendulum using neural networks. *IEEE Control Systems Magazine,* **9,** 31–37.

Anderson, J. A. and E. Rosenfeld. (1988). *Neurocomputing: Foundations of Research.* MIT Press, Cambridge, MA.

Atkenson, C. G. and D. J. Reinkensmeyer. (1989). Using associative content-addressable memories to control robots. In *Proc. of* 1989 *IEEE Int. Conf. on Robotics and Automation,* Scottsdale, Arizona, pp. 1859–1864.

Ballard, D. H. (1988). Cortical connections and parallel processing: Structure and function. In M. Arbib and Hamson (Eds), *Vision, Brain, and Cooperative Computation,* pp. 563–621. MIT Press, Cambridge, MA.

Barron, A. R. (1991). Approximation and estimation bounds for artificial neural networks. In *Proc. 4th Annual Workshop on Computational Learning Theory,* University of California Santa Cruz, U.S.A., pp. 243–249.

Barto, A. G. (1988). An approach to learning control surface by connectionist systems. In M. Arbib and Hanson (Eds), *Vision, Brain and Cooperative Computation,* pp. 665–701. MIT Press, Cambridge, MA.

Barto, A. G. (1990). *Neural Networks for Control,* Chapter 1, pp. 5–58. MIT Press, Cambridge, MA.

Barto, A. G., R. S. Sutton and Ch. W. Anderson (1983). Neuronlike adaptive elements that can solve difficult learning control problems. *IEEE Trans. on System, Man, and Cybernetics,* **13,** 834–846.

Bassi, D. F. (1990). Connectionist dynamic control of robotic manipulators. D. Phil. thesis, University of Southern California.

Bassi, D. F. and G. A. Beckey (1989). Decomposition of neural network model of robot dynamics: a feasibility study. *Simulation and AI,* **220,** 8–13.

Bavarian, B. (1988). Introduction to neural networks for intelligent control. *IEEE Control Systems Magazine,* **8,** 3–7.

Bhat, N. V. and T. J. McAvoy (1990). Use of neural nets for dynamical modelling and control of chemical process systems. *Computers Chem. Engng.,* **14,** 573–5583.

Bhat, N. V., P. A. Minderman, T. McAvoy and N. S. Wang (1990). Modeling chemical process systems via neural computation. *IEEE Control Systems Magazine,* **10,** 24–29.

Billings, S. A. (1985). Introduction to nonlinear system analysis and identification. In K. Godfrey and P. Jones (Eds), *Signal Processing for Control.* Springer-Verlag, Berlin.

Broomhead, D. S. and D. Lowe (1988). Multivariable functional interpolation and adaptive networks. *Complex Systems,* **2,** 321–355.

Bruck, J. (1989) Computing with Networks of Threshold Elements. D. Phil. Thesis, Stanford University.

Bullock, D. (1983). How neural networks factor problems of

sensory motor control. In *Proc. American Control Conf.*, pp. 2271–2275.

Burkill, J. C. and H. Burkill (1970). *A Second Course in Mathematical Analysis*. Cambridge University Press, Cambridge, U.K.

Carrol, M. S. and B. W. Dickinson (1989). Construction of neural nets using the Radon transform. In *Proc. IJCNN*.

Casdagli, M. (1989). Non-linear system prediction of chaotic time series. *Physica D*, **35**, 335–356.

Chen, Fu-Chuang (1989). Back-propagation neural network for nonlinear self-tuning adaptive control. In *IEEE Int. Symposium on Intelligent Control* 1989, pp. 274–279.

Chen, S., S. A. Billings, C. F. Cowan and P. M. Grant (1990a). Practical identification of NARMAX models using radial basis functions. *Int. J. Control*, **52**, 1327–1350.

Chen, S., S. A. Billings and P. M. Grant (1990b). Non-linear system identification using neural networks. *Int. J. Control*, **51**, 1191–1214.

Chen, S., C. F. N. Cowan, S. A. Billings and P. M. Grant (1990c). Parallel recursive prediction error algorithm for training layered neural networks. *Int. J. Control*, **51**, 1215–1228.

Chen, V. C. and Y. H. Pao (1989). Learning control with neural networks. In *Proc. of* 1989 *IEEE Int. Conf. on Robotics and Automation*, Scottsdale, AZ. pp. 1448–1453.

Chen, V. C. (1990). Problem-solving by using reinforcement learning neural nets. In *IEEE Int. Joint Conf. on Neural Networks, IJCNN*'90, pp. 583–586.

Cheok, K. C. and N. J. Huang (1989). Lyapunov stability analysis for self-learning neural model with application to semiactive suspension control system. In *IEEE Int. Symposium on Intelligent Control* 1989, pp. 329–331.

Chester, D. (1990). Why two hidden layers are better than one. In *IEEE Int. Joint Conf. on Neural Networks, IJCNN*'90, pp. 265–268.

Chi, S. R., R. Shoureshi and M. Tenorio (1990). Neural networks for system identification. *IEEE Control Systems Magazine*, **10**, 31–34.

Chua, L. O. and L. Yang (1988a). Cellular neural networks: Applications. *IEEE Trans. on Circuits and Systems*, **35**, 1273–1290.

Chua, L. O. and L. Yang (1988b). Cellular neural networks: Theory. *IEEE Trans. on Circuits and Systems*, **35**, 1257–1272.

Ciliz, M. K. and C. Isik (1989). Time optimal control of mobile robot motion using neural nets. In *IEEE Int. Symposium on Intelligent Control* 1989, pp. 368–373.

Cluett, W. R., S. L. Shah and D. G. Fisher (1988). Robustness analysis of discrete-time adaptive control systems using input–output stability theory: A tutorial. *IEE Proc.*, **135**, 133–141.

Cohen, M. A. and S. Grossberg (1983). Absolute stability of global pattern formation and parallel memory storage by competitive neural networks. *IEEE Trans. on System, Man, and Cybernetics*, **13**, 815–826.

Cybenko, G. (1988). Continuous valued networks with two hidden layers are sufficient. Technical Report, Department of Computer Science, Tufts University.

Cybenko, G. (1989). Approximation by superpositions of a sigmoidal function. *Math. Control Signal Systems*, **2**, 303–314.

Daunicht, W. J. (1990). Defanet—a deterministic approach to function approximation by neural networks. In *IEEE Int. Joint Conf. on Neural Networks, IJCNN*'90, pp. 161–164.

Demircioglu, H. and P. J. Gawthrop (1988). Continuous-time relay self-tuning control. *Int. J. Control*, **47**, 1061–1080.

Ditto, W. L., S. N. Rauseo and M. L. Spano (1990). Experimental control of chaos. *Physical Review Letters*, **65**, 3211.

Dzieliński, A., S. Skoneczny, R. Żbikowski and S. Kukliński (1990). Cellular neural network application to moiré pattern filtering. In *Proc. IEEE CNNA'90 Workshop*, Budapest, Hungary.

Economou, C. G., M. Morari and B. O. Palsson (1986). Internal model control. 5. Extension to nonlinear systems. *Ind. Eng. Chem. Process Des. Dev.*, **25**, 403–411.

Ersü, E. and H. Tolle (1984). A new concept for learning control inspired by brain theory. *Proc. 9th World Congress of IFAC*, **7**, 245–250.

Eykhoff, P. (1974). *System Identification*. Wiley, New York.

Fang, Y. and T. J. Sejnowski (1990). Faster learning for dynamic recurrent backpropagation. *Neural Computation*, **2**, 270–273.

Farmer, D. and J. Sidorowich (1987). Predicting chaotic time series. *Physical Review Letters*, **59**, 845.

Fletcher, R. (1987). *Practical Methods of Optimization*, 2nd ed. Wiley, Chichester.

Franklin, J. A. (1989). Input space representation for refinement learning control. In *IEEE Int. Symposium on Intelligent Control* 1989, pp. 115–122.

Fu, K. S. (1970). Learning control systems—review and outlook. *Trans. IEEE on Aut. Control*, **16**, 210–221.

Funahashi, K. I. (1989). On the approximate realization of continuous mappings by neural networks. *Neural Networks*, **2**, 183–192.

Garcia, C. E. and M. Morari (1982). Internal model control—1. A unifying review and some new results. *Ind. Eng. Chem. Process Des. Dev.*, **21**, 308–323.

Gawthrop, P. J. (1987). Robust stability of a continuous-time self-tuning controller. *Int. J. of Adaptive Control and Signal Processing*, **1**, 31–48.

Gawthrop, P. J. (1990). Robust stability of multi-loop continuous-time self-tuning controllers. *Int. J. of Adaptive Control and Signal Processing*, **4**, 359–382.

Gawthrop, P. J. and D. G. Sbarbaro (1990). Stochastic approximation and multilayer perceptrons: The gain back-propagation algorithm. *Complex System J.*, **4**, 51–74.

Giles, C. L. and T. Maxwell (1987). Learning, invariance, and generalization in high-order neural networks. *Applied Optics*, **26**, 4972–4978.

Girosi, F. and T. Poggio (1989). Representation properties of networks: Kolmogorov's theorem is irrelevant. *Neural Computation*, **1**, 465–469.

Girosi, F. and T. Poggio (1990). Networks and the best approximation property. *Biological Cybernetics*, **63**, 169–176.

Goldberg, K. and B. A. Pearlmutter (1988). Using a neural network to learn the dynamic of the CMU direct-drive arm II. Report CMU-CS-88-160, Carnegie-Mellon University, 1988.

Goodwin, G. C. and K. S. Sin (1984). *Adaptive Filtering Prediction and Control*. Prentice-Hall, Englewood Cliffs, NJ.

Graf, D. H. and W. R. Lalonde (1988). A neural controller for collision-free movement of general robot manipulators. In *IEEE Int. Joint Conf. on Neural Networks, IJCNN*'88, pp. 77–84.

Grant, E. and B. Zhang (1989). A neural net approach to supervised learning of pole balancing. In *IEEE Int. Symposium on Intelligent Control* 1989, pp. 123–129.

Grossberg, S. (1976). Adaptive pattern classification and universal recording: I. parallel development and coding of neural feature detectors. *Biological Cybernetics*, **23**, 121–134.

Grossberg, S. (1988). *Neural Networks and Natural Intelligence*. The MIT Press, Cambridge, MA.

Guez, A., J. L. Elibert and M. Kam (1988). Neural network architecture for control. *IEEE Control Systems Magazine*, **8**, 22–25.

Guez, A. and J. W. Selinsky (1990). A neurocontroller with guaranteed performance for rigid robots. In *IEEE Int. Joint Conf. on Neural Networks, IJCNN*'90, pp. 347–350.

Guhua, A. and A. Mathur (1990). Setpoint control based on reinforcement learning. In *IEEE Int. Joint Conf. on Neural Networks, IJCNN*'90, pp. 511–514.

Haber, R. and Unbehauen, H. (1990). Structure identification of nonlinear dynamic systems—A survey on input/output approaches. *Automatica*, **26**, 651–677.

Hartman, E. and J. D. Keeler (1992). Predicting the future with semi-local units. *Neural Computation*. To appear.

Hebb, D. O. (1949). *The Organization of Behaviour*. Wiley, New York.

Hecht-Nielsen, R. (1987a). Counterpropagation networks. *Applied Optics*, **26**, 4979–4984.
Hecht-Nielsen, R. (1987b). Kolmogorov's mapping neural network existence theorem. In *Proc. IJCNN.*
Hecht-Nielsen, R. (1988). Neurocomputer applications. In R. Eckmiller and Ch. v. d. Malsburg (Eds), *Neural Computers*, pp. 445–453. Springer-Verlag, Berlin.
Helferty, J. J., J. B. Collins and M. Kam (1989). A neural network learning strategy for the control of a one-legged hopping machine. In *Proc. of* 1989 *IEEE Int. Conf. on Robotics and Automation*, Scottsdale, AZ, pp. 1604–1609.
Hopfield, J. J. (1982). Neural networks and physical systems with emergent collective computational abilities. *Proc. of the National Academy of Sciences*, **79**, 2554–2558.
Hopfield, J. J. (1984). Neurons with graded response have collective computational properties like those of two-state neurons. *Proc. of the National Academy of Sciences*, **81**, 3088–3092.
Hornik, K., M. Stinchcombe and H. White (1989). Multilayer feedforward networks are universal approximators. *Neural Networks*, **2**, 359–366.
Hunt, K. J. and D. Sbarbaro (1991). Neural networks for non-linear internal model control. *Proc. IEE Pt. D*, **138**, 431–438.
IEEE (1988). Special issue on neural networks. *IEEE Control Systems Magazine*, **8**.
IEEE (1989). Special issue on neural networks. *IEEE Control Systems Magazine*, **9**.
IEEE (1990). Special issue on neural networks. *IEEE Control Systems Magazine*, **10**.
Johanson, R. (1990). Real-time identification of multilinear systems. In *Preprints of the 11th IFAC World Congress*, Tallin, Estonia, Vol. 3, pp. 119–123.
Johnson, J. L. (1991). Globally stable saturable learning laws. *Neural Networks*, **4**, 47–51.
Jordan, M. I. and D. E. Rumelhart (1991). Forward models: supervised learning with a distal teacher, Occasional Paper #40, Center for Cognitive Science, Massachusetts Institute of Technology (submitted to "Cognitive Science"); 49 pages.
Josin, G. (1988). Neural-space generalization of a topological transformation. *Biological Cybernetics*, **59**, 283–290.
Josin, G. M. (1990). Development of a neural network autopilot model for high performance aircraft. In *IEEE Int. Joint Conf. on Neural Networks, IJCNN'90*, pp. 547–550.
Karsai, G., K. Anderson, G. Cook and K. Ramaswamy (1989). Dynamic modelling and control of nonlinear process using neural network techniques. In *IEEE Int. Symposium on Intelligent Control* 1989, pp. 280–286.
Kawato, M., Y. Uno, M. Isobe and R. Suzuki (1988). Hierarchical neural network model for voluntary movement with application to robotics. *IEEE Control Systems Magazine*, **8**, 8–17.
Keerthi, S. S. and E. G. Gilbert (1986). Moving-horizon approximations for a general class of optimal nonlinear infinite-horizon discrete-time systems. In *Proc. 20th Annual Conf. Information Science and Systems*, Princeton University, pp. 301–306.
Kelly, D. G. (1990). Stability in contractive nonlinear neural networks. *IEEE Trans. on Biomedical Engineering*, **37**, 231–242.
Kohonen, T. (1987). *Self-Organization and Associative Memory*. Springer-Verlag, Berlin.
Kollias, S. and D. Anastassiou (1989). An adaptive least squares for efficient training of artificial neural networks. *IEEE Trans. on Circuits and Systems*, **36**, 1092–1101.
Kolmogorov, A. N. (1957). On the representation of continuous functions of several variables by superposition of continuous functions of one variable and addition. *Dokl. Akad. Nauk SSSR*, **114**, 953–956.
Kosikov, V. S. and A. P. Kurdyukov (1987). Design of a nonsearching self-adjusting system for nonlinear plant. *Automatika i Telemekhanika*, **4**, 58–65.
Kosut, R. L. and B. Friedlander (1985). Robust adaptive control: conditions for global stability. *IEEE Trans. on Aut. Control.* **AC-30**, 610–624.
Kosut, R. L. and C. R. Johnson (1984). An input–output view of robustness in adaptive control. *Automatica*, **20**, 569–582.
Kraft, L. G. and D. P. Campagna (1990). A comparison between CMAC neural network control and two traditional adaptive control systems. *IEEE Control Systems Magazine*, **10**, 36–43.
Kumar, S. and A. Guez (1990). Adaptive pole placement for neurocontrol. In *IEEE Int. Joint Conf. on Neural Networks, IJCNN'90*, pp. 397–400.
Kurkova, V. (1991). Kolmogorov's theorem is relevant. *Neural Computation*, **3**, 617–622.
Lane, S. H., D. A. Handelman and J. J. Gelfand (1991). Higher order CMAC neural networks—theory and practice. In *Proc. American Control Conf.*, Boston, MA. pp. 1579–1585.
Lapedes, A. and R. Farber (1987). Non-linear signal processing using neural networks: prediction and system modelling. Report LA-UR-87-2662, Los Alamos National Laboratory.
Lee, Chuen-Chien and H. R. Berenji (1989). An intelligent controller based on approximate reasoning and reinforcement learning. In *IEEE Int. Symposium on Intelligent Control* 1989, pp. 200–205.
Li, J. H., A. N. Michel and W. Porod (1988). Qualitative analysis and synthesis of a class of neural networks. *IEEE Trans. on Circuits and Systems*, **35**, 976–986.
Lin, J. and R. Unbehauen (1990). Adaptive nonlinear digital filter with canonical piecewise-linear structure. *IEEE Trans. on Circuits and Systems*, **37**, 347–353.
Lippmann, R. P. (1987). An introduction to computing with neural nets. *IEEE ASSP Magazine*, **4**, 4–22.
Ljung, L. (1987). *System Identification—Theory for the User*. Prentice Hall, Englewood Cliffs, NJ.
Ljung, L. and T. Söderström (1983). *Theory and Practice of Recursive Identification*. MIT Press, London.
Lorentz, G. G. (1976). *Mathematical Developments Arising from Hilbert's Problems*, Vol. 2, pp. 419–430. American Mathematical Society.
Martinez, T. M., H. Ritter and K. J. Schulten (1988). Three-dimensional neural net for learning visuomotor-coordination of a robot arm. In *IEEE Int. Joint Conf. on Neural Networks, IJCNN'88*, pp. 351–356.
Mayne, D. Q. and H. Michalska (1990). Receding horizon control of nonlinear systems. *Trans. IEEE on Aut. Control*, **35**, 814–824.
McCulloch, W. S. and W. Pitts (1943). A logical calculus of the ideas immanent in nervous activity. *Bulletin of Mathematical Biophysics*, **9**, 127–147.
Metropolis, N., A. Rosenbluth, M. Rosenbluth, A. Teller and E. Teller (1953). Equation of state calculations by fast computing machines. *J. Chemical Physics*, **21**, 1087–1092.
Michel, A. N., J. A. Farrell and W. Porod (1989). Qualitative analysis of neural networks. *IEEE Trans. on Circuits and Systems*, **36**, 229–243.
Miller, W. T., F. H. Glanz and L. G. Kraft (1987). Application of a general learning algorithm to the control of robotic manipulators. *The Int. J. of Robotic Research*, **6**, 84–98.
Miller, W. T., R. S. Sutton and P. J. Werbos (1990). *Neural Networks for Control*. MIT Press, Cambridge, MA.
Minsky, M. L. and S. A. Papert (1969). *Perceptrons*. The MIT Press, Cambridge, MA.
Minsky, M. L. and S. A. Papert (1988). *Perceptrons (expanded edition)*. The MIT Press, Cambridge, MA.
Moody, J. and C. Darken (1989). Fast-learning in networks of locally-tuned processing units. *Neural Computation*, **1**, 281–294.
Morari, M. and E. Zafiriou (1989). *Robust Process Control*. Prentice-Hall, Englewood Cliffs, NJ.
Narendra, K. S. (1990). *Neural Networks for Control*, Chapter 5, pp. 115–142. MIT Press. Cambridge, MA.
Narendra, K. S. and A. M. Annaswamy (1989). *Stable Adaptive Systems*. Prentice-Hall, Englewood Cliffs, NJ.
Narendra, K. S. and K. Parthasarathy (1990). Identification and control for dynamic systems using neural networks. *IEEE Trans. on Neural Networks*, **1**, 4–27.

Ott, E., C. Grebori and J. A. Yorke (1990). Controlling chaos. *Physical Review Letters*, **64**, 1196.

Park, J. and S. Lee (1990). Neural computation for collision-free path planning. In *Proc. of IEEE Int. Joint Conf. on Neural Networks, IJCNN'90*, Washington, DC.

Pearlmutter, B. A. (1988). Learning state space trajectories in recurrent neural networks. Report CMU-CS-88-191, Carnegie-Mellon University, 1988.

Pearlmutter, B. A. (1989). Learning state space trajectories in recurrent neural networks. *Neural Computation*, **1**, 263–269.

Pearlmutter, B. A. (1990). Dynamic recurrent neural networks. Technical Report CMU-CS-90-196, Carnegie Mellon University, School of Computer Science.

Pineda, F. J. (1987). Generalization of back-propagation to recurrent neural networks. *Physical Review Letters*, **59**, 2229–2232.

Pineda, F. J. (1989). Recurrent backpropagation and the dynamical approach to adaptive neural computation. *Neural Computation*, **1**, 161–172.

Poggio, T. (1982). *Physical and Biological Processing of Images*, pp. 128–153. Springer-Verlag, Berlin.

Poggio, T. and F. Girosi (1990). Networks for approximation and learning. *Proc. of IEEE*, **78**, 1481–1497.

Porcino, D. P. and J. C. Collins (1990). An application of neural networks to the guidance of free-swimming submersibles. In *IEEE Int. Joint Conf. on Neural Networks, IJCNN'90*, pp. 417–420.

Poteryaiko, I. Y. (1991). On Liapunov function for neural networks models with multi-unit interaction. In T. Kohonen, K. Mäkisara, O. Simula and J. Kangas (Eds), *Artificial Neural Networks*, pp. 21–23. Elsevier, North-Holland.

Psaltis, D., A. Sideris and A. A. Yamamura (1988). A multilayered neural network controller. *IEEE Control Systems Magazine*, **8**, 17–21.

Ritter, H. and K. Schulten (1986). Topology conserving mappings for learning motor tasks. In J. S. Denher (Ed.), *Neural Networks for Computing, Aip Conf. Proc.*, **151**, 376–380.

Ritter, H. and K. Schulten, (1988). Extending Kohonen's self-organizing mapping algorithm to learn ballistic movements. In R. Eckmiller and Ch. v. d. Malsburg (Eds), *Neural Computers*, pp. 393–406. Springer-Verlag, Berlin.

Robinson, A. J. (1989). Dynamic error propagation networks. D.Phil. thesis, Cambridge University, U.K.

Robinson, A. J. and F. Fallside (1987). Static and dynamic error propagation networks with application to speech coding. In D. Z. Anderson (Ed.) *Proc. of Neural Information Processing Systems*. American Institute of Physics, 1987.

Rohrs, C. E., L. S. Valavani, M. Athans and G. Stein (1985). Robustness of continuous-time adaptive control in the presence of unmodeled dynamics. *Trans. IEEE*, **AC-30**, 881–889.

Rosenblatt, F. (1958). The perceptron: a probabilistic model for information storage and organization in the brain. *Psychological Review*, **65**, 386–408.

Rudin, W. (1976). *Principles of Mathematical Analysis*, 3rd ed. McGraw-Hill, Auckland.

Rumelhart, D. E., G. E. Hinton and R. J. Williams (1986). Learning internal representations by error propagation. In D. E. Rumelhart and J. L. McClelland (Eds), *Parallel Distributed Processing*. MIT Press, Cambridge, MA.

Rumelhart, D. E. and J. L. McClelland (1986). *Parallel Distributed Processing: Explorations in the Microstructures of Cognition*, Vol. 1: *Foundations*. MIT Press, Cambridge, MA.

Sanner, R. M. and D. L. Akin (1990). Neuromorphic pitch attitude regulation of an underwater telerobot. *IEEE Control Systems Magazine*, **10**, 62–67.

Sastry, S. S. and Bodson (1989). *Adaptive Control: Stability, Convergence, and Robustness*. Prentice-Hall, Englewood Cliffs, NJ.

Sato, M (1990). Real time learning algorithm for recurrent analog neural networks. *Biological Cybernetics*, **62**, 237–241.

Savic, M. and Seow-Hwee Tan (1989). A new class of neural networks suitable for intelligent control. In *IEEE Int. Symposium on Intelligent Control* 1989, pp. 418–423.

Sbarbaro, D. G. and P. J. Gawthrop (1991). Self-organization and adaptation in Gaussian networks. In *9th IFAC/IFORS Symposium on Identification and System Parameter Estimation*, Budapest, Hungary, pp. 454–459.

Scalero, R. S. and N. Tepedelenlioglu (1990). A fast algorithm for neural networks. In *IEEE Int. Joint Conf. on Neural Networks, IJCNN'90*, pp. 77–84.

Sekiguchi, M., S. Nagata and K. Asakawa (1989). Behaviour control for mobile robot by multi-hierachical neural network. In *Proc. of 1989 IEEE Int. Conf. on Robotics and Automation*, Scottsdale, AZ, pp. 1578–1583.

Shamma, S. (1989). Spatial and temporal processing in central auditory networks. In Ch. Koch and I. Segev (Eds), *Methods in Neural Modeling*, pp. 247–289. The MIT Press, Cambridge, MA.

Söderström, T. and P. Stoica (1989). *System Identification*. Prentice-Hall, Hemel Hempstead, U.K.

Sudharsanan, S. I. and M. K. Sundareshan (1990). Equilibrium uniqueness and global exponential stability of a neural network for optimization applications. In *Proc. IEEE Int. Joint Conf. on Neural Networks, IJCNN'90*, Washington, DC.

Tan, S., L. Vandenberghe and J. Vandewalle (1990). Remarks on the stability of asymmetric dynamical neural networks. In *Proc. IEEE Int. Joint Conf. on Neural Networks, IJCNN'90*, San Diego, CA.

Troudet, T. and W. Merril (1989). Neoromorphic learning continuous valued mappings in presence of noise. In *IEEE Int. Symposium on Intelligent Control* 1989, pp. 312–319.

Tsypkin, Ya. Z (1971). *Adaptation and Learning in Automatic Systems*. Academic Press, New York.

Ungar, L. H., B. A. Powell and S. N. Kamens (1990). Adaptive networks for fault diagnosis and process control. *Computers Chem. Engng.*, **14**, 561–572.

Weigand, A. S., B. A. Huberman and D. E. Rumelhart (1990). Predicting the future: a connectionist approach. *Int. J. Neural Systems*, **3**, 193.

Werbos, P. J. (1974). Beyond regression: new tools for prediction and analysis in the behavior sciences. Ph.D. Thesis, Harvard University, Committee on Applied Mathematics.

Werbos, P. J. (1989). Maximizing long-term gas industry profits in two minutes in lotus using neural network methods. *IEEE Trans. on Systems, Man, and Cybernetics*, **19**, 315–333.

Werbos, P. J. (1990). Backpropagation through time: what it does and how to do it? *Proc. of IEEE*, **78**, 1550–1560.

Widrow, B. (1986). Adaptive inverse control. In *Preprints of the 2nd IFAC Workshop on Adaptive Systems in Control and Signal Processing*, Lund, Sweden, pp. 1–5.

Widrow, B. and M. E. Hoff (1960). Adaptive switching circuits. 1960 *IRE WESCON Convention Record, New York: IRE*, pp. 96–104.

Widrow, B. and M. A. Lehr (1990). 30 years of adaptive neural networks: Perceptron, medaline, and backpropagation. *Proc. of the IEEE*, **78**, 1415–1441.

Widrow, B. and S. D. Stearns (1985). *Adaptive Signal Processing*. Prentice-Hall, Englewood Cliffs, NJ.

Widrow, B. and R. Winter (1988). Neural nets for adaptive filtering and adaptive pattern recognition. *IEEE Computer*, **21**, 25–39.

Wiener, N. (1948). *Cybernetics: or Control and Communication in the Animal and the Machine*. MIT Press, Cambridge, MA.

Williams, R. J. and D. Zipser (1989a). Experimental analysis of the real-time recurrent learning algorithm. *Connection Science*, **1**, 87–111.

Williams, R. J. and D. Zipser (1989b). A learning algorithm for continually running fully recurrent neural networks. *Neural Computation*, **1**, 270–280.

Williams, R. J. and D. Zipser (1990). Gradient-based learning algorithms for recurrent connectionist networks. Technical Report NU-CCS-90-9, Northeastern University, Boston, College of Computer Science.

Willis, M. J., C. Di Massimo, G. A. Montague, M. T. Tham and A. J. Morris (1991). On artificial neural networks in process engineering. *Proc. IEE Pt. D,* **138,** 256–266.

Wolfram, S. (1986). *Theory and Applications of Cellular Automata.* Singapore World Scientific, Singapore.

International Workshop. Cellular neural networks and their applications. *Proc. CNNA*'90.

Ydstie, B. E. (1990). Forecasting and control using adaptive connectionist networks. *Computers Chem. Engng.,* **14,** 583–599.

Yeung, D. Y. and G. A. Beckey (1989). Using a context-sensitive learning network for robot arm control. In *Proc. of* 1989 *IEEE Int. Conf. on Robotics and Automation,* Scottsdale, AZ, pp. 1441–1447.

Zak, S. H. (1990). Robust tracking control of dynamic systems with neural networks. In *IEEE Int. Joint Conf. on Neural Networks, IJCNN*'90, pp. 563–566.

Zhao, X. and J. Mendel (1988). An artificial neural minimum-variance estimator. In *IEEE Int. Joint Conf. on Neural Networks, IJCNN*'88, pp. 499–506.

Zipser, D. (1989). A subgrouping strategy that reduces complexity and speeds up learning in recurrent networks. *Neural Computation,* **1,** 552–558.

The Self-Organizing Map

TEUVO KOHONEN, SENIOR MEMBER, IEEE

Invited Paper

Among the architectures and algorithms suggested for artificial neural networks, the Self-Organizing Map has the special property of effectively creating spatially organized "internal representations" of various features of input signals and their abstractions. One novel result is that the self-organization process can also discover semantic relationships in sentences. In this respect the resulting maps very closely resemble the topographically organized maps found in the cortices of the more developed animal brains. After supervised fine tuning of its weight vectors, the Self-Organizing Map has been particularly successful in various pattern recognition tasks involving very noisy signals. In particular, these maps have been used in practical speech recognition, and work is in progress on their application to robotics, process control, telecommunications, etc. This paper contains a survey of several basic facts and results.

I. Introduction

A. On the Role of the Self-Organizing Map Among Neural Network Models

The network architectures and signal processes used to model nervous systems can roughly be divided into three categories, each based on a different philosophy. *Feedforward networks* [94] transform sets of input signals into sets of output signals. The desired input-output transformation is usually determined by external, supervised adjustment of the system parameters. In *feedback networks* [27], the input information defines the initial activity state of a feedback system, and after state transitions the asymptotic final state is identified as the outcome of the computation. In the third category, neighboring cells in a neural network compete in their activities by means of mutual lateral interactions, and develop adaptively into specific detectors of different signal patterns. In this category learning is called *competitive*, *unsupervised*, or *self-organizing*.

The Self-Organizing Map discussed in this paper belongs to the last category. It is a sheet-like artificial neural network, the cells of which become specifically tuned to various input signal patterns or classes of patterns through an unsupervised learning process. In the basic version, only one cell or local group of cells at a time gives the active response to the current input. The locations of the responses tend to become ordered as if some meaningful coordinate system for different input features were being created over the network. The spatial location or coordinates of a cell in the network then correspond to a particular domain of input signal patterns. Each cell or local cell group acts like a separate *decoder* for the same input. It is thus the presence or absence of an active response at that location, and not so much the exact input–output signal transformation or magnitude of the response, that provides an interpretation of the input information.

The Self-Organizing Map was intended as a viable alternative to more traditional neural network architectures. It is possible to ask just how "neural" the map is. Its analytical description has already been developed further in the technical than in the biological direction. But the learning results achieved seem very natural, at least indicating that the adaptive processes themselves at work in the map may be similar to those encountered in the brain. There may therefore be sufficient justification for calling these maps "neural networks" in the same sense as their traditional rivals.

Self-Organizing Maps, or systems consisting of several map modules, have been used for tasks similar to those to which other more traditional neural networks have been applied: pattern recognition, robotics, process control, and even processing of semantic information. The spatial segregation of different responses and their organization into topologically related subsets results in a high degree of efficiency in typical neural network operations.

Although the largest map we have used in practical applications has only contained about 1000 cells, its learning speed, especially when using computational shortcuts, can be increased to orders of magnitude greater than that of many other neural networks. Thus much larger maps than those used so far are quite feasible, although it also seems that practical applications favor hierarchical systems made up of many smaller maps.

It may be appropriate to observe here that if the maps are used for pattern recognition, their classification accuracy can be multiplied if the cells are fine-tuned using supervised learning principles (cf. Sec. III).

Although the Self-Organizing Map principle was introduced in early 1981, no complete review has appeared in compact form, except perhaps in [44], which does not contain the latest results. I have therefore tried to collect a variety of basic material in the present paper.

Manuscript received August 18, 1989; revised March 15, 1990.

The author is with the Department of Computer Science, Helsinki University of Technology, 02150 Espoo, and the Academy of Finland, 00550 Helsinki, Finland.

IEEE Log Number 9038826.

Reprinted from *Proc. IEEE*, vol. 78, no. 9, pp. 1464–1480, Sept. 1990.

B. Brain Maps

As much as a hundred years ago, a quite detailed topographical organization of the brain, and especially of the cerebral cortex, could be deduced from functional deficits and behavioral impairments induced by various kinds of lesion, or by hemorrhages, tumors, or malformations. Different regions in the brain thereby seemed to be dedicated to specific tasks. One modern systematic technique for causing controllable, reversible simulated lesions is to stimulate a particular site with small electric currents, thereby eventually inducing both excitatory and inhibitory effects and disturbing the assumed local function [75]. If such a spatially confined stimulus then disrupts a specific cognitive ability such as naming of objects, it gives at least some indication that this site is essential to that task.

One straightforward method for locating a response is to record the electric potential or train of neural impulses associated with it. Many detailed mappings, especially from the primary sensory and associative areas of the brain, have been made using various electrophysiological recording techniques.

Direct evidence for any localization of brain functions can also be obtained using modern imaging techniques that display the strength and spatial distribution of neural responses simultaneously over a large area, with a spatial resolution of a few millimeters. The two principal methods which use radioactive tracers are positron emission tomography (PET) [80] and autoradiography of the brain through very narrow collimators (gamma camera). PET reveals changes in oxygen uptake and phosphate metabolism. The gamma camera method directly detects changes in cerebral blood flow. Both phenomena correlate with local neural activity, but they are unable to monitor rapid phenomena. In magnetoencephalography (MEG), the low magnetic field caused by electrical neural responses is detected, and by computing its sources, quite rapid neural responses can be directly analyzed, with a spatial resolution of a few millimeters. The main drawback of MEG is that only current dipoles parallel to the surface of the skull are detectable; and since the dipoles are oriented perpendicular to the cortex, only the sulci can be studied with this method. A review of experimental techniques and results relating to these studies can be found in [32].

After a large number of such observations, a fairly detailed organizational view of the brain has evolved [32]. Especially in higher animals, the various cortices in the cell mass seem to contain many kinds of "map" [33], such that a particular location of the neural response in the map often directly corresponds to a specific modality and quality of sensory signal. The field of vision is mapped "quasiconformally" onto the primary visual cortex. Some of the maps, especially those in the primary sensory areas, are ordered according to some feature dimensions of the sensory signals; for instance, in the visual areas, there are line orientation and color maps [11], [116], and in the auditory cortex there are the so-called tonotopic maps [91], [103], [104], which represent pitches of tones in terms of the cortical distance, or other auditory maps [98]. One of the sensory maps is the somatotopic map [29], [30] which contains a representation of the body, i.e., the skin surface. Adjacent to it is a motor map [70] that is topographically almost identically organized. Its cells mediate voluntary control actions on muscles. Similar maps exist in other parts of the brain [97]. On the higher levels the maps are usually unordered, or at most the order is a kind of ultrametric topological order that is not easily interpretable. There are also singular cells that respond to rather complex patterns, such as the human face [9], [93].

Some maps represent quite abstract qualities of sensory and other experiences. For instance, in the word-processing areas, neural responses seem to be organized according to categories and semantic values of words ([6], [15], [65], [67], [106–109], and [114]; cf. also Sec. V).

It thus seems as if the *internal representations* of information in the brain are generally organized *spatially*. Although there is only partial biological evidence for this, enough data are already available to justify further theoretical studies of this principle. Artificial self-organizing maps and brain maps thus have many features in common, and what is even more intriguing, we now fully understand the processes by which such artificial maps can be formed adaptively and completely automatically.

C. Early Work on Competitive Learning

The basic idea underlying what is called competitive learning is roughly as follows: Assume a sequence of statistical samples of a vectorial observable $\mathbf{x} = \mathbf{x}(t) \in \mathbb{R}^n$, where t is the time coordinate, and a set of variable reference vectors $\{\mathbf{m}_i(t): \mathbf{m}_i \in \mathbb{R}^n, i = 1, 2, \cdots, k\}$. Assume that the $\mathbf{m}_i(0)$ have been initialized in some proper way; random selection will often suffice. If $\mathbf{x}(t)$ can somehow be simultaneously compared with each $\mathbf{m}_i(t)$ at each successive instant of time, taken here to be an integer $t = 1, 2, 3, \cdots$, then the best-matching $\mathbf{m}_i(t)$ is to be updated to match even more closely the current $\mathbf{x}(t)$. If the comparison is based on some distance measure $d(\mathbf{x}, \mathbf{m}_i)$, altering $\mathbf{m}_i$ must be such that, if $i = c$ is the index of the best-matching reference vector, then $d(\mathbf{x}, \mathbf{m}_c)$ is decreased, and all the other reference vectors $\mathbf{m}_i$, with $i \neq c$, are left intact. In this way the different reference vectors tend to become specifically "tuned" to different domains of the input variable $\mathbf{x}$. It will be shown below that if p is the probability density function of the samples $\mathbf{x}$, then the $\mathbf{m}_i$ tend to be located in the input space $\mathbb{R}^n$ in such a way that they approximate to p in the sense of some minimal residual error.

Vector Quantization (VQ) (cf., e.g., [19], [54], [58]) is a classical method, that produces an approximation to a continuous probability density function $p(\mathbf{x})$ of the vectorial input variable $\mathbf{x}$ using a finite number of codebook vectors $\mathbf{m}_i$, $i = 1, 2, \cdots, k$. Once the "codebook" is chosen, the approximation of $\mathbf{x}$ involves finding the reference vector $\mathbf{m}_c$ closest to $\mathbf{x}$. One kind of optimal placement of the $\mathbf{m}_i$ minimizes E, the expected rth power of the reconstruction error:

$$E = \int \|\mathbf{x} - \mathbf{m}_c\|^r p(\mathbf{x})\, d\mathbf{x} \tag{1}$$

where $d\mathbf{x}$ is the volume differential in the $\mathbf{x}$ space, and the index $c = c(\mathbf{x})$ of the best-matching codebook vector ("winner") is a function of the input vector $\mathbf{x}$:

$$\|\mathbf{x} - \mathbf{m}_c\| = \min_i \{\|\mathbf{x} - \mathbf{m}_i\|\}. \tag{2}$$

In general, no closed-form solution for the optimal placement of the $\mathbf{m}_i$ is possible, and iterative approximation schemes must be used.

It has been pointed out in [14], [64], and [115] that (1)

defines a placement of the codebook vectors into the signal space such that their point density function is an approximation to $[p(x)]^{n/(n+r)}$, where n is the dimensionality of x and m_i. We usually consider the case $r = 2$. In most practical applications $n \gg r$, and then the optimal VQ can be shown to approximate $p(x)$.

Using the square-error criterion ($r = 2$), it can also be shown that the following stepwise "delta rule," in the discrete-time formalism ($t = 0, 1, 2, \cdots$), defines the optimal values asymptotically. Let $m_c = m_c(t)$ be the closest codebook vector to $x = x(t)$ in the Euclidean metric. The steepest-descent gradient-step optimization of E in the m_c space yields the sequence

$$m_c(t+1) = m_c(t) + \alpha(t)\,[x(t) - m_c(t)],$$
$$m_i(t+1) = m_i(t) \qquad \text{for } i \neq c \tag{3}$$

with $\alpha(t)$ being a suitable, monotonically decreasing sequence of scalar-valued gain coefficients, $0 < \alpha(t) < 1$. This is then the simplest analytical description of competitive learning.

In general, if we express the dissimilarity of x and m_i in terms of a general distance function $d(x, m_i)$, we have first to identify the "winner" m_c such that

$$d(x, m_c) = \min_i \{d(x, m_i)\}. \tag{4}$$

After that an updating rule should be used such that d decreases monotonically: the correction δm_i of m_i must be such that

$$[\text{grad}_{m_i}\, d(x, m_i)]^T \cdot \delta m_i < 0. \tag{5}$$

If (1) is used for signal approximation, it often turns out to be more economical to first observe a number of training samples $x(t)$, which are "classified" (labeled) on the basis of (2) according to the closest codebook vectors m_i, and then to perform the updating operation in a single step. For the new codebook vector m_i, the average is taken of those $x(t)$ that were identified with codebook vector i. This algorithm, termed the *k-means algorithm* is widely used in digital telecommunications engineering [58].

The $m_i(t)$, in the above processes, actually develop into a set of *feature-sensitive detectors*. Feature-sensitive cells are also known to be common in the brain. Neural modelers like Nass and Cooper [72], Pérez *et al.* [79], and Grossberg [21] have been able to suggest how such feature-sensitive cells can emerge from simplified membrane equations of model neurons.

In the above process and its biophysical counterparts, all the cells act independently. Therefore the *order* in which they are assigned to the different domains of input signals is more or less haphazard, most strongly depending on the initial values of the $m_i(0)$. In fact, in 1973, v.d. Malsburg [59] had already published a computer simulation in which he demonstrated *local* ordering of feature-sensitive cells, such that in small subsets of cells roughly corresponding to the so-called columns of the cortex, the cells were tuned more closely than were more remote cells. Later, Amari [1] formulated and analyzed the corresponding system of differential equations, relating them to spatially continuous two-dimensional media. Such continuous layers interacted in the lateral direction; the arrangement was called a *nerve field*. The above studies are of great theoretical importance because they involve a self-organizing tendency. The ordering power they demonstrated was, however, still weak as nerve-field type equations only describe this tendency as a marginal effect. In spite of numerous attempts, no "maps" of practical importance could be produced; ordering was either restricted to a one-dimensional case, or confined to small parcelled areas of the network [60], [77], [78], [99], [100], [110], [111].

Indeed it later transpired that system equations have to involve much stronger, idealized self-organizing effects, and that the organizing effect has to be maximized in every possible way before useful global maps can be created. The present author, in early 1981, was experimenting with various architectures and system equations, and found a process description [34]–[36] that seemed generally to produce globally well-organized maps. Because all the other system models known at that time only yielded results that were significantly more "brittle" with respect to the selection of parameters and to success in achieving the desired results, we may skip them here and concentrate on the computationally optimized algorithm known as the *Self-Organizing Map* algorithm.

II. An Algorithm that Orders Responses Spatially

Readers who are not yet familiar with the Self-Organizing Maps may benefit from a quick look at Figs. 5 and 6 or Fig. 9 to find out what spatial ordering of output responses means.

The Self-Organizing Map algorithm that I shall now describe has evolved during a long series of computer experiments. The background to this research has been expounded in [44]. While the purpose of each detail of the final equations may be clear in concrete simulations, it has proved extremely difficult, in spite of numerous attempts, to express the dynamic properties of this process in mathematical theorems. Strict mathematical analysis only exists for simplified cases. And even they are too lengthy to be reviewed here: cf. [7], [8], [24], [57], [83], [86], [89], [90]. It is therefore hoped that the simulation experiments and practical applications reported below in Secs. II-C, II-E, IV, V, and VI will suffice to convince the reader about the utility of this algorithm.

It may also be necessary to emphasize again that for practical purposes we are trying to extract or explain the self-organizing function in its purest, most effective form, whereas in genuine biological networks this tendency may be more or less disguised by other functions. It may thus be conceivable, as has been verified by numerous simulation experiments, that the two essential effects leading to spatially organized maps are: 1) *spatial concentration* of the network activity on the cell (or its neighborhood) that is *best tuned* to the present input, and 2) further *sensitization or tuning* of the best-matching cell *and its topological neighbors* to the present input.

A. Selection of the Best-Matching Cell

Consider the two-dimensional network of cells depicted in Fig. 1. Their arrangement can be hexagonal, rectangular, etc. Let (in matrix notation) $x = [x_1, x_2, \cdots, x_n]^T \in \mathbb{R}^n$ be the *input vector* that, for simplicity and computational efficiency, is assumed to be connected in parallel to all the neurons i in this network. (We have also shown that subsets of the same input signals can be connected at random to the

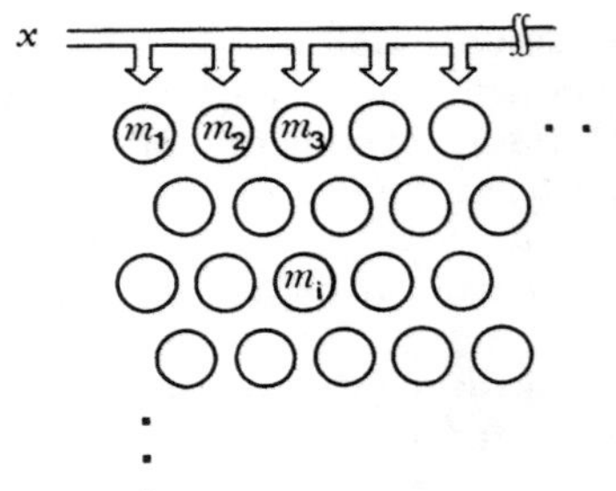

Fig. 1. Cell arrangement for the map and definition of variables.

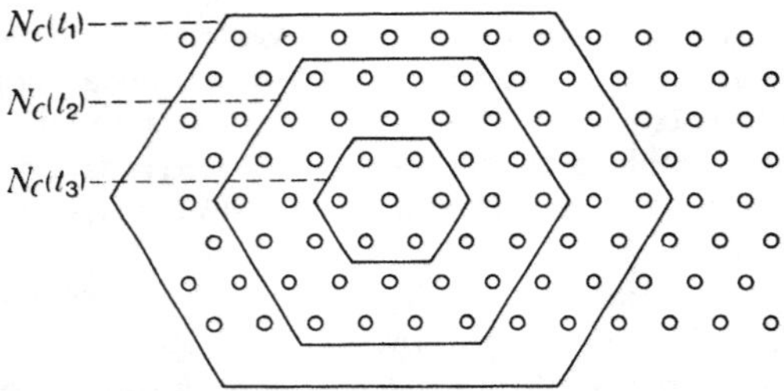

Fig. 2. Examples of topological neighborhood $N_c(t)$, where $t_1 < t_2 < t_3$.

cells; cf. the "tonotopic map" discussed in [35] and [44]. Experiments are in progress in which the input connections to the cells can be made in a cascade.) The *weight vector* of cell i shall henceforth be denoted by $\boldsymbol{m}_i = [m_{i1}, m_{i2}, \cdots, m_{in}]^T \in \mathbb{R}^n$.

The simplest analytical measure for the match of $\boldsymbol{x}$ with the $\boldsymbol{m}_i$ may be the inner product $\boldsymbol{x}^T\boldsymbol{m}_i$. If, however, the self-organizing algorithm is to be used for, say, natural signal patterns relating to metric vector spaces, a better and more convenient (cf. the adaptation law below) matching criterion may be used, based on the *Euclidean distances* between $\boldsymbol{x}$ and $\boldsymbol{m}_i$. The minimum distance defines the "winner" $\boldsymbol{m}_c$ (cf. (2)). A shortcut algorithm to find $\boldsymbol{m}_c$ has been presented in [49].

Comment. Definition of the input vector, $\boldsymbol{x}$, as an ordered set of signal values is only possible if the interrelation between the signals is simple. In many practical problems, such as image analysis (cf. Discussion, Sec. VII) it will generally be necessary to use some kind of preprocessing to extract a set of invariant features for the components of $\boldsymbol{x}$.

B. Adaptation (Updating) of the Weight Vectors

It is crucial to the formation of ordered maps that the cells doing the learning are not affected independently of each other (cf. competitive learning in Sec. I-C), but as *topologically related subsets*, on each of which a similar kind of correction is imposed. During the process, such selected subsets will then encompass different cells. The net corrections at each cell will thus tend to be smoothed out in the long run. An even more intriguing result from this sort of spatially correlated learning is that *the weight vectors tend to attain values that are ordered along the axes of the network.*

In biophysically inspired neural network models, correlated learning by spatially neighboring cells can be implemented using various kinds of lateral feedback connection and other lateral interactions. In the present process we want to enforce lateral interaction directly in a general form, for arbitrary underlying network structures, by defining a *neighborhood set* N_c around cell c. At each learning step, all the cells within N_c are updated, whereas cells outside N_c are left intact. This neighborhood is centered around that cell for which the best match with input $\boldsymbol{x}$ is found:

$$\|\boldsymbol{x} - \boldsymbol{m}_c\| = \min_i \{\|\boldsymbol{x} - \boldsymbol{m}_i\|\}. \tag{2'}$$

The width or radius of N_c can be time-variable; in fact, for good global ordering, it has experimentally turned out to be advantageous to let N_c be very wide in the beginning and shrink monotonically with time (Fig. 2). The explanation for this may be that a wide initial N_c, corresponding to a coarse spatial resolution in the learning process, first induces a rough global order in the $\boldsymbol{m}_i$ values, after which narrowing the N_c improves the spatial resolution of the map; the acquired global order, however, is not destroyed later on. It is even possible to end the process with $N_c = \{c\}$, that is, finally updating the best-matching unit ("winner") only, in which case the process is reduced to simple competitive learning. Before this, however, the "topological order" of the map would have to be formed.

The updating process (in discrete-time notation) may read

$$\boldsymbol{m}_i(t+1) = \begin{cases} \boldsymbol{m}_i(t) + \alpha(t)[\boldsymbol{x}(t) - \boldsymbol{m}_i(t)] & \text{if } i \in N_c(t), \\ \boldsymbol{m}_i(t) & \text{if } i \notin N_c(t), \end{cases} \tag{6}$$

where $\alpha(t)$ is a scalar-valued "adaptation gain" $0 < \alpha(t) < 1$. It is related to a similar gain used in the stochastic approximation processes [49], [92], and as in these methods, $\alpha(t)$ should decrease with time.

An alternative notation is to introduce a scalar "kernel" function $h_{ci} = h_{ci}(t)$,

$$\boldsymbol{m}_i(t+1) = \boldsymbol{m}_i(t) + h_{ci}(t)[\boldsymbol{x}(t) - \boldsymbol{m}_i(t)] \tag{7}$$

whereby, above, $h_{ci}(t) = \alpha(t)$ within N_c, and $h_{ci}(t) = 0$ outside N_c. On the other hand, the definition of h_{ci} can also be more general; a biological lateral interaction often has the shape of a "bell curve". Denoting the coordinates of cells c and i by the vectors $\boldsymbol{r}_c$ and $\boldsymbol{r}_i$, respectively, a proper form for h_{ci} might be

$$h_{ci} = h_0 \exp(-\|\boldsymbol{r}_i - \boldsymbol{r}_c\|^2/\sigma^2), \tag{8}$$

with $h_0 = h_0(t)$ and $\sigma = \sigma(t)$ as suitable decreasing functions of time.

C. Demonstrations of the Ordering Process

The first computer simulations presented here are intended to illustrate the effect that the weight vectors tend to approximate to the density function of the input vectors in an orderly fashion. In these examples, the input vectors were chosen to be two-dimensional for visual display purposes, and their probability density function was arbitrarily selected to be uniform over the area demarcated by the borderlines (square or triangle). Outside the frame the density was zero. The vectors $\boldsymbol{x}(t)$ were drawn from this density function independently and at random, after which they caused adaptive changes in the weight vectors $\boldsymbol{m}_i$.

The $\boldsymbol{m}_i$ vectors appear as points in the same coordinate system as that in which the $\boldsymbol{x}(t)$ are represented; in order to indicate to which unit each $\boldsymbol{m}_i$ value belongs, the points

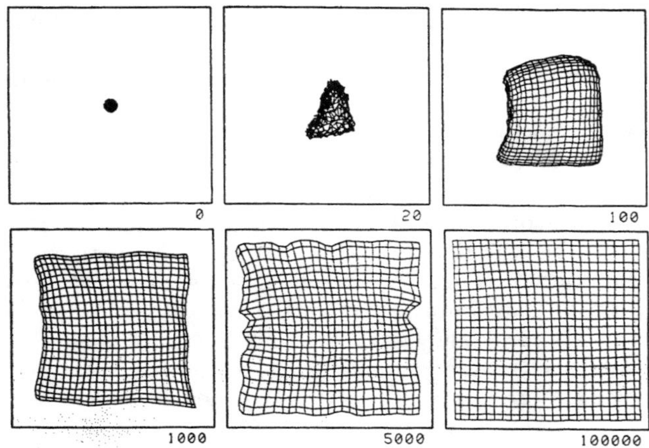

Fig. 3. Weight vectors during the ordering process, two-dimensional array.

corresponding to the m_i vectors have been connected by a lattice of lines conforming to the topology of the processing unit array. In other words, a line connecting two weight vectors m_i and m_j is only used to indicate that the corresponding units i and j are adjacent in the array. In Fig. 3 the arrangement of the cells is rectangular (square), whereas in Fig. 4 the cells are interconnected in a linear chain.

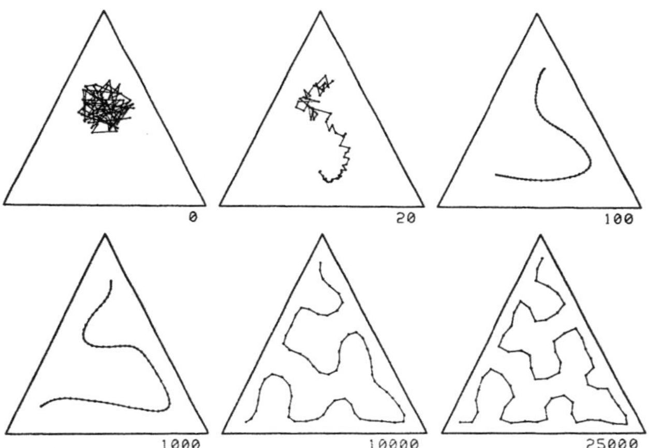

Fig. 4. Weight vectors during the ordering process, one-dimensional array.

Examples of intermediate phases during the self-organizing process are given in Figs. 3 and 4. The initial values $m_i(0)$ were selected at random from a certain domain of values.

As stated above, in Fig. 3 the array was two-dimensional. The results, however, are particularly interesting if the distribution and the array have different dimensionalities: Fig. 4 illustrates a case in which the distribution of x is two-dimensional, but the array is one-dimensional (linear row of cells). The weight vectors of linear arrays tend to approximate to higher-dimensional distributions by Peano curves. A two-dimensional network representing three-dimensional "bodies" (uniform-density function) is shown in Fig. 5.

In practical applications, the input and output weight vectors are usually high-dimensional; e.g., in speech recognition, the dimensionality n may be 15 to 100.

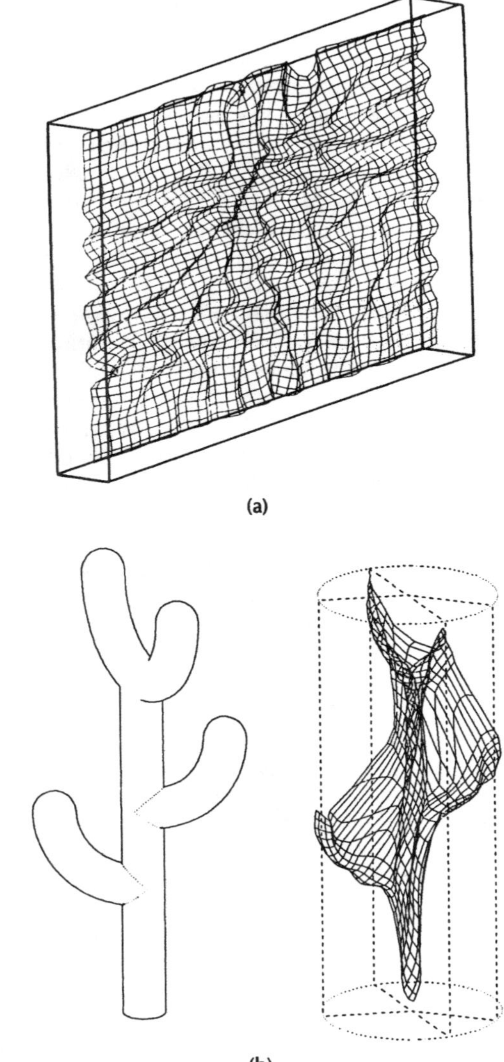

Fig. 5. Representation of three-dimensional (uniform) density functions by two-dimensional maps.

Since no factor present defines a particular orientation in the output map, the latter can be realized in the process in any mirror or point-symmetric inversion, mainly depending on the initial values $m_i(0)$. If a particular orientation is to be favored, the easiest way to achieve this result is by asymmetric choice of the initial values $m_i(0)$.

D. Some Practical Hints for the Application of the Algorithm

When applying the map algorithm, (2) or (2′) and (6) alternate. Input x is usually a random variable with a density function $p(x)$, from which the successive values $x(t)$ are drawn. In real-world observations, such as speech recognition, the $x(t)$ can simply be successive samples of the input observables in their natural order of occurrence.

The process may be started by choosing arbitrary, even random, initial values for the $m_i(0)$, the only restriction being that they should be different.

We shall give numerical examples of efficient process parameters with the simulation examples. It may also be helpful to emphasize the following general conditions.

1) Since learning is a stochastic process, the final statistical accuracy of the mapping depends on the number of steps, which must be reasonably large; there is no way to circumvent this requirement. A "rule of thumb" is that, for good statistical accuracy, the number of steps must be at least 500 times the number of network units. On the other hand, the number of components in $\mathbf{x}$ has no effect on the number of iteration steps, and if hardware neural computers are used, a very high dimensionality of input is allowed. Typically we have used up to 100 000 steps in our simulations, but for "fast learning," e.g., in speech recognition, 10 000 steps and even less may sometimes be enough. Note that the algorithm is computationally extremely light. If only a small number of samples are available, they must be recycled for the desired number of steps.

2) For approximately the first 1000 steps, $\alpha(t)$ should start with a value that is close to unity, thereafter decreasing monotonically. An accurate rule is not important: $\alpha = \alpha(t)$ can be linear, exponential, or inversely proportional to t. For instance, $\alpha(t) = 0.9(1 - t/1000)$ may be a reasonable choice. The *ordering* of the $\mathbf{m}_i$ occurs during this initial period, while the remaining steps are only needed for the fine adjustment of the map. After the ordering phase, $\alpha = \alpha(t)$ should attain small values (e.g., of the order of or less than .01) over a long period. Neither is it crucial whether the law for $\alpha(t)$ decreases linearly or exponentially during the final phase.

3) Special caution is required in the choice of $N_c = N_c(t)$. If the neighborhood is too small to start with, the map will not be ordered globally. Instead various kinds of mosaic-like parcellations of the map are seen, between which the ordering direction changes discontinuously. This phenomenon can be avoided by starting with a fairly wide $N_c = N_c(0)$ and letting it shrink with time. The initial radius of N_c can even be more than half the diameter of the network! During the first 1000 steps or so, when the proper ordering takes place, and $\alpha = \alpha(t)$ is fairly large, the radius of N_c can shrink linearly to, say, one unit; during the fine-adjustment phase, N_c can still contain the nearest neighbors of cell c.

Parallel implementations of the algorithm have been discussed in [25] and [61].

E. Example: Taxonomy (Hierarchical Clustering) of Abstract Data

Although the more practical applications of the Self-Organizing Maps are available, for example, in pattern recognition and robotics, it may be interesting to apply this principle first to abstract data vectors consisting of hypothetical *attributes* or *characteristics*. We will look at an example with implicitly defined (*hierarchical*) structures in the primary data, which the map algorithm is then able to reveal. Although this system is a single-level network, it can produce a hierarchical representation of the relations between the primary data.

The central result in self-organization is that if the input signals have a well-defined probability density function, then the weight vectors of the cells try to imitate it, however complex its form. It is even possible to perform a kind of numerical taxonomy on this model. Because there are no restrictions on the semantic content of the input signals, they can be regarded as arbitrary attributes, with discrete or continuous values. In Table 1, 32 items, each with five hypothetical attributes, are recorded in a data matrix. (This example is completely artificial.) Each of the columns represents one item, and for later inspection the items are labeled "A" through "6", although these labels were not referred to during the learning.

The attribute values $(a_1, a_2, \cdots, a_5)$ constitute the pattern vector $\mathbf{x}$ which acts as a set of signal values at the inputs to the network of the type shown in Fig. 1. During training, the vectors $\mathbf{x}$ were selected from Table 1 at random. Sampling and adaptation were continued iteratively until the asymptotic state could be considered stationary. Such a "learned" network was then calibrated using the items from Table 1 and labeling the best-matching map cells according to the different calibration items. Such a labeled map is shown in Fig. 6. It can be seen that the "images" of different items are related according to a taxonomic graph where the different branches are visible. For comparison, Fig. 7 illus-

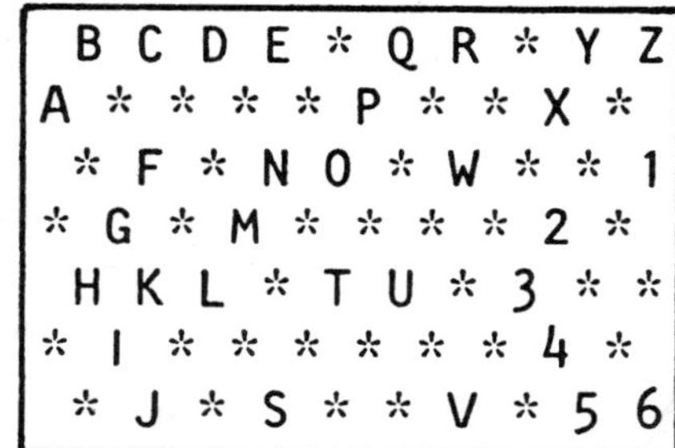

Fig. 6. Self-organized map of the data matrix of Table 1.

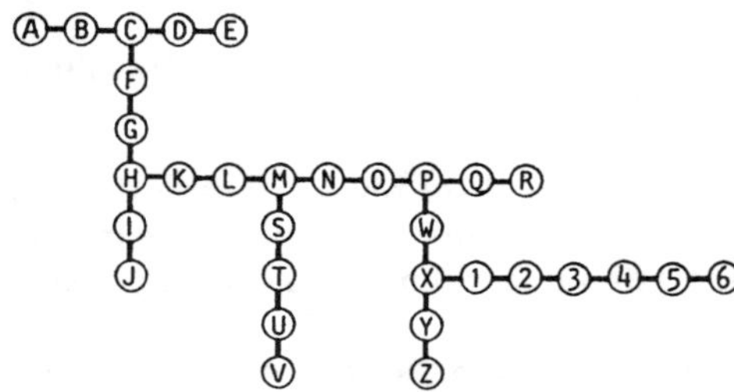

Fig. 7. Minimal spanning tree corresponding to Table 1.

Table 1 Input Data Matrix

Attribute \ Item	A	B	C	D	E	F	G	H	I	J	K	L	M	N	O	P	Q	R	S	T	U	V	W	X	Y	Z	1	2	3	4	5	6
a_1	1	2	3	4	5	3	3	3	3	3	3	3	3	3	3	3	3	3	3	3	3	3	3	3	3	3	3	3	3	3	3	3
a_2	0	0	0	0	0	1	2	3	4	5	3	3	3	3	3	3	3	3	3	3	3	3	3	3	3	3	3	3	3	3	3	3
a_3	0	0	0	0	0	0	0	0	0	0	1	2	3	4	5	6	7	8	3	3	3	3	6	6	6	6	6	6	6	6	6	6
a_4	0	0	0	0	0	0	0	0	0	0	0	0	0	0	0	0	0	0	1	2	3	4	1	2	3	4	2	2	2	2	2	2
a_5	0	0	0	0	0	0	0	0	0	0	0	0	0	0	0	0	0	0	0	0	0	0	0	0	0	0	1	2	3	4	5	6

trates the minimal spanning tree (where the most closely similar pairs of items are linked) describing the similarity relations of the items in Table 1. The system parameters in this process were:

$\alpha = \alpha(t)$: During the first 1000 steps, α decreased linearly with time from .5 to .04 (the initial value could have been closer to unity, say, .9). During the subsequent 10 000 steps, α decreased from .04 to zero linearly with time.

$N_c = N_c(t)$: The lattice was hexagonal, 7 by 10 units, and during the first 1000 steps, the *radius* of N_c decreased from the value six (encompassing the majority of cells in the network) to one (encompassing neuron c and its six neighbors) linearly with time, thereafter keeping the value one.

F. Another Variant of the Algorithm

One further remark may be necessary. It has sometimes been suggested that x be normalized before it is used in the algorithm. Normalization is not necessary in principle, but it may improve numerical accuracy because the resulting reference vectors then tend to have the same dynamic range.

Another aspect, as mentioned above, is that it is also possible to apply a general distance measure in the matching; then, however, the matching and updating laws should be mutually compatible with respect to the same metric. For instance, if the inner-product measure of similarity were applied, the learning equations should read:

$$\mathbf{x}^T(t)\mathbf{m}_c(t) = \max_i \{\mathbf{x}^T(t)\mathbf{m}_i(t)\}, \tag{9}$$

$$\mathbf{m}_i(t+1) = \begin{cases} \dfrac{\mathbf{m}_i(t) + \alpha'(t)\mathbf{x}(t)}{\|\mathbf{m}_i(t) + \alpha'(t)\mathbf{x}(t)\|} & \text{if } i \in N_c(t), \\ \mathbf{m}_i(t) & \text{if } i \notin N_c(t), \end{cases} \tag{10}$$

and $0 < \alpha'(t) < \infty$; for instance, $\alpha'(t) = 100/t$. This process normalizes the reference vectors at each step. The normalization computations slow down the training algorithm significantly. On the other hand, the linear matching criterion applied during recognition is very simple and fast, and amenable to many kinds of simple analog computation, both electronic and optical.

III. Fine Tuning of the Map by Learning Vector Quantization (LVQ) Methods

If the Self-Organizing Map is to be used as a *pattern classifier* in which the cells or their responses are grouped into subsets, each of which corresponds to a discrete class of patterns, then the problem becomes a decision process and must be handled differently. The original Map, like any classical Vector Quantization (VQ) method (cf. Sec. I-D) is mainly intended to approximate input signal values, or their probability density function, by quantized "codebook" vectors that are localized in the input space to minimize a quantization error functional (cf. Sec. III-A below). On the other hand, if the signal sets are to be *classified* into a finite number of categories, then several codebook vectors are usually made to represent each class, and their identity *within* the classes is no longer important. In fact, only decisions made at class borders count. It is then possible, as shown below, to define *effective* values for the codebook vectors such that they directly define *near-optimal decision borders* between the classes, even in the sense of classical Bayesian decision theory. These strategies and learning algorithms were introduced by the present author [38], [43], [45] and called *Learning Vector Quantization (LVQ)*.

A. Type One Learning Vector Quantization (LVQ1)

If several codebook vectors $\mathbf{m}_i$ are assigned to each class, and each of them is labeled with the corresponding class symbol, the class regions in the $\mathbf{x}$ space are defined by simple nearest-neighbor comparison of $\mathbf{x}$ with the $\mathbf{m}_i$; the label of the closest $\mathbf{m}_i$ defines the classification of $\mathbf{x}$.

To define the optimal placement of $\mathbf{m}_i$ in an iterative learning process, initial values for them must first be set using any classical VQ method or by the Self-Organizing Map algorithm. The initial values in both cases roughly correspond to the overall statistical density function $p(\mathbf{x})$ of the input. The next phase is to determine the labels of the codebook vectors, by presenting a number of input vectors with known classification, and assigning the cells to different classes by majority voting, according to the frequency with which each $\mathbf{m}_i$ is closest to the calibration vectors of a particular class.

The classification accuracy is improved if the $\mathbf{m}_i$ are updated according to the following algorithm [41], [43]–[45]. The idea is to pull codebook vectors away from the decision surfaces to demarcate the class borders more accurately. Let $\mathbf{m}_c$ be the codebook vector closest to $\mathbf{x}$ in the Euclidean metric (cf. (2), (2')); this then also defines the classification of $\mathbf{x}$. Apply training vectors $\mathbf{x}$ the classification of which is known. Update the $\mathbf{m}_i = \mathbf{m}_i(t)$ as follows:

$$\mathbf{m}_c(t+1) = \mathbf{m}_c(t) + \alpha(t)[\mathbf{x}(t) - \mathbf{m}_c(t)]$$
$$\text{if } x \text{ is classified correctly,}$$
$$\mathbf{m}_c(t+1) = \mathbf{m}_c(t) - \alpha(t)[\mathbf{x}(t) - \mathbf{m}_c(t)]$$
$$\text{if the classification of } x \text{ is incorrect,}$$
$$\mathbf{m}_i(t+1) = \mathbf{m}_i(t) \text{ for } i \neq c. \tag{11}$$

Here $\alpha(t)$ is a scalar gain $(0 < \alpha(t) < 1)$, which is decreasing monotonically in time, as in earlier formulas. Since this is a fine-tuning method, one should start with a fairly small value, say $\alpha(0) = 0.01$ or 0.02 and let it decrease to zero, say, in 100 000 steps.

This algorithm tends to reduce the point density of the $\mathbf{m}_i$ around the Bayesian decision surfaces. This can be deduced as follows. The minus sign in the second equation may be interpreted as *defining corrections in the same direction as if (10) were used for the class to which $\mathbf{m}_c$ belongs, but with the probability density function of the neighboring (overlapping) class subtracted from that of $\mathbf{m}_c$.* In other words, we would perform a classical Vector Quantization of the function $|p(\mathbf{x}|C_i)P(C_i) - p(\mathbf{x}|C_j)P(C_j)|$ where C_i and C_j are the neighboring classes, $p(\mathbf{x}|C_i)$ is the conditional probability density function of samples $\mathbf{x}$ belonging to class C_i, and $P(C_i)$ is the *a priori* probability of occurrences of the class C_i samples. The difference between the density functions of the neighboring classes, by definition, drops to zero at the Bayes border, inducing the above "depletion layer" of the codebook vectors.

After training, the $\mathbf{m}_i$ will have acquired values such that classification using the "nearest neighbor" principle, by

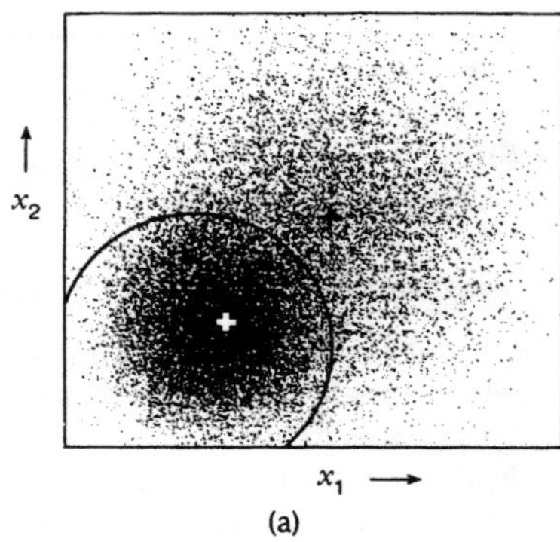

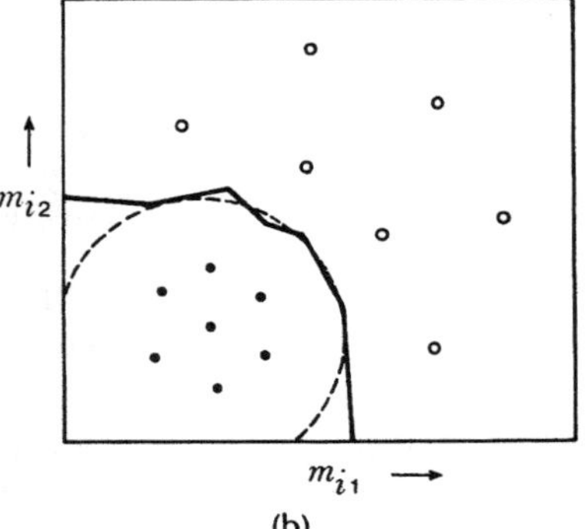

Fig. 8. (a) An illustrative example in which x is two-dimensional and the probability density functions of the classes substantially overlap. (a) The probability density function of $x = [x_1, x_2]^T$ is represented here by its samples, the small dots. The superposition of two symmetric Gaussian density functions corresponding to two different classes C_1 and C_2, with their centroids shown by the white and the dark cross, respectively, is shown. Solid curve: the theoretically optimal Bayes decision surface. (b) Large black dots: reference vectors of class C_1. Open circles: reference vectors of class C_2. Solid curve: decision surface in the Learning Vector Quantization. Broken curve: Bayes decision surface.

comparing of x with the m_i, already rather closely coincides with that of the Bayes classifier. Figure 8 represents an illustrative example in which x is two-dimensional, and the probability density functions of the classes substantially overlap. The decision surface defined by this classifier seems to be near-optimal, although piecewise linear, and the classification accuracy in this rather difficult example is within a fraction of a percent of that achieved with the Bayes classifier. For practical applications of the LVQ1, cf. [12], [76]. A rigorous mathematical discussion of the LVQ1, and suggestions to improve its stability, have been represented in [51].

B. Type Two Learning Vector Quantization (LVQ2)

The previous algorithm can easily be modified to comply even better with Bayes' decisionmaking philosophy [43]–[45]. Assume that two codebook vectors m_i and m_j that belong to different classes and are closest neighbors in the vector space are initially in a wrong position. The (incorrect) discrimination surface, however, is always defined as the midplane of m_i and m_j. Let us define a symmetric *window* of nonzero width around the midplane, and stipulate that *corrections to m_i and m_j shall only be made if x falls into the window on the wrong side of the midplane* (cf. Fig. 9).

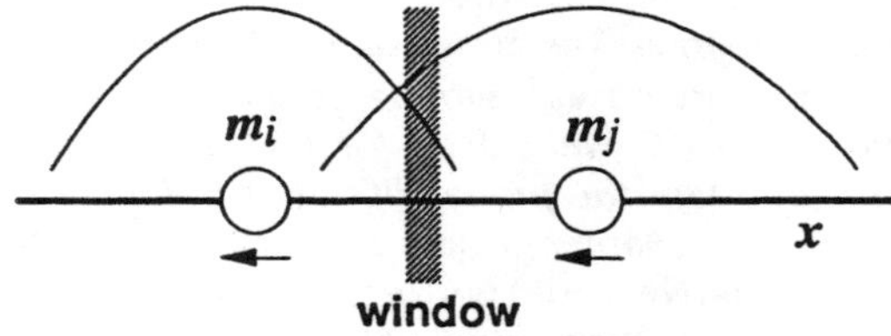

Fig. 9. Illustration of the "window" used in the LVQ2 and LVQ3 algorithms. The curves represent class distributions of x samples.

If the corrections are made according to (12), it is easy to see that for vectors falling into the window, the corrections of both m_i and m_j, on average, have such a direction that the midplane moves towards the crossing surface of the class distributions, and thus asymptotically coincides with the Bayes decision border.

$$m_i(t+1) = m_i(t) - \alpha(t)[x(t) - m_i(t)],$$

$$m_j(t+1) = m_j(t) + \alpha(t)[x(t) - m_j(t)],$$

if C_i is the nearest class, but x belongs to $C_j \neq C_i$ where C_j is the next-to-nearest class ("runner-up"); furthermore x must fall into the "window". In all the other cases,

$$m_k(t+1) = m_k(t). \tag{12}$$

The optimal width of the window must be determined experimentally, and it depends on the number of available samples. With a relatively small number of training samples, a width 10 to 20% of the difference between m_i and m_j seems to be proper.

One question concerns the practical definition of the "window". If we are working in a high-dimensional signal space, it seems reasonable to define the "window" in terms of relative distances d_i and d_j from m_i and m_j, respectively, having constant ratio s. In this way the borders of the "window" are Apollonian hyperspheres. The vector x is defined to lie in the "window" if

$$\min(d_i/d_j, d_j/d_i) > s. \tag{13}$$

If w is the relative width of the window in its narrowest point, then $s = (1 - w)/(1 + w)$. The optimal size of the window depends on the number of available training samples. If we had a large number of samples, a narrow window would guarantee the most accurate location of the border; but for good statistical accuracy the number of samples falling into the window must be sufficient, too, so a 20% window seems a good compromise, at least in the experiments reported below.

For reasons explained in the next section, the classification accuracy of the LVQ2 is first improved when the decision surface is shifted towards the Bayes limit; after that, however, the m_i continue "drifting away". Therefore this algorithm ought to be applied for a relatively short time only, say, starting with $\alpha = 0.02$ and letting it to decrease to zero in at most 10 000 steps.

C. Type Three Learning Vector Quantization (LVQ3)

The LVQ2 algorithm was based on the idea of *differentially* shifting the decision borders towards the Bayes limits, while no attention was paid to what might happen to the location of the m_i in the long run if this process were continued. Thus, although researchers have reported good results, some have had problems, too. It turns out that at least two different kinds of detrimental effect must be taken into account. First, because corrections are proportional to the difference of x and m_i, or x and m_j, the correction on m_j (correct class) is of larger magnitude than that on m_i (wrong class); this results in monotonically decreasing distances $\|m_i - m_j\|$. One remedy is to compensate for this effect, approximately at least, by accepting *all the training vectors from the "window,"* and the only condition is that *one* of m_i and m_j must belong to the *correct* class, and *the other* to the *incorrect class*. The second problem arises from the fact that if the process in (12) is continued, it may lead to another asymptotic equilibrium of m_i that is no longer optimal. Therefore it seems necessary to include corrections that ensure that the m_i continue approximating the class distributions, at least roughly. Combining these ideas, we now obtain an improved algorithm that may be called LVQ3:

$$m_i(t + 1) = m_i(t) - \alpha(t)[x(t) - m_i(t)],$$

$$m_j(t + 1) = m_j(t) + \alpha(t)[x(t) - m_j(t)],$$

where m_i and m_j are the two closest codebook vectors to x, and x and m_j belong to the same class, while x and m_i belong to different classes; furthermore x must fall into the "window";

$$m_k(t + 1) = m_k(t) + \epsilon\alpha(t)[x(t) - m_k(t)]$$

for $k \in \{i, j\}$, if x, m_i, and m_j belong to the same class. (14)

In a series of experiments, applicable values for ϵ between 0.1 and 0.5 were found. The optimal value of ϵ seems to depend on the size of the window, being smaller for narrower windows. This algorithm seems to be self-stabilizing, i.e., the optimal placement of the m_i does not change in continued learning.

Notice that whereas in LVQ1 only one of the m_i values was changed at each step, LVQ2 and LVQ3 change two codebook vectors simultaneously.

IV. Application of the Map to Speech Recognition

When artificial neural networks are to be used for a practical pattern recognition application such as speech recognition, the first task is to make clear whether it is desirable to perform the complete chain of processing operations starting, e.g., from the pre-analysis of the microphone signal and leading on to some form of linguistic encoding of speech using "all-neural" operations, or whether "neural networks" should only be applied at the most critical stage, whereby the rest of the processing operations can be implemented on standard computing equipment. This choice mainly depends on whether the objective is commercial or academic.

Another issue is whether the aim is to demonstrate the ultimate capabilities of "neural networks" in the analysis of dynamical speech information, or whether it is only to replace some of the traditional "spectral" and "vector space" pattern recognition algorithms by highly adaptive, learning "neural network" principles.

In a speech recognizer, a proper place for artificial neural networks is in the *phonemic recognition stage* where exacting statistical analysis is needed. It should be remembered that if phonemes, i.e., classes of different phonological realizations of vowels and consonants, are selected for the basic phonetic units, then account has to be taken of their transformations due to *coarticulation effects*. In other words, the spectral properties of the phonemes are changed in the context or frame of other phonemes. In an "all-neural" speech recognizer it may not be necessary to distinguish or consider phonemes at all, because interpretation of speech is then regarded as an integral, implicit process. Introduction of the phoneme concept already implies that the system must be able to automatically identify them in one form or another and to label the corresponding time interval. Correction of coarticulation effects may then already be implemented in the acoustic analysis itself, by regarding the speech states as Markov processes, and analyzing the state transitions statistically [53]. A different approach altogether is first to apply some vector quantization classification, whereby the speech waveform is only labeled by class symbols of stationary phonemes, as if no coarticulation effects were being taken into account. Corrections can then be made afterwards in a separate postprocessing stage, in symbolic form. We have used the latter approach.

We have implemented a practical "phonetic typewriter" for unlimited speech input using the Self-Organizing Map to spot and recognize phonemes in continuous speech (Finnish and Japanese) [42], [46], [48]. The "network" was fine-tuned for optimal decision accuracy by the Learning Vector Quantization. After that, in the postprocessing stage we applied a self-learning grammar that corrects the majority of coarticulation errors and derives its numerous transformation rules automatically from given examples. This principle, termed *"Dynamically Expanding Context"* [37], [40], actually belongs to the category of learning Artificial Intelligence methods, and thus falls outside the scope of this article (cf. Sec. IV-D below).

A. Acoustic Preprocessing of the Speech Signal

It is known that biological sensory organs such as the inner ear are usually able to adapt to signal transients in a fast, nonlinear way. Nonetheless, we decided to apply conventional frequency analysis to the preprocessing of speech. The main reason for this was that digital Fourier analysis is both accurate and fast, and the fundamentals of digital filtering are well understood. Deviations from physiological reality are not essential since the self-organizing neural network can accept many alternative kinds of preprocessing and can compensate for minor imperfections.

The technical details of the acoustic preprocessing stage are briefly as follows: 1) 5.3-kHz low-pass switched-capacitor filter, 2) 12-bit A/D-converter with 13.02-kHz sampling rate, 3) 256-point FFT formed every 9.83 ms using a Hamming window, 4) logarithmization and smoothing of the power spectrum, 5) combination of spectral channels from the frequency range 200 Hz–5 kHz into a 15-component pattern vector, 6) subtraction of the average from the components, 7) normalization of the pattern vectors. Except for steps 1) and 2), an integrated-circuit signal processor, TMS32010, is used for the computation.

B. Phoneme Map

The simplest type of speech maps formed by self-organization is the static *phoneme map*. There are 21 phonemes in Finnish: /u, o, a, œ, ø, y, e, i, s, m, n, η, l, r, j, v, h, d, k, p, t/. For their representation we used *short-time spectra* as the input patterns $x(t)$. The spectra were evaluated every 9.83 ms. They were computed by the 256-point FFT, from which a 15-component spectral vector was formed by grouping of the channels. In the present study all the spectral samples, even those from the transitory regions, were employed and presented to the algorithm in the natural order of their utterance. During learning, the spectra were not segmented or labeled in any way: any features present in the speech waveform contributed to the self-organized map. After adaptation, the map was calibrated using known stationary phonemes (Fig. 10). The map resembles the well-known *formant maps* used in phonetics; the main difference is that in our maps *complete spectra*, not just two of their resonant frequencies as in formant maps, are used to define the mapping.

Recognition of discrete phonemes is a decision-making process in which the final accuracy only depends on the rate of misclassification errors. It is therefore necessary to try to minimize them using a decision-controlled (supervised) learning scheme, using a training set of speech spectra with known classification.

In practice, for a new speaker, it will be sufficient to dictate 200 to 300 words which are then analyzed by an automatic segmentation method. The latter picks up the training spectra that are applied in the supervised learning algorithm. The finite set of training spectra (of the order of 2000) must be repeated in the algorithm either cyclically or in a random permutation. LVQ1, LVQ2, or LVQ3 can be used as the fine tuning algorithm. A map created for a typical (standard) speaker can then be modified for a new speaker very quickly, using 100 more dictated words, and LVQ fine tuning only.

C. Specific Problems with Transient Phonemes

Generally, the spectral properties of consonants behave more dynamically than those of vowels. Especially in the case of stop consonants, it seems to be better to pay attention to the plosive burst and transient region between the consonant and the subsequent vowel in order to identify the consonant. In our system transient information is coded in additional "satellite" maps (called *transient maps*) and they are trained, using transient spectral samples alone, to describe the dynamic features with higher resolution [48]. Our system was in fact developed in two versions: one for Finnish and one for Japanese. In the Japanese version, four transient maps have been constructed to distinguish the following cases:

1) voiceless stops /k, p, t/ and glottal stop (vowel at the beginning of utterance),
2) voiceless stops /k, p, t/ without comparison with the glottal stop,
3) voiced stops /b, d, g/,
4) nasals /m, n, η/.

Only one transient map has been adopted for the Finnish version, making the distinction between /k, p, t/ and the glottal stop. (/b/ and /g/ do not exist in original Finnish.)

D. Compensation for Coarticulation Effects using the "Dynamically Expanding Context"

Because of coarticulation effects, i.e., transformation of the speech spectra due to neighboring phonemes, systematic errors appear in phonemic transcriptions. For instance, the Finnish word "hauki" (meaning pike) is almost invariably recognized as the phoneme string /haouki/ by our acoustic processor. It may then be suggested that if a transformation rule /aou/ → /au/ is introduced, this error will be corrected. It might also be imagined that it is possible to list and take into account all such variations. However, there may be hundreds of different frames or contexts of neighboring phonemes in which a particular phoneme may occur, and in many cases such empirical rules are contradictory; they are only statistically correct. The frames may

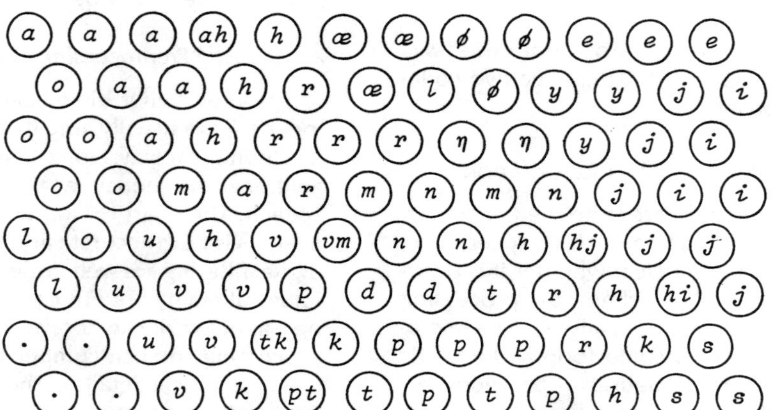

Fig. 10. An example of a *phoneme map*. Natural Finnish speech was processed by a model of the inner ear which performs its frequency analysis. The resulting signals were then connected to an artificial neural network, the cells of which are shown in this picture as circles. The cells were tuned automatically, without any supervision or extra information given, to the acoustic units of speech known as *phonemes*. The cells are labeled by the symbols of those phonemes to which they "learned" to give responses; most cells give a unique answer, whereas the double labels show which cells respond to two phonemes.

also be erroneous. In order to find an optimal and very large system of rules, the *Dynamically Expanding Context* grammar mentioned above was applied [37], [40]. Its rules or productions can be used to transform erroneous symbol strings into correct ones, and even into orthographic text.

Because the correction rules are made accessible in memory using a software content-addressing method (hash coding), they can be applied very quickly, such that the overall operation of the grammar, even with 15 000 rules, is almost in real time. This algorithm is able to correct up to 70% of the errors left by the phoneme map recognizer.

E. Performance of the "Phonetic Typewriter"

In order to get some idea of the accuracy of the map algorithm, we first show a comparative benchmarking of five different methods, namely, classification of manually selected phonemic spectra by the classical parametric Bayes classification, the well-known k-Nearest-Neighbor method (kNN), and LVQ1, LVQ2 and LVQ3.

In this partial experiment, the spectral samples were of Finnish phonemes (divided into 19 classes and using 15 frequency channels for the spectral decomposition). There were 1550 training vectors, and 1550 statistically independent vectors that were only used for testing. The error percentages are given in Table 2.

Table 2 Speech Recognition Experiments with Error Percentages for Independent Test Data

	Parametric Bayes	kNN	LVQ1	LVQ2	LVQ3
Test 1	12.1	12.0	10.2	9.8	9.6
Test 2	13.8	12.1	13.2	12.0	11.5

Note that the parametric Bayes classifier is not even theoretically the best because it assumes that the class samples are normally distributed. We have not been able to find any method, theoretical or heuristic, that classifies speech spectra better than LVQ1, LVQ2 or LVQ3.

In its complete form, the "Phonetic Typewriter" has been tested on several Finnish and Japanese speakers over a long period. To someone familiar with practical speech recognizers it will be clear that it would be meaningless to evaluate and compare different test runs statistically; the results obtained in each isolated test depend so much on the experimental situation and the content of text, the status and tuning of the equipment, as well as on the physical condition of the speaker. The number of tests performed over many years is also too large to be discussed fully here. Let it suffice to mention that the accuracy of spotting and recognizing phonemes in arbitrary continuous speech typically varies between 80 and 90% (depending on the automatic segmentation and recognition of any phoneme), and this figure depends on the speaker and the text. After compensation for coarticulation effects and editing the text into orthographic form, the accuracy, in terms of correctness of any letter, is of the order of 92 to 97%, again depending on the speaker and the text.

The Phonetic Typewriter has already been implemented in several hardware versions using signal processor chips. The latest versions operate in genuine real time with continuous dictation.

It may be of interest here to mention other results, independent of ours. McDermott and Katagiri [28], [66] have carried out experiments on all the Japanese phonemes, and report that LVQ2 gave consistently higher accuracies than Backpropagation Time Delay Neural Networks [105] and was faster in learning.

V. Semantic Map

Demonstrations such as those reported above have indicated that the Self-Organizing Map is indeed able to extract abstract information from multidimensional primary signals, and to represent it as a location, say, in a two-dimensional network. Although this is already a step towards generalization and symbolism, it must be admitted that the extraction of features from geometrically or physically relatable data elements is still a very concrete task, in principle at least.

The operation of the brain at the higher levels relies heavily on abstract concepts, symbolism, and language. It is an old notion that the deepest semantic elements of any language should also be physiologically represented in the neural realms. There is now new physiological evidence for linguistic units being locatable in the human brain [6], [15].

In attempting to devise Neural Network models for linguistic representations, the first difficulty is encountered when trying to find metric distance relations between symbolic items. Unlike with primary sensory signal patterns for which similarity is easily derivable from their mutual distances in the vector spaces in which they are represented, it can not be assumed that encodings of symbols in general have any relationship with the observable characteristics of the corresponding items. How could it then be possible to represent the "logical similarity" of pairs of items, and to map such items topographically? The answer lies in the fact that *the symbol, during the learning process, is presented in context*, i.e., in conjunction with the encodings of a set of other concurrent items. In linguistic representations context might mean a few adjacent words. Similarity between items would then be reflected through the *similarity of the contexts*. Note that for ordered sets of arbitrary encodings, invariant similarity can be expressed, e.g., in terms of the number of items they have in common. On the other hand, it may be evident that the meaning (semantics) of a symbolic encoding is only derivable from the conditional probabilities of its occurrences with other encodings, independent of the type of encoding [68].

However, in the learning process, the literal encodings of the symbols must be memorized, too. Let vector x_s represent the symbolic expression of an item, and x_c the representation of its context. The simplest neural model then assumes that x_s and x_c are connected to the same neural units, i.e., the representation (pattern) vector x of the item is formed as a concatenation of x_s and x_c:

$$x = \begin{bmatrix} x_s \\ x_c \end{bmatrix} = \begin{bmatrix} x_s \\ 0 \end{bmatrix} + \begin{bmatrix} 0 \\ x_c \end{bmatrix}. \tag{15}$$

In other words, the symbol part and the context part form a vectorial sum of two orthogonal components.

The core idea underlying symbol maps is that the two parts are weighted properly such that *the norm of the context part predominates over that of the symbol part during the self-organizing process;* the topographical mapping then mainly reflects the metric relationships of the sets of associated encodings. But since the inputs for symbolic signals are also active all the time, memory traces of them are formed in the corresponding inputs of those cells in the map that have been selected (or actually enforced) by the context part. *If then, during recognition of input information, the context signals are missing or are weaker, the (same) map units are selected solely on the basis of the symbol part. In this way the symbols become encoded into a spatial order reflecting their logical (or semantic) similarities.*

In the following, I shall demonstrate this idea, which was originated by H. Ritter, using a simple language [84]. The simplest definition of the context of a word is to take all those words (together with their serial order) that occur in a certain "window" around the selected word. For simplicity, we shall imagine that the content of each "window" can somehow be presented to the x_c input ports of the neural system. We are not interested here in any particular means for the conversion of, say, temporal signal patterns into parallel ones (this task could be done using paths with different delays, eigenstates that depend on sequences, or any other mechanisms implementable in short-term memory).

The vocabulary used in this experiment is listed in Fig. 11(a) and comprises nouns, verbs, and adverbs. Each word

Bob/Jim/Mary	1
horse/dog/cat	2
beer/water	3
meat/bread	4
runs/walks	5
works/speaks	6
visits/phones	7
buys/sells	8
likes/hates	9
drinks/eats	10/11
much/little	12
fast/slowly	13
often/seldom	14
well/poorly	15

(a)

Sentence Patterns:		
1-5-12	1-9-2	2-5-14
1-5-13	1-9-3	2-9-1
1-5-14	1-9-4	2-9-2
1-6-12	1-10-3	2-9-3
1-6-13	1-11-4	2-9-4
1-6-14	1-10-12	2-10-3
1-6-15	1-10-13	2-10-12
1-7-14	1-10-14	2-10-13
1-8-12	1-11-12	2-10-14
1-8-2	1-11-13	2-11-4
1-8-3	1-11-14	2-11-12
1-8-4	2-5-12	2-11-13
1-9-1	2-5-13	2-11-14

(b)

Mary likes meat
Jim speaks well
Mary likes Jim
Jim eats often
Mary buys meat
dog drinks fast
horse hates meat
Jim eats seldom
Bob buys meat
cat walks slowly
Jim eats bread
cat hates Jim
Bob sells beer
(etc.)

(c)

Fig. 11. Outline of vocabulary used in this experiment. (a) List of used words (nouns, verbs, and adverbs), (b) sentence patterns, and (c) some examples of generated three-word-sentences.

class has further categorial subdivisions, such as names of persons, animals, and inanimate objects. To study semantic relationships in their purest form, it must be stipulated that the semantic meaning be not inferable from any patterns used for the encoding of the individual words, but only from the context in which the words occur (i.e., combinations of words). To this end each word was encoded by a random vector of unit length (here, seven-dimensional).

A sequence of randomly generated meaningful three-word sentences was used as the input data to the self-organizing process. Meaningful sentence patterns had therefore first to be constructed on the basis of word categories (Fig. 11(b)). Each explicit sentence was then constructed by randomly substituting the numbers in a randomly selected sentence pattern from Fig. 11(b) by words with compatible numbering in Fig. 11(a). A total of 498 different three-word sentences are possible, a few of which are exemplified in Fig. 11(c). These sentences were concatenated into a single continuous string, S.

The context of a word in this string was restricted to the pair of words formed by its immediate predecessor and successor in S (ignoring any sentence borders; i.e., words from adjacent sentences in S are uncorrelated, and act like random noise in that field). The code vectors of the predecessor/successor-pair forming the context to a word were concatenated into a single 14-dimensional code vector x_c. In this simple demonstration we thus only took into account the context provided by the immediately adjacent textual environment of each word occurrence. Even this restricted context already contains interesting semantic relationships.

In our computer experiments it turned out that instead of presenting each phrase separately to the algorithm, a much more efficient learning strategy is first to consider each word *in its average context* over a set of possible "windows". The (mean) context of a word was thus first defined as *the average over 10 000 sentences of all code vectors of predecessor/successor-pairs surrounding that particular word.* The resulting thirty 14-dimensional "average word contexts", normalized to unit length, assumed the role of the "context fields" x_c in (14). Each "context field" was combined with a 7-dimensional "symbol field" x_s, consisting of the code vector for the word itself, but scaled to length a. The parameter a determines the relative influence of the symbol part x_s in comparison to the context part x_c and was set to 0.2.

For the simulation, a planar, rectangular lattice of 10 by 15 cells was used. The initial weight vectors of the cells were chosen randomly, so that no initial order was present. Updating was based on (7) and (8). The learning step size was $h_0 = 0.8$ and the radius $\sigma(t)$ of the adjustment zone (cf. (8)) was gradually decreased from an initial value $\sigma_i = 4$ to a final value $\sigma_f = 0.5$ according to the law $\sigma(t) = \sigma_i(\sigma_f/\sigma_i)^{t/t_{max}}$ Here t counts the number of adaptation steps.

After $t_{max} = 2000$ input presentations the responses of the neurons to presentation of the symbol parts alone were tested. In Fig. 12, the symbolic label is written to that site at which the symbol signal $x = [x_s, 0]^T$ gave the maximum response. We clearly see that *the contexts "channel" the word items to memory positions whose arrangements reflects both grammatical and semantic relationships.* Words of same type, i.e., nouns, verbs, and adverbs, are segregated into separate, large domains. Each of these domains is further organized according to similarities on the semantic level. Adverbs with opposite meaning tend to be close to each other, because sentences differing in one word only are regarded as semantically correlated, and the words that are different then usually have the opposite meaning. The groupings of the verbs correspond to differences in the ways they can co-occur with adverbs, persons, animals, and nonanimate objects such as, e.g., food.

It could be argued that the structures resulting in the map were artificially created by a preplanned choice of the sentence patterns allowed as input. This is not the case, however, since it is easy to check that the categorial sentence patterns in Fig. 11(b) almost completely exhaust the possibilities for forming semantically meaningful three-word sentences.

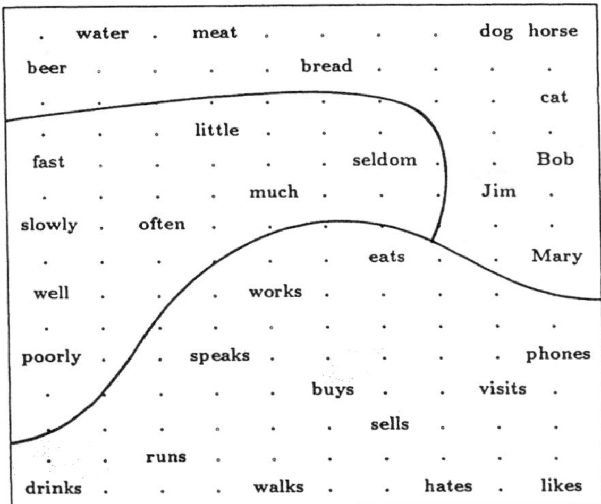

Fig. 12. "Semantic map" obtained on a network of 10 × 15 cells after 2000 presentations of word-context-pairs derived from 10 000 random sentences of the kind shown in Fig. 10(c). Nouns, verbs, and adverbs are segregated into different domains. Within each domain a further grouping according to aspects of meaning is discernible.

VI. Survey of Practical Applications of the Map

In addition to numerous more abstract simulations, theoretical developments, and "toy examples," the following practical problem areas have been suggested for the Self-Organizing Map or the LVQ algorithms. In some of them concrete work is already in progress.

- Statistical pattern recognition, especially recognition of speech [42], [48];
- control of robot arms, and other problems in robotics [17], [18], [63], [85], [87], [88];
- control of industrial processes, especially diffusion processes in the production of semiconductor substrates [62], [102];
- automatic synthesis of digital systems [23];
- adaptive devices for various telecommunications tasks [2], [4], [47];
- image compression [71];
- radar classification of sea-ice [76];
- optimization problems [2];
- sentence understanding [95];
- application of expertise in conceptual domains [96]; and even
- classification of insect courtship songs [73].

Of these, the application to speech recognition has the longest tradition in demonstrating the power of the map method when dealing with difficult stochastic signals. My personal expectations are, however, that the greatest industrial potential of this method may lie in process control and telecommunications.

On the other hand, it is a little surprising that so few applications of the maps to computer vision are being studied. This does not mean that the problems of vision are not important. It is rather that automatic analysis and extraction of visual features, without heuristic or analytical approach, has turned out to be an extremely difficult problem. Biological and artificial vision probably require very complicated hierarchical systems using many stages (e.g., several different maps) [55], [56]. One unclarified problem is how the maps should be interconnected, e.g., whether special nonlinear interfaces are needed [82]; and in hierarchical systems, adaptive normalization of input (cf. [31]) also seems necessary. Only a few isolated problems, such as texture analysis that is under study in our laboratory, might be amenable to the basic method as such.

VII. Discussion

It was stated in Secs. I and III that it is not advisable to use the Self-Organizing Map for classification problems because decision accuracy can be significantly increased if fine tuning such as LVQ is used. Another important notion, that not only concerns the maps but most of the other neural network models as well, is that it would often be absurd to use primary signal elements, such as temporal samples of speech waveform or pixels of an image, for the components of x directly. This is especially true if the input patterns are fine-structured, like line drawings. It is not possible to achieve any *invariances* in perception unless the primary information is first *transformed*, using, e.g., various convolutions with, say, Gabor functions [13], or other, possibly nonlinear functionals of the image field [81], as components of x. Which particular choice of functionals should be used for preprocessing in a particular task is a very difficult and delicate question, and cannot be discussed here.

One question concerns the maximum capacity achievable in the maps. Is it possible to increase their size, to eventually use them for data storage in large knowledge data bases? It can at least be stated that the brain's maps are not particularly extensive; they mainly seem to provide for efficient encoding of a particular subset of signals to enhance the operation and capacity of associative memory [44]. If more extensive systems are required, it might be more efficient to develop hierarchical structures of abstract representations.

The hardware used for the maps has so far only consisted of co-processor boards (cf., e.g., [42], [48]). If the simple algorithm is to be directly built into special hardware, one of its essential operations will be the global extremum selector, for which conventional parallel computing hardware is available [39]. Analog "winner-take-all" circuits can also be used [20], [21], [52]. Another question is whether the learning operations ought to be performed on the board, or whether fixed values for weights could be loaded into the cells. Note that in the latter case the function of the cells can be very simple, like that of the conventional formal neurons. One beneficial property of the maps is that their parameters usually stabilize out into a narrow dynamic range, and the accuracy requirements are then modest. In this case even integer arithmetic operations can provide for sufficient accuracy.

What is the most significant difference between the Self-Organizing Map and other contemporary neural-model approaches? Most of the latter strongly emphasize the aspect of distributed processing, and only consider spatial organization of the processing units as a secondary aspect. The map principle, on the other hand, is in some ways complementary to this idea. The intrinsic potential of this particular self-organizing process for creating a localized, structured arrangement of representations in the basic network module is emphasized.

Actually, we should not talk of the localization of a "function": it is only the response that is localized. I am thus not opposed to the view that neural networks are distributed systems. The massive interconnects that underlie all neural processing are certainly spread over the network; their *effects*, on the other hand, may be "focused" on local sites.

It seems inevitable, however, that any complex processing task requires *organization of information into separate parts*. Distributed processing models in general underrate this issue. Consequently, many models that process features of input data without structuring exhibit slow convergence and poor generalization ability, usually ensuing from ignorance of the localization of the adaptive processes.

On the lower perceptual levels, localization of responses in topographically organized maps has already been demonstrated long time ago, and it is known that such maps need not be prespecified in detail, but can instead organize themselves on the basis of the statistics of the incoming signals. Such maps have already been applied with success in many complex pattern recognition and robot control tasks.

On the higher levels of representation, relationships between items seem to be based on more subtle roles in their occurrence, and are less apparent from their immediate intrinsic properties. Nonetheless it has also been shown recently that even with a simple modeling assumption of semantic roles, topological self-organization of semantic data will take place. To describe the role of an item, it is sufficient that the input data are presented together with a sufficient amount of context. This then controls the adaptation process.

In the practical application that we have studied most carefully, viz. speech recognition, a statistical accuracy of phonemic recognition has been achieved that is clearly equal to or better than the results produced by more conventional methods, even when the latter are based on analysis of signal dynamics [28], [66].

It should be emphasized that the map method is not restricted to using of any particular form of preprocessing, such as amplitude spectra in speech recognition, or even to phonemes as basic phonological units. For instance, analogous maps may be formed for diphones, syllables, or demisyllables, and other spectral representations such as linear prediction coding (LPC) coefficients or cepstra may be used as the input information to the maps.

Although the basic one-level map, as demonstrated in Sec. II-E, has already been shown to be capable of creating hierarchical (ultrametric) representations of structured data distributions, it might be expected that the real potential of the map lies in a genuine hierarchical or otherwise structured system that consists of several interconnected map modules. In a more natural system, such modules might also correspond to contiguous areas in a single large sheet, where each area receives a different kind of external input, as in the different areas in the cortex. In that case, the borders between the modules might be diffuse. The problem of hierarchical maps, however, has turned out to be very difficult. One of the particular difficulties arises if the inputs to a cell come from very different sources; it then seems inevitable that for the comparison of input patterns, an asymmetrical distance function, in which the signal components are provided with adaptive tensorial weights, must be applied [31]. Another aspect concerns the interfaces of modules in a hierarchical map system: the signals merging from different modules may have to be combined nonlinearly [82]. On the other hand, it has already been demonstrated that the map, or the LVQ algorithms, can be used as a preprocessing stage for other models [26], [69], [101]. In the Counterpropagation Network of Hecht-Nielsen [22], competitive learning is neatly integrated into a hierarchical system as a special layer. Combinations of maps have also been studied in [74], [112], and [113].

It should be noted that slightly different self-organization ideas have recently been suggested [5], [10], [16]. They, however, fall outside the scope of this article.

One of the strongest original motives for starting the development of (artificial) neural networks was their use as learning systems that might effectively be able to utilize the vast capacities of active circuits that can be manufactured using semiconductor or optical technologies. It is therefore a little surprising that most of the theoretical research on and simulations of neural networks have been restricted to relatively small networks containing only a few tens to a few thousands of nodes (let alone parallel networks for preprocessing images). The main problem with most circuits seems to be slow convergence of learning, which again indicates that the best learning mechanisms are yet to be found.

References

[1] S.-I. Amari, "Topographic organization of nerve fields," *Bull. Math. Biology*, vol. 42, pp. 339–364, 1980.

[2] B. Angéniol, G. de la Croix Vaubois, and J.-Y. Le Texier, "Self-organizing feature maps and the travelling salesman problem," *Neural Networks*, vol. 1, pp. 289–293, 1988.

[3] N. Ansari and Y. Chen, "Dynamic digital satellite communication network management by self-organization," *Proc. Int. Joint Conf. on Neural Networks, IJCNN-90-WASH-DC* (Washington, DC, 1990) pp. II-567–II-570.

[4] D. S. Bradburn, "Reducing transmission error effects using a self-organizing network," *Proc. Int. Joint Conf. on Neural Networks, IJCNN 89* (Washington, D.C., 1989) pp. II-531–537.

[5] D. J. Burr, "An improved elastic net method for the traveling salesman problem," *Proc. IEEE Int. Conf. on Neural Networks, ICNN-88* (San Diego, Cal., 1988) pp. I-69–I-76.

[6] A. Caramazza, "Some aspects of language processing revealed through the analysis of acquired aphasia: The lexical system," *Ann. Rev. Neurosci.*, vol. 11, pp. 395–421, 1988.

[7] M. Cottrell and J.-C. Fort, "A stochastic model of retinotopy: A self-organizing process," *Biol. Cybern.*, vol. 53, pp. 405–411, 1986.

[8] ——, "Étude d'un processus d'auto-organisation," *Ann. Inst. Henri Poincaré*, vol. 23, pp. 1–20, 1987.

[9] A. R. Damasio, H. Damasio, and G. W. Van Hoesen, "Prosopagnosia: Anatomic basis and behavioral mechanisms," *Neurology*, vol. 24, pp. 89–93, 1975.

[10] R. Durbin and D. Willshaw, "An analogue approach to the travelling salesman problem using an elastic net method," *Nature*, vol. 326, pp. 689–691, 1987.

[11] D. Essen, "Functional organization of primate visual cortex," in *Cerebral Cortex*, vol. 3, A. Peters, E. G. Jones (Eds.). New York: Plenum Press, 1985, pp. 259–329.

[12] J. Fuller and A. Farsaie, "Invariant target recognition using feature extraction," *Proc. Int. Joint. Conf. on Neural Networks, IJCNN-90-WASH-DC* (Washington, DC, 1990) pp. II-595–II-598.

[13] D. Gabor, "Theory of communication," *J.I.E.E.*, vol. 93, pp. 429–459, 1946.

[14] A. Gersho, "On the structure of vector quantizers," *IEEE Trans. Inform. Theory*, vol. IT-25, no. 4, pp. 373–380, July 1979.

[15] H. Goodglass, A. Wingfield, M. R. Hyde, and J. C. Theurkauf, "Category specific dissociations in naming and recognition by aphasic patients," *Cortex*, vol. 22, pp. 87–102, 1986.

[16] I. Grabec, "Self-organization based on the second maximum entropy principle," *First IEE Int. Conf. on Artificial Neural Networks*, Conference Publication No. 313. (London, 1989) pp. 12–16.

[17] D. H. Graf and W. R. LaLonde, "A neural controller for collision-free movement of general robot manipulators," *Proc. IEEE Int. Conf. on Neural Networks, ICNN-88* (San Diego, Cal., 1988) pp. I-77–I-84.

[18] D. H. Graf and W. R. LaLonde, "Neuroplanners for hand/eye coordination," *Proc. Int. Joint Conf. on Neural Networks, IJCNN 89* (Washington, DC, 1989) pp. II-543–II-548.

[19] R. M. Gray, "Vector quantization," *IEEE ASSP Mag.*, vol. 1, pp. 4–29, 1984.

[20] S. Grossberg, "On the development of feature detectors in the visual cortex with applications to learning and reaction-diffusion systems," *Biol. Cybern.*, vol. 21, pp. 145–159, 1976.

[21] —, "Adaptive pattern classification and universal recoding: I. Parallel development and coding of neural feature detectors; II. Feedback, expectation, olfaction, illusions," *Biol. Cybern.*, vol. 23, pp. 121–134 and 187–202, 1976.

[22] R. Hecht-Nielsen, "Applications of counterpropagation network," *Neural Networks*, vol. 1, pp. 131–139, 1988.

[23] A. Hemani and A. Postula, "Scheduling by self organisation," *Proc. Int. Joint. Conf. on Neural Networks, IJCNN-90-WASH-DC* (Washington, DC, 1990) pp. II-543–II-546.

[24] R. E. Hodges and C.-H. Wu, "A method to establish an autonomous self-organizing feature map," *Proc. Int. Joint. Conf. on Neural Networks, IJCNN-90-WASH-DC* (Washington, DC, 1990) pp. I-517–I-520.

[25] R. E. Hodges, C.-H. Wu, and C.-J. Wang, "Parallelizing the self-organizing feature map on multi-processor systems," *Proc. Int. Joint. Conf. on Neural Networks, IJCNN-90-WASH-DC* (Washington, DC, 1990) pp. II-141–II-144.

[26] R. M. Holdaway, "Enhancing supervised learning algorithms via self-organization," *Proc. Int. Joint Conf. on Neural Networks, IJCNN 89* (Washington, D.C., 1989) pp. II-523–II-529.

[27] J. J. Hopfield, "Neural networks and physical systems with emergent collective computational abilities," *Proc. Natl. Acad. Sci. USA*, vol. 79, pp. 2554–2558, 1982.

[28] H. Iwamida, S. Katagiri, E. McDermott, and Y. Tohkura, "A hybrid speech recognition system using HMMs with an LVQ-trained codebook," ATR Technical Report TR-A-0061, ATR Auditory and Visual Perception Research Laboratories, 1989.

[29] J. H. Kaas, R. J. Nelson, M. Sur, C. S. Lin, and M. M. Merzenich, "Multiple representations of the body within the primary somatosensory cortex of primates," *Science*, vol. 204, pp. 521–523, 1979.

[30] J. H. Kaas, M. M. Merzenich, and H. P. Killackey, "The reorganization of somatosensory cortex following peripheral nerve damage in adult and developing mammals," *Annual Rev. Neurosci.*, vol. 6, pp. 325–356, 1983.

[31] J. Kangas, T. Kohonen, and J. Laaksonen, "Variants of self-organizing maps," *IEEE Trans. Neural Networks*, vol. 1, pp. 93–99, 1990.

[32] A. Kertesz, Ed., *Localization in Neuropsychology.* New York, N.Y.: Academic Press, 1983.

[33] E. I. Knudsen, S. du Lac, and S. D. Esterly, "Computational maps in the brain," *Ann. Rev. Neurosci.*, vol. 10, pp. 41–65, 1987.

[34] T. Kohonen, "Automatic formation of topological maps of patterns in a self-organizing system," *Proc. 2nd Scandinavian Conf. on Image Analysis* (Espoo, Finland, 1981) pp. 214–220.

[35] —, "Self-organized formation of topologically correct feature maps," *Biol. Cybern.*, vol. 43, pp. 59–69, 1982.

[36] —, "Clustering, taxonomy, and topological maps of patterns," *Proc. Sixth Int. Conf. on Pattern Recognition* (Munich, Germany, 1982) pp. 114–128.

[37] —, "Dynamically expanding context, with application to the correction of symbol strings in the recognition of continuous speech," *Proc. Eighth Int. Conf. on Pattern Recognition* (Paris, France, 1986) pp. 1148–1151.

[38] —, "Learning Vector Quantization," Helsinki University of Technology, Laboratory of Computer and Information Science, Report TKK-F-A-601, 1986.

[39] —, *Content-Addressable Memories*, 2nd ed. Berlin, Heidelberg, Germany: Springer-Verlag, 1987.

[40] —, "Self-learning inference rules by dynamically expanding context," *Proc. IEEE First Ann. Int. Conf. on Neural Networks* (San Diego, CA, 1987) pp. II-3–II-9.

[41] —, "An introduction to neural networks," *Neural Networks*, vol. 1, pp. 3–16, 1988.

[42] —, "The 'neural' phonetic typewriter," *Computer*, vol. 21, pp. 11–22, March 1988.

[43] —, "Learning vector quantization," *Neural Networks*, vol. 1, suppl. 1, p. 303, 1988.

[44] —, *Self-Organization and Associative Memory*, 3rd ed. Berlin, Heidelberg, Germany: Springer-Verlag, 1989.

[45] T. Kohonen, G. Barna, and R. Chrisley, "Statistical pattern recognition with neural networks: Benchmarking studies," *Proc. IEEE Int. Conf. on Neural Networks, ICNN-88* (San Diego, Cal., 1988) pp. I-61–I-68.

[46] T. Kohonen, K. Mäkisara, and T. Saramäki, "Phonotopic maps—insightful representation of phonological features for speech recognition," *Proc. Seventh Int. Conf. on Pattern Recognition* (Montreal, Canada, 1984) pp. 182–185.

[47] T. Kohonen, K. Raivio, O. Simula, O. Ventä, and J. Henriksson, "An adaptive discrete-signal detector based on self-organizing maps," *Proc. Int. Joint. Conf. on Neural Networks, IJCNN-90-WASH-DC* (Washington, DC, 1990) pp. II-249–II-252.

[48] T. Kohonen, K. Torkkola, M. Shozakai, J. Kangas, and O. Ventä, "Microprocessor implementation of a large vocabulary speech recognizer and phonetic typewriter for Finnish and Japanese," *Proc. European Conference on Speech Technology* (Edinburgh, 1987) pp. 377–380.

[49] H. J. Kushner and D. S. Clark, *Stochastic Approximation Methods for Constrained and Unconstrained Systems.* New York, Berlin: Springer-Verlag, 1978.

[50] J. Lampinen and E. Oja, "Fast self-organization by the probing algorithm," *Proc. Int. Joint Conf. on Neural Networks, IJCNN 89* (Washington, DC, 1989) pp. II-503–II-507.

[51] A. LaVigna, "Nonparametric classification using learning vector quantization," Ph.D. Thesis, University of Maryland, 1989.

[52] J. Lazarro, S. Ryckebusch, M. A. Mahowald, and C. A. Mead, "Winner-take-all network of O(N) complexity," in *Advances in Neural Information Processing Systems I*, D. S. Touretzky, Ed. San Mateo, CA: Morgan Kaufmann Publishers, 1989.

[53] S. E. Levinson, L. R. Rabiner, and M. M. Sondhi, "An introduction to the application of the theory of probabilistic functions of a Markov process to automatic speech recognition," *Bell Syst. Tech. J.*, pp. 1035–1073, Apr. 1983.

[54] Y. Linde, A. Buzo, and R. M. Gray, "An algorithm for vector quantization," *IEEE Trans. Communication*, vol. COM-28, pp. 84–95, 1980.

[55] S. P. Luttrell, "Self-organizing multilayer topographic mappings," *Proc. IEEE Int. Conf. on Neural Networks, ICNN-88* (San Diego, CA, 1988) pp. I-93–I-100.

[56] —, "Hierarchical self-organizing networks," *First IEE Int. Conf. on Artificial Neural Networks*, Conference Publication No. 313. (London, 1989) pp. 2–6.

[57] —, "Self-organization: A derivation from first principles of a class of learning algorithms," *Proc. Int. Joint Conf. on Neural Networks, IJCNN 89* (Washington, DC, 1989) pp. II-495–II-498.

[58] J. Makhoul, S. Roucos, and H. Gish, "Vector quantization in speech coding," *Proc. IEEE*, vol. 73, pp. 1551–1588, 1985.

[59] Ch. v.d Malsburg, "Self-organization of orientation sensitive cells in the striate cortex," *Kybernetik*, vol. 14, pp. 85–100, 1973.

[60] Ch. v.d. Malsburg and D. J. Willshaw, "How to label nerve cells so that they can interconnect in an ordered fashion," *Proc. Natl. Acad. Sci. USA*, vol. 74, pp. 5176–5178, 1977.

[61] R. Mann and S. Haykin, "A parallel implementation of Kohonen feature maps on the Warp systolic computer," *Proc. Int. Joint. Conf. on Neural Networks, IJCNN-90-WASH-DC* (Washington, DC, 1990) pp. II-84–II-87.

[62] K. M. Marks and K. F. Goser, "Analysis of VLSI process data based on self-organizing feature maps," *Proc. Neuro-Nîmes'88* (Nîmes, France, 1988) pp. 337–347.

[63] J. Martinetz, H. J. Ritter, and K. J. Schulten, "Three-dimensional neural net for learning visuomotor coordination of a robot arm," *IEEE Trans. Neural Networks*, vol. 1, pp. 131–136, 1990.

[64] J. Max, "Quantizing for minimum distortion," *IRE Trans. Inform. Theory*, vol. IT-6, no. 2, pp. 7–12, Mar. 1960.

[65] R. A. McCarthy and E. K. Warrington, "Evidence for modality specific meaning systems in the brain," *Nature*, vol. 334, pp. 428–430, 1988.

[66] E. McDermott and S. Katagiri, "Shift-invariant, multi-category phoneme recognition using Kohonen's LVQ2," *Proc. Int. Conf. on Acoustics, Signals, and Speech, ICASSP 89* (Glasgow, Scotland) pp. 81–84.

[67] P. McKenna and E. K. Warrington, "Category-specific naming preservation: A single case study," *J. Neurol. Neurosurg. Psychiatry*, vol. 41, pp. 571–574, 1978.

[68] R. Miikkulainen and M.-G. Dyer, "Forming global representations with extended backpropagation," *Proc. IEEE Int. Conf. on Neural Networks, ICNN 88* (San Diego, CA, 1988) pp. 285–292.

[69] P. Morasso, "Neural models of cursive script handwriting," *Proc. Int. Joint Conf. on Neural Networks, IJCNN 89* (Washington, DC, 1989) pp. II-539–II-542.

[70] J. T. Murphy, H. C. Kwan, W. A. MacKay, and Y. C. Wong, "Spatial organization of precentral cortex in awake primates, III, Input-output coupling," *J. Neurophysiol.*, vol. 41, pp. 1132–1139, 1977.

[71] N. M. Nasrabadi and Y. Feng, "Vector quantization of images based upon the Kohonen Self-Organizing Feature Maps," *Proc. IEEE Int. Conf. on Neural Networks, ICNN-88* (San Diego, CA, 1988) pp. I-101–I-108.

[72] M. M. Nass and L. N. Cooper, "A theory for the development of feature detecting cells in visual cortex," *Biol. Cybern.*, vol. 19, pp. 1–18, 1975.

[73] E. K. Neumann, D. A. Wheeler, J. W. Burnside, A. S. Bernstein, and J. C. Hall, "A technique for the classification and analysis of insect courtship song," *Proc. Int. Joint. Conf. on Neural Networks, IJCNN-90-WASH-DC*, (Washington, DC, 1990) pp. II-257–262.

[74] Y. Nishikawa, H. Kita, and A. Kawamura, "NN/I: a neural network which divides and learns environments," *Proc. Int. Joint. Conf. on Neural Networks, IJCNN-90-WASH-DC*, (Washington, DC, 1990) pp. I-684–I-687.

[75] G. A. Ojemann, "Brain organization for language from the perspective of electrical stimulation mapping," *Behav. Brain Sci.*, vol. 2, pp. 189–230, 1983.

[76] J. Orlando, R. Mann, and S. Haykin, "Radar classification of sea-ice using traditional and neural classifiers," *Proc. Int. Joint. Conf. on Neural Networks, IJCNN-90-WASH-DC*, (Washington, DC, 1990) pp. II-263–II-266.

[77] K. J. Overton and M. A. Arbib, "The branch arrow model of the formation of retinotectal connections," *Biol. Cybern.*, vol. 45, pp. 157–175, 1982.

[78] K. J. Pearson, L. H. Finkel, and G. M. Edelman, "Plasticity in the organization of adult cerebral maps: a computer simulation based on neuronal group selection," *J. Neurosci.*, vol. 12, pp. 4209–4223, 1987.

[79] R. Pérez, L. Glass, and R. J. Shlaer, "Development of specificity in cat visual cortex," *J. Math. Biol.*, vol. 1, pp. 275–288, 1975.

[80] S. E. Petersen, P. T. Fox, M. I. Ponsner, M. Mintun, and M. E. Raichle, "Positron emission tomographic studies of the cortical anatomy of single-word processing," *Nature*, vol. 331, pp. 585–589, 1988.

[81] M. Porat and Y. Y. Zeevi, "The generalized Gabor scheme of image representation in biological and machine vision," *IEEE Trans. Pattern Anal. Machine Intell.*, vol. PAMI-10, pp. 452–468, 1988.

[82] H. Ritter, "Combining self-organizing maps," *Proc. Int. Joint Conf. on Neural Networks, IJCNN 89* (Washington, DC, 1989) pp. II-499–II-502.

[83] —, "Asymptotic level density for a class of vector quantization processes," Helsinki University of Technology, Lab. of Computer and Information Science, Report A9, 1989.

[84] H. Ritter and T. Kohonen, "Self-organizing semantic maps," *Biol. Cybern.*, vol. 61, pp. 241–254, 1989.

[85] H. J. Ritter, T. M. Martinetz, and K. J. Schulten, "Topology conserving maps for learning visuo-motor-coordination," *Neural Networks*, vol. 2, pp. 159–168, 1989.

[86] H. Ritter and K. Schulten, "On the stationary state of Kohonen's self-organizing sensory mapping," *Biol. Cybern.*, vol. 54, pp. 99–106, 1986.

[87] —, "Topology conserving mappings for learning motor tasks," *Proc. Neural Networks for Computing*, AIP Conference (Snowbird, Utah, 1986) pp. 376–380.

[88] —, "Extending Kohonen's self-organizing mapping algorithm to learn ballistic movements," *NATO ASI Series*, vol. F41, pp. 393–406, 1988.

[89] —, "Kohonen's self-organizing maps: exploring their computational capabilities," *Proc. IEEE Int. Conf. on Neural Networks, ICNNN 88* (San Diego, CA, 1988) pp. I-109–I-116.

[90] —, "Convergency properties of Kohonen's topology conserving maps: Fluctuations, stability and dimension selection," *Biol. Cybern.*, vol. 60, pp. 59–71, 1989.

[91] R. A. Reale and T. J. Imig, "Tonotopic organization in auditory cortex of the cat," *J. Comp. Neurol.*, vol. 192, pp. 265–291, 1980.

[92] H. Robbins and S. Monro, "A stochastic approximation method," *Ann. Math. Statist.*, vol. 22, pp. 400–407, 1951.

[93] E. T. Rolls, "Neurons in the cortex of the temporal lobe and in the amygdala of the monkey with responses selective for faces," *Hum. Neurobiol.*, vol. 3, pp. 209–222, 1984.

[94] D. E. Rumelhart, G. E. Hinton, and R. J. Williams, "Learning internal representations by error propagation," in *Parallel Distributed Processing: Explorations in the Microstructure of Cognition. Vol. 1.: Foundations*, D. E. Rumelhart, J. L. McClelland and the PDP research group, Eds. Cambridge, Mass.: MIT Press, 1986, pp. 318–362.

[95] J. K. Samarabandu and O. E. Jakubowicz, "Principles of sequential feature maps in multi-level problems," *Proc. Int. Joint. Conf. on Neural Networks, IJCNN-90-WASH-DC* (Washington, DC, 1990) pp. II-683–II-686.

[96] P. G. Schyns, "Expertise acquisition through concepts refinement in a self-organizing architecture," *Proc. Int. Joint. Conf. on Neural Networks, IJCNN-90-WASH-DC* (Washington, DC, 1990) pp. I-236–I-239.

[97] D. L. Sparks and J. S. Nelson, "Sensory and motor maps in the mammalian superior colliculus," *TINS*, vol. 10, pp. 312–317, 1987.

[98] N. Suga and W. E. O'Neill, "Neural axis representing target range in the auditory cortex of the mustache bat," *Science*, vol. 206, pp. 351–353, 1979.

[99] N. V. Swindale, "A model for the formation of ocular dominance stripes," *Proc. R. Soc.*, vol. B.208, pp. 243–264, 1980.

[100] A. Takeuchi and S. Amari, "Formation of topographic maps and columnar microstructures," *Biol. Cybern.*, vol. 35, pp. 63–72, 1979.

[101] T. Tanaka, M. Naka, and K. Yoshida, "Improved back-propagation combined with LVQ," *Proc. Int. Joint. Conf. on Neural Networks, IJCNN-90-WASH-DC* (Washington, DC, 1990) pp. I-731–734.

[102] V. Tryba, K. M. Marks, U. Rückert, and K. Goser, "Selbstorganisierende Karten als lernende klassifizierende Speicher," *ITG Fachbericht*, vol. 102, pp. 407–419, 1988.

[103] A. R. Tunturi, "Physiological determination of the arrangement of the afferent connections to the middle ectosylvian auditory area in the dog," *Am. J. Physiol.*, vol. 162, pp. 489–502, 1950.

[104] —, "The auditory cortex of the dog," *Am. J. Physiol.*, vol. 168, pp. 712–717, 1952.

[105] A. Waibel, T. Hanazawa, G. Hinton, K. Shikano, and K. J. Lang, "Phoneme recognition using time-delay neural networks," *IEEE Trans. Acoust. Speech and Signal Processing*, vol. ASSP-37, pp. 382–339, 1989.

[106] E. K. Warrington, "The selective impairment of semantic memory," *Q. J. Exp. Psychol.*, vol. 27, pp. 635–657, 1975.

[107] E. K. Warrington and R. A. McGarthy, "Category specific access dysphasia," *Brain*, vol. 106, pp. 859–878, 1983.

[108] —, "Categories of knowledge," *Brain*, vol. 110, pp. 1273–1296, 1987.

[109] E. K. Warrington and T. Shallice, "Category-specific impairments," *Brain*, vol. 107, pp. 829–854, 1984.

[110] D. J. Willshaw and Ch. v.d. Malsburg, "How patterned neural connections can be set up by self-organization," *Proc. R. Soc. London*, vol. B 194, pp. 431–445, 1976.

[111] ——, "A marker induction mechanism for the establishment of ordered neural mappings: its application to the retinotectal problem," *Proc. R. Soc. London*, vol. B 287, pp. 203–243, 1979.

[112] L. Xu and E. Oja, "Vector pair correspondence by a simplified counter-propagation model: a twin topographic map," *Proc. Int. Joint. Conf. on Neural Networks, IJCNN-90-WASH-DC* (Washington, DC, 1990) pp. II-531–534.

[113] ——, "Adding top-down expectation into the learning procedure of self-organizing maps," *Proc. Int. Joint. Conf. on Neural Networks, IJCNN-90-WASH-DC* (Washington, DC, 1990) pp. I-735–738.

[114] A. Yamadori and M. L. Albert, "Word category aphasia," *Cortex*, vol. 9, pp. 112–125, 1973.

[115] P. L. Zador, "Asymptotic quantization error of continuous signals and the quantization dimension," *IEEE Trans. Inform. Theory*, vol. IT-28, pp. 139–149, March 1982.

[116] S. Zeki, "The representation of colours in the cerebral cortex," *Nature*, vol. 284, pp. 412–418, 1980.

Cognitron: A Self-organizing Multilayered Neural Network

Kunihiko Fukushima

NHK Broadcasting Science Research Laboratories, Kinuta, Setagaya, Tokyo, Japan

Received: February 4, 1975

Abstract

A new hypothesis for the organization of synapses between neurons is proposed: "The synapse from neuron x to neuron y is reinforced when x fires provided that no neuron in the vicinity of y is firing stronger than y". By introducing this hypothesis, a new algorithm with which a multilayered neural network is effectively organized can be deduced. A self-organizing multilayered neural network, which is named "cognitron", is constructed following this algorithm, and is simulated on a digital computer. Unlike the organization of a usual brain models such as a three-layered perceptron, the self-organization of a cognitron progresses favorably without having a "teacher" which instructs in all particulars how the individual cells respond. After repetitive presentations of several stimulus patterns, the cognitron is self-organized in such a way that the receptive fields of the cells become relatively larger in a deeper layer. Each cell in the final layer integrates the information from whole parts of the first layer and selectively responds to a specific stimulus pattern or a feature.

1. Introduction

It is thought that the synaptic connections between neurons in the brain are not completely inherited, but are plastically modified by learning or experiences after birth. It is reported, for instance, that early experiences during maturation grossly modify the visual cortex of a cat. In the normal adult cat, neurons of the visual cortex are selectively sensitive to the orientation of the lines and edges in the visual field, and the preferred orientation of different neurons are uniformly distributed in all orientations (Hubel and Wiesel, 1959, 1962 and 1965). Blakemore and Cooper (1970) reared kittens in an abnormal environment consisting entirely of black-and-white stripes of one orientation. These animals had no cortical neurons responding to the orientation perpendicular to the stripes that they saw when they were young. It seems that the visual cortex adjusts itself during maturation to the nature of the visual experience. Perhaps the nervous systems are plastically modified to match the probability of occurrence of features in the visual inputs. Such kind of selforganization of neural networks would be more prominant in the higher center of the brain.

At present, however, the algorithm with which a neural network is self-organized is not known. Although several hypothesis for it have been proposed, none of them has been physiologically substantiated.

The three-layered perceptron proposed by Rosenblatt (1962) is one of the examples of the brain models based on such hypotheses. For a while after the perceptron was proposed, its capability for information processing was greatly expected, and many research works on it have been made. With the progress of the researches, however, it was gradually revealed that the capability of the perceptron is not so large as it had been expected at the beginning.

Although the perceptron consists of only three layers of neurons, it is known that the capability of a layered neural network is greatly enlarged if the number of the neural layers is increased. A model of the mechanism of feature extraction in the visual nervous system proposed by the author (Fukushima, 1970 and 1971) would be one of the examples which shows the capability of a multilayered neural network. In that model, however, the synaptic connections between neurons are fixed, and the plastic modification of the synapses has not been considered. It can be inferred that, even for a neural network with modifiable synapses, the multiplication of the neural layers would increase the capability of the network. Since the algorithm with which a multilayered neural network can be effectively organized has not been known, however, brain models with multilayered structure which have functions of memory or learning have little been reported. Consequently, the selforganizing system or a brain model hitherto reported did not go beyond the confine of a three-layered perceptron, in which only the synapses between the last two layers are modifiable.

In this paper, a new hypothesis for the organization of synapses between neurons is proposed. By introducing this hypothesis, a new algorithm with which a multilayered neural network is effectively organized can be deduced. A self-organizing multilayered neural

Reprinted with permission from *Biological Cybernetics*, vol. 20, pp. 121–136, 1975.

network is constructed following this algorithm, and is named a "cognitron". The performance of a cognitron has been simulated on a digital computer.

In a cognitron, similarly to the animal's brain, synaptic connections between neurons are plastically modified so as to match the nature of its experience. The neurons become selectively responsive to the features which have been frequently presented. Since wasteful functions which are utilized only for the detection of rarely appearing features are not formed, the capability of the neural network can be fully exhibited.

2. Hypothesis on the Synapse Modification

2.1. Hypotheses Hitherto Proposed

According to Marr (1970), the hitherto-proposed hypotheses on the synapse modification are classified into the following three categories as shown in Fig. 1.

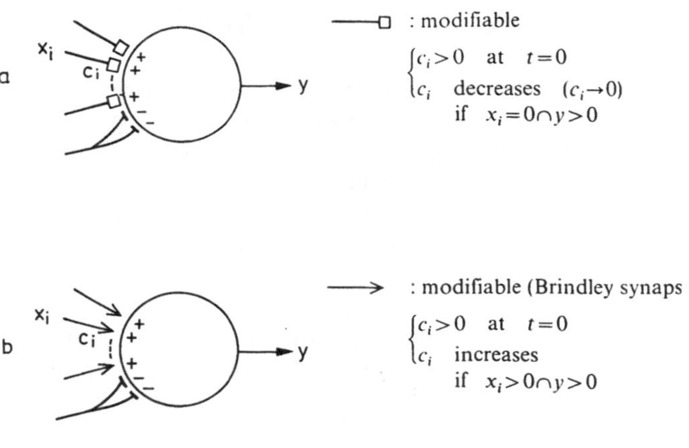

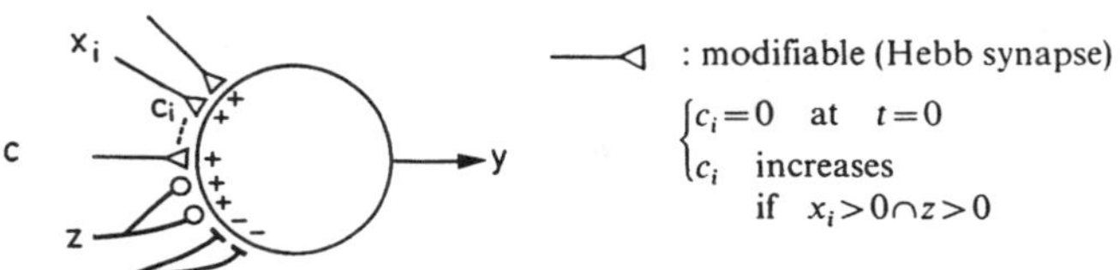

Fig. 1a–c. Hitherto-proposed three hypotheses on the modification of synapses

In the hypothesis of Fig. 1a, a modifiable afferent synapse c_i is excitatory at the beginning. It becomes ineffective, if and only if the postsynaptic cell y firs without presynaptic activity x_i. This hypothesis is based on the idea that the irrelevant synapses become ineffective. Incidentally, the inhibitory afferent synapses shown in the figure control the threshold of the cell y at an appropriate level.

This hypothesis, however, has a fatal disadvantage: once an improper stimulus is given to the neural network, the network would suffer an irrecoverable damage, because the synapses are irreversibly modified only toward extinction.

In the hypothesis of Fig. 1b, a modifiable synapse c_i has a certain amount of excitatory component at the initial state. It will be reinforced if there is a presynaptic activity x_i simultaneously with the firing of the postsynaptic cell y. This hypothesis is based on the idea that only the synapses relevant to the firing of the postsynaptic cell are reinforced. Such synapses are called Brindley Synapses.

In a four-layered perception (Block *et al.*, 1962), the synapses from A^{I}-cells to A^{II}-cells are of this type. Brindley synapses are also used in the model of visual cortex proposed by von der Malsburg (1973), in which orientation sensitive simple cells are self-organized. In these models, Brindley synapses are used only between particular two layers. It is not known whether a multilayered network with only Brindley synapses can be favorably organized or not. Probably, very sophisticated initial connections between neurons would be necessary, in order a multilayered network with only Brindley synapses be favorably organized. The information quantity for such sofisticated initial connections might be too much to be transmitted hereditarily, if they are to be determined by birth not only in the distal system but also in the higher center of the brain.

In the hypothesis of Fig. 1c, the postsynaptic cell y possesses another synaptic input z which controls the reinforcement of its ordinary afferent synapses. An ordinary synapse c_i, which is named a Hebb synapse, is initially ineffective, and is reinforced if there is a presynaptic activity x_i in conjunction with the control signal z. That is, the afferent synapses are reinforced following the state of the input signals at the moment when the control signal z comes. Marr (1969) proposes a hypothesis that the learning in the cerebellum is carried out by the synapses of this type, and that the afferent input from a climbing fiber corresponds to the control signal z. The control signal z is considered to be an instruction from a "teacher" for a "supervised learning". Following this concept, the modifiable synapses in a three-layered perceptron might be classified to this type.

In order to organize a multilayered network composed of Hebb synapses, a "teacher" should give instructions how each individual cell should respond whenever a stimulus is given to the network. These instructions must be given not only to the cells of the final layer, but also to the cells of the intermediate layers. It is difficult to imagine that such a "teacher" exists in the brain.

Although above-mentioned hypotheses have been proposed, none of them has been physiologically substantiated, and they are still a matter of conjecture. Even if these hypotheses be accepted, it seems very difficult to deduce only from these hypotheses an algorithm for a successful organization of a multi-layered neural network.

If the synapses are modified following the new hypothesis proposed below, however, even a multi-layered neural network can be satisfactory organized neither with a special "teacher" nor with sofisticated initial connections.

2.2. A New Hypothesis

A new hypothesis on the modification of synapses is proposed here[1] (Fig. 2):

The synaptic connection from cell x to cell y is reinforced if and only if the following two conditions are simultaneously satisfied.

(i) Presynaptic cell x fires.

(ii) None of the postsynaptic cells situated near the cell y fires stronger (or brisker) than y.

It is physiologically believed that there are two kinds of neurons: excitatory ones and inhibitory ones. The former cells give excitatory effects, and the latter cells give inhibitory effects to the postsynaptic cells. It is assumed that the above hypothesis holds in case where the presynaptic cell x is an inhibitory one as well as an excitatory one. Where, the reinforcement of an inhibitory synapse means that the synapse is made more inhibitory (not less inhibitory).

It is assumed here that a cell does not always have a possibility to have afferent synapses from all the other cells. A cell can have afferent synapses only from a group of cells situated in a particular area predetermined for each cell. This area is named "connectable area" of the cell. The connectable area is determined by the spread of the dendrites of the postsynaptic cell and the spread of the axon terminals of the presynaptic cells.

Condition (ii) in the hypothesis means that, among a group of postsynaptic cells situated in a small area, only one cell has its afferent synapses reinforced. This small area is named a "vicinity area". The neighboring cells generally have approximately the same connectable area. Hence, if condition (ii) were neglected and only condition (i) were imposed, all the neighboring cells would become to have almost the same afferent synapses. As a matter of fact, however, since

[1] This hypothesis has been reported by the author in Japanese (Fukushima, 1974)

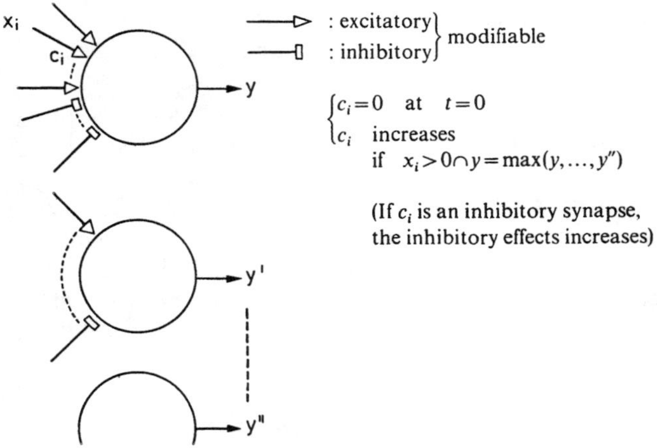

Fig. 2. A new hypothesis on the modification of synapses

condition (ii) is also imposed, only one cell which happens to have yielded a maximum output in the vicinity area is reinforced. The afferent synapses to other cells in the vicinity area, which have almost the same connectable areas, remains unchanged.

This is the case where there is at least one postsynaptic cell firing in the vicinity area. If no postsynaptic cell happens to fire in a vicinity area, however, all the postsynaptic cells in the area are to be reinforced under condition (i). Such a situation might occur, when the neural network is in an initial state where there are no connections between neurons, or when a stimulus unexperienced before is given to the network. Quantitatively speaking, however, as is discussed later, the amounts of the reinforcement of the afferent synapses are smaller when all the cells in the vicinity area are simultaneously reinforced than when only one cell is exclusively reinforced.

Although the connectable areas of the cells in a vicinity area overlap each other, they do not strictly coincide with each other but are slightly different. Therefore, even if all the cells in the vicinity area are once reinforced simultaneously, it does not mean that all of these cells are grown to have identical afferent synapses. Once even a slight difference in characteristics is generated between the cells, it will grow further and further because of condition (ii). Hence, each cell becomes to have its own individual characteristics.

If it is assumed that the synapses grow under these conditions, the neural network would also have a self-repairing function. That is, even if a certain cell is damaged, another cell will substitute for the damaged cell. When a cell which has responded strongly to a certain stimulus becomes nonresponsive because of a damage, another cell which happens to respond stronger than the other cells to this stimulus will be grown so as to substitute for the damaged cell. Until the damage of the first cell, the second cell has been prevented from reinforcement of its afferent synapses.

Here, let us discuss whether the hypothesis is reasonable. Let us suppose, for instance, that the following situation takes place. For the reinforcement of the synapses, a certain kind of chemical substance must be supplied to the postsynaptic cells, and a kind of glia cells participate in the nourishment of this chemical substance. The spread of one glia cell coincides with a vicinity area. The glia cell supplies the nutrient concentratedly to a single cell which has yielded a maximum output within the vicinity area. In case where no cell is firing in the vicinity area, however, all the cells in the area are equally supplied with the nutrient. Since the total amount of the nutrient supplied at a time is limitted, the amount of nutrient given to each individual cell becomes smaller in this case than in case where only one cell is concentratedly nourished. If such a situation is supposed, the hypothesis proposed here would not be so ill-advised.

3. Neural Element

Before discussing the structure of the multilayered neural network "cognitron", let us discuss the characteristics of neural elements employed in a cognitron.

The neural element (which, in this paper, sometimes will be called merely a neuron or a cell) is of analog type with a mechanism of shunting inhibition. The inputs and the output of a neural element take non-negative analog values proportional to the pulse densities (or instantaneous mean frequencies) of the firing of the actual biological neurons. Let $u(1), \ldots, u(N)$ be the inputs from excitatory afferent synapses (that is, the the outputs of the presynaptic cells), and $v(1), \ldots, v(M)$ be the inputs from inhibitory afferent synapses. The output w of this cell is defined by:

$$w = \varphi\left[\frac{1+\sum_{\nu=1}^{N} a(\nu)\cdot u(\nu)}{1+\sum_{\mu=1}^{M} b(\mu)\cdot v(\mu)} - 1\right], \tag{1}$$

where $\varphi[\ ]$ is a function defined by the following equation:

$$\varphi[x] = \begin{cases} x & (x \geqq 0) \\ 0 & (x < 0)\,. \end{cases} \tag{2}$$

The conductances of the excitatory and inhibitory synapses $a(\nu)$ and $b(\mu)$ take non-negative analog values.

The input-to-output characteristics of the cell could be interpreted as follows. The first term in [ ] of Eq. (1) stand for the membrane potential (short-term mean value), which is raised by the inputs from excitatory synapses, and, at the same time, is shunted by the effect of the inputs from inhibitory synapses. It is assumed that the cell fires with a pulse density proportional to the difference between the membrane potential determined in this way and the resting potential indicated by the second term in [ ] of Eq. (1).

Let e be the sum of all the excitatory effects, and h be the sum of all the inhibitory effects. That is,

$$e = \sum_{\nu=1}^{N} a(\nu)\cdot u(\nu)\,, \tag{3}$$

$$h = \sum_{\mu=1}^{M} b(\mu)\cdot v(\mu)\,. \tag{4}$$

With these symbols, Eq. (1) can also be written as

$$w = \varphi\left[\frac{1+e}{1+h} - 1\right] = \varphi\left[\frac{e-h}{1+h}\right]. \tag{5}$$

When the inhibitory input is small ($h \ll 1$), we have $w \doteqdot \varphi[e-h]$, which coincides with the characteristics of the usual analog-threshold-element (Fukushima, 1969). For a system like a cognitron where the synaptic conductances $a(\nu)$ and $b(\mu)$ increase further and further with the progress of learning, the employment of such elements like analog-threshold-elements is improper because their outputs increase boundlessly. In the cell proposed here, however, when the conductances of the input synapses increase and we have $e \gg 1$ and $h \gg 1$, Eq. (5) approximately becomes $w \doteqdot \varphi[e/h-1]$ where the output is determined by the ratio e/h not by the difference of e and h. Therefore, even if the synaptic conductances increase with learning, the output of the cell converges to a certain value without divergence, so long as both the excitatory synaptic conductances $a(\nu)$ and the inhibitory ones $b(\mu)$ increases with the same rate.

Let us look at the input-to-output relation of the cell in case where the excitatory and the inhibitory input increase in proportion. If we write

$$e = \varepsilon x\,, \quad h = \eta x$$

and if $\varepsilon > \eta$ holds, Eq. (5) can be transformed into

$$\begin{aligned} w &= \frac{(\varepsilon-\eta)x}{1+\eta x} \\ &= \frac{\varepsilon-\eta}{2\eta}\left\{1+\tanh\left(\frac{1}{2}\log \eta x\right)\right\}. \end{aligned} \tag{6}$$

This input-to-output relation coincides with a logarithmic relation expressed by Wever-Fechner's law to which a S-shaped saturation expressed by tanh is added. The same expression is often used as an empirical formula in neurophysiology and psychology to approximate the nonlinear input-to-output relations of the sensory receptors (for instance, cones) and the overall sensory systems of animals.

Since the neural element of this type resembles so well in characteristics to the biological neuron, it has a wide application not only for a cognitron but also for various kinds of visual and auditory information processing systems.

4. Structure of a Cognitron

4.1. The Basic Structure

According to the hypothesis discussed in Section 2.2, a self-organizing multilayered neural network, cognitron, is constructed (Fukushima, 1974). At first, the basic idea for the construction of a cognitron is discussed.

The cognitron has a multilayered structure. It consists of a number of neural layers of a similar structure cascaded one after another. The l-th layer U_l consists of excitatory neurons $u_l(\boldsymbol{n})$ and inhibitory neurons $v_l(\boldsymbol{n})$, where $\boldsymbol{n}=(n_x, n_y)$ is a two-dimensional co-ordinates indicating the location of a cell.

An excitatory cell $u_l(\boldsymbol{n})$ receives modifiable synaptic connections from neurons $u_{l-1}(\boldsymbol{n}+\boldsymbol{v})$ $[\boldsymbol{v}\in S_l]$ and $v_{l-1}(\boldsymbol{n})$ in the preceding layer U_{l-1}. If we write the conductances of the synaptic connections as $a_l(\boldsymbol{v}, \boldsymbol{n})$ and $b_l(\boldsymbol{n})$, the output of the cell $u_l(\boldsymbol{n})$ is given by

$$u_l(\boldsymbol{n})=\varphi\left[\frac{1+\sum_{\boldsymbol{v}\in S_l} a_l(\boldsymbol{v}, \boldsymbol{n})\cdot u_{l-1}(\boldsymbol{n}+\boldsymbol{v})}{1+b_l(\boldsymbol{n})\cdot v_{l-1}(\boldsymbol{n})}-1\right] \quad (7)$$

where S_l indicates the connectable area of a cell.

Meanwhile, the inhibitory cell $v_{l-1}(\boldsymbol{n})$ receives fixed excitatory synaptic connections $c_{l-1}(\boldsymbol{v})[\geqq 0]$ from the neighboring excitatory cells $u_{l-1}(\boldsymbol{n}+\boldsymbol{v})$, and yields an output equal to the mean value of the outputs of the neighboring cells:

$$v_{l-1}(\boldsymbol{n})=\sum_{\boldsymbol{v}\in S_l} c_{l-1}(\boldsymbol{v})\cdot u_{l-1}(\boldsymbol{n}+\boldsymbol{v})\,. \quad (8)$$

The values of the fixed synaptic connections are so determined as to satisfies

$$\sum_{\boldsymbol{v}\in S_l} c_{l-1}(\boldsymbol{v})=1\,. \quad (9)$$

As is seen from Eqs. (7) and (8), the connectable area of the cell $u_l(\boldsymbol{n})$ coincides with the connecting area of this inhibitory cell $v_{l-1}(\boldsymbol{n})$. Figure 3 shows how the cells of layers U_{l-1} and U_l are connected.

The reinforcement of the afferent synapses of cell $u_l(\boldsymbol{n})$ takes place only when none of the cells situated in the vicinity of $u_l(\boldsymbol{n})$ is firing stronger than $u_l(\boldsymbol{n})$. Let

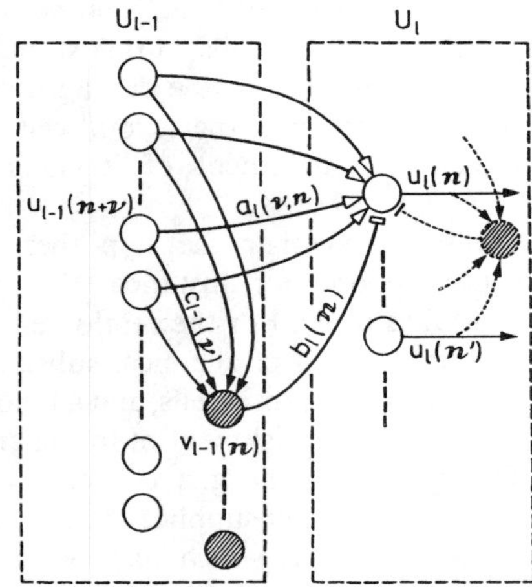

Fig. 3. The basic structure of the cognitron. The interconnections between two adjoining layers are shown

$\delta_l(\boldsymbol{n})$ be a function which takes a value 1 or 0 depending whether the synaptic reinforcement for cell $u_l(\boldsymbol{n})$ is performed or not:

$$\delta_l(\boldsymbol{n})=\begin{cases}1 & \text{if } u_l(\boldsymbol{n})\geqq u_l(\boldsymbol{n}+\boldsymbol{v}) \text{ for every } \boldsymbol{v}\in\Omega_l\\ 0 & \text{otherwise}\end{cases} \quad (10)$$

where Ω_l stand for a vicinity area (for instance, the area in which $|\boldsymbol{v}|<R$ holds).

When $\delta_l(\boldsymbol{n})=1$ holds and the reinforcement is to be performed, the amounts of the synaptic reinforcement $\Delta a_l(\boldsymbol{v}, \boldsymbol{n})$ and $\Delta b_l(\boldsymbol{n})$ change depending whether $u_l(\boldsymbol{n})=0$ or $u_l(\boldsymbol{n})>0$.

When $u_l(\boldsymbol{n})=0$:

$$\Delta a_l(\boldsymbol{v}, \boldsymbol{n})=q_0\cdot c_{l-1}(\boldsymbol{v})\cdot u_{l-1}(\boldsymbol{n}+\boldsymbol{v})\cdot\delta_l(\boldsymbol{n})\,, \quad (11)$$

$$\Delta b_l(\boldsymbol{n})=q_0\cdot v_{l-1}(\boldsymbol{n})\cdot\delta_l(\boldsymbol{n})\,. \quad (12)$$

When $u_l(\boldsymbol{n})>0$:

$$\Delta a_l(\boldsymbol{v}, \boldsymbol{n})=q_1\cdot c_{l-1}(\boldsymbol{v})\cdot u_{l-1}(\boldsymbol{n}+\boldsymbol{v})\cdot\delta_l(\boldsymbol{n})\,, \quad (13)$$

$$\Delta b_l(\boldsymbol{n})=\frac{\sum_{\boldsymbol{v}\in S_l} a_l(\boldsymbol{v}, \boldsymbol{n})\cdot u_{l-1}(\boldsymbol{n}+\boldsymbol{v})}{2v_{l-1}(\boldsymbol{n})}\cdot\delta_l(\boldsymbol{n})$$

$$=\frac{q_1\sum_{\boldsymbol{v}\in S_l} c_{l-1}(\boldsymbol{v})\cdot u_{l-1}^2(\boldsymbol{n}+\boldsymbol{v})}{2v_{l-1}(\boldsymbol{n})}\cdot\delta_l(\boldsymbol{n}) \quad (14)$$

where q_0 and q_1 are positive constants which satisfies

$$q_1>q_0>0\,. \quad (15)$$

4.2. Quantitative Analysis

Let us discuss the implication of this algorithm on reinforcement.

When both $\delta_l(\boldsymbol{n})=1$ and $u_l(\boldsymbol{n})=0$ hold, it is seen from Eq. (10) that there is no neuron firing in the vicinity of $u_l(\boldsymbol{n})$. Consequently, also the other cells in its vicinity are reinforced following the algorithm of Eq. (11) and (12), if they are to be reinforced. As is seen from (15), the amounts of reinforcement are set smaller in this case than in case where $u_l(\boldsymbol{n})$ is the only cell reinforced in its vicinity. The latter case arises when both $\delta_l(\boldsymbol{n})=1$ and $u_l(\boldsymbol{n})>0$ hold.

As shown in Fig. 3, there are two paths of information flow from cell $u_{l-1}(\boldsymbol{n}+\boldsymbol{v})$ in the preceding layer to cell $u_l(\boldsymbol{n})$: the excitatory information through the synapses of conductance $a_l(\boldsymbol{v}, \boldsymbol{n})$, and the inhibitory information transmitted via inhibitory cell $v_{l-1}(\boldsymbol{n})$. As for the inhibitory information flow, the conductance of the synapse from $u_{l-1}(\boldsymbol{n}+\boldsymbol{v})$ to $v_{l-1}(\boldsymbol{n})$ is $c_{l-1}(\boldsymbol{v})$ which is unmodifiable, and the conductance of the inhibitory synapse from $v_{l-1}(\boldsymbol{n})$ to $u_l(\boldsymbol{n})$ is $b_l(\boldsymbol{n})$ which is modifiable. The input-to-output characteristic of cell $v_{l-1}(\boldsymbol{n})$ is linear since it does not have inhibitory inputs. Hence, the overall conductance of the inhibitory path from $u_{l-1}(\boldsymbol{n}+\boldsymbol{v})$ to $u_l(\boldsymbol{n})$ is $c_{l-1}(\boldsymbol{v})\cdot b_l(\boldsymbol{n})$.

Let $r_l(\boldsymbol{v}, \boldsymbol{n})$ be the ratio of the amount of reinforcement of the excitatory conductance $\Delta a_l(\boldsymbol{v}, \boldsymbol{n})$ to that of the over all inhibitory conductance $c_{l-1}(\boldsymbol{v})\cdot \Delta b_l(\boldsymbol{n})$:

$$r_l(\boldsymbol{v}, \boldsymbol{n})=\frac{\Delta a_l(\boldsymbol{v}, \boldsymbol{n})}{c_{l-1}(\boldsymbol{v})\cdot \Delta b_l(\boldsymbol{n})}. \tag{16}$$

That is, $r_l(\boldsymbol{v}, \boldsymbol{n})$ takes a value greater than or less than 1 depending upon which of the excitatory or the inhibitory connection from cell $u_{l-1}(\boldsymbol{n}+\boldsymbol{v})$ to $u_l(\boldsymbol{n})$ be reinforced stronger than the other.

It is reduced from Eqs. (11) and (12), or from Eqs. (13) and (14), that $r_l(\boldsymbol{v}, \boldsymbol{n})>1$ hods if and only if the following conditions are satisfied.

When $u_l(\boldsymbol{n})=0$:

$$u_{l-1}(\boldsymbol{n}+\boldsymbol{v})>v_{l-1}(\boldsymbol{n}). \tag{17}$$

When $u_l(\boldsymbol{n})>0$:

$$u_{l-1}(\boldsymbol{n}+\boldsymbol{v})>\frac{\sum_{\boldsymbol{\mu}\in S_l} c_{l-1}(\boldsymbol{\mu})\cdot u_{l-1}^2(\boldsymbol{n}+\boldsymbol{\mu})}{2v_{l-1}(\boldsymbol{n})}. \tag{18}$$

As is seen from Eqs. (8) and (9), the output of cell $v_{l-1}(\boldsymbol{n})$ is equal to the (weighted) mean value of the outputs of the cells within the connectable area of $u_l(\boldsymbol{n})$. Accordingly, inequality (17) means that, when $u_l(\boldsymbol{n})=0$, the excitatory connections from the cells responding stronger than the mean value are more reinforced than the inhibitory ones. As for the case where $u_l(\boldsymbol{n})>0$, since the right side of inequality (18) is a little complicated, we will consider for a moment a simple case where all the outputs of the cells within the connectable area take value 0 or 1. In this special case, the right-hand side of inequality (18) is equal to 1/2. This means that, when $u_l(\boldsymbol{n})>0$, the excitatory connections from the cells responding stronger than 1/2 are more reinforced than the inhibitory ones.

In both cases, qualitatively speaking, the excitatory connection is more reinforced than the inhibitory one from a cell which is yielding a relatively large response $[r_l(\boldsymbol{v}, \boldsymbol{n})>1]$, and, inversely, the inhibitory connection is more reinforced than the excitatory one from a cell which is yielding a relatively small response $[r_l(\boldsymbol{v}, \boldsymbol{n})<1]$.

Generally, in a cognitron, the number of postsynaptic cells which are reinforced at one time tend to only one within a single vicinity area, after a certain degree of learning has taken place. Hence, at every instance, the distribution of strongly firing cells generally becomes sparse. Therefore, using Eqs. (16) and (11)–(14), we can conclude that the value of $r_l(\boldsymbol{v}, \boldsymbol{n})$ is generally much smaller in case of $u_l(\boldsymbol{n})>0$ than in case of $u_l(\boldsymbol{n})=0$, if the inputs are the same for the two cases. That is, the relative amount of reinforcement of the inhibitory synapses to that of the excitatory ones is much larger in the former case. As the result of such a strong reinforcement of the inhibitory connections, the postsynaptic cells become reluctant to respond to other stimulus patterns than the one to which the cell has been reinforced, and the cognitron acquires the ability to differentiate a pattern from other similar patterns.

The amount of reinforcement for the case of $u_l(\boldsymbol{n})>0$ is so determined that the increment of the total excitatory effect to the postsynaptic cell is just 2 times as large as that of the inhibitory one if the same stimulus pattern is given again. That is,

when $u_l(\boldsymbol{n})>0$:

$$\frac{\sum_{\boldsymbol{v}\in S_l} \Delta a_l(\boldsymbol{v}, \boldsymbol{n})\cdot u_{l-1}(\boldsymbol{n}+\boldsymbol{v})}{\Delta b_l(\boldsymbol{n})\cdot v_{l-1}(\boldsymbol{n})}=2. \tag{19}$$

Therefore, if the same stimulus patterns are given repeatedly, the output of cell $u_l(\boldsymbol{n})$ gradually increases and tend to 1. This conclusion can be deduced from Eqs. (7) and (19).

If the amount of reinforcement is shared in such a way between the excitatory and inhibitory synapses, however, the inhibitory synapses become too strong in case of $u_l(\boldsymbol{n})=0$. Let us consider for a moment what would happen if the synapses were to be reinforced according to Eqs. (13) and (14) even for $u_l(\boldsymbol{n})=0$. In the initial state where the synaptic connections from layer U_{l-1} to layer U_l are not completed, cell $u_l(\boldsymbol{n})$ might be presented with several kinds of different stimuli without yielding a response. If these stimulus patterns are random, the overall inhibitory connection from any of the cells $u_{l-1}(\boldsymbol{n}+\boldsymbol{v})$ would become stronger than that of the excitatory one. Hence, cell $u_l(\boldsymbol{n})$ would become nonresponsive to any of the stimulus patterns, and the organization of the cognitron would stop. Actually, in case of $u_l(\boldsymbol{n})=0$, the amount of reinforcement of the excitatory synapse are kept small so as to satisfies the following equation.

When $u_l(\boldsymbol{n})=0$:

$$\sum_{\boldsymbol{v}\in S_l} \Delta a_l(\boldsymbol{v}, \boldsymbol{n}) = \sum_{\boldsymbol{v}\in S_l} c_{l-1}(\boldsymbol{v}) \cdot \Delta b_l(\boldsymbol{n}) . \qquad (20)$$

Equation (20) means that the synapses are reinforced under the condition that the cell responds neither in an excitatory nor in an inhibitory manner to a uniform pattern (or to a d.c. component of the spatial frequency).

4.3. Lateral Inhibition

In the last two sections, the principles of the construction of a cognitron has been discussed. In the actual construction of the network, however, a little modification is made.

An excitatory cell $u_l(\boldsymbol{n})$ receives lateral inhibition from the neighboring cells. That is, as shown in Fig. 3, there is a mechanism of backward lateral inhibition among the excitatory cells of the same layer.

In the computer simulation mentioned later, however, the backward lateral inhibition is substituted by a forward lateral inhibition in order to save the computation time.

That is, the right-hand side of Eq. (7) is regarded as a intermediate output $u_l'(\boldsymbol{n})$ instead of the final output of the cell $u_l(\boldsymbol{n})$. The fined output $u_l(\boldsymbol{n})$ is obtained by application of lateral inhibition to this intermediate output $u_l'(\boldsymbol{n})$. So, instead of Eq. (7), we have intermediate output $u_l'(\boldsymbol{n})$ by

$$u_l'(\boldsymbol{n}) = \varphi\left[\frac{1+\sum_{\boldsymbol{v}\in S_l} a_l(\boldsymbol{v}, \boldsymbol{n}) \cdot u_{l-1}(\boldsymbol{n}+\boldsymbol{v})}{1+b_l(\boldsymbol{n}) \cdot v_{l-1}(\boldsymbol{n})} - 1\right] \qquad (21)$$

and the final output of the cell is given by

$$u_l(\boldsymbol{n}) = \varphi\left[\frac{1+u_l'(\boldsymbol{n})}{1+\sum_{\boldsymbol{\mu}\in H_l} g_l(\boldsymbol{\mu}) \cdot u_l'(\boldsymbol{n}+\boldsymbol{\mu})} - 1\right]. \qquad (22)$$

In this equation, $g_l(\boldsymbol{\mu})$ represents the conductance of the equivalent forward lateral inhibition transformed from the backward one, and is determined so as to satisfy

$$\sum_{\boldsymbol{\mu}\in H_l} g_l(\boldsymbol{\mu}) = 1 \qquad (23)$$

where H_l represents the spread of this lateral inhibition $g_l(\boldsymbol{\mu})$.

This kind of lateral inhibition is helpful for getting rid of an awkward situation mentioned below. Suppose that the spatial distribution pattern of the outputs of the excitatory cells of U_l happens to be of the shape like, say, Mt. Fuji, and have a single peak surrounded by a wide spread slope. In such a case, any of the cells in the slope is in the state that some other cells in its vicinity are yielding larger outputs than it. Hence, no cell in the slope would be reinforced. It is only the cell at the peak that can be reinforced, and a favorable progress of the self-organization could not be expected.

If the above-mentioned mechanism of lateral inhibition is added to the network, however, the wide spread slope vanishes, and a number of cells in the layer can be reinforced at a time. This results in a successful self-organization.

4.4. Connectable Areas and Branching of Axon Terminals

A cognitron consists of many neural layers of the above-mentioned fundamental structure cascaded one after another. The layers are named U_0, U_1... from the headmost one. Layer U_0 is an input layer to which stimulus patterns are presented. The cells of layer U_0, however, are not necessarily the sensory receptor cells themselves. We suppose that another neural network like a feature extractor is placed in front of layer U_0, and the cells of layer U_0 receive the information already processed to some extent. Here we use the word "receptive field" in a wide sense: The area in layer U_0 (not necessarily the sensory receptor layer) which affects the response of a cell in a deeper layer will be called the receptive field of the cell.

In cascading layers, it is important how to determine the connectable area of each cell. Let us compare three possible methods shown in Fig. 4.

In the method of Fig. 4a, the size of the connectable area of each cell is determined to become equal independent of the layer to which the cell belongs. In this case, however, even if the number of layers are increased, the receptive field of a cell of the last layer does not increase its size so noticeably. In order the receptive field of a cell of the last layer covers the whole U_0 layer, a considerable number of layers should be cascaded.

On the other hand, in the method of Fig. 4b, the size of the connectable area of a cell increases with the depth of the layer in which the cell is situated. In this case, the receptive field of a cell in the last layer can be made to cover the whole U_0 layer with a less number of layers than for the method of Fig. 4a. If the sizes of the connectable areas are determined in such a way, however, all the cells in a deeper layer, especially in the last layer, would become to have almost identical connect-

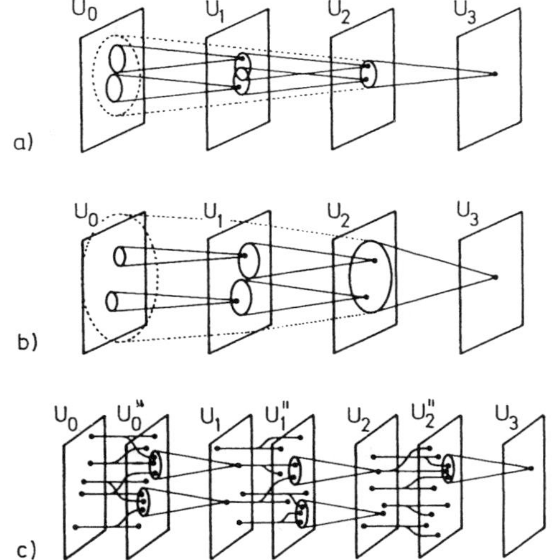

Fig. 4a–c. Three possible methods for interconnecting layers. The connectable area of each cell is differently chosen in these three methods. Method c is adopted for the cognitron discussed in this paper

able area because of too much overlapping. Hence, in order to make each cell have its own individual character, it is necessary to widen the size of the vicinity area Ω_l (which is the area in which the reinforcement of more than two cells is generally prevented). Consequently, we have a situation where only one or a few cell fire at a time in the last layer. This seems to contradict to the situation in the animals brain where it is supposed that many cells would fire for one stimulus pattern, and that the configuration of the firing pattern would correspond to a "concept". Furthermore, no more processing of information could be expected in this case, even if another layer is cascaded after the last layer, since the cells of the newly added layer would normally have only one firing cell at a time in their connectable areas. Judging from these, the method like Fig. 4b does not seem desirable either.

It is true that, however, for some applications, it would be desirable to have a network in which the firing of only one cell correspond to one stimulus. In such a case, it is good enough to design the network having a structure like Fig. 4b.

In Fig. 4c, the axons of the excitatory cells ramifies into a number of branches, and the destination of each branch is probabilistically distributed. The probabilistic distribution here does not mean, however, that the deviation of the destination from the starting position is completely random. The destinations are determined in such a way that branches with large deviation be less numerous.

With this method, the receptive field of each cell of the last layer will cover the whole U_0-layer without widening the connectable areas of postsynaptic cells so much nor increasing the number of layers so much. It is also possible to cascade another layers after the last one, since a number of cells would fire at a time even in the last layer.

In this paper, the method of Fig. 4c is adopted. Suppose that the axon of every excitatory cell $u_l(\boldsymbol{n})$ ramifies into $(K+1)$ branches. Let the spatial pattern representing the firing states of the terminals of these branches be $u_l''(\boldsymbol{n}, k)$ $[k=0, 1, 2, \ldots, K]$. Suppose the branches with $k=0$ are not deviated:

$$u_l''(\boldsymbol{n}, 0) = u_l(\boldsymbol{n}). \tag{24}$$

On the other hand, the destinations of the branches with $k \neq 0$ receive probabilistic permutation. Let $\mathscr{P}_{lk}$ be the operator representing the permutation on $\boldsymbol{n}$. Then we have

$$\{u_l''(\boldsymbol{n}, k)\} = \mathscr{P}_{lk}\{u_l(\boldsymbol{n})\} \quad (k \neq 0). \tag{25}$$

Here the assignment of the permutation of each element is made in such a way that a permutation with small disparity between the positions before and after the permutation occurs more frequently.

In the computer simulation mentioned later, we discuss the case with $K=1$, that is, the case where each axon bifurcates into two branches. One of the two branches goes straight, but the other one receives a probabilistic permutation.

In case where every axons ramifies, Eq. (21) is to be replaced by

$$u_l'(\boldsymbol{n}) = \varphi \cdot \left[\frac{1 + \sum_{k=0}^{K} \sum_{\boldsymbol{v} \in S_l} a_l(\boldsymbol{v}, \boldsymbol{n}, k) \cdot u_{l-1}''(\boldsymbol{n}+\boldsymbol{v}, k)}{1 + b_l(\boldsymbol{n}) \cdot v_{l-1}(\boldsymbol{n})} - 1\right]. \tag{26}$$

Equations (8), (9), and (11)–(14) holds without change, if only the following replacements of the variables are carried out.

$$\begin{aligned} u_{l-1}(\boldsymbol{n}+\boldsymbol{v}) &\to u_{l-1}''(\boldsymbol{n}+\boldsymbol{v}, k) \\ a_l(\boldsymbol{v}, \boldsymbol{n}) &\to a_l(\boldsymbol{v}, \boldsymbol{n}, k) \\ c_{l-1}(\boldsymbol{v}) &\to c_{l-1}(\boldsymbol{v}, k) \\ \textstyle\sum_{\boldsymbol{v} \in S_l} &\to \textstyle\sum_{k=0}^{K} \sum_{\boldsymbol{v} \in S_l}. \end{aligned} \tag{27}$$

5. Computer Simulation

5.1. The Parameters for the Simulation

In the computer simulation, the parameters are chosen in the following way.

The number of the layers are four, and the layers are named $U_0, U_1, \ldots, U_4$ from the front. In each layer, there are $12 \times 12 = 144$ excitatory cells $u_l(\boldsymbol{n})$ and the same number of inhibitory cells $v_l(\boldsymbol{n})$.

The connectable area S_l is a square of $5 \times 5 = 25$ in size for every l. The vicinity area Ω_l in which the reinforcement of more than two cells is generally prevented, is a little smaller and have a rhombic shape whose hight and width are both 5 (13 in area). The conductance $c_l(\boldsymbol{v}, k)$ of the synaptic connection from an axon terminal $u_l''(\boldsymbol{n}+\boldsymbol{v})$ to a cell $v_l(\boldsymbol{n})$ is actually a function of only $\boldsymbol{v}$ and independent of l and k. It is a two-dimensional Gaussian function of $\boldsymbol{v}$.

The spread of the lateral inhibition H_l is a square of 7×7 in size for every l. The value of the interconnection $g_l(\boldsymbol{\mu})$ is independent of l and is a two-dimensional Gaussian function of $\boldsymbol{\mu}$.

The number of the branches of the ramified axon is 2 ($K = 1$). The permutation $\mathscr{P}_{lk}$ ($k = 1$) of the destinations of the axon terminals has been determined in the following way. At first, from a matrix which indicating the starting position of each axon, a pair of horizontally adjoining two elements have been chosen at random and have been permutated to each other. Next, from this new matrix, a pair of vertically adjoing two elements have been chosen at random and also have been permutated. Such an operation has been repeated 576 times each for the horizontal and vertical direction, and a matrix indicating the state of permutation $\mathscr{P}_{lk}$ has been obtained. Different random numbers have been used for $l = 0$, 1, and 2. The r.m.s. values of the disparities between the positions before and after the permutation determined in this way has been 3.28, 3.54, and 4.05 for $l = 0$, 1, and 2, respectively.

The parameters which determine the amount of reinforcement at a time are adjusted to $q_0 = 2.0$ and $q_1 = 16.0$. The initial conductances of all the modifiable synapses $a_l(\boldsymbol{v}, \boldsymbol{n}, k)$ and $b_l(\boldsymbol{n})$ are chosen to be 0.

Here, the method of boundary correction is discussed briefly. Since the size of the neural layers is limited in the computer simulation, the connectable area S_l and the spread of the inhibitory connection H_l of a cell which is situated near the boundary might exceed the boundary of the layer.

In Eqs. (9) and (23), if it is supposed that the summation is taken only within the part of S_l or H_l which is contained within the boundary of the layer, the left-hand sides of these equations would become less than 1 for such a chipped-off summation area.

Hence, in case of $u_l(\boldsymbol{n}) = 0$, if the reinforcement is made following Eqs. (11) and (12), the inhibition would become too small for the cells near the boundary, and Eq. (20) would not be satisfied for such cells. In order to avoid such an effect, the amount of reinforcement given by the right-hand side of Eq. (12) is compensated by dividing it by the value of left-hand side of Eq. (9) calculated in the above-mentioned sense. As for the lateral inhibitions, the values of $g_l(\boldsymbol{\mu})$ themselves are compensated so as to satisfy the equation (23) in the sense mentioned above.

5.2. Responses of the Cells of Each Layer

Five stimulus patterns "0", "1", "2", "3", "4" have been presented repeatedly to layer U_0 in a cyclic manner. These patterns have been selected to become simple figures when directly watched, only for the sake of the convenience for examining the response. These patterns do not necessarily stand for the unprocessed retinal images themselves.

Figure 5 shows how the cells respond to each of the five stimulus patterns at the 20th cycle of presentation. The responses of the cells of each layer are indicated by the whiteness in the photographs. As is seen in Fig. 5, the cognitron has been organized in such a way that each stimulus pattern elicits its own response pattern to every layer. Most of the cells, especially in layer U_3, have become to respond selectively to one stimulus pattern.

In order to show how the permutation of axon branches is made, Fig. 6 exemplifies the responses at the terminals of the permutated axon-branches for the stimulus pattern "4".

5.3. Reverse Reproduction

In order to verify that the synapses have been satisfactorily organized, an experiment by means of reverse reproduction is made. Actually, the information flow through a synapse is unilateral, and, in case of a cognitron, it is always in the direction from the front to the last layer. In the reverse reproduction, however, it is supposed that the direction of information flow through synapses were to be reversed, and the responses of the cells under such condition are obtained by computer simulation. For example, if the reverse reproduction is made from a single cell of layer U_1 only the U_0-cells which have excitatory effect on this U_1-cell would respond. That is, the excitatory part of the receptive field of this U_1-cell can be seen.

To be more exact, in the reverse reproduction, it is assumed that the information flows in the following way. If a cell $u_l(\boldsymbol{n})$ responds with an intensity $u_l(\boldsymbol{n})$, the amount of excitatory effect transmitted to a branched axon $u_{l-1}''(\boldsymbol{n}+\boldsymbol{v}, k)$ of the preceding layer is $a_l(\boldsymbol{v}, \boldsymbol{n}, k) \cdot u_l(\boldsymbol{n})$, and the amount of inhibitory effect is $c_{l-1}(\boldsymbol{v}, k) \cdot b_l(\boldsymbol{n}) \cdot u_l(\boldsymbol{n})$. The inhibitory effect is transmitted via cell

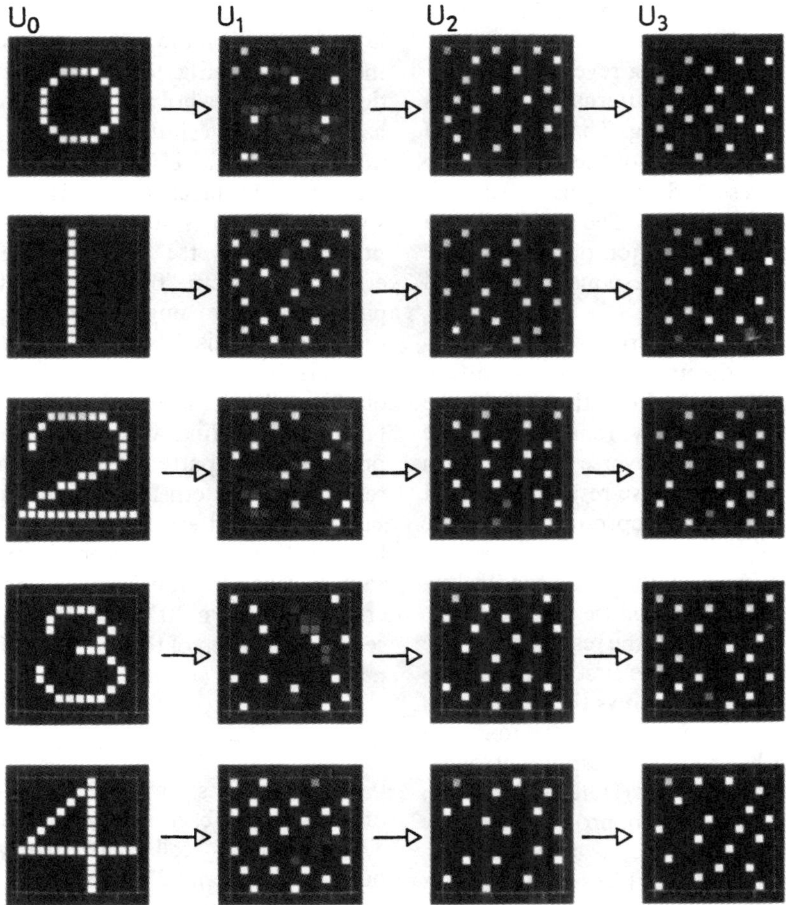

Fig. 5. The responses of the cognitron to five different stimulus patterns. These stimulus patterns have been presented to layer U_0 in a cyclic manner. The responses at the 20th cycle of stimulus presentation are shown

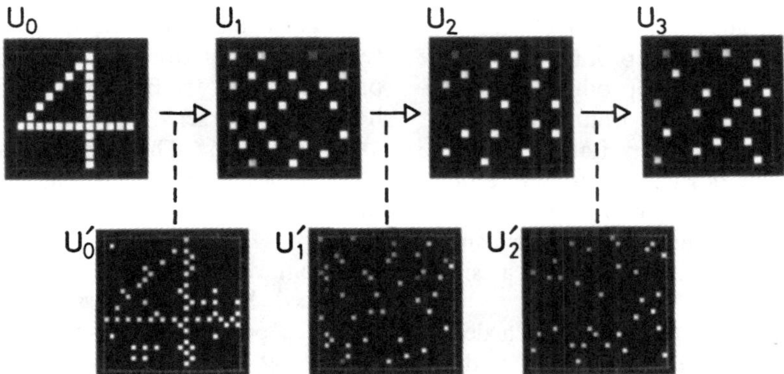

Fig. 6. An example of the responses at the terminals of the permutated axon-branches

$v_{l-1}(\boldsymbol{n})$. It is assumed that the excitatory cells of the preceding layer U_{l-1} has the input-to-output characteristic given by Eq. (5) also for the inputs in the reverse direction. The excitatory input e is assumed to be equal to the sum of the excitatory effects appeared at every branched axon terminals of the cell. The inhibitory input h is also calculated in the same manner. By substituting these values e and h to Eq. (5), the response to the reverse inputs is calculated. It is supposed that the lateral inhibition discussed in Section 4.3 does not work in the process of reverse reproduction.

Figure 7 shows several results of reverse reproduction from the normal response patterns for stimulus "4" after the 20th cycle of learning. That first line of Fig. 7 shows the normal response of the cognitron to stimulus pattern "4" presented to layer U_0 and is identical with the fifth line of Fig. 5. The second line of Fig. 7 shows the reverse reproduction of the response of layer U_0 from the normal response pattern of layer U_1.

In the third line of Fig. 7, the response of layer U_1 is reversely reproduced from the normal response pattern of layer U_2 at first, and, again from this result, the response of layer U_0 is reversely reproduced. The fourth line shows the results of reverse reproduction from layer U_3. As is seen from these results, the input stimulus pattern is correctly reproduced by the reverse reproduction.

In contrast to Fig. 7, in which the reverse reproduction is made from the normal responses of the whole cells of a certain layer, Fig. 8 shows the results of reverse reproduction from single cells. The first line of Fig. 8 is the same as that of Fig. 7, and shows the response of the cognitron in the normal state. In the second line of Fig. 8, a U_1-cell which happened to respond strongly to stimulus "4" is chosen at first. (Here, it is not so important which cell is to be chosen, provided that the cell has yielded a large response to this stimulus. In Fig. 8, the cell whose normal response to this stimulus is maximum among the cells of layer U_1 has been chosen.) From this single cell, the response pattern of layer U_0 is reversely reproduced. Since the connectable area, hence the receptive field, of this cell is situated at the lower right part of layer U_0, only the lower right part of pattern "4" is reproduced. It can be considered that, in the reverse reproduction, excitatory responses are elicited only from the U_0-cells which, in the normal state, give excitatory effects to this U_1-cell. The third line of Fig. 8 shows the result of reverse reproduction from a single U_2-cell which has yielded a maximum output, and the fourth line shows the result of reverse reproduction from a single U_3-cell.

As is seen from these results, the cells in a deeper layer have relatively larger receptive fields, and the receptive field of the chosen U_3-cell covers the whole U_0-layer. That is, the deeper the layer is, the larger becomes the area from which a single cell of the layer integrates the information. In case of Fig. 8, the U_3-cell watches the whole U_0-layer and correctly grasps the information of pattern "4".

Figure 9 sums up the results of reverse reproduction for all the five stimulus patterns. It shows only the responses of U_0-layer and the firing patterns of U_3-layer from which the reverse reproduction are made. In the left half of Fig. 9, the results of reverse reproduction from the whole U_3-layer are shown. The right half of the figure shows the results of reverse reproduction from single U_3-cells whose outputs have been maximum in layer U_3. It is seen that the original stimulus patterns are correctly reproduced by the process of reverse reproduction, except for some errors for pattern "0". The imperfect reproduction of pattern "0" from single U_3-cell seems to be caused by the fact that this U_3-cell happens to be situated near the periphery of U_3-layer, and that the receptive field of this cell has been unable to cover the whole U_0-layer. We can find a U_3-cell, from which a better reproduction of pattern "0" is obtained. The reversely reproduced pattern from such single cell is almost the same as the pattern shown at the left and on the first line in Fig. 9. As is seen from these experiments, reverse reproduction is made correctly except a little error for pattern "0". Then we can conclude that the self-organization of the cognitron has been successfully performed.

5.4. The Time-Course of Synapse Organization

Figure 10 shows the time course of synapse organization observed by the reverse reproduction. That is, the U_3-cell which has yielded a maximum output to pattern "2" at the 20th cycle of presentation is chosen at first. This is the same cell as the one chosen in the experiment shown in the right side on the third line of Fig. 9. In order to observe how the synaptic connections to this cell have been organized, we start again from the initial state. We make the reverse reproduction from this cell after every cycle of presentation of the five stimulus patterns, and observe how the reversely reproduced pattern of U_0-layer varies with time. The numeral letter to the upper left of each pattern in Fig. 10 indicates the time from the initial state measured in cycles of pattern presentations.

From the 1st to the 3rd cycle, no pattern is reproduced by the process of reverse reproduction, since the paths from U_0-cells to this U_3-cell are not formed yet. At the 4th cycle, weak connections begin to be formed between the cells, and a faint pattern begins to appear. From the 4th to the 6th cycle, however, the reproduced pattern does not only contain the components of pattern "2" but also the components of patterns "0" and "1". These results show that, at this stage of development, this U_3-cell has not come to selectively detect pattern "2" yet, and responds also to the components of patterns "0" and "1",

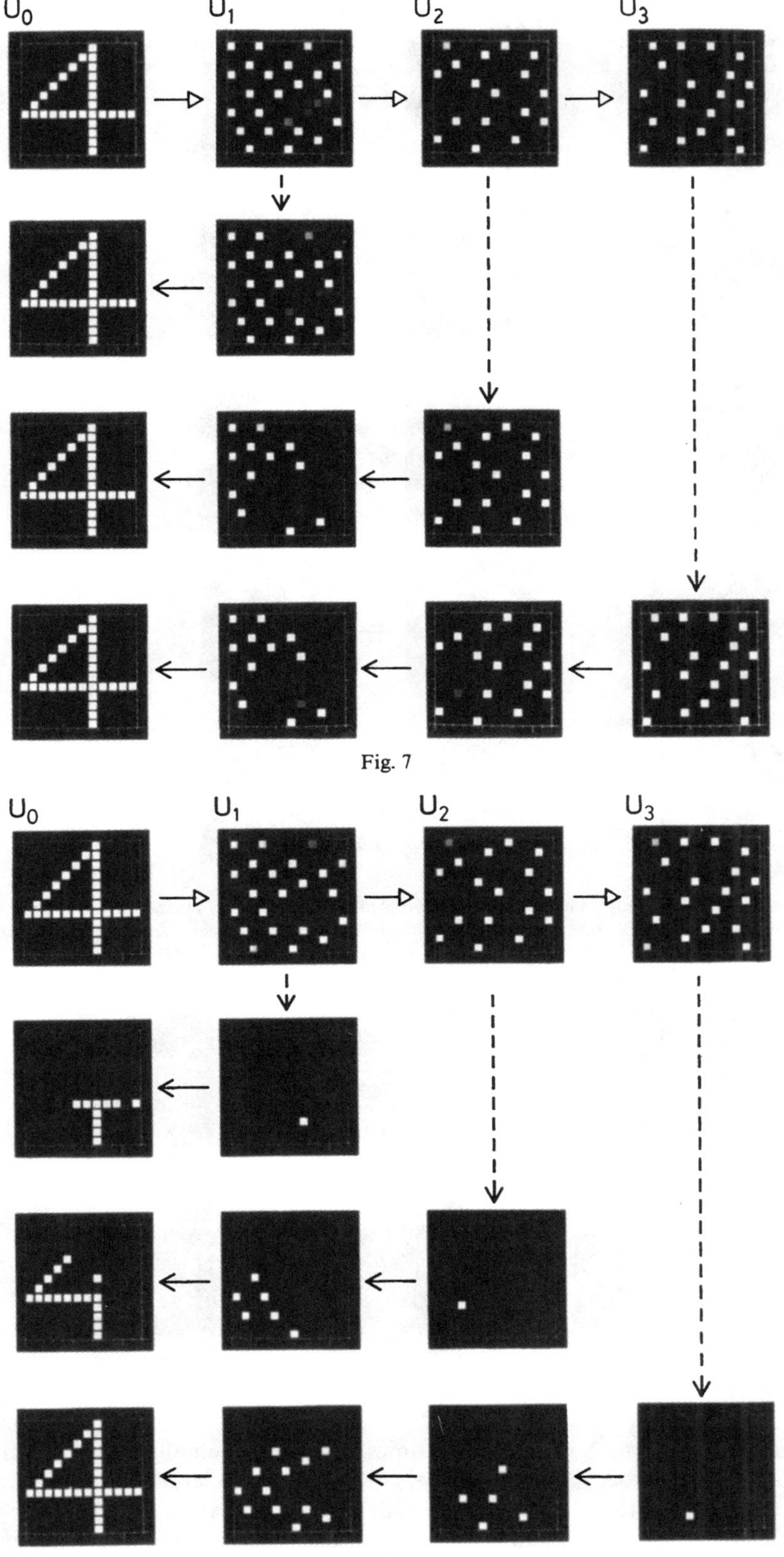

Fig. 7. Reverse reproduction from the normal responses of the whole cells of single layers. The first line shows the normal response to stimulus "4" at the 20th cycle of pattern presentation. The second, the third and the fourth lines show the reverse reproduction from the normal response patterns of layers U_1, U_2, and U_3 respectively

Fig. 8. Reverse reproduction from single cells. The first line shows the normal response to stimulus "4". The second, the third and the fourth lines show the reverse reproduction from the single cells whose normal responses have been maximum among the cells of layers U_1, U_2, and U_3 respectively

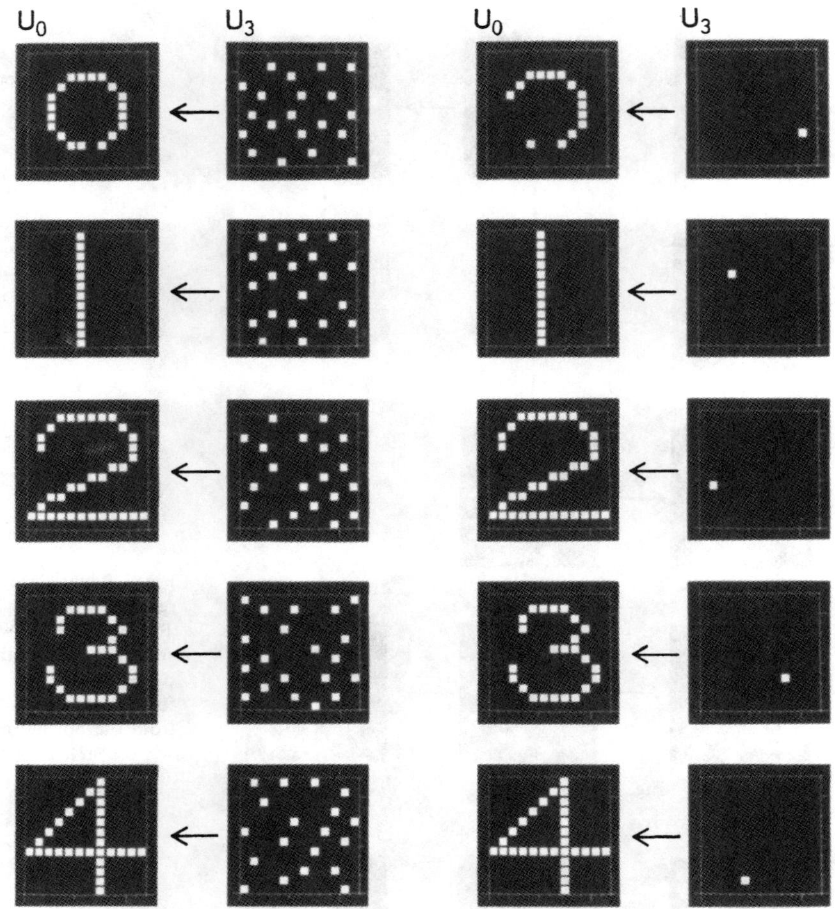

Fig. 9. Summary of the results of reverse reproduction for five stimulus patterns. Left: the reverse reproduction of layer U_0 from the whole U_3 layer. Right: the reverse reproduction of layer U_0 from single U_3-cells whose responses to the stimuli have been maximum in layer U_3

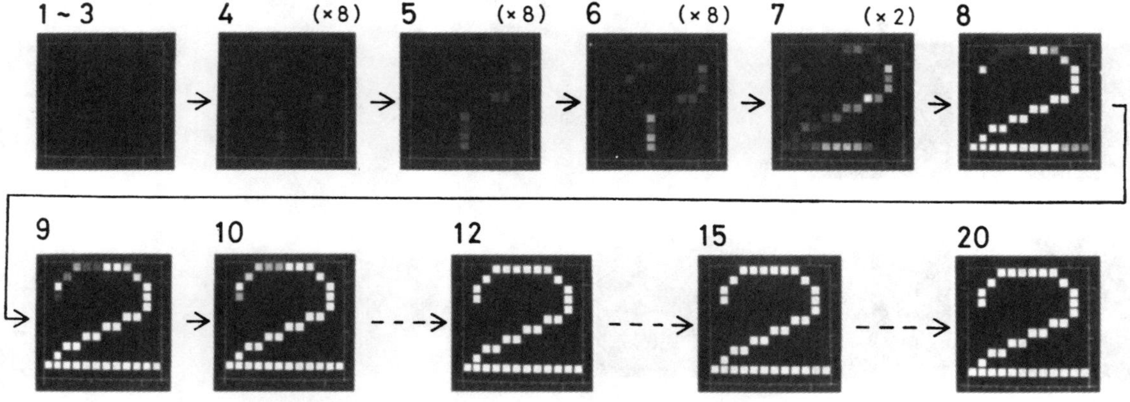

Fig. 10. The time course of synapse organization observed by the reverse reproduction

Since the magnitude of the reproduced patterns from the 4th to the 7th cycle are too small to be displayed with equal level to other patterns, in Fig. 10, the patterns from the 4th to the 6th cycle are displayed with a level 8 times larger than actually is, and the one at the 7th cycle is displayed with a level 2 times larger.

After the 7th cycle, the reproduced patterns gradually become to bear a considerable resemblance

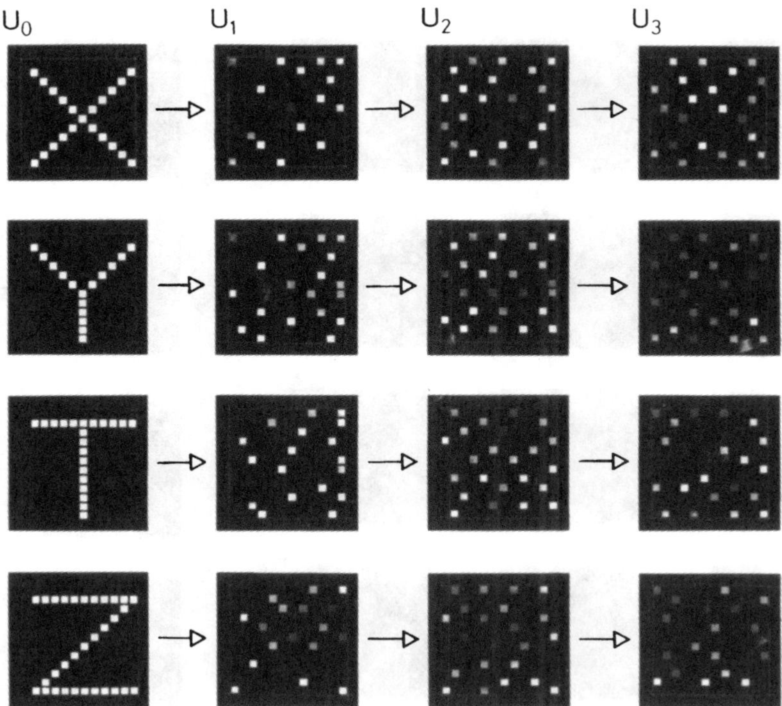

Fig. 11. The response of the cognitron to four stimulus patterns which resembles to one another. These stimulus patterns have been presented to layer U_0 in a cyclic manner. The responses at the 20th cycle of stimulus presentation are shown

to pattern "2", but contain some parts whose levels still remain low. This shows that some of the synapses have not been fully developed yet. After 15th cycle, however, the reproduced patterns reach a steady state, and it is seen that the synaptic connections have been completed.

5.5. Response to Resembling Patterns

Between the stimulus patterns used in the above experiments, there are not strong resemblances. In such a case, the response patterns of U_3-layer also bear little resemblance to each other. In this section, we discuss how the cognitron becomes to respond, if the stimulus patterns have considerable amounts of common components with one another.

The stimulus patterns presented to the cognitron are "*X*", "*Y*", "*T*", and "*Z*", and are shown in the left column of Fig. 11. There are considerable amounts of common components between "*X*" and "*Y*", "*Y*" and "*T*", "*T*" and "*Z*", "*Z*" and "*X*", and also between "*Y*" and "*Z*".

Figure 11 shows how the cells become to respond to these four stimulus patterns after 20 cycles of pattern presentation. In this case, even in layer U_3, there are many cells which respond to more than two different stimulus patterns. Figure 12 shows to which pattern each U_3-cell responds. That is, Fig. 12 indicates, for instance, that the 3rd cell from the left on the 2nd line responds both "*T*" and "*Z*". In this figure, the magnitude of the response of each cell is neglected, and it is only checked whether the cell yields any output to the stimulus or not. In Fig. 12, we can find cells which respond to, for example, both "*X*" and "*Y*", which have several points of similarity to each other. We

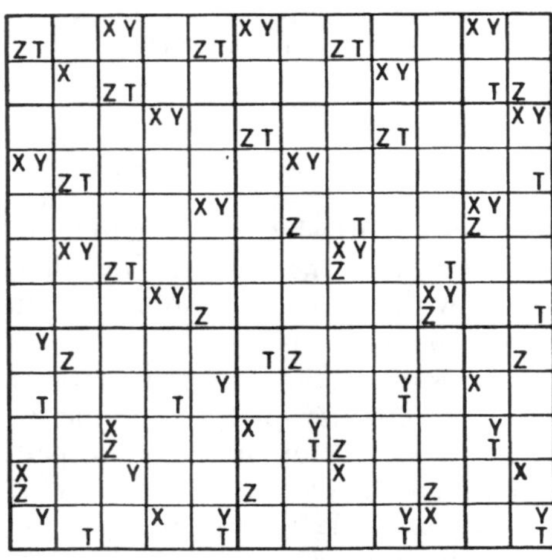

Fig. 12. List of the names of the stimulus patterns to which each U_3-cell responds under the same condition as Fig. 11

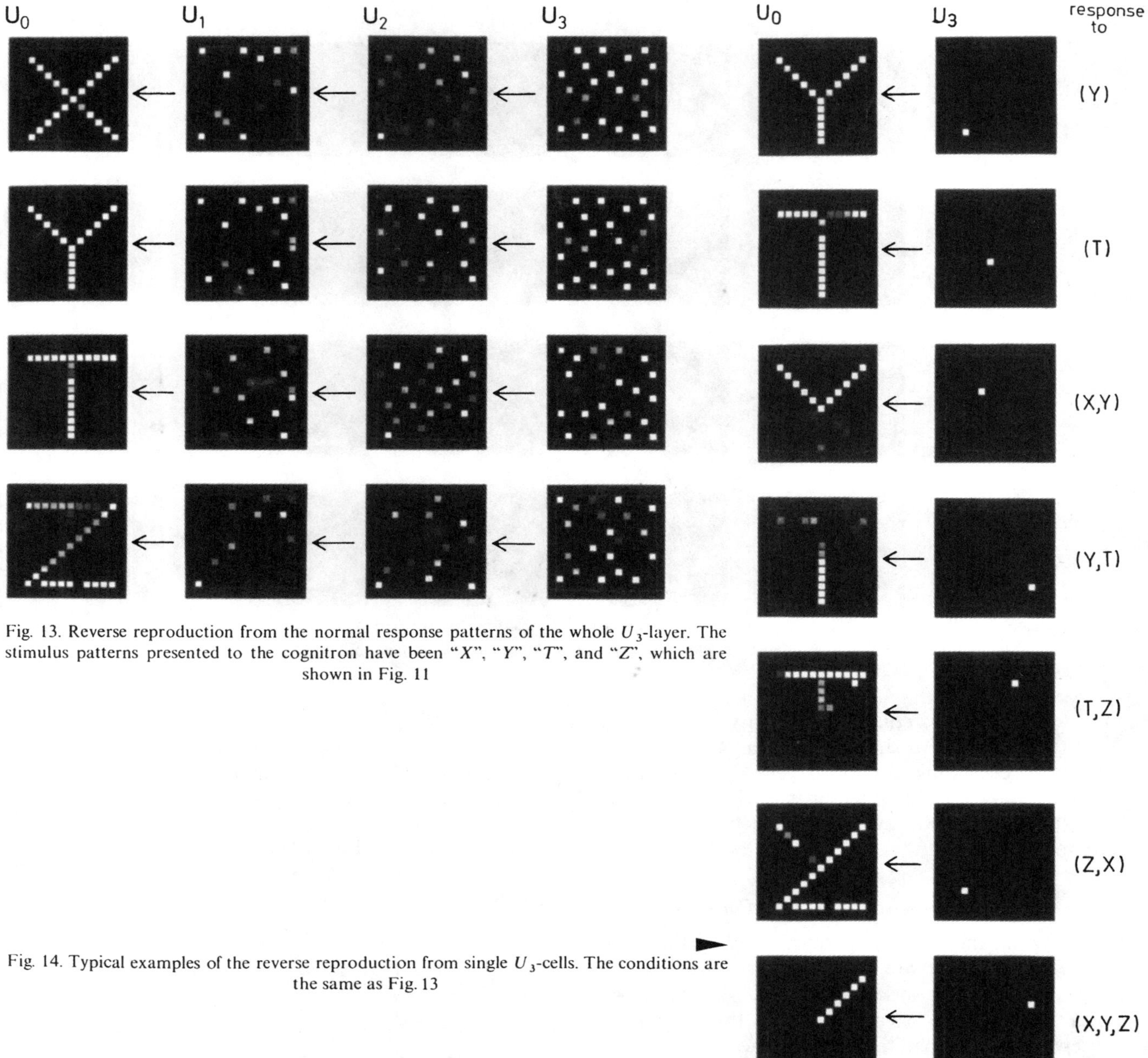

Fig. 13. Reverse reproduction from the normal response patterns of the whole U_3-layer. The stimulus patterns presented to the cognitron have been "*X*", "*Y*", "*T*", and "*Z*", which are shown in Fig. 11

Fig. 14. Typical examples of the reverse reproduction from single U_3-cells. The conditions are the same as Fig. 13

cannot find, however, cells which respond to both "*X*" and "*T*", between which there is little resemblance. These results show that resembling stimulus patterns elicit similar response patterns.

Figure 13 shows the results of reverse reproduction from the normal response patterns of the whole U_3-layer. It is seen that the original stimulus patterns are correctly reproduced except a little error for "*Z*".

Figure 14 shows some typical examples of the reverse reproduction from single U_3-cells. The left side pictures show the reversely reproduced patterns of layer U_0. The right side pictures show the U_3-cells from which reverse reproduction is made. The letter drawn to the right of each of these pictures indicates the names of the stimulus patterns which elicit positive responses to these U_3 cells.

The first line of Fig. 14 shows the reverse reproduction from a single cell which responds only to stimulus "*Y*". It is seen that the pattern "*Y*" is correctly reproduced. Also, in the second line of Fig. 14, the stimulus pattern "*T*" is reproduced fairly well, although the reproduced pattern is missing at one cell. It is seen that the reversely reproduced pattern from a single cell which selectively responds to one stimulus pattern generally coincide with the original stimulus pattern fairly well.

The third line of Fig. 14 shows the reverse reproduction from a cell which responds to both "X" and "Y". The reproduced pattern in this case is a mixture between "X" and "Y". The same situation is observed for the pictures in the following lines of Fig. 14. The U_3-cell shown in the bottom line of Fig. 14 responds to "X", "Y" and "Z". The pattern reversely reproduced from this cell consists of a line component which is the common component among these three stimulus patterns.

In some cases, a complete reproduction of the original pattern is impossible from only one U_3-cell. For example, the U_3-cell shown in the second line of Fig. 14 is the one which gives the best reproduction of pattern "T", but the reversely reproduced pattern contains some error. The reverse reproduction from other U_3-cells only give worse results. If the reverse reproduction is made from the whole cells of U_3-layer, however, a complete reproduction of the original stimulus pattern is performed as has been shown in the third line of Fig. 13. As is seen from these results, the response of a single cell by itself does not always serve for pointing out the presented stimulus pattern correctly, but the configuration of the spatial pattern representing the response of all the U_3-cells serves for the purpose. That is, in the cognitron, even the cells of the last layer generally divide the work among them, rather than only one of the cells works alone for the information processing of each stimulus pattern.

6. Conclusion

A self-organizing multilayered neural network "cognitron" has been constructed following a new hypothesis on the synapse organization. The computer simulation of the cognitron has shown that the cognitron has characteristics similar to that of the animal's brain in many points. Since the cognitron has a multilayered structure, it seems to have a larger capability for information processing than the usual brain models or learning machines proposed before.

The cognitron discussed in this paper is not intended to be a complete system for pattern recognition. If we want to make a pattern recognizer with a cognitron, some other functions must be added to it.

For instance, a feature extractor with a function of normalization of position, size, etc. would be necessary to be added in front of the cognitron, and a decision circuit should be cascaded after the last layer of the cognitron.

In the cognitron proposed here, the neural layers are merely cascaded one after another. It is expected that the neural network would acquire more ability if the structure of the network is modified. For instance, backward couplings between the cells of different layers, or cross couplings between the cells in the same layer would yield a better performance. Even for such cases, the algorithm used for the organization of the cognitron would be successfully applied.

The author conjectures that the hypothesis on the synapse organization proposed here holds not only in the higher center of the brain but also in the distal parts of the sensory systems. This does not mean, however, that this hypothesis is the only one rule which controls the organization of all kinds of synapses. Perhaps, the organization of some kinds of synapses would be controlled by some other rules.

Acknowledgement. The author would like to thank Dr. Kenji Hiwatashi and Dr. Akira Watanabe for their encouragement during this work.

References

Blakemore, C., Cooper, G. F.: Development of the brain depends on the visual environment. Nature (Lond.) **228**, 477–478 (1970)

Block, H. D., Knight, B. W., Rosenblatt, F.: Analysis of four-layer series-coupled perceptron. II. Rev. mod. Phys. **34**, 135–142 (1962)

Fukushima, K.: Visual feature extraction by a multilayered network of analog threshold elements. IEEE Trans. Syst. Sci. Cybernetics SSC-**5**, 322–333 (1969)

Fukushima, K.: A feature extractor for curvilinear patterns: A design suggested by the mammalian visual system. Kybernetik **7**, 153–160 (1970)

Fukushima, K.: A feature extractor for a pattern recognizer. A design suggested by the visual system (in Japanese). NHK Techn. J. **23**, 351–367 (1971)

Fukushima, K.: Self-organizing multilayered neuron network "Cognitron" (in Japanese), Paper of Technical Group on Pattern Recognition and Learning, Inst. Electronics Comm. Engrs. Japan, PRL**74**-25 (1974); and 1974 Nat. Conv. Rec. of Inst. Electronics Commun. Engrs. Japan, No. S9-8 (1974)

Hubel, D. H., Wiesel, T. N.: Receptive fields of single neurones in the cat's visual cortex. J. Physiol. (Lond.) **148**, 574–591 (1959)

Hubel, D. H., Wiesel, T. N.: Receptive fields, binocular interaction and functional architecture in the cat's visual cortex. J. Physiol. (Lond.) **160**, 106–154 (1962)

Hubel, D. H., Wiesel, T. N.: Receptive fields and functional architecture in two nonstriate visual areas (18 and 19) of the cat. J. Neurophysiol. **28**, 229–289 (1965)

Malsburg, C. von der: Self-organization of orientation sensitive cells in the striate cortex. Kybernetik **14**, 85–100 (1973)

Marr, D.: A theory of cerebeller cortex. J. Physiol. (Lond.) **202**, 437–470 (1969)

Marr, D.: A theory of cerebral neocortex. Proc. roy. Soc. Lond. Ser. B, **176**, 161–234 (1970)

Rosenblatt, F.: Principles of Neurodynamics. Washington: Spartan Books 1962

Dr. Kunihiko Fukushima
NHK Broadcasting Science Research Laboratories
1-10-11, Kinuta, Setagaya, Tokyo 157, Japan

Neurons with graded response have collective computational properties like those of two-state neurons

(associative memory/neural network/stability/action potentials)

J. J. HOPFIELD

Divisions of Chemistry and Biology, California Institute of Technology, Pasadena, CA 91125; and Bell Laboratories, Murray Hill, NJ 07974

Contributed by J. J. Hopfield, February 13, 1984

ABSTRACT A model for a large network of "neurons" with a graded response (or sigmoid input–output relation) is studied. This deterministic system has collective properties in very close correspondence with the earlier stochastic model based on McCulloch–Pitts neurons. The content-addressable memory and other emergent collective properties of the original model also are present in the graded response model. The idea that such collective properties are used in biological systems is given added credence by the continued presence of such properties for more nearly biological "neurons." Collective analog electrical circuits of the kind described will certainly function. The collective states of the two models have a simple correspondence. The original model will continue to be useful for simulations, because its connection to graded response systems is established. Equations that include the effect of action potentials in the graded response system are also developed.

Recent papers (1–3) have explored the ability of a system of highly interconnected "neurons" to have useful collective computational properties. These properties emerge spontaneously in a system having a large number of elementary "neurons." Content-addressable memory (CAM) is one of the simplest collective properties of such a system. The mathematical modeling has been based on "neurons" that are different both from real biological neurons and from the realistic functioning of simple electronic circuits. Some of these differences are major enough that neurobiologists and circuit engineers alike have questioned whether real neural or electrical circuits would actually exhibit the kind of behaviors found in the model system even if the "neurons" were connected in the fashion envisioned.

Two major divergences between the model and biological or physical systems stand out. Real neurons (and real physical devices such as operational amplifiers that might mimic them) have continuous input–output relations. (Action potentials are omitted until *Discussion*.) The original modeling used two-state McCulloch–Pitts (4) threshold devices having outputs of 0 or 1 only. Real neurons and real physical circuits have integrative time delays due to capacitance, and the time evolution of the state of such systems should be represented by a differential equation (perhaps with added noise). The original modeling used a stochastic algorithm involving sudden 0–1 or 1–0 changes of states of neurons at random times. This paper shows that the important properties of the original model remain intact when these two simplifications of the modeling are eliminated. Although it is uncertain whether the properties of these new continuous "neurons" are yet close enough to the essential properties of real neurons (and/or their dendritic arborization) to be directly applicable to neurobiology, a major conceptual obstacle has been eliminated. It is certain that a CAM constructed on the basic ideas of the original model (1) but built of operational amplifiers and resistors will function.

Form of the Original Model

The original model used two-state threshold "neurons" that followed a stochastic algorithm. Each model neuron i had two states, characterized by the output V_i of the neuron having the values V_i^0 or V_i^1 (which may often be taken as 0 and 1, respectively). The input of each neuron came from two sources, external inputs I_i and inputs from other neurons. The total input to neuron i is then

$$\text{Input to } i = H_i = \sum_{j \neq i} T_{ij}V_j + I_i. \qquad [1]$$

The element T_{ij} can be biologically viewed as a description of the synaptic interconnection strength from neuron j to neuron i.

CAM and other useful computations in this system involve the change of state of the system with time. The motion of the state of a system of N neurons in state space describes the computation that the set of neurons is performing. A model therefore must describe how the state evolves in time, and the original model describes this in terms of a stochastic evolution. Each neuron samples its input at random times. It changes the value of its output or leaves it fixed according to a threshold rule with thresholds U_i.

$$\begin{aligned} V_i &\rightarrow V_i^0 \text{ if } \sum_{j \neq i} T_{ij}V_j + I_i < U_i \\ &\rightarrow V_i^1 \text{ if } \sum_{j \neq i} T_{ij}V_j + I_i > U_i. \end{aligned} \qquad [2]$$

The interrogation of each neuron is a stochastic process, taking place at a mean rate W for each neuron. The times of interrogation of each neuron are independent of the times at which other neurons are interrogated. The algorithm is thus *asynchronous*, in contrast to the usual kind of processing done with threshold devices. This asynchrony was deliberately introduced to represent a combination of propagation delays, jitter, and noise in real neural systems. Synchronous systems might have additional collective properties (5, 6).

The original model behaves as an associative memory (or CAM) when the state space flow generated by the algorithm is characterized by a set of stable fixed points. If these stable points describe a simple flow in which nearby points in state space tend to remain close during the flow (i.e., a nonmixing flow), then initial states that are close (in Hamming distance) to a particular stable state and far from all others will tend to terminate in that nearby stable state.

The publication costs of this article were defrayed in part by page charge payment. This article must therefore be hereby marked "*advertisement*" in accordance with 18 U.S.C. §1734 solely to indicate this fact.

Abbreviations: CAM, content-addressable memory; RC, resistance–capacitance.

Reprinted with permission from *Proc. of the National Academy of Sciences*, vol. 81, pp. 3088–3092, May 1984.

If the location of a particular stable point in state space is thought of as the information of a particular memory of the system, states near to that particular stable point contain partial information about that memory. From an initial state of partial information about a memory, a final stable state with all the information of the memory is found. The memory is reached not by knowing an address, but rather by supplying in the initial state some subpart of the memory. Any subpart of adequate size will do—the memory is truly addressable by *content* rather than location. A given T matrix contains many memories simultaneously, which are reconstructed individually from partial information in an initial state.

Convergent flow to stable states is the essential feature of this CAM operation. There is a simple mathematical condition which guarantees that the state space flow algorithm converges on stable states. Any symmetric T with zero diagonal elements (i.e., $T_{ij} = T_{ji}$, $T_{ii} = 0$) will produce such a flow. The proof of this property followed from the construction of an appropriate energy function that is always decreased by any state change produced by the algorithm. Consider the function

$$E = -\frac{1}{2}\sum_{i\neq j}\sum T_{ij}V_iV_j - \sum_i I_iV_i + \sum_i U_iV_i. \quad [3]$$

The change ΔE in E due to changing the state of neuron i by ΔV_i is

$$\Delta E = -\left[\sum_{j\neq i} T_{ij}V_j + I_i - U_i\right]\Delta V_i. \quad [4]$$

But according to the algorithm, ΔV_i is positive only when the bracket is positive, and similarly for the negative case. Thus any change in E under the algorithm is negative. E is bounded, so the iteration of the algorithm must lead to stable states that do not further change with time.

A Continuous, Deterministic Model

We now construct a model that is based on continuous variables and responses but retains all the significant behaviors of the original model. Let the output variable V_i for neuron i have the range $V_i^0 \leq V_i \leq V_i^1$ and be a continuous and monotone-increasing function of the instantaneous input u_i to neuron i. The typical input–output relation $g_i(u_i)$ shown in Fig. 1*a* is sigmoid with asymptotes V_i^0 and V_i^1. For neurons exhibiting action potentials, u_i could be thought of as the mean soma potential of a neuron from the total effect of its excitatory and inhibitory inputs. V_i can be viewed as the short-term average of the firing rate of the cell i. Other biological interpretations are possible—for example, nonlinear processing may be done at junctions in a dendritic arbor (7), and the model "neurons" could represent such junctions. In terms of electrical circuits, $g_i(u_i)$ represents the input–output characteristic of a nonlinear amplifier with negligible response time. It is convenient also to define the inverse output–input relation, $g_i^{-1}(V)$.

In a biological system, u_i will lag behind the instantaneous outputs V_j of the other cells because of the input capacitance C of the cell membranes, the transmembrane resistance R, and the finite impedance T_{ij}^{-1} between the output V_j and the cell body of cell i. Thus there is a resistance–capacitance (RC) charging equation that determines the rate of change of u_i.

$$C_i(du_i/dt) = \sum_j T_{ij}V_j - u_i/R_i + I_i$$
$$u_i = g_i^{-1}(V_i). \quad [5]$$

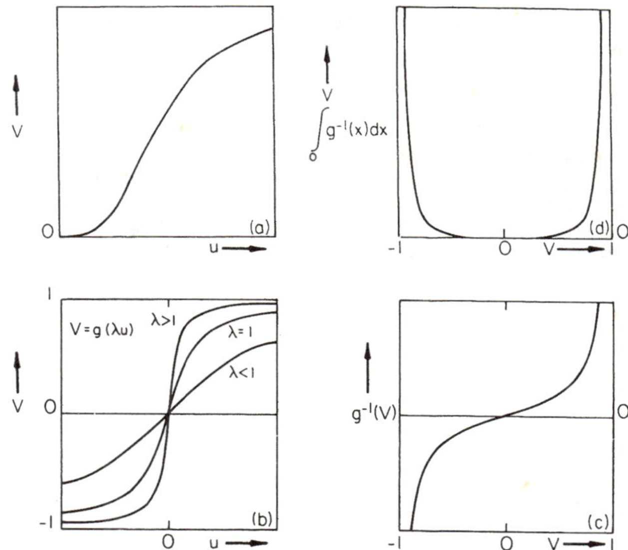

FIG. 1. (*a*) The sigmoid input–output relation for a typical neuron. All the $g(u)$ of this paper have such a form, with possible horizontal and vertical translations. (*b*) The input–output relation $g(\lambda u)$ for the "neurons" of the continuous model for three values of the gain scaling parameter λ. (*c*) The output–input relation $u = g^{-1}(V)$ for the g shown in *b*. (*d*) The contribution of g to the energy of Eq. **5** as a function of V.

$T_{ij}V_j$ represents the electrical current input to cell i due to the present potential of cell j, and T_{ij} is thus the synapse efficacy. Linear summing of inputs is assumed. T_{ij} of both signs should occur. I_i is any other (fixed) input current to neuron i.

The same set of equations represents the resistively connected network of electrical amplifiers sketched in Fig. 2. It appears more complicated than the description of the neural system because the electrical problem of providing inhibition and excitation requires an additional inverting amplifier and a negative signal wire. The magnitude of T_{ij} is $1/R_{ij}$, where R_{ij} is the resistor connecting the output of j to the input line i, while the sign of T_{ij} is determined by the choice of the posi-

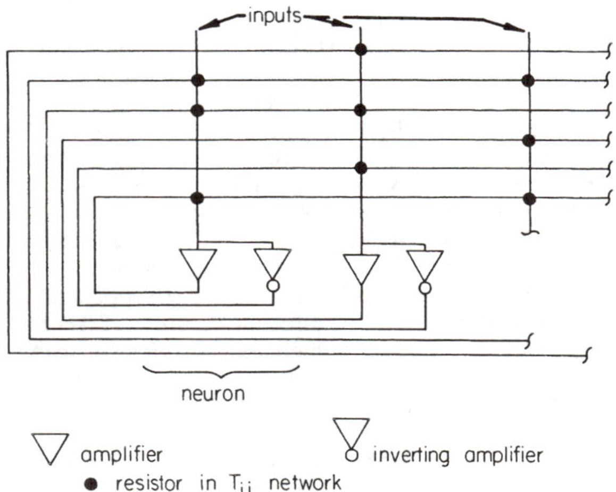

FIG. 2. An electrical circuit that corresponds to Eq. **5** when the amplifiers are fast. The input capacitance and resistances are not drawn. A particularly simple special case can have all positive T_{ij} of the same strength and no negative T_{ij} and replaces the array of negative wires with a single negative feedback amplifier sending a common output to each "neuron."

tive or negative output of amplifier j at the connection site. R_i is now

$$1/R_i = 1/\rho_i + \sum_j 1/R_{ij}, \quad [6]$$

where ρ_i is the input resistance of amplifier i. C_i is the total input capacitance of the amplifier i and its associated input lead. We presume the output impedance of the amplifiers is negligible. These simplifications result in Eq. **5** being appropriate also for the network of Fig. 2.

Consider the quantity

$$E = -\frac{1}{2}\sum_{i,j}\sum T_{ij}V_iV_j + \sum_i (1/R_i)\int_0^{V_i} g_i^{-1}(V)dV + \sum_i I_iV_i. \quad [7]$$

Its time derivative for a symmetric T is

$$dE/dt = -\sum_i dV_i/dt\left(\sum_j T_{ij}V_j - u_i/R_i + I_i\right). \quad [8]$$

The parenthesis is the right-hand side of Eq. **5**, so

$$dE/dT = -\sum C_i(dV_i/dt)(du_i/dt) = -\sum C_i g_i^{-1\prime}(V_i)(dV_i/dt)^2. \quad [9]$$

Since $g_i^{-1}(V_i)$ is a monotone increasing function and C_i is positive, each term in this sum is nonnegative. Therefore

$$dE/dt \leq 0,\ dE/dt = 0 \rightarrow dV_i/dt = 0 \text{ for all } i. \quad [10]$$

Together with the boundedness of E, Eq. **10** shows that the time evolution of the system is a motion in state space that seeks out minima in E and comes to a stop at such points. E is a Liapunov function for the system.

This deterministic model has the same flow properties in its continuous space that the stochastic model does in its discrete space. It can therefore be used in CAM or any other computational task for which an energy function is essential (3). We expect that the qualitative effects of disorganized or organized anti-symmetric parts of T_{ij} should have similar effects on the CAM operation of the new and old system. The new computational behaviors (such as learning sequences) that can be produced by antisymmetric contributions to T_{ij} within the stochastic model will also hold for the deterministic continuous model. Anecdotal support for these assertions comes from unpublished work of John Platt (California Institute of Technology) solving Eq. **5** on a computer with some random T_{ij} removed from an otherwise symmetric T, and from experimental work of John Lambe (Jet Propulsion Laboratory), David Feinstein (California Institute of Technology), and Platt generating sequences of states by using an antisymmetric part of T in a real circuit of a six "neurons" (personal communications).

Relation Between the Stable States of the Two Models

For a given T, the stable states of the continuous system have a simple correspondence with the stable states of the stochastic system. We will work with a slightly simplified instance of the general equations to put a minimum of mathematics in the way of seeing the correspondence. The same basic idea carries over, with more arithmetic, to the general case.

Consider the case in which $V_i^0 < 0 < V_i^1$ for all i. Then the zero of voltage for each V_i can be chosen such that $g_i(0) = 0$ for all i. Because the values of asymptotes are totally unimportant in all that follows, we will simplify notation by taking them as ± 1 for all i. The second simplification is to treat the case in which $I_i = 0$ for all i. Finally, while the continuous case has an energy function with self-connections T_{ii}, the discrete case need not, so $T_{ii} = 0$ will be assumed for the following analysis.

This continuous system has for symmetric T the underlying energy function

$$E = -\frac{1}{2}\sum_{j\neq i}\sum T_{ij}V_iV_j + \sum 1/R_i\int_0^{V_i} g_i^{-1}(V)dV. \quad [11]$$

Where are the maxima and minima of the *first term* of Eq. **11** in the domain of the hypercube $-1 \leq V_i \leq 1$ for all i? In the usual case, all extrema lie at *corners* of the N-dimensional hypercube space. [In the pathological case that T is a positive or negative definite matrix, an extremum is also possible in the interior of the space. This is not the case for information storage matrices of the usual type (1).]

The discrete, stochastic algorithm searches for minimal states at the corners of the hypercube—corners that are lower than adjacent corners. Since E is a linear function of a single V_i along any cube edge, the energy minima (or maxima) of

$$E = -\frac{1}{2}\sum_{i\neq j}\sum T_{ij}V_iV_j \quad [12]$$

for the discrete space $V_i = \pm 1$ are exactly the same corners as the energy maxima and minima for the continuous case $-1 \leq V_i \leq 1$.

The second term in Eq. **11** alters the overall picture somewhat. To understand that alteration most easily, the gain g can be scaled, replacing

$$V_i = g_i(u_i) \text{ by } V_i = g_i(\lambda u_i)$$

and

$$u_i = g_i^{-1}(V_i) \text{ by } u_i = (1/\lambda)g_i^{-1}(V_i). \quad [13]$$

This scaling changes the steepness of the sigmoid gain curve without altering the output asymptotes, as indicated in Fig. 1*b*. $g_i(x)$ now represents a standard form in which the scale factor $\lambda = 1$ corresponds to a standard gain, $\lambda >> 1$ to a system with very high gain and step-like gain curve, and λ small corresponds to a low gain and flat sigmoid curve (Fig. 1*b*). The second term in E is now

$$+\frac{1}{\lambda}\sum_i 1/R_i\int_0^{V_i} g_i^{-1}(V)dV. \quad [14]$$

The integral is zero for $V_i = 0$ and positive otherwise, getting very large as V_i approaches ± 1 because of the slowness with which $g(V)$ approaches its asymptotes (Fig. 1*d*). However, in the high-gain limit $\lambda \rightarrow \infty$ this second term becomes negligible, and the locations of the maxima and minima of the full energy expression become the same as that of Eq. **12** or Eq. **3** in the absence of inputs and zero thresholds. *The only stable points of the very high gain, continuous, deterministic system therefore correspond to the stable points of the stochastic system.*

For large but finite λ, the second term in Eq. **11** begins to contribute. The form of $g_i(V_i)$ leads to a large positive contribution near all surfaces, edges, and corners of the hypercube while it still contributes negligibly far from the surfaces. This leads to an energy surface that still has its maxima at corners but the minima become displaced slightly toward the interior

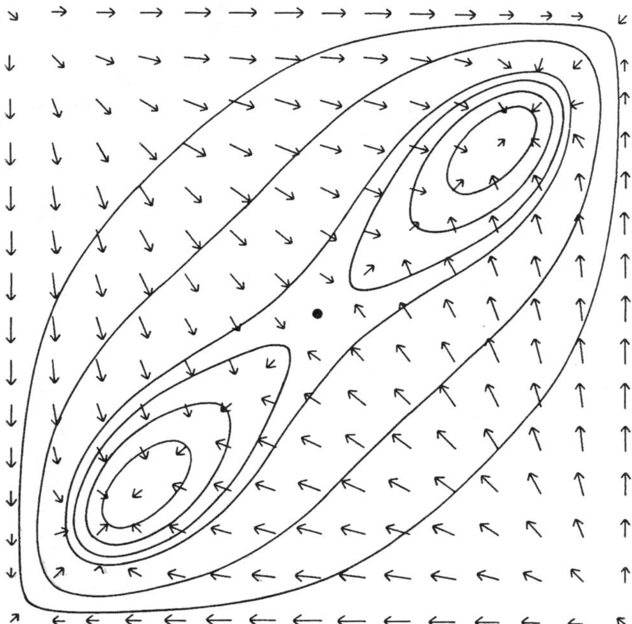

FIG. 3. An energy contour map for a two-neuron, two-stable-state system. The ordinate and abscissa are the outputs of the two neurons. Stable states are located near the lower left and upper right corners, and unstable extrema at the other two corners. The arrows show the motion of the state from Eq. **5**. This motion is not in general perpendicular to the energy contours. The system parameters are $T_{12} = T_{21} = 1$, $\lambda = 1.4$, and $g(u) = (2/\pi)\tan^{-1}(\pi\lambda u/2)$. Energy contours are 0.449, 0.156, 0.017, −0.003, −0.023, and −0.041.

of the space. As λ decreases, each minimum moves further inward. As λ is further decreased, minima disappear one at a time, when the topology of the energy surface makes a minimum and a saddle point coalesce. Ultimately, for very small λ, the second term in Eq. **11** dominates, and the only minimum is at $V_i = 0$. When the gain is large enough that there are many minima, each is associated with a well-defined minimum of the infinite gain case—as the gain is increased, each minimum will move until it reaches a particular cube corner when $\lambda \rightarrow \infty$. The same kind of mapping relation holds in general between the continuous deterministic system with sigmoid response curves and the stochastic model.

An energy contour map for a two-neuron (or two operational amplifier) system with two stable states is illustrated in Fig. 3. The two axes are the outputs of the two amplifiers. The upper left and lower right corners are stable minima for infinite gain, and the minima are displaced inward by the finite gain.

There are many general theorems about stability in networks of differential equations representing chemistry, circuits, and biology (8–12). The importance of this simple symmetric system is not merely its stability, but the fact that the correspondence with a discrete system lends it a special relation to elementary computational devices and concepts.

DISCUSSION

Real neurons and real amplifiers have graded, continuous outputs as a function of their inputs (or sigmoid input–output curves of finite steepness) rather than steplike, two-state response curves. Our original stochastic model of CAM and other collective properties of assemblies of neurons was based on two-state neurons. A continuous, deterministic neuron network of interconnected neurons with graded responses has been analyzed in the previous two sections. It functions as a CAM in precisely the same collective way as did the original stochastic model of CAM. A set of memories can be nonlocally stored in a matrix of synaptic (or resistive) interconnections in such a way that particular memories can be reconstructed from a starting state that gives partial information about one of them.

The convergence of the neuronal state of the continuous, deterministic model to its stable states (memories) is based on the existence of an energy function that directs the flow in state space. Such a function can be constructed in the continuous, deterministic model when T is symmetric, just as was the case for the original stochastic model with two-state neurons. Other interesting uses and interpretations of the behaviors of the original model based on the existence of an underlying energy function will also hold for the continuous ("graded response") model (3).

A direct correspondence between the stable states of the two models was shown. For steep response curves (high gain) there is a 1:1 correspondence between the memories of the two models. When the response is less steep (lower gain) the continuous-response model can have fewer stable states than the stochastic model with the same T matrix, but the existing stable states will still correspond to particular stable states of the stochastic model. This simple correspondence is possible because of the quadratic form of the interaction between different neurons in the energy function. More complicated energy functions, which have occasionally been used in constraint satisfaction problems (13, 14), may have in addition stable states within the interior of the domain of state space in the continuous model which have no correspondence within the discrete two-state model.

This analysis indicates that a real circuit of operational amplifiers, capacitors, and resistors should be able to operate as a CAM, reconstructing the stable states that have been designed into T. As long as T is symmetric and the amplifiers are fast compared with the characteristic RC time of the input network, the system will converge to stable states and cannot oscillate or display chaotic behavior. While the symmetry of the network is essential to the mathematics, a pragmatic view indicates that approximate symmetry will suffice, as was experimentally shown in the stochastic model. Equivalence of the gain curves and input capacitance of the amplifiers is not needed. For high-gain systems, the stable states of the real circuit will be exactly those predicted by the stochastic model.

Neuronal and electromagnetic signals have finite propagation velocities. A neural circuit that is to operate in the mode described must have propagation delays that are considerably shorter than the RC or chemical integration time of the network. The same must be true for the slowness of amplifier response in the case of the electrical circuit.

The continuous model supplements, rather than replaces, the original stochastic description. The important properties of the original model are not due to its simplifications, but come from the general structure lying behind the model. Because the original model is very efficient to simulate on a digital computer, it will often be more practical to develop ideas and simulations on that model even when use on biological neurons or analog circuits is intended. The interesting collective properties transcend the 0–1 stochastic simplifications.

Neurons often communicate through action potentials. The output of such neurons consists of a series of sharp spikes having a mean frequency (when averaged over a short time) that is described by the input–output relation of Fig. 1*a*. In addition, the delivery of transmitter at a synapse is quantized in vesicles. Thus Eq. **5** can be only an equation for the behavior of a neural network neglecting the quantal noise due to action potentials and the releases of discrete vesicles. Because the system operates by moving downhill on an energy surface, the injection of a small amount of quantal noise will not greatly change the minimum-seeking behavior.

Eq. **5** has a generalization to include action potentials. Let all neurons have the same gain curves $g(u)$, input capacitance C, input impedance R, and maximum firing rate F. Let $g(u)$ have asymptotes 0 and 1. When a neuron has an input u, it is presumed to produce action potentials $V_0\delta(t - t_{\text{firing}})$ in a stochastic fashion with a probability $Fg(u)$ of producing an action potential per unit time. This stochastic view preserves the basic idea of the input signal being transformed into a firing rate but does not allow precise timing of individual action potentials. A synapse with strength T_{ij} will deliver a quantal charge V_0T_{ij} to the input capacitance of neuron i when neuron j produces an action potential. Let $P(u_1, u_2, \ldots u_i, \ldots, u_N, t)du_1, du_2, \ldots, du_N$ be the probability that input potential 1 has the value $u_1, \ldots$ The evolution of the state of the network is described by

$$\partial P/\partial t = \sum_i (1/RC)(\partial(u_iP)/\partial u_i)$$

$$+ \sum_j Fg(u_j)[-P + P(u_1 - T_{1j}V_0/C, \ldots, u_i - T_{ij}V_0/C, \ldots)]. \quad [15]$$

If V_0 is small, the term in brackets can be expanded in a Taylor series, yielding

$$\partial P/\partial t = \sum_i (1/RC)(\partial(u_iP)/\partial u_i)$$

$$- \sum_j (\partial P/\partial u_i)(V_0F/C) \sum_i T_{ij}\, g(u_j)$$

$$+ V_0^2F/2C^2 \sum_{i,j,k} g(u_k)T_{ik}T_{jk}\, (\partial^2P/\partial u_i\partial u_j). \quad [16]$$

In the limit as $V_0 \rightarrow 0$, $F \rightarrow \infty$ such that FV_0 = constant, the second derivative term can be omitted. This simplification has the solutions that are identical to those of the continuous, deterministic model, namely

$$P = \prod \delta(u_i - u_i(t)),$$

where $u_i(t)$ obeys Eq. **5**.

In the model, stochastic noise from the action potentials disappears in this limit and the continuous model of Eq. **5** is recovered. The second derivative term in Eq. **16** produces noise in the system in the same fashion that diffusion produces broadening in mobility-diffusion equations. These equations permit the study of the effects of action potential noise on the continuous, deterministic system. Questions such as the duration of stability of nominal stable states of the continuous, deterministic model Eq. **5** in the presence of action potential noise should be directly answerable from analysis or simulations of Eq. **15** or **16**. Unfortunately the steady-state solution of this problem is *not* equivalent to a thermal distribution—while Eq. **15** is a master equation, it does not have detailed balance even in the high-gain limit, and the quantal noise is not characterized by a temperature.

The author thanks David Feinstein, John Lambe, Carver Mead, and John Platt for discussions and permission to mention unpublished work. The work at California Institute of Technology was supported in part by National Science Foundation Grant DMR-8107494. This is contribution no. 6975 from the Division of Chemistry and Chemical Engineering, California Institute of Technology.

1. Hopfield, J. J. (1982) *Proc. Natl. Acad. Sci. USA* **79,** 2554–2558.
2. Hopfield, J. J. (1984) in *Modeling and Analysis in Biomedicine*, ed. Nicolini, C. (World Scientific Publishing, New York), in press.
3. Hinton, G. E. & Sejnowski, T. J. (1983) in *Proceedings of the IEEE Computer Science Conference on Computer Vision and Pattern Recognition* (Washington, DC), pp. 448–453.
4. McCulloch, W. A. & Pitts, W. (1943) *Bull. Math. Biophys.* **5,** 115–133.
5. Little, W. A. (1974) *Math. Biosci.* **19,** 101–120.
6. Little, W. A. & Shaw, G. L. (1978) *Math. Biosci.* **39,** 281–289.
7. Poggio, T. & Torre, V. (1981) in *Theoretical Approaches to Neurobiology*, eds. Reichardt, W. E. & Poggio, T. (MIT Press, Cambridge, MA), pp. 28–38.
8. Glansdorf, P. & Prigogine, R. (1971) in *Thermodynamic Theory of Structure, Stability, and Fluctuations* (Wiley, New York), pp. 61–67.
9. Landauer, R. (1975) *J. Stat. Phys.* **13,** 1–16.
10. Glass, L. & Kauffman, S. A. (1973) *J. Theor. Biol.* **39,** 103–129.
11. Grossberg, S. (1973) *Stud. Appl. Math.* **52,** 213–257.
12. Glass, L. (1975) *J. Chem. Phys.* **63,** 1325–1335.
13. Kirkpatrick, S., Gelatt, C. D. & Vecchi, M. P. (1983) *Science* **220,** 671–680.
14. Geman, S. & Geman, D. (1984) *IEEE Transactions Pat. Anal. Mech. Intell.*, in press.

Article 3.5

NEURONS WITH HYSTERESIS FORM A NETWORK THAT CAN LEARN WITHOUT ANY CHANGES IN SYNAPTIC CONNECTION STRENGTHS

Geoffrey W. Hoffmann
Theoretical Biology and Biophysics, Los Alamos National Laboratory,
Los Alamos, New Mexico 87545, U.S.A.,
and
*Departments of Physics and Microbiology, University of British Columbia,
Vancouver, B.C., Canada V6T 2A6

Maurice W. Benson
Center for Nonlinear Studies, Los Alamos National Laboratory,
Los Alamos, New Mexico 87545, U.S.A.,
and
Department of Mathematics, Lakehead University,
Thunderbay, Ontario, Canada P7B 5E1

ABSTRACT

A neural network concept derived from an analogy between the immune system and the central nervous system is outlined. The theory is based on a neuron that is slightly more complicated than the conventional McCullogh-Pitts type of neuron, in that it exhibits hysteresis at the single cell level. This added complication is compensated by the fact that a network of such neurons is able to learn without the necessity for any changes in synaptic connection strengths. The learning occurs as a natural consequence of interactions between the network and its environment, with environmental stimuli moving the system around in an N-dimensional phase space, until a point in phase space is reached such that the system's responses are appropriate for dealing with the stimuli. Due to the hysteresis associated with each neuron, the system tends to stay in the region of phase space where it is located. The theory includes a role for sleep in learning.

INTRODUCTION

It is difficult to overstate the importance of analogy in the development of theories. An example from physics is the development of the theory of electromagnetism in the last century, which involved the use of complicated mechanical analogies.[1] An example from biology is Burnet's clonal selection theory of the immune response, a theory that is modelled on Darwin's theory of evolution, and describes a survival of the fittest process at the level of single cells within an animal.[2] The precursor of the clonal selection theory was Jerne's natural selection theory of antibody formation[3], which marked the transition from older instructionist ideas to Darwinian selectionist theories. Analogies clearly played a similarly crucial role when Jerne subsequently postulated that the immune system functions as a network; this time he was impressed by analogies between the structures and the functions of the immune system and those of the central nervous system.[4,5]

In order to explore a variety of ideas on how the brain could work, we need to utilize whatever analogies we can think of. Little has formulated a neural network theory[6] based on the analogy between an Ising spin system and a neural network. Klopf[7] and Barto[8] have described neural network models in which they assume that the selfish behaviour of neural networks might be a reflection of analogous selfishness at the level of the neuronal building blocks. It would be interesting to know what analogies with other physical system, if any, aided Hopfield in the formulation of his

*Permanent address

Reprinted with permission from *American Institute of Physics*, pp. 219–225, 1986.

model,[9] which has had such a profound influence on this field during the last four years. His treatment of the effect of symmetric synaptic coefficients, T_{ij}, was probably inspired by familiarity with many physical systems in which symmetric couplings lead to interesting properties.

As mentioned above, similarities between the central nervous system and the immune system led to a new way of viewing the immune system. An exciting possibility is that we can now reverse the process, and apply insights gained in the study of the immune system network to the central nervous system problem. This possibility has been considered also by Edelman and Reeke.[10] The immune system is a complex system, but in the last few years we have learned an enormous amount about how it functions. We now have at least the outline of an immune system network theory that seems to work quite well.[11-16] Since many of the similarities between the immune system and the central nervous system are at a high level,[17] we considered the possibility that the same kind of mathematical model could be applicable to both systems. That idea led to a theory that we here review briefly, and that has been developed in more detail elsewhere.[17,18]

At one level our theory is slightly more complex than other theories, in that we invoke hysteresis at the level of single neurons. This added complexity is compensated by a new simplicity at the level of the network; the network can learn without any changes in the synaptic connection strengths. Learned information is associated solely with a state vector; memory is a consequence of the fact that due to the hysteresis associated with each neuron, the system tends to stay in the region of an N-dimensional phase space to which its experiences have taken it. Its stimulus-response behaviour is determined by its location in that space.

A NEURON WITH HYSTERESIS

Neurophysiologists tell us that we cannot, on the basis of presently available data, exclude the possibility that at least some neurons exhibit hysteresis at the single cell level. A simple, plausible biochemical model that could lead to single cell hysteresis is as follows. Let X be a key neural metabolite that controls the rate of firing of the cell, such that the rate of firing is (say) proportional to the concentration of X. The concentration of X is assumed to be affected by a second substance Y that reflects the net level of inhibitory signals the cell receives. We consider the following simple reaction scheme:

A → X

X → B (slow)

X → B (catalysed by Y)

Excess X inhibits Y

X is produced at a constant rate from A. X is broken down by two processes, one of which is slow and is independent of Y. The second breakdown process involves Y as an enzyme or the activator of an enzyme. High levels of X inhibit the enzymatic breakdown of X by Y. This phenomenon, known to biochemists as substrate inhibition of enzymes, can occur when an enzyme has two binding sites for the substrate. If we denote the concentrations of X and Y in neuron i by x_i and y_i, respectively, an appropriate differential equation for x_i as a function of time can have the form

$$\dot{x}_i = 1 - x_i - \frac{x_i y_i}{1 + \alpha x_i^2} \tag{1}$$

where α is a constant. y_i is given by

$$y_i = \sum_{j=1}^{N} \beta_{ij} x_j \tag{2}$$

β_{ij} is the synaptic connection strength from neuron j to neuron i. The β_{ij}

correspond to the T_{ij} of Hopfield, except that learning can occur with the β_{ij} fixed (see below).

There can be three steady-state values for each x_i, two of which are stable (A and C, Fig. 1), and one of which is unstable (B). With N neurons, each of which can be in either a high x_i or a low x_i steady state, the number of attractors can be almost 2^N. The number of unstable steady states in the system can be close to 3^N.

THE DYNAMICS OF AN N-DIMENSIONAL SYSTEM DISPLAYED ON A PHASE PLANE

It is convenient to consider a simple variant of the system (1) that also has about 2^N attractors:

$$\dot{x}_i = (1 - | x_i |)x_i - y_i \tag{3}$$

When the system (3) is at equilibrium, all the neurons are located on the reverse S-shaped curve

$$y_i = (1 - | x_i |)x_i$$

which is shown in Figure 2. Figure 2a shows trajectories following a small transient stimulus. The same stimulus with a larger amplitude results in a qualitatively identical perturbation of the network (Figure 2b). A still larger stimulus causes the x_i value of one or more neurons to change sign, and the other neurons then relax back to slightly different positions on the equilibrium curve (Figure 2c). The system can make a large number of different responses to a correspondingly large number of different stimuli, and unless the amplitudes of the stimuli exceed a certain threshold level, the system stays in the same region of phase space.

The network has been shown to be capable of producing arbitrary outputs, providing N is sufficiently large.[17]

A particular output is produced if the system has an initial condition that can lead to that output. The hysteresis associated with each neuron tends to keep the system as a whole in a restricted region of the N-dimensional phase space, and this can account for memory. The particular region in that space where the system is located at any instant depends on the set of stimuli to which it has been subjected. Hence, in contrast to conventional neural network models, the formation of memories does not need to involve changes in the synaptic strengths, β_{ij}.

TEACHING

These networks can be taught to exhibit prespecified stimulus-response behavior. We systemmatically perturb the network until it reaches a region of phase space where it gives the desired stimulus-response behavior. Since the system tends to stay in a particular region of the phase space in the absence of systemmatic perturbations, the system is able to retain trained stimulus-response behavior or memories. The idea is that a brain is moved around in the N-dimensional phase space by the stimuli that it receives, until it reaches a point in that phase space such that its responses to the environment are such that it is not strongly perturbed by the environment.

We simply apply stimuli and observe responses. There is no "twiddling" of the synaptic connection strengths. Only the various stimuli themselves are used to move the system around in the N-dimensional phase space until a satisfactory region in that space is found. If at some point a satisfactory response to a stimulus is not obtained, the stimulus is reapplied and can intensify. A wrong response means the system was in an unsatisfactory region of the phase space. Applying the same stimulus again takes the system further away from the region of phase space that did not give the correct response. If a good response is obtained, we pass on to another stimulus. This algorithm presumes to mimic the way real brains learn by experience. In an abstract sense there needs to be some complementarity between the stimulus and the response. If we are exposed to a stimulus and respond to it appropriately, our response can lead to the elimination of, or escape from, the stimulus. If we do not respond appropriately, the stimulus can persist, and possibly intensify, independently of whether the stimulus

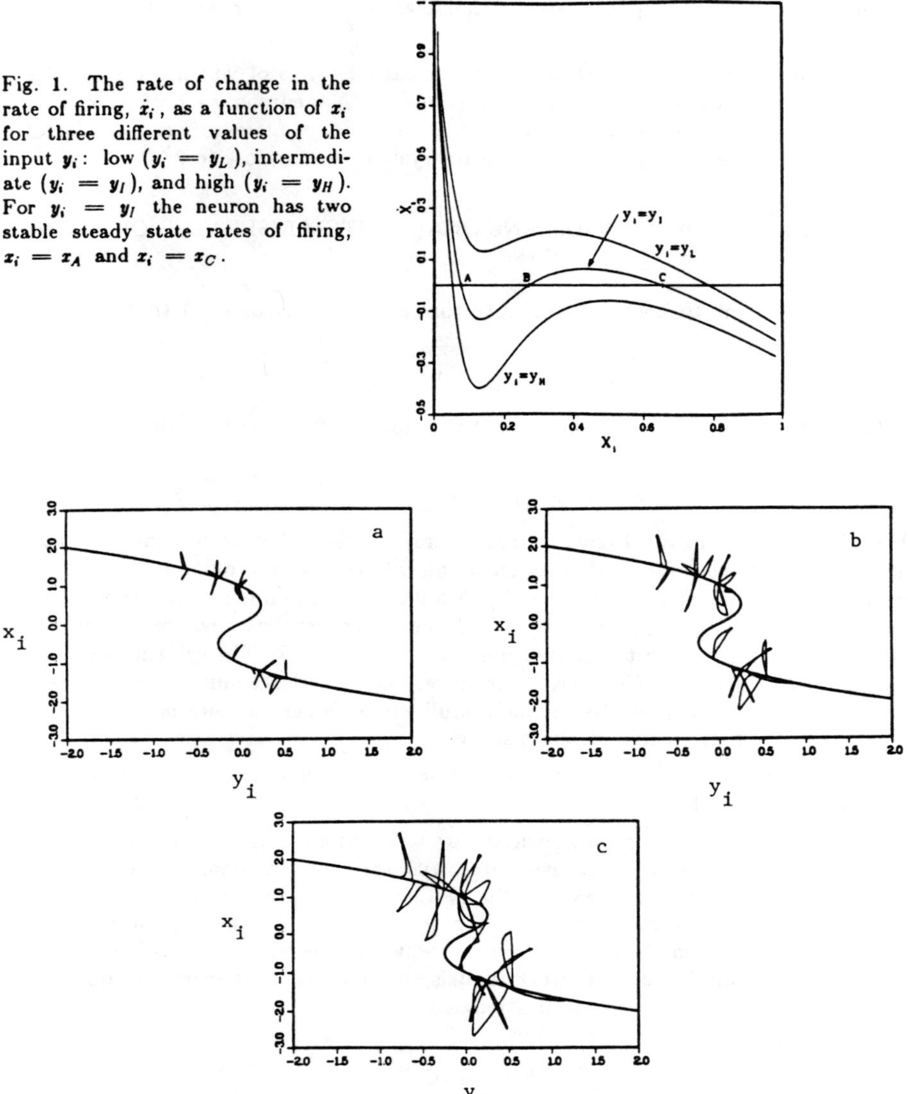

Fig. 1. The rate of change in the rate of firing, $\dot{x}_i$, as a function of x_i for three different values of the input y_i: low ($y_i = y_L$), intermediate ($y_i = y_I$), and high ($y_i = y_H$). For $y_i = y_I$ the neuron has two stable steady state rates of firing, $x_i = x_A$ and $x_i = x_C$.

Fig. 2. The dynamics of a network with 20 neurons portrayed in the y_i, x_i phase plane. The system is at equilibrium when the y_i, x_i coordinates of every neuron is on the reverse S shaped curve. (a) A transient small perturbation causes the neurons to leave the curve, and then relax back to their original positions. (b) A larger stimulus causes a qualitatively similar response. (c) A still larger stimulus causes two neurons to switch from one branch of the curve to the other, and the equilibrium values of each of the other neurons changes somewhat. For these plots we used a network with 20 neurons with a random β_{ij} matrix of connectance 0.5. (The matrix was not symmetric, in contrast to the matrix used for similar figures in reference 18.)

is being provided by a teacher or aspects of the non-human or inanimate environment. In a recent series of experiments we trained networks consisting of 20 neurons to respond to stimuli in prespecified ways using an algorithm based on these ideas.[18] The correct responses were defined in a binary fashion, so that untrained networks had a fifty percent rate of making correct responses. In one series of experiments, we achieved an average accuracy of 73% when we were training for a single stimulus, 70% when there were two stimuli, 66% with 3 stimuli, and 59% when there were 4 stimulus-response pairs being taught. The teaching procedure has not been fine-tuned, so we might be able to do a little better than this. More recent results indicate that networks with completely random β_{ij} can learn just as well as the networks with symmetric β_{ij} that we used initially.

SLEEP

There is an automatic motion of each neuron from the central region (near x_i = 0 in the model (3)) towards one of the attractors (near $x_i = \pm 1$). The system can be strongly influenced by small signals when neurons are in the central region, while much larger stimuli are required when the system is at or near a system attractor. The system might therefore be much more easily taught if there were some means of keeping at least some of the neurons near the central regions. Perhaps during sleep the $\dot{x}_i$ versus x_i characteristic changes to, say,

$$\dot{x}_i = - u x_i$$

where u is a constant. The x_i of the system then relax back towards the origin during sleep, while maintaining their respective relative magnitudes. This part of our theory leads to the prediction that the variance in the rates of firing of neurons involved in memory should increase during waking hours and decrease during sleep.

PROSPECTS

The potential possibilities for systems of the above type has scarcely begun to be explored. In this final section we speculate about the directions that future research might take.

Since the β_{ij} matrix is fixed, special forms of that matrix that permit very rapid computation by digital or optical techniques can be used. As discussed in more detail elsewhere,[18] the use of circulant matrices will permit computations that are up to a factor of $N/\log N$ faster than calculations with random matrices, and such systems are also amenable to the application of extremely rapid analogue optical methods.

The brain is organized largely as groups of neurons that have high internal connectance, with lower connectance between groups. For artificial intelligence applications, it might be profitable to mimic this architecture, and thus obtain reliable behavior from unreliable components. If we couple together groups of our neurons that have been previously trained separately, it should be possible that the larger aggregates will give an improved fraction of "correct" responses than the smaller groups are able to do independently.

A scientific and perhaps philosophical mystery in neural science is the question of the neural bases of pain and joy. The above teaching method is something of a brute force approach, that might be viewed as being based on using only "pain." (If the response is incorrect, we increase the magnitude of the stimulus.) Presumably the efficiency of our teaching could be much greater if we could add something to the algorithm equivalent to joy. A very speculative neural interpretation of pain and joy is in terms of familiar and unfamiliar regions of phase space. Perhaps pain is associated with stimuli that take one much away from familiar regions of phase space (as most completely random strong stimuli would do), and joy results from stimuli that are special in the sense that they move one back towards more familiar regions. When a brain has developed a set of memories, it presumably spends most of its time in a relatively small section of the total phase space. Specific stimuli that it receives periodically (for instance the specific stimuli experienced as a result of consuming food) would be expected to play an important role in keeping it in that limited region of phase space. Stimuli that cause pain are normally not received periodically, and they are likely to be more random in nature. There are many directions in phase space away from any restricted (familiar) region, and only a small number of directions back towards the most familiar region if one is slightly removed from it. Thus this idea would account for the fact that a limited set of specific stimuli can impart joy, while a broad range of relatively nonspecific stimuli can impart pain. New stimuli could move one back towards familiar regions of phase space if they are somehow correlated (in the context of the wiring diagram of our brains) with other stimuli with which we are familiar. If this speculation is correct, we might be able to simulate joy by making a network accustomed to a certain set of stimuli during its "ontogeny." Stimuli that are correlated with the early set might then be used to impart "joy."

The reader might feel that we are being excessively ambitious in trying to account for the phenomena of pain and joy so simplisticly. Such an assessment could well be correct. On the other hand, if we are serious about trying to understand the brain, we need to face up to these more difficult questions, and we have no way of knowing in advance how complex the solutions will be. Furthermore, these ideas can be tested at two levels. We can firstly see whether they do in fact work at the level of our computer model, that is whether they prove useful in the development of a more successful teaching algorithm. Secondly, this concept predicts that the periodic application of a specific stimulus to an animal during its early ontogeny should lead to the animal becoming addicted to that stimulus. That is, if such an animal were then given control over the application of the stimulus, it would be likely to learn to apply the stimulus to itself, and then continue to apply the stimulus with the same periodicity. If both of these predictions can be verified, the hypothesis will become worthy of serious consideration.

ACKNOWLEDGEMENTS

This work was done while we were both on sabbatical leave at the Los Alamos National Laboratory. We thank George Bell and the other members of the Theoretical Biology group and David Campbell of the Center for Nonlinear Studies for their kind hospitality.

This research was financed by the Natural Sciences and Engineering Research Council of Canada, Grant No. A-6729 to G. W. Hoffmann, and the United States Department of Energy.

REFERENCES

1. F. Hund, Geschichte der physikalischen Begriffe (Bibliographisches Institut, Mannheim 1972).
2. F. M. Burnet, The Clonal Selection Theory of Acquired Immunity (Cambridge University Press, Cambridge, 1959).
3. N. K. Jerne, Proc. Nat. Acad. Sci. (USA) 41 849-857 (1955)
4. N. K. Jerne, Sci. Amer. 229 (1), 52-60, (1973).
5. N. K. Jerne, Ann. Immunol. (Inst. Pasteur) 125C 373-389 (1974).
6. W. A. Little, Math. Biosci. 19 101-120 (1974).
7. A. H. Klopf, The Hedonistic Neuron: A Theory of Memory, Learning and Intelligence (Hemisphere, Washington, D.C., 1982).
8. A. G. Barto, Human Neurobiol. 4 229-256 (1985).

Part 4
Theory of Neuronal Approximations

The theory of neural computing should be concerned with the development of neural-like, massively parallel connections in a machine for performing some cognitive tasks—the higher level brain functions.

THE theory of approximation is an important class of problems in both static and dynamic processes. We would like to approximate a given experimental curve by a class of polynomials, or a time evolution of a dynamic process by a difference or differential equation. We would also like to approximate, using the Fourier analysis, a complex periodic (or almost periodic) phenomenon by a series of sinusoids. System identification, estimation of a signal from a noisy measurement, or forecasting stock markets using the past history and current economic trends are just a few examples that have baffled system scientists and economists.

In this part, we describe the theory of neuronal approximations that has captured the attention of neural scientists only recently. It was at the IEEE First International Conference of Neural Networks in 1987 held at San Diego when R. Hecht-Nielsen reiterated the theorem of Kolmogorov (1957) and made some comparisons with the attributes of neural networks. Kolmogorov's theorem states that one can express a continuous multivariable function, on a compact domain, in terms of sums and compositions of single variable functions. Furthermore, the number of single-variable functions required is finite. It implies that there are no nemesis functions that cannot be modeled (approximated) by neural networks.

Indeed, the parallel and layered morphology of the neural systems is responsible for solving a wider class of problems in fields such as system approximation (identification), control, learning, and adaptation. In this part of the book, we present some representative articles that deal with the theory of neuronal approximation. These articles provide a mathematical foundation of neural approximations, and show how layered networks can approximate given static and/or dynamic physical entities to a desired degree of accuracy. Such an approximation, in fact, forms the basis in problems dealing with pattern classification, system identification, control, robotics manipulator problems, and in the design of neural filters and equalizers.

In the first article, (4.1), K. I. Funahashi proves that any continuous mapping can be realized by a multilayer (more than two layers) neural network with at least one hidden layer using sigmoidal activation functions. The theoretical details provided in this article are based on the Kolmogorov–Arnold–Sprecher theorem and will be of great use in the further development of the neural networks in control applications.

In the second article, (4.2), F. Girosi and T. Poggio show that feedforward multilayer neural networks of the backpropagation type do not have the best approximation property for the class of continuous functions defined on a subset of R^n. On the other hand, the authors prove that for methods derived for regularization and, in particular, for radial basis function (RBF) networks, best and unique approximation exists.

K. Hornik, M. Stinchcombe, and H. White, in their article, (4.3), rigorously establish that standard multilayered feedforward networks with as few as one hidden layer, using an arbitrary activation (squashing) function, are capable of approximating any Borel measurable function from one finite dimensional space to another, to any desired degree of accuracy. In this sense, multilayer feedforward networks can be considered as a class of universal approximators.

In article (4.4), G. Cybenko demonstrates that finite linear combinations of compositions of a fixed, invariable function and a set of affine functions can uniformly approximate any continuous function of n-real variables with support in the unit hybercube. The author also discusses approximation properties of other possible types of nonlinearities that might be implemented by multilayered neural network structures.

In article (4.5), N. E. Cotter presents a neural network architecture based on the Stone–Weierstrass theorem. He shows that exponential functions, polynomials, partial fractions, and Boolean functions can be used to create neural approximations for arbitrarily bounded measurable functions. His discussion begins with a theorem and follows with appropriate mathematical developments and a pragmatic description of some abstract terminology.

This set of five articles provides studies of various advance properties of the feedforward neural networks, and none of them deal with the approximation problem of dynamic neural structures. In the final article, (4.6), D. H. Rao and M. M. Gupta propose a dynamic neural unit (DNU) incorporating synaptic and somatic operations. The DNU is comprised of a second-order linear structure with feedforward and feedback weights, thus providing two zeros and two poles within the unit circle. This linear structure is followed by a time-varying nonlinear activation function. It is demonstrated in this article that a network of such DNUs is capable of approximating various nonlinear functions. Such dynamic neural structures also have fast convergence properties for various computations.

This set of six articles represents only a small sample of articles that have appeared during recent years; however, they should motivate readers to further studies in this field.

After having introduced some basic neuro-biological and neuro-mathematical tools, in the next four parts we proceed specifically with the theory of neuro-control and its applications and implementations.

Futher Reading

[1] P. P. Korovkin, *Linear Operators and Approximation Theory.* Delhi: Hindustan Publishing Corp., 1960.

[2] G. A. Watson, *Approximation Theory and Numerical Methods.* New York: John Wiley & Sons, 1980.

[3] L. V. Kantorovich and G. P. Akilov, *Functional Analysis.* Translated by H. L. Silcock, Elmsford, NY: Pergamon Press, 1982.

[4] A. N. Kolmogorov and S. V. Fomin, *Introductory Real Analysis.* Translated by R. A. Silverman, Englewood Cliffs, NJ: Prentice-Hall, 1970.

[5] T. Poggio and F. Girosi, "Networks for approximation and learning," *Proc. IEEE,* vol. 78, no. 9, pp. 1481–1497, Sept. 1990.

[6] G. Cybenko, "Approximation by superpositions of a sigmoidal function," *Math. Contr., Signals, Syst.,* vol. 2, pp. 303–314, 1989.

[7] K. Funahashi, "On the approximate realization of continuous mappings by neural networks," *Neural Networks,* vol. 2, pp. 183–192, 1989.

[8] K. Hornik, M. Stinchecombe and H. White, "Multi-layer feed forward networks are universal approximators," *Neural Networks,* vol. 2, pp. 359–366, 1989.

[9] N. E. Cotter, N. E., "The Stone–Weierstrass theorem and its applications to neural networks," *IEEE Trans. Neural Networks,* vol. 1, no. 4, pp. 290–295, 1990.

[10] E. K. Blum and L. K. Li, "Approximation theory and feedforward networks," *Neural Networks,* vol. 4, pp. 511-515, 1991.

[11] A. R. Gallant and H. White, "There exists a neural network that does not make avoidable mistakes," in *Proc. IEEE on Neural Networks,* vol. I, pp. 657–664, San Diego, 1988.

[12] P. Cardaliaguet and G. Euvard, "Approximation of a function and its derivative with a neural network," *Neural Networks,* vol. 5, pp. 207–220, 1992.

[13] R. Hecht-Nielsen, "Kolmogorov's Mapping Neural Network Existence Theorem," in *Proc. IEEE on Neural Networks,* vol. II, pp. 11–14, San Diego, 1987.

[14] N. E. Cotter and T. J. Gullerm, "The CMAC and a theorem of Kolmogorov," *Neural Networks,* vol. 5, pp. 221–228, 1992.

[15] V. Kurkova, "Kolmogorov's theorem and multilayer neural networks," *Neural Networks,* vol. 5, pp. 501–506, 1992.

[16] F. Girosi and T. Poggio, "Representation of properties of networks: Kolmogorov's theorem is irrelevant," *Neural Computation,* vol. 63, pp. 169–176, 1990.

[17] K. S. Narendra and K. Parthasarthy, "Identification and control of dynamical systems using neural networks," *IEEE Trans. Neural Networks,* pp. 4–27, vol. 1, no. 1, 1990, March.

[18] S. A. Billings, H. B. Jamaludding, and S. Chen, "Properties of neural networks with applications to modeling nonlinear dynamical systems," *Int. J. Control,* vol. 55, no. 1, pp. 193–224, 1992.

[19] K. J. Hunt and D. Sbarbaro, "Neural networks for nonlinear internal model control," *IEE Proc.-D,* vol. 138, no. 5, pp. 431–438, Sept. 1991.

[20] R. D. Jones, Y. C. Lee, C. W. Barnes, G. W. Flake, K. Lee, P. S. Lewis, and S. Qian, "Function approximation and time series prediction with neural networks," in *Proc. Int. Joint Conf. Neural Networks (IJCNN),* pp. 649–665, June 1990.

Article 4.1

On the Approximate Realization of Continuous Mappings by Neural Networks

KEN-ICHI FUNAHASHI

ATR Auditory and Visual Perception Research Laboratories

(*Received 6 May 1988; revised and accepted 14 September 1988*)

Abstract—*In this paper, we prove that any continuous mapping can be approximately realized by Rumelhart-Hinton-Williams' multilayer neural networks with at least one hidden layer whose output functions are sigmoid functions. The starting point of the proof for the one hidden layer case is an integral formula recently proposed by Irie–Miyake and from this, the general case (for any number of hidden layers) can be proved by induction. The two hidden layers case is proved also by using the Kolmogorov-Arnold-Sprecher theorem and this proof also gives non-trivial realizations.*

Keywords—Neural network, Back propagation, Output function, Sigmoid function, Hidden layer, Unit, Realization, Continuous mapping.

1. INTRODUCTION

Since McCulloch-Pitts (1943), there have been many studies of mathematical models of neural networks. Recently, Hopfield, Hinton, Rumelhart, Sejnowski and others have tried many concrete applications such as pattern recognition, and have shown that it is possible to clarify the mechanism of human information processing by the use of these models. In particular, the back propagation algorithm (generalized delta rule) proposed by Rumelhart, Hinton, and Williams (1986) provides a learning rule for multilayer networks. Many applications of this algorithm have been shown recently. However, there has been little theoretical research on the capability of the Rumelhart–Hinton–Williams multilayer network.

On the application to pattern recognition, Lippmann (1987) asserts that arbitrary complex decision regions, including concave regions, can be formed using four-layer networks, but this is only an intuitive assertion. Wieland and Leighton (1987) showed an example of a three-layer network with thresholding units which partitions a space into concave subspaces. Huang and Lippmann (1987) demonstrated by simulations that three-layer networks can form several complex decision regions in pattern recognition application. However, it has been known that any piecewise-linear decision region (which is not necessarily convex) can be realized by a multilayer network (Duda & Fossum, 1966). It's learning algorithm was also proposed (Amari, 1967) based on the same principle as the generalized delta rule. There are also other applications of multilayer networks for forming mappings, such as NETtalk by Sejnowski and Rosenberg (1987).

Hecht-Nielsen (1987) pointed out that Kolmogorov's theorem (Kolmogorov, 1957) and Sprecher's refinement (Sprecher, 1965), which are both known as negative solutions of Hilbert's thirteenth problem, show that any continuous mapping can be represented by a form of four-layer neural network. Uesaka (1971) and Poggio (1983) have also pointed this out. However, the assertion has a problem in that the output function of each unit of this network is not a given sigmoid function.

Irie and Miyake (1988) obtained an integral formula which suggests the realization of functions of several variables by three-layer networks by analogy with the principle of the computerized tomography (CT). But in this integral formula, the output function $\psi(x)$ must satisfy the condition of absolute integrability, so that it cannot be a sigmoid function. Moreover, the function to be realized is given by an integral representation and the formula does not di-

The author wishes to thank Drs. Y. Tohkura, T. Inui and S. Miyake, and Mr. T. Okamoto for their valuable comments on the manuscript. The author also would like to thank anonymous reviewers whose constructive suggestions have improved the quality of this paper.

Requests for reprints should be sent to Ken-ichi Funahashi, ATR Auditory and Visual Perception Research Laboratories, Twin 21 Building, MID Tower 2-1-61 Shiromi, Higashi-ku, Osaka 540, Japan.

Reprinted with permission from *Neural Networks*, vol. 2, K. Funahashi, "On the Approximate Realization of Continuous Mappings by Neural Networks," pp. 183–192, 1989, Pergamon Press Ltd., Oxford, England.

rectly give the realization theorem of functions by networks with finite units.

In neural networks of the feed-forward type by Rumelhart–Hinton–Williams, bounded and monotone increasing differentiable functions such as the sigmoid function $\phi(x) = 1/(1 + e^{-x})$ are used as output functions of units. This is a different point from the McCulloch-Pitts model and perceptron which use heaviside function as output functions of units and is the reason why it is possible to derive a learning algorithm for multilayer networks.

In a feed-forward type network, it's input–output relationship defines a mapping which is called an input–output mapping of the network. We studied the problem of network capabilities from the point of view of input–output mappings.

In this paper, we started from an integral formula recently proposed by Irie and Miyake (1988) and proved the theorem which guarantees the approximate realization of continuous mappings by three-layer (one hidden layer) networks whose output functions for hidden layer are sigmoid, and whose output functions for input and output layers are linear in the sense of uniform topology. It is easy to prove the theorem for k ($\geq$3)-layer networks by using the theorem for a three-layer case. But the proof of the theorem for the case $k > 3$ gives only trivial approximate realization of given mappings. Therefore we show another proof for the four-layer case by using the Kolmogorov–Arnold-Sprecher theorem (Kolmogorov, 1957; Sprecher, 1965).

McCulloch-Pitts showed that any logical circuit can be designed using their model. Correspondingly, our assertion shows that any continuous mapping can be approximately represented by the Rumelhart–Hinton–Williams multilayer network.

2. MULTILAYER NEURAL NETWORKS

The Rumelhart–Hinton–Williams multilayer network that we consider here is a feed-forward type network with connections between adjoining layers only. Networks generally have hidden layers between the input and output layers. Each layer consists of computational units. The input–output relationship of each unit is represented by inputs x_i, output y, connection weights w_i, threshold θ, and differentiable function ϕ as follows:

$$y = \phi\left(\sum_{i=1}^{k} w_i x_i - \theta\right).$$

The learning rule of this network is known as the back propagation algorithm (Rumelhart, Hinton, & Williams, 1986). The back propagation algorithm is an algorithm that uses a gradient descent method to modify weights and thresholds so that the error between the desired output and the output signal of the network is minimized. We generally use a bounded and monotonic increasing differentiable function which is called sigmoid function for each unit's output function.

If a multilayer network has n input units and m output units, then the input–output relationship defines a continuous mapping from n-dimensional Euclidean space to m-dimensional Euclidean space. We call this mapping the input–output mapping of the network. We study the problem of network capabilities from the point of view of input–output mappings. It is observed that for the study of mappings defined by multilayer networks it is sufficient to consider networks whose output functions for hidden layers are the above $\phi(x)$ and whose output functions for input and output layers are linear.

3. APPROXIMATE REALIZATION OF CONTINUOUS MAPPINGS BY NEURAL NETWORKS

We shall consider the possibility of representing continuous mappings by neural networks whose output functions in hidden layers are sigmoid, for example, $\phi(x) = 1/(1 + e^{-x})$. It is simply noted here that general continuous mappings cannot be exactly represented by Rumelhart–Hinton–Williams' networks. For example, if a real analytic output function such as the sigmoid function $\phi(x) = 1/(1 + e^{-x})$ is used, then an input–output mapping of this network is analytic and generally cannot represent all continuous mappings.

Let points of n-dimensional Euclidean space $\mathbf{R}^n$ be denoted by $\mathbf{x} = (x_1, \ldots, x_n)$ and the norm of $\mathbf{x}$ defined by $|\mathbf{x}| = (\Sigma_{i=0}^{n} x_i^2)^{1/2}$.

We prove the following theorems and corollaries in this paper.

Theorem 1.

Let $\phi(x)$ be a nonconstant, bounded and monotone increasing continuous function. Let K be a compact subset (bounded closed subset) of $\mathbf{R}^n$ and $f(x_1, \ldots, x_n)$ be a real valued continuous function on K. Then for an arbitrary $\epsilon > 0$, there exists an integer N and real constants $c_i, \theta_i (i = 1, \ldots, N)$, $w_{ij} (i = 1, \ldots, N, j = 1, \ldots, n)$ such that

$$\tilde{f}(x_1, \ldots, x_n) = \sum_{i=1}^{N} c_i \phi\left(\sum_{j=1}^{n} w_{ij} x_j - \theta_i\right)$$

satisfies $\max_{\mathbf{x} \in K} |f(x_1, \ldots, x_n) - \tilde{f}(x_1, \ldots, x_n)| < \epsilon$. In other words, for an arbitrary $\epsilon > 0$, there exists a three-layer network whose output functions for the hidden layer are $\phi(x)$, whose output functions for input and output layers are linear and which has an input–output function $\tilde{f}(x_1, \ldots, x_n)$ such that

$\max_{\mathbf{x}\in K} |f(x_1, \ldots, x_n) - \tilde{f}(x_1, \ldots, x_n)| < \epsilon$.

The above theorem easily leads to the following general theorem.

Theorem 2.

Let $\phi(x)$ be a nonconstant, bounded and monotone increasing continuous function. Let K be a compact subset (bounded closed subset) of $\mathbf{R}^n$ and fix an integer $k \geq 3$. Then any continuous mapping $f: K \to \mathbf{R}^m$ defined by $\mathbf{x} = (x_1, \ldots, x_n) \to (f_1(\mathbf{x}), \ldots, f_m(\mathbf{x}))$ can be approximated in the sense of uniform topology on K by input–output mappings of k-layer (k-2 hidden layers) networks whose output functions for hidden layers are $\phi(x)$, and whose output functions for input and output layers are linear. In other words, for any continuous mapping $f: K \to \mathbf{R}^m$ and an arbitrary $\epsilon > 0$, there exists a k-layer network whose input–output mapping is given by $\tilde{f}: K \to \mathbf{R}^m$ such that $\max_{\mathbf{x}\in K} d(f(\mathbf{x}), \tilde{f}(\mathbf{x})) < \epsilon$, where $d(,)$ is a metric which induces the usual topology of $\mathbf{R}^m$.

Corollary 1.

Let $\phi(x)$, K be as above and fix an integer $k \geq 3$. Then any mapping $f: \mathbf{x} \in K \to (f_1(\mathbf{x}), \ldots, f_m(\mathbf{x})) \in \mathbf{R}^m$ where $f_i(\mathbf{x})$ $(i = 1, \ldots, m)$ are summable on K, can be approximated in the sense of L^2-topology on K by input–output mappings of k-layer (k-2 hidden layers) networks whose output functions for hidden layers are $\phi(x)$ and whose output functions for input and output layers are linear. In other words, for an arbitrary $\epsilon > 0$, there exists a k-layer network whose input–output mapping is given by $\tilde{f}: \mathbf{x} \in K \to (\tilde{f}_1(\mathbf{x}), \ldots, \tilde{f}_m(\mathbf{x})) \in \mathbf{R}^m$ such that

$$d_{L^2(K)}(f, \tilde{f}) = \left(\sum_{i=1}^{m} \int_K |f_i(x_1, \ldots, x_n) - \tilde{f}_i(x_1, \ldots, x_n)|^2 \, d\mathbf{x} \right)^{1/2} < \epsilon.$$

Corollary 2.

Let K be as above and fix an integer $k \geq 3$. Let $\phi(x)$ be a strictly increasing continuous function such that $\phi((-\infty, \infty)) = (0, 1)$. Then any continuous mapping $f: K \to (0, 1)^m$ can be approximated in the sense of uniform topology on K by input–output mappings of $k(\geq 3)$-layer neural networks whose output functions for hidden and output layers are $\phi(x)$.

Proof. Set $f(\mathbf{x}) = (f_1(\mathbf{x}), \ldots, f_m(\mathbf{x}))$. As $\phi^{-1}: (0, 1) \to (-\infty, \infty)$ is continuous, the theorem 2 is applied to the mapping $\mathbf{x} \to \phi^{-1} f(\mathbf{x}) = (\phi^{-1} f(\mathbf{x}), \ldots, \phi^{-1} f_m(\mathbf{x}))$ and the corollary is obtained easily.

q.e.d.

Remark 1.

Usual output functions such as the sigmoid function $1/(1 + e^{-x})$ used for back-propagation neural networks satisfy the condition of $\phi(x)$ that $\phi(x)$ is a nonconstant, bounded and monotone increasing continuous function.

Remark 2.

Any mapping is approximately realized by a three-layer (one hidden layer) network. However, it should be theoretically studied in the future that the possibility of $k > 3$-layer networks can realize a given mapping with less costs (number of units or connections) than three-layer networks, within error ϵ.

For the application of neural networks to pattern recognition, if m is the number of recognized categories, usually m output units corresponding to these categories are used, and the system is allowed to learn to take values near 1 only for units corresponding to the input categories. Corollaries show that if one uses multilayer networks with hidden layers, any decision region can be formed by a neural network. In particular, a strictly increasing continuous function, as the output function of each unit, can be chosen.

In this paper, we call bounded and monotone increasing continuous functions, sigmoid functions. In particular, a sigmoid function $\phi(x)$ having a weak derivative which is summable has the property that if we set $\phi_\epsilon(x) = \phi(x/\epsilon)(\epsilon > 0)$, then the derivatives $\phi'_\epsilon(x) = (1/\epsilon)\phi'(x/\epsilon)$ converge, in the sense of the generalized function (see, e.g., Gel'fand & Shilov, 1964), to the δ function as $\epsilon \to 0$. That is to say, if $\phi(\infty) - \phi(-\infty) = 1$, then for any smooth function $g(x)$ with compact support,

$$\lim_{\epsilon \to +0} \int_{-\infty}^{\infty} \phi'_\epsilon(x) \cdot g(x) \, dx = g(0).$$

The following examples are included in the class of sigmoid functions considered here.

Example 1. For $\phi(x) = 1/(1 + \exp(-x))$, $\phi'_\epsilon(x) = 1/\epsilon \exp(-x/\epsilon)/(1 + \exp(-x/\epsilon))^2$ and $\phi(x)$ is a sigmoid function.

Example 2. For $\Phi(x) = 1/\sqrt{2\pi} \int_{-\infty}^{x} \exp(-t^2/2)\, dt$, $\Phi'_\epsilon(x) = \epsilon/\sqrt{2\pi} \exp(-x^2/2\epsilon)$ and $\Phi(x)$ is a sigmoid function.

Example 3. For $\phi(x)$ where $\phi(x) = 0 (x < 0)$, $\phi(x) = x (0 < x < 1)$ and $\phi(x) = 1 (x \geq 1)$, $\phi'_\epsilon(x) = 0 (x < 0 \text{ or } x \geq \epsilon)$, $\phi'_\epsilon(x) = 1/\epsilon (0 \leq x < \epsilon)$, and $\phi(x)$ is a sigmoid function.

In the McCulloch-Pitts neural model and perceptron, a threshold function $\phi(x) = 1 (x \geq 0)$, $= 0 (x < 0)$ is used as the output function.

Sigmoid functions $\phi(x)$ where $\phi(-\infty) = 0$ and $\phi(\infty) = 1$ are appropriate as output functions in the neural model because if we set $\phi_\epsilon(x) = \phi(x/\epsilon)(\epsilon > 0)$ then these converge to the threshold function in the McCulloch–Pitts neural model and perceptron as $\epsilon \to +0$.

McCulloch–Pitts shows that one can design any logical circuit using their model. Correspondingly, theorem 2 above shows that any continuous mapping can be approximately represented by multilayer networks with sigmoid output functions.

4. PRELIMINARY 1 (MOLLIFIERS, FOURIER TRANSFORMS)

Fundamental matters used in this paper are reviewed here.

Let $L^p(\mathbf{R}^n)(p \geq 1)$ denote the space of all measurable functions $f(\mathbf{x})$ on $\mathbf{R}^n$ which satisfy

$$\int_{\mathbf{R}^n} |f(\mathbf{x})|^p \, d\mathbf{x} < \infty.$$

The norm of $f \in L^p(\mathbf{R}^n)$ is defined by

$$\|f(\mathbf{x})\|_{L^p} = \left(\int_{\mathbf{R}^n} |f(\mathbf{x})|^p \, d\mathbf{x} \right)^{1/p},$$

and the convergence $f_n(\mathbf{x}) \to f(\mathbf{x})$ in $L^p(\mathbf{R}^n)$ is defined by

$$\lim_{n\to\infty} \|f_n(\mathbf{x}) - f(\mathbf{x})\|_{L^p} = 0.$$

Generally, for any measurable set K, $L^p(K)(p \geq 1)$ is defined similarly.

Let $\rho(\mathbf{x})$ be a function on $\mathbf{R}^n$ which satisfies the following conditions:

(i) $\rho(\mathbf{x}) \geq 0$, $\rho(\mathbf{x})$ has continuous partial derivatives of all orders and the support is contained in the unit sphere $|\mathbf{x}| \leq 1$.
(ii) $\int_{\mathbf{R}^n} \rho(\mathbf{x}) \, d\mathbf{x} = 1$

Then, for $\epsilon > 0$, set $\rho_\epsilon(\mathbf{x}) = (1/\epsilon)^n \rho(\mathbf{x}/\epsilon)$.

If $u(\mathbf{x}) \in L^1_{\text{loc}}$, that is, $u(\mathbf{x})$ is locally summable, consider

$$\rho_\epsilon * u(\mathbf{x}) = \int_{\mathbf{R}^n} \rho_\epsilon(\mathbf{x} - \mathbf{y}) u(\mathbf{y}) \, d\mathbf{y},$$

then the following assertions hold: (a) $\rho_\epsilon * u(\mathbf{x}) \in C^\infty$, that is, $\rho_\epsilon * u(\mathbf{x})$ has continuous partial derivatives of all orders, and the support of $\rho_\epsilon * u(\mathbf{x})$ is contained in the ϵ neighborhood of support of $u(\mathbf{x})$; (b) if $u(\mathbf{x})$ is a continuous function with compact support, then $\rho_\epsilon * u(\mathbf{x}) \to u(\mathbf{x})$ uniformly on $\mathbf{R}^n$ as $\epsilon \to +0$; and (c) if $u(\mathbf{x}) \in L^p(\mathbf{R}^n)(p \geq 1)$ then $\rho_\epsilon * u(\mathbf{x}) \to u(\mathbf{x})$ in L^p as $\epsilon \to +0$. The operator $\rho_\epsilon *$ is called a mollifier.

For $f(\mathbf{x}) \in L^1(\mathbf{R}^n)$, Fourier transform

$$\hat{f}(\boldsymbol{\xi}) = \int_{\mathbf{R}^n} e^{-i\langle \mathbf{x}, \boldsymbol{\xi} \rangle} f(\mathbf{x}) \, d\mathbf{x}, \tag{1}$$

where $\langle \mathbf{x}, \boldsymbol{\xi} \rangle = \sum_{i=1}^n x_i \xi_i$, can be defined and set $\hat{f}(\boldsymbol{\xi}) = \mathscr{F} f(\boldsymbol{\xi})$.

If $f(\mathbf{x})$ satisfies an additional condition that $f(\mathbf{x})$ has continuous partial derivatives of order up to n, then $f(\mathbf{x})$ can be represented at each point by inverse Fourier transform of $\hat{f}(\boldsymbol{\xi})$ as follows.

$$f(\mathbf{x}) = (2\pi)^{-n} \int_{\mathbf{R}^n} e^{i\langle \mathbf{x}, \boldsymbol{\xi} \rangle} \hat{f}(\boldsymbol{\xi}) \, d\boldsymbol{\xi}.$$

The Plancherel theorem especially asserts that $\mathscr{F}$ can be extended one to one onto mapping $\mathscr{F}: L^2(\mathbf{R}^n) \to L^2(\mathbf{R}^n)$ and for $f(\mathbf{x}) \in L^1 \cap L^2(\mathbf{R}^n)$, $\mathscr{F} f(\boldsymbol{\xi})$ is equal to the one defined by (1). Furthermore, for $f(\mathbf{x}) \in L^2(\mathbf{R}^n)$,

$$\left\| \mathscr{F} f(\boldsymbol{\xi}) - \int_{|\mathbf{x}| \leq A} e^{-i\langle \mathbf{x}, \boldsymbol{\xi} \rangle} f(\mathbf{x}) \, d\mathbf{x} \right\|_{L^2} \longrightarrow 0 (A \longrightarrow +\infty)$$

(see e.g., Yosida, 1968).

5. PRELIMINARY 2 (IRIE–MIYAKE'S INTEGRAL FORMULA)

The following theorem is a starting point for proof of Theorem 1.

Theorem (Irie-Miyake)

Let $\psi(x) \in L^1(\mathbf{R})$, that is, let $\psi(x)$ be absolutely integrable and $f(x_1, \ldots, x_n) \in L^2(\mathbf{R}^n)$. Let $\Psi(\xi)$ and $F(w_1, \ldots, w_n)$ be Fourier transforms of $\psi(x)$ and $f(x_1, \ldots, x_n)$ respectively.

If $\Psi(1) \neq 0$, then

$$f(x_1, \ldots, x_n) = \int_{-\infty}^{\infty} \cdots \int_{-\infty}^{\infty} \psi\left(\sum_{i=1}^n x_i w_i - w_0 \right) \frac{1}{(2\pi)^n \Psi(1)} F(w_1, \ldots, w_n) \times \exp(iw_0) \, dw_0 \, dw_1 \cdots dw_n.$$

Remark

This formula precisely asserts that if we set

$$I_{\infty,A}(x_1, \ldots, x_n) = \int_{-A}^{A} \cdots \int_{-A}^{A} \left[\int_{-\infty}^{\infty} \psi\left(\sum_{i=1}^n x_i w_i - w_0 \right) \times \frac{1}{(2\pi)^n \Psi(1)} F(w_1, \ldots, w_n) \times \exp(iw_0) \, dw_0 \right] dw_1 \cdots dw_n$$

then

$$\lim_{A\to\infty} \|I_{\infty,A}(x_1, \ldots, x_n) - f(x_1, \ldots, x_n)\|_{L^2} = 0.$$

Connecting this formula with three-layer networks, Irie and Miyake (1988) assert that arbitrary functions can be represented by a three-layer network with an infinite number of computational units. In this formula, w_0 corresponds to threshold, w_i corresponds to connection weight and $\psi(x)$ corresponds to the output function of the units. However, the sigmoid function $1/(1 + e^{-x})$ does not satisfy the condition of this formula that $\psi(x)$ be absolutely integrable and so the formula does not directly give the realization theorem of functions by networks.

6. PRELIMINARY 3 (SEVERAL LEMMAS)

We prepare several Lemmas for proof of our theorem 1.

Lemma 1.

Let $\phi(x)$ be a nonconstant, bounded and monotone increasing continuous function. For $\alpha > 0$, if we set

$$g(x) = \phi(x + \alpha) - \phi(x - \alpha),$$

then $g(x) \in L^1(\mathbf{R})$, that is,

$$\int_{-\infty}^{\infty} |g(x)|\, dx < \infty.$$

Furthermore, for some $\delta > 0$, if we set

$$g_\delta(x) = \phi(x/\delta + \alpha) - \phi(x/\delta - \alpha),$$

then the value of Fourier transform $G_\delta(\xi)$ of $g_\delta(x)$ at $\xi = 1$ is non-zero.

Proof. Let $|g(x)| \leq M$. For $L > M$,

$$\begin{aligned}\int_{-L}^{L} |g(x)|\, dx &= \int_{-L}^{L} g(x)\, dx = \int_{-L+\alpha}^{L+\alpha} \phi(x)\, dx \\ &\quad - \int_{-L-\alpha}^{L-\alpha} \phi(x)\, dx \\ &= \int_{L-\alpha}^{L+\alpha} \phi(x)\, dx \\ &\quad - \int_{-L-\alpha}^{-L+\alpha} \phi(x)\, dx \leq 4\alpha M.\end{aligned}$$

Therefore,

$$\lim_{L\to\infty} \int_{-L}^{L} |g(x)|\, dx < \infty.$$

We show that for some $\delta > 0$, $G_\delta(1) \neq 0$. If the assertion does not hold, then for any $\delta > 0$,

$$\int_{-\infty}^{\infty} (\phi(x/\delta + \alpha) - \phi(x/\delta - \alpha))e^{-ix}\, dx = 0.$$

By the change of the variable,

$$\int_{-\infty}^{\infty} (\phi(x + \alpha) - \phi(x - \alpha))e^{-ix\delta}\, dx = 0 \quad \text{(for any } \delta > 0). \quad (1)$$

Taking the complex conjugate of the above equation (1),

$$\int_{-\infty}^{\infty} (\phi(x + \alpha) - \phi(x - \alpha))e^{ix\delta}\, dx = 0 \quad \text{(for any } \delta > 0). \quad (2)$$

Since the Fourier transform $G_1(\xi)$ of $g_1(x) = \phi(x + \alpha) - \phi(x - \alpha) \in L^1(\mathbf{R})$ is continuous, so, from (1) and (2), $G_1(\xi)$ is identically zero. Therefore,

$$\phi(x + \alpha) - \phi(x - \alpha) \equiv 0.$$

This is a contradiction, because $\phi(x)$ is not constant. q.e.d.

Remark

Lemma 1 holds for $\phi(x)$ which is locally summable.

Lemma 2

Let $A_i > 0$ $(i = 1, \ldots, m)$, K be a compact subset (bounded closed subset) of $\mathbf{R}^n$ and $h(x_1, \ldots, x_m, t_1, \ldots, t_n)$ be a continuous function on $[-A_1, A_1] \times \cdots \times [-A_m, A_m] \times K$.

Then the function defined by the integral

$$H(\mathbf{t}) = \int_{-A_1}^{A_1} \cdots \int_{-A_m}^{A_m} h(x_1, \ldots, x_m, t_1, \ldots, t_n)\, dx_1 \ldots dx_m$$

can be approximated uniformly on K by the Riemann sum

$$\begin{aligned}H_N(\mathbf{t}) &= \frac{2A_1 \cdots 2A_m}{N^m} \\ &\quad \times \sum_{k_1\cdots k_m=0}^{N-1} h\left(-A_1 + \frac{k_1 \cdot 2A_1}{N}, \ldots, \right. \\ &\quad \left. - A_m + \frac{k_m \cdot 2A_m}{N}, t_1, \ldots, t_n\right).\end{aligned}$$

In other words, for an arbitrary $\epsilon > 0$, there exists a natural number N_0 such that for $N \geq N_0$,

$$\max_{\mathbf{t}\in K} |H(\mathbf{t}) - H_N(\mathbf{t})| < \epsilon.$$

Proof. The function $h(\mathbf{x}, \mathbf{t})$ is continuous on the compact set $[-A_1, A_1] \times \cdots \times [-A_m, A_m] \times K$, so $h(\mathbf{x},$

$\mathbf{t})$ is uniformly continuous. Therefore for any $\epsilon > 0$, we can take the integer N_0 such that if $N \geq N_0$ and

$$\left| x_i - \left(A_i + \frac{k_i \cdot 2A_i}{N} \right) \right| < \frac{2A_i}{N} \ (i = 1, \ldots, m) \text{ then}$$

$$\left| h(x_1, \ldots, x_m, t_1, \ldots, t_n) - h\left(-A_1 + \frac{k_1 \cdot 2A_1}{N}, \ldots, -A_m + \frac{k_m \cdot 2A_m}{N}, t_1, \ldots, t_n \right) \right| < \frac{\epsilon}{2A_1 \cdots 2A_m}.$$

Assertion of the Lemma is obvious from this inequality. q.e.d.

7. PROOF OF THEOREMS

We will prove our theorems in Section 3 under the above preliminaries.

Proof of Theorem 1

Step 1. Because $f(\mathbf{x})(\mathbf{x} = (x_1, \ldots, x_n))$ is a continuous function on a compact subset K of $\mathbf{R}^n$, $f(\mathbf{x})$ can be extended to be a continuous function on $\mathbf{R}^n$ with compact support. We also denote this by $f(\mathbf{x})$.

If we operate the mollifier $\rho_\alpha *$ on $f(\mathbf{x})$, $\rho_\alpha * f(\mathbf{x})$ is C^∞-function with compact support. Furthermore, $\rho_\alpha * f(\mathbf{x}) \to f(\mathbf{x})(\alpha \to +0)$ uniformly on $\mathbf{R}^n$. Therefore we may suppose $f(\mathbf{x})$ is a C^∞-function with compact support for proving Theorem 1. By the Paley–Wiener theorem (see, e.g., Yosida, 1968), the Fourier transform $F(\mathbf{w})(\mathbf{w} = (w_1, \ldots, w_n))$ of $f(\mathbf{x})$ is real analytic and, for any integer N, there exists a constant C_N such that

$$|F(\mathbf{w})| \leq C_N(1 + |\mathbf{w}|)^{-N}. \tag{3}$$

In particular, $F(\mathbf{w}) \in L^1 \cap L^2(\mathbf{R}^n)$.

We define $I_A(x_1, \ldots, x_n)$, $I_{\infty,A}(x_1, \ldots, x_n)$ and $J_A(x_1, \ldots, x_n)$ as follows:

$$I_A(x_1, \ldots, x_n) = \int_{-A}^{A} \cdots \int_{-A}^{A} \psi\left(\sum_{i=1}^{n} x_i w_i - w_0 \right) \cdot \frac{1}{(2\pi)^n \Psi(1)} F(w_1, \ldots, w_n) \times \exp(iw_0)\, dw_0\, dw_1 \cdots dw_n, \tag{4}$$

$$I_{\infty,A}(x_1, \ldots, x_n) = \int_{-A}^{A} \cdots \int_{-A}^{A} \left[\int_{-\infty}^{\infty} \psi\left(\sum_{i=1}^{n} x_i w_i - w_0 \right) \frac{1}{(2\pi)^n \Psi(1)} F(w_1, \ldots, w_n) \times \exp(iw_0)\, dw_0 \right] dw_1 \cdots dw_n, \tag{5}$$

$$J_A(x_1, \ldots, x_n) = \frac{1}{(2\pi)^n} \int_{-A}^{A} \int_{-A}^{A} F(w_1, \ldots, w_n) \exp\left(i \sum_{i=1}^{n} x_i w_i \right) dw_1 \cdots dw_n, \tag{6}$$

where $\psi(x) \in L^1$ is defined by

$$\psi(x) = \phi(x/\delta + \alpha) - \phi(x/\delta - \alpha)$$

for some α and δ so that $\psi(x)$ satisfies Lemma 1 in Section 6.

The essential part of the proof of Irie–Miyake's integral formula is the equality

$$I_{\infty,A}(x_1, \ldots, x_n) = J_A(x_1, \ldots, x_n) \tag{7}$$

and this is derived from

$$\int_{-\infty}^{\infty} \psi\left(\sum_{i=1}^{n} x_i w_i - w_0 \right) \exp(iw_0)\, dw_0 = \exp\left(i \sum_{i=1}^{n} x_i w_i \right) \cdot \Psi(1). \tag{8}$$

In our discussion, using the estimate of $F(\mathbf{w})$, we can prove

$$\lim_{A \to \infty} J_A(x_1, \ldots, x_n) = f(x_1, \ldots, x_n)$$

uniformly on $\mathbf{R}^n$. Therefore

$$\lim_{A \to \infty} I_{\infty,A}(x_1, \ldots, x_n) = f(x_1, \ldots, x_n)$$

uniformly on $\mathbf{R}^n$. That is to say, we can state that for any $\epsilon > 0$ there exists $A > 0$ such that

$$\max_{\mathbf{x} \in \mathbf{R}^n} |I_{\infty,A}(x_1, \ldots, x_n) - f(x_1, \ldots, x_n)| < \epsilon/2. \tag{i}$$

Step 2. We will approximate $I_{\infty,A}$ by finite integrals on K. For $\epsilon > 0$, fix A which satisfies (i).

For $A' > 0$, set

$$I_{A',A}(x_1, \ldots, x_n) = \int_{-A}^{A} \cdots \int_{-A}^{A} \left[\int_{-A'}^{A'} \psi\left(\sum_{i=1}^{n} x_i w_i - w_0 \right) \times \frac{1}{(2\pi)^n \Psi(1)} F(w_1, \ldots, w_n) \times \exp(iw_0)\, dw_0 \right] dw_1 \ldots dw_n.$$

We will show that, for $\epsilon > 0$, we can take $A' > 0$ so that

$$\max_{\mathbf{x} \in K} |I_{A',A}(x_1, \ldots, x_n) - I_{\infty,A}(x_1, \ldots, x_n)| < \epsilon/2. \tag{ii}$$

Using the following equation

$$\int_{-A'}^{A'} \psi\left(\sum_{i=1}^{n} x_i w_i - w_0\right) \exp(iw_0)\, dw_0$$

$$= \int_{\Sigma_{i=1}^{n} x_i w_i - A'}^{\Sigma_{i=1}^{n} x_i w_i + A'} \psi(t) \exp(-it)\, dt \cdot \exp\left(i \sum_{i=1}^{n} x_i w_i\right),$$

the fact $F(\mathbf{x}) \in L^1$ and compactness of $[-A, A]^n \times K$, we can take A' so that

$$\left| \int_{-A'}^{A'} \psi\left(\sum_{i=1}^{n} x_i w_i - w_0\right) \exp(iw_0)\, dw_0 - \int_{-\infty}^{\infty} \psi\left(\sum_{i=1}^{n} x_i w_i - w_0\right) \exp(iw_0)\, dw_0 \right|$$

$$\leq \frac{\epsilon(2\pi)^n |\Psi(1)|}{\left(2 \int_{-\infty}^{\infty} \cdots \int_{-\infty}^{\infty} |F(x)|\, dx + 1\right)} \times \int_{-A}^{A} \cdots \int_{-A}^{A} |F(x)|\, dx \text{ on } K.$$

Therefore,

$$\max_{\mathbf{x} \in K} |I_{A',A}(x_1, \ldots, x_n) - I_{\infty,A}(x_1, \ldots, x_n)|$$

$$\leq \frac{\epsilon}{\left(2 \int_{-\infty}^{\infty} \cdots \int_{-\infty}^{\infty} |F(x)|\, dx + 1\right)} \times \int_{-A}^{A} \cdots \int_{-A}^{A} |F(x)|\, dx < \epsilon/2.$$

Step 3. From (i) and (ii), we can say that for any $\epsilon > 0$, there exist $A, A' > 0$ such that

$$\max_{\mathbf{x} \in K} |f(x_1, \ldots, x_n) - I_{A',A}(x_1, \ldots, x_n)| < \epsilon. \tag{iii}$$

That is to say, $f(\mathbf{x})$ can be approximated by the finite integral $I_{A',A}(\mathbf{x})$ uniformly on K. The integrand of $I_{A',A}(\mathbf{x})$ can be replaced by the real part and is continuous on $[-A', A'] \times \cdots \times [-A, A] \times K$, so by Lemma 2, $I_{A',A}(\mathbf{x})$ can be approximated by the Riemann sum uniformly on K.

Since

$$\psi\left(\sum_{i=1}^{n} x_i w_i - w_0\right) = \phi\left(\sum_{i=1}^{n} w_i x_i/\delta - w_0 + \alpha\right) - \phi\left(\sum_{i=1}^{n} w_i x_i/\delta - w_0 - \alpha\right),$$

the Riemann sum can be represented by a three-layer network. Therefore $f(\mathbf{x})$ can be represented approximately by the three-layer networks. q.e.d.

Proof of Theorem 2

If $k = 3$, set $f:\mathbf{x} = (x_1, \ldots, x_n) \rightarrow (f_1(\mathbf{x}), \ldots, f_m(\mathbf{x}))$ and apply Theorem 1 to each $f_i(\mathbf{x})$.

For the general case, we first remark that a $k(>3)$-layer network can be represented by the composition of k-2 three-layer networks and using the realization of identity mapping by three-layer network,

Proof of Corollary 1

In the expression $f:\mathbf{x} \rightarrow (f_1(\mathbf{x}), \ldots, f_m(\mathbf{x}))$, we extend $f_i(\mathbf{x})$ to functions which take value zero on $\mathbf{R}^n - K$. We also denote these by $f_i(\mathbf{x})$ $(i = 1, \ldots, m)$. We can approximate $f_i(\mathbf{x})(i = 1, \ldots, m)$ by C^∞-functions with compact support by operating mollifier $\rho_\alpha *$ on f_i and apply theorem 2 to $\rho_\alpha * f_i$. q.e.d.

The above proof of the theorem 2 for the case $k > 3$ gives only trivial approximate realizations of given mappings by k-layer networks. Therefore, we shall give a different proof for the case $k = 4$, by using the Kolmogorov–Arnold–Sprecher theorem, which gives nontrivial realizations of continuous mappings.

8. KOLMOGOROV–ARNOLD–SPRECHER'S THEOREM

Let $I = [0, 1]$ denote the closed unit interval, $I^n = [0, 1]^n (n \geq 2)$ the Cartesian product of I.

In his famous thirteenth problem, Hilbert conjectured that there are analytic functions of three variables which cannot be represented as a finite superposition of continuous functions of only two arguments. Kolmogorov (1957) and Arnold refuted this conjecture and proved the following theorem.

Theorem (Kolmogorov)

Any continuous functions $f(x_1, \ldots, x_n)$ of several variables defined on $I^n (n \geq 2)$ can be represented in the form

$$f(\mathbf{x}) = \sum_{j=1}^{2n+1} \chi_j \left(\sum_{i=1}^{n} \psi_{ij}(x_i)\right),$$

where χ_j, ψ_{ij} are continuous functions of one variable and ψ_{ij} are monotone functions which are not dependent on f.

Sprecher (1965) refined the above theorem and obtained the following:

Theorem (Sprecher)

For each integer $n \geq 2$, there exists a real, monotone increasing function $\psi(x)$, $\psi([0, 1]) = [0, 1]$, dependent on n and having the following property:
For each preassigned number $\delta > 0$ there is a rational number ϵ, $0 < \epsilon < \delta$, such that every real continuous

function of n variables, $f(\mathbf{x})$, defined on I^n, can be represented as

$$f(\mathbf{x}) = \sum_{j=1}^{2n+1} \chi\left[\sum_{i=1}^{n} \lambda^i \psi(x_i + \epsilon(j-1)) + j - 1\right],$$

where the function χ is real and continuous and λ is an independent constant of f.

Hecht-Nielsen (1987) pointed out that this theorem means that any continuous mapping $f:\mathbf{x} \in I^n \to (f_1(\mathbf{x}), \ldots, f_m(\mathbf{x})) \in \mathbf{R}^m$ is represented by a form of four-layer neural network with hidden units whose output functions are ψ, $\chi_i (i = 1, \ldots, m)$, where ψ is used for the first hidden layer, χ_i is given by Sprecher's theorem for $f_i(\mathbf{x})$ and $\chi_i (i = 1, \ldots, m)$ are used for the second hidden layer.

9. ALTERNATIVE PROOF OF THEOREM 2 FOR THE CASE $k = 4$

In section 8, we reviewed Kolmogorov's theorem and its refinement from the point of view of neural networks. The Kolmogorov–Arnold–Sprecher theorem and the following proposition are used to prove our theorem 2 for the case $k = 4$. This proposition is a special case (one variable case) of theorem 1 in Section 3.

Proposition

Let $g(x)$ be a continuous function on $\mathbf{R}$ and $\phi(x)$ a bounded and monotone increasing continuous function. For an arbitrary compact subset (bounded closed subset) K of $\mathbf{R}$ and an arbitrary $\epsilon > 0$, there are an integer N and real constants a_i, b_i, $c_i (i = 1, \ldots, N)$ such that

$$\left| g(x) - \sum_{i=1}^{N} c_i \phi(a_i x + b_i) \right| < \epsilon$$

holds on K.

In the appendix, we shall state the direct proof of the above proposition by a different method without using Fourier transforms under the additional condition that $\phi(x)$ has a weak derivative which is summable.

Next we prove theorem 2 for the case $k = 4$ by using the Kolmogorov–Arnold–Sprecher theorem and the above proposition.

Proof. We may suppose that $K = [0, 1]^n$, because $f_p(\mathbf{x}) (p = 1, \ldots, m)$ can be extended continuous functions with compact supports. We apply Sprecher's theorem to $f_p(\mathbf{x}) (p = 1, \ldots, m)$ and represent $f_p(\mathbf{x})$ by the form

$$f_p(\mathbf{x}) = \sum_{j=1}^{2n+1} \chi_p \left[\sum_{i=1}^{n} \lambda^i \psi(x_i + \bar{\epsilon}(j-1)) + j - 1\right]$$

$(p = 1, \ldots, m)$, where λ and $\bar{\epsilon}$ are constants. We apply our proposition to functions χ_p, ψ, and approximate these functions using a sigmoid function ϕ.

Let $K_j (j = 1, \ldots, 2n + 1)$ be the images of $[0, 1]^n$ by mappings

$$\tau_j : \mathbf{x} \longrightarrow \sum_{i=1}^{n} \lambda^i \psi(x_i + \bar{\epsilon}(j-1)) + j - 1 \qquad (j = 1, \ldots, 2n+1)$$

and set $K = \cup K_j$. Take $\delta > 0$ and the closure K_δ of δ neighborhood of K. Continuous functions χ_p $(p = 1, \ldots, m)$ are approximated by

$$\chi_{p,N}(x) = \sum_{i=1}^{N} c_{i,N} \phi(a_{i,N} x + b_{i,N}) \tag{9}$$

so that

$$|\chi_p(x) - \chi_{p,N}(x)| < \epsilon/(4n+2) (p = 1, \ldots, m) \tag{10}$$

on K_δ. As $\chi_{p,N}(x)$ are uniformly continuous on K_δ, sufficiently small η can be taken so that if $|x - y| < \eta (x, y \in K_\delta)$ then $|\chi_{p,N}(x) - \chi_{p,N}(y)| < \epsilon/(4n + 2) (p = 1, \ldots, m)$.

We apply our lemma to τ_j and approximate τ_j on $[0, 1]^n$ by $\tau_{j,N'}$ so that

$$|\tau_j(\mathbf{x}) - \tau_{j,N'}(\mathbf{x})| < \min(\eta, \delta), \tag{11}$$

where $\tau_{j,N'}(x) (j = 1, \ldots, m)$ are defined as follows: We approximate $\psi(x)$ by

$$\psi_{N'}(x) = \sum_{i=1}^{N'} \tilde{c}_i \phi(\tilde{a}_i x + \tilde{b}_i) \tag{12}$$

on $2n\bar{\epsilon}$ neighborhood of $[0, 1]$ and set

$$\tau_{j,N'}(\mathbf{x}) = \sum_{i=1}^{n} \lambda^i \psi_{N'}(x_i + \bar{\epsilon}(j-1)) + j - 1 \tag{13}$$

so that the above inequality (11) is satisfied. Using a transformation

$$\begin{aligned}
&\sum_{j=1}^{2n+1} \chi_p[\tau_j(\mathbf{x})] - \sum_{j=1}^{2n+1} \chi_{p,N}[\tau_{j,N'}(\mathbf{x})] \\
&= \sum_{j=1}^{2n+1} \chi_p[\tau_j(\mathbf{x})] - \sum_{j=1}^{2n+1} \chi_{p,N}[\tau_j(\mathbf{x})] \\
&+ \sum_{j=1}^{2n+1} \chi_{p,N}[\tau_j(\mathbf{x})] - \sum_{j=1}^{2n+1} \chi_{p,N}[\tau_{j,N'}(\mathbf{x})],
\end{aligned}$$

it is seen that $f_p(\mathbf{x}) (p = 1, \ldots, m)$ are approximated by

$$\sum_{j=1}^{2n+1} \chi_{p,N}[\tau_{j,N'}(\mathbf{x})] \quad (p = 1, \ldots, m)$$

on $[0, 1]^n$ so that the errors are less than ϵ. Looking at the form of this approximation, the theorem is obtained. q.e.d.

10. NEURAL NETWORK AND INFORMATION PROCESSING IN THE BRAIN

In the Rumelhart–Hinton–Williams multilayer neural network, input and output values of each unit correspond to pulse-frequencies in a neuron and thus each unit, disregarding time characteristics, is a very simple model of the neuron. When a neural network is implemented for pattern recognition in engineering fields, output units correspond to gnostic cells in the brain.

The approximate realization of continuous mappings using neural networks, which are simple models of the neural system, suggest that there are several gnostic cells in the brain. It also shows the possibility of revealing information processing in the brain through neural network approaches.

11. SUMMARY

We proved the approximate realization theorem of continuous functions by three-layer networks. This theorem leads to the approximate realization theorem of continuous mappings by $k(\geq 3)$-layer networks and we showed that any mapping whose components are summable on compact subset, can be approximately represented by $k(\geq 3)$-layer networks in the sense of L^2-norm. We also showed an alternative proof of the theorem for the case $k = 4$ by using the Kolmogorov–Arnold–Sprecher theorem and a proposition which is a special case of the three-layer case. We consider that one of the problems of analyzing neural network capabilities is solved in the form of the existence theorem of networks which are approximately capable of representing any mapping given.

Presently, for application of neural networks to pattern recognition or related engineering fields, up to four-layer networks are used (Waibel, Hanazawa, Hinton, Shikano, & Lang, 1988; Tamura & Waibel, 1988). The theorems proved here provide that the mathematical base and their use would be fundamental in further discussions of neural network system theory.

REFERENCES

Amari, S. (1967). A theory of adaptive pattern classifiers, *IEEE Transactions on Electronic Computers*, **EC-16**, 299–307.

Duda, R. O., & Fossum, H. (1966). Pattern classification by iteratively determined linear and piecewise linear discriminant functions. *IEEE Transactions on Electronic Computers*, **EC-15**, 220–232.

Gel'fand, I. M., & Shilov, G. E. (1964). *Generalized functions*, (Vol. 1, Chap. 1). New York: Academic Press.

Hecht-Nielsen, R. (1987). Kolmogorov mapping neural network existence theorem. *IEEE First International Conference on Neural Networks*, **3**, 11–13.

Huang, W. Y., & Lippmann, R. P. (1987). Neural net and traditional classifiers. In D. Z. Anderson (Ed.), *Neural information processing Systems, Denver, Colorado*, 1987 (pp. 387–396). New York: American Institute of Physics.

Irie, B., & Miyake, S. (1988). Capabilities of three-layered Perceptrons. *IEEE International Conference on Neural Networks*, **1**, 641–648.

Kolmogorov, A. N. (1957). On the representation of continuous functions of many variables by superposition of continuous functions of one variable and addition. *Doklady Akademii Nauk SSSR*, **144**, 679–681; *American Mathematical Society Translation*, **28**, 55–59 [1963].

Lippmann, R. P. (1987, April). An introduction to computing with neural nets. *IEEE ASSP Magazine*, **4**, pp. 4–22.

McCulloch, W. S., & Pitts, W. (1943). A logical calculus of the idea immanent in nervous activity. *Bulletin of Mathematical Biophysics*, **5**, 115–133.

Poggio, T. (1983). Visual algorithms. In O. J. Braddick & A. C. Sleigh (Eds.), *Physical and biological processing of images* (pp. 128–135). New York: Springer-Verlag.

Rumelhart, D. E., Hinton, G. E., & Williams, R. J. (1986). Learning representations by error propagation. In D. E. Rumelhart, J. L. McClelland and the PDP Research Group (Eds.), *Parallel distributed processing* (Vol. 1, pp. 318–362). Cambridge, MA: MIT Press.

Sejnowski, T. J., & Rosenberg, C. R. (1987). Parallel networks that learn to pronounce English text. *Complex Systems*, **1**, 145–168.

Sprecher, D. A. (1965). On the structure of continuous functions of several variables. *Transactions of the American Mathematical Society*, **115**, 340–355.

Tamura, S. and Waibel, A. (1988). Noise reduction using connectionist models. 1988 *International Conference on Acoustic, Speech, and Signal Processing*, pp. 553–556.

Uesaka, Y. (1971). Analog perceptrons: On additive representation of functions. *Information and Control*, **19**, 41–65.

Waibel, A., Hanazawa, T., Hinton, G., Shikano, K., and Lang, K. (1988). Phoneme recognition: neural networks vs. hidden Markov models. 1988 *International Conference on Acoustic, Speech, and Signal Processing*, pp. 107–110.

Wieland, A., & Leighton, R. (1987). Geometric analysis of neural network capabilities, *IEEE First International Conference on Neural Networks*, **3**, 385–392.

Yosida, K. (1968). *Functional analysis*. New York: Springer-Verlag.

APPENDIX (DIRECT PROOF OF THE PROPOSITION IN SECTION 9 BY A DIFFERENT METHOD)

Proof. There is a continuous function $\tilde{g}(x)$ on **R** which has a compact support such that $g(x) = \tilde{g}(x)$ on K. We may prove the proposition for $\tilde{g}(x)$ and so we may initially suppose that $g(x)$ has a compact support. We may also suppose that $\phi(\infty) - \phi(-\infty) = 1$. For the arbitrary $\epsilon > 0$, we will approximate $g(x)$ on K by a summation of sigmoid functions whose variables are shifted and scaled. Initially, we can approximate $g(x)$ by a simple function (step function) $c(x)$ with compact support so that

$$|g(x) - c(x)| < \epsilon/2 \tag{A.1}$$

on **R** and whose step variances are less than $\epsilon/4$. Here $c(x)$ is represented using the Heaviside function $H(x)$ as follows:

$$c(x) = \sum_{i=1}^{N} c_i H(x - x_i).$$

For a sigmoid function $\phi(x)$, set $\phi_\alpha(x) = \phi(x/\alpha)(\alpha > 0)$. Then $\phi'_\alpha(x) = \dfrac{d}{dx}\phi_\alpha(x)$ converge to the delta function as $\alpha \to 0$. We consider the convolution $c*\phi'_\alpha(x)$ of $c(x)$ and $\phi'_\alpha(x)$. We set $2\epsilon'$ = "minimum width of steps" and obtain

$$c(x) - c*\phi'_\alpha(x) = \int_{-\infty}^{\infty} \phi'_\alpha(y)[c(x) - c(x - y)]\, dy.$$

Divide the integrand of the right term into $(-\infty, -\epsilon')$, $[-\epsilon', \epsilon']$, (ϵ', ∞) and estimate these using the properties of sigmoid functions. For example,

$$\left| \int_{-\epsilon'}^{\epsilon'} \phi'_\alpha(y)[c(x) - c(x - y)]\, dy \right| < \epsilon/4 \int_{-\infty}^{\infty} \phi'_\alpha(y)\, dy = \epsilon/4$$

and other terms will be arbitrarily small for a sufficiently small α. Therefore we obtain

$$|c(x) - c*\phi'_\alpha(x)| < \epsilon/4.$$

As $c*\phi'_\alpha(x) = c'*\phi_\alpha(x)$ and $c'(x)$ is given by

$$c'(x) = \sum_{i=1}^{N} c_i\delta(x - x_i)$$

and so, $c*\phi'_\alpha(x)$ is represented as follows:

$$c*\phi'_\alpha(x) = \sum_{i=1}^{N} c_i\phi_\alpha(x - x_i).$$

That is to say,

$$\left| c(x) - \sum_{i=1}^{N} c_i\phi_\alpha(x - x_i) \right| < \epsilon/2. \tag{A.2}$$

Using (A.1) and (A.2) we obtain

$$\left| g(x) - \sum_{i=1}^{N} c_i\phi_\alpha(x - x_i) \right| < \epsilon.$$

Here $\phi_\alpha(x - x_i) = \phi(x/\alpha - x_i/\alpha)$, so we set $a_i = 1/\alpha$, $b_i = -x_i/\alpha$ and the proposition is proved. q.e.d.

Networks and the Best Approximation Property

F. GIROSI AND T. POGGIO

ARTIFICIAL INTELLIGENCE LABORATORY, CENTER FOR BIOLOGICAL INFORMATION PROCESSING, MASSACHUSETTS INSTITUTE OF TECHNOLOGY, CAMBRIDGE, MA 02139, USA

RECEIVED DECEMBER 28, 1989/ACCEPTED DECEMBER 30, 1989

***Abstract*.—Networks can be considered as approximation schemes. Multilayer networks of the perceptron type can approximate arbitrarily well continuous functions (Cybenko 1988, 1989; Funahashi 1989; Stinchcombe and White 1989). We prove that networks derived from regularization theory and including Radial Basis Functions (Poggio and Girosi 1989), have a similar property. From the point of view of approximation theory, however, the property of approximating theory, however, the property of approximating continuous functions arbitrarily well is not sufficient for characterizing good approximation schemes. More critical is the property of *best approximation*. The main result of this paper is that multilayer perceptron networks, of the type used in backpropagation, do not have the best approximation property. For regularization networks (in particular Radial Basis Function networks) we prove existence and uniqueness of best approximation.**

1. INTRODUCTION

Learning an input-output relation from examples can be considered as the problem of approximating an unknown function $f(x)$ from a set of sparse data points (Poggio and Girosi 1989). From this point of view, feedforward networks are equivalent to a parametric approximating function $f(W, x)$. As an example, consider a feedforward network, of the multilayer perceptron type, with one hidden layer; the vector W corresponds, then, to the two sets of "weights", from the input to the hidden layer, and from the hidden layer to the output. Even before considering the problem of how to find the appropriate values of W for the set of data, the fundamental representational problem must be approached: which class of mappings f can be approximated by F, and how well? The neural network field has recently seen an increasing awareness of this problem. Several results have been published, all showing that multilayer perceptrons of different form and complexity can approximate arbitrarily well a continuous function, provided that an arbitrarily large number of units is available (Cybenko 1988, 1989; Funahashi 1989; Moore and Poggio 1988; Stinchcombe and White 1989; Carrol and Dickinson 1989). This property is shared by algebraic and trigonometric polynomials, as is shown by the classical Weierstrass Theorem, and for this reason we shall refer to it as the Weierstrass property. Results of this type should not be taken to mean that the approximation scheme is a "good" approximation scheme. An indication of the latter point is provided, in the case of multilayer perceptron networks, of the type used for backpropagation, by a closer look at the published results. Taken together, they imply that almost any nonlinearity at the hidden layer and a variety of different architectures (one or more hidden layers, for instance) insures the Weierstrass property (Funahashi 1989; Cybenko 1989; Stinchcombe and White 1989). There is nothing special about sigmoids, and in fact many classical approximation schemes exist that can be represented as a network with a hidden layer and that exhibit the Weierstrass property. In a sense this property is not very useful for characterizing approximation schemes, since many schemes have it. Literature in the field of approximation theory reflects this situation, since it emphasizes other properties in characterizing approximation schemes. In particular, a critical concept is that of *best approximation*. An approximation scheme has the best approximation property if in the set A of approximating functions (for instance the set $F(W, x)$ spanned by parameters W) there is one that has minimum distance from any given function of a larger set Φ (a more formal definition is given later). Several questions can be asked, such as the existence, uniqueness, computability, etc., of the best approximation.

In this paper, we show that feedforward multilayer networks of the backpropagation type (Rumelhart et al. 1986a, b; Sejnowski and Rosenberg 1987) do not have the best approximation property for the class of continuous functions defined on a subset of R^n. On the other hand, we prove that for networks derived from regularization, and in particular for Radial Basis Function networks, best approximation exists and is unique. We also prove that these networks approximate arbitrarily well continuous functions (see Appendix B and C). We have recently shown that Radial Basis Function approximation schemes are a special case of regularization and are therefore equivalent to generalized (radial) splines (Poggio and Girosi

1989). For Radial Basis Function networks we prove existence and uniqueness of best approximation.[1]

The plan of the paper is as follows. We first formalize the previous arguments, then introduce some basic notions from approximation theory. Next, we prove that multilayer networks of the type used for backpropagation do not have the best approximation property, and that networks obtained from regularization theory have this property. In the last section, we discuss the implications of these results and list some open questions. Appendix B proves that the Stone-Weierstrass theorem holds for Gaussian Radial Basis Function networks (with different variances). In appendix C we prove a more general result: regularization networks and, in particular Radial Basis Functions approximate arbitrarily well any continuous function on a compact subset of R^n.

2. Most Networks Approximate Continuous Functions

In recent years there have been attempts to find a mathematical justification for the use of feedforward multilayer networks of the type used for backpropagation. Typical results deal with the possibility, given a network, of approximating any continuous function arbitrarily well. In mathematical terms this means that the set of functions that can be computed by the network is *dense* (see Appendix A) in the space of the continuous functions $C[U]$ defined on some subset U of R^d. The most recent results (Cybenko 1989; Funahashi 1989; Stinchcombe and White 1989) consider networks with just one layer of hidden units, that correspond to the following class of approximating functions:

$$\sum \equiv \{f \in C[U] | f(\mathbf{x}) = \sum_{i=1}^{m} c_i \sigma(\mathbf{x} \cdot \mathbf{w}_i + \theta_i),$$

$$U \subset R^d,\ \mathbf{w}_i \in R^d,\ c_i,\ \theta_i \in R,\ m \in N\} \tag{1}$$

where σ is a continuous function. Depending on σ, the set Σ may or may not be dense in the space of the continuous functions. The set $\mathcal{D}$ of functions σ such that Σ is dense seems to be large. For instance, the *sigmoidal* functions, that is functions such that

$$\lim_{t \to +\infty} \sigma(t) = 1$$

$$\lim_{t \to -\infty} \sigma(t) = 0$$

belong to $\mathcal{D}$ (Cybenko 1989; Funahashi 1989). Many other types of functions in $\mathcal{D}$ can be found in the paper of Cybenko (1989). The set $\mathcal{D}$ has been recently extended by the result of Stinchcombe and White (1989). In fact they prove that it contains all the functions whose mean value is different from zero and whose L_p-norm is finite for $1 \leq p < \infty$.

Other networks can be built, such that the corresponding set of approximating functions is dense in $C[U]$. Consider for example the network in Fig. 1. This is the most general network with one layer of hidden units, and the class of approximating functions corresponding to it is

$$\mathcal{N} \equiv \{f \in C[U] | f(\mathbf{x}) = \sum_{i=l}^{m} c_i H_i(\mathbf{x}),$$

$$U \subset R^d,\ H_i \in C[U],\ m \in N\}. \tag{2}$$

The function H_i are of the form $H_i = H(\mathbf{x}; \mathbf{W}_i)$, where $\mathbf{W}_i$ is a vector of unknown parameters in some multidimensional space and H is a continuous function. If the H_i are appropriately chosen the set $\mathcal{N}$ can be dense in $C[U]$. For example the H_i could be algebraic or trigonometric polynomials, and in this case the denseness of $\mathcal{N}$ would be a trivial consequence of the Stone-Weierstrass theorem (see Appendix B). This theorem allows a significant extension of the set of "basis" functions H_i. Appendix B gives another example, showing how Gaussian functions of radial argument (and different variances) can be used to approximate any continuous functions. Appendix C provides a more powerful result showing that *all networks derived from regularization theory can approximate arbitrarily well continuous functions on a compact subset of* R^n. This result includes, in particular, Radial Basis Functions networks with the radial basis function being the Green's function of a self-adjoint differential operator associated to a Tikhonov stabilizer. Such Green's functions, not necessarily radial, include most of the known approximation schemes, such as the Gaussian and several types of splines and many functions, but not all functions, that satisfy some sufficient conditions given by Micchelli (1986) in order to be interpolating functions.

Since a large number of networks can approximate arbitrarily well any continuous functions, it is natural to ask whether this property is really important from the point of view of approximation theory, and whether other more fundamental properties can be characterized. As we mentioned already, one of the basic properties that an approximating set should have is the *best approximation* property, that guarantees that the approximation problem has a solution. The next section focuses our attention on the relationship between this property and different kind of networks, since this seems to be a more appropriate starting point for a complete analysis of the networks performances from a rigorous mathematical point of view.

3. Basic Facts in Approximation Theory

3.1 The Best Approximation Property

An informal formulation of the approximation problem can be stated as follows: *given a function f belonging to some prescribed set of functions* Φ, *and a given subset A of* Φ, *find the element a of A that is the "closest" to f.*

In order to give this formulation a precise mathematical meaning, some definitions are needed. First of all a notion of "distance" has to be introduced on the set Φ. Since this set is usually assumed to be a normed linear space, with norm

[1] The theory has been extended by introducing the more general schemes of GRBF and HyperBF, which can be considered as the network equivalent of generalized multidimensional splines with free knots.

indicated by $\|\cdot\|$, the distance $d(f,g)$ between two elements f and g of Φ is naturally defined as $\|f - g\|$. Given $f \in \Phi$ and $A \subset \Phi$ we can now define the *distance of f from A as*

$$d(f, A) \equiv \inf_{a \in A} \|f - a\|. \tag{3}$$

If the infimum of $\|f - a\|$ is attained for some element a_0 of A, that is if there exists an $a_0 \in A$ such that $\|f - a_0\| = d(f,A)$, this element is said to be a *best approximation* to *f from A*. A set A is called an *existence set* (*uniqueness set*, resp.) if, to each $f \in \Phi$, there is at least (at most, resp.) one best approximation to f from A. If the set A is an existence set we will also say that it has the best approximation property. A set A is called a *Tchebycheff set* if it is an existence set and a uniqueness set. We are now ready to give a precise formulation of the approximation problem:

Approximation problem: given $f \in \Phi$ and $A \subset \Phi$ find a best approximation to f from A.

From the definition above it is clear that the approximation problem has a solution if and only if A is an existence set, and a large part of approximation theory has been devoted to proving existence theorems, which give sufficient conditions to guarantee existence and possibly uniqueness of closest points. We will only present very simple properties of sets with the best approximations property, and will apply these result to network architectures, in order to understand their properties from the point of view of approximation theory.

We begin with the following observation:

Proposition 3.1 *Every existence set is closed.*

Proof. Let $A \subset \Phi$ be an existence set, and suppose that it is not closed. Then there is a sequence $\{a_n\}$ of elements of A that converges to an element f that is not in A, that is there exists an $f \in \Phi \backslash A$ such that

$$\lim_{n \to \infty} d(f, a_n) = 0.$$

This means that $d(f, A) = 0$, and since A is an existence set there is an element $a_0 \in A$ such that $\|f - a_0\| = 0$. By the properties of the norm this implies that $f = a_0$, which is absurd because $f \notin A$ and $a_0 \in A$. Then A must be closed.

The converse of this proposition is not true, that is closedness is not sufficient for a set to be an existence set. However the stronger condition of compactness is sufficient, as the following theorem shows.

Theorem 3.1 Let A be a compact set in a metric space Φ. Then A is an existence set.

Proof. For each $f \in \Phi$ the distance $d(f, a)$ with $a \in A$, is a continuous real valued function defined on the compact set A. From theorem A.2 of Appendix A it attains its maximum and minimum value on this set and this concludes the proof.

In the next section we apply these simple results to some network architectures.

4. Networks and Approximation Theory

From the point of view of approximation theory a feedforward network is a representation of a set A of parametric functions, and the learning algorithm corresponds to the search of the best approximation to some target function f from A. Since in general a best approximation does not exist unless the set A has some properties (see, for instance, theorem 3.1), it is of interest to understand which classes of networks have these properties.

4.1 Multilayer Networks of the Backpropagation Type do not have the Best Approximation Property

Here we consider the class of networks of the backpropagation type with one layer of hidden units. The space Φ of functions that have to be approximated is chosen to be $C[U]$, the set of continuous functions defined on a subset U of R^d with some unspecified norm. If the number of hidden units is m, the functions that can be computed by such networks belong to the following set σ^m:

$$\sigma^m \equiv \{f \in C[U] | f(x) \sum_{i=1}^{m} c_i \sigma(\mathbf{x} \cdot \mathbf{w}_i + \theta_i),$$
$$\mathbf{w}_i \in R^d,\ c_i,\ \theta_i \in R\} \tag{4}$$

where $\sigma(x)$ is usually a sigmoidal function. We now show that σ^m is not an existence set, and this does not depend on the norm that has been chosen. The result is proved in the case of σ being a sigmoid and for one hidden layer, $\sigma(x) = (1 + e^{-x})^{-1}$, but *it holds for every other non trivial choice of nonlinear function and for networks with more than one hidden layer.*

Proposition 4.1 *The set* σ^m is not an existence set for $m \geq 2$.

Proof. A necessary condition for a set to be an existence set is to be closed. Therefore it is sufficient to show that σ^m is not closed, and this can be done by showing an accumulation point that does not belong to it. Let us consider the following function:

$$f_\delta(\mathbf{x}) = \frac{1}{\delta}\left(\frac{1}{1 + e^{-[\mathbf{w} \cdot \mathbf{x} + \theta]}} - \frac{1}{1 + e^{-[\mathbf{w} \cdot \mathbf{x} + (\theta + \delta)]}}\right).$$

Clearly $f_\delta \in \sigma^m$, $\forall m \geq 2$, but it is easily seen that

$$\lim_{\delta \to 0} f_\delta(\mathbf{x}) \equiv g(\mathbf{x}) = \frac{1}{2(1 + \cosh[\mathbf{w} \cdot \mathbf{x} + \theta])}$$

and $g \notin \sigma^m$, $\forall m \geq 2$. For each $m \geq 2$ the function g is then an accumulation point of σ^m but does not belong to it: σ^m can not be closed and this concludes the proof.

This result reflects a general fact in nonlinear approximation theory: usually the set of approximating functions is not closed, and its closure must be added to it in order to obtain an existence set. This is the case, for instance, for the approximation by *γ-polynomials* in one dimension, that are replaced by the *extended γ-polynomials*, to guarantee the existence of a best approximating element (Braess 1986; Rice 1964, 1969; Hobby and Rice 1967; De Boor 1969).

4.2 Existence and Uniqueness of Best Approximation for Regularization and RBF

One of the possible approaches to the problem of surface reconstruction is given by regularization theory (Tikhonov and Arsenin 1977; Bertero et al. 1988). Poggio and Girosi (1989) have shown that the solution obtained by means of this method maps into a class of networks with one hidden layer (an instance of which are Radial Basis Function networks or RBF). In fact the solution can always be written in the parametric form:

$$f(\mathbf{x}) = \sum_{i=1}^{m} c_i \phi_i(\mathbf{x}) \tag{5}$$

where the c_i are unknown, m is the number of data points and the ϕ_i are fixed, depending on the nature of the problem and on the data points. More precisely the "basis function" ϕ_i is of the form $\phi_i(\mathbf{x}) = G(\mathbf{x}; \mathbf{x}_i)$, where $\mathbf{x}_i$ is a data point and G is the Green's function of some (pseudo) differential operator $\hat{p}$ (a term belonging to the null space of P can also appear, see Appendix C). In the particular case of radial function $G = G(\|\mathbf{x} - \mathbf{x}_i\|)$ the RBF method is recovered, and the solution to the approximation problem is then a linear superposition of radial Green's functions G "centered" on the data points.

Notice that this function can be computed by a network that is a special case of the one represented in Fig. 1. The main difference is that in the general case the functions G_i depend on *unknown* parameters, while in the regularization context only the coefficient c_i are unknown.

Equation 5 means that the approximated solution belongs to the subset T^m of $C[U]$:

$$T^m \equiv \{f \in C[U] | f(\mathbf{x}) = \sum_{i=1}^{m} c_i \phi_i(\mathbf{x}),\ c_i \in R\} \tag{6}$$

Since we have shown that the set of approximating functions associated with networks with one hidden layer of the type used for backpropagation does not have the best approximation property, it is natural to ask whether or not the set T^m has this property.[2] The answer is positive, as is stated in the following proposition:

Proposition 4.2 *The set T^m is an existence set for $m \geq 1$.*

Proof. Let f be a prescribed element of $C[U]$, and let a_0 be an arbitrary point of T^m. We are looking for the closed point to f in T^m. It has to lie in the set

$$\{a \in T^m \mid \|a - f\| \leq \|a_0 - f\|\}.$$

This set is clearly closed and bounded, and by theorem A.1 it is compact. The best approximation property comes from theorem 3.1.

[2] Notice that multilayer perceptrons of the type used for backpropagation cannot be derived from any regularization scheme since they cannot written as the linear superposition of Green's functions of any kind.

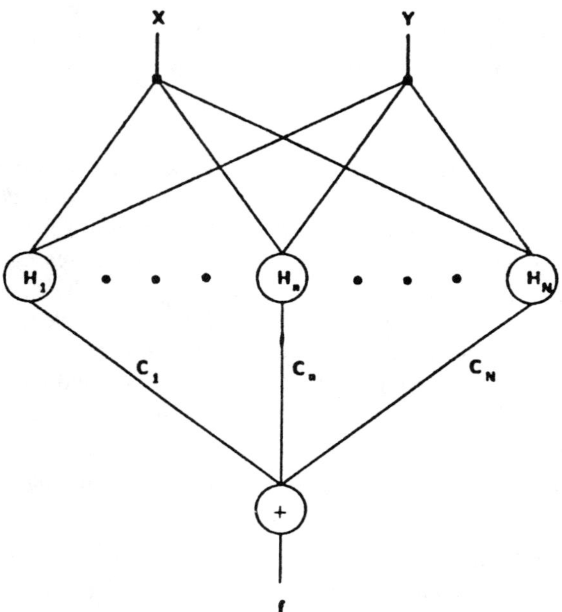

Fig. 1 The most general network with one layer of hidden units. Here we show the two-dimensional case, in which $\mathbf{x} = (x, y)$. Each function H_i can depend on a set of unknown parameters, that are computed during the learning phase, as well as the coefficients c_i. When $H_i = \sigma(\mathbf{x} \cdot \mathbf{w}_i + \theta)$ a network of the backpropagation type is recovered, while $H_i = H(\|\mathbf{x} - \mathbf{t}_i\|)$ corresponds to RBF or GRBF scheme (Broomhead and Lowe 1988; Poggio and Girosi 1989).

From this proposition we can see that when the approximating function is a finite linear combination of basis functions, the set that is spanned by these basis functions is an existence set for $C[U]$. Depending on the norm that is chosen in $C[U]$, the best approximating element can be unique. In fact the following theorem holds (see Appendix A for the definition of *strictly convex*):

Proposition 4.3 *The set T^m, $m \geq 1$ is a Tchebycheff set if the normed space $C[U]$ is strictly convex.*

Proof. The existence has already been proved. Suppose then that there are two best approximating elements f and f' from T^m to a function $g \in C[U]$. Let λ be the distance of g from T^m. Applying the triangular inequality we obtain:

$$\|\tfrac{1}{2}(f + f') - g\| \leq \tfrac{1}{2}\|f - g\| + \tfrac{1}{2}\|f' - g\| = \lambda. \tag{7}$$

Since T^m is a vector space, then $\frac{1}{2}(f + f') \in T^m$ and by definition of λ it follows that $\|\frac{1}{2}(f + f'\| \geq \lambda$. This implies that the equality holds in (7). If $\lambda = 0$ it is clear that $f = f' = g$. If $\lambda \neq 0$, then we can write (7) as

$$\left\|\frac{1}{2}\left[\frac{(f - g)}{\lambda} + \frac{(f' - g)}{\lambda}\right]\right\| = 1. \tag{8}$$

This means that the vectors $(f - g)/\lambda$, $(f' - g)/\lambda$, and their midpoints are all of norm 1, but since strictly convexity holds, then $f = f'$.

Since it is well known that $C[U]$ with the L_p-norms, $l < p < \infty$ is strictly convex (Rice 1964), we have then shown that in most cases regularization theory gives an approximating set with the best approximation property and with a unique best approximating element.

5. Conclusions

5.1 GRBF and Best Approximation

We have recently extended the scheme of (5) to the case in which the number of basis functions is less than the number of data points (Poggio and Girosi 1989; Broomhead and Lowe 1988). The reason for this is that when the number of data points becomes larger the complexity of the network may become too high, being proportional to the number of data points. A solution to the approximation problem is sought of the form:

$$f(\mathbf{x}) = \sum_{i=1}^{n} c_i G(\mathbf{x}; \mathbf{t}_i) \tag{9}$$

where n is smaller than the number of data points and the positions of the "centers" $\mathbf{t}_i$ of the expansion are unknown, having to be found during the learning stage. Does the best approximation property hold for this approximation scheme, that we call Generalized Radial Basis Function (GRBF) method? The answer is no, exactly as for splines with free knots, to which (9) is in fact equivalent. By the same arguments we have used in Section 4.1, we could show that the set G^n of approximating functions generated by (9) (the analogous of the set T^m) is not closed. The scheme, however, has almost the best approximation property in the following sense. The scheme already works satisfactorily if the centers $\mathbf{t}_i$ are fixed to a subset of examples or other positions. In this case G^n is a linear space, and it is an existence set, as well as T^m. We could then have an algorithm in which first the centers are found independently (for instance by the K-means algorithm, see Moody and Darken 1989) and then the c_i are obtained with gradient descent methods (see Poggio and Girosi 1989). In this scheme the best approximation property is preserved, while the computational complexity have been reduced with respect to the exact solution of the regularization problem.

There are other ways to make GRBF a best approximation. The most interesting approach is to follow the theory of γ-polynomials (Braess 1986; Rice 1964, 1969; Hobby and Rice 1967; De Boor 1969) and complete the sets of basis functions with its closure, consisting of an appropriate number of derivatives of the Green's function with respect to its parameters, yielding a best approximation scheme. It seems very difficult to use either of these two approaches for networks of the type used for backpropagation.

5.2 Open Questions

We have not explored the practical consequences of the fact that multilayer networks of the backpropagation type are not best approximation. Intuitively, it seems that the lack of the best approximation property is related to possible practical degeneracies of the solution. In certain situations, because of the fact that the sigmoid, which is asymptomatically constant, contains as an argument one set of parameters (the w_i), the precise values of these parameters may not have any significant effect on the output of the network. The same situation happens for GRBF when the centers inside the Green's function are unknown. In the GRBF case, however, we can freeze the $\mathbf{t}_i$ to reasonable values whereas this is impossible in the backpropagation case.

Other questions remain open as well. The most important questions from the viewpoint of approximation theory are: (1) the computation of the best approximation, i.e., which algorithm to use, (2) a priori bounds on the goodness of the approximation given some generic information on the class of functions to be approximated, and (3) a priori estimates of the complexity of the best approximation, again given generic information on the class of functions to be approximated. In the case of RBF, the latter question is directly related to the size of the required training set, and therefore to the deep issue of sample complexity (see Poggio and Girosi 1989, Sect 9.3). About problems 1) and 2) notice that in practical cases it may be admissible to use a scheme which is not best approximation, *if* it provides an almost as good approximation at a much lower computational cost.

Acknowledgments. We are grateful to G. Palm and E. Grimson for useful suggestions. This report describes research done within the Artificial Intelligence Laboratory and the Center for Biological Information Processing in the Department of Brain and Cognitive Sciences. Support for this research is provided by a grant from ONR, Cognitive and Neural Sciences Division, by the Artificial Intelligence Center of Hughes Aircraft Corporation and by the NATO Scientific Affairs Division (0403/87). Support for the A.I. Laboratory's artificial intelligence research is provided by the Advanced Research Projects Agency of the Department of Defense under Army contract DACA76-85-C-0010 and in part under Office of Naval Research (ONR) contract N00014-85-K-0124. TP is supported by the Uncas and Ellen Whitaker chair.

Appendix A: Definitions and Basic Theorems

We review here some of the definitions that have been used in the paper. Every set will be assumed to have the structure of metric space, unless differently specified, and the concepts of limit point, infimum and supremum are assumed to be known. All these definitions and theorems can be found in any standard text on functional analysis (Yosida 1974; Rudin 1973) and in many books on approximation theory (Braess 1986; Cheney 1981).

An important concept is that of *closure:*

Definition A.1 If $\mathcal{S}$ is a set of elements, then by the closure $[\mathcal{S}]$ of $\mathcal{S}$ we mean the set of all points in $\mathcal{S}$ together with the set of all limit points of $\mathcal{S}$.

We can now define the *closed* sets as following:

Definition A.2 A set $\mathcal{S}$ is closed if it is coincident with its closure $[\mathcal{S}]$.

A closed set then contains all its limit points. Another important definition related to the concept of closure is that of *dense* sets:

Definition A.3 Let $\mathcal{T}$ a subset of the set $\mathcal{S}$. $\mathcal{T}$ is dense in $\mathcal{S}$ if $[\mathcal{T}] = \mathcal{S}$.

If $\mathcal{T}$ is dense in $\mathcal{S}$ then each element of $\mathcal{S}$ can be approximated arbitrarily well by elements of $\mathcal{T}$. As an example we mention the set of rational numbers, that is dense in the set of real numbers, and the set of polynomials that is dense in the space of continuous functions (see Appendix B).

In order to extend some properties of the real valued functions defined on an interval to real valued functions defined on more complex metric spaces it is fundamental to define the compact sets:

Definition A.4 A compact set is one in which every infinite subset contains at least one limit point.

It can be shown that, in finite dimensional metric spaces, there exists a simple characterization of compact sets. In fact the following theorem holds:

Theorem A.1 *Every closed, bounded, finite-dimensional set in a metric linear space is compact.*

The well known Weierstrass theorem on the attainment of the extrema of a continuous function on an interval can now be extended as following:

Theorem A.2 *A continuous real valued function defined on a compact set in a metric space achieves its infimum and supremum on that set.*

A subset of the metric spaces is given by the normed spaces, and among the normed spaces, a special role is played by the strictly convex spaces:

Definition A.5 A normed space is strictly convex if:

$$\|f\| = \|g\| = \|\tfrac{1}{2}(f + g)\| = 1 \Rightarrow f = g$$

The geometrical interpretation of this definition is that a space is strictly convex if the unit sphere does not contain any line segment on its surface.

Appendix B: Gaussian Networks and Stone's Theorem

It has been proved (Cybenko 1989; Funahashi 1989) that a network with a one hidden layer of sigmoidal units can approximate a continuous function arbitrarily well. Here we show that this property, which is well known for algebraic and trigonometric polynomial approximation schemes, is shared by a network with Gaussian hidden units. The proof is a simple application of the Stone-Weierstrass theorem, which is the generalization given by Stone of the Weierstrass approximation theorem (Stone 1937, 1948). Our result was obtained independently from the equivalent proof of Hartman et al. (1989). We first need the definitions of *algebra*.

Definition B.1 An algebra is a set of elements denoted by $\mathcal{Y}$, together with a scalar field $\mathcal{F}$, which is closed under the binary operators of + (addition between elements of $\mathcal{Y}$), $\times$ (multiplication of elements $\mathcal{Y}$), $\times$ (mulitplication of elements of $\mathcal{Y}$), $\cdot$ (multiplication of elements in $\mathcal{Y}$ by elements from the scalar field $\mathcal{F}$), such that

1. $\mathcal{Y}$ together with $\mathcal{F}$, +, and $\cdot$ forms a linear space,
2. if f,g,h are in $\mathcal{Y}$, α is in $\mathcal{F}$, then
   a. $f \times g$ is in $\mathcal{Y}$,
   b. $f \times (g \times h) = (f \times g) \times h$,
   c. $f \times (g + h) = f \times g + f \times h$,
   d. $(f + g) \times h = f \times h + g \times h$,
   e. $\alpha(f \times g) = (\alpha f) \times g = f \times (\alpha g)$.

It is an elementary calculation to show that U is some subset of R^d then $C[U]$ is an algebra with respect to the scalar field R. We can now define a *subalgebra* as following:

Definition B.2 A set $\mathcal{S}$ is a subalgebra of the algebra $\mathcal{Y}$ if

1. $\mathcal{S}$ is a linear subspace of $\mathcal{Y}$,
2. $\mathcal{S}$ is closed under the operation $\times$. That is, if f and g are in $\mathcal{S}$, then $f \times g$ is also in $\mathcal{S}$.

We can now formulate the Stone's theorem:

Theorem B.1 (Stone, 1937) *Let X be a compact metric space, C[X] the set of continuous functions defined on X and A a subalgebra of C[X] with the following two properties:*

1. the function f(x) = 1 belongs to A;

2. for any two distinct points x and y in X there is function $f \in$ a such that $f(x) \neq f(y)$.

Then A is dense in C[X].

As a simple application of this theorem we consider the set of gaussian superpositions, defined as

$$\mathcal{G}_X \equiv \{f \in C[X] \mid f(x) = \sum_{i=1}^{m} c_i e^{-(x - t_i)^2/\sigma_i^2},$$

$$X \subset R^d,\ \mathbf{t}_i \in R^d,\ c_i,\ \sigma_i \in R,\ m \in N\} \tag{10}$$

We can now enunciate the following:

Proposition B.1 The set $\mathcal{G}$ is dense in C[X], where X is a compact subset of R^d.

Proof. In order to use Stone's theorem, we first have to show that $\mathcal{G}_x$ is a subalgebra of $C[X]$, for each compact subset X of R^a. The set $\mathcal{G}_x$ will be subalgebra of C[X] if the product of two of its elements yields another element of $\mathcal{G}_x$. Since $\mathcal{G}_x$ is a linear superposition of Gaussians of different variance and centered on different points it is sufficient to deal with the product of two Gaussians. From the identity below it follows that the product of two gaussians centered on two points $\mathbf{t}_1$ and $\mathbf{t}_2$ is proportional to a Gaussian centered on a point $\mathbf{t}_3$ that is a convex linear combination of $\mathbf{t}_1$ and $\mathbf{t}_2$. In fact we have:

$$\mathbf{e}^{-(\mathbf{x} - \mathbf{t}_1)^2/\sigma_1^2} \cdot \mathbf{e}^{-(\mathbf{x} - \mathbf{t}_2)^2/\sigma_2^2} = c\,\mathbf{e}^{-(\mathbf{x} - \mathbf{t}_3)^2/\sigma_3^2},$$

$$\mathbf{t}_3 = \frac{\sigma_2^2\mathbf{t}_1 + \sigma_1^2\mathbf{t}_2}{\sigma_1^2 + \sigma_2^2}, \qquad \sigma_3^2 = \frac{\sigma_1^2\sigma_2^2}{\sigma_1^2 + \sigma_2^2}, \qquad c = \mathbf{e}^{-(\mathbf{t}_1 - \mathbf{t}_2)^2/\sigma_3^2}.$$

The functions $f(\mathbf{x}) = 1$ belongs to $\mathcal{G}_x$, since it can be considered as Gaussian of infinite variance, and for any distinct points $\mathbf{x}$, $\mathbf{y}$ we can obviously find a function in $\mathcal{G}_x$ such that $f(\mathbf{x}) \neq f(\mathbf{y})$: the conditions of Stone's theorem are then satisfied and $\mathcal{G}_x$ is dense in $C[X]$.

Appendix C: Regularization Networks can Approximate Smooth Functions Arbitrarily Well

In this appendix we briefly describe the regularization method for approximating functions and show that the networks that are derived from a regularization principle can approximate arbitrarily well continuous functions defined on a compact subset of R^n.

Let $S = \{(\mathbf{x}_i, y_i) \in R^n \times R \mid i = 1, \ldots N\}$ be a set of data that we want to approximate by means of a function f. The regularization approach (Tikhonov 1963; Tikhonov and Arsenin 1977; Morozov 1984; Bertero 1986) consists in computing the function f that minimizes the functional

$$H[f] = \sum_{i=1}^{N} (y_i - f(\mathbf{x}_i))^2 + \lambda \|Pf\|^2$$

where P is a constraint operator (usually a differential operator, $\|\cdot\|^2$ is a norm on the function space to whom Pf belongs (usually the L^2 norm) and λ is a positive real number, and so called *regularization parameter.* The structure of the operator P embodies the a priori knowledge about the solution, and therefore depends on the nature of the particular problem that has to be solved. The general form of the solution of this variational problem is given by the following expansion (Poggio and Girosi 1989):

$$f(\mathbf{x}) = \sum_{i=l}^{N} c_i G(\mathbf{x}; \mathbf{x}_i) + p(\mathbf{x}) \tag{11}$$

where G is the Green's function of the differential operator $\hat{P}P$, $\hat{P}$ being the adjoint operator of P, $p(\mathbf{x})$ is a linear combination of functions that span the null space of P, and the coefficients c_i can be found by inverting a matrix that depends on the data points (Poggio and Girosi 1989). We remind the reader that the Green's function of an operator $\hat{P}P$ is the function that satisfies the following differential equation (in the distributions sense):

$$\hat{P}PG(\mathbf{x}; \mathbf{y}) = \delta(\mathbf{x} - \mathbf{y}). \tag{12}$$

It is clear that there is a correspondence between the class of functions that can be written in the form (11) (for any number of data points and for any Green's functions G of a self-adjoint operator) and a subclass of feedforward networks with one layer of hidden units, of the type shown in figure 1. Under mild assumptions on $\hat{P}P$, these networks can approximate continuous functions arbitrarily well, as is stated in the following proposition:

Proposition C.1. *For every continuous function F defined on a compact subset of R^n and every piecewise continuous G which is the Green's function of a self-adjoint differential operator, there exists a function $f^*(\mathbf{x}) = \sum_{i=1}^{N} c_i G(\mathbf{x}; \mathbf{x}_i)$, such that for all $\mathbf{x}$ and any positive ϵ the following inequality holds:*

$$|F(\mathbf{x}) - f^*(\mathbf{x})| < \epsilon$$

Proof. Let F be a continuous function defined on a compact set $D \subset R^n$. Its domain of definition can be extended to all R^n, in such a way that is goes continuously to zero outside D. The resulting function, that we still call F, is a continuous function with bounded support[3]. Consider the space K of "test functions" (Gelfand and Shilov 1964), that consists of real functions $\phi(\mathbf{x})$ with continuous derivatives of all orders and with bounded support (which means that the function and all its derivatives vanish outside of some bounded region). As Gelfand and Shilov show (Appendix 1.1), there always exists a function $\phi(\mathbf{x})$ in K arbitrarily close to F, i.e., such that for all $\mathbf{x}$ and for any $\epsilon > 0$,

$$|F(\mathbf{x}) - \phi| < \epsilon.$$

Thus it is sufficient to show that every function $\phi(\mathbf{x}) \in K$ can be approximated arbitrarily well by a linear superposition of Green's functions (function f^* of proposition C.1).

We start with the identity

$$\phi(x) = \int d\mathbf{y}\,\phi(\mathbf{y})\delta(\mathbf{x} - \mathbf{y}) \tag{13}$$

where the integral is actually taken only over the bounded region in which $\phi(\mathbf{x})$ fails to vanish. By means of (12) we obtain

$$\phi(\mathbf{x}) = \int d\mathbf{y}\,\phi(\mathbf{y})(\hat{P}PG)(\mathbf{x}; \mathbf{y}) \tag{14}$$

and since $\phi(\mathbf{x})$ is in K and $\hat{P}P$ is formally self-adjoint we have

$$\phi(\mathbf{x}) = \int d\mathbf{y}\,G(\mathbf{x}; \mathbf{y})(\hat{P}P\phi)(\mathbf{y}). \tag{15}$$

We can rewrite (15) as

$$\phi(\mathbf{x}) = \int d\mathbf{y}\,G(\mathbf{x}; \mathbf{y})\psi(\mathbf{y}) \tag{16}$$

where $\psi(\mathbf{x}) = \hat{P}P\phi(\mathbf{x})$. Since $G(\mathbf{x}; \mathbf{y})\psi(\mathbf{y})$ is piecewise continuous on a closed domain, this integral exists in the sense of Riemann. By definition of Riemann integral, equation 16 can then be written as

$$\phi(\mathbf{x}) = \Delta^n \sum_{k \in I} \psi(\mathbf{x}_k)G(\mathbf{x}; \mathbf{x}_k) + E_x(\Delta) \tag{17}$$

where $\mathbf{x}_k$ are points of a square grid of spacing Δ, I is the finite set of lattice points where $\psi(\mathbf{x}) \neq 0$, and $E_x(\Delta)$ is the discretization error, with the property

$$\lim_{\Delta \to 0} E_x(\Delta) = 0. \tag{18}$$

[3] The support of a continuous function $F(\mathbf{x})$ is the closure of the set on which $F(\mathbf{x}) \neq 0$.

If we now choose $f^*(\mathbf{x}) = \Delta^n \Sigma_{k \in I} \psi(\mathbf{x}_k) G(\mathbf{x}; \mathbf{x}_k)$, combining (17) and (18) we obtain

$$\lim_{\Delta \to 0} [\phi(\mathbf{x}) - f^*(\mathbf{x})] = 0. \tag{19}$$

Thus every function $\phi \in K$ can be approximated arbitrarily well by a linear superposition of Green's functions G of a self-adjoint operator, and this concludes the proof.

Remark. The conditions of proposition C.1 exclude Green's functions that have singularities in the origin. An example is the Green's function associated with the "membrane" stabilizer $P = \nabla$ in 2 or more dimensions. In 2 dimensions, the membrane Green's function is $G(r) = -\log r$, where $r = \|\mathbf{x} - \mathbf{x}_i\|$ (in 1 dimension $G(x) = |x|$, satisfies the condition of proposition C.1).

Remark. Notice that in order to approximate arbitrarily well any continuous function on a compact domain with functions of the type 11, it is not necessary to include the term p belonging to the null space of P.

References

Bertero M (1986) Regularization methods for linear inverse problems. In: Talenti CG (ed) Inverse problems. Springer, Berlin Heidelberg New York

Bertero M, Poggio T, Torre V (1988) Ill-posed problems in early vision. Proc IEEE 76:869–889

Braess D (1986) Nonlinear approximation theory. Springer, Berlin Heidelberg New York

Broomhead DS, Lowe D (1988) Multivariable functional interpolation and adaptive networks. Complex Syst 2:321–355

Carrol SM, Dickinson BW (1989) Construction of neural nets using the Radon transform. In: Proceedings of the International Joint Conference on Neural Networks, pp I-607–I-611. Washington D.C., June 1989. IEEE TAB Neural Network Committee

Cheney EW (1981) Introduction to approximation theory. Chelsea, New York

Cybenko G (1988) Continuous valued neural networks with two hidden layers are sufficient. Technical report. Department of Computer Sciences, Tufts University, Medford, Mass

Cybenko G (1989) Approximation by superposition of a sigmoidal function. Math Control Syst Signals (in press)

De Boor C (1969) On the approximation by γ-Polynomials. In: Schoenberg IJ (ed) Approximation with special emphasis on spline functions. Academic Press, New York pp 157–183

Funahashi K (1989) On the approximate realization of continuous mappings by neural networks. Neural Networks 2:183–192

Gelfand IM, Shilov GE (1964) Generalized functions, vol 1: Properties and operations. Academic Press, New York

Hartman E, Keeler K, Kowalski JM (1989) Layered neural networks with gaussian hidden units as universal approximations. (submitted for publication)

Hobby CR, Rice JR (1967) Approximation from a curve of functions. Arch Rat Mech Anal 27:91–106

Micchelli CA (1986) Interpolation of scattered data: distance matrices and conditionally positive definite functions. Constn Approx 2:11–22

Moody J, Darken C (1989) Fast learning in networks of locally-tuned processing units. Neural Comput 1:281–294

Moore B, Poggio T (1988) Representations properties of multilayer feed-forward networks. In: Abstracts of the First Annual INNS Meeting. Pergamon Press, New York, p. 502

Morozov VA (1984) Methods for solving incorrectly posed problems. Springer, Berlin Heidelberg New York

Poggio T, Girosi F (1989) A theory of networks for approximation and learning. A.I. Memo No. 1140. Artificial Intelligence Laboratory, Massachusetts Institute of Technology, Cambridge Mass

Rice JR (1964) The approximation of functions, vol. 1 Addison-Wesley, Reading, Mass

Rice JR (1969) The approximation of functions, vol. 2. Addison-Wesley, Reading, Mass

Rudin W (1973) Functional analysis. McGraw-Hill, New York

Rumelhart DE, Hinton GE, Williams RJ (1986a) Learning internal representations by error propagation. In: Parallel distributed processing, chap 8. MIT Press, Cambridge Mass, pp. 318–362

Rumelhart DE, Hinton GE, Williams RJ (1986b) Learning representations by back-propagating errors. Nature 323: 533–536

Sejnowski TJ, Rosenberg CR (1987) Parallel networks that learn to pronounce english text. Complex Syst 1:145–168

Stinchcombe M, White H (1989) Universal approximation using feedforward networks with non-sigmoid hidden layer activation functions. In: Proceedings of the International Joint Conference on Neural Networks. Washington DC, June 1989. IEEE TAB Neural Network Committee, pp I/607–I/611

Stone MH (1937) Applications of the theory of Boolean rings to general topology. AMS Trans 41:375–481

Stone MH (1948) The generalized Weierstrass approximation theorem. Math Mag 21:167–183, 237–254

Tikhonov AN (1963) Solution of incorrectly formulated problems and the regularization method. Soviet Math Dokl 4: 1035–1038

Tikhonov AN, Arsenin VY (1977) Solutions of Ill-posed Problems. Winston, Washington, DC.

Yosida K (1974) Functional analysis. Springer, Berlin Heidelberg New York

Prof. Dr. Tomaso Poggio
Center for Biological Information Processing
Room NE 43–787
Artificial Intelligence Laboratory
Massachusetts Institute of Technology
Cambridge MA 02139
USA

Multilayer Feedforward Networks are Universal Approximators

KURT HORNIK

Technische Universität Wien

MAXWELL STINCHCOMBE AND HALBERT WHITE

University of California, San Diego

(*Received* 16 *September* 1988; *revised and accepted* 9 *March* 1989)

Abstract—*This paper rigorously establishes that standard multilayer feedforward networks with as few as one hidden layer using arbitrary squashing functions are capable of approximating any Borel measurable function from one finite dimensional space to another to any desired degree of accuracy, provided sufficiently many hidden units are available. In this sense, multilayer feedforward networks are a class of universal approximators.*

Keywords—Feedforward networks, Universal approximation, Mapping networks, Network representation capability, Stone-Weierstrass Theorem, Squashing functions, Sigma-Pi networks, Back-propagation networks.

1. INTRODUCTION

It has been nearly twenty years since Minsky and Papert (1969) conclusively demonstrated that the simple two-layer perceptron is incapable of usefully representing or approximating functions outside a very narrow and special class. Although Minsky and Papert left open the possibility that multilayer networks might be capable of better performance, it has only been in the last several years that researchers have begun to explore the ability of multilayer feedforward networks to approximate general mappings from one finite dimensional space to another. Recently, this research has virtually exploded with impressive successes across a wide variety of applications. The scope of these applications is too broad to mention useful specifics here; the interested reader is referred to the proceedings of recent IEEE Conferences on Neural Networks (1987, 1988) for a sampling of examples.

The apparent ability of sufficiently elaborate feedforward networks to approximate quite well nearly any function encountered in applications leads one to wonder about the ultimate capabilities of such networks. Are the successes observed to date reflective of some deep and fundamental approximation capability, or are they merely flukes, resulting from selective reporting and a fortuitous choice of problems? Are multilayer feedforward networks in fact inherently limited to approximating only some fairly special class of functions, albeit a class somewhat larger than the lowly perceptron? The purpose of this paper is to address these issues. We show that multilayer feedforward networks with as few as one hidden layer are indeed capable of universal approximation in a very precise and satisfactory sense.

Advocates of the virtues of multilayer feedforward networks (e.g., Hecht-Nielsen, 1987) often cite Kolmogorov's (1957) superposition theorem or its more recent improvements (e.g., Lorentz, 1976) in support of their capabilities. However, these results require a *different* unknown transformation (g in Lorentz's notation) for each continuous function to be represented, while specifying an exact upper limit to the number of intermediate units needed for the representation. In contrast, quite specific squashing functions (e.g., logistic, hyperbolic tangent) are used in practice, with necessarily little regard for the function being approximated and with the number of hidden units increased *ad libitum* until some desired level of approximation accuracy is reached. Al-

White's participation was supported by a grant from the Guggenheim Foundation and by National Science Foundation Grant SES-8806990. The authors are grateful for helpful suggestions by the referees.

Requests for reprints should be sent to Halbert White, Department of Economics, D-008, UCSD, La Jolla, CA 92093.

Reprinted with permission from *Neural Networks,* vol. 2, K. Hornik, M. Stinchcombe, and H. White, "Multilayer Feedforward Networks are Universal Approximators," pp. 359–366, 1989, Pergamon Press Ltd., Oxford, England.

though Kolmogorov's result provides a theoretically important possibility theorem, it does not and cannot explain the successes achieved in applications.

In previous work, le Cun (1987) and Lapedes and Farber (1988) have shown that adequate approximations to an unknown function using monotone squashing functions can be achieved using two hidden layers. Irie and Miyake (1988) have given a representation result (perfect approximation) using one hidden layer, but with a *continuum* of hidden units. Unfortunately, this sort of result has little practical usefulness, despite its great theoretical utility.

Recently, however, Gallant and White (1988) showed that a particular single hidden layer feedforward network using the monotone "cosine squasher" is capable of embedding as a special case a Fourier network which yields a Fourier series approximation to a given function as its output. Such networks thus possess all the approximation properties of Fourier series representations. In particular, they are capable of approximation to any desired degree of accuracy of any square integrable function on a compact set using a finite number of hidden units. Still, Gallant and White's results do not justify arbitrary multilayer feedforward networks as universal approximators, but only a particular class of single hidden layer networks in a particular (but important) sense. Further related results using the logistic squashing function (and a great deal of useful background) are given by Hecht-Nielsen (1989).

The present paper makes use of the Stone-Weierstrass Theorem and the cosine squasher of Gallant and White to establish that standard multilayer feedforward network architectures using arbitrary squashing functions can approximate virtually any function of interest to any desired degree of accuracy, provided sufficiently many hidden units are available. These results establish multilayer feedforward networks as a class of universal approximators. As such, failures in applications can be attributed to inadequate learning, inadequate numbers of hidden units, or the presence of a stochastic rather than a deterministic relation between input and target. Our results do not address the issue of how many units are needed to attain a given degree of approximation.

The plan of this paper is as follows. In section 2 we present our main results. Section 3 contains a discussion of our results, directions for further research and some concluding remarks. Mathematical proofs are given in an appendix.

2. MAIN RESULTS

We begin with definitions and notation which enable us to speak precisely about the class of multi-layer feedforward networks under consideration.

Definition 2.1

For any $r \in N \equiv \{1, 2, \ldots\}$, $\mathbf{A}^r$ is the set of all affine functions from R^r to R, that is, the set of all functions of the form $A(x) = w \cdot x + b$ where w and x are vectors in R^r, "$\cdot$" denotes the usual dot product of vectors, and $b \in R$ is a scalar. □

In the present context, x corresponds to network input, w corresponds to network weights from input to the intermediate layer, and b corresponds to a bias.

Definition 2.2

For any (Borel) measurable function $G(\cdot)$ mapping R to R and $r \in N$ let $\Sigma^r(G)$ be the class of functions

$$\{f: R^r \to R : f(x) = \sum_{j=1}^{q} \beta_j G(A_j(x)),\ x \in R^r,\ \beta_j \in R,\ A_j \in \mathbf{A}^r,\ q = 1, 2, \ldots\}. \quad \square$$

A leading case occurs when G is a "squashing function," in which case $\Sigma^r(G)$ is the familiar class of output functions for single hidden layer feedforward networks with squashing at the hidden layer and no squashing at the output layer. The scalars β_j correspond to network weights from hidden to output layers.

For convenience, we formally define what we mean by a squashing function.

Definition 2.3

A function $\Psi: R \to [0,1]$ is a squashing function if it is non-decreasing, $\lim_{\lambda\to\infty} \Psi(\lambda) = 1$, and $\lim_{\lambda\to-\infty} \Psi(\lambda) = 0$. □

Because squashing functions have at most countably many discontinuities, they are measurable. Useful examples of squashing functions are the threshold functions, $\Psi(\lambda) = 1_{\{\lambda \geq 0\}}$ (where $1_{\{\cdot\}}$ denotes the indicator function), the ramp function, $\Psi(\lambda) = \lambda 1_{\{0 \leq \lambda \leq 1\}} + 1_{\{\lambda > 1\}}$, and the cosine squasher of Gallant and White (1988), $\Psi(\lambda) = (1 + \cos[\lambda + 3\pi/2])(1/2)\, 1_{\{-\pi/2 \leq \lambda \leq \pi/2\}} + 1_{\{\lambda > \pi/2\}}$.

We define a class of $\Sigma\Pi$ network output functions (Maxwell, Giles, Lee, & Chen, 1986; Williams, 1986) in the following way.

Definition 2.4

For any measurable function $G(\cdot)$ mapping R to R and $r \in N$, let $\Sigma\Pi^r(G)$ be the class of functions

$$\{f: R^r \to R: f(x) = \sum_{j=1}^{q} \beta_j \cdot \prod_{k=1}^{l_j} G(A_{jk}(x)),\ x \in R^r,\ \beta_j \in R,\ A_{jk} \in A^r,\ l_j \in N,\ q = 1, 2, \ldots\}. \quad \square$$

Our general results will be proved first for $\Sigma\Pi$ networks and subsequently extended to Σ networks. The latter are the special case of $\Sigma\Pi$ networks for which $l_j = 1$ for all j.

Notation for the classes of function that we consider approximating is given by the next definition.

Definition 2.5

Let C^r be the set of continuous functions from R^r to R, and let M^r be the set of all Borel measurable functions from R^r to R. We denote the Borel σ-field of R^r as B^r. □

The classes $\Sigma^r(G)$ and $\Sigma\Pi^r(G)$ belong to M^r for any Borel measurable G. When G is continuous, $\Sigma^r(G)$ and $\Sigma\Pi^r(G)$ belong to C^r. The class C^r is a subset of M^r, which in fact contains virtually all functions relevant in applications. Functions that are not Borel measurable exist (e.g., Billingsley, 1979, pp. 36–37) but they are pathological. Our first results concern approximating functions in C^r; we then extend these results to approximating functions in M^r.

Closeness of functions f and g belonging to C^r or M^r is measured by a metric, ρ. Closeness of one class of functions to another class is described by the concept of denseness.

Definition 2.6

A subset S of a metric space (X, ρ) is ρ $-$ *dense* in a subset T if for every $\varepsilon > 0$ and for every $t \in T$ there is an $s \in S$ such that $\rho(s, t) < \varepsilon$. □

In other words, an element of S can approximate an element of T to any desired degree of accuracy. In our theorems below, T and X correspond to C^r or M^r, S corresponds to $\Sigma^r(G)$ or $\Sigma\Pi^r(G)$ for specific choices of G, and ρ is chosen appropriately.

Our first result is stated in terms of the following metrics on C^r.

Definition 2.7

A subset S of C^r is said to be *uniformly dense on compacta in* C^r if for every compact subset $K \subset R^r$ S is ρ_K-dense in C^r, where for $f, g \in C^r$ $\rho_K(f,g) \equiv \sup_{x\in K}|f(x) - g(x)|$. A sequence of functions $\{f_n\}$ *converges to a function f uniformly on compacta* if for all compact $K \subset R^r$ $\rho_K(f_n,f) \to 0$ as $n \to \infty$. □

We may now state our first main result.

Theorem 2.1

Let G be any continuous nonconstant function from R to R. Then $\Sigma\Pi^r(G)$ is uniformly dense on compacta in C^r. □

In other words, $\Sigma\Pi$ feedforward networks are capable of arbitrarily accurate approximation to any real-valued continuous function over a compact set. The compact set requirement holds whenever the possible values of the inputs x are bounded ($x \in K$). An interesting feature of this result is that the activation function G may be *any* continuous nonconstant function. It is not required to be a squashing function, although this is certainly allowed. Another interesting type of activation function allowed by this result behaves like a squashing function for values of $A(x)$ below a given level, but then decreases continuously to zero as $A(x)$ increases beyond this level. Our subsequent results all follow from Theorem 2.1.

In order to interpret the metrics relevant to our subsequent results we introduce the following notion.

Definition 2.8

Let μ be a probability measure on (R^r, B^r). If f and g belong to M^r, we say the are *μ-equivalent* if $\mu\{x \in R^r: f(x) = g(x)\} = 1$ □

Taking μ to be a probability measure (i.e., $\mu(R^r) = 1$) is a matter of convenience; our results actually hold for arbitrary finite measures. The context need not be probabilistic. Regardless, the measure μ describes the relative frequency of occurrence of input "patterns" x. The measure μ is the "input space environment" in the terminology of White (1988a). Functions that are μ-equivalent differ only on a set of patterns occurring with probability (measure) zero, and we are concerned only with distinguishing between classes of equivalent functions.

The metric on classes of μ-equivalent functions relevant for our main results is given by the next definition.

Definition 2.9

Given a probability measure μ on (R^r,B^r) define the metric ρ_μ from $M^r \times M^r$ to R^+ by $\rho_\mu(f,g) = \inf\{\varepsilon > 0: \mu\{x: |f(x) - g(x)| > \varepsilon\} < \varepsilon\}$. □

Two functions are close in this metric if and only if there is only a small probability that they differ significantly. In the extreme case that f and g are μ-equivalent $\rho_\mu(f,g)$ equals zero.

There are many equivalent ways to describe what it means for $\rho_\mu(f_n, f)$ to converge to zero.

Lemma 2.1. All of the following are equivalent.

(a) $\rho_\mu(f_n,f) \to 0$.
(b) For every $\varepsilon > 0$ $\mu\{x: |f_n(x) - f(x)| > \varepsilon\} \to 0$.
(c) $\int \min\{|f_n(x) - f(x)|, 1\}\ \mu(dx) \to 0$. □

From (b) we see that ρ_μ-convergence is equivalent to convergence in probability (or measure). In (b) the Euclidean metric can be replaced by any metric on R generating the Euclidean topology, and the integrand in (c) by any bounded metric on R generating the Euclidean topology. For example $d(a,b) = |a - b|/(1 + |a - b|)$ is a bounded metric generating the Euclidean topology, and (c) is true if and only if $\int d(f_n(x), f(x))\mu(dx) \to 0$.

The following lemma relates uniform convergence on compacta to ρ_μ-convergence.

Lemma 2.2. If $\{f_n\}$ is a sequence of functions in M^r that converges uniformly on compacta to the function f then $\rho_\mu(f_n, f) \to 0$ □

We now state our first result on approximating functions in M^r. It follows from Theorem 2.1 and Lemma 2.2.

Theorem 2.2

For every continuous nonconstant function G, every r, and every probability measure μ on (R^r, B^r), $\Sigma\Pi^r(G)$ is ρ_μ-dense in M^r □

In other words, single hidden layer $\Sigma\Pi$ feedforward networks can approximate any measurable function arbitrarily well, regardless of the continuous nonconstant function G used, regardless of the dimension of the input space r, and regardless of the input space environment μ. In this precise and satisfying sense, $\Sigma\Pi$ networks are universal approximators.

The continuity requirement on G rules out the threshold function $\Psi(\lambda) = 1_{\{\lambda \geq 0\}}$. However, for squashing functions continuity is not necessary.

Theorem 2.3

For every squashing function Ψ, every r, and every probability measure μ or (R^r, B^r), $\Sigma\Pi^r(\Psi)$ is uniformly dense on compacta in C^r and ρ_μ-dense in M^r. □

Because of their simpler structure, it is important to know that the very simplest $\Sigma\Pi$ networks, the Σ networks, have similar approximation capabilities.

Theorem 2.4

For every squashing function Ψ, every r, and every probability measure μ on (R^r, B^r), $\Sigma^r(\Psi)$ is uniformly dense on compacta in C^r and ρ_μ-dense in M^r. □

In other words, standard feedforward networks with only a single hidden layer can approximate any continuous function uniformly on any compact set and any measurable function arbitrarily well in the ρ_μ metric, regardless of the squashing function Ψ (continuous or not), regardless of the dimension of the input space r, and regardless of the input space environment μ. Thus, Σ networks are also universal approximators.

Theorem 2.4 implies Theorem 2.3 and, for squashing functions, Theorem 2.3 implies Theorem 2.2. Stating our results in the given order reflects the natural order of their proofs. Further, deriving Theorem 2.3 as a consequence of Theorem 2.4 obscures its simplicity.

The structure of the proof of Theorem 2.3 (respectively 2.4) reveals that a similar result holds if Ψ is not restricted to be a squashing function, but is any measurable function such that $\Sigma\Pi^1(\Psi)$ (respectively $\Sigma^1(\Psi)$) uniformly approximates some squashing function on compacta. Stinchcombe and White (1989) give a result analogous to Theorem 2.4 for nonsigmoid hidden layer activation functions.

Subsequent to the first appearance of our results (Hornik, Stinchcombe, & White, 1988), Cybenko (1988) independently obtained the uniform approximation result for functions in C^r contained in Theorem 2.4. Cybenko's very different approach makes elegant use of the Hahn-Banach theorem.

A variety of corollaries follows easily from the results above. In all the results to follow, Ψ is a squashing function.

Corollary 2.1

For every function g in M^r there is a compact subset K of R^r and an $f \in \Sigma^r(\Psi)$ such that for any $\varepsilon > 0$ we have $\mu(K) < 1 - \varepsilon$ and for every $x \in K$ we have $|f(x) - g(x)| < \varepsilon$, regardless of Ψ, r, or μ. □

In other words, there is a single hidden layer feedforward network that approximates any measurable function to any desired degree of accuracy on some compact set K of input patterns that to the same degree of accuracy has measure (probability of occurrence) 1. Note the difference between Corollary 2.1 and Theorem 2.1. In Theorem 2.1 g is continuous and K is an arbitrary compact set; in Corollary 2.1 g is measurable and K must be specially chosen.

Our next result pertains to approximation in L_p-spaces. We recall the following definition.

Definition 2.10

$L_p(R^r, \mu)$ (or simply L_p) is the set of $f \in M^r$ such that $\int |f(x)|^p \mu(dx) < \infty$. The L_p norm is defined by $\|f\|_p = [\int |f(x)|^p \mu(dx)]^{1/p}$. The associated metric on L_p is defined by $\rho_p(f, g) = \|f - g\|_p$. □

The L_p approximation result is the following.

Corollary 2.2

If there is a compact subset K *of* R^r such that $\mu(K) = 1$ then $\Sigma^r(\Psi)$ is ρ_p-dense in $L_p(R^r, \mu)$ for every $p \in [1, \infty)$, regardless of Ψ, r, or μ. □

We also immediately obtain the following result.

Corollary 2.3

If μ is a probability measure on $[0,1]^r$ then $\Sigma^r(\Psi)$ is ρ_p-dense in $L_p([0, 1]^r, \mu)$ for every $p \in [1, \infty)$, regardless of Ψ, r, or μ. □

Corollary 2.4

If μ puts mass 1 on a finite set of points, then for every $g \in M^r$ and for every $\varepsilon > 0$ there is an $f \in \Sigma^r(\Psi)$ such that $\mu\{x: |f(x) - g(x)| < \varepsilon\} = 1$. □

Corollary 2.5

For every Boolean function g and every $\varepsilon > 0$ there is an f in $\Sigma^r(\Psi)$ such that $\max_{x \in \{0,1\}^r} |g(x) - f(x)| < \varepsilon$. □

In fact, exact representation of functions with finite support is possible with a single hidden layer.

Theorem 2.5

Let $\{x_1, \ldots, x_n\}$ be a set of distinct points in R^r and let $g : R^r \rightarrow R$ be an arbitrary function. If Ψ achieves 0 and 1, then there is a function $f \in \Sigma^r(\Psi)$ with n hidden units such that $f(x_i) = g(x_i)$, $i \in \{1, \ldots, n\}$. □

With some tedious modifications the proof of this theorem goes through when Ψ is an arbitrary squashing function.

The foregoing results pertain to single output networks. Analogous results are valid for multi-output networks approximating continuous or measurable functions from R^r to R^s, $s \in N$, denoted $C^{r,s}$ and $M^{r,s}$, respectively. We extend Σ^r and $\Sigma\Pi^r$ to $\Sigma^{r,s}$ and $\Sigma\Pi^{r,s}$ respectively be re-interpreting β_j as an $s \times 1$ vector in Definitions 2.2 and 2.4. The function $g: R^r \rightarrow R^s$ has elements g_i, $i = 1, \ldots, s$. We have the following result.

Corollary 2.6

Theorems 2.3, 2.4 and Corollaries 2.1–2.5 remain valid for classes $\Sigma\Pi^{r,s}(\Psi)$ and/or $\Sigma^{r,s}(\Psi)$ approximating functions in $C^{r,s}$ and $M^{r,s}$ with ρ_μ replaced with ρ^s_μ, $\rho^s_\mu(f, g) \equiv \sum_{i=1}^{s} \rho_\mu(f_i, g_i)$ and with ρ_p replaced with its appropriate multivariate generalization. □

Thus, multi-output multilayer feedforward networks are universal approximators of vector-valued functions.

All of the foregoing results are for networks with a single hidden layer. Our final result describes the approximation capabilities of multi-output multilayer networks with multiple hidden layers. For simplicity, we explicitly consider the case of multilayer Σ nets only. We denote the class of output functions for multilayer feedforward nets with l layers (not counting the input layer, but counting the output layer) mapping R^r to R^s using squashing functions Ψ as $\Sigma_l^{r,s}(\Psi)$. (Our previous results thus concerned the case $l = 2$.) The activation rules for the elements of such a network are

$$a_{ki} = G_k(A_i(a_{k-1}))\ i = 1, \ldots, q_k;\ k = 1, \ldots, l,$$

where a_k is a $q_k \times 1$ vector with elements a_{ki}, $a_0 \equiv x$ by convention, $G_1, \ldots, G_{l-1} = \Psi$, G_l is the identity map, $q_0 \equiv r$, and $q_l \equiv s$. We have the following result.

Corollary 2.7

Theorem 2.4 and Corollaries 2.1–2.6 remain valid for multioutput multilayer classes $\Sigma_l^{r,s}(\Psi)$ approximating functions in $C^{r,s}$ and $M^{r,s}$, with ρ_μ and ρ_p replaced as in Corollary 2.6, provided $l \geqslant 2$. □

Thus, $\Sigma_l^{r,s}$ networks are universal approximators of vector valued functions.

We remark that any implementation of a $\Sigma\Pi_l^{r,s}$ network is also a universal approximator as it contains the $\Sigma_l^{r,s}$ networks as a special case. We avoid explicit consideration of these because of their notational complexity.

3. DISCUSSION AND CONCLUDING REMARKS

The results of Section 2 establish that standard multilayer feedforward networks are capable of approximating any measurable function to any desired degree of accuracy, in a very specific and satisfying sense. We have thus established that such "mapping" networks are universal approximators. This implies that any lack of success in applications must arise from inadequate learning, insufficient numbers of hidden units or the lack of a deterministic relationship between input and target.

The results given here also provide a fundamental basis for rigorously establishing the ability of multilayer feedforward networks to learn (i.e., to estimate consistently) the connection strengths that achieve the approximations proven here to be possible. A statistical technique introduced by Grenander (1981) called the "method of sieves" is particularly well suited to this task. White (1988b) establishes such results for learning, using results of White and Woolridge (in press). For this it is necessary to utilize the concept of metric entropy (Kolmogorov & Tinomirov, 1961) for subsets of Σ^r possessing fixed numbers of hidden units. As a natural by-product of the metric entropy results one obtains quite specific rates at which the number of hidden units may grow as the number of training instances increases, while still ensuring the statistical property of consistency (i.e., avoiding overfitting).

An important related area for further research is

the investigation of the rate at which approximations using $\Sigma\Pi$ or Σ networks improve as the number of hidden units increases (the "degree of approximation") when the dimension r of the input space is held fixed. Such results will support rate of convergence results for learning via sieve estimation in multilayer feedforward networks based on the recent approach of Severini and Wong (1987).

Another important area for further investigation that we have neglected completely and that is beyond the scope of our work here is the rate at which the number of hidden units needed to attain a given accuracy of approximation must grow as the dimension r of the input space increases. Investigation of this "scaling up" problem may also be facilitated by consideration of the metric entropy of $\Sigma\Pi^{r,s}$ and $\Sigma^{r,s}$.

The results given here are clearly only one step in a rigorous general investigation of the capabilities and properties of multilayer feedforward networks. Nevertheless, they provide an essential and previously unavailable theoretical foundation establishing that the successes realized to date by such networks in applications are not just flukes, but are instead a reflection of the general universal approximation capabilities of multilayer feedforward networks.

Note added in proof: The authors regret being unaware of the closely related work by Funahashi (this journal, volume 2, pp. 183–192) at the time the revision of this article was submitted. Our Theorem 2.4 and Corollary 2.7 somewhat extend Funahashi's Theorems 1 and 2 by permitting non-continuous activation functions.

MATHEMATICAL APPENDIX

Because of the central role played by the Stone-Weierstrass theorem in obtaining our results, we state it here. Recall that a family $\mathbf{A}$ of real functions defined on a set E is an *algebra* if $\mathbf{A}$ is closed under addition, multiplication, and scalar multiplication. A family $\mathbf{A}$ *separates points* on E if for every x, y in E, $x \neq y$, there exists a function f in $\mathbf{A}$ such that $f(x) \neq f(y)$. The family A *vanishes at no point of* E if for each x in E there exists f in A such that $f(x) \neq 0$. (For further background, see Rudin, 1964, pp. 146–153.)

Stone-Weierstrass Theorem

Let $\mathbf{A}$ be an algebra of real continuous functions on a compact set K. If $\mathbf{A}$ separates points on K and if $\mathbf{A}$ vanishes at no point of K, then the uniform closure $\mathbf{B}$ of $\mathbf{A}$ consists of all real continuous functions on K (i.e., $\mathbf{A}$ is ρ_K-dense in the space of real continuous functions on K).

Proof of Theorem 2.1

We apply the Stone-Weierstrass Theorem. Let $K \subset R^r$ be any compact set. For any G, $\Sigma\Pi^r(G)$ is obviously an algebra on K. If $x, y \in K$, $x \neq y$, then there is an $A \in \mathbf{A}^r$ such that $G(A(x)) \neq G(A(y))$. To see this, pick $a, b \in R$, $a \neq b$ such that $G(a) \neq G(b)$. Pick $A(\cdot)$ to satisfy $A(x) = a$, $A(y) = b$. Then $G(A(x)) \neq G(A(y))$. This ensures that $\Sigma\Pi^r(G)$ is separating on K.

Second, there are $G(A(\cdot))$'s that are constant and not equal to zero. To see this, pick $b \in R$ such that $G(b) \neq 0$ and set $A(x) = 0 \cdot x + b$. For all $x \in K$, $G(A(x)) = G(b)$. This ensures that $\Sigma\Pi^r(G)$ vanishes at no point of K.

The Stone-Weierstrass Theorem thus implies that $\Sigma\Pi^r(G)$ is ρ_K-dense in the space of real continuous functions on K. Because K is arbitrary, the result follows. □

Proof of Lemma 2.1

(a) $\Leftrightarrow$ (b): Immediate.

(b) $\rightarrow$ (c): If $\mu\{x:|f_n(x) - f(x)| > \varepsilon/2\} < \varepsilon/2$ then $\int \min\{|f_n(x) - f(x)|, 1\}\, \mu(dx) < \varepsilon/2 + \varepsilon/2 = \varepsilon$.

(c) $\rightarrow$ (b): This follows from Chebyshev's inequality. □

Proof of Lemma 2.2

Pick an $\varepsilon > 0$. By Lemma 2.1 it is sufficient to find $N \in N$ such that for all $n \geq N$ we have $\int \min\{f_n(x) - f(x), 1\}\, \mu(dx) < \varepsilon$. Without loss of generality, we suppose $\mu(R^r) = 1$. Because R^r is a locally compact metric space, μ is a regular measure (e.g., Halmos, 1974, 52.G, p. 228). Thus there is a compact subset K of R^r with $\mu(K) > 1 - \varepsilon/2$. Pick N such that for all $n \geq N$ $\sup_{x \in K}|f_n(x) - f(x) < \varepsilon/2$. Now $\int_{R^r - K} \min\{|f_n(x) - f(x)|, 1\}\, \mu(dx) + \int_K \min\{|f_n(x) - f(x)|\ 1\}\, \mu(dx) < \varepsilon/2 + \varepsilon/2 = \varepsilon$ for all $n \geq N$. □

Lemma A.1. For any finite measure μ C^r is ρ_μ-dense in M^r.

Proof

Pick an arbitrary $f \in M^r$ and $\varepsilon > 0$. We must find a $g \in C^r$ such that $\rho_\mu(f, g) < \varepsilon$. For sufficiently large M, $\int \min\{|f \cdot 1_{\{|f|<M\}} - f|, 1\} d\mu < \varepsilon/2$. By Halmos (1974, Theorems 55.C and D, p. 241–242), there is a continuous g such that $\int |f \cdot 1_{\{|f|<M\}} - g| d\mu < \varepsilon/2$. Thus $\int \min\{|f - g|, 1\} d\mu < \varepsilon$. □

Proof of Theorem 2.2

Given any continuous nonconstant function, it follows from Theorem 2.1 and Lemma 2.2 that $\Sigma\Pi^r(G)$ is ρ_μ-dense in C^r. Because C^r is ρ_μ-dense in M^r by Lemma A.1, it follows that $\Sigma\Pi^r(G)$ is ρ_μ-dense in M^r (apply the triangle inequality). □

The extension from continuous to arbitrary squashing functions uses the following lemma.

Lemma A.2. Let F be a continuous squashing function and Ψ an arbitrary squashing function. For every $\varepsilon > 0$ there is an element H_ε of $\Sigma^1(\Psi)$ such that $\sup_{\lambda \in \mathbf{R}}|F(\lambda) - H_\varepsilon(\lambda)| < \varepsilon$.

Proof

Pick an arbitrary $\varepsilon > 0$. Without loss of generality, take $\varepsilon < 1$ also. We must find a finite collection of constants, β_j, and affine functions A_j, $j \in \{1, 2, \ldots, Q - 1\}$ such that $\sup_{\lambda \in \mathbf{R}}|F(\lambda) - \Sigma_{j=1}^{Q-1} \beta_j \Psi(A_j(\lambda))| < \varepsilon$.

Pick Q such that $1/Q < \varepsilon/2$. For $j \in \{1, \ldots, Q - 1\}$ set $\beta_j = 1/Q$. Pick $M > 0$ such that $\Psi(-M) < \varepsilon/2Q$ and $\Psi(M) > 1 - \varepsilon/2Q$. Because Ψ is a sqashing function such an M can be found. For $j \in \{1, \ldots, Q - 1\}$ set $r_j = \sup\{\lambda: F(\lambda) = j/Q\}$. Set $r_Q = \sup\{\lambda: F(\lambda) = 1 - 1/2Q)$. Because F is a continuous squashing function such r_j's exist.

For any $r < s$ let $A_{r,s} \in A^1$ be the unique affine function satisfying $A_{r,s}(r) = M$ and $A_{r,s}(s) = -M$. The desired approximation is then $H_\varepsilon(\lambda) = \Sigma_{j=1}^{Q-1} \beta_j \Psi(A_{r_j r_{j+1}}(\lambda))$. It is easy to check that on each of the intervals $(-\infty, r_1]$, $(r_1, r_2]$, $\ldots$, $(r_{Q-1}, r_Q]$, $(r_Q, +\infty)$ we have $|F(\lambda) - H_\varepsilon(\lambda)| < \varepsilon$. □

Proof of Theorem 2.3

By Lemma 2.2 and Theorem 2.2, it is sufficient to show that $\Sigma\Pi^r(\Psi)$ is uniformly dense on compacta in $\Sigma\Pi^r(F)$ for some continuous squashing function F. To show this, it is sufficient to show that every function of the form $\Pi_{k=1}^l F(A_k(\cdot))$ can be uniformly approximated by members of $\Sigma\Pi^r(\Psi)$.

Pick an arbitrary $\varepsilon > 0$. Because multiplication is continuous and $[0,1)^l$ is compact there is a $\delta > 0$ such that $|a_k - b_k| < \delta$ for $0 \leq a_k, b_k \leq 1$, $k \in \{1, \ldots, l\}$ implies $|\Pi_{k=1}^l a_k - \Pi_{k=1}^l b_k| < \varepsilon$.

By Lemma A.2 there is a function $H_\delta(\cdot) = \Sigma_{t=1}^T \beta_t \Psi(A_t^1(\cdot))$

such that $\sup_{\lambda \in R}|F(\lambda) - H_\delta(\lambda)| < \delta$. It follows that

$$\sup_{x \in R^r}\left|\prod_{k=1}^{l} F(A_k(x)) - \prod_{k=1}^{l} H_\delta(A_k(x))\right| < \varepsilon.$$

Because $A_j^l(A_k(\cdot)) \in \boldsymbol{A}^r$, we see that $\Pi_{k=1}^{l} H_\delta(A_k(\cdot)) \in \Sigma\Pi^r(\Psi)$.

Thus $\Pi_{k=1}^{l} F(A_{jk}(\cdot))$ can be uniformly approximated by elements of $\Sigma\Pi^r(\Psi)$. □

The proof of Theorem 2.4 makes use of the following three lemmas.

Lemma A.3. For every squashing function Ψ, every $\varepsilon > 0$, and every $M > 0$ there is a function $\cos_{M,\varepsilon} \in \Sigma^1(\Psi)$ such that

$$\sup_{\lambda \in [-M, +M]}|\cos_{M,\varepsilon}(\lambda) - \cos(\lambda)| < \varepsilon.$$

Proof

Let F be the cosine squasher of Gallant and White (1988) (the third example of squashing functions in Section 2). By adding, subtracting and scaling a finite number of affinely shifted versions of F we can get the cosine function on any interval $[-M, +M]$. The result now follows from Lemma A.2 and the triangle inequality. □

Lemma A.4. Let $g(\cdot) = \Sigma_{j=1}^{Q} \beta_j \cos(A_j(\cdot))$, $A_j \in A^r$. For arbitrary squashing function Ψ, for arbitrary compact $K \subset R^r$, and for arbitrary $\varepsilon > 0$ there is an $f \in \Sigma^r(\Psi)$ such that $\sup_{x \in K}|g(x) - f(x)| < \varepsilon$.

Proof

Pick $M > 0$ such that for $j \in \{1, \ldots, Q\}$ $A_j(K) \subset [-M, +M]$. Because Q is finite, K is compact and the $A_j(\cdot)$ are continuous, such an M can be found. Let $Q' = Q \cdot \Sigma_{j=1}^{Q}|\beta_j|$. By Lemma A. 3 for all $x \in K$ we have $|\Sigma_{j=1}^{Q} \beta_j \cos_{M,\varepsilon/Q'}(A_j(x)) - g(x)| < \varepsilon$. Because $\cos_{M,\varepsilon/Q'} \in \Sigma^1(\Psi)$, we see that $f(\cdot) = \Sigma_{j=1}^{Q} \cos_{M,\varepsilon/Q'}(A_j(\cdot)) \in \Sigma^r(\Psi)$. □

Lemma A.5. For every squashing function Ψ $\Sigma^r(\Psi)$ is uniformly dense on compacta in C^r.

Proof

By Theorem 2.1 the trigonometric polynomials $\{\Sigma_{j=1}^{Q}\beta_j\Pi_{k=1}^{l_j} \cos(A_{jk}(\cdot)) : Q, l_j \in N, \beta_j \in R, A_{jk} \in \boldsymbol{A}^r\}$ are uniformly dense on compacta in C^r. Repeatedly applying the trigonometric identity (cos a) · (cos b) = cos(a + b) − cos (a − b) allows us to rewrite every trigonometric polynomial in the form $\Sigma_{t=1}^{T} \alpha_t \cos(A_t(\cdot))$ where $\alpha_t \in R$ and $A_t \in \boldsymbol{A}^r$. The result now follows from Lemma A.4. □

Proof of Theorem 2.4

By Lemma A.5, $\Sigma^r(\Psi)$ is uniformly dense on compacta in C^r. Thus Lemma 2.2 implies that $\Sigma^r(\Psi)$ is ρ_μ-dense in C^r. The triangle inequality and Lemma A.1 imply that $\Sigma^r(\Psi)$ is ρ_μ-dense in M^r. □

Proof of Corollary 2.1

Fix $\varepsilon > 0$. By Lusin's Theorem (Halmos, 1974, p. 242–243) there is a compact set K^1 such that $\mu(K^1) > 1 - \varepsilon/2$ and $g|K^1$ (g restricted to K^1) is continuous on K^1. By the Tietze extension theorem (Dugundji, 1966, Theorem 5.1) there is a continuous function $g' \in C^r$ such that $g'|K^1 = g|K^1$ and $\sup_{x \in R^r} g'(x) = \sup_{x \in K^1} g|K^1(x)$. By Lemma A.5, $\Sigma^r(\Psi)$ is uniformly dense on compacta in C^r. Pick compact K^2 such that $\mu(K^2) > 1 - \varepsilon/2$. Take $f \in \Sigma^r(\Psi)$ such that $\sup_{x \in K^2}|f(x) - g'(x)| < \varepsilon$. Then $\sup_{x \in K^1 \cap K^2}|f(x) - g(x)| < \varepsilon$ and $\mu(K^1 \cap K^2) > 1 - \varepsilon$. □

Proof of Corollary 2.2

Pick arbitrary $g \in L_p$ and arbitrary $\varepsilon > 0$. We must show the existence of a function $f \in \Sigma^r(\Psi)$ such that $\rho_p(f, g) < \varepsilon$.

It follows from standard theorems (Halmos, 1974, Theorems 55.C and 55.D) that for every bounded function $h \in L_p$ there is a continuous f' such that $\rho_p(h, f') < \varepsilon/3$. For sufficiently large $M \in R$, setting $h = g1_{\{|g| \le M\}}$ gives $\rho_p(g, h) < \varepsilon/3$. Because $\Sigma^r(\Psi)$ is uniformly dense on compacta, there is an $f \in \Sigma^r(\Psi)$ such that $\sup_{x \in K}|f(x) - f'(x)|^p < (\varepsilon/3)^p$. Because $\mu(K) = 1$ by hypothesis we have $\rho_p(f', f) < \varepsilon/3$. Thus $\rho_p(g, f) \le \rho_p(g, h) + \rho_p(h, f') + \rho_p(f', f) < \varepsilon/3 + \varepsilon/3 + \varepsilon/3 = \varepsilon$. □

Proof of Corollary 2.3

Note that $[0, 1]^r$ is compact and apply Corollary 2.2. □

Proof of Corollary 2.4

Let $\underline{\varepsilon} = \min\{\mu(x) : \mu(x) > 0\}$. For all $\varepsilon < \underline{\varepsilon}$ we have that $\rho_\mu(f, g) = \varepsilon$ implies $\mu\{x : |f(x) - g(x)| > \varepsilon\} = 0$. Appealing to Theorem 2.4 finishes the proof. □

Proof of Corollary 2.5

Put mass $1/2^r$ on each point in $\{0, 1\}^r$ and apply Corollary 2.4. □

Proof of Theorem 2.5

There are two steps to this theorem. First, its validity is demonstrated when $\{x_1, \ldots x_n\} \subset R^1$, then the result is extended to R^r.

Step 1: Suppose $\{x_1, \ldots, x_n\} \subset R^1$ and, relabelling if necessary, that $x_1 < x_2 < \ldots < x_{n-1} < x_n$. Pick $M > 0$ such that $\Psi(-M) = 1 - \Psi(M) = 0$. Define A_1 as the constant affine function $A_1 \equiv M_0$ and set $\beta_1 = g(x_1)$. Set $f^1(x) = \beta_1 \cdot \Psi(A_1(x))$. Because $f^1(x) \equiv g(x_1)$ we have $f^1(x_1) = g(x_1)$. Inductively define A_k by $A_k(x_{k-1}) = -M$ and $A_k(x_k) = M$. Define $\beta_k = g(x_k) - g(x_{k-1})$.

Set $f^k(x) = \Sigma_{j=1}^{k} \beta_j\Psi(A_j(x))$. For $i \le k$ $f^k(x_i) = g(x_i)$. f^n is the desired function.

Step 2: Suppose $\{x_1, \ldots, x_n\} \subset R^r$ where $r \ge 2$. Pick $p \in R^r$ such that if $i \ne j$ then $p \cdot (x_i - x_j) \ne 0$. This can be done since $\cup_{i \ne j}\{q : q \cdot (x_i - x_j) = 0\}$ is a finite union of hyperplanes in R^r. Relabelling, if necessary, we can assume that $p \cdot x_1 < p \cdot x_2 < \cdots < p \cdot x_n$. As in the first step find β_j's and A_j's such that $\Sigma_{j=1}^{n} \beta_j\Psi(A_j(p \cdot x_i)) = g(x_i)$. Then $f(x) = \Sigma_{j=1}^{n} \beta_j\Psi(A_j(p \cdot x))$ is the desired function. □

Proof of Corollary 2.6

Using vectors β_i which are 0 except in the ith position we can approximate each g_i to within ε/δ. Adding together δ approximations keeps us within the classes $\Sigma\Pi^{r,s}$ and $\Sigma^{r,s}$. □

The proof of Corollary 2.7 uses the following lemma.

Lemma A.6. Let $\boldsymbol{F}$ (resp. $\boldsymbol{G}$) be a class of functions from R to R (resp. R^r to R) that is uniformly dense on compacta in C^1 (resp. C^r). The class of functions $\boldsymbol{G} \circ \boldsymbol{F} = \{f \circ g : g \in \boldsymbol{G}$ and $f \in \boldsymbol{F}\}$ is uniformly dense on compacta in C^r.

Proof

Pick an arbitrary $h \in C^r$, compact subset K of R^r, and $\varepsilon > 0$. We must show the existence of an $f \in \boldsymbol{F}$ and a $g \in \boldsymbol{G}$ such that $\sup_{x \in K}|f(g(x)) - h(x)| < \varepsilon$.

By hypothesis there is a $g \in \boldsymbol{G}$ such that $\sup_{x \in K}|g(x) - h(x)| < \varepsilon/2$. Because K is compact and h is continuous $\{h(x) : x \in K\}$ is compact. Thus $\{g(x) : x \in K\}$ is bounded. Let S be the necessarily compact closure of $\{g(x) : x \in K\}$.

By hypothesis there is an $f \in \boldsymbol{F}$ such that $\sup_{s \in S}|f(s) - s| < \varepsilon/2$. We see that $f \circ g$ is the desired function, as

$$\begin{aligned}\sup_{x \in K}|f(g(x)) - h(x)| &\le \sup_{x \in K}|f(g(x)) - g(x) + g(x) - h(x)| \\ &\le \sup_{x \in K}|f(g(x)) - g(x)| + \sup_{x \in K}|g(x) - h(x)| \\ &< \varepsilon/2 + \varepsilon/2 = \varepsilon. \quad \square\end{aligned}$$

Proof of Corollary 2.7

We consider only the case where $s = 1$. When $s \ge 2$ apply Corollary 2.6. It is sufficient to show that for every k the class of functions

$$J_k = \left\{ \sum_{i_k} \beta_{i_k} \Psi \left(A_{i_k} \left(\sum_{i_{k-1}} \beta_{i_k i_{k-1}} \times \Psi \left(\ldots \left(\sum_{i_k} \beta_{i_k \ldots i_1} \Psi(A_{i_k \ldots i_1}(x)) \right) \ldots \right) \right) \right) \right\}$$

is uniformly dense on compacta in C^r.

Lemma A.5 proves that this is true when $k = 1$. Induction on k will complete the proof.

Suppose J_k is uniformly dense on compacta in C^r. We must show that J_{k+1} is uniformly dense on compacta in C^r, $J_{k+1} = \{\Sigma_j \beta_j \Psi(A_j(g_j(x))) : g_j \in J_k\}$. Lemma A.5 says that the class of functions $\{\Sigma_j \beta_j \Psi(A_j(\cdot))\}$ is uniformly dense on compacta in C^1. Lemma A.6 and the induction hypothesis complete the proof. □

REFERENCES

Billingsley, P. (1979). *Probability and measure*. New York: Wiley.

Cybenko, G. (1988). *Approximation by superpositions of a sigmoidal function* (Tech. Rep. No. 856). Urbana, IL: University of Illinois Urbana-Champaign Department of Electrical and Computer Engineering.

Dugundji, J. (1966). *Topology*. Boston: Allyn and Bacon, Inc.

Gallant, A. R., & White, H. (1988). There exists a neural network that does not make avoidable mistables. In *IEEE Second International Conference on Neural Networks* (pp. I:657–664). San Diego: SOS Printing.

Grenander, U. (1981). *Abstract inference*. New York: Wiley.

Halmos, P. R. (1974). *Measure theory*. New York: Springer-Verlag.

Hecht-Nielsen, R. (1987). Kolmogorov's mapping neural network existence theorem. In *IEEE First International Conference on Neural Networks* (pp. III:11–14). San Diego: SOS Printing.

Hecht-Nielsen, R. (1989). Theory of the back propagation neural network. In *Proceedings of the International Joint Conference on Neural Networks* (pp. I:593–608). San Diego: SOS Printing.

Hornik, K., Stinchcombe, M., & White, H. (1988). Multilayer feedforward networks are universal approximators (Discussion Paper 88-45). San Diego, CA: Department of Economics, University of California, San Diego.

IEEE First International Conference on Neural Networks (1987). M. Caudill and C. Butler (Eds.). San Diego: SOS Printing.

IEEE Second International Conference on Neural Networks (1988). San Diego: SOS Printing.

Irie, B., & Miyake, S. (1988). Capabilities of three layer perceptrons. In *IEEE Second International Conference on Neural Networks* (pp. I:641–648). San Diego: SOS Printing.

Kolmogorov, A. N. (1957). On the representation of continuous functions of many variables by superposition of continuous functions of one variable and addition. *Doklady Akademii Nauk SSR*, **114**, 953–956.

Kolmogorov, A. N., & Tihomirov, V. M. (1961). ϵ-entropy and ϵ-capacity of sets in functional spaces. *American Mathematical Society Translations*, **2(17)**, 277–364.

Lapedes, A., & Farber, R. (1988). How neural networks work (Tech. Rep. LA-UR-88-418). Los Alamos, NM: Los Alamos National Laboratory.

le Cun, Y. (1987). *Modeles connexionistes de l'apprentissage*. These de Doctorat, Universite Pierre et Marie Curie.

Lorentz, G. G. (1976). The thirteenth problem of Hilbert. In F. E. Browder (Ed.), *Proceedings of Symposia in Pure Mathematics* (Vol. 28, pp. 419–430). Providence, RI: American Mathematical Society.

Maxwell, T., Giles, G. L., Lee, Y. C., & Chen, H. H. (1986). Nonlinear dynamics of artificial neural systems. In J. Denker (Ed.), *Neural networks for computing*. New York: American Institute of Physics.

Minsky, M., & Papert, S. (1969). *Perceptrons*. Cambridge: MIT Press.

Rudin, W. (1964). *Principles of mathematical analysis*. New York: McGraw-Hill.

Severini, J. A., & Wong, W. H. (1987). *Convergence rates of maximum likelihood and related estimates in general parameter spaces* (Working Paper). Chicago, IL: University of Chicago Department of Statistics.

Stinchcombe, M., & White, H. (1989). Universal approximation using feedforward networks with non-sigmoid hidden layer activation functions. In *Proceedings of the International Joint Conference on Neural Networks* (pp. I:613–618). San Diego: SOS Printing.

White, H. (1988a). The case for conceptual and operational separation of network architectures and learning mechanisms (Discussion Paper 88-21). San Diego, CA: Department of Economics, University of California, San Diego.

White, H. (1988b). Multilayer feedforward networks can learn arbitrary mappings: Connectionist nonparametric regression with automatic and semi-automatic determination of network complexity (Discussion Paper). San Diego, CA: Department of Economics, University of California, San Diego.

White, H., & Wooldridge, J. M. (in press). Some results for sieve estimation with dependent observations. In W. Barnett, J. Powell, & G. Tauchen (Eds.), *Nonparametric and semi-parametric methods in econometrics and statistics*. New York: Cambridge University Press.

Williams, R. J. (1986). The logic of activation functions. In D. E. Rumelhart & J. L. McClelland (Eds.), *Parallel distributed processing: Explorations in the microstructures of cognition* (Vol. 1, pp. 423–443). Cambridge: MIT Press.

Approximation by Superpositions of a Sigmoidal Function*

G. Cybenko†

Abstract. In this paper we demonstrate that finite linear combinations of compositions of a fixed, univariate function and a set of affine functionals can uniformly approximate any continuous function of n real variables with support in the unit hypercube; only mild conditions are imposed on the univariate function. Our results settle an open question about representability in the class of single hidden layer neural networks. In particular, we show that arbitrary decision regions can be arbitrarily well approximated by continuous feedforward neural networks with only a single internal, hidden layer and any continuous sigmoidal nonlinearity. The paper discusses approximation properties of other possible types of nonlinearities that might be implemented by artificial neural networks.

Key words. Neural networks, Approximation, Completeness.

1. Introduction

A number of diverse application areas are concerned with the representation of general functions of an n-dimensional real variable, $x \in \mathbb{R}^n$, by finite linear combinations of the form

$$\sum_{j=1}^{N} \alpha_j \sigma(y_j^{\mathrm{T}} x + \theta_j), \tag{1}$$

where $y_j \in \mathbb{R}^n$ and $\alpha_j, \theta \in \mathbb{R}$ are fixed. (y^{T} is the transpose of y so that $y^{\mathrm{T}}x$ is the inner product of y and x.) Here the univariate function σ depends heavily on the context of the application. Our major concern is with so-called sigmoidal σ's:

$$\sigma(t) \to \begin{cases} 1 & \text{as} \quad t \to +\infty, \\ 0 & \text{as} \quad t \to -\infty. \end{cases}$$

Such functions arise naturally in neural network theory as the activation function of a neural node (or *unit* as is becoming the preferred term) [L1], [RHM]. The main result of this paper is a demonstration of the fact that sums of the form (1) are dense in the space of continuous functions on the unit cube if σ is any continuous sigmoidal function. This case is discussed in the most detail, but we state general conditions on other possible σ's that guarantee similar results.

The possible use of artificial neural networks in signal processing and control applications has generated considerable attention recently [B], [G]. Loosely speaking, an artificial neural network is formed from compositions and superpositions of a single, simple nonlinear activation or response function. Accordingly, the output of the network is the value of the function that results from that particular composition and superposition of the nonlinearities. In particular, the simplest nontrivial class of networks are those with one internal layer and they implement the class of functions given by (1). In applications such as pattern classification [L1] and nonlinear prediction of time series [LF], for example, the goal is to select the compositions and superpositions appropriately so that desired network responses

* Date received: October 21, 1988. Date revised: February 17, 1989. This research was supported in part by NSF Grant DCR-8619103, ONR Contract N000-86-G-0202 and DOE Grant DE-FG02-85ER25001.

† Center for Supercomputing Research and Development and Department of Electrical and Computer Engineering, University of Illinois, Urbana, Illinois 61801, U.S.A.

Reprinted with permission from *Mathematics of Control, Signals, and Systems*, vol. 2, pp. 303–314, 1989.

(meant to implement a classifying function or nonlinear predictor, respectively) are achieved.

This leads to the problem of identifying the classes of functions that can be effectively realized by artificial neural networks. Similar problems are quite familiar and well studied in circuit theory and filter design where simple nonlinear devices are used to synthesize or approximate desired transfer functions. Thus, for example, a fundamental result in digital signal processing is the fact that digital filters made from unit delays and constant multipliers can approximate any continuous transfer function arbitrarily well. In this sense, the main result of this paper demonstrates that networks with only one internal layer and an arbitrary continuous sigmoidal nonlinearity enjoy the same kind of universality.

Requiring that finite linear combinations such as (1) exactly represent a given continuous function is asking for too much. In a well-known resolution of Hilbert's 13th problem, Kolmogorov showed that all continuous functions of n variables have an exact representation in terms of finite superpositions and compositions of a small number of functions of one variable [K], [L2]. However, the Kolmogorov representation involves different nonlinear functions. The issue of exact representability has been further explored in [DS] in the context of projection pursuit methods for statistical data analysis [H].

Our interest is in finite linear combinations involving the *same* univariate function. Moreover, we settle for approximations as opposed to exact representations. It is easy to see that in this light, (1) merely generalizes approximations by finite Fourier series. The mathematical tools for demonstrating such completeness properties typically fall into two categories: those that involve algebras of functions (leading to Stone–Weierstrass arguments [A]) and those that involve translation invariant subspaces (leading to Tauberian theorems [R2]). We give examples of each of these cases in this paper.

Our main result settles a long-standing question about the exact class of decision regions that continuous valued, single hidden layer neural networks can implement. Some recent discussions of this question are in [HL1], [HL2], [MSJ], and [WL] while [N] contains one of the early rigorous analyses. In [N] Nilsson showed that any set of M points can be partitioned into two arbitrary subsets by a network with one internal layer. There has been growing evidence through examples and special cases that such networks can implement more general decision regions but a general theory has been missing. In [MSJ] Makhoul *et al.* have made a detailed geometric analysis of some of the decisions regions that can be constructed exactly with a single layer. By contrast, our work here shows that *any* collection of compact, disjoint subsets of $\mathbb{R}^n$ can be discriminated with arbitrary precision. That result is contained in Theorem 3 and the subsequent discussion below.

A number of other current works are devoted to the same kinds of questions addressed in this paper. In [HSW] Hornik *et al.* show that monotonic sigmoidal functions in networks with single layers are complete in the space of continuous functions. Carroll and Dickinson [CD] show that the completeness property can be demonstrated constructively by using Radon transform ideas. Jones [J] outlines a simple constructive demonstration of completeness for arbitrary bounded sigmoidal functions. Funahashi [F] has given a demonstration involving Fourier analysis and Paley–Wiener theory. In earlier work [C], we gave a constructive mathematical proof of the fact that continuous neural networks with two hidden layers can approximate arbitrary continuous functions.

The main techniques that we use are drawn from standard functional analysis. The proof of the main theorem goes as follows. We start by noting that finite summations of the form (1) determine a subspace in the space of all continuous functions on the unit hypercube of $\mathbb{R}^n$. Using the Hahn–Banach and Riesz Representation Theorems, we show that the subspace is annihilated by a finite measure. The measure must also annihilate every term in (1) and this leads to the necessary conditions on σ. All the basic functional analysis that we use can be found in [A], [R2] for example.

The organization of this paper is as follows. In Section 2 we deal with preliminaries, state, and prove the major result of the paper. Most of the technical details of this paper are in Section 2. In Section 3 we specialize to the case of interest in neural network theory and develop the consequences. Section 4 is a discussion of other types of functions, σ, that lead to similar results while Section 5 is a discussion and summary.

2. Main Results

Let I_n denote the n-dimensional unit cube, $[0, 1]^n$. The space of continuous functions on I_n is denoted by $C(I_n)$ and we use $\|f\|$ to denote the supremum (or uniform) norm of an $f \in C(I_n)$. In general we use $\|\cdot\|$ to denote the maximum of a function on its domain. The space of finite, signed regular Borel measures on I_n is denoted by $M(I_n)$. See [R2] for a presentation of these and other functional analysis constructions that we use.

The main goal of this paper is to investigate conditions under which sums of the form

$$G(x) = \sum_{j=1}^{N} \alpha_j \sigma(y_j^T x + \theta_j)$$

are dense in $C(I_n)$ with respect to the supremum norm.

Definition. We say that σ is *discriminatory* if for a measure $\mu \in M(I_n)$

$$\int_{I_n} \sigma(y^T x + \theta)\, d\mu(x) = 0$$

for all $y \in \mathbb{R}^n$ and $\theta \in \mathbb{R}$ implies that $\mu = 0$.

Definition. We say that σ is *sigmoidal* if

$$\sigma(t) \to \begin{cases} 1 & \text{as} \quad t \to +\infty, \\ 0 & \text{as} \quad t \to -\infty. \end{cases}$$

Theorem 1. *Let σ be any continuous discriminatory function. Then finite sums of the form*

$$G(x) = \sum_{j=1}^{N} \alpha_j \sigma(y_j^T x + \theta_j) \tag{2}$$

are dense in $C(I_n)$. In other words, given any $f \in C(I_n)$ and $\varepsilon > 0$, there is a sum, $G(x)$, of the above form, for which

$$|G(x) - f(x)| < \varepsilon \qquad \textit{for all} \quad x \in I_n.$$

Proof. Let $S \subset C(I_n)$ be the set of functions of the form $G(x)$ as in (2). Clearly S is a linear subspace of $C(I_n)$. We claim that the closure of S is all of $C(I_n)$.

Assume that the closure of S is not all of $C(I_n)$. Then the closure of S, say R, is a closed proper subspace of $C(I_n)$. By the Hahn–Banach theorem, there is a bounded linear functional on $C(I_n)$, call it L, with the property that $L \neq 0$ but $L(R) = L(S) = 0$.

By the Riesz Representation Theorem, this bounded linear functional, L, is of the form

$$L(h) = \int_{I_n} h(x)\, d\mu(x)$$

for some $\mu \in M(I_n)$, for all $h \in C(I_n)$. In particular, since $\sigma(y^T x + \theta)$ is in R for all y and θ, we must have that

$$\int_{I_n} \sigma(y^T x + \theta)\, d\mu(x) = 0$$

for all y and θ.

However, we assumed that σ was discriminatory so that this condition implies that $\mu = 0$ contradicting our assumption. Hence, the subspace S must be dense in $C(I_n)$. ■

This demonstrates that sums of the form

$$G(x) = \sum_{j=1}^{N} \alpha_j \sigma(y_j^T x + \theta_j)$$

are dense in $C(I_n)$ providing that σ is continuous and discriminatory. The argument used was quite general and can be applied in other cases as discussed in Section 4. Now, we specialize this result to show that any continuous sigmoidal σ of the form discussed before, namely

$$\sigma(t) \to \begin{cases} 1 & \text{as} \quad t \to +\infty, \\ 0 & \text{as} \quad t \to -\infty, \end{cases}$$

is discriminatory. It is worth noting that, in neural network applications, continuous sigmoidal activation functions are typically taken to be monotonically increasing, but no monotonicity is required in our results.

Lemma 1. *Any bounded, measurable sigmoidal function, σ, is discriminatory. In particular, any continuous sigmoidal function is discriminatory.*

Proof. To demonstrate this, note that for any x, y, θ, φ we have

$$\sigma(\lambda(y^T x + \theta) + \varphi) \begin{cases} \to 1 & \text{for} \quad y^T x + \theta > 0 \quad \text{as} \quad \lambda \to +\infty, \\ \to 0 & \text{for} \quad y^T x + \theta < 0 \quad \text{as} \quad \lambda \to +\infty, \\ = \sigma(\varphi) & \text{for} \quad y^T x + \theta = 0 \quad \text{for all } \lambda. \end{cases}$$

Thus, the functions $\sigma_\lambda(x) = \sigma(\lambda(y^T x + \theta) + \varphi)$ converge pointwise and boundedly to the function

$$\gamma(x) \begin{cases} = 1 & \text{for} \quad y^T x + \theta > 0, \\ = 0 & \text{for} \quad y^T x + \theta < 0, \\ = \sigma(\varphi) & \text{for} \quad y^T x + \theta = 0 \end{cases}$$

as $\lambda \to +\infty$.

Let $\Pi_{y,\theta}$ be the hyperplane defined by $\{x | y^T x + \theta = 0\}$ and let $H_{y,\theta}$ be the open half-space defined by $\{x | y^T x + \theta > 0\}$. Then by the Lesbegue Bounded Convergence Theorem, we have that

$$\begin{aligned} 0 &= \int_{I_n} \sigma_\lambda(x)\, d\mu(x) \\ &= \int_{I_n} \gamma(x)\, d\mu(x) \\ &= \sigma(\varphi)\mu(\Pi_{y,\theta}) + \mu(H_{y,\theta}) \end{aligned}$$

for all φ, θ, y.

We now show that the measure of all half-planes being 0 implies that the measure μ itself must be 0. This would be trivial if μ were a positive measure but here it is not.

Fix y. For a bounded measurable functon, h, define the linear functional, F, according to

$$F(h) = \int_{I_n} h(y^T x)\, d\mu(x)$$

and note that F is a bounded functional on $L^\infty(\mathbb{R})$ since μ is a finite signed measure. Let h be the indicator funcion of the interval $[\theta, \infty)$ (that is, $h(u) = 1$ if $u \geq \theta$ and

$h(u) = 0$ if $u < \theta$) so that

$$F(h) = \int_{I_n} h(y^T x)\, d\mu(x) = \mu(\Pi_{y,-\theta}) + \mu(H_{y,-\theta}) = 0.$$

Similarly, $F(h) = 0$ if h is the indicator function of the open interval (θ, ∞). By linearity, $F(h) = 0$ for the indicator function of any interval and hence for any simple function (that is, sum of indicator functons of intervals). Since simple functons are dense in $L^\infty(\mathbb{R})$ (see p. 90 of [A]) $F = 0$.

In particular, the bounded measurable functions $s(u) = \sin(m \cdot u)$ and $c(u) = \cos(m \cdot u)$ give

$$F(s + ic) = \int_{I_n} \cos(m^T x) + i \sin(m^T x)\, d\mu(x) = \int_{I_n} \exp(im^T x)\, d\mu(x) = 0$$

for all m. Thus, the Fourier transform of μ is 0 and so μ must be zero as well [R2, p. 176]. Hence, σ is discriminatory. ■

3. Application to Artificial Neural Networks

In this section we apply the previous results to the case of most interest in neural network theory. A straightforward combination of Theorem 1 and Lemma 1 shows that networks with one internal layer and an arbitrary continuous sigmoidal function can approximate continuous functions wtih arbitrary precision providing that no constraints are placed on the number of nodes or the size of the weights. This is Theorem 2 below. The consequences of that result for the approximation of decision functions for general decision regions is made afterwards.

Theorem 2. *Let σ be any continuous sigmoidal function. Then finite sums of the form*

$$G(x) = \sum_{j=1}^{N} \alpha_j \sigma(y_j^T x + \theta_j)$$

are dense in $C(I_n)$. In other words, given any $f \in C(I_n)$ and $\varepsilon > 0$, there is a sum, $G(x)$, of the above form, for which

$$|G(x) - f(x)| < \varepsilon \qquad \textit{for all} \quad x \in I_n.$$

Proof. Combine Theorem 1 and Lemma 1, noting that continuous sigmoidals satisfy the conditions of that lemma. ■

We now demonstrate the implications of these results in the context of decision regions. Let m denote Lesbegue measure in I_n. Let $P_1, P_2, \ldots, P_k$ be a partition of I_n into k disjoint, measurable subsets of I_n. Define the decision function, f, according to

$$f(x) = j \qquad \text{if and only if} \quad x \in P_j.$$

This function f can be viewed as a decision function for classification: if $f(x) = j$, then we know that $x \in P_j$ and we can classify x accordingly. The issue is whether such a decision function can be implemented by a network with a single internal layer.

We have the following fundamental result.

Theorem 3. *Let σ be a continuous sigmoidal function. Let f be the decision function for any finite measurable partition of I_n. For any $\varepsilon > 0$, there is a finite sum of the form*

$$G(x) = \sum_{j=1}^{N} \alpha_j \sigma(y_j^T x + \theta_j)$$

and a set $D \subset I_n$, so that $m(D) \geq 1 - \varepsilon$ and

$$|G(x) - f(x)| < \varepsilon \qquad \textit{for} \quad x \in D.$$

Proof. By Lusin's theorem [R1], there is a continuous function, h, and a set D with $m(D) \geq 1 - \varepsilon$ so that $h(x) = f(x)$ for $x \in D$. Now h is continuous and so, by Theorem 2, we can find a summation of the form of G above to satisfy $|G(x) - h(x)| < \varepsilon$ for all $x \in I_n$. Then for $x \in D$, we have

$$|G(x) - f(x)| = |G(x) - h(x)| < \varepsilon. \qquad \blacksquare$$

Because of continuity, we are always in the position of having to make some incorrect decisions about some points. This result states that the total measure of the incorrectly classified points can be made arbitrarily small. In light of this, Thoerem 2 appears to be the strongest possible result of its kind.

We can develop this approximation idea a bit more by considering the decision problem for a single closed set $D \subset I_n$. Then $f(x) = 1$ if $x \in D$ and $f(x) = 0$ otherwise; f is the indicator function of the set $D \subset I_n$. Suppose we wish to find a summation of the form (1) to approximate this decision function. Let

$$\Delta(x, D) = \min\{|x - y|, y \in D\}$$

so that $\Delta(x, D)$ is a continuous function of x. Now set

$$f_\varepsilon(x) = \max\left\{0, \frac{\varepsilon - \Delta(x, D)}{\varepsilon}\right\}$$

so that $f_\varepsilon(x) = 0$ for points x farther than ε away from D while $f_\varepsilon(x) = 1$ for $x \in D$. Moreover, $f_\varepsilon(x)$ is continuous in x.

By Theorem 2, find a $G(x)$ as in (1) so that $|G(x) - f_\varepsilon(x)| < \frac{1}{2}$ and use this G as an approximate decision function: $G(x) < \frac{1}{2}$ guesses that $x \in D^C$ while $G(x) \geq \frac{1}{2}$ guesses that $x \in D$. This decision procedure is correct for all $x \in D$ and for all x at a distance at least ε away from D. If x is within ε distance of D, its classification depends on the particular choice of $G(x)$.

These observations say that points sufficiently far away from and points inside the closed decision region can be classified correctly. In contrast, Theorem 3 says that there is a network that makes the measure of points incorrectly classified as small as desired but does not guarantee their location.

4. Results for Other Activation Functions

In this section we discuss other classes of activation functions that have approximation properties similar to the ones enjoyed by continuous sigmoidals. Since these other examples are of somewhat less practical interest, we only sketch the corresponding proofs.

There is considerable interest in discontinuous sigmoidal functions such as hard limiters ($\sigma(x) = 1$ for $x \geq 0$ and $\sigma(x) = 0$ for $x < 0$). Discontinuous sigmoidal functions are not used as often as continuous ones (because of the lack of good training algorithms) but they are of theoretical interest because of their close relationship to classical perceptrons and Gamba networks [MP].

Assume that σ is a bounded, measurable sigmoidal function. We have an analog of Theorem 2 that goes as follows:

Theorem 4. *Let σ be bounded measurable sigmoidal function. Then finite sums of the form*

$$G(x) = \sum_{j=1}^{N} \alpha_j \sigma(y_j^T x + \theta_j)$$

are dense in $L^1(I_n)$. In other words, given any $f \in L^1(I_n)$ and $\varepsilon > 0$, there is a sum, $G(x)$, of the above form for which

$$\|G - f\|_{L^1} = \int_{I_n} |G(x) - f(x)|\, dx < \varepsilon.$$

The proof follows the proof of Theorems 1 and 2 with obvious changes such as replacing continuous functions by integrable functions and using the fact that $L^\infty(I_n)$ is the dual of $L^1(I_n)$. The notion of being discriminatory accordingly changes to the following: for $h \in L^\infty(I_n)$ the condition that

$$\int_{I_n} \sigma(y^{\mathrm{T}}x + \theta)h(x)\, dx = 0$$

for all y and θ implies that $h(x) = 0$ almost everywhere. General sigmoidal functions are discriminatory in this sense as already seen in Lemma 1 because measures of the form $h(x)\, dx$ belong to $M(I_n)$.

Since convergence in L^1 implies convergence in measure [A], we have an analog of Theorem 3 that goes as follows:

Theorem 5. *Let σ be a general sigmoidal function. Let f be the decision function for any finite measurable partition of I_n. For any $\varepsilon > 0$, there is a finite sum of the form*

$$G(x) = \sum_{j=1}^{N} \alpha_j \sigma(y_j^{\mathrm{T}}x + \theta_j)$$

and a set $D \subset I_n$, so that $m(D) \geq 1 - \varepsilon$ and

$$|G(x) - f(x)| < \varepsilon \qquad \textit{for} \quad x \in D.$$

A number of other possible activation functions can be shown to have approximation properties similar to those in Theorem 1 by simple use of the Stone–Weierstrass theorem [A]. Those include the *sine* and *cosine* functions since linear combinations of $\sin(mt)$ and $\cos(mt)$ generate all finite trigonometric polynomials which are classically known to be complete in $C(I_n)$. Interestingly, the completeness of trigonometric polynomials was implicitly used in Lemma 1 when the Fourier transform's one-to-one mapping property (on distributions) was used. Another classical example is that of exponential functions, $\exp(mt)$, and the proof again follows from direct application of the Stone–Weierstrass theorem. Exponential activation functions were studied by Palm in [P] where their completeness was shown.

A whole other class of possible activation functions have completeness properties in $L^1(I_n)$ as a result of the Wiener Tauberian theorem [R2]. For example, suppose that σ is any $L^1(\mathbb{R})$ function with nonzero integral. Then summations of the form (1) are dense in $L^1(\mathbb{R}^n)$ as the following outline shows.

The analog of Theorem 1 carries through but we change $C(I_n)$ to $L^1(I_n)$ and $M(I_n)$ to the corresponding dual space $L^\infty(I_n)$. The analog of Theorem 3 holds if we can show that an integrable σ with nonzero integral is discriminatory in the sense that

$$\int_{I_n} \sigma(y^{\mathrm{T}}x + \theta)h(x)\, dx = 0 \tag{3}$$

for all y and θ implies that $h = 0$.

To do this we proceed as follows. As in Lemma 1, define the bounded linear functional, F, on $L^1(\mathbb{R})$ by

$$F(g) = \int_{I_n} g(y^{\mathrm{T}}x)h(x)\, dx.$$

(Note that the integral exists since it is over I_n and h is bounded. Specifically, if $g \in L^1(\mathbb{R})$, then $g(y^{\mathrm{T}}x) \in L^1(I_n)$ for any y.)

Letting $g_{\theta,s}(t) = \sigma(st + \theta)$, we see that

$$F(g_{\theta,s}) = \int_{I_n} \sigma((sy)^{\mathrm{T}}x + \theta)h(x)\, dx = 0$$

so that F annihilates every translation and scaling of $g_{0,1}$. Let $\hat{f}$ be the Fourier transform of f. By standard Fourier transform arguments, $\hat{g}_{\theta,s}(z) = \exp(\mathbf{i}z\theta/s)\hat{g}(z/s)/s$. Because of the scaling by s, the only z for which the Fourier transforms of all the $g_{\theta,s}$ can vanish is $z = 0$ but we are assuming that $\int_{\mathbb{R}} \sigma(t)\,dt = \hat{g}_{0,1}(0) \neq 0$. By the Wiener Tauberian theorem [R2], the subspace generated by the functions $g_{\theta,s}$ is dense in $L^1(\mathbb{R})$. Since $F(g_{\theta,s}) = 0$ we must have that $F = 0$. Again, this implies that

$$F(\exp(imt)) = \int_{I_n} \exp(imt)h(t)\,dt = 0$$

for all m and so the Fourier transform of h is 0. Thus h itself is 0. (Note that although the exponential function is not integrable over all of $\mathbb{R}$, it is integrable over bounded regions and since h has support in I_n, that is sufficient.)

The use of the Wiener Tauberian theorem leads to some other rather curious activation functions that have the completeness property in $L^1(I_n)$. Consider the following activation function of n variables: $\sigma(x) = 1$ if x lies inside a finite fixed rectangle with sides parallel to the axes in $\mathbb{R}^n$ and zero otherwise. Let U be an $n \times n$ orthogonal matrix and $y \in \mathbb{R}^n$. Now $\sigma(Ux + y)$ is the indicator funciton of an arbitrarily oriented rectangle. Notice that *no* scaling of the rectangle is allowed—only rigid-body motions in Euclidean space! We then have that summations of the form

$$\sum_{j=1}^{N} \alpha_j \sigma(U_j x + y_j)$$

are dense in $L^1(\mathbb{R}^n)$. This follows from direct application of the Wiener Tauberian theorem [R2] and the observation that the Fourier transform of σ vanishes on a mesh in $\mathbb{R}^n$ that does not include the origin. The intersection of all possible rotations of those meshes is empty and so σ together with its rotations and translations generates a space dense in $L^1(\mathbb{R}^n)$.

This last result is closely related to the classical *Pompeiu Problem* [BST] and using the results of [BST] we speculate that the rectangle in the above paragraph can be replaced by any convex set with a corner as defined in [BST].

5. Summary

We have demonstrated that finite superpositions of a fixed, univariate function that is *discriminatory* can uniformly approximate any continuous function of n real variables with support in the unit hypercube. Continuous sigmoidal functions of the type commonly used in real-valued neural network theory are discriminatory.

This combination of results demonstrates that any continuous function can be uniformly approximated by a continuous neural network having only one internal, hidden layer and with an arbitrary continuous sigmoidal nonlinearity (Theorem 2). Theorem 3 and the subsequent discussion show in a precise way that arbitrary decision functions can be arbitrarily well approximated by a neural network with one internal layer and a continuous sigmoidal nonlinearity.

Table 1 summarizes the various contributions of which we are aware.

While the approximating properties we have described are quite powerful, we have focused only on existence. The important questions that remain to be answered deal with feasibility, namely how many terms in the summation (or equivalently, how many neural nodes) are required to yield an approximation of a given quality? What properties of the function being approximated play a role in determining the number of terms? At this point, we can only say that we suspect quite strongly that the overwhelming majority of approximation problems will require astronomical numbers of terms. This feeling is based on the *curse of dimensionality* that plagues multidimensional approximation theory and statistics. Some recent progress concerned with the relationship between a function being approximated and the number of terms needed for a suitable approximation can be found in [MSJ] and [BH],

Table 1

Function type and transformations	Function space	References
$\sigma(y^T x + \theta)$, σ continuous sigmoidal, $y \in \mathbb{R}^n$, $\theta \in \mathbb{R}$	$C(I_n)$	This paper
$\sigma(y^T x + \theta)$, σ monotonic sigmoidal, $y \in \mathbb{R}^n$, $\theta \in \mathbb{R}$	$C(I_n)$	[F], [HSW]
$\sigma(y^T x + \theta)$, σ sigmoidal, $y \in \mathbb{R}^n$, $\theta \in \mathbb{R}$	$C(I_n)$	[J]
$\sigma(y^T x + \theta)$, $\sigma \in L^1(\mathbb{R})$ $\int \sigma(t)\, dt \neq 0$, $y \in \mathbb{R}^n$, $\theta \in \mathbb{R}$	$L^1(I_n)$	This paper
$\sigma(y^T x + \theta)$, σ continuous sigmoidal, $y \in \mathbb{R}^n$, $\theta \in \mathbb{R}$	$L^2(I_n)$	[CD]
$\sigma(Ux + y)$, $U \in \mathbb{R}^{n \times n}$, $y \in \mathbb{R}^n$, σ indicator of a rectangle	$L^1(I_n)$	This paper
$\sigma(tx + y)$, $t \in \mathbb{R}$, $\sigma \in L^1(\mathbb{R}^n)$ $y \in \mathbb{R}^n$, $\int \sigma(x)\, dx \neq 0$	$L^1(\mathbb{R}^n)$	Wiener Tauberian theorem [R2]

[BEHW], and [V] for related problems. Given the conciseness of the results of this paper, we believe that these avenues of research deserve more attention.

Acknowledgments. The author thanks Brad Dickinson, Christopher Chase, Lee Jones, Todd Quinto, Lee Rubel, John Makhoul, Alex Samarov, Richard Lippmann, and the anonymous referees for comments, additional references, and improvements in the presentation of this material.

References

[A] R. B. Ash, *Real Analysis and Probability*, Academic Press, New York, 1972.

[BH] E. Baum and D. Haussler, What size net gives valid generalization?, *Neural Comput.* (to appear).

[B] B. Bavarian (ed.), Special section on neural networks for systems and control, *IEEE Control Systems Mag.*, **8** (April 1988), 3–31.

[BEHW] A. Blumer, A. Ehrenfeucht, D. Haussler, and M. K. Warmuth, Classifying learnable geometric concepts with the Vapnik–Chervonenkis dimension, *Proceedings of the* 18*th Annual ACM Symposium on Theory of Computing*, Berkeley, CA, 1986, pp. 273–282.

[BST] L. Brown, B. Schreiber, and B. A. Taylor, Spectral synthesis and the Pompeiu problem, *Ann. Inst. Fourier* (*Grenoble*), **23** (1973), 125–154.

[CD] S. M. Carroll and B. W. Dickinson, Construction of neural nets using the Radon transform, preprint, 1989.

[C] G. Cybenko, Continuous Valued Neural Networks with Two Hidden Layers are Sufficient, Technical Report, Department of Computer Science, Tufts University, 1988.

[DS] P. Diaconis and M. Shahshahani, On nonlinear functions of linear combinations, *SIAM J. Sci. Statist. Comput.*, **5** (1984), 175–191.

[F] K. Funahashi, On the approximate realization of continuous mappings by neural networks, *Neural Networks* (to appear).

[G] L. J. Griffiths (ed.), Special section on neural networks, *IEEE Trans. Acoust. Speech Signal Process.*, **36** (1988), 1107–1190.

[HSW] K. Hornik, M. Stinchcombe, and H. White, Multi-layer feedforward networks are universal approximators, preprint, 1988.

[HL1] W. Y. Huang and R. P. Lippmann, Comparisons Between Neural Net and Conventional Classifiers, Technical Report, Lincoln Laboratory, MIT, 1987.

[HL2] W. Y. Huang and R.P. Lippmann, Neural Net and Traditional Classifiers, Technical Report, Lincoln Laboratory, MIT, 1987.

[H] P. J. Huber, Projection pursuit, *Ann. Statist.*, **13** (1985), 435–475.

[J] L. K. Jones, Constructive approximations for neural networks by sigmoidal functions, Technical Report Series, No. 7, Department of Mathematics, University of Lowell, 1988.

[K] A. N. Kolmogorov, On the representation of continuous functions of many variables by superposition of continuous functions of one variable and addition, *Dokl. Akad. Nauk. SSSR*, **114** (1957), 953–956.

[LF] A. Lapedes and R. Farber, Nonlinear Signal Processing Using Neural Networks: Prediction and System Modeling, Technical Report, Theoretical Division, Los Alamos National Laboratory, 1987.

[L1] R. P. Lippmann, An introduction to computing with neural nets, *IEEE ASSP Mag.*, **4** (April 1987), 4–22.

[L2] G. G. Lorentz, The 13th problem of Hilbert, in *Mathematical Developments Arising from Hilbert's Problems* (F. Browder, ed.), vol. 2, pp. 419–430, American Mathematical Society, Providence, RI, 1976.

[MSJ] J. Makhoul, R. Schwartz, and A. El-Jaroudi, Classification capabilities of two-layer neural nets. *Proceedings of the IEEE International Conference on Acoustics, Speech, and Signal Processing*, Glasgow, 1989 (to appear).

[MP] M. Minsky and S. Papert, *Perceptrons*, MIT Press, Cambridge, MA, 1969.

[N] N. J. Nilsson, *Learning Machines*, McGraw-Hill, New York, 1965.

[P] G. Palm, On representation and approximation of nonlinear systems, Part II: Discrete systems, *Biol. Cybernet.*, **34** (1979), 49–52.

[R1] W. Rudin, *Real and Complex Analysis*, McGraw-Hill, New York, 1966.

[R2] W. Rudin, *Functional Analysis*, McGraw-Hill, New York, 1973.

[RHM] D. E. Rumelhart, G. E. Hinton, and J. L. McClelland, A general framework for parallel distributed processing, in *Parallel Distributed Processing: Explorations in the Microstructure of Cognition* (D. E. Rumelhart, J. L. McClelland, and the PDP Research Group, eds.), vol. 1, pp. 45–76, MIT Press, Cambridge, MA, 1986.

[V] L. G. Valiant, A theory of the learnable, *Comm. ACM*, **27** (1984), 1134–1142.

[WL] A Wieland and R. Leighton, Geometric analysis of neural network capabilities, *Proceedings of IEEE First International Conference on Neural Networks*, San Diego, CA, pp. III-385–III-392, 1987.

The Stone–Weierstrass Theorem and Its Application to Neural Networks

NEIL E. COTTER

Abstract—**This paper describes neural network architectures based on the Stone–Weierstrass theorem. Exponential functions, polynomials, partial fractions, and Boolean functions are used to create networks capable of approximating arbitrary bounded measurable functions. A "modified logistic network," satisfying the theorem, is proposed as an alternative to commonly used networks based on logistic squashing functions.**

I. Introduction

In the past several years, researchers have shown that infinitely large neural networks with a single hidden layer are capable of approximating all but the most pathological of functions [1]–[4]. This powerful result assures us that neural networks have two desirable properties: 1) larger networks can produce less error than smaller networks; and 2) there are no nemesis functions that cannot be modeled by neural networks.

The Stone–Weierstrass theorem from classical real analysis can be used to show that certain network architectures possess the universal approximation capability. Hornik *et al.* [4] introduced this theorem as the basis of their proofs and considered existing architectures that fail to satisfy the theorem's hypotheses. In this paper, we focus instead on new architectures satisfying the theorem. This ensures that our networks have the two desirable properties listed above.

By employing the Stone–Weierstrass theorem in the design of our networks, we also guarantee that the networks can compute certain polynomial expressions: if we are given networks exactly computing two functions, f and g, then a larger network can exactly compute a polynomial expression of f and g. We will see how the larger network can be explicitly determined from the smaller ones. Our discussion begins with the theorem and a pragmatic description of some abstract terminology.

II. Stone–Weierstrass Theorem and Terminology

Theorem (Stone–Weierstrass) [5], [6]: Let domain D be a compact space of N dimensions, and let $\mathfrak{F}$ be a set of continuous real-valued functions on D, satisfying the following criteria:

1) Identity Function: The constant function $f(x) = 1$ is in $\mathfrak{F}$.

2) Separability: For any two points $x_1 \neq x_2$ in D, there is an f in $\mathfrak{F}$ such that $f(x_1) \neq f(x_2)$.

3) Algebraic Closure: If f and g are any two functions in $\mathfrak{F}$, then fg and $af + bg$ are in $\mathfrak{F}$ for any two real numbers a and b.

Then $\mathfrak{F}$ is dense in $C(D)$, the set of continuous real-valued functions on D. In other words, for any $\epsilon > 0$ and any function g in $C(D)$, there is a function f in $\mathfrak{F}$ such that $|g(x) - f(x)| < \epsilon$ for all $x \in D$.

A. Compact Spaces

In applications of neural networks, the domain on which we operate is almost always compact. It is a standard result in real analysis that every closed and bounded set in $\Re^N$ is compact [6]. The unit hypercube, that we assume for the domain from now on, is the prototypical example of such a compact space:

$$D = [0, 1]^N. \tag{1}$$

Any region obtained by smoothly distorting (but not tearing) the unit hypercube is also a suitable compact space to which the Stone–Weierstrass theorem can be applied.

B. Measurable Functions and L^p Spaces

The Stone–Weierstrass theorem states conditions that guarantee that a network can approximate continuous functions. Nevertheless, many interesting functions, including step functions, are discontinuous. These functions are members of the set of bounded measurable functions. This set includes continuous functions, bounded functions that have a countable number of discontinuities, and all other bounded functions that are likely to be encountered in practice. Nonmeasurable functions, such as white Gaussian noise, are typified by having discontinuities at uncountably many points in their domain. Such functions are abstractions, however, that are unrealizable in physical systems.

C. Convergence Almost Everywhere

The Stone–Weierstrass theorem can be extended to bounded measurable functions by applying a theorem of Lusin [6].

Theorem (Lusin): If g is a measurable real-valued function that is bounded almost everywhere on a compact domain D, then given $\delta > 0$ there is a continuous real-valued function f on D such that the measure of the set where f is not equal to g is less than δ

$$m\{x: f(x) \neq g(x)\} < \delta. \tag{2}$$

Manuscript received January 24, 1989; revised June 15, 1990. This paper was supported by the National Science Foundation under Grant EET-8810478.

The author is with the Department of Electrical Engineering, University of Utah, Salt Lake City, UT 84112.

IEEE Log Number 9038255.

Reprinted from *IEEE Trans. on Neural Networks*, vol. 1, no. 4, pp. 290–295, Dec. 1990.

In other words, the minimum total volume of open spheres required to cover the set where $f \neq g$ is less than δ.

Results from real analysis allow us to strengthen this result of convergence in measure to convergence almost everywhere, and we conclude that continuous functions can approximate bounded measurable functions.

An alternative formulation is to consider the space $L^p[D]$ consisting of all real measurable Lebesque-integrable functions whose L^p norm is finite on domain D:

$$L^p[D] = \{f: \|f\|_p < \infty\} \tag{3}$$

where the norm for $1 \leq p < \infty$ is

$$\|f\|_p \equiv \left\{\int_D |f|^p\right\}^{1/p}. \tag{4}$$

Because of their practical meanings, it is often convenient to use the L^p spaces: for $p = 1$, we obtain the set of finite area functions on domain D, and for $p = 2$, we obtain the set of finite energy functions on domain D. Because these descriptions are so easily understood, we state our theorems in terms of L^p spaces.

When we say that a set $\mathfrak{F}$ of functions is dense in $L^p[D]$, we will mean that for any f in $L^p[D]$, we can find a sequence f_n of functions in $\mathfrak{F}$ such that

$$\lim_{n\to\infty} \|f - f_n\|_p = 0. \tag{5}$$

It is always possible to find a sequence of continuous functions satisfying this condition. For such a sequence it is also possible to conclude that f_n converges to f almost everywhere.

The practical consequence of the preceding discussion is that an infinitely large neural network can model any $L^p[D]$ function at all but isolated points. A finite network, however, might only accurately model such functions over a subset of the domain.

This kind of problem also occurs with finite Fourier series, manifesting itself as Gibb's phenomenon. The 18% overshoot that occurs at step discontinuities decays away over some interval, causing significant error if too few terms are included in the Fourier series. In the limit of infinite terms, the error is confined to a set of measure zero and might be inconsequential. Users of finite neural networks might observe analogous phenomena. They may be confident, however, that errors can be reduced if they use a larger network and possess a suitable learning algorithm.

D. Advantages of Satisfying the Stone–Weierstrass Theorem

A network that satisfies the Stone–Weierstrass theorem can compute the weighted sum $af + bg$ and the product fg of functions f and g computed by smaller networks. Conversely, we can separate a polynomial expression into smaller terms that can be approximated by neural networks. These networks may be recombined to approximate the original polynomial. We shall describe the procedure for constructing a network that computes sums, and the proof of Theorem 1 doubles as a procedure for constructing a network that computes products. With these procedures, we can explicitly determine synaptic weights for a network approximating a polynomial.

In comparison, networks employing sigmoidal squashing functions, including the popular logistic [7] and hyperbolic tangent functions [8], are unable to exactly compute a product of sigmoids. Using logistic squashing functions, for example, we can trivially calculate logistics along each of two axes

$$f(x_1, x_2) = [1 + \exp(-x_1)]^{-1} \tag{6}$$

and

$$g(x_1, x_2) = [1 + \exp(-x_2)]^{-1}. \tag{7}$$

Nevertheless, no finite-sized "logistic network" can exactly compute fg. Nor does adding layers to the network resolve this problem. In this case, we find it difficult to exploit structural information about the function we wish to approximate.

III. Applying the Stone–Weierstrass Theorem

A. Generic Network Architecture

A tree structure, in which many neurons on one layer feed a single neuron on the next layer, is a generic architecture for networks satisfying the Stone–Weierstrass theorem. An example of a tree-structured network is shown in Fig. 1. The tree structure has one or more hidden layers followed by a linear output neuron. We assume that an arbitrarily large number of neurons are present in each hidden layer.

In hidden layers, we will employ a variety of squashing functions. Since several of the squashing functions are unbounded, we use the term "squashing function" loosely.

Before verifying that the generic architecture satisfies the Stone–Weierstrass theorem, we observe that the approximation capabilities of our networks are unaffected by various modifications to the tree structure.

1) Because tree-structured networks can be embedded in totally connected networks, our theorems extend to totally connected networks.

2) If the range of the function being approximated is appropriately bounded, we may include an invertible continuous squashing function in the output neuron. We obtain this result by observing that, prior to squashing, we have a continuous function and a linear output neuron. (For this argument to be valid, the range of the approximated function must be properly contained in the range of the squashing function.)

3) N copies of a network, placed side by side, can compute continuous N-dimensional vector-valued functions. In other words, single-output networks generalize to multiple outputs.

4) Preprocessing of inputs via any continuous invertible function does not affect the ability of an architecture to approximate continuous functions.

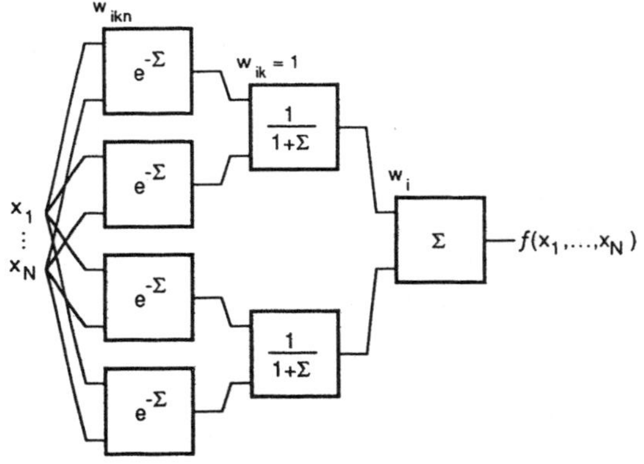

Fig. 1. Tree structured network for modified logistic network. Each box represents a neuron which computes a weighted sum of inputs and passes the result through the function shown in the box.

B. Identity Function

The first hypothesis of the Stone–Weierstrass theorem requires that our neural network be able to compute the identity function $f(x) = 1$. An obvious way to compute this function is set all synaptic weights in the network to zero except for a unitary threshold input to the last summation unit. A more seamless way to compute this function is to use a penultimate-layer squashing function whose output value is one for an input value of zero. Then setting all synaptic weights to zero results in an output value equal to one. This eliminates the need for an extraneous threshold input. With the exception of the modified logistic network, all the networks we discuss are of this type.

C. Separability

The second hypothesis of the Stone–Weierstrass theorem requires that our neural network be able to compute functions that have different values for different points. Without this requirement, the trivial set of functions $\{f: f(x) = n, n \in \Re\}$ would satisfy the Stone–Weierstrass theorem. Separability is satisfied whenever a network can compute strictly monotonic functions of each input variable. All of our networks have this capability.

D. Algebraic Closure—Additive

The third hypothesis of the Stone–Weierstrass theorem requires that our neural network be able to approximate sums and products of functions. If the network can compute either of two functions, f and g, we construct a network that computes $af + bg$ as follows: embed the networks for f and g in a larger network and, at the output neuron, scale the synapses by a and b. Since the output unit is linear, we obtain $af + bg$. The resulting network also has a tree structure as before.

E. Algebraic Closure—Multiplicative

Modeling the product fg of two functions is the last capability we must demonstrate before we can conclude that the Stone–Weierstrass theorem applies to a network. Because the output unit is assumed to be linear, we must represent fg as a sum of functions. Thus, the key to satisfying the Stone–Weierstrass theorem is to find functions that transform multiplication into addition so that products can be written as summations. There are at least three generic functions that accomplish this transformation: exponential functions, partial fractions, and step functions. We use these squashing functions in the next section to construct networks.

IV. Networks Satisfying the Stone–Weierstrass Theorem

A. Decaying-Exponential Networks

Exponential functions are basic to the process of transforming multiplication into addition in several kinds of networks:

$$\exp(x_1)\exp(x_2) = \exp(x_1 + x_2). \tag{8}$$

Decaying-exponential networks use this identity explicitly:

Theorem 1: Let $\mathfrak{F}$ be the set of all functions that can be computed by arbitrarily large decaying-exponential networks on domain $D = [0, 1]^N$:

$$\mathfrak{F} = \Bigg\{ f(x_1, \cdots, x_N) = \sum_{i=1}^{I} w_i \exp\left(-\sum_{n=1}^{N} w_{in}x_n\right): w_i, w_{in} \in \Re \Bigg\}. \tag{9}$$

Then $\mathfrak{F}$ is dense in $L^p[D]$, for $1 \le p < \infty$.

Proof: Let f and g be two functions in $\mathfrak{F}$ where f is as shown in (9) and g is as shown in (10)

$$g(x_1, \cdots, x_N) = \sum_{j=1}^{J} w_j \exp\left(-\sum_{n=1}^{N} w_{jn}x_n\right). \tag{10}$$

Then the product fg may be written as a single summation as follows:

$$fg = \sum_{\iota=1}^{IJ} w_\iota \exp\left(-\sum_{n=1}^{N} w_{\iota n}x_n\right) \tag{11}$$

where $w_\iota = w_i w_j$ and $w_{\iota n} = w_{in} + w_{jn}$.

Hence, fg is in $\mathfrak{F}$, and the network satisfies the Stone–Weierstrass theorem. It follows that $\mathfrak{F}$ is dense in $L^p[D]$, for $1 \le p < \infty$.

Corollary 1: In the decaying-exponential network, hidden-layer weights w_{in} may be restricted to integer or nonnegative integer values. (The latter case yields bounded outputs from the decaying-exponential neurons when inputs are nonnegative. This is the motivation for using negative rather than positive exponentials.) Alternatively, if a threshold input is included in the hidden layer, output layer weights w_i may be restricted to integer values. Restricting weights on both layers to integer values, however, would violate the hypothesis that the network computes all functions $af + bg$ where a and b are real numbers. Note that adding a threshold input on the first layer is equivalent to multiplying w_i by a real constant.

B. Fourier Networks

Fourier networks, introduced by Gallant and White [3], implement Fourier series in network form. Fourier series satisfy the Stone–Weierstrass theorem by the familiar trigonometric identity that transforms multiplication into addition:

$$\cos(A + B) = \cos A \cos B - \sin A \sin B. \quad (12)$$

This identity is merely a disguised form of (8) obtained from complex exponentials. Hence, Fourier networks are also based on exponentials.

A clever twist suggested by Gallant and White is to chop sinusoids into half-cycle sections and add flat tails. The result is a sigmoidal cosine squasher. A sinusoid is obtained by shifting and summing cosine squashers. We shall refer to a slightly modified version of the cosine squasher as a "cosig" function.

By showing that Fourier networks mimic Fourier series, Gallant and White proved an equivalent form of the following theorem and corollary. We note that Fourier series, and hence Fourier networks, satisfy the Stone–Weierstrass theorem. The cosig network, described by Theorem 2, computes a superset of functions computed by Fourier networks.

Theorem 2: Let $\mathfrak{F}$ be the set of all functions that can be computed by arbitrarily large cosig networks, on domain $D = [0, 1]^N$:

$$\mathfrak{F} = \left\{ f(x_1, \cdots, x_N) = \sum_{i=1}^{I} w_i \operatorname{cosig}\left(\sum_{n=1}^{N} w_{in}x_n + \theta_i \right) : w_i, w_{in}, \theta_i \in \Re \right\} \quad (13)$$

where cosig( ) is a cosine squasher

$$\operatorname{cosig}(x) = \begin{cases} 0 & x \leq -\frac{1}{2} \\ \dfrac{1 + \cos(2\pi x)}{2} & -\frac{1}{2} < x < 0 \\ 1 & x \geq 0. \end{cases} \quad (14)$$

Then $\mathfrak{F}$ is dense in $L^p[D]$, for $1 \leq p < \infty$.

Corollary 2: The Fourier network, which is a restricted form of the cosig network, has the same approximation capabilities as the cosig network. In the Fourier network, in order to yield sinusoids corresponding to a Fourier series, first layer weights w_{in} are restricted to fixed integer values, and thresholds θ_i are restricted to values of form $q/4$ where q is an integer.

An interesting feature of a cosig function is that it has zero derivative outside the interval $[-\frac{1}{2}, 0]$. Consequently, a backpropagation learning algorithm would alter only a small subset of neurons for each training pattern presented to the network. This might be advantageous in preventing the learning of new knowledge from corrupting previously stored information.

We also observe that learning in a cosig network might yield something unlike a Fourier series. The first layer might not compute sinusoidal functions. Another possibility is that the network might compute sinusoids but alter their frequencies. This scenario is equivalent to finding a representation for a function in terms of sinusoids that are not harmonically related. These variations on Fourier series might be an efficient way to use computational hardware.

C. Modified Sigma-Pi and Polynomial Networks

A third kind of network based on exponentials is the polynomial or modified sigma-pi network. We observe that powers of x provide another example of (8)

$$x^n x^m = \exp(n \ln x) \exp(m \ln x) = \exp[(n + m) \ln x] = x^{n+m}. \quad (15)$$

Rumelhart *et al.* [7] have described a sigma-pi neuron that computes sums of products of selected inputs. A single layer of such neurons is incapable of approximating a function of the form $f(x_1, \cdots, x_N) = x_n^q$, where q is large, unless the network allows inputs to be multiplied by themselves. The modified sigma-pi or polynomial network, described by the next theorem, explicitly provides the capability of approximating terms of the form $x_1^{q_1} \cdots x_n^{q_N}$ where q_i is any real number. As has been noted by Barron and Barron [9] and Hornik *et al.* [4], this kind of network satisfies the Stone–Weierstrass theorem. In the form of network presented here, synaptic weights serve as exponents for the inputs. We can achieve a similar result by placing a hidden layer of sigma (summation) neurons before the pi (product) neuron.

Theorem 3: Let $\mathfrak{F}$ be the set of all functions that can be computed by arbitrarily large modified sigma-pi networks on domain $D = [0, 1]^N$:

$$\mathfrak{F} = \left\{ f(x_1, \cdots, x_N) = \sum_{i=1}^{I} w_i \prod_{n=1}^{N} x_n^{w_{in}} : w_i, w_{in} \in \Re \right\}. \quad (16)$$

Then $\mathfrak{F}$ is dense in $L^p[D]$, for $1 \leq p < \infty$.

Corollary 3: In the modified sigma-pi or polynomial network, the synaptic weights w_{in} in the hidden layer may be restricted to integer or nonnegative integer values. A standard polynomial network is obtained in this way.

D. Exponentiated-Function Networks

A fourth kind of network based on exponentials is the exponentiated-function network. This network is obtained by preprocessing inputs to a modified sigma-pi network. If g is the preprocessing function, the first layer of the sigma-pi network computes polynomial functions of the form $g(x_1)^{n_1} \cdots g(x_N)^{n_N}$.

Theorem 4: Let $\mathfrak{F}$ be the set of all functions that can be computed by arbitrarily large exponentiated-function networks, on domain $D = [0, 1]^N$:

$$\mathfrak{F} = \left\{ f(x_1, \cdots, x_N) = \sum_{i=1}^{I} w_i \prod_{n=1}^{N} g(x_n)^{w_{in}} : g \in C[0, 1] \text{ invertible}, \; w_i, w_{in} \in \Re \right\}. \tag{17}$$

Then $\mathfrak{F}$ is dense in $L^p[D]$, for $1 \leq p < \infty$.

Corollary 4: In the exponential-function network, the synaptic weights w_{in} in the hidden layer may be restricted to integer or nonnegative integer values.

We note that if f is a logistic function or similar sigmoid, then f^n is also a sigmoid. Thus, if we wish to compute a single-variable function, we may use a network of neurons with special sigmoidal squashing functions and satisfy the Stone–Weierstrass theorem.

E. Partial Fraction Networks

Partial fractions are an example of nonexponential functions that translate multiplication into addition

$$\left(\frac{1}{1 + w_1 x}\right)\left(\frac{1}{1 + w_2 x}\right) = \left(\frac{w_1}{w_1 - w_2}\right)\left(\frac{1}{1 + w_1 x}\right) + \left(\frac{w_2}{w_2 - w_1}\right)\left(\frac{1}{1 + w_2 x}\right). \tag{18}$$

One might suppose that a similar identity holds when $w_i x$ is replaced by $\Sigma_{n=1}^{N} w_{in} x_n$. With this change, however, the product no longer translates into a finite sum. Attempts to adapt the partial fraction method to multiple variables fail, and one may conclude that partial fractions afford a clever way of computing x^n but not $x^n x^m$. Nevertheless, we could use partial fractions in a network that computes functions of single variables

$$f(x) = \sum_{m=1}^{M} \frac{w_m}{1 + mx}. \tag{19}$$

F. Modified Logistic Network

Combining the partial fraction idea with exponentials produces a network that computes arbitrary functions of more than one variable. Furthermore, the combination of partial fractions and exponentials is similar to a logistic function. The resulting modified logistic network, shown in Fig. 1, obeys the Stone–Weierstrass theorem:

Theorem 5: Let $\mathfrak{F}$ be the set of all functions that can be computed by arbitrarily large modified logistic networks, on domain $D = [0, 1]^N$:

$$\mathfrak{F} = \left\{ f(x_1, \cdots, x_N) = \sum_{i=1}^{I} w_i \left[1 + \sum_{k=1}^{K} \exp\left(- \sum_{n=1}^{N} w_{ikn} x_n \right) \right]^{-1} : w_i, w_{ikn} \in \Re \right\}. \tag{20}$$

Then $\mathfrak{F}$ is dense in $L^p[D]$, for $1 \leq p < \infty$.

Corollary 5: In the modified logistic network, the synaptic weights w_{ikn} in the first hidden layer may be restricted to integer or nonnegative integer values. Also, threshold weights may be added in either the output layer or first hidden layer, and nonnegative real or integer weights w_{ik} may be included in the penultimate layer.

The set of functions computed by a modified logistic network (with logistic squashing function appended at the output) is a superset of the set of functions computed by a logistic network. Furthermore, the backward error propagation weight update rules for the modified logistic network are identical to those of a standard logistic network. Hence, the modified logistic network may be substituted for a logistic network when one wishes to take advantage of multiplicative closure.

G. Step Functions and Perceptron Networks

In the 1940's, McCulloch and Pitts [10] showed that a network of neurons with stepped squashing functions, later called perceptrons, could compute any Boolean logic function. Although its input may be real valued, a network of perceptrons computes only binary functions. We may, therefore, use an AND gate to compute the product of two such functions. Since a perceptron with properly chosen threshold and synaptic weights is an AND gate, we can construct a perceptron network that computes fg from networks that compute f and g.

If we add a linear output neuron to a perceptron network, we have a network that computes step functions (meaning stair-step functions with multiple steps). Lippmann [11] offers a heuristic description, paraphrased here, of how such a network with two hidden layers can approximate an arbitrary function f.

First Hidden Layer: Computes hyperplanes, each of which divides the input space into half-spaces assigned values of zero and one. The hyperplanes define boundaries of convex regions in the input space. Neurons are added to the first layer until the value of f is nearly constant over each convex region.

Second Hidden Layer: Computes AND functions that determine which convex region an input point lies in.

Output Layer: Multiplies each output from the second hidden layer by the value of f in the corresponding convex region. In other words, the output of the second hidden layer identifies the convex region, and the synaptic weight in the output layer is the value of f.

In this heuristic argument lies a proof that, if two networks can approximate f and g, we can construct a network that computes fg.

First Hidden Layer: Includes the hyperplanes for both f and g.

Second Hidden Layer: Computes AND functions for every intersection of convex regions from f and g.

Output Layer: Multiplies each output from the second hidden layer by the value of fg in the corresponding convex region.

We cannot apply the Stone–Weierstrass theorem to the aforementioned network because of a mathematical difficulty: the perceptron network computes step functions rather than continuous functions. Step functions, however, are dense in the set of measurable functions by the following lemma [6].

Lemma IV—(Step Functions): If f is an almost everywhere bounded and measurable function on a compact space, then given $\epsilon > 0$, there is a step function h such that

$$|f - h| < \epsilon \tag{21}$$

except on a set of measure less than ϵ.

Intuitively, the idea here is that any continuous function can be accurately approximated by a step function if the steps are sufficiently small. This procedure introduces yet another source of error for finite networks. Similar errors occur in digital-to-analog converters that produce stair-step approximations of bandlimited functions. In this case, suitable low-pass filtering removes the error. The same procedure removes errors for a perceptron network approximating a bandlimited function.

Although the Stone–Weierstrass theorem does not apply to our network, there is an advantage to having shown that we satisfy algebraic closure: we can explicitly construct networks that compute certain polynomial expressions. This is partial compensation for the lack of a multilayer perceptron learning algorithm [12].

V. Conclusion

In the first part of this paper, we observed that errors which approach zero in the limit of large networks may be significant in finite networks. Two possible types of errors are overshoot and slow convergence. A third type of error may arise in two-layer networks: the network may be unable to efficiently approximate the product of two functions.

These difficulties are partially resolved by requiring networks to satisfy the hypotheses of the Stone–Weierstrass theorem, as do the networks presented in this paper. These networks are derived by transforming multiplication into addition. The transformation is accomplished via three types of squashing functions: exponentials, partial fractions, and step functions. Networks based on these functions satisfy the algebraic closure hypotheses of the Stone–Weierstrass theorem, allowing us to explicitly construct networks that compute polynomial expressions of functions computed by smaller networks.

A combination of partial fractions and exponentials produces the modified logistic network. This network is similar to a logistic network but satisfies the Stone–Weierstrass theorem directly. The modified logistic network is suggested as an alternative to the logistic network in situations where one wishes to compute polynomial expressions.

References

[1] G. Cybenko, "Approximations by superpositions of a sigmoidal function," *Math. Contr., Signals, Syst.*, vol. 2, pp. 303-314, 1989.

[2] K. Funahashi, "On the approximate realization of continuous mappings by neural networks," *Neural Networks*, vol. 2, pp. 183-192, 1989.

[3] A. R. Gallant and H. White, "There exists a neural network that does not make avoidable mistakes," in *Proc. IEEE Int. Conf. Neural Networks*, San Diego, CA, July 24-27, 1988, vol. I, pp. 657-664.

[4] K. Hornik, M. Stinchcombe, and H. White, "Multilayer feedforward networks are universal approximators," *Neural Networks*, vol. 2, pp. 359-366, 1989.

[5] L. V. Kantorovich and G. P. Akilov, *Functional Analysis*, 2nd ed. Oxford: Pergamon, 1982.

[6] H. L. Royden, *Real Analysis*, 2nd ed. New York: Macmillan, 1968.

[7] D. E. Rumelhart, G. E. Hinton, and R. J. Williams, *Parallel Distributed Processing: Explorations in the Microstructures of Cognition.* Cambridge, MA: M.I.T. Press, 1986.

[8] J. J. Hopfield, "Neurons with graded response have collective computational properties like those of two-state neurons," in *Proc. Nat. Acad. Sci. USA*, vol. 81, pp. 3088-3092, 1984.

[9] A. Barron and R. Barron, "Statistical learning networks: A unifying view," presented at the Symp. Interface: Statistics Comput. Sci., Reston, VA, 1988.

[10] W. S. McCulloch and W. Pitts, "A logical calculus of the ideas immanent in nervous activity," *Bull. Math. Biophys.*, vol. 5, pp. 115-133, 1943.

[11] R. P. Lippmann, "An introduction to computing with neural nets," *IEEE ASSP Mag.*, pp. 4-22, Apr. 1987.

[12] M. Minsky and S. Papert, *Perceptrons.* Cambridge, MA: M.I.T. Press, 1969.

Dynamic Neural Units and Function Approximation

D.H. Rao and M.M. Gupta

Intelligent Systems Research Laboratory, College of Engineering

University of Saskatchewan, Saskatoon, Canada, S7N 0W0

Abstract- The purpose of this paper is to develop a new *dynamic* model of neuron for applications in robotics and control. The proposed model, called the *dynamic neural unit* (DNU), comprises of two distinct operations: (i) the *synaptic operation*, which determines the optimum feedforward and feedback synaptic weights controlling the dynamics of the neuron, and (ii) the *somatic operation* which determines the optimum gain of the nonlinear activation function for a given task. The function approximation capability of the network of DNUs is described by considering linear and trigonometric operators. Finally, we present a few computer simulation results.

1. Introduction

It is true that the dynamic structure of neurons in the central nervous system (CNS) is extremely complex. However, it has been shown that even a highly simplified model of a neuron, when connected densely in a network, can do significant computations. Over the last decade or so artificial neural networks, or simply neural networks, have been used in a wide range of applications, and have been shown to perform, on a small scale, such higher cognitive functions as memory, recall, learning and adaptation.

In a conventional structure of an artificial neural network, the neuron receives its inputs from a number of other neurons and from the neural sensors. A weighted sum of these inputs constitutes the argument of a 'fixed' nonlinear activation function. The weights correspond to the synapses in a biological neuron while the activation function is analogous to its intercellular current conduction mechanism. The resulting value of the activation function forms the argument of an output function. This output is distributed with weighted connections to other processing units. This is an oversimplified but useful first approximation of a neuron [1]. The neural networks, with this neuronal model, have no feedback connections; that is, connections through weights extending from the outputs of a layer to the inputs of the same or the previous layers. Also, these networks have no memory; their output is solely determined by the current inputs and the values of the weights. The lack of feedback ensures that the networks are conditionally stable [2]. The feedback, also known as recurrent, neural networks, introduced by Hopfield [3], provide an alternative neural network model. This model consists of a single layer network included in a feedback configuration with a time delay. Because feedback networks have feedback paths from their outputs to the inputs, the response of such networks is *dynamic*. In spite of the interesting applications that the feedback neural networks have been used for, the basic architecture of the neuron is static; that is, the neuron simply provides a weighted integration of the synaptic inputs.

In order to emulate, even on a small scale, some of the dynamic and complex functions and to better reflect the dynamics of biological neurons, the neural models must incorporate dynamic elements with continuous feedback. In this paper we describe one such model, called a *dynamic neural unit* (DNU). The architectural details of DNU are presented in the next section. The function approximation is discussed in Section 3. A few function approximation examples are presented at the end of the paper.

2. Architectural Details of the Proposed Dynamic Model

The neural structure described in this paper, called the *dynamic neural unit* (DNU), is motivated by the fact that biological neuronal systems always function with a continuous feedback. One example of such a system is the *reverberating circuit* in the neuronal pool of the central nervous system (CNS) which functions as follows: An incoming signal stimulates the first neuron, which then stimulates the second, third and so forth. However, branches return to the first neuron providing feedback and re stimulates [4].

Using the topology of the reverberating circuit as a basis, the authors proposed a *dynamic neural unit* (DNU) [5] to model biological neuron. The dynamic structure of the DNU is only analogous to that of the reverberating circuit and does not represent any specific anatomical region within the biological nervous system, and output of this structure forms an argument to a time-varying nonlinear activation function. This clearly defines the two basic operations of the DNU: (i) the *synaptic operation* and (ii) the *somatic operation*. The synaptic operation involves the determination of optimum feedforward and feedback synaptic weights which determine the neural dynamic behavior, while the second operation determines the optimum gain (slope) of the nonlinear function.

Reprinted from *IEEE Conf. on Neural Networks*, San Francisco, CA, pp. 743–748, March 28–April 1, 1993.

The dynamic structure of DNU, which enables the synaptic operation, comprises of two unit delay operators with feedforward and feedback weights, $\mathbf{a}_{ff} = [a_0, a_1, a_2]$ and $\mathbf{b}_{fb} = [b_1, b_2]$ respectively. This represents a second-order structure with two poles and two zeros (complex or real). The selection of a nonlinear function, which represents an intercellular current conduction mechanism in biological neuron, needs more attention than what is presently given in neural network paradigm. This is briefly discussed in the following subsection.

2.1 Comments on nonlinear activation function

The proportion of neurons in a neural network that receive inputs greater than the threshold value is given by a nonlinear function of the total applied inputs, say v(k): k representing the discrete-time index. This nonlinear input transformation function, Ψ[v(k)], is related to the distribution of the neural thresholds, g[v(k)], within the neural unit [6]. If the probability distribution of these neural thresholds about an aggregate value θ is given by a *unimodal* distribution, then the nonlinear input transformation may be represented by a *sigmoidal* function. Thus, the proportion of neurons in a neural network receiving inputs greater than the intrinsic threshold is given by the expression [7]

$$u(k) = \Psi[v(k), g_s, \theta] = \int_{-\infty}^{v(k)} g[v(k)]\, dv(k) \qquad (1)$$

where the pair $[g_s, \theta]$ determines the transformational properties of the function Ψ[.]. The parameter g_s is defined as the maximum slope of the sigmoidal relationship at the point of inflection given by the aggregate value θ. In other words, for a particular distribution of neural thresholds, it is possible to determine the proportion of neurons receiving inputs exceeding the threshold by integrating the neural threshold distribution over the total applied input, Eqn. (1).

An important assumption in deriving this sigmoidal function is that each neural unit in a densely connected neural network is comprised of only one 'type' of neuron. This enables the distribution of neural thresholds to be defined as a unimodal function. However, if the network is assumed to be comprised of m different types of neurons then the distribution of thresholds must be redefined as a m-modal function that produce a nonlinearl activation function with m inflection points. In general, a m-modal probability distribution for the neural thresholds is expressed as

$$g(v(k)) = \frac{1}{2m} \sum_{i=1}^{m} g_s^i \operatorname{sech}^2 [g_s^i(v(k) - \theta^i)] \qquad (2)$$

and the corresponding monotonically increasing input transformation function is given by

$$\Psi[v(k)] = \frac{1}{2}\left[1 + \frac{1}{m}\sum_{i=1}^{m} \tanh\left[g_s^i\left(v(k) - \theta^i\right)\right]\right] \qquad (3)$$

where a slope parameter g_s^i and inflection point θ^i exists for each corresponding i-mode. Physiologically, a multimodal distribution would be expected to correspond to the presence of a number of distinct cell type within the population of neurons [6]. In our further discussion, it is assumed that the neural network is comprised of one type of neuron which defines the distribution of thresholds to be a unimodal function. It follows then that the nonlinear function is sigmoidal. Any function Ψ[.] is said to belong to the class of sigmoidal functions, if (a) Ψ[v(k)] is a monotonically increasing function of v(k) in the interval (-∞, ∞), (b) Ψ[v(k)] approaches the asymptotic values 0 and 1 as v(k) approaches -∞ and ∞ receptively, and (c) Ψ[v(k)] has one and only one inflection point. Mathematically, the sigmoidal function is :

$$\lim_{v(k)\to\infty} \Psi[v(k)] = 1, \text{ and } \lim_{v(k)\to-\infty} \Psi[v(k)] = 0.$$

Another important point to be considered in dealing with nonlinear sigmoidal activation function is the parameter which determines the shape of the function. In the biological neuron, the dendrites of each neuron receive pulses at the synapses and convert them to continuously variable dendritic current. The flow of current through the axon membrane modulates the axonal firing rate. For each neuron there is a time-varying nonlinear relationship between the pulse rate at the synapse and the amplitude of the dendritic current [8]. This leads to a possible inference that the main body of the neuron, the soma, may also be changing during neural activities, such as learning, adaptation, and vision perception. Recently, Yamada and Yabuta have considered determining the optimum shape of the sigmoidal function [9]. Independently, it was proposed in [5] that the parameter which controls the shape of the nonlinear function can also be considered as one of the adjustable parameters of the neural structure. This morphological change in neuron during the learning process may be modeled by considering slope of the nonlinear function in the neural structure as one of the adaptable parameters in addition to the synaptic weights. This component of neuronal learning and adaptation may be called *somatic adaptation.*

In view of the above remarks, the dynamic model of neuron, the DNU, described in this paper comprises of a dynamic structure whose output forms an argument to a time-varying sigmoidal activation function.

2.2 Mathematical model of the DNU

The dynamic structure of the DNU may be expressed in difference equation form as

$$v(k) = -b_1\, v(k-1) - b_2\, v(k-2) + a_0\, x(k) + a_1 x(k-1) + a_2\, x(k-2) \quad (4)$$

where $x(k) \in \Re^n$ is the vector of neural inputs, $v(k) \in \Re^1$ is the output of the dynamic structure, and $u(k) \in \Re^1$ is the neural output. The vectors of the signals and adaptable weights of the dynamic neuron may be defined as

$$\gamma(k,v,x) = [v(k-1) \;\; v(k-2) \;\; x(k) \;\; x(k-1) \;\; x(k-2)]^T, \quad (5)$$

$$\zeta_{(a_{ff}, b_{fb})} = [-b_1 \;\; -b_2 \;\; a_0 \;\; a_1 \;\; a_2]^T. \quad (6)$$

Using (5) and (6), Eqn. (4) is rewritten as : $v(k) = \zeta^T_{(a_{ff}, b_{fb})}\, \gamma(k,v,x)$. The nonlinear mapping operation on v(k) yields a neural output u(k) given by

$$u(k) = \Psi[g_s\, v(k)] \quad (7)$$

where $\Psi[.]$ is the sigmoidal function defined earlier and g_s is the somatic gain which controls slope of the sigmoidal function. The sigmoidal function satisfies some algebraic properties on the compact set $\Omega \in [-1,1]^N$.

Lemma 1: The sigmoidal function in Eqn. (7) may be defined as a hyperbolic tangent function which satisfies the following properties: (i) $\Psi[v(k)]$ is strictly increasing; that is $v_1 < v_2$ for each v_1 and $v_2 \in \Re$, then it is true that $\Psi[v_1] < \Psi[v_2]$, (ii) $\Psi[v(k)]$ is uniformly Lipschiz; that is, there exists a constant $C > 0$ such that $|\Psi[v_1] - \Psi[v_2]| \leq C\, |v_1 - v_2|$, $\forall$ $v_1, v_2 \in \Re$, Observe that (i) and (ii) hold iff $0 < \dfrac{\Psi[v_1] - \Psi[v_2]}{v_1 - v_2} \leq C$, $\forall$ v_1 $v_2 \in \Re$, and $v_1 \neq v_2$.

Lemma 2: There exist constants C_1, C_2 such that the neural network estimate $\hat{f}[.]$ on the compact set Ω of $\Re^N$ satisfies

$$0 < C_1 < |\hat{f}[.]| \leq C_2 .$$

3. Function Approximation Using the DNUs

It is known from mathematical analysis that analytic functions can be approximated by means of a power series

$$f(x) = a_0 + a_1\, x + a_2\, x^2 + \ldots.. + a_n\, x^n \quad (8)$$

which converges uniformly to the function f(x) in some interval [-a, a], a > 0. It means that if we put

$$S_n(x) = a_0 + a_1\, x + a_2\, x^2 + \ldots.. + a_n\, x^n , \quad S(0) = a_0 ,$$

then there exists a number $N(\varepsilon)$ for $\varepsilon > 0$ such that the inequality $n > N(\varepsilon)$ would imply the inequality $|f(x) - S_n(x)| < \varepsilon$, $-a \leq x \leq a$, that is, the polynomial $S_n(x)$ differs from the function f(x) very little if the degree n of the polynomial is sufficiently high. It is also known that the equality (8) implies unlimited differentiability of the function f(x) in the interval (- a < x < a), but any continuous function does not by far possess this property [10]. However, Weierstrass has shown that any continuous function approximated by, subjected to chosen, polynomials. It has been demonstrated in [11] that a network of DNUs satisfies the conditions of Stone-Weierstrass theorem in order to approximate arbitrary continuous functions. It is shown in the following theorem that a dense connection of DNUs, for that matter the conventional neural network structure too, can approximate continuous functions when operated in the linear range (a large slope) of the sigmoidal function.

Definition: The function Ψ (.) is said to be a linear operator if it is both additive and homogeneous on Ω [12]. Ψ (.) is said to be additive if: $\Psi(x_1 + x_2) = \Psi(x_1) + \Psi(x_2)$, $(x_1, x_2 \in \Omega)$, and homogeneous if

$$\Psi(\lambda x) = \lambda\, \Psi(x) , \quad (x \in \Omega,\ \lambda \in K) .$$

A necessary condition for a linear function Ψ (.) to be continuous on a space Ω is that Ψ be bounded on every bounded set. The condition is also sufficient if Ω satisfies the first axiom of countability [13].

Lemma 3: If $\Psi_n(1) \rightarrow 1$, $\Psi_n(x) \rightarrow p$, $\Psi_n(x^2) \rightarrow p^2$, then $\Psi_n(g) = 0$, where $g(x) = (x-p)^2$.

Proof: $g(x) = (x-p)^2 = x^2 - 2\,p\,x + p^2$, it follows that $\Psi_n(g)$

$$= \Psi_n(x^2) - 2\,p\, \Psi_n(x) + p^2\, \Psi_n(1) \rightarrow p^2 - 2\,p\,p + p^2 = 0.$$

Theorem 1: If the two conditions $\Psi_n(1) \rightarrow 1$, $\Psi_n(g) \rightarrow 0$ for $n \rightarrow \infty$ where $g(x) = (x-p)^2$ are satisfied for the sequence of linear positive functionals $\Psi_n(f)$, then $\lim_{n \rightarrow \infty} \Psi_n(f) = f(p)$ for any function f(x) continuous at the point x = p and bounded on the real axis.

Proof: In view of boundedness of the function f(x), $-M < f(x) < M$, where $M(f) = \sup_x |f(x)|$, we have

$$-2M < (f(x) - f(p)) < 2M, \ \forall\, x \in \text{on } \Omega. \quad (9a)$$

In view of continuity of this function at the point x = p, we get: $-\varepsilon < (f(x) - f(p)) < \varepsilon$, for $|x-p| < \delta$. (9b)

Equations (9a) and (9b) imply the inequality

$$-\varepsilon - \left(\frac{2M}{\delta^2}\right)g(x) < (f(x) - f(p)) < \varepsilon + \left(\frac{2M}{\delta^2}\right)g(x), \ \forall x. \quad (10)$$

In fact, if $|x-p| < \delta$, then (9) implies (10) since $g(x) = (x-p)^2 \geq 0$, and if $|x-p| \geq \delta$, then $\left(\frac{2M}{\delta^2}\right)g(x) \geq -\left(\frac{2M}{\delta^2}\right)\delta^2 = 2M$, and (10) follows from (8) since $\varepsilon > 0$. Now applying the inequality (10) to the fact that the linear positive functionals are monotonic, we get

$$-\varepsilon \Psi_n(1) - \left(\frac{2M}{\delta^2}\right)\Psi_n(g) \leq \Psi_n(f) - f(p)\Psi_n(1)$$

$$\leq \varepsilon \Psi_n(1) + \left(\frac{2M}{\delta^2}\right)\Psi_n(g). \quad (11)$$

According to the conditions given in the theorem, the right hand side of Eqn. (11) converges to ε, and the left hand side to $-\varepsilon$. Thus, there exists a number $N(\varepsilon)$ such that the inequality $-2\varepsilon < \Psi_n(f) - f(p)\Psi_n(1) < 2\varepsilon$ will be true $\forall\, n > N(\varepsilon)$. Since $\varepsilon > 0$ is arbitrary $\Psi_n(f) - f(p)\Psi_n(1) = \gamma_n \to 0$. Finally, since $\Psi_n(1) \to 1$

$$\Psi_n(f) = f(p)\Psi_n(1) + \gamma_n \to f(p).$$

Hence, the theorem is proved. Sequal to the above theorem, we will now consider a trigonometric polynomial in the following theorem and show that the proposed dynamic model can approximate continuous functions.

Lemma 4: If $\Psi_n(1) \to 1$, $\Psi_n(\cos x) \to \cos p$, $\Psi_n(\sin x) \to \sin p$, then $\Psi_n(g) = 0$, where $g(x) = \sin^2\left\{\frac{(x-p)}{2}\right\}$.

Proof: $g(x) = \sin^2\left\{\frac{(x-p)}{2}\right\} = \frac{1-\cos(x-p)}{2}$

$$= \frac{1}{2}\,[\,1 - \cos p \cos x - \sin p \sin x\,]. \quad (12)$$

$$\Psi_n(g) = \frac{1}{2}\left\{\Psi_n(1) - \cos p\, \Psi_n(\cos x) - \sin p\, \Psi_n(\sin x)\right\}$$

$$\to \frac{1}{2}\left\{1 - \cos^2(p) - \sin^2(p)\right\} = 0. \quad (13)$$

Theorem 2: If the two conditions $\Psi_n(1) \to 1$, $\Psi_n(g) \to 0$ for $n \to \infty$, where $g(x) = \sin^2\left\{\frac{(x-p)}{2}\right\}$, are satisfied for the sequence of linear positive functions $\Psi_n(f)$, then

$$\lim_{n \to \infty} \Psi_n(f) = f(p) \quad (14)$$

for any function f(x) with period 2π, continuous at the point x = p and bounded.

Proof: In view of boundedness of the function f(x), $-M < f(x) < M$, where $M(f) = \sup_x |f(x)|$, the inequalities:

$$-2M < (f(x) - f(p)) < 2M, \ \forall x \in \text{on } \Omega, \text{ and} \quad (15)$$

$$-\varepsilon < (f(x) - f(p)) < \varepsilon \ \text{ for } |x-p| < \delta \quad (16)$$

are valid. Now we take a sub-interval $(p - \delta < x \leq 2\pi + p - \delta)$ of length 2π. The inequality

$$-\varepsilon - \left(\frac{2M}{\sin^2\frac{\delta}{2}}\right)g(x) < f(x) - f(p) < \varepsilon + \left(\frac{2M}{\sin^2\frac{\delta}{2}}\right)g(x), \quad \forall x, \quad (17)$$

which is valid in this sub-interval. In fact, if $|x-p| < \delta$, then the inequality (17) follows from (16), since $g(x) = \sin^2\left\{\frac{(x-p)}{2}\right\} \geq 0$. If $\delta \leq x - p \leq 2\pi - \delta$, then $\frac{\delta}{2} \leq \frac{(x-p)}{2} \leq \pi - \frac{\delta}{2}$, and thus $\sin\frac{(x-p)}{2} \geq \sin\frac{\delta}{2}$, $g(x) = \sin^2\left\{\frac{(x-p)}{2}\right\} \geq \sin^2\frac{\delta}{2}$, $\left(\frac{2M}{\sin^2\frac{\delta}{2}}\right)g(x) \geq 2M$. In order to prove the validity of inequality (17) $\forall$ x we note that the function $g(x) = \sin^2\left\{\frac{(x-p)}{2}\right.$

$= \frac{1-\cos(x-p)}{2}$ has period 2π and according to the conditions of the theorem the function f(x) also has this period, that is,

$$-\varepsilon - \left(\frac{2M}{\sin^2\frac{\delta}{2}}\right) g(x+2k\pi) < f(x\ 2k\pi) - f(p) < \varepsilon + \left(\frac{2M}{\sin^2\frac{\delta}{2}}\right) g(x+2k\pi),\ \forall\ x. \quad (18)$$

If x varies in the sub-interval $(p - \delta, 2\pi + p - \delta)$, then $(x+2\pi)$ will vary in the sub-interval $(2\pi + p - \delta, 4\pi + p - \delta)$, $(x+4\pi)$ in the sub-interval $(4\pi + p - \delta, 6\pi + p - \delta)$, and in general $(x+2k\pi)$ in the sub-interval $(2k\pi + p - \delta, 2k\pi+2\pi + p - \delta)$, $k = 0, \pm 1, \pm 2, \ldots$ Totality of these sub-intervals covers without any gap the whole real axis, and thus the inequality (17), whose validity on every sub-interval follows from (18), is proved for $\forall$ x.

The proof of the theorem is now simple and straightforward. In fact, using the inequality (17) and monotony of the functions $\Psi_n(f)$, we obtain

$$-\varepsilon \Psi_n(1) - \left(\frac{2M}{\sin^2\frac{\delta}{2}}\right)\Psi_n(g) < \Psi_n(f) - f(p)\,\Psi_n(1) < \varepsilon \Psi_n(1) + \left(\frac{2M}{\sin^2\frac{\delta}{2}}\right)\Psi_n(g). \quad (19)$$

In view of the conditions, the right hand side of Eqn. (21) converges to ε, and the left hand side to $-\varepsilon$. Thus, there exists a number $N(\varepsilon)$ such that the inequality $-2\varepsilon < \Psi_n(f) - f(p)\Psi_n(1) < 2\varepsilon$ will be true $\forall\ n > N(\varepsilon)$. Since $\varepsilon > 0$ is arbitrary $\Psi_n(f) - f(p)\Psi_n(1) = \gamma_n \to 0$, $\Psi_n(f) = f(p)\,\Psi_n(1) + \gamma_n \to f(p)$. Hence, the theorem is proved.

Theorem 3: If the function f(x) in the interval $[-\pi, \pi]$ is continuous and even, then there exists an even trigonometric polynomial $\Psi_n(f, x)$ which deviate the least from the function f(x), that is, $d(\Psi, f) = \| f - \Psi_n(f, x) \|$.

Proof: If $\Psi_n(f, x)$ is any polynomial which deviates the least from the function f(x),

$$d(\Psi, f) = \| f - \Psi_n(f, x) \| \quad (20)$$

then replacing x by -x and noting that the function f(x) is even, we obtain

$$d(\Psi, f) = \max_{-\pi \le x \le \pi} \left| f(-x) - \Psi_n(f, -x) \right|$$

$$= \max_{-\pi \le x \le \pi} \left| f(x) - \Psi_n(f, -x) \right| = \| f - \Psi_n(f, -x) \|. \quad (21)$$

Let $Q(x) = \frac{[\Psi_n(f, x) + \Psi_n(f, -x)]}{2}$. The trigonometric polynomial Q(x) is even and has degree not greater than n. Thus,

$$\| f - Q \| = \left\| \left(f - \frac{[\Psi_n(f, x) + \Psi_n(f, -x)]}{2} \right) \right\|$$

$$= \left(\frac{1}{2}\right) \| f - \Psi_n(f, x) + f - \Psi_n(f, -x) \|$$

$$\le \left(\frac{1}{2}\right) \| f - \Psi_n(f, x) \| + \left(\frac{1}{2}\right) \| f - \Psi_n(f, -x) \|$$

$$= \left(\frac{1}{2}\right) d(\Psi, f) + \left(\frac{1}{2}\right) d(\Psi, f) = d(\Psi, f).$$

Hence, it follows that $d(\Psi, f) = \| f - Q \|$, the even polynomial Q(x) deviates the least from the function f(x). Following the above explanation, we will now derive the learning and adaptive algorithm for the adaptable parameters of the DNU.

A few function approximation examples, based on the learning algorithm derived in [5,11], are given at the end of this paper.

4. Conclusions

In this paper we have developed a dynamic model of neuron, called the DNU, which embodies two distinct operations, the synaptic and the somatic. The first operation determines the optimum synaptic weights, while the second determines the optimum shape of sigmoidal function for a given task. The function approximation capability of DNUs has been demonstrated using linear and trigonometric polynomials.

References

[1] J.A. Anderson, "Cognitive and psychological computation with neural models", *IEEE Trans. on Systems, Man, and Cybernetics*, Vol. 13, No. 5, pp. 799-815, Sept./Oct. 1983.

[2] P.D. Wasserman, Neural Computing: theory and practice, *Van Nostrand*, New York, 1989.

[3] J.J. Hopfield, "Neural Networks and physical systems with emergent collective computational abilities", *Proceedings of the National Academy of Sciences*, Vol. 79, pp. 2554 - 2558, 1982.

[4] A.C. Guyton, Text book of medical physiology, *W.B. Saunders Company, Philadelphia*, 1987.

[5] M.M. Gupta and D.H. Rao, "Synaptic and somatic adaptations in dynamic neural networks", *Second Int. Conf. on Fuzzy Logic and Neural Networks*, Fukuoka, Japan, pp. 173-177, July 17-22, 1992.

[6] H.R. Wilson and J.D. Cowan, "Excitatory and inhibitory interactions in localized populations of model neurons", *Biophysical Journal*, Vol. 12, pp. 1-24, 1972.

[7] G.K. Knopf, "Theoretical studies of a dynamic neuro-vision processor: a biologically motivated design", Ph.D. thesis, March 1991, Univ. of Saskatchewan, Saskatoon, Canada.

[8] W.J. Freeman, "Dynamics of image formation by nerve cell assemblies", in E. Basar, H. Flohr, H. Haken and A.J. Mandell (Eds), Synergitcs of the brain, Berlin, *Springer-Verlag*, 1983.

[9] T. Yamada and T. Yabuta, "Neural network controller using autotuning method for nonlinear functions", *IEEE Trans. on Neural Networks*, Vol. 3, No. 4, pp. 595-601, July 1992.

[10] P.P. Korovkin, Linear operators and approximation theory, Hindustan Publishing Corp., Delhi, 1960.

[11] D.H. Rao and M.M. Gupta, "Dynamic neural adaptive control: theoretical development, part I", *The Journal of Mathematical Systems, Estimation and Control*, [Submitted].

[12] L.V. Kantorovich and G.P. Akilov, Functional analysis, Translated by H.L. Silcock, Pergamon Press, NY, 1982.

[13] A.N. Kolmogorov nad S.V. Fomin, Introductory real analysis, Translated by R.A. Silverman, Prentice-Hall, Inc., NJ, 1970.

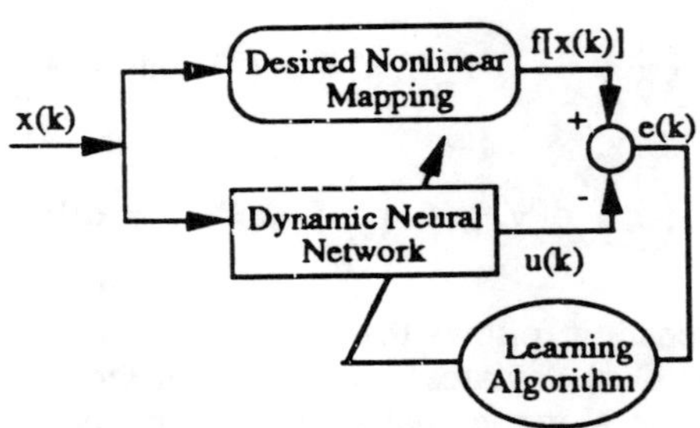

Fig. 1: The learning scheme for a nonlinear function.

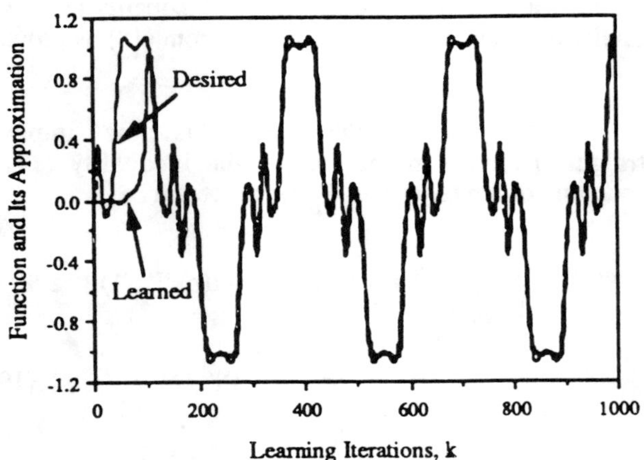

Fig. 2a

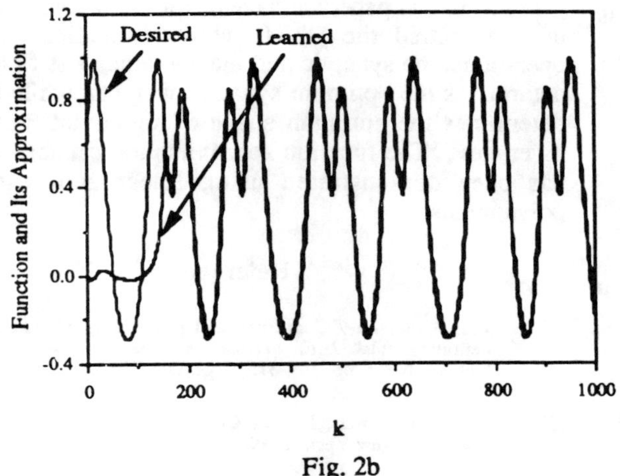

Fig. 2b

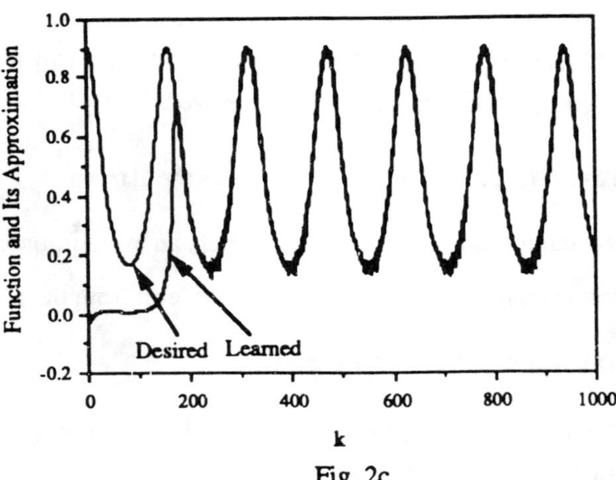

Fig. 2c

Fig. 2: Nonlinear functions and their approximations by the network of dynamic neural units.

Part 5
Neuronal Morphology for Control Systems: Static and Dynamic Structures

The successful operation of an intelligent system depends upon its ability to cope with uncertainties that may arise within the system and the environment. Biology has provided the necessary motivation for neural control methodology with abilities such as learning and adaptivity. Artificial neuro-control promises to provide robust and intelligent systems.

THE basic concepts of learning and adaptation in the field of control systems were introduced in the early sixties and several extensions and advances have been made since then. However, advances in the understanding of the physiology of biological control has spurred the interest of system scientists to explore the field of neuro-control. Biology has certainly provided motivation to the field of neuro-control, as it has to the field of neuro-vision. Recent mathematical models and the architectures of neuro-control systems have generated many theoretical and industrial interests. Recent advances in static (memoryless) and dynamic (with memory) neural networks have created a profound impact on the field of neuro-control. Now, researchers are moving toward the design of intelligent control systems using biological neuro-control as a basis.

In Part Five of the volume, we present an overview of some important advances that have appeared in the literature very recently. This work deals with a diverse group of neuro-control problems, such as the neural architecture for adaptation and control, introduction to backpropagation algorithms, identification, and control problems for a general class of dynamic systems. We also provide an extensive list of research literature in the field of the dynamics and stability of recurrent neural systems, self-learning control, adaptive filters and equalizers, identification and control of nonlinear systems. This set of six articles thus deals with a variety of problems encountered in a typical control situation and provides the basic theoretical foundation of neuro-control systems.

Backpropagation and its variations-based learning algorithms are now most widely used in neural networks. The basic principle of backpropagation algorithm involves the determination of partial derivatives of performance index with respect to the neural weights. In the first article, (5.1), the inventor of the learning algorithm, P. J. Werbos, reviews the basics of backpropagation, which has been found useful in areas like pattern recognition, fault diagnosis, dynamic recurrent systems, and control systems.

Next, we present an article taken from *IEEE Control Systems Magazine,* dealing exclusively with the use of neural networks in control systems. In article (5.2), D. H. Nguyen and B. Widrow show how a neural network can learn of its own accord to control a nonlinear dynamic system. An example of the "truck backer-upper" is given, in which the neural network controller steers a trailer-truck while backing up to a loading dock. The neural-controller is able to guide the truck to the dock from almost any initial position. The technique explored in the article should be applicable to a wide variety of nonlinear control problems.

Mathematically formulated neural network elements, although biologically inspired, represent certain neural models, abstract or computational architectures, for some specific control tasks. In this volume on neural-control, our objective is not to mimic the central nervous system, but to develop some motivations for solving specific problems facing the system scientists. The neural-computational architecture presented in this volume so far has several possible substrates for building intelligent controllers for misbehaved dynamic systems encountered in this complex technological society. Abstract neural network models may be useful in explaining the neurophysiological behavior of the animal kingdom, and may also have some potential for creating intelligent networks for engineering and medical applications. The last four articles in this part deal with various issues in the general field of learning and adaptation in the control of complex dynamic systems.

In article (5.3), L. Jin, P. N. Nikiforuk, and M. M. Gupta address the problems of learning and control of a general class of nonlinear discrete systems using a feedforward multilayered neural network (MNN) architecture. They use an input–output linearization neural approach both for output tracking and model reference purposes. A new learning algorithm with a dead-zone function is presented that can approximate an unknown nonlinear input-output dynamic relationship and its inverse dynamic relationship. The proposed approach has on-line identification and adaptive control abilities.

In article (5.4), K. J. Hunt and D. Sbarbaro present a novel neural network method. The ability of neural networks to model arbitrary nonlinear functions and their inverse is exploited. It is used to directly incorporate networks modeling the plant and its inverse within the control strategy.

In article (5.5), K. S. Narendra and K. Parthasarathy discuss the basic identification and control of dynamic-systems, and provide a neural-network approach for their solution. This article should be of great interest to the reader working in this field. This article is one of the first few articles to present extensive theoretical and application of dynamic neural networks to system identification and control problems.

In the final article, (5.6), M. M. Gupta and D. H. Rao present a novel dynamic neural architecture named the dynamic neural unit (DNU), [see also article (4.6)]. The DNU has a linear dynamics (discrete) followed by a varying sigmoidal-type nonlinear function. A new dynamic learning algorithm is reported that employs a parallel-cascade structure of DNUs to control unknown nonlinear dynamic systems using the inverse-dynamics control approach. The approach presented in this article seems to have great potential for solving some dynamic problems in control, identification, parameter estimation, noise cancellation, equalization, prediction, adaptive echo cancellation, and so on.

Thus, in this part of the volume, we have presented a variety of neural methodologies for the control of complex plants. In the next part, we will present a different methodology—one based upon fuzzy logic.

Further Reading

[1] K. S. Fu, "Learning control systems—Review and outlook," *IEEE Trans. Automat. Contr.*, pp. 210–221, April 1970.

[2] P. J. Antsaklis and K. M. Passino, "Towards intelligent autonomous control systems: Architecture and fundamental issues," *Int. J. Intell. Robotic Syst.*, pp. 315–342, 1989.

[3] M. D. Peek and P. J. Antsaklis, "Parameter learning for performance adaptation," *IEEE Contr. Syst. Mag.*, pp. 3–11, Dec. 1990.

[4] W. L. Baker and J. A. Farrell, "Connectionist learning systems for control," in SPIE Proc. Intell. Robots Comput. Vision IX: Neural, Biological and 3-D Methods, D. P. Casasent, Ed., pp. 181–198, Nov. 7–9, 1990.

[5] K. G. Shin and X. Cui, "Design of a knowledge-based controller for intelligent control systems," *IEEE Trans. Syst., Man, Cybern.*, vol. 21, no. 2, pp. 368–375, March/April 1991.

[6] A. G. Barto, "Connectionist learning for control," in *Neural Networks for Control*, T. Miller, R. S. Sutton, and P. J. Werbos, Eds., Cambridge, MA: MIT Press, pp. 6–58, March/April 1991.

[7] A. Guez, J. L. Eilbert, and M. Kam, "Neural network architecture for control," *IEEE Contr. Syst. Mag.*, pp. 22–25, April 1988.

[8] W. P. Jones and J. Hoskins, "Backpropagation: A generalized delta learning rule," *Byte*, pp. 155–162, Oct. 1987.

[9] Y. Ichikawa and T. Sawa, "Neural network applications for direct feedback controllers," *IEEE Trans. Neural Networks*, vol. 3, no. 2, pp. 224–231, March 1992.

[10] T. Yabuta and T. Yamada, "Neural network controller characteristics with regard to adaptive control," *IEEE Trans. Syst. Man, Cybern.*, vol. 22, no. 1, pp. 170–176, Jan/Feb. 1992.

[11] D. A. Hoskins, J. N. Hwang, and J. Vagners, "Iterative inversion of neural networks and its application to adaptive control," *IEEE Trans. Neural Networks*, vol. 3, no. 2, pp. 292–301, March 1992.

[12] K. S. Narendra and K. Parthasarthy, "Gradient methods for the optimization of dynamical systems containing neural networks," *IEEE Trans. Neural Networks*, vol. 2, no. 2, pp. 4–27, March 1991.

[13] E. Levin, N. Tishby, and S. A. Solla, "A statistical approach to learning and generalization in layered neural networks," *Proc. IEEE*, vol. 78, no. 10, pp. 1568–1574, Oct. 1990.

[14] J. S. Albus, "A new approach to manipulator control: The cerebellar model articulation controller (CMAC)," *J. Dynamic Sys., Meas., Contr.*, pp. 220–227, Sept. 1975.

[15] L. G. Kraft and D. P. Campagna, "A comparison between CMAC neural network control and two traditional adaptive control systems," *IEEE Contr. Syst. Mag.*, pp. 36–43, April 1990.

[16] A. Guez and J. Selinsky, "A trainable neuromorphic controller," *J. of Robotic Syst.*, vol. 5, no. 4, pp. 363–388, 1988.

[17] S. D. Wang and M. S. Yeh, "Self-adaptive neural architectures for control applications," in *Proc. Int. Joint Conf. Neural Networks, (IJCNN)*, pp. 309–314, June 1990.

[18] S. R. Chu and R. Shoureshi, "Applications of neural networks in learning of dynamical systems," *IEEE Trans. Syst., Man, Cybern.*, vol. 22, no. 1, pp. 160–163, Jan./Feb. 1991.

[19] M. Kawato, "Optimization and learning in neural networks for formation and control of coordinated movement," *ATR Technical Report*, Japan, 1990.

[20] C. C. Lee, "A self-learning rule-based controller employing approximate reasoning and neural net concepts," *Int. J. Intell. Syst.*, vol. 6, pp. 71-93, 1991.

[21] F. Delcomyn, "Neural basis of rhythmic behavior in animals," *Science*, vol. 210, pp. 492–498, 1976.

[22] B. Widrow, R. G. Winter, and R. A. Baxter, "Layered neural nets for pattern recognition," *IEEE Trans. Acoust., Speech Signal Process.*, vol. 36, no. 7, pp. 1109–1118, July 1988.

[23] K. S. Narendra and S. Mukhopadhyay, "Associative learning in random environments using neural networks," *IEEE Trans. Neural Networks*, vol. 2, no. 1, pp. 20–31, Jan. 1991.

[24] A. D. Kulkarni, "Solving ill-posed problems with artificial neural networks," *Neural Networks*, vol. 4, pp. 477–484, 1991.

[25] K. Matsuoka, "Stability conditions for nonlinear continuous neural networks with asymmetric connection weights," *Neural Networks*, vol. 5, pp. 495–500, 1992.

[26] J. J. Shynk and S. Roy, "Convergence properties and stationary points of a perceptron learning algorithm," *Proc. IEEE*, vol. 78, no. 10, pp. 1599–1604, Oct. 1990.

[27] T. Yamada and T. Yabuta, "Neural network controller using autotuning method for nonlinear functions," *IEEE Trans. Neural Networks*, vol. 3, no. 4, pp. 595–601, July 1992.

[28] S. Hui and S. Zak, "Analysis of single perceptrons learning capabilities," in *Proc. Amer. Contr. Conf.*, Boston, pp. 809–814, June 26–28, 1991.

[29] E. A. Wan, "Temporal backpropagation for FIR neural networks," in *Proc. Int. Joint Conf. Neural Networks (IJCNN)*, pp. 575–580, June 1990.

[30] G. Tsutsumidani, N. Ohnishi, and N. Sugie, "Properties and learning algorithm of discrete neural network with time delay," in *Proc. Int. Joint Conf. Neural Networks* (IJCNN), pp. 529–534, Nov. 1991.

[31] M. M. Gupta and D. H. Rao, "Synaptic and somatic adaptations in dynamic neural networks," in *Proc. Second Int. Conf. Fuzzy Logic and Neural Networks*, Fukuoka, Japan, pp. 173–177, July 1992.

[32] S. I. Sudharsanan and M. K. Sundareshan, "Training of a three-layer dynamical recurrent neural network for nonlinear input–output mapping," in *Proc. Int. Joint Conf. Neural Networks* (IJCNN), pp. 111–115, Nov. 1991.

[33] S. Chen, C. F. N. Cowan, and P. M. Grant, "Orthogonal least squares learning algorithm for radial basis function networks," *IEEE Trans. Neural Networks*, vol. 2, no. 2, pp. 302–309, March 1991.

[34] Special Issue on Neural Networks, *IEEE Contr. Syst. Mag.*, April 1988.

[35] Special Issue on Neural Networks, *IEEE Contr. Syst. Mag.*, April 1990.

Backpropagation Through Time: What It Does and How to Do It

PAUL J. WERBOS

Backpropagation is now the most widely used tool in the field of artificial neural networks. At the core of backpropagation is a method for calculating derivatives exactly and efficiently in any large system made up of elementary subsystems or calculations which are represented by known, differentiable functions; thus, backpropagation has many applications which do not involve neural networks as such.

This paper first reviews basic backpropagation, a simple method which is now being widely used in areas like pattern recognition and fault diagnosis. Next, it presents the basic equations for backpropagation through time, and discusses applications to areas like pattern recognition involving dynamic systems, systems identification, and control. Finally, it describes further extensions of this method, to deal with systems other than neural networks, systems involving simultaneous equations or true recurrent networks, and other practical issues which arise with this method. Pseudocode is provided to clarify the algorithms. The chain rule for ordered derivatives—the theorem which underlies backpropagation—is briefly discussed.

I. Introduction

Backpropagation through time is a very powerful tool, with applications to pattern recognition, dynamic modeling, sensitivity analysis, and the control of systems over time, among others. It can be applied to neural networks, to econometric models, to fuzzy logic structures, to fluid dynamics models, and to almost any system built up from elementary subsystems or calculations. The one serious constraint is that the elementary subsystems must be represented by functions known to the user, functions which are both continuous and differentiable (i.e., possess derivatives). For example, the first practical application of backpropagation was for estimating a dynamic model to predict nationalism and social communications in 1974 [1].

Unfortunately, the most general formulation of backpropagation can only be used by those who are willing to work out the mathematics of their particular application. This paper will mainly describe a simpler version of backpropagation, which can be translated into computer code and applied directly by neural network users.

Section II will review the simplest and most widely used form of backpropagation, which may be called "basic backpropagation." The concepts here will already be familiar to those who have read the paper by Rumelhart, Hinton, and Williams [2] in the seminal book *Parallel Distributed Processing*, which played a pivotal role in the development of the field. (That book also acknowledged the prior work of Parker [3] and Le Cun [4], and the pivotal role of Charles Smith of the Systems Development Foundation.) This section will use new notation which adds a bit of generality and makes it easier to go on to complex applications in a rigorous manner. (The need for new notation may seem unnecessary to some, but for those who have to *apply* backpropagation to complex systems, it is essential.)

Section III will use the same notation to describe backpropagation through time. Backpropagation through time has been applied to concrete problems by a number of authors, including, at least, Watrous and Shastri [5], Sawai and Waibel *et al.* [6], Nguyen and Widrow [7], Jordan [8], Kawato [9], Elman and Zipser, Narendra [10], and myself [1], [11], [12], [15]. Section IV will discuss what is missing in this simplified discussion, and how to do better.

At its core, backpropagation is simply an efficient and exact method for calculating all the derivatives of a single target quantity (such as pattern classification error) with respect to a large set of input quantities (such as the parameters or weights in a classification rule). Backpropagation through time extends this method so that it applies to dynamic systems. This allows one to calculate the derivatives needed when optimizing an iterative analysis procedure, a neural network with memory, or a control system which maximizes performance over time.

II. Basic Backpropagation

A. The Supervised Learning Problem

Basic backpropagation is current the most popular method for performing the supervised learning task, which is symbolized in Fig. 1.

In supervised learning, we try to adapt an artificial neural network so that its actual outputs ($\hat{Y}$) come close to some target outputs (Y) for a training set which contains T patterns. The goal is to adapt the parameters of the network so that it performs well for patterns from outside the training set.

The main use of supervised learning today lies in pattern

Manuscript received September 12, 1989; revised March 15, 1990.

The author is with the National Science Foundation, 1800 G St. NW, Washington, DC 20550.

IEEE Log Number 9039172.

Reprinted from *Proc. IEEE*, vol. 78, no. 10, pp. 1550–1560, Oct. 1990.

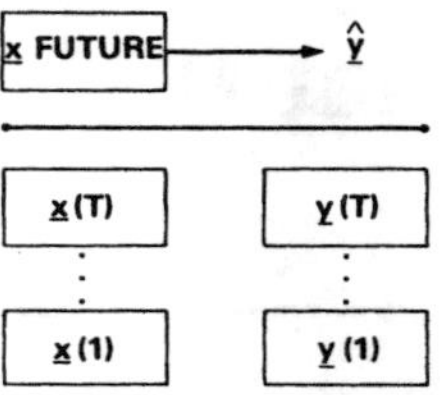

Fig. 1. Schematic of the supervised learning task.

recognition work. For example, suppose that we are trying to build a neural network which can learn to recognize handwritten ZIP codes. (AT&T has actually done this [13], although the details are beyond the scope of this paper.) We assume that we already have a camera and preprocessor which can digitize the image, locate the five digits, and provide a 19 × 20 grid of ones and zeros representing the image of each digit. We want the neural network to input the 19 × 20 image, and output a classification; for example, we might ask the network to output four binary digits which, taken together, identify which decimal digit is being observed.

Before adapting the parameters of the neural network, one must first obtain a training database of actual handwritten digits and correct classifications. Suppose, for example, that this database contains 2000 examples of handwritten digits. In that case, $T = 2000$. We may give each example a label t between 1 and 2000. For each sample t, we have a record of the input pattern and the correct classification. Each input pattern consists of 380 numbers, which may be viewed as a vector with 380 components; we may call this vector $\boldsymbol{X}(t)$. The desired classification consists of four numbers, which may be treated as a vector $\boldsymbol{Y}(t)$. The actual output of the network will be $\hat{\boldsymbol{Y}}(t)$, which may differ from the desired output $\boldsymbol{Y}(t)$, especially in the period *before* the network has been adapted. To solve the supervised learning problem, there are two steps:

- We must specify the "topology" (connections and equations) for a network which inputs $\boldsymbol{X}(t)$ and outputs a four-component vector $\hat{\boldsymbol{Y}}(t)$, an approximation to $\boldsymbol{Y}(t)$. The relation between the inputs and outputs must depend on a set of weights (parameters) $\boldsymbol{W}$ which can be adjusted.
- We must specify a "learning rule"—a procedure for adjusting the weights $\boldsymbol{W}$ so as to make the actual outputs $\hat{\boldsymbol{Y}}(t)$ approximate the desired outputs $\boldsymbol{Y}(t)$.

Basic backpropagation is currently the most popular learning rule used in supervised learning. It is generally used with a very simple network design—to be described in the next section—but the same approach can be used with *any* network of differentiable functions, as will be discussed in Section IV.

Even when we use a simple network design, the vectors $\boldsymbol{X}(t)$ and $\boldsymbol{Y}(t)$ need not be made of ones and zeros. They can be made up of any values which the network is capable of inputting and outputting. Let us denote the components of $\boldsymbol{X}(t)$ as $X_1(t) \cdots X_m(t)$ so that there are m inputs to the network. Let us denote the components of $\boldsymbol{Y}(t)$ as $Y_1(t) \cdots Y_n(t)$ so that we have n outputs. Throughout this paper, the components of a vector will be represented by the *same* letter as the vector itself, in the same case; this convention turns out to be convenient because $\boldsymbol{x}(t)$ will represent a different vector, very closely related to $\boldsymbol{X}(t)$.

Fig. 1 illustrates the supervised learning task in the general case. Given a history of $\boldsymbol{X}(1) \cdots \boldsymbol{X}(T)$ and $\boldsymbol{Y}(1) \cdots \boldsymbol{Y}(T)$, we want to find a mapping from $\boldsymbol{X}$ to $\boldsymbol{Y}$ which will perform well when we encounter *new* vectors $\boldsymbol{X}$ outside the training set. The index "t" may be interpreted either as a time index or as a pattern number index; however, this section will not assume that the order of patterns is meaningful.

B. Simple Feedforward Networks

Before we specify a learning rule, we have to define exactly how the outputs of a neural net depend on its inputs and weights. In basic backpropagation, we assume the following logic:

$$x_i = X_i, \qquad 1 \le i \le m \tag{1}$$

$$\text{net}_i = \sum_{j=1}^{i-1} W_{ij} x_j, \qquad m < i \le N + n \tag{2}$$

$$x_i = s(\text{net}_i), \qquad m < i \le N + n \tag{3}$$

$$Y_i = x_{i+N}, \qquad 1 \le i \le n \tag{4}$$

where the function s in (3) is usually the following sigmoidal function:

$$s(z) = 1/(1 + e^{-z}) \tag{5}$$

and where N is a constant which can be any integer you choose as long as it is no less than m. The value of $N + n$ decides how many neurons are in the network (if we include inputs as neurons). Intuitively, net_i represents the total level of voltage exciting a neuron, and x_i represents the intensity of the resulting output from the neuron. (x_i is sometimes called the "activation level" of the neuron.) It is conventional to assume that there is a threshold or constant weight W_{io} added to the right side of (2); however, we can achieve the same effect by assuming that one of the inputs (such as X_m) is always 1.

The significance of these equations is illustrated in Fig. 2. There are $N + n$ circles, representing all of the neurons

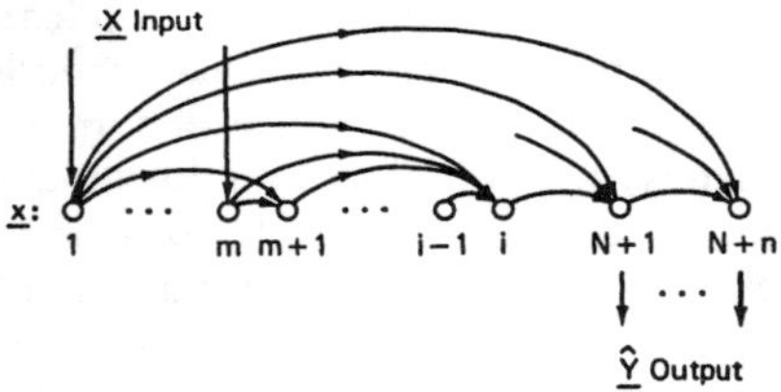

Fig. 2. Network design for basic backpropagation.

in the network, *including* the input neurons. The first m circles are really just copies of the inputs $X_1 \cdots X_m$; they are included as part of the vector $\boldsymbol{x}$ only as a way of simplifying the notation. *Every* other neuron in the network—such as neuron number i, which calculates net_i and x_i—takes input from *every* cell which precedes it in the network. Even the last output cell, which generates $\hat{Y}_n$, takes input from other output cells, such as the one which outputs $\hat{Y}_{n-1}$.

In neural network terminology, this network is "fully connected" in the extreme. As a practical matter, it is usually desirable to limit the connections between neurons. This can be done by simply fixing some of the weights W_{ij} to zero so that they drop out of all calculations. For example, most researchers prefer to use "layered" networks, in which all connection weights W_{ij} are zeroed out, *except* for those going from one "layer" (subset of neurons) to the next layer. In general, one may zero out as many or as few of the weights as one likes, based on one's understanding of individual applications. For those who first begin this work, it is conventional to define only three layers—an input layer, a "hidden" layer, and an output layer. This section will assume the full range of *allowed* connections, simply for the sake of generality.

In computer code, we could represent this network as a Fortran subroutine (assuming a Fortran which distinguishes upper case from lower case):

```
          SUBROUTINE NET(X, W, x, Yhat)
          REAL X(m),W(N+n,N+n),x(N+n),Yhat(n),net
          INTEGER, i,j,m,n,N
C   First insert the inputs, as per equation (1)
          DO 1 i = 1,m
       1  x(i) = X(i)
C   Next implement (2) and (3) together for each value
C     of i
          DO 1000 i = m+1,N+n
C         calculate net_i as a running sum, based on (2)
          net = 0
          DO 10 j = 1,i-1
    10           net = net + W(i,j)*x(j)
C               finally, calculate x_i based on (3) and (5)
  1000          x(i) = 1/(1+exp(-net))
C   Finally, copy over the outputs, as per (4)
          DO 2000 i = 1,n
  2000          Yhat(i) = x(i+N);
```

In the pseudocode, note that X and W are technically the inputs to the subroutine, while x and $Yhat$ are the outputs. $Yhat$ is usually regarded as "the" output of the network, but x may also have its uses outside of the subroutine proper, as will be seen in the next section.

C. Adapting the Network: Approach

In basic backpropagation, we choose the weights W_{ij} so as to minimize square error over the training set:

$$E = \sum_{t=1}^{T} E(t) = \sum_{t=1}^{T} \sum_{i=1}^{n} (1/2)(\hat{Y}_i(t) - Y_i(t))^2. \quad (6)$$

This is simply a special case of the well-known method of least squares, used very often in statistics, econometrics, and engineering; the uniqueness of backpropagation lies in *how* this expression is minimized. The approach used here is illustrated in Fig. 3.

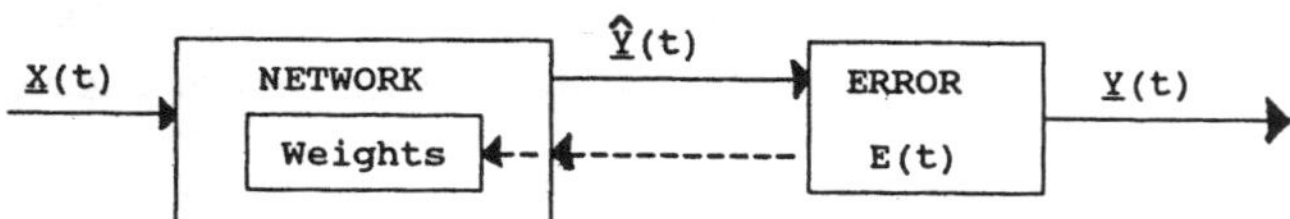

Fig. 3. Basic backpropagation (in pattern learning).

In basic backpropagation, we start with arbitrary values for the weights W. (It is usual to choose random numbers in the range from −0.1 to 0.1, but it may be better to guess the weights based on prior information, in cases where prior information is available.) Next, we calculate the outputs $\hat{Y}(t)$ and the errors $E(t)$ for that set of weights. Then we calculate the derivatives of E with respect to all of the weights; this is indicated by the dotted lines in Fig. 3. If increasing a given weight would lead to more error, we adjust that weight downwards. If increasing a weight leads to less error, we adjust it upwards. After adjusting all the weights up or down, we start all over, and keep on going through this process until the weights and the error settle down. (Some researchers iterate until the error is close to zero; however, if the number of training patterns exceeds the number of weights in the network—as recommended by studies on generalization—it may not be possible for the error to reach zero.) The uniqueness of backpropagation lies in the method used to calculate the derivatives exactly for all of the weights in only one pass through the system.

D. Calculating Derivatives: Theoretical Background

Many papers on backpropagation suggest that we need only use the conventional chain rule for partial derivatives to calculate the derivatives of E with respect to all of the weights. Under certain conditions, this can be a rigorous approach, but its generality is limited, and it requires great care with the side conditions (which are rarely spelled out); calculations of this sort can easily become confused and erroneous when networks and applications grow complex. Even when using (7) below, it is a good idea to test one's gradient calculations using explicit perturbations in order to be sure that there is no bug in one's code.

When the idea of backpropagation was first presented to the Harvard faculty in 1972, they expressed legitimate concern about the validity of the rather complex calculations involved. To deal with this problem, I proved a new chain rule for *ordered* derivatives:

$$\frac{\partial^+ \text{TARGET}}{\partial z_i} = \frac{\partial \text{TARGET}}{\partial z_i} + \sum_{j>i} \frac{\partial^+ \text{TARGET}}{\partial z_j} * \frac{\partial z_j}{\partial z_i} \quad (7)$$

where the derivatives with the superscript represent *ordered* derivatives, and the derivatives without subscripts represent ordinary partial derivatives. This chain rule is valid only for *ordered* systems where the values to be calculated can be calculated one by one (if necessary) in the order z_1, $z_2, \cdots, z_n$, TARGET. The simple partial derivatives represent the *direct* impact of z_i on z_j *through the system equation* which determines z_j. The ordered derivative represents the *total* impact of z_i on TARGET, accounting for both the *direct* and *indirect* effects. For example, suppose that we had a simple system governed by the following two equations, in order:

$$z_2 = 4 * z_1$$

$$z_3 = 3 * z_1 + 5 * z_2.$$

The "simple" partial derivative of z_3 with respect to z_1 (the *direct* effect) is 3; to calculate the simple effect, we *only* look at the equation which determines z_3. However, the ordered derivative of z_3 with respect to z_1 is 23 because of the indirect impact by way of z_2. The simple partial derivative measures what happens when we increase z_1 (e.g., by 1, in this example) and assume that everything else (like z_2) in the equation which determines z_3 remains constant. The ordered derivative measures what happens when we increase z_1, *and also* recalculate all other quantities—like

z_2—which are later than z_1 in the causal ordering we impose on the system.

This chain rule provides a straightforward, plodding, "linear" recipe for how to calculate the derivatives of a given TARGET variable with respect to *all* of the inputs (and parameters) of an ordered differentiable system *in only one pass through the system*. This paper will not explain this chain rule in detail since lengthy tutorials have been published elsewhere [1], [11]. But there is one point worth noting: because we are calculating ordered derivatives of *one* target variable, we can use a simpler notation, a notation which works out to be easier to use in complex practical examples [11]. We can write the ordered derivative of the TARGET with respect to z_i as "F_z_i," which may be described as "the feedback to z_i." In basic backpropagation, the TARGET variable of interest is the error E. This changes the appearance of our chain rule in that case to

$$F\_z_i = \frac{\partial E}{\partial z_i} + \sum_{j>i} F\_z_j * \frac{\partial z_j}{\partial z_i}. \tag{8}$$

For purposes of debugging, one can calculate the true value of any ordered derivative simply by perturbing z_i at the point in the program where z_i is calculated; this is particularly useful when applying backpropagation to a complex network of functions other than neural networks.

E. Adapting the Network: Equations

For a given set of weights $\boldsymbol{W}$, it is easy to use (1)–(6) to calculate $\boldsymbol{Y}(t)$ and $E(t)$ for each pattern t. The trick is in how we then calculate the derivatives.

Let us use the prefix "$F\_$" to indicate the ordered derivative of E with respect to whatever variable the "$F\_$" precedes. Thus, for example,

$$F\_\hat{Y}(t) = \frac{\partial E}{\partial \hat{Y}_i(t)} = \hat{Y}_i(t) - Y_i(t), \tag{9}$$

which follows simply by differentiating (6). By the chain rule for ordered derivatives as expressed in (8),

$$F\_x_i(t) = F\_\hat{Y}_{i-N}(t) + \sum_{j=i+1}^{N+n} W_{ji} * F\_net_j(t),$$

$$i = N + n, \cdots, m + 1 \tag{10}$$

$$F\_net_i(t) = s'(net_i) * F\_x_i(t), \qquad i = N + n, \cdots, m + 1 \tag{11}$$

$$F\_W_{ij} = \sum_{t=1}^{T} F\_net_i(t) * x_j(t) \tag{12}$$

where s' is the derivative of $s(z)$ as defined in (5) and F_Y_k is assumed to be zero for $k \leq 0$. Note how (10) requires us to run *backwards* through the network in order to calculate the derivatives, as illustrated in Fig. 4; this backwards propagation of information is what gives backpropagation its name. A little calculus and algebra, starting from (5), shows us that

$$s'(z) = s(z) * (1 - s(z)), \tag{13}$$

which we can use when we implement (11). Finally, to adapt the weights, the usual method is to set

$$\text{New } W_{ij} = W_{ij} - \text{learning\_rate} * F\_W_{ij} \tag{14}$$

where the learning_rate is some small constant chosen on an ad hoc basis. (The usual procedure is to make it as large as possible, up to 1, until the error starts to diverge; however, there are more analytic procedures available [11].)

F. Adapting the Network: Code

The key part of basic backpropagation—(10)–(13)—may be coded up into a "dual" subroutine, as follows.

```
      SUBROUTINE F_NET(F_Yhat, W, x, F_W)
      REAL F_Yhat(n),W(N+n,N+n),x(N+n),
        F_W(N+n,N+n),F_net(N+n),F_x(N+n)
      INTEGER i,j,n,m,N
C  Initialize equation (10)
      DO 1 i = 1,N
 1       F_x(i) = 0
      DO 2 i = 1,n
 2       F_x(i+N)=F_Yhat(i)
C  RUN THROUGH (10)-(12) AS A SET,
C     FOR i RUNNING BACKWARDS
      DO 1000 i=N+n,m+1,-1
C        complete (10) for the current value of i */
         DO 10 j = i+1,N+n
C        modify "DO 10" if needed to be sure
C          nothing is done if i=N+n
 10        F_x(i) = F_x(i)+W(j,i)*F_net(j)
C        next implement (11), exploiting (13)
         F_net(i) = F_x(i)*x(i)*(1.-x(i))
C        then implement (12) for the current
C          value of i
         DO 12 j = 1,i-1
 12        F_W(i,j)=F_net(i)*x(j)
 1000    CONTINUE
```

Note that the array F_W is the only output of this subroutine.

Equation (14) represented "batch learning," in which weights are adjusted only after *all* T patterns are processed. It is more common to use pattern learning, in which the weights are continually updated after each observation. Pattern learning may be represented as follows:

```
C  PATTERN LEARNING
        DO  1000 pass_number=1, maximum_passes
           DO  100 t =1,T
                CALL NET(X(t), W, x, Yhat)
C               Next Implement equation (9)
                DO  9 i = 1,n
 9                 F_Yhat(i)=Yhat(i)-Y(t,i)
C               Next Implement (10)-(12)
                CALL F_NET(F_Yhat, W, x, F_W)
C               Next Implement (14)
C               Note how weights are updated
C                 within the "DO 100" loop.
```

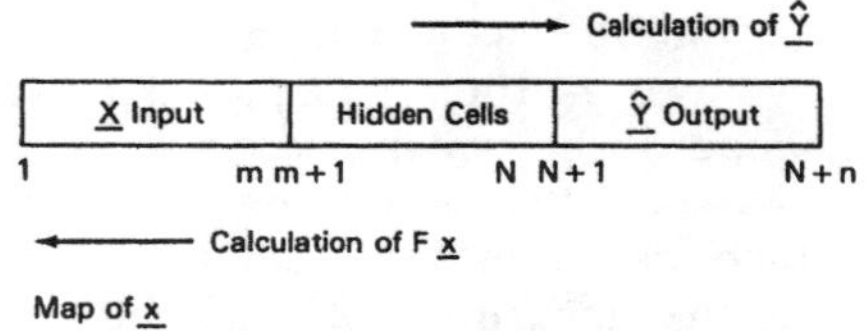

Fig. 4. Backwards flow of derivative calculation.

```
           DO  14 i = m+1,N+n
                  DO  14 j = 1,i-1
14                    W(i,j)=W(i,j)-
                        learning_rate*
                        F_W(i,j)
100        CONTINUE
1000       CONTINUE
```

The key point here is that the weights W are adjusted in response to the *current* vector F_W, which only depends on the current pattern t; the weights are adjusted after each pattern is processed. (In batch learning, by contrast, the weights are adjusted only *after* the "DO 100" loop is completed.)

In practice, maximum_passes is usually set to an enormous number; the loop is exited only when a test of convergence is passed, a test of error size or weight change which can be injected easily into the loop. True real-time learning is like pattern learning, but with only one pass through the data and no memory of earlier times t. (The equations above could be implemented easily enough as a real-time learning scheme; however, this will not be true for backpropagation through time.) The term "on-line learning" is sometimes used to represent a situation which *could* be pattern learning or *could* be real-time learning. Most people using basic backpropagation now use pattern learning rather than real-time learning because, with their data sets, many passes through the data are needed to ensure convergence of the weights.

The reader should be warned that I have not actually tested the code here. It is presented simply as a way of explaining more precisely the preceding ideas. The C implementations which I have worked with have been less transparent, and harder to debug, in part because of the absence of range checking in that language. It is often argued that people "who know what they are doing" do not need range checking and the like; however, people who think they never make mistakes should probably not be writing this kind of code. With neural network code, especially, good diagnostics and tests are very important because bugs can lead to slow convergence and oscillation—problems which are hard to track down, and are easily misattributed to the algorithm in use. If one *must* use a language without range checking, it is extremely important to maintain a version of the code which is highly transparent and safe, however inefficient it may be, for diagnostic purposes.

III. Backpropagation Through Time

A. Background

Backpropagation through time—like basic backpropagation—is used most often in pattern recognition today. Therefore, this section will focus on such applications, using notation like that of the previous section. See Section IV for other applications.

In some applications—such as speech recognition or submarine detection—our classification at time t will be more accurate if we can account for what we saw at earlier times. *Even though* the training set still fits the same format as above, we want to use a more powerful class of networks to do the classification; we want the output of the network at time t to account for variables at earlier times (as in Fig. 5).

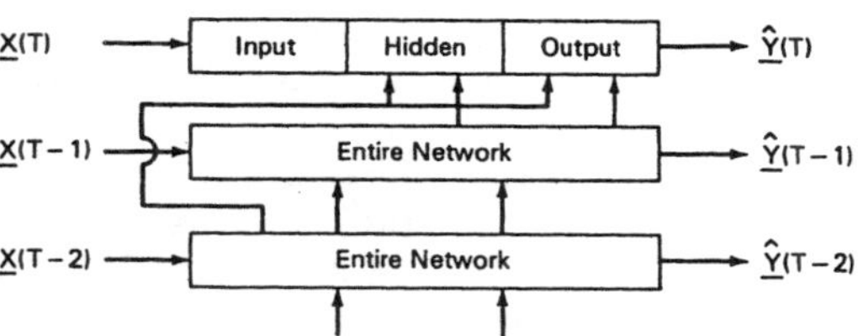

Fig. 5. Generalized network design with time lags.

The Introduction cited a number of examples where such "memory" of previous time periods is very important. For example, it is easier to recognize moving objects if our network accounts for *changes* in the scene from the time $t-1$ to time t, which requires memory of time $t-1$. Many of the best pattern recognition algorithms involve a kind of "relaxation" approach where the representation of the world at time t is based on an *adjustment* of the representation at time $t-1$; this requires memory of the *internal* network variables for time $t-1$. (Even Kalman filtering requires such a representation.)

B. Example of a Recurrent Network

Backpropagation can be applied to any system with a well-defined order of calculations, even if those calculations depend on past calculations within the network itself. For the sake of generality, I will show how this works for the network design shown in Fig. 5 where *every* neuron is potentially allowed to input values from *any* of the neurons at the two previous time periods (including, of course, the input neurons). To avoid excess clutter, Fig. 5 shows the hidden and output sections of the network (parallel to Fig. 2) only for time T, but they are present at other times as well. To translate this network into a mathematical system, we can simply replace (2) above by

$$\text{net}_i(t) = \sum_{j=1}^{i-1} W_{ij}x_j(t) + \sum_{j=1}^{N+n} W'_{ij}x_j(t-1) + \sum_{j=1}^{N+n} W''_{ij}x_j(t-2). \tag{15}$$

Again, we can simply fix some of the weights to be zero, if we so choose, in order to simplify the network. In most applications today, the W'' weights are fixed to zero (i.e., erased from all formulas), and all the W' weights are fixed to zero as well, except for W'_{ii}. This is done in part for the sake of parsimony, and in part for historical reasons. (The "time-delay neural networks" of Watrous and Shastri [5] assumed that special case.) Here, I deliberately include extra terms for the sake of generality. I allow for the fact that *all* active neurons (neurons other than input neurons) can be allowed to input the outputs of *any* other neurons if there is a time lag in the connection. The weights W' and W'' are the weights on those time-lagged connections between neurons. [Lags of more than two periods are also easy to manage; they are treated just as one would expect from seeing how we handle lag-two terms, as a special case of (7).]

These equations could be embodied in a subroutine:

SUBROUTINE NET2($X(t)$, W', W'', $x(t-2)$,

$x(t-1)$, $x(t)$, Yhat),

which is programmed just like the subroutine NET, with the modifications one would expect from (15). The output arrays are $x(t)$ and $Yhat$.

When we call this subroutine for the first time, at $t = 1$, we face a minor technical problem: there is no value for $x(-1)$ or $x(0)$, both of which we need as inputs. In principle, we can use any values we wish to choose; the choice of $x(-1)$ and $x(0)$ is essentially part of the definition of our network. Most people simply set these vectors to zero, and argue that their network will start out with a blank slate in classifying whatever dynamic pattern is at hand, both in the training set and in later applications. (Statisticians have been known to treat these vectors as weights, in effect, to be adapted along with the other weights in the network. This works fine in the training set, but opens up questions of what to do when one applies the network to new data.)

In this section, I will assume that the data run from an initial time $t = 1$ through to a final time $t = T$, which plays a crucial role in the derivative calculations. Section IV will show how this assumption can be relaxed somewhat.

C. Adapting the Network: Equations

To calculate the derivatives of F_W_{ij}, we use the same equations as before, *except* that (10) is replaced by

$$F\_x_i(t) = F\_\hat{Y}_{i-N}(t) + \sum_{j=i+1}^{N+n} W_{ji} * F\_net_j(t) + \sum_{j=m+1}^{N+n} W'_{ji} * F\_net_j(t+1) + \sum_{j=m+1}^{N+n} W''_{ji} * F\_net_j(t+2). \tag{16}$$

Once again, if one wants to fix the W'' terms to zero, one can simply delete the rightmost term.

Notice that this equation makes it impossible for us to calculate $F\_x_i(t)$ and $F\_net_i(t)$ until *after* $F\_net_j(t+1)$ and $F\_net_j(t+2)$ are already known; therefore, we can only use this equation by proceeding *backwards in time*, calculating F_net for time T, and then working our way backwards to time 1.

To adapt this network, of course, we need to calculate $F\_W'_{ij}$ and $F\_W''_{ij}$ as well as F_W_{ij}:

$$F\_W'_{ij} = \sum_{t=1}^{T} F\_net_i(t+1) * x_j(t) \tag{17}$$

$$F\_W''_{ij} = \sum_{t=1}^{T} F\_net_i(t+2) * x_j(t). \tag{18}$$

In all of these calculations, $F\_net(T+1)$ and $F\_net(T+2)$ should be treated as zero. For programming convenience, I will later define quantities like $F\_net'_i(t) = F\_net_i(t+1)$, but this is purely a convenience; the subscript "i" and the time argument are enough to identify which derivative is being represented. (In other words, $net_i(t)$ represents a specific quantity z_i as in (8), and $F\_net_i(t)$ represents the ordered derivative of E with respect to that quantity.)

D. Adapting the Network: Code

To fully understand the meaning and implications of these equations, it may help to run through a simple (hypothetical) implementation.

First, to calculate the derivatives, we need a new subroutine, dual to NET2.

```
      SUBROUTINE F_NET2(F_Yhat, W, W', W", x, F_net,
     F_net', F_net", F_W, F_W', F_W")
        REAL F_Yhat(n), W(N+n,N+n),
          W'(N+n,N+n), W"(N+n,N+n)
        REAL x(N+n),F_net(N+n),F_net'(N+n),
          F_net"(N+n)
        REAL F_W(N+n,N+n),F_W'(N+n,N+n),
          F_W"(N+n,N+n),F_x(N+n)
        INTEGER i,j,n,m,N
C Initialize equation (16)
        DO 1 i = 1,N
 1        F_x(i)=0.
        DO 2 i = 1,n
 2        F_x(i+N)=F_Yhat(i)
C RUN THROUGH (16), (11), AND (12) AS A SET,
C   RUNNING BACKWARDS
        DO 1000 i=N+n,m+1,-1
C         first complete (16)
          DO 161 j = i+1,N+n
 161        F_x(i)=F_x(i)+W(j,i)*F_net(j)
          DO 162 j = m+1,N+n
 162        F_x(i)=F_x(i)+W'(j,i)*F_net'(j)
              +W"(j,i)*F_net"(j)
C         next implement (11)
          F_net(i)=F_x(i)*x(i)*(1-x(i))
C         implement (12), (17), and (18)
            (as running sums)
          DO 12 j = 1,i-1
 12         F_W(i,j)=F_W(i,j)
              +F_net(i)*x(j)
          DO 1718 j =1,N+n
            F_W'(i,j)=F_W'(i,j)
              +F_net'(i)*x(j)
 1718       F_W"(i,j)=F_W"(i,j)
              +F_net" (i)*x(j)
 1000   CONTINUE
```

Notice that the last two DO loops have been set up to perform running sums, to simplify what follows.

Finally, we may adapt the weights as follows, by batch learning, where I use the abbreviation $x(i,)$, to represent the vector formed by $x(i,j)$ across all j.

```
        REAL x(-1:T,N+n), Yhat(T,n)
        DATA x(0,),x(-1,) / (2*(N+n)) * 0.0/
        DO 1000 pass_number=1,maximum_passes
C         First calculate outputs and errors in
C           a forward pass
          DO 100 t=1,T
 100        CALL NET2(X(t),W,W',W",x(t-2),
              x(t-1),x(t,),Yhat(t,))
C         Initialize the running sums to 0 and
C           set F_net(T), F_net(T+1) to 0
          DO 200 i = m+1,N+n
            F_net'(i)=0.
            F_net"(i)=0.
            DO 199 j = 1,N+n
              F_W(i,j)=0.
              F_W'(i,j)=0.
 199          F_W"(i,j)=0.
 200      CONTINUE
```

```
C  NEXT CALCULATE THE DERIVATIVES IN A SWEEP
   BACKWARDS THROUGH TIME
            DO  500 t = T,1,-1
C                 First, calculate the errors at the
C                   current time t
                  DO  410 i = 1,n
  410                F_Yhat(i)=Yhat(t,i)-Y(t,i)
C           Next, calculate F_net(t) for time t and
C             update the F_W running sums
              call F_NET2(F_Yhat,W,W',W''',x(t,),
              F_net,F_net',F_net'',F_W,
              F_W',F_W'')
C           Move F_net(t+1) to F_net(t), in effect,
C             to prepare for a new t value
                  DO  420 i = m+1,N+n
                     F_net''(i)=F_net'(i)
  420                F_net'(i)=F_net(i)
  500             CONTINUE
C  FINALLY, UPDATE THE WEIGHTS BY
   STEEPEST DESCENT
            DO  999 i = m+1,N+n
                  DO  998 j = 1,N+n
                     W(i,j)=W(i,j)-
                       learning_rate*F_W(i,j)
                     W'(i,j)=W'(i,j)-
                       learning_rate*F_W'(i,j)
  998                W''(i,j)-learning_rate*
                       F_W''(i,j);
  999             CONTINUE
 1000       CONTINUE
```

Once again, note that we have to go backwards in time in order to get the required derivatives. (There are ways to do these calculations in forward time, but exact results require the calculation of an entire *Jacobian* matrix, which is far more expensive with large networks.) For backpropagation through time, the natural way to adapt the network is in one big batch. Also note that we need to store a lot of intermediate information (which is inconsistent with real-time adaptation). This storage can be reduced by clever programming if W' and W'' are sparse, but it cannot be eliminated altogether.

In using backpropagation through time, we usually need to use much smaller learning rates than we do in basic backpropagation if we use steepest descent at all. In my experience [20], it may also help to start out by fixing the W' weights to zero (or to 1 when we want to force memory) in an initial phase of adaptation, and slowly free them up.

In some applications, we may not really care about errors in classification at all times t. In speech recognition, for example, we may only care about errors at the *end* of a word or phoneme; we usually do output a preliminary classification before the phoneme has been finished, but we usually do not care about the accuracy of that preliminary classification. In such cases, we may simply set F_Yhat to zero in the times we do not care about. To be more sophisticated, we may replace (6) by a more precise model of what we *do* care about; whatever we choose, it should be simple to replace (9) and the F_Yhat loop accordingly.

IV. Extensions of the Method

Backpropagation through time is a very general method, with many extensions. This section will try to describe the most important of these extensions.

A. Use of Other Networks

The network shown in (1)-(5) is a very simple, basic network. Backpropagation can be used to adapt a wide variety of other networks, including networks representing econometric models, systems of simultaneous equations, etc. Naturally, when one writes computer programs to implement a different kind of network, one must either describe *which* alternative network one chooses or *else* put options into the program to give the user this choice.

In the neural network field, users are often given a choice of network "topology." This simply means that they are asked to declare which *subset* of the possible weights/connections will actually be used. Every weight removed from (15) should be removed from (16) as well, along with (12) and (14) (or whichever apply to that weight); therefore, simplifying the network by removing weights simplifies all the other calculations as well. (Mathematically, this is the same as fixing these weights to zero.) Typically, people will remove an entire block of weights, such that the limits of the sums in our equations are all shrunk.

In a truly brain-like network, each neuron [in (15)] will only receive input from a small number of other cells. Neuroscientists do not agree on how many inputs are typical; some cite numbers on the order of 100 inputs per cell, while others quote 10 000. In any case, all of these estimates are small compared to the billions of cells present. To implement this kind of network efficiently on a conventional computer, one would use a linked list or a list of offsets to represent the connections actually implemented for each cell; the same strategy can be used to implement the backwards calculations and keep the connection costs low. Similar tricks are possible in parallel computers of all types. Many researchers are interested in devising ways to *automatically* make and break connections so that users will not have to specify all this information in advance [20]. The research on topology is hard to summarize since it is a mixture of normal science, sophisticated epistemology, and extensive ad hoc experimentation; however, the paper by Guyon *et al.* [13] is an excellent example of what works in practice.

Even in the neural network field, many programmers try to avoid the calculation of the exponential in (5). Depending on what kind of processor one has available, this calculation can multiply run times by a significant factor.

In the first published paper which discussed backpropagation at length as a way to adapt neural networks [14], I proposed the use of an artificial neuron ("continuous logic unit," CLU) based on

$$s(z) = 0, \qquad z < 0$$

$$s(z) = z, \qquad 0 < z < 1$$

$$s(z) = 1, \qquad z > 1.$$

This leads to a very simple derivative as well. Unfortunately, the *second* derivatives of this function are not well behaved, which can affect the efficiency of some applications. Still, many programmers are now using piecewise linear approximations to (5), along with lookup tables, which can work relatively well in some applications. In earlier experiments, I have also found good uses for a Taylor series approximation:

$$s(z) = 1/(1 - z + 0.5 * z^2), \qquad z < 0$$

$$s(z) = 1 - 1/(1 + z + 0.5 * z^2), \qquad z > 0.$$

In a similar spirit, it is common to speed up learning by "stretching out" $s(z)$ so that it goes from -1 to 1 instead of 0 to 1.

Backpropagation can also be used without using neural networks at all. For example, it can be used to adapt a network consisting entirely of user-specified functions, representing something like an econometric model. In that case, the way one proceeds depends on who one is programming for and what kind of model one has.

If one is programming for oneself and the model consists of a sequence of equations which can be invoked one after the other, then one should consider the tutorial paper [11], which also contains a more rigorous definition of what these "F_x_i" derivatives really mean and a proof of the chain rule for ordered derivatives. If one is developing a tool for others, then one might set it up to look like a standard econometric package (like SAS or Troll) where the user of the system types in the equations of his or her model; the backpropagation would go *inside* the package as a way to speed up these calculations, and would mostly be transparent to the user. If one's model consists of a set of simultaneous equations which need to be solved at each time, then one must use more complicated procedures [15]; in neural network terms, one would call this a "doubly recurrent network." (The methods of Pineda [16] and Almeida [17] are special cases of this situation.)

Pearlmutter [18] and Williams [19] have described alternative methods, designed to achieve results similar to those of backpropagation through time, using a different computational strategy. For example, the Williams–Zipser method is a special case of the "conventional perturbation" equation cited in [14], which rejected this as a neural network method on the grounds that its computational costs scale as the square of the network size; however, the method does yield exact derivatives with a time-forward calculation.

Supervised learning problems or forecasting problems which involve memory can also be translated into control problems [15, p. 352], [20], which allows the use of adaptive critic methods, to be discussed in the next section. Normally, this would yield only an approximate solution (or approximate derivatives), but it would also allow time-forward real-time learning. If the network itself contains calculation noise (due to hardware limitations), the adaptive critic approach might even be more robust than backpropagation through time because it is based on mathematics which allow for the presence of noise.

B. Applications Other Than Supervised Learning

Backpropagation through time can also be used in two other major applications: neuroidentification and neurocontrol. (For applications to sensitivity analysis, see [14] and [15].)

In neuroidentification, we try to do with neural nets what econometricians do with forecasting models. (Engineers would call this the identification problem or the problem of identifying dynamic systems. Statisticians refer to it as the problem of estimating stochastic time-series models.) Our training set consists of vectors $\boldsymbol{X}(t)$ and $\boldsymbol{u}(t)$, not $\boldsymbol{X}(t)$ and $\boldsymbol{Y}(t)$. Usually, $\boldsymbol{X}(t)$ represents a set of observations of the eXternal (sic) world, and $\boldsymbol{u}(t)$ represents a set of actions that we had control over (such as the settings of motors or actuators). The *combination* of $\boldsymbol{X}(t)$ and $\boldsymbol{u}(t)$ is input to the network at each time t. Our target, at time t, is the vector $\boldsymbol{X}(t + 1)$.

We could easily build a network to input these inputs, and aim at these targets. We could simply collect the inputs and targets into the format of Section II, and then use basic backpropagation. But basic backpropagation contains no "memory." The forecast of $\boldsymbol{X}(t + 1)$ would depend on $\boldsymbol{X}(t)$, but *not* on previous time periods. If human beings worked like this, then they would be unable to predict that a ball might roll out the far side of a table after rolling down under the near side; as soon as the ball disappeared from sight [from the current vector $\boldsymbol{X}(t)$], they would have no way of accounting for its existence. (Harold Szu has presented a more interesting example of this same effect: if a tiger chased after such a memoryless person, the person would forget about the tiger after first turning to run away. Natural selection has eliminated such people.) Backpropagation through time permits more powerful networks, which do have a "memory," for use in the same setup.

Even this approach to the neuroidentification problem has its limitations. Like the usual methods of econometrics [15], it may lead to forecasts which hold up poorly over multiple time periods. It does not properly identify where the noise comes from. It does not permit real-time adaptation. In an earlier paper [20], I have described some ideas for overcoming these limitations, but more research is needed. The first phase of Kawato's cascade method [9] for controlling a robot arm is an identification phase, which is more robust over time, and which uses backpropagation through time in a different way; it is a special case of the "pure robust method," which also worked well in the earliest applications which I studied [1], [20].

After we have solved the problem of identifying a dynamic system, we are then ready to move on to controlling that system.

In neurocontrol, we often *start out* with a model or network which describes the system or plant we are trying to control. Our problem is to adapt a *second* network, the *action* network, which inputs $\boldsymbol{X}(t)$ and outputs the control $\boldsymbol{u}(t)$. (In actuality, we can allow the action network to "see" or input the *entire* vector $\boldsymbol{x}(t)$ calculated by the model network; this allows it to *account* for memories such as the recent appearance of a tiger.) Usually, we want to adapt the action network so as to maximize some measure of performance or utility $U(\boldsymbol{X}, t)$ summed over time. Performance measures used in past applications have included everything from the energy used to move a robot arm [8], [9] through to net profits received by the gas industry [11]. Typically, we are given a set of possible initial states $\boldsymbol{x}(1)$, and asked to train the action network so as to maximize the sum of utility from time 1 to a final time T.

To solve this problem using backpropagation through time, we simply calculate the derivatives of our performance measure with respect to all of the weights in the action network. "Backpropagation" refers to how we calculate the derivatives, not to anything involving pattern recognition or error. We then adapt the weights according to these derivatives, as in (12), except that the sign of the adjustment term is now positive (because we are *maximizing* rather than *minimizing*).

The easiest way to implement this approach is to merge

the utility function, the model network, and the action network into one big network. We can then construct the dual to this entire network, as described in 1974 [1] and illustrated in my recent tutorial [11]. However, if we wish to keep the three component networks distinct, then the bookkeeping becomes more complicated. The basic idea is illustrated in Fig. 6, which maps exactly into the approach used by Nguyen and Widrow [7] and by Jordan [8].

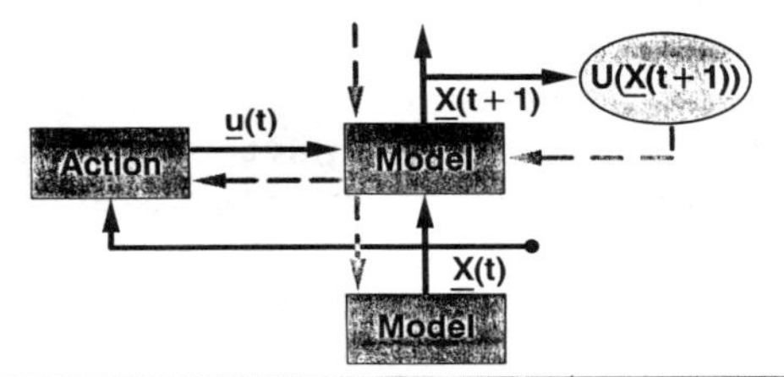

Fig. 6. Backpropagating utility through time. (Dashed lines represent derivative calculations.)

Instead of working with a *single* subroutine, NET, we now need *three* subroutines:

$$\mathrm{UTILITY}(\boldsymbol{X};\ t;\ \boldsymbol{x}'';\ U)$$

$$\mathrm{MODEL}(\boldsymbol{X}(t),\ \boldsymbol{u}(t);\ \boldsymbol{x}(t);\ \boldsymbol{X}(t+1))$$

$$\mathrm{ACTION}(\boldsymbol{x}(t);\ \boldsymbol{W};\ \boldsymbol{x}'(t);\ \boldsymbol{u}(t)).$$

In each of these subroutines, the two arguments on the right are technically outputs, and the argument on the far right is what we usually think of as the output of the network. We need to know the full vector $\boldsymbol{x}$ produced inside the model network so that the action network can "see" important memories. The action network does not need to have its own internal memory, but we need to save its internal state ($\boldsymbol{x}'$) so that we can later calculate derivatives. For simplicity, I will assume that MODEL does not contain any lag-two memory terms (i.e., W'' weights). The primes after the x's indicate that we are looking at the internal states of different networks; they are unrelated to the primes representing lagged values, discussed in Section III, which we will also need in what follows.

To use backpropagation through time, we need to construct dual subroutines for *all* three of these subroutines:

$$F\_\mathrm{UTILITY}(\boldsymbol{x}'';\ t;\ F\_X)$$

$$F\_\mathrm{MODEL}(F\_\mathrm{net}',\ F\_X(t+1);\ \boldsymbol{x}(t);\ F\_\mathrm{net},\ F\_u)$$

$$F\_\mathrm{ACTION}(F\_u;\ \boldsymbol{x}'(t);\ F\_W).$$

The outputs of these subroutines are the arguments on the far right (including *F*_net), which are represented by the broken lines in Fig. 4. The subroutine *F*_UTILITY simply reports out the derivatives of $U(\boldsymbol{x}, t)$ with respect to the variables X_i. The subroutine *F*_MODEL is like the earlier subroutine *F*_NET2, except that we need to output *F*_*u* instead of derivatives to weights. (Again, we are adapting only the action network here.) The subroutine *F*_ACTION is virtually identical to the old subroutine *F*_NET, except that we need to calculate *F*_*W* as a running sum (as we did in *F*_NET2).

Of these three subroutines, *F*_MODEL is by far the most complex. Therefore, it may help to consider some possible code.

```
      SUBROUTINE F_MODEL(F_net', F_X', x, F_net, F_u)
C     The weights inside this subroutine are those
C     used in MODEL, analogous to those in NET2, and are
C     unrelated to the weights in ACTION
      REAL F_net'(N+n),F_X'(n),x(N+n),
     F_net(N+n),F_u(p),F_x(N+n)
      INTEGER i,j,n,m,N,p
      DO 1 i=1,N
1        F_x(i)=0.
      DO 2 i = 1,n
2        F_x(i+N)=F_X(i)
      DO 1000 i = N+n,1,-1
         DO 910 j = i+1,N+n
910         F_x(i)=F_x(i)+W(j,i)*F_net(j)
         DO 920 j = m+1,N+n
920         F_x(i)=F_x(i)+W'(j,i)*F_net'(j)
1000     F_net(i)=F_x(i)*x(i)*(1-x(i));
      DO 2000 i = 1,p
2000     F_u(i)=F_x(n+i)
```

The last small DO loop here assumes that $\boldsymbol{u}(t)$ was part of the input vector to the original subroutine MODEL, inserted into the slots between $\boldsymbol{x}(n+1)$ and $\boldsymbol{x}(m)$. Again, a good programmer could easily compress all this; my goal here is only to illustrate the mathematics.

Finally, in order to adapt the action network, we go through multiple passes, each starting from one of the starting values of $\boldsymbol{x}(1)$. In each pass, we call ACTION and then MODEL, one after the other, until we have built up a stream of forecasts from time 1 up to time T. Then, for each time t going backwards from T to 1, we call the UTILITY subroutine, then *F*_UTILITY, then *F*_MODEL, and then *F*_ACTION. At the end of the pass, we have the correct array of derivatives *F*_*W*, which we can then use to adjust the weights of the action network.

In general, backpropagation through time has the advantage of being relatively quick and exact. That is why I chose it for my natural gas application [11]. However, it cannot account for noise in the process to the controlled. To account for noise in maximizing an arbitrary utility function, we must rely on adaptive critic methods [21]. Adaptive critic methods do *not* require backpropagation through time in any form, and are therefore suitable for true real-time learning. There are other forms of neurocontrol as well [21] which are not based on maximizing a utility function.

C. Handling Strings of Data

In most of the examples above, I assumed that the training data form one lone time series, from t equals 1 to t equals T. Thus, in adapting the weights, I always assumed batch learning (except in the code in Section II); the weights were always adapted *after* a complete set of derivatives was calculated, based on a complete pass through all the data. Mechanically, one could use pattern learning in the backwards pass through time; however, this would lead to a host of problems, and it is difficult to see what it would gain.

Data in the real world are often somewhere between the two extremes represented by Sections II and III. Instead of having a set of unrelated patterns or one continuous time series, we often have a *set* of time series or strings. For example, in speech recognition, our training set may consist of a set of strings, each consisting of one word or one sentence. In robotics, our training set may consist of a set of strings, where each string represents one experiment with a robot.

In these situations, we can apply backpropagation through time to a *single string* of data at a time. For each string, we can calculate complete derivatives and update the weights. Then we can go on to the next string. This is like pattern learning, in that the weights are updated incrementally before the entire data set is studied. It requires intermediate storage for only one string at a time. To speed things up even further, we might adapt the net in stages, initially fixing certain weights (like W_{ii}) to zero or one.

Nevertheless, string learning is not the same thing as real-time learning. To solve problems in neuroidentification and supervised learning, the only consistent way to have internal memory terms *and* to avoid backpropagation through time is to use adaptive critics in a supporting role [15]. That alternative is complex, inexact, and relatively expensive for these applications; it may be unavoidable for true real-time systems like the human brain, but it would probably be better to live with string learning and focus on *other* challenges in neuroidentification for the time being.

D. Speeding Up Convergence

For those who are familiar with numerical analysis and optimization, it goes without saying that steepest descent—as in (12)—is a very inefficient method.

There is a huge literature in the neural network field on how to speed up backpropagation. For example, Fahlman and Touretzky of Carnegie-Mellon have compiled and tested a variety of intuitive insights which can speed up convergence a hundredfold. Their benchmark problems may be very useful in evaluating other methods which claim to do the same. A few authors have copied simple methods from the field of numerical analysis, such as quasi-Newton methods (BFGS) and Polak-Ribiere conjugate gradients; however, the former works only on small problems (a hundred or so weights) [22], while the latter works well only with batch learning and very careful line searches. The need for careful line searches is discussed in the literature [23], but I have found it to be unusually important when working with large problems, including simulated linear mappings.

In my own work, I have used Shanno's more recent conjugate gradient method with batch learning; for a *dense* training set—made up of distinctly different patterns—this method worked better than anything else I tried, including pattern learning methods [12]. Many researchers have used *approximate* Newton's methods, without saying that they are using an approximation; however an *exact* Newton's method can also be implemented in $O(N)$ storage, and has worked reasonably well in early tests [12]. Shanno has reported new breakthroughs in function minimization which may perform still better [24]. Still, there is clearly a lot of room for improvement through further research.

Needless to say, it can be much easier to converge to a set of weights which do not minimize error or which assume a simpler network; methods of that sort are also popular, but are useful only when they clearly fit the application at hand for identifiable reasons.

E. Miscellaneous Issues

Minimizing square error and maximizing likelihood are often taken for granted as fundamental principles in large parts of engineering; however, there is a large literature on alternative approaches [12], both in neural network theory and in robust statistics.

These literatures are beyond the scope of this paper, but a few related points may be worth noting. For example, instead of minimizing square error, we could minimize the 1.5 power of error; all of the operations above still go through. We can minimize E of (5) *plus* some constant k times the sum of squares of the weights; as k goes to infinity and the network is made linear, this converges to Kohonen's pseudoinverse method, a common form of associative memory. Statisticians like Dempster and Efron have argued that the linear form of this approach can be better than the usual least squares methods; their arguments capture the essential insight that people can forecast by analogy to historical precedent, instead of forecasting by a comprehensive model or network. Presumably, an ideal network would bring together both kinds of forecasting [12], [20].

Many authors worry a lot about local minima. In using backpropagation through time in robust estimation, I found it important to keep the "memory" weights near zero at first, and free them up gradually in order to minimize problems. When T is much larger than m—as statisticians recommend for good generalization—local minima are probably a lot less serious than rumor has it. Still, with T larger than m, it is very easy to construct local minima. Consider the example with $m = 2$ shown in Table I.

Table 1 Training Set for Local Minima

t	$X(t)$	$Y(t)$
1	0 1	.1
2	1 0	.1
3	1 1	.9

The error for each of the patterns can be plotted as a contour map as a function of the two weights w_1 and w_2. (For this simple example, no threshold term is assumed.) Each map is made up of straight contours, defining a fairly sharp trough about a central line. The three central lines for the three patterns form a triangle, the vertices of which correspond roughly to the local minima. Even when T is much larger than m, conflicts like this can exist within the training set. Again, however, this may not be an overwhelming problem in practical applications [19].

V. Summary

Backpropagation through time can be applied to many different categories of dynamical systems—neural networks, feedforward systems of equations, systems with time lags, systems with instantaneous feedback between variables (as in ordinary differential equations or simultaneous equation models), and so on. The derivatives which it calculates can be used in pattern recognition, in systems identification, and in stochastic and deterministic control. This paper has presented the key equations of backpropagation, as applied to neural networks of varying degrees of complexity. It has also discussed other papers which elaborate on the extensions of this method to more general applications and some of the tradeoffs involved.

References

[1] P. Werbos, "Beyond regression: New tools for prediction and analysis in the behavioral sciences," Ph.D. dissertation, Committee on Appl. Math., Harvard Univ., Cambridge, MA, Nov. 1974.

[2] D. Rumelhart, D. Hinton, and G. Williams, "Learning internal representations by error propagation," in D. Rumelhart and F. McClelland, eds., *Parallel Distributed Processing, Vol. 1.* Cambridge, MA: M.I.T. Press, 1986.

[3] D. B. Parker, "Learning-logic," M.I.T. Cen. Computational Res. Economics Management Sci., Cambridge, MA, TR-47, 1985.

[4] Y. Le Cun, "Une procedure d'apprentissage pour reseau a seuil assymetrique," in *Proc. Cognitiva '85*, Paris, France, June 1985, pp. 599–604.

[5] R. Watrous and L. Shastri, "Learning phonetic features using connectionist networks: an experiment in speech recognition," in *Proc. 1st IEEE Int. Conf. Neural Networks*, June 1987.

[6] H. Sawai, A. Waibel, P. Haffner, M. Miyatake, and K. Shikano, "Parallelism, hierarchy, scaling in time-delay neural networks for spotting Japanese phonemes/CV-syllables," in *Proc. IEEE Int. Joint Conf. Neural Networks*, June 1989.

[7] D. Nguyen and B. Widrow, "The truck backer-upper: An example of self-learning in neural networks," in W. T. Miller, R. Sutton, and P. Werbos, Eds., *Neural Networks for Robotics and Control.* Cambridge, MA: M.I.T. Press, 1990.

[8] M. Jordan, "Generic constraints on underspecified target trajectories," in *Proc. IEEE Int. Joint Conf. Neural Networks*, June 1989.

[9] M. Kawato, "Computational schemes and neural network models for formation and control of multijoint arm trajectory," in W. T. Miller, R. Sutton, and P. Werbos, Eds., *Neural Networks for Robotics and Control.* Cambridge, MA: M.I.T. Press, 1990.

[10] R. Narendra, "Adaptive control using neural networks," in W. T. Miller, R. Sutton, and P. Werbos, Eds., *Neural Networks for Robotics and Control.* Cambridge, MA: M.I.T. Press, 1990.

[11] P. Werbos, "Maximizing long-term gas industry profits in two minutes in Lotus using neural network methods," *IEEE Trans. Syst., Man, Cybern.*, Mar./Apr. 1989.

[12] —, "Backpropagation: Past and future," in *Proc. 2nd IEEE Int. Conf. Neural Networks*, June 1988. The transcript of the talk and slides, available from the author, are more introductory in nature and more comprehensive in some respects.

[13] I. Guyon, I. Poujaud, L. Personnaz, G. Dreyfus, J. Denker, and Y. Le Cun, "Comparing different neural network architectures for classifying handwritten digits," in *Proc. IEEE Int. Joint Conf. Neural Networks*, June 1989.

[14] P. Werbos, "Applications of advances in nonlinear sensitivity analysis," in R. Drenick and F. Kozin, Eds., *Systems Modeling and Optimization: Proc. 10th IFIP Conf. (1981).* New York: Springer-Verlag, 1982.

[15] —, "Generalization of backpropagation with application to a recurrent gas market model," *Neural Networks*, Oct. 1988.

[16] F. J. Pineda, "Generalization of backpropagation to recurrent and higher order networks," in *Proc. IEEE Conf. Neural Inform. Processing Syst.*, 1987.

[17] L. B. Almeida, "A learning rule for asynchronous perceptrons with feedback in a combinatorial environment," in *Proc. 1st IEEE Int. Conf. Neural Networks*, 1987.

[18] B. A. Pearlmutter, "Learning state space trajectories in recurrent neural networks," in *Proc. Int. Joint Conf. Neural Networks*, June 1989.

[19] R. Williams, "Adaptive state representation and estimation using recurrent connectionist networks," in W. T. Miller, R. Sutton, and P. Werbos, Eds., *Neural Networks for Robotics and Control.* Cambridge, MA: M.I.T. Press, 1990.

[20] P. Werbos, "Learning how the world works: Specifications for predictive networks in robots and brains," in *Proc. 1987 IEEE Int. Conf. Syst., Man, Cybern.*, 1987.

[21] —, "Consistency of HDP applied to a simple reinforcement learning problem," *Neural Networks*, Mar. 1990.

[22] J. Dennis and R. Schnabel, *Numerical Methods for Unconstrained Optimization and Nonlinear Equations.* Englewood Cliffs, NJ: Prentice-Hall, 1983.

[23] D. Shanno, "Conjugate-gradient methods with inexact searches," *Math. Oper. Res.*, vol. 3, Aug. 1978.

[24] —, "Recent advances in numerical techniques for large-scale optimization," in W. T. Miller, R. Sutton, and P. Werbos, Eds., *Neural Networks for Robotics and Control.* Cambridge, MA: M.I.T. Press, 1990.

Article 5.2

Neural Networks for Self-Learning Control Systems

Derrick H. Nguyen and Bernard Widrow

ABSTRACT: Neural networks can be used to solve highly nonlinear control problems. This paper shows how a neural network can learn of its own accord to control a nonlinear dynamic system. An emulator, a multilayered neural network, learns to identify the system's dynamic characteristics. The controller, another multilayered neural network, next learns to control the emulator. The self-trained controller is then used to control the actual dynamic system. The learning process continues as the emulator and controller improve and track the physical process. An example is given to illustrate these ideas. The "truck backer-upper," a neural network controller steering a trailer truck while backing up to a loading dock, is demonstrated. The controller is able to guide the truck to the dock from almost any initial position. The technique explored here should be applicable to a wide variety of nonlinear control problems.

Introduction

This paper addresses the problem of controlling severely nonlinear systems from the standpoint of utilizing neural networks to achieve nonlinear controller design. The methodology shows promise for application to control problems that are so complex that analytical design techniques do not exist and may not exist for sometime to come. Neural networks can be used to implement highly nonlinear controllers with weights or internal parameters that can be determined by a self-learning process.

Neural Networks

A *neural network* is a system with inputs and outputs and is composed of many simple and similar processing elements. The processing elements each have a number of internal parameters called *weights*. Changing the weights of an element will alter the behavior of the element and, therefore, will also alter the behavior of the whole network. The goal here is to choose the weights of the network to achieve a desired input/output relationship. This process is known as *training the network*. The network can be considered memoryless in the sense that, if one keeps the weights constant, the output vector depends only on the current input vector and is independent of past inputs.

The authors are with Information Systems Laboratory, Department of Electrical Engineering, Stanford University, Stanford, CA 94305.

Adalines

The processing element used in the networks in this paper, the Adaline [1], is shown in Fig. 1. It has an input vector $X = \{x_i\}$, which contains n components, a single output y, and a weight vector $W = \{w_i\}$, which also contains n components. The weights are variable coefficients indicated by circles with arrows. The output y equals the sum of inputs multiplied by the weights and then passed through a nonlinear function. (Note: In the early 1960s, Adaline elements utilized sharp quantizers in the form of signum functions. Today both signum and the differentiable sigmoid functions are used.)

$$s(X) = \sum_{i=0}^{n-1} w_i x_i \tag{1}$$

$$y(X) = f(s(X)) \tag{2}$$

The nonlinear function $f(s)$ used here is the sigmoid function

$$f(s) = [1 - \exp(-2s)]/[1 + \exp(-2s)] = \tanh(s) \tag{3}$$

With this nonlinearity, the Adaline behaves similar to a linear filter when its output is small, but saturates to $+1$ or -1 as the output magnitude increases. It should be noted that one of the Adaline's inputs is usually set to $+1$. This provides the Adaline with a way of adding a constant bias to the weighted sum.

The goal here is to train the Adaline to achieve a desired form of behavior. During the training process, the Adaline is presented with an input X, which causes its output to be $y(X)$. We would like the Adaline to output a desired value $d(X)$ instead, and so we adjust the weights to cause the output to be something closer to $d(X)$ the next time X is presented. The value $d(X)$ is called the *desired response*.' Many input, desired-response pairs are used in the training of the weights.

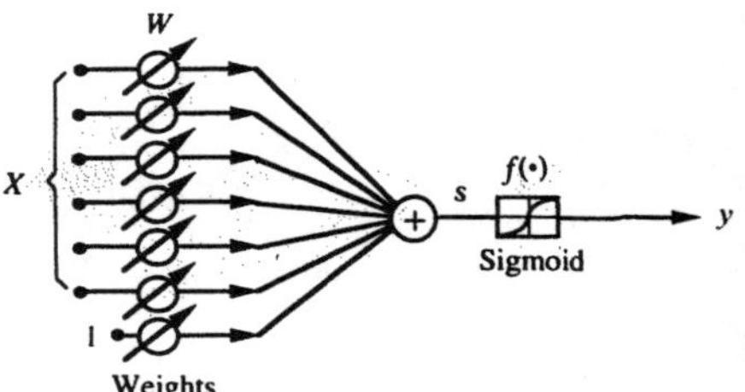

Fig. 1. Adaline with sigmoid.

A good measure of the Adaline's performance is the mean-squared error $\mathbf{J}$, where $E(\cdot)$ denotes an expectation over all available $(X, d(X))$ pairs.

$$\mathbf{J} = E(\text{error}^2) \tag{4}$$

$$= E(d(X) - y(X))^2 \tag{5}$$

$$= E\left(d(x) - f\left(\sum_{i=0}^{n-1} w_i x_i\right)\right)^2 \tag{6}$$

By applying gradient descent [1]–[3], the algorithm to adjust W to minimize $\mathbf{J}$ turns out to be the following, where $f'(s)$ is the derivative of the function $f(s)$.

$$w_{i,\text{new}} = w_{i,\text{old}} + 2\mu\delta x_i \tag{7}$$

$$\delta = (d(X) - y(X))\,f'(s(X)) \tag{8}$$

The designer chooses μ, which affects the speed of convergence and stability of the weights during training. The value δ can be thought of as an "equivalent error" and would be equal to the error $d(X) - y(X)$ if $f(s)$ were the identity function. In this case, Eqs. (7) and (8) would be the same as the 1959 least-mean-squares (LMS) algorithm of Widrow and Hoff [1] and Widrow and Stearns [3].

The preceding algorithm is applied many times with many different $(X, d(X))$ pairs until the weights converge to a minimum of the objective function $\mathbf{J}$.

Back-Propagation Algorithm

In this paper, Adalines are connected together to form what is known as a *layered feedforward neural network*, shown in Fig. 2. A layer of Adalines is created by connecting a number of Adalines to the same input vector. Many layers can then be cascaded, with outputs of one layer connected to the inputs of the next layer, to form a

Reprinted from *IEEE Contr. Syst. Mag.*, pp. 18–23, April 1990.

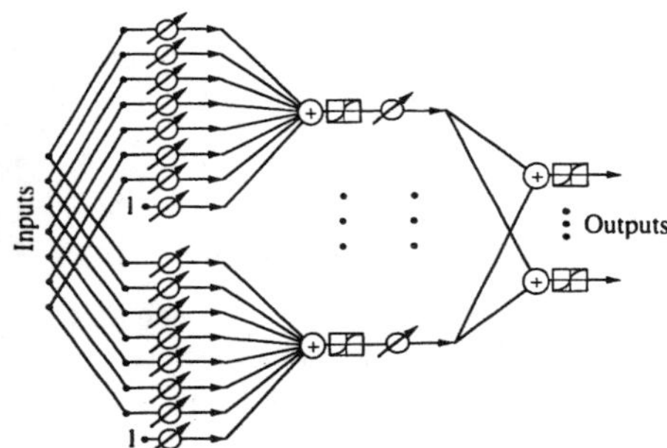

Fig. 2. Two-layer feedforward neural network.

network. It has been proven that a network consisting of only two layers of Adalines can implement any nonlinear function X, $d(X)$ given enough Adalines in the first layer (the layer closest to the input). The idea is that each Adaline in the first layer can take a small piece of the function relating X to $d(X)$ and make a linear approximation to that piece. The second layer then adds the pieces together to form the complete approximation to the desired function. A proof of this is given in [4]. [Note that $d(X)$ can be vector valued since the network can have more than one output.] Despite this theoretical result, networks of many more layers than two are being used. They offer a variety of convergence properties, robustness, and generalization characteristics (an ability to respond correctly to inputs that were not trained in) that can be quite different from those obtainable with a two-layer network.

The algorithm used to train layered neural networks is known as *back-propagation* [2], [5], [6]. This algorithm converges to a set of weights that minimizes the mean-square error

$$\mathbf{J} = E(\|d(X) - y(X)\|^2) \tag{9}$$

where $y(X)$ is the output vector of the last layer of the network. Just as in the case of the single Adaline, it is convenient to define "equivalent error" for each Adaline in the network. For Adaline m in the output layer, the equivalent error is the following, where y_m is the output of Adaline m, δ_j is the equivalent error of the jth Adaline, j indexes the set of all Adalines that have inputs connected to Adaline m's output, and w_{jm} is the weight of the connection from Adaline m's output to Adaline j's input.

$$\delta_m = (d_m(X) - y_m(X))\, f'(s_m(X)) \tag{10}$$

For Adaline m in one of the other layers, the equivalent error is

$$\delta_m = f'(s_m(X)) \sum_j \delta_j w_{jm} \tag{11}$$

Each weight is updated using the same equation as for the single Adaline case, where i ranges over the inputs of Adaline m.

$$w_{mi,\text{new}} = w_{mi,\text{old}} + 2\mu\delta_m x_k \tag{12}$$

Note that this is called the back-propagation algorithm because the equivalent error is computed for the output layer using Eq. (10), and then propagated backward through the layers toward the input layer using Eq. (11). As the equivalent error is computed during the backward propagation, the weights are updated using Eq. (12).

Layered neural networks adapted by means of the back-propagation algorithm are powerful tools for pattern recognition, associative memory, and adaptive filtering. In this paper, adaptive neural networks will be used to solve nonlinear adaptive control problems that are very difficult to solve with conventional methods.

Control Problem

The standard representation of a finite-dimensional discrete-time plant is shown in Fig. 3. The vector u_k represents the inputs to the plant at time k and the vector z_k represents the state of the plant at time k. The function $A(z_k, u_k)$ maps the current inputs and state into the next state. When the plant is linear, the usual state equation holds, where F and G are matrices.

$$z_{k+1} = A(z_k, u_k) = Fz_k + Gu_k \tag{13}$$

The function $A(z_k, u_k)$ would be nonlinear for a nonlinear plant.

A common problem in control is to provide the correct input vector to drive a nonlinear plant from an initial state to a subsequent desired state z_d. The typical approach used in solving this problem involves linearizing the plant around a number of operating points, building linear state-space models of the plant at these operating points, and then building a controller. For nonlinear plants, this approach is usually computationally intensive and requires considerable design effort.

In this paper, the objective is to train a controller—in this case, a neural network—to produce the correct signal u_k to drive the plant to the desired state z_d given the current state of the plant z_k (Fig. 3). Each value of u_k over time plays a part in determining the state of the plant. Knowing the desired state, however, does not easily yield information about the values of u_k that would be required to achieve it.

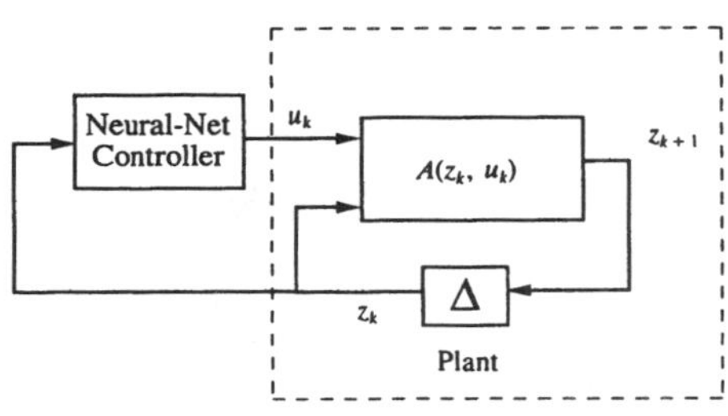

Fig. 3. Plant and controller.

A number of different approaches for training a controller have been described in the literature. They include reinforcement learning [7]–[9], inverse control [10], [11], and optimal control [11]. The architecture and training algorithm presented in this paper are novel in that they require little guidance from the designer to solve the control problem. This approach uses neural networks in optimal control by training the controller to maximize a performance function. The approach is different from [11] in that the plant can be an unknown plant and plant identification is a part of the algorithm. A similar approach has been used by Widrow and Stearns [3], Widrow [10], and Jordan [12].

Training Algorithm

Plant Identification— Training the Plant Emulator

Before training the neural net controller, a separate neural net is trained to behave like the plant. Specifically, the neural net is trained to emulate $A(z_k, u_k)$. Training the emulator is similar to plant identification in control theory, except that the plant identification here (Fig. 4) is done automatically by a neural network capable of modeling nonlinear plants.

In this paper, we assume that the states of the plant are directly observable without noise. A neural net with as many outputs as there are states, and as many inputs as there

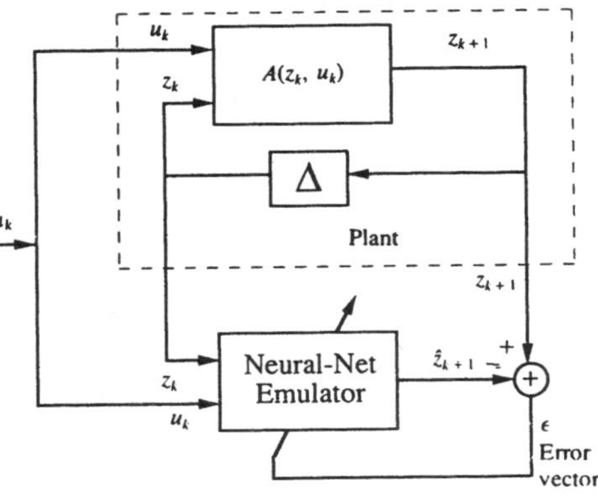

Fig. 4. Training the neural net plant emulator.

are states plus plant inputs, is created. The number of layers in the neural net and the number of nodes in each layer presently are determined empirically since they depend on the degree of nonlinearity of the plant.

In Fig. 4, the training process begins with the plant in an initial state. The plant inputs are generated randomly. At time k, the input of the neural net is set equal to the current state of the plant z_k and the plant input u_k. The neural net is trained by back-propagation [Eqs. (10)–(12)] to predict the next state of the plant, with the value of the next state of the plant z_{k+1} used as the desired response during training. This process is roughly analogous to the steps that would be taken by a human designer to identify the plant. In this case, however, the plant identification is done automatically by a neural network.

Training the Neural Network Controller

Given that the emulator now closely matches the plant dynamics, we use it for the purpose of training the controller. The controller learns to drive the plant emulator from an initial state z_0 to the desired state z_d in K time steps. Learning takes place during many trials or runs, each starting from an initial state and terminating at a final state z_K. The objective of the learning process is to find a set of controller weights that minimizes the error function $\mathbf{J}$, where $\mathbf{J}$ is averaged over the set of initial states z_0.

$$\mathbf{J} = E(\|z_d - z_K\|^2) \qquad (14)$$

The training process for the controller is illustrated in Fig. 5. The training process starts with the neural net plant emulator set in a random initial state z_0. Because the neural net controller initially is untrained, it will output an erroneous control signal u_0 to the plant emulator and to the plant itself. The plant emulator will then move to the next state z_1, and this process continues for K time steps. At this point, the plant is at the state z_K. (Note that the number of time steps K needs to be determined by the designer.)

We now would like to modify the weights in the controller network so that the square error $(z_d - z_K)^2$ will be less at the end of the next run. To train the controller, we need to know the error in the controller output u_k for each time step k. Unfortunately, only the error in the final plant state, $(z_d - z_K)$, is available. However, because the plant emulator is a neural network, we can back-propagate the final plant error $(z_d - z_K)$ through the plant emulator using Eqs. (10) and (11) to get an equivalent error for the controller in the Kth stage. This error then can be used to train the controller by using Eqs. (11) and (12). The emulator in a sense translates the error in the final plant state to the error in the controller output. The real plant cannot be used here because the error cannot be propagated through it. This is why the neural network emulator is needed. The error continues to be back-propagated through all K stages of the run using Eq. (11), and the controller's weight change is computed for each stage. The weight changes from all the stages obtained from the back-propagation algorithm are added together and then added to the controller's weights. This completes the training for one run.

The algorithm described would require saving all the weight changes so that they can be added to the original weights at the end of the run. In practice, for simplicity's sake, the weight changes are added immediately to the weights as they are computed. This does not significantly affect the final result since the weight changes are small and do not affect the controller's weights very much after one run. It is their accumulated effects over a large number of runs that improve the controller's performance.

Figure 5 represents the controller training process. For clarity, the details of error back-propagation are not illustrated there, but are described above and are represented algebraically by Eqs. (10)–(12). Because the training algorithm is essentially an implementation of gradient descent, local minima in the error function may yield suboptimal results. In practice, however, a good solution is almost always achieved by using a large number of Adalines in the hidden layers of the neural networks.

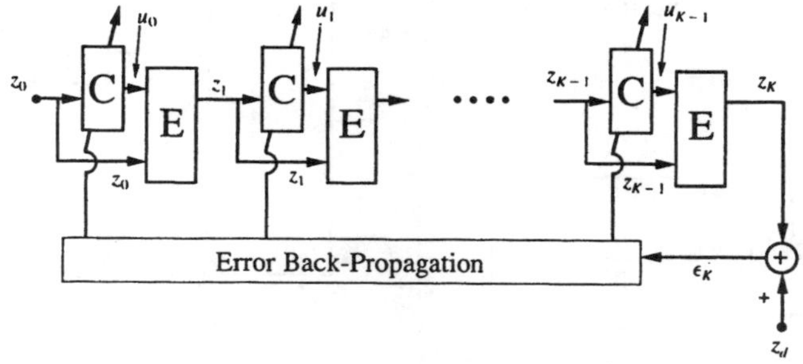

Fig. 5. Training the controller with back-propagation (C = controller, E = emulator).

An Example: Truck Backer-Upper

Backing a trailer truck to a loading dock is a difficult exercise for all but the most skilled truck drivers. Anyone who has tried to back up a house trailer or a boat trailer will realize this. Normal driving instincts lead to erroneous movements, and a great deal of practice is required to develop the requisite skills.

When watching a truck driver backing toward a loading dock, one often observes the driver backing, going forward, backing again, going forward, etc., and finally backing up to the desired position along the dock. The forward and backward movements help to position the trailer for successful backing up to the dock. A more difficult backing up sequence would only allow backing, with no forward movements permitted. The specific problem treated in this example is that of the design by self-learning of a nonlinear controller to control the steering of a trailer truck while backing up to a loading dock from any initial position. Only backing up is allowed. Computer simulation of the truck and its controller has demonstrated that the algorithm described earlier can train a controller to control the truck very well. An experimental two-layer neural controller containing 25 adaptive neural units in the first layer and one unit in the second layer has exhibited exquisite backing up control. The trailer truck can be straight or initially "jack-knifed" and aimed in many different directions, toward and away from the dock, but as long as there is sufficient clearance, the controller appears to be capable of finding a solution.

Figure 6 shows a computer-screen image of the truck, the trailer, and the loading dock. The critical state variables representing the position of the truck and that of the loading dock are θ_{cab}, the angle of the cab, $\theta_{trailer}$, the angle of the trailer, and $x_{trailer}$ and $y_{trailer}$, the Cartesian position of the rear of the center of the trailer. The definition of the state variables is illustrated in Fig. 6.

The truck is placed at some initial position and is backed up while being steered by the controller. The run ends when the truck comes to the dock. The goal is to cause the back of the trailer to be parallel to the loading dock, i.e., to make $\theta_{trailer}$ go to zero and to have the point $(x_{trailer}, y_{trailer})$ be aligned as closely as possible with the point (x_{dock}, y_{dock}). The final cab angle is unimportant.

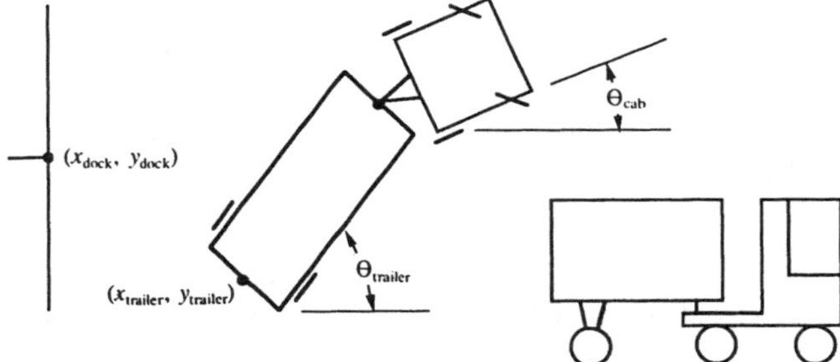

Fig. 6. Truck, trailer, and loading dock.

The controller will learn to achieve these objectives by adapting its weights to minimize the objective function **J**, where **J** is averaged over all training runs.

$$\mathbf{J} = E(\alpha_1(x_{dock} - x_{trailer})^2 + \alpha_2(y_{dock} - y_{trailer})^2 + \alpha_3(0 - \theta_{trailer})^2) \quad (15)$$

The constants α_1, α_2, and α_3 are chosen by the designer to weigh the importance of each error component.

Training

As described in the previous section, the learning process for the truck backer-upper controller involves two stages. The first stage trains a neural network to be an emulator of the truck and trailer kinematics. The second stage enables the neural-network controller to learn to control the truck by using the emulator as a guide. The control process consists of feeding the state vector z_k to the controller, which, in turn, provides a steering signal u_k between -1 (hard right) and $+1$ (hard left) to the truck (k is the time index). At each time step, the truck backs up by a fixed small distance. The next state is determined by the present state and the steering signal, which is fixed during the time step.

The process used to train the emulator is shown in Fig. 4. The emulator used in this example is a two-layer network with 25 Adalines in the first layer and four Adalines in the second layer. A suitable architecture for this network was determined by experiment. There is no theory for this yet. Experience shows that the choice of network architecture is important but a range of variation is permissible. The emulator network has five inputs corresponding to the four state variables x_k and the steering signal u_k, and four outputs corresponding to the next four state variables z_{k+1}.

During training, the truck backs up randomly, going through many cycles with randomly selected steering signals. The emulator learns to generate the next positional state vector when given the present state vector and the steering signal. This is done for a wide variety of positional states and steering angles. The two-layer emulator is adapted by means of the back-propagation algorithm. By this process, the emulator "gets the feel" of how the trailer and truck behave. Once the emulator is trained, then it can be used to train the controller.

Refer to Fig. 7. The identical blocks labeled C represent the controller net. The identical blocks labeled T represent the truck and trailer emulator. Let the weights of C be chosen at random initially. Let the truck back up. The initial state vector z_0 is fed to C, whose output sets the steering angle of the truck. The backing up cycle proceeds with the truck backing a small fixed distance so that the truck and trailer soon arrive at the next state z_1. With C remaining fixed, a new steering angle is computed for state z_1, and the truck backs up a small fixed distance once again. The backing up sequence continues until the truck hits something and stops. The final state z_K is compared with the desired final state (the rear of the trailer parallel to the dock with proper positional alignment) to obtain the final state error vector ϵ_K. (Note that, in reality, there is only one controller C. Figure 7 shows multiple copies of C for the purpose of explanation.) The error vector contains four elements, which are the errors of interest in $z_{trailer}$, $y_{trailer}$, $\theta_{trailer}$, and θ_{cab} and are used to adapt the controller C. The final angle of the cab, θ_{cab}, does not matter and so the element of the error vector due to θ_{cab} is set to zero. Each element of the error vector is also weighted by the corresponding α_i of Eq. (15).

The method of adapting the controller C is illustrated in Fig. 7. The final state error vector ϵ_K is used to adapt the blocks labeled C, which are maintained identical to each other throughout the adaptive process. The controller C is a two-layer neural network. The first layer has the six state variables as inputs, and this layer contains 25 adaptive Adaline units. The second, or output, layer has one adaptive Adaline unit and produces the steering signal as its output. All of the Adaline units have sigmoidal activation functions.

The controller C is adapted as described in the previous section. The weights of C are chosen initially at random. The initial position of the truck is chosen at random. The truck backs up, undergoing many individual back-up moves, until it comes to the dock. The final error is then computed and used by back-propagation to adapt the controller. The error is used to update the weights as it is back-propagated through the network. This way, the controller is adapted to minimize the sum of the squares of the components of the error vector using the method of steepest descent. The entire process is repeated by placing the truck and trailer in another initial position and allowing it to back up until it stops. Once again, the controller weights are adapted. And so on.

The controller and the emulator are two-layered neural networks each containing 25 hidden units. Thus, each stage of Fig. 7 amounts to a four-layer neural network. The entire process of going from an initial state to the final state can be seen from Fig. 7 to be analogous to a neural network having a number of layers equal to four times the

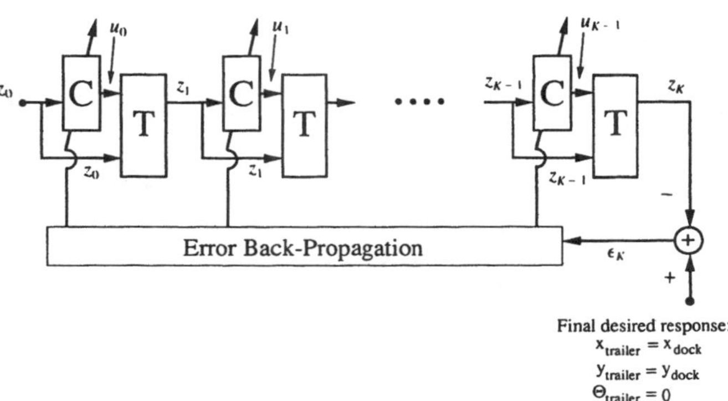

Fig. 7. Training the truck controller with back-propagation (C = controller, T = truck emulator).

number of backing up steps when going from the initial state to the final state. The number of steps K varies, of course, with the initial position of the truck and trailer relative to the position of the target, the loading dock.

The diagram of Fig. 7 was simplified for clarity of presentation. The output error actually back-propagates through the T-blocks and C-blocks. Thus, the error used to adapt each of the C-blocks does originate from the output error ϵ_K, but travels through the proper back-propagation paths. For purposes of back-propagation of the error, the T-blocks are the truck emulator. However, the actual truck kinematics are used when sensing the error ϵ_K itself.

The training of the controller was divided into several "lessons." In the beginning, the controller was trained with the truck initially set to points very near the dock and the trailer pointing at the dock. Once the controller was proficient at working with these initial positions, the problem was made harder by starting the truck farther away from the dock and at increasingly difficult angles. This way, the controller learned to do easy problems first and more difficult problems after it mastered the easy ones. There were 16 lessons in all. In the easiest lesson, the trailer was set about half a truck length from the dock in the x direction pointing at the dock, and the cab at a random angle of ± 30 deg. In the last and most difficult lesson, the rear of the trailer was set randomly between one and two truck lengths from the dock in the x direction and ± 1 truck length from the dock in the y direction. The cab and trailer angles were set to be the same, at a random angle of ± 90 deg. (Note that uniform distributions were used to generate the random parameters.) The controller was trained for about 1000 truck backups per lesson during the early lessons, and 2000 truck backups per lesson during the last few. It took about 20,000 backups to train the controller.

Results

The controller learned to control the truck very well with the preceding training process. Near the end of the last lesson, the root-mean-square (rms) error of y_{trailer} was about 3 percent of a truck length. The rms error of θ_{trailer} was about 7 deg. There is no error in x_{trailer} since a truck backup is stopped when $x_{\text{trailer}} = x_{\text{dock}}$. One may, of course, trade off the error in y_{trailer} with the error in θ_{trailer} by giving them different weights in the objective function during training.

After training, the controller's weights were fixed. The truck and trailer were placed in a variety of initial positions, and backing up was done successfully in each case. A back-up run when using the trained controller is demonstrated in Fig. 8. Initial and final states are shown on the computer screen displays, and the backing up trajectory is illustrated by the time-lapse plot. The trained controller was capable of controlling the truck from initial positions it had never seen. For example, the controller was trained with the cab and trailer placed at angles of ± 90 deg, but was capable of backing up the truck with the cab and trailer placed at any angle provided that there was enough distance between the truck and the dock.

A More Sophisticated Objective Function

The above-described truck controller was trained to minimize only the final state error. One can also train it to minimize total path length or control energy in addition to the final state error. For example, the objective function to minimize control energy is the following, with $\mathbf{J}$ averaged over all training trials.

$$\mathbf{J} = E\left[\alpha_1(x_{\text{dock}} - x_{\text{trailer}})^2 + \alpha_2(y_{\text{dock}} - y_{\text{trailer}})^2 + \alpha_3(\theta_{\text{dock}} - \theta_{\text{trailer}})^2 + \alpha_4 \sum_{k=0}^{K-1} u_k^2\right] \quad (16)$$

A simple change is made to the algorithm to minimize this objective function. In the original algorithm, the equivalent error for the controller at each time step k is computed during the backward pass of the back-propagation algorithm. It is easy to show that control energy can be minimized by adding $-\alpha_4 u_k$ to the equivalent error of the controller at each time step. The modified equivalent error is then back-propagated through the controller to update the controller's weights as earlier. This change makes sense, since using $-\alpha_4 u_k$ as an error in u_k causes the controller to learn to make u_k smaller in magnitude.

Training the controller to minimize control energy would cause it to drive the truck to the dock with as little steering as possible. An example with the controller trained in this manner is shown in Fig. 9. This example uses the same truck and trailer initial position as with the example of Fig. 8. Note that the path of the truck controlled by the new controller contains fewer sharp turns. Of course, the final state error increases somewhat because of the new control objective.

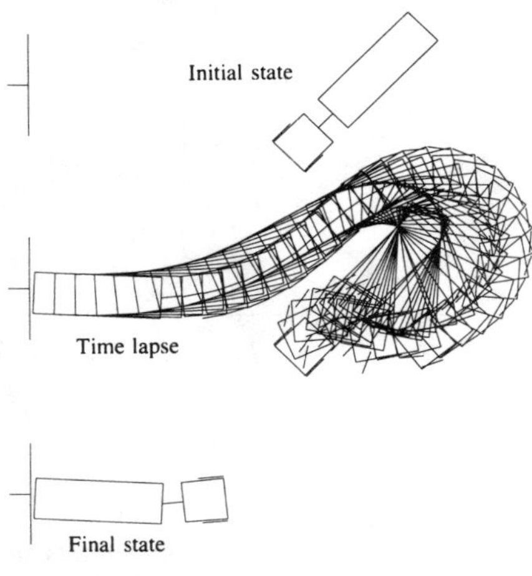

Fig. 8. A backing up example.

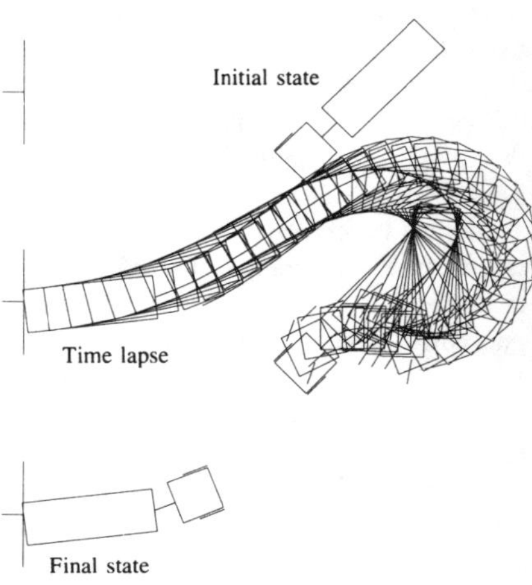

Fig. 9. Backing up while minimizing control energy.

Summary

The truck emulator in the form of a two-layer neural network was able to represent the trailer and truck when jackknifed, in line, or in any condition in between. Nonlinearity in the emulator was essential for accurate modeling of the kinematics. The angle between the truck and the trailer was not small and thus sin θ could not be represented approximately as θ. Controlling the nonlinear kinematics of the truck and trailer required a nonlinear controller, implemented by another two-layer neural network. Self-learning processes were used to determine the parameters of both the emulator and the controller. Thousands of backups were required to train these networks, requiring several hours on a workstation. Without the learning process, however, substantial amounts of human effort and design time would have been required to devise the controller.

The truck "backer-upper" learns to solve sequential decision problems. The control decisions made early in the backing up process have substantial effects on final results. Early moves may not always be in a direction to reduce error, but they position the truck and trailer for ultimate success. In many respects, the truck backer-upper learns a control strategy similar to a dynamic programming problem solution. The learning is done in a layered neural network. Connecting signals from one layer to another corresponds to the idea that the final state of a given backing up cycle is the same as the initial state of the next backing up cycle.

Future research will be concerned with

- Determination of complexity of the emulator as related to the complexity of the system being controlled.
- Determination of the complexity of the controller as related to the complexity of the emulator.
- Determination of the convergence and rate of learning for the emulator and controller.
- Proof of robustness of the control scheme.
- Analytic derivation of the nonlinear controller for the truck backer-upper, and comparison with the self-learned controller.
- Relearning in the presence of movable obstacles.
- Exploration of other areas of application for self-learning neural networks.

Acknowledgments

This research was sponsored by the SDIO Innovative Science and Technology Office and managed by ONR under Contract N00014-86-K-0718, by the Department of the Army Belvoir RD&E Center under Contract DAAK70-89-K-0001, by NASA Ames under Contract NCA2-389, by the Rome Air Development Center under Subcontract E-21-T22-S1 with Georgia Institute of Technology, and by grants from the Thomson CSF Company and the Lockheed Missiles and Space Company.

This material is based on work supported under a National Science Foundation Graduate Fellowship. Any opinions, findings, conclusions, or recommendations expressed in this publication are those of the authors and do not necessarily reflect the views of the National Science Foundation.

References

[1] B. Widrow and M. E. Hoff, Jr., "Adaptive Switching Circuits," in *1960 IRE WESTCON Conv. Record, Part 4*, pp. 96–104, 1960.

[2] D. E. Rumelhart, G. E. Hinton, and R. J. Williams, "Learning Internal Representations by Error Propagation," in D. E. Rumelhart and J. L. McClelland, Eds., *Parallel Distributed Processing*, vol. 1, chap. 8, Cambridge, MA: MIT Press, 1986.

[3] B. Widrow and S. D. Stearns, *Adaptive Signal Processing*, Englewood Cliffs, NJ: Prentice-Hall, 1985.

[4] B. Irie and S. Miyake, "Capabilities of Three-Layered Perceptrons," *Proc. IEEE Intl. Conf. Neural Networks*, pp. I-641, 1988.

[5] D. B. Parker, "Learning Logic," Tech. Rept. TR-47, Center for Comput. Res. Econ. and Manage., Massachusetts Institute of Technology, Cambridge, MA, 1985.

[6] P. Werbos, "Beyond Regression: New Tools for Prediction and Analysis in the Behavioral Sciences," Ph.D. Thesis, Harvard Univ., Cambridge, MA, Aug. 1974.

[7] B. Widrow, N. K. Gupta, and S. Maitra, "Punish/Reward: Learning with a Critic in Adaptive Threshold Systems," *IEEE Trans. Syst., Man, Cybern.*, Sept. 1973.

[8] A. G. Barto, R. S. Sutton, and C. W. Anderson, "Neuronlike Adaptive Elements That Can Solve Difficult Learning Control Problems," *IEEE Trans. Syst., Man, Cybern.*, Sept./Oct. 1983.

[9] C. W. Anderson, "Learning to Control an Inverted Pendulum Using Neural Networks," *IEEE Contr. Syst. Mag.*, vol. 9, no. 3, Apr. 1989.

[10] B. Widrow, "Adaptive Inverse Control," in *Adaptive Systems in Control and Signal Processing 1986*, International Federation of Automatic Control, July 1986.

[11] D. Psaltis, A. Sideris, and A. A. Yamamura, "A Multilayered Neural Network Controller," *IEEE Contr. Syst. Mag.*, vol. 8, Apr. 1988.

[12] M. I. Jordan, "Supervised Learning and Systems with Excess Degrees of Freedom," COINS 88-27, Massachusetts Institute of Technology, Cambridge, MA, 1988.

Article 5.3

Direct adaptive output tracking control using multilayered neural networks

L. Jin
P.N. Nikiforuk
M.M. Gupta

Indexing terms: Adaptive tracking, Neural networks, Nonlinear control systems, Weight learning

Abstract: Multilayered neural networks are used to construct nonlinear learning control systems for a class of unknown nonlinear systems in a canonical form. An adaptive output tracking architecture is proposed using the outputs of the two three-layered neutral networks which are trained to approximate the unknown nonlinear plant to any desired degree of accuracy by using the modified back-propagation technique. A weight-learning algorithm is presented using the gradient descent method with a dead-zone function, and the descent and convergence of the error index during weight learning are shown. The closed-loop system is proved to be stable, with the output tracking error converging to the neighbourhood of the origin. The effectiveness of the proposed control scheme is illustrated through simulations.

1 Introduction

The control of systems with complex, unknown, and nonlinear dynamics has become a topic of considerable importance in the literature. In conventional nonlinear control design, so far, three main approaches have been proposed: adaptive control, Lyapunov-based control and variable structure control. And the feedback linearisation technique of the nonlinear systems are especially appealing from the point view of the nonlinear control design. To achieve the objective of either stabilisation or tracking, however, some strict assumptions were introduced regarding the structure of the uncertainties based on the completely known nonlinear models.

Paper 9524D (C9, C8), first received 2nd December 1991 and in revised form 25th February 1993

The authors are with the Intelligent Systems Research Laboratory, College of Engineering, University of Saskatchewan, Saskatoon, Saskatchewan, Canada S7N 0W0

Advances in the area of artificial neural networks have provided the potential for new approaches to the control of systems with complex, unknown and nonlinear dynamics. The main potentials of the neural networks for control applications can be summarised as: they can be used to approximate any continuous mapping, they perform this approximation through learning, and parallel processing and fault tolerance are easily accomplished. One of the most popular of the neural network architectures for control purposes is the multilayered neural network (MNN) with the error back-propagation (BP) algorithm. It is proved that a three-layered neural network using the back-propagation algorithm can approximate a wide range of nonlinear functions to any desired degree of accuracy [2–4]. To avoid modelling difficulties, a number of multilayered neural network based controllers have been proposed [6–10]. If a control system is regarded as a mapping of control inputs into observation outputs, an appropriate mapping is realised by a MNN which is trained so that the desired response is obtained. For such kinds of adaptive learning control systems using the MNN, the weights of the network need to be updated using the network's output error index, and the learning control law is constructed based on the output of the MNN. Therefore, the main research topics in the field of the neural network control methods are the convergence of the weight-learning schemes and the stability of the closed-loop control systems.

2 Nonlinear control formulation

Assume that a single-input and single-output nonlinear system is given in a canonical form as

$$z^{(n)} + f(z, z^{(1)}, \ldots, z^{(n-1)}) = g(z, z^{(1)}, \ldots, z^{(n-1)})u$$
$$y = z \qquad (1)$$

where $z \in R$ is a physical state, $u \in R$ is the output. Let $x = (x_1, x_2, \ldots, x_n)^T = (z, z^{(1)}, \ldots, z^{(n-1)})^T$ be the state vector, then the system of eqn. 1 can be represented as a state space model

Reprinted with permission from *IEEE Proceedings-D*, vol. 140, no. 6, L. Jin, P. N. Nikiforuk, and M. M. Gupta, "Direct Adaptive Output Tracking Control Using Multilayered Neural Networks," pp. 393–398, Nov. 1993, IEEE.

$$\begin{cases} \dot{x}_1 = x_2 \\ \dot{x}_2 = x_3 \\ \quad\vdots \\ \dot{x}_n = -f(x) + g(x)u \\ y = x_1 \end{cases} \tag{2}$$

Let the desired output $y_d = y_d(t)$ be a continuously differentiable function on $[0, +\infty)$, and its first n derivatives $y_d^{(1)}, \ldots, y_d^{(n)}$ be uniformly bounded. The problem to be addressed consists of finding a control $u(t)$ that will force the output $y(t)$ to track asymptotically the desired output $y_d(t)$; that is

$$\lim_{t\to\infty}(y(t) - y_d(t)) = 0 \tag{3}$$

Since $g(x)$ in the system eqns. 2 is bounded away from zero, its inverse is well defined and the most convenient structure for a nonlinear feedback control is the one in which the input variable u is set to equal [15]

$$u = \frac{-\sum_{k=1}^{n} c_{k-1}x_k + f(x) + v}{g(x)} = \alpha(x) + \beta(x)v \tag{4}$$

where

$$\alpha(x) = \frac{-\sum_{k=1}^{n} c_{k-1}x_k + f(x)}{g(x)} \tag{5}$$

$$\beta(x) = \frac{1}{g(x)} \tag{6}$$

and v is a new control input to be designed for the purpose of output tracking. Since the nonlinear feedback control law (eqn. 4) linearises the state equation (eqn. 2), it can be easily verified that the system of eqn. 2 is transformed into the linear system as follows

$$\begin{cases} \dot{x}_1 = x_2 \\ \dot{x}_2 = x_3 \\ \quad\vdots \\ \dot{x}_n = -\sum_{k=1}^{n} c_{k-1}x_k + v \\ y = x_1 \end{cases} \tag{7}$$

Let $e(t) = y(t) - y_d(t)$ be the output tracking error, and the new control v be chosen as

$$v = \sum_{k=0}^{n} c_k \frac{d^k y_d}{dt^k} = \sum_{k=0}^{n} c_k y_d^{(k)} \tag{8}$$

where $c_n = 1$. The output tracking error $e(t)$ satisfies the following linear error equation

$$e^{(n)} + c_{n-1}e^{(n-1)} + \cdots + c_0 e = 0 \tag{9}$$

If one chooses real numbers $c_0, c_1, \ldots, c_{n-1}$ such that all the roots of the polynomial

$$N(s) = s^n + c_{n-1}s^{n-1} + \cdots + c_0 \tag{10}$$

have a negative real part, then the desired output $y_d(t)$ and its derivatives $y_d^{(1)}(t), \ldots, y_d^{(n-1)}(t)$ are asymptotically tracked. One can set $\Delta x = (\Delta x_1, \Delta x_2, \ldots, \Delta x_n)^T = (e, e^{(1)}, \ldots, e^{(n-1)})^T$. Then, the error equation (eqn. 9) can be rewritten as

$$\Delta\dot{x} = C\,\Delta x \tag{11}$$

where

$$C = \begin{pmatrix} 0 & 1 & 0 & \cdots & 0 \\ 0 & 0 & 1 & \cdots & 0 \\ & & & \ddots & \\ -c_0 & -c_1 & -c_2 & \cdots & -c_{n-1} \end{pmatrix} \tag{12}$$

is a Hurwitz matrix. We define the following Lyapunov function

$$V_0(\Delta x) = \Delta x^T P_0\,\Delta x \tag{13}$$

where P_0 is a symmetric, positive definite matrix solution of the following Lyapunov equation

$$P_0 C + C^T P_0 = 1 \tag{14}$$

Evaluating its time derivative along the state trajectories of the error system (eqn. 11), one obtains

$$\dot{V}_0 = -\Delta x^T\,\Delta x \tag{15}$$

Remark: Another interesting nonlinear control strategy is the so called 'sliding mode' control law [11, 12]

$$v = \sum_{k=1}^{n} c_k \frac{d^k y_d}{dt^k} + c_0 y + \lambda\,\text{sign}\,(S) \tag{16}$$

and

$$u = \frac{1}{g(x)}\left\{\sum_{k=2}^{n} c_{k-1}\left(\frac{d^{k-1}y_d}{dt^{k-1}} - x_k\right) + \left[\frac{d^n y_d}{dt^n} + f(x)\right] + \lambda\,\text{sign}\,(S)\right\} \tag{17}$$

where $\lambda > 0$, and the sliding manifold S is defined as

$$S(e) = \sum_{k=0}^{n-1} c_{k+1}\frac{d^k e}{dt^k} = \sum_{k=0}^{n-1} c_{k+1}e^k \tag{18}$$

The error equation (eqn. 9) is then replaced by

$$e^{(n)} + c_{n-1}e^{(n-1)} + \cdots + \lambda\,\text{sign}\,(S) = 0 \tag{19}$$

Furthermore, if the Lyapunov function is chosen as

$$V_1 = \tfrac{1}{2}SS \tag{20}$$

then by eqns. 16 and 18 one obtains

$$\dot{V}_1 = -\lambda S\,\text{sign}\,(S) < 0 \tag{21}$$

3 Adaptive tracking using multilayered neural networks

Recently, several studies have independently found that a three-layered neural network using the back-propagation algorithm can approximate a wide range of nonlinear functions to any desired degree of accuracy. In this Section, multilayered neural networks are used to construct a nonlinear controller for the purpose of adaptively tracking the desired output $y_d(t)$. Suppose that the continuous functions $f(x)$ and $g(x)$ are unknown. Let the nonlinear system (eqn. 1) be modelled by the neural network

$$
\begin{cases}
\dot{x}_1^* = x_2^* \\
\dot{x}_2^* = x_3^* \\
\vdots \\
\dot{x}_n^* = -\hat{f}(x, w) + \hat{g}(x, l)u \\
y^* = x_1^*
\end{cases}
\tag{22}
$$

In the case that eqn. 22 is a three-layered neural network as shown on Fig. 1, $\hat{f}(x, w)$ and $\hat{g}(x, l)$ can be represented as

$$\hat{f}(x, w) = \sum_{i=1}^{p} w_i H\left(\sum_{j=1}^{n} w_{ij} x_j + \hat{w}_i \right) \tag{23}$$

$$\hat{g}(x, l) = \sum_{i=1}^{q} l_i H\left(\sum_{j=1}^{n} l_{ij} x_j + \hat{l}_i \right) \tag{24}$$

where

$$
\begin{cases}
w = [(w^1)^T, (w^2)^T, (w^3)^T]^T \\
w^1 = [w_1, w_2, \ldots, w_p]^T \\
w^2 = [w_{11}, w_{21}, \ldots, w_{p1}, \ldots, w_{1n}, w_{2n}, \ldots, w_{pn}]^T \\
w^3 = [\hat{w}_1, \hat{w}_2, \ldots, \hat{w}_p]^T
\end{cases}
\tag{25}
$$

and

$$
\begin{cases}
l = [(l^1)^T, (l^2)^T, \ldots, (l^3)^T]^T \\
l^1 = [l_1, l_2, \ldots, l_q]^T \\
l^2 = [l_{11}, l_{21}, \ldots, l_{q1}, \ldots, l_{1n}, l_{2n}, \ldots, l_{qn}]^T \\
l^3 = [\hat{l}_1, \hat{l}_2, \ldots, \hat{l}_q]^T
\end{cases}
\tag{26}
$$

are the weights of the three-layered neural networks shown on Fig. 1. The p and q are the number of nonlinear hidden neurons of the neural networks corresponding

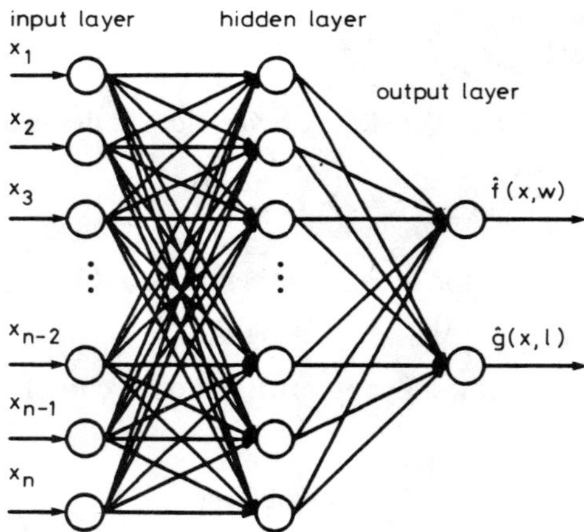

Fig. 1 *Three-layer neural networks*

to $\hat{f}(x, w)$ and $\hat{g}(x, l)$, respectively. The function H in eqns. 23 and 24 is the hyperbolic tangent function, which is similar to a smoothed step function

$$H(x) = \frac{e^x - e^{-x}}{e^x + e^{-x}} \tag{27}$$

Based on the analytic results of Hecht–Nielsen [2] about the capability of layered neural networks to approximate nonlinear functions, one can guarantee that $\hat{f}(x, w)$ and $\hat{g}(x, l)$ are complex enough (that is, they contain enough neurons) to be able to approximate $f(x)$ and $g(x)$ to the desired accuracy. The weights w and l are unknown, and $w(t)$ and $l(t)$ represent the estimates of w and l. To better define the error index relative to the weight learning, the estimated nonlinear system using the neural networks is represented as

$$
\begin{cases}
\dot{x}_1^* = x_2^* \\
\dot{x}_2^* = x_3^* \\
\vdots \\
\dot{x}_n^* = -\hat{f}(x, w(t)) + \hat{g}(x, l(t))u \\
y^* = x_1^*
\end{cases}
\tag{28}
$$

Let the error index of the output of the neural networks be defined as

$$
\begin{aligned}
e^*(t) &= y^{*(n)} - y_d^{(n)} \\
&= -\hat{f}(x, w(t)) + \hat{g}(x, l(t))u - y_d^{(n)}
\end{aligned}
\tag{29}
$$

Then, $w(t)$ and $l(t)$ can be adjusted using a gradient search method such that $e^*(t)$ can be reduced. Let $\pi(t) = [w^T(t), l^T(t)]^T$ be the estimated weight coefficient of the neural

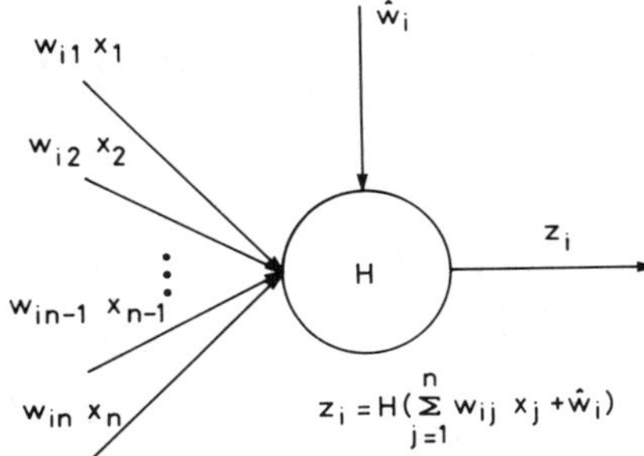

Fig. 2 *i-th neuron of hidden layer*

network model (eqn. 24). Then the updating rule of the weights is easily given as

$$\pi(t + \delta t) = \pi(t) - \eta \frac{\partial e^*}{\partial \pi} \tag{30}$$

where

$$\frac{\partial e^*}{\partial \pi} = \begin{pmatrix} -\partial \hat{f}(x, w(t))/\partial w(t) \\ \partial \hat{g}(x, l(t))/\partial l(t) \end{pmatrix}$$

and η is the step-size parameter which affects the rate of convergence of the weights during learning. On the other hand, it is well known that the major disadvantage of the gradient method is that the rate of convergence of the iterative proceeding near the minimum will be very slow. To improve the rate of convergence near the mininum using a simpler method, one can employ a dead-zone algorithm for updating the weights, proposed by Chen, Khalil [8]. Let the error $e^*(t)$ be applied as the input to a dead-zone function $D(e^*)$

$$
D(e^*) = \begin{cases}
0 & \text{if } |e^*| \leq d_0 \\
e^* - d_0 & \text{if } e^* > d_0 \\
e^* + d_0 & \text{if } e^* < -d_0
\end{cases}
$$

Then, the output of the dead-zone function is used in the following updating rule

$$\pi(t + \delta t) = \pi(t) - \eta D(e^*) \frac{\partial e^*}{\partial \pi} \tag{31}$$

and the estimated nonlinear feedback law as shown in Fig. 3 is designed as

$$\hat{u} = \hat{\alpha}(x, w(t), l(t)) + \hat{\beta}(x, l(t))v \tag{32}$$

where

$$\hat{\alpha}(x, w(t), l(t)) = \frac{-\sum_{k=1}^{n} c_{k-1}x_k + \hat{f}(x, w(t))}{\hat{g}(x, l(t))} \tag{33}$$

$$\hat{\beta}(x, l(t)) = \frac{1}{\hat{g}(x, l(t))} \tag{34}$$

4 Results on convergence of weight learning

For convenience, let $\pi_k = \pi(t + k\delta t)$. The following theorem shows that the error index e^* is descent towards the direction of the origin using the weight iterative proceeding (eqn. 31).

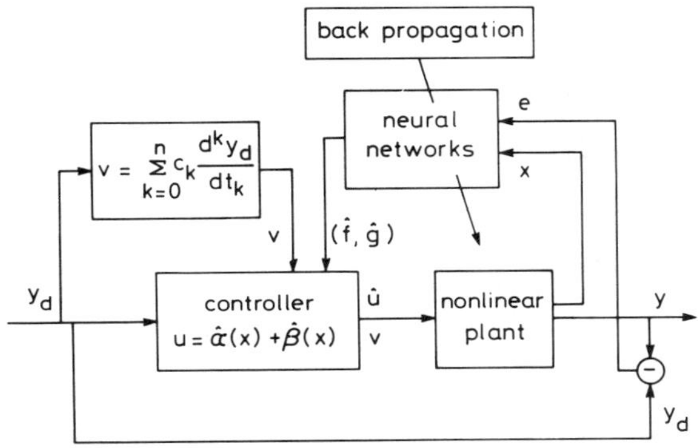

Fig. 3 *Adaptive nonlinear tracking using MNN*

Theorem 1: If there is a weight π_k such that $|e^*(\pi_k)| > d_0$, then, there exists a number $\eta > 0$ such that

$$[e^*(\pi_k + \Delta\pi) - e^*(\pi_k)] \text{ sign } (e^*(\pi_k)) < 0 \tag{35}$$

where $\Delta\pi_k = -\eta D(e^*)\partial e^*/\partial\pi$.

The proof appears in the proof procedure of the following Theorem 2 which provides sufficient condition for the convergence in the case where a $|e^*| \leqslant d_0$ can not be reached in a finite number of learning iterations.

Theorem 2: If the limit of the sequence $\{\pi_k\}$ of eqn. 31 is π^*, then $|e^*(\pi^*)| \leqslant d_0$.

Proof: The proof is by contradiction. Assume that $e^*(\pi^*) > d_0$. Then there exist numbers $\mu > 0$ and $\eta > 0$ such that

$$D(e^*)\left[\frac{\partial e^*(\pi_1)}{\partial\pi}\right]^T\left[\frac{\partial e^*(\pi_2)}{\partial\pi}\right] \geqslant \mu > 0 \tag{36}$$

for all π_1 and π_2 in some neighbourhood of π^*; If π_k is a point in this neighbourhood, then so is π_{k+1}, where

$$\pi_{k+1} = \pi_k - \eta D(e^*)\frac{\partial e^*}{\partial\pi} \tag{37}$$

On the other hand, by the first mean-value theorem, one can imply

$$e^*(\pi_{k+1}) - e^*(\pi_k) = -\eta D(e^*)\left[\frac{\partial e^*(\pi_k)}{\partial\pi}\right]^T\left[\frac{\partial e^*(\pi_k - \theta\Delta\pi_k)}{\partial\pi}\right]$$

$$0 \leqslant \theta \leqslant 1 \tag{38}$$

In this equation, the coefficient of $-\eta$ is at least μ from eqns. 36 and 37. Thus, in moving from π_k to π_{k+1}, the value of $e^*(\pi)$ is decreased by at least $\eta\mu$, but there are an infinite number of terms of the sequence $\{\pi_k\}$ in any neighbourhood of π^*, since

$$\lim_{k\to\infty} \pi_k = \pi^* \tag{39}$$

Hence, by repeating the above argument for successive values of k, one finds that $e^*(\pi^*) \to -\infty$, which contradicts the fact that $\partial e^*(\pi^*)/\partial\pi$ exists. The original assumption that $e^*(\pi^*) > d_0$ is therefore false. By the same proof proceeding, one can show that $e^*(\pi^*) < -d_0$ is false, too, and the theorem is proven. ■

5 Results on feedback stability

To analyse the local stability of the closed-loop system the following assumptions about the nonlinear plant (eqn. 1) and the desired output $y_d(t)$ are required.

Assumption 1: For any $x \in R^n$

$$0 < k_1 \leqslant |g(x)| \tag{40}$$

Assumption 2: For any $t \in [0, +\infty)$, the desired output $y_d(t)$ and its first n derivatives $y_d^{(1)}, \ldots, y_d^n$ are uniformly bounded; that is

$$|y_d^{(i)}(t)| \leqslant m_i \quad i = 0, 1, \ldots, n \tag{41}$$

The following assumption is given based on the analytic results of Hecht-Nielsen [2] about the capability of the MNNs to approximate nonlinear functions using the back-propagation technique.

Assumption 3: There exist weight coefficients w and l such that $\hat{f}(x, w)$ and $\hat{g}(x, l)$ approximate the continuous functions $f(x)$ and $g(x)$ with accuracy ε on Σ, a compact subset of R^n, that is $\exists\, w, l$ such that

$$\max |\hat{f}(x, w) - f(x)| \leqslant \varepsilon \tag{42}$$

$$\max |\hat{g}(x, l) - g(x)| \leqslant \varepsilon \quad \forall x \in \Sigma \tag{43}$$

Lemma 1–3 will show that the MNNs with the hyperbolic tangent function in the hidden layer satisfy some algebraic properties on the compact set Σ.

Lemma 1: The hyperbolic tangent function $H(x)$ satisfies the following properties:

(*a*) $H(x)$ is strictly increasing; for each $x_1, x_2 \in R$ such that $x_1 < x_2$ it is true tha $H(x_1) < H(x_2)$;

(*b*) $H(x)$ is uniformly linear growth; there exists a constant $\beta_1 > 0$ such that $|H(x)| \leqslant \beta_1 |x|$ for all $x \in R$.

Proof: One can easily show that the first part is a true based on the definition of the function $H(x)$. Since $H(x)$ is uniformly Lipschitz [12], there exists a constant $\beta_1 > 0$ such that for all $x_1, x_2 \in R$

$$|H(x_1) - H(x_2)| \leqslant \beta_1 |x_1 - x_2| \tag{44}$$

Hence,

$$|H(x_1)| - |H(x_2)| \leqslant |H(x_1) - H(x_2)| \leqslant \beta_1 |x_1 - x_2| \tag{45}$$

Note that $H(0) = 0$. If one sets $x_2 = 0$, then the second part is implied. ■

Lemma 2: There exist constants $\beta_{1w}, \beta_{2w} > 0$ and $\beta_{1l}, \beta_{2l} > 0$ such that the neural network models (eqns. 23 and 24) on Σ, a compact subset of R^n, satisfy the following conditions

$$|\hat{f}_1(x, w)| \leqslant \beta_{1w}\|x\| + \beta_{2w} \tag{46}$$

$$|\hat{g}_1(x, l)| \leqslant \beta_{1l}\|x\| + \beta_{2l} \quad \forall x \in \Sigma \tag{47}$$

Proof: By Lemma 1,

$$\begin{aligned} |\hat{f}_1(x, w)| &\leqslant \sum_{i=1}^{p} |w_i| \left| H\left(\sum_{i=1}^{n} w_{ij} x_j + \hat{w}_i \right) \right| \\ &\leqslant \sum_{i=1}^{p} |w_i| \left(\beta_1 \left| \sum_{j=1}^{n} w_{ij} x_j x_j + \hat{w}_i \right| \right) \\ &\leqslant \sum_{i=1}^{p} |w_i| (\beta_1 (\|\bar{w}_i\| \|x\| + |\hat{w}_i|) \\ &= \beta_{1w}\|x\| + \beta_{2w} \quad \forall x \in R^n \end{aligned} \tag{48}$$

where $\bar{w}_i = (w_{i1}, w_{i2}, \ldots, w_{in})^T$, and

$$\beta_{1w} = \sum_{i=1}^{p} |w_i| |\beta_1| \|\bar{w}_i\| \tag{49}$$

$$\beta_{2w} = \sum_{i=1}^{p} |w_i| \beta_1 |\hat{w}_i| \tag{50}$$

Hence, the proof is complete. ■

Lemma 3: There exist constants $\hat{k}_1, \hat{k}_2 > 0$ such that the neural network $\hat{g}(x, l)$ on the compact set Σ of R^n satisfies

$$0 < \hat{k}_1 \leqslant |\hat{g}_1(x, l)| \leqslant \hat{k}_2 \quad \forall x \in \Sigma \tag{51}$$

Proof: Note that

$$\hat{g}_1(x, l) \neq 0 \quad \forall x \in \Sigma \tag{52}$$

This implies that $\hat{g}_1(x, l) \gg 0$, or $\hat{g}_1(x, l) \ll 0$ for all $x \in \Sigma$. Since $\hat{g}_1(x, l)$ is the continuous function on the compact set Σ. Hence, there exist the nonzero maximum and minimum of $\hat{g}_1(x, l)$ on Σ, which means that eqn. 51 is true. ■

The following theorem will give the local stability of the adaptive learning control system on the compact set Σ.

Theorem 3: Under the assumptions 1–3. There exists a constant $\delta = \delta(\varepsilon)$ such that the output tracking error of the nonlinear system (eqn. 1) on the compact set Σ using the neural network control law (eqn. 32) is confined to a neighbourhood of the origin defined by $\|\Delta x^*\| \leqslant \delta$.

Proof: The error dynamics of the system under the neural network control law $\hat{u}$ can be obtained as

$$\Delta\dot{x} = C\,\Delta x + b_1(x, w, l) + b_2(x, l)v \tag{53}$$

where $\Delta x = (e, e^{(1)}, \ldots, e^{n-1})^T$, and

$$b_1(x, w, l) = \begin{pmatrix} 0 \\ 0 \\ \vdots \\ g(x)(\hat{\alpha}(x, w, l) - \alpha(x)) \end{pmatrix} \tag{54}$$

$$b_2(x, l) = \begin{pmatrix} 0 \\ 0 \\ \vdots \\ g(x)(\hat{\beta}(x, l) - \beta(x)) \end{pmatrix} \tag{55}$$

Computing the time derivative of Lyapunov function V_0 along the state trajectories of the error dynamics (eqn. 53),

$$\begin{aligned} \dot{V}_0 &= \langle dV_0, C\,\Delta x + b_1(x, w, l) + b_2(x, l)v \rangle \\ &= \langle dV_0, C\,\Delta x \rangle + \langle dV_0, b_1(x, w, l) \rangle \\ &\quad + \langle dV_0, b_2(x, l)v \rangle \\ &\leqslant -\Delta x^T \Delta x \\ &\quad + 2\|P_0\| \|\Delta x\| |g(x)(\hat{\alpha}(x, w, l) - \alpha(x))| \\ &\quad + 2\|P_0\| \|\Delta x\| |g(x)(\hat{\beta}(x, l) - \beta(x))| |v| \end{aligned} \tag{56}$$

On the other hand, by assumption 3 and lemmas 2 and 3,

$$\begin{aligned} &|g(x)(\hat{\alpha}(x, w, l) - \alpha(x))| \\ &= \left| g(x) \left(\frac{-\sum_{k=1}^{n} c_{k-1} x_k + \hat{f}(x, w)}{\hat{g}(x, l)} + \frac{\sum_{k=1}^{n} c_{k-1} x_k - f(x)}{g(x)} \right) \right| \\ &\leqslant \frac{\sum_{k=1}^{n} |c_{k-1} x_k| |\hat{g}(x, l) - g(x)|}{|\hat{g}(x, l)|} \\ &\quad + \frac{|\hat{f}(x, w)| |\hat{g}(x, l) - g(x)|}{|\hat{g}(x, l)|} + \frac{|\hat{g}(x, l)| |\hat{f}(x, w) - f(x)|}{|\hat{g}(x, l)|} \\ &\leqslant \frac{\sum_{k=1}^{n} |c_{k-1} x_k| + |\hat{f}(x, w)| + |\hat{g}(x, l)|}{|\hat{g}(x, l)} \varepsilon \\ &\leqslant \frac{\|c\| \|x\| + \beta_{1w}\|x\| + \beta_{2w} + \beta_{1l}\|x\| + \beta_{2l}}{\hat{k}_1} \varepsilon \\ &\leqslant \|\Delta x\| \delta_1 \varepsilon + \delta_2 \varepsilon \end{aligned} \tag{57}$$

and

$$\begin{aligned} |g(x)(\hat{\beta}(x, l) - \beta(x)| &= \left| g(x) \left(\frac{1}{\hat{g}(x, l)} - \frac{1}{g(x)} \right) \right| \\ &= \frac{|\hat{g}(x, l) - g(x)|}{|\hat{g}(x, l)|} \leqslant \frac{\varepsilon}{\hat{k}_1} \end{aligned} \tag{58}$$

where

$$\delta_1 = \frac{\|c\| + \beta_{1w} + \beta_{1l}}{\hat{k}_1} \tag{59}$$

$$\delta_2 = \delta_1 \|Y_d\| + \frac{\beta_{2w} + \beta_{2l}}{\hat{k}_1} \qquad (60)$$

and

$$c = (c_0, c_1, \ldots, c_{n-1})^T$$

$$Y_d = (y_d, y_d^{(1)}, \ldots, y_d^{(n-1)})^T$$

therefore,

$$\dot{V}_0 \leqslant -\|\Delta x\|(\|\Delta x\|\delta_3 - \delta_4) \qquad (61)$$

where

$$\delta_3 = 1 - 2\|P_0\|\delta_1\varepsilon \qquad (62)$$

$$\delta_4 = 2\|P_0\|\varepsilon\left(\delta_2 + \frac{|v|}{\hat{k}_1}\right) \qquad (63)$$

Thus V_0 can be assured to be nonincreasing whenever $\|\Delta x\| \geqslant \delta \equiv \delta_4/\delta_3$, so that the output tracking error is confined to be a neighbourhood of $\Delta x = 0$ defined by $\|\Delta x\| \leqslant \delta$, which can be arbitrarily small as $\varepsilon \to 0$. ■

6 Simulation results

In this Section, the preceding nonlinear learning control scheme is illustrated using a second order SISO nonlinear plant. Consider a single-link manipulator described by

$$\pi(t) = ml^2\ddot{\theta}(t) + v\dot{\theta}(t) + mgl\cos\theta(t) \qquad (64)$$

where the length, mass and friction coefficient are $l = 1$ m, $m = 2.0$ kg and $v = 1.0$ kg m^2/s, respectively. Let the coefficients of the error equation (eqn. 9) be chosen as $c_1 = 8$ and $c_0 = 30$. The $\hat{f}$ network has 20 hidden neurons, the $\hat{g}$ network has five hidden neurons, the parameters of the weight updating law (eqn. 31) are set at $\eta = 0.005$ and $d_0 = 0.001$, and the initial values of all weights of the $\hat{f}$ and $\hat{g}$ networks are chosen as 0.5. The simulations are performed using a fourth-order, fixed-stepsize Runge–Kutta algorithm with $\Delta t = 0.01$ s. The output tracking trajectories obtained by applying four different desired outputs, step, sine, square, sawtooth, are shown in Fig. 4, where it is seen that the satisfactory output tracking performance has been achieved through the proposed control scheme. Meanwhile, the simulation results shows that the system output response and the convergence of the learning control law are sensitive to the number of the hidden neurons and the initial values of the weights.

7 Concluding remarks

We have used multilayered neural networks to construct a nonlinear learning control law for a class of unknown SISO nonlinear control systems, and it is straightforward to extend the developed MNN controller to the MIMO nonlinear control systems. The two main results given in this paper are:

(*a*) The three-layered neural networks were introduced

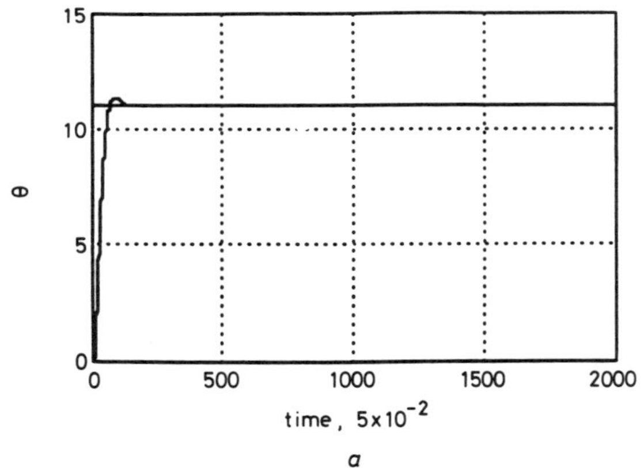

a

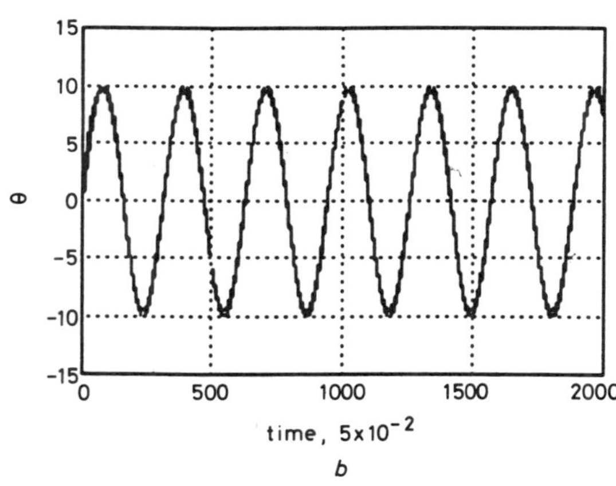

b

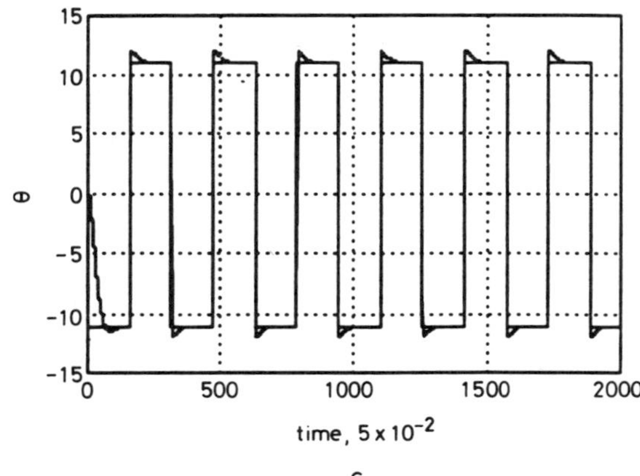

c

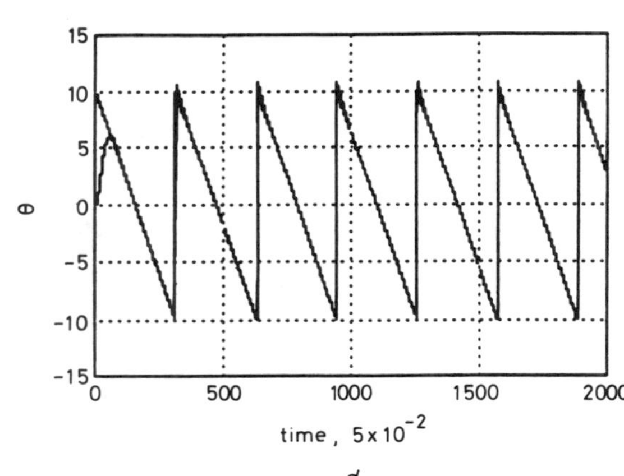

d

Fig. 4 *Simulation results of output tracking control*
a Tracking desired step output *b* Tracking desired sine output *c* Tracking desired square output *d* Tracking desired sawtooth output

to approximate the unknown nonlinear models using the modified error back propagation, and a weight-learning algorithm with a dead-zone function discussed.

(*b*) The convergence of the weight-learning law is guaranteed through the theorems proved in Section 4. The stability of the closed-loop is shown as follows: If there are enough neurons in nonlinear hidden layers of the three-layer neural networks to be able to approximate $f(x)$ and $g(x)$ to the desired accuracy, the output tracking error will converge to a neighbourhood of the origin, which can be arbitrarily small as well as the desired accuracy of the network learning.

8 References

1 RUMELHART, D.E., and McCELLAND, J.L.: 'Learning internal representations by error propagation', *in* 'Parallel distributed processing: explorations in the microstructure of cognition, vol. 1: foundations' (MIT Press, 1986)

2 HECHT-NIELSEN, R.: 'Theory of the back-propagation neural network'. Proceedings of the international joint conference on *Neural Networks*, 1989, pp. I-593-605

3 BLUM, E.K., and LI, L.K.: 'Approximation theory and feedforward networks', *Neural Networks*, 1991, **4**, pp. 511–515

4 SIMPSON, P.K.: 'Artificial neural systems' (Pergamon, 1990)

5 HORNIK, K., STINCHCOMBE, M., and WHITE, H.: 'Multilayer feedforward networks are universal approximators', *Neural Networks*, 1989, **2**, pp. 359–366

6 PSALTIS, D., SIDERIS, A., and YAMAMURA, A.A.: 'A multilayered neural network controller', *IEEE Control Syst. Mag.*, April 1988, pp. 17–21

7 CHEN, F.-C.: 'Back-propagation neural networks for nonlinear self-tuning adaptive control', *IEEE Control Syst. Mag.*, April 1990, pp. 44–48

8 CHEN, F.-C. and KHALIL, H.K.: 'Adaptive control of nonlinear systems using neural networks — a dead-zone approach'. Proceedings of American Control Conference, 1991, pp. 667–672

9 KUMAR, S.S., and GUEZ, A.: 'ART based adaptive pole placement for neurocontrollers', *Neural Networks*, 1991, **4**, pp. 319–335

10 JIN, L., NIKIFORUK, P.N., and GUPTA, M.M.: 'Adaptive tracking of SISO nonlinear systems using multilayered neural networks'. Proceedings of American Control Conference, Chicago, 1992, pp. 56–60

11 SLOTINE, J-J.E.: 'Sliding controller design for nonlinear systems', *Int. J. Control*, 1984, **40**, (2), pp. 421–434

12 SLOTINE, J-J.E.E., and SASTRY, S.S.: 'Tracking control of nonlinear systems using sliding surfaces with application to robot manipulators', *Int. J. Control*, 1983, **38**, (2), pp. 465–492

13 SANNER, R.M., and SLOTINE, J.-J.E.: 'Gaussian networks for direct adaptive control'. Proceedings of American Control Conference, pp. 2153–2159

14 HUI, S., and ZAK, S.H.: 'Analysis of single perceptrons learning capabilities'. Proceedings of American Control Conference, pp. 809–814

15 ISDORI, A.: 'Nonlinear control system' (Springer Verlag, New York, 1989)

16 WALSH, G.R.: 'Methods of optimization' (Wiley, 1975)

Neural networks for nonlinear internal model control

K.J. Hunt
D. Sbarbaro

Indexing terms: Networks, modelling

Abstract: A novel technique, directly using artificial neural networks, is proposed for the adaptive control of nonlinear systems. The ability of neural networks to model arbitrary nonlinear functions and their inverses is exploited. The use of nonlinear function inverses raises questions of the existence of the inverse operators. These are investigated and results are given characterising the invertibility of a class of nonlinear dynamical systems. The control structure used is internal model control. It is used to directly incorporate networks modelling the plant and its inverse within the control strategy. The potential of the proposed method is demonstrated by an example.

1 Introduction

This paper focuses on the use of artificial neural networks for the identification and control of non-linear dynamic systems. Such networks go by other names: parallel distributed processing, or connectionist networks. The term neural networks comes from the similarity of these networks to models of brain activity.

Artificial neural networks provide a distinctive computational paradigm and have proved effective for a range of practical problems where conventional computation techniques have not succeeded.

An artificial neural network consists of many simple processing elements each having a number of inputs and a single output. The output of each element is determined as a nonlinear function of a weighted sum of the inputs. A range of activation functions can be used; in this work we consider Gaussian activation functions because of their function representation properties (discussed in Section 5.2). The basic elements are interconnected using variable strength links, or weights. The strengths of the individual links determine the overall behaviour of the network. The weights of a particular network can be chosen to achieve a desired input-output relationship. The weights are adjusted during a training period when the network is presented with a range of input patterns and its output is compared to the desired output each time. The aim is to successively drive the weights to values which make the network output equal to the target output.

Paper 8183D (C8), first received 17th September 1990 and in revised form 29th April 1991

The authors are with the Control Group, Department of Mechanical Engineering, The University of Glasgow, Glasgow G12 8QQ, United Kingdom

In this work we are primarily concerned with using artificial neural networks for dynamical systems control. The characteristics of neural networks suggest that they may be useful in certain classes of control problems. In particular, the ability of neural networks to represent arbitrary nonlinear mappings encourages the study of neural networks for complex nonlinear control problems. Relevant features of neural networks in the control context are

(i) the ability to represent arbitrary non-linear relations

(ii) adaptation and learning in uncertain systems, provided through both off-line and online weight adaptation

(iii) information is transformed to internal representation allowing data fusion, with both quantitative and qualitative signals

(iv) parallel distributed processing architecture allowing fast processing for large-scale dynamical systems

(v) architecture providing a degree of robustness through fault tolerance and graceful degradation

Each current practical design method based on linear control theory can address a subset of these problems. Although the present emphasis here is on nonlinear control problems, we note that the characteristics of artificial neural networks allow them, potentially, to address all these problems simultaneously.

The ability of artificial neural networks to represent arbitrary non-linear relations is now well established [1, 2]. This has lead to the investigation of non-linear dynamic systems modelling using neural networks [3, 4].

The ability of neural networks to represent nonlinear relations leads to the idea of using networks directly in a model-based control strategy. The idea followed here is based on the possibility of training networks to learn both a system's input/output relationship and the corresponding inverse relationship. A suitable control strategy which directly incorporates the plant model (and the corresponding inverse model) is provided by internal model control (IMC) [5, 6]. The applicability of IMC to nonlinear systems control has been demonstrated by Economou *et al.* [7]. The inverse of the nonlinear operator modelling the plant was shown to play a crucial role in the implementation of nonlinear IMC. These authors studied analytical and numerical methods for the necessary construction of nonlinear operator inverses. In the present work artificial neural networks are used for the construction of plant models and their inverses, and it is intended that they are used directly within the IMC control structure.

Results characterising the invertibility of a class of non-linear dynamic systems are presented. Architectures and algorithms for the training of neural networks for

Reprinted with permission from *IEE Proceedings-D*, vol. 138, no. 5, K. J. Hunt and D. Sbarbaro, "Neural Networks for Nonlinear Internal Model Control," pp. 431–438, Sept. 1991, IEE.

modelling dynamical inverses are described and convergence results for these algorithms are established.

A further novel feature of our approach is the use of radial basis functions, and the Gaussian function in particular, as the nonlinear function in the network hidden units [8]. There is recent theoretical support for the 'best representation' property of such networks [9]. This approach stands in contrast to the widespread use of sigmoidal nonlinearities. To the best of our knowledge Gaussian networks have not yet been used by others in the study of dynamical systems identification and control, although this type of network has been studied in the related problem of time series prediction [10, 11].

The idea of using neural networks for nonlinear IMC has been considered by Bhat and McAvoy [12]. However, they did not investigate invertibility conditions for nonlinear dynamic systems neither did they propose learning algorithms for building the inverse models. Further, they do not appear to have implemented a neural network based nonlinear IMC controller.

In this paper we describe our work relating to these three issues. We have implemented the nonlinear IMC algorithm and performed extensive simulations, with very encouraging results. A representative sample of the results is given below.

2 Internal model control

The IMC structure is now well known and has been shown to underlie a number of control design techniques of apparently different origin [5]. IMC has been shown to have a number of desirable properties; a detailed analysis has been given by Morari and Zafiriou [6]. Here, we briefly summarise these properties.

The nonlinear IMC structure is shown in Fig. 1 (in this Section we follow Economou *et al.* [7]). The nonlinear operators denoted by P, M and C represent the plant, the plant model, and the controller, respectively. The operator F denotes a filter, discussed later in the paper. The double lines used in the block diagram emphasise that the operators are nonlinear and that the usual block diagram manipulations do not hold.

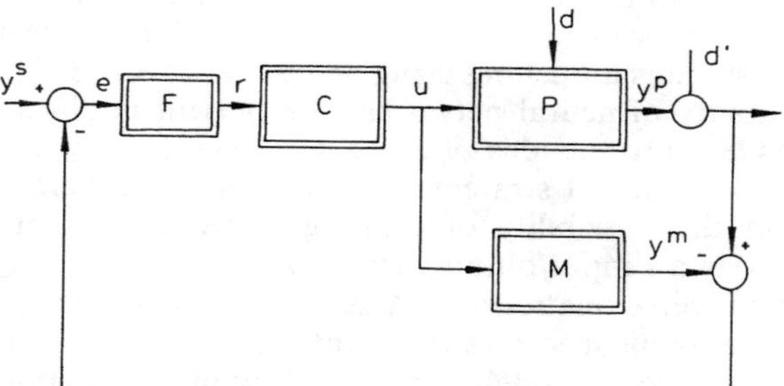

Fig. 1 *Nonlinear IMC structure*

The important characteristics of IMC are summarised with the following properties:

Property P1: Assume that the plant and controller are input–output stable and that the model is a perfect representation of the plant. Then the closed-loop system is input–output stable.

Property P2: Assume that the inverse of the operator describing the plant model exists, that this inverse is used as the controller, and that the closed-loop system is input–output stable with this controller. Then the control will be perfect, i.e. $y = y^s$.

Property P3: Assume that the inverse of the steady state model operator exists, that the steady state controller operator is equal to this, and that the closed-loop system is input–output stable with this controller. Then offset free control is attained for asymptotically constant inputs.

The IMC structure provides a direct method for the design of nonlinear feedback controllers. According to the above properties, if a good model of the plant is available, the closed-loop system will achieve exact set-point following despite unmeasured disturbances acting on the plant.

Discussion so far has considered only the idealised case of a perfect model, leading to perfect control. In practice, however, a perfect model can never be obtained. In addition, the infinite gain required by perfect control would lead to sensitivity problems under model uncertainty. The filter F is introduced to alleviate these problems. By suitable design, the filter can be selected to reduce the gain of the feedback system, thereby moving away from the perfect controller. This introduces robustness into the IMC structure. A full treatment of robustness and filter design for IMC is given by Morari and Zafiriou [6].

The IMC is significant because the stability and robustness properties of the structure can be analysed and manipulated in a transparent manner, even for nonlinear systems. Thus IMC provides a general framework for nonlinear systems control. Such generality is not apparent in alternative approaches to nonlinear control.

A second role of the filter is to project the signal e into the appropriate input space for the controller.

3 Nonlinear IMC using neural networks

We propose a two step procedure for using neural networks directly within the IMC structure. The first step involves training a network to represent the plant response. This network is then used as the plant model operator M in the control structure of Fig. 1. The architecture shown in Fig. 2 provides the method for training a network to represent the plant. Here, the error signal used to adjust the network weights is the difference between the plant output and the network output. Thus, the network is forced towards copying the plant dynamics. Full details of the learning law used here are given in Section 6.1.

Following standard IMC practice (guided by property P2 above) we select the controller as the plant inverse model. The second step in the procedure is to train a second network to represent the inverse of the plant. To do this we use the architecture shown in Fig. 3. Here, for reasons explained in Section 6.2, we employ the plant model (obtained in the first learning step) in the inverse learning architecture rather than the plant itself. For inverse modelling the error signal used to adjust the network is defined as the difference between the (inverse modelling) network input and the plant model output. This tends to force the transfer function between these two signals to unity; i.e. the network being trained is forced to represent the inverse of the plant model. Having obtained the inverse model, this network is then used as the controller block C in the control structure of Fig. 1.

The final IMC architecture incorporating the trained networks is shown in Fig. 4.

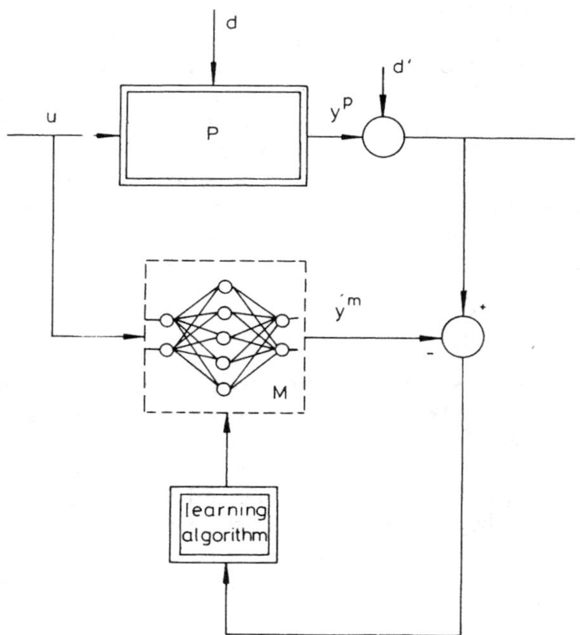

Fig. 2 *Plant identification*

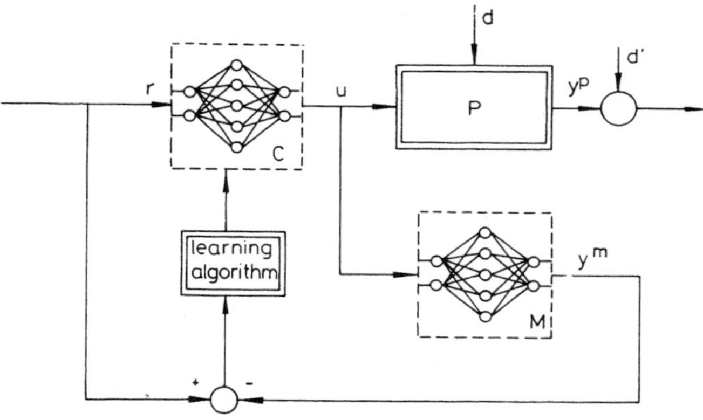

Fig. 3 *Specialised learning structure*

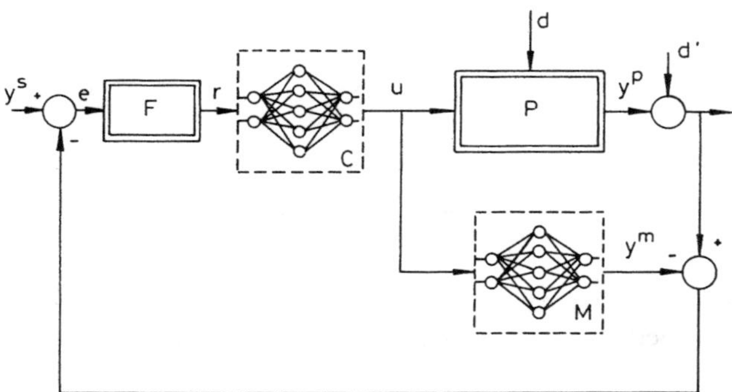

Fig. 4 *General IMC architecture*

4 Invertibility of discrete nonlinear systems

The inversion of nonlinear operators plays a central role in the development of nonlinear IMC. Here we explore the invertibility of nonlinear dynamic systems. Consider a general analytic system Σ governed by the following nonlinear difference equation

$$\Sigma: y^p(k+1) = f(y^p(k), \ldots, y^p(k-n); u(k), \ldots, u(k-m)) \quad y^p \in \mathcal{R} \quad u \in \mathcal{R} \quad (1)$$

Here, $n+1$ is the order of the system, with $m \leqslant n$. For the moment we consider single-input single-output systems; the approach can be generalised to multivariable systems.

The system Σ is called invertible at $[y^p(k), \ldots, y^p(k-n), u(k-1), \ldots, u(k-m)]^T$ if there is a subset A of $\mathcal{R}^{m+n+1}$, such that for $[y^p(k), \ldots, y^p(k-n), u(k-1), \ldots, u(k-m)]^T \in A$

$$f(y^p(k), \ldots, y^p(k-n); u^1(k), \ldots, u(k-m)) \neq f(y^p(k), \ldots, y^p(k-n); u^2(k), \ldots, u(k-m))$$

for any distinct inputs $u^1(k)$, $u^2(k)$.

The system Σ is singular if for any $[y^p(k), \ldots, y^p(k-n), u(k-1), \ldots, u(k-m)]^T \in A$ and for any distinct inputs $u^1(k)$, $u^2(k)$, the resulting outputs are equal:

$$f(y^p(k), \ldots, y^p(k-n); u^1(k), \ldots, u(k-m)) = f(y^p(k), \ldots, y^p(k-n); u^2(k), \ldots, u(k-m))$$

Proposition 1: If $f(y^p(k), \ldots, y^p(k-n); u(k), \ldots, u(k-m))$ is monotonic with respect to $u(k)$ then the system is invertible at $[y^p(k), \ldots, y^p(k-n), u(k-1), \ldots, u(k-m)]^T$.

Proof: If $f(y^p(k), \ldots, y^p(k-n); u(k), \ldots, u(k-m))$ is monotonic and $u^1(k) > u^2(k)$ then for the same point $[y^p(k), \ldots, y^p(k-n), u(k-1), \ldots, u(k-m)]^T$,

$$f(y^p(k), \ldots, y^p(k-n); u^1(k), u(k-1), \ldots, u(k-m)) > f(y^p(k), \ldots, y^p(k-n); u^2(k), u(k-1), \ldots, u(k-m))$$

In the same way,

$$f(y^p(k), \ldots, y^p(k-n); u^1(k), u(k-1), \ldots, u(k-m)) < f(y^p(k), \ldots, y^p(k-n); u^2(k), u(k-1), \ldots, u(k-m))$$

if $u^1(k) < u^2(k)$. The system is therefore invertible.

5 Network architecture

The characteristics of any network are defined by the particular nonlinear activation function of the elements and by the way in which the elements are interconnected. In this Section we define the network architecture used in the present plant and plant inverse modelling studies.

5.1 Basic elements

The simple basic element comprises differently weighted inputs that are processed by a certain nonlinear function, called the activation function. Some of the most frequently used activation functions are shown in Fig. 5. Each

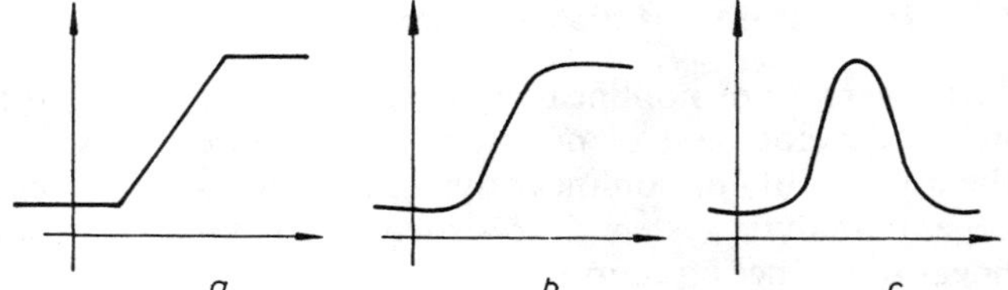

Fig. 5 *Activation functions*
a saturation
b sigmoid
c gaussian

representation of the basic unit has advantages and disadvantages. The choice of a specific unit depends on the problem being studied.

5.2 Gaussian network

The basic elements by themselves are not very powerful in terms of computation or representability but their interconnection allows us to encode relations between the variables, giving different powerful processing capabilities. The connection of several layers, as shown in Fig. 6, gives the possibility of nonlinear mapping between the inputs and the outputs. This capability can be used to

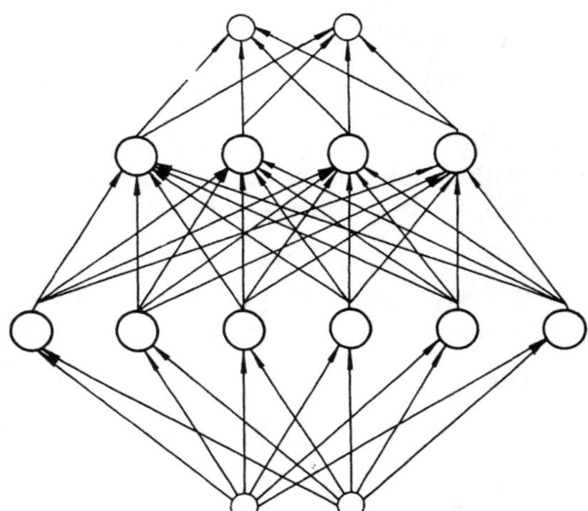

Fig. 6 *Multilayer network*
O processing units
o fan in–out units

represent complex nonlinear relations among the inputs and outputs.

The basic structure having one hidden layer with sigmoid units has been shown to be powerful enough to produce an arbitrary mapping among variables [2]. The standard approach to training this type of network is the back-propagation algorithm. However, this method is known to have drawbacks in terms of speed of convergence and the attainment of a unique global solution to the optimisation problem. These problems arise mainly because the technique involves the solution of a nonlinear problem with local minima.

The use of Gaussian units, on the other hand, has the advantage of simplifying the problem, because in a natural way it produces a partition of the input space. There is theoretical support for representation and consistence problems with these networks [1]. It has been shown that this kind of network has the best approximation property [9].

In this work we now consider networks consisting of many inputs and a single output (by duplication of this structure the approach can easily be generalised to multi-output systems), as shown in Fig. 7. The number of input units n_i corresponds to the dimension of the network input vector, denoted by $x \in \mathscr{R}^{n_i}$. Each linear input unit is given an activation value corresponding to the relevant value of the input vector. The linear output unit is fully connected to the hidden units; the network output y (a nonlinear function, $g(.)$, of the input vector x) is a weighted sum of the activation levels of the N hidden units:

$$y = g(x) = \sum_{i=1}^{N} c_i K_i \tag{2}$$

Here, c_i denotes the connection weight between hidden unit i and the network output. K_i is the output (the activation value) of hidden unit i.

The activation level of a hidden unit in a Gaussian network depends only on the distance between the input vector and the centre of the Gaussian function of that unit. The centre of the hidden unit i is denoted by $x_i \in \mathscr{R}^{n_i}$. More precisely, the activation level K_i of the hidden unit i is defined as

$$K_i = e^{-d_i(x,\, x_i,\, \Delta)} \tag{3}$$

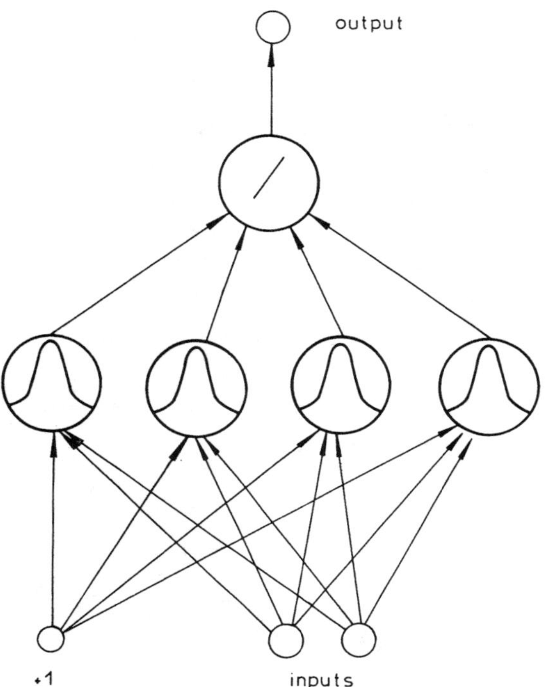

Fig. 7 *Basic Gaussian network*

Here, the distance function d_i for hidden unit i represents the scaled distance between the centre of the function x_i for that unit and the input vector x:

$$d_i(x, x_i, \Delta) = (x - x_i)^T \Delta (x - x_i) \tag{4}$$

The matrix Δ represents the bandwidth of the hidden unit functions. We consider a constant diagonal matrix Δ whose diagonal entries are equal. Having such a Δ constrains the representation but simplifies the learning algorithm. We refer to such a network as a *regular* Gaussian network. Fig. 7 shows the general architecture of this network.

6 Learning algorithms

Section 3 outlined the architectures used for plant modelling and plant inverse modelling. These are necessary in the IMC formulation. The learning laws used for training the networks are considered below.

6.1 Modelling the plant

The plant is modelled using a network described by

$$y^m(k+1) = \sum_{i=1}^{Nm} c_i^m K_i^m \tag{5}$$

where

$$K_i^m = \exp(-d_i^m(x^m(k), x_i^m, \Delta^m)) \tag{6}$$

Here, the m superscript indicates a variable related with the plant model. We choose the structure of the plant model to be the same as that of the plant (eqn. 1), i.e. the model output is a nonlinear function of the present and past plant outputs, and the present and past plant inputs. The model input vector $x^m(k)$ is thus given as

$$x^m(k) = [y^p(k), \ldots, y^p(k-n), u(k), \ldots, u(k-m)]^T$$

We denote the centre of the Gaussian function of hidden unit i as

$$x_i^m = [y_{i,0}, \ldots, y_{i,n}, u_{i,0}, \ldots, u_{i,m}]^T$$

The parameters x_i^m and Δ^m are fixed to meet the interpolation conditions, i.e. the x_i^m are distributed uniformly over the input space and Δ^m is adjusted such that $\sum_{i=1}^{Nm} K_i = 1$ over the input space. There are other possibilities for this [13].

The parameter vector c_i^m is adjusted to minimise the mean square error between the real plant and the model; i.e.

$$c_i^m(k+1) = c_i^m(k) + \alpha K_i(y^p(k+1) - y^m(k+1)) \tag{7}$$

Here, α is a gain parameter. Fig. 2 represents the structure used. Using standard linear systems theory it can be shown that if the plant can be modelled as eqn. 5, the least mean square solution can be found by eqn. 7 [8].

6.2 Inverse model identification

If the model of the plant is invertible then the inverse of the plant can be approximated in a similar way to the plant. This model is then used as the controller. For reasons described in Section 6.2.1 we choose to use the plant model inverse rather than the inverse of the real plant. We utilise a second network described by

$$u(k) = \sum_{i=1}^{NC} c_i^C K_i^C \tag{8}$$

where

$$K_i^C = \exp(-d_i^C(x^C(k), x_i^C, \Delta^C)) \tag{9}$$

Here, the C superscript indicates a variable related with the controller. The inverse of the function f in eqn. 1 (required to obtain $u(k)$) depends on the future plant output value $y^p(k+1)$. In order to obtain a realisable approximation we replace this value by the controller input value r. Finally, since we need to approximate the inverse of the plant model (as opposed to the plant itself), we define the controller network input vector $x^C(k)$ as

$$x^C(k) = [y^m(k), \ldots, y^m(k-n), r(k+1), u(k-1), \ldots, u(k-m)]^T$$

Here, the future value $r(k+1)$ is obtained at time k by suitable definition of the IMC filter F (see Section 7). The centre of the Gaussian function of hidden unit i is given by

$$x_i^C = [y_{i,0}, \ldots, y_{i,n}, r_i, u_{i,1}, \ldots, u_{i,m}]^T$$

6.2.1 Non-iterative methods: To adjust c_i^C the architecture shown in Fig. 3 is used. This architecture is similar to the specialised learning architecture presented by Psaltis *et al.* [14] (the difference being that here we use the plant model, rather than the plant itself). The parameters in c_i^C are adjusted to minimise the mean square error between the output of the model and the input of the controller. This leads to the following learning algorithm:

$$c_i^C(k+1) = c_i^C(k) + \alpha K_i^C(r(k+1) - y^m(k+1)) \frac{\partial y^m(k+1)}{\partial u(k)} \tag{10}$$

Here, if the real plant were used in the learning procedure (as in [14]) then $\partial y(k+1)/\partial u(k)$ would require to be estimated. This can be done using first-order differences [14] changing each input to the plant slightly at the operating point and measuring the change at the output. By using the plant model, however, the derivatives can be calculated explicitly. From eqn. 5 one obtains

$$\frac{\partial y^m(k+1)}{\partial u(k)} = -2\Delta^m \sum_{i=1}^{Nm} c_i^m K_i^m(u(k) - u_{i,0}) \tag{11}$$

Proposition 2: The learning algorithm defined by eqn. (10) converges to a global minimum of the index defined by

$$J = \sum_j (r(j+1) - y^m(j+1))^2 \tag{12}$$

if the system is monotonically increasing with respect to $u(k)$.

Proof: Consider the mean square error defined over all the possible inputs $r(k)$ described by eqn. 12. Its change ΔJ under an adaptation step, eqn. 10 resulting from an input r^* is

$$\Delta J(r^*) = -2 \sum_j (r(j+1) - y^m(j+1)) \sum_i \frac{\partial y^m(j+1)}{\partial u(j)}$$

$$\times K_i^C(x^C(j), x_i^C, \Delta^C)\,\Delta c_i^C$$

Inserting eqn. 10 and averaging over all inputs r_k^*

$$\langle \Delta J \rangle = -2\alpha \sum_l \sum_j (r(j+1)) - y^m(j+1))$$
$$\times (r^*(l+1) - y^{m*}(l+1))$$
$$\times \sum_i \frac{\partial y^m(j+1)}{\partial u(j)} K_i^C(x^C(j), x_i^C, \Delta^C)$$
$$\times \frac{\partial y^{m*}(l+1)}{\partial u(l)} K_i^C(x^{C*}(l), x_i^C, \Delta^C)$$

$$\langle \Delta J \rangle = -2\alpha \sum_i \left(\sum_l \frac{\partial y^m(l+1)}{\partial u(l)} K_i^C(x^C(l), x_i^C, \Delta^C) \right.$$
$$\left. \times (r(l+1) - y^m(l+1)) \right)^2 \qquad (13)$$

Clearly, this quantity can only be negative or zero. For a monotonic function it cannot be zero. From eqn. 13, if the function is not monotonic the algorithm can reach a local minimum when $\partial y/\partial u$ is zero.

Another approach involves the use of a synthetic signal [15]. This leads to the so called general learning architecture [14] as shown in Fig. 8. In this case the adaptation law for the weights does not depend on the derivatives of the plant

$$c_i^C(k+1) = c_i^C(k) + \alpha K_i^C(S_s - u(k)) \qquad (14)$$

Here, S_s is the synthetic signal.

Proposition 3: If the system is invertible the algorithm defined by eqn. 14 converges to the best approximation of the inverse in the least square sense.

Proof: If the system is invertible then there exists an injective mapping which represents the inverse. Thus, from linear systems theory the algorithm defined by eqn. 14 converges to the least squares error [8].

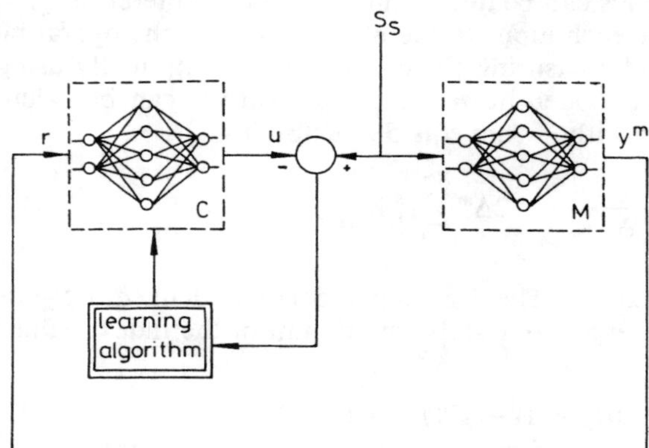

Fig. 8 *Use of synthetic signal, general learning structure*

As Psaltis *et al.* [14] indicated the specialised method allows the training of the inverse network in a region in the expected operational range of the plant. On the other hand, the generalised training procedure produces an inverse over the whole operating space. Psaltis *et al.* suggest a hybrid training procedure in which the specialised and generalised architectures are combined.

6.2.2 Iterative methods:

Iterative methods use a plant model to calculate the inverse. In this case a recursive method is used to find the inverse of the model in each operating point. This method is useful in singular systems which only satisfy the invertibility conditions outlined earlier locally, and not over the whole operating space. This approach can also be used when it is necessary to have small networks due to memory or processing limitations. In this case the restricted accuracy of the trained network can be enhanced by using the network to provide stored initial values for the iterative method, establishing a compromise between speed of convergence and storing capacities.

At time k, the objective is to find an input u which will produce a model output $y^m(k+1)$ equal to $r(k+1)$. It is possible to use the method of successive substitution

$$u^{n+1}(k) = u^n(k) + \gamma(r(k+1) - y^m(k+1))$$

where γ is a weight to be chosen. A functional block diagram is shown in Fig. 9.

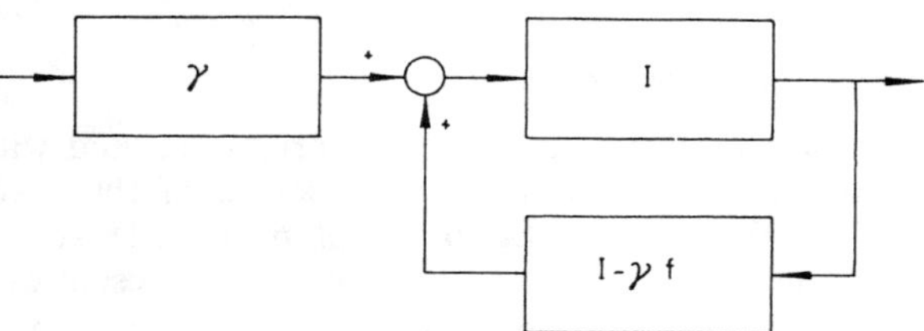

Fig. 9 *Iterative inversion*

According to the small gain theorem [16], the inverse operator is input–output stable if the product of the operator gains in the loop is less than 1

$$\|I\| \; \|I - \gamma f\| \leqslant 1$$

The initial value $u^0(k)$ can be stored in a connectionist network.

7 Examples

The plant to be considered is described by

$$y^p(k+1) = f_1(y^p(k)) + f_2(u(k))$$
$$= \frac{y^p(k)}{1 + y^p(k)^2} + u(k)^3 \qquad (15)$$

This system was previously considered by Narendra and Parthasarathy [3]. The system is monotonic with respect to $u(k)$ and therefore invertible. For all the simulations a simple first-order linear filter F was used. In this case the main objective of the filter was to map the error into the input space defined for the controller.

7.1 Separable case

First, we use the prior information to decompose the system into two parts

$$y^m(k+1) = N_1(y^p(k)) + N_2(u(k))$$

where N_1 and N_2 represent connectionist networks with 40 and 20 units, which represent f_1 and f_2, respectively, over a defined interval. Note that y^p is used for training the network; when training is complete the network can be used independent of the plant with $y^m(k)$ as input to N_1. This approach is followed below.

The structure of the inverse of the model can be deduced from the above equations. We require the input of the controller r to equal the model output

$$r(k+1) = N_1(y^m(k)) + N_2(u(k))$$

Thus, the control signal u is obtained as

$$u(k) = N_2^{-r}(r(k+1) - N_1(y^m(k)))$$

where N_2^{-r} is a network with 40 units, representing the right inverse of N_2. The structure of this system is shown in Fig. 10 [3]. Fig. 11 shows the functions and their estimates after training. The response to reference changes for the IMC structure is shown in Fig. 12. The response to disturbances and references changes is illustrated in Fig. 13, the system in this case is capable of eliminating the step disturbance.

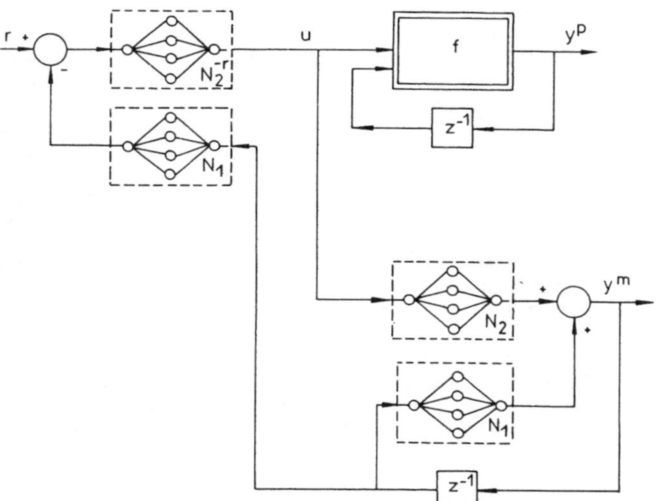

Fig. 10 *Separable system*

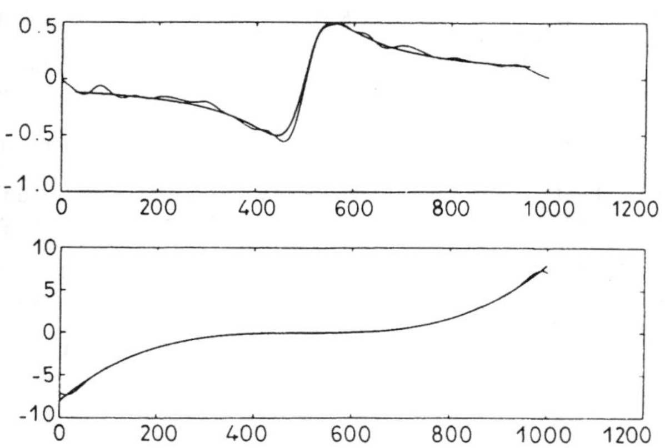

Fig. 11 *Functions and estimates*

7.2 Single network

Alternatively, we may ignore the structure of the nonlinearity. In this case a single connectionist network represents the system

$$y^m(k+1) = N(y^p(k), u(k))$$

The inverse is represented by

$$u(k) = N^{-r}(y^m(k), r(k+1))$$

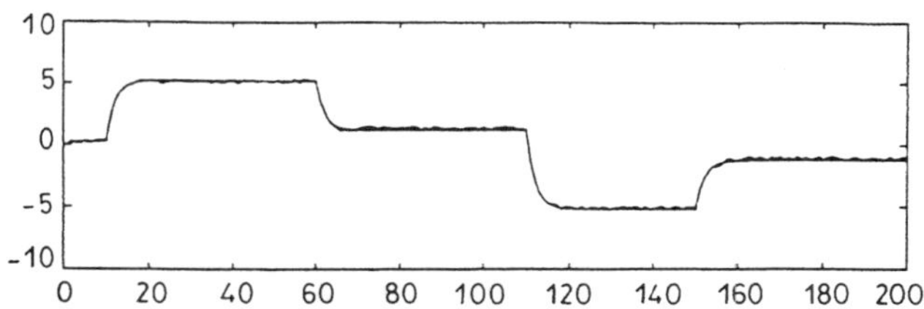

Fig. 12 *Response to changes in reference signal*

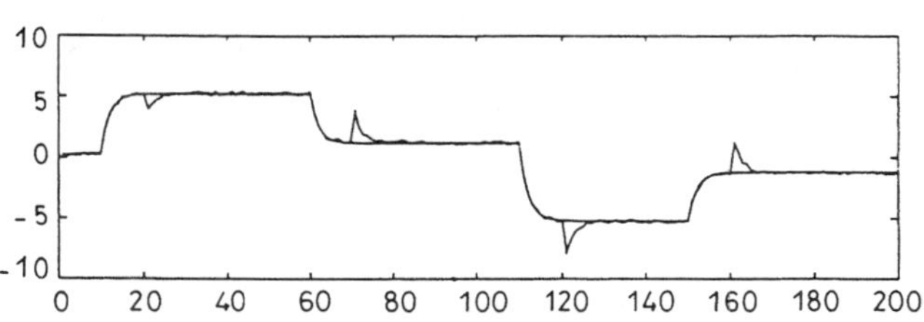

Fig. 13 *Response to changes in reference signal with a perturbation*

This represents the right inverse of N. Fig. 14 shows the response using the IMC structure to changes in the reference signal using a mesh of 100 units, and Fig. 15 shows the relation between the input to the inverse and the output of the model, that is $N[y^m(k), N^{-r}(y^m(k), r(k+1))]$. Fig. 16 shows the response to changes in the reference signal using a mesh of 400 units; in this case the error is much smaller compared to the 100 units case. Finally, Fig. 17 shows the relation between the input to the network that represents the inverse of the model and the output of the model, for this case. Theoretically it is possible to increase the number of units to obtain the desired behaviour.

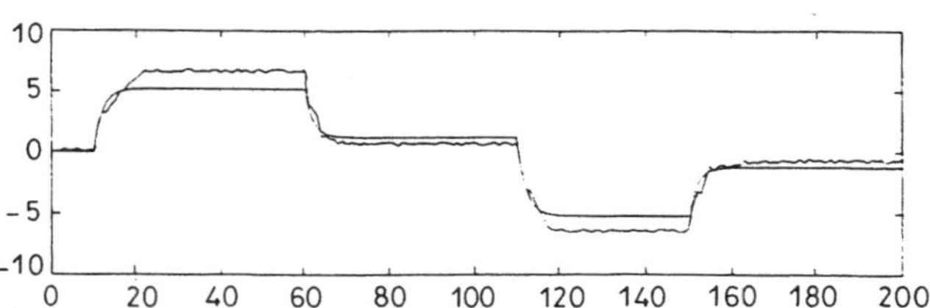

Fig. 14 *Response to changes in reference signal using 100 units*

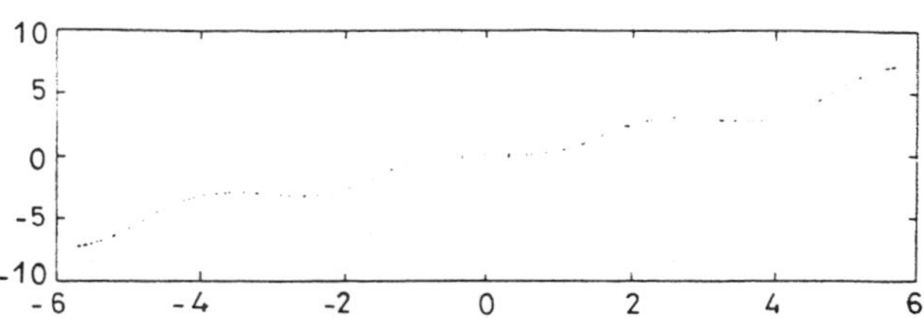

Fig. 15 *Input to controller against output of model*

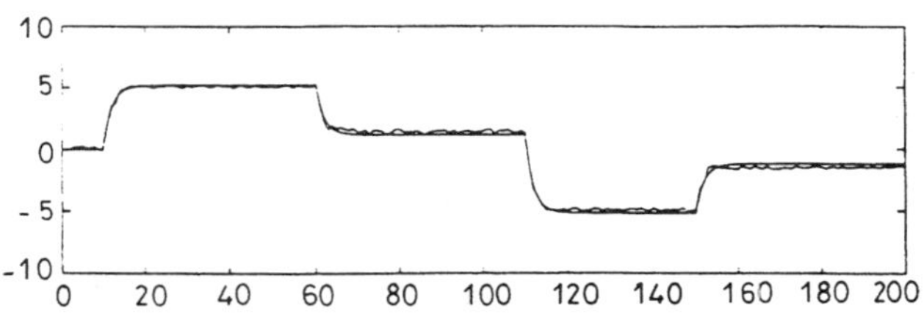

Fig. 16 *Response to changes in reference signal using 400 units*

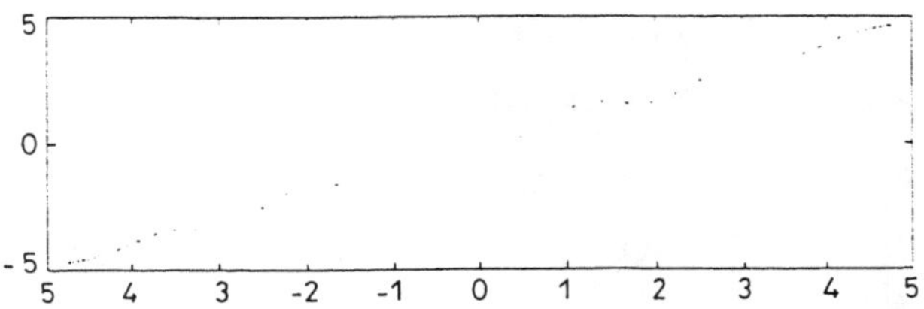

Fig. 17 *Input to controller against output of model*

7.3 Iterative calculation

The fidelity of the approximation is limited by the number of units. In order to increase the accuracy without increasing the number of units it is possible to add a refinement procedure over the first approximation using the iterative procedure outlined in Section 6.2.4.

The method used in the refinement, as described above, is the contraction mapping. Fig. 18 shows the results obtained with this procedure as a complement to a network with 100 units. The increased accuracy as a result of the improvement in the inverse model should be noted. Fig. 19 shows the relation between the input to the inverse and the output of the model for this case.

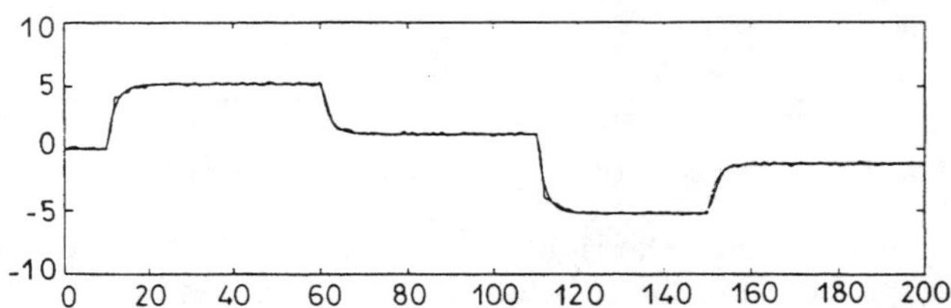

Fig. 18 *Response to changes in reference signal using iterative scheme*

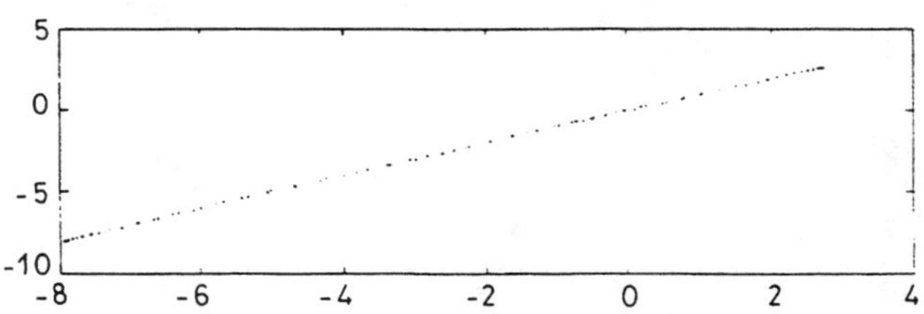

Fig. 19 *Input to controller against output of model*

8 Conclusions

The use of artificial neural networks for non-linear IMC has been explored. In doing so, we studied the invertibility of a class of non-linear dynamical systems and derived an invertibility characterisation.

We proposed architectures and algorithms for training networks to represent nonlinear dynamic relations, and the inverse of these relations. In addition, proof of convergence for the proposed algorithm is provided. The use of neural networks within the IMC structure was demonstrated by example.

In a companion paper [17] we investigate the relationship between IMC and adaptive inverse control [15]. In that work the neural network based IMC is interpreted as a nonlinear adaptive inverse control; the neural networks are viewed as nonlinear adaptive filters.

Future work will focus on the extension of the approach to multivariable systems, and on the robustness of the controller.

9 References

1 GEORGIEV, A.A.: 'Fitting of multivariate functions', *Proc. IEEE*, 1987, **75**, pp. 970–971

2 CYBENKO, G.: 'Approximation by superpositions of a sigmoidal function', *Math. Control Signal Systems*, 1989, **2**, pp. 303–314

3 NARENDRA, K.S., and PARTHASARATHY, K.: 'Identification and control of dynamic systems using neural networks', *IEEE Trans. Neural Networks*, 1990, **1**, pp. 4–27

4 CHEN, S., BILLINGS, S.A., and GRANT, P.M.: 'Non-linear system identification using neural networks', *Int. J. Control*, 1990, **51**, pp. 1191–1214

5 GARCIA, C.E., and MORARI, M.: 'Internal model control — 1. A unifying review and some new results', *Ind. Eng. Chem. Process Des. Dev.*, 1982, **21**, pp. 308–323

6 MORARI, M., and ZAFIRIOU, E.: 'Robust process control' (Prentice-Hall, 1989)

7 ECONOMOU, C.G., MORARI, M., and PALSSON, B.O.: 'Internal model control. 5. extension to nonlinear systems', *Ind. Eng. Chem. Process Des. Dev.*, 1986, **25**, pp. 403–411

8 SBARBARO, D.G., and GAWTHROP, P.J.: 'Self-organization and adaptation in gaussian networks'. 9th IFAC/IFORS symposium on identification and system parameter estimation. Budapest, Hungary, 1991

9 GIROSI, F., and POGGIO, T.: 'Networks and the best approximation property', *Biol. Cybernetics*, 1990, **63**, pp. 169–176

10 WEIGAND, A.S., HUBERMAN, B.A., and RUMELHART, D.E.: 'Predicting the future: A connectionist approach'. Report PARC-SSI-90-20, Stanford University, 1990

11 CASDAGLI, M.: 'Non-linear system prediction of chaotic time series', *Physica D*, 1989, **35**, pp. 335–356

12 BHAT, N., and MCAVOY, T.J.: 'Use of neural nets for dynamical modelling and control of chemical process systems', *Comput. Chem. Eng.*, 1990, **14**, (4–5), pp. 573–583

13 LEE, S., and KIL, R.M.: 'Multilayer feedforward potential function network', *Neural Networks*, 1990

14 PSALTIS, D., SIDERIS, A., and YAMAMURA, A.A.: 'A multilayered neural network controller', *IEEE Control System Mag.*, 1988, **8**, pp. 17–21

15 WIDROW, B.: 'Adaptive inverse control'. Preprints 2nd IFAC workshop on Adaptive Systems in Control and Signal Processing, Lund, Sweden, 1986, pp. 1–5

16 DESOER, C.A., and VIDYASAGAR, M.: 'Feedback systems: input–output properties' (Academic Press, London, 1975)

17 HUNT, K.J., and SBARBARO, D.: 'Adaptive filtering and neural networks for realisation of internal model control' (submitted for publication).

Identification and Control of Dynamical Systems Using Neural Networks

KUMPATI S. NARENDRA FELLOW, IEEE, AND KANNAN PARTHASARATHY

Abstract—**The paper demonstrates that neural networks can be used effectively for the identification and control of nonlinear dynamical systems. The emphasis of the paper is on models for both identification and control. Static and dynamic back-propagation methods for the adjustment of parameters are discussed. In the models that are introduced, multilayer and recurrent networks are interconnected in novel configurations and hence there is a real need to study them in a unified fashion. Simulation results reveal that the identification and adaptive control schemes suggested are practically feasible. Basic concepts and definitions are introduced throughout the paper, and theoretical questions which have to be addressed are also described.**

I. Introduction

MATHEMATICAL systems theory, which has in the past five decades evolved into a powerful scientific discipline of wide applicability, deals with the analysis and synthesis of dynamical systems. The best developed aspect of the theory treats systems defined by linear operators using well established techniques based on linear algebra, complex variable theory, and the theory of ordinary linear differential equations. Since design techniques for dynamical systems are closely related to their stability properties and since necessary and sufficient conditions for the stability of linear time-invariant systems have been generated over the past century, well-known design methods have been established for such systems. In contrast to this, the stability of nonlinear systems can be established for the most part only on a system-by-system basis and hence it is not surprising that design procedures that simultaneously meet the requirements of stability, robustness, and good dynamical response are not currently available for large classes of such systems.

In the past three decades major advances have been made in adaptive identification and control for identifying and controlling linear time-invariant plants with unknown parameters. The choice of the identifier and controller structures is based on well established results in linear systems theory. Stable adaptive laws for the adjustment of parameters in these cases which assure the global stability of the relevant overall systems are also based on properties of linear systems as well as stability results that are well known for such systems [1]. In this paper our interest is in the identification and control of nonlinear dynamic plants using neural networks. Since very few results exist in nonlinear systems theory which can be directly applied, considerable care has to be exercised in the statement of the problems, the choice of the identifier and controller structures, as well as the generation of adaptive laws for the adjustment of the parameters.

Two classes of neural networks which have received considerable attention in the area of artificial neural networks in recent years are: 1) multilayer neural networks and 2) recurrent networks. Multilayer networks have proved extremely successful in pattern recognition problems [2]–[5] while recurrent networks have been used in associative memories as well as for the solution of optimization problems [6]–[9]. From a systems theoretic point of view, multilayer networks represent static nonlinear maps while recurrent networks are represented by nonlinear dynamic feedback systems. In spite of the seeming differences between the two classes of networks, there are compelling reasons to view them in a unified fashion. In fact, it is the conviction of the authors that dynamical elements and feedback will be increasingly used in the future, resulting in complex systems containing both types of networks. This, in turn, will necessitate a unified treatment of such networks. In Section III of this paper this viewpoint is elaborated further.

This paper is written with three principal objectives. This first and most important objective is to suggest identification as well as controller structures using neural networks for the adaptive control of unknown nonlinear dynamical systems. While major advances have been made in the design of adaptive controllers for linear systems with unknown parameters, such controllers cannot be used for the global control of nonlinear systems. The models suggested consequently represent a first step in this direction. A second objective is to present a prescriptive method for the dynamic adjustment of the parameters based on back propagation. The term dynamic back propagation is introduced in this context. The third and final objective is to state clearly the many theoretical assumptions that have to be made to have well posed problems. Block diagram representations of systems commonly used in systems theory, as well as computer simulations, are included throughout the paper to illustrate the various concepts introduced. The paper is organized as follows:

Manuscript received August 25, 1989; revised November 15, 1989. This work was supported by the National Science Foundation under grant EET-8814747 and by Sandia National Laboratories under Contract 84-1791.

The authors are with the Department of Electrical Engineering, Yale University, New Haven, CT 06520.

IEEE Log Number 8933588.

Reprinted from *IEEE Trans. on Neural Networks*, vol. 1, no. 1, pp. 4–27, March 1990.

Section II deals with basic concepts and notational details used throughout the paper. In Section III, multilayer and recurrent networks are treated in a unified fashion. Section IV deals with static and dynamic methods for the adjustment of parameters of neural networks. Identification models are introduced in Section V while Section VI deals with the problem of adaptive control. Finally, in Section VII, some directions are given for future work.

II. Preliminaries, Basic Concepts, and Notation

In this section, many concepts related to the problem of identification and control are collected and presented for easy reference. While only some of them are directly used in the procedures discussed in Sections V and VI, all of them are relevant for a broad understanding of the role of neural networks in dynamical systems.

A. Characterization and Identification of Systems

System characterization and identification are fundamental problems in systems theory. The problem of characterization is concerned with the mathematical representation of a system; a model of a system is expressed as an operator P from an input space $\mathcal{U}$ into an output space $\mathcal{Y}$ and the objective is to characterize the class $\mathcal{P}$ to which P belongs. Given a class $\mathcal{P}$ and the fact that $P \in \mathcal{P}$, the problem of identification is to determine a class $\hat{\mathcal{P}} \subset \mathcal{P}$ and an element $\hat{P} \in \hat{\mathcal{P}}$ so that $\hat{P}$ approximates P in some desired sense. In static systems, the spaces $\mathcal{U}$ and $\mathcal{Y}$ are subsets of $\mathbb{R}^n$ and $\mathbb{R}^m$, respectively, while in dynamical systems they are generally assumed to be bounded Lebesgue integrable functions on the interval $[0, T]$ or $[0, \infty)$. In both cases, the operator P is defined implicitly by the specified input–output pairs. The choice of the class of identification models $\hat{\mathcal{P}}$, as well as the specific method used to determine $\hat{P}$, depends upon a variety of factors which are related to the accuracy desired, as well as analytical tractability. These include the adequacy of the model $\hat{P}$ to represent P, its simplicity, the ease with which it can be identified, how readily it can be extended if it does not satisfy specifications, and finally whether the $\hat{P}$ chosen is to be used off line or on line. In practical applications many of these decisions naturally depend upon the prior information that is available concerning the plant to be identified.

1. Identification of Static and Dynamic Systems: The problem of pattern recognition is a typical example of identification of static systems. Compact sets $U_i \subset \mathbb{R}^n$ are mapped into elements $y_i \in \mathbb{R}^m; (i = 1, 2, \cdots,)$ in the output space by a decision function P. The elements of U_i denote the pattern vectors corresponding to class y_i. In dynamical systems, the operator P defining a given plant is implicitly defined by the input–output pairs of time functions $u(t), y(t), t \in [0, T]$. In both cases the objective is to determine $\hat{P}$ so that

$$\|\hat{y} - y\| = \|\hat{P}(u) - P(u)\| \leq \epsilon, \qquad u \in \mathcal{U} \tag{1}$$

for some desired $\epsilon > 0$ and a suitably defined norm (denoted by $\|.\|$) on the output space. In (1), $\hat{P}(u) = \hat{y}$ denotes the output of the identification model and hence $\hat{y} - y \triangleq e$ is the error between the output generated by $\hat{P}$ and the observed output y. A more detailed statement of the identification problem of dynamical systems is given in Section II-C.

2. The Weierstrass Theorem and the Stone–Weierstrass Theorem: Let $C([a, b])$ denote the space of continuous real valued functions defined on the interval $[a, b]$ with the norm of $f \in C([a, b])$ defined by

$$\|f\| = \sup_t \left\{|f(t)| : t \in [a, b]\right\}.$$

The famous approximation theorem of Weierstrass states that any function in $C([a, b])$ can be approximated arbitrarily closely by a polynomial. Alternately, the set of polynomials is dense in $C([a, b])$. Naturally, Weierstrass's theorem and its generalization to multiple dimensions finds wide application in the approximation of continuous functions $f: \mathbb{R}^n \rightarrow \mathbb{R}^m$ using polynomials (e.g., pattern recognition). A generalization of Weierstrass's theorem due to Stone, called the Stone–Weierstrass theorem can be used as the starting point for all the approximation procedures for dynamical systems.

Theorem: (Stone–Weierstrass [10]): Let $\mathcal{U}$ be a compact metric space. If $\hat{\mathcal{P}}$ is a subalgebra of $C(\mathcal{U}, \mathbb{R})$ which contains the constant functions and separates points of $\mathcal{U}$ then $\hat{\mathcal{P}}$ is dense in $C(\mathcal{U}, \mathbb{R})$.

In the problems of interest to us we shall assume that the plant P to be identified belongs to the space $\mathcal{P}$ of bounded, continuous, time-invariant and causal operators [11]. By the Stone–Weierstrass theorem, if $\hat{\mathcal{P}}$ satisfies the conditions of the theorem, a model belonging to $\hat{\mathcal{P}}$ can be chosen which approximates any specified operator $P \in \mathcal{P}$.

A vast literature exists on the characterization of nonlinear functionals and includes the classic works of Volterra, Wiener, Barret, and Urysohn. Using the Stone–Weierstrass theorem it can be shown that a given nonlinear functional under certain conditions can be represented by a corresponding series such as the Volterra series or the Wiener series. In spite of the impressive theoretical work that these represent, very few have found wide application in the identification of large classes of practical nonlinear systems. In this paper our interest is mainly on representations which permit on-line identification and control of dynamic systems in terms of finite dimensional nonlinear difference (or differential) equations. Such nonlinear models are well known in the systems literature and are considered in the following subsection.

B. Input-State-Output Representation of Systems

The method of representing dynamical systems by vector differential or difference equations is currently well established in systems theory and applies to a fairly large class of systems. For example, the differential equations

$$\begin{aligned} \frac{dx(t)}{dt} &\triangleq \dot{x}(t) = \Phi[x(t), u(t)] \qquad t \in \mathbb{R}^+ \\ y(t) &= \Psi[x(t)] \end{aligned} \tag{2}$$

where $x(t) \triangleq [x_1(t), x_2(t), \cdots, x_n(t)]^T$, $u(t) \triangleq [u_1(t), u_2(t), \cdots, u_p(t)]^T$ and $y(t) \triangleq [y_1(t), y_2(t), \cdots, y_m(t)]^T$ represent a p input m output system of order n with $u_i(t)$ representing the inputs, $x_i(t)$ the state variables, and $y_i(t)$ the outputs of the system. Φ and Ψ are static nonlinear maps defined as $\Phi: \mathbb{R}^n \times \mathbb{R}^p \rightarrow \mathbb{R}^n$ and $\Psi: \mathbb{R}^n \rightarrow \mathbb{R}^m$. The vector $x(t)$ denotes the state of the system at time t and is determined by the state at time $t_0 < t$ and the input u defined over the interval $[t_0, t)$. The output $y(t)$ is determined completely by the state of the system at time t. Equation (2) is referred to as the input-state-output representation of the system. In this paper we will be concerned with discrete-time systems which can be represented by difference equations corresponding to the differential equations given in (2). These take the form

$$x(k+1) = \Phi[x(k), u(k)]$$
$$y(k) = \Psi[x(k)] \qquad (3)$$

where $u(.)$, $x(.)$, and $y(.)$ are discrete time sequences. Most of the results presented can, however, be extended to continuous time systems as well. If the system described by (3) is assumed to be linear and time invariant, the equations governing its behavior can be expressed as

$$x(k+1) = Ax(k), + Bu(k)$$
$$y(k) = Cx(k) \qquad (4)$$

where A, B, and C are $(n \times n)$, $(n \times p)$, and $(m \times n)$ matrices, respectively. The system is then parameterized by the triple $\{C, A, B\}$. The theory of linear time-invariant systems, when C, A, and B are known, is very well developed and concepts such as controllability, stability, and observability of such systems have been studied extensively in the past three decades. Methods for determining the control input $u(.)$ to optimize a performance criterion are also well known. The tractability of these different problems may be ultimately traced to the fact that they can be reduced to the solution of n linear equations in n unknowns. In contrast to this, the problems involving nonlinear equations of the form (3), where the functions Φ and Ψ are known, result in nonlinear algebraic equations for the solution of which similar powerful methods do not exist. Consequently, as shown in the following sections, several assumptions have to be made to make the problems analytically tractable.

C. Identification and Control

1. Identification: When the functions Φ and Ψ in (3), or the matrices A, B, and C in (4), are unknown, the problem of identification of the unknown system (referred to as the plant in the following sections) arises [12]. This can be formally stated as follows [1]:

The input and output of a time-invariant, causal discrete-time dynamical plant are $u(.)$ and $y_p(.)$, respectively, where $u(.)$ is a uniformly bounded function of time. The plant is assumed to be stable with a known parameterization but with unknown values of the parameters. The objective is to construct a suitable identification model (Fig. 1(a)) which when subjected to the same input $u(k)$ as the plant, produces an output $\hat{y}_p(k)$ which approximates $y_p(k)$ in the sense described by (1).

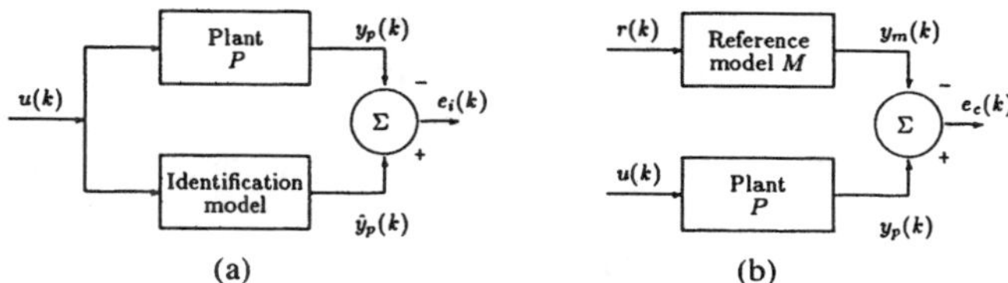

Fig. 1. (a) Identification. (b) Model reference adaptive control.

2. Control: Control theory deals with the analysis and synthesis of dynamical systems in which one or more variables are kept within prescribed limits. If the functions Φ and Ψ in (3) are known, the problem of control is to design a controller which generates the desired control input $u(k)$ based on all the information available at that instant k. While a vast body of frequency and time-domain techniques exist for the synthesis of controllers for linear systems of the form described in (4) with A, B, and C known, similar methods do not exist for nonlinear systems, even when the functions $\Phi(.,.)$ and $\Psi(.)$ are specified. In the last three decades there has been a great deal of interest in the control of plants when uncertainty exists regarding the dynamics of the plant [1]. To assure mathematical tractability, most of the effort has been directed towards the adaptive control of linear time-invariant plants with unknown parameters. Our interest in this paper lies primarily in the identification and control of unknown nonlinear dynamical systems.

Adaptive systems which make explicit use of models for control have been studied extensively. Such systems are commonly referred to as model reference adaptive control (MRAC) systems. The implicit assumption in the formulation of the MRAC problem is that the designer is sufficiently familiar with the plant under consideration so that he can specify the desired behavior of the plant in terms of the output of a reference model. The MRAC problem can be qualitatively stated as follows (Fig. 1(b)).

a. Model reference adaptive control: A plant P with an input–output pair $\{u(k), y_p(k)\}$ is given. A stable reference model M is specified by its input–output pair $\{r(k), y_m(k)\}$ where $r: \mathrm{N} \rightarrow \mathbb{R}$ is a bounded function. The output $y_m(k)$ is the desired output of the plant. The aim is to determine the control input $u(k)$ for all $k \geq k_0$ so that

$$\lim_{k \to \infty} \left| y_p(k) - y_m(k) \right| \leq \epsilon$$

for some specified constant $\epsilon \geq 0$.

As described earlier, the choice of the identification model (i.e., its parameterization) and the method of adjusting its parameters based on the identification error $e_i(k)$ constitute the two principal parts of the identification problem. Determining the controller structure, and adjusting its parameters to minimize the error between the output of the plant and the desired output, represent the corresponding parts of the control problem. In Section II-C-3, some well-known methods for setting up an identification model and a controller structure for a linear plant as well as the adjustment of identification and control parameters are described. Following this, in Section II-C-4, the problems encountered in the identification and control of nonlinear dynamical systems are briefly presented.

3. Linear Systems: For linear time-invariant plants with unknown parameters, the generation of identification models are currently well known. For a single-input single-output (SISO) controllable and observable plant, the matrix A and the vectors B and C in (4) can be chosen in such a fashion that the plant equation can be written as

$$y_p(k+1) = \sum_{i=0}^{n-1} \alpha_i y_p(k-i) + \sum_{j=0}^{m-1} \beta_j u(k-j) \quad (5)$$

where α_i and β_j are constant unknown parameters. A similar representation is also possible for the multi-input multi-output (MIMO) case. This implies that the output at time $k+1$ is a linear combination of the past values of both the input and the output. Equation (5) motivates the choice of the following identification models:

$$\hat{y}_p(k+1) = \sum_{i=0}^{n-1} \hat{\alpha}_i(k)\hat{y}_p(k-i) + \sum_{j=0}^{m-1} \hat{\beta}_j(k)u(k-j) \quad (6)$$

(Parallel model)

$$\hat{y}_p(k+1) = \sum_{i=0}^{n-1} \hat{\alpha}_i(k)y_p(k-i) + \sum_{j=0}^{m-1} \hat{\beta}_j(k)u(k-j)$$

(Series—parallel model) (7)

where $\hat{\alpha}_i(i = 0, 1, \cdots, n-1)$ and $\hat{\beta}_j(j = 0, 1, \cdots, m-1)$ are adjustable parameters. The output of the parallel identification model (6) at time $k+1$ is $\hat{y}_p(k+1)$ and is a linear combination of its past values as well as those of the input. In the series-parallel model, $\hat{y}_p(k+1)$ is a linear combination of the past values of the input and output of the plant. To generate stable adaptive laws, the series-parallel model is found to be preferable. In such a case, a typical adaptive algorithm has the form

$$\hat{\alpha}_i(k+1) = \hat{\alpha}_i(k) - \eta \cdot \frac{e(k+1)y_p(k-i)}{1 + \sum_{i=0}^{n-1} y_p^2(k-i) + \sum_{j=0}^{m-1} u^2(k-j)} \quad (8)$$

where $\eta > 0$ determines the step size. In the following discussions, the constant vector of plant parameters $[\alpha_0, \cdots, \alpha_{n-1}, \beta_0, \cdots, \beta_{m-1}]^T$ will be denoted by p and that of the identification model $[\hat{\alpha}_0, \cdots, \hat{\alpha}_{n-1}, \hat{\beta}_0, \cdots, \hat{\beta}_{m-1}]^T$ by $\hat{p}$.

Linear time-invariant plants which are controllable can be shown to be stabilizable by linear state feedback. This fact has been used to design adaptive controllers for such plants. For example, if an upper bound on the order of the plant is known, the control input can be generated as a linear combination of the past values of the input and output respectively. If $\theta(k)$ represents the control parameter vector, it can be shown that a constant vector θ^* exists such that when $\theta(k) \equiv \theta^*$ the plant together with the controller has the same input–output characteristics as the reference model. Adaptive algorithms for adjusting $\theta(k)$ in a stable fashion are now well known and have the general form shown in (8).

4. Nonlinear Systems: The importance of controllability and observability in the formulation of the identification and control problems for linear systems is evident from the discussion in Section II-C-3. Other well-known results in linear systems theory are also called upon to choose a reference model as well as a suitable parameterization of the plant and to assure the existence of a desired controller. In recent years a number of authors have addressed issues such as controllability, observability, feedback stabilization, and observer design for nonlinear systems [13]–[16]. In spite of such attempts constructive procedures, similar to those available for linear systems, do not exist for nonlinear systems. Hence, the choice of identification and controller models for nonlinear plants is a formidable problem and successful identification and control has to depend upon several strong assumptions regarding the input–output behavior of the plant. For example, if a SISO system is represented by the equation (3), we shall assume that the state of the system can be reconstructed from n measurements of the input and output. More precisely, $y_p(k) = \Psi[x(k)]$, $y_p(k+1) = \Psi[\Phi[x(k), u(k)]]$, $\cdots$, $y_p(k+n-1) = \Psi[\Phi[\cdots \Phi[\Phi[x(k), u(k)], u(k+1)], \cdots, u(k+n-2)]]$ yield n nonlinear equations in n unknowns $x(k)$ if $u(k), \cdots, u(k+n-2)$, $y_p(k), \cdots y_p(k+n-1)$ are specified and we shall assume that for any set of values of $u(k)$ in a compact region in $\mathcal{U}$, a unique solution to the above problem exists. This permits identification procedures to be proposed for nonlinear systems along lines similar to those in the linear case.

Even when the function Φ is known in (3) and the state vector is accessible, the determination of $u(.)$ for the plant to have a desired trajectory is an equally difficult problem. Hence, for the generation of the control input, the existence of suitable inverse operators have to be assumed. If a controller structure is assumed to generate the input $u(.)$, further assumptions have to be made to assure the existence of a constant control parameter vector to achieve the desired objective. All these indicate that considerable progress in nonlinear control theory will be needed to obtain rigorous solutions to the identification and control problems.

In spite of the above comments, the linear models described in Section II-C-3 motivate the choice of structures for identifiers and controllers in the nonlinear case. It is in these structures that we shall incorporate neural networks as described in Sections V and VI. A variety of considerations discussed in Section III reveal that both multilayer neural networks as well as recurrent networks, which are currently being extensively studied, will feature as subsystems in the design of identifiers and controllers for nonlinear dynamical systems.

III. Multilayer and Recurrent Networks

The assumptions that have to be made to assure well posed problems using models suggested in Sections V and VI are closely related to the properties of multilayer and recurrent networks. In this section, we describe briefly the two classes of neural networks and indicate why a unified treatment of the two may be warranted to deal with more complex systems in the future.

A. Multilayer Networks

A typical multilayer network with an input layer, an output layer, and two hidden layers is shown in Fig. 2. For convenience we denote this in block diagram form as shown in Fig. 3 with three weight matrices W^1, W^2, and W^3 and a diagonal nonlinear operator Γ with identical sigmoidal elements γ [i.e., $\gamma(x) = 1 - e^{-x}/1 + e^{-x}$] following each of the weight matrices. Each layer of the network can then be represented by the operator

$$N_i[u] = \Gamma[W^i u] \tag{9}$$

and the input–output mapping of the multilayer network can be represented by

$$y = N[u] = \Gamma[W^3\Gamma[W^2\Gamma[W^1u]]] = N_3N_2N_1[u]. \tag{10}$$

In practice, multilayer networks have been used successfully in pattern recognition problems [2]–[5]. The weights of the network W^1, W^2, and W^3 are adjusted as described in Section IV to minimize a suitable function of the error e between the output y of the network and a desired output y_d. This results in the mapping function $N[u]$ realized by the network, mapping vectors into corresponding output classes. Generally a discontinuous mapping such as a nearest neighbor rule is used at the last stage to map the input sets into points in the range space corresponding to output classes. From a systems theoretic point of view, multilayer networks can be considered as versatile nonlinear maps with the elements of the weight matrices as parameters. In the following sections we shall use the terms "weights" and "parameters" interchangeably.

B. Recurrent Networks

Recurrent networks, introduced in the works of Hopfield [6] and discussed quite extensively in the literature, provide an alternative approach to pattern recognition. One version of the network suggested by Hopfield consists of a single layer network N_1, included in feedback configuration, with a time delay (Figs. 4 and 5). Such a network represents a discrete-time dynamical system and can be described by

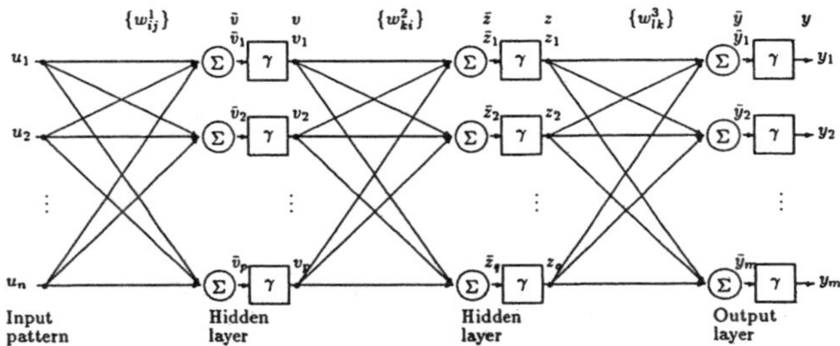

Fig. 2. A three layer neural network.

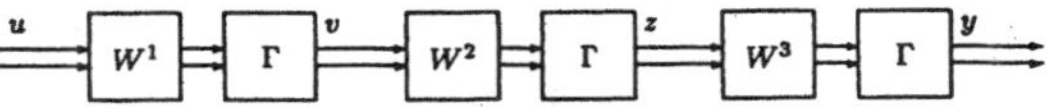

Fig. 3. A block diagram representation of a three layer network.

$$x(k+1) = N_1[x(k)], \quad x(0) = x_0.$$

Given an initial value x_0, the dynamical system evolves to an equilibrium state if N_1 is suitably chosen. The set of initial conditions in the neighborhood of x_0 which converge to the same equilibrium state is then identified with that state. The term "associative memory" is used to describe such systems. Recently, both continuous-time and discrete-time recurrent networks have been studied with constant inputs [17]. The inputs rather than the initial conditions represent the patterns to be classified in this case. In the continuous-time case, the dynamic system in the feedback path has a diagonal transfer matrix with identical elements $1/(s + \alpha)$ along the diagonal. The system is then represented by the equation

$$\dot{x} = -\alpha x + N_1[x] + I \tag{11}$$

so that $x(t) \in \mathbb{R}^n$ is the state of the system at time t, and the constant vector $I \in \mathbb{R}^n$ is the input.

C. A Unified Approach

In spite of the seeming differences between the two approaches to pattern recognition using neural networks, it is clear that a close relation exists between them. Recurrent networks with or without constant inputs are merely nonlinear dynamical systems and the asymptotic behavior of such systems depends both on the initial conditions as well as the specific input used. In both cases, this depends critically on the nonlinear map represented by the neural network used in the feedback loop. For example, when no input is used, the equilibrium state of the recurrent network in the discrete case is merely the fixed point of the mapping N_1. Thus the existence of a fixed point, the conditions under which it is unique, the maximum number of fixed points that can be achieved in a given network are all relevant to both multilayer and recurrent networks. Much of the current literature deals with such problems [18] and for mathematical tractability most of them as-

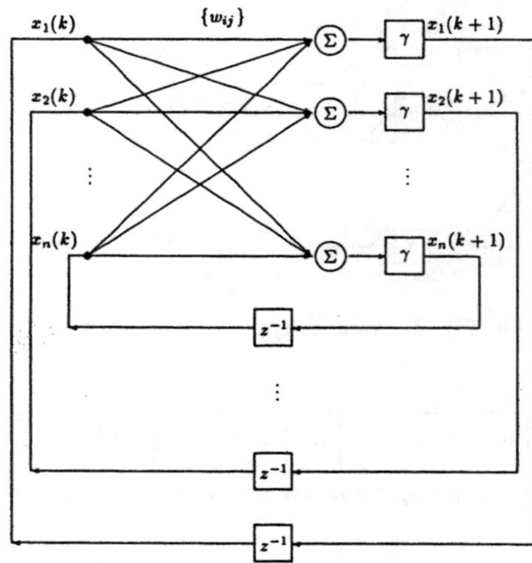

Fig. 4. The Hopfield network.

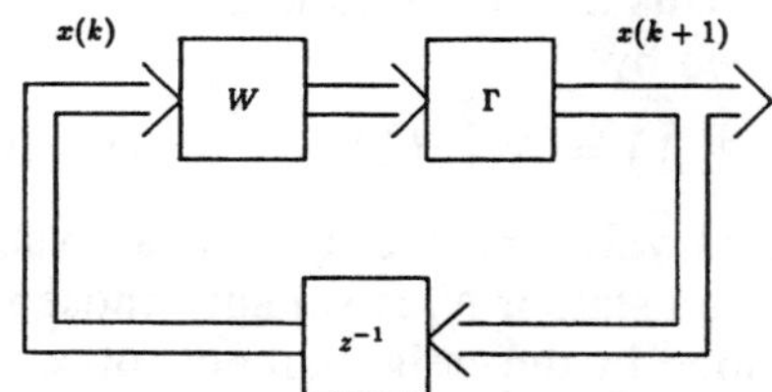

Fig. 5. Block diagram representation of the Hopfield network.

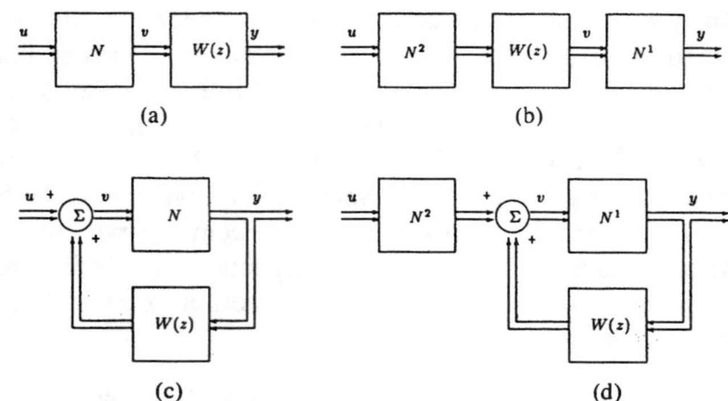

Fig. 6. (a) Representation 1. (b) Representation 2. (c) Representation 3. (d) Representation 4.

sume that recurrent networks contain only single layer networks (i.e., $N_1[.]$). As mentioned earlier, inputs when they exist are assumed to be constant. Recently, two layer recurrent networks have also been considered [19] and more general forms of recurrent networks can be constructed by including multilayer networks in the feedback loop [20]. In spite of the interesting ideas that have been presented in these papers, our understanding of such systems is still far from complete. In the identification and control problems considered in Sections V and VI, multilayer networks are used in cascade and feedback configurations and the inputs to such models are functions of time.

D. Generalized Neural Networks

From the above discussion, it follows that the basic elements in a multilayer network is the mapping $N_1[.] = \Gamma[W^1.]$, while the addition of the time delay element z^{-1} in the feedback path (Fig. 5) results in a recurrent network. In fact, general recurrent networks can be constructed composed of only the basic operations of 1) delay, 2) summation, and 3) the nonlinear operator $N_i[.]$. In continuous-time networks, the delay operator is replaced by an integrator. In some cases (as in (11)) multiplication by a constant is also allowed. Hence such networks are nonlinear feedback systems which consist only of elements $N_1[.]$, in addition to the usual operations found in linear systems.

Since arbitrary linear time-invariant dynamical systems can be constructed using the operations of summation, multiplication by a constant and time delay, the class of nonlinear dynamical systems that can be generated using generalized neural networks can be represented in terms of transfer matrices of linear systems [i.e., $W(z)$] and nonlinear operators $N[.]$. Fig. 6 shows these operators connected in cascade and feedback in four configurations which represent the building blocks for more complex systems. The superscript notation N^i is used in the figures to distinguish between different multilayer networks in any specific representation.

From the discussion of generalized neural networks, it follows that the mapping properties of $N_i[.]$ and consequently $N[.]$ (as defined in (10)) play a central role in all analytical studies of such networks. It has recently been shown in [21], using the Stone–Weierstrass theorem, that a two layer network with an arbitrarily large number of nodes in the hidden layer can approximate any continuous function $f \in C(\mathbb{R}^n, \mathbb{R}^m)$ over a compact subset of $\mathbb{R}^n$. This provides the motivation to assume that the class of generalized networks described is adequate to deal with a large class of problems in nonlinear systems theory. In fact, all the structures used in Section V and VI for the construction of identification and controller models are generalized neural networks and are closely related to the configurations shown in Fig. 6. For ease of discussion in the rest of the paper, we shall denote the class of functions generated by a network containing N layers by the symbol $\mathfrak{N}^N_{i_1, i_2, \cdots, i_{N+1}}$. Such a network has i_1 inputs, i_{N+1} outputs and $(N-1)$ sets of nodes in the hidden layers, each containing $i_2, i_3, \cdots, i_N$ nodes, respectively.

IV. Back Propagation in Static and Dynamic Systems

In both static identification (e.g., pattern recognition) and dynamic system identification of the type treated in this paper, if neural networks are used, the objective is to determine an adaptive algorithm or rule which adjusts the parameters of the network based on a given set of input–output pairs. If the weights of the networks are considered as elements of a parameter vector θ, the learning process involves the determination of the vector θ^* which optimizes a performance function J based on the output error. Back propagation is the most commonly used method for this purpose in static contexts. The gradient of the performance function with respect to θ is computed as $\nabla_\theta J$ and θ is adjusted along the negative gradient as

$$\theta = \theta_{\text{nom}} - \eta \nabla_\theta J |_{\theta = \theta_{\text{nom}}}$$

where η, the step size, is a suitably chosen constant and θ_{nom} denotes the nominal value of θ at which the gradient is computed. In this section, a diagrammatic representation of back propagation is first introduced. Following this, a method of extending this concept to dynamical systems is described and the term dynamic back propagation is defined. Prescriptive methods for the adjustment of weight vectors are suggested which can be used in the identification and control problems of the type discussed in Sections V and VI.

In the early 1960's, when the adaptive identification and control of linear dynamical systems were extensively studied, sensitivity models were developed to generate the partial derivatives of the performance criteria with respect to the adjustable parameters of the system. These models were the first to use sensitivity methods for dynamical systems and provided a great deal of insight into the necessary adaptive system structure [22]–[25]. Since conceptually the above problem is identical to that of determining the parameters of neural networks in identification and control problems, it is clear that back-propagation can be extended to dynamical systems as well.

A. A Diagrammatic Representation of Back Propagation

In this section we introduce a diagrammatic representation of back propagation. While the diagrammatic and algorithmic representations are informationally equivalent, their computational efficiency is different since the former preserves information about topological and geometric relations. In particular, the diagrammatic representation provides a better visual understanding of the entire process of back propagation, lends itself to modifications which are computationally more efficient and suggests novel modifications of the existing structure to include other functional extensions.

In the three layered network shown in Fig. 2, $u^T \triangleq [u_1, u_2, \cdots, u_n]$ denotes the input pattern vector while $y^T \triangleq [y_1, y_2, \cdots, y_m]$ is the output vector. $v^T \triangleq [v_1, v_2, \cdots, v_p]$ and $z^T \triangleq [z_1, z_2, \cdots, z_q]$ are the outputs at the first and the second hidden layers, respectively. $\{w_{ij}^1\}_{p\times n}$, $\{w_{ki}^2\}_{q\times p}$ and $\{w_{lk}^3\}_{m\times q}$ are the weight matrices associated with the three layers as shown in Fig. 2. The vectors $\bar{v} \in \mathbb{R}^p$, $\bar{z} \in \mathbb{R}^q$ and $\bar{y} \in \mathbb{R}^m$ are as shown in Fig. 2 with $\gamma(\bar{v}_i) = v_i$, $\gamma(\bar{z}_k) = z_k$ and $\gamma(\bar{y}_l) = y_l$ where $\bar{v}_i$, $\bar{z}_k$, and $\bar{y}_l$ are elements of $\bar{v}$, $\bar{z}$ and $\bar{y}$ respectively. If $y_d^T = [y_{d1}, y_{d2}, \cdots, y_{dm}]$ is the desired output vector, the output error vector for a given input pattern u is defined as $e \triangleq y - y_d$. The performance criterion J is then defined as

$$J = \sum_S \|e\|^2$$

where the summation is carried out over all patterns in a given set S. If the input patterns are assumed to be presented at each instant of time, the performance criterion J may be interpreted as the sum squared error over an interval of time. It is this interpretation which is found to be relevant in dynamic systems. In the latter case, the inputs and outputs are time sequences and the performance criterion J has the form $(1/T) \sum_{i=k-T+1}^{k} e^2(i)$, where T is a suitably chosen integer.

While strictly speaking the adjustment of the parameters should be carried out by determining the gradient of J in parameter space, the procedure commonly followed is to adjust it at every instant based on the error at that instant and a small step size η. If θ_j represents a typical parameter, $\partial e/\partial\theta_j$ has to be determined to compute the gradient as $e^T(\partial e/\partial\theta_j)$. The back propagation method is a convenient method of determining this gradient.

Fig. 7 shows the diagrammatic representation of back propagation for the three layer network shown in Fig. 2. The analytical method of deriving the gradient is well known in the literature and will not be repeated here. Fig. 7 merely shows how the various components of the gradient are realized. In our example, it is seen that signals u, v, and z and $\gamma'(\bar{v})$, $\gamma'(\bar{z})$, and $\gamma'(\bar{y})$, as well as the error vector, are used in the computation of the gradient (where $\gamma'(x)$ is the derivative of $\gamma(x)$ with respect to x). qm, pq, and np multiplications are needed to compute the partial derivatives with respect to the elements of W^3, W^2, and W^1, respectively. The structure of the weight matrices in the network used to compute the derivatives is seen to be identical to that in the original network while the signal flow is in the opposite direction, justifying the use of the term "back propagation." For further details regarding the diagrammatic representation, the reader is referred to [26] and [27]. The advantages of the diagrammatic representation mentioned earlier are evident from Fig. 7. More relevant to our purpose is that the same representation can be readily modified for the dynamic case. In fact, the diagrammatic representation was used extensively in all the simulation studies described in Sections V and VI.

B. Dynamic Back Propagation

In a causal dynamical system the change in a parameter at time k will produce a change in the output $y(t)$ for all $t \geq k$. For example, given a nonlinear dynamical system $x(k+1) = \Phi[x(k), u(k), \theta]$; $y(k) = \Psi[x(k)]$ where θ is a parameter, u is the input and x is the state vector defined in (3), the partial derivative of $y(k)$ with respect to θ can be obtained by solving the linear state equations

$$z(k+1) = A(k)z(k) + v(k), \; z(k_0) = 0$$

$$w(k) = C(k)z(k) \tag{12}$$

where $z(k) = \partial x(k)/\partial\theta \in \mathbb{R}^n$, $A(k) = \Phi_x(k) \in \mathbb{R}^{n\times n}$, $v(k) = \Phi_\theta(k) \in \mathbb{R}^n$, $w(k) = \partial y(k)/\partial\theta \in \mathbb{R}^m$ and $C(k)$

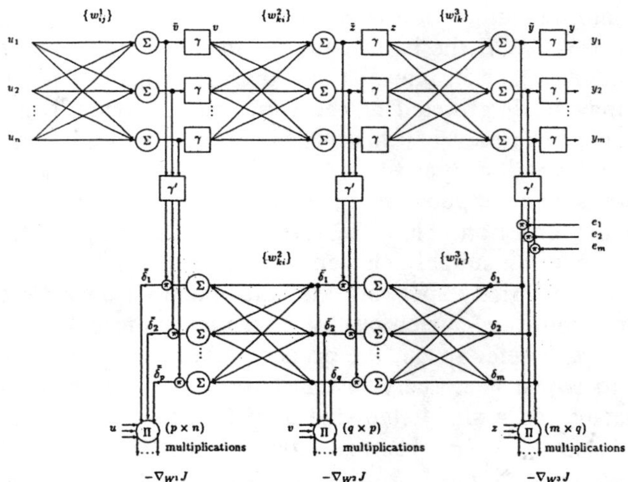

Fig. 7. Architecture for back propagation.

$= \Psi_x(k) \in \mathbb{R}^{m \times n}$. Φ_x and Ψ_x are Jacobian matrices and the vector Φ_θ represents the partial derivative of Φ with respect to θ. Equation (12) represents the linearized equations of the nonlinear system around the nominal trajectory and input. If $A(k)$, $v(k)$, and $C(k)$ can be computed, $w(k)$, the partial derivative of y with respect to θ can be obtained as the output of a dynamic sensitivity model.

In the previous section, generalized neural networks were defined and four representations of such networks with dynamical systems and multilayer neural networks connected in series and feedback were shown in Fig. 6. Since complex dynamical systems can be expressed in terms of these four representations, the back-propagation method can be extended to such systems if the partial derivative of the outputs with respect to the parameters can be determined for each of the representations. In the following we indicate briefly how (12) can be specialized to these four cases. In all cases it is assumed that the partial derivative of the output of a multilayer neural network with respect to one of the parameters can be computed using static back propagation and can be realized as the output of the network in Fig. 7.

In representation 1, the desired output $y_d(k)$ as well as the error $e(k) \triangleq y(k) - y_d(k)$ are functions of time. Representation 1 is the simplest situation that can arise in dynamical systems. This is because

$$\frac{\partial e(k)}{\partial \theta_j} = \frac{\partial y(k)}{\partial \theta_j} = W(z) \frac{\partial v}{\partial \theta_j}$$

where θ_j is a typical parameter of the network N. Since $\partial v/\partial \theta_j$ can be computed at every instant using static back propagation, $\partial e(k)/\partial \theta_j$ can be realized as the output of a dynamical system $W(z)$ whose inputs are the partial derivatives generated.

In representation 2, the determination of the gradient is rendered more complex by the presence of neural network N^1. If θ_j is a typical parameter of N^1, the partial derivative $\partial e(k)/\partial \theta_j$ is computed by static back propagation. However, if θ_j is a typical parameter of N^2

$$\frac{\partial y_i}{\partial \theta_j} = \sum_l \frac{\partial y_i}{\partial v_l} \frac{\partial v_l}{\partial \theta_j}.$$

Since $\partial v/\partial \theta_j$ can be computed using the method described in representation 1 and $\partial y_i/\partial v$ can be obtained by static back propagation, the product of the two yield the partial derivative of the signal y_i with respect to the parameter θ_j.

Representation 3 shows a neural network connected in feedback with a transfer matrix $W(z)$. The input to the nonlinear feedback system is a vector $u(k)$. If θ_j is a typical parameter of the neural network, the aim is to determine the derivatives $\overline{\partial} y_i(k)/\overline{\partial} \theta_j$ for $i = 1, 2, \cdots, m$ and all $k \geq 0$. We observe here for the first time a situation not encountered earlier, in that $\overline{\partial} y_i(k)/\overline{\partial} \theta_j$ is the solution of a difference equation, i.e., $\overline{\partial} y_i(k)/\overline{\partial} \theta_j$ is affected by its own past values

$$\frac{\overline{\partial} y}{\overline{\partial} \theta_j} = \frac{\partial N[v]}{\partial v} W(z) \frac{\overline{\partial} y}{\overline{\partial} \theta_j} + \frac{\partial N[v]}{\partial \theta_j}. \tag{13}$$

In (13), $\overline{\partial} y/\overline{\partial} \theta_j$ is a vector and $\partial N[v]/\partial v$ and $\partial N[v]/\partial \theta_j$ are the Jacobian matrix and a vector, respectively, which are evaluated around the nominal trajectory. Hence it represents a linearized difference equation in the variables $\overline{\partial} y_i/\overline{\partial} \theta_j$. Since $\partial N[v]/\partial v$ and $\partial N[v]/\partial \theta_j$ can be computed at every instant of time, the desired partial derivatives can be generated as the output of a dynamical system shown in Fig. 8(a) (the bar notation $\overline{\partial} y/\overline{\partial} \theta_j$ is used in (13) to distinguish between $\partial y/\partial \theta_j$ and $\partial N[v]/\partial \theta_j$).

In the final representation, the feedback system is preceded by a neural network N^2. The presence of N^2 does not affect the computation of the partial derivatives of the output with respect to the parameters of N^1. However, if θ_j is a typical parameter of N^2, it can be shown that $\partial y/\partial \theta_j$ can be obtained as

$$\frac{\partial y}{\partial \theta_j} = \frac{\partial N^1[v]}{\partial v} \left[\frac{\partial N^2[u]}{\partial \theta_j} + W(z) \frac{\partial y}{\partial \theta_j} \right]$$

or alternately it can be represented as the output of the dynamical system shown in Fig. 8(b) whose inputs can be computed at every instant of time.

In all the problems of identification and control that we will be concerned with in the following sections, the matrix $W(z)$ is diagonal and consists only of elements of the form z^{-d_i} (i.e., a delay of d_i units). Further since dynamic back propagation is considerably more involved than static back propagation, the structure of the identification models is chosen, wherever possible, so that the latter can be used. The models of back propagation developed here can be applied to general control problems where neural networks and linear dynamical systems are interconnected in arbitrary configurations and where static back propagation cannot be justified. For further details the reader is

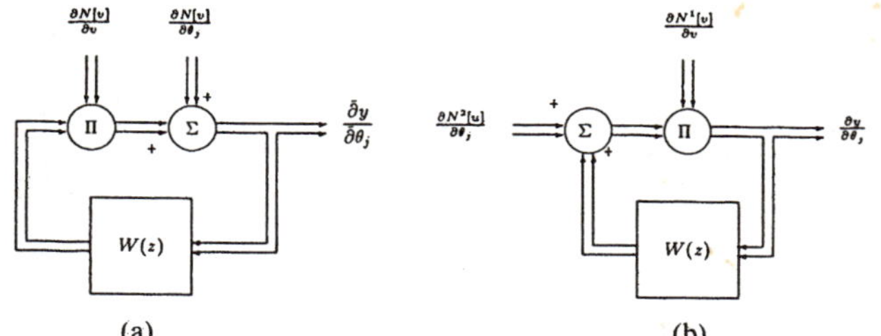

Fig. 8. (a) Generation of gradient in representation 3. (b) Generation of gradient in representation 4.

referred to [27]. A paper based on [27] but providing details concerning the implementation of the algorithms in practical applications is currently under preparation.

V. Identification

As mentioned in Section III, the ability of neural networks to approximate large classes of nonlinear functions sufficiently accurately make them prime candidates for use in dynamic models for the representation of nonlinear plants. The fact that static and dynamic back-propagation methods, as described in Section IV, can be used for the adjustment of their parameters also makes them attractive in identifiers and controllers. In this section four models for the representation of SISO plants are introduced which can also be generalized to the multivariable case. Following this, identification models are suggested containing multilayer neural networks as subsystems. These models are motivated by the models which have been used in the adaptive systems literature for the identification and control of linear systems and can be considered as their generalization to nonlinear systems.

A. Characterization

The four models of discrete-time plants introduced here can be described by the following nonlinear difference equations:

Model I: $y_p(k+1)$

$$= \sum_{i=0}^{n-1} \alpha_i y_p(k-i) + g[u(k), u(k-1), \cdots, u(k-m+1)]$$

Model II: $y_p(k+1)$

$$= f[y_p(k), y_p(k-1), \cdots, y_p(k-n+1)] + \sum_{i=0}^{m-1} \beta_i u(k-i)$$

Model III: $y_p(k+1)$

$$= f[y_p(k), y_p(k-1), \cdots, y_p(k-n+1)] + g[u(k), u(k-1), \cdots, u(k-m+1)]$$

Model IV: $y_p(k+1)$

$$= f[y_p(k), y_p(k-1), \cdots, y_p(k-n+1); u(k), u(k-1), \cdots, u(k-m+1)] \tag{14}$$

where $[u(k), y_p(k)]$ represents the input-output pair of the SISO plant at time k, and $m \leq n$. The block diagram representation of the various models are shown in Fig. 9. The functions $f: \mathbb{R}^n \to \mathbb{R}$ in Models II and III and $f: \mathbb{R}^{n+m} \to \mathbb{R}$ in Model IV, and $g: \mathbb{R}^m \to \mathbb{R}$ in (14) are assumed to be differentiable functions of their arguments. In all the four models, the output of the plant at the time $k+1$ depends both on its past n values $y_p(k-i)$ $(i = 0, 1, \cdots, n-1)$ as well as the past m values of the input $u(k-j)$ $(j = 0, 1, \cdots, m-1)$. The dependence on the past values $y_p(k-i)$ is linear in Model I while in Model II the dependence on the past values of the input $u(k-j)$ is assumed to be linear. In Model III, the nonlinear dependence of $y_p(k+1)$ on $y_p(k-i)$ and $u(k-j)$ is assumed to be separable. It is evident that Model IV in which $y_p(k+1)$ is a nonlinear function of $y_p(k-i)$ and $u(k-j)$ subsumes Models I–III. If a general nonlinear SISO plant can be described by an equation of the form (3) and satisfies the stringent observability condition discussed in Section II-C-4, it can be represented by such a model. In spite of its generality, Model IV is, however, analytically the least tractable and hence for practical applications some of the other models are found to be more attractive. For example, as will be apparent in the following section, Model II is particularly suited for the control problem.

From the results given in Section III, it follows that under fairly weak conditions on the function f and/or g in (14), multilayer neural networks can be constructed to approximate such mappings over compact sets. We shall assume for convenience that f and/or g belong to a known class $\mathfrak{N}^N_{i_1, i_2, \cdots, i_{N+1}}$ in the domain of interest, so that the plant can be represented by a generalized neural network as discussed in Section III. This assumption motivates the choice of the identification models and allows the statement of well posed identification problems. In particular, the identification models have the same structure as the plant but contain neural networks with adjustable parameters.

Let a nonlinear dynamic plant be represented by one of the four models described in (14). If such a plant is to be identified using input-output data, it must be further assumed that it has bounded outputs for the class of permissible inputs. This implies that the model chosen to represent the plant also enjoys this property. In the case of Model I, this implies that the roots of the characteristic equation $z^n - \alpha_0 z^{n-1} - \cdots - \alpha_{n-2} z - \alpha_{n-1} = 0$ lie in the interior of the unit circle. In the other three cases no such simple algebraic conditions exist. Hence the study of the stability properties of recurrent networks containing multilayer networks represents an important area of research.

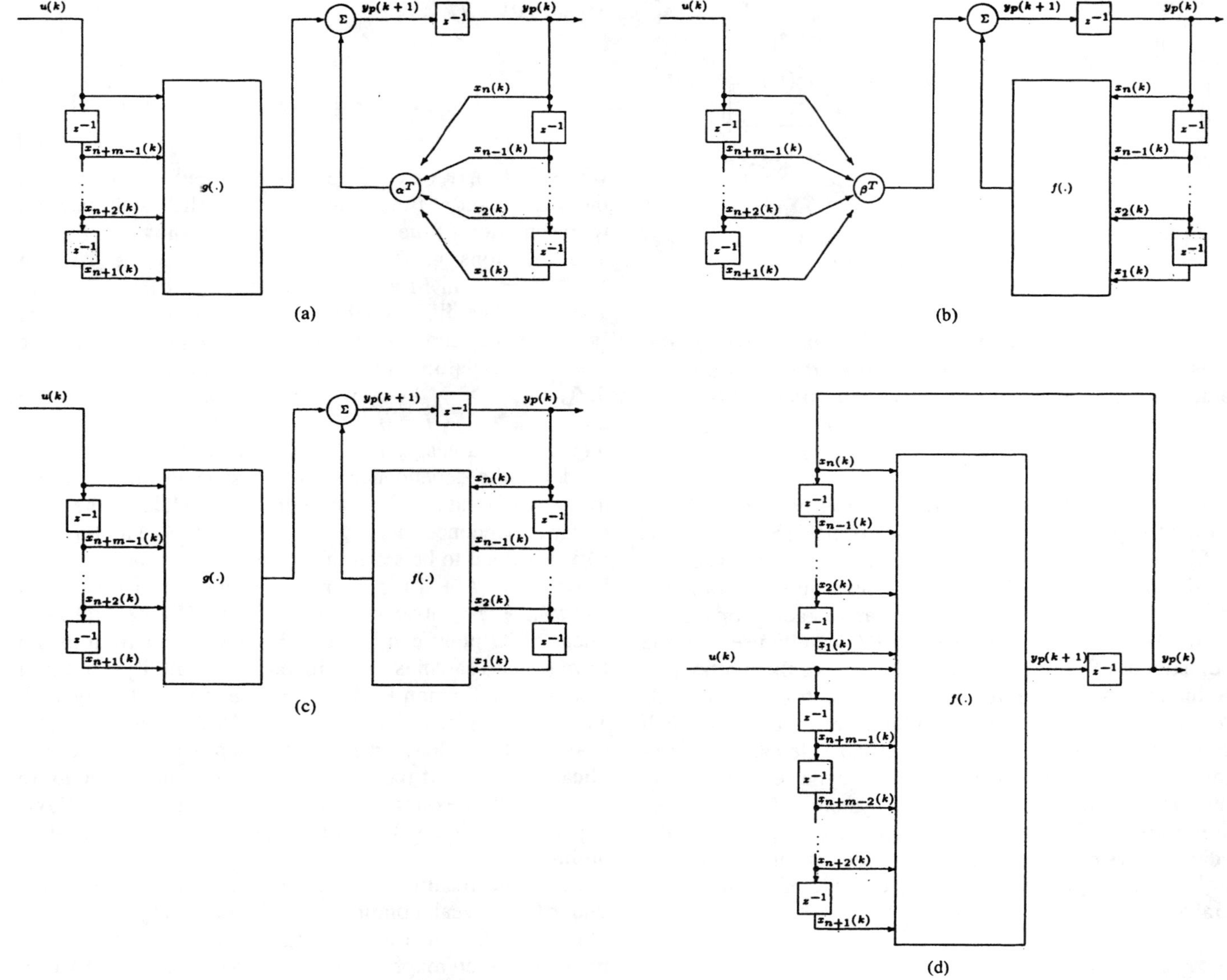

Fig. 9. Representation of SISO plants. (a) Model I. (b) Model II. (c) Model III. (d) Model IV.

The models described thus far are for the representation of discrete-time plants. Continuous-time analogs of these models can be described by differential equations, as stated in Section II. While we shall deal exclusively with discrete-time systems, the same methods also carry over to the continuous time case.

B. Identification

The problem of identification, as described in Section II-C, consists of setting up a suitably parameterized identification model and adjusting the parameters of the model to optimize a performance function based on the error between the plant and the identification model outputs. Since the nonlinear functions in the representation of the plant are assumed to belong to a known class $\mathfrak{N}^{N}_{i_1, i_2, \cdots, i_{N+1}}$ in the domain of interest, the structure of the identification model is chosen to be identical to that of the plant. By assumption, weight matrices of the neural networks in the identification model exist so that, for the same initial conditions, both plant and model have the same output for any specified input. Hence the identification procedure consists in adjusting the parameters of the neural networks in the model using the method described in Section IV based on the error between the plant and model outputs. However, as shown in what follows, suitable precautions have to be taken to ensure that the procedure results in convergence of the identification model parameters to their desired values.

1. Parallel Identification Model: Fig. 10(a) shows a plant which can be represented by Model I with $n = 2$ and $m = 1$. To identify the plant one can assume the structure of the identification model shown in Fig. 10(a) and described by the equation

$$\hat{y}_p(k + 1) = \hat{\alpha}_0 \hat{y}_p(k) + \hat{\alpha}_1 \hat{y}_p(k - 1) + N[u(k)].$$

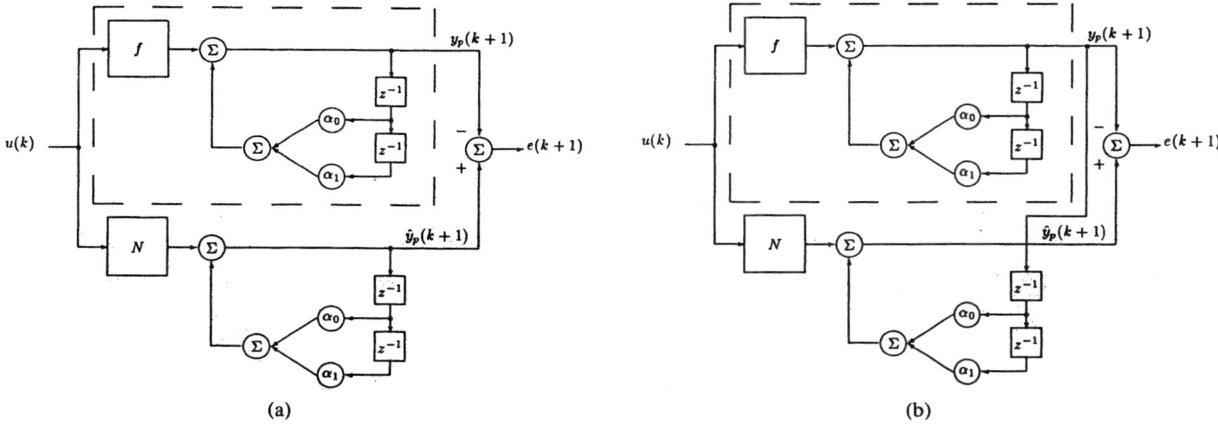

Fig. 10. (a) Parallel identification model. (b) Series-parallel identification model.

As mentioned in Section II-C-3, this is referred to as a parallel model. Identification then involves the estimation of the parameters $\hat{\alpha}_i$ as well as the weights of the neural network using dynamic back propagation based on the error $e(k)$ between the model output $\hat{y}_p(k)$ and the actual output $y_p(k)$.

From the assumptions made earlier, the plant is bounded-input bounded-output (BIBO) stable in the presence of an input (in the assumed class). Hence, all the signals in the plant are uniformly bounded. In contrast to this, the stability of the identification model as described here with a neural network cannot be assured and has to be proved. Hence if a parallel model is used, there is no guarantee that the parameters will converge or that the output error will tend to zero. In spite of two decades of work, conditions under which the parallel model parameters will converge even in the linear case are at present unknown. Hence, for plant representations using Models I–IV, the following identification model, known as the series-parallel model, is used.

2. *Series-Parallel Model:* In contrast to the parallel model described above, in the series-parallel model the output of the plant (rather than the identification model) is fed back into the identification model as shown in Fig. 10(b). This implies that in this case the identification model has the form

$$\hat{y}_p(k+1) = \hat{\alpha}_0 y_p(k) + \hat{\alpha}_1 y_p(k-1) + N[u(k)].$$

We shall use the same procedure with all the four models described earlier. The series-parallel identification model corresponding to a plant represented by Model IV has the form shown in Fig. 11. TDL in Fig. 11 denotes a tapped delay line whose output vector has for its elements the delayed values of the input signal. Hence the past values of the input and the output of the plant form the input vector to a neural network whose output $\hat{y}_p(k)$ corresponds to the estimate of the plant output at any instant of time k. The series-parallel model enjoys several advantages over the parallel model. Since the plant is assumed to be BIBO stable, all the signals used in the identification procedure (i.e., inputs to the neural networks) are bounded. Further, since no feedback loop exists in the model, static back propagation can be used to adjust the parameters reducing the computational overhead substantially. Finally, assuming that the output error tends to a small value asymptotically so that $y_p(k) \approx \hat{y}_p(k)$, the series-parallel model may be replaced by a parallel model without serious consequences. This has practical implications if the identification model is to be used off line. In view of the above considerations the series-parallel model is used in all the simulations in this paper.

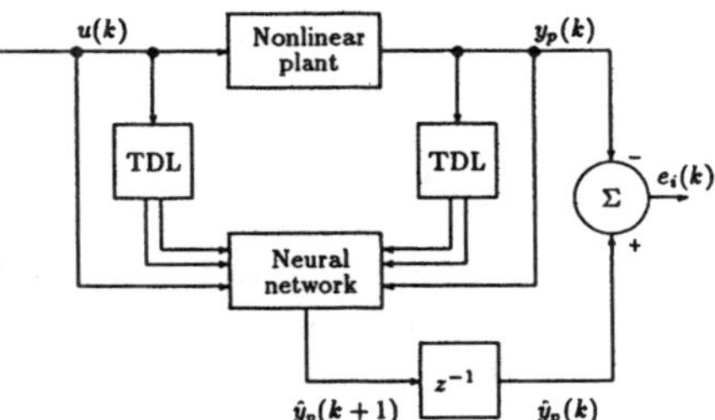

Fig. 11. Identification of nonlinear plants using neural networks.

C. Simulation Results

In this section simulation results of nonlinear plant identification using the models suggested earlier are presented. Six examples are presented where the prior information available dictates the choice of one of the Models I–IV. Each example is chosen to emphasize a specific point. In the first five examples, the series-parallel model is used to identify the given plant and static back-propagation is used to adjust parameters of the neural networks. A final example is used to indicate how dynamic back propagation may be used in identification problems. Due to space limitations, only the principal results are presented here. The reader interested in further details is referred to [27]–[29].

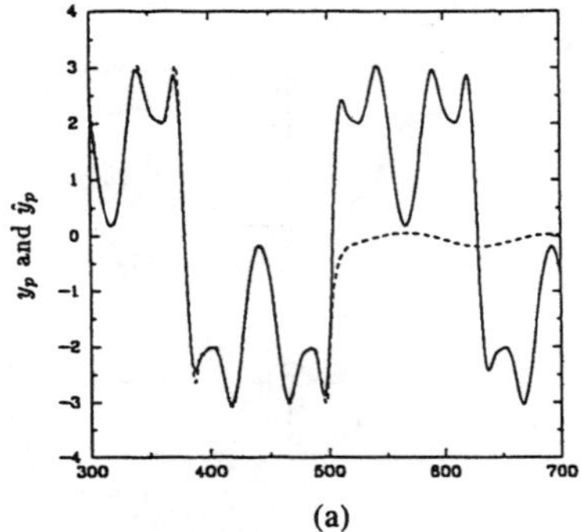

(a)

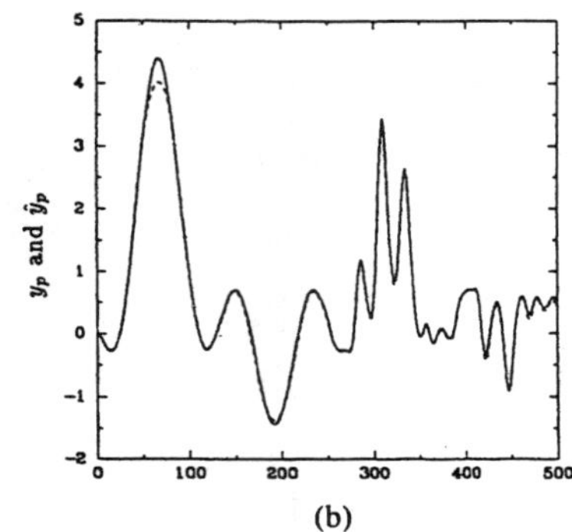

(b)

Fig. 12. Example 1: (a) Outputs of the plant and identification model when adaptation stops at $k = 500$. (b) Response of plant and identification model after identification using a random input.

1. Example 1: The plant to be identified is governed by the difference equation

$$y_p(k+1) = 0.3y_p(k) + 0.6y_p(k-1) + f[u(k)] \tag{15}$$

where the unknown function has the form $f(u) = 0.6 \sin(\pi u) + 0.3 \sin(3\pi u) + 0.1 \sin(5\pi u)$. From (15), it is clear that the unforced linear system is asymptotically stable and hence any bounded input results in a bounded output. In order to identify the plant, a series-parallel model governed by the difference equation

$$\hat{y}_p(k+1) = 0.3y_p(k) + 0.6y_p(k-1) + N[u(k)]$$

was used. The weights in the neural network were adjusted at every instant of time ($T_i = 1$) using static back propagation. The neural network belonged to the class $\mathfrak{N}^3_{1,20,10,1}$ and the gradient method employed a step size of $\eta = 0.25$. The input to the plant and the model was a sinusoid $u(k) = \sin(2\pi k/250)$. As seen from Fig. 12(a), the output of the model follows the output of the plant almost immediately but fails to do so when the adaptation process is stopped at $k = 500$, indicating that the identification of the plant is not complete. Hence the identification procedure was continued for 50 000 time steps using a random input whose amplitude was uniformly distributed in the interval $[-1, 1]$ at the end of which the adaptation was terminated. Fig. 12(b) shows the outputs of the plant and the trained model. The nonlinear function in the plant in this case is $f[u] = u^3 + 0.3u^2 - 0.4u$. As can be seen from the figure, the identification error is small even when the input is changed to a sum of two sinusoids $u(k) = \sin(2\pi k/250) + \sin(2\pi k/25)$ at $k = 250$.

2. Example 2: The plant to be identified is described by the second-order difference equation

$$y_p(k+1) = f[y_p(k), y_p(k-1)] + u(k)$$

where

$$f[y_p(k), y_p(k-1)] = \frac{y_p(k)\, y_p(k-1)[y_p(k) + 2.5]}{1 + y_p^2(k) + y_p^2(k-1)}. \tag{16}$$

This corresponds to Model II. A series-parallel identifier of the type discussed earlier is used to identify the plant from input–output data and is described by the equation

$$\hat{y}_p(k+1) = N[y_p(k), y_p(k-1)] + u(k) \tag{17}$$

where N is a neural network with $N \in \mathfrak{N}^3_{2,20,10,1}$. The identification process involves the adjustment of the weights of N using back propagation.

Some prior information concerning the input–output behavior of the plant is needed before identification can be undertaken. This includes the number of equilibrium states of the unforced system and their stability properties, the compact set $\mathcal{U}$ to which the input belongs and whether the plant output is bounded for this class of inputs. Also, it is assumed that the mapping N can approximate f over the desired domain.

a. Equilibrium states of the unforced system: The equilibrium states of the unforced system $y_p(k+1) = f[y_p(k), y_p(k-1)]$ with f as defined in (16) are $(0, 0)$ and $(2, 2)$, respectively, in the state space. This implies that while in equilibrium without an input, the output of the plant is either the sequence $\{0\}$ or the sequence $\{2\}$. Further, for any input $|u(k)| \leq 5$, the output of the plant is uniformly bounded for initial conditions $(0, 0)$ and $(2, 2)$ and satisfies the inequality $|y_p(k)| \leq 13$.

Assuming different initial conditions in the state space and with zero input, the weights of the neural network were adjusted so that the error $e(k+1) = y_p(k+1) - N[y_p(k), y_p(k-1)]$ is minimized. When the weights converged to constant values, the equation $\hat{y}_p(k+1) = N[\hat{y}_p(k), \hat{y}_p(k-1)]$ was simulated for initial conditions within a radius of 4. The identified system was found to have the same trajectories as the plant for the same initial conditions. The behavior of the plant and the identified model for different initial conditions are shown in Fig. 13. It must be emphasized here that in practice the initial conditions of the plant cannot be chosen at the discretion of the designer and must be realized only by using different inputs to the plant.

b. Identification: While the neural network realized above can be used in the identification model, a separate

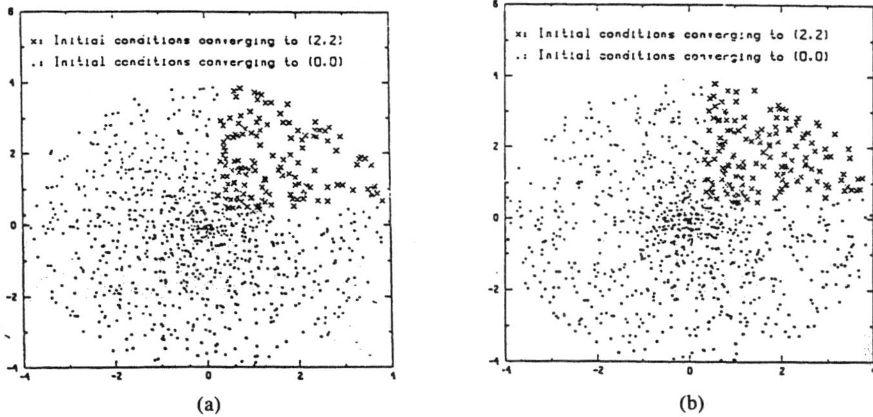

Fig. 13. Example 2: Behavior of the unforced system. (a) Actual plant. (b) Identified model of the plant.

simulation was carried out using both inputs and outputs and a series-parallel model. The input $u(k)$ was assumed to be an i.i.d. random signal uniformly distributed in the interval $[-2, 2]$ and a step size of $\eta = 0.25$ was used in the gradient method. The weights in the neural network were adjusted at intervals of five steps using the gradient of $\Sigma_{i=k-4}^{k} e^2(i)$. Fig. 14 shows the outputs of the plant and the model after the identification procedure was terminated at $k = 100\,000$.

3. Example 3: In Example 2, the input is seen to occur linearly in the difference equation describing the plant. In this example the plant is described by Model III and has the form

$$y_p(k+1) = \frac{y_p(k)}{1 + y_p(k)^2} + u^3(k)$$

which corresponds to $f[y_p(k)] = y_p(k)/(1 + y_p(k)^2)$ and $g[u(k)] = u^3(k)$ in (14). A series-parallel identification model consists of two neural networks N_f and N_g belonging to $\mathfrak{N}^3_{1,20,10,1}$ and can be described by the difference equation

$$\hat{y}_p(k+1) = N_f[y_p(k)] + N_g[u(k)].$$

The estimates $\hat{f}$ and $\hat{g}$ are obtained by using neural networks N_f and N_g. The weights in the neural networks were adjusted at every instant of time using a step size of $\eta = 0.1$ and was continued for 100 000 time steps. Since the input was a random input in interval $[-2, 2]$, $\hat{g}$ approximates g only over this interval. Since this in turn results in the variation of y_p over the interval $[-10, 10]$, $\hat{f}$ approximates f over the latter interval. The functions $\hat{f}$ and $\hat{g}$ as well as f and g over their respective domains are shown in Fig. 15(a) and (b). In Fig. 15(c), the outputs of the plant as well as the identification model for an input $u(k) = \sin(2\pi k/25) + \sin(2\pi k/10)$ are shown and are seen to be indistinguishable.

4. Example 4: The same methods used for identification of plants in examples 1-3 can be used when the unknown plants are known to belong to Model IV. In this

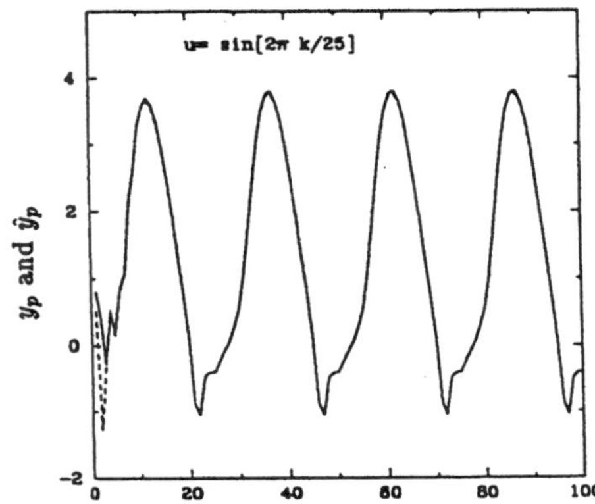

Fig. 14. Example 2: Outputs of the plant and the identification model.

example, the plant is assumed to be of the form

$$y_p(k+1) = f[y_p(k), y_p(k-1), y_p(k-2), u(k), u(k-1)]$$

where the unknown function f has the form

$$f[x_1, x_2, x_3, x_4, x_5] = \frac{x_1 x_2 x_3 x_5 (x_3 - 1) + x_4}{1 + x_3^2 + x_2^2}.$$

In the identification model, a neural network N belonging to the class $\mathfrak{N}^3_{5,20,10,1}$ is used to approximate the function f. Fig. 16 shows the output of the plant and the model when the identification procedure was carried out for 100 000 steps using a random input signal uniformly distributed in the interval $[-1, 1]$ and a step size of $\eta = 0.25$. As mentioned earlier, during the identification process a series-parallel model is used, but after the identification process is terminated the performance of the model is studied using a parallel model. In Fig. 16, the input to the plant and the identified model is given by $u(k) = \sin(2\pi k/250)$ for $k \leq 500$ and $u(k) = 0.8 \sin(2\pi k/250) + 0.2 \sin(2\pi k/25)$ for $k > 500$.

5. Example 5: In this example, it is shown that the same methods used to identify SISO plants can be used to

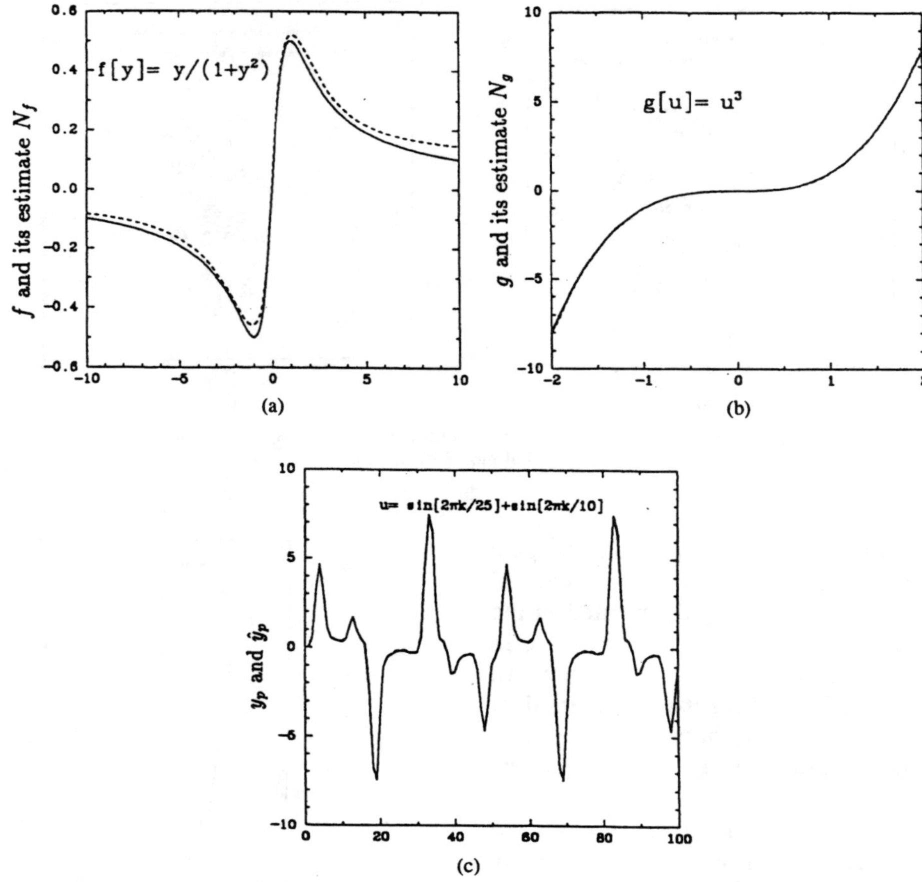

Fig. 15. Example 3: (a) Plots of the functions f and $\hat{f}$. (b) Plots of the functions g and $\hat{g}$. (c) Outputs of the plant and the identification model.

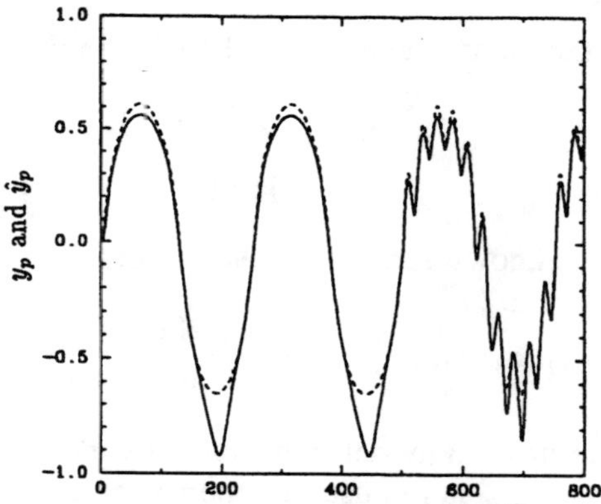

Fig. 16. Example 4: Identification of Model IV.

identify MIMO plants as well. The plant is described by the equations

$$\begin{bmatrix} y_{p1}(k+1) \\ y_{p2}(k+1) \end{bmatrix} = \begin{bmatrix} \dfrac{y_{p1}(k)}{1 + y_{p2}^2(k)} \\ \dfrac{y_{p1}(k)y_{p2}(k)}{1 + y_{p2}^2(k)} \end{bmatrix} + \begin{bmatrix} u_1(k) \\ u_2(k) \end{bmatrix}. \quad (18)$$

This corresponds to the multivariable version of Model II. The series-parallel identification model consists of two neural networks N^1 and N^2 and is described by the equation

$$\begin{bmatrix} \hat{y}_{p1}(k+1) \\ \hat{y}_{p2}(k+1) \end{bmatrix} = \begin{bmatrix} N^1[y_{p1}(k), y_{p2}(k)] \\ N^2[y_{p1}(k), y_{p2}(k)] \end{bmatrix} + \begin{bmatrix} u_1(k) \\ u_2(k) \end{bmatrix}.$$

The identification procedure was carried out for 100 000 time steps using a step size of $\eta = 0.1$ with random inputs $u_1(k)$ and $u_2(k)$ uniformly distributed in the interval $[-1, 1]$. The responses of the plant and the identification model for a vector input $[\sin(2\pi k/25), \cos(2\pi k/25)]^T$ are shown in Fig. 17.

Comment: In examples 1, 3, 4, and 5 the adjustment of the parameters was carried out by computing the gradient of $e^2(k)$ at instant k while in example 2 adjustments were based on the gradient of an error function evaluated over an interval of length 5. While from a theoretical point of view it is preferable to use a larger interval to define the error function, very little improvement was observed in the simulations. This accounts for the fact that in examples 3, 4, and 5 adjustments were based on the instantaneous rather than an average error signal.

6. Example 6: In examples 1–5, a series-parallel identification model was used and hence the parameters of the neural networks were adjusted using the static back-propagation method. In this example, we consider a simple

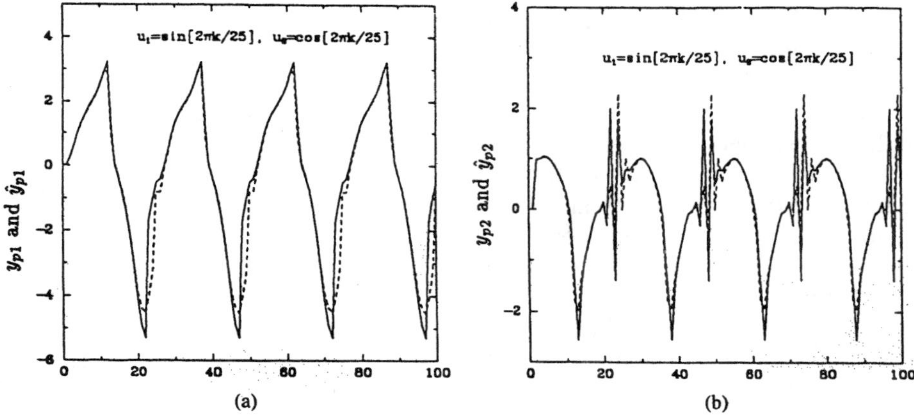

(a) (b)

Fig. 17. Example 5: Responses of the plant and the identification model.

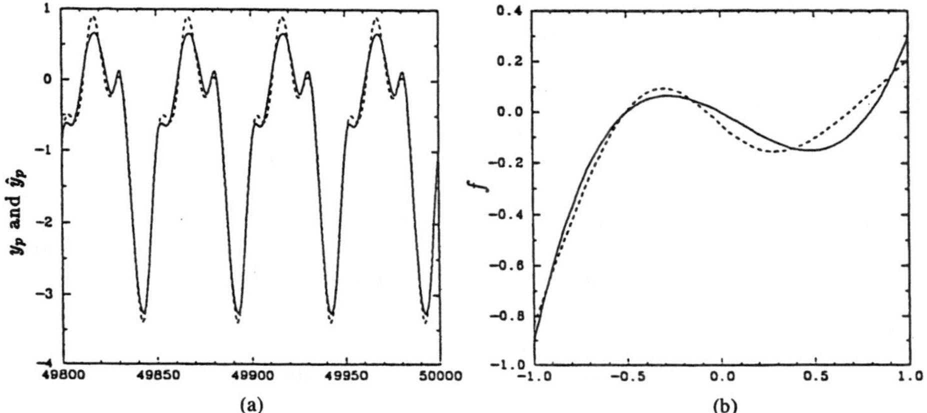

(a) (b)

Fig. 18. Example 6: (a) Outputs of plant and identification model. (b) $f[u]$ and $N[u]$ for $u \in [-1, 1]$.

first order nonlinear system which is identified using the dynamic back-propagation method discussed in Section IV. The nonlinear plant is described by the difference equation

$$y_p(k+1) = 0.8y_p(k) + f[u(k)]$$

where the function $f[u] = (u - 0.8)u(u + 0.5)$ is unknown. However, it is assumed that f can be approximated to the desired degree of accuracy by a multilayer neural network.

The identification model used is described by the difference equation

$$\hat{y}_p(k+1) = 0.8\hat{y}_p(k) + N[u(k)]$$

and the neural network belonged to the class $\mathfrak{N}^3_{1,20,10,1}$. The model chosen corresponds to representation 1 in Section IV (refer to Fig. 6(a)). The objective is to adjust a total of 261 weights in the neural network so that $e(k) \triangleq \hat{y}_p(k) - y_p(k) \to 0$ asymptotically. Defining the performance criterion to be minimized as $J = (1/2T) \Sigma_{i=k-T+1}^{k} e^2(i)$, the partial derivative of J with respect to a weight θ_j in the neural network can be computed as $(\partial J/\partial\theta_j) = (1/T) \Sigma_{i=k-T+1}^{k} e(i)(\partial e(i)/\partial\theta_j)$. The quantity $(\partial e(i)/\partial\theta_j)$ can be computed in a dynamic fashion using the method discussed in Section IV and used in the following rule to update θ:

$$k - T + 1 = \theta k - T + 1 - \eta\left[\frac{1}{T}\sum_{i=\theta(k+1)}^{k} e(i)\frac{\partial e(i)}{\partial\theta}\right]$$

where η is the step size in the gradient procedure.

Fig. 18(a) shows the outputs of the plant and the identification model when the weights in the neural network were adjusted after an interval of 10 time steps using a step size of $\eta = 0.01$. The input to the plant (and the model) was $u(k) = \sin(2\pi k/25)$. In Fig. 18(b), the function $f(u) = (u - 0.8)u(u + 0.5)$, as well as the function realized by the three layer neural network after 50 000 steps for $u \in [-1, 1]$, are shown. As seen from the figure, the neural network approximates the given function quite accurately.

VI. Control of Dynamical Systems

As mentioned in Section II, for the sake of mathematical tractability most of the effort during the past two decades in the model reference adaptive control theory has been directed towards the control of linear time-invariant plants with unknown parameters. Much of the theoretical work in the late 1970's was aimed at determining adaptive laws for the adjustment of the control parameter vector $\theta(k)$ which would result in stable overall systems. In 1980 [30]–[33], it was conclusively shown that for both discrete-time and continuous-time systems, such stable adaptive laws could be determined provided that some prior information concerning the plant transfer function was available. Since that time much of the research in the area has been directed towards determining conditions which assure the robustness of the overall system under different types of perturbations.

In contrast to the above, very little work has been reported on the global adaptive control of plants described by nonlinear difference or differential equations. It is in the control of such systems that we are primarily interested in this section. Since, in most problems, very little theory exists to guide the analysis, one of the aims is to indicate precisely how the nonlinear control problem differs from the linear one and the nature of the theoretical questions that have to be answered.

Algebraic and Analytic Parts of Adaptive Control Problems: In conventional adaptive control theory, two stages are generally distinguished in the adaptive process. In the first, referred to as the *algebraic part*, it is first shown that the controller has the necessary degrees of freedom to achieve the desired objective. More precisely, if some prior information regarding the plant is given, it is shown that a controller parameter vector θ^* exists for every value of the plant parameter vector p, so that the output of the controlled plant together with the controller approaches the output of the reference model asymptotically. The *analytic part* of the problem is then to determine stable adaptive laws for adjusting $\theta(k)$ so that $\lim_{k \to \infty} \theta(k) = \theta^*$ and the output error tends to zero.

Direct and Indirect Control: For over 20 years, two distinct approaches have been used [1] to control a plant adaptively. These are 1) *direct control* and 2) *indirect control.* In direct control, the parameters of the controller are directly adjusted to reduce some norm of the output error. In indirect control, the parameters of the plant are estimated as the elements of a vector $\hat{p}(k)$ at any instant k and the parameter vector $\theta(k)$ of the controller is chosen assuming that $\hat{p}(k)$ represents the true value p of the plant parameter vector. Even when the plant is assumed to be linear and time invariant, both direct and indirect adaptive control result in overall nonlinear systems. Figs. 19 and 20 represent the structure of the overall adaptive system using the two methods for the adaptive control of a linear time-invariant plant [1].

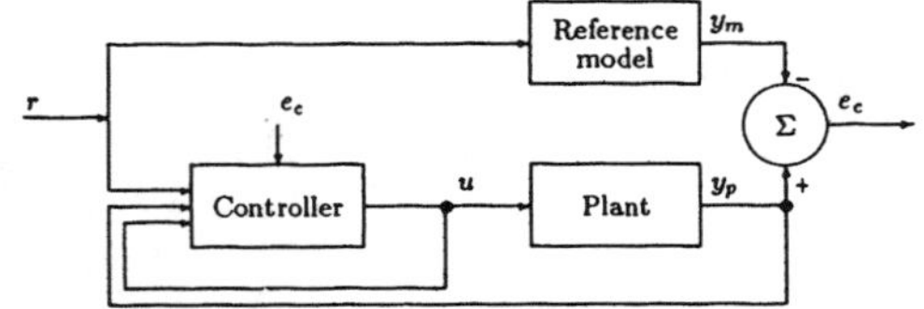

Fig. 19. Direct adaptive control.

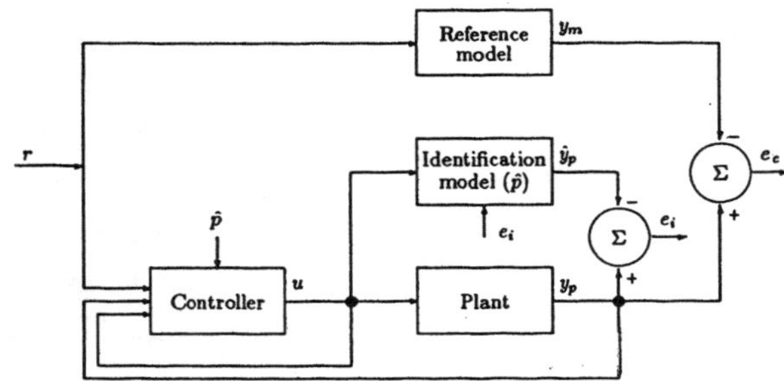

Fig. 20. Indirect adaptive control.

A. Adaptive Control of Nonlinear Systems Using Neural Networks

For a detailed treatment of direct and indirect control systems the reader is referred to [1]. The same approaches which have proved successful for linear plants can also be attempted when nonlinear plants have to be adaptively controlled. The structure used for the identification model as well as the controller are strongly motivated by those used in the linear case. However, in place of the linear gains, nonlinear neural networks are used.

Methods for identifying nonlinear plants using delayed values of both plant input and output were discussed in the previous section and Fig. 11 shows a general identification model. Fig. 21 shows a controller whose output is the control input to the plant and whose inputs are the delayed values of the plant input and output, respectively.

1. Indirect Control: At present, methods for directly adjusting the control parameters based on the output error (between the plant and the reference model outputs) are not available. This is because the unknown nonlinear plant in Fig. 21 lies between the controller and the output error e_c. Hence, until such methods are developed, adaptive control of nonlinear plants has to be carried out using indirect methods. This implies that the methods described in Section V have to be first used on line to identify the input–output behavior of the plant. Using the resulting identification model, which contains neural networks and linear dynamical elements as subsystems, the parameters of the controller are adjusted. This is shown in Fig. 22. It is this procedure of identification followed by control that is adopted in this section. Dynamic back propagation through a system consisting of only neural networks and linear dynamic elements was discussed in Section IV to determine the gradient of a performance index with respect to the adjustable parameters of a system. Since identification of the unknown plant is carried out using only neural networks and tapped delay lines, the identification model can be used to compute the partial derivatives of a performance index with respect to the controller parameters.

B. Simulation Results

The procedure adopted to adaptively control a nonlinear plant depends largely on the prior information available regarding the unknown plant. This includes knowl-

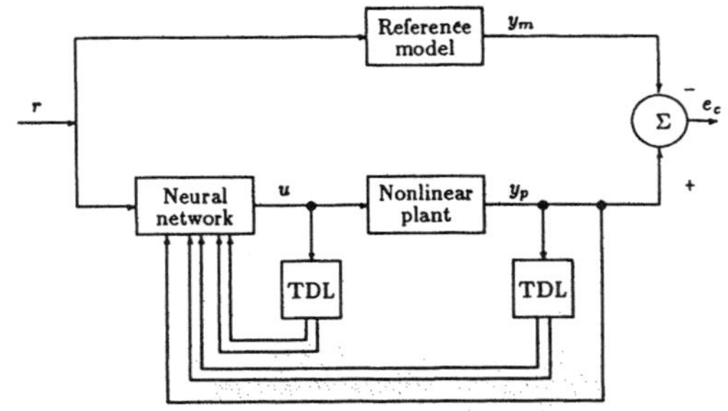

Fig. 21. Direct adaptive control of nonlinear plants using neural networks.

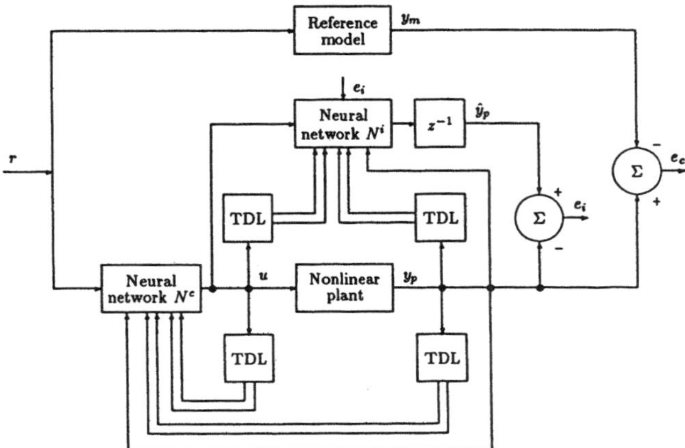

Fig. 22. Indirect adaptive control using neural networks.

edge of the number of equilibrium states of the unforced system, their stability properties, as well as the amplitude of the input for which the output is also bounded. For example, if the plant is known to have a bounded output for all inputs u belonging to some compact set $\mathcal{U}$, then the plant can be identified off line using the methods outlined in Section V. During identification, the weights in the identification model can be adjusted at every instant of time ($T_i = 1$) or at discrete time intervals ($T_i > 1$). Once the plant has been identified to the desired level of accuracy, control action can be initiated so that the output of the plant follows the output of a stable reference model. It must be emphasized that even if the plant has bounded outputs for bounded inputs, feedback control may result in unbounded solutions. Hence, for on-line control, identification and control must proceed simultaneously. The time intervals T_i and T_c, respectively, over which the identification and control parameters are to be updated have to be judiciously chosen in such a case.

Five examples, in which nonlinear plants are adaptively controlled, are included below and illustrate the ideas discussed earlier. As in the previous section, each example is chosen to emphasize a specific point.

1. Example 7: We consider here the problem of controlling the plant discussed in example 2 which is described by the difference equation

$$y_p(k+1) = f[y_p(k), y_p(k-1)] + u(k)$$

where the function

$$f[y_p(k), y_p(k-1)] = \frac{y_p(k)y_p(k-1)[y_p(k)+2.5]}{1+y_p^2(k)+y_p^2(k-1)} \quad (19)$$

is assumed to be unknown. A reference model is described by the second-order difference equation

$$y_m(k+1) = 0.6y_m(k) + 0.2y_m(k-1) + r(k)$$

where $r(k)$ is a bounded reference input. If the output error $e_c(k)$ is defined as $e_c(k) \triangleq y_p(k) - y_m(k)$, the aim of control is to determine a bounded control input $u(k)$ such that $\lim_{k\to\infty} e_c(k) = 0$. If the function $f[.]$ in (19) is known, it follows directly that at stage k, $u(k)$ can be computed from a knowledge of $y_p(k)$ and its past values as

$$u(k) = -f[y_p(k), y_p(k-1)] + 0.6y_p(k) + 0.2y_p(k-1) + r(k) \quad (20)$$

resulting in the error difference equation $e_c(k+1) = 0.6e_c(k) + 0.2e_c(k-1)$. Since the reference model is asymptotically stable, it follows that $\lim_{k\to\infty} e_c(k) = 0$ for arbitrary initial conditions. However, since $f[.]$ is unknown, it is estimated on line as $\hat{f}$ as discussed in example 2 using a neural network N and the series-parallel method.

The control input to the plant at any instant k is computed using $N[.]$ in place of f as

$$u(k) = -N[y_p(k), y_p(k-1)] + 0.6y_p(k) + 0.2y_p(k-1) + r(k). \quad (21)$$

This results in the nonlinear difference equation

$$y_p(k+1) = f[y_p(k), y_p(k-1)] - N[y_p(k), y_p(k-1)] + 0.6y_p(k) + 0.2y_p(k-1) + r(k) \quad (22)$$

governing the behavior of the plant. The structure of the overall system is shown in Fig. 23.

In the first stage, the unknown plant was identified off line using random inputs as described in example 2. Following this, (21) was used to generate the control input. The response of the controlled system with a reference input $r(k) = \sin(2\pi k/25)$ is shown in Fig. 24(b).

In the second stage, both identification and control were implemented simultaneously using different values of T_i and T_c. The asymptotic response of the system when identification and control start at $k = 0$ with $T_i = T_c = 1$ is shown in Fig. 25(a). Since it is desirable to adjust the control parameters at a slower rate than the identification parameters, the experiment was repeated with $T_i = 1$ and $T_c = 3$ and is shown in Fig. 25(b). Since the identification process is not complete for small values of k, the control can be theoretically unstable. However, this was not observed in the simulations. If the control is initiated at time

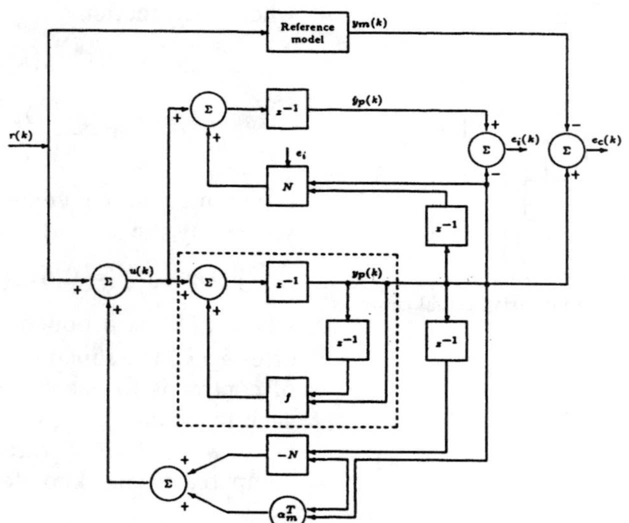

Fig. 23. Example 7: Structure of the overall system.

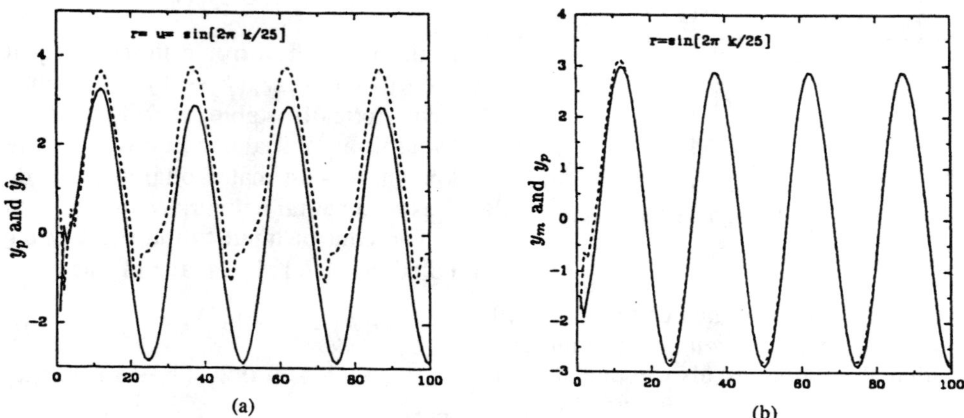

Fig. 24. Example 7: (a) Response for no control action. (b) Response for $r = \sin(2\pi k/25)$ with control.

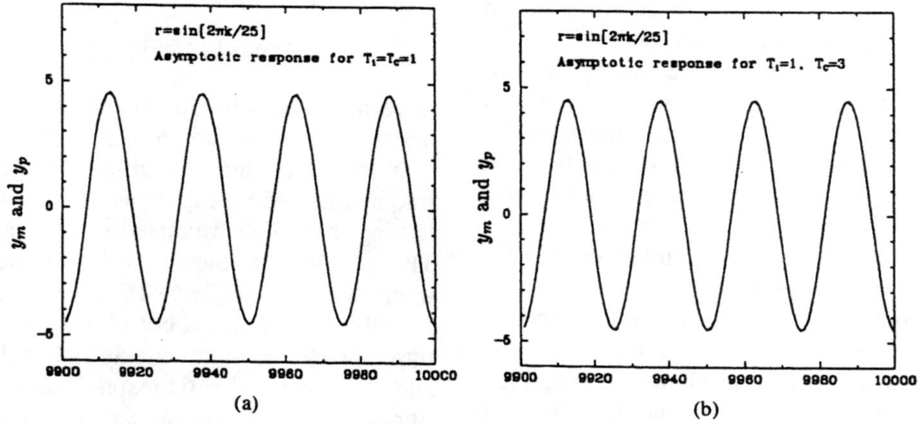

Fig. 25. Example 7: (a) Response when control is initiated at $k = 0$ with $T_i = T_c = 1$. (b) Response when control is initiated at $k = 0$ and $T_i = 1$ and $T_c = 3$.

$k = 0$ using nominal values of the parameters of the neural network with $T_i = T_c = 10$, the output of the plant was seen to increase in an unbounded fashion as shown in Fig. 26.

The simulations reported above indicate that for stable and efficient on-line control, the identification must be sufficiently accurate before control action is initiated and hence T_i and T_c should be chosen with care.

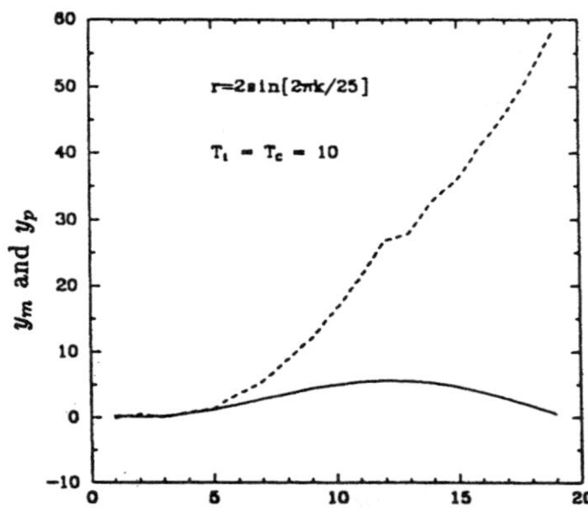

Fig. 26. Example 7: Response when control is initiated at $k = 0$ with $T_i = T_c = 10$.

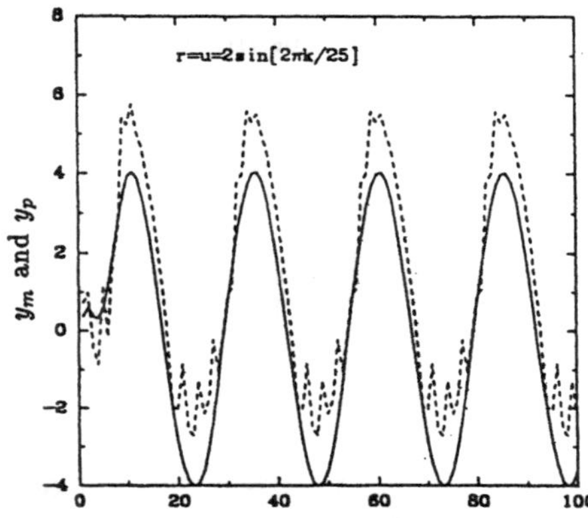

Fig. 27. Example 8: Responses of the reference model and the plant when no control action is taken.

2. *Example 8:* The unknown plant in this case corresponds to Model II and can be described by a difference equation of the form

$$y_p(k+1) = f[y_p(k), y_p(k-1), \cdots, y_p(k-n+1)] + \sum_{j=0}^{m-1} \beta_j u(k-j), \qquad m \leq n.$$

It is assumed that the parameters β_j ($j = 0, 1, \cdots, m-1$) are unknown, but that β_0 is nonzero with a known sign. The specific plant used in the simulation study was

$$y_p(k+1) = \frac{5y_p(k)y_p(k-1)}{1 + y_p^2(k) + y_p^2(k-1) + y_p^2(k-2)} + u(k) + 0.8u(k-1). \tag{23}$$

The output of the stable reference model is described by

$$y_m(k+1) = 0.32y_m(k) + 0.64y_m(k-1) - 0.5y_m(k-2) + r(k)$$

where r is the uniformly bounded reference input. The responses of the reference model and the plant when $r(k) = u(k) = \sin(2\pi k/25)$ are shown in Fig. 27. While the output of the reference model is also a sinusoid of the same frequency, the response of the plant is seen to contain higher harmonics. It is assumed that sgn $\beta_0 = +1$ and that $\beta_0 \geq 0.1$. This enables a projection type algorithm to be used in the identification procedure so that the estimate $\hat{\beta}_0$ of β_0 satisfies the inequality $\hat{\beta}_0 \geq 0.1$. The control input any instant of time k is generated as

$$u(k) = \frac{1}{\hat{\beta}_0}[-\hat{f}_k[y_p(k), y_p(k-1), y_p(k-2)] - \hat{\beta}_1 u(k-1) + 0.32y_p(k) + 0.64y_p(k-1) - 0.5y_p(k-2) + r(k)]. \tag{24}$$

In Fig. 28, the plant is identified over a period of 50 000 time steps using an input which is random and distributed uniformly over the interval $[-2, 2]$. At the end of this interval, the control is implemented as given in (24). The response of the plant as well as the reference model are shown in Fig. 28. In Fig. 28(a) the reference input is $r(k) = \sin(2\pi k/25)$, while in Fig. 28(b) the reference input is $r(k) = \sin(2\pi k/25) + \sin(2\pi k/10)$. In both cases the control system is found to perform satisfactorily. Since the plant is identified over a sufficiently long time with a general input, the parameters $\hat{\beta}_0$ and $\hat{\beta}_1$ are found to converge to 1.005 and 0.8023, respectively, which are close to the true values of 1 and 0.8.

In Fig. 29 the response of the controlled plant to a reference input $r(k) = \sin(2\pi k/25)$ is shown, when identification and control are initiated at $k = 0$. Since the input is not sufficiently general, $\hat{\beta}_0(k)$ and $\hat{\beta}_1(k)$ tend to values 4.71 and 3.59 so that the asymptotic values of the parameter errors are large. In spite of this, the output error is seen to tend to zero for values of k greater than 9900. This example reveals that good control may be possible without good parameter identification.

3. *Example 9:* In this case, the plant is described by the same equation as in (23) with $0.8u(k-1)$ replaced by $1.1u(k-1)$ and the same procedure is adopted as in example 8 to generate the control input. It is found that the output error is bounded and even tends to zero while the control input grows in an unbounded fashion (Fig. 30). This is a phenomenon which is well known in adaptive control theory and arises due to the presence of zeros of the plant transfer function lying outside the unit circle. In the present context $u(k) + 1.1u(k-1)$ can be zero even as $u(k) = (-1.1)^k$ tends to ∞ in an oscillatory fashion. The same phenomenon can also occur in systems where the dependence of y_p on u in nonlinear.

4. *Example 10:* The control of the nonlinear multivariable plant with two inputs and two outputs, discussed in example 5, is considered in this example and the plant is described by (18). The reference model is linear and is described by the difference equations

$$\begin{bmatrix} y_{m1}(k+1) \\ y_{m2}(k+1) \end{bmatrix} = \begin{bmatrix} 0.6 & 0.2 \\ 0.1 & -0.8 \end{bmatrix} \begin{bmatrix} y_{m1}(k) \\ y_{m2}(k) \end{bmatrix} + \begin{bmatrix} r_1(k) \\ r_2(k) \end{bmatrix}$$

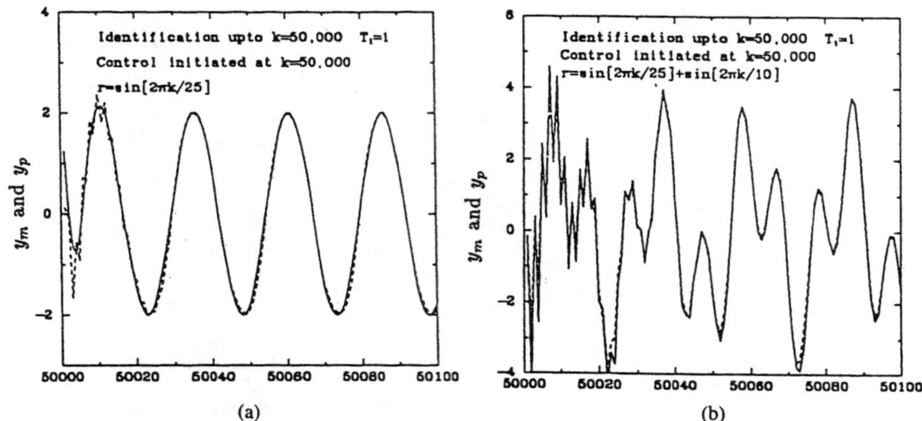

Fig. 28. Example 8: Identification followed by control.

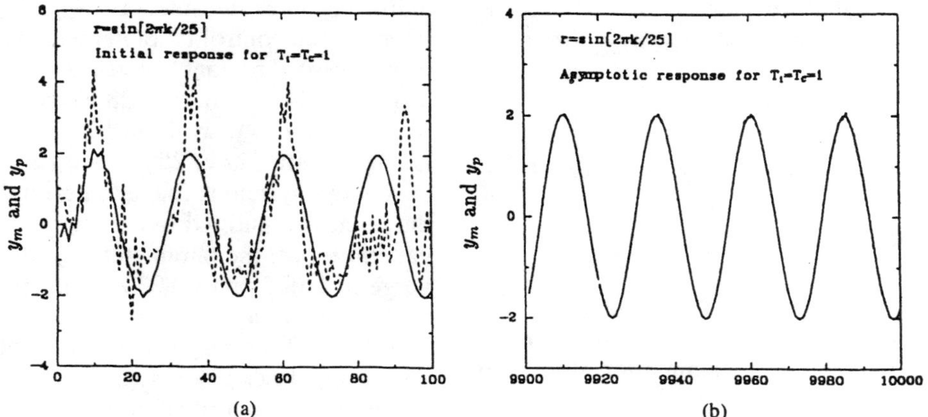

Fig. 29. Example 8: Initial response when control action is taken at $k = 0$ with $T_i = T_c = 1$. (b) Asymptotic response with $T_i = T_c = 1$.

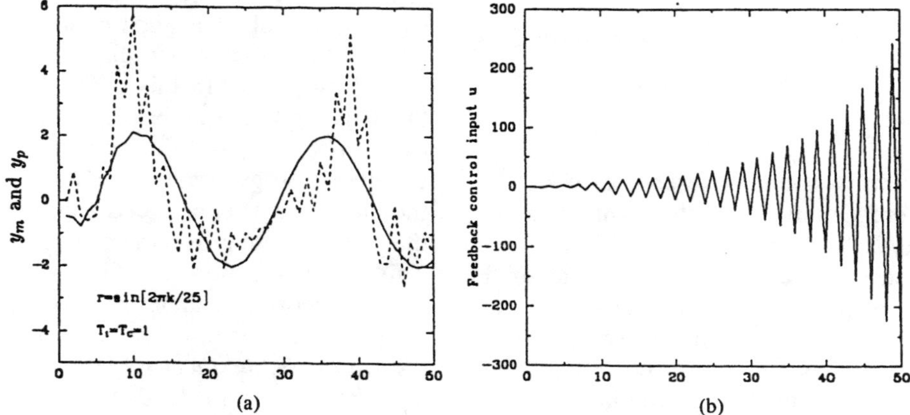

Fig. 30. Example 9: (a) Outputs of the reference model and the plant when control is initiated at $k = 0$. (b) The feedback control input u.

where r_1 and r_2 are bounded reference inputs. The plant is identified as in example 5 and control is initiated after the identification process is complete. The responses of the plant as compared to the reference model for the same inputs are shown in Fig. 31. The improvement in the responses, when the neural networks in the identification model are used to generate the control input to the plant are evident from the figure. The outputs of the controlled plant and the reference model are shown and indicate that the output error is almost zero.

5. Example 11: In examples 7–10, the output of the plant depends linearly on the control input. This makes the computation of the latter relatively straightforward. In this example the plant is described by Model III and has

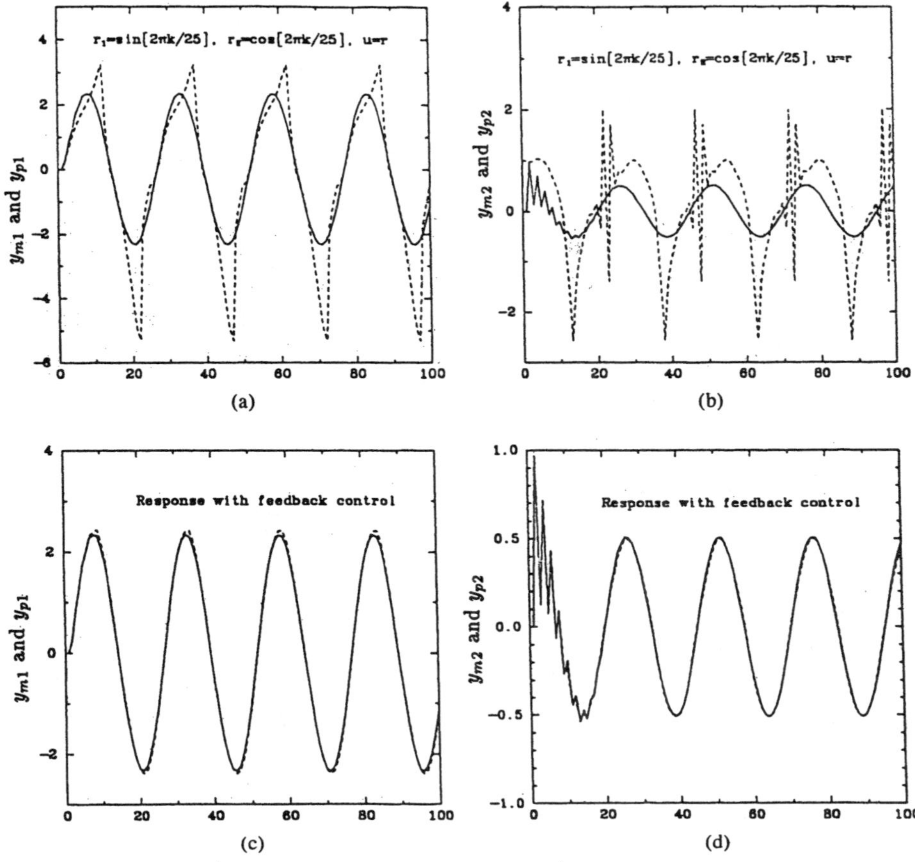

Fig. 31. Example 10: (a), (b) Outputs of the reference model and the plant when no control action is taken. (c), (d) Outputs of the reference model and the plant with feedback control.

the form

$$y_p(k+1) = \frac{y_p(k)}{1 + y_p(k)^2} + u^3(k)$$

which was identified successfully in example 3. Choosing the reference model as

$$y_m(k+1) = 0.6y_m(k) + r(k)$$

the aim once again is to choose $u(k)$ so that $\lim_{k\to\infty} | y_p(k) - y_m(k)| = 0$. If $f[y_p] = y_p/(1 + y_p^2)$ and $g[u] = u^3$, the control input in this case is chosen as

$$u(k) = \widehat{g^{-1}}[-\hat{f}[y_p(k)] + 0.6y_p(k) + r(k)] \quad (25)$$

where $\hat{f}$ and $\widehat{g^{-1}}$ are the estimates of f and g^{-1}, respectively. The estimates $\hat{f}$ and $\hat{g}$ are obtained as described earlier using neural networks N_f and N_g. Since $\hat{g}[u]$ has been realized as the output of a neural network N_g, the weights of a neural network $N_c \in \mathfrak{N}^3_{1,20,10,1}$ (shown in Fig. 32) can be adjusted so that $N_g[N_c(r)] \approx r$ as $r(k)$ varies over the interval $[-4, 4]$. The range $[-4, 4]$ was chosen for $r(k)$ since this assures that the input to the identification model varies over the same range for which the estimates $\hat{f}$ and $\hat{g}$ are valid. In Fig. 33 $N_g[N_c(r)]$ is plotted against r and is seen to be unity over the entire range.

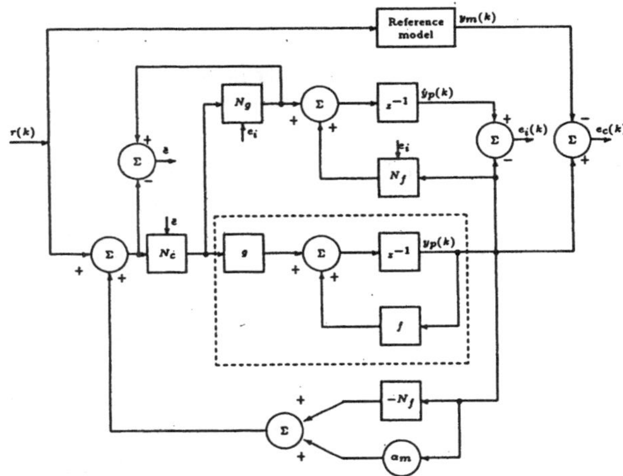

Fig. 32. Example 11: Structure of the overall system.

The determination of N_c was carried out over 25 000 time steps using a random input uniformly distributed in the interval $[-4, 4]$ and a step size of $\eta = 0.01$. Since the plant nonlinearities f and g as well as g^{-1} have been estimated using neural networks N_f, N_g, and N_c, respectively, the control input to the plant can be determined using (25). The output of the plant to a reference input $r(k) = \sin(2\pi k/25) + \sin(2\pi k/10)$ is shown in Fig.

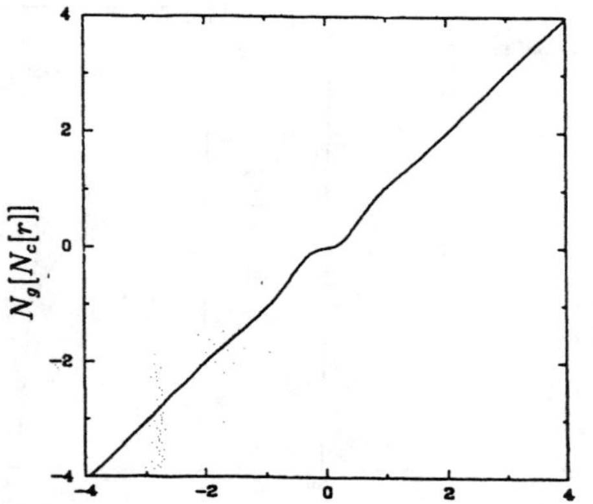

Fig. 33. Example 11: Plot of the function $N_g[N_c(r)]$.

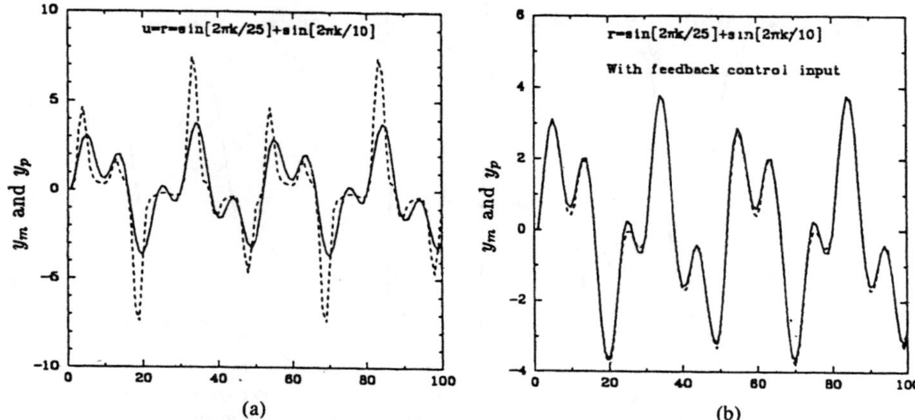

Fig. 34. Example 11: (a) Outputs of the reference model and plant without a feedback controller. (b) Outputs of the reference model and plant with a feedback controller.

34(a) when a feedback controller is not used; the response with a controller is shown in Fig. 34(b). The response in Fig. 34(b) is identical to that of the reference model and is almost indistinguishable from it. Hence, from this example we conclude that it may be possible in some cases to generate a control input to an unknown plant so that almost perfect model following is achieved.

VII. Comments and Conclusions

In this paper models for the identification and control of nonlinear dynamic systems are suggested. These models, which include multilayer neural networks as well as linear dynamics, can be viewed as generalized neural networks. In the specific models given, the delayed values of relevant signals in the system are used as inputs to multilayer neural networks. Methods for the adjustment of parameters in generalized neural networks are treated and the concept of dynamic back propagation is introduced in this context to generate partial derivatives of a performance index with respect to adjustable parameters on line. However, in many identifiers and controllers it is shown that by using a series-parallel model, the gradient can be obtained with the simpler static back-propagation method.

The simulation studies on low order nonlinear dynamic systems reveal that identification and control using the methods suggested can be very effective. There is every reason to believe that the same methods can also be used successfully for the identification and control of multivariable systems of higher dimensions. Hence, they should find wide application in many areas of applied science.

Several assumptions were made concerning the plant characteristics in the simulation studies to achieve satisfactory identification and control. For example, in all cases the plant was assumed to have bounded outputs for the class of inputs specified. An obvious and important extension of the methods in the future will be to the control of unstable systems in some compact domain in the state space. All the plants were also assumed to be of relative degree unity (i.e., input at k affects the output at $k + 1$), minimum phase (i.e., no unbounded input lies in the null space of the operator representing the plant) and Models II and III used in control problems assumed that inverses of operators existed and could be approximated. Future work will attempt to relax some or all of these assumptions. Further, in all cases the gradient method is used exclusively for the adjustment of the parameters of the plant. Since it is well known that such methods can lead to instability for large values of the step size η, it is essential that efforts be directed towards determining stable adaptive laws for adjusting the parameters. Such work is currently in progress.

A number of assumptions were made throughout the paper regarding the plant to be controlled for the methods to prove successful. These include stability properties of recurrent networks with multilayer neural networks in the forward path, controllability, observability, and identifiability of the models suggested as well as the existence of nonlinear controllers to match the response of the reference model. At the present stage of development of nonlinear control theory, few constructive methods exist for checking the validity of these assumptions in the context of general nonlinear systems. However, the fact that we are dealing with special classes of systems represented by generalized neural networks should make the development of such methods more tractable. Hence, concurrent theoretical research in these areas is needed to justify the models suggested in this paper.

Acknowledgment

The authors would like to thank the reviewers and the associate editor for their careful reading of the paper and their helpful comments.

References

[1] K. S. Narendra and A. M. Annaswamy, *Stable Adaptive Systems.* Englewood Cliffs, NJ: Prentice-Hall, 1989.

[2] D. J. Burr, "Experiments on neural net recognition of spoken and written text," *IEEE Trans. Acoust., Speech, Signal Processing*, vol. 36, no. 7, pp. 1162–1168, July 1988.

[3] R. P. Gorman and T. J. Sejnowski, "Learned classification of sonar targets using a massively parallel network," *IEEE Trans. Acoust., Speech, Signal Processing*, vol. 36, no. 7, pp. 1135–1140, July 1988.

[4] T. J. Sejnowski and C. R. Rosenberg, "Parallel networks that learn to pronounce English text," *Complex Syst.*, vol. 1, pp. 145–168, 1987.

[5] B. Widrow, R. G. Winter, and R. A. Baxter, "Layered neural nets for pattern recognition," *IEEE Trans. Acoust., Speech, Signal Processing*, vol. 36, no. 7, pp. 1109–1118, July 1988.

[6] J. J. Hopfield, "Neural networks and physical systems with emergent collective computational abilities," *Proc. Nat. Acad. Sci., U.S.*, vol. 79, pp. 2554–2558, Apr. 1982.

[7] J. J. Hopfield and D. W. Tank, "Neural computation of decisions in optimization problems," *Biolog. Cybern.*, vol. 52, pp. 141–152, 1985.

[8] D. W. Tank and J. J. Hopfield, "Simple 'neural' optimization networks: An A/D converter, signal decision circuit, and a linear programming circuit," *IEEE Trans. Syst., Man, Cybern.*, vol. CAS-33, no. 5, pp. 533–541, May 1986.

[9] H. Rauch and T. Winarske, "Neural networks for routing communication traffic," *IEEE Control Syst. Mag.*, vol. 8, no. 2, pp. 26–30, Apr. 1988.

[10] N. B. Haaser and J. A. Sullivan, *Real Analysis.* New York: Van Nostrand Reinhold, 1971.

[11] P. G. Gallman and K. S. Narendra, "Identification of nonlinear systems using a Uryson model," Becton Center, Yale University, New Haven, CT, tech. rep. CT-38, Apr. 1971; also *Automatica*, Nov. 1976.

[12] L. Ljung and T. Söderstrom, *Theory and Practice of Recursive Identification.* Cambridge, MA: M.I.T. Press, 1985.

[13] S. N. Singh and W. J. Rugh, "Decoupling in a class of nonlinear systems by state variable feedback," *Trans. ASME*, vol. 94, pp. 323–324, 1972.

[14] E. Freund, "The structure of decoupled nonlinear systems," *Int. J. Contr.*, vol. 21, pp. 651–654, 1975.

[15] A. Isidori, A. J. Krener, C. Gori Giorgi, and S. Monaco, "Nonlinear decoupling via feedback: A differential geometric approach," *IEEE Trans. Automat. Contr.*, vol. AC-26, pp. 331–345, 1981.

[16] S. S. Sastry and A. Isidori, "Adaptive control of linearizable systems," *IEEE Trans. Automat. Contr.*, vol. 34, no. 11, pp. 1123–1131, Nov. 1989.

[17] F. J. Pineda, "Generalization of back propagation to recurrent networks," *Phys. Rev. Lett.*, vol. 59, no. 19, pp. 2229–2232, Nov. 1987.

[18] J. H. Li, A. N. Michel and W. Porod, "Qualitative analysis and synthesis of a class of neural networks," *IEEE Trans. Circuits Syst.*, vol. 35, no. 8, pp. 976–986, Aug. 1988.

[19] B. Kosko, "Bidirectional associative memories," *IEEE Trans. Syst., Man, Cybern.*, vol. 18, no. 1, pp. 49–60, Jan./Feb. 1988.

[20] K. S. Narendra and K. Parthasarathy, "Neural networks and dynamical systems. Part I: A gradient approach to Hopfield networks," Center Syst. Sci., Dept. Electrical Eng., Yale University, New Haven, CT, tech. rep. 8820, Oct. 1988.

[21] K. Hornik, M. Stinchcombe, and H. White, "Multilayer feedforward networks are universal approximators," Dept. Economics, University of California, San Diego, CA, discussion pap., Dept. Economics, June 1988.

[22] K. S. Narendra and L. E. McBride, Jr., "Multiparameter self-optimizing system using correlation techniques," *IEEE Trans. Automat. Contr.*, vol. AC-9, pp. 31–38, 1964.

[23] P. V. Kokotovic, "Method of sensitivity points in the investigation and optimization of linear control systems," *Automat. Remote Contr.*, vol. 25, pp. 1512–1518, 1964.

[24] L. E. McBridge, Jr. and K. S. Narendra, "Optimization of time-varying systems," *IEEE Trans. Automat. Contr.*, vol. AC-10, no. 3, pp. 289–294, 1965.

[25] J. B. Cruz, Jr., Ed., *System Sensitivity Analysis, Benchmark Papers in Electrical Engineering and Computer Science.* Stroudsburg, PA: Dowden, Hutchinson and Ross, 1973.

[26] K. S. Narendra and K. Parthasarathy, "A diagrammatic representation of back propagation," Center for Syst. Sci., Dept. of Electrical Eng., Yale University, New Haven, CT, tech. rep. 8815, Aug. 1988.

[27] K. S. Narendra and K. Parthasarathy, "Back propagation in dynamical systems containing neural networks," Center Syst. Sci., Dept. Electrical Eng., Yale University, New Haven, CT, tech. rep. 8905, Mar. 1989.

[28] K. S. Narendra and K. Parthasarathy, "Neural networks and dynamical systems. Part II: Identification," Center for Syst. Sci., Dept. of Electrical Eng., Yale University, New Haven, CT, tech. rep. 8902, Feb. 1989.

[29] K. S. Narendra and K. Parthasarathy, "Neural networks and dynamical systems. Part III: Control," Center Syst. Sci., Dept. Electrical Eng., Yale University, New Haven, CT, tech. rep. 8909, May 1989.

[30] K. S. Narendra, Y. H. Lin, and L. S. Valavani, "Stable adaptive controller design—Part II: Proof of stability," *IEEE Trans. Automat. Contr.*, vol. 25, pp. 440–448, June 1980.

[31] A. S. Morse, "Global stability of parameter adaptive systems," *IEEE Trans. Automat. Contr.*, vol. 25, pp. 433–439, June 1980.

[32] G. C. Goodwin, P. J. Ramadge, and P. E. Caines, "Discrete time multivariable adaptive control," *IEEE Trans. Automat. Contr.*, vol. 25, pp. 449–456, June 1980.

[33] K. S. Narendra and Y. H. Lin, "Stable discrete adaptive control," *IEEE Trans. Automat. Contr.*, vol. 25, vol. 456–461, June 1980.

Dynamic Neural Units with Applications to the Control of Unknown Nonlinear Systems

M.M. Gupta
D.H. Rao
Intelligent Systems Research Laboratory
College of Engineering
University of Saskatchewan
Saskatoon S7N 0W0, Canada

Abstract

To emulate higher cognitive functions, such as sensory information processing and the attendant complexities of learning, memory storage, control, and vision, the biological neuron has to be modeled based upon feedback networks. The authors proposed such a neuronal model, called a "dynamic neural unit" (DNU) based upon the topology of a reverberating circuit in a neuronal pool of the central nervous system. The DNU architecture embodies delay elements, feedforward and feedback synaptic weights, and a nonlinear activation function. In this article, the concept of DNU is extended by incorporating both synaptic and somatic operations. The synaptic operation provides the optimum feedforward and feedback weights, while the somatic operation determines the optimum gain (slope) of the nonlinear activation function for a given task. The architectural details and the learning and adaptive algorithm to modify adjustable parameters, namely, feedforward and feedback synaptic weights, and slope of nonlinear function of the DNU are presented. The algorithm implementation of an isolated DNU is given. Considering DNU as the basic processing element, a three-stage dynamic neural network is developed. The effectiveness of the proposed dynamic neural network architecture as applied to the control of unknown nonlinear systems is discussed and extensive simulation results are presented. © 1993 John Wiley and Sons, Inc.

1. Introduction

The adaptive control of linear systems has been successfully developed during the last decade or two and the detailed descriptions of such systems may be found in refs. 1–3. While adaptive control has shown potential in controlling complex systems and offers good disturbance rejection, the region of operation of the control system is restricted. This is because the adjustable parameters of the adaptive controller are typically based upon convergence and stability analysis, which may limit the performance of the compensated system.[4] The kind of stability that is proven is only asymptotic tracking and global boundedness of all signals—no uniform convergence or Lyapunov stability is established. Perhaps the most unrealistic among the conditions are the assumptions that there are no disturbances and that the plant order is not higher than the order of the model. Simulated evidence showed later that "mild" violation to the latter assumptions can cause most of the adaptive control algorithms to go unstable.[5]

These limitations can be seen as restrictions on the acceptable operating region of a controller. By enhancing a conventional controller with learning, one can effectively expand the region of operation of the controller and create a more robust controller. The control system can then compensate for larger changes in the plant and its environment.[4] The goal of machine learning in this case is to broaden the region of operation of an adaptive control system by allowing the controller parameters to better adapt to different plant and environmental conditions.

Neural networks with their massive parallelism and ability to learn offer exciting possibilities the design of adaptive controllers for systems with complex, unknown, and nonlinear dynamics. This is an advantage when controlling systems that cannot be modeled accurately. The potential benefits of such networks are: (1) They could be used to approximate any continuous linear or

Reprinted with permission from *Journal of Intelligent and Fuzzy Systems*, vol. 1, no. 1, M. M. Gupta and D. H. Rao, "Dynamic Neural Units with Applications to the Control of Unknown Linear Systems," pp. 73–92, Copyright 1993. John Wiley & Sons, Inc.

nonlinear mapping functions; (2) they perform this approximation through learning attributes; and (3) parallel processing and fault tolerance are easy to accomplish.

The most commonly used neural network architecture in control applications is the multilayered neural network (MNN) with error backpropagation (BP) learning algorithms. Layered neural networks are nonlinear parametric models that can approximate any continuous input–output relation.[6–8] The quality of approximation depends upon the architecture of the network used, as well as upon the complexity of the system.[9] The problem of finding a suitable set of parameters that approximate an unknown relation, F, is usually solved using learning algorithms. For networks with certain architectural restrictions (for example, static networks with feedforward connections only), the procedure is employed by propagating the error backward from the output nodes through the hidden layers to adjust the weights using a gradient descent approach. Extension of the BP training to more general network environments, specifically with dynamic elements, has been of interest.[10] Feedback (recurrent) neural networks have been recently used for the identification and control of nonlinear dynamic systems.[11]

Although a feedback neural network comprises of delays, summation, and nonlinear operators, the basic architecture of the neuron simply provides a weighted integration of the synaptic inputs over a period of time. This model of neuron is static in the sense that there are no dynamic elements embodied in the model architecture. To emulate some of the intriguing dynamic functions, such as learning and adaptation, and better reflect the dynamics of the biologic neuron, it may be essential to model the artificial neuron based upon a feedback network.[14,15]

The authors proposed[12] a new architecture to model the biologic neuron, named the *dynamic neural unit* (DNU), whose structure is analogous to that of the reverberating circuit in a neuronal pool of the central nervous system. The topology of the DNU embodies delay elements, feedforward and feedback synaptic weights, and a nonlinear operator. A three-stage dynamic neural network, with DNU as the basic computing element, has been used successfully for controlling unknown linear and nonlinear dynamic systems by considering only the synaptic weights as the adjustable parameters of the network.[12]

In this article, we extend the learning algorithm of the DNU by considering the idea that the main body of the neuron, the soma, may also be changing during the process of learning and adaptation.[13] A simple model of somatic learning may be represented by considering the slope of the nonlinear operator as an adjustable parameter of the DNU.

This article is organized as follows. In Section 2, we describe the architecture of DNU. A simple model of somatic learning is proposed. The adaptive algorithm for the adjustable parameters of the DNU is developed in this section. In Section 3, a three-stage dynamic neural network is developed by considering DNU as the basic processing element. This is followed, in the fourth section, by several computer simulation results for nonlinear systems. Finally, in Section 5 we present the conclusions. The definitions of certain terms used in this article are given in the appendix.

2. Dynamics of an Isolated DNU

2.1. Architectural Details

It is currently understood that the biological neuron provides two distinct mathematical operations distributed over the *synapse*, the junction point between an axon and the dendrite, and the *soma*, the main body of the neuron. These two neuronal mathematical operations may be called (1) the *synaptic operation* and (2) the *somatic operation*. From the biological point of view, these two operations are physically separate, but in the modeling of a biological neuron these operations have been combined (for example, thresholding in the soma is transferred to the synaptic operation). In this section, we present the mathematical details of an isolated DNU, the basic computing (processing) element in a multistage dynamic neural network, which embodies the synaptic and somatic operations separately.

The DNU proposed in ref. 12 considers the synaptic weights of a neuron as the adaptable parameters, which leads only to synaptic learning. A modified structure of DNU, which accounts for both synaptic and somatic learning, is shown in Figure 1.

The dynamic structure of DNU, as shown in Figure 1, consists of two delay elements and two feedforward and feedback paths weighted by the synaptic weights $\mathbf{a}_{ff}$ and $\mathbf{b}_{fb}$ respectively. This is a second-order dynamic structure that can be de-

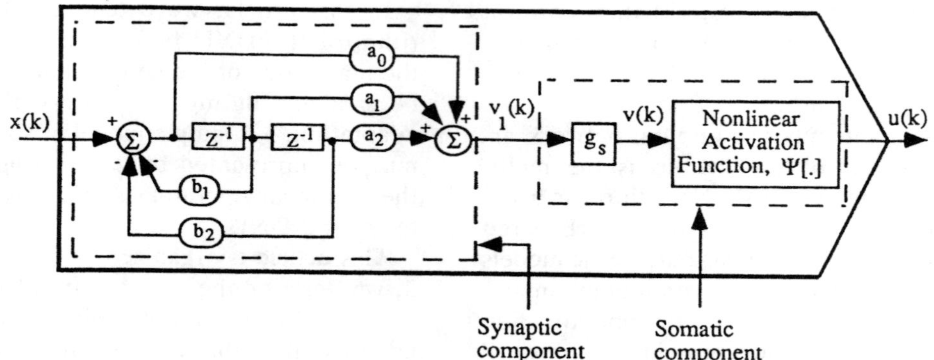

Figure 1. Basic structure of DNU, which consists of synaptic and somatic components. z^{-1} is the unit delay operator, k is the discrete-time index, and $\mathbf{a}_{ff} = [a_0, a_1, a_2]$ and $\mathbf{b}_{fb} = [b_1, b_2]$ are the adjustable feedforward and feedback weights contributing to the forward and feedback signals, respectively. The output of this dynamic structure, $v_1(k)$, is passed through the somatic gain (slope), g_s, and a nonlinear activation function $\Psi[.]$ that produces the neuron output $u(k) \in R^1$ in response to an input signal $x(k) \in R^n$.

scribed by the following difference equation:

$$v_1(k) = -b_1 v_1(k-1) - b_2 v_1(k-2) + a_0 x(k) + a_1 x(k-1) + a_2 x(k-2) \tag{1}$$

where $x(k) \in R^n$ is the neural input, $v_1(k) \in R^1$ is the output of the dynamic structure, $u(k) \in R^1$ is the neural output, k is the discrete-time index, and $\mathbf{a}_{ff} = [a_0, a_1, a_2]$ and $\mathbf{b}_{fb} = [b_1, b_2]$ are the vectors of adaptable feedforward and feedback weights, respectively.

Alternatively, eq. (1) may be written in the transfer function form as

$$w(k, \mathbf{a}_{ff}, \mathbf{b}_{fb}) = \frac{v_1(k)}{x(k)} = \frac{[a_0 + a_1 z^{-1} + a_2 z^{-2}]}{[1 + b_1 z^{-1} + b_2 z^{-2}]} \tag{2}$$

The vectors of signals and adaptable weights of the dynamic neuron are defined now as follows:

$$\gamma(k, v_1, x) = [v_1(k-1)\ v_1(k-2)\ x(k)\ x(k-1)\ x(k-2)]^T \tag{3}$$

and

$$\zeta_{(\mathbf{a}_{ff}, \mathbf{b}_{fb})} = [-b_1\ -b_2\ a_0\ a_1\ a_2]^T \tag{4}$$

where the superscript T denotes transpose

Using (3) and (4), eq. (1) is rewritten as

$$v_1(k) = \gamma^T(k, v_1, x)\zeta_{\mathbf{a}_{ff}, \mathbf{b}_{fb}} = \left\{ \begin{bmatrix} (v_1(k-1)) \\ (v_1(k-2)) \\ (x(k)) \\ (x(k-1)) \\ (x(k-2)) \end{bmatrix} [-b_1\ -b_2\ a_0\ a_1\ a_2] \right\} \tag{5}$$

The nonlinear mapping operation on $v_1(k)$ yields a neural output $u(k)$ given by

$$u(k) = \Psi[g_s v_1(k)] \tag{6}$$

where $\Psi[\cdot]$ is some nonlinear activation function. The selection of a nonlinear function is dependent upon the following assumption:

Assumption 1. *The neural network is comprised by only one "type" of neuron.*

This assumption enables the probability distribution of neural thresholds about an aggregate value Θ to be defined as a *unimodal* function, as shown in Figure 2a. It follows then that the nonlinear transformation is *sigmoidal*, as shown in Figure 2b. However, if the neural unit is assumed to be comprised of m different types of neurons then the distribution of the thresholds can be redefined in m-modal functions that produce a nonlinear input transformation with m-inflection points.[16,17]

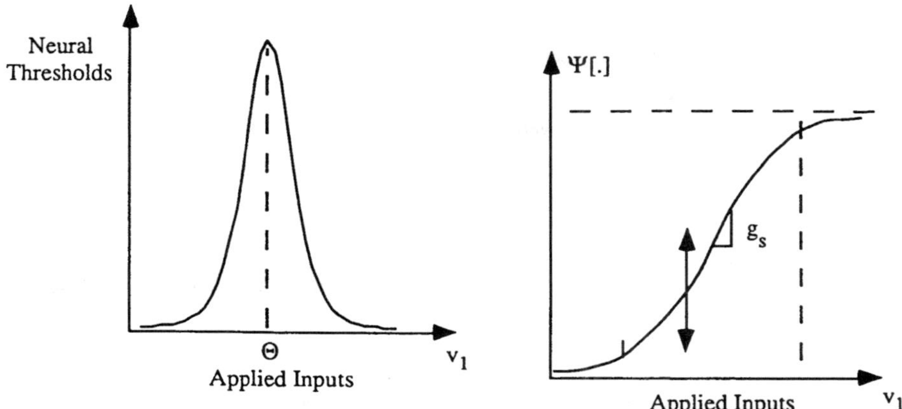

Figure 2. (a) *Unimodal* probability distribution function of the neural thresholds Θ. (b) Nonlinear *sigmoidal* function, with somatic gain (slope) g_S, arising from the neural threshold distribution given in (a).

2.2. Approximation of a Function by a Neural Network

If a neural network model is employed to represent a dynamic system, it is important to establish the form of expansion that the network provides. The investigation of this property consists of expanding the nonlinear function of the neural network.[18] Assuming, for example, that the nonlinear model is a *sigmoidal* function (Fig. 2b) expressed as

$$\Psi[v_1(k)] = \frac{1}{1+e^{-[v_1(k)]}} = \frac{1}{2} + \frac{1}{4} v_1(k) - \frac{1}{48} v_1^3(k) + \frac{1}{480} v_1^5(k) - \cdots + \cdots \quad (7)$$

The input–output relationship for an isolated DNU is then given by

$$\begin{aligned} u(k) &= \frac{1}{2} + \frac{1}{4}[\gamma^T(k, v_1, x)\zeta_{(\mathbf{a}_{ff},\mathbf{b}_{fb})}] - \frac{1}{48}[\gamma^T(k, v_1, x)\zeta_{(\mathbf{a}_{ff},\mathbf{b}_{fb})}]^3 + \cdots \\ &= c_0 + c_1 v_1(k-1) + c_2 v_1(k-2) + c_3 x(k) \\ &\quad + c_4 x(k-1) + c_5 x(k-2) + c_6 v_1^3(k-1) \\ &\quad + c_7 v_1(k-1)x(k) + c_8 v_1(k-1)v_1(k-2) \\ &\quad + c_9 v_1^2(k-1) + c_6 v_1^3(k-1)x(k-1) + \cdots \end{aligned} \quad (8)$$

where the c_i are the coefficients that are functions of the $x(k)$s and $v_1(k)$s. It is easy to see how this result can be generalized to higher-order network approximations and other nonlinear functions. Because c_0, a constant, exists in eq. (8), this shows that multistage neural networks include a mean value in the system representation.[18] If the mathematical operations are extended to both the positive and negative neural outputs, and, thus, the neural activity is expanded for both the excitatory and inhibitory inputs, the activation (sigmoidal) function then can be defined as a hyperbolic tangent function given by

$$\Psi[v(k)] = \left[\frac{e^{(g_s v_1)} - e^{-(g_s v_1)}}{e^{(g_s v_1)} + e^{-(g_s v_1)}}\right] = \tan h[g_s v_1] = \tan h[v] \quad (9)$$

where $v = g_s v_1$ and g_s is the somatic gain, the parameter that controls the slope of the activation function. In the limit as $g_s \to \infty$, the sigmoidal function tends to become the sign (binary) function with an infinite slope at $v = 0$ and a zero slope for $v \neq 0$. Figure 3 shows $\Psi[v]$ and its derivative $\Psi'[v]$ for an increasing value of g_s.

The concept of varying the gain g_s, as introduced in this article, during the process of learning and adaptation is biologically motivated as explained below. In the biological neuron, at the macroscopic level the dendrites of each neuron receive pulses at the synapses and convert them to continuously variable dendritic current. The flow of current through the axon membrane modulates the axonal firing rate. For each neuron, there is a time-varying nonlinear relationship between the pulse rate at the synapse and the amplitude of the dendritic current.[19] This morphological change of the neuron during the learning process may be modeled by considering the slope of the nonlinear function in the neural

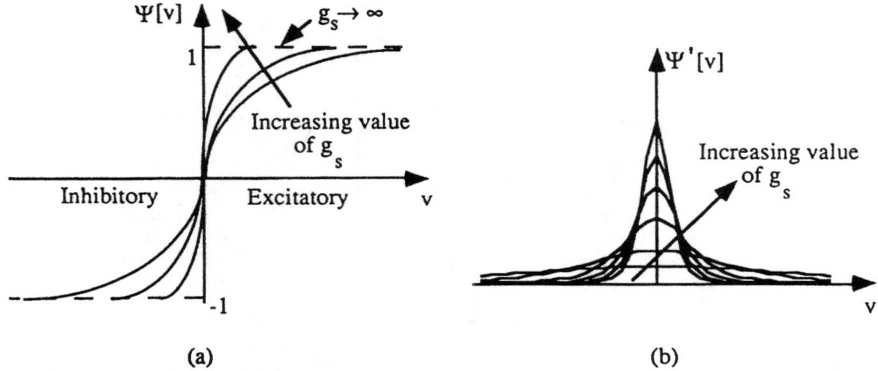

Figure 3. (a) Sigmoid function $\Psi[v(k)] = \tan h[g_s v_1]$. (b) Derivative, $\Psi'[v]$, of sigmoidal function that tends to become a sign function as $\tan h[g_s v_1] \mid_{g_s \to \infty}$, that is, the slope $\Psi'[v]$ tends to become narrow with an increasing value of g_s. This also represents the axonal gain in biological neuron.

structure as one of the adaptable parameters in addition to the synaptic weights. This component of neuronal learning may be called *somatic learning*.

In the following discussion, we assumed a hyperbolic tangent function with slope g_s as the time-varying nonlinear relationship between the synaptic and somatic components.

The following lemmas show that a DNU with the hyperbolic tangent function satisfies some algebraic properties on the compact set Σ.

Lemma 1. *The hyperbolic tangent function satisfies the following properties*:

1. $\Psi[v(k)]$ is strictly increasing, that is, for $v_a < v_b$ in R

$$\Psi[v_a] < \Psi[v_b] \tag{10}$$

2. $\Psi[v(k)]$ is uniformly Lipschiz, that is, there exists a constant $C > 0$ such that

$$|\Psi[v_a] - \Psi[v_b]| \leq C|v_a - v_b| \quad \forall v_a, v_b \in R \tag{10b}$$

Observe that (1) and (2) are valid iff

$$0 < \frac{\Psi[v_a] - \Psi[v_b]}{v_a - v_b} \leq C \quad \forall v_a, v_b \in R, \text{and}$$
$$v_a \neq v_b \tag{10c}$$

3. $\Psi[v(k)]$ has one and only one inflection point (because of Assumption 1).

Lemma 2. *There exist constants C_1, C_2 such that the neural network $\hat{f}[\cdot]$ on the compact set Σ of R^n satisfies*

$$0 < C_1 < |\hat{f}[\cdot]| \leq C_2 \quad \forall q \in \text{on } \Sigma \tag{11}$$

Note that $\hat{f}[\cdot] \neq 0 \; \forall q \in$ on Σ. Hence, this implies that $\hat{f}[\cdot] > 0$ or $\hat{f}[\cdot] < 0$ for all $q \in$ on Σ, and, therefore, there exist nonzero maximum and minimum, which means that eq. (11) is proved.

Based upon the above assumptions and lemmas, we will now derive the learning and adaptive algorithm for the adaptable parameters of the DNU.

2.3. Learning and Adaptive Algorithm

If neural networks are used in both pattern recognition (static identification) and dynamic system identification and control, the objective of the algorithm is to adjust the parameters of the network based on a given set of input–output pairs. If the weighs of the network are considered as the elements of a parameter vector $\phi_{(\mathbf{a}_{ff}, \mathbf{b}_{fb})}$, the learning process involves the determination of the vector $\phi^*_{(\mathbf{a}_{ff}, \mathbf{b}_{fb})}$ that optimizes a performance index J based upon the output error. The performance index J may be defined as an even function of the error, that is,

$$J = \frac{1}{2} E\{e^2(k; \phi_{(\mathbf{a}_{ff}, \mathbf{b}_{fb})})\} \tag{12}$$

where E is the expectation operator and e is the error signal defined as the difference between the desired response $y_d(k)$ and the actual response $u(k)$:

$$e(k) = y_d(k) - u(k) \tag{13}$$

The gradient of performance index J with respect to $\phi_{(\mathbf{a}_{ff},\mathbf{b}_{fb})}$ is computed as $\nabla_{\phi_{(\mathbf{a}_{ff},\mathbf{b}_{fb})}} J$ and $\phi_{(\mathbf{a}_{ff},\mathbf{b}_{fb})}$ is adjusted as the negative gradient of performance index, that is,

$$\phi_{(\mathbf{a}_{ff},\mathbf{b}_{fb})}(k+1) = \phi_{(\mathbf{a}_{ff},\mathbf{b}_{fb})}(k) - \mathrm{dia}[\mu]\nabla_{\phi_{(\mathbf{a}_{ff},\mathbf{b}_{fb})}} J|_{\phi_{(\mathbf{a}_{ff},\mathbf{b}_{fb})}=\phi_{(\mathbf{a}_{ff},\mathbf{b}_{fb})}(k)} \tag{14}$$

where dia$[\mu]$ is the diagonal matrix of the adaptive gains, $\phi_{(\mathbf{a}_{ff},\mathbf{b}_{fb})}(k+1)$ and $\phi_{(\mathbf{a}_{ff},\mathbf{b}_{fb})}(k)$ are the values of parameter vector at $(k+1)$th and kth instance, respectively, and $\nabla_{\phi_{(\mathbf{a}_{ff},\mathbf{b}_{fb})}} J$ is the gradient of performance function (index) J evaluated at $\phi_{(\mathbf{a}_{ff},\mathbf{b}_{fb})}(k)$, which is written as $[\partial J / \partial\phi_{(\mathbf{a}_{ff},\mathbf{b}_{fb})}(k)]$. In eq. (14), the constraint on the parameter vector $\phi_{(\mathbf{a}_{ff},\mathbf{b}_{fb})}(k)$ is

$$\|\phi_{(\mathbf{a}_{ff},\mathbf{b}_{fb})}(k)\| = \left|\sum_{i=0}^{n} \sqrt{\phi^2_{i(\mathbf{a}_{ff},\mathbf{b}_{fb})}(k)}\right| = 1 \tag{15a}$$

and dia$[\mu]$, the matrix of adaptive gains, is given by

$$\mathrm{dia}[\mu] = \begin{bmatrix} \mu_{a_i} & 0 \\ 0 & \mu_{b_j} \end{bmatrix} \tag{15b}$$

where μ_{a_i}, $i=0,1,2$, and μ_{b_j}, $j=1,2$, are the individual gains of adaptable parameters of DNU, which determine the stability and speed of convergence to optimal values, and $\|.\|$ represents the norm or the supremum of $\phi_{(\mathbf{a}_{ff},\mathbf{b}_{fb})}(k)$.

Thus, the feedforward parameters a_{ff_i}, $i=0,1,2$, and the feedback parameters b_{fb_j}, $j=1,2$, are adapted according to the following set of equations:

$$a_{ff_i}(k+1) = a_{ff_i}(k) - \mu_{a_i}\frac{\partial J}{\partial a_{ff_i}(k)} \quad i=0,1,2 \tag{16a}$$

$$b_{fb_j}(k+1) = b_{fb_j}(k) - \mu_{b_j}\frac{\partial J}{\partial b_{fb_j}(k)} \quad j=1,2 \tag{16b}$$

Equation (14) can be extended to the other adjustable parameter of DNU, the slope g_s of the activation function, as

$$g_s(k+1) = g_s(k) - \mu_{g_s}\nabla_{g_s} J|_{g_s=g_s(k)} \tag{17}$$

From the definitions of the performance index, eq. (12), and the error signal, eq. (13), the gradient of performance index is given by

$$\begin{aligned}
\frac{\partial J}{\partial\phi_{(\mathbf{a}_{ff},\mathbf{b}_{fb})}} &= \frac{1}{2}E\left[\frac{\partial[y_d(k)-u(k)]^2}{\partial\phi_{(\mathbf{a}_{ff},\mathbf{b}_{fb})}}\right] \\
&= E\left\{e(k)\left[-\frac{\partial u(k)}{\partial\phi_{(\mathbf{a}_{ff},\mathbf{b}_{fb})}}\right]\right\} = E\left\{e(k)\left[-\frac{\partial\Psi(v)}{\partial\phi_{(\mathbf{a}_{ff},\mathbf{b}_{fb})}}\right]\right\} \\
&= E\left\{-e(k)\left[\frac{\partial\Psi(v)}{\partial v}\frac{\partial v}{\partial\phi_{(\mathbf{a}_{ff},\mathbf{b}_{fb})}}\right]\right\} \text{ (by chain rule)} \\
&= E\left\{-e(k)\left[g_s\frac{4}{[e^{(g_s v_1)}-e^{-(g_s v_1)}]^2}\frac{\partial v}{\partial\phi_{(\mathbf{a}_{ff},\mathbf{b}_{fb})}}\right]\right\} \\
&= E\{-e(k)[g_s\,\mathrm{sech}^2[v_1(k)]\mathbf{s}_{\phi_{(\mathbf{a}_{ff},\mathbf{b}_{fb})}}(k)]\}
\end{aligned} \tag{18}$$

where the term $\mathbf{s}_{\phi_{(\mathbf{a}_{ff},\mathbf{b}_{fb})}}(k) = [\partial v/\partial\phi_{(\mathbf{a}_{ff},\mathbf{b}_{fb})}] = g_s[\partial v_1/\partial\phi_{(\mathbf{a}_{ff},\mathbf{b}_{fb})}]$ is named as a vector of parameter-state (or sensitivity) signals.

Theorem 1. *The parameter-state signals for the feedforward weights are given by the relation*

$$\mathbf{s}_{\phi_{\mathbf{a}_{ff_i}}}(k) = g_s[x(k-i)] \quad i=0,1,2 \tag{19}$$

Proof: From eq. (5), we can write

$$\begin{aligned}
\mathbf{s}_{\phi_{\mathbf{a}_{ff_i}}}(k) &= g_s\frac{\partial}{\partial a_{ff_i}(k)}\left\{\begin{bmatrix}(v_1(k-1))\\(v_1(k-2))\\(x(k))\\(x(k-1))\\(x(k-2))\end{bmatrix}[-b_1\ \ -b_2\ \ a_0\ \ a_1\ \ a_2]\right\} \\
&= g_s\frac{\partial}{\partial a_{ff_i}(k)}\left\{\begin{bmatrix}(x(k))\\(x(k-1))\\(x(k-2))\end{bmatrix}[a_0\ \ a_1\ \ a_2]\right\} \quad i=0,1,2
\end{aligned} \tag{20}$$

Thus, the individual parameter-state signals for the feedforward weights are:

For $i=0$

$$s_{a_0}(k) = g_s[x(k)] \tag{21a}$$

For $i = 1$

$$s_{a_1}(k) = g_s[x(k-1)] \tag{21b}$$

For $i = 2$

$$s_{a_2}(k) = g_s[x(k-2)] \tag{21c}$$

Therefore, the parameter-state signals for the feedforward weights are

$$\mathbf{s}_{\phi_{\mathbf{a}_{ffi}}}(k) = g_s[x(k-i)] \qquad i = 0, 1, 2 \qquad \blacksquare$$

Theorem 2. *The parameter-state signals for the feedback weights are given by the relation*

$$\mathbf{s}_{\phi_{\mathbf{b}_{fb_j}}}(k) = -g_s[v_1(k-j)] \qquad j = 1, 2 \tag{22}$$

Proof:

$$\mathbf{s}_{\phi_{\mathbf{b}_{fb_j}}}(k) = g_s \frac{\partial}{\partial b_{ff_j}(k)} \left\{ \begin{bmatrix} (v_1(k-1)) \\ (v_1(k-2)) \\ (x(k)) \\ (x(k-1)) \\ (x(k-2)) \end{bmatrix} [-b_1 \; -b_2 \; a_0 \; a_1 \; a_2] \right\}$$

Thus, the individual parameter state signals for the feedback weights are:
For $j = 1$

$$s_{b_1}(k) = -g_s[v_1(k-1)] \tag{23a}$$

For $j = 2$

$$s_{b_2} = -g_s[v_1(k-2)] \tag{23b}$$

Therefore, the parameter state signals for the feedback weights may be written as

$$\mathbf{s}_{\phi_{\mathbf{b}_{fb_j}}}(k) = -g_s[v_1(k-j)], \quad j = 1, 2 \qquad \blacksquare$$

From eqs. (21) and (23), the parameter-state signals or sensitivity model of the feedforward and feedback weights of the DNU is developed as shown in Figure 4.

From eqs. (16) and (18), the algorithm for updating the components of the weighting vector is

$$a_{ff_i}(k+1) = a_{ff_i}(k) + \mu_{a_i} E \times \{e(k) g_s \sec h^2[v_1(k)] \mathbf{s}_{\phi_{\mathbf{a}_{ffi}}}(k)\} \qquad i = 0, 1, 2 \tag{24a}$$

and

$$b_{fb_j}(k+1) = b_{fb_j}(k) + \mu_{bj} E \times \{e(k) g_s \sec h^2[v_1(k)] \mathbf{s}_{\phi_{\mathbf{a}_{ffi}}}(k)\} \qquad j = 1, 2 \tag{24b}$$

Similarly, it can be shown that

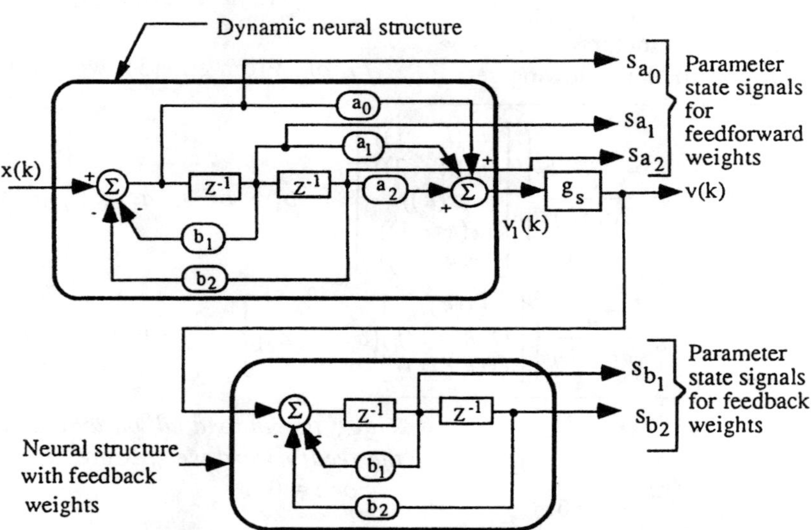

Figure 4. Parameter-state signals or the sensitivity model of an isolated DNU.

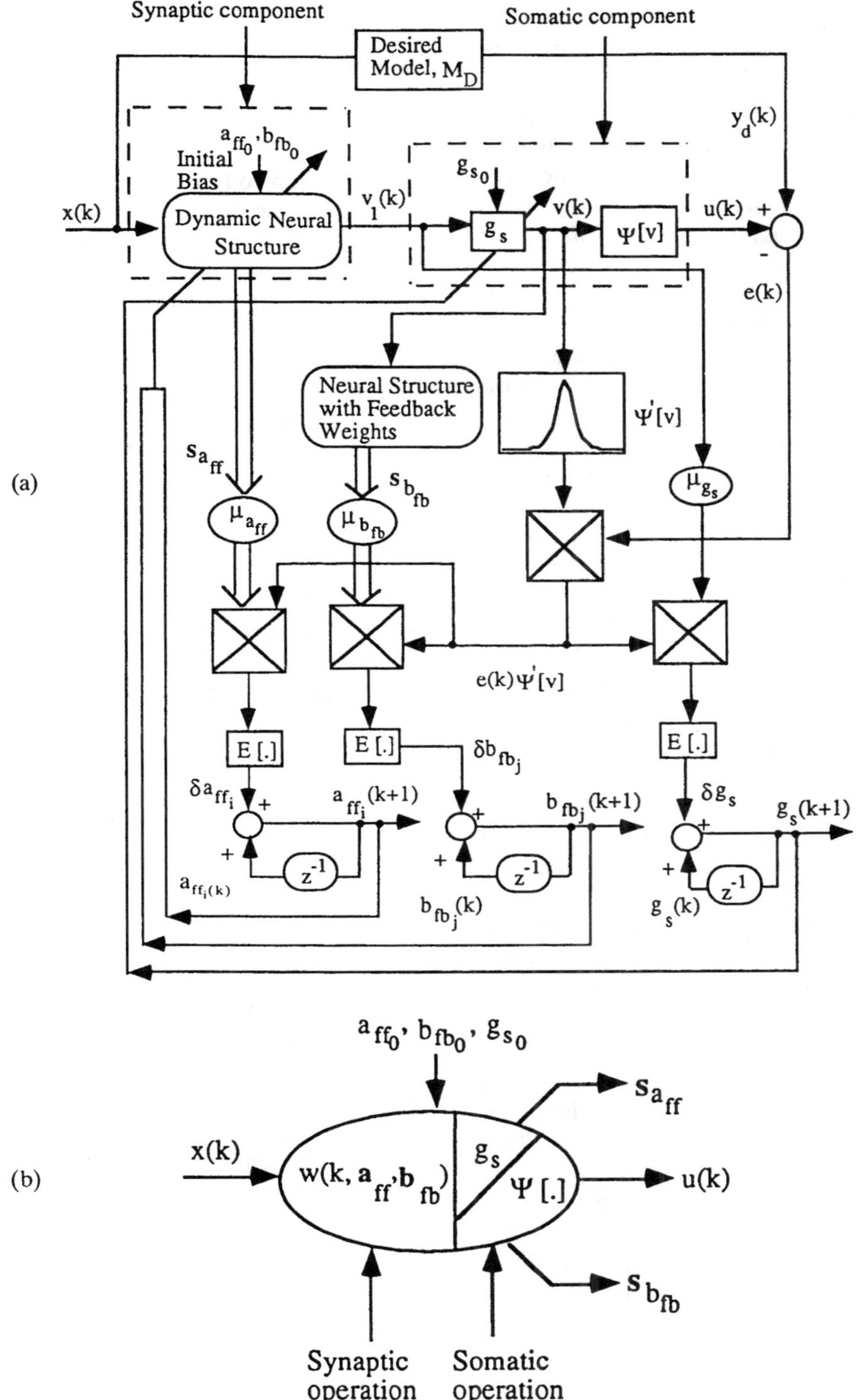

Figure 5. (a) Implementation of the learning and adaptive algorithm for an isolated DNU. z^{-1} represents the unit delay, δa_{ff_i}, δb_{fb_j}, and δg_s are the adaptive terms of the feedforward and feedback weights of the dynamic structure of DNU, and the slope of the nonlinear function, respectively. (b) Symbolic representation of DNU. $w(k, \mathbf{a}_{ff}, \mathbf{b}_{fb})$ represents the dynamic structure of the neuron and $\mathbf{a}_{ff}$ and $\mathbf{b}_{fb}$ are the adaptable feedforward and feedback weights with the corresponding parameter state signals $\mathbf{s}_{\mathbf{a}_{ff}}$ and $\mathbf{s}_{\mathbf{b}_{fb}}$, respectively. The parameter g_s represents slope (the somatic gain) of the nonlinear function that contributes to the somatic adaptation and $\Psi[\cdot]$ represents the nonlinear activation function. $\mathbf{a}_{ff0}$, $\mathbf{b}_{fb_0}$, and g_{s_0} represent the initial values of the adjustable parameters.

$$g_s(k+1) = g_s(k)[1 + \mu_{g_s} E\{e(k) \times \sec h^2[v_1(k)]v_1(k)\}] \quad (24c)$$

Equations (24a) and (24b) correspond to synaptic adaptation and eq. (24c) to somatic adaptation of the DNU. The implementation scheme of eq. (24) is shown in Figure 5a and the symbolic representation of the DNU in Figure 5b.

The following remarks are in order with reference to the deduction of the algorithm for the adjustable parameters of the DNU:

1. The desired model, M_D, in Figure 5a is an entity representing a physical reality (in the case of system identification, for example) or model in the designer's mind (in the case of pattern classification problems, for example).
2. The somatic gain, g_s, is equivalent to increasing the strengths of the components of the weighting vector by a factor of g_s. However, due to the constraint on $\phi_{(\mathbf{a}_{ff},\mathbf{b}_{fb})}$ [eq. (15a)] the somatic gain can be treated independently for somatic adaptation. Further, it provides more flexibility in the learning and adaptation process in the case of fuzzy neural networks.
3. The expectation of a random process x with probability density function $p(x)$ is defined as $E[x] = \int_{-\infty}^{\infty} xp(x)\,dx$. Thus, $E[x]$ is an averaging process and, for an ergodic process, can be approximated by the time average $E[x(t)] = \text{Lim}_{T\to\infty} (1/2T) \int_{-T}^{T} x(t)\,dt$. For a discrete case, $E[x(k)] = 1/K \sum_{i=t+1-K}^{t} x(i)$.
4. Equations (24a)–(24c) make use of the inner product of the error signal, $e(k)$, and the derivative of the nonlinear function, $\Psi'(v)$. This term is significant in the sense that during the learning process if $|v|$ is small (in the neighborhood of zero) then it provides a large weight to the error, thus making a large change in the weighting vector $\phi_{(\mathbf{a}_{ff},\mathbf{b}_{fb})}$. On the other hand, if $|v|$ is large $\Psi'(v)$ is small, thus providing a little weight to the error.

3. Multistage Dynamic Neural Networks

Although a single neuron can control unknown linear systems[12] and perform certain simple pattern detection functions,[6] the power of neural computation comes from the neurons connected in a network structure. Larger networks in general offer greater computational capabilities. Arranging neurons in layers or stages is supposed to mimic the layered structure of a certain portion of the brain.[18] These multilayered networks have been proven to have capabilities beyond those of a single layer. Several independent studies have established that a three-layered neural network can approximate any nonlinear function of interest to any degree of accuracy.[7,8] In general, multistage (multilayered) neural networks can be considered versatile nonlinear maps with the elements of the weight matrices as parameters.[11] In this section, we develop a three-stage dynamic neural network using DNU as the basic processing element.

Consider a single-stage neural system with "p" DNUs as shown in Figure 6a

The input–output mapping of a single-stage network with DNUs configured in sigma mode can be expressed as

$$\begin{aligned} u(k) &= [\Psi^{(1)}\{g_s^{(1)}(w^{(1)}(k, \mathbf{a}_{ff}^1, \mathbf{b}_{fb}^1)x^{(1)}(k))\}] + \\ &+ [\Psi^{(i)}\{g_s^{(i)}(w^{(i)}(k, \mathbf{a}_{ff}^i, \mathbf{b}_{fb}^i)x^{(i)}(k))\}] + \\ &+ [\Psi^{(p)}\{g_s^{(p)}(w^{(p)}(k, \mathbf{a}_{ff}^p, \mathbf{b}_{fb}^p)x^{(p)}(k))\}] \\ &= \omega_1(x) + \cdots + \omega_i(x) + \cdots + \omega_p(x) \\ &= \sum_{n=1}^{p} \omega_n(x) \qquad n = 1, \ldots, i, \ldots, p \end{aligned} \quad (25)$$

where $\omega(x) = [\Psi\{g_s(w(k, \mathbf{a}_{ff}, \mathbf{b}_{fb}))x(k)\}$ represents the output of an isolated DNU.

The dynamic neural system shown in Figure 6 and described by eq. (25) maps an n-dimensional input vector $x(k) \in R^n$ into a p-dimensional neural output vector $u(k) \in R^p$ where the mapping operation of an ith neuron is given by:

1. The linear mapping operation (synaptic operation)

$$v_{li}(k) = w^{(i)}(k, \mathbf{a}_{ff}^i, \mathbf{b}_{fb}^i)x^{(i)}(k) \quad (26a)$$

2. and the nonlinear mapping operation (somatic operation)

$$u_i(k) = \Psi^{(i)}\{g_s^{(i)}(v_{li}(k))\} \qquad i = 1, 2, \ldots, p \quad (26b)$$

A multistage dynamic neural network can be formed by cascading a group of single-stage

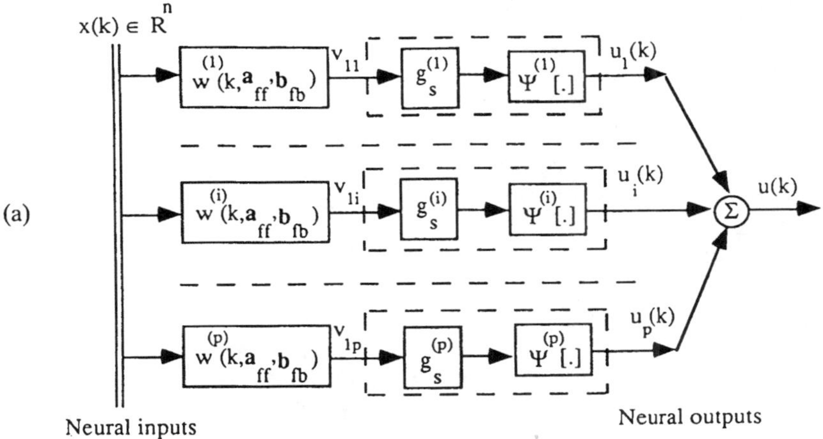

Figure 6. (a) Sigma connection of "p" DNUs to form a single-stage dynamic neural network. (b) Equivalent representation of one-stage dynamic neural network with "p" DNUs in parallel (in sigma mode).

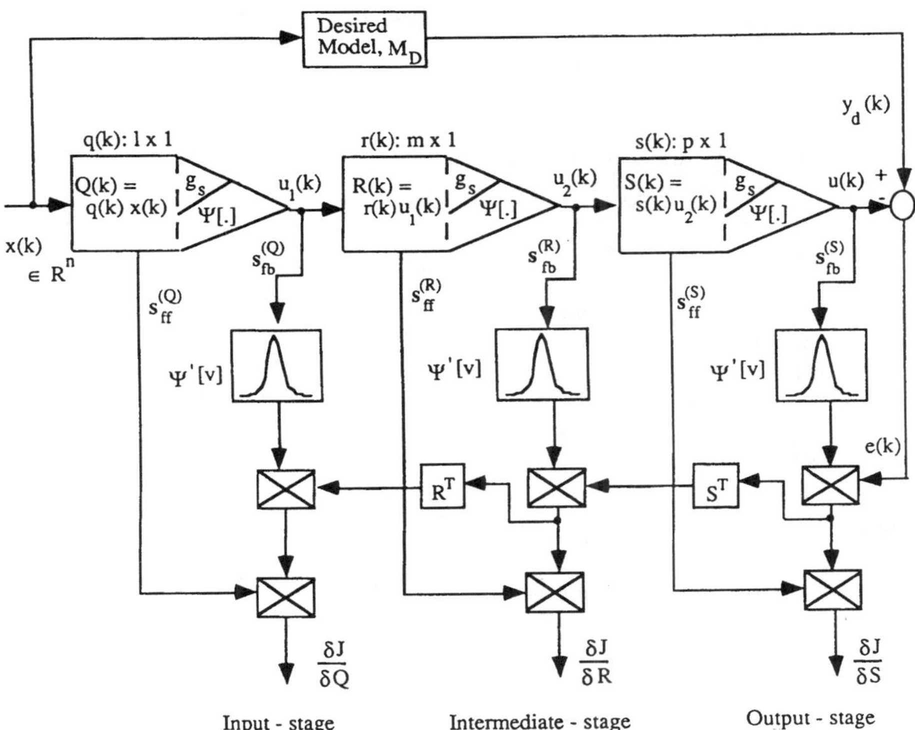

Figure 7. Morphology of a three-stage dynamic neural network and the parameter-state model. The dynamic neural network consists of input (Q)-, intermediate (R)-, and output (S)-stages. $\mathbf{s}_{ff}^{(Q)}$, $\mathbf{s}_{ff}^{(R)}$, and $\mathbf{s}_{ff}^{(S)}$ represent the feedforward parameter-state signals of the Q, R, and S dynamic neural stages (layers), respectively. Similarly, $\mathbf{s}_{fb}^{(Q)}$, $\mathbf{s}_{fb}^{(R)}$, and $\mathbf{s}_{fb}^{(S)}$ represent the feedback parameter-state signals of the Q, R, and S dynamic neural stages, respectively.

DNUs while the output of one stage provides the input to the subsequent stages. Using the above concepts, a three-stage dynamic neural network consisting of an input stage, an intermediate stage, and an output stage is formed. Figure 7 shows the morphology of a three-stage neural network and the corresponding parameter-state (sensitivity) model for the implementation of the learning and adaptive algorithm derived in the preceding section.

In the following section, we consider the formulation of nonlinear control problem.

3.1. Formulation of Nonlinear Control Problem

Assume that a single-input–single-output (SISO) nonlinear discrete system is given in the form

$$q(k+1) = f[q(k), u(k)] \quad \text{(state equation)}$$
$$y(k) = g[q(k)] \quad \text{(output equation)} \tag{27}$$

where $q \in R^n$ are the state variables, $u(k) \in R^1$ is the control input, $f[.]$ and $g[.]$ are the nonlinear maps on R^n, $f[.]$ is bounded away from zero, and $y(k) \in R^1$ is the plant output.

If the system described by eq. (27) is linear and time invariant, the well-developed concepts of controllability, observability, and stability can be applied to determine the control input $u(k)$ to optimize a performance function. The tractability of these different problems may be ultimately traced to the fact that they can be reduced to the solution of n linear equations in n unknowns. In contrast to this, problems involving nonlinear equations of the form (27), where the functions $f[.]$ and $g[.]$ are known, result in nonlinear algebraic equations for which similar powerful methods do not exist.[11] As a consequence, several assumptions have to be made to make the problems analytically tractable.

The problem to be addressed consists of finding a control signal $u(k)$ that will force the output $y(k)$ to track asymptotically the desired output $y_d(k)$, that is

$$\lim_{k \to \infty} [y_d(k) - y(k)] = 0 \tag{28}$$

To achieve the above, the following assumptions about the nonlinear plant are required:

Assumption 2. *For any $q \in R^n$*

$$0 < k_1 \leq |f[.]| \tag{29a}$$

Assumption 3. *For any $k \in [0, \infty]$ the desired output $y_d(k)$ and its n-derivatives $y_d^{(1)}(k)$, $y_d^{(2)}(k), \ldots, y_d^{(n)}(k)$, are uniformly bounded, that is,*

$$|y_d^{(i)}(k)| \leq m_i \qquad i = 0, 1, 2, \ldots, n \tag{29b}$$

Assumption 4. *There exist coefficients $\mathbf{a}_{ff}$ and $\mathbf{b}_{fb}$ such that* $\hat{f}[.]$ and $\hat{g}[.]$ *approximate the nonlinear functions $f[.]$ and $g[.]$, respectively, with an accuracy ε on Σ, a compact subset of R^n, that is,*

$$\max|f[.] - \hat{f}[.]| \leq \varepsilon \tag{29c}$$
$$\max|g[.] - \hat{g}[.]| \leq \varepsilon \qquad \forall q \in \text{ on } \Sigma \tag{29d}$$

A nonlinear system can be represented by one of the four discrete-time models as suggested in ref. 11. These models are described by the following nonlinear difference equations:

Model I

$$y(k+1) = \sum_{i=0}^{n-1} \alpha_i y(k-i) + g[u(k), u(k-1), \ldots, u(k-m+1)] \tag{30a}$$

Model II

$$y(k+1) = f[y(k), y(k-1), \ldots, y(k-n+1)] + \sum_{j=0}^{m-1} \beta_j u(k-j) \tag{30b}$$

Model III

$$y(k+1) = f[y(k), y(k-1), \ldots, y(k-n+1)] + g[u(k), u(k-1), \ldots, u(k-m+1)] \tag{30c}$$

Model IV

$$y(k+1) = f[y(k), y(k-1), \ldots, y(k-n+1), u(k), u(k-1), \ldots, u(k-m+1)] \tag{30d}$$

where $[u(k), y(k)]$ represents the input–output pair of a SISO plant at time k. In all four models, the output of the plant at time $(k+1)$ depends both upon on its past n values as well as the past m values of the input (output of the neural net). In the above equations, $f[.]$ and $g[.]$ are nonlinear functions that may take different forms. In Model I, the plant output $y(k+1)$ is a linear function of the past values $y(k-1)$, while in Model II the relation between $y(k+1)$ and the past values of the control input $u(k-j)$ is assumed linear. In Model III, the nonlinear relation of $y(k+1)$ with $y(k-1)$ and $u(k-j)$ is assumed to be separable, while Model IV, in which $y(k+1)$ is a nonlinear function of $y(k-1)$ and $u(k-1)$ subsumes Models I–III, is analytically least tractable.[11]

The performance of the dynamic neural network with only synaptic adaptation has been discussed extensively for the above four models in ref. 12. A study of the performance of the proposed dynamic neural network for a set of nonlinear systems represented by Model IV [eq. 30(d)] is reported in this article from the adaptive control point of view.

4. Computer Simulation Results

In the computer simulation studies discussed in this section, a SISO nonlinear dynamic system of the form

$$y(k+1) = f[y(k), y(k-1), \ldots, y(k-n+1), u(k), u(k-1), \ldots, u(k-m+1)]$$

was cascaded with the dynamic neural network presented in the earlier section. This control configuration is shown in Figure 8.

If the error is defined as $e(k) = y_d(k) - y(k)$, where $y_d(k)$ is the desired output and $y(k)$ is the actual output of the plant under control, the aim of control is to determine a bounded control input $u(k)$ such that

$$\lim_{k\to\infty} [y_d(k) - y(k) = e(k)] = 0$$

In our further discussion, the input signal $x(k)$ is considered the desired response for the unknown nonlinear plant to follow, thereby avoiding the necessity for obtaining the desired model.

The neural network used for computer simulation studies consisted of three stages with two DNUs in each stage. In this section, we present four simulation examples, each signifying a particular control objective. In Example 1, we consider a nonlinear dynamic plant governed by the difference equation (30d). The parameters of the plant, β_{ff} and α_{fb}, and the nonlinear function $f[.]$ are assumed unknown. As the neural network is trained, the plant response follows the desired command signal closely. In Example 2, the study of robustness of the dynamic neural network is carried out by making the plant undergo different nonlinear functions at different instants of the control process.

Example 1. *Control of an unknown nonlinear plant.*

The plant to be controlled is governed by the difference equation

$$y(k) = f[y(k-1), y(k-2), u(k), u(k-1), u(k-2)] \qquad (31)$$

with an arbitrary unknown function $f[.] = [2 + \cos\{7\pi(y^2(k-1) + y^2(k-2))\}] + e^{-u(k)}/1 + u^2(k-1) + u^2(k-2)$, *and the plant parameters are* $\beta_{ff} = [1.2, 1, 0.8]$ *and* $\alpha_{fb} = [1.3, 0.9, 0.7]$.

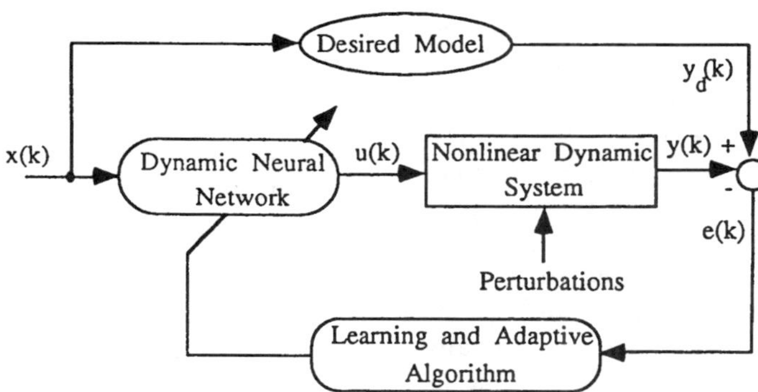

Figure 8. Control scheme for a nonlinear dynamic system using dynamic neural network.

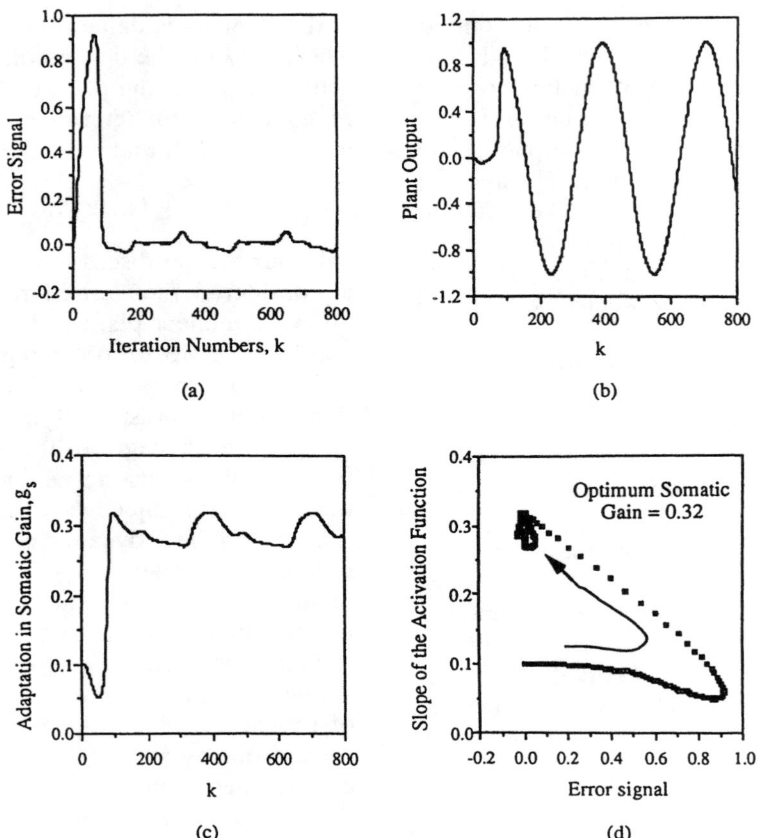

Figure 9. Example 1. (a) Error response. (b) Plant output. (c) Adaptation in the slope of the activation function. (d) Relationship between the slope of the activation function and the error signal.

The input to the system was $x(k) = \sin(2\pi k/250)$ *in the interval* $[-1, 1]$. *The simulation results obtained for this case are shown in Figure 9. Figures 9a and 9b show the error and plant responses, respectively. From these figures, it is observed that except for an initial error the plant response is close to that of the desired signal. Figures 9c and 9d show the adaptation in the somatic gain (slope),* g_s, *of the activation function with respect to iteration numbers and error signal, respectively.*

It may be observed from Figures 9a and 9b that the training of the neural network was slow during the initial period, which resulted in a large initial error. However, this depends upon the initial settings of the adjustable parameters ($\mathbf{a}_{ff}$, $\mathbf{b}_{fb}$, g_s) of the DNU. As the training proceeds, the nonlinear plant followed the command input closely, which resulted in a small error. The adaptation in somatic gain, g_s, as observed from Figures 9c and 9d, was initially large and settled down to an optimum value as the error signal converged to the preset tolerance limit. The optimum somatic gain was found to be 0.32.

Example 2. *Control of a nonlinear plant with variations in nonlinear characteristics.*

In this example, two cases are considered. In the first, the nonlinear function $f[.]$ *was arbitrarily changed to some other function at certain instant of the control process. In the second case, three different nonlinear functions were considered. To investigate the robustness of the dynamic neural network, the nonlinear functions were changed from one function to another during the control process.*

Case (i): The nonlinear plant to be controlled was the same as in Example 1 with the following nonlinearity:

$$f[.] = \frac{[\sin\{\pi(y^2(k-2)+0.5)\}] + 0.3\sin(2\pi u(k))}{1+u^2(k-1)+u^2(k-2)} \quad \text{for } k < 250 \qquad (32a)$$

During the control process, the nonlinearity $f[.]$ was changed to the following:

$$f[.] = [\sin\{\pi(y^2(k-2))\} + \sqrt{|\{u^2(k) + u^2(k-1) + u^2(k-2)\}|}] \quad \text{for } k \geq 250 \quad (32b)$$

This change in the characteristics of the nonlinear function is shown in Figure 10a, and the corresponding error and plant responses are shown in Figures 10b and 10c, respectively. It can be observed from the error and output responses that the dynamic neural network was able to drive the plant toward the desired response even in the presence of changing nonlinear dynamic characteristics.

Case (ii): The nonlinear plant considered in this case was the same as in Case (i). The objective of this simulation was to demonstrate the robustness of the dynamic neural network. Initially, the neural network was trained to a particular nonlinear function $f[.]$. While the control operation was in progress, the functions used to model the nonlinear system were changed. The neural network is then presented with these different nonlinear functions in a random sequence. It was observed that the effect of introducing different nonlinear functions on system performance was less in subsequent occurrences of variations in nonlinearity characteristics. The three nonlinear functions used in this simulation example were as follows:

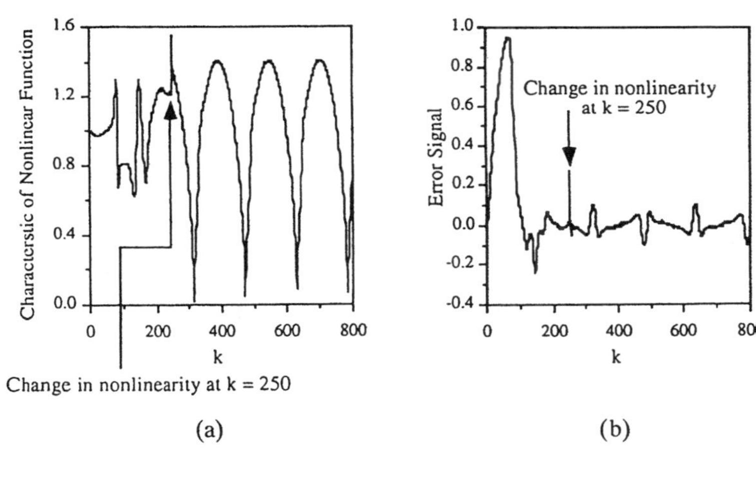

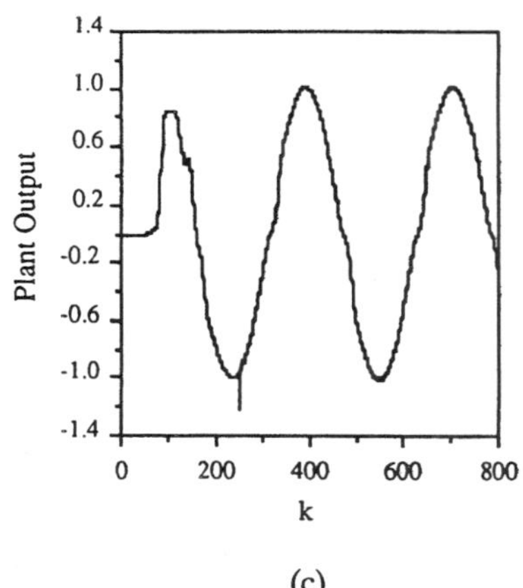

Figure 10. Example 2, Case (i). (a) Characteristic of the nonlinear function $f[.]$. (b) Error response. (c) Plant output.

$$f[\cdot] = \frac{[\sin\{\pi(y^2(k-2)+0.5)\}] + 0.3\sin(2\pi u(k))}{1+u^2(k-1)+u^2(k-2)}$$

for $0 \le k < 350$ and $2000 \le k < 2500$ (33a)

$$f[\cdot] = \frac{[2+\cos\{7\pi(y^2(k-1)+y^2(k2))\}] + e^{-u(k)}}{1+u^2(k-1)+u^2(k-2)} \tag{33b}$$

for $350 \le k < 800$, $1200 \le k < 1600$, and $k > 2500$

$$f[\cdot] = e^{(y^2(k-1)+y^2(k-2))} + \sqrt{|\{u^2(k)+u^2(k-1)+u^2(k-2)\}|} \tag{33c}$$

for $800 \le k < 1200$ and $1600 \le k < 2000$

The simulation results obtained for this case are shown in Figure 11.

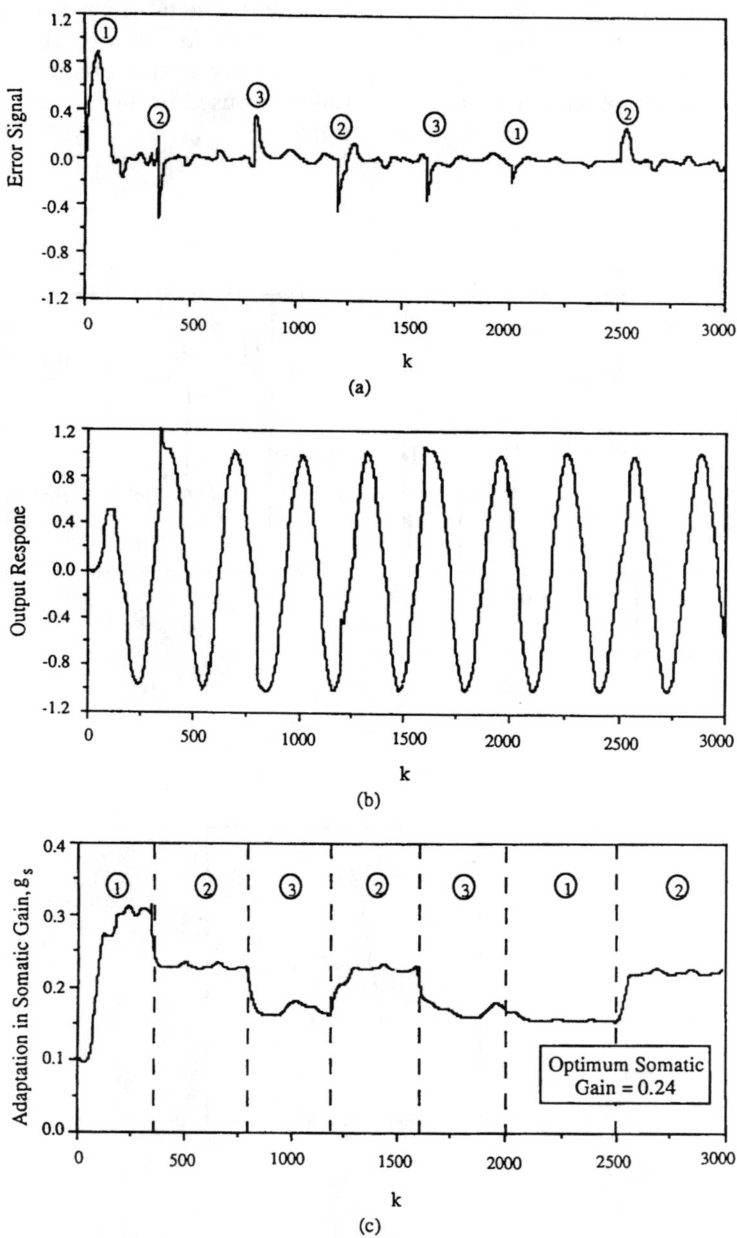

Figure 11. Example 2, Case (ii). (a) Error response to the changes in nonlinearity characteristics. Numbers in circles, 1, 2, and 3 represent different nonlinearity characteristics as described in eqs. (33a), (33b), and (33c), respectively. (b) Output response. (c) Adaptation in somatic gain.

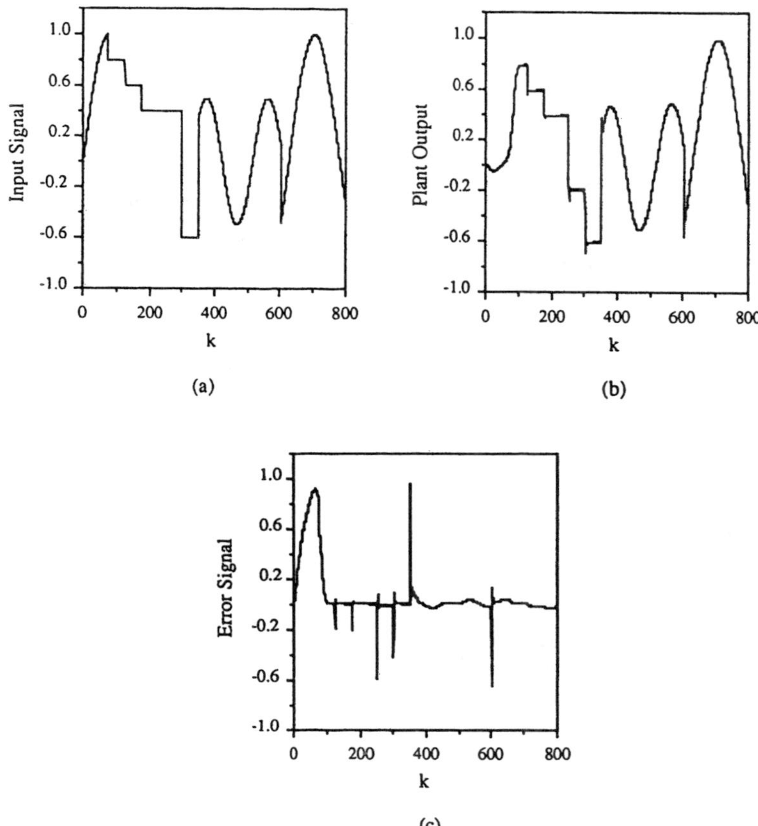

Figure 12. Example 3. (a) Varying input signal. (b) Plant output. (c) Error signal.

Example 3. *Control of a nonlinear plant with variations in input signal.*

The goal of this simulation was to demonstrate the input signal adaptation of the dynamic neural network. The plant and the nonlinear function $f[.]$ *in this example were the same as in Example* 1; *however, in this example the input signal* $x(k)$ *was varied in the interval* $[-1, 1]$ *as follows*:

$$
\begin{aligned}
x(k) &= \sin(2\pi k/250) && 0 \le k < 75 \\
x(k) &= 0.8 && \text{for } 75 \le k < 125 \\
x(k) &= 0.6 && \text{for } 125 \le k < 175 \\
x(k) &= 0.4 && \text{for } 175 \le k < 300 \\
x(k) &= -0.6 && \text{for } 300 \le k < 350 \\
x(k) &= 0.5\cos(2\pi k/150) && \text{for } 350 \le k < 600 \\
x(k) &= \sin(2\pi k/250) && \text{for } k \ge 600
\end{aligned}
\tag{34}
$$

The input variations are shown in Figure 12a. The corresponding plant and error responses are shown in Figures 12b and 12c, respectively. This example demonstrates the ability of the dynamic neural network that makes the plant adaptively follow the varying input signal.

Example 4. *Control of a nonlinear plant with parameter perturbations.*

The purpose of this simulation example was to study the effect of the plant parameter perturbations on the performance of the dynamic neural network. The nonlinear plant considered in this example was of the same form as in the above examples with the following parameter values and the nonlinearity characteristics:

$$\beta_{ff} = [1.4, 1.2, 0] \text{ and } \alpha_{fb} = [1.2, 1, 0]$$

$$f[.] = \frac{[\sin\{\pi(y^2(k-2))\} + \sqrt{|\{u^2(k) + u^2(k-1)\} + u^2(k-2)|}]}{1 + u^3(k)}$$

During the control process, the plant parameters β_2 *and* α_2 *were perturbed as follows*:

$$\begin{rcases} \beta_2 = \beta_2 + 0.3 \\ \alpha_2 = \alpha_2 + 0.3 \end{rcases} \quad \text{for } 175 \leq k < 350$$

$$\begin{rcases} \beta_2 = \beta_2 + 0.5 \\ \alpha_2 = \alpha_2 + 0.7 \end{rcases} \quad \text{for } 350 \leq k < 750 \tag{35}$$

$$\begin{rcases} \beta_2 = \beta_2 + 0.2 \\ \alpha_2 = \alpha_2 - 0.4 \end{rcases} \quad \text{for } k \geq 750$$

These perturbations in the plant parameters are shown in Figures 13a and 13b. The characteristic of the nonlinear function $f[.]$ is shown in Figure 13c, and the corresponding error and plant responses are shown in Figures 13d and 13e, respectively. The adaptation of the somatic gain is depicted in Figure 13f.

As may be seen from the simulation studies reported in Figure 13, the plant was able to follow the input signal fairly well even when there were perturbations in the plant parame-

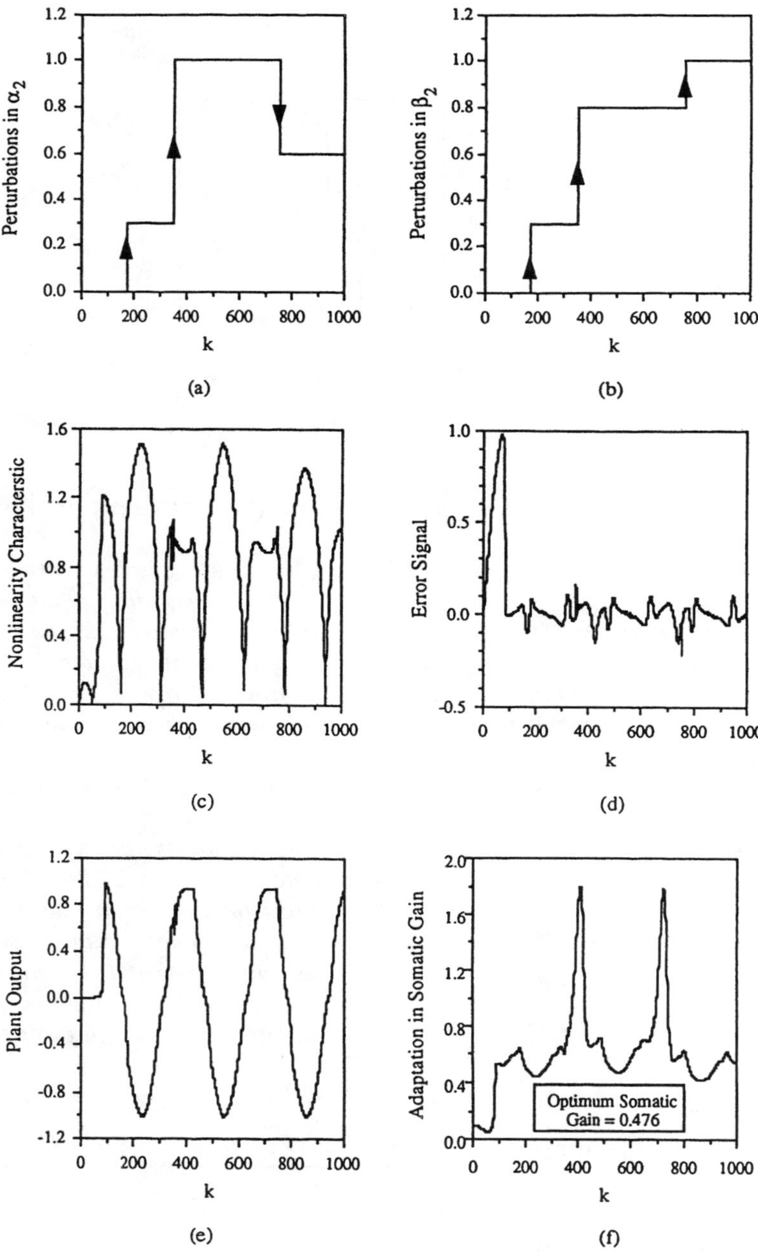

Figure 13. Example 4. (a) Perturbations in α_2. (b) Perturbations in β_2. (c) Characteristic of the nonlinear function $f[.]$. (d) Error signal. (e) Plant output. (f) Adaptation in the somatic gain.

ters, thereby demonstrating the parameter adaptation capability of the dynamic neural network. The adaptation in the somatic gain was found to be significant. This could lead to instability when the magnitude of perturbations is large. To prevent instability, it may be necessary to reduce the adaptive gain μ_{g_s} in the somatic gain algorithm [eq. 24(c)].

5. Conclusions

A new architecture of an artificial neuron, the DNU, has been presented that enables both the synaptic and somatic adaptations. The DNU does not necessarily represent any specific anatomical region within the biological nervous system. Rather, it represents a reasonable mathematical model of the functional dynamics of the reverberating circuit in the neuronal pool of the central nervous system. The algorithm for adaptively modifying the parameters of the DNU, namely, the feedforward and feedback synaptic weights, and the slope of the nonlinear function has been derived.

A multistage dynamic neural network has been developed by considering the DNU as a basic processing element. The morphology of the three-stage dynamic neural network and the corresponding parameter-state (the sensitivity) model have been presented. The effectiveness of the proposed neural structure and the adaptive algorithm have been demonstrated for a class of unknown nonlinear systems. From the simulation results presented, it can be inferred that the three-stage dynamic neural network is capable of adaptively controlling an unknown nonlinear control system. This ability of the dynamic neural network was shown for changes in the characteristics of nonlinear function, input signal variations, and plant parameter perturbations.

The neural network architecture proposed in this article provides fast learning in comparison with many conventional neural network architectures used for control applications. This is due to the fact that the model of neuron, the DNU, is basically dynamic in nature, unlike in conventional structures. It has been shown by the authors in ref. 20 that the proposed dynamic neural network learns the inverse kinematics transformations of two- and three-linked robots faster than the conventional multilayered feedforward neural network.

The neural network discussed in this article is based upon the DNU, which promises to provide robust control characteristics for a class of unknown nonlinear and time-varying plants. Further work is under investigation on the use of DNU for the control of flexible structures in time-varying and unknown environments.

The research reported in this article was partially supported by an NSERC (Canada) research grant. The authors are grateful to the reviewers for their useful comments.

References

1. M. M. Gupta, Ed., Adaptive Methods for Control System Design, IEEE Press, New York, 1986.
2. K. J. Astrom, "Adaptive Feedback Control." In *Proceedings of the IEEE*, Vol. 75, No. 2, 1987, pp. 185–217.
3. K. S. Narendra and A.M. Annaswamy, "Stable Adaptive Systems." Prentice Hall, Englewood, Cliffs, NJ, 1989.
4. M. D. Peek and P. J. Antsaklis, "Parameter Learning for Performance Adaptation." *IEEE Control Syst. Mag*. Vol. 10, No. 3 (1990).
5. R. Ortega and Y. Tang, "Robustness of Adaptive Controllers – A Survey." *Automatica*, **25**, 651 (1989).
6. R. P. Lipmann, "An Introduction to Computing with Neural Nets". *ASSP Mag*., **4**(2) 4 (1987).
7. G. Cybenko, "Approximation by Superpositions of a Sigmoidal Function". *Math. Control, Signals, Syst*. **2**, 303 (1989).
8. K. Hornik, M. Stinchecombe, and H. White, "Multi-Layer Feed Forward Networks Are Universal Approximators." *Neural Networks*, **2**, 359 (1989).
9. E. Levin, N. Tishby, and S. A. Solla, In "A Statistical Approach to Learning and Generalization in Layered Neural Networks." *Proceedings of the IEEE*, Vol. 78, No. 10, 1990, pp. 1568–1574.
10. S. I. Sudharsanan and M. K. Sundareshan, "Training of a Three-Layer Dynamical Recurrent Neural Network for Nonlinear Input-Output Mapping." *IJCNN Sing*. **2**, 111 (1991).
11. K. S. Narendra and K. Parthasarthy, "Identification and Control of Dynamical Systems Using Neural Networks." *IEEE Trans. Neural Networks*, **1**, 4 (1990).
12. M. M. Gupta and D.H. Rao, "Dynamic Neural Units in the Control of Linear and Nonlinear Systems." In *Proceedings of the International Joint Conf. on Neural Networks*, June 1992, pp. 100–105.
13. M. M. Gupta and D. H. Rao, "Synaptic and Somatic Adaptations in Dynamnic Neural Networks." In *Proceedings of the Second International*

Conf. on Fuzzy Logic and Neural Networks, 1992, pp. 173–177.
14. S. Hui and S. Zak, "Analysis of Single Perceptrons Learning Capabilities." In *Proceedings of the American Control Conf.* 1991, pp. 809–814.
15. S. Chen, S. A. Billings, and P. M. Grant, "Nonlinear System Identification Using Neural Networks." *Int. J. Control*, **51**, 1191 (1990).
16. H. R. Wilson and J. D. Cowan, "Excitatory and Inhibitory Interactions in Localized Populations of Model Neurons." *Biophys. J.*, **12** (1972).
17. M. M. Gupta and G. K. Knopf, "A Multitask Visual Information Processor with a Biologically Motivated Design." *J. Visual Comm. Image Represent.*, **3**, 230 (1992).
18. S. A. Billings, H. B. Jamaludding, and S. Chen, "Properties of Neural Networks with Applications to Modeling Nonlinear Dynamical Systems." *Int. J. Control*, **55** (1) 193 (1992).
19. W. J. Freeman, "Dynamnics of Image Formation by Nerve Cell Assemblies." In *Synergistics of the Brain*, E. Basar, H. Flohr, H. Haken, and A. J. Mandell, Eds., Berlin, Springer-Verlag, pp. 102–121, 1983.
20. M. M. Gupta and D.H. Rao, "Neural Learning of Robot Inverse Kinematics Transformations." In *Neural and Fuzzy Systems: The Emerging Science of Intelligent Computing*, S. Mitra, W. Kraske, and M. M. Gupta, Eds. SPIE Institute Series Editor, Bellingham, Washington 1993 (to appear).

Appendix: Definitions of Certain Key Terms

Action potential: The pulse of electric potential generated across the membrane of a neuron following the application of a stimulus greater than the threshold value.

Axon: The output node of a neuron that carries the action potential to other neurons in the network.

Dendrite: The input part of the neuron that carries a temporal summation of action potentials to the soma.

Excitatory neuron: A neuron that transmits an action potential that has a positive or excitatory influence on the recipient nerve cells.

Inhibitory neuron: A neuron that transmits an action potential that has a negative or inhibitory influence on the recipient nerve cells.

Neural activity: A neuron is assumed to be active if it's firing a sequence of action potentials and the action potential is greater than the rest potential (−90 mV).

Neuron: The basic nerve cell or computing unit for biologic information processing.

Reverberation: The process whereby neural activity circulates at a constant level within a nervous layer even after the stimulus (input) is removed.

Soma: The body of a neuron that processes the dendritic inputs.

Somatic Gain: The parameter that changes the slope of nonlinear activation function, in this article the sigmoidal function, used in the architecture of neuron.

Synapse: The junction point between the axon of a presynaptic neuron and the dendrite of a post-synaptic neuron.

Synaptic Learning: It is that component of learning which pertains to the determination of optimum synaptic weights based on the minimization of certain performance index of error. Only the term 'learning' is used in conventional neural network architectures, for

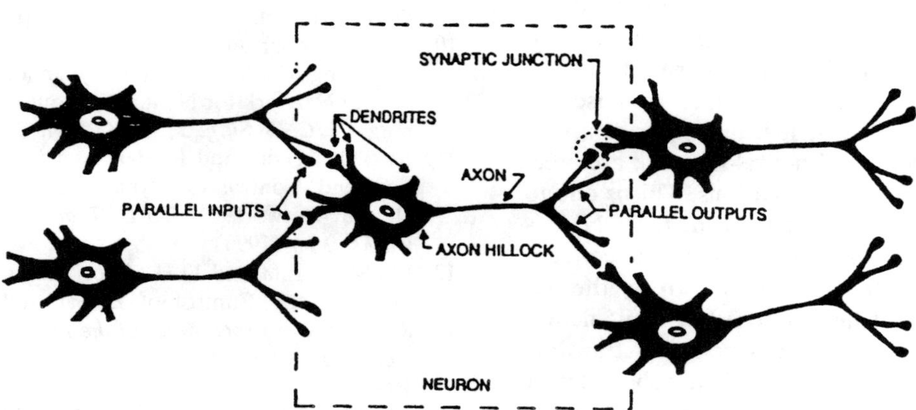

Figure A1. Schematic view of biological neuron. Each neuron receives parallel inputs through its dendrites and transmits a common output via the axon to other neurons.

the nonlinear activation function is kept constant. However, in DNU architecture, proposed in this article, the slope (somatic gain) of sigmoidal function is also adapted in addition to synaptic weights, and hence the terms synaptic and somatic learning or operations are used.

A schematic diagram of a biological neuron is shown in Figure A1.

Part 6
Fuzzy–Neural Control Systems

The faculty of cognition and perception in humans is very complex, but it possesses a very efficient mechanism for information processing and expression. "The weather is just beautiful for playing a game of golf," conveys a large amount of information in just a brief sentence. This is the attribute of graded membership in our natural language and thinking processes that makes our knowledge base and information processing abilities so efficient. This notion of graded membership has been adopted in the calculus of fuzzy logic due to Lotfi A. Zadeh. Now, this notion promises to provide some robust algorithms for intelligent robotic systems.

HUMAN cognition, which embodies all our thinking and logic processes, and our actions and decisions, is full of imprecision and uncertainties. The natural world in which we live is a world of imprecision. Imprecision arises from physical phenomena and it is inherent in our cognitive process. The human brain has evolved around cognitive uncertainties, and through the process of learning, it is able to extract important information and make decisions. It is a robust and, to a large measure, a fault-tolerant system. A mystery, however, shrouds the working and mathematics of this cognitive process. This mystery has existed for years because of the lack of the right mathematical tools for understanding and modeling this mysterious cognitive process. The conventional mathematical tools and artificial logical tools are nothing but paradoxes even in simple human logical tasks. It was in the early sixties when Professor Lotfi A. Zadeh pondered over these paradoxes and the mathematical models of human logic and reasoning. He carefully considered the relevance and validity of the precision of our decisions and binary logic (yes or no, white or black, 0 or 1) under cognitive uncertainties.

Zadeh's earlier works, prior to the mid-sixties, were devoted to the use of "precise" mathematical tools in systems theory and decision processes. No doubt it was a great contribution to the field, but his seminal work on fuzzy logic, which appeared in an article in 1965 (reproduced in this volume), renewed a widespread interest in the fields of systems, decision analysis, cognitive uncertainty, and modeling of human cognition (process of learning, thinking, reasoning, and adaptation). Since the publication of that work, a worldwide community of scholars and scientists have made many important contributions to the field—leading to some important commercial, industrial, and domestic applications. In growing numbers, investigators in a wide variety of fields—ranging from psychology, sociology, philosophy, and economics to natural sciences, engineering, and computer sciences—are exploring this new path (which seems to be very bright and promising) to the understanding of human reasoning and cognition. They are also developing novel methods for dealing with systems and processes that are too complex to be analyzed by conventional quantitative techniques.

In retrospect, it is evident that the trend toward the use of fuzzy logic—a logic that is much closer in spirit to human cognition and language than the conventional logical systems—could have been anticipated during the past century. What held back the development of fuzzy logic were the attitudes from the mechanistic era of the nineteenth century and, more recently, the habits of programmatic reasoning fostered by the rapidly widening use of digital machine computation.

Fuzzy logic rests on the notion that the key elements in human cognition are based not on precise numbers but on a class of numbers (objects) in which the transition from membership to nonmembership is gradual rather than abrupt.

In Part Six of this volume, we present a set of five representative articles that deal exclusively with the theory of fuzzy sets and fuzzy logic, and that deal with the combined field of fuzzy logic and neural networks—the fuzzy–neural networks. This newly developing field of fuzzy–neural networks encompasses many features of the human brain—the cognitive aspects in fuzzy logic, and the learning and adaptation strengths of neural networks. This new field, combined with the genetic aspects of humans, promises to provide many theoretical and applied advances.

The articles collected in this part are intended to provide the reader with a broad view of the fuzzy–neural field. More specifically, article (6.1) is the seminal work of Lotfi A. Zadeh, which appeared in 1965. Professor Zadeh defined a fuzzy set as a class of objects with a continuum of grades of membership in the interval [0, 1]. The notions of inclusion, union, intersection, complement, relation, convexity, and so forth, for fuzzy sets are introduced and the various properties of these notions are established. In 1965, no one knew that this article would stir the interests of so many scholars, scientists, and applied engineers.

Another important work due to Zadeh is included that specially addresses the use of this new theory in control systems and decision processes—the subject of this volume. In article (6.2), Zadeh presents an approach that is substantially different from the conventional qualitative techniques of systems analysis. This approach has three main distinguishing features: 1) use of so-called linguistic variables in place of, or in addition to, numerical variables; 2) characterization of relations between variables by fuzzy conditional statements; and 3) characterization of complex relations by fuzzy algorithms. This approach provides an approximate, yet effective, means of describing the behavior of systems that are too complex or too ill-defined to admit to precise mathematical analysis. Since the publication of this article, many extensions of the approach have appeared in the literature.

After having presented some inspiring works on fuzzy sets and fuzzy logic, the next three articles are works related to the combined field of fuzzy logic and neural networks.

In article (6.3), M. M. Gupta provides some basic principles of fuzzy–neural computing using synaptic (confluence) and somatic (aggregation, thresholding, and activation) operations. He then provides a detailed neuronal morphology based upon fuzzy logic and its generalization using T-operators (generalized OR and AND operations), including the learning and adaptation algorithm.

Article (6.4), by J. R. Jang, presents a generalized control strategy that enhances fuzzy controllers with self-learning capability. This methodology, termed temporal backpropagation, is model-insensitive. The inverted pendulum system is used as a test-bed to demonstrate the effectiveness of this new fuzzy control approach.

Advances in fuzzy logic and neuro-control have introduced many innovations in the design and architecture of new VLSI implementations of fuzzy-neurons. In article (6.5), T. Yamakawa—the pioneer in the field of fuzzy–neural technology—and his research colleagues, E. Uchino, T. Miki, and H. Kusanagi, present a structure and a learning algorithm of a new fuzzy neuron. The synaptic operations are nonlinear, and are characterized by IF–THEN rules. This neo-fuzzy neuron has been used for nonlinear system identification and for the prediction of chaotic behavior.

Further Reading

[1] L. A. Zadeh, "Fuzzy sets," *Inform. Contr.*, vol. 8, pp. 338–353, 1965.

[2] P. K. Simpson, "Fuzzy min-max neural networks—Part I: Classification," *IEEE Trans. Neural Networks*, vol. 3, no. 5, pp. 776–786, Sept. 1992.

[3] M. M. Gupta and G. K. Knopf, "Fuzzy neural network approach to control systems," in *Proc. First Int. Symp. Uncertainty Modeling Anal.*, Maryland, pp. 483–488, Dec. 3–5, 1990.

[4] M. M. Gupta, "Uncertainty and information: The emerging paradigms," *Int. Neuro Mass-Parallel Comput. Inform. Syst.*, vol. 2, pp. 65–70, 1991.

[5] A. Kaufmann and M. M. Gupta, *Introduction to Fuzzy Arithmetic: Theory and Applications.* New York: Van Nostrand Reinhold, Second Edition, 1991.

[6] S. K. Pal and S. Mitra, "Multilayer Perceptron, fuzzy sets, and classification," *IEEE Trans. Neural Networks*, vol. 3, no. 5, pp. 683–697, Sept. 1992.

[7] T. Yamakawa, E. Uchino, T. Miki, and H. Kusanagi, "A neo fuzzy neuron and its application to system identification and prediction of the system behavior," in *Proc. 2nd Int. Conf. Fuzzy Logic and Neural Networks (IIZUKA-92)*, Iizuka, Japan, pp. 477–483, July 17–22, 1992.

[8] M. M. Gupta and D. H. Rao, "Virtual cognitive systems (VCS): Neural–Fuzzy logic approach," *IFAC Conf.*, vol. 8, pp. 323–330, Sydney, Australia, July 19–23, 1993.

[9] Special Issue on Fuzzy Logic and Neural Networks, *IEEE Trans. Neural Networks*, vol. 3, no. 5, Sept. 1992.

[10] M. M. Gupta, G. N. Saridis, and B. R. Gaines, Eds., *Fuzzy Automata and Decision Processes*. New York: North-Holland, 1977.

[11] M. M. Gupta, R. K. Ragade, and R. R. Yager, Eds., *Advances in Fuzzy Set Theory and Applications.* New York: North-Holland, 1979.

[12] M. M. Gupta and E. Sanchez, Eds., *Fuzzy Information and Decision Processes.* New York: North-Holland, 1982.

[13] M. M. Gupta and E. Sanchez, Eds., *Approximate Reasoning in Decision Analysis*. New York: North-Holland, 1982.

[14] M. M. Gupta, A. Kandel, and W. Bandler, Eds., Approximate Reasoning in Expert Systems, New York: North-Holland, 1985.

[15] A. Kaufmann and M. M. Gupta, *Introduction to Fuzzy Arithmetic: Theory and Applications.* 2nd Ed. New York: Van Nostrand Reinhold, 1991. Japanese translation, Ohmsha Ltd., Tokyo, 1992

[16] M. M. Gupta and T. Yamakawa, Eds., *Fuzzy Computing: Theory, Hardware and Applications.* New York: North-Holland, Oxford, 1988.

[17] M. M. Gupta, and T. Yamakawa, Eds., *Fuzzy Logic in Knowledge-Based Systems*, Decision and Control, New York: North-Holland, 1988.

[18] A. Kaufmann and M.M. Gupta, *Fuzzy Mathematical Models in Engineering and Management Science*. New York: North-Holland 1988. Japanese translation by M. Matsuka, Ohmsha Ltd. Tokyo, 1992.

[19] B. M. Ayyub, M. M.Gupta, and L. N. Kanal, Eds., *Analysis and Management of Uncertainty: Theory and Application.* The Netherlands: Kluver Academic Publishers, 1992.

[20] B. Kosko, *Neural Networks and Fuzzy Systems.* Englewood Cliffs, NJ: Prentice Hall, 1992.

Article 6.1
Fuzzy Sets*

L. A. Zadeh

Department of Electrical Engineering and Electronics Research Laboratory, University of California, Berkeley, California

A fuzzy set is a class of objects with a continuum of grades of membership. Such a set is characterized by a membership (characteristic) function which assigns to each object a grade of membership ranging between zero and one. The notions of inclusion, union, intersection, complement, relation, convexity, etc., are extended to such sets, and various properties of these notions in the context of fuzzy sets are established. In particular, a separation theorem for convex fuzzy sets is proved without requiring that the fuzzy sets be disjoint.

I. INTRODUCTION

More often than not, the classes of objects encountered in the real physical world do not have precisely defined criteria of membership. For example, the class of animals clearly includes dogs, horses, birds, etc. as its members, and clearly excludes such objects as rocks, fluids, plants, etc. However, such objects as starfish, bacteria, etc. have an ambiguous status with respect to the class of animals. The same kind of ambiguity arises in the case of a number such as 10 in relation to the "class" of all real numbers which are much greater than 1.

Clearly, the "class of all real numbers which are much greater than 1," or "the class of beautiful women," or "the class of tall men," do not constitute classes or sets in the usual mathematical sense of these terms. Yet, the fact remains that such imprecisely defined "classes" play an important role in human thinking, particularly in the domains of pattern recognition, communication of information, and abstraction.

The purpose of this note is to explore in a preliminary way some of the basic properties and implications of a concept which may be of use in dealing with "classes" of the type cited above. The concept in question is that of a *fuzzy set*,[1] that is, a "class" with a continuum of grades of membership. As will be seen in the sequel, the notion of a fuzzy set provides a convenient point of departure for the construction of a conceptual framework which parallels in many respects the framework used in the case of ordinary sets, but is more general than the latter and, potentially, may prove to have a much wider scope of applicability, particularly in the fields of pattern classification and information processing. Essentially, such a framework provides a natural way of dealing with problems in which the source of imprecision is the absence of sharply defined criteria of class membership rather than the presence of random variables.

We begin the discussion of fuzzy sets with several basic definitions.

II. DEFINITIONS

Let X be a space of points (objects), with a generic element of X denoted by x. Thus, $X = \{x\}$.

* This work was supported in part by the Joint Services Electronics Program (U.S. Army, U.S. Navy and U.S. Air Force) under Grant No. AF-AFOSR-139-64 and by the National Science Foundation under Grant GP-2413.

[1] An application of this concept to the formulation of a class of problems in pattern classification is described in RAND Memorandum RM-4307-PR, "Abstraction and Pattern Classification," by R. Bellman, R. Kalaba and L. A. Zadeh, October, 1964.

Reprinted from *Information and Control,* vol. 8, pp. 338–353, 1965.

A *fuzzy set (class)* A in X is characterized by a *membership (characteristic) function* $f_A(x)$ which associates with each point[2] in X a real number in the interval [0, 1],[3] with the value of $f_A(x)$ at x representing the "grade of membership" of x in A. Thus, the nearer the value of $f_A(x)$ to unity, the higher the grade of membership of x in A. When A is a set in the ordinary sense of the term, its membership function can take on only two values 0 and 1, with $f_A(x) = 1$ or 0 according as x does or does not belong to A. Thus, in this case $f_A(x)$ reduces to the familiar characteristic function of a set A. (When there is a need to differentiate between such sets and fuzzy sets, the sets with two-valued characteristic functions will be referred to as *ordinary sets* or simply *sets*.)

Example. Let X be the real line R^1 and let A be a fuzzy set of numbers which are much greater than 1. Then, one can give a precise, albeit subjective, characterization of A by specifying $f_A(x)$ as a function on R^1. Representative values of such a function might be: $f_A(0) = 0$; $f_A(1) = 0$; $f_A(5) = 0.01$; $f_A(10) = 0.2$; $f_A(100) = 0.95$; $f_A(500) = 1$.

It should be noted that, although the membership function of a fuzzy set has some resemblance to a probability function when X is a countable set (or a probability density function when X is a continuum), there are essential differences between these concepts which will become clearer in the sequel once the rules of combination of membership functions and their basic properties have been established. In fact, the notion of a fuzzy set is completely nonstatistical in nature.

We begin with several definitions involving fuzzy sets which are obvious extensions of the corresponding definitions for ordinary sets.

A fuzzy set is *empty* if and only if its membership function is identically zero on X.

Two fuzzy sets A and B are *equal*, written as $A = B$, if and only if $f_A(x) = f_B(x)$ for all x in X. (In the sequel, instead of writing $f_A(x) = f_B(x)$ for all x in X, we shall write more simply $f_A = f_B$.)

The *complement* of a fuzzy set A is denoted by A' and is defined by

$$f_{A'} = 1 - f_A . \tag{1}$$

As in the case of ordinary sets, the notion of containment plays a central role in the case of fuzzy sets. This notion and the related notions of union and intersection are defined as follows.

Containment. A is *contained in* B (or, equivalently, A is a *subset of* B, or A is *smaller than or equal to* B) if and only if $f_A \leqq f_B$. In symbols

$$A \subset B \Leftrightarrow f_A \leqq f_B . \tag{2}$$

Union. The *union* of two fuzzy sets A and B with respective membership functions $f_A(x)$ and $f_B(x)$ is a fuzzy set C, written as $C = A \cup B$, whose membership function is related to those of A and B by

$$f_C(x) = \text{Max}\,[f_A(x), f_B(x)], \qquad x \in X \tag{3}$$

or, in abbreviated form

$$f_C = f_A \vee f_B . \tag{4}$$

Note that $\cup$ has the associative property, that is, $A \cup (B \cup C) = (A \cup B) \cup C$.

Comment. A more intuitively appealing way of defining the union is

[2] More generally, the domain of definition of $f_A(x)$ may be restricted to a subset of X.

[3] In a more general setting, the range of the membership function can be taken to be a suitable partially ordered set P. For our purposes, it is convenient and sufficient to restrict the range of f to the unit interval. If the values of $f_A(x)$ are interpreted as truth values, the latter case corresponds to a multivalued logic with a continuum of truth values in the interval [0, 1].

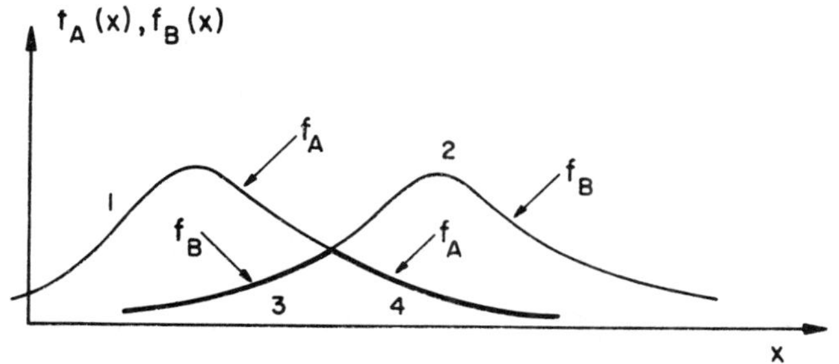

FIG. 1. Illustration of the union and intersection of fuzzy sets in R^1

the following: The union of A and B is the smallest fuzzy set containing both A and B. More precisely, if D is any fuzzy set which contains both A and B, then it also contains the union of A and B.

To show that this definition is equivalent to (3), we note, first, that C as defined by (3) contains both A and B, since

$$\text{Max}\,[f_A, f_B] \geqq f_A$$

and

$$\text{Max}\,[f_A, f_B] \geqq f_B\,.$$

Furthermore, if D is any fuzzy set containing both A and B, then

$$f_D \geqq f_A$$

$$f_D \geqq f_B$$

and hence

$$f_D \geqq \text{Max}\,[f_A, f_B] = f_C$$

which implies that $C \subset D$. Q.E.D.

The notion of an intersection of fuzzy sets can be defined in an analogous manner. Specifically:

Intersection. The *intersection* of two fuzzy sets A and B with respective membership functions $f_A(x)$ and $f_B(x)$ is a fuzzy set C, written as $C = A \cap B$, whose membership function is related to those of A and B by

$$f_C(x) = \text{Min}\,[f_A(x), f_B(x)], \qquad x \in X, \tag{5}$$

or, in abbreviated form

$$f_C = f_A \wedge f_B\,. \tag{6}$$

As in the case of the union, it is easy to show that the intersection of A and B is the *largest* fuzzy set which is contained in both A and B. As in the case of ordinary sets, A and B are *disjoint* if $A \cap B$ is empty. Note that $\cap$, like $\cup$, has the associative property.

The intersection and union of two fuzzy sets in R^1 are illustrated in Fig. 1. The membership function of the union is comprised of curve segments 1 and 2; that of the intersection is comprised of segments 3 and 4 (heavy lines).

Comment. Note that the notion of "belonging," which plays a fundamental role in the case of ordinary sets, does not have the same role in the case of fuzzy sets. Thus, it is not meaningful to speak of a point x "belonging" to a fuzzy set A except in the trivial sense of $f_A(x)$ being positive. Less trivially, one can introduce two levels α and β ($0 < \alpha < 1$, $0 < \beta < 1$, $\alpha > \beta$) and agree to say that (1) "x belongs to A" if $f_A(x) \geqq \alpha$; (2) "x does not belong to A" if $f_A(x) \leqq \beta$; and (3) "x has an indeterminate status relative to A" if $\beta < f_A(x) < \alpha$. This leads to a three-valued logic (Kleene, 1952) with three truth values: T ($f_A(x) \geqq \alpha$), F ($f_A(x) \leqq \beta$), and U ($\beta < f_A(x) < \alpha$).

III. SOME PROPERTIES OF $\cup$, $\cap$, AND COMPLEMENTATION

With the operations of union, intersection, and complementation defined as in (3), (5), and (1), it is easy to extend many of the basic identities which hold for ordinary sets to fuzzy sets. As examples, we have

$$\left.\begin{aligned}(A \cup B)' &= A' \cap B' \\ (A \cap B)' &= A' \cup B'\end{aligned}\right\}\text{De Morgan's laws} \qquad \begin{aligned}(7)\\(8)\end{aligned}$$

$$C \cap (A \cup B) = (C \cap A) \cup (C \cap B) \qquad \text{Distributive laws.} \qquad (9)$$

$$C \cup (A \cap B) = (C \cup A) \cap (C \cup B) \qquad (10)$$

These and similar equalities can readily be established by showing that the corresponding relations for the membership functions of A, B, and C are identities. For example, in the case of (7), we have

$$1 - \text{Max}\,[f_A, f_B] = \text{Min}\,[1 - f_A, 1 - f_B] \qquad (11)$$

which can be easily verified to be an identity by testing it for the two possible cases: $f_A(x) > f_B(x)$ and $f_A(x) < f_B(x)$.

Similarly, in the case of (10), the corresponding relation in terms of f_A, f_B, and f_C is:

$$\text{Max}\,[f_C, \text{Min}\,[f_A, f_B]] = \text{Min}\,[\text{Max}\,[f_C, f_A], \text{Max}\,[f_C, f_B]] \qquad (12)$$

which can be verified to be an identity by considering the six cases:

$$f_A(x) > f_B(x) > f_C(x), f_A(x) > f_C(x) > f_B(x), f_B(x) > f_A(x) > f_C(x),$$

$$f_B(x) > f_C(x) > f_A(x), f_C(x) > f_A(x) > f_B(x), f_C(x) > f_B(x) > f_A(x).$$

Essentially, fuzzy sets in X constitute a distributive lattice with a 0 and 1 (Birkhoff, 1948).

AN INTERPRETATION FOR UNIONS AND INTERSECTIONS

In the case of ordinary sets, a set C which is expressed in terms of a family of sets $A_1, \cdots, A_i, \cdots, A_n$ through the connectives $\cup$ and $\cap$, can be represented as a network of switches $\alpha_1, \cdots, \alpha_n$, with $A_i \cap A_j$ and $A_i \cup A_j$ corresponding, respectively, to series and parallel combinations of α_i and α_j. In the case of fuzzy sets, one can give an analogous interpretation in terms of sieves. Specifically, let $f_i(x)$, $i = 1, \cdots, n$, denote the value of the membership function of A_i at x. Associate with $f_i(x)$ a sieve $S_i(x)$ whose meshes are of size $f_i(x)$. Then, $f_i(x) \vee f_j(x)$ and $f_i(x) \wedge f_j(x)$ correspond, respectively, to parallel and series combinations of $S_i(x)$ and $S_j(x)$, as shown in Fig. 2.

More generally, a well-formed expression involving $A_1, \cdots, A_n$, $\cup$, and $\cap$ corresponds to a network of sieves $S_1(x), \cdots, S_n(x)$ which can be found by the conventional synthesis techniques for switching circuits. As a very simple example,

$$C = [(A_1 \cup A_2) \cap A_3] \cup A_4 \qquad (13)$$

corresponds to the network shown in Fig. 3.

Note that the mesh sizes of the sieves in the network depend on x and that the network as a whole is equivalent to a single sieve whose meshes are of size $f_C(x)$.

IV. ALGEBRAIC OPERATIONS ON FUZZY SETS

In addition to the operations of union and intersection, one can define a number of other ways of forming combinations of fuzzy sets and relating them to one another. Among the more important of these are the following.

Algebraic product. The *algebraic product* of A and B is denoted by AB

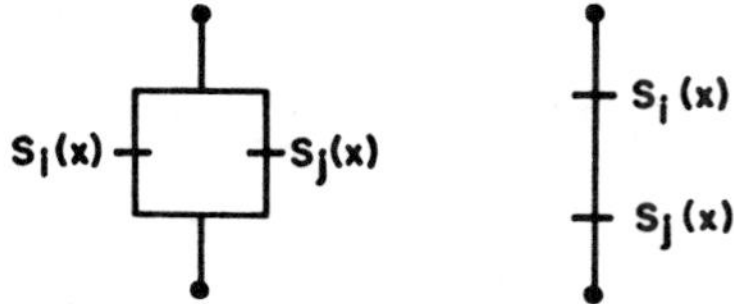

FIG. 2. Parallel and series connection of sieves simultating ⋃ and ⋂

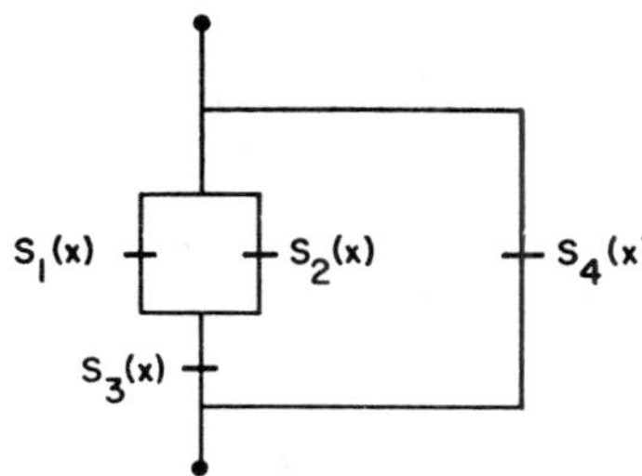

FIG. 3. A network of sieves simultating $\{[f_1(x) \vee f_2(x)] \wedge f_3(x)\} \vee f_4(x)$

and is defined in terms of the membership functions of A and B by the relation

$$f_{AB} = f_A f_B . \tag{14}$$

Clearly,

$$AB \subset A \cap B. \tag{15}$$

Algebraic sum.[4] The *algebraic sum* of A and B is denoted by $A + B$ and is defined by

$$f_{A+B} = f_A + f_B \tag{16}$$

provided the sum $f_A + f_B$ is less than or equal to unity. Thus, unlike the algebraic product, the algebraic sum is meaningful only when the condition $f_A(x) + f_B(x) \leqq 1$ is satisfied for all x.

Absolute difference. The *absolute difference* of A and B is denoted by $|A - B|$ and is defined by

$$f_{|A-B|} = |f_A - f_B|.$$

Note that in the case of ordinary sets $|A - B|$ reduces to the relative complement of $A \cap B$ in $A \cup B$.

Convex combination. By a convex combination of two vectors f and g is usually meant a linear combination of f and g of the form $\lambda f + (1 - \lambda)g$, in which $0 \leqq \lambda \leqq 1$. This mode of combining f and g can be generalized to fuzzy sets in the following manner.

Let A, B, and Λ be arbitrary fuzzy sets. The *convex combination of* A, B, *and* Λ is denoted by $(A, B; \Lambda)$ and is defined by the relation

$$(A, B; \Lambda) = \Lambda A + \Lambda' B \tag{17}$$

where Λ' is the complement of Λ. Written out in terms of membership functions, (17) reads

$$f_{(A,B;\Lambda)}(x) = f_\Lambda(x) f_A(x) + [1 - f_\Lambda(x)] f_B(x), \qquad x \in X. \tag{18}$$

A basic property of the convex combination of A, B, and Λ is expressed by

$$A \cap B \subset (A, B; \Lambda) \subset A \cup B \qquad \text{for all } \Lambda. \tag{19}$$

[4] The dual of the algebraic product is the *sum* $A \oplus B = (A'B')' = A + B - AB$. (This was pointed out by T. Cover.) Note that for ordinary sets ⋂ and the algebraic product are equivalent operations, as are ⋃ and ⊕.

This property is an immediate consequence of the inequalities

$$\text{Min}\,[f_A(x), f_B(x)] \leqq \lambda f_A(x) + (1 - \lambda) f_B(x)$$
$$\leqq \text{Max}\,[f_A(x), f_B(x)], \qquad x \in X \quad (20)$$

which hold for all λ in [0, 1]. It is of interest to observe that, given any fuzzy set C satisfying $A \cap B \subset C \subset A \cup B$, one can always find a fuzzy set Λ such that $C = (A, B; \Lambda)$. The membership function of this set is given by

$$f_\Lambda(x) = \frac{f_C(x) - f_B(x)}{f_A(x) - f_B(x)}, \qquad x \in X. \quad (21)$$

Fuzzy relation. The concept of a *relation* (which is a generalization of that of a *function*) has a natural extension to fuzzy sets and plays an important role in the theory of such sets and their applications—just as it does in the case of ordinary sets. In the sequel, we shall merely define the notion of a fuzzy relation and touch upon a few related concepts.

Ordinarily, a relation is defined as a set of ordered pairs (Halmos, 1960); e.g., the set of all ordered pairs of real numbers x and y such that $x \geqq y$. In the context of fuzzy sets, a *fuzzy relation in X* is a fuzzy set in the product space $X \times X$. For example, the relation denoted by $x \gg y$, $x, y \in R^1$, may be regarded as a fuzzy set A in R^2, with the membership function of A, $f_A(x, y)$, having the following (subjective) representative values: $f_A(10, 5) = 0$; $f_A(100, 10) = 0.7$; $f_A(100, 1) = 1$; etc.

More generally, one can define an *n-ary fuzzy relation* in X as a fuzzy set A in the product space $X \times X \times \cdots \times X$. For such relations, the membership function is of the form $f_A(x_1, \cdots, x_n)$, where $x_i \in X$, $i = 1, \cdots, n$.

In the case of binary fuzzy relations, the *composition* of two fuzzy relations A and B is denoted by $B \circ A$ and is defined as a fuzzy relation in X whose membership function is related to those of A and B by

$$f_{B \circ A}(x, y) = \text{Sup}_v\, \text{Min}\,[f_A(x, v), f_B(v, y)].$$

Note that the operation of composition has the associative property

$$A \circ (B \circ C) = (A \circ B) \circ C.$$

Fuzzy sets induced by mappings. Let T be a mapping from X to a space Y. Let B be a fuzzy set in Y with membership function $f_B(y)$. The inverse mapping T^{-1} induces a fuzzy set A in X whose membership function is defined by

$$f_A(x) = f_B(y), \qquad y \in Y \quad (22)$$

for all x in X which are mapped by T into y.

Consider now a converse problem in which A is a given fuzzy set in X, and T, as before, is a mapping from X to Y. The question is: What is the membership function for the fuzzy set B in Y which is induced by this mapping?

If T is not one-one, then an ambiguity arises when two or more distinct points in X, say x_1 and x_2, with different grades of membership in A, are mapped into the same point y in Y. In this case, the question is: What grade of membership in B should be assigned to y?

To resolve this ambiguity, we agree to assign the larger of the two grades of membership to y. More generally, the membership function for B will be defined by

$$f_B(y) = \text{Max}_{x \in T^{-1}(y)} f_A(x), \qquad y \in Y \quad (23)$$

where $T^{-1}(y)$ is the set of points in X which are mapped into y by T.

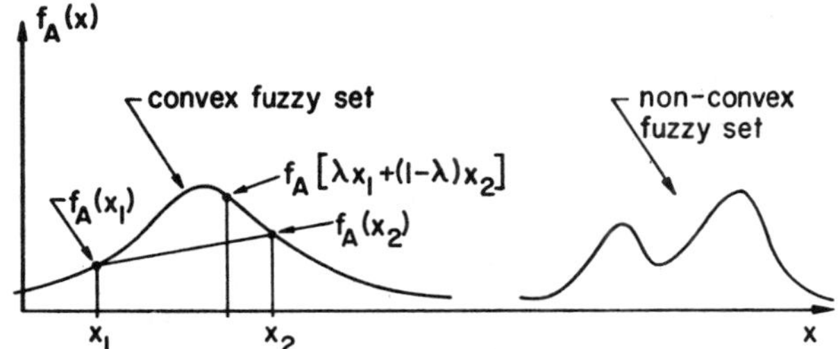

FIG. 4. Convex and nonconvex fuzzy sets in E^1

V. CONVEXITY

As will be seen in the sequel, the notion of convexity can readily be extended to fuzzy sets in such a way as to preserve many of the properties which it has in the context of ordinary sets. This notion appears to be particularly useful in applications involving pattern classification, optimization and related problems.

In what follows, we assume for concreteness that X is a real Euclidean space E^n.

DEFINITIONS

Convexity. A fuzzy set A is *convex* if and only if the sets Γ_α defined by

$$\Gamma_\alpha = \{x \mid f_A(x) \geqq \alpha\} \tag{24}$$

are convex for all α in the interval (0, 1].

An alternative and more direct definition of convexity is the following[5]: A is *convex* if and only if

$$f_A[\lambda x_1 + (1 - \lambda)x_2] \geqq \text{Min}\,[f_A(x_1), f_A(x_2)] \tag{25}$$

for all x_1 and x_2 in X and all λ in [0, 1]. Note that this definition does not imply that $f_A(x)$ must be a convex function of x. This is illustrated in Fig. 4 for $n = 1$.

To show the equivalence between the above definitions note that if A is convex in the sense of the first definition and $\alpha = f_A(x_1) \leqq f_A(x_2)$, then $x_2 \in \Gamma_\alpha$ and $\lambda x_1 + (1 - \lambda)x_2 \in \Gamma_\alpha$ by the convexity of Γ_α. Hence

$$f_A[\lambda x_1 + (1 - \lambda)x_2] \geqq \alpha = f_A(x_1) = \text{Min}\,[f_A(x_1), f_A(x_2)].$$

Conversely, if A is convex in the sense of the second definition and $\alpha = f_A(x_1)$, then Γ_α may be regarded as the set of all points x_2 for which $f_A(x_2) \geqq f_A(x_1)$. In virtue of (25), every point of the form $\lambda x_1 + (1 - \lambda)x_2$, $0 \leqq \lambda \leqq 1$, is also in Γ_α and hence Γ_α is a convex set. Q.E.D.

A basic property of convex fuzzy sets is expressed by the

THEOREM. *If A and B are convex, so is their intersection.*

Proof: Let $C = A \cap B$. Then

$$f_C[\lambda x_1 + (1 - \lambda)x_2] = \text{Min}\,[f_A[\lambda x_1 + (1 - \lambda)x_2], f_B[\lambda x_1 + (1 - \lambda)x_2]]. \tag{26}$$

Now, since A and B are convex

$$\begin{aligned} f_A[\lambda x_1 + (1 - \lambda)x_2] &\geqq \text{Min}\,[f_A(x_1), f_A(x_2)] \\ f_B[\lambda x_1 + (1 - \lambda)x_2] &\geqq \text{Min}\,[f_B(x_1), f_B(x_2)] \end{aligned} \tag{27}$$

and hence

$$f_C[\lambda x_1 + (1 - \lambda)x_2] \geqq \text{Min}\,[\text{Min}\,[f_A(x_1), f_A(x_2)], \text{Min}\,[f_B(x_1), f_B(x_2)]] \tag{28}$$

[5] This way of expressing convexity was suggested to the writer by his colleague, E. Berlekamp.

or equivalently

$$f_C[\lambda x_1 + (1-\lambda)x_2] \geqq \text{Min}\,[\text{Min}\,[f_A(x_1), f_B(x_1)], \text{Min}\,[f_A(x_2), f_B(x_2)]] \tag{29}$$

and thus

$$f_C[\lambda x_1 + (1-\lambda)x_2] \geqq \text{Min}\,[f_C(x_1), f_C(x_2)]. \quad \text{Q. E. D.} \tag{30}$$

Boundedness. A fuzzy set A is *bounded* if and only if the sets $\Gamma_\alpha = \{x \mid f_A(x) \geqq \alpha\}$ are bounded for all $\alpha > 0$; that is, for every $\alpha > 0$ there exists a finite $R(\alpha)$ such that $\| x \| \leqq R(\alpha)$ for all x in Γ_α.

If A is a bounded set, then for each $\epsilon > 0$ then exists a hyperplane H such that $f_A(x) \leqq \epsilon$ for all x on the side of H which does not contain the origin. For, consider the set $\Gamma_\epsilon = \{x \mid f_A(x) \geqq \epsilon\}$. By hypothesis, this set is contained in a sphere S of radius $R(\epsilon)$. Let H be any hyperplane supporting S. Then, all points on the side of H which does not contain the origin lie outside or on S, and hence for all such points $f_A(x) \leqq \epsilon$.

LEMMA. *Let A be a bounded fuzzy set and let $M = \text{Sup}_x f_A(x)$. (M will be referred to as the* maximal grade *in A.) Then there is at least one point x_0 at which M is essentially attained in the sense that, for each $\epsilon > 0$, every spherical neighborhood of x_0 contains points in the set $Q(\epsilon) = \{x \mid f_A(x) \geqq M - \epsilon\}$.*

Proof.[6] Consider a nested sequence of bounded sets $\Gamma_1, \Gamma_2, \cdots$, where $\Gamma_n = \{x \mid f_A(x) \geqq M - M/(n+1)\}$, $n = 1, 2, \cdots$. Note that Γ_n is nonempty for all finite n as a consequence of the definition of M as $M = \text{Sup}_x f_A(x)$. (We assume that $M > 0$.)

Let x_n be an arbitrarily chosen point in Γ_n, $n = 1, 2, \cdots$. Then, $x_1, x_2, \cdots$, is a sequence of points in a closed bounded set Γ_1. By the Bolzano-Weierstrass theorem, this sequence must have at least one limit point, say x_0, in Γ_1. Consequently, every spherical neighborhood of x_0 will contain infinitely many points from the sequence $x_1, x_2, \cdots$, and, more particularly, from the subsequence $x_{N+1}, x_{N+2}, \cdots$, where $N \geqq M/\epsilon$. Since the points of this subsequence fall within the set $Q(\epsilon) = \{x \mid f_A(x) \geqq M - \epsilon\}$, the lemma is proved.

Strict and strong convexity. A fuzzy set A is *strictly convex* if the sets Γ_α, $0 < \alpha \leqq 1$ are strictly convex (that is, if the midpoint of any two distinct points in Γ_α lies in the interior of Γ_α). Note that this definition reduces to that of strict convexity for ordinary sets when A is such a set.

A fuzzy set A is *strongly convex* if, for any two distinct points x_1 and x_2, and any λ in the open interval $(0, 1)$

$$f_A[\lambda x_1 + (1-\lambda)x_2] > \text{Min}\,[f_A(x_1), f_A(x_2)].$$

Note that strong convexity does not imply strict convexity or vice-versa. Note also that if A and B are bounded, so is their union and intersection. Similarly, if A and B are strictly (strongly) convex, their intersection is strictly (strongly) convex.

Let A be a convex fuzzy set and let $M = \text{Sup}_x f_A(x)$. If A is bounded, then, as shown above, either M is attained for some x, say x_0, or there is at least one point x_0 at which M is essentially attained in the sense that, for each $\epsilon > 0$, every spherical neighborhood of x_0 contains points in the set $Q(\epsilon) = \{x \mid M - f_A(x) \leqq \epsilon\}$. In particular, if A is strongly convex and x_0 is attained, then x_0 is unique. For, if $M = f_A(x_0)$ and $M = f_A(x_1)$, with $x_1 \neq x_0$, then $f_A(x) > M$ for $x = 0.5x_0 + 0.5x_1$, which contradicts $M = \text{Max}_x f_A(x)$.

More generally, let $C(A)$ be the set of all points in X at which M is essentially attained. This set will be referred to as the *core* of A. In the case of convex fuzzy sets, we can assert the following property of $C(A)$.

[6] This proof was suggested by A. J. Thomasian.

THEOREM. *If A is a convex fuzzy set, then its core is a convex set.*

Proof: It will suffice to show that if M is essentially attained at x_0 and x_1, $x_1 \neq x_0$, then it is also essentially attained at all x of the form $x = \lambda x_0 + (1 - \lambda)x_1$, $0 \leqq \lambda \leqq 1$.

To the end, let P be a cylinder of radius ϵ with the line passing through x_0 and x_1 as its axis. Let x_0' be a point in a sphere of radius ϵ centering on x_0 and x_1' be a point in a sphere of radius ϵ centering on x_1 such that $f_A(x_0') \geqq M - \epsilon$ and $f_A(x_1') \geqq M - \epsilon$. Then, by the convexity of A, for any point u on the segment $x_0'x_1'$, we have $f_A(u) \geqq M - \epsilon$. Furthermore, by the convexity of P, all points on $x_0'x_1'$ will lie in P.

Now let x be any point in the segment x_0x_1. The distance of this point from the segment $x_0'x_1'$ must be less than or equal to ϵ, since $x_0'x_1'$ lies in P. Consequently, a sphere of radius ϵ centering on x will contain at least one point of the segment $x_0'x_1'$ and hence will contain at least one point, say w, at which $f_A(w) \geqq M - \epsilon$. This establishes that M is essentially attained at x and thus proves the theorem.

COROLLARY. *If $X = E^1$ and A is strongly convex, then the point at which M is essentially attained is unique.*

Shadow of a fuzzy set. Let A be a fuzzy set in E^n with membership function $f_A(x) = f_A(x_1, \cdots, x_n)$. For notational simplicity, the notion of the *shadow* (projection) of A on a hyperplane H will be defined below for the special case where H is a coordinate hyperplane, e.g., $H = \{x \mid x_1 = 0\}$.

Specifically, the *shadow* of A on $H = \{x \mid x_1 = 0\}$ is defined to be a fuzzy set $S_H(A)$ in E^{n-1} with $f_{S_H(A)}(x)$ given by

$$f_{S_H(A)}(x) = f_{S_H(A)}(x_2, \cdots, x_n) = \operatorname{Sup}_{x_1} f_A(x_1, \cdots, x_n).$$

Note that this definition is consistent with (23).

When A is a convex fuzzy set, the following property of $S_H(A)$ is an immediate consequence of the above definition: If A is a convex fuzzy set, then its shadow on any hyperplane is also a convex fuzzy set.

An interesting property of the shadows of two convex fuzzy sets is expressed by the following implication

$$S_H(A) = S_H(B) \text{ for all } H \Rightarrow A = B.$$

To prove this assertion,[7] it is sufficient to show that if there exists a point, say x_0, such that $f_A(x_0) \neq f_B(x_0)$, then their exists a hyperplane H such that $f_{S_H(A)}(x_0^*) \neq f_{S_H(B)}(x_0^*)$, where x_0^* is the projection of x_0 on H.

Suppose that $f_A(x_0) = \alpha > f_B(x_0) = \beta$. Since B is a convex fuzzy set, the set $\Gamma_\beta = \{x \mid f_B(x) > \beta\}$ is convex, and hence there exists a hyperplane F supporting Γ_β and passing through x_0. Let H be a hyperplane orthogonal to F, and let x_0^* be the projection of x_0 on H. Then, since $f_B(x) \leqq \beta$ for all x on F, we have $f_{S_H(B)}(x_0^*) \leqq \beta$. On the other hand, $f_{S_H(A)}(x_0^*) \geqq \alpha$. Consequently, $f_{S_H(B)}(x_0^*) \neq f_{S_H(A)}(x_0^*)$, and similarly for the case where $\alpha < \beta$.

A somewhat more general form of the above assertion is the following: Let A, but not necessarily B, be a convex fuzzy set, and let $S_H(A) = S_H(B)$ for all H. Then $A = \operatorname{conv} B$, where $\operatorname{conv} B$ is the convex hull of B, that is, the smallest convex set containing B. More generally, $S_H(A) = S_H(B)$ for all H implies $\operatorname{conv} A = \operatorname{conv} B$.

Separation of convex fuzzy sets. The classical separation theorem for ordinary convex sets states, in essence, that if A and B are disjoint convex sets, then there exists a separating hyperplane H such that A is on one side of H and B is on the other side.

It is natural to inquire if this theorem can be extended to convex fuzzy

[7] This proof is based on an idea suggested by G. Dantzig for the case where A and B are ordinary convex sets.

sets, without requiring that A and B be disjoint, since the condition of disjointness is much too restrictive in the case of fuzzy sets. It turns out, as will be seen in the sequel, that the answer to this question is in the affirmative.

As a preliminary, we shall have to make a few definitions. Specifically, let A and B be two bounded fuzzy sets and let H be a hypersurface in E^n defined by an equation $h(x) = 0$, with all points for which $h(x) \geqq 0$ being on one side of H and all points for which $h(x) \leqq 0$ being on the other side.[8] Let K_H be a number dependent on H such that $f_A(x) \leqq K_H$ on one side of H and $f_B(x) \leqq K_H$ on the other side. Let M_H be Inf K_H. The number $D_H = 1 - M_H$ will be called the *degree of separation of A and B by H.*

In general, one is concerned not with a given hypersurface H, but with a family of hypersurfaces $\{H_\lambda\}$, with λ ranging over, say, E^m. The problem, then, is to find a member of this family which realizes the highest possible degree of separation.

A special case of this problem is one where the H_λ are hyperplanes in E^n, with λ ranging over E^n. In this case, we define the *degree of separability* of A and B by the relation

$$D = 1 - \bar{M} \tag{31}$$

where

$$\bar{M} = \mathrm{Inf}_H M_H \tag{32}$$

with the subscript λ omitted for simplicity.

Among the various assertions that can be made concerning D, the following statement[9] is, in effect, an extension of the separation theorem to convex fuzzy sets.

THEOREM. *Let A and B be bounded convex fuzzy sets in E^n, with maximal grades M_A and M_B, respectively* [$M_A = \mathrm{Sup}_x f_A(x)$, $M_B = \mathrm{Sup}_x f_B(x)$]. *Let M be the maximal grade for the intersection $A \cap B$ ($M = \mathrm{Sup}_x \mathrm{Min} \cdot [f_A(x), f_B(x)]$). Then $D = 1 - M$.*

Comment. In plain words, the theorem states that the highest degree of separation of two convex fuzzy sets A and B that can be achieved with a hyperplane in E^n is one minus the maximal grade in the intersection $A \cap B$. This is illustrated in Fig. 5 for $n = 1$.

Proof: It is convenient to consider separately the following two cases: (1) $M = \mathrm{Min}\,(M_A, M_B)$ and (2) $M < \mathrm{Min}\,(M_A, M_B)$. Note that the latter case rules out $A \subset B$ or $B \subset A$.

Case 1. For concreteness, assume that $M_A < M_B$, so that $M = M_A$. Then, by the property of bounded sets already stated there exists a hyperplane H such that $f_B(x) \leqq M$ for all x on one side of H. On the other side of H, $f_A(x) \leqq M$ because $f_A(x) \leqq M_A = M$ for all x.

It remains to be shown that there do not exist an $M' < M$ and a hyperplane H' such that $f_A(x) \leqq M'$ on one side of H' and $f_B(x) \leqq M'$ on the other side.

This follows at once from the following observation. Suppose that such H' and M' exist, and assume for concreteness that the core of A (that is, the set of points at which $M_A = M$ is essentially attained) is on the plus side of H'. This rules out the possibility that $f_A(x) \leqq M'$ for all x on the plus side of H', and hence necessitates that $f_A(x) \leqq M'$ for all x on the minus side of H', and $f_B(x) \leqq M'$ for all x on the plus side of H'. Consequently, over all x on the plus side of H'

$$\mathrm{Sup}_x \mathrm{Min}\,[f_A(x), f_B(x)] \leqq M'$$

and likewise for all x on the minus side of H'. This implies that, over all

[8] Note that the sets in question have H in common.

[9] This statement is based on a suggestion of E. Berlekamp.

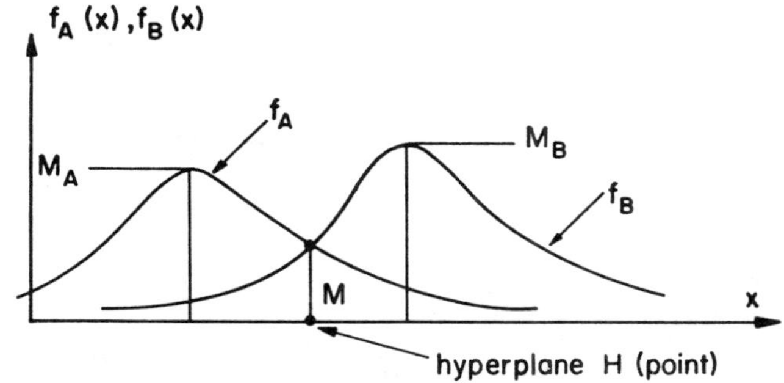

FIG. 5. Illustration of the separation theorem for fuzzy sets in E^1

x in X, $\text{Sup}_x \text{ Min } [f_A(x), f_B(x)] \leqq M'$, which contradicts the assumption that $\text{Sup}_x \text{ Min } [f_A(x), f_B(x)] = M > M'$.

Case 2. Consider the convex sets $\Gamma_A = \{x \mid f_A(x) > M\}$ and $\Gamma_B = \{x \mid f_B(x) > M\}$. These sets are nonempty and disjoint, for if they were not there would be a point, say u, such that $f_A(u) > M$ and $f_B(u) > M$, and hence $f_{A \cap B}(u) > M$, which contradicts the assumption that $M = \text{Sup}_x f_{A \cap B}(x)$.

Since Γ_A and Γ_B are disjoint, by the separation theorem for ordinary convex sets there exists a hyperplane H such that Γ_A is on one side of H (say, the plus side) and Γ_B is on the other side (the minus side). Furthermore, by the definitions of Γ_A and Γ_B, for all points on the minus side of $H, f_A(x) \leqq M$, and for all points on the plus side of $H, f_B(x) \leqq M$.

Thus, we have shown that there exists a hyperplane H which realizes $1 - M$ as the degree of separation of A and B. The conclusion that a higher degree of separation of A and B cannot be realized follows from the argument given in Case 1. This concludes the proof of the theorem.

The separation theorem for convex fuzzy sets appears to be of particular relevance to the problem of pattern discrimination. Its application to this class of problems as well as to problems of optimization will be explored in subsequent notes on fuzzy sets and their properties.

RECEIVED: November 30, 1964

REFERENCES

BIRKHOFF, G. (1948), "Lattice Theory," Am. Math. Soc. Colloq. Publ., Vol. 25, New York.

HALMOS, P. R. (1960), "Naive Set Theory." Van Nostrand, New York.

KLEENE, S. C. (1952), "Introduction to Metamathematics," p. 334. Van Nostrand, New York.

Outline of a New Approach to the Analysis of Complex Systems and Decision Processes

LOTFI A. ZADEH

***Abstract*—The approach described in this paper represents a substantive departure from the conventional quantitative techniques of system analysis. It has three main distinguishing features: 1) use of so-called "linguistic" variables in place of or in addition to numerical variables; 2) characterization of simple relations between variables by fuzzy conditional statements; and 3) characterization of complex relations by fuzzy algorithms.**

A *linguistic variable* is defined as a variable whose values are sentences in a natural or artificial language. Thus, if *tall, not tall, very tall, very very tall,* etc. are values of *height*, then *height* is a linguistic variable. *Fuzzy conditional statements* are expressions of the form IF A THEN B, where A and B have fuzzy meaning, e.g., IF x is *small* THEN y is *large*, where *small* and *large* are viewed as labels of fuzzy sets. A *fuzzy algorithm* is an ordered sequence of instructions which may contain fuzzy assignment and conditional statements, e.g., $x =$ *very small*, IF x is *small* THEN y is *large*. The execution of such instructions is governed by the *compositional rule of inference* and the *rule of the preponderant alternative*.

By relying on the use of linguistic variables and fuzzy algorithms, the approach provides an approximate and yet effective means of describing the behavior of systems which are too complex or too ill-defined to admit of precise mathematical analysis. Its main applications lie in economics, management science, artificial intelligence, psychology, linguistics, information retrieval, medicine, biology, and other fields in which the dominant role is played by the animate rather than inanimate behavior of system constituents.

I. INTRODUCTION

THE ADVENT of the computer age has stimulated a rapid expansion in the use of quantitative techniques for the analysis of economic, urban, social, biological, and other types of systems in which it is the animate rather than inanimate behavior of system constituents that plays a dominant role. At present, most of the techniques employed for the analysis of *humanistic*, i.e., human-centered, systems are adaptations of the methods that have been developed over a long period of time for dealing with *mechanistic* systems, i.e., physical systems governed in the main by the laws of mechanics, electromagnetism, and thermodynamics. The remarkable successes of these methods in unraveling the secrets of nature and enabling us to build better and better machines have inspired a widely held belief that the same or similar techniques can be applied with comparable effectiveness to the analysis of humanistic systems. As a case in point, the successes of modern control theory in the design of highly accurate space navigation systems have stimulated its use in the theoretical analyses of economic and biological systems. Similarly, the effectiveness of computer simulation techniques in the macroscopic analyses of physical systems has brought into vogue the use of computer-based econometric models for purposes of forecasting, economic planning, and management.

Given the deeply entrenched tradition of scientific thinking which equates the understanding of a phenomenon with the ability to analyze it in quantitative terms, one is certain to strike a dissonant note by questioning the growing tendency to analyze the behavior of humanistic systems as if they were mechanistic systems governed by difference, differential, or integral equations. Such a note is struck in the present paper.

Essentially, our contention is that the conventional quantitative techniques of system analysis are intrinsically unsuited for dealing with humanistic systems or, for that matter, any system whose complexity is comparable to that of humanistic systems. The basis for this contention rests on what might be called the *principle of incompatibility*. Stated informally, the essence of this principle is that as the complexity of a system increases, our ability to make precise and yet significant statements about its behavior diminishes until a threshold is reached beyond which precision and significance (or relevance) become almost mutually exclusive characteristics.[1] It is in this sense that precise quantitative analyses of the behavior of humanistic systems are not likely to have much relevance to the real-world societal, political, economic, and other types of problems which involve humans either as individuals or in groups.

An alternative approach outlined in this paper is based on the premise that the key elements in human thinking are not numbers, but labels of fuzzy sets, that is, classes of objects in which the transition from membership to nonmembership is gradual rather than abrupt. Indeed, the pervasiveness of fuzziness in human thought processes suggests that much of the logic behind human reasoning is not the traditional two-valued or even multivalued logic, but a logic with fuzzy truths, fuzzy connectives, and fuzzy rules of inference. In our view, it is this fuzzy, and as yet not well-understood, logic that plays a basic role in what may well be one of the most important facets of human thinking, namely, the ability to *summarize* information—to extract from the collections of masses of data impinging

Manuscript received August 1, 1972; revised August 13, 1972. This work was supported by the Navy Electronic Systems Command under Contract N00039-71-C-0255, the Army Research Office, Durham, N.C., under Grant DA-ARO-D-31-124-71-G174, and NASA under Grant NGL-05-003-016-VP3.

The author is with the Department of Electrical Engineering and Computer Sciences and Electronics Research Laboratory, University of California, Berkeley, Calif. 94720.

[1] A corollary principle may be stated succinctly as, "The closer one looks at a real-world problem, the fuzzier becomes its solution."

Reprinted from *IEEE Trans. Syst., Man., Cybern.*, vol. SMC-3, no. 1, pp. 28–44, Jan. 1973.

upon the human brain those and only those subcollections which are relevant to the performance of the task at hand.

By its nature, a summary is an approximation to what it summarizes. For many purposes, a very approximate characterization of a collection of data is sufficient because most of the basic tasks performed by humans do not require a high degree of precision in their execution. The human brain takes advantage of this tolerance for imprecision by encoding the "task-relevant" (or "decision-relevant") information into labels of fuzzy sets which bear an approximate relation to the primary data. In this way, the stream of information reaching the brain via the visual, auditory, tactile, and other senses is eventually reduced to the trickle that is needed to perform a specified task with a minimal degree of precision. Thus, the ability to manipulate fuzzy sets and the consequent summarizing capability constitute one of the most important assets of the human mind as well as a fundamental characteristic that distinguishes human intelligence from the type of machine intelligence that is embodied in present-day digital computers.

Viewed in this perspective, the traditional techniques of system analysis are not well suited for dealing with humanistic systems because they fail to come to grips with the reality of the fuzziness of human thinking and behavior. Thus, to deal with such systems realistically, we need approaches which do not make a fetish of precision, rigor, and mathematical formalism, and which employ instead a methodological framework which is tolerant of imprecision and partial truths. The approach described in the sequel is a step—but not necessarily a definitive step—in this direction.

The approach in question has three main distinguishing features: 1) use of so-called "linguistic" variables in place of or in addition to numerical variables; 2) characterization of simple relations between variables by conditional fuzzy statements; and 3) characterization of complex relations by fuzzy algorithms. Before proceeding to a detailed discussion of our approach, it will be helpful to sketch the principal ideas behind these features. We begin with a brief explanation of the notion of a linguistic variable.

1) *Linguistic and Fuzzy Variables:* As already pointed out, the ability to summarize information plays an essential role in the characterization of complex phenomena. In the case of humans, the ability to summarize information finds its most pronounced manifestation in the use of natural languages. Thus, each word x in a natural language L may be viewed as a summarized description of a fuzzy subset $M(x)$ of a universe of discourse U, with $M(x)$ representing the meaning of x. In this sense, the language as a whole may be regarded as a system for assigning atomic and composite labels (i.e., words, phrases, and sentences) to the fuzzy subsets of U. (This point of view is discussed in greater detail in [4] and [5].) For example, if the meaning of the noun *flower* is a fuzzy subset $M(\textit{flower})$, and the meaning of the adjective *red* is a fuzzy subset $M(\textit{red})$, then the meaning of the noun phrase *red flower* is given by the intersection of $M(\textit{red})$ and $M(\textit{flower})$.

If we regard the color of an object as a variable, then its values, *red*, *blue*, *yellow*, *green*, etc., may be interpreted as labels of fuzzy subsets of a universe of objects. In this sense, the attribute *color* is a *fuzzy variable*, that is, a variable whose values are labels of fuzzy sets. It is important to note that the characterization of a value of the variable *color* by a natural label such as *red* is much less precise than the numerical value of the wavelength of a particular color.

In the preceding example, the values of the variable *color* are atomic terms like *red*, *blue*, *yellow*, etc. More generally, the values may be sentences in a specified language, in which case we say that the variable is *linguistic*. To illustrate, the values of the fuzzy variable *height* might be expressible as *tall*, *not tall*, *somewhat tall*, *very tall*, *not very tall*, *very very tall*, *tall but not very tall*, *quite tall*, *more or less tall*. Thus, the values in question are sentences formed from the label *tall*, the negation *not*, the connectives *and* and *but*, and the hedges *very*, *somewhat*, *quite*, and *more or less*. In this sense, the variable *height* as defined above is a linguistic variable.

As will be seen in Section III, the main function of linguistic variables is to provide a systematic means for an approximate characterization of complex or ill-defined phenomena. In essence, by moving away from the use of quantified variables and toward the use of the type of linguistic descriptions employed by humans, we acquire a capability to deal with systems which are much too complex to be susceptible to analysis in conventional mathematical terms.

2) *Characterization of Simple Relations Between Fuzzy Variables by Conditional Statements:* In quantitative approaches to system analysis, a dependence between two numerically valued variables x and y is usually characterized by a table which, in words, may be expressed as a set of conditional statements, e.g., IF x is 5 THEN y is 10, IF x is 6 THEN y is 14, etc.

The same technique is employed in our approach, except that x and y are allowed to be fuzzy variables. In particular, if x and y are linguistic variables, the conditional statements describing the dependence of y on x might read (the following italicized words represent the values of fuzzy variables):

IF x is *small* THEN y is *very large*
IF x is *not very small* THEN y is *very very large*
IF x is *not small and not large* THEN y is *not very large*

and so forth.

Fuzzy conditional statements of the form IF A THEN B, where A and B are terms with a fuzzy meaning, e.g., "IF John is *nice* to you THEN you should be *kind* to him," are used routinely in everyday discourse. However, the meaning of such statements when used in communication between humans is poorly defined. As will be shown in Section V, the conditional statement IF A THEN B can be given a precise meaning even when A and B are fuzzy rather than nonfuzzy sets, provided the meanings of A and B are defined precisely as specified subsets of the universe of discourse.

In the preceding example, the relation between two fuzzy variables x and y is *simple* in the sense that it can be characterized as a set of conditional statements of the form IF A THEN B, where A and B are labels of fuzzy sets representing the values of x and y, respectively. In the case of more complex relations, the characterization of the dependence of y on x may require the use of a fuzzy algorithm. As indicated below, and discussed in greater detail in Section VI, the notion of a fuzzy algorithm plays a basic role in providing a means of approximate characterization of fuzzy concepts and their interrelations.

3) *Fuzzy-Algorithmic Characterization of Functions and Relations:* The definition of a fuzzy function through the use of fuzzy conditional statements is analogous to the definition of a nonfuzzy function f by a table of pairs $(x,f(x))$, in which x is a generic value of the argument of f and $f(x)$ is the value of the function. Just as a nonfuzzy function can be defined algorithmically (e.g., by a program) rather than by a table, so a fuzzy function can be defined by a fuzzy algorithm rather than as a collection of fuzzy conditional statements. The same applies to the definition of sets, relations, and other constructs which are fuzzy in nature.

Essentially, a fuzzy algorithm [6] is an ordered sequence of instructions (like a computer program) in which some of the instructions may contain labels of fuzzy sets, e.g.:

Reduce x *slightly* if y is *large*
Increase x *very slightly* if y is *not very large and not very small*
If x is *small* then stop; otherwise increase x by 2.

By allowing an algorithm to contain instructions of this type, it becomes possible to give an approximate fuzzy-algorithmic characterization of a wide variety of complex phenomena. The important feature of such characterizations is that, though imprecise in nature, they may be perfectly adequate for the purposes of a specified task. In this way, fuzzy algorithms can provide an effective means of approximate description of objective functions, constraints, system performance, strategies, etc.

In what follows, we shall elaborate on some of the basic aspects of linguistic variables, fuzzy conditional statements, and fuzzy algorithms. However, we shall not attempt to present a definitive exposition of our approach and its applications. Thus, the present paper should be viewed primarily as an introductory outline of a method which departs from the tradition of precision and rigor in scientific analysis—a method whose approximate nature mirrors the fuzziness of human behavior and thereby offers a promise of providing a more realistic basis for the analysis of humanistic systems.

As will be seen in the following sections, the theoretical foundation of our approach is actually quite precise and rather mathematical in spirit. Thus, the source of imprecision in the approach is not the underlying theory, but the manner in which linguistic variables and fuzzy algorithms are applied to the formulation and solution of real-world problems. In effect, the level of precision in a particular application can be adjusted to fit the needs of the task and the accuracy of the available data. This flexibility constitutes one of the important features of the method that will be described.

II. Fuzzy Sets: A Summary of Relevant Properties

In order to make our exposition self-contained, we shall summarize in this section those properties of fuzzy sets which will be needed in later sections. (More detailed discussions of topics in the theory of fuzzy sets which are relevant to the subject of the present paper may be found in [1]–[17].)

Notation and Terminology

A fuzzy subset A of a universe of discourse U is characterized by a membership function $\mu_A: U \to [0,1]$ which associates with each element y of U a number $\mu_A(y)$ in the interval $[0,1]$ which represents the grade of membership of y in A. The *support* of A is the set of points in U at which $\mu_A(y)$ is positive. A *crossover point* in A is an element of U whose grade of membership in A is 0.5. A *fuzzy singleton* is a fuzzy set whose support is a single point in U. If A is a fuzzy singleton whose support is the point y, we write

$$A = \mu/y \tag{2.1}$$

where μ is the grade of membership of y in A. To be consistent with this notation, a nonfuzzy singleton will be denoted by $1/y$.

A fuzzy set A may be viewed as the union (see (2.27)) of its constituent singletons. On this basis, A may be represented in the form

$$A = \int_U \mu_A(y)/y \tag{2.2}$$

where the integral sign stands for the union of the fuzzy singletons $\mu_A(y)/y$. If A has a finite support $\{y_1,y_2,\cdots,y_n\}$, then (2.2) may be replaced by the summation

$$A = \mu_1/y_1 + \cdots + \mu_n/y_n \tag{2.3}$$

or

$$A = \sum_{i=1}^{n} \mu_i/y_i \tag{2.4}$$

in which μ_i, $i = 1,\cdots,n$, is the grade of membership of y_i in A. It should be noted that the + sign in (2.3) denotes the union (see (2.27)) rather than the arithmetic sum. In this sense of +, a finite universe of discourse $U = \{y_1,y_2,\cdots, y_n\}$ may be represented simply by the summation

$$U = y_1 + y_2 + \cdots + y_n \tag{2.5}$$

or

$$U = \sum_{i=1}^{n} y_i \tag{2.6}$$

although, strictly, we should write (2.5) and (2.6) as

$$U = 1/y_1 + 1/y_2 + \cdots + 1/y_n \tag{2.7}$$

and

$$U = \sum_{i=1}^{n} 1/y_i. \tag{2.8}$$

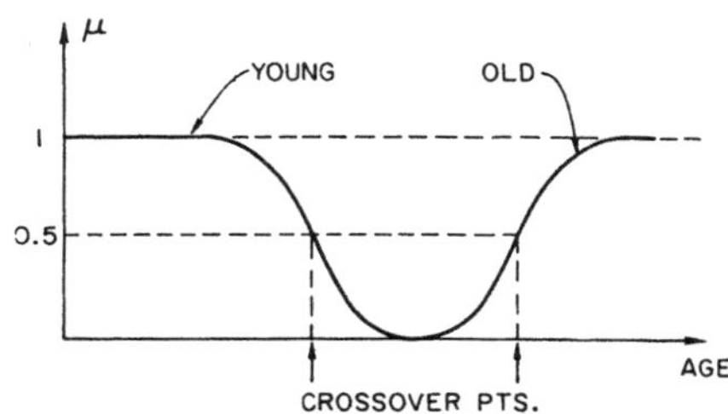

Fig. 1. Diagrammatic representation of *young* and *old*.

As an illustration, suppose that

$$U = 1 + 2 + \cdots + 10. \tag{2.9}$$

Then a fuzzy subset[2] of U labeled *several* may be expressed as (the symbol $\triangleq$ stands for "equal by definition," or "is defined to be," or "denotes")

$$several \triangleq 0.5/3 + 0.8/4 + 1/5 + 1/6 + 0.8/7 + 0.5/8. \tag{2.10}$$

Similarly, if U is the interval [0,100], with $y \triangleq age$, then the fuzzy subsets of U labeled *young* and *old* may be represented as (here and elsewhere in this paper we do not differentiate between a fuzzy set and its label)

$$young = \int_0^{25} 1/y + \int_{25}^{100} \left(1 + \left(\frac{y-25}{5}\right)^2\right)^{-1} /y \tag{2.11}$$

$$old = \int_{50}^{100} \left(1 + \left(\frac{y-50}{5}\right)^{-2}\right)^{-1} /y. \tag{2.12}$$

(see Fig. 1).

The grade of membership in a fuzzy set may itself be a fuzzy set. For example, if

$$U = \text{TOM} + \text{JIM} + \text{DICK} + \text{BOB} \tag{2.13}$$

and A is the fuzzy subset labeled *agile*, then we may have

$$agile = medium/\text{TOM} + low/\text{JIM} + low/\text{DICK} + high/\text{BOB}. \tag{2.14}$$

In this representation, the fuzzy grades of membership *low*, *medium*, and *high* are fuzzy subsets of the universe V

$$V = 0 + 0.1 + 0.2 + \cdots + 0.9 + 1 \tag{2.15}$$

which are defined by

$$low = 0.5/0.2 + 0.7/0.3 + 1/0.4 + 0.7/0.5 + 0.5/0.6 \tag{2.16}$$

$$medium = 0.5/0.4 + 0.7/0.5 + 1/0.6 + 0.7/0.7 + 0.5/0.8 \tag{2.17}$$

$$high = 0.5/0.7 + 0.7/0.8 + 0.9/0.9 + 1/1. \tag{2.18}$$

[2] A is a subset of B, written $A \subset B$, if and only if $\mu_A(y) \le \mu_B(y)$, for all y in U. For example, the fuzzy set $A = 0.6/1 + 0.3/2$ is a subset of $B = 0.8/1 + 0.5/2 + 0.6/3$.

Fuzzy Relations

A *fuzzy relation* R from a set X to a set Y is a fuzzy subset of the Cartesian product $X \times Y$. ($X \times Y$ is the collection of ordered pairs (x,y), $x \in X$, $y \in Y$). R is characterized by a bivariate membership function $\mu_R(x,y)$ and is expressed

$$R \triangleq \int_{X \times Y} \mu_R(x,y)/(x,y). \tag{2.19}$$

More generally, for an *n*ary fuzzy relation R which is a fuzzy subset of $X_1 \times X_2 \times \cdots \times X_n$, we have

$$R \triangleq \int_{X_1 \times \cdots \times X_n} \mu_R(x_1,\cdots,x_n)/(x_1,\cdots,x_n), \quad x_i \in X_i, \quad i = 1,\cdots,n. \tag{2.20}$$

As an illustration, if

$$X = \{\text{TOM, DICK}\} \quad \text{and} \quad Y = \{\text{JOHN, JIM}\}$$

then a binary fuzzy relation of *resemblance* between members of X and Y might be expressed as

$$resemblance = 0.8/(\text{TOM, JOHN}) + 0.6/(\text{TOM, JIM}) + 0.2/(\text{DICK, JOHN}) + 0.9/(\text{DICK, JIM}).$$

Alternatively, this relation may be represented as a *relation matrix*

$$\begin{array}{c} \\ \text{TOM} \\ \text{DICK} \end{array} \begin{array}{c} \text{JOHN} \quad \text{JIM} \\ \begin{bmatrix} 0.8 & 0.6 \\ 0.2 & 0.9 \end{bmatrix} \end{array} \tag{2.21}$$

in which the (i,j)th element is the value of $\mu_R(x,y)$ for the ith value of x and the jth value of y.

If R is a relation from X to Y and S is a relation from Y to Z, then the *composition* of R and S is a fuzzy relation denoted by $R \circ S$ and defined by

$$R \circ S \triangleq \int_{X \times Z} \vee_y (\mu_R(x,y) \wedge \mu_s(y,z))/(x,z) \tag{2.22}$$

where $\vee$ and $\wedge$ denote, respectively, max and min.[3] Thus, for real a,b,

$$a \vee b = \max(a,b) \triangleq \begin{cases} a, & \text{if } a \ge b \\ b, & \text{if } a < b \end{cases} \tag{2.23}$$

$$a \wedge b = \min(a,b) \triangleq \begin{cases} a, & \text{if } a \le b \\ b, & \text{if } a > b \end{cases} \tag{2.24}$$

and $\vee_y$ is the supremum over the domain of y.

If the domains of the variables x, y, and z are finite sets, then the relation matrix for $R \circ S$ is the max–min product[4] of the relation matrices for R and S. For example, the max–min product of the relation matrices on the left-hand side of (2.25) results in the relation matrix $R \circ S$ shown on the

[3] Equation (2.22) defines the max–min composition of R and S. Max–product composition is defined similarly, except that $\wedge$ is replaced by the arithmetic product. A more detailed discussion of these compositions may be found in [2].

[4] In the max–min matrix product, the operations of addition and multiplication are replaced by $\vee$ and $\wedge$, respectively.

right-hand side of

$$\overset{R}{\begin{bmatrix} 0.3 & 0.8 \\ 0.6 & 0.9 \end{bmatrix}} \circ \overset{S}{\begin{bmatrix} 0.5 & 0.9 \\ 0.4 & 1 \end{bmatrix}} = \overset{R \circ S}{\begin{bmatrix} 0.4 & 0.8 \\ 0.5 & 0.9 \end{bmatrix}}. \tag{2.25}$$

Operations on Fuzzy Sets

The negation *not*, the connectives *and* and *or*, the hedges *very*, *highly*, *more or less*, and other terms which enter in the representation of values of linguistic variables may be viewed as labels of various operations defined on the fuzzy subsets of U. The more basic of these operations will be summarized.

The *complement* of A is denoted $\neg A$ and is defined by

$$\neg A \triangleq \int_U (1 - \mu_A(y))/y. \tag{2.26}$$

The operation of complementation corresponds to negation. Thus, if x is a label for a fuzzy set, then *not* x should be interpreted as $\neg x$. (Strictly speaking, $\neg$ operates on fuzzy sets, whereas *not* operates on their labels. With this understanding, we shall use $\neg$ and *not* interchangeably.)

The *union* of fuzzy sets A and B is denoted $A + B$ and is defined by

$$A + B \triangleq \int_U (\mu_A(y) \vee \mu_B(y))/y. \tag{2.27}$$

The union corresponds to the connective *or*. Thus, if u and v are labels of fuzzy sets, then

$$u \text{ or } v \triangleq u + v \tag{2.28}$$

The *intersection* of A and B is denoted $A \cap B$ and is defined by

$$A \cap B \triangleq \int_U (\mu_A(y) \wedge \mu_B(y))/y. \tag{2.29}$$

The intersection corresponds to the connective *and*; thus

$$u \text{ and } v \triangleq u \cap v. \tag{2.30}$$

As an illustration, if

$$U = 1 + 2 + \cdots + 10 \tag{2.31}$$

$$u = 0.8/3 + 1/5 + 0.6/6 \tag{2.32}$$

$$v = 0.7/3 + 1/4 + 0.5/6 \tag{2.33}$$

then

$$u \textit{ or } v = 0.8/3 + 1/4 + 1/5 + 0.6/6 \tag{2.34}$$

$$u \textit{ and } v = 0.7/3 + 0.5/6. \tag{2.35}$$

The *product* of A and B is denoted AB and is defined by

$$AB \triangleq \int_U \mu_A(y)\mu_B(y)/y. \tag{2.36}$$

Thus, if

$$A = 0.8/2 + 0.9/5 \tag{2.37}$$

$$B = 0.6/2 + 0.8/3 + 0.6/5 \tag{2.38}$$

then

$$AB = 0.48/2 + 0.54/5. \tag{2.39}$$

Based on (2.36), A^α, where α is any positive number, is defined by

$$A^\alpha \triangleq \int_U (\mu_A(y))^\alpha/y. \tag{2.40}$$

Similarly, if α is a nonnegative real number, then

$$\alpha A \triangleq \int_U \alpha\mu_A(y)/y. \tag{2.41}$$

As an illustration, if A is expressed by (2.37), then

$$A^2 = 0.64/2 + 0.81/5 \tag{2.42}$$

$$0.5A = 0.4/2 + 0.45/5. \tag{2.43}$$

In addition to the basic operations just defined, there are other operations that are of use in the representation of linguistic hedges. Some of these will be briefly defined. (A more detailed discussion of these operations may be found in [15].)

The operation of *concentration* is defined by

$$\text{CON}(A) \triangleq A^2. \tag{2.44}$$

Applying this operation to A results in a fuzzy subset of A such that the reduction in the magnitude of the grade of membership of y in A is relatively small for those y which have a high grade of membership in A and relatively large for the y with low membership.

The operation of *dilation* is defined by

$$\text{DIL}(A) \triangleq A^{0.5}. \tag{2.45}$$

The effect of this operation is the opposite of that of concentration.

The operation of *contrast intensification* is defined by

$$\text{INT}(A) \triangleq \begin{cases} 2A^2, & \text{for } 0 \le \mu_A(y) \le 0.5 \\ \neg 2(\neg A)^2, & \text{for } 0.5 \le \mu_A(y) \le 1. \end{cases} \tag{2.46}$$

This operation differs from concentration in that it increases the values of $\mu_A(y)$ which are above 0.5 and diminishes those which are below this point. Thus, contrast intensification has the effect of reducing the fuzziness of A. (An entropy-like measure of fuzziness of a fuzzy set is defined in [16].)

As its name implies, the operation of *fuzzification* (or, more specifically, *support fuzzification*) has the effect of transforming a nonfuzzy set into a fuzzy set or increasing the fuzziness of a fuzzy set. The result of application of a fuzzification to A will be denoted by $F(A)$ or $\tilde{A}$, with the wavy overbar referred to as a *fuzzifier*. Thus $x \approx 3$ means "x is approximately equal to 3," while $x = \tilde{3}$ means "x is a fuzzy set which approximates to 3." A fuzzifier F is characterized by its *kernel* $K(y)$, which is the fuzzy set resulting from the application of F to a singleton $1/y$. Thus

$$K(y) \triangleq \widetilde{1/y}. \tag{2.47}$$

In terms of K, the result of applying F to a fuzzy set A is given by

$$F(A; K) \triangleq \int_U \mu_A(y)K(y) \tag{2.48}$$

where $\mu_A(y)K(y)$ represents the product (in the sense of (2.41)) of the scalar $\mu_A(y)$ and the fuzzy set $K(y)$, and $\int_U$ should be interpreted as the union of the family of fuzzy sets $\mu_A(y)K(y)$, $y \in U$. Thus (2.48) is analogous to the integral representation of a linear operator, with $K(y)$ playing the role of impulse response.

As an illustration of (2.48), assume that U, A, and $K(y)$ are defined by

$$U = 1 + 2 + 3 + 4 \qquad (2.49)$$

$$A = 0.8/1 + 0.6/2 \qquad (2.50)$$

$$K(1) = 1/1 + 0.4/2 \qquad (2.51)$$

$$K(2) = 1/2 + 0.4/1 + 0.4/3.$$

Then, the result of applying F to A is given by

$$\begin{aligned} F(A;K) &= 0.8(1/1 + 0.4/2) + 0.6(1/2 + 0.4/1 + 0.4/3) \\ &= 0.8/1 + 0.32/2 + 0.6/2 + 0.24/1 + 0.24/3 \\ &= 0.8/1 + 0.6/2 + 0.24/3. \end{aligned} \qquad (2.52)$$

The operation of fuzzification plays an important role in the definition of linguistic hedges such as *more or less*, *slightly*, *much*, etc. Examples of its uses are given in [15].

Language and Meaning

As was indicated in Section I, the values of a linguistic variable are fuzzy sets whose labels are sentences in a natural or artificial language. For our purposes, a language L may be viewed as a correspondence between a set of terms T and a universe of discourse U. (This point of view is described in greater detail in [4] and [5]. For simplicity, we assume that T is a nonfuzzy set.) This correspondence may be assumed to be characterized by a fuzzy *naming relation* N from T to U, which associates with each term x in T and each object y in U the degree $\mu_N(x,y)$ to which x applies to y. For example, if $x = young$ and $y = 23$ years, then $\mu_N(young, 23)$ might be 0.9. A term may be atomic, e.g., $x =$ *tall*, or composite, in which case it is a concatenation of atomic terms, e.g., $x =$ *very tall man*.

For a fixed x, the membership function $\mu_N(x,y)$ defines a fuzzy subset $M(x)$ of U whose membership function is given by

$$\mu_{M(x)}(y) \triangleq \mu_N(x,y), \qquad x \in T, \quad y \in U. \qquad (2.53)$$

This fuzzy subset is defined to be the *meaning* of x. Thus, the meaning of a term x is the fuzzy subset $M(x)$ of U for which x serves as a label. Although x and $M(x)$ are different entities (x is an element of T, whereas $M(x)$ is a fuzzy subset of U), we shall write x for $M(x)$, except where there is a need for differentiation between them. To illustrate, suppose that the meaning of the term *young* is defined by

$$\mu_N(young,y) = \begin{cases} 1, & \text{for } y \leq 25 \\ \left(1 + \left(\dfrac{y-25}{5}\right)^2\right)^{-1}, & \text{for } y > 25. \end{cases} \qquad (2.54)$$

Then we can represent the fuzzy subset of U labeled *young* as (see (2.11))

$$young = \int_0^{25} 1/y + \int_{25}^{100} \left(1 + \left(\frac{y-25}{5}\right)^2\right)^{-1} \Big/ y \qquad (2.55)$$

with the right-hand member of (2.55) representing the meaning of *young*.

Linguistic hedges such as *very*, *much*, *more or less*, etc., make it possible to modify the meaning of atomic as well as composite terms and thus serve to increase the range of values of a linguistic variable. The use of linguistic hedges for this purpose is discussed in the following section.

III. Linguistic Hedges

As stated in Section II, the values of a linguistic variable are labels of fuzzy subsets of U which have the form of phrases or sentences in a natural or artificial language. For example, if U is the collection of integers

$$U = 0 + 1 + 2 + \cdots + 100 \qquad (3.1)$$

and *age* is a linguistic variable labeled x, then the values of x might be *young*, *not young*, *very young*, *not very young*, *old and not old*, *not very old*, *not young and not old*, etc.

In general, a value of a linguistic variable is a composite term $x = x_1x_2 \cdots x_n$, which is a concatenation of atomic terms $x_1,\cdots,x_n$. These atomic terms may be divided into four categories:

1) *primary terms*, which are labels of specified fuzzy subsets of the universe of discourse (e.g., *young* and *old* in the preceding example);
2) the negation *not* and the connectives *and* and *or*;
3) *hedges*, such as *very*, *much*, *slightly*, *more or less* (although *more or less* is comprised of three words, it is regarded as an atomic term), etc.;
4) *markers*, such as parentheses.

A basic problem P_l which arises in connection with the use of linguistic variables is the following: Given the meaning of each atomic term x_i, $i = 1,\cdots,n$, in a composite term $x = x_1 \cdots x_n$ which represents a value of a linguistic variable, compute the meaning of x in the sense of (2.53). This problem is an instance of a central problem in quantitative fuzzy semantics [4], namely, the computation of the meaning of a composite term. P_l is a special case of the latter problem because the composite terms representing the values of a linguistic variable have a relatively simple grammatical structure which is restricted to the four categories of atomic terms 1)–4).

As a preliminary to describing a general approach to the solution of P_l, it will be helpful to consider a subproblem of P_l which involves the computation of the meaning of a composite term of the form $x = hu$, where h is a hedge and u is a term with a specified meaning; e.g., $x =$ *very tall man*, where $h =$ *very* and $u =$ *tall man*.

Taking the point of view described in [15], a hedge h may be regarded as an operator which transforms the fuzzy set $M(u)$, representing the meaning of u, into the fuzzy set $M(hu)$. As stated already, the hedges serve the function of

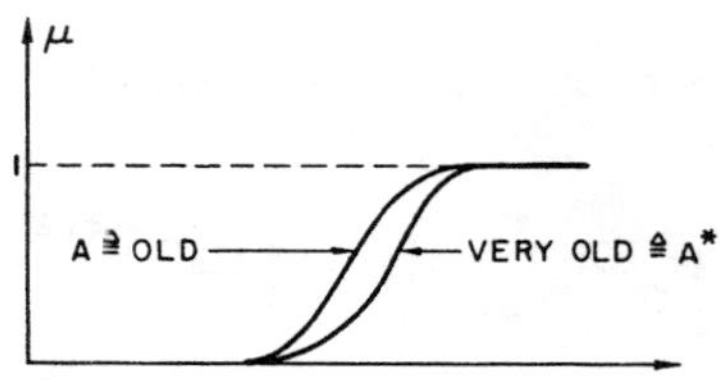

Fig. 2. Effect of hedge *very*.

generating a larger set of values for a linguistic variable from a small collection of primary terms. For example, by using the hedge *very* in conjunction with *not*, *and*, and the primary term *tall*, we can generate the fuzzy sets *very tall*, *very very tall*, *not very tall*, *tall and not very tall*, etc. To define a hedge h as an operator, it is convenient to employ some of the basic operations defined in Section II, especially concentration, dilation, and fuzzification. In what follows, we shall indicate the manner in which this can be done for the natural hedge *very* and the artificial hedges *plus* and *minus*. Characterizations of such hedges as *more or less*, *much*, *slightly*, *sort of*, and *essentially* may be found in [15].

Although in its everyday use the hedge *very* does not have a well-defined meaning, in essence it acts as an intensifier, generating a subset of the set on which it operates. A simple operation which has this property is that of concentration (see (2.44)). This suggests that *very x*, where x is a term, be defined as the square of x, that is

$$\textit{very } x \triangleq x^2 \tag{3.2}$$

or, more explicitly

$$\textit{very } x \triangleq \int_U \mu_x^2(y)/y. \tag{3.3}$$

For example, if (see Fig. 2)

$$x = \textit{old men} \triangleq \int_{50}^{100} \left(1 + \left(\frac{y-50}{5}\right)^{-2}\right)^{-1} \Big/ y \tag{3.4}$$

then

$$x^2 = \textit{very old men} = \int_{50}^{100} \left(1 + \left(\frac{y-50}{5}\right)^{-2}\right)^{-2} \Big/ y. \tag{3.5}$$

Thus, if the grade of membership of JOHN in the class of *old men* is 0.8, then his grade of membership in the class of *very old men* is 0.64. As another simple example, if

$$U = 1 + 2 + 3 + 4 + 5 \tag{3.6}$$

and

$$\textit{small} = 1/1 + 0.8/2 + 0.6/3 + 0.4/4 + 0.2/5 \tag{3.7}$$

then

$$\textit{very small} = 1/1 + 0.64/2 + 0.36/3 + 0.16/4 + 0.04/5. \tag{3.8}$$

Viewed as an operator, *very* can be composed with itself. Thus

$$\textit{very very } x = (\textit{very } x)^2 = x^4. \tag{3.9}$$

For example, applying (3.9) to (3.7), we obtain (neglecting small terms)

$$\textit{very very small} = 1/1 + 0.4/2 + 0.1/3. \tag{3.10}$$

In some instances, to identify the operand of *very* we have to use parentheses or replace a composite term by an atomic one. For example, it is not grammatical to write

$$x = \textit{very not exact} \tag{3.11}$$

but if *not exact* is replaced by the atomic term *inexact*, then

$$x = \textit{very inexact} \tag{3.12}$$

is grammatically correct and we can write

$$x = (\neg \textit{exact})^2. \tag{3.13}$$

Note that

$$\textit{not very exact} = \neg(\text{very exact}) = \neg(\textit{exact}^2) \tag{3.14}$$

is not the same as (3.13).

The artificial hedges *plus* and *minus* serve the purpose of providing milder degrees of concentration and dilation than those associated with the operations CON and DIL (see (2.44), (2.45)). Thus, as operators acting on a fuzzy set labeled x, *plus* and *minus* are defined by

$$\textit{plus } x \triangleq x^{1.25} \tag{3.15}$$

$$\textit{minus } x \triangleq x^{0.75}. \tag{3.16}$$

In consequence of (3.15) and (3.16), we have the approximate identity

$$\textit{plus plus } x = \textit{minus very } x. \tag{3.17}$$

As an illustration, if the hedge *highly* is defined as

$$\textit{highly} = \textit{minus very very} \tag{3.18}$$

then, equivalently,

$$\textit{highly} = \textit{plus plus very}. \tag{3.19}$$

As was stated earlier, the computation of the meaning of composite terms of the form *hu* is a preliminary to the problem of computing the meaning of values of a linguistic variable. We are now in a position to turn our attention to this problem.

IV. Computation of the Meaning of Values of a Linguistic Variable

Once we know how to compute the meaning of a composite term of the form *hu*, the computation of the meaning of a more complex composite term, which may involve the terms *not*, *or*, and *and* in addition to terms of the form *hu*, becomes a relatively simple problem which is quite similar to that of the computation of the value of a Boolean expression. As a simple illustration, consider the computation of the meaning of the composite term

$$x = \textit{not very small} \tag{4.1}$$

where the primary term *small* is defined as

$$\textit{small} = 1/1 + 0.8/2 + 0.6/3 + 0.4/4 + 0.2/5 \tag{4.2}$$

with the universe of discourse being

$$U = 1 + 2 + 3 + 4 + 5. \tag{4.3}$$

By (3.8), the operation of *very* on *small* yields

$$\textit{very small} = 1/1 + 0.64/2 + 0.36/3 + 0.16/4 + 0.04/5 \tag{4.4}$$

and, by (2.26),

$$\textit{not very small} = \neg(\textit{very small}) = 0.36/2 + 0.64/3 + 0.84/4 + 0.96/5 \approx 0.4/2 + 0.6/3 + 0.8/4 + 1/5. \tag{4.5}$$

As a slightly more complicated example, consider the composite term

$$x = \textit{not very small and not very very large} \tag{4.6}$$

where *large* is defined by

$$\textit{large} = 0.2/1 + 0.4/2 + 0.6/3 + 0.8/4 + 1/5. \tag{4.7}$$

In this case,

$$\textit{very large} = \textit{large}^2 = 0.04/1 + 0.16/2 + 0.36/3 + 0.64/4 + 1/5 \tag{4.8}$$

$$\textit{very very large} = (\textit{large}^2)^2 \approx 0.1/3 + 0.4/4 + 1/5 \tag{4.9}$$

$$\textit{not very very large} \approx 1/1 + 1/2 + 0.9/3 + 0.6/4 \tag{4.10}$$

and hence

not very small and not very very large

$$\approx (0.4/2 + 0.6/3 + 0.8/4 + 1/5) \cap (1/1 + 1/2 + 0.9/3 + 0.6/4) \approx 0.4/2 + 0.6/3 + 0.6/4. \tag{4.11}$$

An example of a different nature is provided by the values of a linguistic variable labeled *likelihood.* In this case, we assume that the universe of discourse is given by

$$U = 0 + 0.1 + 0.2 + 0.3 + 0.4 + 0.5 + 0.6 + 0.7 + 0.8 + 0.9 + 1 \tag{4.12}$$

in which the elements of U represent probabilities. Suppose that we wish to compute the meaning of the value

$$x = \textit{highly unlikely} \tag{4.13}$$

in which *highly* is defined as (see (3.18))

$$\textit{highly} = \textit{minus very very} \tag{4.14}$$

and

$$\textit{unlikely} = \textit{not likely} \tag{4.15}$$

with the meaning of the primary term *likely* given by

$$\textit{likely} = 1/1 + 1/0.9 + 1/0.8 + 0.8/0.7 + 0.6/0.6 + 0.5/0.5 + 0.3/0.4 + 0.2/0.3. \tag{4.16}$$

Using (4.15), we obtain

$$\textit{unlikely} = 1/0 + 1/0.1 + 1/0.2 + 0.8/0.3 + 0.7/0.4 + 0.5/0.5 + 0.4/0.6 + 0.2/0.7 \tag{4.17}$$

and hence

very very unlikely

$$= (\textit{unlikely})^4 \approx 1/0 + 1/0.1 + 1/0.2 + 0.4/0.3 + 0.2/0.4. \tag{4.18}$$

Finally, by (4.14)

highly unlikely

$$= \textit{minus very very unlikely} \approx (1/0 + 1/0.1 + 1/0.2 + 0.4/0.3 + 0.2/0.4)^{0.75} \approx 1/0 + 1/0.1 + 1/0.2 + 0.5/0.3 + 0.3/0.4. \tag{4.19}$$

It should be noted that in computing the meaning of composite terms in the preceding examples we have made implicit use of the usual precedence rules governing the evaluation of Boolean expressions. With the addition of hedges, these precedence rules may be expressed as follows.

Precedence	Operation
First	*h, not*
Second	*and*
Third	*or*

As usual, parentheses may be used to change the precedence order and ambiguities may be resolved by the use of association to the right. Thus *plus very minus very tall* should be interpreted as

plus (*very* (*minus* (*very* (*tall*)))).

The technique that was employed for the computation of the meaning of a composite term is a special case of a more general approach which is described in [4] and [5]. The approach in question can be applied to the computation of the meaning of values of a linguistic variable provided the composite terms representing these values can be generated by a context-free grammar. As an illustration, consider a linguistic variable x whose values are exemplified by *small, not small, large, not large, very small, not very small, small or not very very large, small and* (*large or not small*), *not very very small and not very very large*, etc.

The values in question can be generated by a context-free grammar $G = (V_T, V_N, S, P)$ in which the set of terminals V_T comprises the atomic terms *small, large, not, and, or, very*, etc.; the nonterminals are denoted S, A, B, C, D, and E; and the production system is given by

$$\begin{array}{ll} S \rightarrow A & C \rightarrow D \\ S \rightarrow S \text{ or } A & C \rightarrow E \\ A \rightarrow B & D \rightarrow \text{very } D \\ A \rightarrow A \text{ and } B & E \rightarrow \text{very } E \\ B \rightarrow C & D \rightarrow \text{small} \\ B \rightarrow \text{not } C & E \rightarrow \text{large} \\ C \rightarrow (S). & \end{array} \tag{4.20}$$

Each production in (4.20) gives rise to a relation between the fuzzy sets labeled by the corresponding terminal and nonterminal symbols. In the case of (4.20), these relations are (we omit the productions which have no effect on the associated fuzzy sets)

$$S \to S \text{ or } A \Rightarrow S_L = S_R + A_R$$

$$A \to A \text{ and } B \Rightarrow A_L = A_R \cap B_R$$

$$B \to \text{not } C \Rightarrow B_L = \neg C_R$$

$$D \to \text{very } D \Rightarrow D_L = {D_R}^2$$

$$E \to \text{very } E \Rightarrow E_L = {E_R}^2$$

$$D \to \text{small} \Rightarrow D_L = \text{small}$$

$$E \to \text{large} \Rightarrow E_L = \text{large} \tag{4.21}$$

in which the subscripts L and R are used to differentiate between the symbols on the left- and right-hand sides of a production.

To compute the meaning of a composite term x, it is necessary to perform a syntactical analysis of x in terms of the specified grammar G. Then, knowing the syntax tree of x, one can employ the relations given in (4.21) to derive a set of equations (in triangular form) which upon solution yield the meaning of x. For example, in the case of the composite term

$$x = \text{not very small and not very very large}$$

the solution of these equations yields

$$x = (\neg \text{small}^2) \cap (\neg \text{large}^4) \tag{4.22}$$

which agrees with (4.11). Details of this solution may be found in [4] and [5].

The ability to compute the meaning of values of a linguistic variable is a prerequisite to the computation of the meaning of fuzzy conditional statements of the form IF A THEN B, e.g., IF x is *not very small* THEN y is *very very large*. This problem is considered in the following section.

V. Fuzzy Conditional Statements and Compositional Rule of Inference

In classical propositional calculus,[5] the expression IF A THEN B, where A and B are propositional variables, is written as $A \Rightarrow B$, with the implication $\Rightarrow$ regarded as a connective which is defined by the truth table.

A	B	$A \Rightarrow B$
T	T	T
T	F	F
F	T	T
F	F	T

Thus,

$$A \Rightarrow B \equiv \neg A \vee B \tag{5.1}$$

[5] A detailed discussion of the significance of implication and its role in modal logic may be found in [18].

in the sense that the propositional expressions $A \Rightarrow B$ (A *implies* B) and $\neg A \vee B$ (*not* A *or* B) have identical truth tables.

A more general concept, which plays an important role in our approach, is a *fuzzy conditional statement*: IF A THEN B or, for short, $A \Rightarrow B$, in which A (the antecedent) and B (the consequent) are fuzzy sets rather than propositional variables. The following are typical examples of such statements:

IF *large* THEN *small*
IF *slippery* THEN *dangerous*

which are abbreviations of the statements

IF x is *large* THEN y is *small*
IF the road is *slippery* THEN driving is *dangerous*.

In essence, statements of this form describe a relation between two fuzzy variables. This suggests that a fuzzy conditional statement be defined as a fuzzy relation in the sense of (2.19) rather than as a connective in the sense of (5.1).

To this end, it is expedient to define first the *Cartesian product* of two fuzzy sets. Specifically, let A be a fuzzy subset of a universe of discourse U, and let B be a fuzzy subset of a possibly different universe of discourse V. Then, the Cartesian product of A and B is denoted by $A \times B$ and is defined by

$$A \times B \triangleq \int_{U \times V} \mu_A(u) \wedge \mu_B(v)/(u,v) \tag{5.2}$$

where $U \times V$ denotes the Cartesian product of the nonfuzzy sets U and V; that is,

$$U \times V \triangleq \{(u,v) \mid u \in U, v \in V\}.$$

Note that when A and B are nonfuzzy, (5.2) reduces to the conventional definition of the Cartesian product of nonfuzzy sets. In words, (5.2) means that $A \times B$ is a fuzzy set of ordered pairs (u,v), $u \in U$, $v \in V$, with the grade of membership of (u,v) in $A \times B$ given by $\mu_A(u) \wedge \mu_B(v)$. In this sense, $A \times B$ is a fuzzy relation from U to V.

As a very simple example, suppose that

$$U = 1 + 2 \tag{5.3}$$

$$V = 1 + 2 + 3 \tag{5.4}$$

$$A = 1/1 + 0.8/2 \tag{5.5}$$

$$B = 0.6/1 + 0.9/2 + 1/3. \tag{5.6}$$

Then

$$A \times B = 0.6/(1,1) + 0.9/(1,2) + 1/(1,3) + 0.6/(2,1) + 0.8/(2,2) + 0.8/(2,3). \tag{5.7}$$

The relation defined by (5.7) may be conveniently represented by the relation matrix

$$\begin{array}{c c} & \begin{array}{ccc} 1 & 2 & 3 \end{array} \\ \begin{array}{c} 1 \\ 2 \end{array} & \begin{bmatrix} 0.6 & 0.9 & 1 \\ 0.6 & 0.8 & 0.8 \end{bmatrix}. \end{array} \tag{5.8}$$

The significance of a fuzzy conditional statement of the form IF A THEN B is made clearer by regarding it as a special case of the conditional expression IF A THEN B ELSE C, where A and (B and C) are fuzzy subsets of possibly different universes U and V, respectively. In terms of the Cartesian product, the latter statement is defined as follows:

$$\text{IF } A \text{ THEN } B \text{ ELSE } C \triangleq A \times B + (\neg A \times C) \quad (5.9)$$

in which $+$ stands for the union of the fuzzy relations $A \times B$ and $(\neg A \times C)$.

More generally, if $A_1,\cdots,A_n$ are fuzzy subsets of U, and $B_1,\cdots,B_n$ are fuzzy subsets of V, then[6]

$$\text{IF } A_1 \text{ THEN } B_1 \text{ ELSE IF } A_2 \text{ THEN } B_2 \cdots \text{ ELSE IF } A_n \text{ THEN } B_n$$

$$\triangleq A_1 \times B_1 + A_2 \times B_2 + \cdots + A_n \times B_n. \quad (5.10)$$

Note that (5.10) reduces to (5.9) if IF A THEN B ELSE C is interpreted as IF A THEN B ELSE IF $\neg A$ THEN C. It should also be noted that by repeated application of (5.9) we obtain

$$\text{IF } A \text{ THEN (IF } B \text{ THEN } C \text{ ELSE } D) \text{ ELSE } E$$

$$= A \times B \times C + A \times \neg B \times D + \neg A \times E. \quad (5.11)$$

If we regard IF A THEN B as IF A THEN B ELSE C with unspecified C, then, depending on the assumption made about C, various interpretations of IF A THEN B will result. In particular, if we assume that $C = V$, then IF A THEN B (or $A \Rightarrow B$) becomes[7]

$$A \Rightarrow B \triangleq \text{IF } A \text{ THEN } B \triangleq A \times B + (\neg A \times V). \quad (5.12)$$

If, in addition, we set $A = U$ in (5.12), we obtain as an alternative definition

$$A \Rightarrow B \triangleq U \times B + (\neg A \times V). \quad (5.13)$$

In the sequel, we shall assume that $C = V$, and hence that $A \Rightarrow B$ is defined by (5.12). In effect, the assumption that $C = V$ implies that, in the absence of an indication to the contrary, the consequent of $\neg A \Rightarrow C$ can be any fuzzy subset of the universe of discourse. As a very simple illustration of (5.12), suppose that A and B are defined by (5.5) and (5.6). Then, on substituting (5.8) in (5.12), the relation matrix for $A \Rightarrow B$ is found to be

$$A \Rightarrow B = \begin{bmatrix} 0.6 & 0.9 & 1 \\ 0.6 & 0.8 & 0.8 \end{bmatrix}.$$

It should be observed that when A, B, and C are nonfuzzy sets, we have the identity

$$\text{IF } A \text{ THEN } B \text{ ELSE } C = (\text{IF } A \text{ THEN } B) \cap (\text{IF } \neg A \text{ THEN } C) \quad (5.14)$$

which holds only approximately for fuzzy A, B, and C. This indicates that, in relation to (5.15), the definitions of IF A THEN B ELSE C and IF A THEN B, as expressed by (5.9) and (5.12), are not exactly consistent for fuzzy A, B, and C. It should also be noted that if 1) $U = V$, 2) $x = y$, and 3) $A \Rightarrow B$ holds for all points in U, then, by (5.12),

$$A \Rightarrow B \text{ implies and is implied by } A \subset B \quad (5.15)$$

exactly if A and B are nonfuzzy and approximately otherwise.

As will be seen in Section VI, fuzzy conditional statements play a basic role in fuzzy algorithms. More specifically, a typical problem which is encountered in the course of execution of such algorithms is the following. We have a fuzzy relation, say, R, from U to V which is defined by a fuzzy conditional statement. Then, we are given a fuzzy subset of U, say, x, and have to determine the fuzzy subset of V, say, y, which is induced in V by x. For example, we may have the following two statements.

1) x is *very small*
2) IF x is *small* THEN y is *large* ELSE y is *not very large*

of which the second defines by (5.9) a fuzzy relation R. The question, then, is as follows: What will be the value of y if x is *very small*? The answer to this question is provided by the following rule of inference, which may be regarded as an extension of the familiar rule of *modus ponens*.

Compositional Rule of Inference: If R is a fuzzy relation from U to V, and x is a fuzzy subset of U, then the fuzzy subset y of V which is induced by x is given by the composition (see (2.22)) of R and x; that is,

$$y = x \circ R \quad (5.16)$$

in which x plays the role of a unary relation.[8]

As a simple illustration of (5.16), suppose that R and x are defined by the relation matrices in (5.17). Then y is given by the max–min product of x and R:

$$\overset{x}{[0.2 \;\; 1 \;\; 0.3]} \circ \overset{R}{\begin{bmatrix} 0.8 & 0.9 & 0.2 \\ 0.6 & 1 & 0.4 \\ 0.5 & 0.8 & 1 \end{bmatrix}} = \overset{y}{[0.6 \;\; 1 \;\; 0.4]}. \quad (5.17)$$

As for the question raised before, suppose that, as in (4.3), we have

$$U = 1 + 2 + 3 + 4 + 5 \quad (5.18)$$

with *small* and *large* defined by (4.2) and (4.7), respectively. Then, substituting *small* for A, *large* for B and *not very large* for C in (5.9), we obtain the relation matrix R for the fuzzy conditional statement IF *small* THEN *large* ELSE *not very large*. The result of the composition of R with x = *very*

[6] It should be noted that, in the sense used in ALGOL, the right-hand side of (5.10) would be expressed as $A_1 \times B_1 + (\neg A_1 \cap A_2) \times B_2 + \cdots + (\neg A_1 \cap \cdots \cap \neg A_{n-1} \cap A_n) \times B_n$ when the A_i and B_i, $i = 1,\cdots,n$, are nonfuzzy sets.

[7] This definition should be viewed as tentative in nature.

[8] If R is visualized as a fuzzy graph, then (5.16) may be viewed as the expression for the fuzzy ordinate y corresponding to a fuzzy abscissa x.

small is

$$\overset{x}{[1 \quad 0.64 \quad 0.36 \quad 0.16 \quad 0.04]} \circ \overset{R}{\begin{bmatrix} 0.2 & 0.4 & 0.6 & 0.8 & 1 \\ 0.2 & 0.4 & 0.6 & 0.8 & 0.8 \\ 0.4 & 0.4 & 0.6 & 0.6 & 0.6 \\ 0.6 & 0.6 & 0.6 & 0.4 & 0.4 \\ 0.8 & 0.8 & 0.64 & 0.36 & 0.2 \end{bmatrix}}$$

$$= \overset{y}{[0.36 \quad 0.4 \quad 0.6 \quad 0.8 \quad 1]}. \qquad (5.19)$$

There are several aspects of (5.16) that are in need of comment. First, it should be noted that when $R = A \Rightarrow B$ and $x = A$ we obtain

$$y = A \circ (A \Rightarrow B) = B \qquad (5.20)$$

as an exact identity, when A, B, and C are nonfuzzy, and an approximate one, when A, B, and C are fuzzy. It is in this sense that the compositional inference rule (5.16) may be viewed as an approximate extension of *modus ponens*. (Note that in consequence of the way in which $A \Rightarrow B$ is defined in (5.12), the more different x is from A, the less sharply defined is y.)

Second, (5.16) is analogous to the expression for the marginal probability in terms of the conditional probability function; that is

$$r_j = \sum_i q_i p_{ij} \qquad (5.21)$$

where

$$q_i = \Pr\{X = x_i\}$$

$$r_j = \Pr\{Y = y_j\}$$

$$p_{ij} = \Pr\{Y = y_j \mid X = x_i\}$$

and X and Y are random variables with values $x_1, x_2, \cdots$ and $y_1, y_2, \cdots$, respectively. However, this analogy does not imply that (5.16) is a relation between probabilities.

Third, it should be noted that because of the use of the max–min matrix product in (5.16), the relation between x and y is not continuous. Thus, in general, a small change in x would produce no change in y until a certain threshold is exceeded. This would not be the case if the composition of x with R were defined as max–product composition.

Fourth, in the computation of $x \circ R$ one may take advantage of the distributivity of composition over the union of fuzzy sets. Thus, if

$$x = u \text{ or } v \qquad (5.22)$$

where u and v are labels of fuzzy sets, then

$$(u \text{ or } v) \circ R = u \circ R \text{ or } v \circ R. \qquad (5.23)$$

For example, if x is *small or medium*, and $R = A \Rightarrow B$ reads IF x is *not small and not large* THEN y is *very small*, then we can write

(*small or medium*) ∘ (*not small and not large* ⇒ *very small*)
= *small* ∘ (*not small and not large* ⇒ *very small*) *or medium* ∘ (*not small and not large* ⇒ *very small*). (5.24)

As a final comment, it is important to realize that in practical applications of fuzzy conditional statements to the description of complex or ill-defined relations, the computations involved in (5.9), (5.10), and (5.16) would, in general, be performed in a highly approximate fashion. Furthermore, an additional source of imprecision would be the result of representing a fuzzy set as a value of a linguistic variable. For example, suppose that a relation between fuzzy variables x and y is described by the fuzzy conditional statement IF *small* THEN *large* ELSE IF *medium* THEN *medium* ELSE IF large THEN *very small*.

Typically, we would assign different linguistic values to x and compute the corresponding values of y by the use of (5.16). Then, on approximating to the computed values of y by linguistic labels, we would arrive at a table having the form shown below:

Given		Inferred	
A	B	x	y
small	*large*	*not small*	*not very large*
medium	*medium*	*very small*	*very very large*
large	*very small*	*very very small*	*very very large*
		not very large	*small or medium*

Such a table constitutes an approximate linguistic characterization of the relation between x and y which is inferred from the given fuzzy conditional statement. As was stated earlier, fuzzy conditional statements play a basic role in the description and execution of fuzzy algorithms. We turn to this subject in the following section.

VI. Fuzzy Algorithms

Roughly speaking, a fuzzy algorithm is an ordered set of fuzzy instructions which upon execution yield an approximate solution to a specified problem. In one form or another, fuzzy algorithms pervade much of what we do. Thus, we employ fuzzy algorithms both consciously and subconsciously when we walk, drive a car, search for an object, tie a knot, park a car, cook a meal, find a number in a telephone directory, etc. Furthermore, there are many instances of uses of what, in effect, are fuzzy algorithms in a wide variety of fields, especially in programming, operations research, psychology, management science, and medical diagnosis.

The notion of a fuzzy set and, in particular, the concept of a fuzzy conditional statement provide a basis for using fuzzy algorithms in a more systematic and hence more effective ways than was possible in the past. Thus, fuzzy algorithms could become an important tool for an approximate analysis of systems and decision processes which are much too complex for the application of conventional mathematical techniques.

A formal characterization of the concept of a fuzzy algorithm can be given in terms of the notion of a fuzzy Turing machine or a fuzzy Markoff algorithm [6]–[8]. In this section, the main aim of our discussion is to relate the concept of a fuzzy algorithm to the notions introduced in the preceding sections and illustrate by simple examples some of the uses of such algorithms.

The instructions in a fuzzy algorithm fall into the following three classes.

1) *Assignment Statements:* e.g.,

$x \approx 5$
$x = small$
x is *large*
x is *not large and not very small.*

2) *Fuzzy Conditional Statements:* e.g.,

IF x is *small* THEN y is *large* ELSE y is *not large*
IF x is positive THEN decrease y *slightly*
IF x is *much greater* than 5 THEN stop
IF x is *very small* THEN go to 7.

Note that in such statements either the antecedent or the consequent or both may be labels of fuzzy sets.

3) *Unconditional Action Statements:* e.g.,

multiply x by x
decrease x *slightly*
delete the first *few* occurrences of 1
go to 7
print x
stop.

Note that some of these instructions are fuzzy and some are not.

The combination of an assignment statement and a fuzzy conditional statement is executed in accordance with the compositional rule (5.16). For example, if at some point in the execution of a fuzzy algorithm we encounter the instructions

1) $x = very\ small$
2) IF x is *small* THEN y is *large* ELSE y is *not very large*

where *small* and *large* are defined by (4.2) and (4.7), then the result of the execution of 1) and 2) will be the value of y given by (5.19), that is,

$$y = 0.36/1 + 0.4/2 + 0.64/3 + 0.8/4 + 1/5. \quad (6.1)$$

An unconditional but fuzzy action statement is executed similarly. For example, the instruction

$$\text{multiply } x \text{ by itself a } \textit{few} \text{ times} \quad (6.2)$$

with *few* defined as

$$few = 1/1 + 0.8/2 + 0.6/3 + 0.4/4 \quad (6.3)$$

would yield upon execution the fuzzy set

$$y = 1/x^2 + 0.8/x^3 + 0.6/x^4 + 0.4/x^5. \quad (6.4)$$

It is important to observe that, in both (6.1) and (6.4), the result of execution is a fuzzy set rather than a single number. However, when a human subject is presented with a fuzzy instruction such as "take *several* steps," with *several* defined by (see (2.10))

$$several = 0.5/3 + 0.8/4 + 1/5 + 1/6 + 0.8/7 + 0.5/8 \quad (6.5)$$

the result of execution must be a single number between 3 and 8. On what basis will such a number be chosen?

As pointed out in [6], it is reasonable to assume that the result of execution will be that element of the fuzzy set which has the highest grade of membership in it. If such an element is not unique, as is true of (6.5), then a random or arbitrary choice can be made among the elements having the highest grade of membership. Alternatively, an external criterion can be introduced which linearly orders those elements of the fuzzy set which have the highest membership, and thus generates a unique greatest element. For example, in the case of (6.5), if the external criterion is to minimize the number of steps that have to be taken, then the subject will pick 5 from the elements with the highest grade of membership.

An analogous question arises in situations in which a human subject has to give a "yes" or "no" answer to a fuzzy question. For example, suppose that a subject is presented with the instruction

$$\text{IF } x \text{ is } \textit{small} \text{ THEN stop ELSE go to 7} \quad (6.6)$$

in which *small* is defined by (4.2). Now assume that $x = 3$, which has the grade of membership of 0.6 in *small.* Should the subject execute "stop" or "go to 7"? We shall assume that in situations of this kind the subject will pick that alternative which is more true than untrue, e.g., "x is *small*" over "x is *not small*," since in our example the degree of truth of the statement "3 is *small*" is 0.6, which is greater than that of the statement "3 is *not small.*" If both alternatives have more or less equal truth values, the choice can be made arbitrarily. For convenience, we shall refer to this rule of deciding between two alternatives as the *rule of the preponderant alternative.*

It is very important to understand that the questions just discussed arise only in those situations in which the result of execution of a fuzzy instruction is required to be a single element (e.g., a number) rather than a fuzzy set. Thus, if we allowed the result of execution of (6.6) to be fuzzy, then for $x = 3$ we would obtain the fuzzy set

$$0.6/\text{stop} + 0.4/\text{go to 7}$$

which implies that the execution is carried out in parallel. The assumption of parallelism is implicit in the compositional rule of inference and is basic to the understanding of fuzzy algorithms and their execution by humans and machines.

In what follows, we shall present several examples of fuzzy algorithms in the light of the concepts discussed in the preceding sections. It should be stressed that these examples are intended primarily to illustrate the basic aspects of fuzzy algorithms rather than demonstrate their effectiveness in the solution of practical problems.

It is convenient to classify fuzzy algorithms into several basic categories, each corresponding to a particular type of application: definitional and identificational algorithms; generational algorithms; relational and behavioral algorithms; and decisional algorithms. (It should be noted that an algorithm of a particular type can include algorithms of other types as subalgorithms. For example, a definitional algorithm may contain relational and decisional subalgorithms.) We begin with an example of a definitional algorithm.

Fuzzy Definitional Algorithms

One of the basic areas of application for fuzzy algorithms lies in the definition of complex, ill-defined or fuzzy concepts in terms of simpler or less fuzzy concepts. The following are examples of such fuzzy concepts: sparseness of matrices; handwritten characters; measures of complexity; measures of proximity or resemblance; degrees of clustering; criteria of performance; soft constraints; rules of various kinds, e.g., zoning regulations; legal criteria, e.g., criteria for insanity, obscenity, etc.; and fuzzy diseases such as arthritis, arteriosclerosis, schizophrenia.

Since a fuzzy concept may be viewed as a label for a fuzzy set, a *fuzzy definitional algorithm* is, in effect, a finite set of possibly fuzzy instructions which define a fuzzy set in terms of other fuzzy sets (and possibly itself, i.e., recursively) or constitute a procedure for computing the grade of membership of any element of the universe of discourse in the set under definition. In the latter case, the definational algorithm plays the role of an *identificational algorithm*, that is, an algorithm which identifies whether or not an element belongs to a set or, more generally, determines its grade of membership. An example of such an algorithm is provided by the procedure (see [5]) for computing the grade of membership of a string in a fuzzy language generated by a context-free grammar.

As a very simple example of a fuzzy definitional algorithm, we shall consider the fuzzy concept *oval*. It should be emphasized again that the oversimplified definition that will be given is intended only for illustrative purposes and has no pretense at being an accurate definition of the concept *oval*. The instructions comprising the algorithm OVAL are listed here. The symbol T in these instructions stands for the object under test. The term CALL CONVEX represents a call on a subalgorithm labeled CONVEX, which is a definitional algorithm for testing whether or not T is convex. An instruction of the form IF A THEN B should be interpreted as IF A THEN B ELSE go to next instruction.

Algorithm OVAL:

1) IF T is not closed THEN T is not *oval*; stop.
2) IF T is self-intersecting THEN T is not *oval*; stop.
3) IF T is not CALL CONVEX THEN T is not *oval*; stop.
4) IF T does not have two *more or less* orthogonal axes of symmetry THEN T is not *oval*; stop.
5) IF the major axis of T is not *much* longer than the minor axis THEN T is not *oval*; stop.
6) T is *oval*; stop.

Subalgorithm CONVEX: Basically, this subalgorithm involves a check on whether the curvature of T at each point maintains the same sign as one moves along T in some initially chosen direction.

1) $x = a$ (some initial point on T).
2) Choose a direction of movement along T.
3) $t \approx$ direction of tangent to T at x.
4) $x' \approx x + 1$ (move from x to a neighboring point).
5) $t' \approx$ direction of tangent to T at x'.
6) $\alpha \approx$ angle between t' and t.
7) $x \approx x'$.
8) $t \approx$ direction of tangent to T at x.
9) $x' \approx x + 1$.
10) $t' \approx$ direction of tangent to T at x'.
11) $\beta \approx$ angle between t' and t.
12) IF β does not have the same sign as α THEN T is not convex; return.
13) IF $x' \approx a$ THEN T is convex; return.
14) Go to 7).

Comment: It should be noted that the first three instructions in OVAL are nonfuzzy. As for instructions 4) and 5), they involve definitions of concepts such as "*more or less* orthogonal," and "*much* longer," which, though fuzzy, are less complex and better understood than the concept of *oval*. This exemplifies the main function of a fuzzy definitional algorithm, namely, to reduce a new or complex fuzzy concept to simpler or better understood fuzzy concepts. In a more elaborate version of the algorithm OVAL, the answers to 4) and 5) could be the degrees to which the conditions in these instructions are satisfied. The final result of the algorithm, then, would be the grade of membership of T in the fuzzy set of oval objects.

In this connection, it should be noted that, in virtue of (5.15), the algorithm OVAL as stated is approximately equivalent to the expression

$$\begin{aligned} \textit{oval} = {} & \text{closed} \cap \text{non-self-intersecting} \cap \text{convex} \\ & \cap\ \textit{more or less} \text{ orthogonal axes of symmetry} \\ & \cap \text{ major axis } \textit{much} \text{ larger than minor axis} \end{aligned} \tag{6.7}$$

which defines the fuzzy set *oval* as the intersection of the fuzzy and nonfuzzy sets whose labels appear on the right-hand side of (6.7). However, one significant difference is that the algorithm not only defines the right-hand side of (6.7), but also specifies the order in which the computations implicit in (6.7) are to be performed.

Fuzzy Generational Algorithms

As its designation implies, a fuzzy generational algorithm serves to generate rather than define a fuzzy set. Possible applications of generational algorithms include: generation of handwritten characters and patterns of various kinds; cooking recipes; generation of music; generation of sentences in a natural language; generation of speech.

As a simple illustration of the notion of a generational algorithm, we shall consider an algorithm for generating the letter **P**, with the height h and the base b of **P** constituting the parameters of the algorithm. For simplicity, **P** will be generated as a dotted pattern, with eight dots lying on the vertical line.

Algorithm **P**(h,b)*:*

1) $i = 1$.
2) $X(i) = b$ (first dot at base).

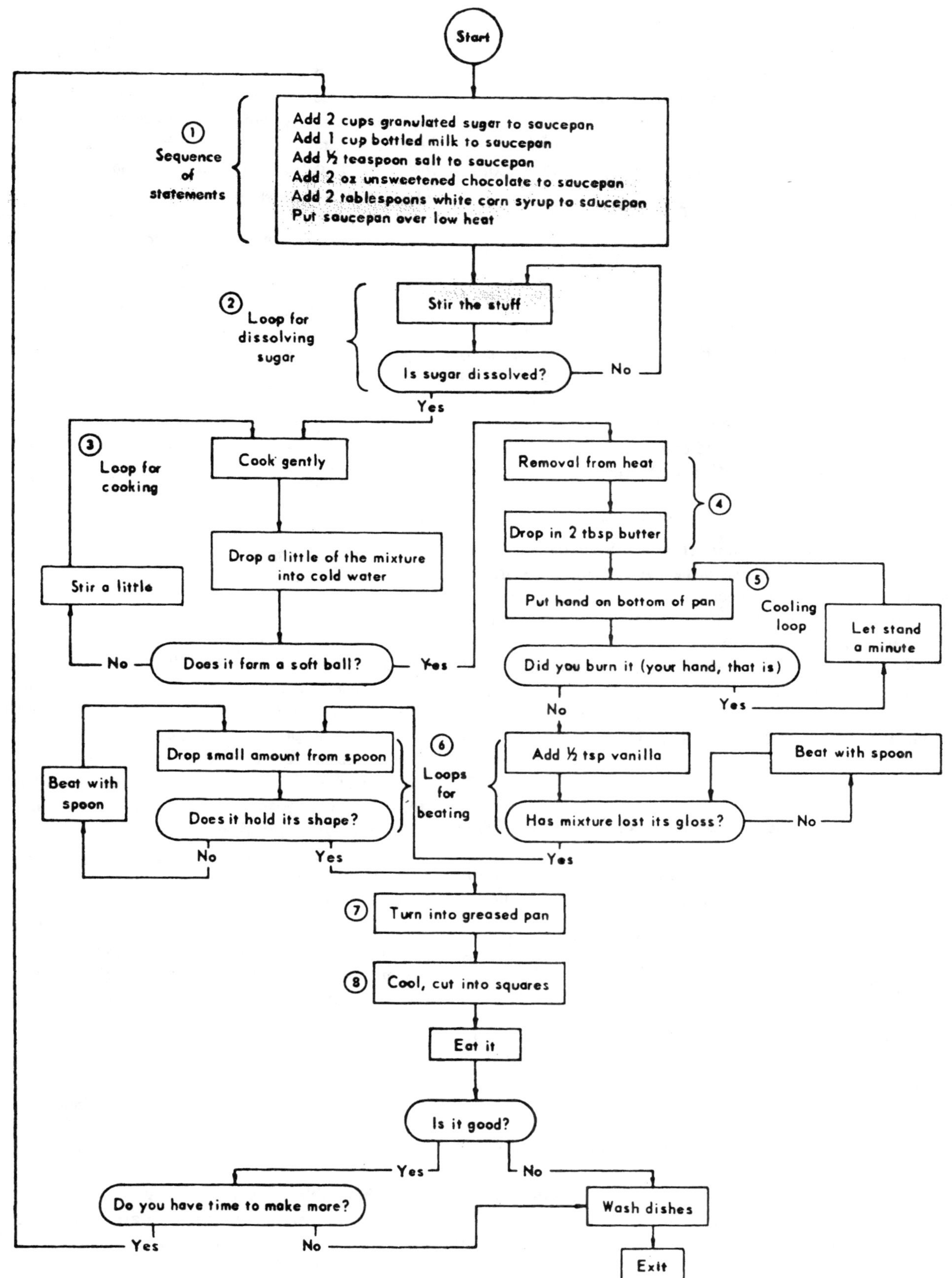

Fig. 3. Recipe for chocolate fudge (from [19]).

3) $X(i+1) \approx X(i) + h/6$ (put dot *approximately* $h/6$ units of distance above $X(i)$).
4) $i = i + 1$.
5) IF $i = 7$ THEN make right turn and go to 7).
6) Go to 3.
7) Move by $h/6$ units; put a dot.
8) Turn by 45°; move by $h/6$ units; put a dot.
9) Turn by 45°; move by $h/6$ units; put a dot.
10) Turn by 45°; move by $h/6$ units; put a dot.
11) Turn by 45°; move by $h/6$ units; put a dot; stop.

The algorithm as stated is of open-loop type in the sense that it does not incorporate any feedback. To make the algorithm less sensitive to errors in execution, we could introduce fuzzy feedback by conditioning the termination of the algorithm on an approximate satisfaction of a specified test. For example, if the last point in step 11) does not fall on the vertical part of **P**, we could return to step 8) and either reduce or increase the angle of turn in steps 8)–11) to correct for the terminal error. The flowchart of a cooking recipe for chocolate fudge (Fig. 3), which is reproduced from [19], is a good example of what, in effect, is a fuzzy generational algorithm with feedback.

Fuzzy Relational and Behavioral Algorithms

A *fuzzy relational algorithm* serves to describe a relation or relations between fuzzy variables. A relational algorithm which is used for the specific purpose of approximate description of the behavior of a system will be referred to as a *fuzzy behavioral algorithm.*

A simple example of a relational algorithm labeled R which involves three parameters x, y, and z is given. This algorithm defines a fuzzy ternary relation R in the universe of discourse $U = 1 + 2 + 3 + 4 + 5$ with *small* and *large* defined by (4.2) and (4.7).

Algorithm $R(x,y,z)$:

1) IF x is *small* and y is *large* THEN z is *very small* ELSE z is *not small.*
2) IF x is *large* THEN (IF y is *small* THEN z is *very large* ELSE z is *small*) ELSE z and y are *very very small.*

If needed, the meaning of these conditional statements can be computed by using (5.9) and (5.11). The relation R, then, will be the intersection of the relations defined by instructions 1) and 2).

Another simple example of a relational fuzzy algorithm $F(x,y)$ which illustrates a different aspect of such algorithms is the following.

Algorithm $F(x,y)$:

1) IF x is *small* and x is increased *slightly* THEN y will increase *slightly.*
2) IF x is *small* and x is increased *substantially* THEN y will increase *substantially.*
3) IF x is *large* and x is increased *slightly* THEN y will increase *moderately.*
4) IF x is *large* and x is increased *substantially* THEN y will increase *very substantially.*

As in the case of the previous example, the meaning of the fuzzy conditional statements in this algorithm can be computed by the use of the methods discussed in Sections IV and V if one is given the definitions of the primary terms *large* and *small* as well as the hedges *slightly*, *substantially*, and *moderately*.

As a simple example of a behavioral algorithm, suppose that we have a system S with two nonfuzzy states (see [3]) labeled q_1 and q_2, two fuzzy input values labeled *low* and *high*, and two fuzzy output values labeled *large* and *small.* The universe of discourse for the input and output values is assumed to be the real line. We assume further that the behavior of S can be characterized in an approximate fashion by the algorithm that will be given. However, to represent the relations between the inputs, states, and outputs, we use the conventional state transition tables instead of conditional statements.

Algorithm BEHAVIOR:

u_t \ x_t	x_{t+1}: q_1	x_{t+1}: q_2	y_t: q_1	y_t: q_2
low	q_2	q_1	*large*	*small*
high	q_1	q_1	*small*	*large*

where

u_t input at time t
y_t output at time t
x_t state at time t.

On the surface, this table appears to define a conventional nonfuzzy finite-state system. What is important to recognize, however, is that in the case of the system under consideration the inputs and outputs are fuzzy subsets of the real line. Thus we could pose the question: What would be the output of S if it is in state q_1 and the applied input is *very low*? In the case of S, this question can be answered by an application of the compositional inference rule (5.16). On the other hand, the same question would not be a meaningful one if S is assumed to be a nonfuzzy finite-state system characterized by the preceding table.

Behavioral fuzzy algorithms can also be used to describe the more complex forms of behavior resulting from the presence of random elements in a system. For example, the presence of random elements in S might result in the following fuzzy-probabilistic characterization of its behavior:

u_t \ x_t	x_{t+1}: q_1	x_{t+1}: q_2	y_t: q_1	y_t: q_2
low	q_2 *likely*	q_1 *likely*	*large likely*	*small likely*^{2}
high	q_1 *likely*^{2}	q_1 *unlikely*^{2}	*small likely*^{2}	*large unlikely*^{2}

In this table, the term *likely* and its modifications by *very* and *not* serve to provide an approximate characterization of probabilities. For example, IF the input is *low* and the present state is q_1, THEN the next state is *likely* to be q_2. Similarly, IF the input is *high* and the present state is q_2 THEN the output is *very unlikely* to be *large.* If the meaning of *likely* is defined by (see (4.16))

$$likely = 1/1 + 1/0.9 + 1/0.8 + 0.8/0.7 + 0.6/0.6 + 0.5/0.5 + 0.3/0.4 + 0.2/0.3 \quad (6.8)$$

then

$$unlikely = 0.2/0.7 + 0.4/0.6 + 0.5/0.5 + 0.7/0.4 + 0.8/0.3 + 1/0.2 + 1/0.1 + 1/0 \quad (6.9)$$

$$very\ likely \approx 1/1 + 1/0.9 + 1/0.8 + 0.6/0.7 + 0.4/0.6 + 0.3/0.5 + 0.1/0.4 \quad (6.10)$$

$$very\ unlikely \approx 0.2/0.6 + 0.3/0.5 + 0.5/0.4 + 0.6/0.3 + 1/0.2 + 1/0.1 + 1/0. \quad (6.11)$$

Fuzzy Decisional Algorithms

A *fuzzy decisional algorithm* is a fuzzy algorithm which serves to provide an approximate description of a strategy or decision rule. Commonplace examples of such algorithms, which we use for the most part on a subconscious level, are the algorithms for parking a car, crossing an intersection, transferring an object, buying a house, etc.

To illustrate the notion of a fuzzy decisional algorithm, we shall consider two simple examples drawn from our everyday experiences.

Example—Crossing a traffic intersection: It is convenient to break down the algorithm in question into several subalgorithms, each of which applies to a particular type of intersection. For our purposes, it will be sufficient to describe only one of these subalgorithms, namely, the subalgorithm SIGN, which is used when the intersection has a stop sign. As in the case of other examples in this section, we shall make a number of simplifying assumptions in order to shorten the description of the algorithm.

Algorithm INTERSECTION:

1) IF signal lights THEN CALL SIGNAL ELSE IF stop sign THEN CALL SIGN ELSE IF blinking light THEN CALL BLINKING ELSE CALL UNCONTROLLED.

Subalgorithm SIGN:

1) IF no stop sign on your side THEN IF no cars in the intersection THEN cross at *normal* speed ELSE wait for cars to leave the intersection and then cross.
2) IF not *close* to intersection THEN continue approaching at normal speed for a *few* seconds; go to 2).
3) *Slow down.*
4) IF in a *great* hurry and no police cars in sight and no cars in the intersection or its *vicinity* THEN cross the intersection at *slow* speed.
5) IF *very close* to intersection THEN stop; go to 7).
6) Continue *approaching* at *very slow* speed; go to 5).
7) IF no cars *approaching* or in the intersection THEN cross.
8) Wait a *few* seconds; go to 7).

It hardly needs saying that a realistic version of this algorithm would be considerably more complex. The important point of the example is that such an algorithm could be constructed along the same lines as the highly simplified version just described. Furthermore, it shows that a fuzzy algorithm could serve as an effective means of communicating know-how and experience.

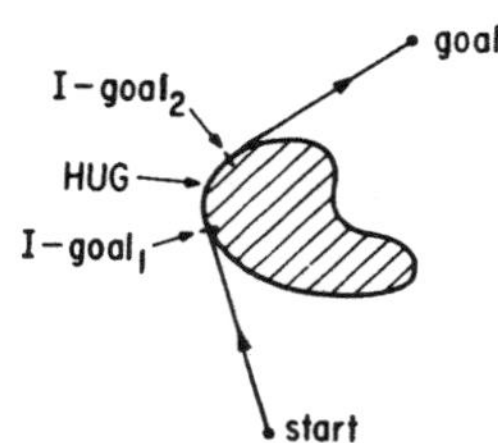

Fig. 4. Problem of transferring blindfolded subject from *start* to *goal*.

As a final example, we consider a decisional algorithm for transferring a blindfolded subject H from an initial position *start* to a final position *goal* under the assumption that there may be an obstacle lying between *start* and *goal* (see Fig. 4). (Highly sophisticated nonfuzzy algorithms of this type for use by robots are incorporated in Shakey, the robot built by the Artificial Intelligence Group at Stanford Research Institute. A description of this robot is given in [20].)

The algorithm, labeled OBSTACLE, is assumed to be used by a human controller C who can observe the way in which H executes his instructions. This fuzzy feedback plays an essential role in making it possible for C to direct H to *goal* in spite of the fuzziness of instructions as well as the errors in their execution by H. The algorithm OBSTACLE consists of three subalgorithms: ALIGN, HUG, and STRAIGHT. The function of STRAIGHT is to transfer H from *start* to an intermediate goal $I\text{-}goal_1$, and then from $I\text{-}goal_2$ to *goal*. (See Fig. 4.) The function of ALIGN is to orient H in a desired direction; the function of HUG is to guide H along the boundary of the obstacle until the goal is no longer obstructed.

Instead of describing these subalgorithms in terms of fuzzy conditional statements as we have done in previous examples, it is instructive to convey the same information by flowcharts, as shown in Figs. 5–7. In the flowchart of ALIGN, ε denotes the error in alignment, and we assume for simplicity that ε has a constant sign. The flowcharts of HUG and STRAIGHT are self-explanatory. Expressed in terms of fuzzy conditional statements, the flowchart of STRAIGHT, for example, translates into the following instructions.

Subalgorithm STRAIGHT:

1) IF not *close* THEN take a step; go to 1).
2) IF not *very close* THEN take a *small* step; go to 2).
3) IF not *very very close* THEN take a *very small* step; go to 3).
4) Stop.

VII. Concluding Remarks

In this and the preceding sections of this paper, we have attempted to develop a conceptual framework for dealing

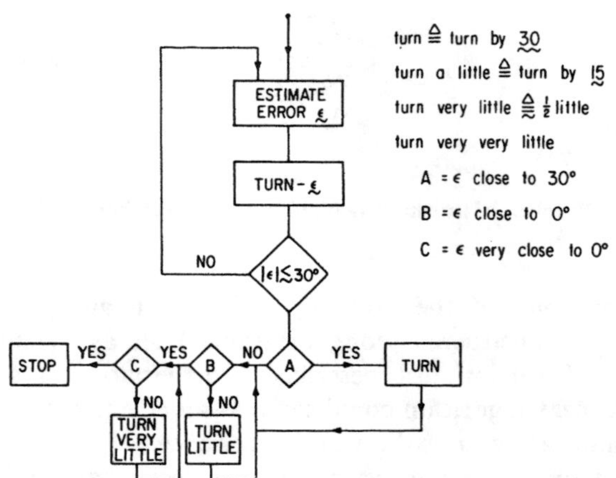

Fig. 5. Subalgorithm ALIGN.

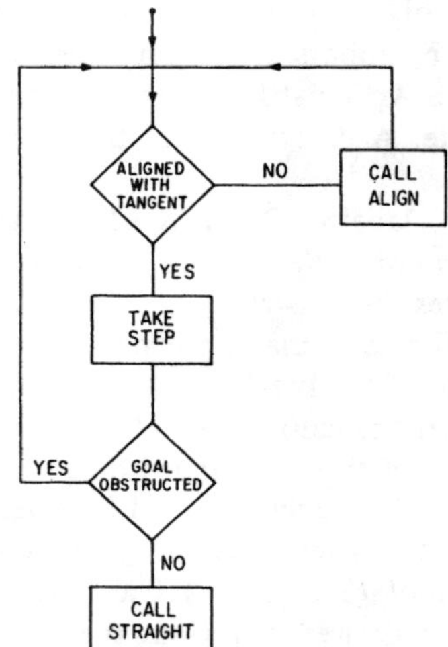

Fig. 6. Subalgorithm HUG.

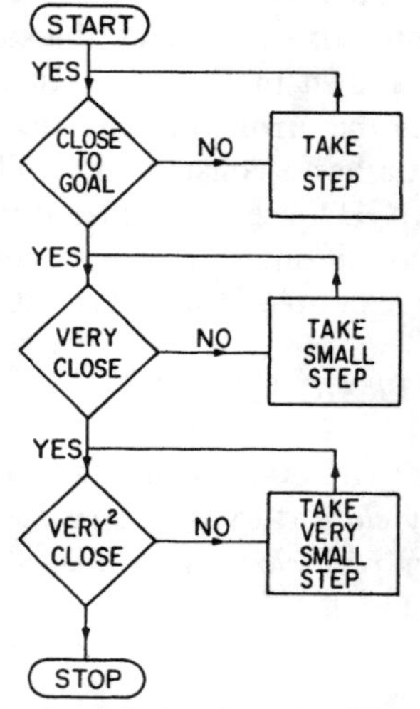

Fig. 7. Subalgorithm STRAIGHT.

with systems which are too complex or too ill-defined to admit of precise quantitative analysis. What we have done should be viewed, of course, as merely a first tentative step in this direction. Clearly, there are many basic as well as detailed aspects of our approach which we have treated incompletely, if at all. Among these are questions relating to the role of fuzzy feedback in: the execution of fuzzy algorithms; the execution of fuzzy algorithms by humans; the conjunction of fuzzy instructions; the assessment of the goodness of fuzzy algorithms; the implications of the compositional rule of inference and the rule of the preponderant alternative; and the interplay between fuzziness and probability in the behavior of humanistic systems.

Nevertheless, even at its present stage of development, the method described in this paper can be applied rather effectively to the formulation and approximate solution of a wide variety of practical problems, particularly in such fields as economics, management science, psychology, linguistics, taxonomy, artificial intelligence, information retrieval, medicine, and biology. This is particularly true of those problem areas in these fields in which fuzzy algorithms can be drawn upon to provide a means of description of ill-defined concepts, relations, and decision rules.

References

[1] L. A. Zadeh, "Fuzzy sets," *Inform. Contr.*, vol. 8, pp. 338–353, 1965.

[2] ——, "Similarity relations and fuzzy orderings," *Inform. Sci.*, vol. 3, pp. 177–200, 1971.

[3] ——, "Toward a theory of fuzzy systems," in *Aspects of Network and System Theory*, R. E. Kalman and N. DeClaris, Eds. New York: Holt, Rinehart and Winston, 1971.

[4] ——, "Quantitative fuzzy semantics," *Inform. Sci.*, vol. 3, pp. 159–176, 1971.

[5] ——, "Fuzzy languages and their relation to human and machine intelligence," in *Proc. Conf. Man and Computer*, 1970; also Electron. Res. Lab., Univ. California, Berkeley, Memo. M-302, 1971.

[6] ——, "Fuzzy algorithms," *Inform. Contr.*, vol. 12, pp. 94–102, 1968.

[7] E. Santos, "Fuzzy algorithms," *Inform. Contr.*, vol. 17, pp. 326–339, 1970.

[8] L. A. Zadeh, "On fuzzy algorithms," Electron. Res. Lab., Univ. California, Berkeley, Memo. M-325, 1971.

[9] S.-K. Chang, "On the execution of fuzzy programs using finite-state machines," *IEEE Trans. Comput.*, vol. C-21, pp. 241–253, Mar. 1972.

[10] S. S. L. Chang and L. A. Zadeh, "Fuzzy mapping and control," *IEEE Trans. Syst., Man, Cybern.*, vol. SMC-2, pp. 30–34, Jan. 1962.

[11] R. E. Bellman and L. A. Zadeh, "Decision-making in a fuzzy environment," *Management Sci.*, vol. 17, pp. B-141–B-164, 1970.

[12] J. A. Goguen, "The logic of inexact concepts," *Syn.*, vol. 19, pp. 325–373, 1969.

[13] G. Lakoff, "Hedges: a study in meaning criteria and the logic of fuzzy concepts," in *Proc. 8th Reg. Meet. Chicago Linguist. Soc.*, 1972.

[14] L. A. Zadeh, "A system-theoretic view of behavior modification," Electron. Res. Lab., Univ. California, Berkeley, Memo. M-320, 1972.

[15] ——, "A fuzzy-set-theoretic interpretation of hedges," Electron. Res. Lab., Univ. California, Berkeley, Memo. M-335, 1972.

[16] A. De Luca and S. Termini, "A definition of a non-probabilistic entropy in the setting of fuzzy sets theory," *Inform. Contr.*, vol. 20, pp. 301–312, 1972.

[17] R. C. T. Lee, "Fuzzy logic and the resolution principle," *J. Ass. Comput. Mach.*, vol. 19, pp. 109–119, 1972.

[18] G. E. Hughes and M. J. Cresswell, *An Introduction to Modal Logic*. London: Methuen, 1968.

[19] R. S. Ledley, *Fortran IV Programming*. New York: McGraw-Hill, 1966.

[20] B. Raphael, R. Duda, R. E. Fikes, P. E. Hart, N. Nilsson, P. W. Thorndyke, and B. M. Wilbur, "Research and applications—artificial intelligence," Stanford Res. Inst., Menlo Park, Calif., Final Rep., Oct. 1971.

Article 6.3

FUZZY LOGIC AND NEURAL NETWORKS

Madan M. Gupta
Intelligent Systems Research Laboratory
College of Engineering
University of Saskatchewan
Phone (306) 966-5451, Fax (306) 966-8710

ABSTRACT

In this paper we give some basic principles of fuzzy neural computing using synaptic and somatic operations. We first briefly review the neural systems based upon conventional algebraic synaptic (confluence) and somatic (aggregation) operations. Then we provide a detailed neuronal morphology based upon fuzzy logic and its generalization in the form of T-operators. For such fuzzy logic based neurons, we then develop the learning and adaptation algorithm.

Keywords: Fuzzy systems, neural systems, synaptic and somatic operations, fuzzy logic.

1. Introduction

In this paper we elucidate the basic principle of neural computing - the neuronal morphology of the brain, and an artificial neural computing systems [1-3]. In the conventional neural computing systems, the main mathematical tools employed are based upon algebraic and differential calculus. In the development of the cognitive machine we employ the learning and adaptation strengths of neural networks and the cognitive attributes found in fuzzy logic and its generalization in the form of T-operators. Then, using the generalized form of fuzzy logic, we develop the fuzzy neural computing paradigms and extend the neural operations to the synaptic (confluence) and somatic (aggregation) operations.

The theory presented in this brief exposition follows a pedagogical style, and starts from the very basic notion of neural computing and fuzzy logic learning to the advanced theory of cognitive neural computing.

The contents of the paper are: a brief introduction to the biological cognitive faculty and the biological neurons - the basic computing elements, artificial neurons, introduction to fuzzy logic based neurons with learning and adaptation algorithm.

2. The Neuron: Synaptic and Somatic Neural Operations

Nature has developed a very complex neural structure in most biological species [2-3]. The biological neurons, over one hundred billion, in the central nervous systems (CNS) of humans play a very important role in various complex sensory, control and cognitive aspects of information processing and decision making [3-5]. In neuronal information processing there are a variety of complex mathematical operations and mapping functions involved which synergically act in a parallel-cascade structure forming a complex pattern of neuronal layers evolving into a sort of pyramidical pattern. The information flows from one neuronal layer to another in the forward direction with continuous feedback evolving into a dynamic pyramidical structure. The pyramidical structure is in the sense of the extraction and convergence of information at each point in the forward direction. Biological neuronal morphology (structure), indeed, provides a clue to as well as a challenge in the design of a realistic cognitive computing machine. In this paper we will provide only some basic insights into neuronal mathematical operations decimated into (i) *synaptic* (confluence) operations and (ii) *somatic* (aggregation and nonlinear mapping) operations. We will consider only a single neuron and study some of its intrinsic properties.

Reprinted with permission from *Proceedings of the Tenth International Conference on Multiple Criteria Decision Making* (*TAIPEI '92*), vol. 3, pp. 281–294, July 19–24, 1992, Japan.

From the neuro-biological, as well as the neuro-mathematical point of view, we identify the two key neuronal elements in a biological neuron which are responsible for providing neuronal attributes such as learning, adaptation, knowledge (storage or memory to past experience), aggregation and nonlinear mapping operations on neuronal information [1-4].

(i) *Synapse*: Synapse is a storage element of the past experience (knowledge); it learns from the neuronal environment and continuously adapts its strength. The past experience appears in the form of synaptic strength, what we call synaptic weight. There are over one hundred billion neurons in our central nervous systems and, on the average, there are over 1000 synapses per neuron. Mathematically, a synapse provides a confluence operation between the new neuronal inputs and the past experience, and sends signals to the dendrite - the input to the main body (soma) of the neuron. We will refer to the mathematical operation in synapse as '*synaptic operation*' or '*synaptic confluence operation*'.

(ii) *Soma*: Soma (neural cell) refers to the main body of the neuron. It receives signals from a very few to over 1000 synapses through its dendrites and provides an aggregation operation. If the aggregated value of the dendritic inputs exceeds a certain threshold, it fires, providing an axonal (output) signal which is a sort of nonlinear function of the aggregated value. We will refer to the mathematical operations in soma as '*somatic operation*'. In the somatic operation, one can identify three distinct mathematical operations: (i) aggregation, (ii) thresholding, and (iii) nonlinear transformation (mapping).

Keeping in view the synaptic and somatic operations briefly described above, we depict a neuron as a mathematical processor which receives an n - dimensional input signal vector $\mathbf{x}(t) \in R^n$ and yields a scalar axonic output $y(t) \in R^1$. Mathematically, the neural process *Ne* can be depicted as a mapping operation from the neural input vector $\mathbf{x}(t) \in R^n$ to the scalar neural output $y(t) \in R$ as

$$Ne: \mathbf{x}(t) \in R^n \rightarrow y(t) \in R^1 \qquad (1)$$

where, $\mathbf{x}(t) = [x_1(t), x_2(t) \ldots x_i(t) \ldots x_n(t)]^T \in R^n$.

Alternatively, we write

$$y(t) = Ne\,[\mathbf{x}(t) \in R^n] \in R^1 \qquad (2)$$

Figure 2 shows a typical neuron depicting the mathematical operations of Equations (1) and (2).

A biological neuron, as we have described above, provides two distinct mathematical operations distributed over the *synapse* (the junction point between an axon and the dendrite) and the *soma*, the main body of the neuron. We will call these two neuronal mathematical operations; (i) *synaptic operations*, and (ii) *somatic operations*. These two neuronal operations, which play two distinct mathematical functions in a biological neuron, are shown in Figure 3. From the biological point of view, these two operations are physically separate. However, we will show shortly that from a mathematical point of view, certain attributes such as thresholding in soma can be transferred to the synaptic operation. Also, the mathematics of the synaptic and somatic operations can be customized, subject to certain rules of course, according to the task in hand. Now, in the following paragraphs, we will provide a general mathematical description of synaptic and somatic operations.

(i) *Synaptic Operation*: The *synaptic weighting vector* $\mathbf{w}(t) \in R^n$ at the junction point between the neural input and the dendrite provides a storage (memory) to the past experience (knowledge-base). Thus, the *synaptic weight*, $w_i(t)$, may be viewed as a representation of the past experience but which has the ability to adapt to the new experience (*learning attribute*). The synaptic operation provides a *confluence* operation between the past experience $\mathbf{w}(t) \in R^n$ and the neural inputs $\mathbf{x}(t) \in R^n$: Thus, the *synaptic confluence operation*, or just the *synaptic operation*, assigns a relative weight (significance) to each incoming neural signal component $x_i(t)$ according to the past experience (knowledge) stored in $w_i(t)$. The weighted synaptic output (dendrite signal) can be written as

$$z_i(t) = w_i(t) \circledcirc x_i(t), i = 1, 2 \ldots n \quad (3)$$

where $\circledcirc$ is the synaptic confluence, operation.

The confluence operator, $\circledcirc$, as will be explained in the following sections, can be modeled by mathematical operations such as **product** and logical generalized **AND** operations. We define the synaptic operations for non-fuzzy and fuzzy signals as follows.

(a) For non-fuzzy signals, $x_i(t)$, $w_i(t) \in (-\infty, \infty)$, and we define the confluence operation $\circledcirc$ by the product operation. Thus,

$$z_i(t) = w_i(t) \cdot x_i(t), i = 1, 2 \ldots n. \quad (4)$$

(b) For signals (fuzzy signals) bounded by the graded membership over the unit interval [0, 1], we define the confluence operation by the generalized **AND** operation [7-11]. The generalized **AND** operation can be expressed using the notion of triangular norms (T-norms) [9-11]. Thus, we define the logical synaptic confluence operation as

$$z_i(t) = w_i(t) \textbf{ AND } x_i(t), i = 1, 2 \ldots n, \in [0, 1]. \quad (5)$$

For such signals (binary or fuzzy)* , the signal vector $\mathbf{x}(t) \in R^n$ as well as the weighting vector $\mathbf{w}(t) \in R^n$ each is defined over the unit hypercube $[0, 1]^n$.

(ii) ***Somatic Operation***: Figure 2.

The somatic operation is carried out in the neural cell, the main body of the neuron, on the weighted neural input signals $z_i(t)$, $i = 1, 2 \ldots n$, and is a two step process as explained below.

(a) ***Somatic Aggregation operation***: Figure 2.

The first somatic operation is the aggregation operation on the dendritic input signals $z_i(t)$, $i = 1, 2 \ldots n$, which essentially maps an nth order vector $z(t) \in R^n$ into a scalar signal $u(t) \in R^n$.

For this somatic operation we introduce the generalized aggregation operation ϕ . Thus, we have

$$u(t) = \mathop{\phi}_{i=1}^{n} z_i(t) \quad (6a)$$

or

$$u(t) = \mathop{\phi}_{i=1}^{n} w_i(t) \circledcirc x_i(t) \quad (6b)$$

The aggregation operation $\phi[\cdot]$ can be modeled by mathematical operations such as summation and logical generalized **OR** operations. We define this operation for non-fuzzy and fuzzy signals as follows:

(i) For non-fuzzy signals

$x_i(t)$, $w_i(t) \in (-\infty, \infty)$, we define aggregation operation as

$$u(t) = \sum_{i=1}^{n} z_i(t) \in R^1 \quad (7a)$$

---

* Binary logic is a subset of fuzzy logic.

or

$$u(t) = \sum_{i=1}^{n} w_i(t) \cdot x_i(t) \in R^1. \tag{7b}$$

One can view this aggregation operation as a linear mapping from n-dimensional dendritic inputs $\{z_i(t)\}$ to one dimensional space, Eqn. (7a). This somatic linear mapping operation can be combined with the synaptic confluence operation yielding a linear weighted mapping from n-dimensional neural input $\mathbf{x}(t) \in R^n$ to one dimensional space $u(t) \in R^1$, (7b). Alternatively, expressing this neuronal operation as a scalar product of two vectors, we write,

$$u(t) = \mathbf{w}^T(t) \cdot \mathbf{x}(t) \in R^1 \tag{7c}$$

where

$$\mathbf{w}(t) = [w_1(t), w_2(t) \ldots w_n(t)]^T \in R^n$$
$$= \text{vector of synaptic weights,}$$

and

$$\mathbf{x}(t) = [x_1(t), x_2(t) \ldots x_n(t)]^T \in R^n$$
$$= \text{vector of neural inputs.}$$

Thus, the combined synaptic weighting and somatic aggregation operations provide a linear mapping of neural inputs $\mathbf{x}(t) \in R^n$ to $u(t) \in R^1$.

(ii) For fuzzy signals bounded by the graded membership over the unit interval [0, 1], we define the aggregation operation by the generalized **OR** operation [9-11]. The generalized **OR** operation can be expressed using the notion of triangular conorm (T-conorm). Thus, we define the logical somatic aggregation operation as,

$$u(t) = \underset{i=1}{\overset{n}{\mathbf{OR}}} \; [z_i(t)] \in [0, 1] \tag{8a}$$

$$= \underset{i=1}{\overset{n}{\mathbf{OR}}} \; [w_i(t) \; \mathbf{AND} \; x_i(t)] \in [0, 1]. \tag{8b}$$

Again, one can view this aggregation operation as a logical mapping from an n-dimensional neural input $\mathbf{x}(t) \in [0, 1]^n$, to one dimensional space $u(t) \in [0, 1]$ as shown in (8b).

Expressing this neural operation as a scalar logical **AND** of two vectors, the neural input vector and the synaptic weighting vector, we write

$$u(t) = \mathbf{w}^T(t) \; \mathbf{AND} \; \mathbf{x}(t) \tag{8c}$$

where $\mathbf{w}(t), \mathbf{x}(t) \in [0, 1]^n$, and the vector **AND**ed operation is combined with individual **OR**ed operation for $i = 1, 2, \ldots n$.

(ii) *Nonlinear Somatic Operation with Thresholding*:

The biological neuronal processes generate some interesting mathematical mapping properties because of their nonlinear operations combined with a thresholding in the soma. As a matter of fact, if the neuron would carry out only linear operations, the mathematical attractiveness as well as the robustness in many applications will disappear. The purpose of this section is to briefly explore these properties for their applications to neural computing systems.

We will consider this important somatic operation again for two different situations, non-fuzzy signals and fuzzy signals.

(i) For non-fuzzy signals, $u(t) \in (-\infty, \infty)$, we define the somatic nonlinear mapping operation on $u(t)$ yielding a neural output $y(t)$ as

$$y(t) = f[u(t), w_o] \in R \tag{9}$$

where the nonlinear function $f[\cdot]$ with thresholding w_o is shown in Figure 4. It should be noted, Figure 3, that the neural output $y(t)$ is zero if the weighted aggregate $u(t)$ of the neural input signal $\mathbf{x}(t) \in R^n$ is less than the threshold value w_o. That is, the neuron will fire (will provide an output) only if the weighted aggregate of $\mathbf{x}(t) \in R^n$ exceeds the threshold w_o. If $u(t)$ exceeds w_o, the neural output $y(t)$ increases monotonically with increasing $u(t)$ to a saturation value, say 1. Depending upon the mapping properties (shape) of the nonlinear function $f[\cdot]$, the value of $y(t)$ is distributed over the interval [0, 1].

(ii) For fuzzy signals with bounded graded membership $u(t)$ over the unit interval [0, 1], we define the nonlinear mapping with thresholding $w_o \in [0, 1]$ as follows:

$$v(t) = u(t) \textbf{ OR } w_o, \; v(t) \in [0, 1] \tag{10a}$$

and, $$y(t) = v^{\alpha}(t) \tag{10b}$$

where α is a positive constant. For $0 < \alpha \leq 1$, the operation (10b) provides a contrast dilation to the membership $v(t)$, whereas for $\alpha > 1$ it provides a concentration operation.

The contrast dilation, has the property of increasing the membership value, whereas the concentration, decreases it. The contrast intensification decreases the membership value for $0 < v \leq 0.5$, and increases it for $0.5 < v \leq 1$. The blurring operation increases the membership value for $0 < v \leq 0.5$ and decreases it for $0.5 < v \leq 1$.

Here, we have provided a few useful nonlinear operations on the graded membership v, the nonlinear operations can be expanded to some other nonlinear and or logical operations.

In summary, a single neuron can be viewed as a nonlinear mathematical operator or logical operator providing a mapping from many neural inputs, $\mathbf{x}(t) \in R^n$, to a single neural output, $y(t) \in R^1$. Following the analogy of a biological neuron, we have divided the neural mathematical operations into two parts: (i) *the synaptic confluence operation*, and (ii) *the somatic aggregation and nonlinear mapping with thresholding (bias).* We have defined these operations for signals over the interval $(-\infty, \infty)$, and for fuzzy signals with graded membership over the unit interval [0, 1]. It is to be noted again that the neural processor modeled in this section may differ in many ways with its biological counterpart, however, they still retain many attributes for the confluence operation in the synapse and aggregation, thresholding and nonlinear operations of the somas. We will combine some of these operations, from the mathematical point of view, in order to produce a generalized mathematical framework for a neural processor [3].

3. Generalized Mathematical Model of a Neuron

In Section 2, we developed a model providing synaptic and somatic operations using a biological plausibility. In this section, we will generalize this mathematical model for real-valued non-fuzzy neural inputs and will devote the next section exclusively to fuzzy signals.

Recalling Equation (9) and Figure 3 we define a new variable $v(t)$ as

$$v(t) = u(t) - w_o \tag{11}$$

where $u(t)$ is the weighted aggregate value of neural inputs defined in (7), and w_o is the threshold (bias). Thus, if the weighted output $u(t)$ is less than w_o, the neural output $y(t)$ is zero. This implies that the neuron will fire only when the weighted aggregate value exceeds the threshold w_o. Thus, we redefine $y(t)$ as

$$y(t) = f[u(t),\ w_o] = \phi[v(t)] \tag{12}$$

and using Equation (7) and (11) we write the neural output y(t) as,

$$v(t) = \underset{i=0}{\overset{n}{\phi}} \quad w_i(t) \circledcirc x_i(t) \tag{13a}$$

$$y(t) = \phi[v(t)] \tag{13b}$$

where $\phi[v(t)]$ is shown in Figure 4 with the threshold shifted to the origin, and © and ϕ represent the generalized confluence and aggregation operations defined in (6).

As a special case, representing the confluence operation by product, and aggregation operation by summation, we rewrite (12) and (13) as

$$v(t) = \sum_{i=0}^{n} w_i(t) \cdot x_i(t)$$

$$= \mathbf{w}_a^T(t) \cdot \mathbf{x}_a(t) \in R^1 \tag{13c}$$

and

$$y(t) = \phi[v(t)] \in R^1 \tag{13d}$$

where $\phi[\cdot]$ is a sigmoidal nonlinear mapping function, and $\mathbf{x}_a(t)$ is the augmented vectors defined as

$$\mathbf{w}_a(t) = [w_o, w_1, w_2 \ldots w_2]^T \in R^{n+1}$$

= augmented vector of synaptic weights including the threshold w_o

$$\mathbf{x}_a(t) = [x_o, x_1, x_2 \ldots x_2]^T \in R^{n+1}, x_o = 1$$

= augmented vector of neural inputs, where $x_o = 1$ accounts for the threshold (bias) term.

A generalized neural model with threshold shifted to synapse is shown in Figure 5.

4. Fuzzy Logic Based Neural Morphology

In Section 2 we developed basic synaptic and somatic neuronal operations using a biological plausibility. In Section 3, we then described a generalized mathematical model of a neuron. This mathematical model was developed in terms of a generalized confluence operation, ©, and generalized aggregation operation, ϕ. In this section, we will use these generalized mathematical operations to develop a fuzzy logic based neural morphology [3]. This treatment is valid for signals, not necessarily fuzzy, but which are bounded over the interval [0, 1] or [-1, 1].

4.1 Fuzzy Logic: Brief introduction [5-10]

The theory of fuzzy logic is developed in order to capture the uncertainties associated with human cognitive processes such as in thinking, reasoning, perception, etc. "Driving conditions in Saskatoon in January, when it is *extremely cold,* are not so *bad* compared to what they are in *late* February or *early* March, especially when there is a *large* temperature variation between day time (*snow is melting*) and night time (*freezing cold*)" In this statement, the words in italics express various vague notions, which can be modeled using the notion of graded membership inherent in fuzzy set theory [5-10]. The theory of fuzzy logic, unlike that of the binary logic, is developed on the notion of graded membership distributed over the closed unit interval [0, 1].

If we express our neural input signal in terms of their membership functions each over the interval [0, 1], rather than in their absolute amplitudes, then we can write the augmented vector of neural inputs as

$$x_a(t) = [x_o(t), x_1(t) \ldots x_i(t) \ldots x_n(t)]^T \in [0, 1]^{n+1}$$

where these neural signal (including the bias term, x_o) are bounded by the (n + 1) dimensional hybercube $[0, 1]^{n+1}$.

Similarly, the augmented synaptic weighting vector $w_a(t)$ can be expressed over the unit hybercube $[0, 1]^{n+1}$.

We perform mathematical operations on these signals using logical operations (connectives) such as **OR**, **AND** (or their generalized form based upon triangular norm or T-operators) and **negation.**

Let us express x_1 and x_2 over [0, 1], then we define, **AND** (T-operation) as a **T** mapping function [8-10]:

T: $[0, 1] \times [0, 1] \rightarrow [0, 1]$ given by

$$y_1 = [x_1 \textbf{ AND } x_2] \triangleq [x_1 \textbf{ T } x_2] = \textbf{T}[x_1, x_2] \quad (15)$$

Similarly, we define the generalized **OR**, (T-conorm) as a **S** mapping function

S: $[0, 1] \times [0, 1] \rightarrow [0, 1]$ given by

$$y_2 = [x_1 \textbf{ OR } x_2 \triangleq [x_1 \textbf{ S } x_2] = \textbf{S}[x_1, x_2] \quad (16)$$

Negation **N** on $x_1 \in [0, 1]$ is defined as a mapping:

N: $[0, 1] \rightarrow [0, 1]$ with the following properties:

$$y_3 = \textbf{N}[x_1] = 1 - x_1 \quad (17)$$

Thus, N(0) = 1, N(1) = 1,
and N(N(x) = x.

Now, we give some important properties of the **T** and **S** operators.

$$\textbf{T}(0, 0) = 0, \quad \textbf{T}(1, 1) = 1, \quad \textbf{T}(1, x) = x, \quad \textbf{T}(x, y) = \textbf{T}(y, x) \quad (18)$$

$$\textbf{S}(0, 0) = 0, \quad \textbf{S}(1, 1) = 1, \quad \textbf{S}(0, x) = x, \quad \textbf{S}(x, y) = \textbf{S}(y, x) \quad (19)$$

Also, De'Morgan's Theorems are stated as follows:

$$\textbf{T}(x_1, x_2) = 1 - \textbf{S}(1 - x_1, 1 - x_2) \text{ and, } \textbf{S}(x_1, x_2) = 1 - \textbf{T}(1 - x_1, 1 - x_2) \quad (20)$$

4.2 Fuzzy Logic Neurons

In the development of fuzzy logic based neural morphology, we will use the following combined synaptic and somatic operations:

Let the augmented vector of neural inputs and synaptic weights be represented by

$$x_a(t) \in [0, 1]^{n+1}, \text{ and } w_a(t) \in [0, 1]^{n+1}$$

respectively, Then, in (13a), by replacing the © - operation by the **T** - operation, and the ϕ - operation by the **S** - operation, we get

$$v(t) = \mathop{\textbf{S}}_{i=0}^{n} [w_i(t) \textbf{ T } x_i(t)) \in [0, 1] \quad (21a)$$

and

$$y(t) = \phi[v(t)] \in [0, 1] \quad (21b)$$

where the nonlinear mapping $\phi[\cdot]$ may be one of the functions defined in (10b), or some other equivalent functions.

5. Unipolar to Bipolar Transformation

The logical operations defined in the preceding section are unipolar signals over the positive unit interval [0, 1]. Such logical operations provide only the neural state corresponding to the excitatory (positive) interactions. In order to consider both the excitatory (positive) and the inhibitory (negative) interactions, of the neural input vector, we must consider both $\mathbf{x}_a(t)$ and its negated values $N[\mathbf{x}_a(t)]$, thus making the neural inputs of dimensions (2n + 2).

Alternatively, we may express the neural inputs and synaptic weights as bipolar signals and weights over the interval [-1, 1] and redefine the logical operations over this interval. We will provide a brief description of this transformation for unipolar, [0, 1], to bipolar [-1, 1], and of the definition of logical operations over the interval [-1, 1].

Let $x(t) \in [0, 1]$ be a unipolar signal. The corresponding bipolar signal z(t) is defined as

$$z(t) = 2\,x(t) - 1 \tag{22}$$

The negation is defined as

$N[x] = 1 - x$, for unipolar signal,

$N[z] = -z$, for bipolar signal.

In Table 1, we give a summary of logical **T** and **S** operations for both unipolar and bipolar signals. Also in Table 2, we define some important logical functions and operations such as Godel's implication, degree of equality using Godel's implication, degree of equality using Lukasiewicz conjuction, and degree of error (inequality) for two bipolar signals z_1 and $z_2 \in [-1, 1]$.

In the next section, we will present a model of the fuzzy neurons with learning and adaptation algorithm.

Table 1: **Summary of Logical Operations (T-Operations) on Unipolar and Bipolar Signals**

Unipolar Signals $x \in [0,1]$	Bipolar Signal $z \in [-1,1]$
(i) **Bipolar to unipolar transformation** $x = \frac{z+1}{2}$	(i) **U nipolar to bipolar transformation** $z = 2x - 1$
(ii) **Negation:** $N[x] = \bar{x} = 1 - x$	(ii) **Negation:** $N[z] = \bar{z} = -z$
(iii) **Boundary Conditions** (a) **T- operator** (generalized **AND**) $T(0, 0) = 0$; $T(1, 1) = 1$ $T(1, x) = x$; $T(x_1, x_2) = T(x_2, x_1)$ (b) **S-Operator** (generalized **OR**) $S(0, 0) = 0$; $S(1, 1) = 1$ $S(0, x) = x$; $S(x_1, x_2) = S(x_2, x_1)$	(iii) **Boundary Conditions** (a) **T- operator** (generalized **AND**) $T(-1, -1) = -1$; $T(1, 1) = 1$ $T(1, z) = z$; $T(z_1, z_2) = T(z_2, z_1)$ (b) **S-Operator** (generalized **OR**) $S(-1, -1) = -1$; $S(1, 1) = 1$ $S(-1, z) = z$; $S(z_1, z_2) = S(z_2, z_1)$
(iv) **Generalized De'Morgan's Theorem** $T(x_1, x_2) = 1 - S\,(1 - x_1, 1 - x_2)$ $S(x_1, x_2) = 1 - T\,(1 - x_1, 1 - x_2)$	(iv) **Generalized De'Morgan's Theorem** $T(z_1, z_2) = -S\,(-z_1, -z_2)$ $S(z_1, z_2) = -T\,(-z_1, -z_2)$

Table 2: **Summary of Some Important Logical Functions and Operations**

(These logical functions and operations can be defined on both the unipolar, [0, 1] and bipolar, [-1, 1], signals, but here, we will consider only the bipolar signals).

(a) **Godel's implication** : Ⓖ

Godel's implication $[z_1 Ⓖ z_2]$ (read as, z_1 implies z_2) is defined as

$$[z_1 Ⓖ z_2] = [z_1 \rightarrow z_2] = \begin{cases} 1, & z_1 \leq z_2 \\ z_2, & z_1 > z_2 \end{cases}$$

(b) **Degree of equality** (using Godel's implication) $\eta(z_1, z_2)$

Given z_1 and z_2 over [-1, 1], *towhat degree they are equal* is defined as

$$\eta(z_1, z_2) = \frac{1}{2}[(\{z_1 Ⓖ z_2\} \mathbf{T} \{z_2 Ⓖ z_1\} + \{\bar{z}_1 Ⓖ \bar{z}_2\} \mathbf{T} \{\bar{z}_2 Ⓖ \bar{z}_1\}] \in [-1, 1]$$

where $\bar{z} = -z$.

(c) **Degree of equality** (using Lukasiewicz's conjunction): $\eta\ (z_1, z_2)$

Again, given z_1 and z_2 over [-1, 1], *to what degree they are equal* is defined as

$$\eta(z_1, z_2) = [1 - |z_1 - z_2|] \in [-1, 1]$$

(d) **Degree of inequality** (degree of error): $E(z_1, z_2) \in [-1, 1]$

Given z_1 and z_2 over [-1, 1], in order to find the degree of difference or degree of inequality, we define

$E[z_1, z_2]$ as the negation on the degree of equality; that is,

$$E[z_1, z_2] = \mathbf{N}[\eta(z_1, z_2)] = -\eta(z_1, z_2).$$

Thus, the degree of error, using Lukasiewicz conjunction can be defined simply as

$$E(z_1, z_2) = |z_1 - z_2| - 1,$$

6. Learning and Adaptation in Fuzzy Logic Neuron

Figure 6 shows a fuzzy logic neuron for bipolar signals and bipolar synaptic weights.

The augmented neural input signals $\mathbf{x}_a(t)$ are defined over the unit hypercube $[0, 1]^{n+1}$. Using the transformation given in (22), we transform the unipolar neural inputs $\mathbf{x}_a(t)$ into bipolar signals $\mathbf{z}_a(t) \in [-1, 1]^{n+1}$.

The logical operation of this neuron is summarized as follows

$$v(t) = \mathop{\mathbf{S}}_{i=0}^{n} [w_i(t)\ \mathbf{T}\ z_i(t)] \tag{23a}$$

which is equivalent to

$$v(t) = \mathbf{w}_a^T(t)\ \mathbf{AND}\ \mathbf{z}_a(t) \in [-1, 1] \tag{23b}$$

(a logical scalar product operation), where $w_o(t)$ and $z_o(t)$ correspond to the bias terms and $z_o = 1$.

The neural output is defined as

$$y = \phi(v) \in [-1, 1] \tag{24a}$$

where $\phi[v]$ is defined as

$$\phi[v] = |v|^{\alpha} \cdot \text{sgn}[v], \alpha > 0. \tag{24b}$$

Let us define an error signal with respect to the desired neural output, $y_d(t) \in [-1, 1]$, as

$$e(t) = y_d(t) - y(t) \in [-1, 1]. \tag{25}$$

The degree of equality and degree of error (inequality) between $y_d(t)$ and $y(t)$ are defined in Table 2.

The objective of learning and adaptation in neural networks is to adapt the parameters of the neural structures, in this case $\mathbf{w}_a(t)$ and α in Eqns. (24a) and (24b), in order to minimize an error function.

We give the adaptive rules for $\mathbf{w}_a(t)$ and $\alpha(t)$ as follows:

$$\mathbf{w}_a(t+1) = \mathbf{w}_a(t) \ \mathbf{OR} \ \Delta\mathbf{w}_a(t) \tag{26a}$$

and

$$\alpha(t+1) = \alpha(t) \ \mathbf{OR} \ \Delta\alpha(t) \tag{26b}$$

where

$$\Delta\mathbf{w}_a(t) = z_a(t) \ \mathbf{AND} \ E(y_d, y) \tag{27a}$$

and

$$\Delta\alpha(t) = v(t) \ \mathbf{AND} \ E(y_d, y) \tag{27b}$$

where $E(y_d, y)$ is the degree (grade) of error between $y_d(t)$ and $y(t)$ as defined in Table 2.

7. Conclusions

In this paper we have developed a generalized neural model using the biological plausibility of synaptic and somatic operations. This generalized model can be specialized for nonfuzzy signals defined over $(-\infty, \infty)$ using the conventional mathematics as widely reported in the literature. However, the main interest of this paper was to develop a basic neuronal morphology using fuzzy logic operations rather than conventional mathematical operations. We also introduced the notion of bipolar signals over the interval [-1, 1]. Such signals may represent a special class of signals or they may just represent the graded membership of fuzzy signals.

The approach followed is rather non-conventional, and we have given the learning and adaptation laws again using logical operations. The next papers in this series will describe the operation of dynamic fuzzy neurons and their applications in signal processing, control and vision.

References

1. Amari, S.I., (1977), "A Mathematical Approach to Neural Systems", in J. Metzler (Ed.), *Systems Neuroscience*, Academic Press, New York, pp. 67-117.
2. Hecht-Nielsen, R., (1990), "Neurocomputing", Addison-Wesley, Reading, MA.
3. Gupta, M.M., [1992], "Introduction to Neural Computing Systems with Applications to Control and Vision", Class Notes, Intelligent Systems Research Laboratory, University of Saskatchewan, Canada.
4. Gupta, M.M., [1991], "Uncertainty and Information The Emerging Paradigms", *International Journal of Neuro and Mass-Parallel Computing and Information Systems*, Vol. 2, pp. 65-70.
5. Bellman, R.E. and Zadeh, L.A., (1970), "Decision Making in a Fuzzy Environment", *Management Science*, Vol. 17, B.141-B, p. 164.

6. Zadeh, L.A., (1973), "Outline of a New Approach to the Analysis of Complex Systems and Decision Processes", *IEEE Trans. Systems Man Cybernetics,* Vol. 3, pp. 28-44.

7 Alsina, C., Trillas, E. and Valverde, L., [1983], "On Some Logical Connectives for Fuzzy Set Theory", *J. Math Anal. Appl.*, Vol. 93, pp. 15-26.

8. Weber, S., (1983), "A General Concept of Fuzzy Connectives, Negations, and Implications Based on t-norms and t-conorms", Fuzzy Sets and Systems, North-Holland, Vol 11, pp. 115-134.

9. Gupta, M.M. and Qi, J. [1991], "Connectives (**AND, OR, NOT**) and T-operators" in Fuzzy Reasoning in Condition Logic in Expert Systems", I.R. Goodman, M.M. Gupta, et al (Editors), Elsevier Science Publishers, pp. 211-234.

10. Gupta, M.M. and Qi, J. [1991], "Theory of T-norms and Fuzzy Inference Methods", Fuzzy Sets and Systems, North-Holland, Vol 40, pp. 431-450.

11. Kaufmann, A. and Gupta, M.M., (1988), "Fuzzy Mathematical Models in Engineering and Management Science", North Holland, Amsterdam, New York, (Also, Japanese Translation, Ohmsha Publication, Tokyo, 1992).

12. Zadeh, L.A., (1965), "Fuzzy Sets, *Information and Control,* Vol. 8, pp. 338-353.

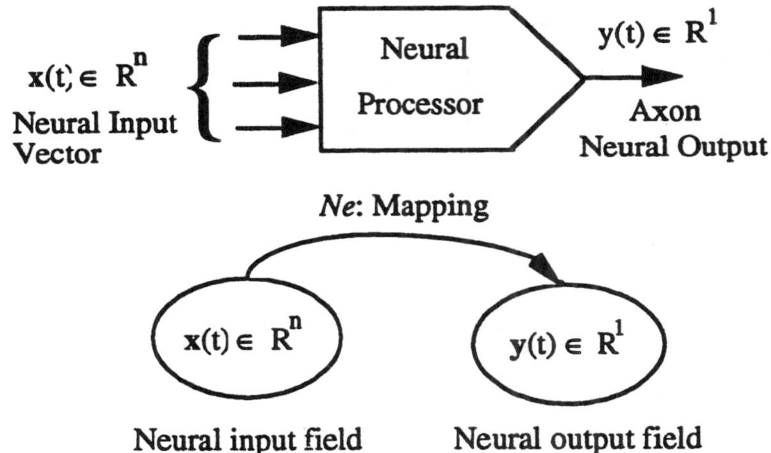

Figure 1. Neuron as a mathematical processor with the mapping function Ne

$$Ne: \mathbf{x}(t) \in R^n \rightarrow y(t) \in R^1.$$

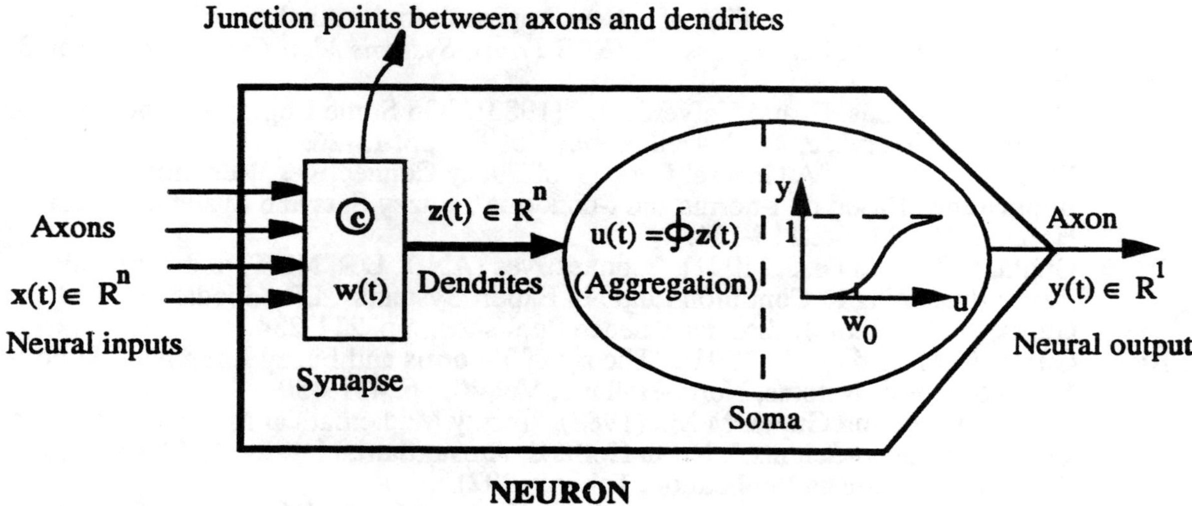

Figure 2. Neuronal *Synaptic* and *Somatic* mathematical operations.

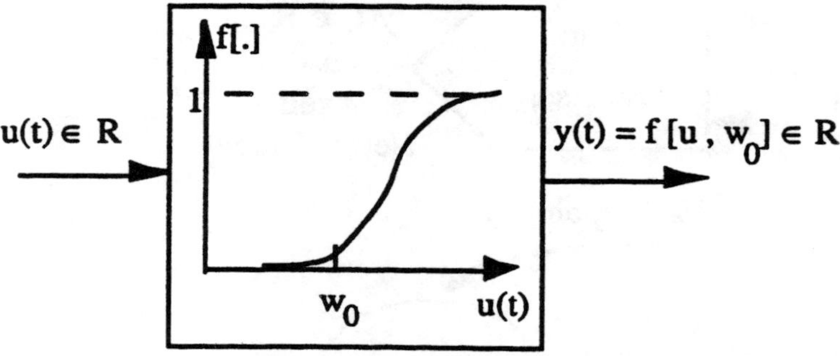

Figure 3. Somatic nonlinear mapping $f[u, w_0]$ with thresholding w_0.

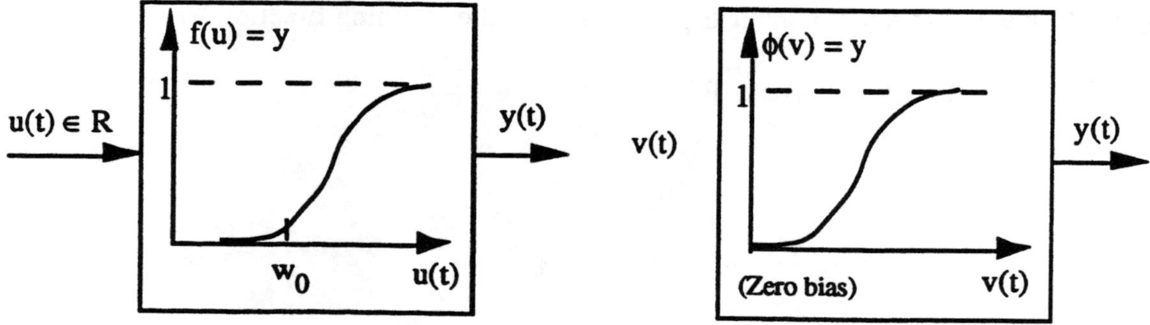

Figure 4. Shifting the threshold w_0 to the origin, $y(t) = f[u, w_0] = \phi[v]$.

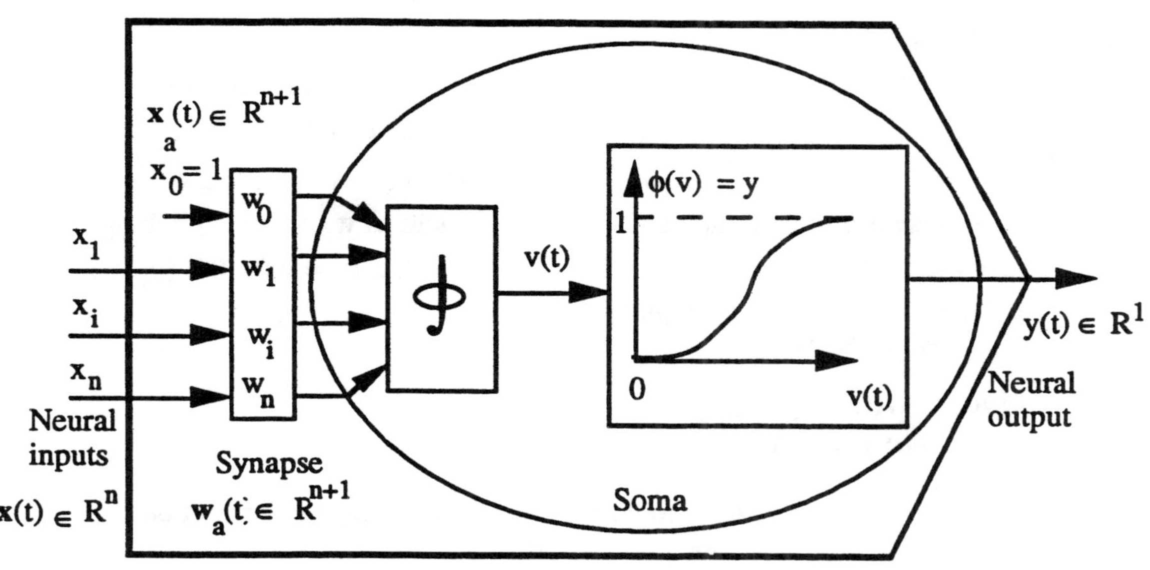

Figure 5. Generalized neural model with augmented vectors $\mathbf{w}_a(t)$, $\mathbf{x}_a(t)$:

$$v(t) = \oint_{i=0}^{n} w_i(t) \circledcirc x_i(t), \quad y(t) = \phi[v(t)].$$

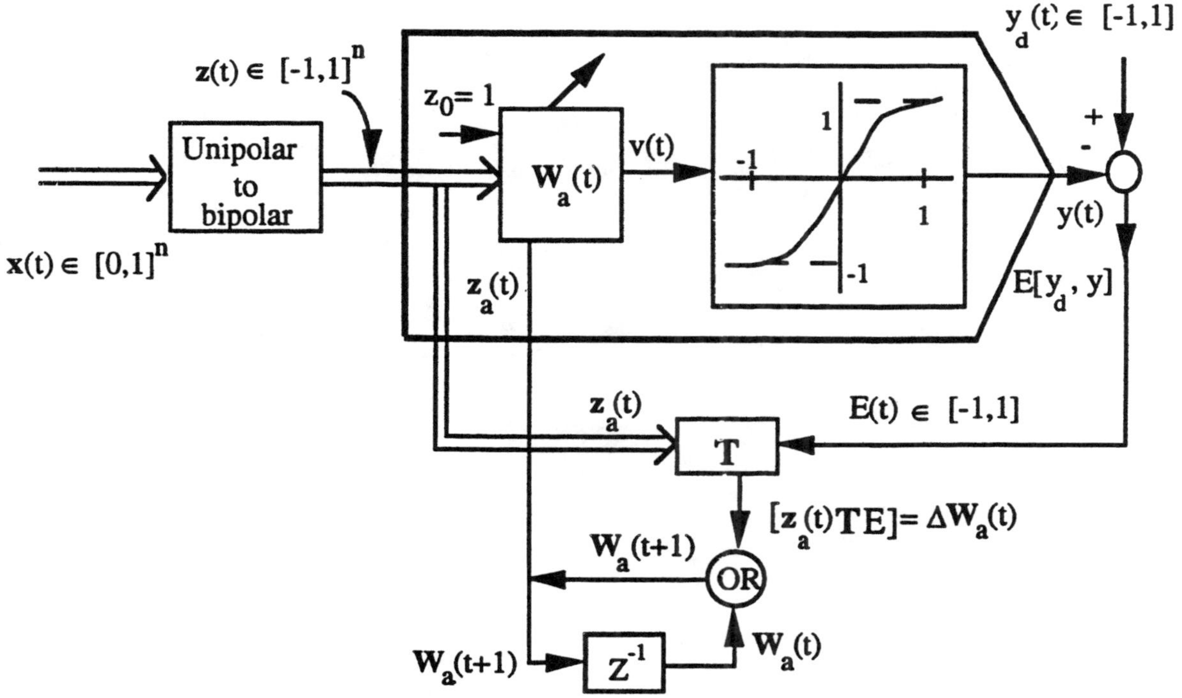

$\mathbf{x}(t) \in [0, 1]^n$

$\mathbf{z}(t) = (2x - 1) \in [-1, 1]^n$

$\mathbf{z}_a(t) = [z_o, \mathbf{z}]^T \; \varepsilon[-1,1]^{n+1}$

(augmented vector of neural inputs)

$\mathbf{w}_a(t) = \in [-1, 1]^{n+1}$

(augmented vector of synaptic weights)

$v(t) = \mathbf{w}_a(t)$ **AND** $\mathbf{x}_a(t)$

$y(t) = \phi[v(t)] = |v(t)|^{\alpha}\, \mathrm{sgn}[v]$,

$\alpha > 0,\; y(t) \in [-1, 1]$

Learning and Adaptation Law

$\mathbf{w}_a(t+1) = \mathbf{w}_a(t)$ **OR** $\Delta\, \mathbf{w}_a(t)$,

$\alpha\,(t+1) = \alpha\,(t)$ **OR** $\Delta\, \alpha\,(t)$,

where, $\Delta\, \mathbf{w}_a(t) = \mathbf{z}_a(t)$ **T** $E(y_d, y)$

and, $\Delta\, \alpha\,(t) = v\,(t)$ **T** $E\,(y_d, y)$

Figure 6. Fuzzy logic neuron with learning and adaptation.

Article 6.4

Self-Learning Fuzzy Controllers Based on Temporal Back Propagation

Jyh-Shing R. Jang, *Student Member, IEEE*

Abstract—**This paper presents a generalized control strategy that enhances fuzzy controllers with self-learning capability for achieving prescribed control objectives in a near-optimal manner. This methodology, termed temporal back propagation, is model-insensitive in the sense that it can deal with plants that can be represented in a piecewise-differentiable format, such as difference equations, neural networks, GMDH structures, and fuzzy models. Regardless of the numbers of inputs and outputs of the plants under consideration, the proposed approach can either refine the fuzzy if–then rules if human experts, or automatically derive the fuzzy if–then rules obtained from human experts are not available. The inverted pendulum system is employed as a test-bed to demonstrate the effectiveness of the proposed control scheme and the robustness of the acquired fuzzy controller.**

I. Introduction

Fuzzy controllers (FC's) have recently found various applications in industry as well as in household appliances. For complex and/or ill-defined systems that are not easily controlled by conventional control schemes, FC's provides a feasible alternative since they can easily capture the approximate, qualitative aspects of human knowledge and reasoning. However, the performance of FC's relies on two important factors: the soundness of knowledge acquisition techniques and the availability of domain (human) experts. These two factors substantially restrict the application domains of FC's.

We have proposed the ANFIS (adaptive-network-based fuzzy inference system) architecture to solve the first problem concerning the automatic elicitation of knowledge in the form of fuzzy if–then rules. The proposed architecture can identify the near-optimal membership functions and other parameters of a rule base for achieving a desired input–output mapping. The basics of the ANFIS architecture are introduced in the next section.

This paper addresses the second problem: how does one control a system through a self-learning FC. In other words, without resorting to human experts, we want to construct an FC that can perform a prescribed control task. The learning aspects of FC's have always been an interesting topic, and recent developments are mostly based on reinforcement learning [2], [9], [10]. Our learning method is based on a special form of gradient descent (called back propagation), which is used for training artificial neural networks [13], [15]. To control the plant's trajectory, we apply the back-propagation-type gradient descent method to propagate the error signals through different time stages. This is called TBP (*temporal back propagation*) and it is explained in Section III.

The proposed control strategy is quite general and can be used to control plants with diverse characteristics. Moreover, the *a priori* knowledge that we have about the plant can be applied in an auxiliary manner to speed up the learning process. In our simulation described in Section IV, we successfully employ the TBP to construct a fuzzy controller with only four fuzzy if–then rules for balancing an inverted pendulum system. The discussion and conclusions are given in Section V.

Manuscript received June 13, 1991; revised November 30, 1991. This work was supported by NASA under Grant NCC-2-275, by LLNL under Grant ISCR 89-12, by the MICRO State Program under Award 90-191, by the MICRO industry, and by Rockwell under Grant B02302532.

The author is with the Department of Electrical Engineering and Computer Sciences, University of California, Berkeley, CA 94720.

IEEE Log Number 9201467.

II. Basics of ANFIS

When employed as a controller, a *fuzzy inference system* is often called a fuzzy controller; therefore we will use the terms fuzzy inference system and fuzzy controller interchangeably throughout this paper. In this section, we describe the basics of *adaptive networks,* which include artificial neural networks as a special case. A special configuration of adaptive networks, ANFIS, is discussed in detail since it is functionally equivalent to a fuzzy controller.

A. Adaptive Networks

An adaptive network (Fig. 1) is a multilayer feedforward network in which each node performs a particular function (*node function*) on incoming signals using a set of parameters specific to this node. The form of node functions may vary from node to node, and the choice of each node function depends on the overall function which the adaptive network is designed to implement.

To reflect different adaptive capabilities, we use both circle and square nodes in an adaptive network. A square node (adaptive node) has modifiable parameters while a circle node (fixed node) has none. The parameter set of an adaptive network is the union of the parameter sets of each adaptive node. In order to achieve a desired input–output mapping, these parameters are updated according to given training data and a gradient-based update procedure described below.

Suppose that a given adaptive network has L layers and the kth layer has $\#(k)$ nodes. We can denote the node in the ith position of the kth layer by (k, i) and its node function (or node output) by O_i^k. Since a node output depends on its

Reprinted from *IEEE Trans. on Neural Networks*, vol. 3, no. 5, pp. 714–723, Sept. 1992.

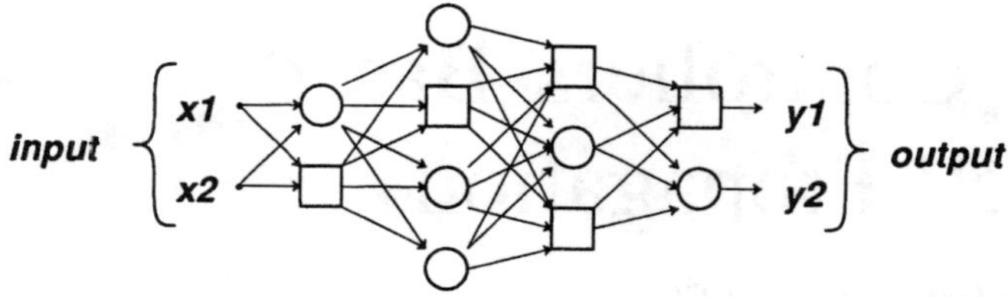

Fig. 1. An adaptive network.

incoming signals and its parameter set, we have

$$O_i^k = O_i^k\left(O_i^{k-1}, \cdots, O_{\#(k-1)}^{k-1}, a, b, c, \cdots\right), \quad (1)$$

where $a, b, c,$, etc. are the parameters pertaining to this node. (Note that we use O_i^k as both the node output and the node function.)

Assuming the given training data set has P entries, we can define the *error measure* for the pth ($1 < p < P$) entry of training data entry as the sum of squared errors:

$$E_p = \sum_{m=1}^{\#(L)} \left(T_{m,p} - O_{m,p}^L\right)^2, \quad (2)$$

where $T_{m,p}$ is the mth component of pth target output vector, and $O_{m,p}^L$ is the mth component of an actual output vector produced by the presentation of the pth input vector. Hence the overall error measure is $E = \sum_{p=1}^{P} E_p$.

In order to develop an update procedure that implements gradient descent in E over the parameter space, first we have to calculate the *error rate,* $\partial E_p / \partial O$, for the pth training data and for each node output O. The error rate for the output node at (L, i) can be calculated readily from (2):

$$\frac{\partial E_p}{\partial O_{i,p}^L} = -2(T_{i,p} - O_{i,p}^L). \quad (3)$$

For the internal node at (k, i) the error rate can be derived by the chain rule:

$$\frac{\partial E_p}{\partial O_{i,p}^k} = \sum_{m=1}^{\#(k+1)} \frac{\partial E_p}{\partial O_{m,p}^{k+1}} \frac{\partial O_{m,p}^{k+1}}{\partial O_{i,p}^k}, \quad (4)$$

where $1 \le k \le L-1$. That is, the error rate of an internal node can be expressed as the linear combination of the error rates of the nodes in the next layer. Therefore for all $1 \le k \le L$ and $1 \le i \le \#(k)$, we can find $\partial E_p / \partial O_{i,p}^k$ by (3) and (4).

Now if α is a parameter of the given adaptive network, we have

$$\frac{\partial E_p}{\partial \alpha} = \sum_{O^* \in S} \frac{\partial E_p}{\partial O^*} \frac{\partial O^*}{\partial \alpha}, \quad (5)$$

where S is the set of nodes whose outputs depend on α directly. Then the derivative of the overall error measure E with respect to α is

$$\frac{\partial E}{\partial \alpha} = \sum_{p=1}^{P} \frac{\partial E_p}{\partial \alpha}. \quad (6)$$

Accordingly, the update formula for the generic parameter α is

$$\Delta\alpha = -\eta \frac{\partial E}{\partial \alpha}, \quad (7)$$

in which η is a update rate (or learning rate) of parameter α. Usually η is further expressed as

$$\eta = \frac{S}{\sqrt{\sum_\alpha \left(\frac{\partial E}{\partial \alpha}\right)^2}}, \quad (8)$$

where S is the *step size,* the length of each gradient transition in parameter space. By a proper selection of S, we can vary the speed of convergence.

B. ANFIS: Adaptive-Network-Based Fuzzy Inference System

For simplicity, we assume the fuzzy inference system under consideration has two inputs x and y, one output z, and the rule base contains two fuzzy if–then rules of Takagi and Sugeno's type [14]. The corresponding ANFIS architecture is shown in Fig. 2, where node functions in the same layer are of the same type, as described below:

Layer 1: Every node i in this layer is a square node with a node function

$$O_i^1 = \mu_{A_i}(x), \quad (9)$$

where x is the input to node i, and A_i is the linguistic label (*small, large,* etc.) associated with this node function. In other words, O_i^1 is the membership function of A_i and it specifies the degree to which the given x satisfies the quantifier A_i. Usually we choose $\mu_{A_i}(x)$ to be bell shaped with maximum equal to 1 and minimum equal to 0, such as

$$\mu_{A_i}(x) = \frac{1}{1 + [(\frac{x-c_i}{a_i})^2]^{b_i}}, \quad (10)$$

or

$$\mu_{A_i}(x) = \exp\left\{-\left[\left(\frac{x-c_i}{a_i}\right)^2\right]^{b_i}\right\}, \quad (11)$$

where $\{a_i, b_i, c_i\}$ is the parameter set. As the values of these parameters change, the bell-shaped functions vary accordingly, thus exhibiting various forms of membership functions on linguistic label A_i. In fact, any continuous and piecewise-differentiable functions, such as commonly used trapezoidal or trangular-shaped membership functions, are also qualified candidates for node functions in this layer. Parameters in this layer are referred to as *premise parameters.*

Layer 2: Every node in this layer is a circle node labeled Π which multiplies the incoming signals and sends the product out. Each node output represents the *firing strength* (or *weight*) of a rule. (In fact, other *T-norm* operators can be used as the node function for generalized AND function.)

Layer 3: Every node in this layer is a circle node labeled N. The ith node calculates the ratio of the i rule's firing strength to the sum of all rules' firing strengths. In other words, nodes in this layer compute the *normalized firing strength* of each rule.

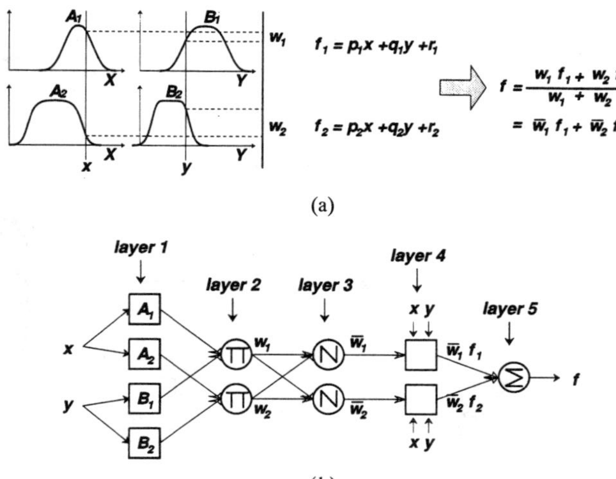

Fig. 2. (a) Fuzzy inference system. (b) Equivalent ANFIS.

Layer 4: Every node i in this layer is a square node with a node function

$$O_i^4 = w_n(p_i x + q_i y + r_i), \tag{12}$$

where w_n is the output of layer 3, and $\{p_i, q_i, r_i\}$ is the parameter set. Parameters in this layer are referred to as *consequent parameters.*

Layer 5: This is a circle node labeled $\sum$ that sums all incoming signals.

Thus we have constructed an adaptive network which is functionally equivalent to a fuzzy inference system. This ANFIS architecture then can update its parameters according to the gradient descent update procedure mentioned before. Other ANFIS's corresponding to different type of fuzzy if–then rules and defuzzification mechanisms, plus a fast update procedure combining the gradient method and a sequential least square estimate, can be found in [6] and [7].

III. Constructing Self-Learning Fuzzy Controllers

In this section, we propose a generalized control scheme which can construct a fuzzy controller through *temporal back propagation,* such that the state variables can follow a given desired trajectory as closely as possible. The basic idea is to implement both the controller and the plant at each time stage as a *stage adaptive network,* and cascade these stage adaptive networks into a *trajectory adaptive network* to facilitate the temporal back propagation learning process.

A. Stage Adaptive Network

Fig. 3 is a block diagram of a feedback control system consisting of a fuzzy controller and a plant. We assume the delay through the controller is small and the state variables are accessible with accuracy. Moreover, the plant block is viewed as a static system since the dependency of the next state on the present state is shown explicitly. Before finding a controller to control the plant state, we have to find mathematical expressions for both the controller block and the plant block. This step is referred to as the *implementation.* In our case, we are going to implement both blocks as adaptive networks.

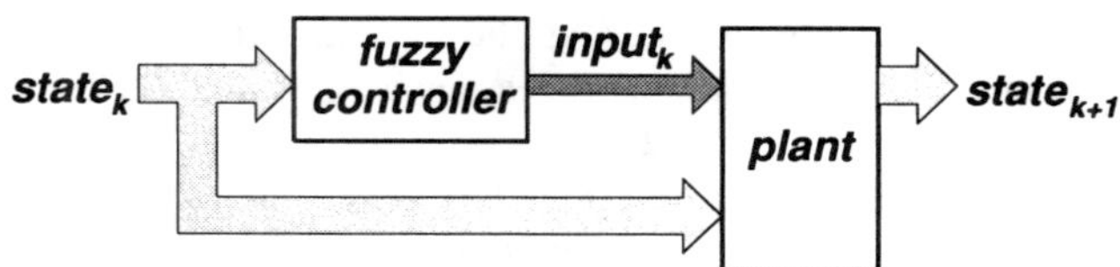

Fig. 3. Block diagram of a fuzzy controller and a plant. Also a stage adaptive network at time stage k.

An obvious candidate for implementing the FC block in Fig. 3 is the ANFIS architecture, since it has exactly the same function as a fuzzy controller, as shown in Fig. 2. If we have p inputs to the plant, then the FC block can be implemented either as p ANFIS's or as an ANFIS that has rules with multiple consequents.

Suppose that we have a human expert who knows how to control the plant. Then the domain knowledge can be transformed into fuzzy if–then rules and the corresponding parameters (which characterize membership functions) can be used as the initial parameters of the FC block in the learning process. As a result, the domain knowledge can guide the TBP learning process to get started from a point in the parameter space that is not far from the optimal one, and the TBP can fine-tune the domain knowledge for achieving a better performance. This cooperative relation between the domain knowledge and the TBP learning process is not always present in other types of controllers.

On the other hand, if we do not have *a priori* knowledge about controlling the plant, then the number of fuzzy if–then rules has to be decided more or less by trial and error. Fortunately, as a consequence of ANFIS's remarkable representational power [6], [7], we usually do not need many rules to construct the desired mapping from state variables to control action.

As for the implementation of the plant block, we can choose whatever function approximators that can best represent the input–output behavior of the plant. This model-insensitive attribute is mostly due to the flexibility of adaptive networks, which allows us to choose either conventional models (difference or differential equations, transfer functions, etc.) or unconventional ones (ANFIS, neural networks [13], radial basis function networks [11], GMDH structure [5], etc.) to implement the plant block.

In the case where the plant can be modeled as a set of n (= number of state variables) first-order difference equations, then the plant block can be replaced with n nodes, each of which uses one difference equation to obtain the state variable at the next time step. Furthermore, if the state equations of the plant are a set of first-order differential equations:

$$\dot{\vec{x}}(t) = \vec{f}\Big(\vec{x}(t), \vec{in}(t), t\Big), \tag{13}$$

where $\vec{x}(t)$ is a vector consisting of state variables at time t and $\vec{in}(t)$ is the input vector to the plant, then we can just employ a linear approximation to get the difference equations as below:

$$\vec{x}(h*k+h) = h*\vec{f}\Big(\vec{x}(h*k), \vec{in}(h*k), h*k\Big) + \vec{x}(h*k), \tag{14}$$

where k is an integer and h is the sampling time. Therefore the plant block still has n nodes, each of which performs a component function of (14).

When the sampling time h is too big or the plant has fast dynamics, the linear approximation may not be a reasonable estimate of the next state. In this case, we can utilize a large body of numerical analysis techniques to obtain a more precise estimate, for instance, the second-order Runge–Kutta method:

$$\begin{cases} \vec{a} = h * \vec{f}\Big(\vec{x}(h*k), \vec{in}(h*k), h*k\Big), \\ \vec{b} = h * \vec{f}\Big(\vec{x}(h*k) + \vec{a}, \vec{in}(h*k), h*k+h\Big) \\ \vec{x}(h*k+h) = \vec{x}(h*k) + 0.5 * \Big(\vec{a} + \vec{b}\Big). \end{cases} \tag{15}$$

However, implementing the above equations as an adaptive network (without modifiable parameters) is more complex and some intermediate nodes would be present in the resulting network for the intermediate variable vectors $\vec{a}$ and $\vec{b}$. Higher order Runge–Kutta formulas may be used to implement the plant block as well, but the increased complexity of the adaptive network could slow down the learning process even more.

Consequently, the block diagram of Fig. 3 can also be viewed as an adaptive network containing two subnetworks, the FC block (ANFIS) and the plant block. Subsequently, we refer to the adaptive network of Fig. 3 as SAN_k, representing the *stage adaptive network* at time stage k.

B. Trajectory Adaptive Network

Given the state of the plant at time $t = k * h$, the FC will generate an input to the plant and the plant will evolve to the next state at time $(k+1)*h$. By repeating this process starting from $t = 0$, we obtain a plant state trajectory determined by the initial state and the parameters of the FC. The state transition from $t = 0$ to $m * h$ is shown conceptually in Fig. 4, which again, is an adaptive network consisting of m SAN_k's, $k = 0$ to $m-1$. Accordingly we can still apply the back-propagation gradient descent to minimize the differences between adaptive network outputs and desired outputs. In order to make the inputs and outputs more explicit, we redraw Fig. 4 to get the *trajectory adaptive network* shown in Fig. 5, where the inputs to the network are the initial state of the plant at time = 0; the outputs of the network are the state trajectory from $t = h$ to $m * h$; and the adjustable parameters all pertain to the FC block implemented as an ANFIS. Hence each entry of the training data is of the form

$$(initial\ state;\ desired\ trajectory), \tag{16}$$

and the corresponding error measure to be minimized is

$$E = \sum_{k=1}^{m} ||\vec{x}(h*k) - \vec{x}_d(h*k)||^2, \tag{17}$$

where $\vec{x}_d(h*k)$ is the desired trajectory at $t = h*k$. With some minor modifications of Fig. 5, the above error measure

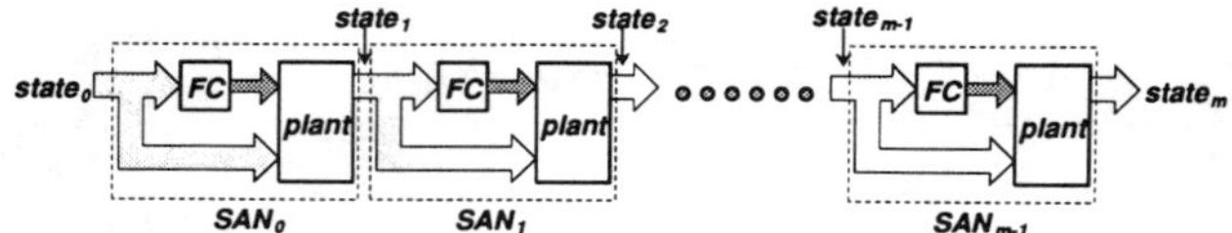

Fig. 4. State transition diagram.

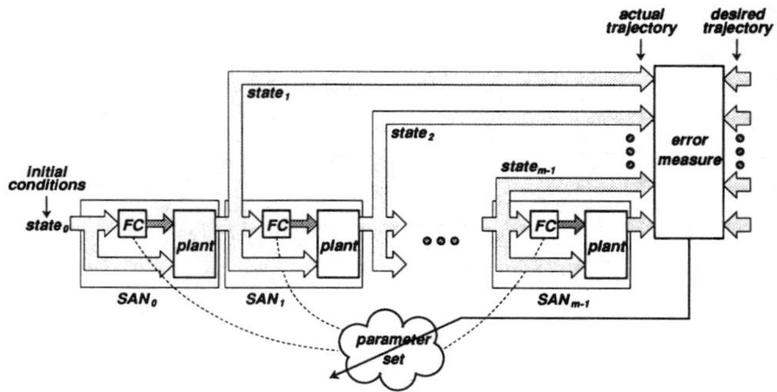

Fig. 5. A trajectory adaptive network for control application.

can be revised as

$$E = \sum_{k=1}^{m} ||\vec{x}(h*k) - \vec{x}_d(h*k)||^2 + \lambda * \sum_{k=0}^{m-1} ||\vec{in}(h*k)||^2, \tag{18}$$

where $\vec{in}(h*k)$ is the controller's output at time $h*k$. By a proper selection of λ, a compromise between trajectory error and control effort can be obtained.

Since the error signals of back propagation can propagate through different time stages, this control methodology is called *temporal back propagation*, or simply TBP. As a matter of fact, the basic idea of TBP is similar to Nguyen and Widrow's approach [12] to construct a self-learning neural controller, which was called back propagation through time by Werbos [4]. We generalize the idea to a much more flexible building block, the adaptive network, which has two advantages over neural networks:

1) accommodation of *a priori* knowledge from a human operator, in the form of fuzzy if–then rules;
2) no need to remodel the plant in neural networks if we already have other existing models for it, such as difference equations.

In the trajectory adaptive network shown in Fig. 5, though there are m FC blocks, all of them refer to the same fuzzy controller at different time stages. That is, there is only one parameter set which belongs to all m FC blocks at different time stages. For clarity, this parameter set is shown explicitly in Fig. 5 and it is updated according to the output of the error measure block.

IV. Application to the Inverted Pendulum System

The proposed control scheme is quite general and it can be applied to a variety of control problems. In this section, we demonstrate the effectiveness of the TBP by applying it to a benchmark problem in intelligent control—the inverted pendulum system.

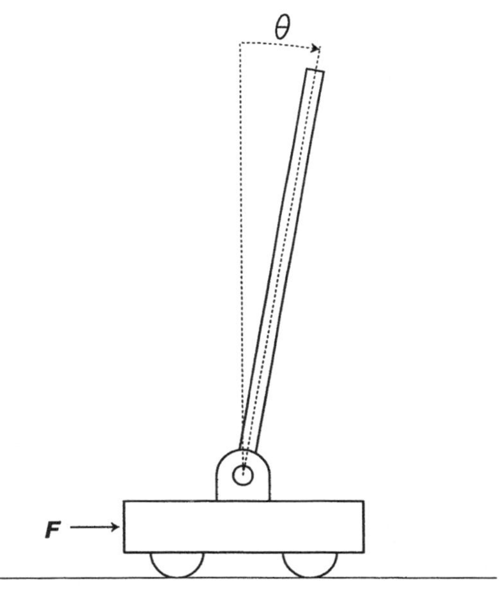

Fig. 6. The inverted pendulum system.

A. The Inverted Pendulum System

The inverted pendulum system (Fig. 6) is composed of a rigid pole and a cart on which the pole is hinged. The cart moves on the rail tracks to its right or left, depending on the force exerted on the cart. The pole is hinged to the cart through a frictionless free joint such that it has only one degree of freedom. The control goal is to balance the pole starting from nonzero conditions by supplying appropriate force to the cart.

The dynamics of the inverted pendulum system are characterized by four state variables: θ (angle of the pole with respect to the vertical axis), $\dot{\theta}$ (angular velocity of the pole), z (position of the cart on the track), and $\dot{z}$ (velocity of the cart). The behavior of these four state variables is governed by the following two second-order differential equations [1], [3]:

$$\ddot{\theta} = \frac{g * \sin\theta + \cos\theta * \left(\frac{-F - m*l*\dot{\theta}^2*\sin\theta}{m_c+m}\right)}{1 * \left(\frac{4}{3} - \frac{m*\cos^2\theta}{m_c+m}\right)} \tag{19}$$

$$\ddot{z} = \frac{F + m * l * \left(\dot{\theta}^2 * \sin\theta - \ddot{\theta} * \cos\theta\right)}{m_c + m} \tag{20}$$

where g (acceleration due to gravity) is 9.8 $\mathrm{m/s^2}$, m_c (mass of cart) is 1.0 kg, m (mass of pole) is 0.1 kg, l (half-length of pole) is 0.5 m, and F is the applied force in newtons. Our control goal here is to balance the pole without regard to the cart's position and velocity; hence only (19) is relevant in our simulation.

B. Simulation Settings

Fig. 7 shows the stage adaptive network used in our simulation. Both the controller and the plant block, together with the learning rule, are explained below.

1) Plant Block: As mentioned earlier, there are several ways to implement the plant block depending on how well we know

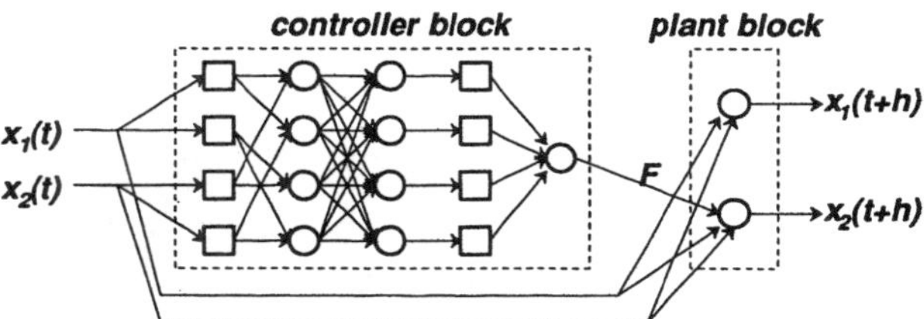

Fig. 7. The implementation of a stage adaptive network.

the plant. In this case, the plant is a deterministic nonlinear dynamical system with precisely defined differential equations, so we can use just two nodes to calculate the state variables at the next time step by linear approximation:

$$\begin{cases} x_1(t+h) = h\dot{x}_1(t) + x_1(t) \\ x_2(t+h) = h\dot{x}_2(t) + x_2(t), \end{cases} \tag{21}$$

where $x_1(\cdot) = \theta(\cdot)$, and $x_2(\cdot) = \dot{\theta}(\cdot)$. These two equations are the node functions of the plant block in Fig. 7.

2) Controller Block: We assume that no domain knowledge (from a human operator's point of view) about the inverted pendulum system is available. The controller block in Fig. 3 is implemented as an ANFIS with two inputs, each of which is assigned two membership functions, so it is a fuzzy controller with four fuzzy if–then rules of Takagi and Sugeno's type [14]. See the controller block in Fig. 7. (Though the number of fuzzy rules can be more than four, the simulation indicates four rules are enough for balancing the pole.)

Without any domain knowledge, we have to set the initial parameters subjectively. The consequent parameters of the FC are all set at zero, which means the control action is zero initially, as shown in Fig. 9. As a conventional way of setting membership functions in a fuzzy controller, the premise parameters are set in such a way that the membership functions can cover the domain interval (or universe of discourse) completely with sufficient overlapping of each other. Parts (a) and (b) of Fig. 8 illustrate the initial membership functions in the form of (10); the domain interval for θ (degrees) and $\dot{\theta}$ (degrees/s) are assumed to be [–20, 20] and [–50, 50], respectively.

3) Temporal Back Propagation: We employ 100 stage adaptive networks to construct the trajectory adaptive network, and each stage adaptive network corresponds to the time transition of 10 ms. That is, the time step (h) used is 10 ms, and the trajectory adaptive network corresponds to a time interval from $t = 0$ to $t = 1$ s. If h is too small, a large network has to be built to cover the same time span, which increases the signal propagation time and thus delays the whole learning process. On the other hand, if h is too big, then the linear approximation of the plant behavior may not be precise enough and a higher order approximation has to be used instead.

The training data set contains desired input–output pairs of the format

$$(\textit{initial condition};\ \textit{desired trajectory}), \tag{22}$$

where the initial condition is a two-element vector which specifies the initial condition of the pole; the desired trajectory

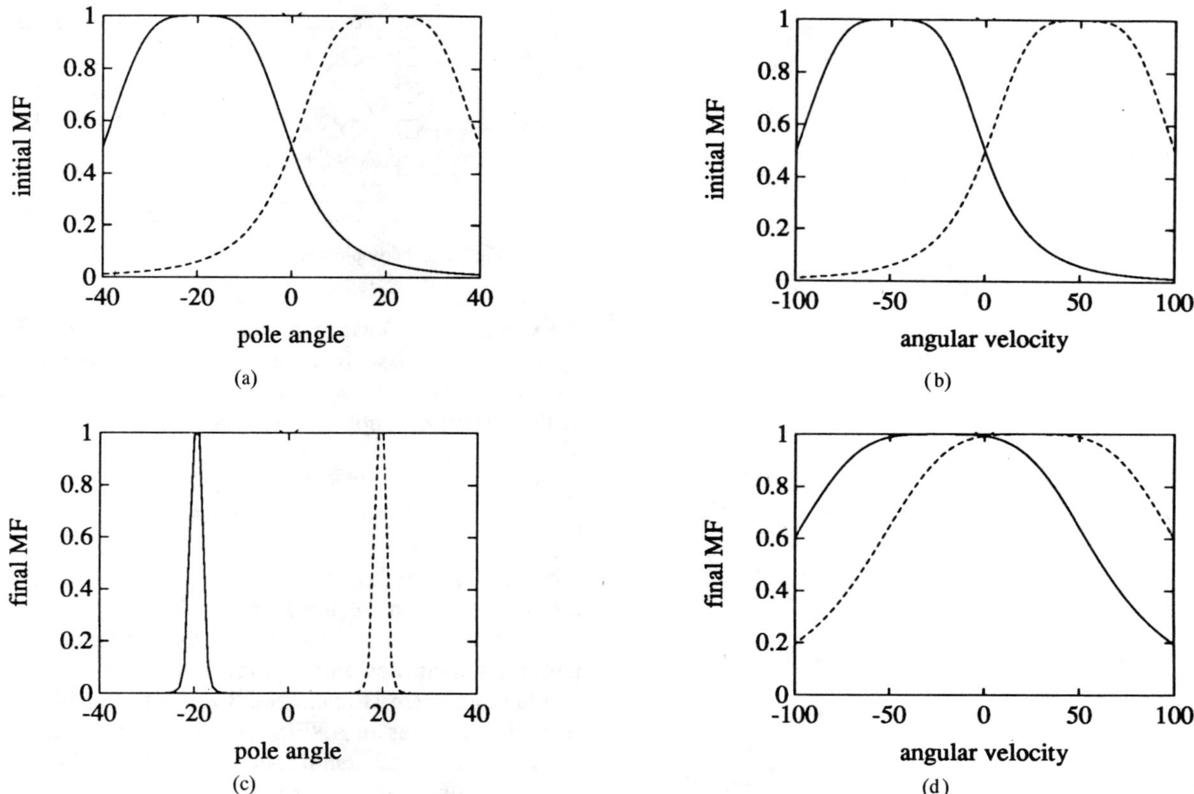

Fig. 8. (a), (b) Initial membership functions; (c), (d) final membership functions.

is a 100-element vector which contains the desired pole angle at each time step. In our simulation, only two entries of training data are used: the initial conditions are $(10, 0)$ and $(-10, 0)$, respectively, and the desired trajectory is always a zero vector. In short, we expect that the trajectory adaptive network not only can learn to balance the pole from an initial pole angle of $+10°$ or $-10°$, but also can achieve the control goal in an near-optimal manner which minimizes the error measure

$$E = \sum_{k=1}^{100} \theta^2(0.01 * k) + \lambda * \sum_{k=0}^{99} f^2(0.01 * k), \tag{23}$$

where $f(0.01 * k)$ is the controller's output force and $\lambda(= 10)$ accounts for the relative unit cost of control effort.

To speed up the convergence, we follow a strict gradient descent in the sense that each transition of the parameters will lead to a smaller error measure. If the error measure increases after parameter update, we back up to the original point in the parameter space and decrease the current step size by half. This process is repeated until the weight update leads to a smaller error measure. However, this step size update rule tends to use a small step size if the error measure surface encountered in the first few updates is not smooth. Therefore we multiply the step size by 4 after observing three consecutive transitions without any backup actions. The initial step size in the simulation is 20 and the learning process stops whenever the number of transitions in parameter space (which is equal to the number of reductions in error measure) reaches 10.

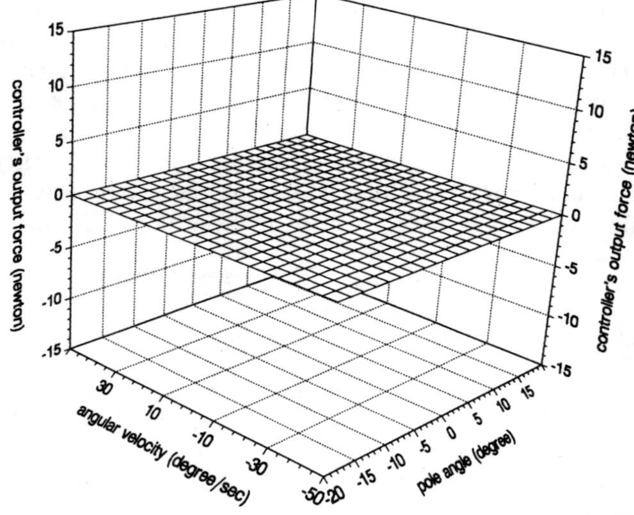

Fig. 9. Initial control action surface.

C. Simulation Results

All the simulation settings mentioned above are referred to as the *reference setting;* other simulations are based on this setting with minor changes. In the learning task with the reference setting, it is amazing to observe that the FC is able to balance the pole right after the first parameter transition, and it keeps on refining the controller (minimizing the error measure) until the tenth parameter transition is done. Parts (a) and (b) of Fig. 8 show the initial membership functions on pole angle and angular velocity; parts (c) and (d) show the final membership functions. Fig. 9 is the initial control action surface, and Fig. 10 is the final control action surface after the tenth parameter transition. A listing of the fuzzy rules with numerical parameters can be found in the Appendix.

Fig. 8 indicates that the final membership functions for θ are quite different from the initial membership functions. Visually there are no membership functions covering the interval $[-10, 10]$ of θ, making the linguistic interpretation of the fuzzy rules difficult. However, since we are utilizing the FC as a functional approximator that can generate the required nonlinear mapping, linguistically desired features (such as enough overlapping between membership functions and total coverage of the domain interval) do not have to be one of the FC's attributes in this case. If we want to keep those desired features, we can either impose certain constraints on the premise parameters or simply increase the number of parameters of the fuzzy controller to endow it with more degree of freedom. Fig. 11 shows the membership functions of a nine-rule fuzzy controller which has about the same performance as the four-rule fuzzy controller. Because of its larger number of degrees of freedom, the premise parameters of the nine-rule fuzzy controller do not have to change much to minimize the error measure; therefore the final membership functions can cover all the domain intervals with desired overlapping.

Solid curves in Fig. 12 demonstrate the state variable trajectories under the reference setting: parts (a), (b) and (d) show the pole angle (degree), angular velocity (degrees/s) and control actions (newtons) from $t = 0$ to $t = 2s$; (c) is the state space plot which reveals how the trajectory approaches the origin from the initial point $(10, 0)$. The dashed and dotted curves in Fig. 12 correspond to λ equal to 40 and 100, respectively. From (a), it is observed that a smaller λ (solid curve) achieves the control goal faster since the controller can apply a larger force to balance the pole. For a large λ (dotted curve), the controller's output has to be kept small, thus slowing the approach to the goal.

To demonstrate how the fuzzy controller can survive substantial changes of plant parameters, we use poles of different lengths to test the controller obtained from the reference setting. The results are shown in Fig. 13, where solid, dashed, and dotted curves correspond to a pole half lengths of 0.5 (reference setting), 0.25, and 0.125 m respectively. It is remarkable to note how the controller can handle the shorter pole easily and gracefully.

In the learning phase, we supply only two training data, corresponding to initial conditions $(10, 0)$ and $(-10, 0)$ of the pole. Now it would be interesting to know how the FC (obtained from the reference setting) deals with other initial conditions. In this part, we monitor the pole behavior starting from other initial conditions which make the control goal even harder. Fig. 14 shows the results, where the solid, dashed, and dotted curves correspond to the initial conditions $(10, 20)$, $(15, 30)$, and $(20, 40)$, respectively. Again, the same fuzzy controller can perform the control task starting from the unseen initial conditions. Fig. 14 together with Fig. 13 reveals the robustness and fault tolerance of the fuzzy controller obtained from the TBP.

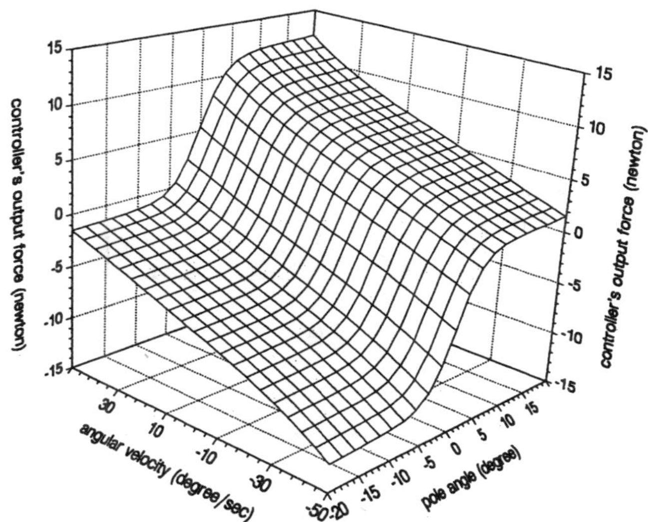

Fig. 10. Final control action surface.

V. Conclusions

We have proposed a generalized controller design methodology, called temporal back-propagation (TBP), for constructing self-learning fuzzy controllers. This methodology employs the adaptive network as a building block and the back-propagation gradient method as the update procedure to minimize the difference between an actual trajectory and a given desired trajectory. Because of the flexibility of this methodology, we can easily customize it for a wide range of control applications. The inverted pendulum is used as a test-bed to verify the effectiveness of the TBP and to exhibit the robustness and fault tolerance of the resulting fuzzy controller.

Best of all, the TBP is not tailored for fuzzy controllers only. Almost all nonpathogenic mathematical representations or equations (such as difference equations, ANFIS architecture, feedforward neural networks, radial function basis networks, and the GMDH structure) can be used to implement the plant block as well as the controller block. This provides us with plenty of freedom in choosing accurate and efficient models.

Appendix

Each linguistic label used in the FC is characterized by three parameters, as described in (10). If θ is in degrees, and $\dot{\theta}$ in degrees/s, the initial fuzzy if–then rules are

$$\begin{cases} \text{If } \theta \text{ is } A_1 \text{ and } \dot{\theta} \text{ is } B_1, \text{ then force } = 0 \\ \text{If } \theta \text{ is } A_1 \text{ and } \dot{\theta} \text{ is } B_2, \text{ then force } = 0 \\ \text{If } \theta \text{ is } A_2 \text{ and } \dot{\theta} \text{ is } B_1, \text{ then force } = 0 \\ \text{If } \theta \text{ is } A_2 \text{ and } \dot{\theta} \text{ is } B_2, \text{ then force } = 0 \end{cases}, \tag{A1}$$

where A_1, A_2, B_1 and B_2 are the linguistic labels characterized by $\{20, 2, -20\}$, $\{20, 2, 20\}$, $\{50, 2, -50\}$, and $\{50, 2, 50\}$, respectively. Fig. 9 is the initial control action surface.

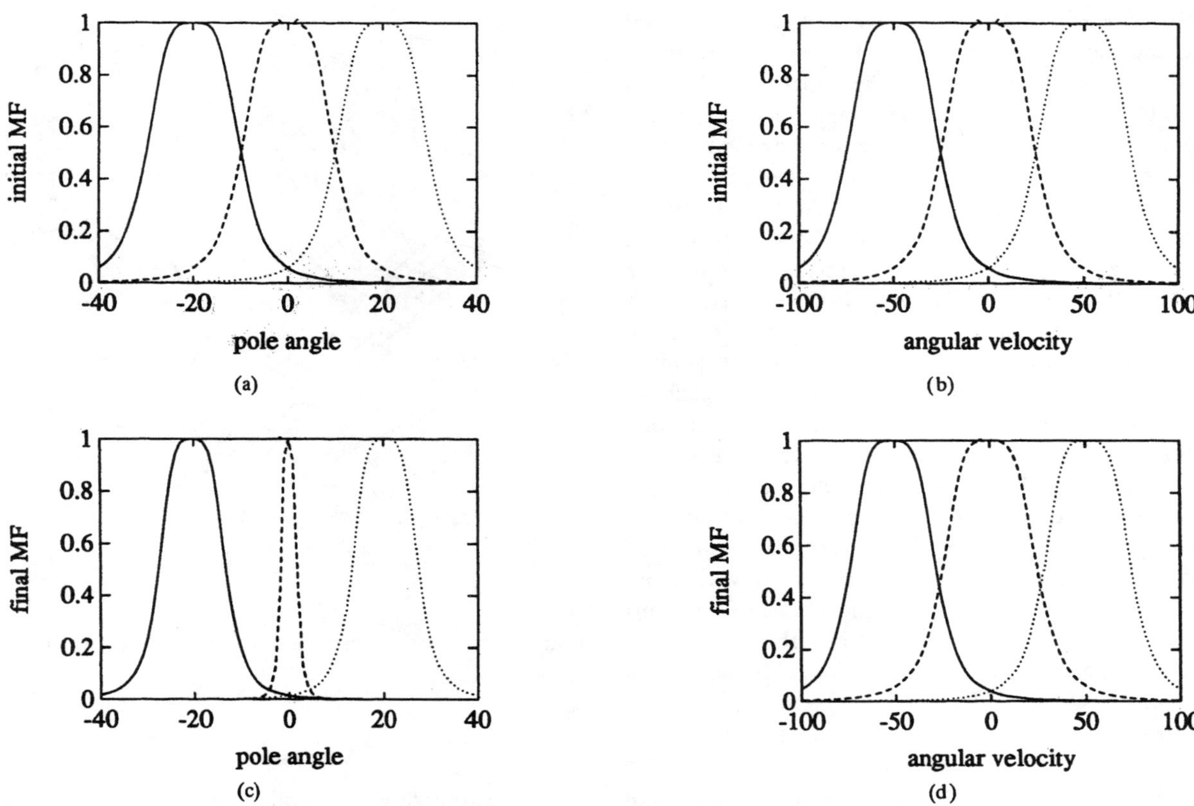

Fig. 11. (a), (b) Initial membership functions and (c), (d) final membership functions of a nine-rule fuzzy controller.

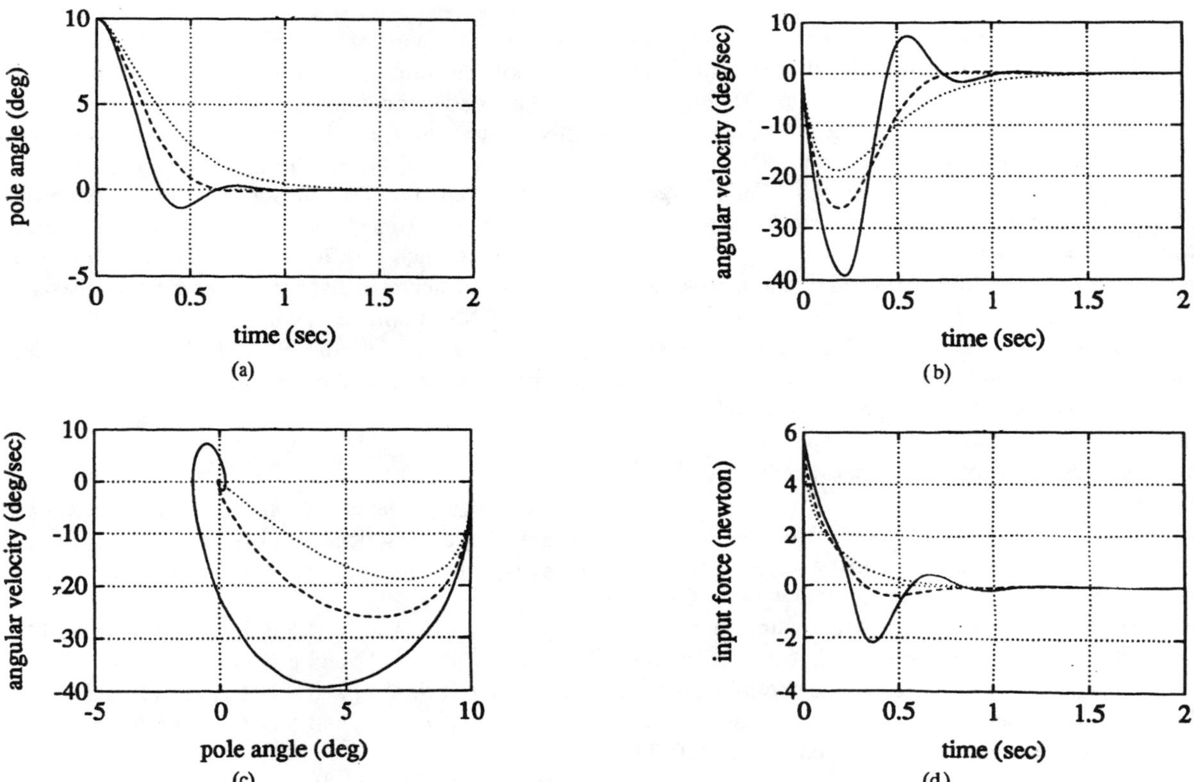

Fig. 12. (a) Pole angle, (b) pole angular velocity, (c) state space, and (d) input force. (Solid, dashed, and dotted curves correspond to $\lambda = 10$, 40, and 100, respectively.)

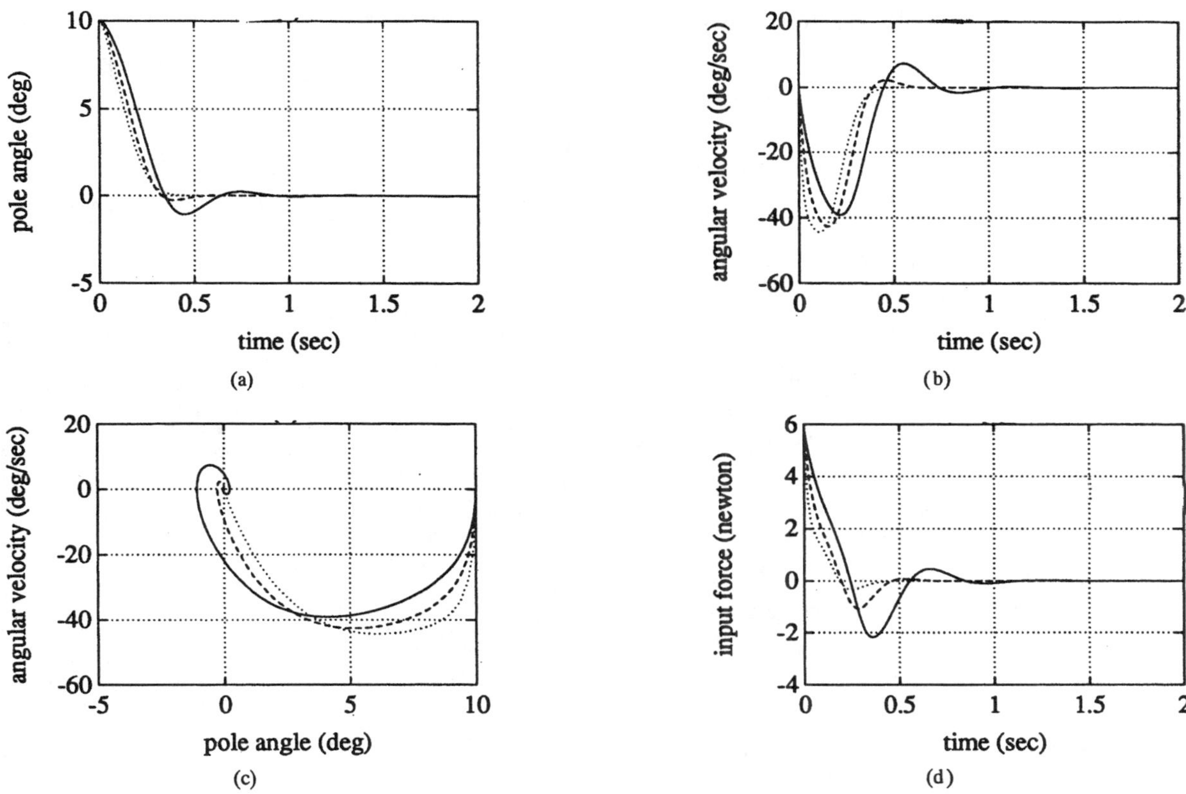

Fig. 13. (a) Pole angle, (b) pole angular velocity, (c) state space, and (d) input force. (Solid, dashed, and dotted curves correspond to half-pole lengths of 0.5, 0.25, and 0.125 m, respectively.)

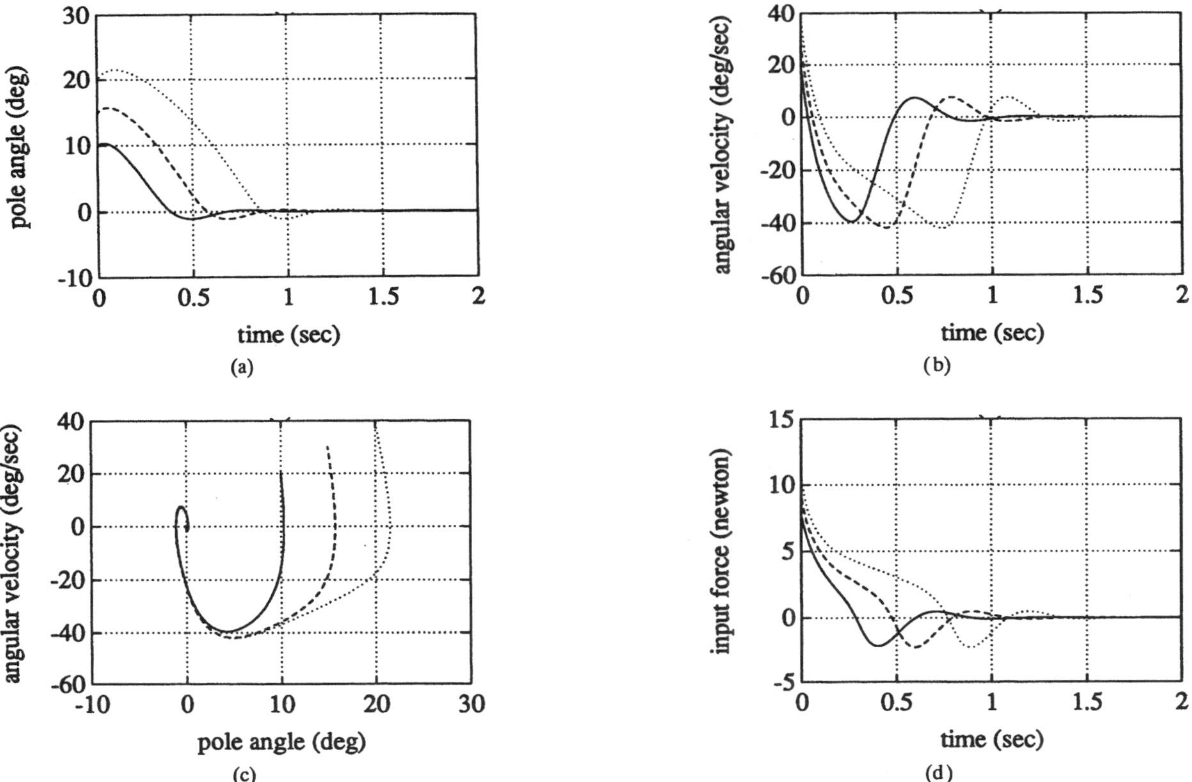

Fig. 14. (a) Pole angle, (b) pole angular velocity, (c) state space, and (d) input force. (Solid, dashed, and dotted curves correspond to initial conditions (10, 20), (15, 30), and (20, 40), respectively.)

The final fuzzy if–then rules derived from the reference settings are

$$\begin{cases} \text{If } \theta \text{ is } A_1 \text{ and } \dot{\theta} \text{ is } B_1, \text{ then force } = 0.0502 * \theta \\ \qquad + 0.1646 * \dot{\theta} - 10.09 \\ \text{If } \theta \text{ is } A_1 \text{ and } \dot{\theta} \text{ is } B_2, \text{ then force } = 0.0083 * \theta \\ \qquad + 0.0119 * \dot{\theta} - 1.09 \\ \text{If } \theta \text{ is } A_2 \text{ and } \dot{\theta} \text{ is } B_1, \text{ then force } = 0.0083 * \theta \\ \qquad + 0.0119 * \dot{\theta} + 1.09 \\ \text{If } \theta \text{ is } A_2 \text{ and } \dot{\theta} \text{ is } B_2, \text{ then force } = 0.0502 * \theta \\ \qquad + 0.1646 * \dot{\theta} + 10.09, \end{cases} \tag{A2}$$

where A_1, A_2, B_1, B_2 are the linguistic labels characterized by {–1.59, 2.34, –19.49}, {–1.59, 2.34, 19.49}, {85.51, 1.94, –23.21}, and {85.51, 1.94, 23.21}, respectively. Fig. 10 is the final control action surface.

ACKNOWLEDGMENT

The guidance and help of Prof. L. A. Zadeh and other members of the "fuzzy group" at UC Berkeley are gratefully acknowledged.

REFERENCES

[1] A. G. Barto, R. S. Sutton, and C. W. Anderson, "Neuronlike adaptive elements that can solve difficult learning control problems," *IEEE Trans. Syst. Man, Cybern.*, vol. SMC-13, no. 5, pp. 834–846, 1983.

[2] H. R. Berenji, "Refinement of approximate reasoning-based controllers by reinforcement learning," in *Proc. Eighth Int. Workshop of Machine Learning,* (Evanston, IL), June 1991.

[3] R. H. Cannon, *Dynamics of Physical Systems.* New York: McGraw-Hill, 1967.

[4] W. T. Miller III, R. S. Sutton, and P. J. Werbos, Eds., *Neural Networks for Control.* Cambridge, MA: MIT Press, 1990.

[5] A. G. Ivakhnenko, "Polynomial theory of complex systems, "*IEEE Trans. Syst. Man, Cyber.*, vol. SMC-1, pp. 364–378, Oct. 1971.

[6] J. S. Jang, "ANFIS; Adaptive-network-based fuzzy inference systems," submitted to *IEEE Trans. Syst., Man, Cybern.*

[7] J. S. Jang, "Fuzzy modeling using generalized neural networks and Kalman filter algorithm," in *Proc. Ninth National Conf. Artificial Intelligence (AAAI-91),* July 1991, pp. 762–767.

[8] J. S. Jang, "Rule extraction using generalized neural networks, "in *Proc. 4th IFSA World Congress,* July 1991.

[9] C.-C. Lee, "Intelligent control based on fuzzy logic and neural network theory," in *Proc. Int. Conf. Fuzzy Logic and Neural Networks,* (Iizuka), 1990, pp. 759–764.

[10] C.-C. Lee, "A self-learning rule-based controller employing approximate reasoning and neural net concepts," *Int. J. Intelligent Systems,* vol. 5, no. 3, pp. 71–93, 1991.

[11] J. Moody and C. Darken, "Fast learning in networks of locally-tuned processing units," *Neural Computation,* vol. 1, pp. 281–294, 1989.

[12] D. H. Nguyen and b. Widrow, "Neural networks for self-learning control systems," *IEEE Control Systems Magazine,* pp. 18–23, Apr. 1990.

[13] D. E. Rumelhart, G. E. Hinton, and R. J. Williams, "Learning internal representations by error propagation," in *Parallel Distributed Processing: Explorations in the Microstructure of Cognition,* D. E. Rumelhart and James L. McClelland, Eds., vol. 1. Cambridge, MA: MIT Press, 1986, ch. 8, pp. 318–362.

[14] T. Takagi and M. Sugeno, "Derivation of fuzzy control rules from human operator's control actions," in *Proc. IFAC Symp. Fuzzy Information, Knowledge Representation and Decision Analysis,* July 1983, pp. 55–60.

[15] P. Werbos, "Beyond Regression: New tools for prediction and analysis in the behavioral sciences," Ph.D. thesis, Harvard University, 1974.

Article 6.5

A Neo Fuzzy Neuron and Its Applications to System Identification and Prediction of the System Behavior

Takeshi Yamakawa*, Eiji Uchino*, Tsutomu Miki* and Hiroaki Kusanagi
Kyushu Institute of Technology
Iizuka, Fukuoka 820, Japan
Phone : +81-948-29-7712 Fax : +81-948-29-7742
E-mail : yamakawa@ces.kyutech.ac.jp

Key words : neo fuzzy neuron, nonlinear synapse neuron, high-speed learning, delay element, system identification, prediction of behavior, complementary membership function, fuzzy segmentation, sequential prediction, adaptive prediction

Abstract
This paper describes a new model of fuzzy neuron named *neo fuzzy neuron*, the synaptic transfer characteristics of which is nonlinear and the soma of which is simply modeled by a summation. Its nonlinear synapse is characterized by a set of *complementary membership functions* and corresponding weights. The learning algorithm of this neo fuzzy neuron was described in detail and the computer simulation illustrates the effective learning in a short learning cycle. The attempt was made to apply this neo fuzzy neuron to system identification of nonlinear dynamical systems. Furthermore, it was applied to predict the chaotic behavior of the system under test.

1. INTRODUCTION

The attempt was made to modify a conventional neuron model by fuzzy concept in order to obtain a novel neuron model better adapted to the study of the behavior of systems which are imprecisely defined by virtue of their high degree of complexity. S. C. Lee and E. T. Lee modified the "all-or-none" activity of the MacCulloch-Pitts model to have the output equal to zero in non-firing state and some positive number μ_i in firing, where $0<\mu_i\leq 1$ [1]. And they presented a general synthesis procedure for realizing any fuzzy automaton using fuzzy neurons. Takeshi Yamakawa presented another type of fuzzy neuron, in which linear synaptic connections are replaced with a nonlinearity characterized by a membership function labeled as "tightly connected", "loosely connected",etc.,and excitatory connections and inhibitory connections are represented by fuzzy logic intersections and fuzzy logic complements followed by fuzzy logic intersections, respectively [2, 3]. This fuzzy neuron was employed to construct a hardware system which achieves hand-written character recognition within 1 microsecond [4,5]. A design algorithm of a fuzzy neuron for pattern recognition was established by example-based learning [6].

This paper describes a new type of fuzzy neuron model , each *nonlinear synapse* of which is characterized by a set of fuzzy implication rules with singleton weights in consequents. The learning algorithm to assign the weights is also touched upon and its high speed learning is discussed.

2. STRUCTURE AND LEARNING ALGORITHM OF A NEO FUZZY NEURON

* Takeshi Yamakawa, Eiji Uchino and Tsutomu miki also work with a Japanese foundation named Fuzzy Logic Systems Institute (FLSI) in Iizuka, Fukuoka 820, Japan.

The structure of the neo fuzzy neuron is shown in Fig.1(a), where the characteristics of each synapse is represented by a nonlinear function f_i and the soma doesn't exhibit a sigmoidal function at all. Aggregation of synaptic signals is achieved by an algebraic sum. Thus the output of this neo fuzzy neuron can be represented by the following equation;

$$\hat{y} = f_1(x_1) + f_2(x_2) + \cdots + f_m(x_m) = \sum_{i=1}^{m} f_i(x_i) \quad (1)$$

The structure of the nonlinear synapse is shown in Fig.1(b). The input space x_i is divided into several *fuzzy segments* which are characterized by membership functions $\mu_{i1}, \mu_{i2},..., \mu_{ij},..., \mu_{in}$ within the range between x_{min} and x_{max} as shown in Fig.1(c) 1, 2, ..., j, ...,n are numbers assigned to labels of fuzzy segments. The membership functions are followed by variable weights $w_{i1}, w_{i2}, ..., w_{ij}, ..., w_{in}$.

Mapping from x_i to $f_i(x_i)$ is determined by fuzzy inferences and a defuzzification as shown in Fig.2. The fuzzy inference adopted here is that of a singleton consequent, that is, each weight w_{ij} is a deterministic value such as 0.3. It should be emphasized that each membership function in antecedent is triangular and assigned to be *complementary* (so called by the authors) with neighbouring ones. In other words, an input signal x_i activates only two membership functions simultaneously and the sum of grades of these two neighbouring membership functions labelled by k and k+1 is always equal to 1, that is, $\mu_{i,k}(x_i) + \mu_{i,k+1}(x_i) = 1$ So that the defuzzification taking a center of gravity doesn't need a division and the output of the neo fuzzy neuron can be represented by the following simple equation.

$$f_i(x_i) = \frac{\sum_{j=1}^{n} \mu_{ij}(x_i)\cdot w_{ij}}{\sum_{j=1}^{n} \mu_{ij}(x_i)} = \frac{\mu_{ik}(x_i)\cdot w_{ik} + \mu_{i,k+1}(x_i)\cdot w_{i,k+1}}{\mu_{ik}(x_i) + \mu_{i,k+1}(x_i)} = \mu_{ik}(x_i)\cdot w_{ik} + \mu_{i,k+1}(x_i)\cdot w_{i,k+1}. \quad (2)$$

This equation can be realized by the architecture shown in

Reprinted from *Proc. of the 2nd Int. Conf. on Fuzzy Logic and Neural Networks,* Iizuka, Japan, July 17–22, T. Yamakawa, E. Uchino, T. Miki, and H. Kusanagin, "A Neo Fuzzy Neuron and Its Applications to System Identification and Prediction of the System Behavior," pp. 477–483, © 1992, Pergamon Press Ltd., Oxford, England.

Fig.1(b)

The weights w_{ij} are assigned by learning, the rule of which is described by n if-then rules as shown in the following;

[Rule No.j]

If x_i lies in the fuzzy segment μ_{ij}, then the corresponding weight w_{ij} should be increased directly proportional to the output error $(y - \hat{y})$, because the error is caused by the weight.

This proposition can be represented by the following equation;

$$W_{ij}(t+1) = W_{ij}(t) + \alpha_i \mu_{ij}(x_i)(y - \hat{y}). \quad (3)$$

where α_i is a learning coefficient for the i-th nonlinear synapse. This is the learning algorithm of this neo fuzzy neuron. After the learning, this neo fuzzy neuron facilitates extrapolation of data.

3. IDENTIFICATION OF NONLINEAR DYNAMICAL SYSTEMS BY A NEO FUZZY NEURON

An m-input neo fuzzy neuron can be combined with a series of delay elements to obtain a one-input and one-output signal processing system as shown in Fig.3. This system can achieve the identification of dynamical systems even though they are chaotic.

In order to examine the feasibility of system identification of this neo fuzzy neuron with delay elements, the test signal is generated which is very chaotic and also very difficult to predict its behavior. The test signal is generated by the dynamical system defined by the following recurrence formula;

$$y_{n+1} = f_1(x_1) + f_2(x_2) + f_3(x_3)$$

$$= f_1(y_n) + f_2(y_{n-1}) + f_3(y_{n-2})$$

$$= \frac{5y_n}{1 + y_n^2} - 0.5y_n - 0.5y_{n-1} + 0.5y_{n-2}. \quad (4)$$

with initial values of $y_0 = 0.733$, $y_1 = 0.234$ and $y_2 = 0.973$. The example is shown in Fig.4, which is quitely chaotic and thus so difficult to predict its time sequential behavior.

This dynamical system is identified by leaning of a neo fuzzy neuron with delay elements. Although the system is a three-dimensional dynamical system as represented by Eq.(4), a 1-synapse, 2-synapse, 3-synapse, 4-synapse and 5-synapse neo fuzzy neuron were examined in order to study how precisely they can identify the system of interest.

The number of fuzzy segments in each input space x_i is 12 and thus twelve membership functions are assigned to each synapse. Each weight corresponding to the membership function are changed according to the following equation.

$$W_{ij}(t+1) = W_{ij}(t) + \alpha_i \mu_{ij}(x_i)(y_{n+1} - \widehat{y_{n+1}}) \quad (5)$$

The learning is achieved as follows. All initial values of weights are zero. The 50 time series $y_0, y_1, y_2, y_3, \ldots\ldots, y_{49}$ (Fig.4) are sequentially applied to the input terminal y_{n+1} (Fig.3) and the data is shifted down one by one through the delay chain. In this sequence, all the weights are changed by Eq.(5), where the learning coefficient α_i is 0.2. One cycle of learning is completed by 45 sets of five data applied from delay elements to synapse inputs x_1, x_2, x_3, x_4, x_5 in case of 5-synapse neo fuzzy neuron. After 100 learning cycles, a root-mean-square value of error was measured for every number of synapses. The experimental results are shown in Table 1 which suggests us that a 3-dimensional dynamical system is not necessarily identified by a 3-synapse neo fuzzy neuron, but is done more precisely by a neo fuzzy neuron of more synapses.

The effect of the learning coefficient α_i on learning speed was examined, the experimental results of which are shown in Fig.5. When α_i is small, the learning speed is low. On the other hand, a big α_i causes an oscillatory learning. The optimum value for α_i could be obtained to be 0.2 for all synapses ($\alpha_1=\alpha_2=\ldots=\alpha_5=\alpha=0.2$) and in this case, the learning of 20 cycles is enough to get a reasonably small error.

After the learning of 100 cycles, the 3-dimensional dynamical system was approximately identified by a 5-synapse neo fuzzy neuron. The experimental results are shown in Fig.6 by dotted lines. Solid lines in Fig.6(a), (b) and (c) represents the true nonlinear functions in the given dynamical system. A fuzzy inference with singletons in consequents exhibits a linear interpolation and thus any nonlinear functions are approximately represented by a piecewise linear function. So that some error is essentially produced. However, this error can be compensated by fourth and fifth synapses. Thus the more the number of synapses increases, the smaller the error becomes as shown in Table 1.

4. BEHAVIOR PREDICTION OF NONLINEAR DYNAMICAL SYSTEMS BY A NEO FUZZY NEURON

Once the learning of the neo fuzzy neuron with a delay chain has been completed by data obtained from a nonlinear dynamical system under test, the neo fuzzy neuron facilitates the prediction of its behavior, even though it is quite chaotic. There are a several modes of prediction, which are described in detail in the following.

4.1 SINGLE-STEP PREDICTION

The learning is made in the system configuration shown in Fig.3. In other words, the data obtained from the unknown dynamical system is applied to the input terminal of the delay chain and after several tens of learning, all weights in each synapse of the fuzzy neuron are determined. If the time series is assigned in the delay chain so that the last data is allocated at the input x_1 of the neo fuzzy neuron, then the output $\widehat{y_{n+1}}$ of the neo fuzzy neuron provides the prediction of one step ahead.

4.2 MULTI-STEP PREDICTION

When we want to predict the behavior of several steps ahead, the system configuration with a neo fuzzy neuron should be a little bit modified as shown in Fig.7, that is, in order to predict the behavior of p steps ahead p-1 delay elements in series should be added to the input of the system as shown in Fig.7. After the learning, the output provides the prediction, if the time series is assigned in the delay chain so that the last data is allocated at the input x_1 of the neo fuzzy neuron.

4.3 SEQUENTIAL PREDICTION

By feeding the output of the neo fuzzy neuron back to the input of the system as shown in Fig.8, sequential prediction can be achieved. All the parameters w_{ij} of 5-synapse neo fuzzy neuron are zero at the initial state and assigned after 100 learning cycle. One learning cycle consists of 45 sets of five data sampled from the test signal shown in Fig.4.and if the time series is assigned in the delay chain so that the last data is allocated at the input x_1 of the neo fuzzy neuron, then the output of the neo fuzzy neuron provides the prediction of one step ahead. And it is fed back to the input of the delay chain. A clock signal forwards this signal to x_1 and simultaneously produces the next step prediction. Thus, the continuous and *sequential prediction* is obtained. The experimental results is shown in Fig.9. Roughly speaking, about 20 step prediction is reasonable from Fig.9. However, even a very small deviation causes a large error, so that this prediction is not suitable for that of long period.

4.4 ADAPTIVE PREDICTION

In order to predict the behavior of the system under test continuously, it is necessary to retrain or relearn by using a new data. In Fig.3 (5-synapse in this case), the output of the neo fuzzy neuron provides with one-step prediction, when y_{49}, y_{48}, ...,y_{45} are allocated at x_1, x_2, ..., x_5. This one set of five data has not yet been utilzed for learning to assign weights. One step later, we can get the true signal and compare it with the predicted one. Learning with new data is achieved to revise all the weights. 45 sets of data (y_1, y_2, ...y_5), (y_2, y_3, ..., y_6), (y_3, ..., y_7), ...(y_{45}, ..., y_{49}) are used for revising weights by 50 learning cycles in accordance with the change of the circumstances. Thus all the weights are revised at every step by 50 learning cycles and the adaptation of the parameters to the cahnge of the circumstances are accomplished. Therefore, the authors call this prediction to be *adaptive prediction*.

5. CONCLUSION

A structure and a learning algorithm of a neo fuzzy neuron was described. It is one kind of nonlinear synapse neuron and the nonlinearity is characterized by fuzzy linguistic if-then rules. Membership functions in antecedents should be complementary to avoid dividers in the defuzzifier.

This neo fuzzy neuron can be applied to identify the nonlinear dynamical system in good accuracy after a short learning cycle, even though it exhibits chaotic behavior.

The neo fuzzy neuron can be applied to predict the behavior of the system under test. Adaptive prediction is the most promising application.

REFERENCES

[1]S.C. Lee and E. T. Lee, "Fuzzy Neurons and Automata," Proc. 4th Princeton Conf. on Information Science and Systems, pp.381-385 (1970).

[2]Takeshi Yamakawa, Japanese Patent Application, No.TOKU-GAN-HEI 1-133690, (May 1989).

[3]Takeshi Yamakawa and Shin-ya Tomoda, "A Fuzzy Neuron and Its Application to Pattern Recognition", Proc. the Third IFSA Congress, Seattle, Washington , August 6-11 (1989).

[4]Takeshi Yamakawa, "Pattern Recognition Hardware System Employing a Fuzzy Neuron", Proc. the International Conference on Fuzzy Logic & Neural Networks, Iizuka, Japan, July 20-24 (1990).

[5]Takeshi Yamakawa et al, Japanese Patent Applications, No.TOKU-GAN-HEI 3-101392, 3-101393, 3-102757, 3-102758, (May 1991).

[6]Takeshi Yamakawa and Masuo Furukawa, "A Design Algorithm of Membership Functions for a Fuzzy Neuron Using Example-Based Learning", Proc. IEEE International Conference on Fuzzy Systems 1992,San Diego,March 8-12 (1992).

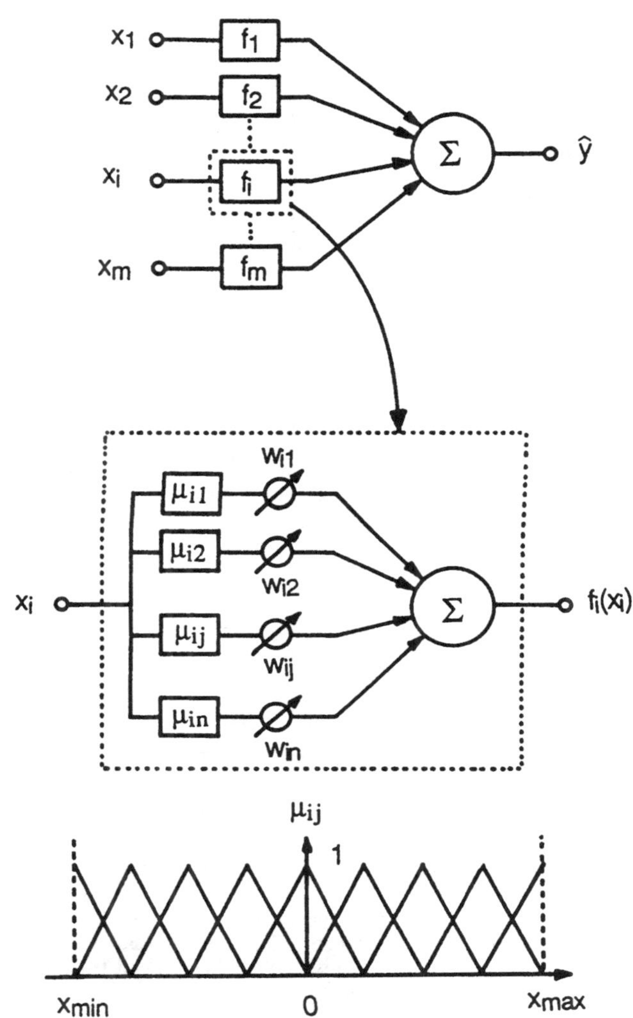

Fig.1 (a)Structure of the neo fuzzy neuron, each synaptic characteristics of which is represented by a nonlinear function f_i. (b)Structure of the nonlinear synapse which is described with a set of if-then rulesincluding singletons in consequents. (c)Triangular and complementary membership functions assigned for the fuzzy segments in the input space.

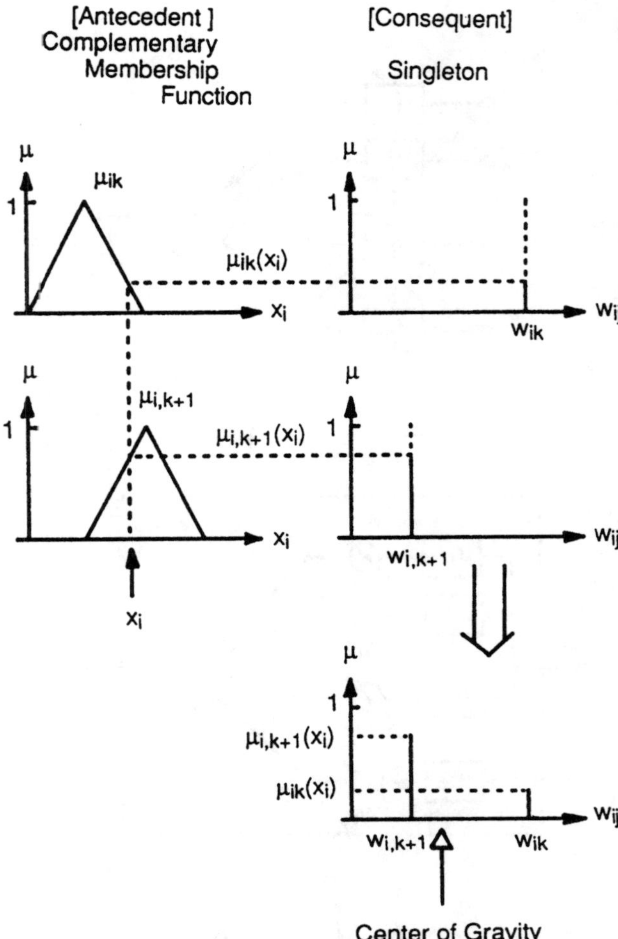

Fig.2 The nonlinear synapse is characterized by a very simple fuzzy inference. The input signal activates only two rules, because antecedents of all rules are described by complementary membership functions. Furthermore, all the consequents of all rules are described by singletons, so that the defuzzification is simply achieved without arithmetic division.

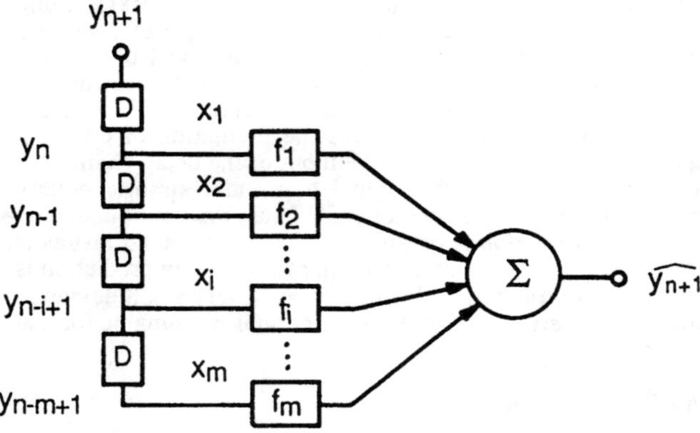

Fig.3 Combination of the neo fuzzy neuron and a series of delay elements D which facilitates system identification and prediction of its behavior.

Table 1 Root-mean-square errors for 1-, 2-, 3-, 4- and 5-synapse.

number of synapsises (dimensions)	1	2	3	4	5
$\sqrt{\overline{error^2}}$	8.59×10^{-1}	5.24×10^{-1}	3.86×10^{-2}	3.23×10^{-2}	1.87×10^{-2}

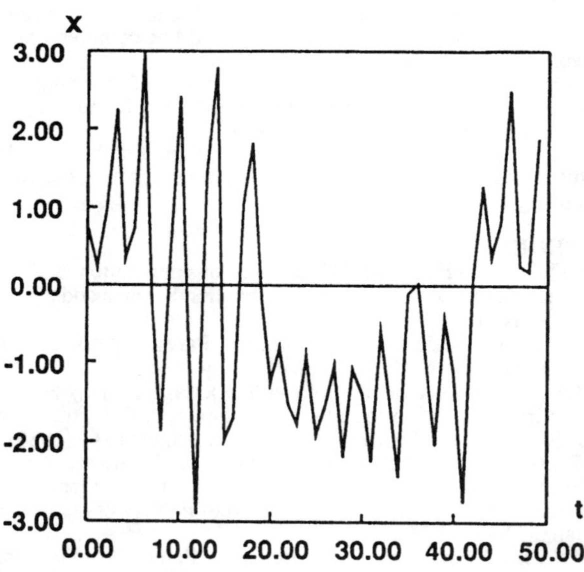

Fig.4 A chaotic test signal generated by the 3-dimensional dynamical system defined by Eq.(4) and the initial values of $y_0 = 0.733$, $y_1 = 0.234$, $y_3 = 0.973$.

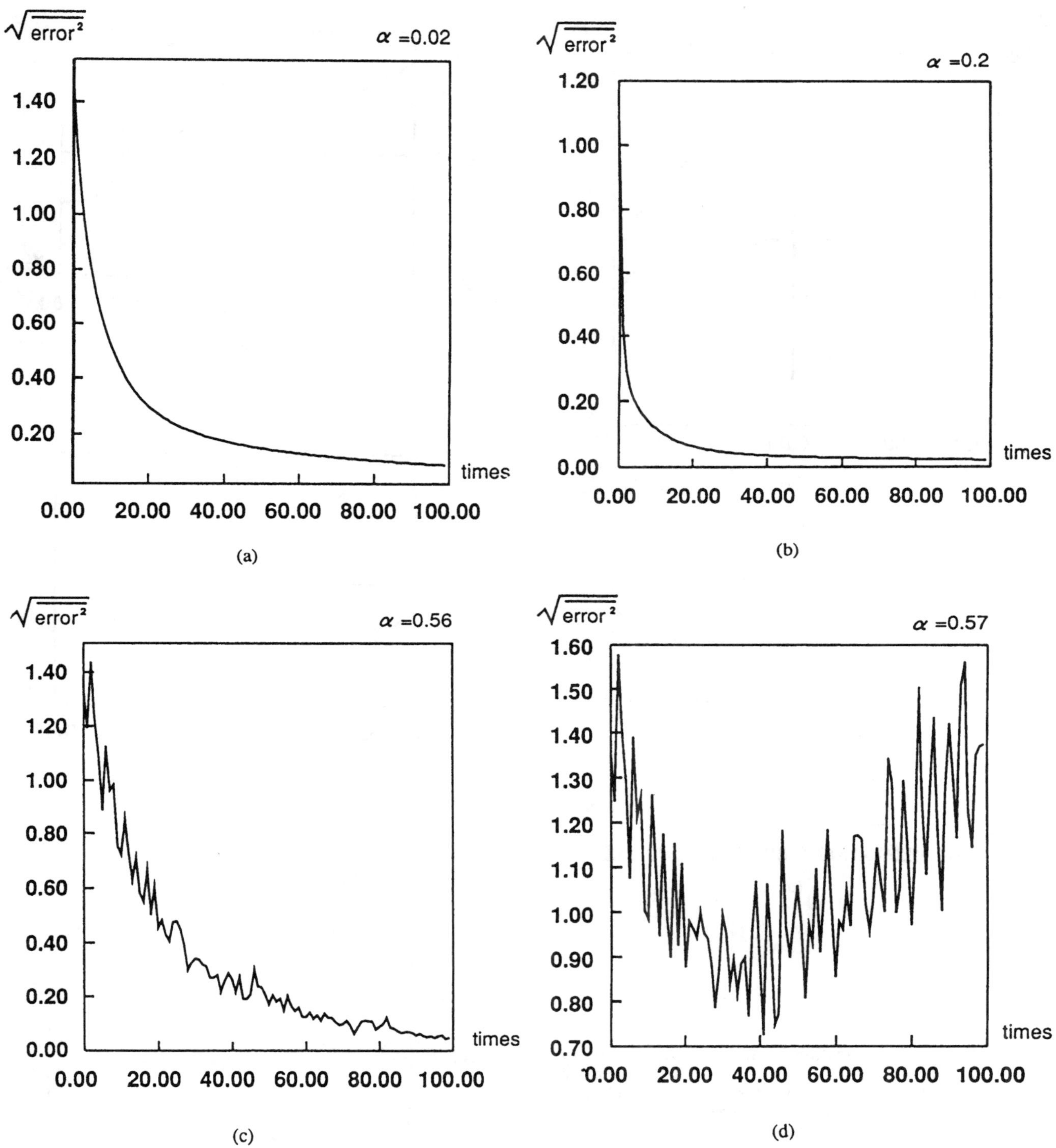

Fig.5 The effect of the learning coefficient α_i on learning speed, where all α_i's are assigned to be equal to α. (a) $\alpha = 0.02$, (b) $\alpha = 0.2$, (c) $\alpha = 0.56$, (d) $\alpha = 0.57$.

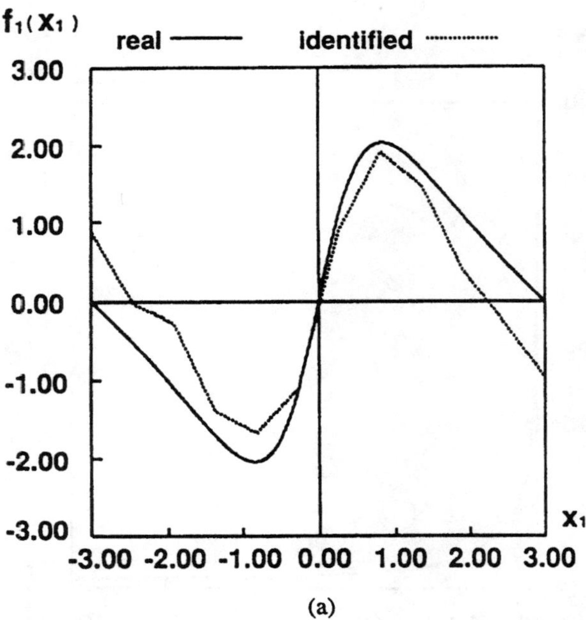

(a)

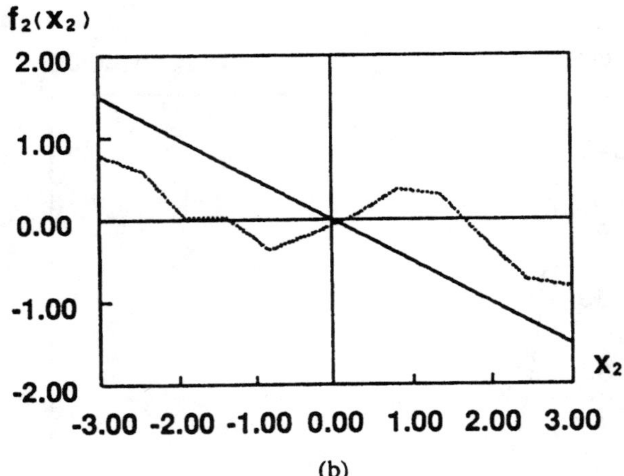

(b)

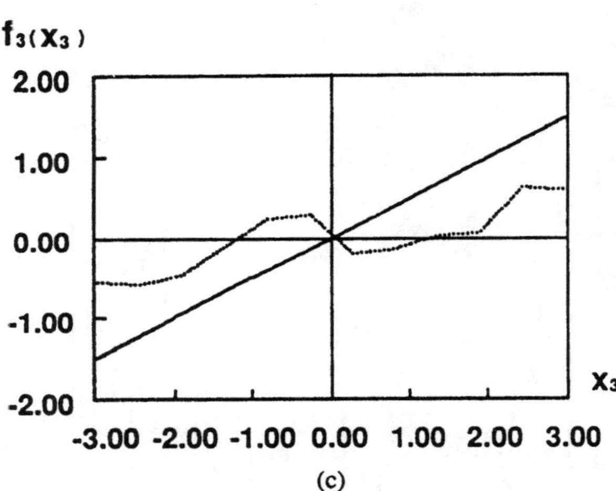

(c)

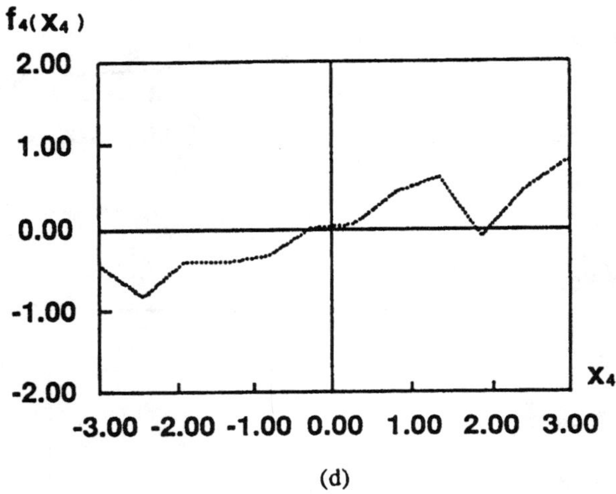

(d)

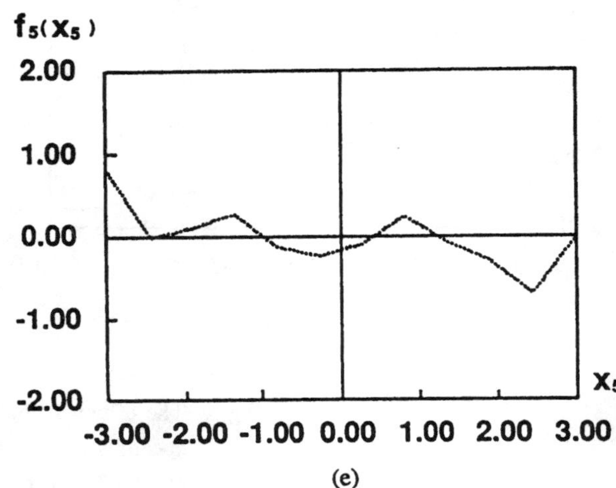

(e)

Fig.6 System identification by a 5-synapse neo fuzzy neuron (dotted lines). The 3-dimensional dynamical system to be obtained is illustrated with solid lines. (a)first dimensional ; $f_1(x_1)$, (b)second dimensional ; $f_2(x_2)$, (c)third dimensional ; $f_3(x_3)$, (d)fourth dimensional ; $f_4(x_4)$, (e)fifth dimensional ; $f_5(x_5)$.

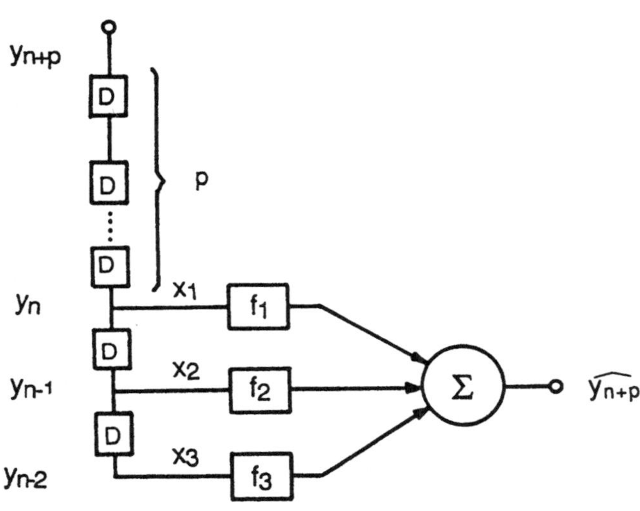

Fig.7 System configuration to predict p steps ahead.

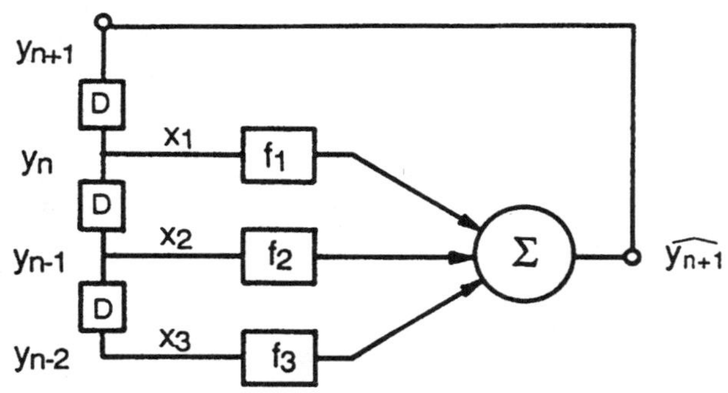

Fig.8 System configuration to achieve sequential prediction.

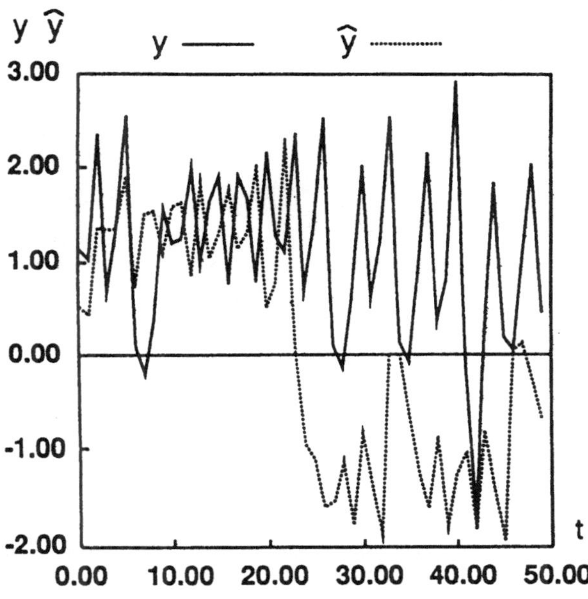

Fig.9 Sequential prediction of a chaotic behavior by the neo fuzzy neuron.

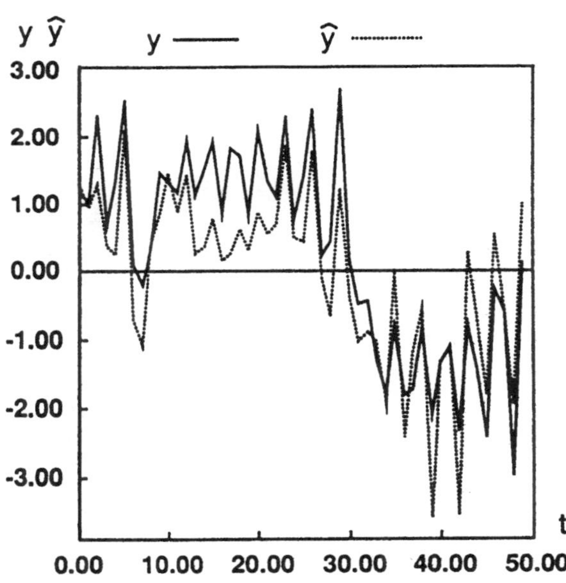

Fig.10 Adaptive prediction of a chaotic behavior by the neo fuzzy neuron.

Part 7
Neuro-Control Systems: Applications

One of the last frontiers of science, perhaps its ultimate challenge, is the understanding of the biological basis of mentation and cognition. Can this process be described by the known physical laws, or does it lie within the domain of metaphysical framework?

IT was in the early sixties that the theoretical foundations of adaptive control systems were laid. Since then, the theoretical developments have very successfully been applied to industrial control problems, such as aircraft control, autopilots for steering ships, process control, robotics, and power systems control. For detailed developments of this field, readers are directed to the IEEE Press reprint book, *Adaptive Methods for Control Systems Design,* 1986, edited by Madan M. Gupta. Since 1985, we have extended the field of adaptive control under new names, such as intelligent control systems, neuro-control systems, and others. The neurons—the basic building blocks of the CNS—play many important functions in the sensory, locomotion, and cognitive aspects of the CNS. Neurons in the higher cortical level provide some sort of cognition; that is, intelligence—the powers to reason, think, learn and adapt. Therefore, we can give an alternative name to ***neuro-control systems,*** the subject of this volume, as ***neo-adaptive control systems.*** The subject of adaptive control systems, with various names such as neo-adaptive control systems, intelligent control systems, cognitive control systems, and neuro-control systems, therefore, falls within the domain of control of complex industrial and robotic systems with reasoning, learning, and adaptive abilities.

In recent years, novel neural morphologies with learning and adaptive capabilities have infused new control power into the control of complex dynamic systems. Theoretical developments in the field are evolving, and many new applications are springing up. After having presented a large amount of theoretical literature in the field of neuro-control systems in the last six parts, we present some applications of various neuro-control morphologies.

The articles presented in this part cannot be classified into the category of pure applications-oriented articles. Since the field is still evolving, these articles have a mixed flavor of theory and some applications, although most of these applications have only a computer simulation basis.

In the first five articles, we present robotics-oriented neuro-control applications. In article (7.1), D. H. Rao and M. M. Gupta present a new dynamic neural architecture called a *dynamic neural processor,* which consists of two dynamic neural units coupled as excitatory and inhibitory neurons, (see also Article 5.6). This neural model is inspired by the collective computational properties of the biological neural subpopulations. It is demonstrated in this article that the proposed neural architecture can be implemented to compute inverse-kinematics transformation of two-linked robots. Simulation studies indicate the robustness and fast convergence of this neural approach compared to conventional feedforward neural networks.

Coordinated actions of multiple systems is a challenging control problem. In article (7.2), X. Cui and K. G. Shin present a knowledge-based intelligent coordinator using neural networks. At the coordination level, a hierarchical structure of the knowledge-based coordinator (KBC) is presented that coordinates the control actions of various low-level controllers. This article also presents some simulation studies of the coordination of two two-linked robots holding a single object.

Although researchers in the field of conventional adaptive control and artificial intelligence have made many advances, neither ideology seems capable of realizing autonomous operation. That is, neither approach can produce machines that can interact with uncertain environmental situations as do humans or higher animals. In article (7.3), B. L. Digney and M. M. Gupta offer a method that combines reinforcement learning with the behavior-based control system. Behaviors are impossible to embed in control functions. The approach presented in this article is based upon the distributed adaptive control systems (DACS), which essentially serve as the robot's artificial nervous system. Several simulation studies are presented on the gait coordination of a quadruped mobile robot for a graceful motion control. This article, thus, addresses many interesting problems associated with robotics and other complex systems.

As demonstrated in articles (7.2) and (7.3), a combination of low-level control action and cognitive control tends to yield a superior class of control performance. To some extent, cognitive tasks can be performed by storing the knowledge in neural networks. D. A. Handelman, S. H. Lane, and J. J. Gelfand, in article (7.4), present a methodology for integrating both computational paradigms for the purpose of robotic control and patterning the integration after the acquisition of models of human motor skills. The authors are highly inspired and motivated by the features of biological control systems. The designer of an intelligent control scheme can learn a great deal by studying the structural (morphological), functional, and behavioral aspects of biological systems. This article shows how some of the biological paradigms can be used for improving the control performance. It provides integrated forms of computation that enable robotic skill acquisition and control.

The control of the movement of a multijointed manipulator, in general, includes highly nonlinear operations. The control algorithms have to cope with additional problems when the manipulator has more joints than necessary for a given task; that is, when it has extra degrees of freedom. The human arm provides such a redundancy in manipulating a complex task. The model of the human arm with three joints (shoulder, elbow, and wrist) moving in a two-dimensional horizontal plane can be considered as a manipulator with one extra degree of freedom, for only two joints are sufficient to reach a desired point in the given work space. The existence of the third joint produces an additional degree of freedom, producing an infinite number of possible solutions. In order to solve the dynamic control problem of the joints to reach a given (x–y) coordinate, a neuronal approach is used in article (7.5), by M. Bruwer and H. Cruse. This neuronal approach has several advantages over the algorithmic approach used conventionally. The results presented in the article qualitatively correspond to that obtained from human subjects.

The next four articles introduce different applications of neural networks.

In article (7.6), Q. H. Wu, B. W. Hogg, and G. W. Irwin present a neuro-controller for a nonlinear, multivariable turbogenerator. The neuro-controller has a compact structure that can easily be extended to cater to more complex dynamic systems.

Beamforming is one of the main functions of a passive phased-array processing system. The main objective of this process is to suppress unwanted jamming interferences and produce the optimal beamformer responses with minimum noise. In article (7.7), P-R. Chang, W-H. Yang, and K-K. Chan propose a cost-effective analogy solution for computing the minimum variance distortion response (MVDR) beamforming problem using the Kennedy–Chua canonical neural network. This canonical feature is able to guarantee the stability and robustness of the solution, and the convergence time can be characterized by the dominant time constant of the network. This approach seems to have many other practical implications.

Next, we present two articles that deal with the neural control of chemical processes. In article (7.8), M. J. Willis et al. consider the use of neural networks for solving some complex process control problems. Simulation studies and the industrial process data are presented. The authors also consider the possibility of neural control models in model-based control strategies.

In article (7.9), N. V. Bhat, et al. discuss a neural-modeling, nonlinear chemical process. The authors' conclusion is that neural nets hold great promise for cost-effective modeling of chemical processes.

Finally, in article (7.10), W. T. Miller III, F. H. Glanz, and L. G. Kraft III describe a neural-network structure called a CMAC (cerebellar model arithmetic computer), which is an alternative to the commonly used backpropagation-based multilayered network. In this article, the authors explain the motivation, mathematical properties, and functioning of the CMAC, and its several potential applications. The CMAC has the advantage that it is fast in software and can be realized in high-speed hardware. It also can learn a large variety of nonlinear functions, reducing the need to use slow backpropagation-based neural networks. The CMAC neural network seems to have great potential in further theoretical and hardware implementations, and possible applications in control and vision problems.

In this part of the book, we have presented a small sample of representative applications in the domain of robotics and industrial process control. In the next part, we will deal with the hardware aspects of neural networks.

Further Reading

[1] A. N. Michel and J. A. Farrell, "Associative memories via artificial neural networks," *IEEE Contr. Syst. Mag.*, pp. 6–17, April 1990.

[2] C. W. Anderson, "Learning to control an inverted pendulum using neural networks," *IEEE Contr. Syst. Mag.*, pp. 31–37, April 1986.

[3] M. Kawato, Y. Uno, M. Isobe, and R. Suzuki, "Hierarchical neural network model for voluntary movement with application to robotics," *IEEE Contr. Syst. Mag.*, pp. 8–15, April 1988.

[4] A. Waibel, T. Hanazawa, G. Hinton, K. Shikano, and K.J. Lang, "Phoneme recognition using time-delay neural networks," *IEEE Trans. Acoust., Speech Signal Process.*, vol. 37, no. 3, pp. 328–339, March 1989.

[5] A. F. Murray, "Silicon implementations of neural networks," *IEE Proc.-F,* vol. 138, no. 1, pp. 3–12, Feb. 1991.

[6] H. C. Card, C. R. Schneider, and W. R. Moore, "Hebbian plasticity in MOS synapses," *IEE Proc.-F,* vol. 138, no. 1, pp. 13–16, Feb. 1991.

[7] L. Tarassenko, J. N. Tombs, and J. H. Reynolds, "Neural network architectures for content-addressable memory," *IEE Proc.-F,* vol. 138, no. 1, pp. 33–39, Feb. 1991.

[8] L. Udapa and S. S. Udapa, "Neural networks for the classification of nondestructive evaluation signals," *IEE Proc.-F,* vol. 138, no. 1, pp. 41–45, Feb. 1991.

[9] W. J. Daunicht, "Control of manipulators by neural networks," *IEE Proc.*, vol. 136, pt. E, no. 5, pp. 395–399, Sept. 1989.

[10] H. Hashomoto, T. Kubota, M. Kudou, and F. Harashima, "Self-organizing visual servo system based on neural networks," *IEEE Contr. Syst. Mag.*, pp. 31–36, April 1992.

[11] L. C. Rabelo and X. J. R. Avula, "Hierarchical neurocontroller architecture for robotic manipulation," *IEEE Contr. Syst. Mag.*, pp. 37–41, April 1992.

[12] B. P. Yuhas, M. H. Goldstein, Jr., T. J. Sejnowski, and R. E. Jenkins, "Neural network models of sensory integration for improved vowel recognition," *Proc. IEEE,* vol. 78, no. 10, pp. 1658–1668, Oct. 1990.

[13] A. D. Handelman, H. L. Stephen, and J. J. Gelfand, "Integrating neural networks and knowledge-based systems for intelligent robotic control," *IEEE Contr. Syst. Mag.*, pp. 77–87, April 1990.

[14] T. Sejnowski and C. R. Rosenberg, "NETtalk: A neural network that learns to read aloud," Tech. Rep. JHU/EECS-86/01, John Hopkins University, 1986.

[15] D. J. Burr, "Experiments on neural net recognition of spoken and written text," *IEEE Trans. Acoust., Speech Signal Process.*, vol. 36, no. 7, pp. 1162–1168, July 1988.

[16] R. P Gorman and T. J. Sejnowski, "Learned classification of sonar targets using a massively parallel network," *IEEE Trans. Acoust., Speech Signal Process.*, vol. 36, no. 7, pp. 1135–1140, July 1988.

[17] F. C. Hoppensteadt, "Signal processing by model neural networks," *SIAM Rev.*, vol. 34, no. 3, pp. 426–444, Sept. 1992.

[18] A. Guez and Z. Ahmad, "Solution to the inverse kinematics problem in robotics by neural networks," in *Proc. IEEE Int. Conf. Neural Networks*, San Diego, Calif., pp. 617–624, March 1988.

[19] J. Barhen, S. Gulati, and M. Zak, "Neural learning of constrained nonlinear transformations," *IEEE Computer*, pp. 67–76, June 1989.

[20] M. M. Gupta, D. H. Rao, and P. N. Nikiforuk, "Dynamic neural network based inverse-kinematics transformation of two- and three-linked robots," IFAC Conf., vol. 3, pp. 289–296, Sydney, Australia, July 19–23, 1993.

[21] M. J. Willis, G. A. Montague, C. D. Massimo, M. T. Tham, and A. J. Morris, "Artificial neural networks in process estimation and control," *Automatica*, vol. 28, no. 6, pp. 1181–1187, 1992.

[22] H. Miyamota, M. Kawato, T. Setoyama, and R. Suzuki, "Feedback-error-learning neural network for trajectory control of a robotic manipulator," *Neural Networks*, vol. 1, pp. 251–265, 1988.

[23] M. M. Gupta, *Adaptive Methods for Control System Design*. New York: IEEE Press, 1986.

General Learning Scheme for Robot Coordinate Transformations Using Dynamic Neural Network

M.M. Gupta and D.H. Rao

Intelligent Systems Research Laboratory, College of Engineering
University of Saskatchewan, Saskatoon, Canada, S7N 0W0

ABSTRACT

By virtue of their functional approximation, learning and adaptive capabilities, the computational neural networks can be suitably employed for learning robot coordinate transformations. The major drawback of conventional static feedforward neural networks based on back-propagation learning algorithm is in their very large convergence time for a given task. Any attempts to accelerate the learning process by increasing the values of learning constants in the algorithm often result in unstable systems. The intent of this paper is to describe a neural network structure called dynamic neural processor (DNP), and examine briefly how it can be used in developing a learning scheme for computing robot inverse kinematic transformations. The architecture and learning algorithm of the proposed dynamic neural network structure, the DNP, are described. Computer simulations are provided to demonstrate the effectiveness of the proposed learning scheme using the DNP.

1. Introduction

The accuracy of a robotic manipulator depends upon its ability to move in a given particular task-space to a desired set of Cartesian locations. As a consequence, the inverse kinematic problem becomes important as it must be solved in real-time to compute joint angles for a desired end-effector position. The inverse kinematic problem can be solved either by algebraic, geometric or iterative techniques. The common approach is to use a closed form solution to the inverse kinematic problem. However, these solutions are manipulator dependent and still often too difficult to solve in a closed form [1].

Advances in the area of neural networks have given a different direction to robotic control. By virtue of their functional mapping and dynamic iterative capabilities, neural networks can be employed for learning coordinate transformations [1 - 6]. Because of their parallel and distributed computations, neural networks have the ability to learn associations between example patterns. These patterns could represent, for example, the task space coordinates and the corresponding joint angles of the model leg. The association between these two sets of patterns basically amounts to the inverse kinematic computations in the robotic paradigm. The advantage of using neural approach, over the conventional inverse kinematic algorithms, is that neural networks can avoid time consuming calculations. Furthermore, in a manner that is typical of neural networks, it would be very easy to modify the learned associations upon changes in the structure of the mechanism. It is advantageous, therefore, to employ neural networks for learning inverse kinematic transformations of robotic systems.

The methodology normally employed in the above task is to train the neural network off-line for possible data patterns within the robot task space. Because of the generalization capability, the neural networks can learn the associated patterns very fast and recall the learned patterns almost instantaneously. The trained network is then used to achieve the desired voluntary movements. This technique, therefore, involves two modes of operation, namely the training phase and the performing phase. Furthermore, the major drawback of this technique, is a very long training procedure, besides being the fact that the static neural networks based on back-propagation learning algorithms do take a very large convergence time for any given task.

The emergence of dynamic neural computing has made it possible to develop learning schemes which can be used to arrive at feasible solutions fast to complex problems, such as the inverse kinematics problem in robotics. The intent of this paper is to develop such a learning scheme using a dynamic neural network. The neural network structure discussed in this paper, called the dynamic neural processor (DNP), is developed based on the neuro-physiological evidence that neural activities of any complexity are a result of interactions between excitatory (positive) and inhibitory (negative) neural subpopulations. The architectural details of the DNP are presented in the next section. The learning algorithm for the adaptable weights of the neural network is also presented in this section, followed by the description of the learning scheme in Section 3. Computer simulation studies are also elucidated in this section, followed by conclusions in the last section.

Reprinted with permission from *SPIE's Conference on Intelligent Robots and Computer Vision XI: Algorithms and Techniques,* vol. *Proc. SPIE* 2055, pp. 524–535, M. M. Gupta and D. H. Rao, "General Learning Scheme for Robot Coordinate Transformations Using Dynamic Neural Network," Boston, 1993.

2. The Dynamic Neural Processor (DNP) Based on Neural Subpopulations

2.1 The Motivation

The artificial or computational neural network structures described in the existing literature often consider the behavior of a single neuron as the basic computing unit for describing neural information processing operations. Each computing unit in the network is based on the concept of an *idealized* neuron. An ideal neuron is assumed to respond optimally to the applied inputs. However, experimental studies in neuro-physiology show that the response of a biological neuron appear random [8, 9], and only by averaging many observations is it possible to obtain predictable results. In general, the behavior of a biological neuron is an unpredictable mechanism for processing information. However, mathematical analysis has shown that these random cells can transmit reliable information if they are sufficiently redundant in numbers. It is postulated [9], therefore, that the collective activity generated by large numbers of locally redundant neurons is more significant in a computational context than the activity generated by a single neuron .

The total neural activity results from a collective assembly of cells called *neural population*, or *neural mass*. The neural population comprises of neurons, and its properties have a generic resemblance to those of individual neurons. But it is not identical to them, and its properties can not be predicted from measurements on single neurons. This is due to the fact that the properties of neural population depend on various parameters of individual neurons and also depend upon the interconnections between neurons [10]. The study of neural networks based on single-neuron analysis precludes the above two facets of biological neural structures.

Each neural population may be further divided into several coexisting *subpopulations*. A subpopulation contains a large class of similar neurons that lie in close spatial proximity. The neurons in each subpopulation are assumed to receive a common set of inputs and provide corresponding outputs. The individual synaptic connections within any subpopulation are random, but dense enough to ensure at least one mutual connection between any two neurons. The most common neural mass is the mixture of *excitatory* (positive) and *inhibitory* (negative) subpopulations of neurons. The excitatory neural subpopulation increases the electro-chemical potential of the post-synaptic neuron, while the inhibitory subpopulation reduces the electro-chemical potential.

The minimum topology of such a neural mass contains excitatory (positive), inhibitory (negative), excitatory - inhibitory (synaptic connection from excitatory to inhibitory), and inhibitory - excitatory (synaptic connection from inhibitory to excitatory) feedback loops. One of the most important attributes of a neural mass is that some of its fundamental characteristics may be described interms of linear systems analysis [10]. This property implies that the operations in a neural mass, within appropriate limits of amplitude, conform to the principle of superposition.

In view of the above remarks, it is envisaged that the neural models developed based upon the concept of neural population may present a better representation of biological neural systems. Based on this hypothesis, the authors proposed *dynamic neural processor* (DNP) for robotics and control applications [11, 12]. The strength of synaptic connections between the two subpopulations of neurons were fixed and were chosen arbitrarily. A modified DNP structure with adaptable synaptic strengths, developed in this paper, is described in the following sub-section.

2.2 The Architecture and Mathematical Model of DNP

The basic functional node of the DNP is the *dynamic neural unit* (DNU) proposed in [13]. A brief description of DNU is given below.

(i) Dynamic Neural Unit (DNU)

The DNU comprises of memory elements (delay operators), and feedforward and feedback synaptic weights as shown in Fig. 1. The output of this dynamic structure constitutes the argument to a time-varying nonlinear activation function. The DNU performs two distinct operations: (i) the *synaptic operation* and (ii) the *somatic operation*. The first operation corresponds to the adaptation of feedforward and feedback synaptic weights, while the second to the adaptation of gain (shape) of the nonlinear activation function. The DNU consists of delay elements, feedforward and feedback paths weighted by the synaptic weights $\mathbf{a}_{ff}$ and $\mathbf{b}_{fb}$ respectively representing a second-order structure followed by a nonlinear activation function. The linear part of the DNU structure can be expressed by the following difference equation

$$v_1(k) = - b_1 v_1(k-1) - b_2 v_1(k-2) + a_0 s(k) + a_1 s(k-1) + a_2 s(k-2) \quad (1)$$

where $s(k) \in \Re^n$ is the neural input vector, $v_1(k) \in \Re^1$ is the output of the dynamic structure, $u(k) \in \Re^1$ is the neural output, k is the discrete-time index, z^{-1} is the unit delay operator, and $\mathbf{a}_{ff} = [a_0, a_1, a_2]$ and $\mathbf{b}_{fb} = [b_1, b_2]$ are the vectors of adaptable feedforward and feedback weights respectively. The vectors of signals and adaptable weights of DNU are defined as follows

$$\Gamma^T(k,v_1,s) = \left[v_1(k-1) \;\; v_1(k-2) \;\; s(k) \;\; s(k-1) \;\; s(k-2)\right], \text{ and} \quad (2)$$

$$\zeta^T_{(\mathbf{a}_{ff},\mathbf{b}_{fb})} = \left[-b_1 \;\; -b_2 \;\; a_0 \;\; a_1 \;\; a_2\right] \text{ (the superscript T denotes transpose).} \quad (3)$$

Using (2) and (3), Eqn. (1) is rewritten as

$$v_1(k) = \Gamma(k,v_1,s)\, \zeta^T_{(\mathbf{a}_{ff},\mathbf{b}_{fb})} . \quad (4)$$

The nonlinear mapping operation on $v_1(k)$ yields a neural output u(k) given by

$$u(k) = \Psi\left[g_S\, v_1(k) - \theta\right] \quad (5)$$

where $\Psi[.]$ is some nonlinear activation function, usually the sigmoidal function and g_S is the somatic gain which controls the slope of the activation function, and θ is a threshold which has to be exceeded for the neuron to fire. Many different forms of mathematical functions can be used to model the nonlinear behavior of the biological neuron. We use a bounded and monotonically increasing and differentiable sigmoidal activation function for each unit's output function. To extend the mathematical operations to both the excitatory and inhibitory inputs, the activation (sigmoidal) function is defined over [-1,1] as

$$\Psi[v(k)] = \tanh\left[g_S\, v_1(k) - \theta\right] = \tanh\,[v(k)] \quad (6)$$

where $v(k) = g_S\, v_1(k) - \theta$. Now we describe the morphology of dynamic neural processor (DNP) in the following paragraphs.

(ii) Dynamic Neural Processor (DNP)

The dynamic neural processor comprises of two DNUs coupled in excitatory and inhibitory modes as depicted in Fig. 2. In this structure, $s_\lambda(k)$ and $u_\lambda(k)$ represent respectively the stimulus (input) and state response (output) of the neural computing unit where the subscript λ indicates either an excitatory, E, or inhibitory, I, state. $s_{t\lambda}(k)$ denotes the total input to the neural unit, $w_{\lambda\lambda}$ represent the strength of self-synaptic connections (w_{EE}, w_{II} in Fig. 2), and $w_{\lambda\lambda'}$ repsent the strength of inter-synaptic connections from one neural unit to another (w_{IE}, w_{EI} in Fig. 2).

The functional dynamics exhibited by a neural computing unit, the DNU, is defined by a second-order difference equation as represented by Eqn. (1). The state variables $u_E(k+1)$ and $u_I(k+1)$ generated at time (k+1) by the excitatory and inhibitory neural units of the proposed neural processor are modeled as:

$$u_E(k+1) = \mathbf{E}\,[u_E(k)\,,\; v_E(k)], \text{ and } u_I(k+1) = \mathbf{I}\,[u_I(k)\,,\; v_I(k)] \quad (7)$$

where $v_E(k)$ and $v_I(k)$ represent the proportion of neurons in the neural unit that receive inputs greater than an intrinsic threshold, and **E** and **I** represent the excitatory and inhibitory actions of the neurons. The neurons that receive inputs greater than a threshold value is given by a nonlinear function of $v_\lambda(k)$. The total inputs incident on the excitatory and inhibitory neural units are respectively

$$s_{tE}(k) = w_E s_E(k) + w_{EE} u_E(k-1) - w_{IE} u_I(k-1) - \theta_E \quad \text{, and} \tag{8}$$

$$s_{tI}(k) = w_I s_I(k) - w_{II} u_I(k-1) + w_{EI} u_E(k-1) - \theta_I \tag{9}$$

where w_E and w_I are scaling factors of the excitatory and inhibitory neural inputs respectively, w_{EE} and w_{II} represent the self-synaptic connection strengths, w_{IE} and w_{EI} represent the inter-neuron synaptic strengths, and θ_E and θ_I represent the thresholds of excitatory and inhibitory neurons respectively. The above equations may be written in the matrix form as:

A direct analytical solution for determining the steady-state and temporal behavior exhibited by the DNP is not possible because of the inherent nonlinearities in Eqns. (8) and (9). However, these nonlinear equations can be analyzed qualitatively by obtaining the phase trajectories in the u_E - u_I phase plane [9, 14]. These trajectories enable the system characteristics to be observed without solving the nonlinear equations. The locus of points where the phase trajectories have a given slope is called an *isocline curve* [15]. The steady-state activity exhibited by DNUs of the neural processor can be investigated by determining the isocline curves corresponding to $u_E(k+1) = 0$ and $u_I(k+1) = 0$. The average activity of excitatory and inhibitory groups of neurons evolve according to:

$$u_E(k+1) = u_E(k) + \left(1 - r_E u_E(k)\right) \Psi_E \left[s_{tE}(k)\right] \quad \text{: Excitatory neuron} \tag{10a}$$

$$u_I(k+1) = u_I(k) + \left(1 - r_I u_I(k)\right) \Psi_I \left[s_{tI}(k)\right] \quad \text{: Inhibitory neuron} \tag{10b}$$

where r_E and r_I represent the absolute refractory periods (during which the neurons can not refire) of excitatory and inhibitory neurons respectively. From Eqns. (8) and (10), the equations for isocline curves may be written as

$$u_I(k) = \frac{1}{w_{IE}}\left[\left(w_E s_E(k) - \theta_E\right) + \Psi_E^{-1}\left[\frac{u_E(k)}{\left(1 - r_E u_E(k)\right)}\right] + w_{EE} u_E(k)\right] \tag{11a}$$

for $u_E(k+1) = 0$,

$$u_E(k) = \frac{1}{w_{EI}}\left[\left(- w_I s_I(k) - \theta_I\right) - \Psi_I^{-1}\left[\frac{u_I(k)}{\left(1 - r_I u_I(k)\right)}\right] + w_I u_I(k)\right] \tag{11b}$$

for $u_I(k+1) = 0$.

From the mathematical properties of sigmoid functions, it is evident that Ψ_E and Ψ_I have unique inverses which monotonically increase over the range $[-\infty, \infty]$. Therefore, u_I as defined by Eqn. (11a) will always be a monotonically increasing function of u_E. On the other hand, because of the negative sign before Ψ_I^{-1} in Eqn. (11b), u_E will be a generally decreasing function of u_I. This qualitative difference between the two isoclines is a direct manifestation of the antisymmetry between excitation and inhibition [16]. The synaptic weights w_{EE} and w_{II}, in Eqns. (11a) and (11b), must always be nonvanishing for the isoclines to be nontrivial, thus making feedback between the subpopulations an essential feature of the DNP. A typical plot of the isoclines, Eqns. (11a) and (11b), for $s_E(k) = 0$, $s_I(k) = 0$ is shown in Fig. 3. In this case there is one steady-state solution corresponding to the one intersection of the two curves. Depending upon the strength of synaptic connections chosen there may be more than one steady-state solutions, and the solution may be stable (+) or unstable (-) depending upon the position of intersection of the two isoclines.

Due to the fact that the functional operation of a neural population can be approximated by linear systems theory, the response of the DNP, u(k), is the superposition of the individual responses, $u_\lambda(k)$, of excitatory and inhibitory neurons in the neuronal subpopulation, and is given by:

$$u(k) = u_E(k) + u_I(k)\,. \tag{12}$$

The interpretation of Eqn. (12) is that total activity of the neural population is the summation of excitatory and inhibitory postsynaptic responses. This describes the steady-state behavior of the DNP.

The above description provides an insight into the nonlinear behavior of the neural processor and, therefore, can be employed to approximate nonlinear functions and to model dynamic systems such as robots. In order to achieve the functional approximation, learning and adaptation capabilities of DNP, it is necessary to develop a learning algorithm to update the processor's weights. This algorithm is developed in the following sub-section.

2.3 The Learning Algorithm

The learning process involves the adaptation of feedforward and feedback weights, and somatic gain which minimize the error function. In an iterative learning scheme, the control sequence is modified in each learning iteration to cause the neural output u(k) to approach the desired state $u_d(k)$. If the error, e(k), can be reduced to an infinitesimally small value as the number of learning iterations increase, the learning scheme is said to be convergent; that is $u(k) \rightarrow u_d(k)$ as $k \rightarrow \infty$ or,

$$\lim_{k \rightarrow \infty} \left[u_d(k) - u(k) = e(k)\right] \rightarrow 0 \tag{13}$$

for an arbitrary set of initial conditions. Error e(k) and the components of the parameter vector $\Omega_{(\mathbf{a}_{ff}, \mathbf{b}_{fb}, g_s, w_{\lambda\lambda'})}$ vary with every learning trial k. To obtain $\Omega_{(\mathbf{a}_{ff}, \mathbf{b}_{fb}, g_s, w_{\lambda\lambda'})}(k+1)$ requires only the information set $\{e(k-m), e(k), \Omega_{(\mathbf{a}_{ff}, \mathbf{b}_{fb}, g_s, w_{\lambda\lambda'})}(k)\}$, where m = 1,2, ... which determines the size of the window. As the number of learning trials increase, the information set reduces to only $\{\Omega^*_{(\mathbf{a}_{ff}, \mathbf{b}_{fb}, g_s, w_{\lambda\lambda'})}(k), e^*(k)\}$ which indicates that the DNU parameters and the error have converged to the optimal values. To achieve this, a performance index, which has to be optimized with respect to the parameter vector, is defined as

$$J = E\left\{ F\left[e(k;\Omega_{(\mathbf{a}_{ff}, \mathbf{b}_{fb}, g_s, w_{\lambda\lambda'})})\right]\right\} \tag{14}$$

where E is the expectation operator. A commonly used form of $F\left[e(k;\Omega_{(\mathbf{a}_{ff}, \mathbf{b}_{fb}, g_s, w_{\lambda\lambda'})})\right]$ in Eqn. (14) is an even function of the error; that is,

$$J = \frac{1}{2} E\left\{ e^2\left(k;\Omega_{(\mathbf{a}_{ff}, \mathbf{b}_{fb}, g_s, w_{\lambda\lambda'})}\right)\right\} \tag{15}$$

where E is the expectation operator and e(k) is the error signal defined as the difference between the desired signal $u_d(k)$ and the actual signal u(k). Each component of the vector $\Omega_{(\mathbf{a}_{ff}, \mathbf{b}_{fb}, g_s, w_{\lambda\lambda'})}$ is adapted in such a way so as to minimize J using the steepest-descent algorithm. This adaptation algorithm may be written as

$$\Omega_{(\mathbf{a}_{ff}, \mathbf{b}_{fb}, g_s, w_{\lambda\lambda'})}(k+1) = \Omega_{(\mathbf{a}_{ff}, \mathbf{b}_{fb}, g_s, w_{\lambda\lambda'})}(k) + \delta\Omega_{(\mathbf{a}_{ff}, \mathbf{b}_{fb}, g_s, w_{\lambda\lambda'})}(k) \tag{16}$$

where $\Omega_{(\mathbf{a}_{ff}, \mathbf{b}_{fb}, g_s, w_{\lambda\lambda'})}(k+1)$ is the new parameter vector, $\Omega_{(\mathbf{a}_{ff}, \mathbf{b}_{fb}, g_s, w_{\lambda\lambda'})}(k)$ is the present parameter vector, and $\delta\Omega_{(\mathbf{a}_{ff}, \mathbf{b}_{fb}, g_s, w_{\lambda\lambda'})}(k)$ is an adaptive adjustment in the parameter vector. In the steepest-descent method, the adjustment of the parameter vector is made proportional to the negative of the gradient of the performance index J, that is,

$$\delta\Omega_{(\mathbf{a}_{ff},\mathbf{b}_{fb},\, g_s,\, w_{\lambda\lambda'})}(k) \propto (-\nabla J), \text{ where } \nabla J = \frac{\partial J}{\partial \Omega_{(\mathbf{a}_{ff},\mathbf{b}_{fb},\, g_s,\, w_{\lambda\lambda'})}}. \text{ Thus}$$

$$\delta\Omega_{(\mathbf{a}_{ff},\mathbf{b}_{fb},\, g_s,\, w_{\lambda\lambda'})}(k) = -\,\mathrm{dia}[\mu]\frac{\partial J}{\partial \Omega_{(\mathbf{a}_{ff},\mathbf{b}_{fb},\, g_s,\, w_{\lambda\lambda'})}} = -\,\mathrm{dia}[\mu]\,\nabla J \tag{17}$$

where dia[μ] is the matrix of individual adaptive gains. In the above equation, the dia[μ] is

$$\mathrm{dia}[\mu] = \begin{bmatrix} \mu_{a_i} & 0 & 0 & 0 \\ 0 & \mu_{b_j} & 0 & 0 \\ 0 & 0 & \mu_{g_s} & 0 \\ 0 & 0 & 0 & \mu_{\lambda\lambda'} \end{bmatrix} \tag{18}$$

where μ_{a_i}, $i = 0,1,2$, μ_{b_j}, $j = 1,2$, μ_{g_s} are the individual learning gains of the adaptable parameters of the DNU, and $\mu_{\lambda\lambda'}$ denote the learning gains for the self- and inter-neuron synaptic connections. Representing the DNU's synaptic weight vector as $\phi_{(\mathbf{a}_{ff},\mathbf{b}_{fb})}$, the gradient of performance index with respect to $\phi_{(\mathbf{a}_{ff},\mathbf{b}_{fb})}$ is obtained as

$$\frac{\partial J}{\partial \phi_{(\mathbf{a}_{ff},\mathbf{b}_{fb})}} = \frac{1}{2}E\left[\frac{\partial[u_d(k) - u(k)]^2}{\partial \phi_{(\mathbf{a}_{ff},\mathbf{b}_{fb})}}\right] = E\left[e(k)\left\{-\frac{\partial \Psi(v)}{\partial \phi_{(\mathbf{a}_{ff},\mathbf{b}_{fb})}}\right\}\right]$$

$$= E\left[-e(k)\left\{\frac{\partial \Psi(v)}{\partial v}\frac{\partial v}{\partial \phi_{(\mathbf{a}_{ff},\mathbf{b}_{fb})}}\right\}\right] = E\left[-e(k)\left\{\mathrm{sech}^2[v(k)]\,\mathbf{P}\phi_{(\mathbf{a}_{ff},\mathbf{b}_{fb})}(k)\right\}\right] \tag{19}$$

where $\mathbf{P}\phi_{(\mathbf{a}_{ff},\mathbf{b}_{fb})}(k) = \dfrac{\partial v(k)}{\partial \phi_{(\mathbf{a}_{ff},\mathbf{b}_{fb})}} = g_s\dfrac{\partial v_1(k)}{\partial \phi_{(\mathbf{a}_{ff},\mathbf{b}_{fb})}}$ represents a vector of parameter-state (or sensitivity) signals. The parameter-state signals for feedforward and feedback weights are respectively given by the relations [13]:

$$\mathbf{P}\phi_{a_{ff_i}}(k) = g_s\,[s\,(k-i)],\quad i = 0, 1, 2,\quad \mathbf{P}\phi_{b_{fb_j}}(k) = -\,g_s\,[v_1\,(k-j)],\quad j = 1,2. \tag{20}$$

Similarly, the gradient of performance index with respect to somatic gain g_s is given by

$$\frac{\partial J}{\partial g_s} = \frac{1}{2}E\left[\frac{\partial[u_d(k) - u(k)]^2}{\partial g_s}\right] = E\left[-e(k)\left\{\mathrm{sech}^2[v(k)]\,v_1(k)\right\}\right]. \tag{21}$$

The adaptation in self- and inter-neuron synaptic connections may be obtained as follows:

$$\frac{\partial J}{\partial w_{\lambda\lambda'}} = \frac{1}{2}E\left[\frac{\partial[u_d(k) - u(k)]^2}{\partial w_{\lambda\lambda'}}\right] = E\left[-e(k)\left\{\frac{\partial \Psi(v)}{\partial v}\frac{\partial v}{\partial w_{\lambda\lambda'}}\right\}\right]$$

$$= E\left[-e(k)\left\{\mathrm{sech}^2[v(k)]\,g_s\,u_\lambda(k-1)\right\}\right]. \tag{22}$$

From the above equations, the algorithm to update the DNP parameters can be written as

$$a_{ff_i}(k+1) = a_{ff_i}(k) + \mu_{a_i} E\left[e(k)\, \text{sech}^2[v(k)]\, P\phi_{a_{ff_i}}(k)\right], \; i = 0,1,2, \tag{23a}$$

$$b_{fb_j}(k+1) = b_{fb_j}(k) + \mu_{b_j} E\left[e(k)\, \text{sech}^2[v(k)]\, P\phi_{b_{fb_j}}(k)\right], \; j = 1,2, \tag{23b}$$

$$g_s(k+1) = g_s(k) + \mu_{g_s} E\left[e(k)\, \text{sech}^2[v(k)]\, v_1(k)\right], \text{ and} \tag{23c}$$

$$w_{\lambda\lambda'}(k+1) = w_{\lambda\lambda'}(k) + \mu_{\lambda\lambda'} E\left[-e(k)\left\{\text{sech}^2[v(k)]\, g_s\, u_\lambda(k-1)\right\}\right]. \tag{23d}$$

For clarity, Eqn. (23d) is written for individual synaptic weights as:

$$w_{EE}(k+1) = w_{EE}(k) + \mu_{EE} E\left[-e(k)\left\{\text{sech}^2[v(k)]\, g_s\, u_E(k-1)\right\}\right]$$

$$w_{IE}(k+1) = w_{IE}(k) + \mu_{IE} E\left[-e(k)\left\{\text{sech}^2[v(k)]\, g_s\, u_I(k-1)\right\}\right]$$

$$w_{EI}(k+1) = w_{EI}(k) + \mu_{EI} E\left[-e(k)\left\{\text{sech}^2[v(k)]\, g_s\, u_E(k-1)\right\}\right]$$

$$w_{II}(k+1) = w_{II}(k) + \mu_{II} E\left[-e(k)\left\{\text{sech}^2[v(k)]\, g_s\, u_I(k-1)\right\}\right]$$

Equations (23a) and (23b) provide adaptation in synaptic weights, while Eqn. (23c) in sigmoidal gain of the DNU, and Eqn. (23d) provides adaptation in external synaptic weights. Theoretical development of functional approximation using DNP is described in [16, 17]. This approximation feature of the neural processor is exploited in the on-line learning of inverse kinematic transformations of a two-linked robot which is discussed in the following section.

3. On-Line Learning of Inverse Kinematic Transformations

3.1 The Learning Scheme

The proposed learning scheme, shown in Fig. 4, uses the DNP to determine joint angles for a given set of desired Cartesian coordinates. These estimated joint angles, which act as inputs to the forward kinematics, are checked against the pre-defined robot task space. This additional level of control makes the robot operate within the work-space. Additional rules and inferences may be incorporated in to the first-level thereby making it a knowledge-based system.

As depicted in this figure, the first level would set up the robot task space which in turn configures the neural structure. That is, it defines the number of parallel layers in the dynamic neural network. For example, if the robot task-space specifications (interms of joint angle constraints, link lengths) are for a two-linked robot, then the neural processor will comprise of one layer (one pair of excitatory and inhibitory neurons). The estimated joint angles which are the outputs of the neural processor are always checked against the specifications so that the robot movements are confined to the pre-defined task-space. Inputs to DNP are the desired Cartesian locations, and the outputs are the estimated joint angles which become, with appropriate scaling, the inputs to the robot forward kinematics. The error signals are calculated as the difference between the targeted and the observed positions. The neural weights are adapted based on the learning algorithm developed in the preceding section.

In this paper a two-linked structure, as a model of human leg, is considered to demonstrate the effectiveness of the proposed neural processor and the learning scheme. The joints at which rotary motion occurs (within limits) are analogous to the hip and the knee joints of the human leg. The free tip of the second link, also called the 'end point', describes the trajectories based on a Cartesian coordinate system. The origin of the coordinate system is the first (hip) joint, which is assumed to be fixed in space, while the end point coordinates (x, y) are located with respect to the two perpendicular axes, X, Y. The hip joint is considered as the anchor (fixed) point. The position of the leg can also be restricted using the angles formed at the two joints with the reference axes as shown. The relationship between these two angles, defined as θ_1, θ_2 and the end point coordinates, x and y, form the kinematic equations of the two-link leg as described by the following equations:

$$x = L_1 \cos(\theta_1) + L_2 \cos(\theta_1 + \theta_2), \tag{24a}$$

$$y = L_1 \sin(\theta_1) + L_2 \cos(\theta_1 + \theta_2). \tag{24b}$$

One way of obtaining the solution to inverse kinematics problem is to constrain the movement of the two links to certain convenient angular ranges which will usually avoid the occurrence of multiple solutions. Incidentally, the structure of the human leg, with its hinge like joint at the knee, only permits a constrained motion of the shank. Taking into consideration the above mentioned constraints, the two joints of the two-link leg model were constrained to move within the specific angular ranges: $-30° < \theta_1 < 180°$, and $0° < \theta_2 < 180°$.

3.2 Computer Simulation Studies

Example 1: In this example, the coordinates x and y of the end-effector were selected at random and applied to the inputs of DNP. The neural weights were adjusted until the output error decreased to a pre-determined value of 0.05. Figure 5 shows the actual and the learned x-y coordinates of a two-linked robot.

Example 2: In sequel to Example 1, different end-effector positions were presented to the neural processor. The results obtained are shown in Fig. 6. Also shown in this figure are the positions P_1 and P_2 which are outside the robot task-space. The neural processor could not learn these positions because of the pre-defined specifications, but as may be observed from the results that the link orientations were in that direction.

Example 3: The successful operation of an intelligent robot depends upon its ability to cope with perturbations that may cause dynamic changes in the robotic structure. In this example we consider such a case where one of the links, L_2, of the robot undergoes stretching effect, for example in a telescopic robot, during the learning process. Due to this dynamic perturbation, the observed end-effector position may not match with that of the desired position necessitating the readjustments in the neural weights. The error trajectories of robot links are shown in Fig. 7a and 7b. From these results it may be seen that the DNP could adapt to the change in robot dynamics, thereby demonstrating the robustness of the learning scheme.

Example 4: In this example, performance comparison of the proposed dynamic neural network with that of a simplified DNP architecture is discussed. The simplified neural processor comprises of two conventional neurons configured in excitatory and inhibitory modes. In the conventional neural structure, inputs to the neuron are just summed up and the summated signal is passed through a 'fixed' sigmoidal function. To compare the performance of dynamic neural and the simplified structures, different end-effector positions were presented to both the networks. The results obtained are tabulated in Table 1. From these results, it is clear that the proposed dynamic neural network could learn the desired patterns must faster than its simplified counterpart. Furthermore, it could not learn some of the patterns marked '*' in the table, may be due to improper selection of initial values of the weights and values of the learning constants in the algorithm. From our initial findings it may be inferred that dynamic networks could converge faster compared to the neural structure with a static neuron.

4. Conclusions

A neural network structure called the dynamic neural processor (DNP) which emphasizes the aggregate dynamic properties of a neural population has been presented. The temporal behavior of this neural model has been briefly discussed. The learning algorithm to update the DNU parameters and self- and inter-synaptic weights of the two subpopulations has been developed. An on-line learning scheme comprising of two hierarchical levels has been proposed. This scheme, taking the desired Cartesian coordinates as inputs, could determine the robot joint angles, and make the robot move in a pre-defined task-space. Robustness of the learning scheme has been demonstrated for perturbations in the robot dynamics. However, the trajectory control aspect is not considered in this paper. It is also interesting to extend the concepts developed in this paper to a flexible structure, and to other complex robotic systems.

5. References

[1] W. Golnazarian, R. Shell and E. Hall, "Robot Control Using Networks With Adaptive Learning Steps", *SPIE's Conf. on Intelligent Robots and Computer Vision XI*, Vol. 1826, pp. 122-129, Nov. 1992.

[2] M. Kawato, K. Furukawa and R. Suzuki, "A Hierarchical Neural-Network Model for Control and Learning of Voluntary Movement", *Biological Cybernetics*, 57, pp. 169-185, 1987.

[3] A. Guez and Z. Ahmad, "Solution to the Inverse Kinematics Problem in Robotics by Neural Networks", *IEEE Int. Conf. on Neural Networks*, San Diego, Calif., pp. 617-624, March 1988.

[4] J. Barhen, S. Gulati and M. Zak, "Neural Learning of Constrained Nonlinear Transformations", *IEEE Computer*, pp. 67-76, June 1989.

[5] L. Nguyen and R.V. Patel, "A Neural Network Based Strategy for the Inverse Kinematics Problem in Robotics", *Proceedings of the International Symposia on Robotics and Manufacturing*, Vol. 3, pp. 995-1000, 1990.

[6] M.M. Gupta and D.H. Rao, "Neural Learning Robot of Inverse Kinematics Transformations ", in Neural and Fuzzy Systems: The Emerging Science of Intelligent Computing, *SPIE Press Series*, Eds. S. Mitra, M.M. Gupta and W. Kraske, (In Press).

[7] A.D. Handelman, H.L. Stephen and J.J. Gelfand, " Integrating Neural Networks and Knowledge -Based Systems for Intelligent Robotic Control ", *IEEE Control System Magazine*, pp. 77-87, April 1990.

[8] W.J. Freeman, Mass Action in the nervous System, *Academic Press*, New York, 1975.

[9] H.R. Wilson and J.D. Cowan, "Excitatory and Inhibitory Interactions in Localized Populations of Model Neurons", *Biophysical Journal*, Vol. 12, pp. 1-24, 1972.

[10] W.J. Freeman, "Linear Analysis of the Dynamics of Neural Masses", *Biophysical Journal*, Vol. 1, *Ann. R. Biophy.*, pp. 225-256, 1972.

[11] D.H. Rao and M.M. Gupta, "A Multi-Functional Dynamic Neural Processor for Control Applications", *American Control Conference*, San Francisco, June 2-4, 1993.

[12] D.H. Rao and M.M. Gupta, "A Neural Processor for Coordinating Multiple Systems with Dynamic Uncertainties", *International Symposium on Uncertainty and Management* (ISUMA), Maryland, pp. 633-640, April 25-28, 1993.

[13] M.M. Gupta and D.H. Rao, "Dynamic Neural Units With Applications to the Control of Unknown Nonlinear Systems", *The Journal of Intelligent and Fuzzy Systems*, Vol. 1, No. 1, pp. 73-92, Jan. 1993.

[14] M.M. Gupta and G.K. Knopf, "A Multitask Visual Information Processor with a Biologically Motivated Design", *Journal of Visual Communication and Image Representation*, 3, No. 3, pp. 230-246, Sept. 1992.

[15] J.J.E. Slotine and W. Li, Applied Nonlinear Control, *Prentice Hall*, New Jersey, 1991.

[16] D.H. Rao and M.M. Gupta, "Dynamic Neural Units and Function Approximation", *IEEE Conf. on Neural Networks*, San Francisco, pp. 743-748, March 28-April 1, 1993.

[17] M.M. Gupta and D.H. Rao, "Adaptive Control of unknown Nonlinear Systems Using Multi-Stage Dynamic Neural Networks", *SPIE 's Conference on Intelligent Robots and Computer Vision XI*, Boston, pp. 130-142, Nov. 18-22, 1992.

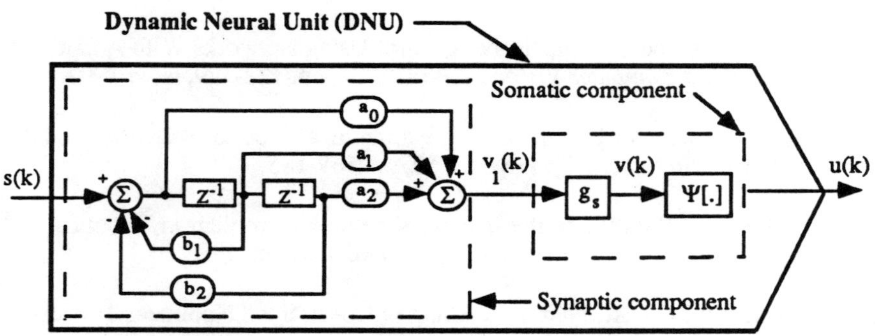

Fig. 1: The structure of dynamic neural unit (DNU).

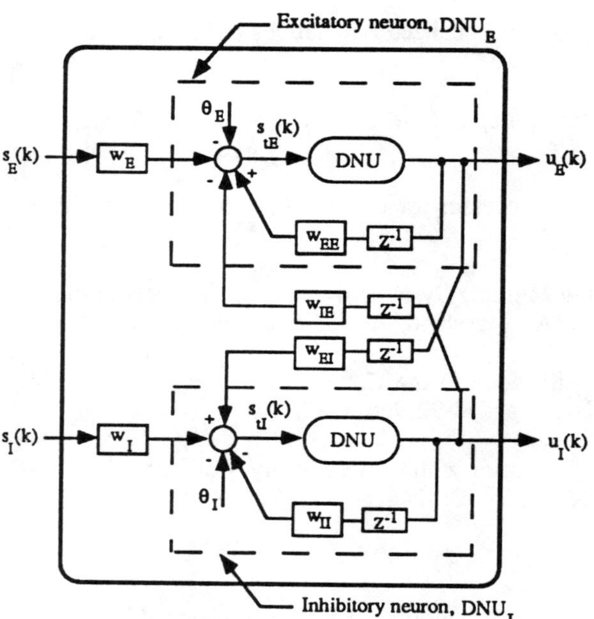

Fig. 2: The dynamic neural processor (DNP) with two dynamic neural units coupled as excitatory and inhibitory neurons represented as DNU_E and DNU_I respectively. w_{EE}, w_{II} represent the strength of self-synaptic connections, while w_{IE}, w_{EI} represent the inter-neuron connections. w_E and w_I are the scaling factors for excitatory and inhibitory neural inputs. $u_E(k)$ and $u_I(k)$ represent the responses of the excitatory and inhibitory neurons to input signals $s_E(k)$ and $s_I(k)$ respectively. θ_E and θ_I represent the thresholds of excitatory and inhibitory neurons respectively.

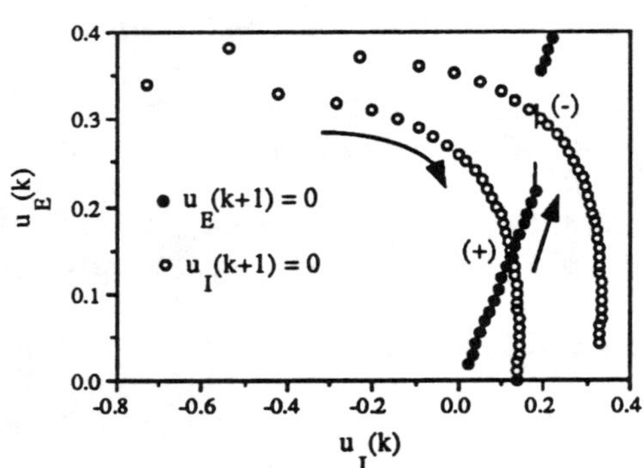

Fig. 3: The isocline curves, Eqns. (11a) and (11b), for $s_E(k) = 0$, $s_I(k) = 0$, and (+) denotes stability and (-) instability of steady-state. The weight parameters for these curves are: $w_{EE} = 20$, $w_{IE} = 5$, $w_{EI} = 8$, $w_{II} = 10$, $\theta_E = \theta_I = 0.5$.

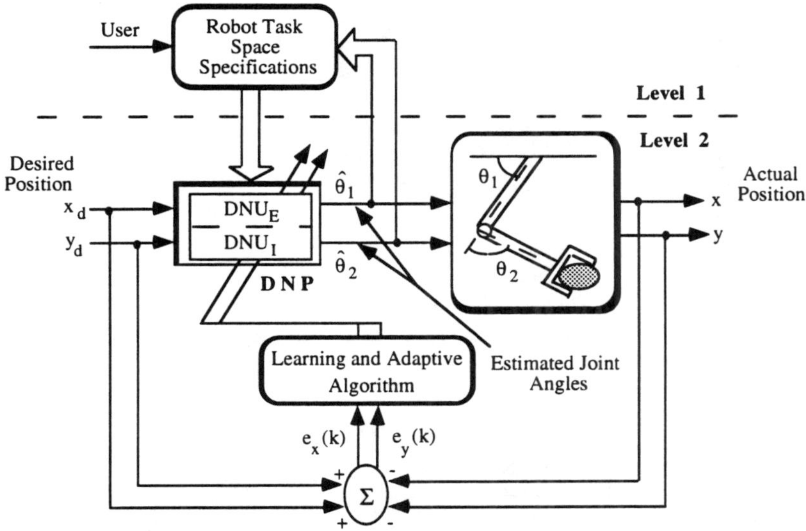

Fig. 4: The proposed learning scheme, with two hierarchical levels, for on-learning of inverse kinematic transformations of a two-linked robot.

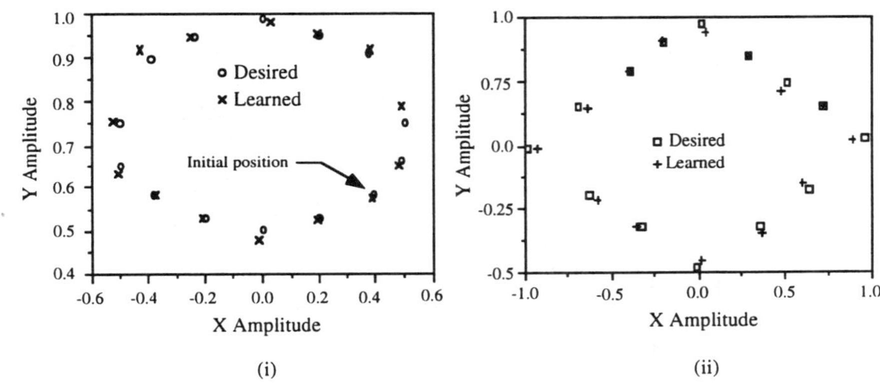

Fig. 5: Illustration of the actual and the learned positions of different trajectories; Example 1.

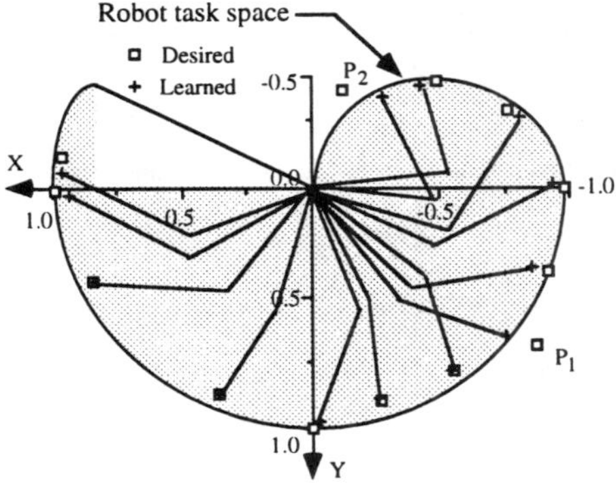

Fig. 6: Representation of the actual and the learned positions in- and out-side of the task-space in a sagattial plane; Example 2.

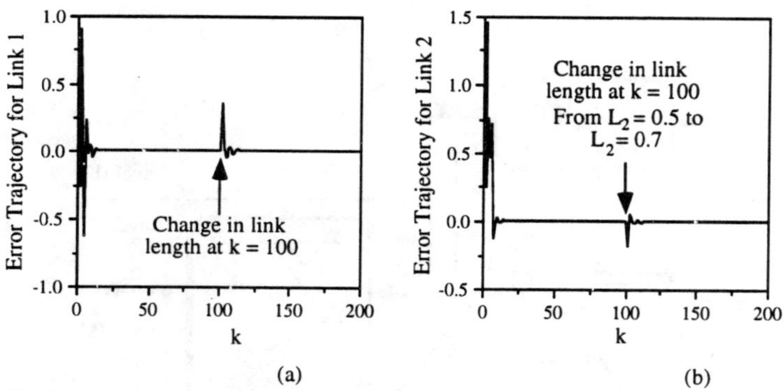

Fig. 7: Error trajectories of robot links when the length of link L_2 was changed from 0.5 to 0.7 units at time instant k = 100; Example 3.

Table 1: Performance comparison

Desired Coordinates		**Dynamic Neural Processor**			**Simplified version of DNP**		
x_d	y_d	x	y	Learning Iterations	x	y	Learning Iterations
- 0.6	0.5	- 0.602	0.51	422	- 0.58	0.508	1512
- 0.5	0.75	- 0.504	0.753	104	- 0.493	0.742	173
- 0.38	0.58	- 0.377	0.579	101	- 0.39	0.588	201
- 0.2	0.53	- 0.209	0.533	80	- 0.19	0. 539	598
0.0	0.5	0.07	0.499	144	0.09	0.495	157
0.5	0. 75	0.492	0.744	319	0.49	0.748	1033
0.3	0.85	0.302	0.86	199	0. 29	0.85	3047
0.2	0.95	0.192	0.952	174	0.19	0.942	782
0.0	0.99	0.09	0.993	164	0.009	0.982	687
- 0.25	0.9	- 0.258	0.903	36	- 0.26	0.894	104
- 0.3	0.8	- 0.309	0.793	31	- 0.31	0.809	191
- 0.6	0.6	- 0.601	0.59	237	- 0.592	0.592	667
- 0.2	0.8	- 0.207	0.812	36	- 0.192	0. 802	143
0.38	0.58	0.378	0.579	31	*	*	
- 0.8	- 0.2	- 0.792	- 0.195	69	*	*	
- 0.3	- 0.45	- 0.312	- 0.45	382	*	*	
0.5	0.4	0.492	0.404	53	*	*	
0.3	0.7	0.297	0.694	24	*	*	
0.0	0.96	0.08	0.961	60	*	*	

Performance comparison between DNP and its simplified version. The sign '*' indicates that the simplified neural structure could not learn these patterns; Example 4.

Intelligent Coordination of Multiple Systems with Neural Networks

Xianzhong Cui, *Student Member, IEEE*, and Kang G. Shin, *Senior Member, IEEE*

***Abstract*— Many control applications require cooperation of two or more independently designed, separately located, but mutually affecting, subsystems. In addition to the good behavior of each subsystem, effective coordination of these subsystems is very important to achieve the desired overall system performance. However, such coordination is very difficult to accomplish due mainly to the lack of precise system models and/or dynamic parameters as well as the lack of efficient tools for system analysis, design, and real-time computation of optimal solutions. A new multiple-system coordinator that combines the techniques of intelligent control and neural networks, and forms the high-level coordinator in a hierarchical structure, is proposed. The basic idea is to estimate the effects of the control commands to subsystems using a predictor and modify these commands using a knowledge-based coordinator so as to achieve the desired performance. The predictor is designed for multiple-input, multiple-output systems using neural networks. The knowledge-based coordinator is responsible for a goal-oriented search in its knowledge base and the overall system stability. Because the internal structure and parameters of the low level are not affected by using the proposed method, some commercially designed servo controllers for single systems can be coordinated to perform more sophisticated tasks for multiple systems than originally intended.**

I. Introduction

ALTHOUGH some basic principles in coordinating multiple systems were developed in early 80s [1], most related publications addressed only conceptual interpretation, and very few of them dealt with actual applications. The main difficulty in coordinating multiple systems comes from the lack of precise system models and parameters as well as the lack of efficient tools for system analysis, design, and real-time computation of optimal solutions. New methods for analysis and design are thus required for the closed-loop coordination of multiple systems.

Since intelligent control does not depend only on mathematical analyses and manipulations, it is an attractive candidate to deal with complex system control problems. An intelligent controller achieves the desired performance by searching for a goal in its knowledge base. There are three basic structures for intelligent control: *performance-adaptive*, *parameter-adaptive*, and *hierarchical structure*. The performance-adaptive structure is motivated by human expert control and/or human cognition ability, and attempts to control a system directly with an intelligent controller. Several examples of this structure are given in [2]–[5]. On the other hand, in a parameter-adaptive structure, the intelligent controller works as an on-line tuner of a conventional (usually PID) controller [6]–[8]. In a hierarchical structure, the intelligent controller [9] is a high-level controller, which attempts to modify only the reference input to the low level. The low-level subsystem could be a servo control system, and its internal structure and parameters are not affected by adding this high-level controller. One of the main tasks associated with an intelligent controller is to design a knowledge base. An inference engine will then conduct a goal-oriented search in the knowledge base according to the characteristics of system performance. The error and/or error increment of system output, and the quality of a step response are commonly used to evaluate system performance. Other additional characteristics were also suggested. For example, the estimated, dominant pole location of a closed-loop system was suggested to express system performance in [5], though no knowledge base was built on it. In [4], the output error and its derivative were arranged into a phase plane divided into 48 areas, on the basis of which rules were designed. The goal was to control the system to reach the origin of this plane. In [9], the multiple-step prediction of system output was used to characterize system performance, and the knowledge to control the system was then simply represented by a decision tree.

However, all the results reported in the literature were intended for single systems. Most of the system characteristics mentioned above may not be suitable for coordinating multiple systems, because system performance may not be easily defined and related to the measured data and control inputs. In fact, for a complex multiple-system, even human's knowledge on how to coordinate it to achieve the desired performance is limited and incomplete. So, it is difficult to design a complete knowledge base for such a system. Addition of a coordinator (not necessarily an intelligent one) leads the problem of coordinating multiple systems to form a hierarchical structure. Such an addition should not interfere with the internal structure and parameters of low-level subsystems, making the structure of performance- or parameter- adaptive intelligent controllers unsuitable for multiple-system coordination. The internal structure and/or parameters of low-level subsystems are usually not known to the coordinator. Moreover, stability analysis becomes very important, due mainly to the uncertain low-level structure and/or parameters, incomplete knowledge of the coordination and system characteristics. We should therefore answer the following questions when designing an intelligent coordinator:

Manuscript received September 12, 1990; revised March 17, 1991. This work was supported in part by the National Science Foundation under Grant No. DMC-8721492.

The authors are with the Real-Time Computing Laboratory, Department of Electrical Engineering and Computer Science, The University of Michigan, Ann Arbor, MI 48109-2122.

IEEE Log Number 9100408.

Reprinted from *IEEE Trans. Syst., Man., Cybern.*, vol. 21, no. 6, pp. 1488–1497, Nov./Dec. 1991.

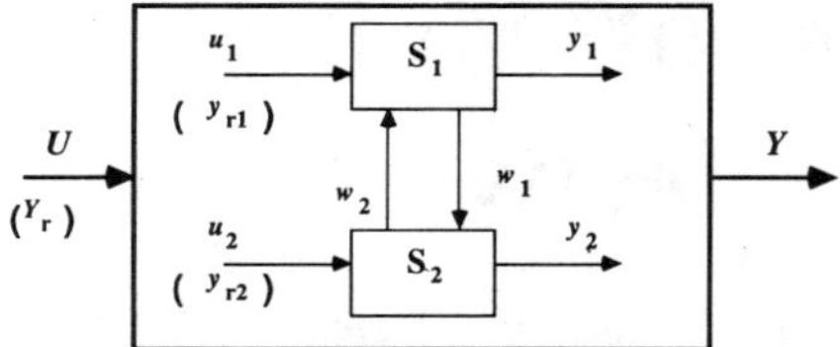

Fig. 1. Interaction of two systems.

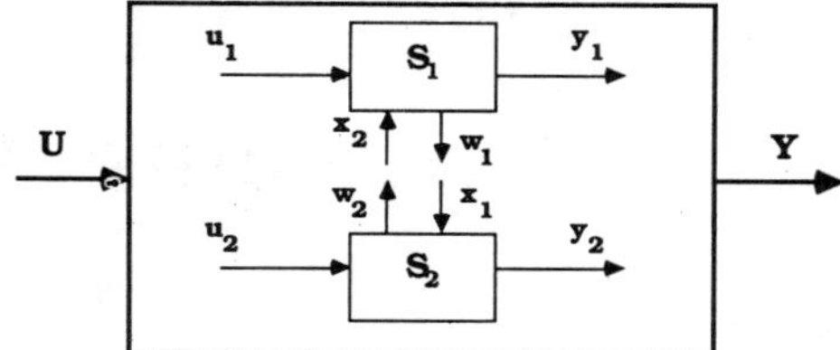

Fig. 2. Goal coordination of two systems.

1) What are the strategy and the structure to coordinate multiple systems?
2) What are the characteristics of multiple-system performance?
3) What is the knowledge necessary for coordination?
4) How should knowledge be represented?
5) How can the qualitative knowledge be extracted from sensor data?
6) How can the result of qualitative reasoning be changed into the quantitative control signals of actuators?
7) How can system stability be analyzed and guaranteed?

We propose a knowledge-based coordinator (KBC) for multiple systems by combining the techniques of intelligent control and neural networks (NN's). The KBC is a high-level coordinator within a hierarchical structure. The detailed structure and/or parameters of low-level subsystems are not required by the KBC, thus allowing individual subsystems to be designed independently. This implies that some commercially designed controllers can be coordinated to perform more sophisticated tasks than originally intended. In Section II, the problem of multiple-system coordination is stated, and some basic principles of multiple-system coordination are reviewed. The proposed scheme and the assumptions used are described in Section III. Section IV addresses the design of a KBC, including the knowledge representation, solution existence, and system stability. Section V deals with the design of an NN-based predictor with multiple-input multiple-output (MIMO). The basic structure of the NN-based predictor, a vector version of the back propagation algorithm, and the updating problem will be discussed there. As an example, the coordination of two 2-link robots holding a single object is discussed in Section VI. The paper concludes with Section VII.

II. The Problem and Principles of Multiple-System Coordination

Fig. 1 describes two interacting systems, and this description can be easily generalized to the case of more than two systems. The system dynamics are described by

$$\boldsymbol{S}_1(\boldsymbol{u}_1, \boldsymbol{y}_1, \boldsymbol{w}_2) = 0 \quad \text{and} \quad \boldsymbol{S}_2(\boldsymbol{u}_2, \boldsymbol{y}_2, \boldsymbol{w}_1) = 0$$

where $\boldsymbol{u}_i \in \boldsymbol{R}^{q_i}$, $\boldsymbol{w}_i \in \boldsymbol{R}^{m_i}$, and $\boldsymbol{y}_i \in \boldsymbol{R}^{p_i}$ for $i = 1, 2$. Let $p = p_1 + p_2$, $q = q_1 + q_2$ and $m = m_1 + m_2$, then the constraints are expressed by $\boldsymbol{S}_0 = \{(\boldsymbol{U}, \boldsymbol{Y}, \boldsymbol{W}) : \boldsymbol{S}_1 = 0, \boldsymbol{S}_2 = 0\}$, where $\boldsymbol{U} = \left[\boldsymbol{u}_1{}^T, \boldsymbol{u}_2^T\right]^T \in \boldsymbol{R}^q$ is the augmented control input vector, $\boldsymbol{Y} = \left[\boldsymbol{y}_1{}^T, \boldsymbol{y}_2{}^T\right]^T \in \boldsymbol{R}^p$ the augmented system output vector, and $\boldsymbol{W} = \left[\boldsymbol{w}_1{}^T, \boldsymbol{w}_2{}^T\right]^T \in \boldsymbol{R}^m$ the vector representing interactions between the two systems.

Usually, the cost function of a multiple-system is the sum of the cost functions of all component systems:

$$J(\boldsymbol{U}, \boldsymbol{Y}, \boldsymbol{W}) \equiv J_1(\boldsymbol{u}_1, \boldsymbol{y}_1, \boldsymbol{w}_2) + J_2(\boldsymbol{u}_2, \boldsymbol{y}_2, \boldsymbol{w}_1). \quad (1)$$

The problem of coordinating multiple systems can be stated as an optimization problem: minimize the cost function J subject to the constraint $\boldsymbol{S}_0$.

Though there are no general approaches to solving this problem for a complex multiple-system, some conceptual methods and basic principles have been suggested in [1]. One of these methods is called model coordination. Under this method, the problem is divided into two-level optimization problems. First, suppose the interaction $\boldsymbol{W}$ is fixed at $\boldsymbol{Z}$, then compute

$$\boldsymbol{H}(\boldsymbol{Z}) = \min_{(\boldsymbol{U}, \boldsymbol{Y}, \boldsymbol{Z}) \in \boldsymbol{S}_0} J(\boldsymbol{U}, \boldsymbol{Y}, \boldsymbol{Z}).$$

where $\boldsymbol{H}(\boldsymbol{Z})$ is then minimized over all allowable values of $\boldsymbol{Z}$. This two-level optimization problem is solved iteratively until the desired performance is achieved. Another method is called *goal coordination*, in which the system is represented as in Fig. 2. Suppose $\boldsymbol{w}_i$ is not necessarily equal to $\boldsymbol{x}_i$. The overall optimality is achieved by sequentially optimizing two subsystems, while treating $\boldsymbol{w}_i$ as an ordinary input variable of each corresponding subsystem. This requires $\boldsymbol{x}_i$ and $\boldsymbol{w}_i$ to be equal, which is called the *interaction balance principle.* Similar to the process of model coordination, the optimality is achieved iteratively. Another basic principle of coordination, called the *interaction prediction principle*, is stated as follows. Let

$$\hat{\boldsymbol{W}} = \left[\hat{\boldsymbol{w}}_1^T, \hat{\boldsymbol{w}}_2^T\right]^T$$

be the predicted interaction and $\boldsymbol{W} = \left[\boldsymbol{w}_1{}^T, \boldsymbol{w}_2{}^T\right]^T$ be the actual interaction under the control $\boldsymbol{U}$. Then the overall optimum will be achieved if the prediction gives the true value, that is: $\hat{\boldsymbol{W}} = \boldsymbol{W}$.

Obviously, solving these optimization problems largely depends on the knowledge of the structure and/or dynamic parameters of low-level subsystems and mathematical synthesis. Moreover, in a hierarchical system it is desirable that adding a high-level coordinator should not affect the internal structure and/or parameters of the low-level subsystems, and should only give appropriate coordination commands to them, so that each level can be designed independently of other levels. That is, the higher the level is, the more intelligence it has, and the less precise its knowledge about the low levels becomes. These requirements motivated us to design a knowledge-based coordinator (KBC).

To design a coordinator, we first need to define a system performance index. It should be chosen to express the desired system performance and should also be amenable to some optimization methods. For example, the performance index defined in (1) is suitable for the concepts of model coordination and goal coordination. To design a KBC, one needs an index to explicitly express system performance, and such an index will henceforth be called the *principal output.* The overall system performance index may not necessarily be the simple summation of the performance indices of all component systems. Because only the system constraints are important for coordination, one may not even be able to define subsystem performance indices. Moreover, we want to relate the principal output directly to the coordination commands. The coordination commands are defined as the reference inputs to subsystems. From the following sections, one can see that both the explicit expression for system performance and the direct relationship between the principal output and the coordination commands will simplify the design of the knowledge base and the goal-oriented search.

III. Description of the Principal Output Prediction Scheme

In a hierarchical structure, each level can be viewed as a mapping from its reference input to the output. The servo controller of each subsystem is usually designed separately from, and independently of, the others. In order not to interfere with the internal structure and/or parameters of the lower level, the only effective control variable is the reference input to the lower level. The reference inputs are a set of predesigned commands, which represent the desired overall behavior of the multiple systems. For example, when multiple robots work in a common workspace, the reference input is the desired trajectory of each robot generated without considering the presence of other robots. The purpose of a high-level coordinator is to modify the desired trajectories to avoid collision among the robots. From a high-level coordinator's points of view, the following conditions are assumed.

C1: Each subsystem is a stable, closed-loop control system.

C2: Each subsystem has a linear response to its reference input.

C3: Each subsystem will remain stable even during its interaction with other subsystems.

C4: System performance can be described explicitly by the principal output.

In Fig. 1, let $\boldsymbol{Y}$ be the principal output vector of the multiple-system, $\boldsymbol{Y}_r = [\boldsymbol{Y}_{r1}^T, \boldsymbol{Y}_{r2}^T]^T$ be the vector of reference input to the low level. Note that the components of $\boldsymbol{Y}$ may not be simply the outputs of subsystems, but could be a function of these outputs:

$$\boldsymbol{Y} = \boldsymbol{F}_0(\boldsymbol{y}_1, \boldsymbol{y}_2), \text{ where } \boldsymbol{F}_0 : \boldsymbol{R}^{p_1} \times \boldsymbol{R}^{p_2} \rightarrow \boldsymbol{R}^p.$$

Because each subsystem is a closed-loop control system, $\boldsymbol{y}_i$ can be represented as

$$\boldsymbol{y}_i = \boldsymbol{f}_i(\boldsymbol{y}_{ri}, \boldsymbol{w}_j), \text{where } i, j = 1, 2, j \neq i, \text{ and } \boldsymbol{f}_i : \boldsymbol{R}^{n_i} \times \boldsymbol{R}^{m_j} \rightarrow \boldsymbol{R}^{p_i}.$$

Then $\boldsymbol{Y}$ can be represented as

$$\boldsymbol{Y} = \boldsymbol{F}(\boldsymbol{y}_{r1}, \boldsymbol{y}_{r2}, \boldsymbol{w}_1, \boldsymbol{w}_2) \tag{2}$$

where $\boldsymbol{F} : \boldsymbol{R}^{n_1} \times \boldsymbol{R}^{n_2} \times \boldsymbol{R}^{m_1} \times \boldsymbol{R}^{m_2} \rightarrow \boldsymbol{R}^p$. The principal-output vector $\boldsymbol{Y}$ in (2) establishes an explicit relationship between the overall system performance and the reference input. Let $\hat{\boldsymbol{Y}}(k+d/k)$ and $\boldsymbol{Y}_d(k+d)$ be the $d - step$ ahead prediction and the desired value of the principal output $\boldsymbol{Y}(k)$ at time $k + d$, respectively. Then, the performance index of the overall system can be defined as

$$J(k) = \left[\boldsymbol{Y}_d(k+d) - \hat{\boldsymbol{Y}}(k+d/k)\right]^T \cdot \left[\boldsymbol{Y}_d(k+d) - \hat{\boldsymbol{Y}}(k+d/k)\right].$$

The purpose of using a coordinator is to choose a suitable reference input vector $\boldsymbol{Y}_r(k)$ so as to minimize $J(k)$ at time k subject to a set of constraints.

Suppose the prediction of the principal output corresponding to each choice of $\boldsymbol{Y}_r(k)$ is available, and the constraints can be expressed with a set of production rules. Then, in each sampling interval, the desired performance can be obtained by iteratively trying different reference inputs and adjusting them according to the principal output prediction. For example, we propose the following algorithm to coordinate two subsystems, where the superscript i denotes the iteration count.

1) Compute the principal output prediction $\hat{\boldsymbol{Y}}^0(k+d/k)$ for given reference inputs $\boldsymbol{y}_{r1}^0(k)$ and $\boldsymbol{y}_{r2}^0(k)$.
2) Using $\hat{\boldsymbol{Y}}^i(k+d/k)$, modify the reference inputs of subsystem 1, $\boldsymbol{y}_{r1}^i(k)$, $i = 0, 1, 2, \ldots$.
3) Compute $\hat{\boldsymbol{Y}}^{i+1}s(k+d/k)$ for given reference inputs $\boldsymbol{y}_{r1}^i(k)$ and $\boldsymbol{y}^0{}_r2(k)$.
4) Set $i \leftarrow i + 1$ and repeat steps 2) and 3) until $\hat{\boldsymbol{Y}}^{i+1}(k + d/k)$ cannot be improved any further with $\boldsymbol{y}_{r1}^i(k)$ due to the constraints.
5) Set $i \leftarrow 0$.
6) Using $\hat{\boldsymbol{Y}}^i(k+d/k)$. modify the reference inputs of subsystem 2, $\boldsymbol{y}_{r2}^i(k)$, $i = 0, 1, 2, \ldots$.
7) Compute $\hat{\boldsymbol{Y}}^{i+1}(k+d/k)$ for given reference inputs $\boldsymbol{y}_{r1}^0(k)$ and $\boldsymbol{y}_{r2}^i(k)$.
8) Set $i \leftarrow i + 1$ and repeat steps (6) and (7) until $\hat{\boldsymbol{Y}}^{i+1}(k+d/k)$ cannot be improved any further with $\boldsymbol{y}_{r2}^i(k)$ due to the constraints.
9) Set $i \leftarrow 0$ and repeat steps 2)–8) until $\hat{\boldsymbol{Y}}^i(k+d/k)$ reaches its desired value.

The conceptual structure of this scheme is given in Fig. 3. Obviously, this scheme needs a multiple-step predictor to compute $\hat{\boldsymbol{Y}}^i(k+d/k)$, and a KBC for the modification process of the reference inputs. By using this principal output predictor to characterize system performance, the knowledge to coordinate multiple systems becomes clear, thus simplifying the design of a knowledge base. We now need to address the following two problems: 1) Given the principal output prediction, how can we design this KBC? This will be discussed in the next section. 2) How can we design such a principal output predictor? In Section V, an MIMO

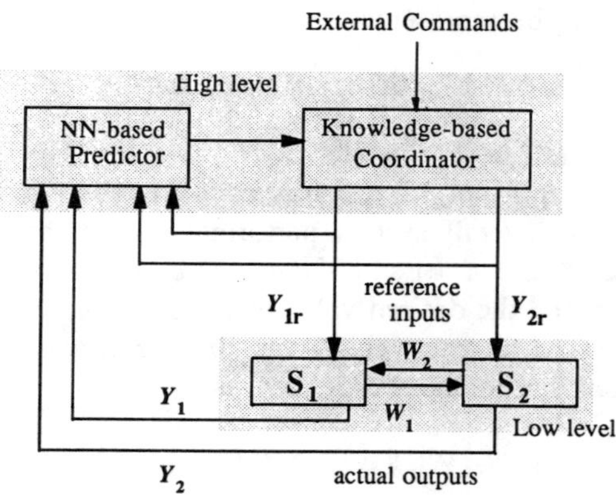

Fig. 3. Conceptual structure of the knowledge-based coordination system.

predictor is designed using NN's. With the ability of learning an input-output (I/O) mapping from experience, an NN can be used to track the variation of the mapping. However, an NN alone cannot form an intelligent coordination/control system. As a general method of representing systems with learning ability, NN's lack the ability of logical reasoning and decision making, interpretation of environmental changes, and quick response to unexpected situations. Therefore, a KBC is needed. Despite its drawbacks, the NN-based predictor establishes an explicit relationship between the principal output and the reference inputs to subsystems. Hence, the knowledge base is simplified. One can also add easily to the knowledge base such rules as the constraints of subsystems, operation monitoring, system protection, and switching of the coordination schemes. The KBC will emphasize system coordination but not data interpretation, while the ability of learning will rely mainly on the NN, that is, the NN will adapt itself to the model/parameter uncertainties, disturbances, component failures, and so on.

IV. Design of the Knowledge-Based Coordinator

A multiple-system with the KBC forms a hierarchical structure, and the low-level subsystems are viewed as a mapping from their reference input to the principal output. The goal is to modify the reference input so that the principal output reaches its desired value. For a given multiple-system we must define the principal output. Note that knowledge-based coordination is not strictly a mathematical optimization problem. The principal output must 1) have an explicit relation to the reference inputs, and 2) be measurable or computable from measured data. Because a multiple-system is designed to perform a common task(s) among the component systems, such a principal output is usually defined to express the situation of the common task(s), though it may not explicitly reflect some of generally used optimization criteria, such as energy or time.

As an example, consider the coordinated control of two robots. The two robots' operations may be tightly coupled or loosely coupled. They are tightly coupled, for example, when they hold a single object rigidly and are coordinated to move the object. On the other hand, they are loosely coupled, when they work in a common workspace and are coordinated to avoid collision. Suppose each robot is equipped with a servo controller that was originally designed for a single robot. The two robots are coordinated by modifying each robot's reference input. For the tightly coupled case, the principal output can be defined as the object's position error or the internal/external force exerted on the object. For the loosely coupled case, on the other hand, the positions and/or velocities of the robots' end-effectors can be used to represent the status of collision avoidance, and thus, they are qualified to be the principal output. For both cases, an explicit relationship between the system performance and the reference input is established by defining the appropriate principal output.

As stated in the previous sections, we want to use the principal output predictor to see where each reference input of the subsystem will lead to. In this section, it is assumed that such a predictor is well-designed and gives the true value of the principal output. (The design of such a predictor is treated in the next section.)

Given the principal output prediction, the simplified knowledge on how to coordinate a multiple-system can be stated in two steps:

1) Modify the reference input and feed the modified input to the predictor.
2) **IF** the principal output prediction yield good performance **THEN** feed the reference input to individual subsystems **ELSE** remodify it.

Since only one reference input is modified at each time, the remaining problems are then in which direction the reference input is modified (increase or decrease), how much it should be modified, and what its limits are. For a single-system, we have already developed such a KBC in [9]. For a multiple-system, the modification process of each reference input is similar to that of a single-system, so only the related results of [9] are summarized below.

A. Knowledge Representation

Using a predictor, the performance of a multiple-system is characterized by the predicted tracking error in its principal output that results from the application of the current reference input. Thus, the space of predicted tracking errors $\boldsymbol{E}$ forms the input space of the KBC's knowledge base. The goal of the KBC is then to implement the modification process discussed thus far. It is not difficult to express this process in terms of a set of production rules. The possible actions that the KBC can take include: increase the reference inputs, decrease the reference inputs, or keep them unchanged. Because our scheme is based on the modification of the reference input according to the corresponding principal output prediction, the internal structure and parameters of low-level subsystems are not affected. To simplify the design of production rules, the predicted output error is considered as the system characteristics or the input of the knowledge base. For each element of the reference input, the basic modification process can be represented by a decision tree as shown in Fig. 4. The ijth node in the tree is represented by $([a_j^i, b_j^i], c_j^i)$, where c_j^i is the

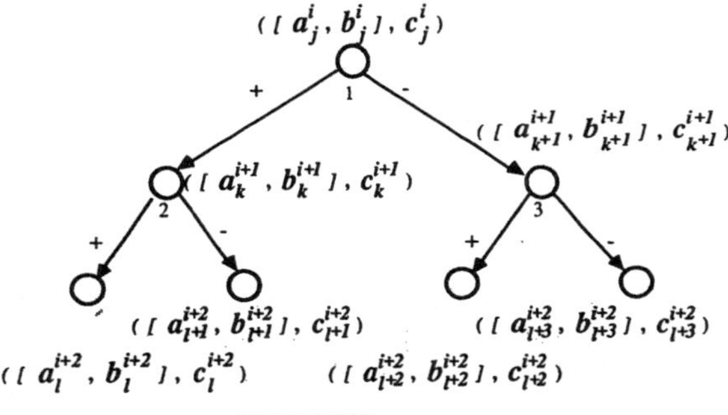

Fig. 4. Decision tree.

quantity added to the reference input,[1]

$$y_r^{i+1}(k) = y_r^o(k) + c_j^i,$$

where $y_r^o(k)$ is an element of the original reference input vector to one of the subsystems at time k, y_r^{i+1} is its modified value after the ith iteration, and $[a_j^i, b_j^i]$ is the interval to be searched, where $a_j^i < c_j^i < b_j^i$ for all i, j. By giving the reference input $y_r^i(k)$, at any node $([a_j^i, b_j^i], c_j^i)$, the interval $[a_j^i, b_j^i]$ will be split into two subintervals $[a_k^{i+1}, b_k^{i+1}] \equiv [a_j^i, c_j^i]$ and $[a_{k+1}^{i+1}, b_{k+1}^{i+1}] \equiv [c_j^i, b_j^i]$, which form two successive nodes. At the ith iteration and at the ijth node, let the predicted tracking error resulting from $y_r^i(k)$ be denoted as

$$e_j^i(k) = \hat{y}^i(k + d/k) - Y_d(k + d)$$

where y_d is an element of $\boldsymbol{Y}_d$ and $\hat{y}^i(k + d/k)$ is the corresponding element of $\hat{\boldsymbol{Y}}^i(k + d/k)$.

Then, c_j^i is computed as:

$$c_j^i = \begin{cases} b_j^i - (b_j^i - a_j^i)K, & \text{if } e_j^i(k) < 0 \\ a_j^i + (b_j^i - a_j^i)K, & \text{if } e_j^i(k) > 0 \end{cases}$$

and $0 < K < 1$ is a weighting coefficient that determines the step size of the iterative operation. a_0^0 and b_0^0 are the predesigned lower and upper bounds for the amount of reference input modification, and usually $c_0^0 = 0$, that is, at the beginning, the reference input is not modified.

The structure of this decision tree shows that the simplest inference process is similar to forward chaining, starting from the root node. However, after a period of operation, we may learn that a positive augment c_j^i is always needed. Then, the inference process may start at any node with $c_j^i > 0$ and go forward or backward according to the sign of predicted tracking error. Note that this backward search does not mean a reverse search, but rather intends to find a suitable node to start the forward search. As soon as the forward search begins, the process is not reversible.

[1] Because only the reference input to one subsystem is modified at a time, to simplify the notation, subsystem 1 and 2 will not be distinguished within this section, that is, $y_r(k)$ will represent one element of either $\boldsymbol{y}_{r1}(k)$ or $\boldsymbol{y}_{r2}(k)$.

B. Solution Existence and Stability Analysis

The basic forms of production rules are as follows.

IF $((e_j^i(k) < 0)$ **AND** $(|e_j^i(k)| > \epsilon)$
THEN increase c_j^i **AND** compute $y_r^{i+1}(k) = y_r^0(k) + c_j^i$;
IF $(e_j^i(k) > 0)$ **AND** $(|e_j^i(k)| > \epsilon)$
THEN decrease c_j^i **AND** compute $y_r^{i+1}(k) = y_r^0(k) + c_j^i$;
IF $|e_j^i(k)| \leqslant \epsilon$
THEN set $y_r^{i+1}(k) = y_r^i(k)$ **AND** stop the iterative operation.

$\epsilon > 0$ is a prespecified error tolerance. Because the amount of modification to the reference input is bounded, or $a_0^0 < c_j^i < b_0^0$ for all i, j, there may be a case where $|e_j^i(k)| > \epsilon$ for all c_j^i. To avoid such a case, the desired trajectory needs to be designed carefully. For example, when the desired trajectory is a step function, for a second order system, at $k = 0$ the system response cannot have a jump no matter how large the reference input is. A reasonable choice of ϵ is another way to prevent this problem. This existence problem can be monitored by adding, for example, the following rule into the knowledge base:

IF $((|c_j^i - b_0^0| < \delta)$ **OR** $(|c_j^i - a_0^0| < \delta))$ **AND** $(|e_j^i(k)| > \epsilon$
THEN (change a_0^0 or b_0^0 automatically and continue the search)
OR (ask the operator for an adjustment)
OR (stop the iterative operation and choose c_j^i with the smallest $e_j^i(k)$ as the best output).

$\delta > 0$ is a prespecified tolerance.

Suppose the weighting factor K is set too small or too large, then the search for c_j^i may take a long time. This would not be acceptable if the required computation cannot be completed within one sampling interval. The case of the computation/search time exceeding one sampling interval is equivalent to having no solution. This case is monitored by

IF (the search time $> T_{\max}$) **AND**
$(|e_j^i(k)| > \epsilon)$ **THEN** (stop the iterative operation)
AND (choose c_j^i with the smallest $e_j^i(k)$ as the best output) **AND** (modify the weighting coefficient K).

$T_{\max}$ is the prespecified maximum search time.

Suppose the prediction gives the true principal output, and let us consider the KBC and the closed-loop subsystem. The KBC can then be viewed as a map $\boldsymbol{M}_0 : \boldsymbol{E} \to \boldsymbol{Y}_R$, specified by all the production rules, where $\boldsymbol{E}$ is the space of predicted principal output tracking error and $\boldsymbol{Y}_R$ the reference input space. The low-level closed-loop subsystem is also a map, $\boldsymbol{L}:\boldsymbol{Y}_R \to \boldsymbol{E}$, which is specified by the desired dynamic properties of the servo controller. Because $\boldsymbol{L}$ represents a well-designed controller and there exists a reference input at time k, $\boldsymbol{Y}_r^i(k) \in \boldsymbol{Y}_R$ such that $e_j^i(k) = 0$. Thus, it is reasonable to assume that $\boldsymbol{L}$ is a linear map. The properties of the map $\boldsymbol{M} \equiv \boldsymbol{L}\boldsymbol{M}_0 : \boldsymbol{E} \to \boldsymbol{E}$ depends mainly on the properties of the map $\boldsymbol{M}_0$. In fact, all the antecedents of production rules are established based on the prediction of principal output. If the predictor gives the true principal output, then the properties of the invariant map $\boldsymbol{M}:\boldsymbol{E} \to \boldsymbol{E}$ is determined solely by the knowledge base. For system stability, all production rules in the knowledge base must form a contraction map. More formally, we give the following theorem without proof. (See [9] for its proof.)

Theorem: Suppose 1) the principal output prediction of a multiple-system is computable and the predictor gives the true principal output, and 2) $\boldsymbol{L}{:}\boldsymbol{Y}_R \rightarrow \boldsymbol{E}$ of the low-level closed-loop subsystem is a linear map. If the map $\boldsymbol{M}_0{:}\boldsymbol{E} \rightarrow \boldsymbol{Y}_R$ is given by a decision tree, then the composite map $\boldsymbol{M} \equiv \boldsymbol{L}\boldsymbol{M}_0{:}\boldsymbol{E} \rightarrow \boldsymbol{E}$ is a contraction map.

At each node of the decision tree, the iterative learning process is performed and the rules always keep the search direction pointed to the node where the tracking error decreases. This implies that the iterative learning process decreases the tracking error. As mentioned in Section III, the inference process is not reversible, and thus, it is impossible to have an unstable system response.

V. Design of an NN-Based Predictor

Though it is assumed that the principal output $\boldsymbol{Y}$ is measurable or computable from the measured data, it may be very difficult to derive a closed-form expression for (2). Therefore, it is almost impractical to design such a principal output predictor with mathematical synthesis alone, even if such a closed form exists. The development of NN's suggested that an I/O mapping can be approximated by a multilayer perceptron [10], [11]. With the ability of learning from examples, an NN can be trained to retain the dynamical property of an I/O mapping. Typically, a set of I/O pairs is arranged as $(u_1, y_1), (u_2, y_2), \ldots$, where $y_i = f(u_i)$ is a mapping. Using these training data, the connection weights within the NN are reorganized so as to represent the mapping relation. One of the most popular NN structures is the multilayer perceptron with the back propagation (BP) algorithm [12], [13]. The computation of BP includes two steps: 1) compute the NN's output forward from its INPUT to OUTPUT layers, and 2) modify the connection weights backward from its OUTPUT to INPUT layers. In what follows, an MIMO predictor is designed using an NN. The BP algorithm is extended to a vector form, and the problem of tracking a time-varying system is also addressed.

Referring to (2), the *d-step* ahead prediction of $\boldsymbol{Y}$ can be represented by

$$\hat{\boldsymbol{Y}}(k+d/k) = \boldsymbol{F}_p(\bar{\boldsymbol{Y}}_{r1}, \bar{\boldsymbol{Y}}_{r2}, \bar{\boldsymbol{Y}}) \tag{3}$$

where

$$\bar{\boldsymbol{Y}}_{r1} = \Big(\boldsymbol{Y}_{r1}(k+i_1), \ldots, \boldsymbol{Y}_{r1}(k), \boldsymbol{Y}_{r1}(k-1), \ldots, \boldsymbol{Y}_{r1}(k-i_2)\Big)$$

$$\bar{\boldsymbol{Y}}_{r2} = \Big(\boldsymbol{Y}_{r2}(k+j_1), \ldots, \boldsymbol{Y}_{r2}(k), \boldsymbol{Y}_{r2}(k-1), \ldots, \boldsymbol{Y}_{r2}(k-j_2)\Big)$$

$$\bar{\boldsymbol{Y}} = \Big(\boldsymbol{Y}(k), \boldsymbol{Y}(k-1), \ldots, \boldsymbol{Y}(k-i)\Big)$$

$$\boldsymbol{F}_p : \boldsymbol{R}^{n_1} \times \boldsymbol{R}^{n_2} \times \boldsymbol{R}^p \rightarrow \boldsymbol{R}^p$$

where i, i_1, i_2, j_1 and j_2 are constant integers. The interaction effects among subsystems are implicitly included in the historical data of $\bar{\boldsymbol{Y}}$. In (3), the principal output prediction is directly represented as a mapping of the reference inputs and the historical data of $\bar{\boldsymbol{Y}}$. A three-layer (with one hidden layer) perceptron is designed to learn the relationship of (3), and forms the backbone of the NN-based predictor. Fig. 5 shows the structure of the predictor, where the reference inputs $\bar{\boldsymbol{Y}}_{r1}$ and $\bar{\boldsymbol{Y}}_{r2}$, and the historical data $\bar{\boldsymbol{Y}}$ are fed to the nodes at the INPUT layer. When the NN becomes well-trained, the predictions $\hat{\boldsymbol{Y}}(k+d/k)$ for $d = 1, 2, \ldots$ are then produced from the OUTPUT nodes. Two problems should be solved before implementing this NN-based predictor:

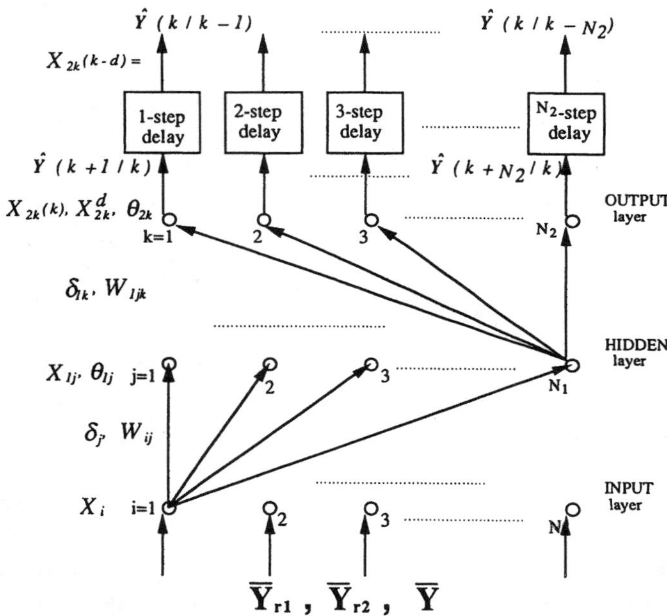

Fig. 5. Structure of the NN-based predictor.

1) How to efficiently represent and compute an MIMO mapping with the NN ?
2) How to track a time-varying mapping ?

These problems have been treated in [14], and thus, only the key points are summarized below for completeness.

A. Multidimensional Back Propagation Algorithm

Traditionally, each node of a multilayer perceptron is designed only to perform scalar operations, and the number of nodes or layers is increased to represent a more complex I/O mapping. However, if each node handles only one element of a vector, then a multilayer mesh is required to learn a complex MIMO mapping, such as (3). Training such a mesh is impractical due to the lack of a systematic algorithm. Therefore, we design a multilayer perceptron with the ability of vector operations in order to easily specify some known coupling relations within the predicted system, and to get an easier (thus more intuitive) form of the training algorithm. Referring to Fig. 5, the corresponding multi-dimensional BP algorithm is summarized below.

All inputs and outputs of this NN are vectors, $\boldsymbol{X}_i \in \boldsymbol{R}^{n\times 1}, \boldsymbol{X}_{1j} \in \boldsymbol{R}^{m\times 1}$, and $\boldsymbol{X}_{2k} \in \boldsymbol{R}^{p\times 1}$ are the output of INPUT, HIDDEN and OUTPUT layers, respectively, for $1 \le i \le N, 1 \le j \le N_1$, and $1 \le k \le N_2$.

1) Compute the Output of the HIDDEN Layer $\boldsymbol{X}_{1j}$:

$$\boldsymbol{X}_{1j} \equiv [x_{1j1}, \ldots, x_{1jm}]^T = \boldsymbol{f}_j(\boldsymbol{O}_{1j}) = \left[\frac{1}{1+\exp(-(o_{1j1}+\theta_{1j1}))}, \ldots, \frac{1}{1+\exp(-(o_{1jm}+\theta_{1jm}))}\right]^T$$

where $\boldsymbol{O}_{1j} = \sum_{i=1}^{N} \boldsymbol{W}_{ij}\boldsymbol{X}_i, \ j = 1, 2, \ldots, N_1$, and $\boldsymbol{W}_{ij} \in \boldsymbol{R}^{m \times n}$ is the weights from the INPUT to the HIDDEN layer, $\boldsymbol{\theta}_{1j} = [\theta_{1j1}, \ldots, \theta_{1jm}]^T$ the threshold at the HIDDEN layer.

2) Compute the Output of the OUTPUT Layer

$$\boldsymbol{X}_{2k} \equiv [x_{2k1}, \ldots, x_{2kp}]^T = \boldsymbol{f}_k(\boldsymbol{O}_{2k}) = \left[\frac{1}{1+\exp(-(o_{2k1}+\theta_{2k1}))}, \ldots, \frac{1}{1+\exp(-(o_{2kp}+\theta_{2kp}))}\right]^T$$

where $\boldsymbol{O}_{2k} = \sum_{j=1}^{N_1} \boldsymbol{W}_{1jk}\boldsymbol{X}_{1j}, k = 1, 2, \ldots, N_2$, and $\boldsymbol{W}_{1jk} \in \boldsymbol{R}^{p \times m}$ is the weights from the HIDDEN to the OUTPUT layer, $\boldsymbol{\theta}_{2k} = [\theta_{2k1}, \ldots, \theta_{2kp}]^T$ the threshold at the OUTPUT layer.

3) Update the Weights from the HIDDEN to the OUTPUT Layer $\boldsymbol{W}_{1jk}$:

$$\boldsymbol{W}_{1jk}(k+1) = \boldsymbol{W}_{1jk}(k) + \Delta\boldsymbol{W}_{1jk}$$

where

$$\Delta\boldsymbol{W}_{1jk} = \eta_1[\delta_{1k}\boldsymbol{T}]^T$$

$$\delta_{1k} = \left[\boldsymbol{X}_{2k}^d - \boldsymbol{X}_{2k}\right]^T diag\,[x_{2k1}(1-x_{2k1}), x_{2k2}(1-x_{2k2}), \cdots x_{2kp}(1-x_{2kp})]$$

and $\boldsymbol{T}_1$ is a $p \times m \times p$ tensor, with the lth matrix as

$$\boldsymbol{T}_{1l} = \begin{bmatrix} 0 \\ \cdots \\ (\boldsymbol{X}_{1j})^T \\ 0 \\ \cdots \end{bmatrix}^T \text{—at the } l\text{th row, } l=1,2,\cdots,p.$$

4) Update the Weights from the INPUT to the Hidden Layer $\boldsymbol{W}_{ij}$:

$$\boldsymbol{W}_{ij}(k+1) = \boldsymbol{W}_{ij}(k) + \Delta\boldsymbol{W}_{ij}$$

where

$$\boldsymbol{W}_{ij} = \eta[\delta_j\boldsymbol{T}_2]^T$$

$$\delta_j = \left(\sum_{k=1}^{N_2} \delta_{1k}\boldsymbol{W}_{1jk}\right) diag\,[x_{1j1}(1-x_{1j1}), x_{1j2}(1-x_{1j2}), \cdots, x_{1jm}(1-x_{1jm})]$$

and $\boldsymbol{T}_2$ is a $m \times n \times m$ tensor, with the lth matrix as

$$\boldsymbol{T}_{2l} = \begin{bmatrix} 0 \\ \cdots \\ (\boldsymbol{X}_i)^T \\ 0 \\ \cdots \end{bmatrix}^T \text{—at the } l\text{th row , } l=1,2,\cdots,m.$$

5) Update the Thresholds at the OUTPUT and the HIDDEN Layers $\boldsymbol{\theta}_{2k}$, $\boldsymbol{\theta}_{1j}$:

$$\boldsymbol{\theta}_{2k}(k+1) = \boldsymbol{\theta}_{2k}(k)\eta_{1\theta}[\delta_{1k}]^T$$
$$\boldsymbol{\theta}_{1j}(k+1) = \boldsymbol{\theta}_{1j}(k) + \eta_\theta[\delta_{1k}]^T$$

where $\eta_1, \eta, \eta_{1\theta}$, and $\eta_\theta > 0$ are the gain factors.

B. Tracking a Time-Varying System

In the above algorithm, the NN is trained by using the prediction error

$$\boldsymbol{E}_k(k) = \boldsymbol{X}_{2k}^d(k) - \boldsymbol{X}_{2k}(k) = \boldsymbol{Y}(k+d) - \hat{\boldsymbol{Y}}(k+d/k), \quad (4)$$

where $\boldsymbol{X}_{2k}^d(k)$ is the desired value of $\boldsymbol{X}_{2k}^d(k)$ at time k. This implies that the NN should be trained by using the system's future output $\boldsymbol{Y}(k+d)$, which are unknown. A set of training data can be acquired beforehand, and used to train the NN. After the NN is "well trained," its weights will no longer be modified. Then, the NN produces the correct outputs, whenever the inputs are present at the INPUT layer. However, for time-varying mappings, it is meaningless to say that an NN is "well trained". Moreover, in a real-time control system, it is desirable to always operate the system in closed-loop. This means that the NN-based predictor should be "updated" (rather than trained) so as to track a time-varying system. To update the NN-based predictor on-line, the basic idea is to modify the weights of the NN using the *a posteriori* prediction error:

$$\boldsymbol{E}_k(k-d) = \boldsymbol{X}_{2k}^d(k-d) - \boldsymbol{X}_{2k}(k-d) = \boldsymbol{Y}(k) - \hat{\boldsymbol{Y}}(k/t-d). \quad (5)$$

It is shown in [14] that one can use (5) instead of (4) in the algorithm, and keep all other formulas unchanged. The scaling problem and error analysis have also been addressed in [14], and concluded that the accuracy of the NN-based predictor depends only on the accuracy of the NN's approximation of the actual mapping.

VI. Coordinated Control of Two 2-Link Robots

To demonstrate how to apply the proposed scheme for solving real life problems, we consider the problem of coordinating two 2-link robots holding a rigid object. The low-level subsystems include two robots each with a separately designed servo controllers. The basic configuration of this example is given in Fig. 6. The Cartesian frame is fixed at the base of robot 1, and the trajectories of the object and the robots' end-effectors are specified relative to this frame. The task is to move the object forward and then backward in X direction while keeping the height in Y direction constant. The desired trajectory of the object is selected by a high-level planner as the reference input to the low level. If the two robots hold the object firmly, then the dynamics of the system are modeled as follows.

Dynamics of the Object: Let $\boldsymbol{f}_i = (f_{ix}, f_{iy})^T$ be the force exerted by the end-effector of robot i on the object in Cartesian space. Then the motion of the object is described by

$$m\ddot{\boldsymbol{P}} + m\boldsymbol{g} = \boldsymbol{f}, \quad \boldsymbol{f} = \boldsymbol{W}\boldsymbol{F} \equiv [\boldsymbol{I}_2, \boldsymbol{I}_2]\begin{bmatrix}\boldsymbol{f}_1 \\ \boldsymbol{f}_2\end{bmatrix} \quad (6)$$

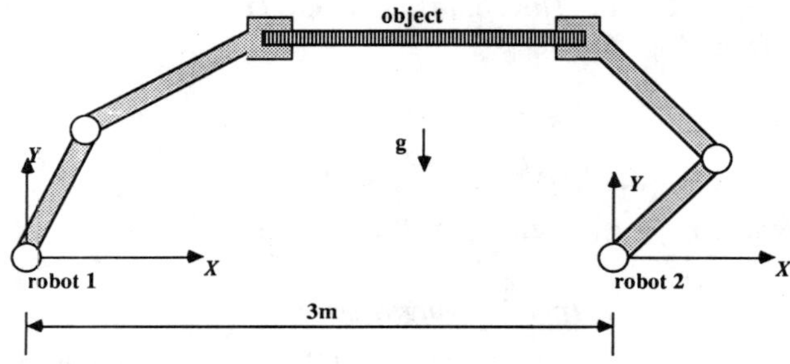

Robots	Link 1	Link 2
Length	1 m	1 m
Mass Center	0.5 m	0.5 m
Mass	20 kg	10 kg
Moment of inertia	0.8 kg ms^2	0.2 kg ms^2
Object: mass=5 kg; length=1 m.		

Fig. 6. Two 2–link robots holding an object.

where m is the mass of the object, $\boldsymbol{P}$ the position of the object in Cartesian space, $\boldsymbol{g}$ the gravitational acceleration, $\boldsymbol{f}$ the external force exerted on the object by the two robots, and $\boldsymbol{I}_2$ is a 2×2 unit matrix. From (6), one can see that, to achieve the object's specified acceleration, the combination of forces shared by the two robots is not unique.

Dynamics of Each Robot with Servo Controller: Suppose two robots have an identical mechanical configuration, then the force-constrained dynamic equation of robot i in joint space is given by [15]:

$$\boldsymbol{H}\ddot{\boldsymbol{q}}_i + \boldsymbol{h}(\boldsymbol{q}_i, \dot{\boldsymbol{q}}_i) + \boldsymbol{J}_i^T \boldsymbol{f}_i = \boldsymbol{\tau}_i, \qquad i = 1, 2$$

where $\boldsymbol{q}_i$ is the vector of the robot's joint positions, $\boldsymbol{H}$ is the inertia matrix, $\boldsymbol{h}$ is the centrifugal, Coriolis, and gravitational forces, $\boldsymbol{J}_i$ is the Jacobian matrix, and $\boldsymbol{\tau}_i$ is the vector of joint torques. Suppose both robots are position-controlled with the computed torque algorithm. That is, the control input to robot i is

$$\boldsymbol{\tau}_i = \hat{\boldsymbol{H}}(\ddot{\boldsymbol{q}}_{id} - \boldsymbol{K}_{Di}(\dot{\boldsymbol{q}}_i - \dot{\boldsymbol{q}}_{id}) - \boldsymbol{K}_{pi}(\boldsymbol{q}_i - \boldsymbol{q}_{id})) + \hat{\boldsymbol{h}} \qquad (7)$$

where $\hat{\boldsymbol{H}}$ and $\hat{\boldsymbol{h}}$ are the estimated values of $\boldsymbol{H}$ and $\boldsymbol{h}$, $\boldsymbol{q}_{id}$ is the desired value of $\boldsymbol{q}_i$, $\boldsymbol{K}_{Di}$ and $\boldsymbol{K}_{pi}$ are the controllers' gains. The reference input to the system is the desired trajectory of the object specified by $\boldsymbol{P}_d, \dot{\boldsymbol{P}}_d$ and $\ddot{\boldsymbol{P}}_d$, which will be transformed into the desired trajectories of the end-effector and the joints of each robot.

Problem Statement: Suppose the object is a rigid body and there is no relative motion between the end-effectors and the object. For (6), let $\boldsymbol{f}_d$ and $\boldsymbol{F}_d$ be the desired values of $\boldsymbol{f}$ and $\boldsymbol{F}$, respectively. Then, we have

$$\boldsymbol{F}_d = \boldsymbol{F}_{Md} + \boldsymbol{F}_{Id} \equiv \boldsymbol{W}^* \boldsymbol{f}_d + (\boldsymbol{I}_4 - \boldsymbol{W}^* \boldsymbol{W})\boldsymbol{y}_0 \qquad (8)$$

where $\boldsymbol{W}^* \in \boldsymbol{R}^{4\times 2}$ is the pseudo-inverse of $\boldsymbol{W}$, $\boldsymbol{I}_4$ is a 4×4 unit matrix, and $\boldsymbol{y}_0 \in \boldsymbol{R}^{4\times 1}$ an arbitrary vector in the null space of $\boldsymbol{W}$. Therefore, the forces exerted by the end-effectors consist of two parts:

$$\boldsymbol{F}_{Md} \equiv \begin{bmatrix} \boldsymbol{F}_{M1d} \\ \boldsymbol{F}_{M2d} \end{bmatrix} \in \boldsymbol{R}^{4\times 1}$$

is the force to move the object and

$$\boldsymbol{F}_{Id} \equiv \begin{bmatrix} \boldsymbol{F}_{I1d} \\ \boldsymbol{F}_{I2d} \end{bmatrix} \in \boldsymbol{R}^{4\times 1}$$

is the internal force. The following two problems arise: 1) sharing the moving force $\boldsymbol{F}_{Md}$ by two robots, 2) changing the internal force so as to satisfy a set of constraints, such as joint torque limits or energy capacity.

In (8), $\boldsymbol{F}_{Md}$ can be specified by the desired trajectory. $\boldsymbol{F}_{Id}$ is given as the desired internal force, for example, $\boldsymbol{F}_{Id} = 0$ for the least energy consumption. Because $\boldsymbol{W}^*$ is a constant matrix and both $\boldsymbol{f}_d$ and $\boldsymbol{F}_{Id}$ are specified, the desired force $\boldsymbol{F}_d$ is determined uniquely. However, this ideal situation of load sharing may not be achieved due to the force and trajectory tracking errors. These errors may be caused by modeling/parameter errors, control performance tradeoff, and/or disturbances. It is therefore necessary to share the load by, or reassign the load to, each robot dynamically. Our goal is to design a KBC to coordinate the two robots moving the object while minimizing the internal force.

Principal Output and Its NN-Based Predictor: The reference inputs to the low-level subsystems are the desired acceleration $\ddot{\boldsymbol{P}}_{id}$, velocity $\dot{\boldsymbol{P}}_{id}$ and position $\boldsymbol{P}_{id}$ of robot i's end-effector, $i = 1, 2$. The internal force can be used to evaluate system performance, and has an explicit relation to the reference inputs. So, the internal force is defined as the principal output. Because the force exerted by each robot to achieve a specified acceleration of the object is not unique, it is possible to adjust the internal force by modifying the reference inputs. Since the position tracking error needs to be kept small and the desired acceleration has an explicit relationship to the force exerted on the object, only the desired acceleration is modified so as to reduce the internal force. Then, the desired acceleration issued to each robot is $\ddot{\boldsymbol{P}}_{idm}$ – the modified value of $\ddot{\boldsymbol{P}}_{id}$, $i = 1, 2$. An NN-based predictor is designed to predict the force exerted on the object, which corresponds to each reference input. The predicted internal force (that is, the principal output) is then computed. The NN-based predictor has eight nodes at the INPUT layer, and the inputs are $\boldsymbol{P}_{1d}(k), \boldsymbol{P}_{1d}(k-1), \boldsymbol{P}_{2d}(k)$, $\boldsymbol{P}_{2d}(k-1)$, $\ddot{\boldsymbol{P}}_{1dm}(k)$, $\ddot{\boldsymbol{P}}_{1dm}(k-1), \ddot{\boldsymbol{P}}_{2dm}(k)$, and $\ddot{\boldsymbol{P}}_{2dm}(k-1)$. There are five HIDDEN nodes and six OUTPUT nodes with outputs $\hat{\boldsymbol{f}}_i(k+d/k)$, for $\subset = 1, 2, d = 1, 2, 3$.

Simulations Results: The two robots move the object in X direction from the initial position to the final position over one-meter distance in five seconds, and then move back to the initial position. The sampling interval is $10ms$. Force predictions are used for the modification process, and position tracking is achieved by the position controllers. The 1-step ahead predictions $\hat{\boldsymbol{f}}_i(k+1/k), i = 1, 2$ are used in the KBC. The desired internal force is set to zero. Without the KBC, the internal force error in X direction is plotted in Fig. 7. After adding the KBC, the root-mean-square error of the internal force in X direction is reduced by 63% as shown in Fig. 8. Moreover, both the external force error and the position tracking error are kept almost the same as those without the KBC. The detailed results are summarized in Table I. Since there is no motion in Y direction, the internal force error in that direction is small enough not to require the KBC.

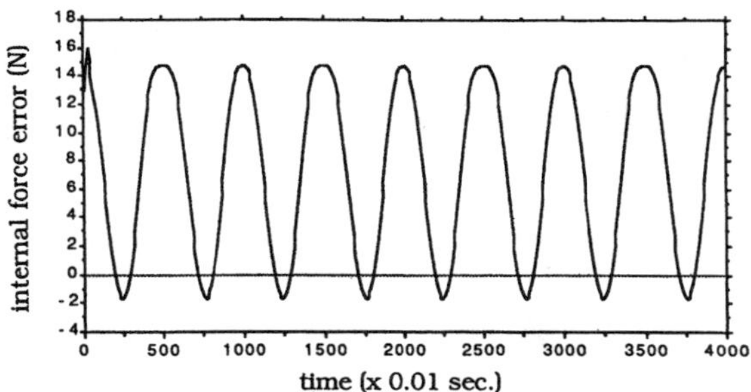

Fig. 7. Internal force error in X direction without the KBC.

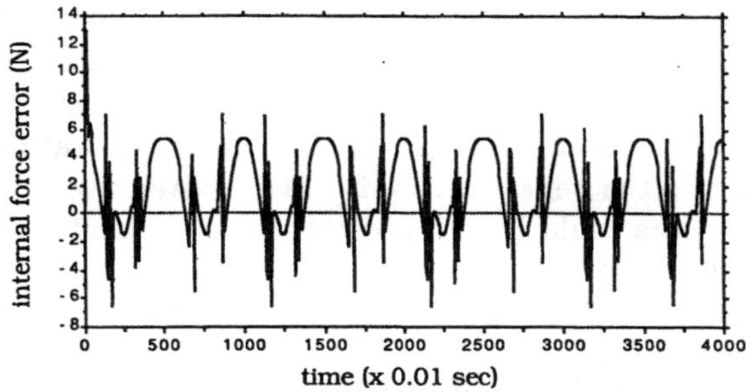

Fig. 8. Internal force error in X direction with the KBC.

TABLE I
RMS ERRORS OF INTERNAL FORCES, EXTERNAL FORCES AND OBJECT'S POSITION TRACKING

Sample Interval for Statistics		RMS Errors of Internal Forces (N)	
		without KBC	with KBC
0 – 1000	in X direction	9.58447	3.85020
	in Y direction	0.93141	0.53177
1001 – 2000	in X direction	9.57130	3.53340
	in Y direction	0.92339	0.49949
2001 – 3000	in X direction	9.57097	3.53688
	in Y direction	0.92339	0.49956

Sample Interval for Statistics		RMS Errors of External Forces (N)	
		without KBC	with KBC
0 – 1000	in X direction	0.72359	0.95822
	in Y direction	2.54853	2.54345
1001 – 2000	in X direction	0.34883	0.70199
	in Y direction	0.01436	0.03345
2001 – 3000	in X direction	0.34883	0.70346
	in Y direction	0.01436	0.03355

Sample Interval for Statistics		RMS Tracking Errors of Object's Positions (m)	
		without KBC	with KBC
0 – 1000	in X direction	0.03509	0.03529
	in Y direction	0.05733	0.05784
1001 – 2000	in X direction	0.03509	0.03530
	in Y direction	0.05759	0.05813
2001 – 3000	in X direction	0.03509	0.03530
	in Y direction	0.05759	0.05813

VII. CONCLUSION

Focusing on the problem of coordinating multiple systems, a knowledge-based coordinator is designed using the techniques of both intelligent control and neural networks. As the high-level coordinator in a hierarchical structure, its basic principle is to modify the reference inputs of low-level subsystems according to the principal output prediction so as to achieve the desired performance. By adding the proposed KBC, the internal structure and parameters of the low-level subsystems are not affected. Hence, each servo controller of the low-level subsystems can be designed separately from, and independently of, the others; no constraints need to be imposed on the design of low-level controllers. This implies that some commercially designed servo controllers for a single system can be coordinated to work for a multiple-system.

Using the principal output and its prediction, and a structure of decision tree for knowledge representation, the knowledge base necessary to coordinate multiple systems is greatly simplified while guaranteeing system stability. By using a predictor, the negative effects of system time delay is eliminated and each reference input is analyzed before putting it in operation. The unknown parameters and/or time-varying properties of a multiple-system are handled by the NN-based predictor, while leaving the logical reasoning and decision making on the coordination to the KBC.

To test this new scheme, the coordination problem for two 2-link robots holding a rigid object is simulated. By modifying the reference input of each robot, the internal force exerted on the object is reduced by 63%, indicating the scheme's potential for the effective coordination of multiple robots.

REFERENCES

[1] R. E. Larson, P. L. McEntire, and J. G. O'Reilly, Eds. "Tutorial: Distributed control (2nd ed.)," *IEEE Computer Society*, 1982.

[2] S. Lee and M. H. Kim, "Cognitive control of dynamic systems," in *Proc. IEEE 2nd Int. Symp. Intelligent Contr.*, 1987, pp. 455–460.

[3] Z. Geng and M. Jamshidi, "Expert self-learning controller for Robot manipulator," in *Proc. 1988 IEEE Int. Conf. Decision and Contr.*, pp. 1090–1095.

[4] A. J. Krijgsman, H. M. T. Broeders, H. B. Verbruggen, and P. M. Bruijn, "Knowledge-based control," in *Proc. 1988 IEEE Int. Conf. Decision and Contr.*, pp. 570–574.

[5] J. Jiang and R. Doraiswami, "Information acquisition in expert control system design using adaptive filters," in *Proc. IEEE The 2nd Int. Symp. Intelligent Contr.*, 1987, pp. 165–170.

[6] K. L. Anderson, G. L. Blankenship, and L. G. Lebow, "A rule-based adaptive PID controller," in *Proc. 1988 IEEE Int. Conf. Decision and Contr.*, pp. 564–569.

[7] B. Porter, A. H. Jones, and C. B. McKeown, "Real-time expert controller for plants with actuator non-linearities," in *Proc. IEEE The 2nd Int. Symp. Intelligent Contr.*, 1987, pp. 171–177.

[8] G. K. H. Pang, "A blackboard control architecture for real-time control," in *Proc. 1988 Amer. Contr. Conf.*, pp. 221–226.

[9] K. G. Shin and X. Cui, "Design of a knowledge-based controller for intelligent control systems," *IEEE Trans. Syst., Man, Cybern.*, vol. 21, pp. 368–375, Mar./Apr. 1991.

[10] G. Cybenko, "Approximation by superpositions of a sigmoidal function," *Mathematics of Control, Signals and Systems*, vol. 2, no. 4, pp. 303–314, 1989.

[11] V. Vemuri (reprint edited by) "Artificial neural networks: Theoretical concepts," *Press of IEEE Computer Society*, 1988.

[12] P. J. Werbos, "Back propagation: Past and future," *Proc. 1988 Int. Conf. Neural Networks*, vol. 1, pp. I343-I353.

[13] D. E. Rumelhart and J. L. McCelland, "Learning internal representations by error propagation," *Parallel Distributed Processing: Explorations in the Microstructure of Cognition, Vol. 1: Foundations.* Cambridge, MA: MIT Press, 1986.

[14] K. G. Shin and X. Cui, "Design of a general-purpose MIMO predictor with neural networks," in *Proc. 13th IMACS World Congress on Computation and Applied Mathematics*, Dublin, Ireland, July 1991.

[15] H. Asada and J.-J. E. Slotine, *Robot Analysis and Control.* New York: Wiley-Interscience, 1986.

Article 7.3

A Distributed Adaptive Control System for a Quadruped Mobile Robot

Bruce L. Digney and M. M. Gupta
Intelligent Systems Research Laboratory, College of Engineering
University of Saskatchewan, Saskatoon, Sask. CANADA S7N 0W0
E_mail: digney@dvinci.usask.ca

Abstract— **In this research, a method by which reinforcement learning can be combined into a behavior based control system is presented. Behaviors which are impossible or impractical to embed as predetermined responses are learned through self-exploration and self-organization using a temporal difference reinforcement learning technique. This results in what is referred to as a distributed adaptive control system (DACS); in effect the robot's artificial nervous system. A DACS is developed for a simulated quadruped mobile robot and the locomotion behavior level is isolated and evaluated. At the locomotion level the proper actuator sequences were learned for all possible gaits and eventually graceful gait transitions were also learned. When confronted with an actuator malfunction, all gaits and transitions were adapted resulting in new *limping* gaits for the quadruped.**

I. Introduction

Although conventional control and artificial intelligence researchers have made many advances, neither ideology seems capable of realizing autonomous operation. That is, neither can produce machines which can interact with the world with an ease comparable to humans or at least higher animals. In responding to such limitations, many researchers have looked to biological/physiological based systems as the motivation to design artificial systems. As an example are the behavior based systems of Brooks [1] and Beer [2]. Behavior based control systems consist of a hierarchical structure of simple behavior modules. Each module is responsible for the sensory motor responses of a particular level of behavior. The overall effect is that higher level behaviors are recursively built upon lower ones and the resulting system operates in a self-organizing manner. Both Brooks and Beer's systems were loosely based upon the nervous systems of insects. These artificial insects operated in a hardwired manner and exhibited an interesting repertoire of simple behaviors. By *hardwired* it is meant that each behavior module had its responses predetermined and was simply programmed externally. Although this approach is successful with simple behaviors, it is obvious that many situations exist where predetermined solutions are impossible or impractical to obtain. It is subsequently proposed that by incorporating learning into the behavior based control system, these difficult behaviors could be acquired through self-exploration and self-learning.

Complex behaviors are usually characterized by a sequence of actions with success or failure only known at the end of that sequence. Also, the critical error signal is only an indication of the success or failure of the system and no information regarding error gradients can be determined, as in the case of continuous valued error feedback. Thus the required learning mechanism must be capable of both reinforcement learning as well as temporal credit assignment. Incremental dynamic programming techniques such as Barto's [3] temporal difference (TD) appear to be well suited to such tasks. Based upon Barto's previous adaptive heuristic critic [4], TD employs adaptive state and action evaluation functions to incrementally improve its action policy until successful operation is attained. The incorporation of TD learning into behavior based control results in a framework of adaptive (ABMs) and non-adaptive behavior modules which is referred to here as a distributed adaptive control system (DACS). The remainder of this report will be concerned with a brief description of the DACS and ABMs, and implementing of the locomotion level ABM within the DACS of a simulated quadruped mobile robot. This level is considered appropriate because the actuator sequences for quadruped locomotion are not intuitively obvious and difficult to determine. Other levels such as global navigation, task planning and task coordination are implemented and discussed by Digney [5].

II. Distributed Adaptive Control Systems

The DACS shown in Figure 1 is comprised of various adaptive and non-adaptive behavior modules. Non-adaptive

Reprinted from *IEEE Conf. on Neural Networks*, San Francisco, pp. 144–149, March 1993.

modules are present as inherent knowledge and are used where adaptive solutions are not required. All modules receive sensory inputs and respond with actions in an attempt to perform a command specified by a higher level. The performance of commands in most cases will require a sequence of actions by the lower level system and possibly the cooperation of many lower level systems. The coupling between ABMs is shown in Figure 2. In this configuration, the action from level $l+1$ becomes the command for level l. Level $l+1$ also supplies goal based reinforcement, r_g, to drive level l towards successful completion of that command. Level l in turn issues actions to level $l-1$ and receives environmental based reinforcement, r_e, from level $l-1$. This environment based reinforcement is representative of the difficulty or cost incurred while performing the requested actions and is included to drive level l to a cost effective solution. While operating, level l may enter a state which is in some way damaging or dangerous. To drive the system away from such a state, sensor based reinforcement, r_s, is used. Sensor based reinforcement is supplied from sensors at level l. It is analogous to *pain or fear* and will ensure that level l operates in a safe manner. These three reinforcements are combined into a total reinforcement signal, r_t, according to Equation 1.

$$r_t = \alpha_e \cdot r_e + \alpha_g \cdot r_g + \alpha_s \cdot r_s \tag{1}$$

where: α_e, α_g and α_s are the relative importance of the reinforcements.

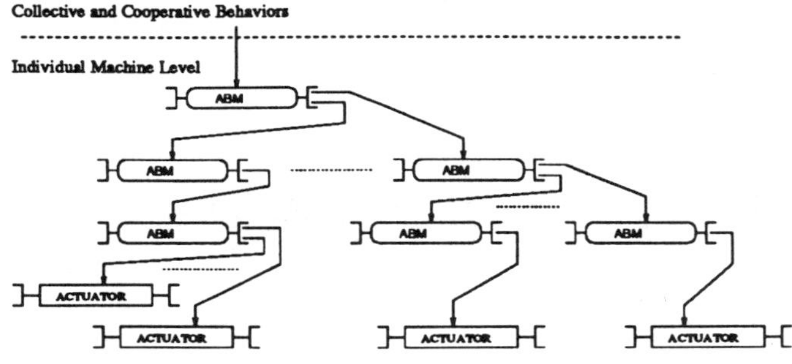

Figure 1: Schematic of DACS

It can be seen from Figure 2 that the flow of environmental and sensor based reinforcement is in the upward direction. This will result in lower level skills and behaviors being learned first, then other higher level behaviors, converging in a recursive manner toward the highest level. Figure 1 shows this highest level as existing within a single physical machine. However, in the case of multiple machines operating in a collective, higher abstract behavior levels are possible. Within the context of this paper, only behaviors relevant to individual machines will be discussed. In the absence of higher collective behaviors controlling individual machines, the purpose or task of the machine is embedded within the DACS as an *instinct or drive*. This instinct is the high level action which results in a feeling of accomplishment or positive reinforcement within the DACS. It is then the responsibility of the adaptive behavior modules within the DACS to learn the skills and behaviors necessary to fulfill this drive. This concept as well as the self-organizing characteristics that result from such interactions are further discussed by Digney [5].

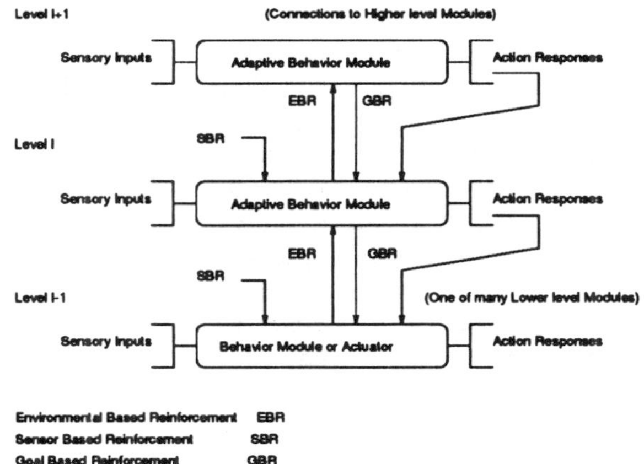

Figure 2: Hierarchy of Three ABMs

The ABM is the primary adaptive building block for the DACS. Within it exist computational mechanisms for state classification, learning and the combination of reinforcement signals. Figure 3 shows a schematic of an ABM complete with incoming command, sensory and reinforcement signals. For clarity the outgoing reinforcement signals have been removed. For any particular level, say l, the ABM observes the relevant system states through appropriate sensors. For a perception system consisting of N sensors, the state S_l, is defined as

$$S_l = \begin{bmatrix} s_0 & \cdots & s_n & \cdots & s_N \end{bmatrix}^T \tag{2}$$

where: s_n is the individual sensor reading, $0 < n < N$

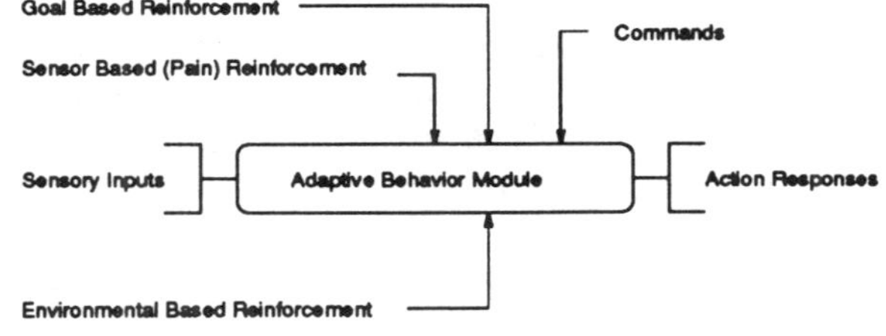

Figure 3: Single ABM

State transitions are detected and the resulting states are classified using an idealized neural classification scheme. This classification embodies the macroscopic operating principles of unsupervised neural networks such as ART2-A [6] and will be assumed adequate in the context of these simulations. The Temporal Difference (TD) algorithm as developed by Barton [7] learns by adjusting state and action evaluation functions then uses these evaluations to choose an optimum action policy. It can be shown that these two evaluation functions can be combined into a single action dependent evaluation function, say $Q_{s,u}$, similar to that described by Barto [7]. Given the system at state s, the action taken, u^*, is the action which satisfies

$$\max_u\{Q_{s,u} + \zeta\} \Rightarrow u^* \tag{3}$$

where: ζ is a random valued function.

In Equation 3, $Q_{s,u}$ and ζ can be thought of as the goal driven and exploration driven components of the action policy respectively. Taking the action u^* results in the transition from state s to state v and the incurring of a total reinforcement signal r_t. The action dependent evaluation function error is obtained by modifying the TD error equation and is

$$e = \gamma \cdot Q_{virtual} - Q_{s,u^*} + r_t. \tag{4}$$

where: $Q_{virtual}$ is the virtual state evaluation value of the next state v and γ is the temporal discount factor.

If action, u^*, does not achieve the desired goal, the virtual state evaluation is,

$$Q_{virtual} = \max_u\{Q_{v,u}\}. \tag{5}$$

It is easily seen that $Q_{virtual}$ becomes the minimum action dependent evaluation function of the new state, v, (remember the evaluation functions are negative in sign) and in effect corresponds to the action most likely to be taken when the system leaves state v.

If the action, u^*, achieves the desired goal, the virtual state evaluation is,

$$Q_{virtual} = 0. \tag{6}$$

This provides relative state evaluations and allows for opended or cyclic goal states. This is illustrated by considering that for cyclic goals it is the dynamic transitions between states that constitutes a goal state and not simplely the arrival at a static system state(s).

This error is used to adapt the evaluation functions according to LMS rules as follows

$$Q_{s,u=u^*}(k+1) = Q_{s,u=u^*}(k) + \eta \cdot e \tag{7}$$

$$Q_{s,u\neq u^*}(k+1) = Q_{s,u\neq u^*}(k) \tag{8}$$

where: η is the rate of adaption and k is the index of adaption.

As the evaluation function converges, the goal driven component begins to dominate over the exploration driven component. The resulting action policy will perform the command in a successful and efficient manner. Generally, an ABM will be capable of performing more than a single command. For an ABM capable of c_{max} commands, the vector of the evaluation functions is defined as:

$$Eval_l = [\ Q_{s,u,0}, \ \cdots \ Q_{s,u,c}, \ \cdots, \ Q_{s,u,c_{max}}\ ]^T \tag{9}$$

where: $Q_{s,u}$ is the evaluation function and c the particular command $0 < c < c_{max}$.

III. DACS for a Quadruped Moble Robot

To evaluate the DACS, the simulated quadruped shown in Figure 4 was used. This mobile robot was placed inside a simulated three dimensional landscape where it is left to develop skills and behaviors as it interacts with its environment. This world is made up of ramps, plateaus, cliffs and walls, as well as various substances of interest. In the absence of any predetermined knowledge it is the responsibility of the DACS and in particular the ABMs to acquire the skills and behaviors for successful operation.

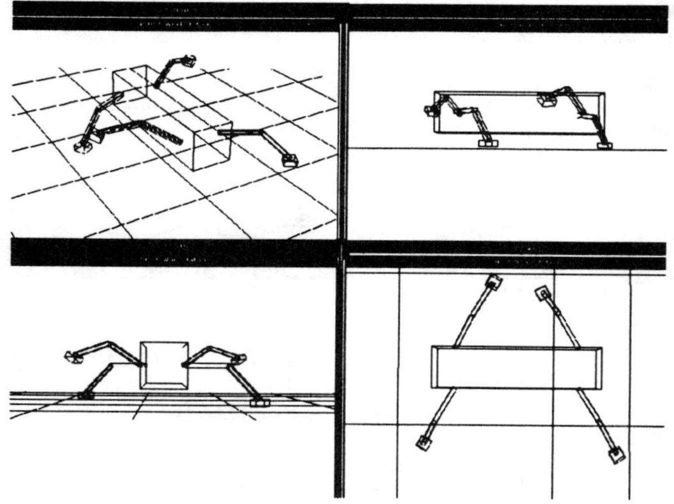

Figure 4: Simulated Quadruped

Although not the most efficient method of locomotion, the learning of quadruped walking provides interesting and challenging problems. Involved is the learning of complex actuator sequences in the midst of numerous false goal states and modes of failure. Figure 5 shows the locomotion ABM with the appropriate sensory, reinforcement and motor action connections.

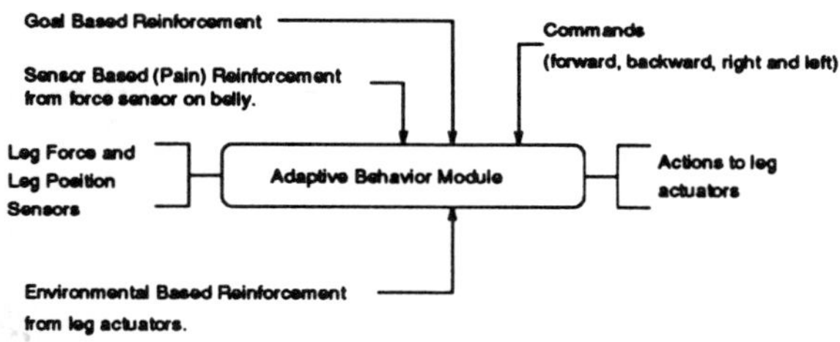

Figure 5: Locomotion ABM

The commands, $c_{locomotion}$, are issued from the ABM above and are dependent upon the possible sensory states of that module. In this case these sensors are capable of detecting all realizable modes of body motion. The commands for the locomotion level are defined in Equation 10.

$$c_{locomotion} = \begin{cases} 0 & \text{forward} \\ 1 & \text{left turn} \\ \vdots & \vdots \\ c_{max} & \text{all possible modes} \end{cases} \tag{10}$$

For any specific command the locomotion ABM will issue action responses, $u_{locomotion}$, to the actuators driving the legs in the horizontal, h, and vertical, v, directions. Within this action vector are the individual actuator commands to extend, ex, or retract, rt, as shown in Equation 11 and 12.

$$u_{locomotion} = \begin{bmatrix} c_{leg_0}, & c_{leg_1}, & c_{leg_2}, & c_{leg_3} \end{bmatrix}^T \tag{11}$$

where

$$c_{leg} = \begin{cases} h & \text{hold} \\ v_{ex} & \text{extend vertical} \\ v_{rt} & \text{retract vertical} \\ h_{ex} & \text{extend horizontal} \\ h_{rt} & \text{retract horizontal} \end{cases} \tag{12}$$

Each leg is equipped with sensors for measuring the forces on each foot and the positions of each leg. The forces on the foot are are biased such that $-f_{max} < f_{leg} < f_{max}$, where f_{max} is the highest force magnitude expected. Similarly the position sensors are biased such that $-l_{max} < l_{leg} < l_{max}$, where l_{max} is half the stroke of the linear actuator. For an arbitrary leg, leg_n, and direction, d, the force and position descriptions of state are

$$s_{force,d,leg_n} = \begin{cases} -1 & \text{if } f_{leg_n,d} < 0 \\ 0 & \text{if } f_{leg_n,d} = 0 \\ +1 & \text{if } f_{leg_n,d} > 0 \end{cases} \tag{13}$$

$$s_{pos,d,leg_n} = \begin{cases} -1 & \text{if } l_{leg_n,d} < 0 \\ +1 & \text{if } l_{leg_n,d} > 0 \end{cases} \tag{14}$$

The reinforcement signals are defined as

$$r_g = \begin{cases} 0 & \text{if } c_{locomotion} \text{ is performed,} \\ -R_g & \text{otherwise} \end{cases} \tag{15}$$

$$r_e = \begin{cases} -R_{low} & \text{easy operation} \\ -R_{high} & \text{difficult operation} \end{cases} \tag{16}$$

$$r_s = \begin{cases} -R_s & \text{if } f_{belly} > 0, \\ 0 & \text{otherwise} \end{cases} \tag{17}$$

where: f_{belly} is a force sensor on the bottom of the robot and R_* are appropriate positive values.

The net effect of these reinforcements is to drive the locomotion system to learn efficient actuator sequences that will perform the specified gaits while maintaining the quadruped's balance. At this level of abstraction the quadruped is said to lose balance when the line drawn between any two legs in contact with the ground does not pass through the quadruped's center of gravity.

Equations 13 and 14, when combined into Equation 18, define the complete state of the locomotion system. For the transition between any pair of past and present states, the total reinforcement can be determined by combining Equations 15, 16 and 17 into Equation 1. The total reinforcement is then used to adapt the command specific evaluation functions of Equation 19 according to Equations 4, 7 and 8. As these evaluation functions converge, all realizable gaits should be achieved.

$$S_{locomotion} = \begin{bmatrix} s_{force,d,leg_0} \\ \vdots \\ s_{pos,d,leg_0} \end{bmatrix} \tag{18}$$

$$Eval_{locomotion} = \begin{bmatrix} Q_{s,u,forward} \\ Q_{s,u,backward} \\ Q_{s,u,rightturn} \\ Q_{s,u,leftturn} \end{bmatrix} \tag{19}$$

IV. Simulation and Results

Using the reinforcement scheme described above, the locomotion ABM was simulated and its abilities to learn in an initially unknown environment and adapt to changes and malfunctions were evaluated. In the following section results from the locomotion tests are presented. Note that the temporal discount factor and the adaption rate chosen for these tests was 0.9 and 0.5 respectively. The reinforcements R_g, R_{low}, R_{high} and R_s were set to 1.0, 1.0, 2.0 and 4.0 respectively.

All possible gaits, including forward, backward, right turn and left turn were discovered and eventually mastered. Figure 6 shows the quadruped's performance while learning the four gaits. Shown on the vertical axis is the cumulative negative reinforcement per step, which can be thought of as the difficulty encountered while performing a single step. The number of steps taken is shown on the horizontal axis. Improvement is obvious as the quadruped initially functions poorly, then eventually learns the optimum actions necessary to perform and maintain a particular gait. In the simulation, the quadruped alternates between all possible gaits, making it possible to evaluate the quadruped's ability to learn graceful transitions between the gaits. In the performance graph of Figure 6, a zone of slow improvement is evident. It is here where the graceful gait transitions are being learned. A mechanical malfunction was simulated by deactivating the horizontal actuator on a single leg. This effectively made the leg able to support only a vertical load and unable to apply any horizontal force to the body of the quadruped. Figure 7 shows the performance of the quadruped under both intact and crippled conditions. Initially, the intact quadruped learns the four standard gaits. Once crippling occurs, a recovery period is required as the quadruped relearns all gaits, this time with a limp. As before, graceful transitions are eventually learned for the new limping gaits. For an intact quadruped Figures 8 and 9 show the individual leg movements that comprise forward and right turning gaits respectively. The corresponding leg commands and reinforcement signals are shown in Tables 1 and 2 for these forward and right turning gaits respectively. In these simulations the end of the quadruped closest to the reader is the front and leg_0,leg_1,leg_2, and leg_3 are considered the front right, front left, rear left and rear right legs respectively. Also, the actuator extensions v_{ex} and h_{ex} cause the leg to move downward and towards the front. The actuator retractions v_{rt} and h_{rt} cause the leg to move upward and towards the rear.

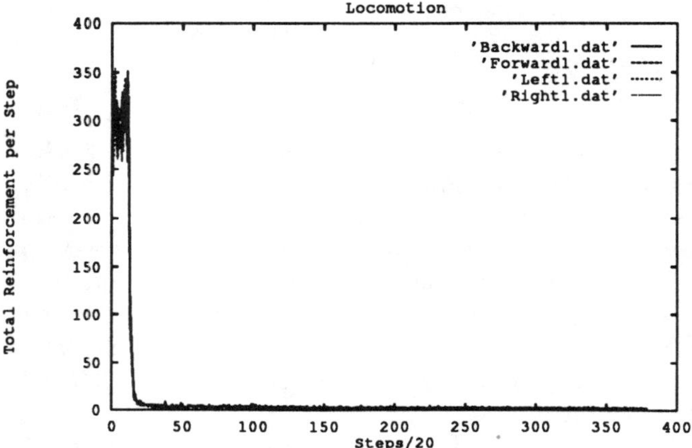

Figure 6: Performance of Intact Quadruped

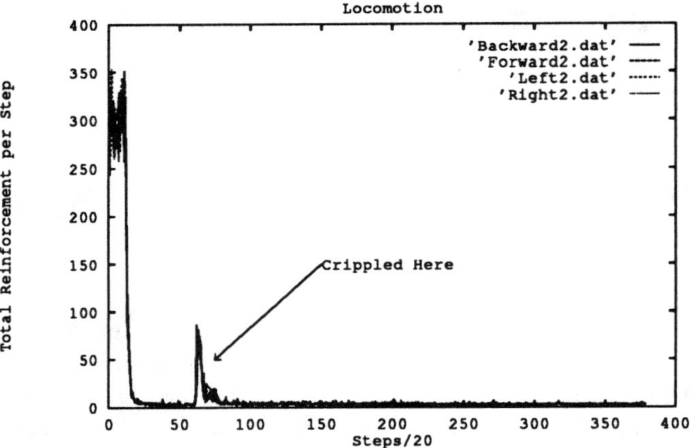

Figure 7: Performance of Crippled Quadruped

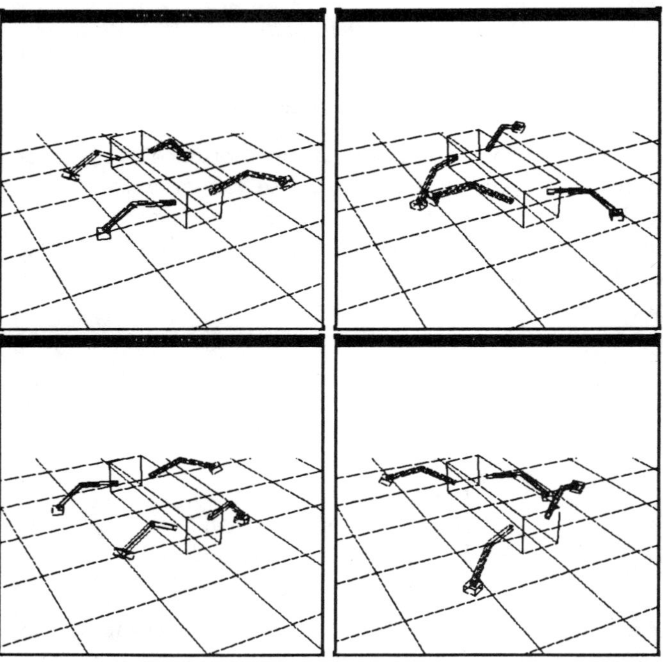

Figure 8: Forward gait(motion towards the reader). Note frame sequence: top left (frame 1) → top right (frame 2) → bottom left (frame 3) → bottom right (frame 4) → top left (frame 1).

Table 1: Command Sequence: Forward Gait

Frame	Commands				Reinforcements		
k	leg_0	leg_1	leg_2	leg_3	r_e	r_g	r_s
$1 \rightarrow 2$	v_{rt}	v_{ex}	v_{rt}	v_{ex}	R_{low}	R_g	0
$2 \rightarrow 3$	h_{ex}	h_{rt}	h_{ex}	h_{rt}	R_{low}	0	0
$3 \rightarrow 4$	v_{ex}	v_{rt}	v_{ex}	v_{rt}	R_{low}	R_g	0
$4 \rightarrow 1$	h_{rt}	h_{ex}	h_{rt}	h_{ex}	R_{low}	0	0

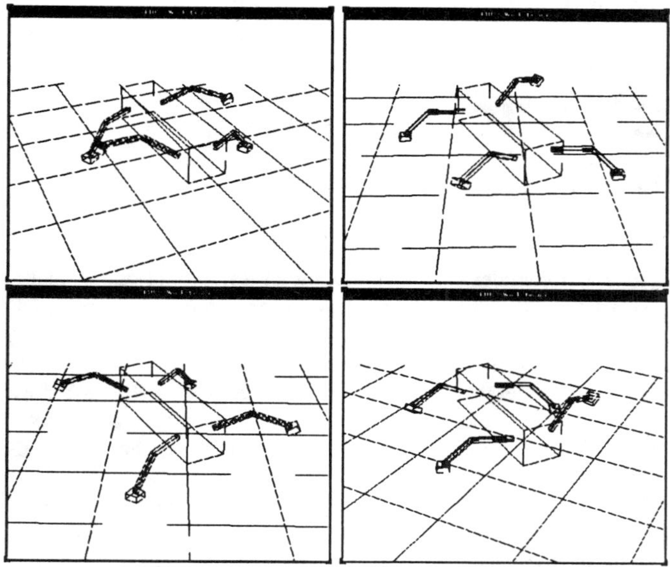

Figure 9: Right Turn Gait(rotation is clockwise as viewed from top) Note frame sequence: top left (frame 1) → top right (frame 2) → bottom left (frame 3) → bottom right (frame 4) → top left (frame 1).

Table 2: Command Sequence: Right Turn Gait

Frame	Commands				Reinforcements		
k	leg_0	leg_1	leg_2	leg_3	r_e	r_g	r_s
1 → 2	h_{ex}	h_{ex}	h_{rt}	h_{rt}	R_{low}	0	0
2 → 3	v_{ex}	v_{rt}	v_{ex}	v_{rt}	R_{low}	R_g	0
3 → 4	h_{rt}	h_{rt}	h_{ex}	h_{ex}	R_{low}	0	0
4 → 1	v_{ex}	v_{ex}	v_{rt}	v_{ex}	R_{low}	R_g	0

V. Discussion

The results achieved in the locomotion tests showed initial poor performance followed by rapid improvement and eventual convergence to an optimum solution. It is during the initial poor performance that the ABM explores the state space, discovering all the obtainable states of the system. For the locomotion ABM, these states are the leg positions and the force on each foot. During this exploration, the learning system refined the initially random action policy to a policy that achieves the specified command. These commands themselves are not supplied to the system as predetermined knowledge but are discovered as realizable commands by the system above. If the quadruped had been placed in water, then it would learn to swim or had the physical configuration allowed it to perform other gaits then these gaits would be learned also.

In response to actuator malfunctions the ABM, after a period of recovery, adapted to the changes and converged to another optimum action policy, in this case a limping leg sequences. This malfunction is labeled as *non-severe* because the adaption required for the DACS to recover is confined to a single ABM. If the malfunction was such that some gaits became impossible to perform, then adaption of higher ABMs would be required for recovery, and what is called a *severe* malfunction would have occurred.

VI. Conclusions

The ABM for locomotion successfully learned the complex action sequences for all realizable gaits. The locomotion ABM was able to successfully recover from a non-severe malfunction. The success of a single ABM can not be extrapolated to imply the success of the entire DACS. Work is currently being performed in the simulation of other behaviors such as body coordination, local navigation, global navigation, task planning, and task coordination. Once these other levels can be incorporated into the DACS, the concepts of self-organization and self-configuration can be explored.

References

[1] Brooks, R (1991) Intelligence Without Reason, *AI Memo No. 1293*, MIT Artificial Intelligence Lab.

[2] Beer, R.D., Chiel, H.J., and Sterling, L.S. (1990) A biological perspective on autonomous agent design, *Robotics and Autonomous Systems 6*, pp 169-186.

[3] Barto, A.G., R.S. Sutton and C.H. Watkins (1989) Learning and Sequential Decision Making, *COINS Technical Report*

[4] Barto, A.G., Sutton, R.S., and Anderson, C.W. (1983) Neuronlike adaptive elements that can solve difficult learning control problems, *IEEE Transactions on Systems, Man, and Cybernetics SMC-13*, pp 834-846.

[5] Digney, B.L. (1992) Emergent Intelligence in a Distributed Adaptive Control System, *Ph.D. Thesis* circulated in draft, University of Saskatchewan, Saskatoon, Saskatchewan.

[6] Carpenter, G.A., S. Grossberg, and D., Rosen, (1990) ART2-A: An adaptive resonance Algorithm for rapid category learning and recognition, *IJCNN 1991 Seattle Washington*, Vol II pp 151-156.

[7] Barto, A.G., S.J. Bradtke and S.P. Singh (1991) Real-time Learning and Control Using Asynchronous Dynamic Programming *University of Amherst Technical Report 91 - 57*

Integrating Neural Networks and Knowledge-Based Systems for Intelligent Robotic Control

David A. Handelman, Stephen H. Lane, and Jack J. Gelfand

ABSTRACT: Artificial neural networks and knowledge-based systems offer very different capabilities concerning control system design, implementation, and performance. This paper presents a methodology for integrating both computational paradigms for the purpose of robotic control, patterning the integration after models of human motor skill acquisition. The initial control task chosen to demonstrate the integration technique involves teaching a two-link manipulator how to make a specific type of swing. A three-level task hierarchy is defined consisting of low-level reflexes, reflex modulators, and an execution monitor. The rule-based execution monitor first determines how to make a successful swing using rules alone. It then teaches Cerebellar Model Articulation Controller (CMAC) neural networks how to accomplish the task by having them observe rule-based task execution. Following initial training, the execution monitor continuously evaluates neural network performance and re-engages swing-maneuver rules whenever changes in the manipulator or its operating environment necessitate retraining of the networks. Simulation results show the interaction between rule-based and network-based system components during various phases of training and supervision.

Introduction

Conventional automatic control system design methods involve the construction of a mathematical model describing the dynamic system to be controlled and the application of analytical techniques to this model to derive a control law. These conventional techniques break down, however, when a representative model is difficult to obtain due to uncertainty or sheer complexity, or when the model produced violates the underlying assumptions of the control law synthesis techniques. Biological control systems, on the other hand, are quite successful at dealing with uncertainty, complexity, and strong nonlinearities. They smoothly coordinate many degrees of freedom during the execution of dexterous manipulative tasks within unstructured environments, solving complex planning and redundancy management problems with apparent ease. If intelligent controllers are to emulate these capabilities found in nature, then a great deal can be learned by directly studying the structural, functional, and behavioral aspects of biological systems.

Features of biological control systems include the following: (1) a hierarchical and modular neural network processing architecture, (2) distributed computation among the various levels of the hierarchy, and (3) the utilization of tightly integrated, yet distinct, forms of sensorimotor processing during the acquisition of motor skills. The goal of our research is to improve the performance and adaptability of intelligent control systems by developing efficient processing architectures and learning procedures based on these biological paradigms. This paper focuses on the third feature mentioned, integrated forms of computation that enable robotic skill acquisition.

Features of Human Skill Acquisition

It is well known that humans pass through various levels of competence as motor skills are acquired. Brooks [1] describes motor skill as the optimal use of programmed movements, and motor control as a hierarchy of plans, programs, and subprograms that finally exit into nonlearned automatic ("reflex") adjustments. Fitts and Posner [2] define three phases of skill learning: (1) the Cognitive (Early) Phase, wherein a beginner tries to understand the task; (2) the Associative (Intermediate) Phase, where patterns of response emerge and gross errors are eliminated; and (3) the Autonomous (Final) Phase, when task execution requires little cognitive control. Adams [3] defines two stages of motor learning: (1) the Verbal-Motor Stage, where corrections are based on verbal descriptions of how well the task is being accomplished, and (2) the Motor Stage, where "conscious" behavior eventually becomes "automatic," and attentional mechanisms pick out only those channels of information relevant to learning.

Our goal is to investigate how robots might benefit from similar learning characteristics. We begin by focusing on representations of knowledge that support the behavioral models of motor control and skill acquisition mentioned here. One scheme used to classify memory and learning distinguishes between *declarative* and *reflexive* mechanisms [4]. Motions indicative of declarative memory and learning require conscious effort, are characterized by inference, comparison, and evaluation, and provide insight into not only how something is done, but why. Motions involving reflexive mechanisms relate specific responses to specific stimuli, are automatic, and require little or no thought. Tasks initially learned declaratively often become reflexive through repetition. Conversely, when familiar tasks are attempted in novel situations, reflexive knowledge must be converted back into declarative form to become useful. For example, although one may become adroit at tying one's own necktie, tying someone else's necktie requires some thought due to the change in perspective. This shifting of task-specific knowledge between declarative and reflexive forms plays a fundamental role in skill acquisition, affecting computational resource allocation, the focusing of attention, and the ability to adapt [5], [6].

In Ref. [7], Anderson attempts to model the transition between the cognitive and associative stages of Fitts and Posner as a knowledge compilation process whereby a general declarative form of knowledge is converted into a more specific procedural representation. During the transition from declarative to procedural processing, operational production rules are "chunked," becoming more condensed and task specific. In Anderson's model, the final shift to the most autonomous form of processing is ac-

An early version of this paper was presented at the 1989 IEEE International Conference on Robotics and Automation, Scottsdale, Arizona, May 14–19, 1989. David A. Handelman and Stephen H. Lane are with Robicon Systems, Inc., 301 North Harrison St., Suite 242, Princeton, NJ 08540. Jack J. Gelfand is the Coordinator of Interdisciplinary Research for the Human Information Processing Group, Department of Psychology, Princeton University, Princeton, NJ 08544.

Reprinted from *IEEE Contr. Syst. Mag.*, pp. 77–87, April 1990.

complished by fine-tuning the selectivity of the problem-solving search procedures.

A quasineural model for the development of automatic processing accompanying skill acquisition has been proposed by Schneider [8] and Schneider and Detweiler [9]. In this model, a shift in the control mechanism changes the gating of information vectors between visual, lexical, semantic, and motor processing units. The initial stage of cognitive processing involves direct control of information between units. During practice, associative learning enables direct association of input/output vector pairs, and priority learning determines the transmission strength of these vectors between processing modules. A gradual transition from controlled to automatic processing occurs with practice. In the final and most automatic stage of skill acquisition within this model, communication between processing units is not gated by the controller, but is determined solely by the transmission strengths learned during the priority learning process.

In the model of cognitive skill acquisition presented by Anderson, the representation of knowledge and functional architecture remain constant as the system goes through various stages of learning. Similarly, in Schneider's model, the interconnections and processing modules remain the same, but a change in the control of information flow between units causes the execution to shift from a serial to a more parallel processing mode. Although, in the future, we intend to consider incorporating features of these models into our architecture for intelligent robotic control, this paper concentrates on exploring the properties of a system in which both control and learning shift between two different functional architectures encoding unique representations of a control law.

Integration of Declarative and Reflexive Processing

The utility of distinguishing between declarative and reflexive processing for automatic control was first proposed by Handelman and Stengel [10]. The current approach provides robotic controllers with both forms of processing and studies how their interaction might enable high levels of dexterity and adaptability. From a functional standpoint, we want to give robots the capability to learn, and strategies for learning. Ideally, the process would parallel the training of an athlete by a coach, whereby the robot learns through trial and error how to perfect a task initially specified by a designer in a high-level task language. The analogy to athletics is intentional. In professional baseball, for example, a batter's bat often comes into contact with a pitched ball less than 0.5 sec after the ball is released by the pitcher, leaving little time for computationally intensive sensorimotor processing. Despite variations in pitchers and batter strength due to fatigue and injury, good players maintain adequate task performance. Such skill, adaptability, and efficiency contrast sharply with the first childhood swings of these athletes. We want to endow robots with a similar learning capability.

We implement declarative and reflexive forms of processing using *knowledge-based systems* and *artificial neural networks*, respectively. Knowledge-based systems have demonstrated the ability to encode and exercise expert knowledge within limited domains, providing a hierarchical software organization and explainable solutions [11], [12]. The inferencing capabilities of these systems effectively implement declarative task knowledge. Neural networks exhibit characteristics of associative memory, pattern matching, generalization, and learning by example [13], [14]. Insofar as they directly and efficiently compute specific outputs in response to specific inputs, neural networks can be viewed as implementing reflexive task knowledge.

Knowledge-based systems and neural network techniques have only recently found application in the field of control. Most control applications of knowledge-based systems involve high-level monitoring, diagnosis, or planning [15]–[17], whereas neural networks have been used to learn reflexlike responses and manipulator inverse dynamics and inverse kinematics [18]–[24]. Although hierarchical and modular intelligent control system architectures have been proposed [25], few research efforts, control-oriented and otherwise, have combined both computational paradigms in a single integrated problem-solver [26]–[29].

Knowledge-based systems provide a control system designer with a convenient mechanism for automated complex decision making. Task-specific knowledge may be defined explicitly. Neural networks, on the other hand, encode knowledge implicitly, adjusting internal weights such that their input/output relationships remain consistent with observed training data. Our approach to robotic control is to utilize the strengths of each processing technique where appropriate. In general, we first decompose a control problem into a clearly defined hierarchy of functional tasks (following guidelines such as those proposed by Sadiris [25]). We then classify tasks according to the most appropriate form of knowledge representation: declarative, reflexive, or hybrid (amenable to both representations). Declarative and reflexive tasks are implemented using knowledge-based systems and neural networks, respectively. Hybrid tasks utilize both, with an initial knowledge-based form ultimately yielding to a neural-network-based form.

The focus of this paper concerns the shifting of knowledge and control between declarative and reflexive forms within hybrid tasks. A schematic of our approach is illustrated in Fig. 1, depicting three phases of skill acquisition within a control system that contains both knowledge-based systems and neural networks.

Knowledge-based system components first determine how to accomplish a given control objective using rules and algorithms within the knowledge base. They then teach neural networks how to accomplish the same task by having them observe and generalize on knowledge-based task execution. Following initial training, the neural networks gradually assume control responsibility. Finally, neural network task execution is optimized through reinforcement learning. The knowledge-based system components continuously evaluate neural network performance and reengage rule-based control whenever errors due to changes in the dynamic system or its operating environment necessitate retraining of the networks. The knowledge-based subsystems thereby ensure proper task completion while neural network relearning takes place. The interaction between the rules and the networks during various phases of training and supervision gives the resulting controller unique adaptive capabilities.

Neural networks are intended to form pieces of a learned world model as well. Consider, for example, the issues involved in teaching a robot how to hit a ball successfully with a bat. In this task, there exist causal relationships within and between the robot and its environment that can be addressed independently. The manner in which the ball approaches the robot during the pitch will depend on how and where the ball was released by the pitcher, the shape, size, weight, and texture of the ball, and the prevailing atmospheric conditions. The manner in which the robot's bat approaches the ball during the swing will depend on the characteristics of the bat, including its initial position, and on the robot's strength, agility, and skill. The position where the ball ends up after being hit by the robot's bat depends on similar factors, including the elasticity of the ball. Although analytical dynamic models can be utilized in such circumstances, consider the ease with which artificial neural networks can be made to encode such causality.

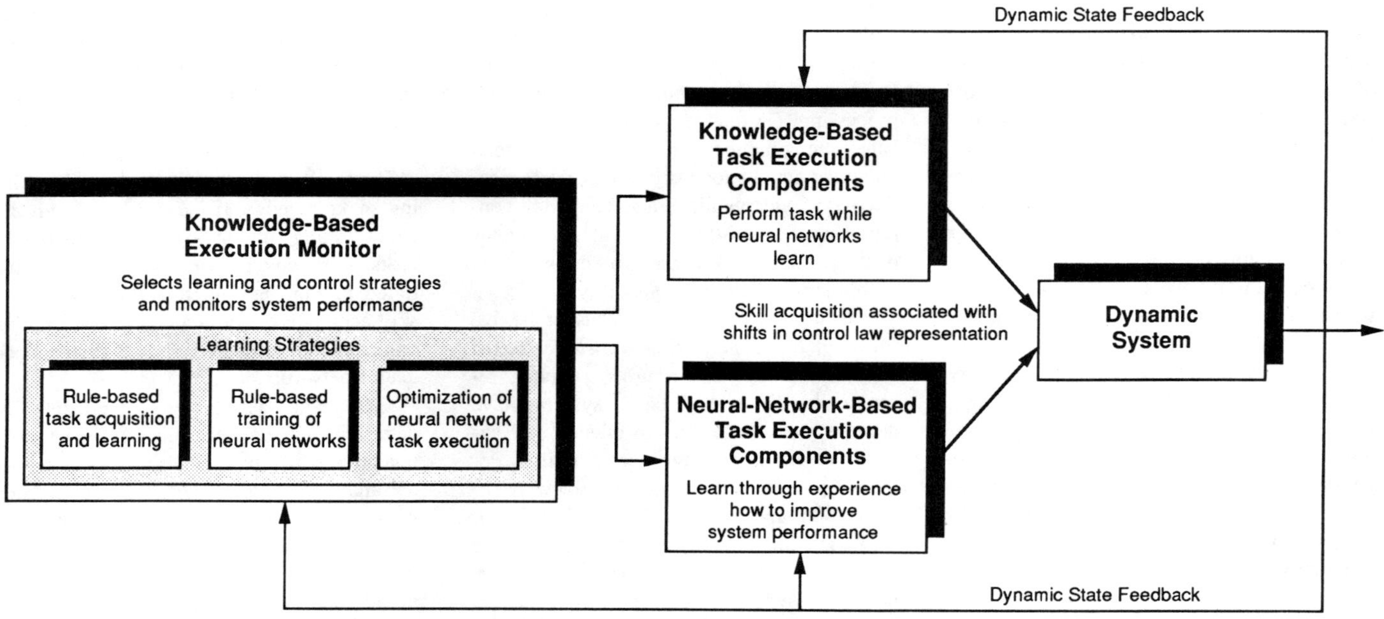

Fig. 1. Integration of knowledge-based systems and neural networks for intelligent robotic control.

By simply watching practice pitches (gathering data), the robot might train a network to predict how the ball will arrive based on how it leaves the pitcher's hand. Similarly, with no ball involved, the robot might handle and swing its bat, building up a network-based model relating time-dependent joint torques to resulting bat trajectories (getting a "feel" for the bat). By piecing together these networks representing the flight of the pitched ball and the flight of the swung bat, the robot should be able to hit the ball. Given a competent learning strategy, the robot could eventually train, through trial and error, a larger network capable of placing the ball at a desired location. If task performance suddenly degrades, individual pieces of the causal puzzle could be examined separately. Did the ball, pitcher, or atmospheric conditions change? Stop swinging the bat and observe more pitches, updating the relevant network if necessary. Did robot strength diminish? Swing the bat and relearn if necessary.

Our approach to control is designed to enable this type of adaptive skill acquisition, with the knowledge-based system components providing the learning strategies, and the neural networks the learning capability. The intent is convenience in design (utilizing knowledge-based programming techniques and applicable algorithms and analytical models), with efficiency in implementation (utilizing the network's convenient function approximation capabilities). Note that, in this presentation of our control approach, we do not attempt to optimize the design or performance of the individual knowledge-based system or neural network components, but use simplified versions of both to highlight the properties of the integrated system. Part of our ongoing research reported elsewhere involves optimization of network-based control performance through reinforcement learning [30], [31].

Rule-Based Supervision of Neural Network Training

The first control task chosen to investigate rule-based supervision of neural network training involves an analogy to one piece of the baseball hitting scenario mentioned earlier, teaching a two-link manipulator how to make a swing. The swing is intended to become part of a higher-level task involving the striking of a ball. The research presented in this paper, however, considers only the swing. The simulated manipulator, shown in Fig. 2, is composed of two cylindrical links, each with a length of 1 m and a radius of 0.1 m. The inner link (located between the shoulder and elbow joints) has a mass of 1 kg, and the outer-link mass varies between 0.5 and 1 kg in the simulations to be described. Rotation of the shoulder and elbow joints occurs in a vertical plane, against gravity. The overall task of the control system is to vary joint torques over time such that the outer link moves from its resting position (pointing straight down) through an imaginary contact point. The outer link is to strike the contact point at the link's midpoint and at a specified angle (slope). A swing is

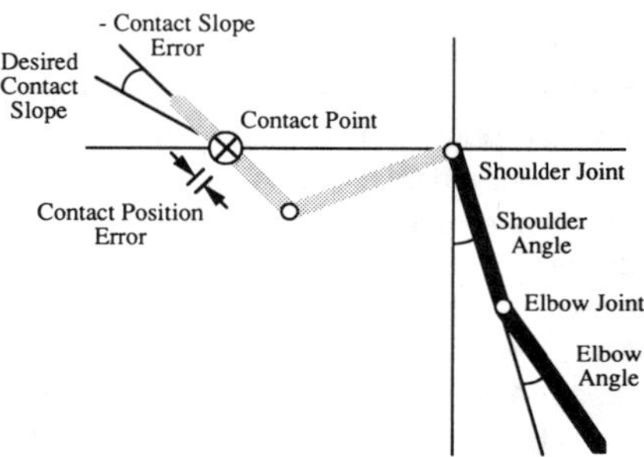

Fig. 2. Simulated planar two-link manipulator.

deemed successful when the two swing-maneuver performance measures, *contact position error* and *contact slope error*, fall within specified bounds.

Figure 3 shows a block diagram of the controller used to accomplish this task. Three levels of a controller task hierarchy are defined: low-level reflexes, reflex modulators, and an execution monitor. On the lowest level (the innermost loop in the block diagram), each joint is endowed with simple reflexes that command torque as a function of angular position and velocity. The term "reflex" is borrowed from human motor control and is meant to represent either a simple position or force servo system. It is believed that many human voluntary motions involve the modulation of reflexes [32], [33]. Consequently, the middle loop of Fig. 3 is used to provide the low-level reflex loops with time-varying gains and commands needed to carry out the swing maneuver. Fi-

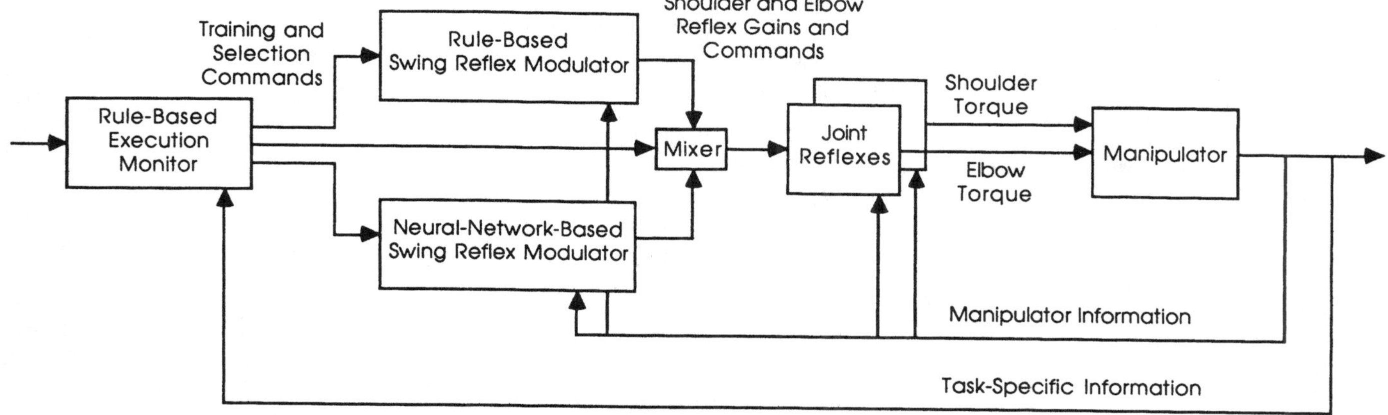

Fig. 3. Block diagram of manipulator control system.

nally, the execution monitor [11] supervises various phases of reflex modulation learning. These levels of the controller hierarchy are described in detail next.

Joint Reflexes

On the lowest level, each joint reflex acts as a hybrid position/torque servomechanism. The two available reflexes are (1) springlike position control with joint angle and angular velocity feedback and (2) constant torque control. The commanded torque for each joint is represented by an equation of the form

$$\tau = k_1\tau_0 - k_2[c_1(\theta - \theta_0) + c_2\dot{\theta}] \quad (1)$$

where k_1 is the *reflex torque* gain, k_2 the *reflex stiffness* gain, τ_0 the *base torque* command, and θ_0 the *reflex angle* command. θ is the joint angle, $\dot{\theta}$ is the joint velocity, and the parameters c_1 and c_2 are used to scale angle and velocity feedback, respectively. Joint motor torque saturation is modeled as a limiting function of ± 25 Nm for the shoulder and ± 15 Nm for the elbow.

Reflex gain and command modulation is used to generate the manipulator swinging motion. Implemented in the middle feedback loop of Fig. 3, modulation is accomplished using either rules, a neural network, or both, depending on the mode of operation.

Table 1
Rule-Based Swing-Maneuver Steps

Swing Step 1

Stiffen shoulder about 20 deg with gain 0.5 and
stiffen elbow about 0 deg with gain 0.1
until shoulder angle >= 10 deg

Swing Step 2

Torque shoulder with gain -1.0
until shoulder angle <= -65 deg

Swing Step 3

Torque shoulder with gain 0.5 and
stiffen elbow about -60 deg with gain 0.2
until shoulder angle >= -45 deg

Swing Step 4

Torque shoulder with gain -1.3 and
stiffen elbow about 0 deg with gain 0.1
until shoulder and elbow velocities are below 50 deg/sec

Swing Step 5

Stiffen shoulder and elbow about 0 deg with gain 1.0
until magnitudes of shoulder and elbow velocities are
below 20 deg/sec

Rule-Based Swing Reflex Modulator

Table 1 characterizes some of the rules used to execute a rule-based swing. The rules are intended to emulate the vertical swinging of a baseball bat and were obtained (by the control system designer) through trial and error. Consequently, they do not depend on knowledge of an explicit dynamic model of the manipulator. The rule-based components of the controller were implemented with a system originally developed as part of a research program in applications of artificial intelligence to fault-tolerant flight control [34], [35]. The same technique used previously for the automatic execution of aircraft emergency procedures is used here to execute a rule-based swing maneuver.

The Princeton Rule-Based Controller embodies a control technique whereby control actions occur as a consequence of *search* through a knowledge base of *parameters*, *rules*, and *procedures*. Parameters, each with an assigned value, collectively represent a partial description of the "state of the world" pertinent to a given control objective. Search implies the attempt to increase the system's knowledge of the world by inferring additional parameter values, most of which are assumed unknown when the search begins. Information expressing relationships and dependencies between parameters is contained in rules, each of which contains a *premise* and an *action*. If a rule premise is true when tested, its action is executed, possibly causing the inference of additional parameter values. Rule testing is guided by an *inference engine*.

Control actions are intended to occur as by-products of search. Buried within the premises and actions of rules are procedures (or calls to procedures) that invoke time-critical control tasks. These procedures are treated as building blocks upon which higher-level control actions are built using rules. In the present case, the low-level joint reflexes of Fig. 3 represent analytical building blocks, and rules are used to provide coordinated reflex modulation.

Reflex modulation is made to occur as a

by-product of search through the rules and parameters. The inference engine executes on the highest level a repetitive cycle involving *knowledge base initialization* and *backward-chaining* (*goal-directed*) *search* on a parameter named SWING_SEARCH_COMPLETED. This is represented as

```
repeat
  initialize_knowledge_
  base;
  determine_value_of
  (SWING_SEARCH_COMPLETED)
until done;
```

Knowledge base initialization assigns the value of UNKNOWN to most parameters. Backward-chaining search on parameter SWING_SEARCH_COMPLETED results in the inference engine testing rules capable of supplying a value for this parameter in their action, such as the following:

```
[RULE_SWING_1
 [PREMISE
    '($OR ($EQ SWING_
           REQUIRED          'FALSE)
           ($EQ SWING_
           STEP_CHECKED
           'TRUE]
 [ACTION
    ($SETQ SWING_SEARCH_
    COMPLETED               'TRUE]]
```

Note that the premise of this LISP-like rule depends on other parameters whose values also are as yet unknown. When tested, the premise first searches for a value for parameter SWING_REQUIRED. A value of FALSE implies that a rule-based swing is not required, and is determined by rules within the execution monitor. Other rules manage swing procedure invocation.

As indicated by Table 1, the rule-based swing maneuver is implemented as a series of steps. Steps are executed one at a time, in order. Cyclic search on parameter SWING_SEARCH_COMPLETED results in cyclic search on parameter SWING_STEP_CHECKED (assuming that SWING_REQUIRED is TRUE). In effect, the control system continually asks the question, "Has the appropriate swing step been checked?" Separate rule groups perform distinct intrastep functions, monitoring the "state" of the swing and quickly performing a piece of the active step if required. Hence, execution of the entire swing maneuver requires many search cycles, some involving nothing more than the monitoring of the swing state.

The result of using a declarative rule-based representation for control actions is an organized and easily maintained software hierarchy. This same technique is used to implement the execution monitor described later.

Network-Based Swing Reflex Modulator

Neural-network-based reflex modulation depicted in Fig. 3 is performed with Cerebellar Model Articulation Controller (CMAC) modules. Originally conceived by Albus [18], the CMAC is a perceptronlike table look-up technique for reproducing nonlinear functions with multiple-input/multiple-output variables over particular regions of the function space. CMAC can learn by example, and its mapping scheme provides automatic generalization (interpolation) between input states, so that similar inputs produce similar outputs. CMAC neural networks were chosen over other network models due to their ability to rapidly learn multidimensional functions on traditional computing architectures. In the present case, CMAC modules are used to encode implicitly, through learning, the sensorimotor relationships represented explicitly by the swing-maneuver rules summarized in Table 1.

Figure 4 shows a functional schematic of a three-input, single-output CMAC module, where the inputs are represented by x_1, x_2, and x_3, and the output is y. To compute an output, the CMAC algorithm quantizes its inputs, generates a list of active memory addresses, and sums the weights located at these addresses. As shown in Fig. 4, it uses a series of mappings to transform input values into output values:

$$q_i = Q(x_i, x_{i_{\min}}, x_{i_{\max}}, q_{i_{\max}}), \quad i = 1, \ldots, m \qquad (2)$$

$$v_{ji} = V(q_i, g, j), \quad j = 1, \ldots, g \qquad (3)$$

$$v_j = \text{concat}(v_{j1}, v_{j2}, \ldots, v_{jm}) \qquad (4)$$

$$p_j = P(v_j) \qquad (5)$$

$$w_j = W[p_j] \qquad (6)$$

$$y = \sum_{j=1}^{g} w_j \qquad (7)$$

First, the values of each input are quantized [Eq. (2)], where m is the number of inputs. For the example in Fig. 4, the mapping Q produces q_1, q_2, and q_3, quantized versions of the three inputs x_1, x_2, and x_3. The resolution of this quantization depends on expected minimum and maximum input values, $x_{i_{\min}}$ and $x_{i_{\max}}$, and on the number of quantization levels, $q_{i_{\max}}$, as shown in Fig. 5. The next mapping, V [Eq. (3)], computes segments of addresses that, when concatenated [Eq. (4)], form virtual weight addresses. In Fig. 4, each of the four virtual addresses shown (v_1, v_2, v_3, and v_4) is composed of three segments, one from each input. The quantization and segment mappings give the CMAC the ability to generalize (produce similar outputs in response to similar inputs). Continuous variations in input values translate into discrete variations in input quantization levels. If an input quantization level changes by 1, only one of its virtual address segments changes by 1. Assume that the value of input x_2 produces quantization $q_2 = 4$. In this case, Fig. 5 shows that the virtual address segments associated with this input would be $v_{12} = 5$, $v_{22} = 6$, $v_{32} = 7$, and $v_{42} = 4$. If q_2 shifts from 4 to 5, all virtual address segments remain the same except v_{42}, which shifts from 4 to 8. Consequently, for the network of Fig. 4, outputs associated with neighboring input quantization levels will have three of four

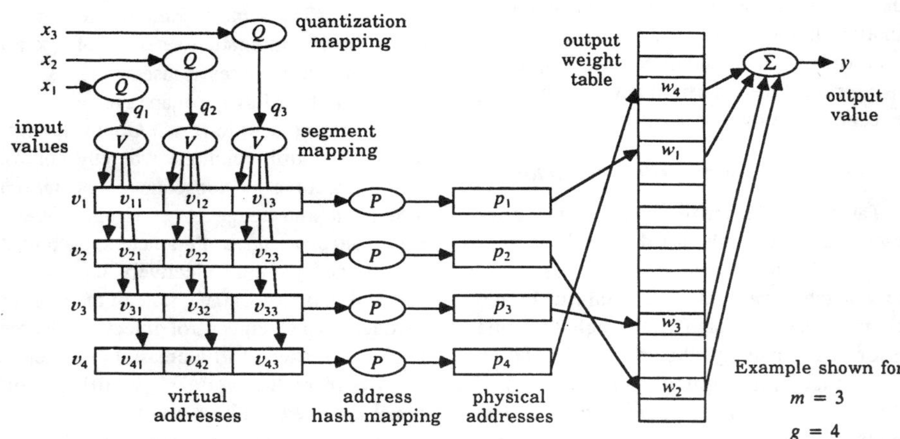

Fig. 4. Functional schematic of Cerebellar Model Articulation Controller (CMAC).

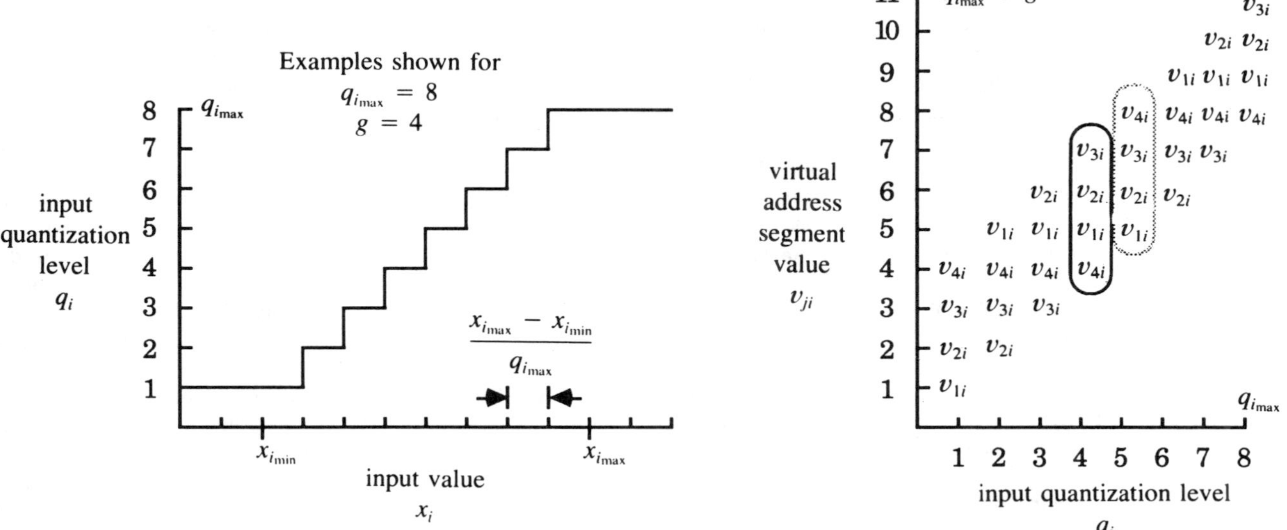

Fig. 5. CMAC input quantization and segment mapping functions.

virtual weight addresses in common because only one address will have changed (assuming the other two input levels remain constant).

The number of weights summed to obtain an output is referred to here as the amount of network generalization, g. For a given input quantization mapping, an increase in g results in an increase in the amount of shared weights between neighboring input/output pairs. An increase in the number of quantization levels, $q_{i_{max}}$, results in higher input resolution, but concurrently increases the size of the virtual address space. To implement CMAC modules using a realistic amount of memory, physical memory addresses are generated by passing the virtual addresses through the many-to-one uniform random mapping P of Eq. (5). As described in Ref. [18], implementation of this mapping using a form of hash coding compresses the huge virtual address space into a compact amount of memory and minimizes the probability of physical address collisions (where two virtual addresses share the same physical address).

The final mappings [Eqs. (6) and (7)] compute the output y by summing the weights w_j located at the physical memory addresses. During training, if the output of the CMAC does not match a desired output for a given input state, the weights pointed to by the physical addresses are updated using the following simple steepest-descent update rule:

$$w_j \leftarrow w_j + \beta \frac{(y_d - y)}{g} \tag{8}$$

In this equation, y_d is the desired network output, y the actual output, and β the learning factor. With $\beta = 1$, y_d can be learned in one iteration. With $\beta < 1$, weights are adjusted to reduce the output error over many iterations. Values of β too close to 1 can produce unstable learning behavior in certain situations [13]. Thus, the state of a CMAC can be characterized by the manner in which its inputs are quantized, the generalization and physical memory used, the learning factor, and the presentation order and content of the training data.

The CMAC processing architecture enables a powerful, straightforward, and efficient form of learning applicable to robotics. CMAC modules are utilized within the manipulator controller of Fig. 3 to implement a reflexive swing maneuver. These modules are depicted in Fig. 6. As used here (and originally by Albus), the CMAC modules implement what is termed a "chained response" in the biological motor control literature. With the CMAC properly trained, input values representing the manipulator state cause the CMAC to output gains and commands to the low-level reflexes that ultimately move the manipulator into the next task-dependent area of the manipulator state space. The resultant manipulator state then defines new input values which, in turn, produce CMAC outputs that again "push" the manipulator into the next desired state. Thus, after the CMAC network has been trained, rules initiate a network-based swing by "throwing" the manipulator into an area of the manipulator state space recognized by the network. The network then reflexively executes the swing maneuver. When the manipulator finally enters a configuration recognized by rules as concluding the swing maneuver, the rules disengage the CMAC network and once again take over reflex modulation responsibility. Training and su-

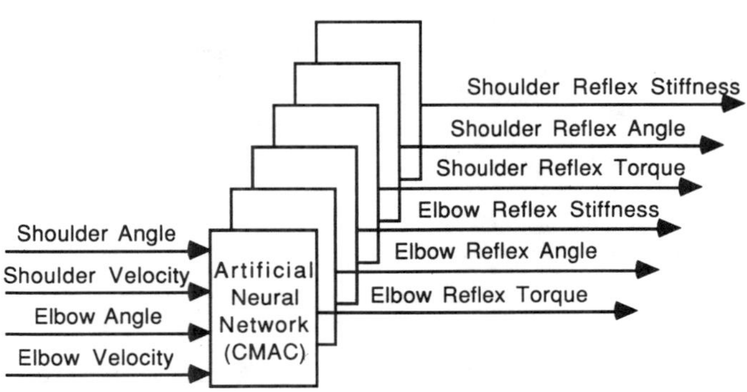

Fig. 6. CMAC modules used for neural-network-based swing.

pervision of CMAC reflex modulation are performed by the execution monitor.

Rule-Based Execution Monitor

The execution monitor is entirely rule-based and orchestrates three *swing modes*: *rule-based*, *rule-and-network-based*, and *network-based*. Under the Rule-Based Swing Mode, proper reflex modulation is acquired and carried out by rules only. Cyclic backward-chaining searches are synchronized with a real-time clock. After an adequate swing has been obtained, the CMAC network modules begin duplicating the rule-based reflex modulation commands, learning by example after each rule-based search cycle. Under the Rule-and-Network-Based Swing Mode, both rule-based and CMAC-based reflex modulation suggestions are generated, and CMAC learning continues. If at any point in time during a swing the CMAC-based outputs are deemed close enough to the rule-based command suggestions, they are fed directly to the reflexes. Finally, under the Network-Based Swing Mode, CMAC learning ceases, and CMAC outputs directly modulate shoulder and elbow joint reflexes throughout the entire swing maneuver. Switching between swing modes is dictated by rules characterized in Table 2.

Simulation Results

Operation of the integrated swing-maneuver controller of Fig. 3 was simulated on a personal computer equipped with a 10-MHz 80286 processor, an 8-MHz 80287 math coprocessor, and 640 KBytes of random-access memory (RAM). Source code for the CMAC modules was written in Pascal, whereas the rule-based system components (including 77 rules) were translated automatically from LISP to Pascal as described in Refs. [34] and [35]. CMAC weights were initialized to zero. Following a discussion of simulation results detailing the operation of the controller, real-time performance issues will be addressed.

Figure 7 depicts the activity of the rule-based system components during various phases of the swing maneuver. Each data point in the top plot corresponds to the number of rules tested by the controller inference engine during a backward-chaining search cycle. Adjacent data points are separated by the amount of time required to complete a search cycle, in this case forced to coincide with a controller sample rate of 20 per second. At the start of the simulation, the manipulator outer-link mass is set to 0.5 kg, and the base torque command is set to 3 Nm.

The controller performs swings at 5-sec intervals. It begins swinging at $t = 5$ sec while in the Rule-Based Swing Mode. Peaks and valleys in the number of rules tested reflect execution of the various swing steps of Table 1. As shown by the plot of swing contact error in Fig. 7, errors associated with the first three swings remain high. The outer link never reaches the contact point because the applied torques do not overcome gravity by a sufficient amount, prompting successive increases in base torque. By $t = 20$ sec, a base torque command of 9 Nm results in acceptable contact error, enabling training of the CMAC neural network modules.

During the 15 swings between $t = 25$ and 100 sec, the rule-based system components teach the CMAC network how to modulate reflexes to accomplish a successful swing. The first five swings are performed in the Rule-Based Swing Mode, with the CMAC network learning, but not contributing to, modulation of the shoulder and elbow joint reflexes. By $t = 50$ sec, CMAC command suggestion errors have dropped low enough for the controller to enter the Rule-and-Network-Based Swing Mode.

Although the manipulator does not come to rest after each swing, it slows considerably. Despite slight variations in initial conditions, the manipulator sweeps out relatively consistent trajectories in joint space in response to rule-based reflex modulation. The Rule-and-Network-Based Swing Mode is designed to admit perturbations off these nominal trajectories by allowing "premature" CMAC suggestions to modulate reflexes when they are deemed close enough to the rule-based suggestions. While in this mode, continued training on rule-based suggestions results in CMAC outputs accounting for an increasing percentage of the 100

Table 2
Sample of Execution Monitor Rules

Rule-Based Swing Mode	Rule-and-Network-Based Swing Mode	Network-Based Swing Mode
If swing mode was rule-based and required swing torque has not been found and adequate contact was not made then adjust swing torque	If swing mode was rule-and-network-based and adequate contact was not made then set swing mode to rule-based and mark required swing torque as not found and stop network training	If swing mode was network-based and adequate contact was not made then set swing mode to rule-and-network-based
If swing mode was rule-based and required swing torque has not been found and adequate contact was made then mark swing torque as found and begin network training	If swing mode was rule-and-network-based and adequate contact was made but network-based command suggestions were not consistent with rule-based commands then make no swing mode change	If swing mode was network-based and adequate contact was made then make no swing mode change
If swing mode was rule-based and required swing torque has been found and adequate contact was made but network-based command suggestions were not consistent with rule-based commands then make no swing mode change	If swing mode was rule-and-network-based and adequate contact was made and network-based command suggestions were consistent with rule-based commands then set swing mode to network-based	
If swing mode was rule-based and required swing torque has been found and adequate contact was made and network-based command suggestions were consistent with rule-based commands then set swing mode to rule-and-network-based		

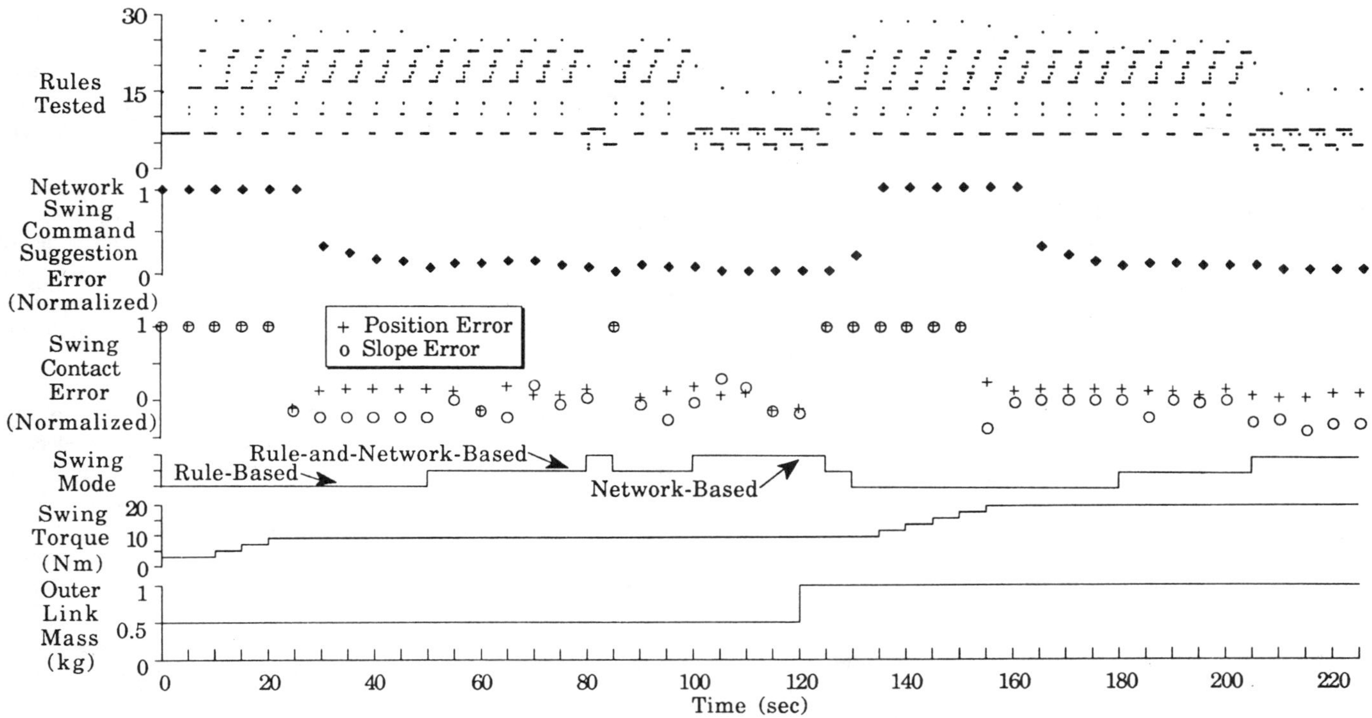

Fig. 7. Rule-based system activity during integrated rule/network-based swing maneuver.

joint torque updates performed during each swing (20 samples/sec × 5 sec/swing).

At $t = 80$ sec, the controller switches to the Network-Based Swing Mode, but the resulting swing falls short of the contact point. Evidently, by wandering into an area of the state space with which the CMAC network had little familiarity, the manipulator received inadequate CMAC reflex modulation suggestions. The controller momentarily switches back to the Rule-and-Network-Based Swing Mode to re-engage network training. Finally, by $t = 100$ sec, four consecutive network-based swings can be accomplished. Figure 8 compares controller performance for swings at $t = 25$, 50, and 100 sec. Note in Fig. 7 that the number of rules tested during network-based swings drops considerably, reflecting a shift from a declarative to a reflexive form of processing.

To demonstrate the inherent adaptability of the integrated controller, the outer-link mass is increased from 0.5 kg to 1.0 kg at $t = 120$ sec. As shown in Fig. 7, insufficient

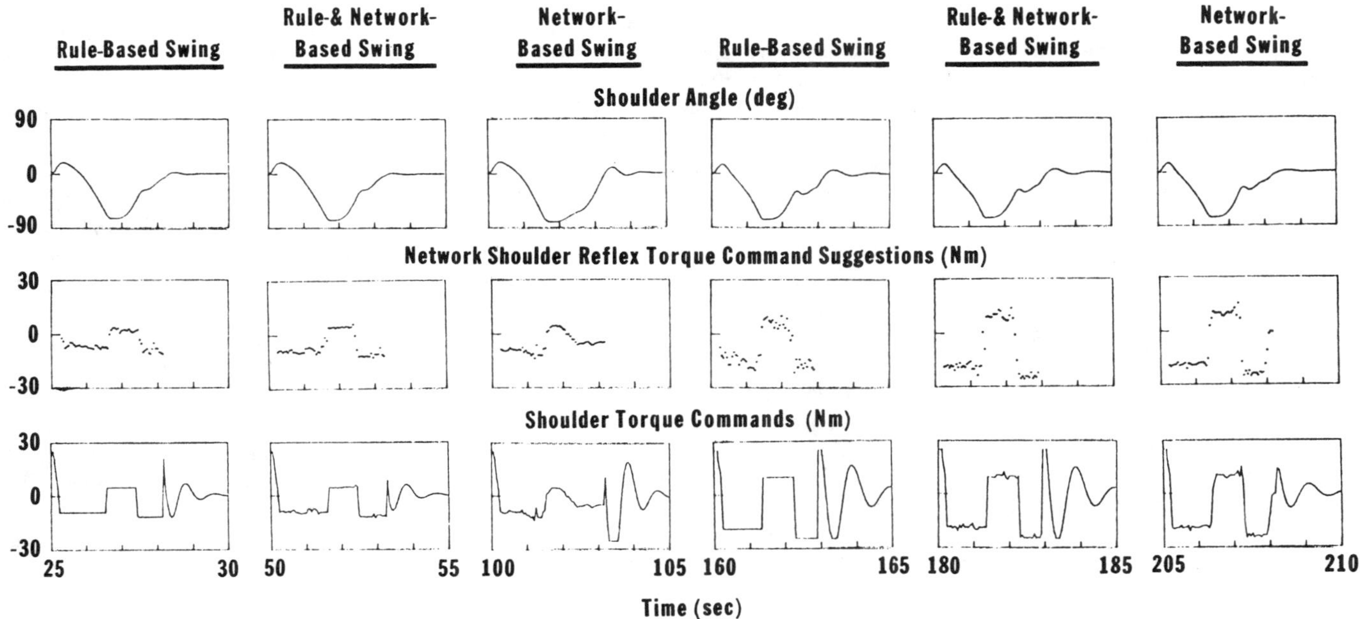

Fig. 8. Comparison of swing performance during various swing modes.

base torque causes the manipulator to miss the contact point over the next six swings. During this time, the swing mode regresses from network-based, to rule-and-network-based, to rule-based (shifting from a reflexive form of processing back to a declarative form), and the base torque command is steadily increased. By t = 155 sec, a base torque of 19 Nm results in a successful rule-based swing. Between t = 160 and 205 sec, the CMAC network relearns by rule-based example how to modulate properly the low-level joint reflex gains. By t = 205 sec, four successful network-based swings once again can be performed. Figure 8 compares controller performance for swings at t = 160, 180, and 205 sec.

Utilizing the personal computer hardware and software described earlier, processing requirements for the various controller operations included the following. While in the Rule-Based Swing Mode with no CMAC learning, a typical controller backward-chaining search cycle testing 23 rules required 0.014 sec. The CMAC network (four inputs, six outputs) required approximately 64 KBytes of memory (RAM). The reading (recall) of a single set of CMAC network outputs required 0.023 sec, and the updating (learning) of a set of CMAC input/output points required 0.047 sec. Although the controller hardware used here cannot support the simulated sample rate of 20 per second, implementation of the integrated controller on one or more state-of-the-art microprocessor boards would achieve real-time performance.

Conclusions

There are a number of problems to be overcome before artificial neural networks become useful components of truly autonomous learning systems. Although neural networks have been shown to be very efficient at learning from experience, in general training data must be supplied by an outside operator who must closely supervise the learning process. In addition, if the system containing the network is required to perform its task during learning, a provision must be made to control its operation during this period. We are developing an automated approach to solving these problems in which knowledge-based systems supervise the training of neural networks and control the operation of the system during the learning process.

Knowledge-based systems and neural networks are being used to emulate shifts between declarative and reflexive mechanisms believed present in biological motor control. The goal is to build robotic systems that can be trained like athletes how to accomplish complex tasks through instruction and practice. In addition to encoding plans and optimized motor control sequences, neural networks also are intended to represent causal pieces of the environment, with the knowledge-based system providing initial rough-cut task execution, network training data, and network optimization and performance monitoring. Although the specific swing-maneuver control problem addressed in this paper can be solved analytically (assuming an accurate dynamic model of the manipulator and its environment), the application in its entirety—the ball-hitting task—was chosen to represent a simplified version of a more complex problem wherein important information about the world may not be available prior to system operation.

The simulation demonstrates one instance in which a neural network could be made to "replace," in a functional sense, the responsibilities of a group of rules within a predominately rule-based control system. Shifts between declarative rule-based processing and reflexive network-based processing within the swing-maneuver controller are traced by variations in the number of rules being tested over time and the errors associated with network outputs during learning. One may ask, however, if rule-based system components can execute a task well, why use neural networks? From a practical point of view, the intent is to replace a computationally intensive form of processing with a more efficient one. Unfortunately, the execution times presented in this paper seem to indicate the opposite effect; the rule-based components required less time per measurement sample than the network components did to perform the same task. This occurred in part because the rule-based controller software utilities (products of previous research) were meticulously optimized, whereas the Cerebellar Model Articulation Controller (CMAC) neural network utilities were not. Highly parallel hardware architectures for CMACs have been proposed by others [21], and their use would vastly improve the network performance. Note that although the rule-based system would benefit from parallel hardware as well, the levels of dependency implied by backward-chaining search through a hierarchical knowledge base imply that parallelization would benefit the CMACs more than the rule-based system used here. Finally, the task chosen in this paper to introduce and demonstrate our declarative/reflexive approach to control was not computationally intensive.

We are presently investigating a more complex problem in robotic skill acquisition involving the satisfaction of multiple goals and constraints for a redundant six-link planar manipulator [31], [36]. The rule-based and algorithmic processing required to perform simultaneous end-point positioning, balance, posture control, and obstacle avoidance are expected to require significantly more computation than the ultimate functionally equivalent neural-network-based representation, with its efficient form of function approximation (a simple summing of weights) and its ability to generalize on relatively sparse trajectory exemplars. We intend to have neural networks functionally replace most of the knowledge-based controller during system operation, and then have them improve on task execution. During the final phase of skill acquisition, the network will optimize its operation (minimizing energy or jerk, time-scaling maneuvers, etc.) through reinforcement learning. In effect, the knowledge-based system will provide the reinforcement learning optimization process with a good "first guess" of the control law, thereby improving its overall rate of convergence.

Important issues related to the integration of knowledge-based systems and neural networks for control still remain and need to be addressed. These include the following:

- How should optimal learning strategies used by the execution monitor be generated?
- Given that a group of neural networks are assigned to perform a complex, multifaceted task, how should the execution monitor knowledge base be set up to optimally allocate knowledge among the neural networks and determine their training order?
- In the preliminary example investigated, control was transferred smoothly between the knowledge-based system and the neural network. What stability problems might arise in more complex systems?
- Can the transfer of knowledge between system components be bidirectional, i.e., can the knowledge-based system infer new rules or modify existing ones based on neural network performance?

Many robotic tasks might benefit from the use of trial-and-error learning despite its consequent failures, such as the preliminary dropping of objects by a pair of dexterous manipulators, or the stumbling of a biped learning how to walk. It is hoped that by coupling multiple representations of knowledge within such learning strategies, robots may eventually tackle manipulative tasks once thought to reside exclusively within the domain of human achievement.

Acknowledgments

The authors acknowledge many helpful discussions with Gregory Clark, Carl Olson, and Paul Thagard of the Princeton University Department of Psychology. This work was supported by the James S. McDonnell Foundation.

References

[1] V. Brooks, *The Neural Basis of Motor Control*, New York: Oxford University Press, p. 12, 1986.

[2] P. Fitts and M. Posner, *Human Performance*, Belmont, CA: Brooks/Cole Pub. Co., p. 8, 1967.

[3] J. Adams, "A Closed-Loop Theory of Motor Learning," *J. Motor Behavior*, vol. 3, no. 2, pp. 111-149, 1971.

[4] I. Kupfermann, "Learning," in E. Kandel and J. Schwartz, Eds., *Principles of Neural Science*, New York: Elsevier, p. 810, 1985.

[5] G. Stelmach and B. Hughes, "Does Motor Skill Automation Require a Theory of Attention?," in R. Magill, Ed., *Memory and Control of Action*, North-Holland, 1983.

[6] W. Schneider and A. Fisk, "Attention Theory and Mechanisms for Skilled Performance," in R. Magill, Ed., *Memory and Control of Action*, North-Holland, 1983.

[7] J. Anderson, "Acquisition of Cognitive Skill," *Psychological Review*, vol. 89, pp. 369-406, 1982.

[8] W. Schneider, "Toward a Model of Attention and the Development of Automatic Processing," in M. Posner and O. Marin, Eds., *Attention and Performance XI*, Hillsdale, NJ: Lawrence Erlbaum Assoc., 1985.

[9] W. Schneider and M. Detweiler, "A Connectionist/Control Architecture for Working Memory," in G. Bauer, Ed., *The Psychology of Learning and Motivation*, vol. 21, New York: Academic Press, 1987.

[10] D. Handelman and R. Stengel, "Rule-Based Mechanisms of Learning for Intelligent Adaptive Flight Control," *Proc. 1988 Amer. Contr. Conf.*, Atlanta, pp. 208-213, June 1988.

[11] E. Charniak and D. McDermott, *Introduction to Artificial Intelligence*, Reading, MA: Addison-Wesley, 1985.

[12] B. Buchanan and E. Shortliffe, *Rule-Based Expert Systems: The MYCIN Experiments of the Stanford Heuristic Programming Project*, Reading, MA: Addison-Wesley, 1984.

[13] D. Rumelhart and J. McClelland, Eds., *Parallel Distributed Processing: Explorations in the Microstructure of Cognition, Volume 1: Foundations*, Cambridge, MA: MIT Press, 1986.

[14] S. Fahlman and G. Hinton, "Connectionist Architectures for Artificial Intelligence," *IEEE Computer*, vol. 20, no. 1, pp. 100-109, Jan. 1987.

[15] J. James and G. Suski, "A Survey of Some Implementations of Knowledge-Based Systems for Real-Time Control," *Proc. 27th Conf. Decision and Control*, Austin, TX, pp. 580-585, Dec. 1988.

[16] D. Handelman and R. Stengel, "Combining Expert System and Analytical Redundancy Concepts for Fault-Tolerant Flight Control," *J. Guid., Contr., Dynam.*, vol. 12, no. 1, pp. 39-45, Jan.-Feb. 1989.

[17] D. Handelman and R. Stengel, "Perspectives on the Use of Rule-Based Control," in M. G. Rodd and G. J. Suski, Eds., *Artificial Intelligence in Real-Time Control: Proceedings of the IFAC Workshop*, New York: Pergamon Press, pp. 27-32, 1989.

[18] J. Albus, "A New Approach to Manipulator Control: The Cerebellar Model Articulation Controller (CMAC)," *J. Dynam. Syst., Meas., Contr.*, vol. 97, pp. 270-277, 1975.

[19] M. Ito, *The Cerebellum and Neural Control*, New York: Raven Press, 1984.

[20] M. Kawato, K. Furukawa, and R. Suzuki, "A Hierarchical Neural-Network Model for Control and Learning of Voluntary Movement," *Biolog. Cybern.*, vol. 57, pp. 169-185, 1987.

[21] W. Miller, III, "Sensor-Based Control of Robotic Manipulators Using a General Learning Algorithm," *IEEE J. Robotics and Automation*, vol. RA-3, no. 2, pp. 157-165, Apr. 1987.

[22] M. Kuperstein, "Generalized Neural Model for Adaptive Sensory-Motor Control of Single Postures," *Proc. 1988 IEEE Intl. Conf. Robotics and Automation*, Philadelphia, pp. 140-143, Apr. 1988.

[23] C. Anderson, "Learning to Control an Inverted Pendulum with Connectionist Networks," *Proc. 1988 Amer. Contr. Conf.*, Atlanta, vol. 3, pp. 2294-2304, June 1988.

[24] S. Lane, D. Handelman, and J. Gelfand, "A Neural Network Computational Map Approach to Reflexive Motor Control," *Proc. IEEE Intl. Symp. Intell. Contr.*, Arlington, VA, Aug. 1988.

[25] G. Saridis, "Intelligent Robotic Control," *IEEE Trans. Automat. Contr.*, vol. AC-28, no. 5, pp. 547-557, May 1983.

[26] W. Hutchison and K. Stephens, "Integration of Distributed and Symbolic Knowledge Representations," *Proc. 1st Intl. Conf. Neural Networks*, San Diego, vol. 2, pp. 395-398, June 1987.

[27] D. Day, "Towards Integrating Automatic and Controlled Problem Solving," *Proc. 1st Intl. Conf. Neural Networks*, San Diego, vol. 2, pp. 661-670, June 1987.

[28] S. Gallant, "Connectionist Expert Systems," *Comm. ACM*, vol. 31, no. 2, pp. 152-169, Feb. 1988.

[29] W. Oliver and W. Schneider, "Using Rules and Task Division to Augment Connectionist Learning," *Proc. 10th Cog. Sci. Soc. Conf.*, Montreal, pp. 55-61, Aug. 1988.

[30] S. Lane, D. Handelman, and J. Gelfand, "Development of Adaptive B-Splines Using CMAC Neural Networks," *Proc. Intl. Joint Conf. Neural Networks*, Washington, DC, June 1989, New York: IEEE Press, 1989.

[31] S. Lane, D. Handelman, and J. Gelfand, "Sensorimotor Skill Acquisition and Reinforcement Learning in Intelligent Robotic Control Systems," submitted to *Biological Cybernetics*.

[32] C. Gallistel, *The Organization of Action: A New Synthesis*, Hillsdale, NJ: Lawrence Erlbaum Assoc., 1980.

[33] R. Stein and C. Capaday, "The Modulation of Human Reflexes During Functional Motor Tasks," *Trends in Neural Science*, vol. 11, no. 7, pp. 328-332, 1988.

[34] D. Handelman and R. Stengel, "An Architecture for Real-Time Rule-Based Control," *Proc. 1987 Amer. Contr. Conf.*, Minneapolis, MN, pp. 1636-1642, June 1987.

[35] D. Handelman, *A Rule-Based Paradigm for Intelligent Adaptive Flight Control*, Ph.D. Dissertation, Mechanical and Aerospace Engineering Dept., Princeton Univ., Princeton, NJ, Apr. 1989.

[36] D. Handelman, S. Lane, and J. Gelfand, "Rule-Based Training of Neural Networks for Goal-Directed Control of Redundant Robots," submitted to *IEEE Trans. Syst., Man, Cybern.*

A Network Model for the Control of the Movement of a Redundant Manipulator

M. Brüwer and H. Cruse

Fachbereich Biologie der Universität, Postfach 8640, D-4800 Bielefeld 1, Federal Republic of Germany

Abstract. In an earlier investigation (Cruse and Brüwer 1987) an algorithmic model was proposed which describes targeting movements of a human arm when restricted to a horizontal plane. As three joints at shoulder, elbow and wrist are allowed to move, the system is redundant. Two models are discussed here which replace this algorithmic model by a network model. Both networks solve the static problem, i.e. they provide the joint angles which the arm has to adopt in order to reach a given point in the workspace. In the first model the position of this point is given in the form of $x-y$ coordinates, the second model obtains this information by means of a retina-like input layer. The second model is expanded by a simple procedure to describe movements from a start to an end point. The results qualitatively correspond to those obtained from human subjects. The advantages of the network models in comparison to the algorithmic model are discussed.

Introduction

The control of the movement of a multijointed manipulator in general includes strongly non-linear operations (Benati et al. 1980). The control algorithm has to cope with additional problems when the manipulator has more joints than necessary for a given task, i.e. when it has extra degrees of freedom. The human arm provides such a redundant manipulator.

To obtain the simplest version of a redundant arm, in earlier experimental investigations (Cruse 1986, Cruse and Brüwer 1987) the number of degrees of freedom of the human arm was reduced to three, namely shoulder, elbow and wrist joint, while allowing the arm to move in a two-dimensional, horizontal plane. The axes of rotation of the three joints were perpendicular to the plane. The position of the endeffector, the tip of a pointer attached to the palm, is determined by two cartesian coordinates in the horizontal plane. Therefore, two joints were sufficient to move the endeffector in the workspace. The existence of the third joint produced an additional degree of freedom and therefore made the system redundant (Fig. 1a). Thus a given point in the two dimensional workspace can be reached by a number of different combinations of joint angles of the manipulator.

The question arises of how the control system selects one of this infinite number of possible positions when trying to reach a given point. To solve this static problem, the following hypothesis was proposed (Cruse 1986). To each joint a cost function is attached which defines a cost value for each joint angle. The cost functions show a minimum at about the middle of the angle range of the joint and the cost values increase to either of the extreme angles. The total cost of a manipulator position is described as the sum of the actual cost values of all joints. When reaching to a given point in the workspace, according to this hypothesis, that manipulator position is selected out of the geometrically possible positions which shows the minimum total cost value. In this way the number of degrees of freedom of the system is reduced and thus the redundancy problem can be solved.

To solve the kinema problem, i.e. the question of how the joints are controlled during the movement from a start to an end or target point, a hypothesis was formulated which represents a compromise between the minimum cost strategy described above and two other strategies (Cruse and Brüwer 1987; Cruse 1989) which will be explained later in more detail. This hypothesis was formulated as an algorithmic model which at least qualitatively behaves like a human subject. This algorithmic model has however some disadvantages (see Discussion) which might be overcome when replacing the algorithmic model by an neuronal network model (Cruse 1989). As a first step

Reprinted with permission from *Biological Cybernetics*, vol. 62, pp. 549–555, 1990.

towards this aim two network models will be presented in this paper which solve the static problem. For the second model in addition a way will be shown as to how the kinematic experiments can be simulated.

Two Network Models Solving the Static Problem

The first model consists of three layers of neurons (Fig. 1b), the input layer, an intermediate "hidden" layer and the output layer. The network models were simulated on a HP 9000/320 using the software of McClelland and Rumelhart (1988). The input layer consists of two neurons the excitation of which corresponds to the value of the cartesian workspace coordinates x and y (see Fig. 1a). The intermediate layer consists of 20 hidden units. The output layer contains three units which give the angle values of the three joints. Only feedforward connections and only connections between neurons of directly neighbouring layers are allowed. The excitation of each neuron was between 0 and 1 with a resting value of 0.5. To train the network we followed the error back propagation rule with a learning rate of 0.05 and a momentum of 0.9 (McClelland and Rumelhart 1983). For one learning cycle (epoch) 32 input patterns were used which correspond to the $x-y$ coordinates of the 32 points in the workspace as shown in Fig. 1a. The corresponding training patterns for the output, i.e. the angle tripels, were taken from the experimental results of one subject which made targeting movements to each of the 32 points (the results were from the same subject as were the data shown in Figs. 2 to 8 in Cruse and Brüwer, 1987).

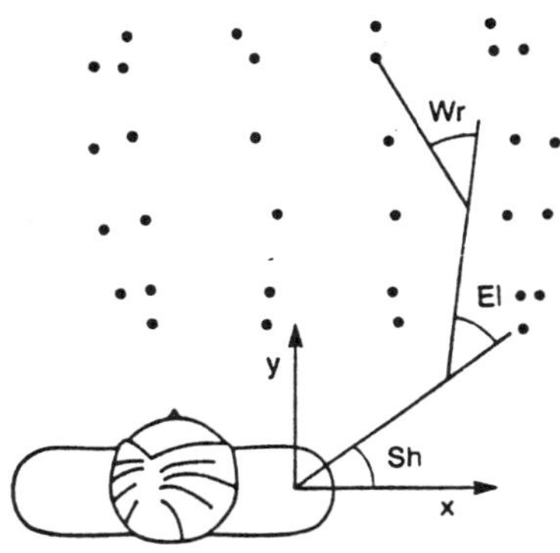

a

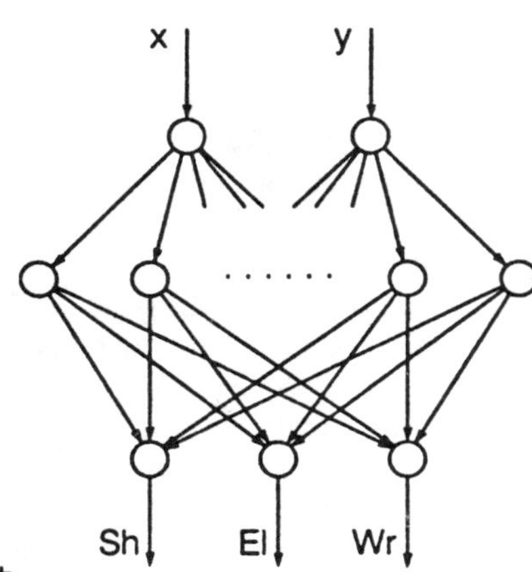

b

Fig. 1. a Top view of the experimental arrangement. The arm is moved in a horizontal plane with the coordinates x and y. The origin is at the shoulder joint. The three joint angles at shoulder (Sh), elbow (El) and wrist (Wr) determine the position of the tip of the arm. The dots show the position of the 32 target points. **b** A simple network consisting of two input units which obtain the coordinate values x and y of the target point, 20 hidden units and three output units which give the angle values of the three joints at shoulder, elbow and wrist

Figure 2a shows the progress of learning the network made during the presentation of 100000 cycles. The mean deviation between angles values calculated by the network and those assumed by the subject was then 1.4 degrees per joint. In Fig. 2b the $x-y$ coordinates of these points are compared with the positions the network actually pointed to. As the network calculates angles and not $x-y$ coordinates, the latter were determined from the angle values by a separate calculation using the corresponding trigonometric formulas. The mean Euclidean distance was 1.1 (± 0.8) cm for the 32 trained points (mean $\pm$S.D.). Using less than 20 hidden units led to slower learning and to larger mistakes.

In order to test the capacity of the network to respond to untrained target points, the response of the network to 117 new target points was calculated. These untrained points lie on a 9×13 grid in the same area as the 32 trained points. As was shown in Fig. 2b for the 32 target points, in Fig. 2c the deviation for the 117 untrained points is shown. The mean Euclidean distance is 0.9 (± 0.5) cm for the untrained points.

In this first model the input values are specified in the form of the $x-y$ coordinates of the target points. However, the position of the target points might also be measured by a "retina-like" sensory system, i.e. a sensory system which locates a point in the workspace not by two coordinates values – for example the cartesian $x-y$ coordinates – but by means of its position within a topologically ordered map. To simulate such a system the following simple network was investigated. The network consists of only two layers, which means that this second model does not contain a hidden layer. As in the first model the output layer consists of three neurons the excitation of which represents the value of the three joint angles. The input layer consists of a two-dimensional array of 10×10 neurons which can be considered as a form of a primitive "retina". A target point in the workspace is

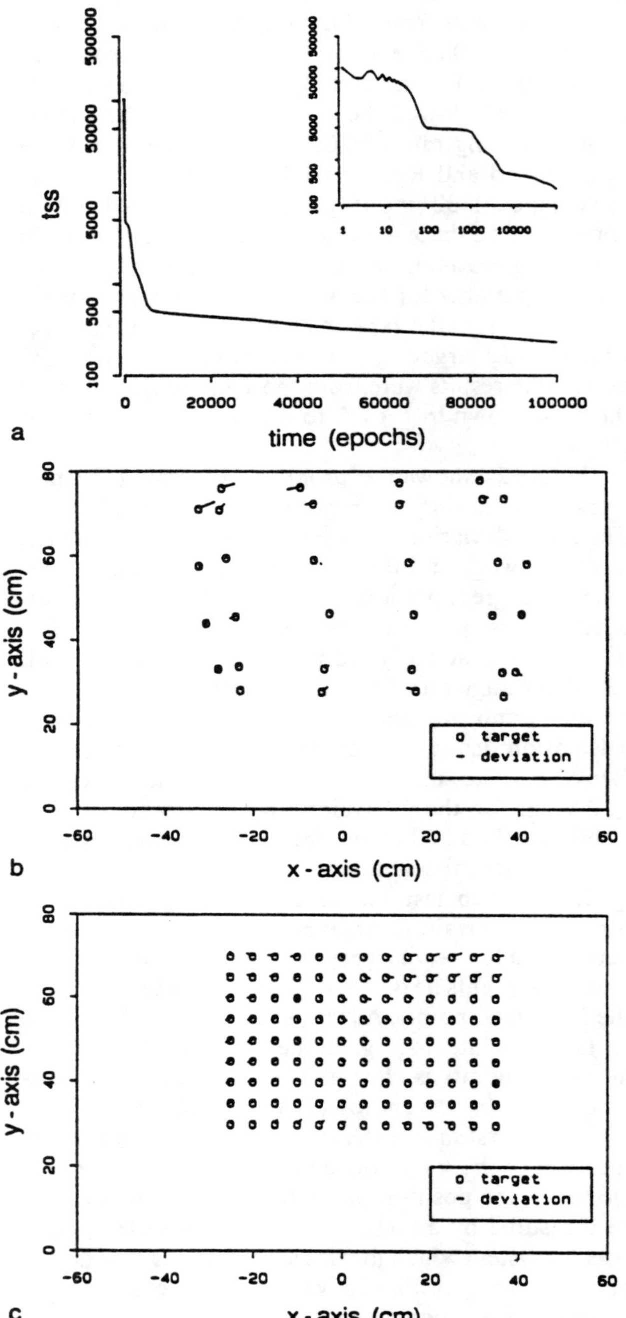

Fig. 2. **a** Learning curve for the network shown in Fig. 1b. Abscissa is the number of learning cycles (epochs). In one epoch each of the 32 target points was presented once. Ordinate is the total sum of squared errors (tss) in one epoch. This error is the deviation of the angle values produced by the network from the angle values obtained during experiments with the human subject. The inset figure shows the same learning curve with a logarithmic abscissa. **b** The mean deviation from each of the 32 target points after 100000 epochs. The circles show the position of the target points in workspace coordinates. The length and direction of the attached bar shows the mean deviation. **c** The deviation for 117 untrained target points shown in the same way

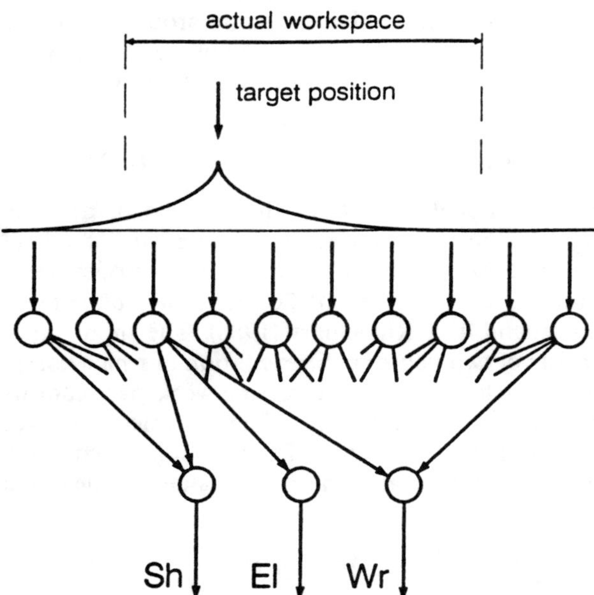

Fig. 3. A network using a two dimensional, "retina-like" input layer. The input consists of a 10×10 units input layer. Only a cross section along the x-axis of the workspace is shown. Around the target position an excitation function with exponential decay is calculated. These values serve as input to the 10×10 layer. All 10×10 units are connected to the three output units which give the angle values of the three joints at shoulder (Sh), elbow (El), and wrist (Wr)

projected onto this layer in a kind of blurred projection. The target point produces an excitation which is highest at the position of the target point and decays exponentially with the Euclidean distance in all directions (Fig. 3). Thus, the input layer obtains a volcano-like excitation function defined on the $x-y$ coordinates of the workspace decreasing with a decay constant of 10 cm. The individual unit of the input layer measures only that value which appears at its coordinate position. In order to avoid cutting parts of this excitation function when the target point is near the margin of the workspace, the application of these "receptive fields" required the introduction of units in the input layer which have no direct corresponding points in the workspace but lie outside the actual workspace. They only serve as acceptors for the outside parts of the excitation functions. Two rows of neurons were used for this purpose. Thus, only the central 6×6 neurons correspond to that range of the workspace within which the 32 target points lie. This means that the projection of the workspace onto the input layer has a rather coarse spatial resolution as the distance between two neurons projected onto the workspace was about 12 cm in the direction of the x-axis and 8 cm in the direction of the y-axis. However, because of the spread excitation function the receptive

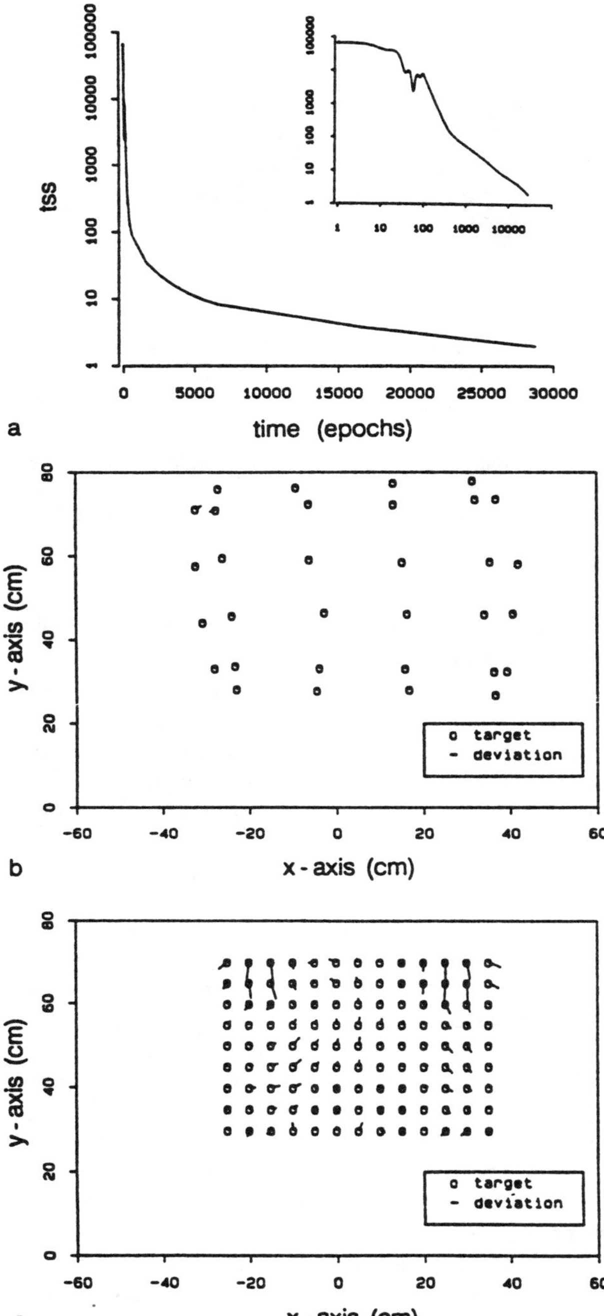

Fig. 4. a Learning curve for the network shown in Fig. 3. Abscissa is the number of learning cycles (epochs). In one epoch each of the 32 target points was presented once. Ordinate is the total sum of squared errors (tss) in one learning cycle. This error is the deviation of the angle values produced by the network from the angle values obtained during experiments with the human subject. The inset figure shows the same learning curve with a logarithmic abscissa. **b** The mean deviation from each of the 32 target points after 29000 epochs. The circles show the position of the target points in workspace coordinates. The length and direction of the attached bar shows the mean deviation. **c** The deviation for 117 untrained target points shown in the same way

fields of the neighbouring units overlap and each input neuron obtained stimuli from a range of an area of about 23×19 cm in the workspace.

The calculation of this excitation function could have been done by using an parallel network. For this purpose an additional preprocessing layer with a fine grid had to be used with feedforward connections to the input layer and fixed weighting factors (Reichardt and Mac Ginitie 1962). However, in order to use as few neurons as possible to save computing time, in this simulation the excitation values were calculated algorithmically and then given as input values onto the 10×10 input layer. Thus by the training procedure only the weighting factors from each of the 10×10 neurons to the 3 output neurons were changed. This was done as was described for the first model. Figure 4a and b shows the learning process and its result in the same way as was done in Fig. 2a, b for the earlier model (learning rate 0.01, momentum 0.98). Correspondingly, Fig. 4c presents the deviation for a set of 117 unlearned target points in $x-y$ coordinates. The mean Euclidean distance was 0.2 (± 0.4) cm for the 32 trained points (Fig. 4b) and 1.4 (± 1.1) cm for the untrained points (Fig. 4c). The higher deviations in the latter case were mainly due to the deviations in the upper right and left corners as shown in Fig. 4c. The mean deviation given in angular degrees was 0.1 deg per joint.

Kinematics

The task of the models described up to now was to provide the angle values for shoulder, elbow and wrist joint so that the arm points to a position arbitrarily given within the workspace, i.e. to solve the static problem. To model the movement from a given start to a given end or target point, the system needs at least the information on the position of these two points. Furthermore the question is open along which path the tip of the arm moves between start and end point. Most earlier authors proposed that in general the tip (endeffector) followed a straight line between both points. This was however shown not to be true in all cases (Atkeson and Hollerbach 1985; Cruse and Brüwer 1987; Flash 1987). If we for a moment leave aside the problem of how the mechanism producing the form of the path is constructed, for a given path a very simple solution for this kinematik problem would be the following. Using the second model, this unknown mechanism could be thought to move an imaginary target point along the path from the start to the end point. For each position of the imaginary target point both models described above could then provide the corresponding angle values.

However, it is not necessary to introduce such a mechanism which moves an imaginary point. Different

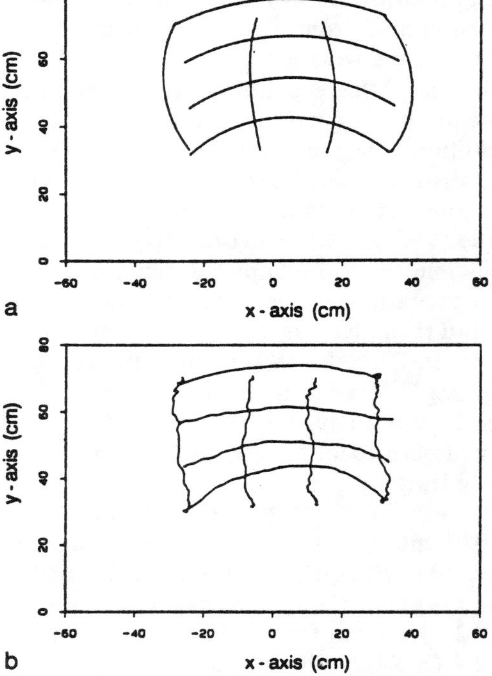

Fig. 5a and b. The path of the tip of the arm in workspace coordinates of eight movements. In four cases the line connecting start and end point are parallel to the x-axis, in the other four cases parallel to the y-axis. Direction of movement is from left to right or from top to bottom. **a** Movement calculated by the network shown in Fig. 3. **b** Movement of the human subject. The results are from that subject, the static data of which were used to train both networks

other solutions are feasible to produce a movement between start and end point without explicitely determining the path. A simple approach using the second model is as follows. To bring the arm to the starting position, the trained network is presented with the coordinates of the start point or, to be more exact, with the corresponding excitation function as was described above for the static situation. Then, to introduce the position of the end point, the excitation function belonging to the end point is also additively given to the input layer. However, at the very beginning the whole excitation function of the end point was multiplied with a scaling factor Fe = 0. This means that the occurrence of the end point does not affect the behaviour of the network. To produce a movement of the arm, this scaling factor Fe was increased while a corresponding scaling factor Fs attached to the excitation function of the start point was decreased following the condition Fs = 1 − Fe. The parameter time is introduced in such a way that the scaling factor Fe is increased in steps of 0.01 and altogether 100 steps were used for the movement from start to end point.

This simple procedure actually produced movements of the arm and some examples are shown in

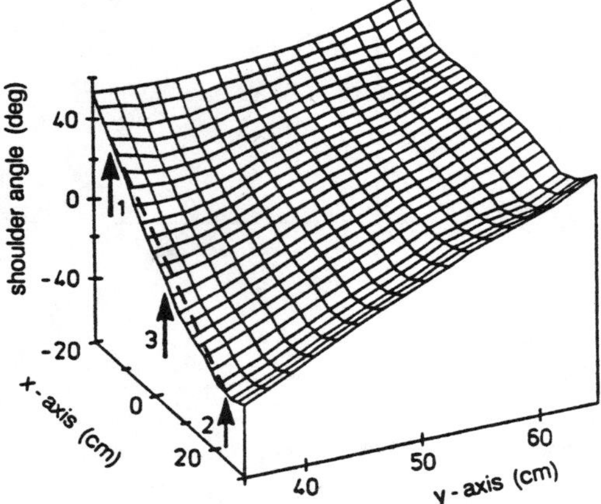

Fig. 6. The values of the shoulder angle (ordinate) calculated by the network shown in Fig. 3 for a grid of 25 × 17 target points plotted over their $x-y$ coordinate values of the workspace. For further explanation see text

Fig. 5a. The results are presented in workspace coordinates and show the movement of the tip of the arm. As can be seen in Fig. 5a the path in general does not follow a straight line. Remarkably these results qualitatively show similar properties as do those of the human subjects. For comparison movements of a human subject are shown in Fig. 5b (see also Cruse and Brüwer 1987). The curvature of the path produced by the network is however somewhat stronger. This is true in particular of the top down paths in Fig. 5.

Why do these curved paths appear? Figure 6 shows the angle values for the shoulder joint produced by the network over the workspace coordinates. As an example let us consider the movement from the start point to the end point which in Fig. 6 are marked by 1 and 2, respectively. This corresponds to a movement parallel to the x-axis and at small y values ($y = 35$ cm, see also Fig. 5). When during the movement from point 1 to point 2 both points are excited equally (Fs = Fe), this should result in a position at about the middle between the points 1 and 2, which is marked as point 3 in Fig. 6. However, as a comparison with the interrupted line connecting both values qualitatively shows, the actual angle value obtained in the static situation is smaller than the mean value of point 1 and point 2 which are given into the network. This means that the movement deviates to higher values of shoulder angle which produces curves paths in the way shown in Fig. 5a. A similar effect is found when considering the angles of elbow and wrist joint.

Discussion

Models on the basis of neuronal networks for the control of a manipulator have already been presented by several authors. Kohonen (1982a) and Ritter and Schulten (1987) and Ritter et al. (1989) provide systems which are based on the Kohonen algorithm (Kohonen 1982b). Another model given by Josin (1988) uses the error back propagation procedure of McClelland and Rumelhart (1988). All these models concern the non redundant arm. Here two simple models are proposed which are able to solve the redundancy problem for the static situation. The first model obtains $x-y$ coordinate values as input as does the model of Josin (1988). The second model obtains the input information via a retina-like input layer which is similar to the approach of Kohonen (1982a, b) and Ritter and Schulten (1987).

The redundancy problem was solved by an earlier algorithmic model by the application of cost functions attached to each joint (see Introduction). Both models proposed here solve the problem on the basis of distributed networks. This occurs in such a way that during the training session the weighting factors in the network connections were learned. However, no obvious explicit representation of the cost functions was found in the values of the weighting factors. The network was trained using the data obtained from a human subject and thus used an external teacher. However, as the data of this subject could be well described by three cost functions, the network could have also be trained with practically the same result by implementing these cost functions into the training program.

The second network model was extended by a simple procedure to be able to describe movements of the redundant arm from a start to an end point within the workspace. The system showed curved paths qualitatively similar to those found in humans. In the algorithmic model (Cruse and Brüwer 1987) the curved paths were produced by superimposing to the minimum cost strategy a second strategy called mass-spring strategy. For this strategy the angles of the end point have to be calculated before the movement starts and each joint moves independently of the other joints to its final angle value. The extension proposed for the network model follows a similar logic as by means of the excitation function the final angles are also determined. However, it was not immediately clear to us that the system shows these similarities also during the movement.

The network in its actual form cannot deal with three properties found in the human experiments. First, one type of experiment performed by Cruse and Brüwer (1987) was to start the movement with the arm in an "uncomfortable" position, i.e. a position which does not obey the minimum cost principle. These experiments cannot be modelled by this network. For this purpose the network had to be enlarged by input units which monitor the actual value of the three joint angles. Second, our earlier algorithmic model in addition to the minimum cost strategy and the mass-spring strategy contains a third strategy which also is not yet implemented in the network model. This so-called pseudoinverse control has the effect that during the movement the incremental changes of the three angles dS, dE, and dW obey the rule that the sum $dS^2+dE^2+dW^2$ assumes a minimum. This strategy is also important when modelling movements starting with uncomfortable arm positions. Third, both the experimental investigations and the algorithmic model showed the following qualitative property which was not found in the network. The actual movements show some kind of hysteresis so that the exact form of the path between two points depends on the direction of the movement, i.e. which of the points is the start and which is the end point. In the algorithmic model this property arises from the implementation of the pseudoinverse control.

However, two disadvantages of the algorithmic model are not found in the network model. In the algorithmic model the necessary linearisation required an iterative method of calculation. Higher exactness then requires smaller iteration steps and thus larger computing time. In the parallel network computing time is extremely short when implemented on a real parallel system. Higher exactness is only a question of the resolution of the individual units or could be increased by increasing the number of units but does not necessarily influence the computing time. Another disadvantage of the algorithmic model is the appearence of so-called singularities. As described by Cruse (1989) a singularity occurs in the algorithmic model when the wrist angle obtains a value of zero. In this case it is not possible for the model to calculate the incremental angle values for the subsequent movement step. For the network model this problem does not exist because the angle values are not calculated but represented in the combination of weighting factors of the different synapses and thus correspond to a kind of distributed look-up table. This property is also of practical importance for the control of artificial manipulators as these also have to deal with the problem of singularities.

Finally a more technical problem concerning the second network model should be mentioned. As described above, the grid of the input layer is rather coarse. Nevertheless, because of the intensive overlapping of the receptive fields – which can be considered as a sort of spatial low pass filter (v. Seelen 1968) – the network is able to interpolate and to develop a continuous projection (Fig. 6) (see also Baldi and Heiligenberg 1988). In addition, during training because of this overlap a stimulus also influences neighbouring units which therefore has a similar effect as has the Kohonen algorithm (1982b).

Just after we finished this work, a paper appeared by Massone and Bizzi (1989) in which the control of a three joint arm moving in a horizontal plane is simulated by means of a three layered network. This model differs from our second model in the following respects. Whereas our model simulates the results of human experiments, the model of Massone and Bizzi adopts a control algorithm proposed by Mussa Ivaldi et al. (1988). It provides a very elegant method to learn and control movements of a redundant manipulator whereby the movements show a bell-shaped velocity profile. As presented, the model can perform movements in the horizontal plane in all directions but all of them start from the same arm position. In contrast, in our model each point of the workspace can be used as start or end point. However, no effort was made to produce a bell-shaped velocity profile in our model nor was the movement simulated at the level of individual muscles as was the case in the model of Massone and Bizzi (1989). Although our model uses only two layers – there is no hidden layer in the second model – it provides a finer spatial resolution which is not limited to a given grid of pixels. Another difference is that no explicit formulation of the path of the endeffector is used, whereas this path was part of the training procedure in the model of Massone and Bizzi. As was described above, our model in its present form is not able to simulate movements starting from an uncomfortable arm position. As the model of Massone and Bizzi contains feedback from the output values to an upper layer, it contains the basic architecture necessary to solve this problem. Therefore only minor changes of the latter model might be necessary to also fulfill the requirements which are needed to simulate movements starting from uncomfortable positions.

References

Atkeson CG, Hollerbach JM (1985) Kinematic features of unrestrained vertical arm movements. J Neurosci 5:2318–2330

Baldi P, Heiligenberg W (1988) How sensory maps could enhance resolution through ordered arrangements of broadly tuned receivers. Biol Cybern 59:313–318

Benati M, Gaglia S, Morasso P, Tagliasco V, Zaccaria R (1980) Anthropomorphic robotics. I. Representing mechanical complexity. Biol Cybern 38:125–140

Cruse H (1986) Constraints for joint angle control of the human arm. Biol Cybern 54:125–132

Cruse H (1989) The control of path and joint angles in a human arm. In: Personnaz L, Dreyfus G (eds) Neural networks from models to applications. 71–77. IDSET, Paris

Cruse H, Brüwer M (1987) The human arm as a redundant manipulator: the control of path and joint angles. Biol Cybern 57:137–144

Flash T (1987) The control of hand equilibrium trajectories in multi-joint arm movements. Biol Cybern 57:257–274

Josin G (1988) Neural-space generalization of a topological transformation. Biol Cybern 59:283–290

Kohonen T (1982a) Clustering, taxonomy, and topological maps of patterns. Proceedings of the Sixth International Conference on Pattern Recognition. IEEE Computer Society Press, Silver Springs, pp 114–128

Kohonen T (1982b) Self-organized formation of topologically correct feature maps. Biol Cybern 43:59–69

Massone L, Bizzi E (1989) A neural network for limb trajectory formation. Biol Cybern 61:417–425

McClelland JL, Rumelhart DE (1988) Explorations in parallel distributed processing. MIT Press, Cambridge, Mass

Mussa Ivaldi FA, Morasso P, Zaccaria R (1988) Kinematic networks – a distributed model for representing and regularizing motor redundancy. Biol Cybern 60:1–16

Reichardt W, Mac Ginitie G (1962) Zur Theorie der lateralen Inhibition. Kybernetik 1:155–165

Ritter HJ, Schulten KJ (1987) Extending Kohonen's self-organizing mapping algorithm to learn ballistic movements. In: Eckmiller R, von der Malsburg C (eds) Neural computers. Springer, Berlin Heidelberg New York, pp 393–406

Ritter HJ, Martinez TM, Schulten KJ (1989) Topology-conserving maps for learning visuo-motor coordination. Neural Net 2:159–168

Seelen W von (1968) Informationsverarbeitung in homogenen Netzen von Neuronenmodellen. Kybernetik 5:133–148

Received: October 26, 1989
Accepted in revised form: December 11, 1989

Prof. Dr. Holk Cruse
Fachbereich Biologie
Universität Bielefeld
Postfach 8640
D-4800 Bielefeld 1
Federal Republic of Germany

A Neural Network Regulator for Turbogenerators

Q. H. Wu, B. W. Hogg, and G. W. Irwin

***Abstract*—This paper presents a neural network (NN) based regulator for nonlinear, multivariable turbogenerator control. A hierarchical architecture of an NN is proposed for regulator design, consisting of two subnetworks, which are used for input–output (I–O) mapping and control respectively, based on the back-propagation (BP) algorithm. The regulator has the flexibility for accepting more sensory information to cater for multi-input, multioutput systems. Its operation does not require a reference model or inverse system model and it can produce more acceptable control signals than are obtained by using sign of plant errors during training. I–O mapping of turbogenerator systems using NN's has been investigated and the regulator has been implemented on a complex turbogenerator system model. Simulation results show satisfactory control performance and illustrate the potential of the NN regulator in comparison with an existing adaptive controller.**

I. Introduction

BOTH the theory and the practice of adaptive control have been developed over the past three decades [1]. Most adaptive control algorithms are based on linear models of the plant, so that classical techniques of system analysis and synthesis can be used in practical applications. However, industrial systems are nonlinear; consequently, unmodeled dynamics and robustness problems arise in practical applications, and supervisory control is required [1], [2].

Improved learning algorithms, coupled with advances in microelectronics, have stimulated considerable renewed interest in neural networks (NN's) across a spectrum of research areas. For control engineering, NN's are attractive in several aspects. They possess the capacity for nonlinear plant modeling, can handle large amounts of sensory information, perform collective processing and learning, and offer the potential for highly parallel computation.

There has been some research on using NN's for the control of dynamic systems. Psaltis *et al.* [3] studied a number of interesting techniques for using back-propagation networks to control plants. Saerens and Soquet [4] presented a neural controller without a specific learning stage. Nguyen and Widrow [5] applied NN's to self-learning of backing a trailer truck into a loading dock. An NN forward model was used by Jordan and Jacobs [6] for system learning and control. The least-squares method was mixed with the back-propagation (BP) algorithm by Sbarbaro and Gawthrop [7] to enhance the ability of the algorithm to learn complex mappings under stochastic conditions; the method is used with an inverse model for robot control. Chen [8] developed an NN-based self-tuning controller with a traditional model structure. Identification and control of dynamic systems using NN's were expounded by Narendra and Parthasarathy [9], who outlined their applications in MRAC systems. However, there is an absence of reported work on practical applications.

An approach to the design of an NN regulator for turbogenerator system control is presented in this paper. A turbogenerator is a highly nonlinear, fast-acting multivariable system with a wide range of time constants in different loops. Interconnected in a power system, turbogenerators operate over a wide range of operating conditions, from lagging to leading phase, and are subjected to different types of disturbances, for example changes in terminal voltage and tap positions of transformers, adjustments of operating conditions by changing inputs of governor and excitation systems, and short circuits on transmission lines. The characteristics of turbogenerators vary as conditions change, but the outputs have to be coordinated so as to satisfy the requirements of power system operation. Most turbogenerator adaptive controllers [10] are designed on the basis of a linear model and are subject to the problem of unequal numbers of system inputs and outputs. This complicates the design of adaptive controllers and degrades practical operation and robustness of designed controllers. Use of NN's may deal with these difficulties.

The NN regulator for turbogenerators introduced in this paper avoids the use of reference model, as discussed in [9], owing to difficulties in choosing a reference model for such a complex system. The use of an inverse model, as employed in [7], is impossible because of the high-gain loops between the turbogenerator and actuators. Since the turbogenerator is a multivariable system, the NN-based traditional model structure, studied in [8], is not suitable. To satisfy the requirements of turbogenerator control, a hierarchical architecture is proposed for NN regulator design. A similar approach was used in [5] and [6]. The NN regulator consists of two subnetworks, which are used for I–O mapping and control respectively. This solves the problem of using the signs of plant errors during the back-propagation procedure, which is used in [3] and [4]. The paper reports simulation studies on the modeling and adaptive control of a turbogenerator system. The performance of the NN regulator under different operating conditions and varying disturbances is investigated and comparisons are made with an existing adaptive controller.

II. Neural Network Regulator

A turbogenerator consists of a synchronous generator, an exciter with an automatic voltage regulator, and a turbine and

Manuscript received August 20, 1990; revised May 13, 1991.
The authors are with the Department of Electrical Engineering, Queen's University of Belfast, Belfast BT9 5AH, United Kingdom.
IEEE Log Number 9102399.

Reprinted from IEEE Trans. on *Neural Networks*, vol. 3, no. 1, pp. 95–100, January 1992.

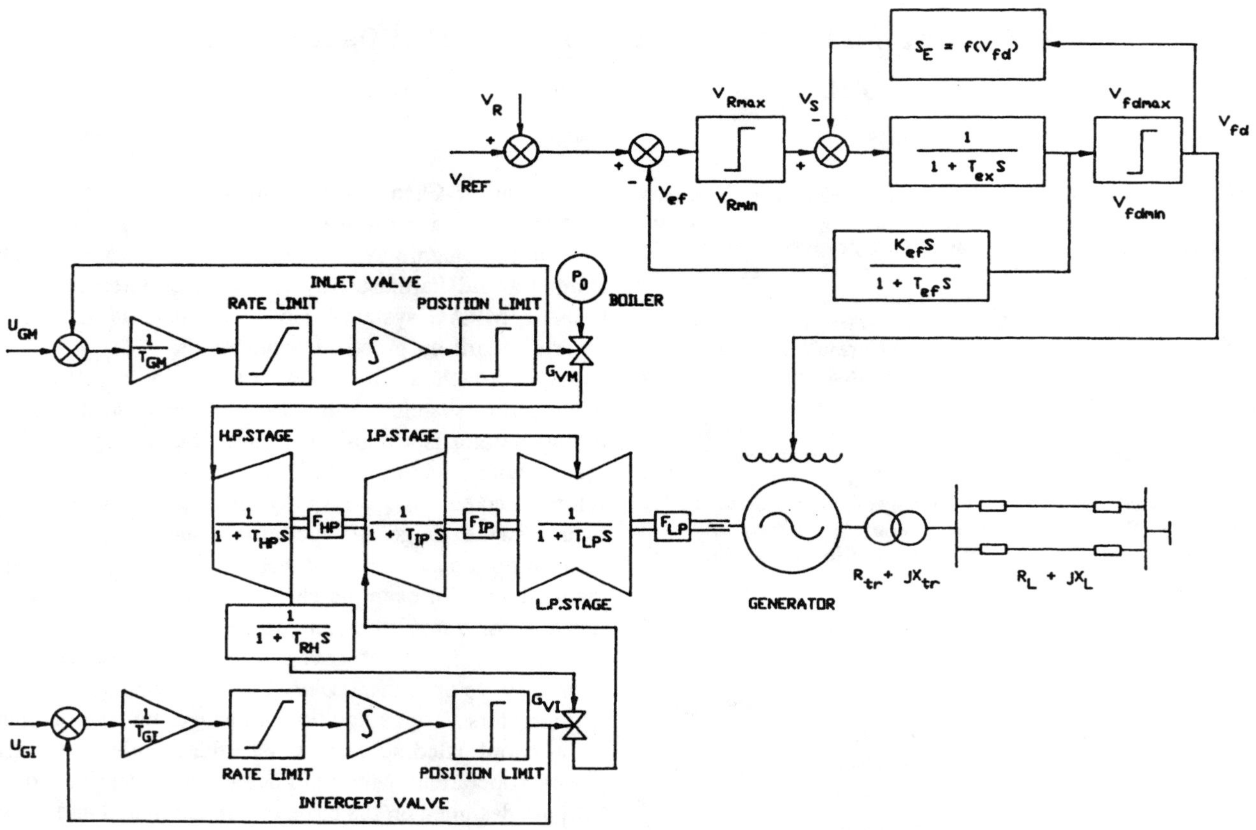

Fig. 1. A turbogenerator unit connected to a power system.

governor system. The synchronous generator is represented by seventh-order nonlinear differential equations. The excitation system is described as a first-order system to represent a static exciter, with a small time constant and limits on the positive and negative ceilings of the field voltage. A three-stage turbine with reheat is employed to drive the synchronous generator. The steam is produced by a conventional coal-fired or oil-fired boiler, which is assumed to be a constant steam source. An outline of a turbogenerator connected to infinite bus power system is illustrated in Fig. 1. The simulation and parameters of the synchronous generator, exciter, turbine, and governor systems are presented in [10]. Further information on such models may be found in work by Hogg [11].

For digital controller design, the turbogenerator may be described as a discrete-time nonlinear system as follows:

$$Y(t+k) = F[Y(t), Y(t-1), \cdots, Y(t-p), U(t), U(t-1), \cdots, U(t-q)]. \quad (1)$$

Here $Y(t) \in R^n$ is the output vector, $U(t) \in R^m$ is the input vector, k represents the system time delay, and p and q are the orders of the time series $\{Y(t)\}$ and $\{U(t)\}$. Feedback control may be defined as

$$U(t) = G[Y(t), Y(t-1), \cdots, Y(t-s)]. \quad (2)$$

This is chosen to minimize the cost function:

$$J_C = \frac{1}{2} \sum_j r_j [y_j(t+k) - d_j(t+k)]^2 \quad (3)$$

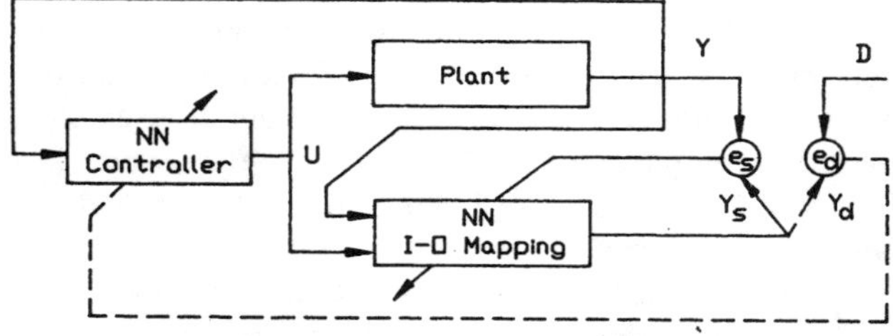

Fig. 2. An NN regulator for adaptive control.

where y_j and d_j denote the jth elements of the system output, $Y(t)$, and the desired track, $D(t)$, respectively. The penalty coefficient, r_j, is applied on the jth output so as to coordinate the performance of the system outputs.

Usually, identification techniques are employed to find the system function, F, and the control law, G, is determined on the basis of stability considerations. However, F and G are usually chosen to be linear; otherwise classical methods of system analysis and synthesis cannot be easily applied. Here the nonlinear, vector mapping functions F and G are represented using multilayered NN's without restriction on their structures. The BP algorithm [12] is employed for NN training, and such training occurs on-line. The result is the integrated NN illustration in Fig. 2.

The NN regulator consists of two subnetworks: an NN mapper (NNM) and an NN controller (NNC). The former is for I–O mapping while the latter performs as an adaptive

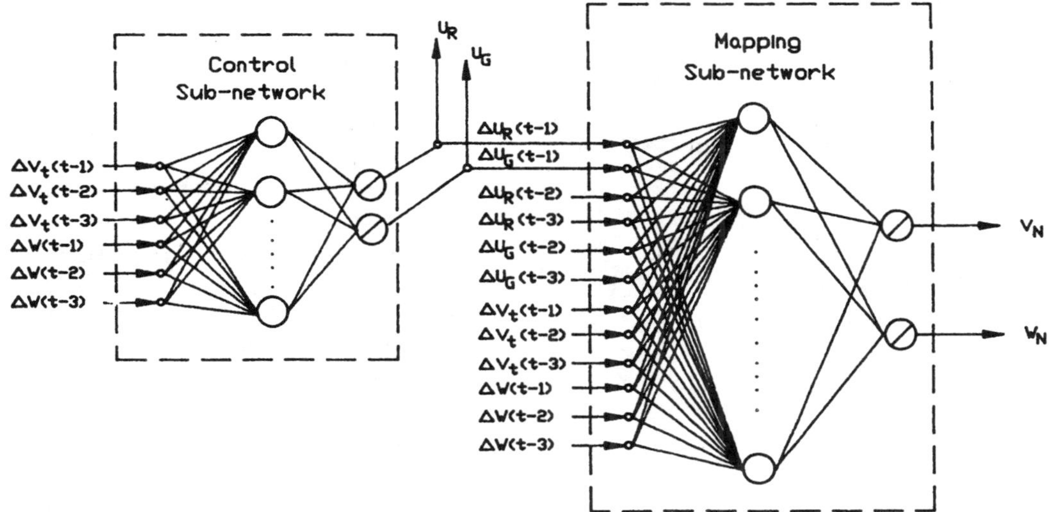

Fig. 3. A hierarchical architecture of an NN regulator for multivariable turbogenerator control.

controller. The two subnetworks operate with different time schedules during a sample interval. At the first time instant, the NNM is employed, using errors between the plant and the network outputs, denoted by e_s in Fig. 2. The performance index

$$J_M = \frac{1}{2}\sum_j r_j[y_j(t) - \hat{y}_j(t)]^2 \tag{4}$$

is used in training, where $\hat{y}_j$ is the jth output of the subnetwork. The weights in the NNM are updated as follows:

$$W_M(t) = W_M(t-1) - \lambda\nabla E_M(W_M) \tag{5}$$

where the subscript M refers to the NNM, and $W_M(t)$ is the weight vector consisting of all weights in the subnetwork at time t. The quantity $\nabla E_m(W_M)$ is the gradient vector, consisting of the derivatives of J_M with respect to each weight in the NNM, and λ is the step length for weight updating.

At the second time instant, the weights in the NNC are modified as adaptive parameters, according to errors e_d between the desired system outputs and the outputs of the NNM, as shown in Fig. 2. The updating of the weights in the NNC is achieved as follows:

$$W_C(t) = W_C(t-1) - \lambda\nabla E_C(W_C) \tag{6}$$

where the subscript C refers to the NNC, and $W_C(t)$ is the weight vector consisting of all weights in the subnetwork NNC at time t. At this stage the NNM functions as a channel for error back-propagation and its weights are unchanged. The outputs of the control subnetwork are sent to control the plant and are also taken as inputs to the NNM for the next mapping stage. The weights in the NNC are updated to minimize the cost function in (3), with $\hat{y}$ used instead of y for two reasons. Future information is involved in the cost function, and also the updating of controller parameters with the gradient vector is directly related to the outputs of the NNM. After proper training of the NNM, the error between $\hat{y}$ and y tends to zero, as will be seen in the next section.

To control the major outputs of the turbogenerator, the terminal voltage, V_t, and the generator speed, ω, are measured on-line and used as feedback variables to yield the excitation control signal, V_R, and the governor input, U_G. In (1) and (2), the system inputs and outputs are then chosen as

$$Y(t) = [\Delta V_t(t), \Delta\omega(t)]^T \qquad U(t) = [\Delta V_R(t), \Delta U_G(t)]^T$$

where $\Delta V_t(t)$ is the deviation from the setpoint and $\Delta\omega$ is the deviation from synchronous speed. The quantities $\Delta V_R(t)$ and $\Delta U_G(t)$ are the changes in excitation and governor inputs respectively.

Fig. 3 illustrates the NN regulator for turbogenerator control. The orders p, q, and s in (1) and (2) are all chosen to be 2 based on experience with the identification of turbogenerator ARMA models [10]. Empirical observations show that higher orders will not improve system dynamic performance. The neurons in the hidden layers were assigned nonlinear functions, those producing the control signals $\Delta V_R(t)$ and $\Delta U_G(t)$ being linear. The mapping and control subnetworks are connected in series and trained hierarchically by the BP algorithm, as discussed earlier. The results of simulation studies on I–O mapping and adaptive control are reported in the next two sections.

III. Turbogenerator I–O Mapping Results

In this section attention is focused on the NNM subnetwork in Fig. 3. A 12–14–2 network was employed with the nonlinear function $y = (1-e^{-x})/(1+e^{-x})$ in the hidden units, and the time delay, k, is chosen to be 1. Two PRBS's were employed to excite the simulated turbogenerator system [10] with amplitudes set at 30% and 20% of the given inputs of the excitation and governor systems respectively. Note that with these levels of perturbation, linear modeling is not appropriate.

Samples of the system outputs, taken every 20 ms, were used to train the NNM using back-propagation errors for weight adjustment, with λ in (5) set at 0.2. The weights were

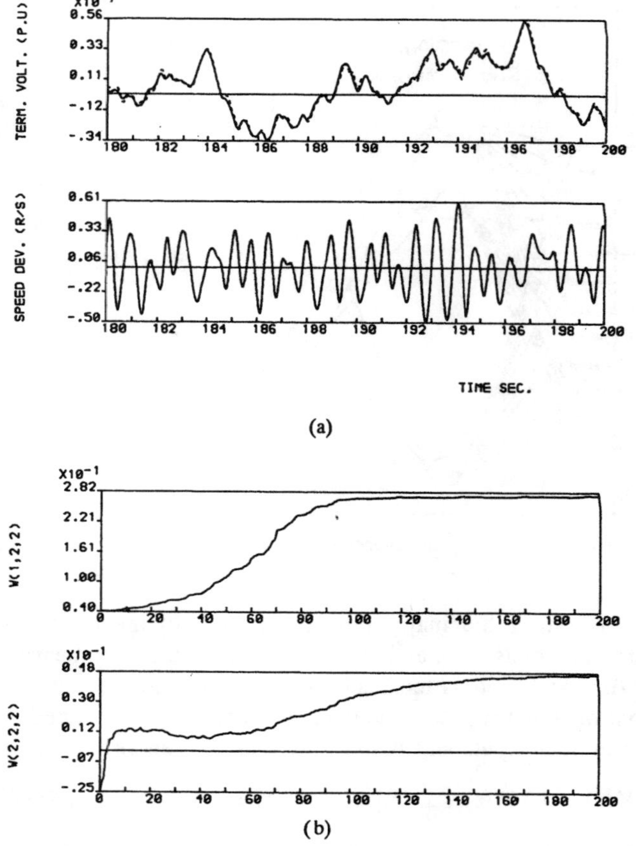

Fig. 4. Turbogenerator I–O mapping using NN's (P_t = 0.8 p.u., Q_t = 0.2 p.u.). (a) NN responses to PRBS inputs. (b) Variation of NN weights.

initialized as random values in the range ±0.1 and the penalty coefficients applied on the voltage and speed were 20 and 1 respectively.

Fig. 4 illustrates the training behavior, with the solid lines in Fig. 4(a) denoting the turbogenerator output responses to the PRBS inputs and the outputs of the NNM denoted by the dashed lines. Fig. 4(a) shows the I–O mapping after 200 s of training and Fig. 4(b) illustrates the variation of weights during the training.

Fig. 4 shows that NN's can be used for I–O mapping of turbogenerator systems with good mapping quality under large disturbances which cause the system to operator well outside the linear range. Therefore NN's can realize complex nonlinear turbogenerator dynamics whose significant modes are excited by the PRBS's.

IV. Turbogenerator Control Results

The NN regulator shown in Fig. 3 was implemented on a simulated model of the turbogenerator [10], [11]. The weights in the NNC are updated by the errors between the outputs of the NNM and a predictor, discussed in [10]. The NN regulator, described in Section II, operates on-line as follows:

1) $Y(t)$ and $D(t)$ are sampled, and the vectors of time series shifted.
2) The patterns $\{Y(t-1)\}$ and $\{U(t-1)\}$ are input to the NN regulator.
3) The weights in the NNM are updated to minimize the cost function J_M.
4) The patterns $\{Y(t-1)\}$ and $\{U(t-1)\}$ are input again to the NN regulator.
5) The weights in the NNC are updated to minimize the cost function J_C.
6) The control input vector $U(t)$ is applied to the turbogenerator system.

The weights of the NN regulator were initialized randomly to lie within the range ±0.1. The NNC was a 6–8–2 network with the same neuron function as in the NNM. The step length, λ, used in training the NNC was set at 0.1, half that used in the NNM. This use of different step sizes in the two subnetworks yielded larger back-propagated signals for weight updating in the NNC, which speeds convergence and makes the regulator more robust.

The performance of the NN regulator has been evaluated by introducing changes of setpoints and also by simulating large disturbances from short circuits on the transmission lines, after 100 s of training with PRBS's. The results are shown in Fig. 5, where the initial operation condition is $P_t = 0.8$ p.u. and $Q_t = 0.2$ p.u. Fig. 5(a) shows the response to changes in the terminal voltage setpoint of ±5%, at 3 s intervals, in the presence of additive random noise on the terminal voltage. The response is well damped, and follows the demanded changes, without offsets in terminal voltage or speed. Fig. 5(b) shows the response to a three-phase short circuit to earth at the sending end of one transmission line, which is cleared after 120 ms. This constitutes a major transient disturbance, during which the system operates in a highly nonlinear mode. The responses are well damped and most satisfactory. The effective control action is illustrated by the exciter and governor inputs.

The NN regulator is also evaluated at other operating points and compared with an adaptive controller [10]. Figs. 6 and 7 illustrate two controller performances with the initial operation condition $P_t = 0.5$ p.u. and $Q_t = -0.1$ p.u., where the turbogenerator system is operating with a leading phase. Figs. 6(a) and 7(a) confirm that the NN regulator can provide satisfactory control performance under different operating conditions. Although the adaptive controller possesses the ability to adapt to condition change, this is limited by the use of an assumed linear model. For an operating condition which is far from the normal, or under different disturbances, the controller may not provide the same qualitative performance, as shown by Figs. 6(b) and 7(b), unless an expert control is involved [1]. The better control performance of the NN regulator under different operating conditions can be ascribed to its nonlinear structure, which has the flexibility to cater for more sensory information than the existing adaptive controller, in which a linear combination of outputs is required to improve system damping.

V. Conclusions

The paper has presented an NN regulator for adaptive, multivariable control of a turbogenerator. This consists of two subnetworks for modeling and adaptive control respectively. The controller avoids the use of sign of plant errors during

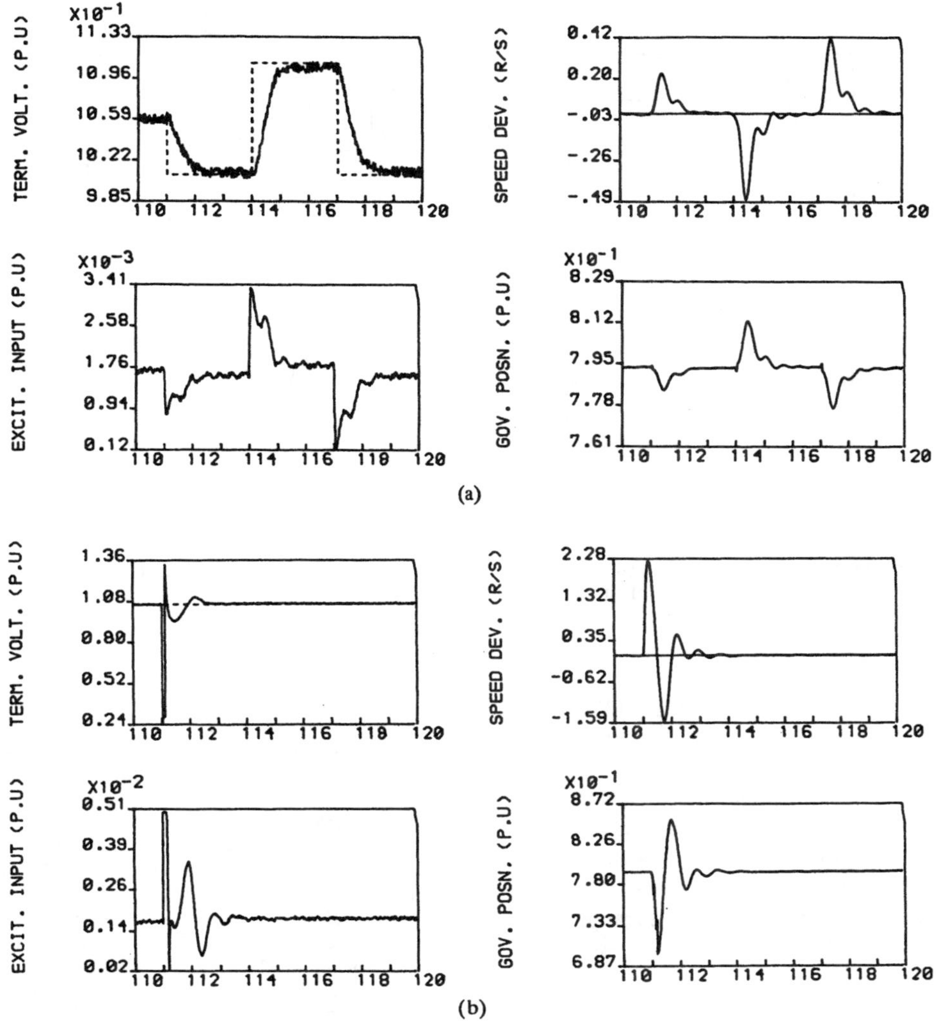

Fig. 5. Responses to small and large disturbances under NN regulator control ($P_t = 0.8$ p.u., $Q_t = 0.2$ p.u.). (a) Responses to voltage setpoint changes. (b) Responses to three-phase short circuit.

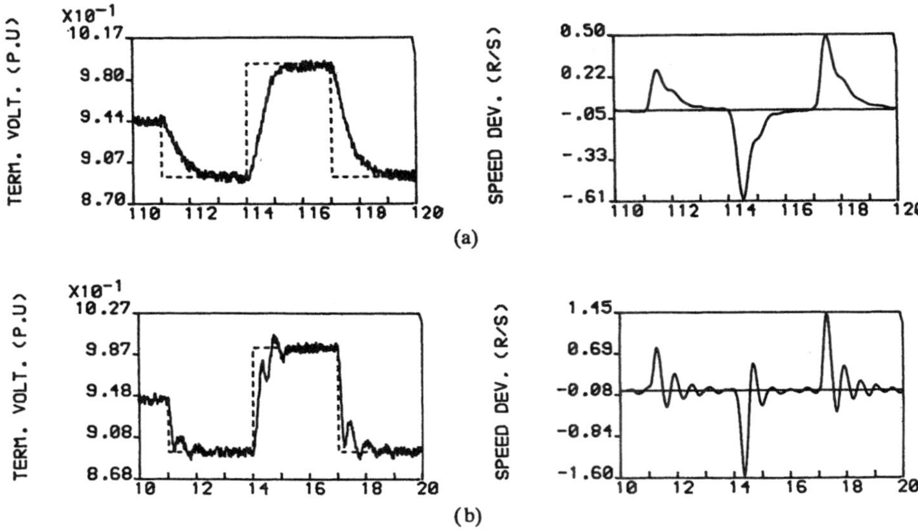

Fig. 6. Comparison of an NN regulator and an adaptive controller with voltage setpoint changes ($P_t = 0.5$ p.u., $Q_t = -0.1$ p.u.). (a) Controlled by NN regulator. (b) Controlled by adaptive controller.

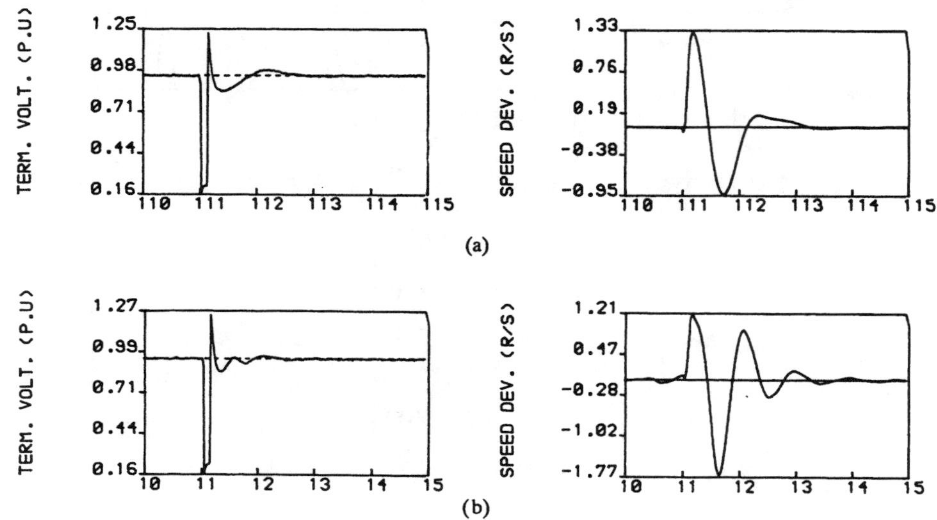

Fig. 7. Comparison of an NN regulator and an adaptive controller with three-phase short circuit ($P_t = 0.5$ p.u., $Q_t = -0.1$ p.u.). (a) Controlled by NN regulator. (b) Controlled by adaptive controller.

back-propagation, does not require a reference model or inverse system model, and avoids the use of probing signals.

The NN regulator has a compact structure, which can easily be extended to cater for more complex dynamic systems or additional control loops. The modeling and adaptive control performance have been evaluated by simulation studies on a detailed nonlinear model of a turbogenerator system. The NN regulator compares well with an existing adaptive controller under different operating conditions and a range of disturbances.

References

[1] K. J. Astrom and B. Wittenmark, *Adaptive control.* Reading, MA: Addison-Wesley, 1989.

[2] Q. H. Wu and B. W. Hogg, "Robust self-tuning regulator for a synchronous generator," *Proc. Inst. Elec. Eng.*, vol. 135, pt. D, no. 6, pp. 463–473, 1988.

[3] D. Psaltis, A. Sideris, and A. Yamamura, "Neural controllers," in *Proc. First Int. Conf. Neural Networks* (San Diego, CA), vol. 4, 1987, pp. 551–558.

[4] M. Saerens and A. Soquet, "A neural controller," in *Proc. First IEE Int. Conf. Artificial Neural Networks,* Oct. 1989, pp. 211–215.

[5] D. Nguyen and B. Widrow, "The truck backer-upper: An example of self-learning in neural networks," in *Proc. Int. Joint Conf. Neural Networks* (Washington, DC), vol. 2, June 1989, pp. II357–II363.

[6] M. I. Jordan and R. A. Jacobs, "Learning to control an unstable system with forward modelling," in *Advances in Neural Information Processing Systems,* vol. 2, D. S. Touretzky, Ed. San Mateo, CA: Morgan Kaufmann, 1990, pp. 324–331.

[7] D. Sbarbaro and P. J. Gawthrop, "Learning complex mappings by stochastic approximation," presented at IJCNN-90, Washington DC, Jan. 1990.

[8] F.-C. Chen, "Back-propagation neural network for nonlinear self-tuning adaptive control," in *Proc. IEEE Int. Symp. Intelligent Control* (Albany, NY), September 25–26, 1989, pp. 274–279.

[9] K. S. Narendra and K. Parthasarathy, "Identification and control of dynamical systems using neural networks," *IEEE Trans. Neural Networks,* vol. 1, pp. 4–27, Mar. 1990.

[10] Q. H. Wu and B. W. Hogg, "Adaptive controller for a turbogenerator system," *Proc. Inst. Elec. Eng.*, vol. 135, pt. D, no. 1, pp. 35–42, 1988.

[11] B. W. Hogg, "Representation and control of turbogenerators in electrical power systems," in *Modelling of Dynamic Systems,* vol. 2, H. Nicholson, Ed. London: Peter Peregrinus, 1981, ch. 5.

[12] D. E. Rumelhart, G. E. Hinton, and R. J. Williams, "Learning internal representations by error propagation," in *Parallel Distributed Processing,* vol. 1, D. E. Rumelhart and J. L. McClelland, Eds. Cambridge, MA: MIT Press, 1986.

Article 7.7

A Neural Network Approach to MVDR Beamforming Problem

Po-Rong Chang, *Member, IEEE,* Wen-Hao Yang, and Kuan-Kin Chan, *Member, IEEE*

***Abstract*—A Hopfield-type neural network approach which leads to an analog circuit for implementing the real-time adaptive antenna array is presented. An optimal pattern of the array can be steered by updating the weights across the array in order to maximize the output signal-to-noise ratio (SNR). Furthermore, it is shown that the problem of adjusting the array weights can be characterized as a constrained quadratic nonlinear programming. Practically, the adjustment of settings is required to respond to a rapid time-varying environment. Many numerical algorithms have been developed for solving such problems using digital computers. The main disadvantage of these algorithms is that they generally converge slowly. To tackle this difficulty, a neural analog circuit solution is particularly attractive in real-time applications with minimization of a cost function subject to constraints. A novel Hopfield-type neural net with a number of graded-response neurons designed to perform the constrained quadratic nonlinear programming would lead to such a solution in a time determined by RC time constants, not by algorithmic time complexity. The constrained quadratic programming neural net has associated it with an energy function which the net always seeks to minimize. A fourth-order Runge–Kutta simulation is conducted to verify the performance of the proposed analog circuit. It shows that the circuit operates at a much higher speed than conventional techniques and the computation time of solving a linear array of 10 elements is about 0.1 ns for RC = 5×10^{-9}.**

I. Introduction

BEAMFORMING is one of the main functions of a passive phased-array processing system. It involves forming multiple beams through applying appropriate delay and weighting elements to the signals received by the sensors. The purpose is to suppress unwanted jamming interferences and to produce the optimal beamformer response which contains minimal contributions due to noise. The most commonly employed technique for deriving the adaptive weights uses a closed loop gradient descent algorithm where the weight updates are derived from estimates of the correlation between the signal in each channel and the summed output of the array. This process can be implemented in an analog fashion using correlation loops [1] or digitally in the form of the Widrow least mean square (LMS) algorithm [2]. The fundamental limitation for this technique is one of poor convergence for a broad dynamic range signal environment. Several different approaches for choosing optimum beamformer weights are summarized in [3]. In many applications none of those approaches is satisfactory. The desired signal may be of unknown strength and may always be present, resulting in signal cancellation with the multiple sidelobe canceller and preventing estimation of signal and noise covariance matrices in the maximum signal-to-noise (SNR) processor. These limitations can be overcome through the application of linear constraints to the weights. The basic concept of linearly constrained minimum variance (LCMV) beamforming is to constrain the response of the beamformer such that the desired signals are passed with specified gain and phase. The weights are chosen to minimize output power subject to the response constraint. When the beamformer has unity response in the look direction, the LCMV problem would become the minimum variance distortionless response (MVDR) beamformer problem, which is a very general approach employed to control beamformer response.

The weights of an MVDR-based beamformer should be updated in real time in order to respond to the rapid time-varying environment. Meanwhile, the evaluation of weights is computationally intensive and can hardly meet the real-time requirement. Systolic implementations of optimum beamformers have been studied to improve the computational speed by a number of investigators. McWhirter and Shephered [5] showed how a triangular systolic array of the type proposed by Gentleman and Kung [6] can be applied to the problem of linearly constrained minimum variance problem, subject to one or more simultaneous linear equality constraints. Their fully pipelined implementation requires $O(p^2 + kp)$ arithmetic operations per cycle time where p is the number of antenna elements and k is the number of look direction constraints. As an alternative to the digital approach, an analog approach based on Hopfield-type neural networks could operate at much higher speed and requires less hardware than digital implementation.

Tank and Hopfield [7] have shown how a class of neural networks with symmetric connections between neurons presents a dynamics that leads to the optimization of a quadratic functional. Recently Chua and Lin [8] and Kennedy and Chua [9], [10] extended the design of Hopfield network and introduced a canonical nonlinear programming circuit which is able to handle more general optimization problems. They showed that a canonical neural network assigned to solve the optimization problem would reach a solution in a time determined by RC time constant, not by algorithmic time complexity. Therefore, the converge speed of reaching the optimal solution is dramatically improved. Experiments show that a MVDR-based neural analog circuit is quite robust and independent of signal power level and the computation time

Manuscript received June 10, 1991; revised November 7, 1991. This work was supported in part by the National Science Council, Republic of China, under Contract NSC81-0404-E009-580.

The authors are with the Department of Communication Engineering, National Chiao Tung University, Hsinchu, Taiwan, Republic of China.

IEEE Log Number 9106609.

Reprinted from *IEEE Trans. Antennas Propagat.*, vol. 40, no. 3, pp. 313–322, March 1992.

of solving a linear array of 10 elements is about 0.1 ns for $RC = 5 \times 10^{-9}$.

II. Problem Formulation

Considering a linear array composed of L isotropic antenna elements which receive signals from sources of variation frequency f_0 located far from the array, $x_l(t)$ is defined as a complex output of the lth element at the sampling time t, and can be expressed as [4]

$$x_l(t) = m(t)e^{j2\pi f_0(t+\tau_l(\theta,\phi))} + n_l(t) + x_{Il}(t) \quad (1)$$

where

$$\tau_l(\theta,\phi) = \frac{\mathbf{r}_l \cdot \hat{\mathbf{s}}(\theta,\phi)}{c} \quad (2)$$

is the time delay of the lth element relative to a reference point chosen at origin. $\mathbf{r}_l$ is the position vector of the lth element. $\hat{\mathbf{s}}(\theta, \phi)$ is an unit vector in the direction (θ, ϕ) of the source, and c is the propagation speed of the plane wave in free space.

The source amplitude $m(t)$ is characterized statistically by

$$E[m(t)] = 0 \quad (3)$$

$$E[m(t)m^*(t)] = p_s \quad (4)$$

where $E[\cdot]$ is the expectation operator, p_s is the power of the source, and the asterisk denotes the complex conjugate. $x_{Il}(t)$ is the component of the directional interferences received by the lth element and possess the same statistics as the source. In addition, $n_l(t)$ is a white random noise with properties

$$E[n_l(t)] = 0, \qquad l = 1,2,\cdots,L \quad (5)$$

$$E[n_l(t)n_k^*(t)] = \sigma_n^2\delta_{lk}, \qquad l,k = 1,2,\cdots,L. \quad (6)$$

Let the signal waveforms derived from the L elements of a beamformer be represented by an L-dimensional complex vector

$$\mathbf{X} \stackrel{\text{def}}{=} [x_1, x_2, \cdots, x_L]^T \quad (7)$$

and the weights of element outputs be represented by L-dimensional complex vector $\mathbf{W}$,

$$\mathbf{W} \stackrel{\text{def}}{=} [w_1, w_2, \cdots, w_L]^T \quad (8)$$

where T denotes the transpose of the vector. Then the output of the beamformer can be written as

$$y(t) = \sum_{l=1}^{L} w_l^* x_l(t) = \mathbf{W}^H\mathbf{X}(t) \quad (9)$$

where H denotes the complex conjugate transpose of a vector.

Since each component of $\mathbf{X}(t)$ is modeled as a zero mean stationary process, the mean output power of the beamformer is given by

$$P(\mathbf{W}) = E[y(t)y^*(t)] = \mathbf{W}^H\mathbf{R}\mathbf{W} \quad (10)$$

where $\mathbf{R}$ is the array correlation matrix.

In order to achieve the optimal utilization of the mean output power of the beamformer, the weights are chosen based on the statistics of the data received at the array such that the output contains minimal influence due to noise as well as interference signals arriving from other directions. Different criteria exist for choosing optimum beamformer weights, which are summarized in [4]. A general approach is to constrain the response of the beamformer so that the desired signals are passed with specified gain and phase. The weights are chosen to minimize the output power subject to the required constraints. This has the effect of preserving the desired signal while minimizing contributions due to noise and interfering signals arriving from directions other than the direction of interest. Based on the above concept, determination of weights with linear constraints to the weight vector is called the linearly constrained minimum variance beamforming problem, which is usually formulated as

$$\min_{\mathbf{W}} \quad \phi(\mathbf{W}) = \mathbf{W}^H\mathbf{R}\mathbf{W} \quad (11)$$

$$\text{subject to} \quad \mathbf{W}^H\mathbf{S}_0 = r \quad (12)$$

where r is a complex constant, $\mathbf{S}_0$ is the steering vector associated with the look direction and is given by

$$\mathbf{S}_0 = \left[1, \exp\left(j\frac{2\pi d}{\lambda_0}\cos\theta_0\right), \cdots, \exp\left(j\frac{2\pi d}{\lambda_0}(L-1)\cos\theta_0\right)\right]^T \quad (13)$$

where d is the element spacing, λ_0 is the wavelength of the plane wave in free space, and θ_0 is the look direction angle (the angle between the axis of the linear array and the direction of the desired signal source).

The method of Lagrange multipliers can be used to solve (11) resulting in

$$\hat{\mathbf{W}} = r\frac{\mathbf{R}^{-1}\mathbf{S}_0}{\mathbf{S}_0^H\mathbf{R}^{-1}\mathbf{S}_0}. \quad (14)$$

Note that in practice the presence of uncorrelated noise ensures that $\mathbf{R}$ is invertible. If $r = 1$, then (11) is often termed the minimum variance distortionless response beamformer.

The MVDR beamforming problem defined in both (11) and (12) is indeed a complex-value constrained quadratic programming problem, which cannot be solved by neural network directly. In order to meet the requirement of neural-based optimizer, one should convert it into a real-value constrained quadratic programming formulation. To achieve this goal, the complex vectors $\mathbf{W}$, $\mathbf{W}_0$, and matrix $\mathbf{R}$ should first be decomposed into their real and imaginary constituents, or

$$\mathbf{W} = \mathbf{W}_r + j\mathbf{W}_i$$

$$\mathbf{R} = \mathbf{R}_r + j\mathbf{R}_i \quad (15)$$

and

$$\mathbf{S}_0 = \mathbf{S}_{0_r} + j\mathbf{S}_{0i} \quad (16)$$

where $\mathbf{W}_r$, $\mathbf{R}_r$, $\mathbf{S}_{0r}$ and $\mathbf{W}_i$, $\mathbf{R}_i$, $\mathbf{S}_{0i}$ are the real parts and imaginary parts of $\mathbf{W}$, $\mathbf{R}$, and $\mathbf{S}_0$, respectively.

Next, substituting (15) into (10), the mean output power becomes

$$\begin{aligned}\mathbf{W}^H\mathbf{R}\mathbf{W} &= (\mathbf{W}_r + j\mathbf{W}_i)^H(\mathbf{R}_r + j\mathbf{R}_i)(\mathbf{W}_r + j\mathbf{W}_i)\\ &= \mathbf{W}_r^T\mathbf{R}_r\mathbf{W}_r - \mathbf{W}_r^T\mathbf{R}_i\mathbf{W}_i + \mathbf{W}_i^T\mathbf{R}_i\mathbf{W}_r\\ &\quad + \mathbf{W}_i^T\mathbf{R}_r\mathbf{W}_i + j[\mathbf{W}_r^T\mathbf{R}_r\mathbf{W}_i + \mathbf{W}_r^T\mathbf{R}_i\mathbf{W}_r\\ &\quad - \mathbf{W}_i^T\mathbf{R}_r\mathbf{W}_r + \mathbf{W}_i^T\mathbf{R}_i\mathbf{W}_i]. \qquad (17)\end{aligned}$$

Since $\mathbf{R}$ is a positive-definite Hermitian matrix, both $\mathbf{R}_r$ and $\mathbf{R}_i$ are identified as the symmetric and skew-symmetric matrices, respectively. Employing the above fact, it can be shown that the imaginary part of $\mathbf{W}^H\mathbf{R}\mathbf{W}$ vanishes and consequently

$$\begin{aligned}\mathbf{W}^H\mathbf{R}\mathbf{W} &= \mathbf{W}_r\mathbf{R}_r\mathbf{W}_r - \mathbf{W}_r^T\mathbf{R}_i\mathbf{W}_i + \mathbf{W}_i^T\mathbf{R}_i\mathbf{W}_r + \mathbf{W}_i^T\mathbf{R}_r\mathbf{W}_i\\ &= \mathbf{v}^T\begin{bmatrix}\mathbf{R}_r & -\mathbf{R}_i\\ \mathbf{R}_i & \mathbf{R}_r\end{bmatrix}\mathbf{v} \qquad (18)\end{aligned}$$

where $\mathbf{v}$ is a $2L$-dimensional real weight vector defined by

$$\mathbf{v} = \begin{bmatrix}\mathbf{W}_r\\ \mathbf{W}_i\end{bmatrix}. \qquad (19)$$

Similarly, the linear constraint can be written as

$$\begin{bmatrix}\mathbf{S}_{0_r}^T & \mathbf{S}_{0_i}^T\\ -\mathbf{S}_{0_i}^T & \mathbf{S}_{0_r}^T\end{bmatrix}\mathbf{v} = \begin{bmatrix}1\\0\end{bmatrix}. \qquad (20)$$

For clarity, we let

$$\mathbf{G} = \begin{bmatrix}2\mathbf{R}_r & -2\mathbf{R}_i\\ 2\mathbf{R}_i & 2\mathbf{R}_r\end{bmatrix} \qquad (21)$$

$$\mathbf{B} = \begin{bmatrix}\mathbf{S}_{0_r}^T & \mathbf{S}_{0_i}^T\\ -\mathbf{S}_{0_i}^T & \mathbf{S}_{o_r}^T\end{bmatrix} \qquad (22)$$

and

$$\mathbf{e} = \begin{bmatrix}1\\0\end{bmatrix} \qquad (23)$$

where $\mathbf{G}$, $\mathbf{B}$, and $\mathbf{e}$ are $(2L) \times (2L)$ symmetric, positive-definite matrix, $2 \times (2L)$ matrix, and (2×1) column vector, respectively. Therefore, the complex-value MVDR problem becomes the following equivalent real canonical quadratic programming problem with linear equality constraints

$$\min_{\mathbf{v}} \qquad \phi(\mathbf{v}) = \tfrac{1}{2}\mathbf{v}^T\mathbf{G}\mathbf{v} \qquad (24)$$

$$\text{subject to} \qquad \mathbf{f}(\mathbf{v}) = \mathbf{B}\mathbf{v} - \mathbf{e} = 0 \qquad (25)$$

where $\mathbf{f}(\mathbf{v})$ is a 2×1 column vector.

III. A Neural-Based Canonical Nonlinear Programming Circuit

To allow a beamformer to respond to a rapid time-varying environment, the weights should be adaptively controlled to satisfy (14) in real time. However, it is computationally intensive and is very costly to implement using discrete components. Digital systolic implementations of optimal beamformers have been studied by a number of investigators [5]. They are usually designed to both compute and implement the adaptive weights. As an alternative to the digital approach, an analog approach based on Hopfield-type neural networks could operate at a much higher speed and requires less hardware than digital implementation. It is shown that Hopfield-type neural networks can solve a number of difficult optimization problems [7]–[10] in a time determined by the system RC time constants, not by algorithmic time complexity. Based on this fact, the neural-based analog circuits are suggested to be one of the favorable choices for the real-time implementation for solving the MVDR problem.

Artificial neural networks contain a large number of identical computing elements or neurons with specific interconnection strengths between neuron pairs. The massively parallel processing power of neural network in solving difficult problems lies in the cooperation of highly interconnected computing elements. Tank and Hopfield [7] demostrated that their networks have the real-time capability in solving various optimization problems, especially, the linear programming and signal decomposition/decision problems, by the programming of synaptic weights stored as a conductance matrix. Recently, Chua and Lin [8] and Kennedy and Chua [10] have extended the method to deal with more general nonlinear programming problems. They proposed a canonical nonlinear programming circuit, which includes dc voltage and current sources, multiport transformers and a network of conductances. Chua and Lin [8] also showed that the linear programming network of Tank and Hopfield is, in fact, a special case of the canonical nonlinear programming circuit. Since the risk of instability in the network is ever presented, Kennedy and Chua [10] introduced their modified canonical design which can guarantee the stability of the network solution. In order to obtain a robust and stable solution of MVDR problem in real time, the circuit proposed by Kennedy and Chua is particularly considered in the design of our implementation in this paper.

The general nonlinear programming problem can be stated as the attempt to minimize a scalar cost function

$$\phi(v_1, v_2, \cdots, v_n). \qquad (26)$$

This minimization is to be accomplished subject to a set of m inequality (or equality) constraints

$$f_j(v_1, v_2, \cdots, v_n) \geq 0 (\text{or} = 0), \qquad 1 \leq j \leq m \quad (27)$$

where m and n are two independent integers.

The canonical nonlinear programming circuit shown in Fig. 1 consists of controlled current and voltage sources, nonlinear resistors, and linear capacitors. The voltages v_i across the capacitors on the right-hand side of Fig. 1 represent the values of the variables involved in the nonlinear programming. The currents i_j through the voltage-controlled nonlinear resistors $g_j(f_j(\mathbf{v}))$ represent constraint satisfaction,

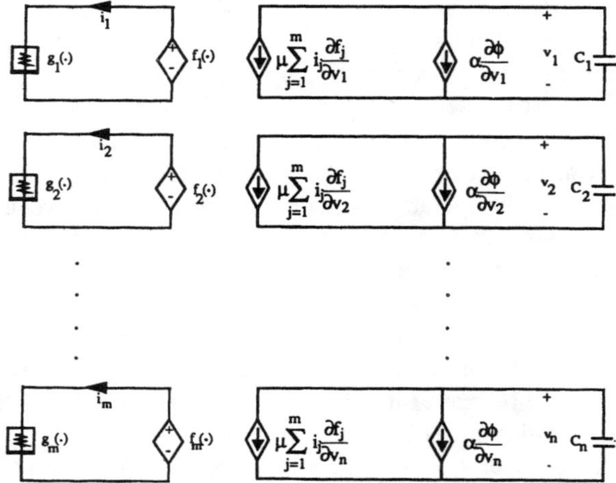

Fig. 1. Canonical nonlinear programming circuit.

where $\mathbf{v} = [v_1, v_2, \cdots, v_n]^T$. Note that the nonlinear functions $g_j(\cdot)$ are used to impose the constraints in the circuit realization. By reading directly from Fig. 1, the circuit equations for the network are given by

$$C\frac{dv_k}{dt} = -\alpha\frac{\partial \phi}{\partial v_k} - \mu \sum_{j=1}^{m} i_j \frac{\partial f_j}{\partial v_k}, \qquad k = 1, 2, \cdots, n \tag{28}$$

where $i_j = g_j(f_j(\mathbf{v}))$, and both α and μ are positive scaling factors.

Next, we would like to show that the circuit equation of (28) converges to a minimum of the cost function $\phi(\mathbf{v})$ subject to a set of constraints. Before discussing this critical issue, several considerations involved in the constrained problem should be identified. The constraints in (27) define a subspace of the multidimensional parameter space called the feasibility region. Solving a constrained problems is, hence, the process of finding that point inside the corresponding feasibility region (including the boundary) where the value of the cost function $\phi(\mathbf{v})$ is the minimum one. To solve a constrained problem defined in (26) and (27), we convert it in an equivalent unconstrained problem. The way to do this is to define a pseudo-cost function $E(\mathbf{v})$ as follows:

$$E(\mathbf{v}) = \alpha\phi(\mathbf{v}) + \mu P(\mathbf{v}) \tag{29}$$

where $\phi(\mathbf{v})$ is the original cost function, $P(\mathbf{v})$ is referred to as the penalty function, and α and μ are called the acceleration factor and the penalty multiplier, respectively.

Different penalty function alternatives can be used in practice. Owing to the considerations in [8], [9], it can be concluded that for a function to qualify as a valid penalty function it must monotonically increase as the $f_j(\mathbf{v})$ deviates from the satisfaction of those constraints. In particular, either the absolute value or the square operator fulfills this requisite, resulting in the following expressions for $P(\mathbf{v})$:

$$P(\mathbf{v}) = \begin{cases} \sum_{j=1}^{m} |g_j(f_j(\mathbf{v}))|, & \text{for absolute value} \\ \frac{1}{2}\sum_{j=1}^{m} g_j^2(f_j(\mathbf{v})), & \text{for square operator.} \end{cases} \tag{30}$$

Moreover, the output of the jth constraint amplifier $g_j(f_j(\mathbf{v}))$ can be defined as follows:

$$g_j(f_j(\mathbf{v})) = \begin{cases} f_j(\mathbf{v}), & \text{for equality constraint} \\ U(-f_j(\mathbf{v}))f_j(\mathbf{v}), & \text{for inequality constraint} \end{cases} \tag{31}$$

where $U(\cdot)$ is the unit step function.

It is interesting to note that the pseudo-cost function $E(\mathbf{v})$ can be identified as energy function for the system of circuit equations (28) and the system is ensured to be completely stable. By complete stability, the system will not oscillate, but will converge to a stable equilibrium state. For simplicitly of analysis, it is assumed the first-order time derivative of $E(\mathbf{v})$ exists and is continuous. In order to make the pseudo-cost function $E(\mathbf{v})$ be differentiable, the square operator would be particularly considered in the penalty function. By using the fact that $i_j = g_j(f_j(\mathbf{v}))$, the time derivative of E becomes

$$\begin{aligned} \frac{dE}{dt} &= \alpha\sum_{k=1}^{n}\frac{\partial\phi}{\partial v_k}\frac{dv_k}{dt} + \mu\sum_{j=1}^{m}\sum_{k=1}^{n} g_j(f_j)\frac{df_j}{\partial v_k}\frac{dv_k}{dt} \\ &= \sum_{k=1}^{n}\left[\alpha\frac{\partial\phi}{\partial v_k} + \mu\sum_{j=1}^{m} g_j(f_j)\frac{\partial f_j}{\partial v_k}\right]\frac{dv_k}{dt} \\ &= -\sum_{k=1}^{n} C_k\left(\frac{dv_k}{dt}\right)^2 \\ &\le 0. \end{aligned} \tag{32}$$

Since each C_k is strictly positive and $(dv_k/dt)^2 \ge 0$ for all k, the time derivative of the pseudo-cost function $E(\mathbf{v})$ is always less than zero. This implies that the circuit will force $E(\mathbf{v})$ to be monotonically decreased except at the equilibrium points where the time derivative vanishes. Nevertheless, the equilibrium points may be either local minimum or inflection points of $E(\mathbf{v})$. The second-order conditions, which are defined in terms of the Hessian matrix $\nabla^2 E(\mathbf{v})$ of second partial derivative of $E(\mathbf{v})$, must be derived in order to determine the status of those equilibrium points. It is shown in [12] that the equilibrium point is a local minimum if the Hessian matrix of $E(\mathbf{v})$ is positive-definite. Usually, the Hessian matrix $\nabla^2 E(\mathbf{v})$ is an $n \times n$ symmetric matrix defined by

$$\nabla^2 E(\mathbf{v}) = \left[\frac{\partial^2 E(\mathbf{v})}{\partial v_i v_j}\right] \tag{33}$$

where v_i is the ith component of $\mathbf{v}$.

By substituting (24) and (25) into (29), we have

$$E(\mathbf{v}) = \frac{\alpha}{2}\mathbf{v}^T\mathbf{G}\mathbf{v} + \mu(\mathbf{B}\mathbf{v} - \mathbf{e})^T(\mathbf{B}\mathbf{v} - \mathbf{e}). \tag{34}$$

As a consequence, its Hassian matrix becomes

$$\nabla^2 E(\mathbf{v}) = \alpha\mathbf{G} + 2\mu\mathbf{B}^T\mathbf{B}. \tag{35}$$

Since G and $\mathbf{B}^T\mathbf{B}$ are positive-definite and positive semidefinite, respectively, the Hassian matrix $\nabla^2 E(\mathbf{v})$ which is linear sum of $\mathbf{G}$ and $\mathbf{B}^T\mathbf{B}$ is positive definite. Therefore the equilibrium point is also the local minimum point. Furthermore, the feasible region over which the constraints of (25) are satisfied can be shown to be a convex set. Since the Hassian matrix of (35) is positive definite throughout the feasible region, it is shown in [12] that any local minimum of $E(\mathbf{v})$ is a global minimum over this feasible region. As a result, the circuit solution of the Hopfield-type network tends to a global minimum of the original cost function $\phi(\mathbf{v})$ within the region over which the constraints are satisfied, when $dE/dt = 0$.

IV. A Neural-Based Circuit Implementation for the MVDR Beamforming Problem

Basically, the circuit shown in Fig. 1 is used to solve a general nonlinear programming. In the case of MVDR-based constrained quadratic problem, a more compact neural-based circuit realization using the existing solid-state devices is possible. Usually, the circuit would include two particular modules. The first module is called the variable amplifier, which can perform the integral of a sum of $(m + 1)$ input currents $(-\alpha\partial\phi/\partial v_k)$ and $(-\mu i_j \partial f_j/\partial v_k)$ and then produces the desired output variable v_k. Fig. 2(a) shows the circuit implementation of the variable amplifier consisting of an integrator and a unity gain inverting amplifier. Op amp 1 produces a voltage v which is in proportion to the integral of the total input current I. The inverting amplifier including op amp 2 and resistor R reproduces this voltage, but with opposite sign. The second module is called the constraint amplifier which is used to perform the constraint satisfaction function $g_j(\cdot)$. Since the MVDR problem has two equality constraints, the output of each $g_j(\cdot)$ would be identical to its input. Therefore the circuit realization of $g_j(\cdot)$ is particularly simple and shown in Fig. 2(b). Without loss of generality, the penalty multiplier μ may be included in $g_j(\cdot)$. Thus, the circuit yields the input–output relation: $O = -\mu I$, where μ represents the magnitude of the resistance, and O and I are an output voltage and an input current, respectively. If the input current I is equal to $-f_j(\mathbf{v})$, then $O = -\mu(-f_j(\mathbf{v})) = \mu f_j(\mathbf{v}) = \mu g_j(f_j(\mathbf{v})) = \mu i_j$.

It should be noted that the canonical nonlinear programming circuit model uses both current i_j and voltage v_k as variables. However, both i_j and v_k are represented as "voltages" in the neural network implementation [10]. By employing the virtual short circuit property of the op amp [10], these voltages would be converted to currents which are suitable for performing the weighted sum operation. Looking upon the circuit system dynamics of (28), the controlled

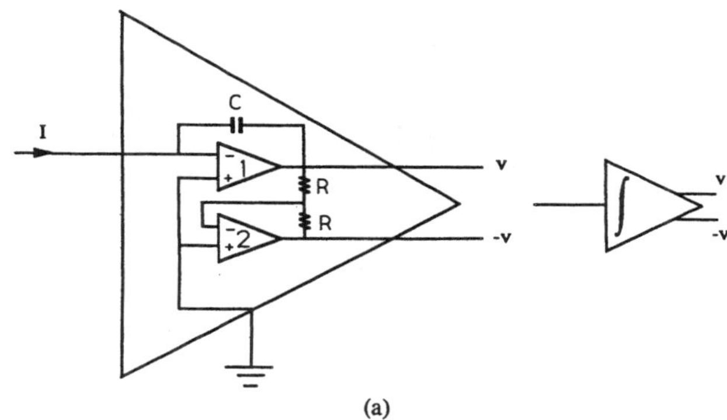

(a)

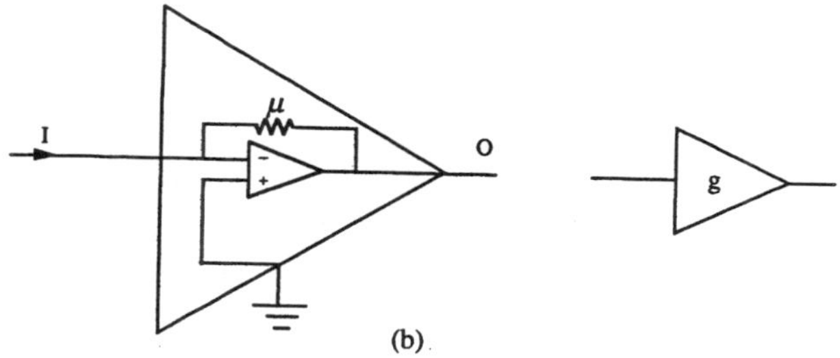

(b)

Fig. 2. Basic blocks of neural circuit. (a) Variable amplifier. (b) Constraint amplifier.

current sources $\partial\phi/\partial v_k$ and $\partial f_j/\partial v_k$ should be determined prior to the implementation. From (24) and (25), it follows that

$$\frac{\partial\phi(\mathbf{v})}{\partial v_k} = \sum_{i=1}^{2L} g_{ki} v_i \tag{36}$$

and

$$\frac{\partial f_j(\mathbf{v})}{\partial v_k} = b_{jk} \tag{37}$$

where g_{ki} and b_{jk} are the (k, i) and (j, k) entries of $\mathbf{G}$ and $\mathbf{B}$, respectively.

Equation (36) shows that the input current $\partial\phi/\partial v_k$ is a linear sum if the v_k weighted by conductances g_{ki}. In the case of linear constraints, the weights $\partial f_j(\mathbf{v})/\partial v_k$ are constants and so may be implemented directly as conductances. Combining (28), (36), and (37), one would obtain the state equations to the circuit implementation as

$$i_j = \mu g_j(f_j(\mathbf{v})) = \mu\left(\sum_{i=1}^{2L} b_{ji} v_i - e_j\right) \tag{38}$$

and

$$\begin{aligned} C\frac{dv_k}{dt} &= -\sum_{i=1}^{2L} (\alpha g_{ki}) v_i - \sum_{j=1}^{2} i_j b_{jk} \\ &= -\sum_{i=1}^{2L} g'_{ki} v_i - \sum_{j=1}^{2} i_j b_{jk}. \end{aligned} \tag{39}$$

Note that the acceleration factor α is included in g_{ki} and then g'_{ki} is defined as (αg_{ki}).

According to (38) and (39), a circuit realization is shown in Fig. 3. It should be noted that the elements of the $\mathbf{e}$, $\mathbf{G}'$,

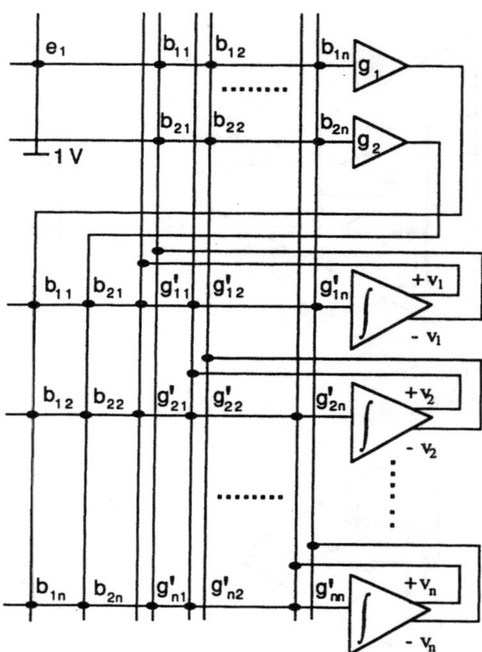

Fig. 3. Schematic diagram of neural circuit for MVDR-based beam forming problem.

and **B** matrices are realized directly as resistive connections and that their associated matrix entries correspond to conductance values. Considering the jth row of the upper constraint block, the input current I_j to the jth constraint amplifier is given by

$$I_j = e_j + \sum_{i=1}^{2L} b_{ji}(-v_i) = e_j - \sum_{i=1}^{2L} b_{ji}v_i = -f_j(\mathbf{v}) \tag{40}$$

where $e_1 = 1$ and $e_2 = 0$. Note that the resistor b_{ji} is connected to the negative output terminal of the ith variable amplifier.

Thus, we would obtain the output voltage of the jth constraint amplifier as

$$i_j = O = -\mu I_j = \mu f_j(\mathbf{v}). \tag{41}$$

Since g'_{ki} and b_{ji} are connected to the output terminal of the ith variable amplifier and the output of the jth constraint amplifier respectively, the current I_k flowing into the kth variable amplifier is obtained by

$$I_k = \sum_{i=1}^{2L} g'_{ki}v_i + \sum_{j=1}^{2} i_j b_{jk}. \tag{42}$$

By using the fact that $C dv_k/dt = -I_k$ for the kth variable amplifier, it yields

$$C\frac{dv_k}{dt} = -\sum_{i=1}^{2L} g'_{ki}v_i - \sum_{j=1}^{2} i_j b_{jk}. \tag{43}$$

As discussed in (40)–(43), it has been proven that the circuit schematic shown in Fig. 3 precisely implements the dynamic equation of (38) and (39) and then is applied to carrying out the optimal solution of the MVDR problem.

V. Illustrated Examples

To verify the effectiveness of MVDR-based neural analog circuit implementation, a linear array of ten elements with half-wavelength spacing is considered in the following examples. The variance of white noise present on each element, i.e, σ_n^2 is assumed to be equal to 0.1. In addition, there are two interference sources which fall in the main lobe and the first sidelobe of the conventional array pattern, respectively. The first interference makes an angle 72° ($\theta_1 = 72°$) with the line of the array and has the power which is taken to be 30 dB more than the white noise power. And, the second interference makes an angle of 98°($\theta_2 = 98°$) and has the power which is 10 dB more than the white noise power. Based on the above assumptions, it is concluded that the power levels for these two interferences would be identified as 100 and 1, that is, $p_1 = 100$ and $p_2 = 1$, respectively. The look direction of signal is assumed to be orthogonal to the array. Three signal powers varied from 0 to 20 dB above the white noise power are employed in our example.

The parameters g'_{ki} and b_{jk} involved in the proposed circuit for this particular example could be obtained according to the value of each component in $\mathbf{G}'$ and $\mathbf{B}$. The $2 \times 20(L = 10)$ matrix $\mathbf{B}$ includes two 10×1 vectors associated with the look direction that are $\mathbf{S}_{0r}$ and $\mathbf{S}_{0i}$ shown in Table I. Another 20×20 symmetric positive-definite matrix $\mathbf{G}'$ consists of both $\mathbf{R}_r$ and $\mathbf{R}_i$ which are 10×10 matrices in terms of the signal power parameter p_s and the acceleration factor α. More details about the expression of $\mathbf{G}'$ are shown in Table I. In addition, the penalty multiplier μ in each constraint amplifier and the acceleration factor α are taken to be 0.4 and 10^{-3}, respectively.

We have simulated the MVDR-based neural analog circuit described by (38) and (39) with the initial guess $\mathbf{v}(0) = \mathbf{0}$, using the simultaneous differential equation solver (DVERK in the IMSL). This routine solves a set of nonlinear differential equations based on the fourth-order Runge–Kutta method. The capacitance C involved in each variable amplifier of the circuit is assumed to be 1 pF. Figs. 4 and 5 show the time evolutions of three output noise powers and the resulting output SNR's corresponding to their associated signal power levels 0.1, 1, and 10, respectively. It is worth observing that the converge time of each curve in Figs. 4 and 5 is almost independent of p_s and equal to 0.1 ns. Since the converge time is characterized by the time constant of system, one may use the dominant equivalent time constant τ, which is defined in Appendix I to verify this particular time behavior. By using the $\mathbf{G}'$ and $\mathbf{B}$ given in Table I, these dominant time constants are found to be also independent of p_s and equal to 5×10^{-9} s. It has been shown that the converge time of each curve is bounded by the dominant time constant. In addition, while reaching the equilibrium state, those resulting solution weights do not satisfy both equality constraints exactly. Appendix II shows that these two steady-state constraint viola-

TABLE I

$$
\mathbf{R}_{rn} = \begin{bmatrix}
101.10 & 57.37 & -35.60 & -97.13 & -73.91 & 13.54 & 88.81 & 86.16 & 7.81 & -77.98 \\
57.37 & 101.10 & 57.37 & -35.60 & -97.13 & -73.91 & 13.54 & 88.81 & 86.16 & 7.81 \\
-35.60 & 57.37 & 101.10 & 57.37 & -35.60 & -97.13 & -73.91 & 13.54 & 88.81 & 86.16 \\
-97.13 & -35.60 & 57.37 & 101.10 & 57.37 & -35.60 & -97.13 & -73.91 & 13.54 & 88.81 \\
-73.91 & -97.13 & -35.60 & 57.37 & 101.10 & 57.37 & -35.60 & -97.13 & -73.91 & 13.54 \\
13.54 & -73.91 & -97.13 & -35.60 & 57.37 & 101.10 & 57.37 & -35.60 & -97.13 & -73.91 \\
88.81 & 13.54 & -73.91 & -97.13 & -35.60 & 57.37 & 101.10 & 57.37 & -35.60 & -97.13 \\
86.16 & 88.81 & 13.54 & -73.91 & -97.13 & -35.60 & 57.37 & 101.10 & 57.37 & -35.60 \\
7.81 & 86.16 & 88.81 & 13.54 & -73.91 & -97.13 & -35.60 & 57.37 & 101.10 & 57.37 \\
-77.98 & 7.81 & 86.16 & 88.81 & 13.54 & -73.91 & -97.13 & -35.60 & 57.37 & 101.10
\end{bmatrix}
$$

$$
\mathbf{R}_{in} = \begin{bmatrix}
0.00 & -82.11 & -92.44 & -21.75 & 68.53 & 99.82 & 44.74 & -48.95 & -99.97 & -64.18 \\
82.11 & 0.00 & -82.11 & -92.44 & -21.75 & 68.53 & 99.82 & 44.74 & -48.95 & -99.97 \\
92.44 & 82.11 & 0.00 & -82.11 & -92.44 & -21.75 & 68.53 & 99.82 & 44.74 & -48.95 \\
21.75 & 92.44 & 82.11 & 0.00 & -82.11 & -92.44 & -21.75 & 68.53 & 99.82 & 44.74 \\
-68.53 & 21.75 & 92.44 & 82.11 & 0.00 & -82.11 & -92.44 & -21.75 & 68.53 & 99.82 \\
-99.82 & -68.53 & 21.75 & 92.44 & 82.11 & 0.00 & -82.11 & -92.44 & -21.75 & 68.53 \\
-44.74 & -99.82 & -68.53 & 21.75 & 92.44 & 82.11 & 0.00 & -82.11 & -92.44 & -21.75 \\
48.95 & -44.74 & -99.82 & -68.53 & 21.75 & 92.44 & 82.11 & 0.00 & -82.11 & -92.44 \\
99.97 & 48.95 & -44.74 & -99.82 & -68.53 & 21.75 & 92.44 & 82.11 & 0.00 & -82.11 \\
64.18 & 99.97 & 48.95 & -44.74 & -99.82 & -68.53 & 21.75 & 92.44 & 82.11 & 0.00
\end{bmatrix}
$$

$$\mathbf{R}_{rs} = [1]_{10\times 10} \qquad \mathbf{R}_{is} = [0]_{10\times 10}$$

$$\mathbf{R}_r = \alpha(p_s\mathbf{R}_{rs} + \mathbf{R}_{rn}) \qquad \mathbf{R}_i = \alpha(p_s\mathbf{R}_{is} + \mathbf{R}_{in})$$

$$\mathbf{S}_{0r} = [1]^T_{1\times 10} \qquad \mathbf{S}_{0_i} = [0]^T_{1\times 10}$$

$$\mathbf{G}' = \begin{bmatrix} 2\mathbf{R}_r & -2\mathbf{R}_i \\ 2\mathbf{R}_i & 2\mathbf{R}_r \end{bmatrix} \qquad \mathbf{B} = \begin{bmatrix} \mathbf{S}^T_{0_r} & \mathbf{S}^T_{0_i} \\ -\mathbf{S}^T_{0_i} & \mathbf{S}^T_{0_r} \end{bmatrix}$$

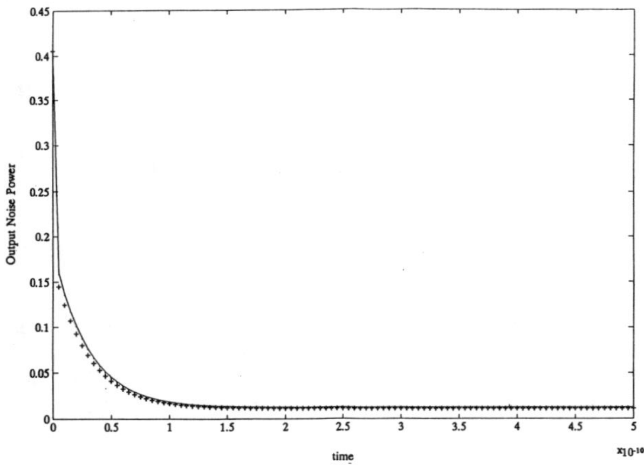

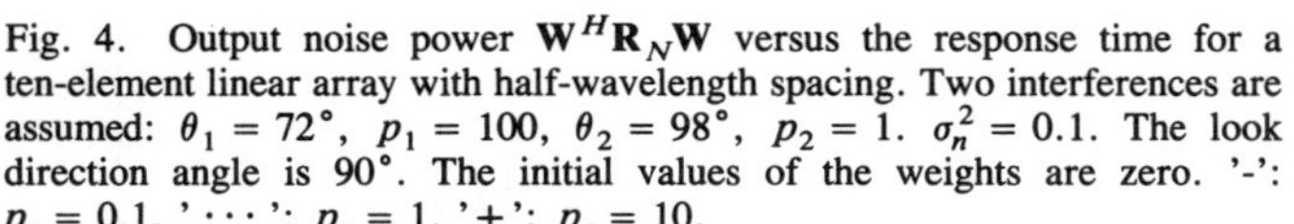

Fig. 4. Output noise power $\mathbf{W}^H\mathbf{R}_N\mathbf{W}$ versus the response time for a ten-element linear array with half-wavelength spacing. Two interferences are assumed: $\theta_1 = 72°$, $p_1 = 100$, $\theta_2 = 98°$, $p_2 = 1$. $\sigma_n^2 = 0.1$. The look direction angle is 90°. The initial values of the weights are zero. '-': $p_s = 0.1$, '$\cdots$': $p_s = 1$, '+': $p_s = 10$.

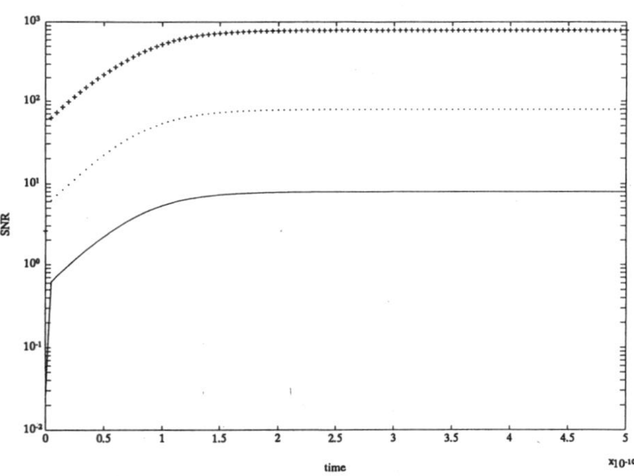

Fig. 5. Output signal-to-noise ratio versus the response time for a ten-element linear array with half-wavelength spacing. Two interferences are assumed: $\theta_1 = 72°$, $p_1 = 100$, $\theta_2 = 98°$, $p_2 = 1$. $\sigma_n^2 = 0.1$. The look direction angle is 90°. The initial values of weights are zero. '-': $p_s = 0.1$, '$\cdots$': $p_s = 1$, '+': $p_s = 10$.

tion errors could be estimated by $\hat{\text{err}}_1 = 2\alpha p_s/(\mu + \alpha p_s)$ and $\hat{\text{err}}_2 = 0$. As a result, the penalty multiplier should be chosen large enough to make $\hat{\text{err}}_1$ sufficiently small. The final results are illustrated in the following table.

Signal Power	Theoretical SNR Value	Simulation Results $\mathbf{v}(0) = \mathbf{0}$	Weighting α	Weighting μ	Constraint Error err_1	Constraint Error err_2	Error Estimates $\hat{\text{err}}_1$	Error Estimates $\hat{\text{err}}_2$
0.1	7.956	7.956	0.001	0.4	0.00056	0	0.0005	0
1	79.56	79.56	0.001	0.4	0.005	0	0.005	0
10	795.6	795.6	0.001	0.4	0.048	0	0.0488	0

It is worth simulating the case of large jammers by varying the INR's (interference-to-noise ratio) for both interfering signals from 30–70 dB simultaneously. Fig. 6 shows that the resulting output SNR of each curve is almost independent of INR and reaches the same steady state level. Since the feedback current into the variable amplifier is approximately proportional to the INR for the case of large jammers, it is expected that the converge time improves as INR increases. However, the acceleration factor is adjusted to smaller value in view of stability consideration and the dominant time constant in such case is about 10^{-5} s. Another situation of particular interest is the performance evaluation of the beamformer under the influence of broad-band jammers. This is given by way of examples which demonstrate how the resulting SNR varies as the element spacing changed from 0.35 λ_0

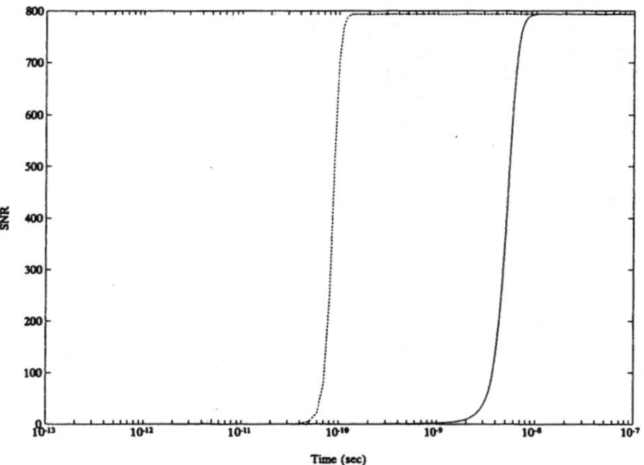

Fig. 6. Output signal-to-noise ratio versus the response time for 10 element linear array with half-wavelength spacing. Two interferences are assumed: $\theta_1 = 72°$, $\theta_2 = 98°$. $\sigma_n^2 = 0.1$. The look direction angle is 90°, $p_s = 10$. The initial values of the weights are zero. '-': INR = 30 dB, '--': INR = 50 dB, '···': INR = 70 dB.

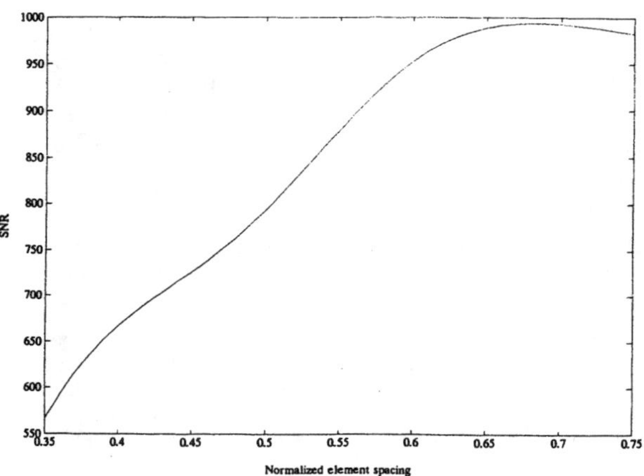

Fig. 7. Output signal-to-noise ratio versus the element spacing for a ten-element linear array. Two interferences are assumed: $\theta_1 = 72°$, $p_1 = 100$, $\theta_2 = 98°$, $p_2 = 100$. $\sigma_n^2 = 0.1$. The look direction angle is 90°, $p_s = 10$. The initial values of the weights are zero.

to 0.75 λ_0, where λ_0 is the wavelength and the signal source corresponds to the half-wavelength spacing. For comparison, all the parameter settings are taken to be the same as used in the previous case. One observes from Fig. 7 that the SNR starts to increase as the element spacing increases. It reaches its maximum level when the element spacing is equal to 0.68 λ_0. Beyond that the SNR level drops slightly. The reason is that the larger element spacing causes the grating lobes to appear in the array and thus degrades the overall output SNR. Finally, a comprehensive example is conducted to test the network performance for the case of closely spaced jammers. The phases of the interfering signals are assumed such that the angles of arrival for the two jammers are as close as $\theta_1 = 72°$ and $\theta_2 = 74°$, respectively. The powers of both interferences are taken to be of the same level and equal to 100. The signal power, look direction angle and element spacing are suggested to be 10, 90° and 0.5 λ_0 respectively. The resultant array output SNR is shown in Fig. 8 which reaches its maximum value of 950 rapidly. Fig. 9 compares the power patterns of the resultant adaptive array pattern with that of the conventional uniform array pattern. One observes clearly that two sharp nulls presented in the pattern correspond to the directions of arrival of the interferences. As a result, the interferences are suppressed and the output SNR of the adaptive array reaches the optimal value.

VI. Conclusion

In this paper, we have proposed a cost-effective analog circuit implementation for computing the MVDR beamforming problem based on Kennedy and Chua's cannonical neural network. Their novel neural-based optimizer is able to guarantee the stability and robustness of the solution, the converge time can be characterized by the dominant time constant of the network. It turns out that the speed of reaching a steady-state solution depends essentially on the time constant, not on the algorithmic time complexity. Finally, a linear array of 10 elements with three signal levels is constructed accordingly to verify the performance of the proposed cir-

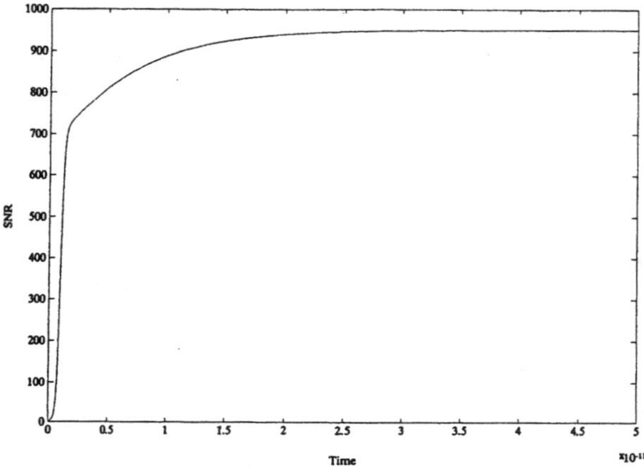

Fig. 8. Output signal-to-noise ratio versus the response time for a ten-element linear array with half-wavelength spacing. Two interferences are assumed: $\theta_1 = 72°$, $p_1 = 100$, $\theta_2 = 74°$, $p_2 = 100$. $\sigma_n^2 = 0.1$. The look direction angle is 90°, $p_s = 10$. The initial values of the weights are zero.

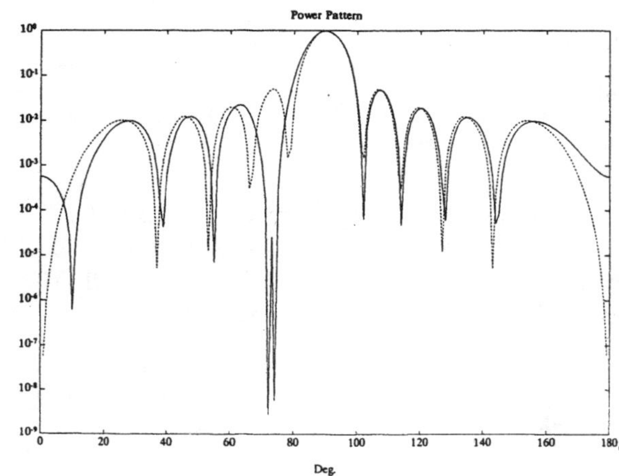

Fig. 9. The conventional uniform array pattern '--' and the resultant adaptive array pattern '-' for a 10-element linear array with half-wavelength spacing. Two interferences are assumed: $\theta_1 = 72°$, $p_1 = 100$, $\theta_2 = 74°$, $p_2 = 100$. $\sigma_n^2 = 0.1$. The look direction angle is 90°, $p_s = 10$. The initial values of the weights are zero.

cuit. It shows that the MVDR-based neural circuit is able to quickly attain its optimal performance in 0.1 ns when the dominant time constant is 5×10^{-9} s and work satisfactorily under the stringent environment of strong jammers as well as closely spaced jammers.

Appendix I
The Dominant Time Constant for an MVDR-Based Neural Circuit

Substituting (38) into (39), one may obtain the system of first-order differential equations in matrix form and given by

$$\frac{d\mathbf{v}}{dt} = -\frac{1}{C}\left[(\mathbf{G}' + \mu \mathbf{B}^T\mathbf{B})\mathbf{v} + \mu \mathbf{B}^T\mathbf{e}\right]$$
$$= -\mathbf{M}\mathbf{v} + \mathbf{N}\mathbf{e} \tag{44}$$

where $\mathbf{M} = (\mathbf{G}' + \mu\mathbf{B}^T\mathbf{B})/C$ and $\mathbf{N} = -\mu\mathbf{B}^T/C$.

[11] showed that the solution of (44) can be expressed as a linear combination of modes $t^k e^{-\lambda_i t}$ where λ_i is the ith eigenvalue of $\mathbf{M}$. It is known that the time behavior of system (44) would be dominated by the minimum eigenvalue $\lambda_{\min}(\mathbf{M}) = \min_i(\lambda_i)$. Consequently, a dominant time constant for the MVDR-based neural circuit is defined by

$$\tau = \frac{1}{\lambda_{\min}(\mathbf{M})}. \tag{45}$$

Appendix II
Estimation of Steady-State Constraint Violation Errors

Let the steady-state constraint violation errors for both equality constraints defined in (25) be represented by err_1 and err_2, respectively. While reaching the equilibrium state, these two constraints become

$$\sum_{k=1}^{2L} B_{1k} v_k = (1 - \text{err}_1) \tag{46}$$

and

$$\sum_{k=1}^{2L} B_{2k} v_k = \text{err}_2. \tag{47}$$

Since the noise power p_{noise} is assumed to be insignificant to the signal power, the objective function ϕ is given by

$$\phi = p_s\left[(1 - \text{err}_1)^2 + \text{err}_2^2\right] + p_{\text{noise}}$$
$$\approx p_s\left[(1 - \text{err}_1)^2 + \text{err}_2^2\right] \tag{48}$$

and the penalty function is calculated as

$$P = \tfrac{1}{2}(\text{err}_1^2 + \text{err}_2^2). \tag{49}$$

Therefore, the energy function becomes

$$E\big|_{\text{steady state}} = \alpha\phi + \mu P$$
$$= \tfrac{1}{2}\{2\alpha p_s(1 - \text{err}_1)^2 + \mu \text{err}_1^2 + (2\alpha p_s + \mu)\text{err}_2^2\}. \tag{50}$$

By the chain rule and the fact that $dE/dt = 0$ at the equilibrium state, the time derivative of E becomes

$$\left.\frac{dE}{dt}\right|_{\text{steady state}} = \frac{\partial E}{\partial \text{err}_1}\cdot\frac{d\text{err}_1}{dt} + \frac{\partial E}{\partial \text{err}_2}\cdot\frac{d\text{err}_2}{dt} = 0. \tag{51}$$

Equation (51) implies that both partial derivatives of E with respect to err_1 and err_2 should vanish respectively, i.e.,

$$\left.\frac{\partial E}{\partial \text{err}_1}\right|_{\text{err}_1 = \hat{\text{err}}_1} = -2\alpha p_s(1 - \hat{\text{err}}_1) + \mu \hat{\text{err}}_1 = 0 \tag{52}$$

and

$$\left.\frac{\partial E}{\partial \text{err}_2}\right|_{\text{err}_2 = \hat{\text{err}}_2} = (2\alpha p_s + \mu)\hat{\text{err}}_2 = 0 \tag{53}$$

where $\hat{\text{err}}_1$ and $\hat{\text{err}}_2$ are the estimates of err_1 and err_2, respectively. Then, we obtain

$$\hat{\text{err}}_1 = \frac{2\alpha p_s}{\mu + \alpha p_s} \tag{54}$$

and

$$\hat{\text{err}}_2 = 0. \tag{55}$$

References

[1] S. P. Applebaum, "Adaptive array," *IEEE Trans. Antennas Propagat.*, vol. AP-24, pp. 585–598, Sept. 1976.

[2] B. Widrow and S. Stearns, *Adaptive Signal Processing*. Englewood Cliffs, NJ: Prentice-Hall; 1985.

[3] B. D. Van Veen and K. M. Buckley, "Beamforming: A versatile approach to spatial filtering," *IEEE ASSP Mag.*, pp. 4–24, Apr. 1988.

[4] L. C. Godara, "Error analysis of the optimal antenna array processors," *IEEE Trans. Aerospace Electron. Syst.*, vol. AES-22, pp. 395–409, July 1986.

[5] J. G. McWhirter and T. J. Shepherd, "Systolic array processor for MVDR beamforming," *Proc. Inst. Elec. Eng.*, vol. 136, pt. F, no. 2, Apr. 1989.

[6] W. M. Gentleman and H. T. Kung, "Matrix triangularisation by systolic array," *Proc. SPIE*, Real time signal processing IV, 1981, p.298.

[7] D. W. Tank and J. J. Hopfield, "Simple 'neural' optimization networks: An A/D converter, signal decision circuit, and a linear programming circuit," *IEEE Trans. Circuit Syst.*, vol. CAS-33, pp. 533–541, May 1986.

[8] L. O. Chua and G. N. Lin, "Nonlinear programming without computation," *IEEE Trans. Circuit Syst.*, vol. CAS-31, pp. 182–188, Feb. 1984.

[9] M.P. Kennedy and L. O. Chua, "Unifying the Tank and Hopfield linear programming network and the canonical nonlinear programming circuit of Chua and Lin," *IEEE Trans. Circuit Syst.*, vol. CAS-34, pp. 210–214, Feb. 1987.

[10] ——, "Neural networks for nonlinear programming," *IEEE Trans. Circuit Syst.*, vol. 35, pp. 554–562, May 1988.

[11] C. T. Chen, *Linear System Theory and Design*. New York: Holt, Rinehart and Winston, 1984.

[12] D. G. Luenberger, *Linear and Nonlinear Programming, 2nd ed.* Reading, MA: Addison-Wesley, 1984.

Artificial neural networks in process engineering

M.J. Willis, MEng
C. Di Massimo, Ing. ICPI
G.A. Montague, PhD
M.T. Tham, PhD
Prof. A.J. Morris, PhD, CEng, MIEE

Indexing terms: Algorithms, Modelling, Process control

Abstract: Artificial neural networks are made up of highly interconnected layers of simple 'neuron-like' nodes. The neurons act as nonlinear processing elements within the network. An attractive property of artificial neural networks is that, given the appropriate network topology, they are capable of characterising nonlinear functional relationships. Furthermore, the structure of the resulting neural network based process model may be considered generic, in the sense that little prior process knowledge is required in its determination. The methodology therefore provides a cost efficient and reliable process modelling technique. The concepts involved in the formulation of artificial neural networks are introduced. Their suitability for solving some process engineering problems is discussed and illustrated using results obtained from both simulation studies and recent applications to industrial process data. In the latter, neural network models were used to provide estimates of biomass concentration in industrial fermentation systems and of top product composition of an industrial distillation tower. Measurements from established instruments such as off-gas carbon dioxide in the fermenter and overheads temperature in the distillation column were used as the secondary variables for the respective processes. The advantage of using these estimates for feedback control is demonstrated. The possibility of using neural network models directly in model based control strategies is also considered. The range of applications is an indication of the utility of artificial neural network methodologies within a process engineering environment.

Paper 7704D (C9), first received 7th June and in revised form 17th September 1990

The authors are with the Department of Chemical and Process Engineering, Merz Court, The University of Newcastle upon Tyne, Newcastle upon Tyne NE1 7RU, United Kingdom

1 Introduction

The idea of using artificial neural networks (ANNs) as a potential solution strategy for problems which require complex data analysis is not new. Over the last 40 to 50 years, scientists have been attempting to realise the 'real' neural structure of the human brain, and to develop an algorithmic equivalent of the human learning process. The principal motivation behind this research is the desire to achieve the sophisticated level of information processing that the brain is capable of. However, the structure of the human brain is extremely complex, with approximately 10^{11} neurons and between 10^{14} and 10^{15} synapses (the connections between neurons). Whilst the function of single neurons is relatively well understood, their collective role within the conglomeration of cerebrum elements is less clear and a subject of avid postulations. Consequently, the architecture of an ANN is based upon a primitive understanding of the functions of the biological neural system. Even if neurophysiology could untangle the complexities of the brain, due to the limitations of current hardware technology it will be extremely difficult, if not impossible, to emulate exactly its immensely distributed structure. Thus, rather than accurately model the intricacies of the human cerebral functions, ANNs attempt to capture and utilise the connectionist philosophy on a more modest and manageable scale.

Within the area of process engineering, process design and simulation, process supervision, control and estimation, and process fault detection and diagnosis rely on the effective processing of unpredictable and imprecise information. To tackle such tasks, current approaches tend to be based on some 'model' of the process in question. The model can be based on qualitative knowledge derived from experience, or quantified in terms of an analytical (usually linear) process model, or a loosely integrated combination of both. Although the resultant procedures can provide acceptable solutions, there are many situations in which they are prone to failure because of the uncertainties and the nonlinearities intrinsic to many process systems. These are, however, exactly the problems that a well-trained human decision process excels in

Reprinted with permission from *IEE Proceedings-D*, vol. 138, no. 3, M. J. Willis, C. D. Massimo, G. A. Montague, M. T. Tham, and A. J. Morris, "Artificial Neural Networks in Process Engineering," pp. 256–266, May 1990, IEE.

solving. Thus, if ANNs fulfil their projected promise, they may form the basis of improved alternatives to current engineering practice. Indeed, applications of artificial neural networks to solve process engineering problems have already been reported.

One of the major obstacles to the widespread use of advanced modelling and control techniques is the cost of model development and validation. The utility of neural networks in providing viable process models was demonstrated in Bhat *et al.* [1, 2], where the technique was used to characterise successfully two nonlinear chemical systems as well as to interpret biosensor data. The application of ANNs to provide cost efficient and reliable process models therefore appears to be highly promising.

In adjunct area, the use of neural network based models (NNMs) for the online estimation of process variables was considered by Montague *et al.* [3]. Lant *et al.* [4] further discussed the relative mertis of process estimation using an adaptive linear estimator [5] and a estimator based on an NNM. In addition, the applicability of neural networks for improving process operability was investigated by Di Massimo *et al.* [6]. In some situations, techniques based on the use of an NNM may offer significant advantages over conventional model based techniques. For instance, if the NNM is sufficiently accurate, it could theoretically be used in place of an online analyser. Indeed, such a philosophy may be used to provide more frequent measurements than could be achieved by hardware instrumentation. This is advantageous from the control viewpoint as the feedback signals will not be subject to measurement delays. Consequently, significant improvements in control performance can be expected. A tentative exposition of a neural model based inferential controller was also presented in Montague *et al.* [3]. Here, an NNM was used to provide estimates of 'difficult-to-measure' controlled variables by inference from other easily measured outputs. These estimates were then used for feedback control.

If an accurate NNM is available, then it could obviously also be directly applicable within a model based control strategy. The particularly attractive feature is the potential to handle nonlinear systems. Psaltis *et al.* [7] and Narendra and Parthasarathy [8] investigated the use of a multilayered neural network processor for plant control. A novel learning architecture was proposed to ensure that the inputs to the plant gave the desired response. Willis [9], on the other hand, proposed the use of NNMs in the synthesis of cost-function based controllers. Here, an online optimisation algorithm is used to determine the future inputs that will minimise the deviations between setpoints and the predicted outputs obtained from an NNM.

Another promising area for ANNs is in fault diagnosis and the development of intelligent control systems [10]. Here, a robust control system must be designed to accommodate highly complex, unquantifiable data. Hoskins and Himmelblau [11] described the desirable characteristics of neural networks for knowledge representation in chemical engineering processes. Their paper illustrated how an artificial neural network could be used to learn and discriminate successfully amongst faults. Birky and McAvoy [12] presented an application of a neural network to learn the design of control configurations for distillation columns. They were able to demonstrate that their approach was an effective and efficient means to extract process knowledge.

2 Process modelling via the use of artificial neural networks

As would have been noted in almost all of the work cited above, an NNM is used in place of conventional models. The accuracy of the NNM model may be influenced by altering the topology (structure of the network). It is the topology of the network, together with the neuron processing function, which impart to an ANN its powerful signal processing capabilities. Although a number of ANN architectures have been proposed (see Reference 13), the 'feedforward' ANN (FANN) is by far the most widely applied. Indeed, a number of authors [14, 15] have recently claimed that any continuous function can be approximated arbitrarily well on a compact set by an FANN; although the topology of the network to achieve this approximation is still open to question. However, in all the claims no more than two hidden layers have been considered necessary. This result essentially states that an FANN could be confidently used to model a wide range of nonlinear relationships. From previous discussions, the implications of this statement are therefore considerable. In view of this, subsequent discussions will therefore be restricted to FANNs.

2.1 Feedforward artificial neural networks

The architecture of a typical FANN is shown in Fig. 1. The nodes in the different layers of the network represent

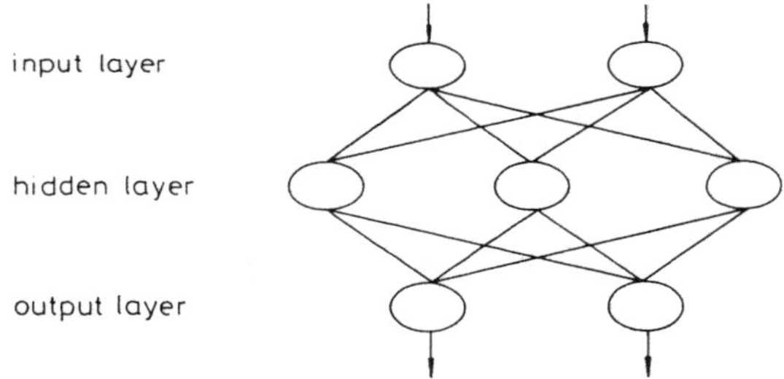

Fig. 1 *Schematic of feedforward neural network architecture*

'neuron-like' processing elements. There is always an input and an output layer. The number of neurons in both these layers depends on the respective number of inputs and outputs being considered. In contrast, hidden layers may vary from zero to any finite number, depending on specification. The number of neurons in each hidden layer is also a user specification. It is the hidden layer structures which essentially defines the topology of an FANN.

The neurons in the input layer do not perform data processing functions; they merely provide a means by which scaled data is introduced into the network. These signals are then 'fed forward' through the network via the connections, through hidden layers, and eventually to the final output layer. Each interconnection has associated with it a weight which modifies the strength of the signal flowing along that path. Thus, with the exception of the neurons in the input layer, the input to each neuron is a weighted sum of the outputs from neurons in the previous layer. For example, if the information from the ith neuron in the $j-1$th layer, to the kth neuron in the jth layer is $I_{j-1,i}$, then the total input to the kth neuron in the jth layer is given by

$$\alpha_{j,k} = d_{j,k} + \sum_{i=1}^{n} w_{j-1,i,k} I_{j-1,i} \tag{1}$$

where $d_{j,k}$ is a bias term which is associated with each interconnection. The output of each node is obtained by passing the weighted sum $\alpha_{j,k}$, through a nonlinear operator. This is typically a sigmoidal function, the simplest of which has the mathematical description

$$I_{j,k} = 1/(1 + \exp(-\alpha_{j,k})) \tag{2}$$

and the response characteristics shown in Fig. 2. Although the function given by eqn. 2 has been widely adopted, in principle any function with a bounded derivative could be employed [16]. It is, however, interesting to note that a sigmoidal nonlinearity has also been observed in human neuron behaviour [17]. Within an ANN, this function provides the network with the ability to represent nonlinear relationships. Additionally, note that the magnitude of bias term in eqn. 1 effectively determines the coordinate space of the nonlinearity. This implies that the network is also capable of characterising the structure of the nonlinearities: a highly desirable feature.

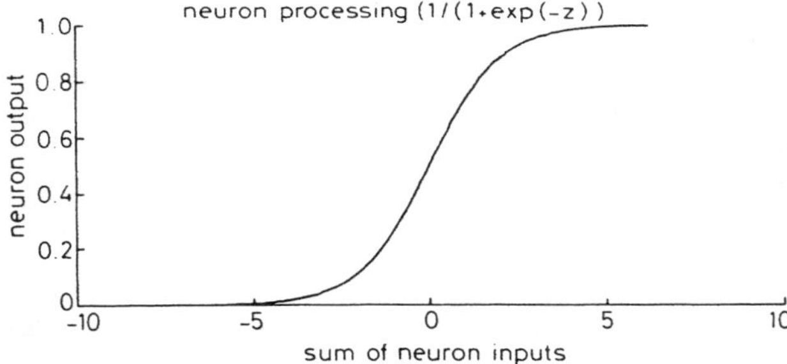

Fig. 2 *Sigmoidal function*

To develop a process model using the neural network approach, the topology of the network must first be declared. The convention used in referring to a network with a specific topology follows that adopted by Bremmerman and Anderson [18]. For example, an FANN with three input neurons, two hidden layers with five and nine neurons, respectively, and two neurons in the output layer will be referred to as a (3-5-9-2) network. A (2-10-1) network thus refers to an FANN with two neurons in the input layer, ten neurons in one hidden layer, and one neuron in the output layer.

2.2 *Algorithms for network training (weight selection)*

On having specified the network topology, a set of input-output data is used to 'train' the network; i.e. to determine appropriate values for the weights (including the bias terms) associated with each interconnection. The data are propagated forward through the network to produce an output which is compared with the corresponding output in the data set, hence generating an error. This error is minimised by making changes to the weights, and may involve many passes through the training data set. When no further decrease in error is possible, the network is assumed to have 'converged', and the last set of weights is retained as the parameters of the NNM. Process modelling using ANNs is therefore very similar to identifying the coefficients of a parametric model of specified order. Loosely speaking, specifying the topology of an ANN is similar to specifying the 'order' of the process model. For a given topology, the magnitudes of the weights define the characteristics of the network. However, unlike conventional parametric model forms, which have an *a priori* assigned structure, the weights of an ANN also define the structural properties of the model. Thus, an ANN has the capability to represent complex systems whose structural properties are unknown.

A numerical search technique is usually applied to determine the weights as this task is usually not amenable to analytical solution. Clearly, determining the weights of the network can be regarded as a nonlinear optimisation problem. The objective function for the optimisation is written as

$$V(\Theta, t) = \tfrac{1}{2}\Sigma E(\Theta, t)^2 \tag{3}$$

where Θ is a vector of network weights, E is the output prediction error, and t is time. The simplest optimisation technique makes use of the Jacobian of the objective function to determine the search direction, and can be generalised by

$$\Theta^{t+1} = \Theta^t + \delta^t S \nabla V(\Theta, t) \tag{4}$$

where δ^t is the 'learn' rate which influences the rate of weight adjustments, and S is the identity matrix. Eqn. 4 was used by Rumelhart and McClelland [19] as the basis for their 'back-error propagation' algorithm, which is a distributed gradient descent technique. In this approach, weights in the jth layer are adjusted by making use of locally available information, and a quantity which is 'back-propagated' from neurons in the $j+1$th layer. However, it is well known that steepest descent methods may be inefficient, especially when the search approaches the minima. Therefore, in most neural network applications, a 'momentum' term is added to eqn. 4 as follows:

$$\Theta^{t+1} = \Theta^t + \delta^t S \nabla V(\Theta, t) + \beta(\Theta^t - \Theta^{t-1}) \qquad 0 < \beta < 1 \tag{5}$$

i.e. the current change in weight is forced to be dependent upon the previous weight change. Here, β is a factor which is used to influence the degree of this dependence. Although this modification does yield improved performances, in training networks where there are numerous weights, gradient methods have to perform exhaustive searches and are also rather prone to failure. A potentially more appealing method would be a Newton-like algorithm which includes second derivative information. In this case, the S matrix in eqn. 4 would be the inverse of the Hessian. Recently, Sbarbaro [20] approximated the Hessian using an approach similar to that utilised in 'classical' recursive system identification algorithms (e.g. Reference 21), and showed that faster network convergence could be achieved. The increase in computational overheads, due to the matrix inversion, is balanced by improved convergence speeds.

An alternative approach was proposed by Bremermann and Anderson [18]. Postulating that weight adjustments occur in a random manner, and that weight changes follow a multivariate Gaussian distribution with zero mean, their algorithm adjusts weights by adding Gaussian distributed random values to old weights. The new weights are accepted if the resulting prediction error is smaller than that recorded using the previous set of weights. This procedure is repeated until the reduction in error is negligible. They claimed that the proposed technique was 'neurobiologically more plausible'. Indeed, parallels have been drawn to bacterial 'chemotaxis', and thus the procedure is referred to as the chemotaxis algorithm.

In this contribution, attention will be confined to the back-error propagation and the chemotaxis algorithm: the former because of its popularity, and the latter because of its flexibility and ease of use. The respective techniques are summarised in Appendixes 7.1 and 7.2.

2.3 Comparison of back-error propagation and the chemotaxis algorithm

The complexity of the weight adjustment 'rule' has been a major criticism of the back-error propagation philosophy [22]. On the other hand, the chemotaxis algorithm, which was proposed as an alternative learning mechanism, appears to adjust network weights in an arbitrary manner. However, algorithms that are formulated within a 'random walk' framework have long been preferred in particle physics (e.g. see References 23 and 24). In Reference 24 it was shown that there was a useful connection between multivariable optimisation and statistical mechanics (the behaviour of systems with many degrees of freedom in thermal equilibrium at a finite temperature). Indeed, this analogy provided the basis for an increasingly popular optimisation technique known as 'simulated-annealing' [24]. In applications to very large scale problems, this method has been found to be faster and more accurate than other combinatorial minimisation procedures.

The chemotaxis approach is analogous to the simulated-annealing technique, and should therefore possess similar accelerated convergence properties. In fact, modifying the chemotaxis algorithm to that of simulated-annealing is relatively easy. The basic differences between the two are that simulated-annealing generates changes in search directions according to the Boltzmann probability distribution, and that it has the ability to accept values that cause an increase in the objective function. Whilst this algorithm is marginally more complex, it has a greater potential for achieving global minimum. Although Bremermann and Anderson [18] did attempt to incorporate the essence of simulated-annealing within their chemotaxis approach, they concluded that the modification was detrimental to their algorithm.

As may be appreciated, it is in the application to complex systems where ANNs are likely to achieve their maximum benefits. Under these circumstances, it is possible that a complex network architecture is also necessary, and it is here that the back-error propagation becomes extremely cumbersome. Since the chemotaxis algorithm presents a more efficient method for determining the weights of large networks, it is potentially a more viable technique. To compare the two approaches, a (1-2-1) and a (1-10-10-1) network with known weights were used to generate two sets of 'standard' test data. These sets were then used to train corresponding 'empty' networks using both the chemotaxis algorithm and the back-error propagation. The ideal technique would result in a set of weights equivalent to those used to generate the data in the respective networks.

The performances of the two network paradigms were quantified in terms of the integral of squared error (ISE). Figs. 3 and 4 show the ISE, recorded as a function of the number of iterations in the training circuit, for the (1-2-1) and (1-10-10-1) networks, respectively. In both figures, the ISE recorded using the back-error propagation showed a smoother trajectory. The 'erratic' ISE traces relate to the chemotaxis approach, and are a direct consequence of the random manner in which the weights of the network were being adjusted. In this comparison, an attempt was made to achieve the best possible results from both techniques; i.e. to achieve minimum ISE using the least number of iterations.

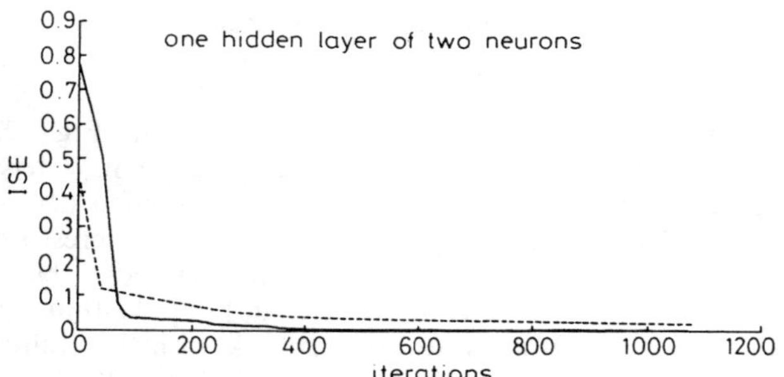

Fig. 3 *ISE against iterations using a (1-2-1) network*
------ back-error propagation
——— chemotaxis algorithm

From Figs. 3 and 4, the chemotaxis approach appears to produce superior results. For the (1-2-1) network, this contradicts the results of Bremmerman and Anderson [18], who showed that back-error propagation approach was superior for simple network structures, whereas the chemotaxis method would be better for larger networks. Nevertheless, the objective of this study was not to prove the superiority of one technique over the other. Indeed, it is possible that given the correct choice of experimental conditions, back-error propagation could be shown to be superior for all cases. The issue here is to demonstrate

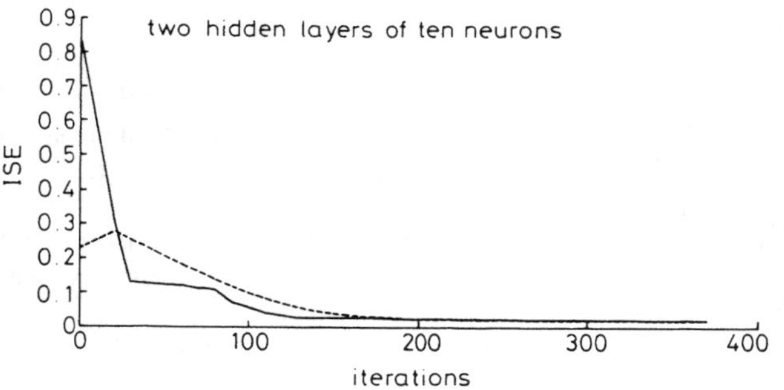

Fig. 4 *ISE against iterations using a (1-10-10-1) network*
------ back-error propagation
——— chemotaxis algorithm

that the chemotaxis algorithm is not only a feasible approach, it is also a more alluring technique due to its neurobiological appeal; the potential to avoid local minima; ease of computation and its implementation flexibility. A good example of the latter feature, is the ease with which dynamics may be incorporated into an FANN.

2.4 Dynamic neural networks

The ANNs discussed above merely perform a nonlinear mapping between inputs and outputs. Dynamics are not inherently included within their structures, whereas in many practical situations dynamic relationships exist between inputs and outputs. As such, the ANNs will fail to capture the essential characteristics of the system. Although dynamics can be introduced in a rather inelegant manner by making use of time histories of the data, a rather more attractive approach is inspired by analogies with biological systems. Studies by Holden [17] suggest that dynamic behaviour is an essential element of the neural processing function. It has also been suggested that a first-order lowpass filter may provide the appropriate representation of the dynamic characteristics [25]. The introduction of these filters is relatively straightforward, with the output of the neuron being transformed in the following manner:

$$y^f(t) = \Omega y^f(t-1) + (1-\Omega)y(t) \quad 0 \leqslant \Omega \leqslant 1 \tag{6}$$

Suitable values of filter time constants cannot be specified *a priori*, and thus the problem becomes one of determining Ω in conjunction with the network weights. An appealing feature of the chemotaxis approach is that the algorithm does not require modification to enable incorporation of filter dynamics: the filter time constants are determined in the same manner as network weights.

A particular instance where the use of a time history would still be appropriate is when uncertainty can exist over system time delays. The use of input data over the range of expected dead-times could serve to compensate for this uncertainty. In this situation a 'limited' time history together with neuron dynamics may be appropriate.

3 Potential application areas

In applying the ANN approach to process modelling, it is necessary to specify the topology of the network. Whereas two hidden layers have been claimed to be sufficient to approximate any nonlinear function, in any given data set one hidden layer may be adequate to describe the system behaviour. Therefore the approach adopted was to commence with a relatively small network and progressively increase both the amount of neurons and hidden layers until no significant improvement in fit was observed on the test data. Whilst recognising this as an *ad hoc* procedure, present theory does not allow a more rigorous mathematical approach. However, recent work [26] has indicated that is is possible to develop procedures which aid in the topology specification.

3.1 Inferential estimation

There are many industrial situations where infrequent sampling of the 'controlled' process output can present potential operability problems. Linear adaptive estimators have been employed to provide 'fast' inferences of variables that are 'difficult to measure' [27]. Although results from industrial evaluations have been promising, it is suggested that, due to their ability to capture nonlinear characteristics, the use of NNMs may provide improved estimation performances. The following sections present the results of some evaluation studies.

3.1.1 Biomass estimation in continuous mycelial fermentation: Due to industrial confidentiality, a thorough description of the fermentation process is not permissible. Nevertheless, note that biomass concentration within the fermenter is the primary control variable. However, with existing sensor technology, online measurements are not possible. In this example, biomass concentrations are determined from laboratory analysis and only made available to the process operators at four-hourly intervals. However, this frequency is inadequate for effective control. The control problem would be alleviated if the process operators were provided with more frequent information. Fortunately, a number of other (secondary) process measurements, such as dilution rate (the manipulative input), carbon dioxide evolution rate (CER), oxygen uptake rate (OUR), and alkali addition rate, provide useful information pertaining to biomass concentration in the fermenter. If the complex relationships between these variables and biomass concentration can be established via an ANN, then the resulting NNM could be used to infer biomass concentrations at a frequency suitable for control purposes. Although other variables, e.g. pH, temperature etc., affect the nonlinear relationship, tight environmental regulation maintains these at an approximate steady state.

These preliminary neural network studies considered a time history of inputs in order to incorporate dynamics. To model the process, a (6-4-1) network was specified, and 'trained' on appropriate process data. Although a number of input variables could be used to train the ANN, any duplication of information in the network could be detrimental to the quality of estimates. This problem is similar to that due to multicollinearities in multivariate regression. Inputs to the network should therefore be chosen with care. Two variables have already been identified as being 'critical' in previous estimation work [3], namely CER and fermenter dilution rate. Thus, at each pass of the training circuit, inputs to the network were the current values of CER and dilution rate values in the data base. Additionally, two previous respective data pairs were also used. The data for comparison with network output for weight adjustment purposes was the biomass concentration corresponding to the first input data pair. Therefore, six neurons were specified in the input layer, and one in the output layer. This serves to demonstrate that time histories of the data can be used to incorporate dynamics. However, the use of dynamic neural networks as described in Section 2.4 is a more elegant technique and will therefore be demonstrated in later examples.

Two sets of fermentation data were used to assess the goodness of fit produced by the neural net model. Fig. 5 demonstrates the ability of the neural network to fit the first data set, even when there was a major change in operating conditions. Here, a step change in the fermenter dilution rate has been applied at approximately 200 hours. The performance of the NNM when applied to the second data set is shown in Fig. 6. It should be

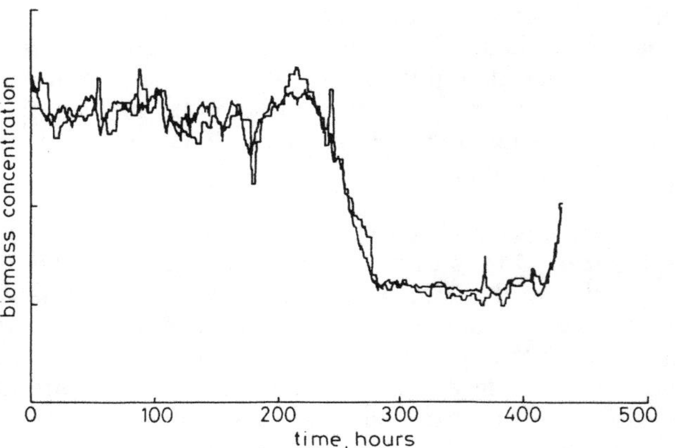

Fig. 5 *Application of neural network estimator to a continuous mycelial fermentation (data set 1)*

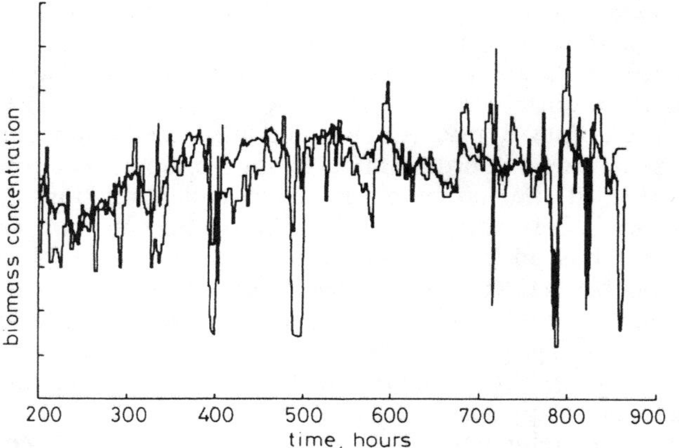

Fig. 6 *Application of neural network estimator to a continuous mycelial fermentation (data set 2)*

noted that, at around 400 and 500 hours, the CO_2 analyser failed, resulting in poor biomass estimates. Nevertheless, the overall ability of the neural network to fit the fermentation data in the presence of significant process disturbances was not compromised. The integrity of networks in the presence of such sensor failures is a subject of current study.

3.1.2 Biomass estimation in fed-batch penicillin fermentation: The measurement difficulties encountered in the control of penicillin fermentation are similar to those experienced in the continuous fermentation system described above. The growth rate of the biomass has to be controlled to a low level in order to optimise penicillin production. Whilst the growth rate can be influenced by manipulating the rate of substrate (feed) addition, it is not possible to measure the rate of growth online. The aim is, therefore, to develop a model which relates an online measurement to the rate of biomass growth. In this case, the OUR was used as the online measured variable, although the CER could have provided the same information. This is because the ratio of CER to OUR, known as the respiratory quotient (RQ), remains constant after approximately 20 hours into the fermentation.

The continuous fermentation process considered previously normally operates around a steady state. Thus a linear estimator may prove sufficient in the vicinity of normal process operating conditions. On the other hand, the fed-batch penicillin fermentation presents a more difficult modelling problem since the system passes through a wide spectrum of dynamic characteristics, never achieving a steady state. The results of applying an NNM to predict fermentation behaviour should therefore provide an indication of the viability of the technique for modelling complex processes. Data from two fermentation batches were used to train the neural network. The current online measurement of OUR was one of the inputs to the neural network. Unlike the previous case, fermenter feed rate was not used as an input variable since it is essentially constant from batch to batch at any point of each fermentation. Thus, the information it contains would not contribute towards the prediction of variations in biomass levels between fermentations. However, since the characteristics of the fermentation are also a function of time, the batch time was considered a pertinent input. The specified network therefore had a (2-3-1) toplogy. Figs. 7 and 8 show the performance of the

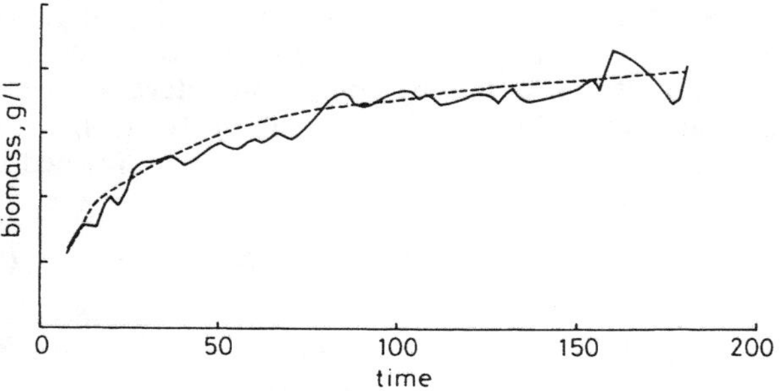

Fig. 7 *Biomass estimation of penicillin process (data set 1)*
——— actual
------- estimate

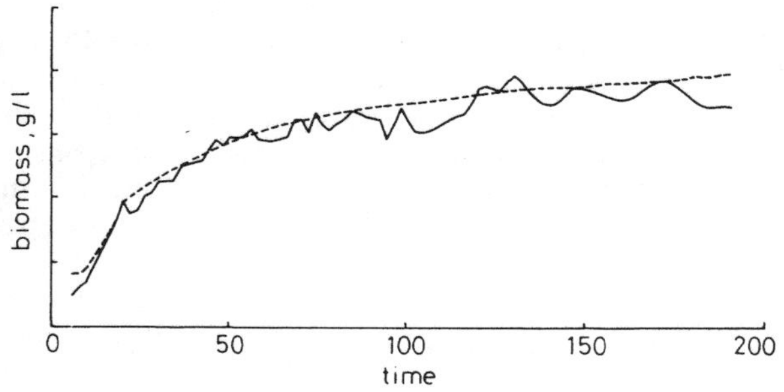

Fig. 8 *Biomass estimation of penicillin process (data set 2)*
——— actual
------- estimate

neural estimator when applied to the two training data sets. As expected, good estimates of biomass were achieved.

OUR data from another fermentation batch were then introduced to the NNM, resulting in the estimates shown in Fig. 9. It can be observed that the estimates produced by the NNM is very acceptable, and almost as good as those observed in Figs. 7 and 8. These results are very encouraging, since work on the development of a nonlinear observer to estimate the biomass concentration of the penicillin fermentation has not been as successful [28]. Compared to the mechanistic model based observer, relatively good estimates have been achieved without the need for corrective action from offline biomass assays. Nevertheless, the possibility of introducing offline

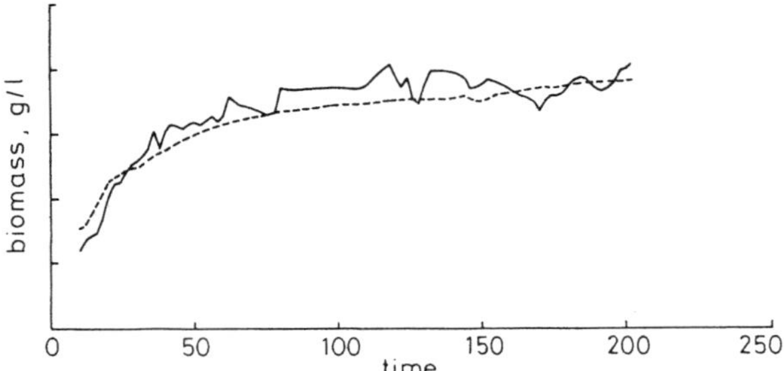

Fig. 9 *Biomass estimation of penicillin process (data set 3)*
——— actual
------- estimate

biomass data to improve the performance of the neural estimator is presently under investigation.

3.1.3 Estimation of product composition in a high purity distillation: Distillation columns, especially those with high purity products, can exhibit highly nonlinear dynamics, resulting in controllability problems. These difficulties may be exacerbated by the measurement delays due to the use of online composition analysers. The column being considered here is an industrial demethaniser, where the control objective is to regulate the composition of the high purity top product stream by varying reflux flow rate. Apart from top product composition, which is subject to an analyser delay of 20 minutes, all other process variables are available at 5 minute intervals. As for the bioreactor systems discussed above, the aim is to use fast secondary measurements, here column overheads vapour temperature and reflux flow rate, to provide an estimate of a slow primary measurement (top product composition) every 5 minutes. Since column vapour temperature can be overly sensitive to disturbances which do not necessarily affect product composition, the use of a tray liquid temperature would be preferable. However, this choice of secondary variable was dictated by its availability at the time of the tests.

The neural estimator was based on a (2-9-9-1) network. First, the weights of a network consisting of neurons without dynamics were determined. It can be observed from Fig. 10 that the estimates were rather poor and estimator performance was inconsistent. The addition of first-order lowpass filters to the output of each neuron, as described previously, were used to incorporate dynamics into the network. The filter constants were determined simultaneously with network weights. The resulting NNM produced the results shown in Fig. 11. Clearly, much improved estimates have been achieved, thus highlighting the advantage of introducing dynamics into the network.

3.2 ANN applications in control

3.2.1 Inferential control via ANNs: With the availability of 'fast' and accurate product quality estimates, the option of closed loop 'inferential' control instantly becomes feasiable. The effectiveness of such a strategy has been demonstrated by Guilandoust *et al.* [5, 29] where adaptive linear algorithms were used to provide inferred estimates of the controlled output for feedback control. Here, the practicality of an NNM based inferential control scheme is explored via nonlinear simulation.

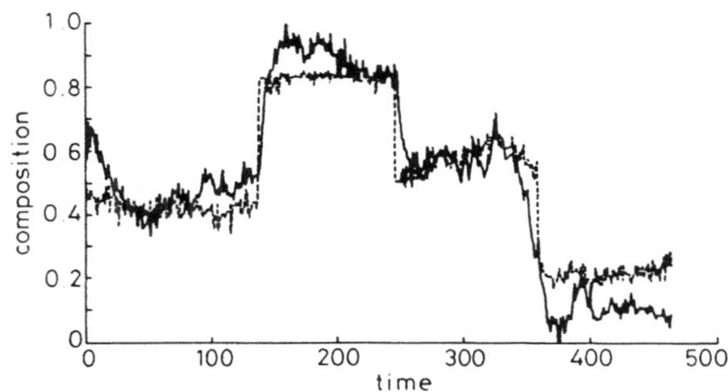

Fig. 10 *Neural network estimator applied to high purity column (back-error propagation)*
------- neural estimation
——— measure

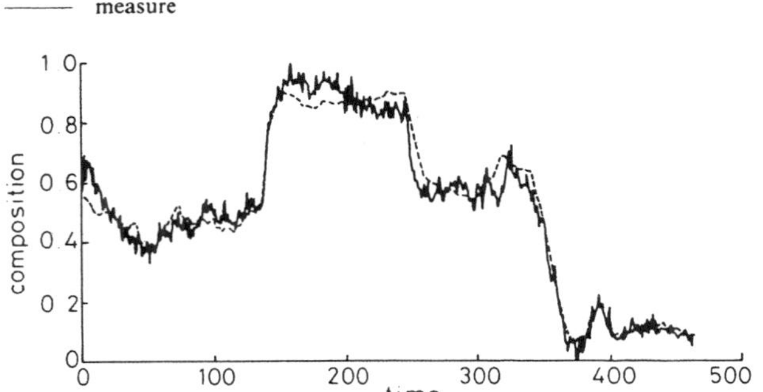

Fig. 11 *Neural network estimator applied to high purity column (chemotaxis algorithm)*
------- neural estimation
——— measure

The process under consideration is a ten stage pilot plant column, installed at the University of Alberta, Canada (see Fig. 12). It separates a 50-50 wt.% methanol-water feed mixture which is introduced at a rate of 18.23 g/s into the column on the fourth tray. Bottom product composition is measured by a gas chromatograph which has cycle time of 3 minutes. The control objective is to use steam flow rate to regulate bottom product composition to 5 wt.% methanol, when the column is disturbed by step changes in feed flow rate. The column is modelled by a comprehensive set of dynamic heat and mass balance relationships [30]. Both the pilot plant and the nonlinear model have been widely used by many investigators to study different advanced control schemes (e.g. [31]), including adaptive inferential control [5, 29].

The training data set consists of paired values of the input variable, i.e. temperature of the liquid on tray 2 (the second tray from the reboiler), and the output variable, i.e. the corresponding bottom product composition. The data was collected at 3 minute intervals, this being the cycle time of the gas chromatograph. The chemotaxis algorithm was used to determine both the weights and filter time constants of a (2-4-4-1) network.

Fixed parameter PI controllers were used to effect composition regulation, and their initial settings were determined using the techniques described by Willis [9]. The controllers were then fine-tuned to obtain responses with minimum integral of absolute error. Comparative performances of PI feedback control and NNM based inferential control under major disturbance conditions, assessed by subjecting the column to a sequence of step disturbances in feed flow rate, namely at 10% decrease

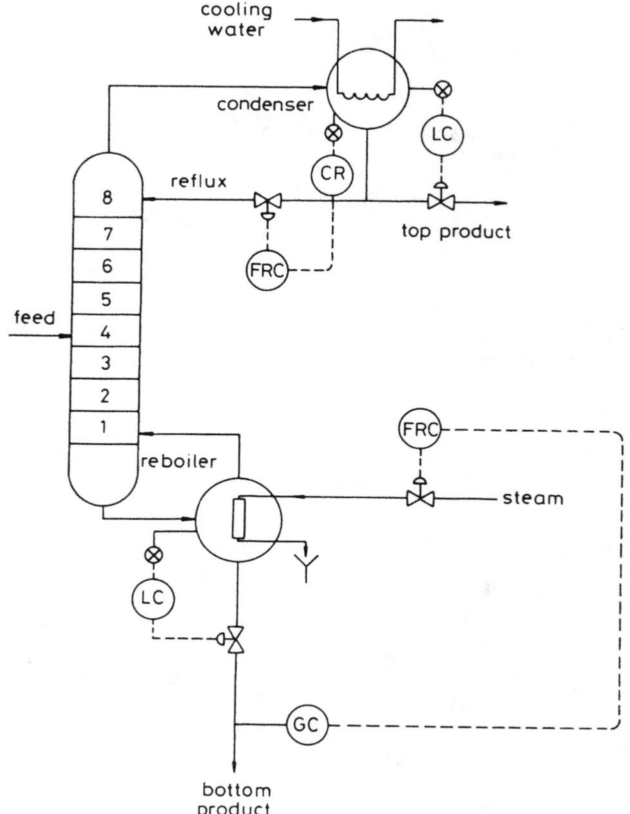

Fig. 12 *Schematic of pilot plant distillation column*

CR = analyser recorder
FRC = flow recorder/controller
GC = gas chromatograph
LC = level controller

from steady state followed by a 10% increase back to steady state. As reported in References 5 and 29, the responses of bottom product composition and tray liquid temperatures exhibit quite different gains and time constants, depending on the direction of inputs. Thus, the NNM must be able to characterise these nonlinear effects in order to provide accurate composition estimates. Moreover, for the inferential control technique to be practicable, the NNM is required to provide composition estimates at a rate faster than that obtainable from the gas chromatograph. This is achieved by introducing into the NNM, tray 2 liquid temperatures sampled every 0.5 minutes. Although the NNM had been trained on data collected at 3 minute intervals, the use of data obtained at a faster rate is permissible because all that is required is a consistent data set.

The inferential controlled response, using a PI controller ($K_c = 2.2$ and $T_I = 20$), together with composition estimates, are shown in Fig. 13. It can be observed that, although the estimate and actual composition compare favourably, there is a difference between the two values. This is because the neural estimator is operating in the open loop; i.e. the estimates are not corrected by feedback of measured composition. Thus, if the estimates are used for control, offsets between the actual process output and the desired value will occur. However, offset free inferential control can be achieved by treating the elimination of estimation errors as a control problem. Here, whenever a new composition value is available, i.e. every 3 minutes, the estimation error is calculated and

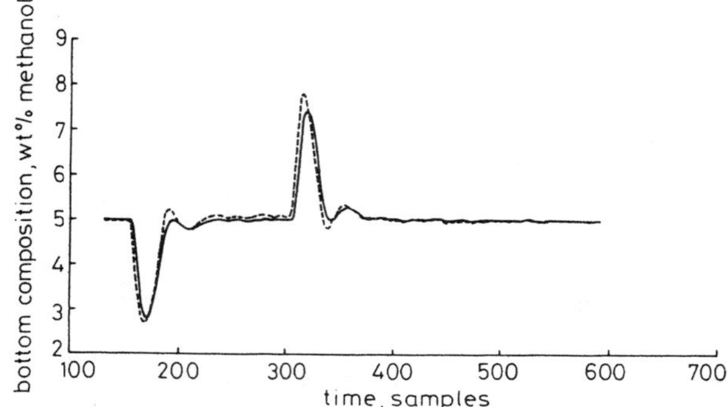

Fig. 13 *Inferential control using a neural network*

——— actual
------- estimate

used to correct subsequent estimates. This technique is based on the internal model control (IMC) strategy of Garcia and Morari [32]. The success of this modification is demonstrated in Fig. 13, where it may be observed that it is the actual composition that is at setpoint.

Fig. 14 demonstrates the effect of using the actual (analyser delayed) composition values for control. The

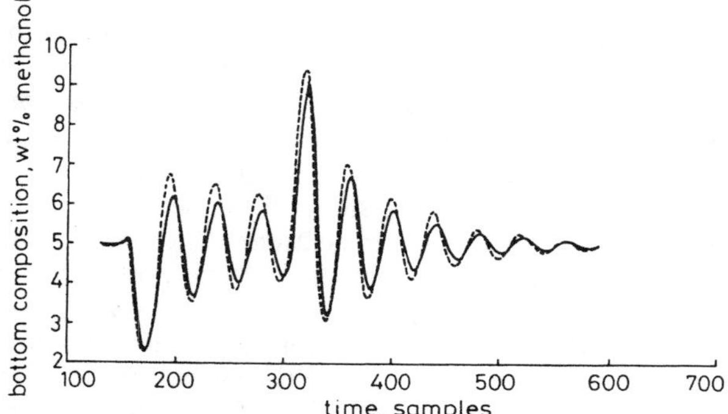

Fig. 14 *Control using analyser delayed value*

——— actual
------- estimate

oscillatory responses observed are a direct consequence of using the same control settings as in the previous case. It is quite clear that these parameters are unsuitable for a feedback signal which suffers from analyser delays. Reducing the proportional gain to 1.57 resulted in the response shown in Fig. 15, where the response shown in Fig. 13 has also been superimposed. Although a much improved performance has been achieved, comparison with the inferential controlled response will show that there is an increase in peak overshoot of more 1.0 wt.% methanol.

The performance of the NNM based inferential control scheme is clearly better. Using the output of the neural estimator for feedback control may be regarded as implicitly providing dead-time compensation in the closed loop. This then permits the use of higher gain control. Moreover, as changes in feed flow rate affect tray 2 temperature before bottom product composition, the estimates will contain information about the impending effects of the disturbance. As a result, disturbance rejection responses are superior to that obtained using conventional feedback control.

3.2.2 Neural predictive control: If the NNM is of sufficient accuracy, then it should also be possible to employ

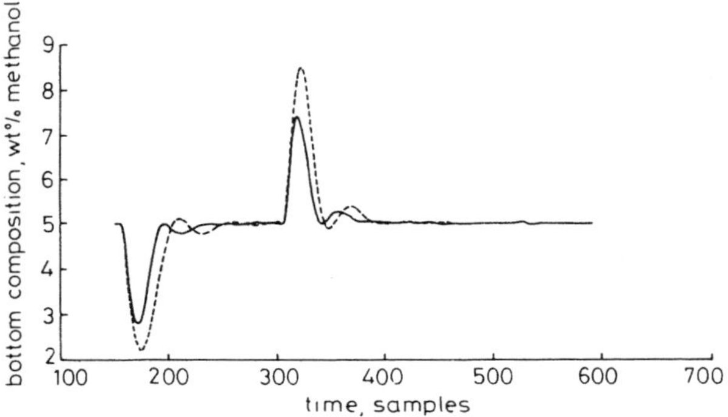

Fig. 15 *Detuned control using analyser delayed value*

------- PI control
——— FANN

the NNM directly within a model based control strategy. A potentially useful algorithm would be one which minimises future output deviations from setpoint, whilst taking into account the control action necessary to achieve the objective. Being common to predictive control strategies, this concept is not new. However, the attraction of using a NNM instead of other model forms is its ability to effectively represent complex nonlinear systems. Thus the resulting controller should prove to be more robust in practical situations, where the structure of the nonlinearity is usually unknown: a common cause of failures in the current approaches to nonlinear control system design. The proposed neural predictive controller (NPC) minimises the following loss function:

$$J(N_1, N_2, N_u) = \sum (Y_{t+j} - W_{t+j})^2 + \sum \mu \, \Delta U_{t+k-1}^2 \qquad (7)$$

i.e. the sum of squares of future setpoint tracking errors is minimised. N_1 is the minimum output prediction horizon; N_2 is the maximum output prediction horizon; N_u is the control horizon; and μ is the move suppression factor. The prediction horizons specifies the range of future outputs which are being considered in the minimisation, while the control horizon defines the number of control moves required to reach the desired objective. If μ is nonzero, excessive control effort will also be penalised. The role of the NNM is to provide the outputs predictions (Y_{t+j}, $j = N_1, \ldots, N_2$). If these are accurate, then, knowing the future setpoints, minimisation of eqn. 7 allows the determination of the necessary future control moves. Depending on the control horizon specified, a range of future control moves may have been calculated at each sample instant. However, control is implemented in a 'receding horizon' manner; i.e. only the first of the calculated control sequence is applied, so as to avoid uncertainties associated with extrapolation. The structure of such a controller is shown in Fig. 16. As with IMC strategy, the primary function of the outer loop is to eliminate prediction error. The proposed NPC algorithm is summarised in Appendix 7.3.

To assess the viability of the NPC, it was applied to the nonlinear exothermic continuous stirred tank reactor (CSTR) example studied by Economou and Morari [33]. The objective of control is to enable the reactor to operate at near maximum feed conversion conditions and, at the same time, prevent thermal runaway. The manipulated variable is the inlet feed stream temperature, and the output concentration is the controlled variable. Feed conversion has a well defined maximum with respect to temperature. As a result, deviations either side of the maximum cause the system gain to change sign. Systems of this type are not integral controllable [33]; i.e. they will not remain stable over the entire operating region when integrating controllers are employed. To

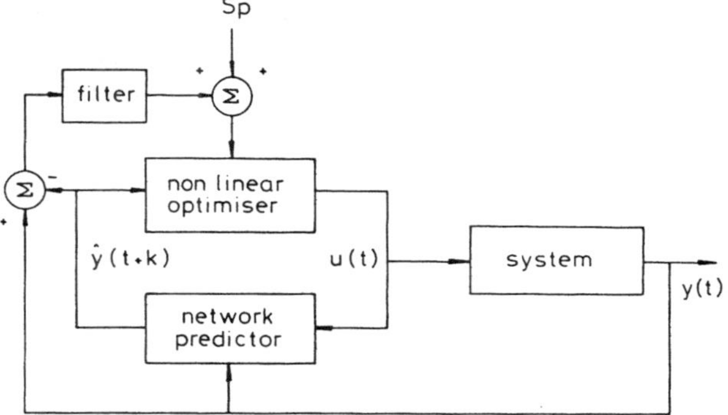

Fig. 16 *Block diagram of neural predictive controller*

maintain stability, proportional controllers must be used instead. However, this results in offsets from the desired value.

The NPC algorithm, with $N_1 = N_2 = N_u = 1$, and $\mu = 0.1$, was applied to the CSTR. The resulting controller could therefore be interpreted as a variant of the generalised minimum variance (GMV) algorithm of Clarke and Gawthrop [34]. For clarity of exposition, it was assumed that there is no analyser dead-time (or equivalently, a 'perfect' inference of composition is possible). This allows attention to be focussed on the effectiveness of the proposed strategy in controlling a process with severe nonlinear characteristics. A (1-9-9-1) network was trained on a set of feed flow rate and output concentration data obtained by subjecting the CSTR to a series of setpoint changes whilst under proportional control. Once the NNM was considered to have converged, the network weights were fixed, and the NNM used within the NPC scheme to provide predictions of output concentrations. The Nelder-Mead algorithm [35] was then employed to determine the control which minimised the deviations of the predicted outputs from setpoint. During implementation, 'hard' limits on the control signal were also imposed to avoid network saturation. This should not limit the applicability of the technique since such constraints are commonly applied in practice.

Fig. 17 shows the system responding to a series of setpoint demands. It can be observed that an offset occurred

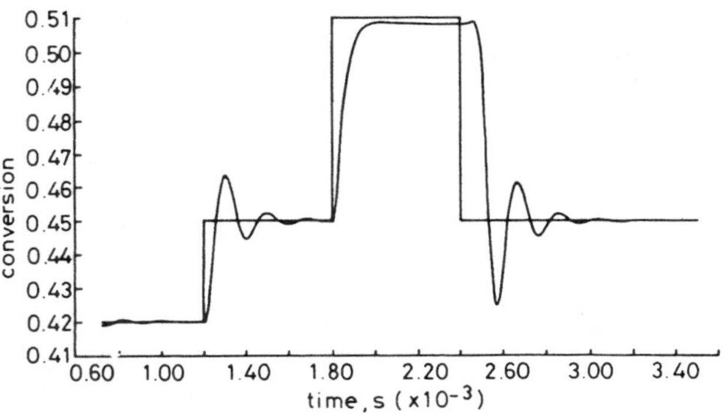

Fig. 17 *Reactor conversion*

during the second setpoint change. This is because the output concentration is already at the maximum possible, a result of limitations in reaction kinetics. This in turn caused the manipulated input to saturate at its lower allowable limit, as shown in Fig. 18. The consequence of this is reflected in the response to the next setpoint demand: the output concentration responds only after a long delay (Fig. 17). However, setpoint tracking was quickly restored once the saturation effects have dissipated. Noting that the CSTR is a severe test of controller

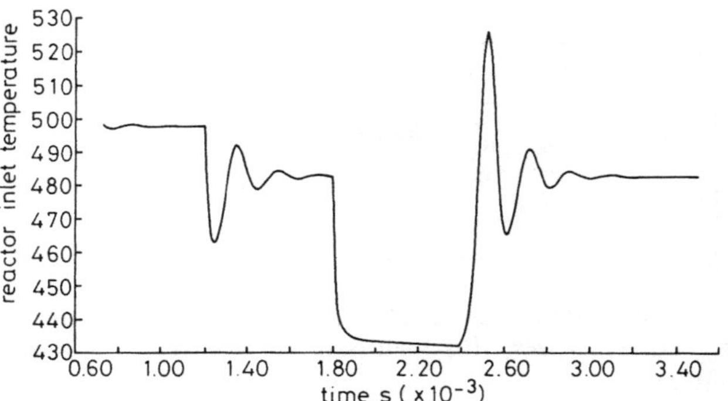

Fig. 18 *Reactor inlet temperature*

capabilities, the controlled response can be considered to be quite adequate. The ability of the NPC to recover gracefully from control saturation effects is particularly encouraging. Although the results are by no means conclusive, they highlight the potential of using NNMs in nonlinear controller synthesis.

4 Concluding remarks

In this contribution, the concept of artificial neural networks was introduced. Additionally, two different neural network paradigms were discussed and their performances evaluated. Compared to the popular back-error propagation, the chemotaxis algorithm was found to be more flexible and easier to implement. The applicability of neural networks to solving some, currently difficult, process engineering problems was then explored.

First, NNMs were used as process estimators. Of particular interest is the ability of the neural estimators to provide 'fast' inferences of important, but 'difficult-to-measure', process outputs, from other easily measured variables. Applications to data obtained from industrial processes reveal that, given an appropriate topology, the network could be trained to characterise the behaviour of all the systems considered: namely industrial continuous and fed-batch fermenters, and a commercial scale high purity distillation column. ANNs therefore exhibit potential as 'soft-sensors', i.e. sensors based on software rather than hardware [27]. Next, it was demonstrated that significant improvements in process regulation can be achieved, if the estimates produced by the neural estimators were used as feedback signals for control. This is possible because the use of secondary variables means that the effects of load disturbances could be anticipated in a feedforward sense. In principle, an NNM can also be employed directly within a model based control strategy. A tentative strategy was described, where the ability of the NNM to represent nonlinear systems was exploited to provide nonlinear predictive control. The results of applications to a highly nonlinear exothermic reaction system were highly encouraging.

The results presented are evidence that ANNs can be a valuable tool for alleviating many current process engineering problems. However, it is stressed that the field is still very much in its infancy and many questions still have to be answered. Determining the 'optimum' network topology is one example. Currently, *ad hoc* procedures based on 'gut-feeling' are used. This arbitrary facet of an otherwise promising philosophy is a potential area of active research. A formalised technique for choosing the appropriate network topology is desirable. There also appears to be no established methodology for determining the stability of ANNs. This is perhaps the most important issue that has to be addressed before the full potentials of ANN methodologies can be realised online. Nevertheless, given the resources and effort that are currently being infused into both academic and commercial research in this area, it is predicted that, within the decade, neural networks will have established themselves as valuable tools for solving process engineering problems.

5 Acknowledgments

The authors gratefully acknowledge the support of the Department of Chemical and Process Engineering, University of Newcastle upon Tyne; Smith Kline Beechams, Irvine; ICI Engineering; ICI Bioproducts; and Dr. David Peel Dr. C. Kambhampati for their critical comments and kind assistance.

6 References

1 BHAT, N., MINDERMAN, P., McAVOY, T., and WANG, N.: 'Modelling chemical process systems via neural computation'. Proc. 3rd Int. Symp. 'Control for Profit', Newcastle-upon-Tyne, 1989

2 BHAT, N., MINDERMAN, P., and McAVOY, T.J.: 'Use of neural nets for modelling of chemical process systems'. Preprints IFAC Symp. Dycord + 89, Maastricht, The Netherlands, Aug. 21–23, 1989, pp. 147–153

3 MONTAGUE, G.A., HOFLAND, A.G., LANT, P.A., DI MASSIMO, C., SAUNDERS, A., THAM, M.T., and MORRIS, A.J.: 'Model based estimation and control: adaptive filtering, nonlinear observers and neural networks'. Proc. 3rd Int. Symp. 'Control for profit', Newcastle-upon-Tyne, 1989

4 LANT, P.A., WILLIS, M.J., MONTAGUE, G.A., THAM, M.T., and MORRIS, A.J.: 'A comparison of adaptive estimation with neural based techniques for bioprocess application'. Preprints ACC, San Diego, 1990, pp. 2173–2178

5 GUILANDOUST, M.T., MORRIS, A.J., and THAM, M.T.: 'Adaptive inferential control', *IEE Proc. D, Control Theory & Appl.*, 1987, **134**, pp. 171–179

6 DI MASSIMO, C., WILLIS, M.J., MONTAGUE, G.A., KAMBHAMPATI, C., HOFLAND, A.G., THAM, M.T., and MORRIS, A.J.: 'On the applicability of neural networks in chemical process control', to be presented at AIChE Annual Meeting, Chicago, 1990

7 PSALTIS, D., SIDERIS, A., and YAMAMURA, A.A.: 'A multilayered neural network controller', *IEEE Control Syst. Mag.*, April 1988, pp. 17–21

8 NARENDRA, K.S., and PARTHASARATHY, K.: 'Identification and control of dynamical systems using neural networks', *IEEE Trans.*, 1990, **NN-1**, (1), pp. 4–27

9 WILLIS, M.J.: 'Adaptive systems in chemical process control'. PhD thesis (in preparation)

10 BAVARIAN, B.: 'Introduction to neural networks for intelligent control', *IEEE Control Syst. Mag.*, April 1988, pp. 3–7

11 HOSKINS, J.C., and HIMMELBLAU: 'Artificial neural network models of knowledge representation in chemical engineering', *Comput. Chem. Engng.*, 1988, **12**, (9/10), pp. 881–890
12 BIRKY, G.J., and McAVOY, T.J.: 'A neural net to learn the design of distillation controls'. Preprints IFAC Symp. Dycord + 89, Maastricht, The Netherlands, Aug. 21–23, 1989, pp. 205–213
13 LIPPMANN, R.P.: 'An introduction to computing with neural nets', *IEEE ASSP Mag.*, April, 1987
14 CYBENKO, G.: 'Approximation by superpositions of sigmoidal function', *Math. Control Signal Syst.*, 1989, **2**, pp. 303–314
15 HECHT-NIELSON, R.: 'Neurocomputing' (Addison-Wesley, 1990)
16 RUMELHART, D.E., HINTON, G.E., and WILLIAMS, R.J.: 'Learning representations by back-propagating errors', *Nature*, 1986, **323**, pp. 533–536
17 HOLDEN, A.V.: 'Models of the stochastic activity of neurons' (Springer Verlag, 1976)
18 BREMERMANN, H.J., and ANDERSON, R.W.: 'An alternative to back-propagation: a simple rule for synaptic modification for neural net training and memory'. Internal Report, Department of Mathematics, University of California, Berkeley, 1989
19 RUMELHART, D.E., and McCLELLAND, J.L.: 'Parallel distributed processing: explorations in the microstructure of cognition. Vol. 1: Foundations' (MIT Press, 1986)
20 SBARBARO, D.: 'Neural nets and nonlinear system identification'. Control Engineering Report 89.9, University of Glasgow, 1989
21 LJUNG, L., and SODERSTROM, T.: 'Theory and practice of recursive identification' (MIT Press, 1983)
22 CRICK, F.: 'The recent excitement about neural networks', *Nature*, 1989, **337**, (12), pp. 129–132
23 METROPOLIS, N., ROSENBLUTH, A., ROSENBLUTH, M., TELLER, A., and TELLER, E.: *J. Chem. Phys.*, 1953, **21**, p. 1087
24 KIRKPATRICK, S., GELATT, C.D., and VECCHI, M.P.: 'Optimisation by simulated annealing', *Science*, 1983, **220**, pp. 671–690
25 TERZUOLO, C.A, McKEEN, T.A., POPPELE, R.E., and ROSENTHAL, N.P.: 'Impulse trains, coding and decoding', *in* TERZUOLO, C.A. (Ed.): 'Systems analysis to neurophysiological problems' (University of Minnesota, Minneapolis, 1969), pp. 86–91
26 WANG, J.: 'Multi-layered neural networks: approximated canonical decomposition of non-linearity'. Internal report, Department of Chemical and Process Engineering, University of Newcastle-upon-Tyne, 1990
27 THAM, M.T., MORRIS, A.J., and MONTAGUE, G.A.: 'Soft sensing: a solution to the problem of measurement delays', *Chem. Eng. Res. Des.*, 1989, **67**, (6), pp. 547–554
28 DI MASSIMO, C., SAUNDERS, A.C.G., MORRIS, A.J., and MONTAGUE, G.A.: 'Non-linear estimation and control of mycelial fermentations'. ACC, Pittsburgh, USA, 1989, pp. 1994–1999
29 GUILANDOUST, M.T., MORRIS, A.J., and THAM, M.T.: 'An adaptive estimation algorithm for inferential control', *Ind. Eng. Chem. Res.*, 1988, **27**, pp. 1658–1664
30 SIMONSMEIR, U.F.: 'Modelling of a nonlinear binary distillation column'. MSc Thesis, University of Alberta, Edmonton, Canada, 1977
31 MORRIS, A.J., NAZER, Y., WOOD, R.K., and LIEUSON, H.: 'Evaluation of self-tuning controllers for distillation column control'. IFAC Conf. 'Digital computer app. to process control', eds. Isermann and Haltenecker, Dusseldorf, Germany, 1981, pp. 345–354
32 GARCIA, C., and MORARI, M.: 'Internal model control, pt. 1, a unifying review and some new results', *Ind. Eng. Chem. Process Des. Dev.*, 1982, **21**, p. 308
33 ECONOMOU, C.G., and MORARI, M.: 'Internal model control, pt 5, extension to nonlinear systems', *Ind. Eng. Chem. Process Des. Dev.*, 1986, **25**, pp. 403–411
34 CLARKE, D.W., and GAWTHROP, P.J.: 'Self-tuning control', *Proc. IEE*, 1979, **126**, (6), pp. 633–640
35 PRESS, W.H., FLANNERY, B.P., TEUKOLSKY, S.A, and VETTERLING, W.T.: 'Numerical recipes' (Cambridge University Press, 1988)

7 Appendixes

7.1 Summary of the back-error propagation [19]

Step I: Initialise weights with small random values.

Step II: Set δ (the learn rate) and β (the momentum term) in eqn. 5.

Step III: Present a single set of inputs, and propagate data forward to obtain the predicted outputs.

Step IV: Update the weights according to eqn. 5.
For hidden to output weights, $V(\Theta, t)$ is given by

$$V(\Theta, t)_{ok} = I_{o,k}(1 - I_{ok})(y_k - I_{o,j})$$

where y_k is the actual process output and the subscipt 'o' denotes an output layer. For input to hidden weights, $V(\Theta, t)$ is given by

$$V(\Theta, t)_{jk} = \sigma_{j,k}^{t+1} I_{j-1,k}$$

where

$$\sigma_{j,k}^{t+1} = I_{j,k}(1 - I_{j,k})\Gamma_{j,k}^{t+1}$$

$$\Gamma_{j,k}^{t+1} = \sum_{k=1}^{n} \Gamma_{j+1,k}^{t} w_{j,k,i}$$

Step V: Return to Step III and proceed with the next data record.

Steps III to V are repeated for successive sweeps through the data set until convergence is deemed to have been achieved. Whereas an orderly sequence of input/output presentation is usually successful, in some cases superior results may be achieved by random choice of the data sequence (T. McAvoy, personal communication, 1990).

7.2 Summary of the chemotaxis algorithm [18]

Step I: Initialise weights with small random values.

Step II: Present the inputs, and propagate data forward to obtain the predicted outputs.

Step III: Determine the cost of the objective function E_1 over the whole data set.

Step IV: Generate a Gaussian distributed random vector.

Step V: Increment the weights with random vector.

Step VI: Calculate the objective function E_2, based on the new weights.

Step VII: If E_2 is smaller than E_1, retain the modified weights, set E_1 equal to E_2, and go to Step V. If E_2 is larger than E_1, go to Step IV.

Note, that, during the minimisation, the allowable variance of the increments may be adjusted to assist network convergence.

7.3 Summary of prototype neural predictive controller

Step I: Obtain the future setpoint sequence

$$W(t + j) \quad j = N_1, \ldots, N_2$$

Step II: Using the NNM, generate predictions of process outputs

$$Y(t + j \mid t) \quad j = N_1, \ldots, N_2$$

correcting for possible errors based upon knowledge of past estimation errors.

Step III: Calculate the future deviations from setpoint

$$e(t + j) = W(t + j) - Y(t + j \mid t) \quad j = N_1, \ldots, N_2$$

Step IV: Minimise the cost function, eqn. 7, to obtain the optimal sequence of controls:

$$U(t + j) \quad j = 0, 1, \ldots, N_u$$

Step V: Implement $U(t)$. Go to Step I.

Modeling Chemical Process Systems via Neural Computation

Naveen V. Bhat, Peter A. Minderman, Jr., Thomas McAvoy, and Nam Sun Wang

ABSTRACT: This paper discusses the use of neural nets for modeling nonlinear chemical systems. Three cases are considered: a steady-state reactor, a dynamic pH stirred tank system, and interpretation of biosensor data. In all cases, a back-propagation net is used successfully to model the system. One advantage of neural nets is that they are inherently parallel and, as a result, can solve problems much faster than a serial digital computer. Furthermore, neural nets have the ability to "learn." Rather than programming neural computers, one presents them with a series of examples, and from these examples the nets learn the governing relationships involved in the training data base.

Introduction

One reason for the recent interest in neural computation is that neural nets hold the promise of solving problems that have heretofore proven extremely difficult for traditional digital computers. Not only are neural nets parallel computing devices and, therefore, fast, but they are capable of learning by example. Currently, there are a wide variety of neural nets being studied or used in applications. Some of the important ones are the adaptive resonance networks, counterpropagation networks, neocognitrons, Hopfield networks, etc., each of which has its own special applications. However, the most widely used learning neural net is back-propagation [1].

Back-propagation is an example of a mapping network that learns an approximation to a function, y equal to $f(x)$, from sample x, y pairs. The fact that the function to be learned is nonlinear presents no problem to a back-propagation net. Representative applications of back-propagation include speech synthesis and recognition, visual pattern recognition, analysis of sonar signals, defense applications, medical diagnosis, and learning in control systems. We have used back-propagation successfully on a number of problems typical of those found in the chemical/petroleum industry, including sensor interpretation [2], dynamic modeling [3], [4], and in learning how to design distillation control systems. This paper draws together the authors' results in these areas and presents some new results on steady-state modeling of a nonlinear chemical reactor.

Figure 1 schematically shows an example of a back-propagation neural net used herein. The boxes and circles are neurons, and the lines between the neurons are called *interconnects*. As can be seen, a back-propagation net has three layers: input, hidden, and output. Cybenko [5] showed that a continuous neural network with two hidden layers and any fixed continuous sigmoidal nonlinearity, i.e., a back-propagation net, can approximate any continuous function arbitrarily well on a compact set. Although Cybenko's results do not give any insight into just how large a back-propagation net is required, they do show that the fundamental structure of back-propagation is such that it can model any continuous nonlinear function.

One of the chief barriers to the more widespread use of advanced modeling and control techniques in the chemical/petroleum industry is the cost of model development and validation. Often modeling costs account for over 75 percent of the expenditures in an advanced control project. Since neural nets can learn by example, they may offer a cost-effective method of developing useful process models.

This paper discusses the use of neural nets for modeling nonlinear chemical process systems. Both a steady-state reactor and a dynamic pH continuously stirred tank reactor (CSTR) are treated, as well as the use of back-propagation for interpreting biosensor data. It is shown that a back-propagation net is capable of learning the underlying relationships. Once a back-propagation model is available, then it can be used directly online, even now. When parallel chips are widely available, back-propagation will be even more attractive. Indeed, the back-propagation algorithm can be simulated in a standard digital computer so that no special hardware is required for its implementation. After discussing the back-propagation algorithm, the various examples are treated.

Back-Propagation

The governing equations for a back-propagation net, such as shown in Fig. 1, have

The authors are with the Department of Chemical Engineering, University of Maryland, College Park, MD 20742.

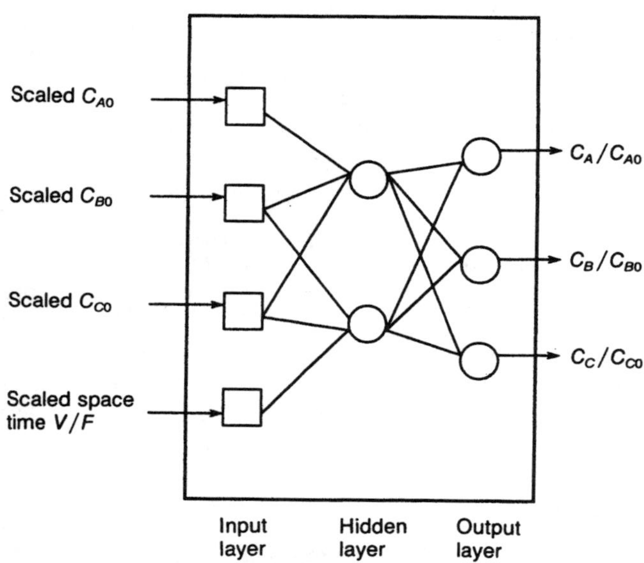

Fig. 1. BPN for steady-state modeling application.

Reprinted from *IEEE Contr. Syst. Mag.*, pp. 24–29, April 1990.

been derived by Rumelhart and McClelland [6] and are briefly reviewed here. The neurons in the input layer simply store the input values. The hidden layer and output layer neurons each carry out two calculations. First, they multiply all inputs and a bias (equal to a constant value of 1) by a weight and then they sum the result as

$$S_j = \sum_{i=1}^{N} w_{ij} x_i \tag{1}$$

Second, the output of the neuron, O_j, is calculated as the sigma function of S_j as

$$O_j = \sigma(S_j) \tag{2}$$

where

$$\sigma(z) = [1 + \exp(-z)]^{-1} \tag{3}$$

Note that not all of the nets use the sigmoidal function as given in Eq. (3), but the ones presented here do. A back-propagation net learns by making changes in its weights in a direction to minimize the sum of squared errors between its predictions and a training data set. The minimization is done using the steepest descent algorithm, which is known to have two main drawbacks. First, the rates of convergence of this algorithm are extremely slow and the improvement per iteration falls sharply. Other minimization techniques, such as the conjugate gradient or Newton's method, have better convergence properties and can be tried in the future. A second drawback is that this algorithm often leads to a local minima, but this fact has not posed a problem in the actual functioning of the net. Assume that there are R input/output pairs, $x^{(r)}y^{(r)}$, available for training the net. After presentation of pair r, the weights are changed as follows:

$$w_{uv}^{(r)} = w_{uv}^{(r-1)} + \Delta w_{uv}^{(r)} \tag{4}$$

with $\Delta w_{uv}^{(r)}$ given by:

Hidden to output weights:

$$\Delta w_{jk}^{(r)} = \sigma(S_k)\,[y_k^{(r)} - O_k^{(r)}]\,O_j \tag{5}$$

Input to hidden weights:

$$\Delta w_{ij}^{(r)} = \sigma(S_j) \left[\sum_{k=1}^{P} [\sigma'(S_k)] \cdot [y_k^{(r)} - O_k^{(r)}]\, w_{jk}^{(r-1)} \right] x_i^{(r)} \tag{6}$$

and

$$\sigma'(S_k) = \sigma(S_k)\,[1 - \sigma(S_k)] \tag{7}$$

After presentation of the first input/output pair, one proceeds with the second pair, and so on. The weights are changed with each presentation. One might assume that setting all the weights to zero may be an acceptable starting point. If all the weights start with equal values and the solution requires unequal weights to be developed, then the system can never learn. The reason is because the error is propagated back through the neurons in proportion to the value of the weights, as shown by Eq. (6). All the error signals to the hidden nodes remain identical, and the system starts out at a local minima and remains there. This problem is counteracted by starting the system with a set of randomized weights distributed uniformly between -0.5 and $+0.5$.

The chosen activation function has a special feature. The function cannot reach its final values of 0 and 1 without infinitely large inputs. The useful region of the activation function is approximately between 0.1 and 0.9, and the variables's input to the net is scaled within this range.

Steady-State Example

Model

To illustrate the use of back-propagation, the following isothermal CSTR reaction sequence is considered, where constituent A goes to constituent B with reaction constant k_1 while constituent B goes to constituent C with reaction constant k_2.

$$A \xrightarrow{k_1} B \xrightarrow{k_2} C$$

The volume of the reactor is V, the feed flow rate is F, and C_A and C_B are the concentrations of A and B, respectively. At steady state, the mole fractions of A, B, and C in the reactor are given by the following expressions:

$$C_A/C_{A0} = b_1 \tag{8}$$

$$C_B/C_{B0} = b_2(C_{B0}/C_{A0}) + (k_1 V/F) b_1 b_2 \tag{9}$$

$$C_C/C_{A0} = C_{C0}/C_{A0} + (k_2 V/F) b_2 C_{B0}/C_{A0} + k_1 k_2 (V/F)^2/(b_1 b_2) \tag{10}$$

where $b_i = [1 + k_i V/F]^{-1}$ for $i = 1, 2$.

It is assumed that historical data on product concentrations are available at the different operating conditions shown in Table 1. These data are used to train the net so that it can be used to optimize reactor yield. To generate a data set, k_1 is taken as 0.16 min^{-1} and k_2 as 0.06 min^{-1}. Equations (8)-(10) show that the product concentrations are a function of both the reactor feed composition and the space time, V/F. Furthermore, although the reaction mechanisms are linear, the kinetic parameters appear in the yield relationships in a nonlinear manner. In using neural nets, the underlying reactor kinetics are assumed to be unknown. In generating historical data, six different values of V/F and two different values of feed composition were assumed. Noise (10 percent Gaussian) was added to these data.

Back-Propagation Net Used

The back-propagation net used is schematically shown in Fig. 1. This net was simulated on a MicroVAX computer. The net inputs are the scaled feed composition and reactor space time. The outputs are the dimensionless product concentrations. One of the issues that has to be decided in configuring a back-propagation net is the number of hidden elements to be used. This number was arbitrarily set at nine for this illustration. The assumed historical data were fed to the net until convergence was achieved. The rate of convergence of the net is slow in this case. Figure 2 shows a plot of the root-mean-square (rms) error between actual and predicted concentrations as a function of the number of times that a data set was presented to the net. The rms error is the sum of the errors of the A, B, and C concentrations. As can be seen, it takes 10,000 presentations before convergence is achieved. The final net predictions of the concentrations for the low-purity feedstock A versus space time are shown in Fig. 3 together with the actual model values. Other concentrations of B and C exhibit similar behavior. As can be seen, the back-propagation net does an excellent job in learning the underlying governing CSTR model.

Optimizing Reactor Performance

In the historical data base, two different feed compositions were used. Here it is as-

Table 1
Process Data

	Feed Composition (A/B/C)	Space Time (−) 0.02	0.04	0.08	0.16	0.32	0.64
Learning set	(99%, 0%, 1%) (80%, 5%, 15%)	(A/B/C) data generated from Eqs. (8)-(10) and corrupted with 10 percent Gaussian noise					
Test set	(90%, 2%, 8%)						

Note: Nominal feed composition and space times were also corrupted with 10 percent noise in the learning set.

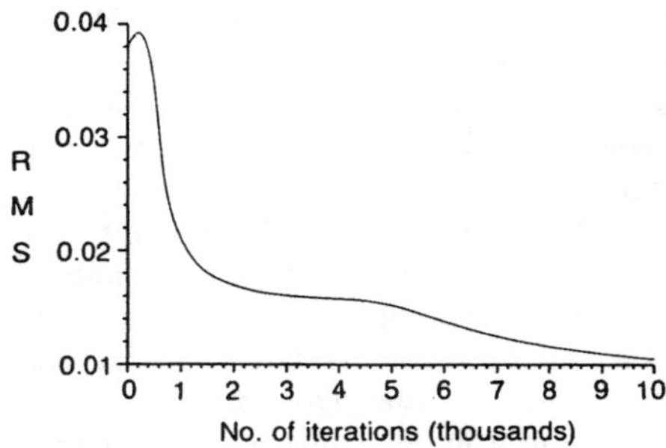

Fig. 2. Rate of convergence.

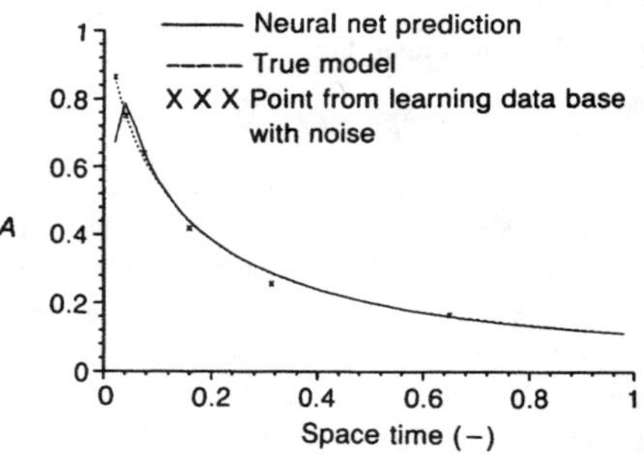

Fig. 3. Comparison of neural net and model predictions (low-purity feedstock).

sumed that a new feedstock is used, where the composition is 90 percent A, 2 percent B, and 8 percent C. The trained net is asked to predict the space time to use to maximize the yield of component B. Figure 4 gives a plot of the exit composition of B versus V/F as predicted by the net. Also shown is the model prediction. The optimum values of V/F are 10.0 min as predicted by the net, and the true value is also 10.0 min. The actual yield at this space time is 61.3 percent. As these results show, the back-propagation net is able to generate essentially optimum operating conditions in this simple example. It should be emphasized that only historical data were used to train the net. However, these data did span the region over which the net was asked to make a prediction. Back-propagation nets do well when asked to interpolate, but their performance is poorer when asked to extrapolate. This point is illustrated in Fig. 4 at low space times. As V/F approaches zero, the highest errors between the net results and the actual model results exist.

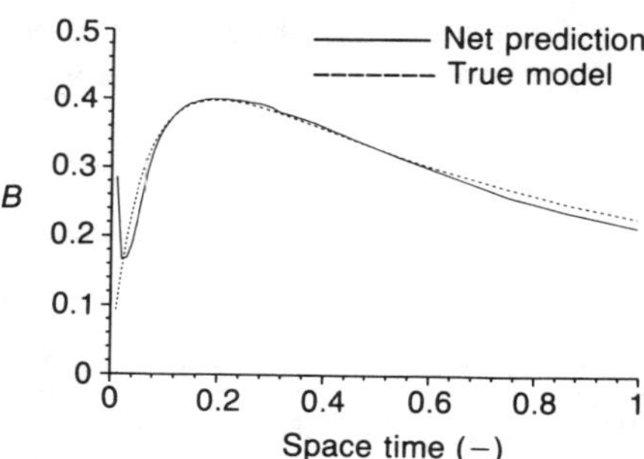

Fig. 4. Comparison of neural net and model predictions (medium-purity feedstock, test case).

Although this example is simple, it illustrates the potential power of neural computation. By using only readily available historical information, a back-propagation net is able to model the reactor without any knowledge of the detailed kinetic mechanisms involved. Such an approach, if successful on a commercial reactor, would be extremely cost-effective. It is felt that there are many chemical process systems where the approach outlined here could be used successfully. Bioprocesses would appear to be particularly attractive candidates because they are so difficult to model from first principles. The next section discusses the use of back-propagation nets to learn dynamic models from input/output data.

Dynamic Example

Our results on using back-propagation to develop nonlinear dynamic models have been discussed in detail in Refs. [3] and [4]. A summary of these results is given here. To illustrate the neural net methodology, the same pH CSTR as discussed in [3], [4] is treated. Recently, we have applied the same modeling approach to data taken from an industrial distillation tower [7] that had seven inputs and four outputs. Our results on the industrial data were comparable to those that follow. It is assumed that an historical data base of inputs and outputs is available. After describing the system, the back-propagation methodology is presented.

System Considered

We have studied the dynamic response of pH in a stirred tank reactor. The CSTR is shown schematically in Fig. 5. The CSTR has two input streams, one containing sodium hydroxide and the other acetic acid. A dynamic model for the pH in the tank can be obtained using the approach presented by McAvoy et al. [8]. By writing material balances on Na^+ and total acetate ($HAC + AC^{-1}$), one gets:

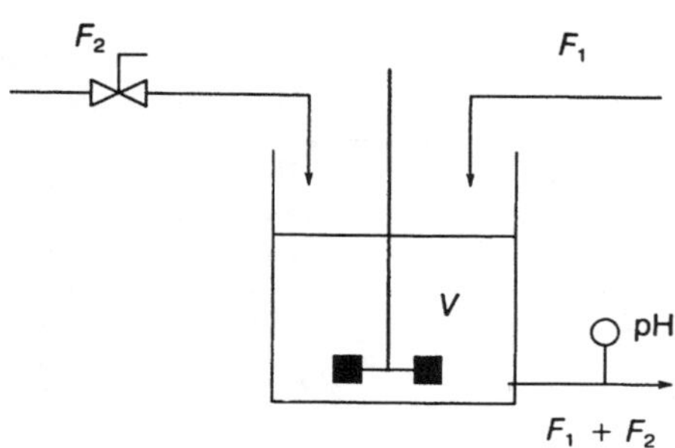

Fig. 5. pH in a CSTR.

Total acetate balance:

$$F_1C_1 - (F_1 + F_2) = Vd\xi/dt \quad (11)$$

Sodium ion balance:

$$F_2C_2 - (F_1 + F_2) = Vd\zeta/dt \quad (12)$$

where

$$\zeta = [Na^+]$$
$$\xi = [HAC] + [AC^-]$$

Combining the HAC and H_2O equilibrium relationships with the requirement for electroneutrality gives

$$\zeta + [H^+] = K_w/[H^+] + K_a[HAC]/[H^+] \quad (13)$$

Parameters for the CSTR considered are given in Table 2. A training data base was developed by forcing the F_2 stream with a 2 percent PRBS (pseudorandom binary signal) superimposed on its steady-state value. The F_2 and pH responses used for training are shown in Fig. 6. In the pH model, Eqs. (11) and (12) are essentially linear differential equations. Equation (13) is a highly nonlinear algebraic equation. Thus, the model has highly nonlinear steady-state characteristics and essentially linear dynamic characteristics.

Back-Propagation Dynamic Modeling (BDM)

To model the pH response shown in Fig. 6, a back-propagation net was used. The input to the net consisted of a moving window of pH and F_2 values, as illustrated in Fig. 7. The center of the window is taken as the current time, t_0. Past and present values of pH and F_2, as well as future values of F_2, are fed to the net. The output from the net is the pH in the future. Thus, the back-propagation modeling approach is similar to traditional ARMA (autoregressive moving average) modeling. Various sizes for the input layer were tried. For the results that follow, the input layer had 15 neurons and the hidden and output layers five neurons. The pH output was predicted one to five time steps into the future. A time step, Δt, of 0.2 min is used.

At the beginning of the training process, the window is placed at the beginning of the data base. The first 5 pH and 10 F_2 values are input to the net. The desired net outputs are the pH values at $t_0 = 0.5$ plus one through five Δt in the future. After the first data presentation, the window moves Δt

Table 2

CSTR Parameters Used	Value
Volume of tank	1000 liters
Flow rate of HAC	81 liters/min
Steady-state flow rate of NaOH	515 liters/min
Steady-state pH	9
Concentration of HAC in F_1	0.3178 mol./liter
Concentration of NaOH in F_2	0.05 mol./liter

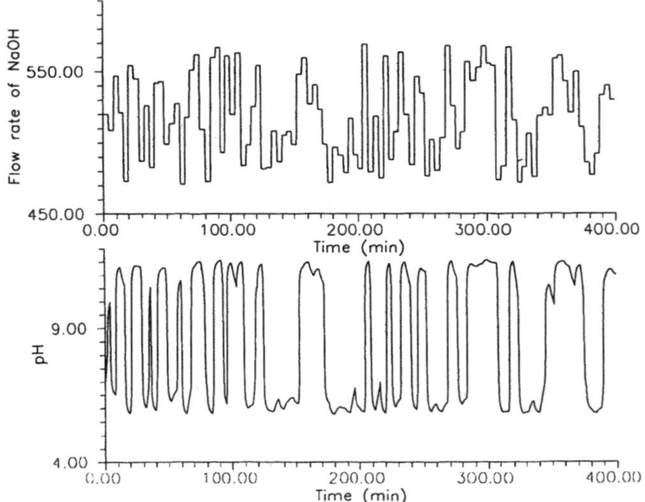

Fig. 6. Data used for the dynamic CSTR example.

down the data base. Again the current and past four pH and F_2 values and the future five F_2 values are input to the net. This process is continued until the end of the data base is reached, and then the process is repeated by starting at the beginning of the data base until convergence is achieved.

Results

After convergence, the net gives excellent pH predictions, as the results in Fig. 8 show. Although the net predicts pH at one to five times steps into the future, only predictions for the first time step are shown for simplicity. The agreement between predicted and actual pH at the other four time steps is as good as that at one time step. Figure 9 illustrates the ability of the BDM model to learn the nonlinear process characteristics.

The net weights were fixed at the values obtained from the training illustrated in Fig. 8. For this training, the pH would settle out at 9 if the PRBS signal were stopped. To generate a test set, the steady-state hydroxide flow, F_2, was lowered slightly so that the steady-state pH in the CSTR was 7. Then a 1 percent PRBS signal was introduced into F_2. As can be seen, the net's response is close to that of the system. In [3] a comparison between a traditional ARMA model and back-propagation is given for this same forcing around pH equal to 7. It is shown that the net does better than an ARMA model in learning the nonlinear process characteristics.

Once a trained back-propagation model is available, then it can be used in a straightforward manner for process control. Figure 10 presents a schematic diagram showing how the neural model is used. The approach is essentially the same as that used in dynamic matrix control [9], except that the nonlinear neural net model is used in place of the linear convolution model. Basically, the optimizer calculates future control moves to minimize a performance index, e.g., a sum of squared errors between desired and predicted future pH values. Alternatively, one can train a net to model the inverse process dynamics and then use this inverse in an IMC (internal model control) configuration [3], [4].

Interpreting Biosensor Data

Overview

Currently, process industries face the constant challenge of holding down the prices of products vis-à-vis the inflationary trend of raw materials. To meet this challenge effectively, plants must be made more efficient and productivity must increase. Although computer-based methods have been developed to meet these goals, often the required measurements are lacking. One area that holds great promise to achieve these goals is new sensor technology [10]. Traditionally, the process industries have relied on standard measurements such as flow, pressure, and temperature. Today, it is becoming increasingly important to measure composition, since this is the key variable in terms of economic importance. Presently, there is a great

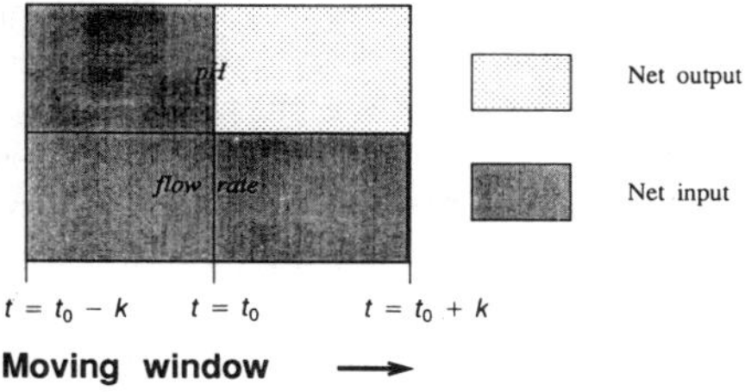

Fig. 7. Window of data.

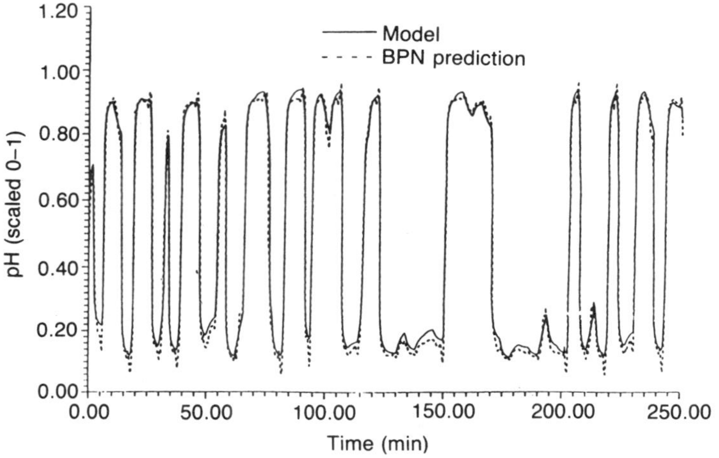

Fig. 8. Prediction of training data (steady-state pH = 9).

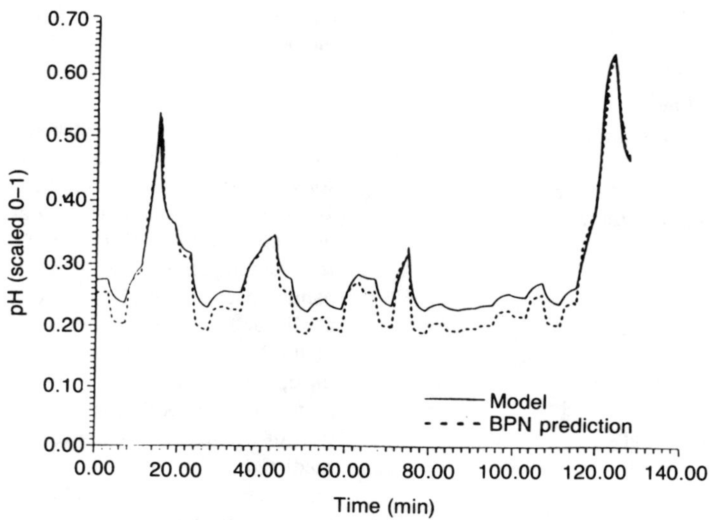

Fig. 9. Prediction of test data (steady-state pH = 7).

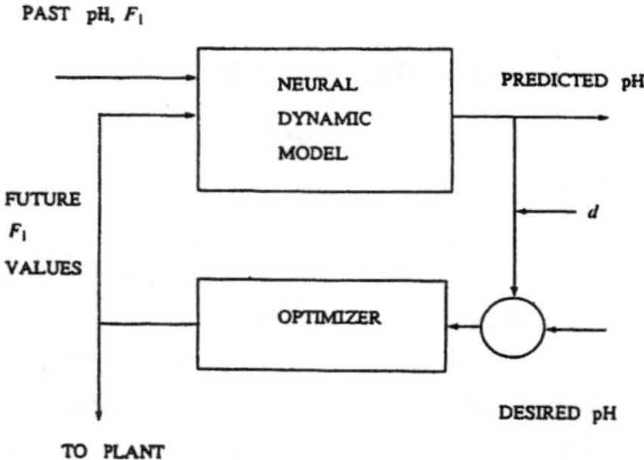

Fig. 10. Neural-net-based predictive control.

deal of activity in developing new methods for sensing composition. Most of these new methods, however, have not been applied on-line. It is critically important that promising laboratory techniques be tried on real problems so that measurement science can be transferred to engineering practice. For this transition to occur, reliable and accurate methods for interpreting sensor signals must be available.

We have studied [2] how to interpret signals from light-induced fluorescence spectra for estimating bioconcentrations. Specifically, mixtures of tryptophan and tyrosine are considered. The input to the net is the measured fluorescence intensity at various wavelengths. The output from the net is the concentration of the fluorophores present. Nonlinear models describing the fluorescence spectra have been developed [11] that show that a linear least-squares approach to deconvoluting the spectral data gives poor results. The nonlinear characteristics of the system must be included. Back-propagation is capable of handling the nonlinearities in the spectral data. Both simulated data from the nonlinear model and actual experimental data have been studied [2]. The actual data are discussed here. Although a specific sensor system is treated, the approach taken is general, and it can be applied to other sensor systems as well.

For our experimental work, 33 samples containing binary mixtures of tryptophan and tyrosine were considered. The composition and mole fraction of these samples are such that the total composition ranges over three decades from 10^{-4} to 10^{-6} molar. To handle this variation, we took the logarithm of total concentration and then scaled it to the range 0 to 1 before presentation to the back-propagation net. Thus the net outputs were mole fraction and the scaled logarithm of total concentration. Finally, the data were split into a training set and a test set by random choice. The test set had 11 elements and the training set 22.

Back-Propagation Results

The inputs to the back-propagation net were the fluorescence spectra at 30 wavelengths. A number of approaches to scaling this intensity data were tried. Taking the logarithm of the intensity produced the best results. Thus a scaling logarithm of intensity was used as the net input. The number of hidden elements in the net was varied from 2 to 10. After convergence, the average sum of squared error in predicting the test set data (PRESS) was used as a means of choosing the optimum number of hidden elements. Figure 11 gives a plot of the PRESS versus the number of hidden elements. It was determined that eight hidden elements were optimum. A comparison between the actual test set data and the back-propagation results for eight hidden elements shows excellent agreement is achieved. The average absolute percent errors compared with the actual values are 5.18 percent for mole fraction and 9.07 percent for total concentration. The maximum absolute percent errors are 25.6 and 25.3 percent for each variable, respectively.

To determine just how well back-propagation does in deconvoluting the spectral data, a benchmark is needed. Recently, Carey et al. [12] have compared several chemometric methods of analyzing data such as fluorescence spectra. The methods evaluated include: standard least squares, principal component regression, and partial least squares (PLS). They concluded that the latter two methods are significantly superior to standard least squares. The best technique appears to be PLS [13]–[15], and this is the method used for comparison here. With PLS, a matrix calibration model of the following form is assumed:

$$Y = XC + E \tag{14}$$

where Y is the measured spectra, C is a calibration matrix, X is the matrix of variables to be predicted, and E is assumed to be a matrix of random noise. The PLS method is used to determine an approximation to the inverse of CC^T from a set of calibration samples. Once the inverse of CC^T is known, it is straightforward to calculate X from spectral measurements on an unknown sample. A key aspect of PLS is that it is essentially a linear technique. As a result, even though it is widely used, it may prove to be inferior to back-propagation. In PLS, one has to decide how many factors to keep in the approximation of the inverse of CC^T. This decision is made using cross-validation in which the calibration data are divided into a training set and a calibration set. To compare PLS with back-propagation, the same 11-member test-set sample was used for each technique. A plot of the PRESS for PLS is also shown in Fig. 11, where it can be seen that seven factors are optimal. The PLS results for seven factors show that the average absolute percent errors are 11.68 percent for mole fraction and 13.73 percent for total composition. The maximum absolute percent errors are 50.99 and 20.70 percent for each variable, respectively. As can be seen, back-propagation does a much better job in predicting the test data than does PLS. This result is very encouraging because PLS appears to have wide applicability for sensor interpretation.

It should be emphasized again that, although a specific sensor is considered, the neural net approach discussed is generally applicable. The only requirement is that one

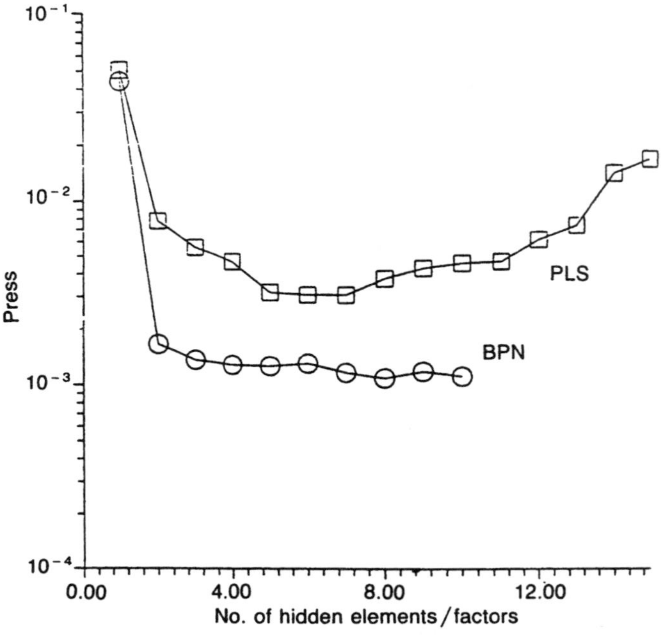

Fig. 11. Comparison of PLS and BPN.

have a calibration set with both inputs and outputs available. The advantage of back-propagation for sensor analysis is that one no longer has to be as concerned with linearity of response. Rather, reproducibility of response becomes a key issue. The characteristics of back-propagation allow one to consider sensor methods that heretofore were rejected because of their nonlinear nature.

Conclusions

This paper has discussed the use of back-propagation neural nets for learning nonlinear models from plant input/output data and for interpreting biosensor data. Because of its ability to identify nonlinear relationships and because it is fundamentally a parallel technique, back-propagation is a particularly promising neural algorithm for application in the chemical/petroleum industries. The back-propagation modeling approach was illustrated on two reactor examples and on fluorescence spectra interpretation. The neural net is able to learn the underlying governing relationships in all cases. It is concluded that neural nets hold great promise for cost-effective modeling of chemical process systems.

Acknowledgments

This research has been supported by the National Science Foundation under Grant EET 87-20046 and through its support of the Systems Research Center at the University of Maryland. Experimental data were taken at the laboratories of the Chemical Process Metrology Division of the National Institute of Standards and Technology. Collaboration with the Hecht Nielsen Company and the use of its neural coprocessor board are gratefully acknowledged.

References

[1] P. Werbos, "Beyond Regression: New Tools for Prediction and Analysis in Behavioral Sciences," Ph.D. Thesis, Harvard Univ., 1974.

[2] T. McAvoy, N. Wang, S. Naidu, and N. Bhat, "Use of Neural Nets for Interpreting Biosensor Data," *Proc. Joint Conf. Neural Networks*, Washington, DC, pp. I227–I233, June 1989.

[3] N. Bhat and T. McAvoy, "Use of Neural Nets for Dynamic Modeling and Control of Chemical Process Systems," to appear in *Computers and Chemical Engineering*, 1989.

[4] N. Bhat and T. McAvoy, "Use of Neural Nets for Dynamic Modeling and Control of Chemical Process Systems," *Proc. 1989 Amer. Automat. Contr. Conf.*, Pittsburgh, PA, pp. 1342–1348, June 1989.

[5] G. Cybenko, "Continuous Value Neural Networks with Two Hidden Layers Are Sufficient," *Math Contr. Signal & Sys.* vol. 2, pp. 303–314, 1989.

[6] D. Rumelhart and J. McClelland, *Parallel Distributed Processing: Explorations in the Microstructure of Cognition*, vol. I, chap. 8, Cambridge, MA: MIT Press, 1987.

[7] N. Bhat and T. McAvoy, "Dynamic Modeling Via Neural Computing," to be presented at Annual AIChE Meeting, San Francisco, Nov. 1989.

[8] T. McAvoy, E. Hsu, and S. Lowenthal, "Dynamics of pH in CSTR's," *Ind. Eng. Chem. Process Des. Develop.*, vol. 11, pp. 68–70, 1972.

[9] C. Cutler and B. Ramaker, "Dynamic Matrix Control—A Computer Control Algorithm," AIChE 86th National Meeting, Houston, TX, Apr. 1979; also, in *Joint Automat. Contr. Conf. Proc.*, San Francisco, 1980.

[10] N. Amundsen, "Frontiers of Chemical Engineering: Needs and Opportunities," Report to National Research Council, 1987.

[11] P. Rinaudo, "Analysis of Laser Induced Fluorescence Spectra," M.S. Thesis, Univ. of Maryland, College Park, chap. 5, 1987.

[12] W. Carey, K. Beebe, E. Sanchez, P. Geladi, and B. Kowalski, "Chemometric Analysis of Multisensor Arrays," *Sensors and Actuators*, vol. 9, pp. 223–234, 1986.

[13] H. Wold, *Festschrift Jerzy Neyman*, New York: Wiley, pp. 411–444, 1966.

[14] K. Joreskog and H. Wold, Eds., *Systems Under Indirect Observation*, pts. I and II, Amsterdam: North Holland, 1982.

[15] P. Geladi and B. Kowalski, "Partial Least Squares Regression: A Tutorial," *Anal. Chim. Acta*, vol. 185, pp. 1–17, 1986.

Article 7.10

CMAC: An Associative Neural Network Alternative to Backpropagation

W. THOMAS MILLER, III, MEMBER, IEEE, FILSON H. GLANZ, MEMBER, IEEE, AND L. GORDON KRAFT, III

The CMAC neural network, an alternative to backpropagated multilayer networks, is described. CMAC has the advantages of the following properties: local generalization, rapid algorithmic computation based on LMS training, incremental training, functional representation, output superposition, and a fast practical hardware realization, all of which are discussed. A geometrical explanation of how CMAC works is provided, and brief descriptions of applications in robot control, pattern recognition, and signal processing are given. Possible disadvantages of CMAC are that it does not have global generalization and that it can have noise due to hash coding. Care must be exercised, as with all neural networks, to assure that a low error solution will be learned.

I. Introduction and Background

Within the last eight years there has been increased interest in systems that learn, using models of biological neurons. As technology has developed in diverse application areas, there has been an increased need for controlling complex systems, with nonlinearities and many degrees of freedom, leading to a search for new ways of handling the large amounts of computation and numbers of variables found in these problems. Many of these systems are so complex that the physical principles and/or the overwhelming size makes writing meaningful model equations impossible.

In this paper we describe a neural network called CMAC (Cerebellar Model Arithmetic Computer) [1], [2]. At the University of New Hampshire, we have been using CMAC in real-time control of an industrial robot and in other applications, typically employing neural networks with hundreds of thousands of adjustable weights that can be trained to approximate nonlinearities which are not explicitly written out or even known. CMAC can learn nonlinear relationships from a very broad category of functions. Furthermore, the learning algorithm generally converges in a small number of iterations.

Manuscript received September 14, 1989; revised March 16, 1990. The work summarized in this manuscript was supported in part by the National Science Foundation under grant IRI-8813225, by the Office of Naval Research and the National Institute of Standards and Technology under ONR grant N00014-89-J-1686, and by the Defense Advanced Research Projects Agency under ONR grant N00014-89-J-3100.

The authors are with Department of Electrical and Computer Engineering, University of New Hampshire, Durham, New Hampshire 03824, USA.

IEEE Log Number 9039173.

Neural networks of one form or another have been "modeled" for a number of decades. Some of the better known models are the Perceptron of Rosenblatt [3], the adaline of Widrow [4], [5], and the recurrent networks of Hopfield [6], [7]. Much of the past effort was devoted to systems with binary inputs and outputs. Because it was recognized in the 1960s that not all binary functions can be realized with a single "linear" threshold element, there has been considerable interest in multilayer networks and the question of how to train them. In the last decade backpropagation has been developed as a method of training multilayer networks, and most recent application papers use this approach. Nonbinary inputs and outputs have also been considered.

The CMAC neural network described in this paper is an alternative to the well known backpropagation-trained analog multilayer neural network. An alternative is useful since backpropagation has the disadvantages of: requiring many iterations to converge and therefore inappropriate for online real-time learning (this imposes the need to use small networks, and therefore many difficult problems of current interest cannot be solved); requiring an large number of computations per iteration so that the algorithm runs slowly unless implemented in expensive custom hardware; having an error surface which can have relative minima, a hazard to backpropagation training which is based on gradient search techniques; not allowing successful incremental learning (if one has finite time to train) in the sense that to achieve reasonable convergence all inputs must be seen before any weight change can take place.

In the 1970s James Albus reported the work he had been doing on CMAC [1], [2], [8], [9]. Albus used CMAC to do rote learning of movements of an artificial arm. After a number of years of being referred to in the robot control literature as being impractical, it was demonstrated that CMAC was not only practical, but that it could be used to learn general state space dependent control responses [10]. Since then the Robotics Laboratory at the University of New Hampshire has been investigating and using CMAC with considerable success [10]–[22]. Other groups working with CMAC include Ersu *et al.* [23]–[26], and Moody at Yale [27]. We are

Reprinted from *Proc. IEEE*, vol. 78, no. 10, pp. 1561–1567, Oct. 1990.

also aware of several industrial research groups using CMAC.

CMAC is an associative neural network in that only a small subset of the network influences any instantaneous output, and that subset is determined by the input to the network. The associative mapping built into CMAC assures local generalization: similar inputs produce similar outputs while distant inputs produce nearly independent outputs. As the result of the built-in associative properties, we have found that the number of training passes required for network convergence is orders of magnitude smaller with CMAC than with backpropagation on real problems.

II. Description of CMAC

In this section we describe CMAC in a mostly geometrical way: the input space, the outputs, and the mechanism of generalization.

Figure 1 shows a set-oriented overview of CMAC. An input vector is the collection of N appropriate sensors of the real

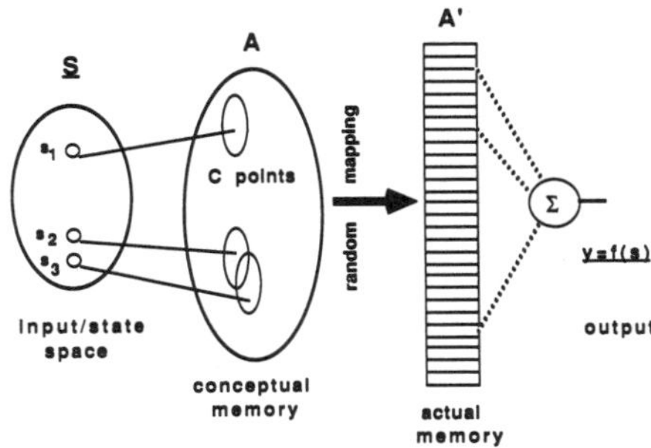

Fig. 1. Set-oriented block diagram of CMAC. The input/state space and the conceptual memory A are N-dimensional. The actual memory A' has as many dimensions as there are output components.

world and/or measures of the desired goal. The input space consists of the set of all possible input vectors. The number N of input vector components and the number of outputs is arbitrary within some practical limits. The CMAC algorithm maps any input it receives into a set of C points in a large "conceptual" memory (A in Fig. 1) in such a way that two inputs that are "close" in input space will have their C points overlap in the A memory, with more overlap for closer inputs. If two inputs are far apart in the input space there will be no overlap in their C-element sets in the A memory, and therefore no generalization.

For practical systems the input space is extremely large. For example, a system with 10 inputs, each of which can take on 100 different values, would have $100^{10} = 10^{20}$ points in its input space, requiring a correspondingly large number of locations in the memory A. Since most learning problems do not involve all of the input space, the memory requirement is reduced by mapping the A memory onto a much smaller physical memory A'. As a result, any input presented to CMAC will generate C real memory locations, the contents of which will be added in order to obtain an output. Notice that the nonlinearity that all neural networks must have is in the associative input mapping, not in the sigmoid/threshold function normally at the output of each neuron.

Figure 2 is a network diagram of a two-input CMAC as implemented in our laboratory. Each variable in the input

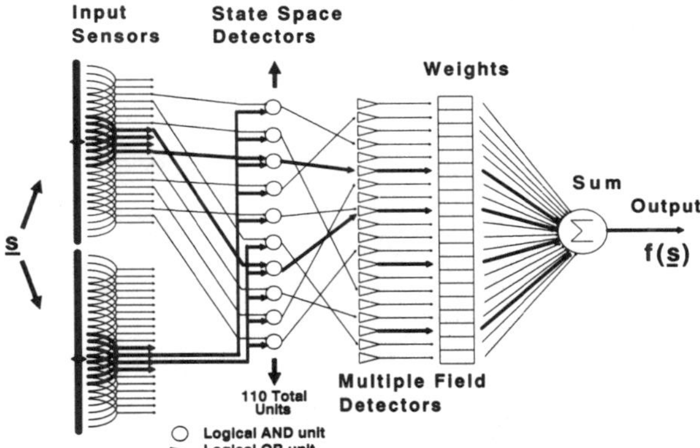

Fig. 2. A simple example of a CMAC neural network with two inputs and one output. The generalization parameter C has the value 4. Note that only a partial set of the state space detectors is shown.

state vector s is fed to a series of input sensors with overlapping receptive fields. Each input sensor produces a binary output which is ON if the input falls within its receptive field and is OFF otherwise. The width of the receptive field of each sensor produces input generalization, while the offset of the adjacent fields produces input quantization. Each input variable excites exactly C input sensors, where C is the ratio of generalization width to quantization width ($C = 4$ in Fig. 2, $C = 32$ to 256 in typical implementations).

The binary outputs of the input sensors are combined in a series of threshold logic units (called state-space detectors) with thresholds adjusted to produce logical AND functions (the output is ON only if all inputs are ON). Each of these units receives one input from the group of sensors for each input variable, and thus its input receptive field is the interior of a hypercube in the input hyperspace (the interior of a square in the two-dimensional input space of Fig. 2). The state space detectors of Fig. 2 correspond logically to the individual memory locations of the A memory in Fig. 1.

If the input sensors were fully interconnected, there would be a very large number of state-space detectors, and a large subset of these detectors would be excited for each possible input. The input sensors are interconnected in a sparse and regular fashion, however, in such a way that each input vector excites exactly C state-space detectors. Figure 3 depicts the organization of the receptive fields in the input space. The total collection of state-space detectors is divided into C subsets. The receptive fields of the units in each of the subsets are organized so as to span the input space without overlap. Each input vector excites one state-space detector from each subset, for a total of C excited detectors for any input. There are many ways to organize the receptive fields of the individual subsets which produce similar results. In our implementation, each of the subsets of state-space detectors is identical in organization, but each subset is offset relative to the others along hyperdiagonals in the input hyperspace (adjacent subsets are offset by the quantization level of each input variable).

The organization of the receptive fields of the state-space detectors guarantees that a fixed number C of detectors is excited by any input. However, the total number of state-space detectors can still be large for many practical problems. On the other hand, it is unlikely that the entire input state-space of a large system would be visited in solving a

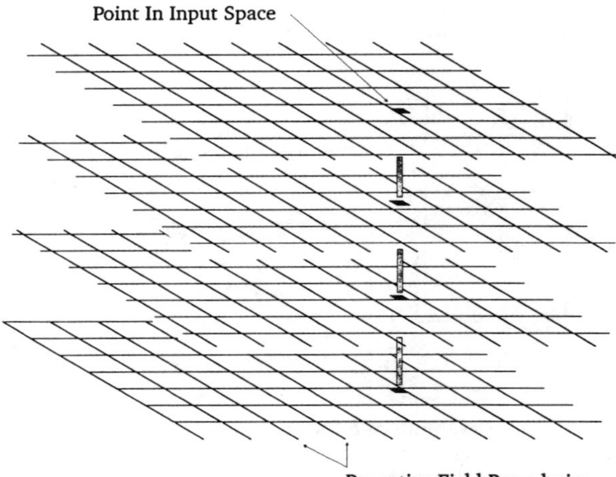

Fig. 3. The organization of the receptive fields for the simple two-input CMAC in Fig. 2, with $C = 4$. The state space detectors are organized into four subsets, offset relative to each other in the input space. The boundaries of the receptive fields are shown for each subset. The dark square in each plane indicates the same region of the input space projected onto the receptive fields of each of the four subsets.

specific problem (most of the possible input vectors would never be experienced). Thus it is not necessary to store unique information for each receptive field. Following this logic, the outputs of the state-space detectors are connected randomly to a smaller set of threshold logic units (called multiple field detectors in Fig. 2) with thresholds adjusted such that the output will be ON if any input is ON (a logical OR function). The receptive field of each of these units is thus the union of the fields of many of the state-space detectors. Since exactly C state-space detectors are excited by any input, at most C multiple field detectors will be excited by any input. The converging connections between the large set of state-space detectors and the smaller set of multiple field detectors are referred to as "collisions." In practice, the converging connections are implemented by assigning a virtual address to each of the state-space detectors, and passing the addresses of the active state-space detectors through a random hashing function.

Finally, the output of each multiple field detector is connected, through an adjustable weight, to an output summing unit. The output for a given input is thus the sum of the weights selected by the excited multiple field detectors. The set of all weights in Fig. 2 corresponds to the A' memory of Fig. 1. Note that while the number of input sensors and state-space detectors is determined by the number of inputs, their dynamic ranges, the level of quantization, and the degree of generalization, the number of multiple field detectors and adjustable weights is an independent design parameter. The necessary size of the weight memory is related to the size of the subset of the input space that is likely to be visited in solving a particular problem, rather than to the total size of the input space. Thus, simple problems in a many dimensional input space require only small weight memories, even though the size of the input space may be huge.

Ideally, the associative mapping within the CMAC network assures that nearby points in the input space generalize while distant points do not generalize. The effect of the converging connections between the state-space detectors and the multiple field detectors, however, is to create randomly distributed low-magnitude generalization with distant points in the space. Figure 4 illustrates this effect

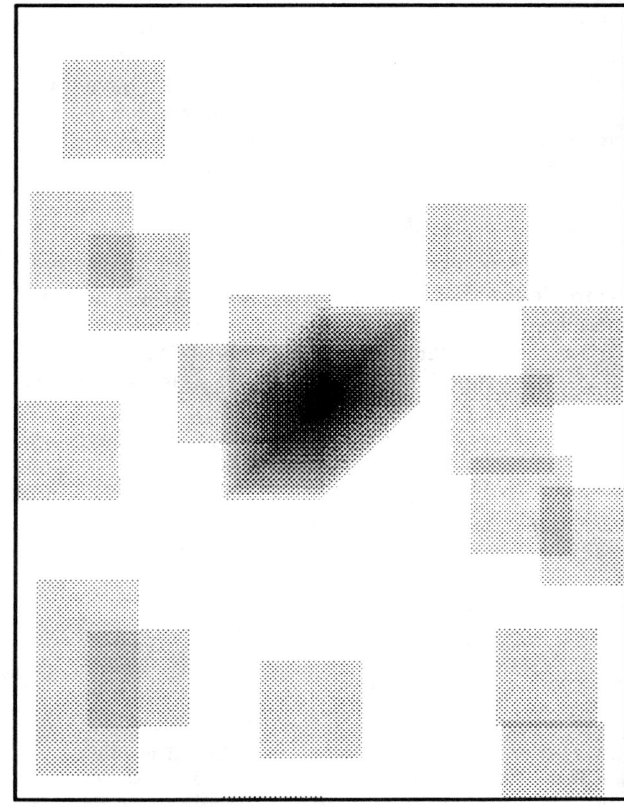

Fig. 4. Generalization of a single point within a 128×96 input space, using a CMAC with $C = 16$ and 1000 memory locations. Greater generalization is indicated by darker shading.

for a simple two-input CMAC. The size of the input space in this example is 128 points along the horizontal axis and 96 points along the vertical axis, for a total of 12288 points in the input space. The figure shows the degree to which a central point in the space generalizes with all other points in the space, for a CMAC with 1000 adjustable weights and $c = 16$. Note that even when using this small memory (relative to the size of the input space), the characteristic of local generalization in the space is largely preserved. Note also that the generalization region is not symmetrical about the horizontal or vertical axes. The asymmetry is related to the distribution of the state-space detector receptive fields in the input space (Fig. 3) and varies somewhat for different points within the input space. Other distributions for the detectors are possible, which would preserve the associative mapping characteristics of CMAC but would modify the shapes of the generalization regions. The effects of receptive field distribution on CMAC performance has not yet been systematically studied.

Network training is typically based on observed training data pairs $\mathbf{s}_o$ and f_o (supervised learning), where f_o is the desired network output in response to the input $\mathbf{s}_o$, using the least mean square (LMS) training rule [28]. Although the CMAC network can be trained to represent arbitrary nonlinear relationships between the input vector **s** and the output $f(\mathbf{s})$, convergence of the weights during training can be demonstrated by comparing the CMAC structure with well-known linear adaptive elements [4], [5]. Let **x** represent the vector of binary outputs of the multiple field detectors in Fig. 2, and let **w** represent the vector of adjustable weights. For each training input $\mathbf{s}_0$ there is a corresponding pattern of multiple field detector outputs $\mathbf{x}_0$ which is dependent on the fixed pattern of the CMAC network interconnections, but is independent of the values of the weights. The weight update described above can then be written as

$$\mathbf{d}\boldsymbol{w} = (\beta/C)(f_o - \boldsymbol{w}^T\mathbf{x}_0)\mathbf{x}_0$$

which is the well-known LMS adaptation rule for linear adaptive elements.

The nonlinear nature of the CMAC network is embodied in the interconnections of the input sensors, state-space detectors, and multiple field detectors, which perform a fixed nonlinear mapping of the continuous valued input vector **s** to a many-dimensional binary valued vector x (which has tens or hundreds of thousands of dimensions in typical implementations). the adaptation problem is linear in this many-dimensional space, and all of the convergence theorems for linear adaptive elements apply [16].

Because CMAC uses a linear output stage, superposition holds in the weight domain. For example, if the weight set $\mathbf{W}_1$ produces the nonlinear function $f_1(\mathbf{s})$ and the weight set $\mathbf{W}_2$ produces the nonlinear function $f_2(\mathbf{s})$, then the weight set $\mathbf{W}_1 + \mathbf{W}_2$ will produce the function $f_1(\mathbf{s}) + f_2(\mathbf{s})$. It is thus possible to characterize the ability of a CMAC network of given dimension, quantization, and generalization to reproduce arbitrary functions by examining its ability to produce orthogonal basis functions, such as multidimensional sinusoids.

III. Properties of CMAC

This section summarizes the main properties of CMAC and thus serves as a basis for understanding the concepts and details introduced in the previous section.

1) CMAC accepts real inputs and gives real outputs. The input components are quantized, but the number of levels can be as large as desired so that any degree of accuracy is achievable.

2) CMAC has a built-in local generalization, meaning that input vectors that are "close" in the input (state) space will give outputs that are close, even if the input has not be trained on, as long as there has been training in that region of the state-space. The measure of "closeness" is not Euclidean distance but "city block distance," a generalization of Hamming distance where the distance is the sum, over all components, of the absolute value of the differences of each component. Locally generalizing networks have less learning interference than globally generalizing networks such as a multilayer perceptron.

3) CMAC has the property that large networks can be used and trained in practical time, even with the software version of the system. This is because there is a small number of calculations per output even though there is a large number of weights. In our CMAC realizations at UNH we typically have tens or hundreds of thousands of weights and 10 to 128 additions per output; in both hardware and software this is a small amount of computation in comparison to that required by an equivalent multilayer perceptron. As an example, in a pattern recognition problem we have trained both CMAC and a two-layer backpropagated perceptron network. We found that CMAC took about 50 iterations while the multilayer network took about 12,000 iterations. These results are in agreement with Moody's chaotic sequences results [27]. Notice that the results were iterations, not time units. Since backpropagation takes so many operations for one iteration, it is very much slower than CMAC per iteration.

4) CMAC uses the LMS adaptation rule of Widrow and Hoff [4]. This least squares algorithm is equivalent to a gradient search of a surface which is quadratic and therefore has a unique minimum. In contrast, backpropagation is known to be a gradient search of a surface which may have relative minima. Of course, when using LMS training, the well-known precautions should be observed [28].

5) CMAC can learn a wide variety of functions. It is easy to show, for example, that a one-input CMAC can learn any discrete one-dimensional single-valued function, given a few mild conditions on the parameters of the CMAC.

6) CMAC obeys superposition in the output space, which means that a multidimensional discrete Fourier series can be used to show what functions can be learned. For example, if multidimensional sinusoids of various spatial frequencies (harmonics) can be learned, a whole class of functions is also learnable. Notice that the functions learned are still nonlinear, since the superposition is in the output space only. Notice also that in spite of the spatial frequency constraints implied by the superposition discussed above, there is no such constraint in time response if consecutive inputs to CMAC are not neighbors in the state-space as would be the case when learning a function whose input is the state of a shift register.

7) CMAC has a practical hardware realization using logic cell arrays [29]. VLSI versions are possible with learning cycles in the microsecond region.

IV. Example Applications of CMAC

One of the advantages of CMAC at this stage of neural network development is that its realization in software is sufficiently fast and efficient for large networks that many applications are possible. We have used CMAC in a number of real-time robotic [10]–[15], [22], pattern recognition [17], [21], signal processing [18], [19], and speech processing demonstrations.

In this section we briefly describe three applications in which we have used CMAC. Readers who would like more information on a particular application are referred to the appropriate papers in the literature. We describe a robot control application, a character recognition application, and a deconvolution/inverse filtering application.

A. Robot Control

One robot tracking problem which we have used to demonstrate neural network control involved the control of a five-axis industrial robot with a video camera attached to the fifth axis in the place of a gripper [16]. The camera looked at an object on a conveyor belt and the control problem was to move the robot in such a way as to keep the object at a fixed orientation and size on the video monitor screen, and simultaneously have the centroid of the object follow a trajectory defined in the video image space rectangular coordinate system. No kinematics of the robot were known to the system, no height measurements other than the length of the object were found by the image processing, and no camera–screen calibrations were given to the system. All of these had to be learned by the neural network system. Figure 5 shows the physical setup of the system. The object in this case was a white disposable razor.

Because of the slowness of the image processing, updates of the object image were too infrequent. A second CMAC was used to predict the object position in between image processing results. This CMAC learned to do its prediction with no initial knowledge of the system. The acquisition and orientation of the object and the trajectory following of the object centroid were all done as the object passed along the five-foot-long conveyor belt, having been placed (within limits) at a random orientation and position on the belt. The trajectories to be followed were either repeated or random, and were described by specifying both video image X and Y coordinate velocities and accelerations.

Training was done at the end of each control cycle based on the error between the desired position and the actual position of the object in the image. In order to initiate training, a simple fixed gain control loop was included in parallel in the system in order to keep the camera close enough to the object initially that the image could be found and the error determined for training. As the system learned more, the fixed gain controller had less and less influence on the control and the learning had more and more. The input vector had 12 components which describe the robot joint positions, the predicted image parameters, and the desired image parameter changes. There were four outputs from the inverse model, one to drive each of the four robot joint motors. The fifth motor was not driven in order to keep the camera vertical. The image prediction CMAC has as inputs the latest image parameters, the latest joint angles, and the latest motor voltages. The outputs of this CMAC were the predicted changes in image parameters. Each network consisted of 16 384 × 4 16-bit weights, and the generalization parameter was $C = 64$.

The RMS control error for each of the four image parameters decreased each trial and generally converged to the sensor sensitivity of one pixel within 15 trials in the repeated trajectory case and about 50 trials in the random trajectory case. The average error was always below the error of the fixed gain controller without learning.

B. Pattern Recognition

A second application of CMAC shows its use in pattern recognition. In this example the first five letters of the alphabet, each in lower case and upper case, and each in 72 rotations 5° apart around the circle, were trained into CMAC. The images were obtained by putting each letter on a calibrated rotating platform in front of a camera which was interfaced through a frame buffer to an 80386-based personal computer. In order to obtain a reasonably sized feature vector, the computer averaged all pixels in each box of an 8 × 8 array centered on the letters. The rows of these averages were summed, as were the columns, giving the 16

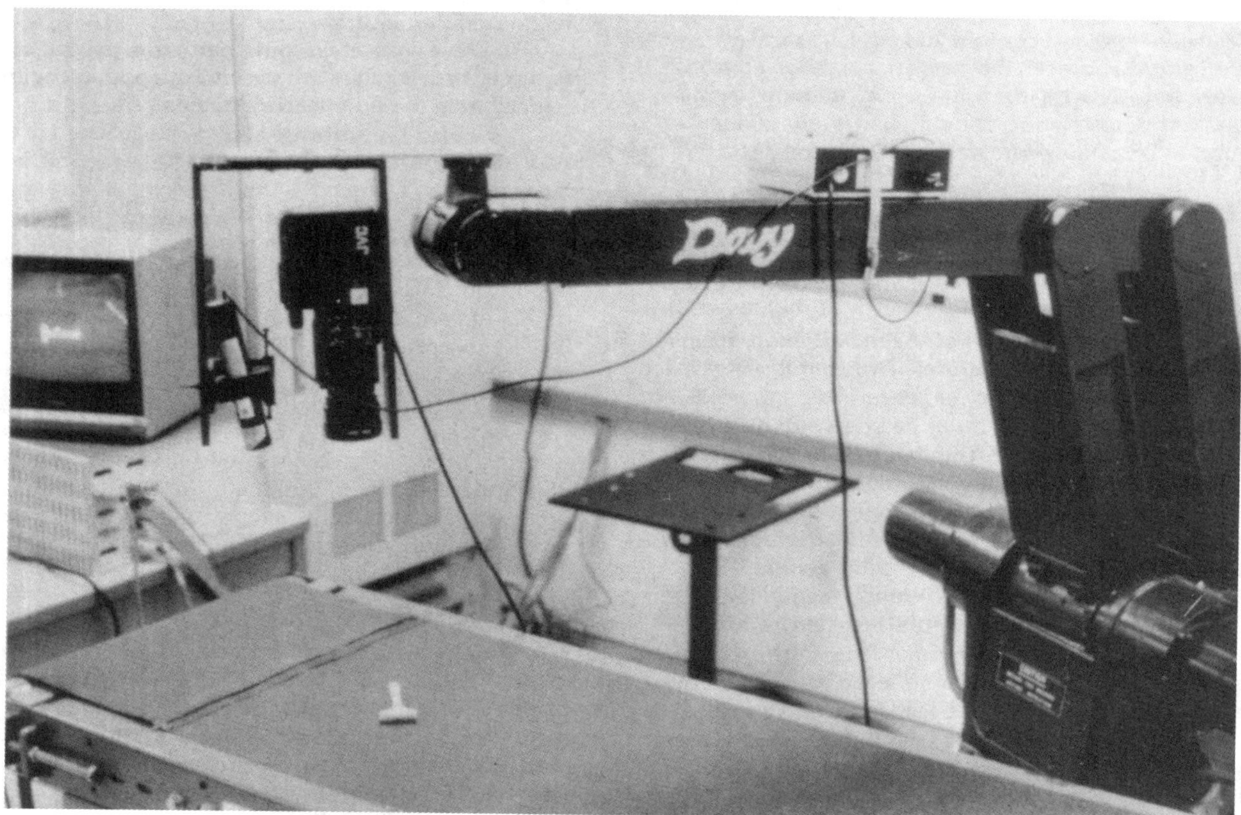

Fig. 5. Robot control system: robot, conveyer belt, and object on belt and in video image.

features used in the pattern recognition. Files of these features for each letter (case and rotation) were made to use in the training. A separate file was made of new images to act as a test set after training.

Table 1 summarizes the results. Five other classifiers were included in order to compare the CMAC results with standard techniques. The table briefly lists these and also gives the notation used. The "random" letters were placed at purely random rotations with no attention paid to the 5° increments used for the training data. Note that, although the nearest neighbor technique did as well as CMAC, test sample computation for the nearest neighbor approach increases linearly with the number of training samples whereas CMAC test sample computation is independent of the number of training samples.

Table 1 Character Recognition Results

	Percent Correct on Test Set[a]					
EXPT[b]	CMAC	Temp	F-WT	R-Coord	Hyper	NNBR
U/U	100.0	78.3	75.0	100.0	82.8	100.0
U/RU	100.0	71.1	68.3	99.4	77.8	100.0
UL/U	100.0	56.4	63.3	93.3	70.3	100.0
UL/RU	100.0	47.2	57.8	93.3	71.7	100.0
L/L	100.0	85.3	60.0	99.2	73.6	100.0
L/RL	100.0	83.9	61.1	100.0	73.9	100.0
UL/L	100.0	38.6	46.9	83.3	36.1	100.0
UL/RL	100.0	38.3	50.0	83.3	35.6	100.0

[a]Classification techniques shown:

CMAC: A CMAC based classifier.
Temp: Standard template matching.
F-Wt: Feature weighting method.
R-Coord: Rotated coordinate method.
Hyper: "Optimal" hyperplane separation.
NNBR: Nearest neighbor method.

[b]Experiments indicated by *Train/Test* where:

U: 72 upper case examples, 5° rotations.
RU: 36 random upper case examples, random rotations.
L: 72 lower case examples, 5° rotations.
RL: 36 random lower case examples, random rotations.

C. Signal Processing

A third application involves a signal processing problem [19]: Given the output of a nonlinear channel with memory, learn to generate the original input. Such a problem arises, for example, in equalization of a nonlinear communication channel [30]. The nonlinearity used in this example is

$$y(n) = \arctan [x(n)] + B \arctan [x(n - 1)]$$

where $B = 0.1$. The memory of the system in this case is one. In this simulation, $y(n)$ is created by generating uniformly distributed pseudorandom noise $x(n)$ and feeding it into the equation above, along with the last input value $x(n - 1)$.

The input to CMAC has two components: $y(n)$ and $y(n - 1)$. The value of $x(n)$ is used as the desired output. On each presentation of the input 2-vector, training is done until the squared error is appropriately small. Note that there is no repeating of the input vectors (except by chance) since they are generated from the $x(n)$s which are not repeated, that is, there is no fixed training set. Figure 6 shows the input sequence $y(n)$ to CMAC, the desired output (goal) sequence $x(n)$, and the CMAC output. As can be seen from the figure, the CMAC output is almost exactly the same as the goal; that is, CMAC has learned to invert the nonlinear output in order to obtain the input sequence (the goal) $x(n)$. If a new random number sequence is used to generate a test sequence, which is run through the trained CMAC, the output is of similar high quality.

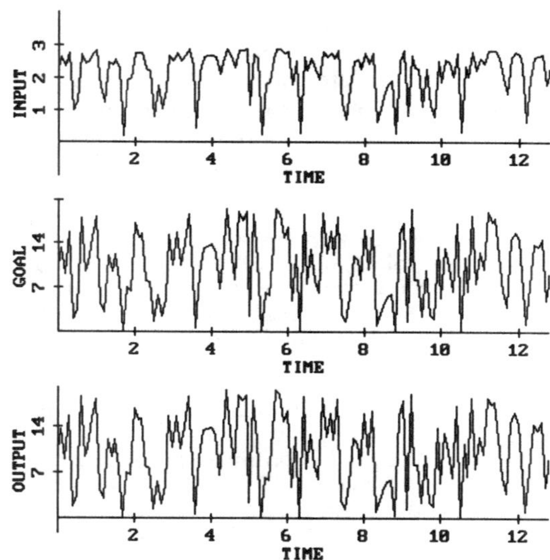

Fig. 6. Nonlinear inverse filtering showing CMAC output time series and the time series it has learned. The input time series is generated from the goal time series by the arctangent nonlinearity given in the text equation.

V. Conclusions

In the foregoing we have explained the properties of a neural network called CMAC, described the mechanism of CMAC, and given several examples of how we have applied CMAC to solve a variety of problems. This form of neural network has advantages and disadvantages in comparison to other forms of neural networks. CMAC has the advantages that it is fast in software and can be realized in high-speed hardware, thereby leading to the practical use of more weights and larger systems in solving problems. Furthermore, CMAC is able to learn a large variety of nonlinear functions, reducing the need to use slow backpropagation-based networks. And CMAC has a local generalization which may have advantages over global generalization: There is little or no learning interference due to recent learning in remote parts of the input space, and local generalization usually requires a smaller number of additions, therefore giving fast computation speeds. Furthermore, CMAC/LMS training takes fewer iterations than backpropagation. The speed and network size advantage of CMAC means that real systems and real problems can be undertaken with results beyond the quality of other standard methods. On the other hand, CMAC has some disadvantages: The generalization is not global, so we will not find mysterious properties "emerging" from the training, if that is an advantage; collisions due to the hash coding necessary to reduce the memory size to something realizable can cause a noise or interference if care is not taken in design to prevent that happening; and some design care must be exercised in order to be assured that a low error solution will be learned in a specific application.

In conclusion, the CMAC neural network is a viable alternative to the multilayer backpropagated network for learning situations with analog inputs and outputs, especially where speed of convergence and speed of computation are important, and where large numbers of weights are needed. CMAC has proven able to learn unknown nonlinear functions quickly and generalize on inputs it has never seen.

References

[1] J. S. Albus, "A theory of cerebellar functions," Mathematical Biosciences, vol. 10, pp. 25–61, 1971.

[2] —, "Theoretical and experimental aspects of a cerebellar model," Ph.D. dissertation, Univ. of Maryland, 1972.

[3] F. Rosenblatt, *The Principles of Neurodynamics.* New York: Spartan, 1962.

[4] B. Widrow and M. E. Hoff, "Adaptive switching circuits," *Proc. IRE Western Electronic Show and Conv.*, vol. 4, pp. 96–104, 1960.

[5] B. Widrow, "Generalization and information storage in networks of adaline 'neurons,' " *Self-Organizing Systems*, M. C. Yovits, Ed. Wash., DC: Spartan, 1962, pp. 435–461.

[6] J. J. Hopfield, "Neurons with graded response have collective computational properties like those of two-state neurons," *Proc. Natl. Acad. Sci. USA*, vol. 81, pp. 3088–3092, May 1984.

[7] —, "Neural networks and physical systems with emergent collective computational abilities," *Proc. Natl. Acad. Sci. USA*, vol. 79, pp. 2554–2558, Apr. 1982.

[8] J. S. Albus, "Data storage in the cerebellar model articulation controller," *J. Dynamic systems, Measurement and Control*, pp. 228–233, Sept. 1975.

[9] —, "A new approach to manipulator control: the cerebellar model articulation controller (CMAC)," pp. 220–227, Sept. 1975.

[10] W. T. Miller, "A nonlinear learning controller for robotic manipulators," *Proc. SPIE, Intelligent Robots and Computer Vision*, vol. 726, pp. 416–423, Oct. 1986.

[11] —, "A learning controller for nonrepetitive robotic operations," *Proc. Workshop on Space Telerobotics*, publication 87-13, vol. II, Pasadena, CA, pp. 273–281, Jan. 10–22, 1987.

[12] —, "Sensor based control of robotic manipulators using a general learning algorithm," *IEEE Trans. Robotics Automat.*, vol. RA-3, pp. 157–165, Apr. 1987.

[13] W. T. Miller, F. H. Glanz, and L. G. Kraft, "Application of a general learning algorithm to the control of robotic manipulators," *Internat. J. Robotics Research*, vol. 6, pp. 84–98, Summer 1987.

[14] W. T. Miller and R. P. Hewes, "Real time experiments in neural network based learning during high speed nonrepetitive robotic operations," *Proc. 3rd IEEE Int. Symp. on Intelligent Control*, pp. 513–518, Aug. 24–26, 1988.

[15] W. T. Miller, "Real time learned sensor processing and motor control for a robot with vision," *1st Annual Conf. of the Intl. Neural Network Society*, p. 347, Sept. 1988.

[16] —, "Real time application of neural networks for sensor-based control of robots with vision," *IEEE SMC*, vol. 19, pp. 825–831, July–Aug. 1989.

[17] F. H. Glanz and W. T. Miller, "Shape recognition using a CMAC based learning system," *Proc. SPIE Conf. on Robotics and Intelligent Systems*, vol. 848, pp. 294–298, Nov. 1987.

[18] —, "Deconvolution using a CMAC neural network," *Proc. 1st Annual Conf. of the Intl. Neural Network Society*, p. 440, Sept. 1988.

[19] —, "Deconvolution and nonlinear inverse filtering using a neural network," *Intl. Conf. on Acoustics and Signal Processing*, vol. 4, pp. 2349–2352, May 23–29, 1989.

[20] L. G. Kraft and D. P. Campagna, "A comparison of CMAC neural network and traditional adaptive control systems," *Proc. 1989 American Control Conf.*, vol. 1, pp. 884–889, June 21–23, 1989.

[21] D. Herold, W. T. Miller, L. G. Kraft, and F. H. Glanz, "Pattern recognition using a CMAC based learning system," *Proc. SPIE, Automated Inspection and High Speed Vision Architectures II*, vol. 1004, pp. 84–90, Nov. 10–11, 1989.

[22] W. T. Miller, R. P. Hewes, F. H. Glanz, and L. G. Kraft, "Real-time dynamic control of an industrial manipulator using a neural-network-based learning controller," *IEEE Trans. Robotics Automat.*, vol. 6, pp. 1–9, Feb. 1990.

[23] E. Ersu and H. Tolle, "Hierarchical learning control—An approach with neuron-like associative memories," *Proc. IEEE Conf. on Neural Information Processing Systems*, Nov. 1988.

[24] —, "A new concept for learning control inspired by brain theory," *Proc. FAC 9th World Congress*, July 2–6, 1984.

[25] E. Ersu and J. Militzer, "Real-time implementation of an associative memory-based learning control scheme for non-linear multivariable processes," *Proc. 1st Measurements and Control Symp. on Applications of Multivariable Systems Techniques*, pp. 109–119, 1984.

[26] E. Ersu and X. Mao, "Control of pH using a self-organizing control concept with associative memories," *Proc. Intl. IASTED Conf. on Applied Control and Identification*, 1983.

[27] J. Moody, "Fast learning in multi-resolution hierarchies," in *Advances in Neural Information Processing*, D. Touretzky, Ed. Morgan Kaufmann, 1989.

[28] B. Widrow and S. D. Stearns, *Adaptive Signal Processing.* Englewood Cliffs, NJ: Prentice-Hall, 1985.

[29] W. T. Miller, B. A. Box, and E. C. Whitney, "Design and implementation of a high speed CMAC neural network using programmable CMOS logic cell arrays," Univ. of New Hampshire, Rept. no. ECE.IS.90.01, Feb. 6, 1990.

[30] K. Hardwicke, "On the applicability of piecewise affine models in nonlinear adaptive equalization," MS thesis, University of Texas at Austin, 1987.

Part 8
Neural Hardware: Architectures and Implementations

Modification of cognitive behavior by experience is in the process of learning and adaptation. One class of learning is through the modification of synaptic and somatic behavior of a neuron in the cognitive faculty.

OVER the last three decades, biology has provided a great impetus to theoretical developments in the field of neural computing. These theoretical developments have led to some interesting applied work, especially in the areas of neuro-vision and neuro-control systems. Because the conventional algorithmic-based computer did not serve all of the purposes of the industrial community, various novel neural architectures and implementations have started to spring up. These architectures were motivated by the synaptic parallelism and somatic processing power in the CNS, and have enjoyed the developments of various technologies: digital, analog, and optical. The new technology based on molecular computing will have a large impact on the field of neural networks in the coming years.

In order to provide some biological fidelity, all of the existing theoretical developments in the field of neural computing adopt an information-processing, storage, and retrieval process associated with changing the connectivity pattern of synapses. The processing strength is associated with the dynamics, thresholding, and sigmoidal mapping operations of the soma.

What makes the biological information processing so efficient and effective compared to the information processing using the digital technology of today? C. Mead, in article (8.1), argues that the biological information processing systems operate on completely different computational principles from those with which engineers are familiar with. For many problems, particularly those in which the input data are ill-conditioned and the computation can be specified in a *relative* manner, the biological solutions are many orders of magnitude more effective compared to what we have been able to implement using digital methods. This advantage can be attributed principally to the use of elementary physical phenomena, such as computational primitives, and to the representation of information by the relative values of analog signals, rather than by the abso-lute values of digital signals. Though the author of this article does not mention it, it is worth pointing out to the reader that biological information processing is based upon relative grades inherent to fuzzy logic. Perhaps this is the reason why humans may take several seconds in numerical computations (for example, $315.618 \times 5.245 = ?$), but do very well in pattern recognition. This article lays a basic philosophical background for the hardware aspects of neuronal computing.

In article (8.2), J. J. Hopfield and D. W. Tank provide a new conceptual framework and minimization principle for the understanding of computations in neural circuits. The celebrated work of Hopfield and Tank consists of nonlinear graded responses organized into networks with effectively symmetric synaptic connections. This neural architecture attempts to retain certain important biological computational features. The authors show that certain complex optimization problems can be analyzed and understood without the need to follow the circuit dynamics in detail. The basic conceptual details provided in this article may lead to various other neuronal architectural morphologies.

Current computational models of neurons have somewhat merged in the silicon-based environment of analog–VLSI circuits. One typical network design utilizes transconductor amplifiers as the variable synaptic weights, both excitatory and inhibitory. The soma of this neuron can be constructed using operational amplifiers acting as compactors. In article (8.3), M. R. DeYong, R. L. Findley, and C. Fields offer an alternative approach to the implementation of the VLSI model of neurons and an alternative motivation for modeling the neuron behavior in general. In particular, the authors present the design of a hybrid analog–digital processing element (PE) that exhibits the temporal and waveform characteristics of biological neurons. The hybrid PE operates at the nanosecond time scale, which enables it to produce real-time solutions to complex spatio–temporal problems found in high-speed signal-vision processing applications. The VLSI prototype chips, which have been designed, fabricated, and tested, have yielded encouraging results and seem to have many potential applications.

The power of neural computing in robotics and other industrial applications hinges upon their distinct features of robust processing, and learning and adaptive capabilities in uncertain and noisy environments.

In article (8.4), S-Y. Kung and J-N. Hwang propose a ring VLSI systolic architecture for implementing neural networks with applications to robotics. The variety of neural configurations considered include a single-layer feedback neural network, competitive learning networks, and multilayer feedforward networks. The authors show that such an architecture is suitable for robotic tasks such as task planning, path planning, and path control.

In this final article, (8.5), T. Yamakawa presents, "a tutorial to enlighten outsiders or beginners on a utility of a fuzzy system by providing a broad scope overview, especially analog mode hardware, with the author's original work. At first, the difference between deterministic words and fuzzy words is explained as well as fuzzy logic. The description of the system by using mathematical equations, linguistic rules and parameter distribution (neural networks) are discussed. The algorithm of fuzzy inference and defuzzification is presented and their hardware implementation is discussed in detail together with an advanced one. The fuzzy logic controller was applied to stabilize a glass with wine and a mouse moving around the plate on the tip of the inverted pendulum."

This set of five representative articles published in recent years discuss the architectural implementation and fault tolerance aspects of neural networks and provide only a limited view of a vast field.

Let us more closely peer into present day progress—the neural field, the vast amount of literature, and the emerging technologies: all these lead us to believe that, before the end of this century, we will be carrying a large amount of computing power (not counting our own cognitive faculty), which will be, of course, a fraction of that of human cognitive power. We will have neural calculators in our pockets, neural power books in our briefcases, and neural workstations in our workplaces. Then we have to ponder—what to do with such enormous neural computing power?

Further Reading

[1] C. Mead and M. Ismail, Eds., *Analog VLSI Implementation of Neural Systems.* Boston: Kluwer Academic Publishers, 1989.

[2] B. E. Boser, E. Sackinger, J. Bromely, Y. L. Cun, and L. D. Jackel, "Hardware requirements for neural network pattern classifiers: A case study and implementation," *IEEE Micro,* pp. 32–40, Feb. 1992.

[3] Special Issue on Neural-Networks Hardware, *IEEE Trans. Neural Networks,* vol. 3, no. 3, May 1992.

[4] A. F. Murray, "Silicon implementations of neural networks," *IEE Proc.-F,* vol. 138, no. 1, pp. 3–12, Feb. 1991.

[5] H. C. Card, C. R. Schneider, and W. R. Moore, "Hebbian plasticity in MOS synapses," *IEE Proc.-F,* vol. 138, no. 1, pp. 13–16, Feb. 1991.

[6] S. Y. Kung and J. N. Hwang, "A unified systolic architecture for artificial neural networks," *J. Parallel Distributed Comput.,* vol. 6, no. 2, pp. 358–387, April 1989.

[7] Y. S. Abu-Mostafa and D. Psaltis, "Optical neural computers," *Scient. Amer.,* pp. 88–95, March 1982.

[8] A. T. Smith and J. F. Walkup, "Optical implementations of the alternating projection neural network," *Optical Eng.,* vol. 3, no. 10, pp. 1522–1528, Oct. 1991.

Article 8.1

Neuromorphic Electronic Systems

CARVER MEAD

Invited Paper

Biological information-processing systems operate on completely different principles from those with which most engineers are familiar. For many problems, particularly those in which the input data are ill-conditioned and the computation can be specified in a relative manner, biological solutions are many orders of magnitude more effective than those we have been able to implement using digital methods. This advantage can be attributed principally to the use of elementary physical phenomena as computational primitives, and to the representation of information by the relative values of analog signals, rather than by the absolute values of digital signals. This approach requires adaptive techniques to mitigate the effects of component differences. This kind of adaptation leads naturally to systems that learn about their environment. Large-scale adaptive analog systems are more robust to component degredation and failure than are more conventional systems, and they use far less power. For this reason, adaptive analog technology can be expected to utilize the full potential of wafer-scale silicon fabrication.

Two Technologies

Historically, the cost of computation has been directly related to the energy used in that computation. Today's electronic wristwatch does far more computation than the Eniac did when it was built. It is not the computation itself that costs—it is the energy consumed, and the system overhead required to supply that energy and to get rid of the heat: the boxes, the connectors, the circuit boards, the power supply, the fans, all of the superstructure that makes the system work. As the technology has evolved, it has always moved in the direction of lower energy per unit computation. That trend took us from vacuum tubes to transisitors, and from transistors to integrated circuits. It was the force behind the transition from n-MOS to CMOS technology that happened less than ten years ago. Today, it still is pushing us down to submicron sizes in semiconductor technology.

So it pays to look at just how much capability the nervous system has in computation. There is a myth that the nervous system is slow, is built out of slimy stuff, uses ions instead of electrons, and is therefore ineffective. When the Whirlwind computer was first built back at M.I.T., they made a movie about it, which was called "Faster than Thought." The Whirwind did less computation than your wristwatch does. We have evolved by a factor of about 10 million in the cost of computation since the Whirlwind. Yet we still cannot begin to do the simplest computations that can be done by the brains of insects, let alone handle the tasks routinely performed by the brains of humans. So we have finally come to the point where we can see what is difficult and what is easy. Multiplying numbers to balance a bank account is not that difficult. What is difficult is processing the poorly conditioned sensory information that comes in through the lens of an eye or through the eardrum.

A typical microprocessor does about 10 million operations/s, and uses about 1 W. In round numbers, it cost us about 10^{-7} J to do one operation, the way we do it today, on a single chip. If we go off the chip to the box level, a whole computer uses about 10^{-5} J/operation. A whole computer is thus about two orders of magnitude less efficient than is a single chip.

Back in the late 1960's we analyzed what would limit the electronic device technology as we know it; those calculations have held up quite well to the present [1]. The standard integrated-circuit fabrication processes available today allow us to build transistors that have minimum dimensions of about 1 μ (10^{-6} m). By ten years from now, we will have reduced these dimensions by another factor of 10, and we will be getting close to the fundamental physical limits: if we make the devices any smaller, they will stop working. It is conceiveable that a whole new class of devices will be invented—devices that are not subject to the same limitations. But certainly the ones we have thought of up to now—including the superconducting ones—will not make our circuits more than about two orders of magnitude more dense than those we have today. The factor of 100 in density translates rather directly into a similar factor in computation efficiency. So the ultimate silicon technology that we can envision today will dissipate on the order of 10^{-9} J of energy for each operation at the single chip level, and will consume a factor of 100–1000 more energy at the box level.

We can compare these numbers to the energy requirements of computing in the brain. There are about 10^{16} synapases in the brain. A nerve pulse arrives at each synapse about ten times/s, on average. So in rough numbers, the brain accomplishes 10^{16} complex operations/s. The power dissipation of the brain is a few watts, so each operation costs only 10^{6} J. The brain is a factor of 1 billion more efficient than our present digital technology, and a factor of

Manuscript received February 1, 1990; revised March 23, 1990.
The author is with the Department of Computer Science, California Institute of Technology, Pasadena, CA 91125.
IEEE Log Number 9039181.

Reprinted from *Proc. IEEE*, vol. 78, no. 10, pp. 1629–1636, Oct. 1990.

10 million more efficient than the best digital technology that we can imagine.

From the first integrated circuit in 1959 until today, the cost of computation has improved by a factor about 1 million. We can count on an additional factor of 100 before fundamental limitations are encountered. At that point, a state-of-the-art digital system will still require 10 MW to process information at the rate that it is processed by a single human brain. The unavoidable conclusion, which I reached about ten years ago, is that we have something fundamental to learn from the brain about a new and much more effective form of computation. Even the simplest brains of the simplest animals are awesome computational instruments. They do computations we do not know how to do, in ways we do not understand.

We might think that this big disparity in the effectiveness of computation has to do with the fact that, down at the device level, the nerve membrane is actually working with single molecules. Perhaps manipulating single molecules is fundamentally more efficient than is using the continuum physics with which we build transistors. If that conjecture were true, we would have no hope that our silicon technology would ever compete with the nervous system. In fact, however, the conjecture is false. Nerve membranes use *populations* of channels, rather than individual channels, to change their conductances, in much the same way that transistors use populations of electrons rather than single electrons. It is certainly true that a single channel can exhibit much more complex behaviors than can a single electron in the active region of a transistor, but these channels are used in large populations, not in isolation.

We can compare the two technologies by asking how much energy is dissipated in charging up the gate of a transistor from a 0 to a 1. We might imagine that a transistor would compute a function that is loosely comparable to synaptic operation. In today's technology, it takes about 10^{-13} J to charge up the gate of a single minimum-size transistor. In ten years, the number will be about 10^{-15} J—within shooting range of the kind of efficiency realized by nervous systems. So the disparity between the efficiency of computation in the nervous system and that in a computer is primarily attributable not to the individual device requirements, but rather to the way the devices are used in the system.

Where Did the Energy Go?

Where did all the energy go? There is a factor of 1 million unaccounted for between what it costs to make a transistor work and what is required to do an operation the way we do it in a digital computer. There are two primary causes of energy waste in the digital systems we build today.

1) We lose a factor of about 100 because, the way we build digital hardware, the capacitance of the gate is only a very small fraction of capacitance of the node. The node is mostly wire, so we spend most of our energy charging up the wires and not the gate.

2) We use far more than one transistor to do an operation; in a typical implementation, we switch about 10 000 transistors to do one operation.

So altogether it costs 1 million times as much energy to make what we call an operation in a digital machine as it costs to operate a single transistor.

I do not believe that there is any magic in the nervous system—that there is a mysterious fluid in there that is not defined, some phenomenon that is orders of magnitude more effective than anything we can ever imagine. There is nothing that is done in the nervous system that we cannot emulate with electronics if we understand the principles of neural information processing. I have spent the last decade trying to understand enough about how it works to be able to build systems that work in a similar way; I have had modest success, as I shall describe.

So there are two big opportunities. The first factor-of-100 opportunity, which can be done with either digital or analog technology, is to make alogrithms more local, so that we do not have to ship the data all over the place. That is a big win—we have built digital chips that way, and have achieved a factor of between 10 and 100 reduction in power dissipation. That still leaves the factor of 10^4, which is the difference between making a digital operation out of bunches of AND and OR gates, and using the physics of the device to do the operation.

Evolution has made a lot of inventions, as it evolved the nervous system. I think of systems as divided into three somewhat arbitrarily levels. There is at the bottom the *elementary functions*, then the *representation of information*, and at the top the *organizing principles*. All three levels must work together; all three are very different from those we use in human-engineered systems. Furthermore, the nervous system is not accompanied by a manual explaining the principles of operation. The blueprints and the early prototypes were thrown away a long time ago. Now we are stuck with an artifact, so we must try to reverse engineer it.

Let us consider the primitive operations and representations in the nervous system, and contrast them with their counterparts in a digital system. As we think back, many of us remember being confused when we were first learning about digital design. First, we decide on the information representation. There is only one kind of information, and that is the bit: It is either a 1 or a 0. We also decide the elementary operations we allow, usually AND, OR, and NOT or their equivalents. We start by confining ourselves to an incredibly impoverished world, and out of that, we try to build something that makes sense. The miracle is that we can do it! But we pay the factor of 10^4 for taking all the beautiful phyics that is built into those transistors, mashing it down into a 1 or a 0, and then painfully building it back up, with AND and OR gates to reinvent the multiply. We then string together those multiplications and additions to get more complex operations—those that are useful in a system we wish to build.

Computation Primitives

What kind of computation primitives are implemented by the device physics we have available in nervous tissue or in a silicon integrated circuit? In both cases, the state variables are analog, represented by an electrical charge. In the nervous system, there are state variables represented by chemical concentrations as well. To build a nervous system or a computer, we must be able to make specific connections. A particular output is connected to certain inputs and not to others. To achieve that kind of specificity, we must be able to isolate one signal on a single electrical node, with minimum coupling to other nodes. In both electronics and the nervous system, that isolation is achieved by building an energy barrier, so that we can put some charge on an electrical node somewhere, and it does not leak over to some other node nearby. In the nervous system, that energy barrier is built by the difference in the dielectric constant between fat and aqueous solutions. In electonics, it is built by the difference in the bandgap between silicon and silicon dioxide.

We do basic aggregation of information using the conservation of change. We can dump current onto an electrical node at any location, and it all ends up as charge on the node. Kirchhoff's law implements a distributed addition, and the capacitance of the node integrates the current into the node with respect to time.

In nervous tissue, ions are in thermal equilibrium with their surroundings, and hence their energies are Boltzmann distributed. This distribution, together with the presence of energy barriers, computes a current that is an exponential function of the barrier energy. If we modulate the barrier with an applied voltage, the current will be an exponential function of that voltage. That principle is used to create active devices (those that produce gain or amplification in signal level), both in the nervous system and in electronics. In addition to providing gain, an individual transistor computes a complex nonlinear function of its control and channel voltages. That function is not directly comparable to the functions that synapses evaluate using their presynaptic and postsynaptic potentials, but a few transistors can be connected strategically to compute remarkably competent synaptic functions.

Fig. 1(a) and (b) shows the current through a nerve membrane as a function of the voltage across the membrane. A plot of the current out of a synapse as the function of the voltage across the presynaptic membrane is shown in (c). The nervous system uses, as its basic operation, a current that increases exponentially with voltage. The channel current in a transistor as a function of the gate voltage is shown in (d). The current increases exponentially over many orders of magnitude, and then becomes limited by space charge, which reduces the dependence to the familiar quadratic. Note that this curve is hauntingly similar to others in the same figure. What class of computations can be implemented efficiently using expontential functions as primitives? Analog electronic circuits are an ideal way to explore this question.

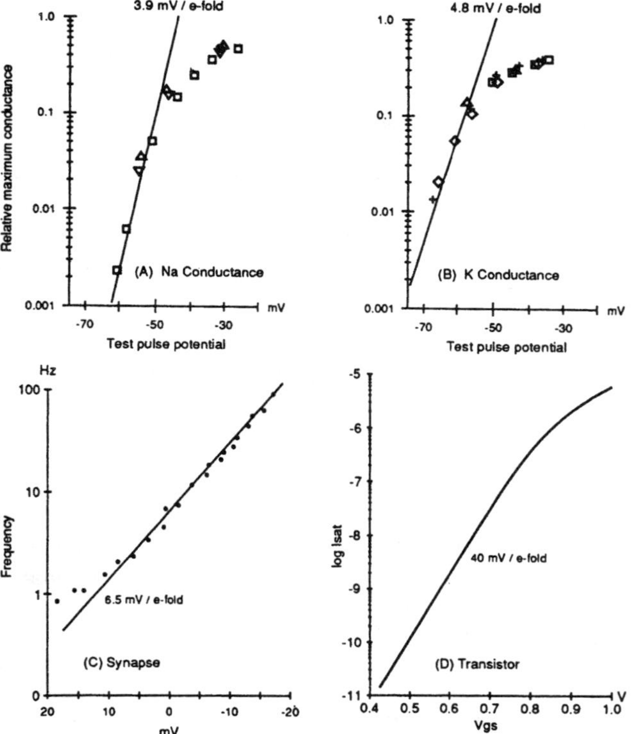

Fig. 1. Current-voltage plots for several important devices, each showing the ubiquitous exponential characteristic. Curves *A* and *B* show the behavior of populations of active ion channels in nerve membrane. Curve *C* illustrates the exponential dependence of the arrival rate of packets of the neurotransmitter at the postsynaptic membrane on the presynaptic membrane potential. Curve *D* shows the saturation current of a MOS transistor as a function of gate voltage.

Most important, the nervous system contains mechanisms for long-term learning and memory. All higher animals undergo permanent changes in their brains as a result of life experiences. Neurobiologists have identified at least one mechanism for these permanent changes, and are actively pursuing others. In microelectronics, we can store a certain quantity of charge on a floating polysilicon node, and that charge will be retained indefinitely. The floating node is completely surrounded by high-quality silicon dioxide—the world's most effective known insulator. We can sense the charge by making the floating node the gate of an ordinary MOS transistor. This mechanism has been used since 1971 for storing digital information in EPROM's and similar devices, but there is nothing inherently digital about the charge itself. Analog memory comes as a natural consequence of this near-perfect charge-storage mechanism. A silicon retina that does a rudimentary form of learning and long-term memory is described in the next section [2]. This system uses ultraviolet light to move charge through the oxide, onto or off the floating node. Tunneling to and from the floating node is used in commercial EEPROM devices. Several hot-electron mechanisms also have been employed to transfer charge through the oxide. The ability to learn and retain analog information for long periods is thus a natural consequence of the structures created by modern silicon processing technology.

The fact that we can build devices that implement the same basic operations as those the nervous system uses leads to the inevitable conclusion that we should be able to build entire systems based on the organizing principles used by the nervous system. I will refer to these systems generically as *neuromorphic systems*. We start by letting the device physics define our elementary operations. These functions provide a rich set of computational primitives, each a direct result of fundamental physical principles. They are not the operations out of which we are accustomed to building computers, but in many ways, they are much more interesting. They are more interesting than AND and OR. They are more interesting than multiplication and addition. But they are very different. If we try to fight them, to turn them into something with which we are familiar, we end up making a mess. So the real trick is to invent a representation that takes advantage of the inherent capabilities of the medium, such as the abilities to generate exponentials, to do integration with respect to time, and to implement a zero-cost addition using Kirchhoff's law. These are powerful primitives; using the nervous system as a guide, we will attempt to find a natural way to integrate them into an overall system-design strategy.

Retinal Computation

I shall use two examples from the evolution of silicon retinas to illustrate a number of physical principles that can be used to implement computation primitives. These examples also serve to introduce general principles of neural computation, and to show how these principles can be applied to realize effective systems in analog electronic integrated-circuit technology.

In 1868, Ernst Mach [3] described the operation performed by the retina in the following terms.

The illumination of a retinal point will, in proportion to the difference between this illumination and the average of the illumination on neighboring points, appear brighter or darker, respectively, depending on whether the illumination of it is above or below the average. The weight of the retinal points in this average is to be thought of as rapidly decreasing with distance from the particular point considered.

For many years, biologists have assembled evidence about the detailed mechanism by which this computation is accomplished. The neural machinery that performs this first step in the chain of visual processing is located in the outer plexiform layer of the retina, just under the photoreceptors. The lateral spread of information at the outer plexiform layer is mediated by a two-dimensional network of cells coupled by resistive connections. The voltage at every point in the network represents a spatially weighted average of the photoreceptor inputs. The farther away an input is from a point in the network, the less weight it is given. The weighting function decreases in a generally exponential manner with distance.

Using this biological evidence as a guide, Mahowald [4], [5] reported a silicon model of the computation described by Mach. In the silicon retina, each node in the network is linked to its six neighbors with resistive elements to form a hexagonal array, as shown in Fig. 2. A single bias circuit

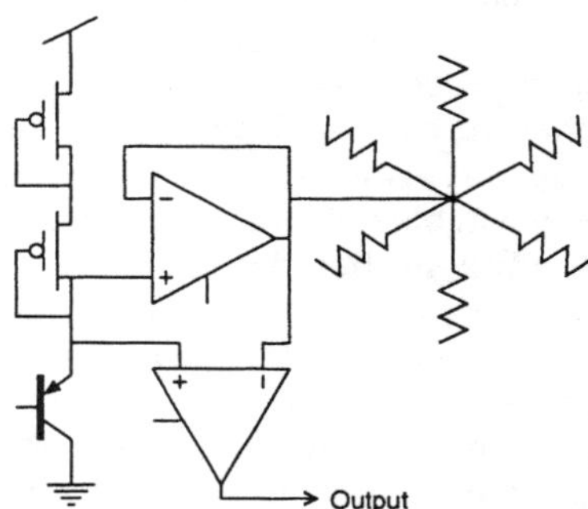

Fig. 2. Schematic of pixel from the Mahowald retina. The output is the difference between the potential of the local receptor and that of the resistive network. The network computes a weighted average over neighboring pixels.

associated with each node controls the strength of the six associated resistive connections. Each photoreceptor acts as a voltage input that drives the corresponding node of the resistive network through a conductance. A transconductance amplifier is used to implement a unidirectional conductance so the photoreceptor acts an effective voltage source. No current can be drawn from the output node of the photoreceptor because the amplifier input is connected to only the gate of a transistor.

The resistive network computes a spatially weighted average of photoreceptor inputs. The spatial scale of the weighting function is determined by the product of the lateral resistance and the conductance coupling the photoreceptors into the network. Varying the conductance of the transconductance amplifier or the strength of the resistors changes the space constant of the network, and thus changes the effective area over which signals are averaged.

From an engineering point of view, the primary function of the computation performed by a silicon retina is to provide an automatic gain control that extends the useful operating range of the system. It is essential that a sensory system be sensitive to changes in its input, no matter what the viewing conditions. The structure executing this level-normalization operation performs many other functions as well, such as computing the contrast ratio and enhancing edges in the image. Thus, the mechanisms responsible for keeping the system operating over an enormous range of image intensity have important consequences with regard to the representation of data.

The image enhancement performed by the retina was also described by Mach.

Let us call the intensity of illumunation $u = f(x, y)$. The brightness sensation v of the corresponding retinal point is given by

$$v = u - m\left(\frac{d^2u}{dx^2} + \frac{d^2u}{dy^2}\right)$$

where m is a constant. If the expression in parentheses is positive, then the sensation of brightness is reduced; in the opposite case, it is increased. Thus, v is not only influenced by u, but also its second differential quotients.

The image-enhancement property described by Mach is a result of the receptive field of the retinal computation, which shows an antagonistic center-surround response. This behavior is a result of the interaction of the photoreceptors, the resistive network, and the output amplifier. A transconductance amplifier provides a conductance through which the resistive network is driven towards the photoreceptor potential. A second amplifier senses the voltage difference across that conductance, and generates an output proportional to the difference between the photoreceptor potential and the network potential at that location. The output thus represents the difference between a center intensity and a weighted average of the intensities of surrounding points in the image.

The center-surround computation sometimes is referred to as a Laplacian filter, which has been used widely in computer vision systems. This computation, which can be approximated by a difference in Gaussians, has been used to help computers localize objects; this kind of enhancement is effective because discontinuities in intensity frequently correspond to object edges. Both of these mathematical forms express, in an analytically tractable way, the computation that occurs as a natural result of an efficient physical implementation of local normalization of the signal level.

In addition to its role in gain control and spatial filtering, the retina sharpens the time response of the system as an intrinsic part of its analog computation. Effective temporal processing requires that the time scale of the computation be matched to the time scale of external events. The temporal response of the silicon retina depends on the properties of the horizontal network. The voltage stored on the capacitance of the resistive network is the temporally as well as spatially averaged output of the photoreceptors. Because the capacitance of the horizontal network is driven by a finite conductance, its response weights its input by an amount that decreases exponentially into the past. The time constant of integration is set by the bias voltages of the wide-range amplifier and of the resistors. The time constant can be varied independently of the space constant, which depends on only the difference between these bias voltages, rather than on their absolute magnitude. The output of the retinal computation is thus the difference between the immediate local intensity and the spatially and temporally smoothed image. It therefore enhances both the first temporal and second spatial derivatives of the image.

Adaptive Retina

The Mahowald retina has given us a very realistic real-time model that shows essentially all of the perceptually interesting properties of early vision systems, including several well-known optical illusions such as Mach bands. One problem with the circuit is its sensitivity to transistor offset voltages. Under uniform illumination, the output is a random pattern reflecting the properties of individual transistors, no two of which are the same. Of course, biological retinas have precisely the same problem. No two receptors have the same sensitivity, and no two synapses have the same strength. The problem in wetware is even more acute than it is in silicon. It is also clear that biological systems use adaptive mechanisms to compensate for their lack of precision. The resulting system performance is well beyond that of our most advanced engineering marvels. Once we understand the principles of adaptation, we can incorporate them into our silicon retina.

All of our analog chips are fabricated in silicon-gate CMOS technology [6]. If no metal contact is made to the gate of a particular transistor, that gate will be completely surrounded by silicon dioxide. Any change parked on such a floating gate will remain for eons. The first floating-gate experiments of which I am aware were performed at Fairchild Research Laboratories in the mid-1960's. The first product to represent data by charges stored on a floating gate was reported in 1971 [7]. In this device, which today is called an EPROM, electrons are placed on the gate by an avalanche breakdown of the drain junction of the transistor. This injection can be done selectively, one junction at a time. Electrons can be removed by ultraviolet light incident on the chip. This so-called erase operation is performed on all devices simultaneously. In 1985, Glasser reported a circuit in which either a binary **1** or a binary **0** could be stored selectively in each location of a floating-gate digital memory [8]. The essential insight contributed by Glasser's work was that there is no fundamental asymetry to the current flowing through a thin layer of oxide. Electrons are excited into the conduction band of the oxide from both electrodes. The direction of current flow is determined primarily by the direction of the electric field in the oxide. In other words, the application of ultraviolet illumination to a capacitor with a silicon-dioxide dielectric has the effect of shunting the capacitor with a very small leakage conductance. With no illumination, the leakage conductance is effectively zero. The leakage conductance present during ultraviolet illumination thus provides a mechanism for adapting the charge on a float gate.

Frank Werblin suggested that the Mahowald retina might benefit from the known feedback connections from the resistive network to the photoreceptor circuit. A pixel incorporating a simplified version of this suggestion is shown in Fig. 3 [2]. In this circuit, the output node is the

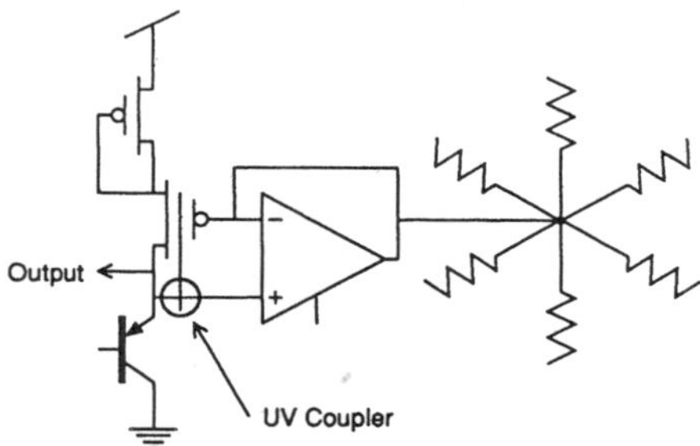

Fig. 3. Schematic of a pixel that performs a function similar to that of the Mahowald retina, but can be adapted with ultraviolet light to correct for output variations among pixels. This form of adaptation is the simplest form of learning. More sophisticated learning paradigms can be evolved directly from this structure.

emitter of the phototransistor. The current out of this node is thus set by the local incident-light intensity. The current into the output node is set by the potential on the resistive network, and hence by the weighted average of the light intensity in the neighborhood. The difference between these two currents is converted into a voltage by the effective resistance of the output node, determined primarily by the Early effect. The advantage of this circuit is that small differences between center intensity and surround intensity are translated into large output voltages, but the large dynamic range of operation is preserved. Retinas fabricated with this pixel show high gain, and operate properly over many orders of magnitude in illumination. The transconductance amplifier has a hyperbolic-tangent relationship between the output current and the input differential voltage. For proper operation, the conductance formed by this amplifier must be considerably smaller than that of the resistive network node. For that reason, when a local output node voltage is very different from the local network voltage, the amplifier saturates and supplies a fixed current to the node. The arrangement thus creates a center-surround response only slightly different from that of the Mahowald retina.

To reduce the effect of transistor offset voltages, we make use of ultraviolet adaptation to the floating gate that has been interposed between the resistive network and the pull-up transistor for the output node. The network is capacitively coupled to the floating node. The current into the output node is thus controlled by the voltage on the network, with an offset determined by the charge stored on the floating node. There is a region where the floating node overlaps the emitter of the phototransistor, shown inside the dark circle in Fig. 3. The entire chip is covered by second-level metal, except for openings over the phototransistors. The only way in which ultraviolet light can affect the floating gate is by interchanging electrons with the output node. If the output node is high, the floating gate will be charged high, thereby decreasing the current into the output node. If the output node is low, the floating gate will be charged low, thereby increasing the current into the output node. The feedback occasioned by ultraviolet illumination is thus negative, driving all output nodes toward the same potential.

Adaptation and Learning

The adaptive retina is a simple example of a general computation paradigm. We can view the function of a particular part of the nervous system as making a prediction about the spatial and temporal porperties of the world. In the case of the retina, these predictions are the simple assertions that the image has no second spatial derivative and no first temporal derivative. If the image does not conform to these predictions, the difference between expectation and experience is sent upward to be processed at higher levels. A block diagram of the essential structure is shown in Fig. 4.

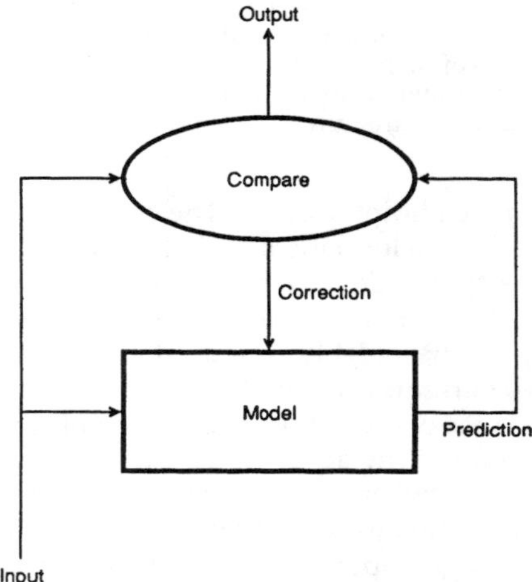

Fig. 4. Conceptual arrangement of a single level of a neural processing system. The computation consists of a prediction of the input, and a comparison of that prediction to the actual input. When the model accurately predicts the input, no information is passed to the next level, and no correction is made to the model. When the model fails to predict the input, the difference is used to correct the model. Random differences will cause a continued small "random walk" of the model parameters around that required for correct prediction. Systematic differences will cause the model to center itself over the true behavior of the input. Most routine events are filtered out to low level, reserving the capabilities of higher centers for genuinely interesting events.

The box labeled "model" is a predictor, perhaps a crude one; in the case of the retina, the model is the resistive network. We give the predictor the input over time, and it computes what is likely to happen next, just before the actual input arrives. Then, when that input materializes, it is compared to the prediction. If the two values are the same, no new information is produced; the system already knew what was about to happen. What happened is what was expected; therefore, no information is sent up to the next level of processing. But when something unexpected has occurred, there is a difference, and that difference is transferred on up to the next level to be interpreted. If we repeat this operation at each level of the nervous system, the information will be of higher quality at each subsequent level because we process only the information that could not be predicted at lower levels.

Learning in this kind of system is provided by the adaptation feedback from the comparator to the model. If the model is making predictions that are systematically different from what happens in nature, the ongoing corrections based on the individual differences will cause the model to learn what actually happens, as well as can be captured at its level of representation. It is only those events that are truly random, or that cannot be predicted from this level and therefore appear random, that will cancel out over all experience. The system parameters will undergo a local random walk, but will stay nearly centered on the average of what nature is providing as input. The retina is presented with a wide variety of scenes; it sees white edges and black edges. But every pixel in the retina sees the same intensity, averaged over time. Corrections towards this average constantly correct differences in photoreceptor sensitivity and variation in the properties of individual neurons and synapses. All other information is passed up to higher levels. Even this simple level of prediction removes a great deal of meaningless detail from the image, and provides a higher level of representation for the next level of discrimination.

That a system composed of many levels organized along the lines of Fig. 4 can compute truly awesome results is perhaps not surprising: each level is equipped with a model of the world, as represented by the information passed up from lower levels. All lower level processing may, from the point of view of a given level, be considered preprocessing. The most important property of this kind of system is that the same mechanism that adapts out errors and mismatches in its individual components also enables the system to build its own models through continued exposure to information coming in from the world. Although this particular example of the adaptive retina learns only a simple model, it illustrates a much more general principle: this kind of system is *self-organizing* in the most profound sense.

Nueral Silicon

Over the past eight years, we have designed, fabricated, and evaluated hundreds of test chips and several dozen complete system-level designs. All these adaptive analog chips were fabricated using standard, commercially available CMOS processing, provided to us under the auspices of DARPA's MOSIS fabrication service. These designs include control systems, motor-pattern generators, retina chips that track bright spots in an image, retina chips that focus images on themselves, and retina chips that perform gain control, motion sensing, and image enhancement. We have made multiscale retinas that give several levels of resolution, stereo-vision chips that see depth, and chips that segment images. A wide variety of systems has been designed to process auditory input; most of them are based on a biologically sensible model of the cochlea. There are monaural chips that decompose sound into its component features, binaural chips that compute horizontal and vertical localization of sound sources, and Seehear chips that convert a visual image into an auditory image—one where moving objects produce sound localized in the direction of the object.

This variety of experiments gives us a feeling for how far we have progressed on the quest for the nine order of magnitude biological advantage. The retina described in the preceding section is a typical example; it contains about 10^5 devices, performs the equivalent of about 10^8 operations/s, and consumes about 10^{-3} W of power. This and other chips using the same techniques thus perform each operation at a cost of only about 10^{-11} J compared to about 10^{-7} J/operation for a digital design using the same technology, and with 10^6 J/operation for the brain. We are still five orders of magnitude away from the efficiency of the brain, but four orders of magnitude ahead of that realized with digital techniques. The real question is how well the adaptive analog approach can take advantage of future advances in silicon fabrication. My prediction is that adaptive analog techniques can utilize the potential of advanced silicon fabrication more fully than can any other approach that has been proposed. Today (1990), a typical 6 in diameter wafer contains about 10^8 devices, partitioned into several hundred chips. After fabrication, the chips are cut apart and are put into packages. Several hundred of these packages are placed on a circuit board, which forms interconnections among them.

Why not just interconnect the chips on the wafer where they started, and dispense with all the extra fuss, bother, and expense? Many attempts by many groups to make a digital wafer-scale technology have met with abysmal failure. There are two basic reasons why wafer-scale integration is very difficult. First, a typical digital chip will fail if even a single transistor or wire on the chip is defective. Second, the power dissipated by several hundred chips of circuitry is over 100 W, and getting rid of all that heat is a major packaging problem. Together, these two problems have prevented even the largest computer companies from deploying wafer-scale systems successfully. The low-power dissipation of adaptive analog systems eliminates the packaging problem; wafers can be mounted on edge, and normal air convection will adequately remove the few hundred milliwatts of heat dissipated per wafer. Due to the robustness of the neural representation, the failure of a few components per square centimeter will not materially affect the performance of the system: its adaptive nature will allow the system simply to learn to ignore these inputs because they convey no information. In one or two decades, I believe we will have 10^{10} devices on a wafer, connected as a complete adaptive analog system. We will be able to extract information from connections made around the periphery of the wafer, while processing takes place in massively parallel form over the entire surface of the wafer. Each wafer operating in this manner will be capable of approximately 10^{13} operations/s. At that time, we will still not understand nearly as much about the brain as we do about the technology.

Scaling Laws

The possibility of wafer-scale integration naturally raises the question of the relative advantage conveyed by a three-dimensional neural structure over a two-dimensional one. Both approaches have been pursued in the evolution of animal brains so the question is of great interest in biology as well. Let us take the point of view that whatever we are going to build will be a space-filling structure. If it is a sheet, it will have neurons throughout the whole plane; if it is a volume, neurons will occupy the whole volume. If we allow every wire from every neuron to be as long as the dimensions of the entire structure, we will obviously get an explosion in the size of the structure as the number of neurons increases. The brain has not done that. If we compare our brain to a rat brain, we are not noticeably less efficient in our use of wiring resources. So the brain has evolved a *mostly local* wiring strategy to keep the scaling from getting out of hand. What are the requirements of a structure that keep the fraction of its resources devoted to wire from exploding as it is made larger? If the structure did not scale, a large brain would be all wire and would have no room for the computation.

First, let us consider the two-dimensional case. For the purpose of analysis, we can imagine that the width W of each wire is independent of the wire's length L, and that the probability that a wire of length between L and $L + dL$ is dedicated to each neuron is $p(L)\,dL$. The expected area of such a wire is the $WL\,p(L)\,dL$. The entire plane, of length and width L_{max}, is covered with neurons, such that there is one neuron per area A. Although the wires from many neurons overlap, the total wire from any given neuron must fit in area A. We can integrate the areas of the wires of all lengths associated with a given neuron, assuming that the shortest wire is of unit length:

$$\int_1^{L_{max}} WL\,p(L)\,dL = A.$$

The question is then: What are the bounds on the form of $p(L)$ such that the area A required for each neuron does not grow explosively as L_{max} becomes large? We can easily see that if $p(L) = 1/L^2$, the area A grows as the logarithm of L_{max}—a quite reasonable behavior. If $p(L)$ did not decrease at least this fast with increasing L_{max}, the human brain would be much more dominated by wire than it is, compared to the brain of a rat or a bat. From this argument, I conclude that the nervous system is organized such that, on the average, the number of wires decreases no more slowly than the inverse square of the wire's length.

We can repeat the analysis for a three-dimensional neural structure of extent L_{max}, in which each neuron occupies volume V. Each wire has a cross-sectional area S, and thus has an expected volume $SL\,p(L)$. As before, the total wire associated with each neuron must fit in volume V:

$$\int_1^{L_{max}} SL\,p(L)\,dL = V.$$

So the three-dimensional structure must follow the same scaling law as its two-dimensional counterpart. If we build a space-filling structure, the third dimension allows us to contact more neurons, but it does not change the basic scaling rule. The number of wires must decrease with wire length in the same way in both two and three dimensions.

The cortex of the human brain, if it is stretched out, is about 1 m/side, and 1 mm thick. About half of that millimeter is wire (white matter), and the other half is computing machinery (gray matter). This basically two-dimensional strategy won out over the three-dimensional strategies used by more primitive animals, apparently because it could evolve more easily: new areas of cortex could arise in the natural course of evolution, and some of them would be retained in the genome if they conveyed a competitive advantage on their owners. This result gives us hope that a neural structure comprising many two-dimensional areas, such as those we can make on silicon wafers, can be made into a truly useful, massively parallel, adaptive computing system.

Conclusion

Biological information-processing systems operate on completely different principles from those with which engineers are familiar. For many problems, particularly those in which the input data are ill-conditioned and the computation can be specified in a relative manner, biological solutions are many orders of magnitude more effective than those we have been able to implement using digital methods. I have shown that this advantage can be attributed principally to the use of elementary physical phenomena as computational primitives, and to the representation of information by the relative values of analog signals, rather than by the absolute values of digital signals. I have argued that this approach requires adaptive techniques to correct for differences between nominally identical components, and that this adaptive capability leads naturally to systems that learn about their environment. Although the adaptive analog systems build up to the present time are rudimentary, they have demonstrated important principles as a prerequisite to undertaking projects of much larger scope. Perhaps the most intriguing result of these experiments has been the suggestion that adaptive analog systems are 100 times more efficient in their use of silicon, and they use 10 000 times less power than comparable digital systems. It is also clear that these systems are more robust to component degradation and failure than are more conventional systems. I have also argued that the basic two-dimensional limitation of silicon technology is not a serious limitation in exploiting the potential of neuromorphic systems. For these reasons, I expect large-scale adaptive analog technology to permit the full utilization of the enormous, heretofore unrealized, potential of wafer-scale silicon fabrication.

References

[1] B. Hoeneisen and C. A. Mead, "Fundamental limitations in microelectronics—I. MOS technology," *Solid-State Electron.*, vol. 15, pp. 819–829, 1972.

[2] C. Mead, "Adaptive retina," in *Analog VLSI Implementation of Neural Systems*, C. Mead and M. Ismail, Eds. Boston, MA: Kluwer, 1989, pp. 239–246.

[3] F. Ratliff, *Mach Bands: Quantitative Studies on Neural Networks in the Retina*. San Francisco, CA: Holden-Day, 1965, pp. 253–332.

[4] M. A. Mahowald and C. A. Mead, "A silicon model of early visual processing," *Neural Networks*, vol. 1, pp. 91–97, 1988.

[5] —, "Silicon retina," in C. A. Mead, *Analog VLSI and Neural Systems*. Reading, MA: Addison-Wesley, 1989, pp. 257–278.

[6] C. A. Mead, *Analog VLSI and Neural Systems*. Reading, MA: Addison-Wesley, 1989.

[7] D. Frohman-Bentchkowsky, "Memory behavior in a floating-gate avalanche-injection MOS (FAMOS) structure," *Appl. Phys. Lett.*, vol. 18, pp. 332–334, Apr. 1971.

[8] L. A. Glasser, "A UV write-enabled PROM," in *Proc. 1985 Chapel Hill Conf. VLSI*, H. Fuchs, Ed. Rockville, MD: Computer Science Press, 1985, pp. 61–65.

Computing with Neural Circuits: A Model

JOHN J. HOPFIELD AND DAVID W. TANK

A new conceptual framework and a minimization principle together provide an understanding of computation in model neural circuits. The circuits consist of nonlinear graded-response model neurons organized into networks with effectively symmetric synaptic connections. The neurons represent an approximation to biological neurons in which a simplified set of important computational properties is retained. Complex circuits solving problems similar to those essential in biology can be analyzed and understood without the need to follow the circuit dynamics in detail. Implementation of the model with electronic devices will provide a class of electronic circuits of novel form and function.

A COMPLETE UNDERSTANDING OF HOW A NERVOUS SYSTEM computes requires comprehension at several different levels. Marr (*1*) noted that the computational problem the system is attempting to solve (the problem of stereopsis in vision, for example) must be characterized. An understanding at this level requires determining the input data, the solution, and the transformations necessary to compute the desired solution from the input. The goal of computational neurobiology is to understand what these transformations are and how they take place. Intermediate computational results are represented in a pattern of neural activity. These representations are a second, and system-specific, level of understanding. It is important to understand how algorithms—transformations between representations—can be carried out by neural hardware. This understanding requires that one comprehend how the properties of individual neurons, their synaptic connections, and the dynamics of a neural circuit result in the implementation of a particular algorithm. Recent theoretical and experimental work attempting to model computation in neural circuits has provided insight into how algorithms can be implemented. Here we define and review one class of network models—nonlinear graded-response neurons organized into networks with effectively symmetric synaptic connections—and illustrate how they can implement algorithms for an interesting class of problems (*2*).

Early attempts to understand biological computation were stimulated by McCulloch and Pitts, who described (*3*) a "logical calculus of the ideas immanent in nervous activity." In these early theoretical studies, biological neurons were modeled as logical decision elements described by a two-valued state variable (on-off), which were organized into logical decision networks that could compute simple Boolean functions. The timing of the logical operations was controlled by a system clock. In studies of the "perceptron" by Rosenblatt (*4*), simple pattern recognition problems were solved by logical decision networks that used a system of feed-forward synaptic connectivity and a simple learning algorithm. Several reviews of McCulloch and Pitts and perceptron work are available (*5*). More recent studies have used model neurons having less contrived properties, with continuous dynamics and without the computerlike clocked dynamics. For example, Hartline *et al.* (*6*) showed that simple linear models with continuous variables could explain how lateral inhibition between adjacent photoreceptor cells enhanced the detection of edges in the compound eye of *Limulus*. Continuous variables and dynamics have been widely used in simulating membrane currents and synaptic integration in single neurons (*7*) and in simulating biological circuits, including central pattern generators (*8*) and cortical structures (*9*). Both two-state (*10, 11*) and continuous-valued nonlinear models (*12*) have been extensively studied in networks organized to implement algorithms for associative memories and associative tasks (*13*).

The recent work being reviewed here has been directed toward an understanding of how particular computations can be performed by selecting appropriate patterns of synaptic connectivity in a simple dynamical model system. Circuits can be designed to provide solutions to a rich repertoire of problems. Early work (*10*) was designed to examine the computational power of a model system of two-state neurons operating with organized symmetric connections. The inclusion of feedback connectivity in these networks distinguished them from perceptron-like networks, which emphasized feed-forward connectivity. Graded-response neurons described by continuous dynamics were combined with the synaptic organization described by earlier work to generate a more biologically accurate model (*14*) whose computational properties include those of the earlier model. General principles for designing circuits to solve specific optimization problems were subsequently developed (*15–17*). These networks demonstrated the power and speed of circuits that were based on the graded-response model. Unexpectedly, new computational properties resulted (*15*) from the use of nonlinear graded-response neurons instead of the two-state neurons of the earlier models. The problems that could be posed and solved on these neural circuits included signal decision and decoding problems, pattern recognition problems, and other optimization problems having combinatorial complexity (*15–20*).

One lesson learned from the study of these model circuits is that a detailed description of synaptic connectivity or a random sampling of neural activity is generally insufficient to determine how the circuit computes and what it is computing. As an introduction to the circuits we review, this analysis problem is illustrated on a simple and well-understood model neural circuit. We next define and discuss the simple dynamical model system and the underlying assumptions and simplifications that relate this model to biological neural circuits. A conceptual framework and minimization principle applied to the model provide an understanding of how these circuits compute, specifically, how they compute solutions to optimization problems. The design and architecture of circuits for two specific problems are presented, including the formerly enigmatic circuit used earlier to illustrate the analysis problem.

J. J. Hopfield is with the Divisions of Chemistry and Biology, California Institute of Technology, Pasadena, CA 91125. D. W. Tank and J. J. Hopfield are with the Molecular Biophysics Research Department, AT&T Bell Laboratories, Murray Hill, NJ 07974.

Reprinted with permission from *Science*, J. J. Hopfield and D. W. Tank, "Computing with Neural Circuits: A Model," vol. 233, pp. 625–633, Aug. 1986.

Understanding Computation in a Simple Neural Circuit

Let us analyze the hypothetical neural circuit shown in Fig. 1 with simulation experiments based on the tools and methods of neurophysiology and anatomy. The analysis will show that the usual available neurobiological measures and descriptions are insufficient to explain how even small circuits of modest complexity compute. The seven-neuron circuit in Fig. 1 is designed to compute in a specific way that will later be described. From a neurobiological viewpoint, the basic anatomy of the circuit contains four principal neurons (*21*), identified in the drawing as P_0, P_1, P_2, and P_3. Each neuron has an axon leaving the circuit near the bottom of the figure. The computational results of the circuit must be evident in the activity of these neurons. The one input pathway, from a neuron external to the circuit, is provided by axon Q. Neurons IN_1, IN_2, and IN_3 are intrinsic interneurons in the circuit.

In attempting to understand the circuit's operation, we simultaneously monitor the activity (computer simulated) in each of the seven neurons while providing for a controllable level of impulse activity in the input axon Q. Results from this experiment on the hypothetical circuit for several fixed levels of input activity are shown in Fig. 2A. The top trace represents our controlled activity in Q. In each time epoch this activity is progressively larger, as illustrated by the increasing number of action potentials per unit time. Although the activity of IN_3 is steadily rising as the activity in Q increases, the activities of the other neurons in the circuit are not simply related to this input. From these results we know what the output patterns of activity on the principal neurons are for specific levels of impulse activity on the input axon Q, but we cannot explain what computation the circuit is computing. Furthermore, we do not know how the structure and organization of the circuit has provided these particular patterns of neural activities for the different input intensities.

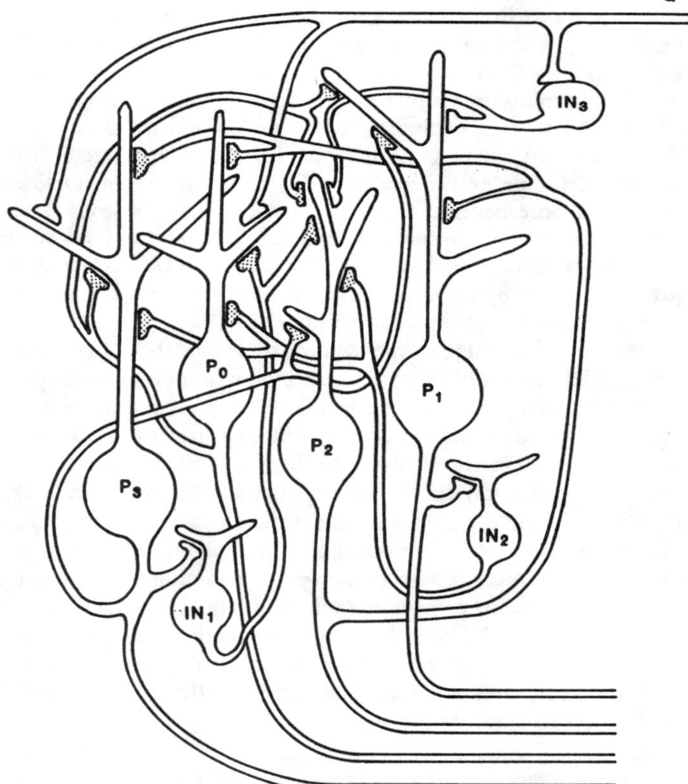

Fig. 1. "Anatomy" of a simple model neural circuit. Input axon Q has excitatory synapses (direct or effective) on each of the principal neurons P_0 through P_3. Each of these principal neurons has inhibitory synapses (direct or indirect) with all other principal neurons. Inhibitory synapses are shaded. IN_1 to IN_3, intrinsic interneurons.

Study of the synaptic organization of the connections between the neurons by electrophysiological or ultrastructural techniques could provide the numerical description of synaptic strengths shown in Table 1. The results of these experiments would show that each individual principal neuron P_i inhibits the other three principal neurons (P_j). There is either monosynaptic inhibition from P_i to P_j or polysynaptic inhibition by an excitatory synapse from P_i to an interneuron (IN_k), which then forms an inhibitory synapse with P_j (for example, the P_1-to-P_2 pathway in Fig. 1). This synaptic organization provides an "effective" inhibitory synapse between any two principal neurons; an action potential elicited in one principal neuron always contributes to inhibition of each of the others. Similar experiments measuring the strengths of the synaptic connections between the input axon Q and the P_i would show effective excitatory connections (Table 1). While the organization between principal neurons could be described classically as "lateral inhibition," the output patterns of activity in the P_i, shown in Fig. 2A for different input intensities, cannot be explained by this qualitative description.

Given the synaptic strengths in Table 1 and an appropriate mathematical description of the neurons, we can simulate the model neural circuit and produce the output activity patterns for the different inputs. Such detailed simulations can also be done for real neural circuits if the required parameters are known. In general, an ability to correctly predict a complex result that relies solely on simulation of the system provides a test of the simulation model, but does not provide an understanding of the result. Thus, despite our classical analysis of the simple neural circuit in Fig. 1, we still have no understanding of *why* these particular synaptic strengths (Table 1) provide these particular relations between input and output activity. Computation in the circuit shown in Fig. 1 can, however, be defined and understood within the conceptual framework provided by an analysis of dynamics in the simple neural circuit model we now discuss.

The Model Circuits and Their Relation to Biology

Neurons are continuous, dynamical systems, and neuron models must be able to describe smooth, continuous quantities such as graded transmitter release and time-averaged pulse intensity. In McCulloch-Pitts models, neurons were logical decision elements described by a two-valued state variable (on-off) and received synaptic input from a small number of other neurons. In general, McCulloch-Pitts models do not capture two important aspects of biological neurons and circuits: analog processing and high interconnectivity. While avoiding these limitations, we still want to model individual neurons simply. In the absence of appropriate simplifications, the complexities of the individual neurons will loom so large that it will be impossible to see the effects of organized synaptic interactions. A simplified model must describe a neuron's effective output, input, internal state, and the relation between its input and output.

In the face of the staggering diversity of neuronal properties, the goal of compressing their complicated characteristics is especially difficult. For the present, let us consider a prototypical biological neuron having inputs onto its dendritic arborization from other neurons and outputs to other neurons from synapses on its axon.

Action potentials initiate near the soma and propagate along the axon, activating synapses. Although we *could* model the detailed synaptic, integrative, and spike-initiating biophysics of this neuron, following, for example, the ideas of Rall (*7*), the first simplification we make in our description of the neuron is to neglect electrical effects attributable to the shape of dendrites and axon. (The axon and dendrite space-constants are assumed to be very large.) Our model neuron has the capacitances and conductances of the arborization added directly to those of the soma. The input currents from all synaptic channels are simply additive; more complex interactions between input currents are ignored. Membrane potential changes are assumed to arrive at the presynaptic side of synapses at the same time as they are initated at the soma. The second simplification is to deal only with "fast" synaptic events. When a potential fluctuation occurs in the presynaptic terminal of a chemical synapse, a change in the concentration of neurotransmitter is followed (with a slight delay) by a current in the postsynaptic cell. In our model neurons we presume this delay is much shorter than the membrane time constant of the neuron.

These two suppositions on time scale mean that when a change in potential is initiated at the soma of cell j, it introduces an effectively instantaneous conductance change in a postsynaptic cell i. The amount of the conductance change depends on the nature and strength of the synapse from cell j to cell i.

Biological neurons that produce action potentials do so (in steady state) at a rate determined by the net synaptic input current. This current acts indirectly by charging the soma and changing the cell potential. A characteristic charging or discharging time constant is determined by the cell capacitance C and membrane resistance R. The input current is "integrated" by the cell RC time constant to determine a value of an effective "input-potential," u. Conceptually, this potential u is the cell membrane potential after deletion of the action potentials. Action potentials (and postsynaptic responses in follower cells) are then generated at a rate dependent on the value of u. Dependencies of firing rates on input currents (and hence u) vary greatly, but have a generally sigmoid and monotonic form (Fig. 3A), rising continuously between zero and some maximum value (*22*). The firing rate of cell i can be described by the function $f_i(u_i)$. For processing in which individual action potentials are not synchronized or highly significant, a model that suppresses the details of action potentials should be adequate. In such a limiting case, two variables describe the state of neuron i: the effective input potential u_i and the output firing rate $f_i(u_i)$. The strength of the synaptic current into a postsynaptic neuron j due to a presynaptic neuron i is proportional to the product of the presynaptic cell's output $[f_i(u_i)]$

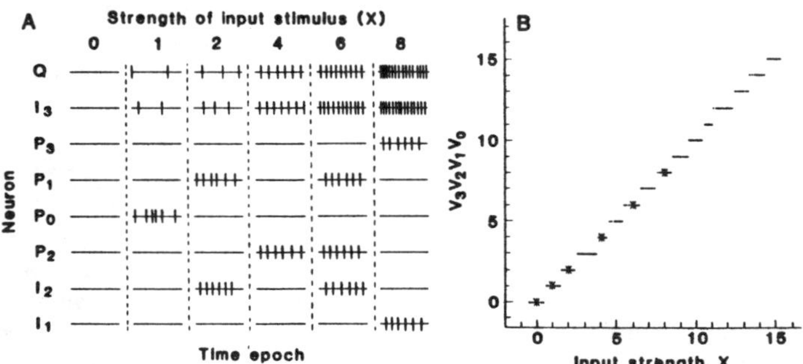

Fig. 2. (A) Results of an experiment in which the activity in each neuron in the circuit of Fig. 1 was simultaneously recorded (by simulation) as a function of the strength of the input stimulus on axon Q. The strength of the input stimulus is indicated by the numbers above each time epoch. (B) A selective rearrangement of the data in (A) illustrating the analog-binary computation being performed by the circuit. The digital word $V_3V_2V_1V_0$ is calculated from the records.

Table 1. Effective synaptic strengths for the circuit in Fig. 1.

Postsynaptic neuron	Presynaptic neuron				
	P_0	P_1	P_2	P_3	Q
P_0		−2	−4	−8	+1
P_1	−2		−8	−16	+2
P_2	−4	−8		−32	+4
P_3	−8	−16	−32		+8

and the strength of the synapse from i to j. In our model, the strength of this synapse is represented by the parameter T_{ij}, so that the postsynaptic current is given by $T_{ij}\,f_j(u_j)$. The net result of our description is that action potentials have their effects represented by continuous variables, just as the usual equations describing the behavior of electrical circuits replace discrete electrons by continuous charge and current variables.

Many neurons, both central and peripheral, show a graded response and do not normally produce action potentials (*23*). The presynaptic terminals of these graded-response neurons secrete neurotransmitters, and hence induce postsynaptic currents, at a rate dependent on the presynaptic cell potential. The effective output of such cells is also a monotonic sigmoid function of the net synaptic input. Thus the model treats both neurons with graded responses and those exhibiting action potentials with the same mathematics.

We can now describe the dynamics of an interacting system of N neurons. The following set of coupled nonlinear differential equations results from our simplifications and describes how the state variables of the neurons (u_i; $i = 1,\ldots, N$) will change with time under the influence of synaptic currents from other neurons in the circuit.

$$C_i\frac{du_i}{dt} = \sum_{j=1}^{N} T_{i,j}\,f_j(u_j) - \frac{u_i}{R_i} + I_i \qquad (i = 1,\ldots, N) \tag{1}$$

These equations might be thought of as a description of "classical" neurodynamics (*12, 14*). They express the net input current charging the input capacitance C_i of neuron i to potential u_i as the sum of three sources: (i) postsynaptic currents induced in i by presynaptic activity in neuron j, (ii) leakage current due to the finite input resistance R_i of neuron i, and (iii) input currents I_i from other neurons external to the circuit. The time evolution of any hypothetical circuit, defined by specific values of T_{ij}, I_i, f_i, C_i, and R_i, can be simulated by numerical integration of these equations.

Some intuitive feeling for how a model neural circuit might behave can be provided by considering the electrical circuit shown in Fig. 3B, which obeys the same differential equation (Eq. 1). The "neurons" consist of amplifiers in conjunction with feedback circuits composed of wires, resistors, and capacitors organized to represent axons, dendritic arborization, and synapses connecting the neurons. The firing rate function of our model neurons $[f_i(u_i)]$ is replaced in the circuit by the output voltage V_i of amplifier i. This output is $V_i = V_i^{max}\,g_i(u_i)$, where the dimensionless function $g_i(u_i)$ has the same sigmoid monotonic shape (Fig. 3A) as $f_i(u_i)$ and a maximum value of 1. V_i^{max} is the electrical circuit equivalent of the maximum firing rate of cell i. The input impedance of our model neuron is represented in the circuit by an equivalent resistor ρ_j and an input capacitor C_j connected in parallel from the amplifier input to ground. These components define the time constants of the neurons and provide for the integrative analog summation of the synaptic input currents from other neurons in the network. To provide for both excitatory and inhibitory synaptic connections between neurons while using conventional electrical components, each amplifier is given two outputs—a normal (+) output and an inverted (−) output of the same magnitude but opposite in sign. A synapse between two neurons is defined by a conductance T_{ij}, which connects one of the two outputs of amplifier j to the input of amplifier i. This connection is made with a resistor of value $R_{ij} = 1/|T_{ij}|$. If the synapse is excitatory ($T_{ij} > 0$), this resistor is connected to the normal (+) output of amplifier j. For an inhibitory synapse ($T_{ij} < 0$), it is connected to the inverted (−) output of amplifier j. Thus, the normal and inverted outputs for each neuron allow for the construction of both excitatory and inhibitory connections through the use of normal (positive valued) resistors. The circuits include a wire providing an externally supplied input current I_i for each neuron (Fig. 3B). These inputs can be used to set the general level of excitability of the network through constant biases, which effectively shift the input-output relation along the u_i axis, or to provide direct parallel inputs to drive specific neurons. As in Eq. 1, the net input current to any neuron is the sum of the synaptic currents (flowing through the set of resistors connecting its input to the outputs of the other neurons), externally provided currents, and leakage current.

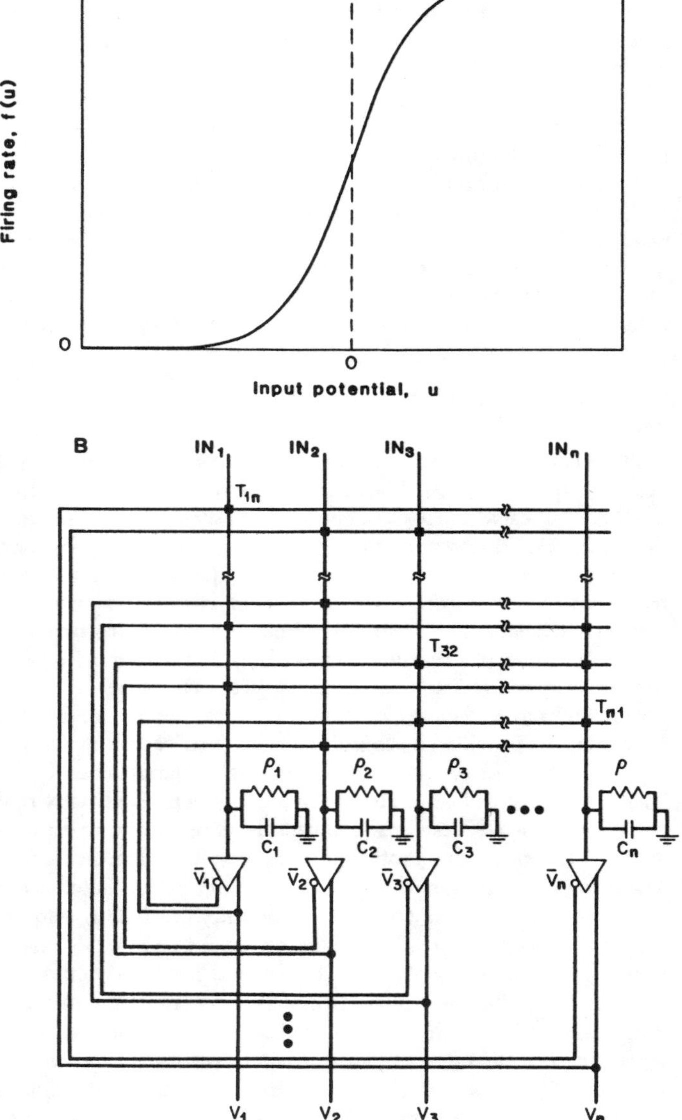

Fig. 3. (A) The sigmoid monotonic input-output relation used for the model neurons. (B) The model neural circuit in electrical components. The output of any neuron can potentially be connected to the input of any other neuron. Black squares at intersections represent resistive connections (with conductance T_{ij}) between outputs and inputs. Connections between inverted outputs (represented by the circles on the amplifiers) and inputs represent negative (inhibitory) connections.

In the model represented by Eq. 1 and Fig. 3, the properties of individual model neurons have been oversimplified, in comparison with biological neurons, to obtain a simple system and set of equations. However, essential features that have been retained include the idea of a neuron as transducer of input to output, with a smooth sigmoid response up to a maximum level of output; the integrative behavior of the cell membrane; large numbers of excitatory and inhibitory connections; the reentrant or fedback nature of the connections; and the ability to work with both graded-response neurons and neurons that produce action potentials. None of these features was the *result* of approximations. Their inclusion in a simplified model emphasizes features of the biological system we believe important for computation. The model retains the two important aspects for computation: dynamics and nonlinearity.

The model of Eq. 1 and Fig. 3 has immense computing power, achieved through organized synaptic interactions between the neurons. The model neurons lack many complex features that give biological neurons, taken individually, greater computational capabilities. It seems an appropriate model for the study of how the cooperative effects of neuronal interactions can achieve computational power.

A New Concept for Understanding the Dynamics of Neural Circuitry

A specific circuit of the general form described by Eq. 1 and Fig. 3 is defined by the values of the synapses (T_{ij}) and input currents (I_i). Given this architecture, the state of the system of neurons is defined by the values of the outputs V_i (or, equivalently, the inputs u_i) of each neuron. The circuit computes by changing this state with time. In a geometric space with a Cartesian axis for each neural output V_i, the instantaneous state of the system is represented by a point. A given circuit has dynamics that can be pictured as a time history or motion in this state space. For a circuit having arbitrarily chosen values for the synaptic connections, these motions can be very complex, and no simplifying description has been found. A broad class of simplified circuits, however, has a unifying principle of behavior while remaining capable of powerful computation. These circuits are literally or effectively symmetric.

A symmetric circuit is defined as having synaptic strength and sign (excitation or inhibition) of the connection from neuron i to j the same as from j to i. The two neurons need not, however, have the same input-output relation, threshold, or capacitance. Our model circuit (Fig. 3B) is symmetric if, for all i and j, T_{ij} is equal to T_{ji}. This symmetry refers only to connections between neurons in the circuit. It specifically excludes the input connections (represented in Fig. 3B as the input currents I_i) and any output connections from the circuit.

Symmetry of the connections results in a powerful theorem about the behavior of the system. The only additional conditions necessary are that the input-output relation of the model neurons be monotonic and bounded and that the external inputs I_i (if any) should change only slowly over the time of the computation. The theorem shows that a mathematical quantity E, which might be thought of as the "computational energy," decreases during the change in neural state with time described by Eq. 1. Started in any initial state, the system will move in a generally "downhill" direction of the E function, reach a state in which E is a local minimum, and stop changing with time. The system cannot oscillate. This concept can be illustrated graphically by a flow map in a state-space diagram. Each line corresponds to a possible time-history of the system, with

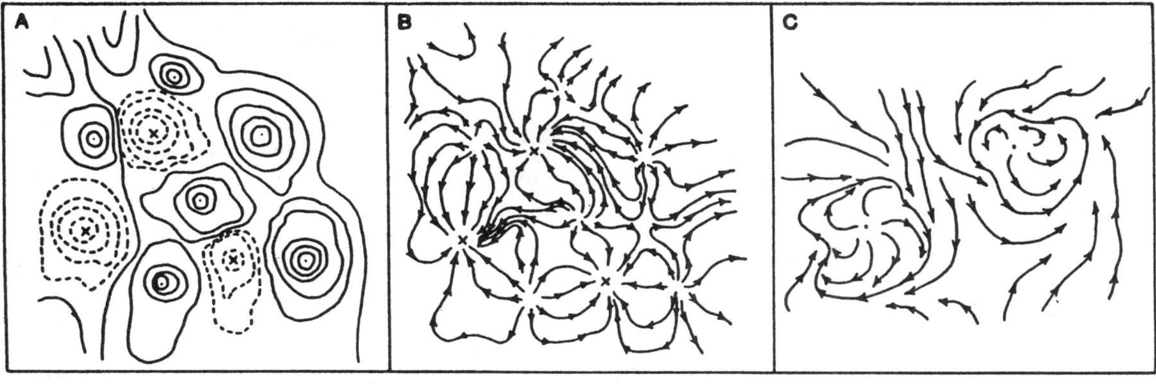

Fig. 4. (A) Energy-terrain contour map for the flow map shown in (B). (B) Typical flow map of neural dynamics for the circuit of Fig. 3 for symmetric connections ($T_{ij} = T_{ji}$). (C) More complicated dynamics that can occur for unrestricted (T_{ij}). Limit cycles are possible.

arrows showing the direction of motion. The structure imposed on the flow map for a circuit with symmetry is illustrated for a two-dimensional state space in Fig. 4. With symmetric connections, the flow map of the neural dynamics resembles Fig. 4B. Such a flow, in which each trajectory goes to stable points and stops, results from always going "downhill" on an "energy-terrain," coming to the bottom of a local valley, and stopping. The contour map of an E function that matches the flow in Fig. 4B is shown in Fig. 4A; it shows separated hills and valleys. The valleys are located where the trajectories in Fig. 4B stop. For a nonsymmetric circuit, the complications illustrated in the flow map in Fig. 4C can occur. This flow map has trajectories corresponding to complicated oscillatory behaviors. Such trajectories are undoubtedly important in neural computations, but as yet we lack the mathematical tools to manipulate and understand them at a computational level. The motion of a neural circuit comprising N neurons must be pictured in a space of N dimensions rather than the two dimensions of Fig. 4, but the qualitative picture of the effects of symmetric synaptic strengths is exactly the same.

The computational energy is a global quantity not felt by an individual neuron. The states of individual neurons simply obey the neural equations of motion (Eq. 1). The computational energy is our way of understanding why the system behaves as it does. A similar situation occurs in the concept of entropy in a simple gas. We understand that when a nonequilibrium state is set up with all the air molecules in one corner of the room, a uniform distribution will rapidly result. We explain that fact by the tendency of the entropy of isolated systems to increase whenever possible, but the individual molecules know nothing of entropy. They simply follow their Newtonian equations of motion.

Symmetric chemical synapses are observed in neural systems (*24*). Nonrectifying electrical synapses are intrinsically symmetric synapses of positive sign (*25*). Lateral inhibition in the visual system of *Limulus* is implemented with symmetric inhibitory synapses (*6*). An asymmetric network can also behave as though it were symmetric. In the olfactory bulb, the local circuit of mitral cell to granule cell to mitral cell provides an equivalent symmetric inhibitory connection between the pair of mitral cells (*26*). A similar situation occurs in the circuit shown in Fig. 1, where a direct equivalence between a neural circuit which is manifestly not symmetric and one which is effectively symmetric can be made if the inhibitory interneurons (IN_1, IN_2) are faster than other neurons.

The requirement of symmetry for this theorem can also be weakened. We have proven stability for a wide class of circuits having organized asymmetry between two sets of neurons with different time constants (*16*). (A neurobiological example would be the existence, in mammalian systems, of fast inhibitory interneurons that could provide effective symmetric inhibitory connections between neurons that are otherwise excitatory.) In one potentially useful example (*16*), stability could be guaranteed even though the sign of T_{ij} was always opposite that of T_{ji}. Also, there is a family of transformations by which a broader class of synaptic organizations can be made equivalent to symmetric ones (*27*). From an empirical viewpoint, moderate disorganized asymmetry (for example, having a random set of connections missing in an otherwise symmetric associative memory circuit) has little experimental effect on dynamic stability (*28*). Because the general features of symmetric circuits persist in circuits that are only equivalently symmetric, and real neural circuits can often be so viewed (except for inputs and outputs), the behavior of symmetric circuit models should be of direct use in trying to understand how neural computation is done in biology.

In general, systems having organized asymmetry can exhibit oscillation and chaos (*29*). In some neural systems like central pattern generators (*8*), coordinated oscillation *is* the desired computation of the circuit. Processing in the olfactory bulb also seems to make explicit use of oscillatory patterns (*30*). In such a case, proper combinations of symmetric synapses can enforce chosen phase relationships between different oscillators, an effect similar to those presented above.

Hard Problems Naturally Solved by Model Neural Circuits

In thinking about how difficult computational problems can be done on such networks, it is useful to recall the simple problem of associative memory, which these networks implement in a "natural" fashion (*10, 13*). This naturalness has two aspects. (i) The symmetry of the networks is natural because, in simple associations, if A is associated with B, B is symmetrically associated with A. (ii) If the desired memories can be made the stable states of a network, the desired computation (given partial information as input, find the memory that most resembles it) can be directly visualized as a motion toward the nearest stable state whose position is the recalled memory. Finally, the way the connection strengths must be chosen for a given set of memories can be easily implemented by learning rules (*13*) such as the one proposed by Hebb (*31*).

To what extent can more difficult computations—for example, those relevant to object recognition or speech perception—be carried out naturally on these model neural circuits? One of the characteristics of such computations seems to be a combinatorial explosion—the huge number of possible answers that must be considered. The desired computation (for example, matching a set of words to a sound pattern) can often be stated as an optimization. Although it is not yet known how to map most biological problems onto model circuits, it is now possible to design model circuits to solve nonbiological problems having combinatorial complexity.

Because well-defined problems have been used, the effectiveness of the neural circuit computation can be quantified. We will review two circuit examples.

The idea of most algorithms or procedures for optimization is to move in a space of possible configurations representing solutions, progress in a direction that tends to decrease the cost function being minimized, and hope that the space and method of moving are smooth enough that a good solution will ultimately be reached. Such ideas lie behind conventional computer optimization algorithms and the recent work in simulated annealing (*32*) and Bayesian image analysis (*33*). In our approach (*15–17*), the optimization problem is mapped onto a neural network in such a way that the network configurations correspond to possible solutions to the problem. An E function appropriate to the problem is then constructed. The form of the E function is chosen so that at configurations representing possible solutions, E is proportional to the cost function of the problem. Since, in general, E is minimized as the circuit computes, the dynamics produce a path through the space that tends to minimize the energy and therefore the cost function. Eventually, a stable-state configuration is reached that corresponds to a local minimum in the E function. The solution to the problem is then decoded from this configuration.

It is particularly easy to construct appropriate E functions when the sigmoid input-output relation is steep, because in this "high-gain" limit, each neuron will be either very near 0 output or very near its maximal output when the system is in a low E stable state (*14*). In the high-gain case, the energy function is

$$E = -\frac{1}{2}\sum_{i,j} T_{ij} V_i V_j - \sum_j I_j V_j \qquad (2)$$

When lower gain is considered, terms containing the function $g_i(u_i)$ must be included in E (*14*). The following two examples make use of this high-gain limit.

The simple seven-neuron circuit described in Fig. 1 was designed according to this conceptual framework to be a four-bit analog-to-binary (A-B) converter. Given an analog input to the circuit represented by the time-averaged impulse activity in the input axon Q, the neural circuit is organized to adjust the firing rates in the principal neurons so that they can be interpreted as the binary number numerically equal to the time-averaged input activity. Reorganization of the data in Fig. 2A will illustrate this computation. In each time epoch in Fig. 2A, assign the value 0 or 1 to the variable V_i representing the output of P_i; if P_i is firing strongly, $V_i = 1$; if it is quiescent, $V_i = 0$. Represent the activity in axon Q by a continuous variable X. The value of the binary word interpreted from the ordered list of numbers ($V_3V_2V_1V_0$) is plotted in Fig. 2B for each of the different values of input strength X. The data points (asterisks) lie on a staircase function (dotted line) characteristic of an A-B converter. (Although not shown, the outputs computed for any other input would also lie on this curve.)

Through the consideration of a specific energy function in the high-gain limit and the synaptic strengths and inputs listed in Table 1, the behavior of the neural circuit can be predicted and understood. We decide in advance that outputs $V_3V_2V_1V_0$ of P_3 through P_0 are interpreted as a computed binary word. The problem to be solved is stated as an optimization: Given analog input X, which binary word (set of outputs) best represents the numerical value of X? The solution is provided when the following E is minimized (*16*):

$$E = -\frac{1}{2}\left(X - \sum_{j=0}^{3} 2^j V_j\right)^2 + \sum_{j=0}^{3} (2^{2j-1})\,[V_j(1 - V_j)] \qquad (3)$$

The second term in E is minimized (and numerically equal to 0) when all V_j are either close to 0 or close to 1. Since E is minimized as the circuit converges, stable states having the correct "syntax" tend to develop. Since the first term in E is a minimum when the expression in the parentheses vanishes, this term biases the circuit towards the states closest, in the least-squares sense, to the analog value of X. The E in Eq. 3 is like that in Eq. 2, a quadratic in the V_i. Rearranging Eq. 3 and comparing it with this general form yields values for T_{ij} and I_i for a circuit of the form in Fig. 3B that can be deduced within a common scale factor as

$$T_{ij} = -2^{(i+j)};\ I_i = (-2^{(2i-1)} + 2^iX) \qquad (4)$$

The coefficient of X in I_i is the synaptic strength from the input axon Q to the principal neurons. These specific values are equal to the strengths of the "effective" synapses tabulated in Table 1. Knowledge that E is minimized as the circuit computes provides an *understanding* of how this synaptic organization both enforces the necessary syntax and biases the network to choose the optimum solution.

Our second example is a neural circuit that computes solutions to the traveling salesman problem (TSP) (*15*). In this frequently studied optimization problem (*34*), a salesman is required to visit in some sequence each of n cities; the problem is to determine the shortest closed tour in which every city is visited only once. Specific problems are defined by the distances (d_{ij}) between pairs of cities (i, j). Assigning letters to the cities in a TSP permits a solution to be specified by an ordered list of letters. For example, the list *CAFGB* is interpreted as "visit C, then A, then F, then G, then B, and finally return to C." For an n-city TSP, this list can be decoded from the outputs of $N = n^2$ neurons if we let a single neuron correspond to the hypothesis that one of the n cities is in a particular one of the n possible positions in the final tour list. This rule suggests the arrangement illustrated in Fig. 5 for displaying the neural output states. The output of a neuron (V_i) is graphically illustrated by shading; a filled square represents a neuron which is "on" and firing strongly. An empty square represents a neuron that is not firing. The output states of the n neurons in each row are interpreted as information about the location of a particular city in the tour solution. The output states of the n neurons in each column are interpreted as information about what cities are to be associated with a particular position in the tour. If the neuron from column 5 in row C is "on," the hypothesis that city C is in position 5 in the final tour solution is true.

Hypothetically, each of the n cities could indicate its position in any one of the n possible tour locations. Therefore, 2^N possible "neural states" could conceivably be represented by these outputs. However, only a subset of these actually correspond to valid solutions to the TSP (valid tours): a city must be in one, and only one, position in a valid tour, and any position must be filled by one and only one city. This constraint implies that only output states in which only one neuron is "on" in every row and in every column are of the correct "syntax" to represent valid solutions to the TSP. A TSP circuit that is to operate correctly must have synapses favoring this subset of states. Simple lateral inhibition between neurons within each row and column will provide this bias. For example, if $V_{B,2}$ (representing city B in position 2) is "on," all other neurons in row B and column 2 should be inhibited. This can be provided by the inhibitory connections from neuron $V_{B,2}$ drawn in Fig. 5 (red lines). Similar row and column inhibitory connections are drawn for neuron $V_{D,5}$. A complex "topology" of syntax-enforcing connections is generated. We can also think of these connections as contributing a term to the E function for the circuit. For example, a term $+A\ V_{X,i}\ V_{Y,i}$ in E makes a contribution $-A$ to the synaptic strength $T_{X,i;Y,i}$ and represents a mutual lateral inhibition between neurons (X,i) and (Y,i). The term is positive (higher E) when both of these neurons are "on," but contributes nothing if only one of the two is "on." The proper combination of similar terms in an E function can specify the synapses that coordinate correct syntax.

In a syntactically correct state representing a valid solution (tour), if neurons $V_{X,i}$ and $V_{Y,i+1}$ are both "on," the salesman travels from city X directly to city Y. Therefore, the distance $d_{X,Y}$ between these two cities is included in the total tour length for that solution. A term of the form $+d_{X,Y} V_{X,i} V_{Y,i+1}$ in the E function provides a "distance" contribution of d_{xy} to the value of E when these neurons are "on." Similar terms, properly summed, will add to E a value equal to the length of the tour. Since the circuit minimizes E, the final state will be biased toward those valid solutions representing short tours. Such inhibitory connections are drawn in Fig. 5 with blue lines for neurons $V_{B,2}$ and $V_{D,5}$. In TSP and in the earlier example, the rules of syntax are expressed in inhibitory connections. It seems easier to define what these systems should not do (by inhibitory connections), and to define what they should do by default, rather than to define what they should do by writing syntax in excitatory connections.

The inhibitory synapses define the computational connections for the TSP circuit. With a common sigmoid gain curve, R, and C for each neuron, the description of the circuit is complete. The gain curve is chosen so that with zero input, a neuron has a nonzero but modest output. This circuit can rapidly compute good solutions to a TSP problem (*15*). When started from an initial "noise" state favoring no particular tour, the network rapidly converges to a steady state describing a very short tour. The state of the circuit at several time points in a typical convergence is illustrated in Fig. 6. In a 30-city problem, there are about 10^{30} possible tours—the combinatorial problem has gotten completely out of hand. But the circuit of 900 neurons can find one of the best 10^7 solutions in a single convergence—a few time constants of the circuit. It selects good answers and rejects bad ones by a factor of 10^{23}.

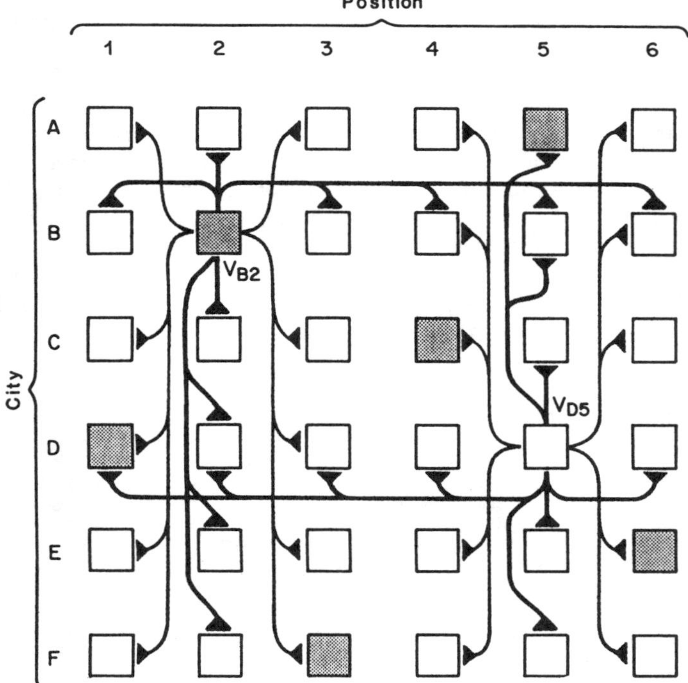

Fig. 5. A stylized picture of the syntax and connections of the TSP neural circuit. Each neuron is symbolically indicated by a square. The neurons are arranged in an n by n array. Each city is associated with n neurons in a row, and each position in the final tour is associated with n neurons in a column. A given neuron ($V_{X,j}$) represents the hypothesis that city X is in position j in the solution. The patterns of synaptic connection for two different neurons are indicated.

The continuous response characteristic of the analog neurons in the TSP circuit represents partial knowledge or belief. A value for $V_{X,j}$ between 0 and 1 represents the "strength" of the hypothesis that city X is in position j of the tour. During an analog convergence, several conflicting solutions or propositions can be simultaneously considered through the continuous variables. It is as though the logical operations of a calculation could be given continuous values between "true" and "false" and evolve toward certainty only near the end of the calculation. This is evident during the TSP convergence process (Fig. 6) and is important for finding good solutions to this problem (*15*). If the gain is greatly increased, the output of any given neuron will usually be either 1 or 0, and the potential analog character of the network will not be utilized. When operated in this mode, the paths found are little better than random. The analog nature of the neural response is in this problem essential to its computational effectiveness. This use of a continuous variable between true and false is similar to the theory of fuzzy sets (*35*) and to the use of evidence voting for the truth of competing propositions in Bayesian inference and connectionist modeling in cognitive psychology (*36*). Two-state neurons do not capture this computational feature.

Discussion

The work reviewed here has shown that a simple model of nonlinear neurons organized into networks with effectively symmetric connections has a "natural" capacity for solving optimization problems. The general behavior can be readily adapted to particular problems by appropriately selecting the synaptic connections. Optimization problems are ubiquitous where goals are attempted in the presence of conflicting constraints, and they arise in problems of perception (What three-dimensional shape "best" describes a given shading pattern in a two-dimensional image?), behavioral choice, and motor control (What is the optimum trajectory to move an appendage to minimize internal stresses?). Hence circuits consistent with this model could efficiently solve problems important in biological information processing.

Biologically relevant problems in vision have already been formulated in terms of optimization problems. Edge-detection, stereopsis, and motion detection can be described as "ill-posed" problems, and solutions can be found by minimizing appropriate quadratic functionals (*37*). The emphasis in these formulations has been simple convex problems with a single minimum in the energy. Networks solving these problems can be implemented by linear circuits having local connections. The nonlinear circuits described here can implement solutions to much more complex problems and have recently been used to solve the object-discontinuity problem in early vision (*18*).

The concept of an energy function and its use in circuit design provide an understanding of *how* model neural circuits rapidly compute solutions to optimization problems. The state of each neuron changes in time in a simple way determined by the state of neurons to which it is connected, but the organization of the synapses results in collective dynamics that minimize an E function relevant to the optimization problem. Knowledge of this E function helps us understand the collective dynamics. The two circuit examples reviewed here, the A-B converter and TSP circuit, were "forward-engineered." Given the optimization problem, a representation of hypothetical solutions to the problem as a particular set of neural states was constructed. Synaptic connections in the operating circuit move the neural state toward these solution states and bias

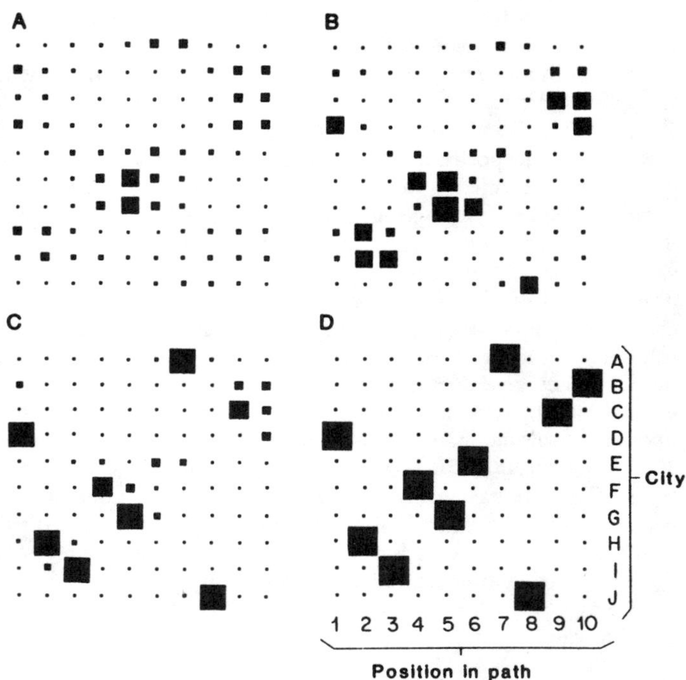

Fig. 6. The convergence of a ten-city analog circuit to a tour. The linear dimension of each square is proportional to the value of $V_{X,i}$. (A to C) Intermediate times. (D) The final state. Indices illustrate how the final state is decoded into a tour (solution of the TSP).

this motion toward the best solution. The values of these synaptic strengths were summarized in the single algebraic statement of the *E* function. [The two problems illustrate different ways in which "data" modulate the circuit parameters: as input currents in the A-B converter or as changes in the connection strengths in the TSP circuit (*17*).] Forward-engineered examples of model neural circuits add to the known repertoire of computational circuits that seem neurobiologically plausible. The general problem of neurobiology is "reverse engineering"—to understand the operation of a complex biological circuit with unknown design principles and internal representations. In general, the set of neural circuits whose functioning is understood provides an information base for hypothesizing function in biological neural circuits in the same way that the study of understood electrical circuits aids the attempt to understand or reverse engineer an unfamiliar electrical circuit diagram.

When a problem falls naturally onto a neural circuit, its convergence to a collective analog decision in a few time constants represents immense computation for the amount of hardware involved. For example, the 30-city TSP can be done on a network of 900 neurons. When that kind of combinatorial problem occurs in perception and pattern recognition, the input to the system will occur in parallel and take little time. A biological neural network of this structure would converge to an answer in a few neural time constants, thus in about 0.1 second. An electronic circuit of the same structure would converge in about 1 μsec. A comparably good solution to this problem, with conventional algorithms used for the TSP, can be found in about 0.1 second on a typical microcomputer having 10^4 times as many devices. The effectiveness of the neural system on the basis of computations per device per time constant is great in comparison with the usual general-purpose digital machine. The ability of the model networks to compute effectively is based on large connectivity, analog response, and reciprocal or reentrant connections. The computations are qualitatively different from those performed by Boolean logic.

Other specific circuit designs have been studied. Many problems in signal processing can be described as the attempt to detect the presence or absence of a waveform having a known stereotyped shape in the presence of other waveforms and noise. (The recognition of phonemes in a stream of speech is conceptually similar, but fraught with large problems of variability from the stereotype form.) We have described the general organization of neural circuits that could solve this task (*16*). Energy functions have been described for other combinatorial optimization problems, including graph coloring (*17*), the Euclidean-match problem (*17*), and the transposition code problem (*15*). Circuits that relax the restriction on a symmetric connection matrix (as biology does) have also been studied. A circuit designed to provide solutions to linear programming problems (*16*) functions without oscillation when the characteristic times of these elements are properly specified, even though its computing elements have antisymmetric connection strengths. The associative memory originally discussed (*10*) and used in a model of learning in a simple invertebrate (*38*) can be described as an optimization problem (*15*). The same conceptual framework can seemingly be applied to a large number of different problems.

Because the basic idea of the model neural circuit can be expressed as an electrical circuit, there have been efforts to build such hardware. Associative memories of 32 neurons (amplifiers) have been built in conventional electrical circuit technology (*39*). A 22-neuron circuit has been successfully microfabricated on a single silicon chip (*40*). Shrinking this kind of network to a compact size seems possible (*41*). The most compact and useful form of such a device would involve an electrically writable resistance change in a two-terminal device, which would function approximately as a Hebbian (*31*) synapse. Examples of such material fabrications exist (*42*). A 32-neuron system has been fabricated that uses optics to implement connections (*43*). Technological questions have so far focused chiefly on associative memory. Similar circuits could be used to solve problems in signal detection and analysis, such as artificial visual systems, in which there tends to be immense data overload and where concurrent distributed processing is desired.

In both biological neural systems and man-made computing structures, hierarchy and rhythmic or timed behaviors are important. The addition of rhythms, adaptation, and timing provides a mechanism for moving from one aspect of a computation to another and for dealing with time-dependent inputs and will lead to new computational abilities even in small networks. Hierarchy is necessary to keep the number of synaptic connections to a reasonable level. To extend the present ideas from neural circuit to neural system, such notions will be essential.

REFERENCES AND NOTES

1. D. Marr, *Vision* (Freeman, San Francisco, 1982).
2. Implementations different from those reviewed here are discussed in: T. Poggio and C. Koch, *Artificial Intelligence Lab. Memo 773* (Massachusetts Institute of Technology, Cambridge, 1984); S. E. Fahlman, G. E. Hinton, T. J. Sejnowski, *Proceedings of National Conference on Artificial Intelligence* (1983), p. 109; S. Geman and D. Geman, *IEEE Trans. Pattern Anal. Mech. Intell.* **6**, 721 (1984).
3. W. S. McCulloch and W. Pitts, *Bull. Math. Biophys.* **5**, 115 (1943).
4. F. Rosenblatt, *Principles of Neurodynamics* (Spartan, Washington, DC, 1961).
5. W. S. McCulloch, *Embodiments of Mind* (MIT Press, Cambridge, MA, 1965); M. Minsky and S. Papert, *Perceptrons* (MIT Press, Cambridge, MA, 1969); A. C. Scott, *J. Math. Psychol.* **15**, 1 (1977).
6. H. K. Hartline, H. G. Wagner, F. Ratliff, *J. Gen. Physiol.* **39**, 651 (1956).
7. For a summary of appropriate mathematical methods, see D. Noble, J. J. B. Jack, and R. Tsien [*Electric Current Flow in Excitable Cells* (Clarendon, Oxford, 1974)].
8. D. K. Hartline, *Biol. Cybern.* **33**, 223 (1979); D. H. Perkel and B. Mulloney, *Science* **185**, 181 (1974); A. I. Selverston, *Behav. Brain Sci.* **3**, 535 (1980); W. O. Freisen and G. S. Stent, *Biol. Cybern.* **28**, 27 (1977).
9. See, for example: R. D. Traub and R. K. S. Wong, *Science* **216**, 745 (1982).
10. J. J. Hopfield, *Proc. Natl. Acad. Sci. U.S.A.* **79**, 2554 (1982).
11. While space does not permit the review of modern developments using two-state neurons, the following references provide an introduction to this literature: W. A. Little, *Math. Biosci.* **19**, 101 (1974); ——— and G. L. Shaw, *ibid.* **39**, 281 (1978);

G. L. Shaw, D. J. Silverman, J. C. Pearson, *Proc. Natl. Acad. Sci. U.S.A.* **82**, 3364 (1985); M. Y. Choi and B. A. Huberman, *Phys. Rev. A* **28**, 1204 (1983); T. Hogg and B. A. Huberman, *Proc. Natl. Acad. Sci. U.S.A.* **81**, 6871 (1984); M. A. Cohen and S. Grossberg, *IEEE Trans. Syst. Man Cybern.* **SMC-13**, 815 (1983); K. Nakano, *ibid.* **SMC-2** (no. 3) (1972). For recent work on stochastic models applied to two states, see G. E. Hinton, T. J. Sejnowski, and D. H. Ackley [*Technical Report CMU-CS-84-119* (Carnegie-Mellon University, Pittsburgh, 1984)].

12. We refer to the dynamics as "classical" because we are ignoring propagation time delays and the quantal nature of action potentials, in analogy to classical mechanics. Similar equations have been described: T. J. Sejnowski, in *Parallel Models of Associative Memory*, G. E. Hinton and J. A. Anderson, Eds. (Erlbaum, Hillsdale, NJ, 1981), p. 189.
13. T. Kohonen, *Self-Organization and Associative Memory* (Springer-Verlag, Berlin, 1984); T. Kohonen *et al.*, *Neuroscience* **2**, 1065 (1977); T. Kohonen, *Biol. Cybern.* **43**, 59 (1982); L. N. Cooper, F. Liberman, E. Oja, *ibid.* **33**, 9 (1979); D. Ackley, G. E. Hinton, T. J. Sejnowski, *Cognit. Sci.* **9**, 147 (1985); D. d'Humieres and B. A. Huberman, *J. Stat. Phys.* **34**, 361 (1984); K. Fukushima, *Biol. Cybern.* **36**, 193 (1980); A. G. Barto, R. S. Sutton, P. S. Brouwer, *ibid.* **40**, 201 (1981); *Parallel Models of Associative Memory*, G. E. Hinton and J. A. Anderson, Eds. (Erlbaum, Hillsdale, NJ, 1981). The capacity of associative memories constructed from networks of two-state neurons is discussed by P. Peretto [*Biol. Cybern.* **50**, 51 (1984)], D. J. Amit, H. Gutfreund, H. Sompolinsky [*Phys. Rev. Lett.* **55**, 1530 (1985)], R. J. McEliece and E. C. Posner [*JPL Telecommunications and Data Acquisition Progress Report 42-83* (1985), p. 209], Y.-S. Abu-Mostafa and J. St. Jacques [*IEEE Trans. Inf. Theory* **IT-31**, 461 (1985)], and L. Personnaz, I. Guyon, and G. Dreyfus [*J. Physique Lett.* **46**, L-359 (1985)].
14. J. J. Hopfield, *Proc. Natl. Acad. Sci. U.S.A.* **81**, 3088 (1984).
15. ______ and D. W. Tank, *Biol. Cybern.* **52**, 141 (1985).
16. D. W. Tank and J. J. Hopfield, *IEEE Circuits Syst.* **CAS-33**, 533 (1986).
17. J. J. Hopfield and D. W. Tank, in *Disordered Systems and Biological Organization* (Springer-Verlag, Berlin, 1986).
18. C. Koch, J. Marroquin, A. Yuille, *Proc. Natl. Acad. Sci. U.S.A.* **83**, 4263 (1986).
19. E. Mjolsness, *CalTech Computer Science Dept. Publication 5153:DF* (1984).
20. T. J. Sejnowski and G. E. Hinton, in *Vision, Brain, and Cooperative Computation*, M. Arbib and A. R. Hanson, Eds. (MIT Press, Cambridge, MA, 1985).
21. G. M. Shepherd, *The Synaptic Organization of the Brain* (Oxford Univ. Press, New York, 1979), p. 3.
22. For a review of observed input-output relations and their ionic basis, see W. E Crill and P. C. Schwindt [*Trends NeuroSci* **6**, 236 (1983)].
23. K. G. Pearson, in *Simpler Networks and Behavior*, J. C. Fentress, Ed. (Sinauer, Sunderland, MA, 1976).
24. J. McCarragher and R. Chase, *J. Neurosci* **16**, 69 (1985).
25. M. V. L. Bennett and D. A. Goodenough, *Neurosci. Res. Program Bull.* **16**, 371 (1978).
26. W. Rall and G. M. Shepherd, *J. Neurophysiol.* **31**, 884 (1968).
27. C. Koch and A. Yuille, personal communication.
28. J. Platt, personal communication.
29. E. Harth, *IEEE Trans. Syst. Man Cybern.* **SMC-13**, 782 (1983).
30. W. J. Freeman, *Mass Action in the Nervous System* (Academic Press, New York, 1975).
31. D. O. Hebb, *The Organization of Behavior* (Wiley, New York, 1949).
32. S. Kirkpatrick, C. D. Gelatt, M. P. Vecchi, *Science* **220**, 671 (1983).
33. S. Geeman and D. Geeman, *IEEE Trans. Pattern Anal.* **6**, 721 (1984).
34. M. R. Garey and D. S. Johnson, *Computers and Intractability* (Freeman, New York, 1979).
35. L. A. Zadeh, *IEEE Trans. Syst. Man Cybern.* **SMC-3**, 38 (1974).
36. D. H. Ballard, *Pattern Recognition* **13**, 111 (1981); J. A. Feldman and D. H. Ballard, *Cognit. Sci.* **6**, 205 (1982); D. L. Waltz and J. B. Pollack, *ibid.* **9**, 51 (1985); J. B. Pollack and D. L. Pollack, *Byte Magazine* **11** (no. 2), 189 (1986).
37. T. Poggio, V. Torre and C. Koch, *Nature (London)* **317**, 314 (1985).
38. A. E. Gelperin, J. J. Hopfield, D. W. Tank, in *Model Neural. Networks and Behavior*, A. Selverston, Ed. (Plenum, New York, 1985).
39. J. Lambe, A. Moopenn, A. P. Thakoor, *Proceedings of the AIAA: Fifth Conference on Computers in Space*, in press; J. Lambe, A. Moopenn, A. P. Thakoor, *Jet Propulsion Lab Publication* **85-69** (Jet Propulsion Laboratory, Pasadena, CA, 1985).
40. M. Sivilotti, M. Emmerling, C. Mead, *1985 Conference on Very Large Scale Integration*, H. Fuchs, Ed. (Computer Science, Rockville, MD, 1985), p. 329.
41. L. D. Jackel, R. E. Howard, H. P. Graf, B. Straughn, J. S. Denker, *J. Vac. Sci. Tech.*, in press.
42. A. E. Owen, P. G. Le Comber, G. Sarrabayrouse, W. E. Spear, *IEE Proc.* **129** (part 1, no. 2), 51 (1982).
43. D. Psaltis and N. Farhat, *Optics Lett.* **10**, 98 (1985).

44. Supported in part by NSF grant PCM-8406049.

Article 8.3

The Design, Fabrication, and Test of a New VLSI Hybrid Analog–Digital Neural Processing Element

Mark R. DeYong, *Student Member, IEEE,* Randall L. Findley, *Member, IEEE,* and Chris Fields

Abstract—**A hybrid analog–digital neural processing element with the time-dependent behavior of biological neurons has been developed. The hybrid processing element is designed for VLSI implementation and offers the best attributes of both analog and digital computation. Custom VLSI layout reduces the layout area of the processing element, which in turn increases the expected network density. The hybrid processing element operates at the nanosecond time scale, which enables it to produce real-time solutions to complex spatiotemporal problems found in high-speed signal processing applications. VLSI prototype chips have been designed, fabricated, and tested with encouraging results. Systems utilizing the time-dependent behavior of the hybrid processing element have been simulated and are currently in the fabrication process. Future applications are also discussed.**

I. Introduction

CURRENT physical and computational models of neurons have somewhat merged in the silicon-based environment of analog very large scale integration (VLSI) circuitry [1]–[4]. One typical network design utilizes transconductance amplifiers as the variable conductances that represent the excitatory and inhibitory synaptic inputs to the neuron. The body of the neuron (soma and axon hillock) is modeled by an operational amplifier (OA) acting as a comparator. Signals between neurons are modeled by slowly varying levels [5] or by frequency-modulated streams of square pulses [3], [6].

Here we offer an alternative approach to the implementation of VLSI models of neurons and an alternative motivation for the modeling of neuron behavior in general. This paper presents the design of a hybrid analog–digital processing element (PE) that models the temporal and waveform characteristics of biological neurons [7]–[9]. The approach utilized in the design of the new PE differs from the standard approach [4] in two ways: first, the new PE is based not on OA building blocks but rather on custom circuitry; second, the time and voltage ranges of the neural signals (ms and mV) are scaled to those typically found in VLSI networks (ns and V).

Manuscript received July 10, 1991; revised October 7, 1991. This work was supported by the NASA Innovative Research Program (Grant NAGW 1592) and by an NSF grant to the Computing Research Laboratory of New Mexico State University.

M. R. DeYong is with the Department of Electrical and Computer Engineering and the Computing Research Laboratory, New Mexico State University, Las Cruces, NM 88003.

R. L. Findley was with the Department of Electrical and Computer Engineering, New Mexico State University, Las Cruces. He is now with the Advanced Workstation Division, IBM, Austin, TX.

C. Fields is with the Computing Research Laboratory, New Mexico State University, Las Cruces, NM 88003.

IEEE Log Number 9105644.

The differences in the two methodologies are due to differences in motivation. The models developed by Mead *et al.* are meant to be accurate models of the computational behavior of biological neural systems and are intended to offer insight into their functionality. The motivation behind the new PE is toward applications in high-speed analog signal processing systems; therefore, while maintaining biological detail provides a way of increasing functionality, the speed and size of the PE are also a concern. Custom circuit design minimizes the layout area of the networks. Operating at the several volt amplitude scale eliminates the problems associated with subthreshold operation (mV scale) such as noise susceptibility, slower operation speeds, and fan-in/fan-out limitations. Operating at the ns time scale offers compensation for the reduced degree of parallelism available in current integrated technologies, compared with that found in biological neural networks. The new PE is a hybrid device offering the best characteristics of both analog and digital computation: the speed and lower hardware overhead associated with analog systems, and the noise immunity and ease of design inherent in digital systems.

To investigate the new PE, the Simulation Program with Integrated Circuit Emphasis (SPICE) was used to simulate circuit models that mimic the input/output responses of neurons. These new models exhibit behavior qualitatively similar to the electrophysiological behavior of neurons. The resulting hybrid PE is well suited for VLSI implementation. Its behavior is controlled by external bias voltages, and is therefore adaptable. Test chips containing prototypes of the new PE have been fabricated and tested. A variety of systems ranging from logic gates to winner-take-all (WTA) networks based on the hybrid PE have also been simulated.

II. Conventional Artificial Neural Networks Do Not Model the Intrinsic Dependence of Neural Computation

While artificial neural network (ANN) research has achieved some notable successes, there is considerable reason to believe that the current processing elements could be improved upon. As stated by Miall [10], "the current forms of neural networks, while suitable for some computational tasks, have an impoverished temporal repertoire and so are unsuited to many time dependent operations faced by animals." Current models replace the intrinsic time dependence of the postsynaptic potentials (PSP's) generated by synaptic activity with the summation of weighted input levels, and the streams of action potentials (AP's) are replaced with the output of the PE. All

Reprinted from *IEEE Trans. on Neural Networks,* vol. 3, no. 3, pp. 363–374, May 1992.

computation is assumed to occur within a short increment:

$$y(t + \tau) = \sigma(\Sigma \alpha_i x_i(t)), \quad (1)$$

where y is the output, x_i are the inputs. α_i are the input weights, and τ is the time increment.

This model has two main shortcomings. First, although the signals are time-dependent the processing is not, which limits its application, and second, it provides no insight into computation in the biological network at the scale of the individual cell. This model can process time-varying signals, but it cannot be directly applied to the class of distributed decision making (DDM) problems involving the temporal relationships between PE's (e.g., temporal winner take all (Temporal WTA) problems, as in Barnden [11]. It is now reasonably clear that such relationships are used ubiquitously to solve real-time signal processing problems, such as visual object identification and tracking (e.g. [12] and [13]).

Although modeling the details of biochemistry down to the ion channels is in all likelihood unnecessary, we expect that a PE that does not lump the influences of the axon, the synapse, and the dendrite into a single weight on a single time scale will better reproduce several computational characteristics of biological neurons. Thus the first goal in the design of the new PE was to accommodate the biological detail deemed critical to capturing the intrinsic time dependence of neurons. This not only aids in the study of biological neural networks, but also provides a means for investigating computational mechanisms clearly not available in the standard model.

The development of biologically realistic artificial neuron models has been of interest for the last four decades and has diverged in several directions ranging from the purely mathematical parallel conductance model introduced in [14] to the symbolic network simulator GENESIS in [15]. These models, however, concentrate solely on biological realism and tend to lose sight of their hardware implementation and application to real-world engineering problems. The artificial neuron models developed here are designed for VLSI implementation and model biological detail at the level relevant to real-world problem solving.

III. Structure of the New PE

The overall structure of the new PE is illustrated in Fig. 1. Input to the PE is provided by three types of chemical synapses, which supply signals (referenced to a resting potential) to the axon hillock. The excitatory synapse supplies current, the inhibitory synapse draws current, and the shunting synapse either supplies or draws current, depending upon the soma voltage. These synaptic waveforms are generated in response to an AP from another cell arriving at the input of the synapse. The axon hillock produces spike trains whenever the soma voltage exceeds a threshold level. The spikes are separated by a minimum refractory period. The amplitude and duration of the synapses' outputs are adjustable, as are the threshold, pulse width, and refractory period parameters of the axon hillock. All circuitry has been implemented using devices that are available in the 2 μm n-well CMOS processes

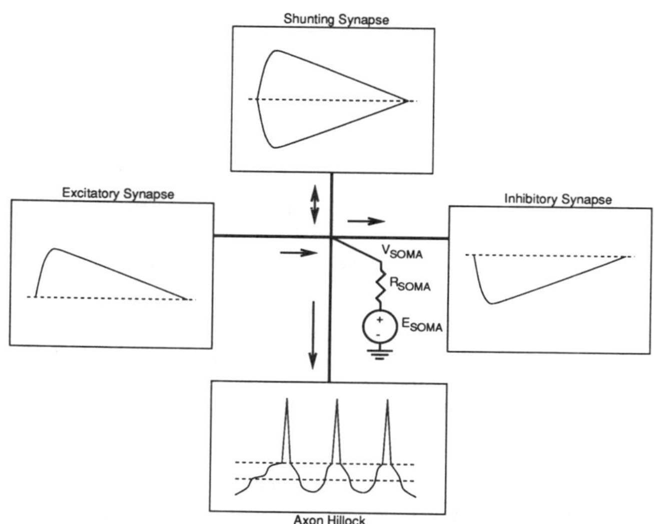

Fig. 1. Processing element structure. The outputs of three types of chemical synapses provide the input to the axon hillock. The synaptic outputs are referenced from the cell resting potential. The axon hillock conveys the cell state in the form of action potential (AP) streams. This figure represents the structure of the complete PE. Note that any combination of synapses may be connected to the SOMA node.

supported by the Metal Oxide Semiconductor Implementation System (MOSIS).

A. Chemical Synapse Models

As an AP enters the presynaptic region from an axon collateral it causes the conductance of the Ca^{2+} channels in the cell membrane to increase, allowing calcium ions to flow into the presynaptic terminal, owing to the potential gradient across the presynaptic region membrane and the concentration gradient of calcium [16]. The amount of calcium influx determines the neurotransmitter released at the junction. Neurotransmitter receptors on the postsynaptic side of the junction receive the chemical signal from the presynaptic region. The bonding of a neurotransmitter and a receptor causes the receptor to induce a chemical reaction which alters the conductance of the ion channels. If the particular postsynaptic region is excitatory, the receptors open the Na^+ channels, allowing sodium ions to flow into the postsynaptic region in response to the concentration gradient. This positive ionic current flow causes a change in the membrane potential of the postsynaptic region in the form of an excitatory postsynaptic potential (EPSP). In an inhibitory synapse the receptors open the K^+ channels, allowing an outflux of potassium ions in response to the concentration gradient, which results in an inhibitory postsynaptic potential (IPSP).

The synaptic sites in a biological neuron are located throughout an elaborate dendritic tree. Inside the dendritic tree and soma, a complex, nonlinear spatiotemporal integration of the input signals is performed in continuous time. The instantaneous firing rate of the neuron depends on the value of the convolved input, and on the recent firing history of the hillock. This type of signal processing and the temporal relations of incoming signals are lost with the traditional synapse model. The inclusion of the time-varying synaptic current in our

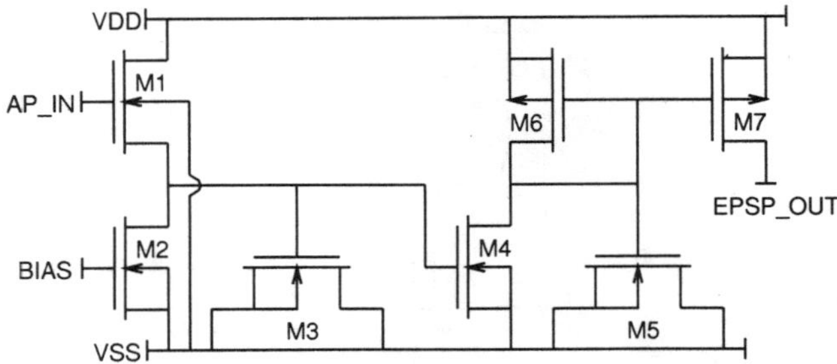

Fig. 2. Excitatory synapse model (ESYN) schematic diagram. Capacitors $M3$ and $M5$ control the fall time and rise time, respectively. The output is driven by a current mirror in order to isolate the ESYN circuitry from the effects of loading from other devices. Note that in all schematic diagrams crossover points are nodes, except those formed by arcs.

PE not only is more biologically realistic; it also provides a computational function not represented in the conventional ANN model. Therefore our strategy in the development of the new PE synapse models was to design VLSI circuits with the time-dependent I/O behavior of biological synapses.

Since electronic current in VLSI circuitry operates much faster than ionic current in neurons, the timing has been linearly scaled to take advantage of this speed while maintaining the relative shapes and timing of the signals. VLSI circuitry is much more reliable when it operates in the saturation region than when it operates in the subthreshold region; therefore the voltage scale has also been scaled from the millivolt range of neural signals to the volt range of saturated MOS devices. The operating range of a typical neuron is roughly -80 mV, just below the Nernst potential for K^+ ions, to roughly $+60$ mV, just above the Nernst potential of Na^+ ions. This range has been scaled linearly to the $0 \rightarrow 5$ V range of the MOS devices. This maps the resting potential of approximately -60 mV to 1 V and the typical threshold voltage of -40 mV to 1.7 V. The operating range for temporal relations in neurons, which is in the ms range, has also been scaled to the ns range of the VLSI devices (approximately 1 ns per 10 ms in our current PE).

1) Excitatory Synapse Model (ESYN): The amount of current supplied to the dendrites by the synapse is a time-varying function with an exponential rise and decay. Changes in the current result from changes in the conductance of the synapse. To the first order, this change in conductance can be written as

$$G(t) = At \exp - t/\tau, \tag{2}$$

where A is the maximum synapse conductance, t is time, and τ is the time constant of the synapse. This is commonly called [15] an alpha function.

Fig. 2 shows the circuitry that produces the desired positive alpha function in response to an action potential at the input. This circuitry has been named ESYN since it models the behavior of the excitatory synapse. The two capacitors (M3 and M5) control the fall and rise times, respectively. The desired output current is created in M6 and is mirrored to the actual output through M7. A typical EPSP is shown in Fig. 3.

2) Inhibitory Synapse Model (ISYN) The second type of synapse that was to be modeled is the inhibitory synapse (ISYN). The function of the inhibitory synapse is to cancel the depolarizing effects of the EPSP's, thus preventing the EPSP's from depolarizing the soma sufficiently to produce an action potential. Fig. 4 shows the circuitry used in this model. Note that there is a large amount of similarity to the circuitry in the ESYN. This is because the waveform to be created by the ISYN is simply a negative version (with respect to the resting potential) of that created by the ESYN. Again, capacitors M3 and M5 establish the fall and rise times, respectively. For the ISYN, transistor M4 is now a current source instead of a current sink, and current mirror M6–M7 is now created with n-channel devices instead of p-channel devices. The ISYN model is scalable in both timing and amplitude, as is the ESYN.

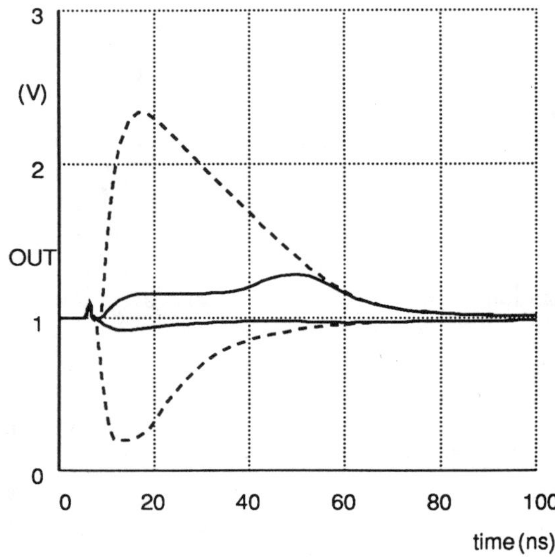

Fig. 3. PSP's generated by the ESYN, ISYN, and SSYN models. The dashed curves represent the EPSP (positive going) and IPSP (negative going) waveforms. The solid curves show the effect of the SSYN on the EPSP (top) and IPSP (bottom) waveforms.

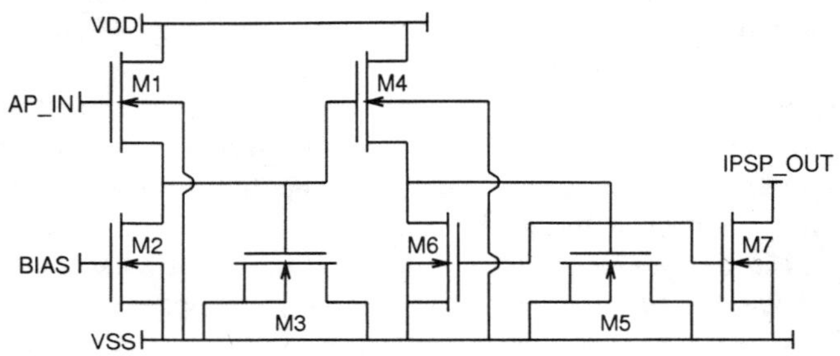

Fig. 4. Inhibitory synapse model (ISYN) schematic diagram. This circuit consists only of n-channel devices. When a pulse is received at the input, a time-varying current is drawn from the soma potential, which serves to inhibit depolarization.

IPSP's created with this model are negative exponential waveforms with respect to the resting potential. The ISYN pulls the potential toward ground, which has the effect of drawing a current from E_{SOMA}. A typical IPSP is also shown in Fig. 3. Note that the flat spot in the waveform is due to the fact that $v_{DS7} < v_{GS7}$, which pulls the device out of the saturation region. This causes the current mirror to display nonideal characteristics.

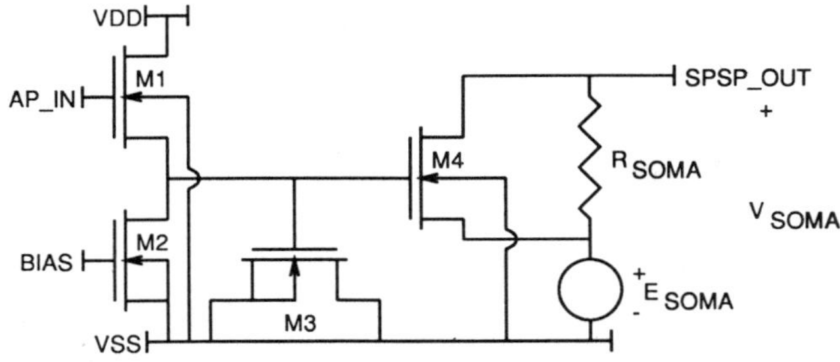

Fig. 5. Shunting synapse model (SSYN) schematic diagram. The circuitry modeling shunting inhibition opens a resistive path to the resting potential source. Thus, hyperpolarized potentials will tend to be pulled up and depolarized potentials will tend to be pulled down.

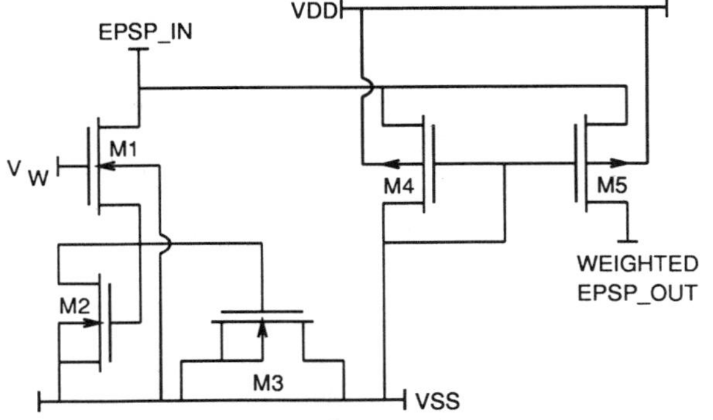

Fig. 6. Schematic of synaptic weighting circuitry. This circuit consists of an RC network that is used to draw a time-varying current from the ESYN and ISYN outputs. The current mirror (M4–M5) isolates the weighting circuitry from the load placed on the synapse. An n-channel current mirror is used in the ISYN weighting circuitry.

3) Shunting Synapse Model (SSYN) Shunting inhibition is due to the influx of Cl^- ions whose Nernst potential is approximately at the resting potential. Thus the SSYN must hold the soma potential (V_{SOMA}) at the resting potential regardless of whether V_{SOMA} is currently positive or negative. This is done by placing a time-varying resistance in series with the resting potential source (E_{SOMA}). Thus, a positive V_{SOMA} will cause current to flow into the source whereas a negative V_{SOMA} will cause current to be supplied by the source. The circuitry of the shunting synapse is shown in Fig. 5.

Fig. 3 shows the effects of the SSYN on the ESYN and ISYN as simulated by SPICE. In these simulations, the SSYN was fired at precisely the same time as the other synapse. It can be seen that the SSYN will draw nearly as much current as is supplied by the ESYN and provide nearly as much current as is drawn by the ISYN. The SSYN does not perform the job perfectly, but produces adequate results. Thus the effects of depolarization and hyperpolarization are both nearly canceled by the shunting function.

B. Synapse Weighting

One method of synapse weighting is to simply place a variable resistor in parallel with the output. Thus a constant current, dependent solely upon the value of the resistance, would be subtracted from the output. The effects of this would be pure attenuation. It may be more interesting to

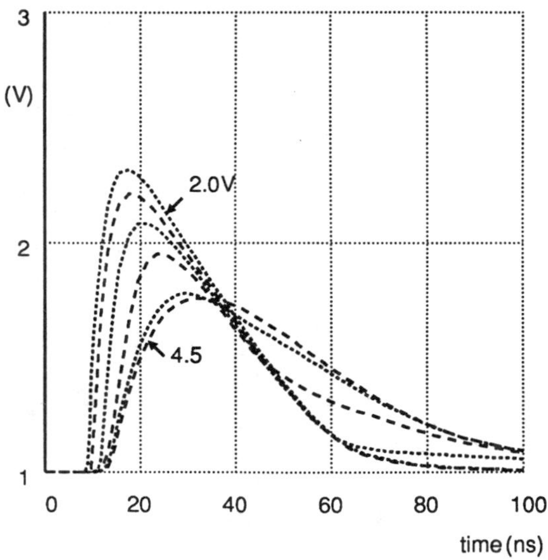

Fig. 7. Effect of weighting circuitry on EPSP. This figure shows the variations in the efficacy of the EPSP's achievable with the weighted ESYN model. Plots are shown for $V_W = 2.0, 2.5, 3.0, 3.5, 4.0$, and 4.5 V.

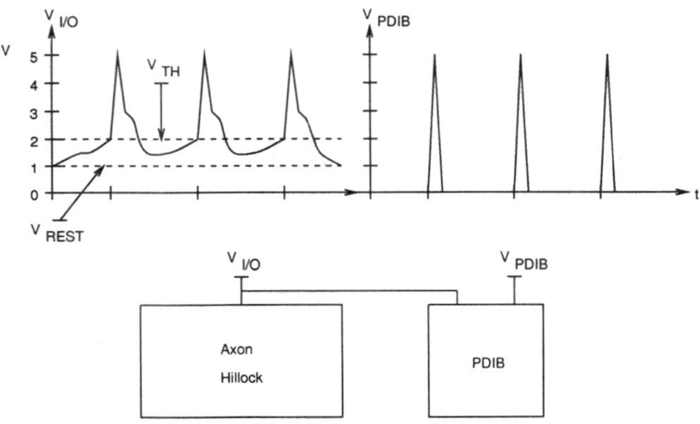

Fig. 8. Axon hillock black box diagram. The action potentials (AP's) generated at the axon hillock I/O node are referenced from the soma potential. The PDIB circuit removes the dc component of the AP; thus the AP's at the PDIB output node are referenced from ground. This allows the AP's from the PDIB output to be directly input to the synapse circuits.

also delay the signal as it is attenuated, as is done by the dendrites in the neuron. This will have the effect of lumping together the synapse's location in the dendritic tree and the synaptic efficacy. This is the primary assumption made in this model. The benefit of doing so is to save the excessive amount of space required to build compartmental models of the dendrites (cf. [17]) at the cost of losing the separability of synapse and dendrite. It is important to mention, however, that the effects of both the synapse and the dendrites are maintained. The time-varying synaptic alpha functions are still maintained with the ability to increase their efficacy, as is the attenuating and spreading function of the dendrites. The difference is that as the efficacy of the synapse is increased; it will also appear as though the synapse has moved closer to the soma (i.e., the dendrites have been shortened). This synaptic weighting scheme in conjunction with a model of some learning mechanism allows for the dynamic modification of network behavior.

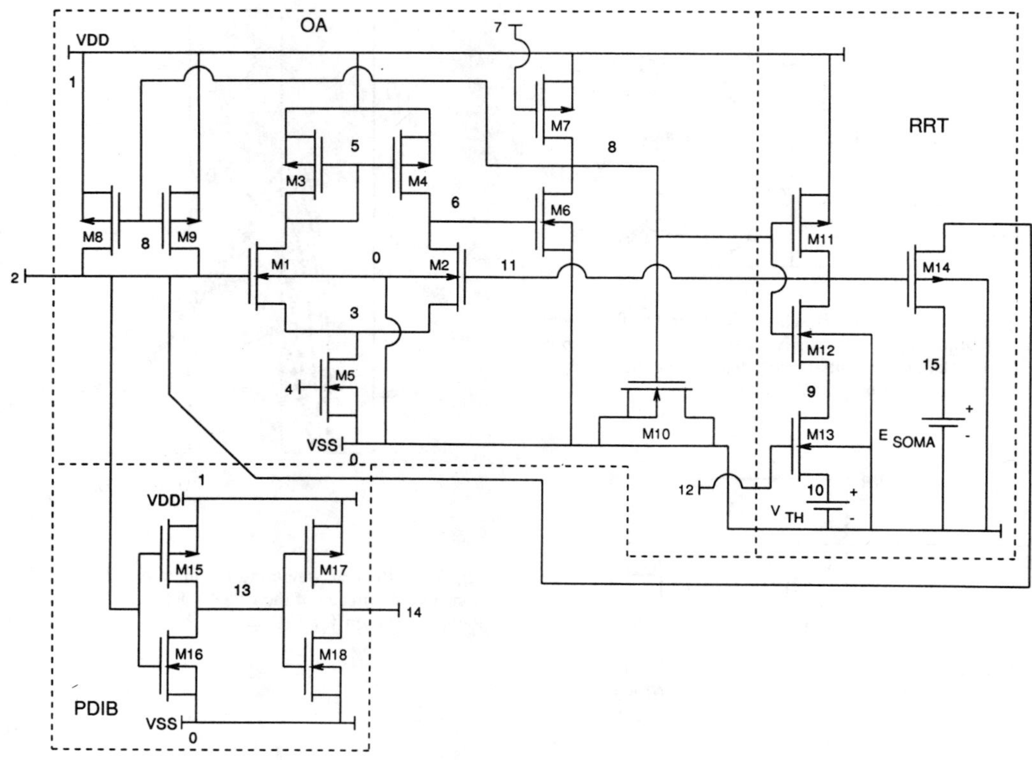

Fig. 9. Axon hillock schematic diagram. The axon hillock circuitry is based on a differential amplifier and employs positive feedback to produce the high voltage gain needed to generate an action potential. Negative feedback is used to reset the circuit and enforce the refractory period.

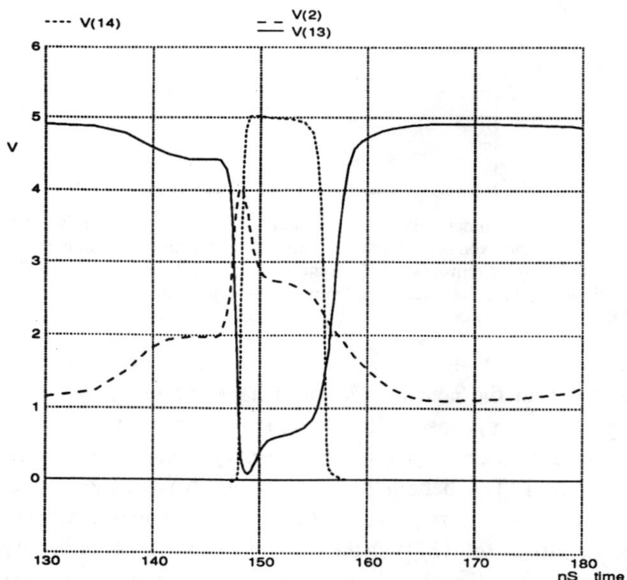

Fig. 10. PDIB effect on action potential. The function of the PDIB subcircuit is to remove the dc offset of the AP's generated by the axon hillock. This is achieved by passing the AP's through two unbalanced CMOS inverters.

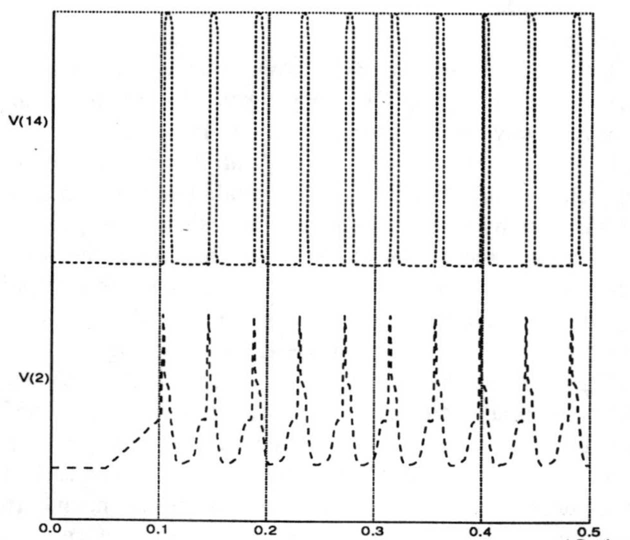

Fig. 11. Typical action potential stream. This figure shows a stream of AP's both before and after the PDIB subcircuit. The pulse width and refractory period of the AP's at node 14 are approximately equal to 10 and 40 ns, respectively.

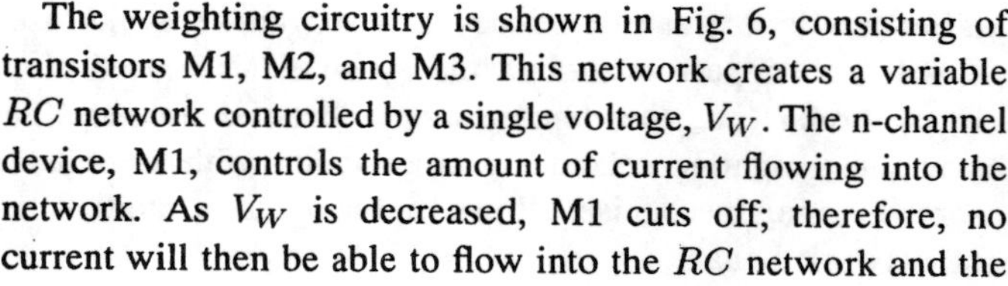

The weighting circuitry is shown in Fig. 6, consisting of transistors M1, M2, and M3. This network creates a variable RC network controlled by a single voltage, V_W. The n-channel device, M1, controls the amount of current flowing into the network. As V_W is decreased, M1 cuts off; therefore, no current will then be able to flow into the RC network and the output waveform will look very much like no RC network is present. As V_W is increased, more current will be allowed to flow into the network; thus the RC network will have the effect of attenuating and delaying the output waveform.

The final current mirror at the output (M4–M5) is added to isolate the weight from other synapses. The output current of this mirror is equal to the output current of the ESYN circuit (Fig. 2) minus the current in the weighting network times the mirror gain.

The sources of M4 and M5 must be shown in Fig. 6 rather than connected to VDD in order to ensure $v_{GS4} = v_{GS5}$. This causes $v_{SB} \neq 0$, and V_T will, therefore, depend on the source potential. Thus the performance of this current mirror will not be ideal. Fig. 7 shows the effect of the weighting circuitry on the ESYN output.

C. Axon Hillock Model

The action potential initiation zone (axon hillock) monitors the soma potential to determine the state of the cell. If the cell membrane becomes depolarized to the threshold voltage, the sodium channels in the hillock open completely, causing a high influx of Na^+ ions. This inward surge of positive ions pulls the membrane potential toward the Nernst potential for sodium, approximately +80 mV, generating the rising edge of an AP. At this point the potassium channels open, which prevents the membrane potential from reaching damaging levels. The K^+ ions are pushed out of the cell by both the potential and concentration gradients, thus causing the falling edge of the AP. There is a minimum period of time needed for the cell to recover from the generation of an AP, called the refractory period [16], [18]. If the soma potential remains above the threshold value for this refractory period, another AP is fired. The length of the refractory period is inversely proportional to the soma activity. The AP streams move along the lossless axon and collaterals to the presynaptic regions on the dendritic trees of neighboring cells.

Fig. 8 shows a black box diagram of axon hillock model. A single node serves as both the input and the output node. The voltage of the I/O node is monitored for the condition $V_{\text{SOMA}} - V_{\text{TH}} > 0$ (where V_{TH} is the axon hillock threshold), at which time an AP is generated at the I/O node. The AP is referenced from the I/O node potential. The presynaptic region removes the dc component of the AP and isolates the axon hillock from the input capacitances of the synapses connected to the axon, as does the presynaptic region of the neuron. Thus the AP's at the output of the presynapse model are referenced from ground. The axon hillock threshold, pulse width, and refractory period parameters are adaptable and only slightly interdependent.

To examine the functionality of the axon hillock model, shown schematically in Fig. 9, the circuitry may be broken down into three blocks:

1) a two-stage OA with positive feedback;
2) the reset, refractory, and threshold circuit (RRT);
3) the presynapse double inverter buffer (PDIB).

The OA stage produces the rising edge of an AP in response to a difference in voltage between its two input terminals. This difference is amplified and fed back into the OA input, which produces the high gain needed to generate the AP. The rising edge activates the RRT circuitry. The first function of the RRT is to generate the falling edge of the AP; it does this by reversing the polarity of the voltage difference between the OA input terminals, allowing the OA to reset. Its second function

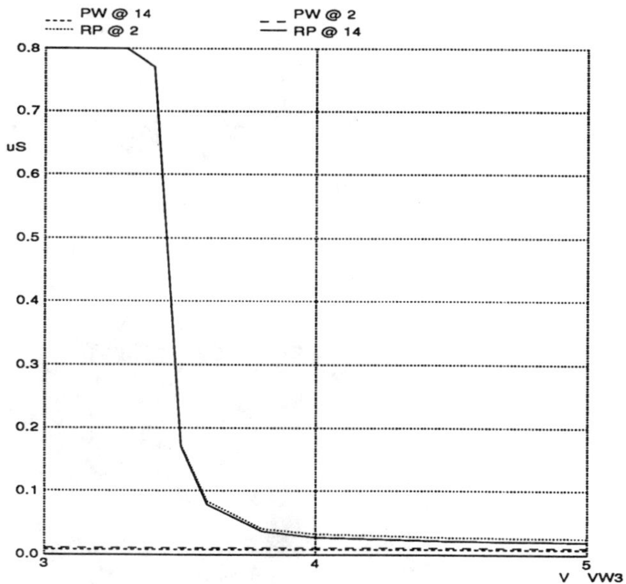

Fig. 12. Nonlinear behavior of axon hillock w.r.t. V_{W3}. The behavior of the refractory period w.r.t. V_{W3} is extremely nonlinear, which allows for wide operation range. Note that the pulse width is virtually independent of V_{W3}.

is to enforce the refractory period, which is determined by how long the input polarity reversal is maintained. The threshold value of the axon hillock is the quiescent voltage of the RRT subcircuit.

The OA stage is centered on an n-channel input differential amplifier with a current mirror load. The differential amplifier amplifies the difference between V_{G1} and V_{G2} over some input range. This range is determined by the amount of current sunk (ISS) by the current sink (M5). ISS is constant if M5 is in saturation since I_D is independent of V_{DS} in the saturation mode. The output range, gain, and speed of operation are also dependent on ISS. The output of the differential amplifier is then passed through the second stage of the OA, an inverting amplifier with a variable current source load. The output of the inverting amplifier drives two p-channel devices, M8 and M9, whose output is fed back into the OA input. This positive feedback causes the OA to *latch* high and thus generate the rising edge of an AP.

The RRT is responsible for the generation of the falling edge of the AP. It performs this task in the following manner. As the input voltage of the RRT falls, M11 turns on and M12 turns off. M11 pulls V_{11} from its resting value of V_{TH} toward VDD; thus reduces v_{id} to zero or possibly a negative level. At the same time, the rising V_{11} turns M14 on. M14 pulls V_2 toward E_{SOMA}. This fast reversal of the v_{id} polarity resets the differential amplifier. The inverting amplifier stage then begins to reset. As V_8 rises, M8 and M9 turn off, allowing V_2 to be pulled closer to E_{SOMA} by M14. The V_8 rise also turns M11 off and M12 on. The V_{11} voltage falls toward V_{TH} at the rate set by M13. As V_{11} falls to its quiescent value, M11 turns off, allowing V_2 to return to its uninfluenced level. If the input activity at the I/O node is sufficient to raise it back to V_{TH}, another AP will be fired.

Fig. 13. ESYN VLSI masks. This figure shows the VLSI masks of the ESYN model without the weighting circuitry isolation mirror. The layout is approximately 105 μm by 70 μm. Notice that the most distinct layout features are the timing capacitors.

The function of the PDIB is to remove the dc component of the AP generated at the I/O node, isolate the axon hillock circuitry from the parasitic load on the PE's output, and increase the PE's drive capabilities. The PDIB changes three of the AP characteristics; dc offset, amplitude, and waveform shape. Since M14 is always on to some degree, V_2 never reaches VDD. The PDIB boosts the AP amplitude to VDD. The change in dc offset and waveform shape are directly related. The AP's generated at the I/O node ride on a dc offset approximately equal to $V_{\rm TH}$. To remove a portion of the offset, the switching point of the first CMOS inverter is set above $VDD/2$. To remove the remainder of the offset the switching point of the second inverter is then set below $VDD/2$.

Fig. 10 shows a typical AP and the changes it experiences as it passes through the PDIB, V_2, V_{13}, and V_{14}. As the AP passes through the first CMOS inverter a large portion of the dc offset is removed and its amplitude is increased by almost 1V. The second inverter removes the remainder of the offset and pushes the AP all the way to VDD. If the switching point of the first inverter is moved even higher above $VDD/2$, the pulse width of the AP at node 14 will approach the pulse width of the AP at node 2 at the 3 V level, approximately 2.0E^{-9} s. The pulse width of the AP at node 14 in Fig. 10 is 8.0E^{-9} s. Fig. 11 shows a typical stream of AP's both before and after the PDIB.

The operating point, pulse width, and refractory period parameters are controlled by V_{W1}, V_{W2}, and V_{W3}, respectively. In Fig. 9, the bias voltages V_{W1}, V_{W2}, and $W3$ correspond to nodes 4, 7, 12, respectively. The nonlinear behaviors of the pulse width and refractory period with respect V_{W3} are shown in Fig. 12.

The soma resistor, $R_{\rm SOMA}$ (shown in Fig. 1), can be implemented in a variety of ways in VLSI. For our simulation we assumed that $R_{\rm SOMA}$ was linear. The simplest method for obtaining a linear resistor in VLSI is to form it out of polysilicon. This method, however, is very expensive in terms of layout area. In the prototype fabrications space was not an issue; therefore the polysilicon implementation of $R_{\rm SOMA}$ was used. Current simulations are investigating the use of a single n-channel device to act as $R_{\rm SOMA}$. This produces a nonlinear $R_{\rm SOMA}$, but owing to the function of $R_{\rm SOMA}$, its nonlinearity is of little consequence. The soma potential, $E_{\rm SOMA}$, is simply a voltage source that is external to the chip, such as VDD.

IV. Fabrication and Testing

A set of test chips has been fabricated in a 2 μm, n-well, double metal, single poly, scalable CMOS process supported by MOSIS. The TINYCHIP package offered by MOSIS provides approximately 2 mm by 2 mm of workspace. This is much smaller than the dies utilized in industry, but is more than adequate for the simple test circuitry needed to verify the functionality of the excitatory synapse and axon hillock models. The layout masks were generated using the VLSI CAD system MAGIC. MAGIC is equipped with an on-line design rule checker and a Caltech intermediate form (CIF) extraction tool. The mask geometries are limited to Manhattan shapes in order to simplify design rule violation checking. Figs. 13 and 14 show the mask layers of the ESYN and axon hillock, respectively.

Unbuffered I/O pads were used for all of the pins since our application is concerned with analog signals. The pins do, however, contain electrostatic discharge (ESD) protection to protect the internal devices from large voltage fluctuations. These pads produce a large parasitic capacitance that must be driven by the internal devices in order to communicate with

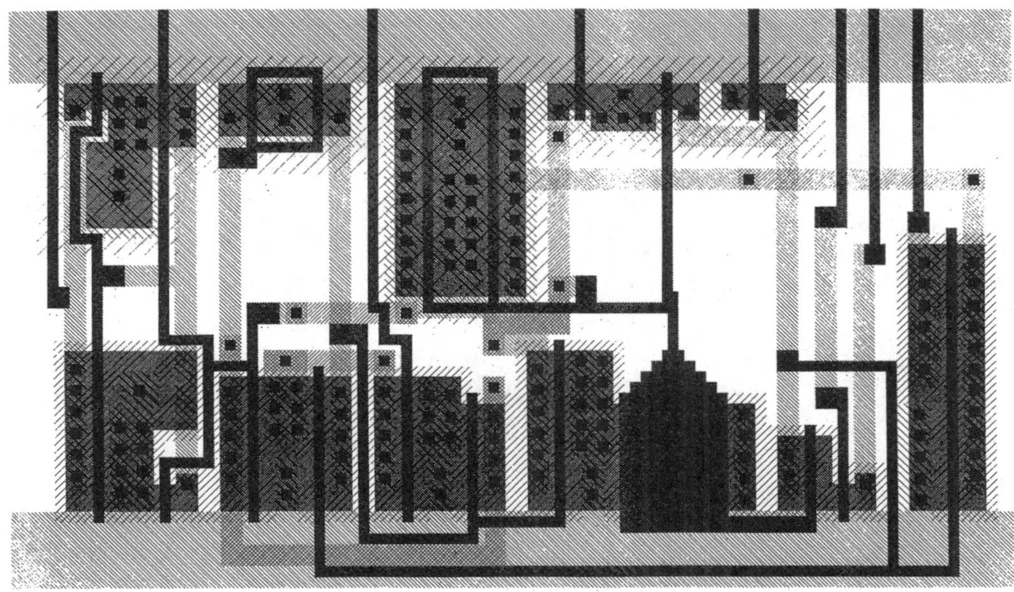

Fig. 14. Axon hillock VLSI masks. The dimensions of the axon hillock layout are approximately 180 μm by 110 μm. Once again the timing capacitor is the most noticeable feature. Note the axon hillock layout is only about 2.5 times the area of the ESYN layout.

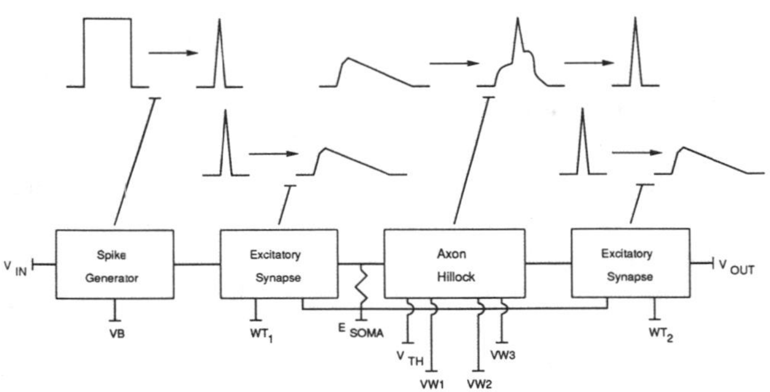

Fig. 15. PE test loop configuration. The test loop configuration is designed as an indirect test of the axon hillock functionality and a direct test of the ESYN functionality. The spike generator circuit produces an AP-like waveform from the rising edge of a square-wave input.

the outside world. In addition, the probes of the test equipment have a load impedance.

The axon hillock model is designed for VLSI implementation and according to simulations operates at very high speeds. This poses a problem when it comes to the hardware testing. To directly view the output of the axon hillock with an oscilloscope would require that it drive the combined capacitance associated with the pin of the chip package and the oscilloscope probe, which is on the order of $20E^{-12}$ F. This load is very large by VLSI standards. Another problem that arises when trying to pass high-frequency signals into and out of the chip is the cross-talk between the pins. To overcome these two obstacles the test loop shown in Fig. 15 is constructed as an indirect test of the axon hillock functionality.

The on-chip spike generator produces a single AP in response to the rising edge of an input pulse. The pulse can be made slow enough as to avoid the high-frequency pin problem. The AP's produced by the spike generator are indistinguishable from those generated by the axon hillock model. The output of the spike generator is fed into an excitatory synapse. This synapse drives the axon hillock, which in turn drives another excitatory synapse. The second synapse drives an output pin. Since the PSP's are much slower than the AP's, the large capacitances of the pin and probe do not present such a loading problem to the synapse.

The test-loop configuration is simplistic and contains devices other than the excitatory synapse and axon hillock which have not been tested; therefore it only provides proof of basic functionality and offers no insight into the more detailed aspects of the model's operation.

The procedure utilized in the testing of the test loop structure is simple and straightforward; square waves ranging in frequency from 100 kHz to 1 MHz in 10 kHz steps are input to the spike generator; at each frequency step the weight of the output excitatory synapse is swept through its operational range, (as determined from SPICE simulations), and the output is recorded. Fig. 16 shows the output response of the test-loop compared with that obtained from simulation (for a 200 kHz square wave input). It can be seen that the test and simulation results are quite similar. The differences between the magnitude of the test and simulation EPSP's can be attributed to two things: first, the pulse width of the AP fed into the output synapse may be larger than determined through simulation; second, the load on the output synapse may not be as large as expected. Both will cause the actual output of the excitatory synapse to be greater than that obtained from simulation. Fig. 17 shows the output of the test loop for a variety of biases on the output ESYN. The biases correspond to the first four settings used in the ESYN simulation plot (Fig. 7). Further increase in V_W of the output ESYN produced no significant change in the EPSP shape and/or duration.

It was stated earlier that the test-loop was to serve as an indirect test of the hillock's functionality; therefore, no direct measurements of the hillock output are presented. From

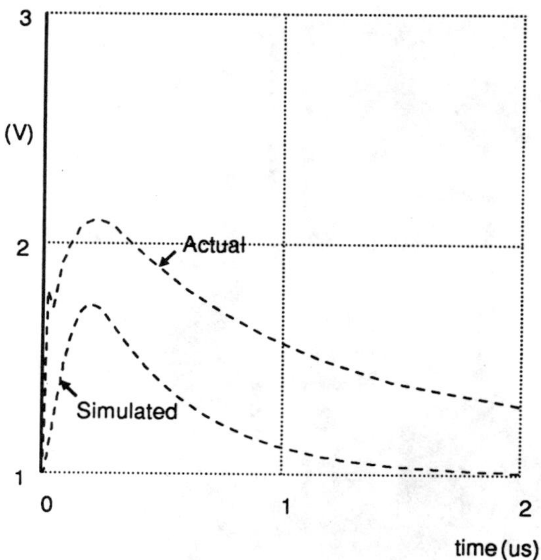

Fig. 16. Comparison of actual and simulated EPSP's. Note the actual EPSP is larger and not as delayed as the simulated EPSP. This is possibly a result of overestimating the capacitive load of the output pin and oscilloscope probe.

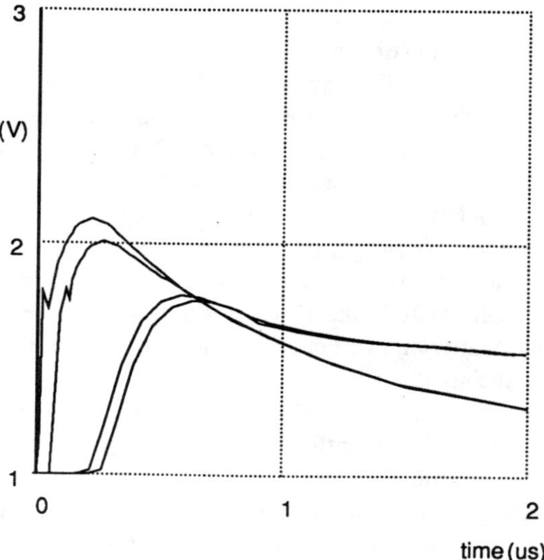

Fig. 17. Actual EPSP's for various V_W. The waveforms correspond to the first four biases of Fig. 7, ($V_W = 2.0, 2.5, 3.0,$ and 3.5 V, from top to bottom). Any increase beyond 3.5 V did not produce any noticeable changes.

the output of the test-loop, however, proof of the hillock functionality and its response to variations in V_{W1} and V_{W2} can be observed. First, note that since the test-loop output does not saturate we know that the axon hillock model is indeed firing distinct AP's and that the reset circuitry of the hillock is functional. If the hillock "latched on," it would cause the output ESYN to saturate. Second, as V_{W2} of the hillock was decreased the time duration of the output EPSP increased. This is an indication that the pulse width of the AP produced by the hillock was increasing as V_{W2} was decreased. This is the behavior predicted by the hillock simulations w.r.t. V_{W2}. And third, it was noticed that as the V_{W1} bias of the hillock was increased, the time duration and amplitude of the output EPSP decreased. In simulation it was seen that as V_{W1} was increased, ISS increased. This increase in ISS increased the overall operation speed of the hillock and thus shortened the AP pulse width. This would correspond to a decrease in the output EPSP time duration and amplitude. Currently a configuration that offers a more complete test of the axon hillock behavior is in fabrication. This test configuration will allow the response of the hillock model w.r.t. the input and bias voltages to be measured directly.

V. Example Computation Using the New PE

WTA mechanisms have been used in a variety of applications, from speech recognition and processing to selective attention in computer vision systems to resource allocation in large distributed computer networks [11], [20]. WTA mechanisms are defined in [21] as networks that choose a single winner or k winners from a set of contending inputs. The choice of the winner is based on a criterion that must be true for a contender to be chosen a winner. The criteria may be earliness, highest activation, arbitrary choice, etc. Most WTA mechanisms perform the task of choosing the winner or winners through the use of lateral inhibition between contending input units. WTA algorithms are typically designed for implementation in parallel processing systems. These algorithms, when implemented as software on serial machines, often lose their efficiency and questions arise as to their stability and convergence properties.

Since the output of the conventional PE is a scalar value, the criterion used by the WTA network is typically the highest activation. In [11], however, it was shown that there is a one-to-one transformation between activation level and earliness. Thus any activation-based problem can be transformed into a time-based problem. It is also shown that many problems can be processed much more efficiently by the time-based WTA algorithms.

The new PE is a hybrid analog–digital device. Analog PSP's and discrete digital AP's are combined to offer a more biologically realistic model neuron in which computation is a function of temporal as well as spatial relationships. As an example of this type of computation, consider the WTA network presented in [8]. This single-layer mutual inhibition WTA network, shown in Fig. 18, utilizes the new PE to pick a single winner from a group of contenders based on the temporal properties of the contending input signals. This particular network has two distinct methods for picking a winner, the choice of which is a function of the relationship between the input frequencies and the time constant of the inhibitory feedback.

The WTA network shown in Fig. 18 functions in the following manner. When an AP enters one of the input lines, it fires the excitatory synapse, which in turn causes the hillock to become active. The output activity is negatively fed back to all of the other PE inputs through an inhibitory synapse, which sets up a period of inhibition by pulling the input potential toward VSS. During this period of inhibition any EPSP's generated by the excitatory synapse are *absorbed* and cause no output activity. The duration of this inhibition period is a function of the IPSP time constant.

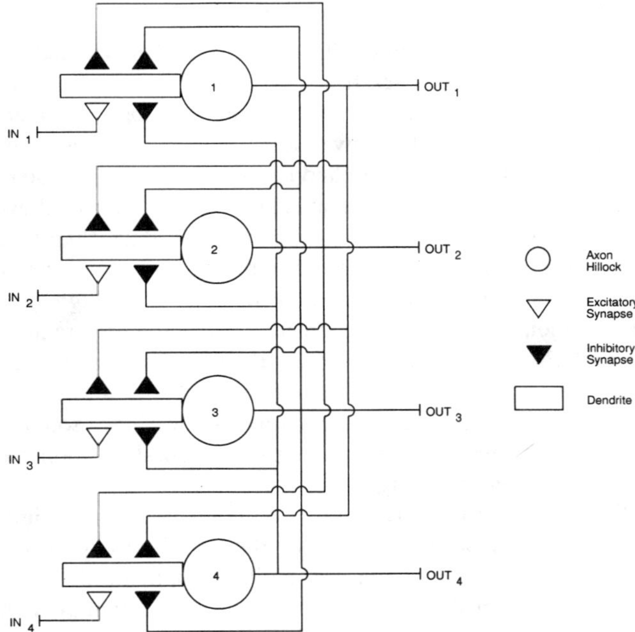

Fig. 18. Bifunctional-single-layer-lateral-inhibition WTA network. This WTA network uses lateral inhibition to choose a winner from four contending inputs. The mechanism by which a winner is chosen is a function of the period of the incoming spike trains and the time duration of the IPSP feedback.

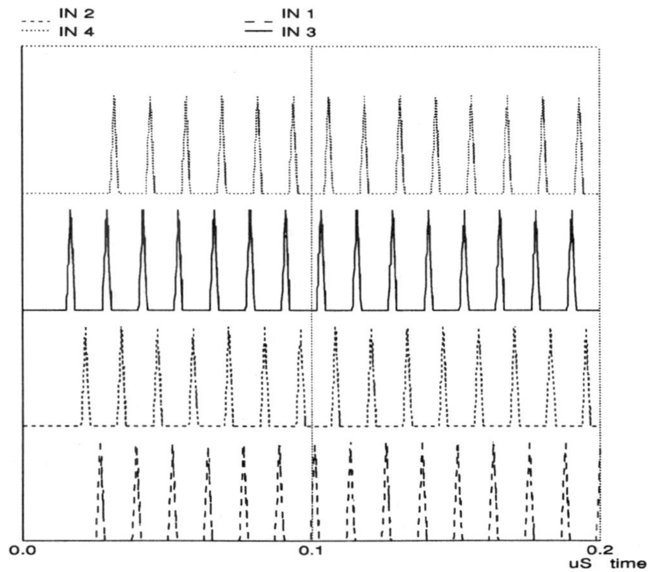

Fig. 19 Temporal WTA input waveforms. The input waveforms to the temporal WTA network are all of equal frequency, with a period less than the time duration of the IPSP feedback.

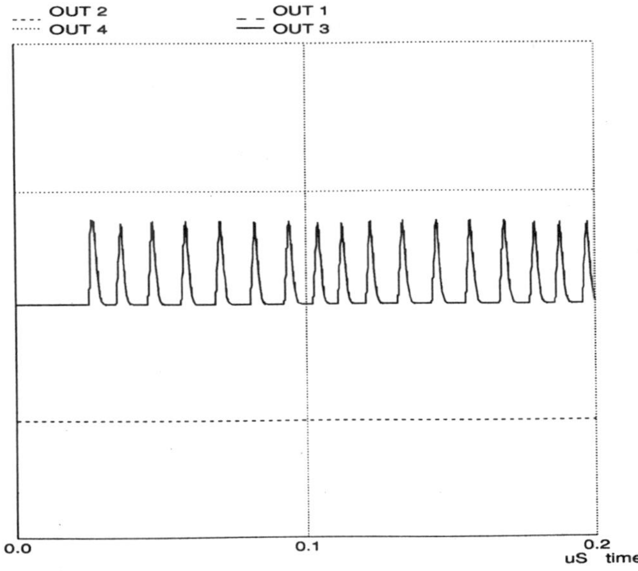

Fig. 20. Temporal WTA output waveforms. Comparison of the temporal WTA input and output waveforms reveals that the network converges to the correct solution in approximately 8 ns.

If the period of the slowest incoming signal is less than the time constant of the IPSP's, the periods of inhibition overlap; therefore, only the first PE to receive input becomes active. In this case the network behaves as a temporal WTA, basing the choice of a winner on the input arrival time. The second distinct behavior of the system occurs when the period of the fastest incoming signal is more than the time constant of the IPSP's. In this situation the inhibition periods on any particular dendrite have gaps during which a different PE may become active. The probability of an inactive PE becoming active during a gap is directly related to the input frequency on that line; the higher the frequency the higher the probability of becoming active. Because of this characteristic, the PE with the highest frequency eventually dominates the others and remains the only active element. The network behaves as a maximum frequency network (MAXfNET) under these conditions. Figs. 19–22 show the input and output waveforms for both the temporal WTA and MAXfNET behaviors.

Computation within the WTA network is a result of the temporal differences between the network and its input, and the spatial characteristics of the network. This simplistic spatiotemporal behavior can be expanded to larger, more intricate networks which make WTA-type decisions based on neighborhoods of mutual excitation and inhibition. This WTA network may be applied to any problem that the conventional WTA algorithms can solve and can often produce a solution in a more computationally efficient manner. This network exploits the parallel nature of the WTA mechanisms.

VI. Future Applications

An example of an application in which neighborhood activity is of interest is in the area of amperometric electrode arrays, as reviewed by Wang *et al.* [22]. The paper discusses the use of electrode arrays in the analysis of chemical reactions. The electrodes in the array are coated with various agents with diverse transport properties. The coatings are partially selective, with permeabilities based on molecular weight, polarity, charge, etc. The set of steady-state current measurements taken from the electrodes represents a fingerprint of the specific chemical reaction. These fingerprints vary as the concentrations of the various mixture components change with time. The temporal characteristics of the new PE allow it to be used in the classification of these dynamic fingerprints.

Another area of application for the new PE is robotic control. The intelligent movement of a robot through its

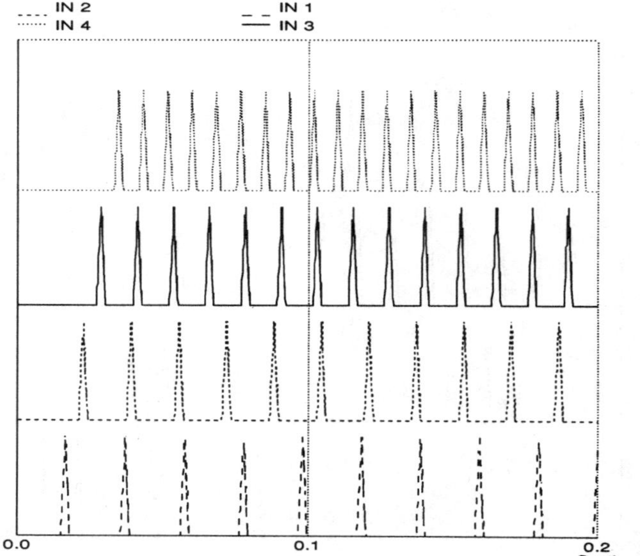

Fig. 21. MAXfNET input waveforms. The input waveforms to the MAXfNET network are of varying frequencies, with periods greater than the time duration of the IPSP feedback.

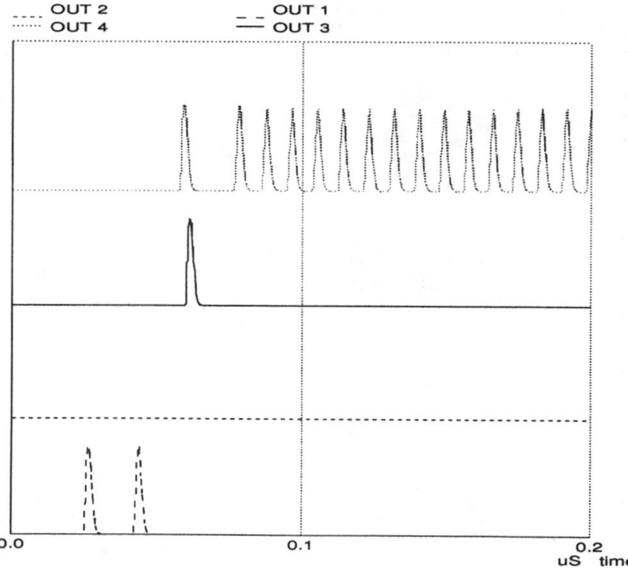

Fig. 22. MAXfNET output waveforms. Comparison of the MAXfNET input and output waveforms reveals that the network converges to the correct solution approximately 40 ns after the high-frequency input appears. Note that the correct input was chosen even though it was last to arrive. Transient temporal WTA behavior is observed on the output of PE 1.

workspace, for example, requires the analysis of enormous amounts of data from external sensors. The coincidence and anticoincidence between these sensor readings are used to make movement decisions such as direction, speed, and distance. Conventional methods for performing the comparisons have high hardware overheads and slow operation speeds. Neighborhood networks of the new PE on the other hand offer a real-time, low-hardware solution; the coincidence and anticoincidence of sensor signals can be determined via the excitatory and inhibitory feedback between and within the sensor neighborhoods.

VII. Conclusions

A hybrid analog–digital PE has been developed and implemented in VLSI circuitry. The new PE adds back the spatiotemporal integration of signals found in biological neural networks, but lost in the conventional artificial neuron model. This allows the PE to be applied to a class of DDM problems not directly solvable by the conventional model. It was shown through both simulation and experimental data that these signals can be scaled back from the millisecond time scale to a nanosecond time scale to make them suitable for high-speed computation. Custom layouts greatly reduce the area required for each PE and, therefore, increase the overall layout density. Thus, it is our hypothesis that networks of the new PE can be utilized as high-speed, low-hardware overhead solutions to problems involving complex spatiotemporal relationships between time-varying signals.

Since the new PE performs analog computation and digital communication, we discussed how this could be exploited to offer the best of both analog and digital computation. Computation is performed with the speed of analog computation and the adaptability expected in neural networks. The results of each PE are transmitted through a digital pulse stream, which is less susceptible to signal degradation, and is suitable for off-chip communication. The application of the new PE to time-dependent signal processing applications is the subject of future investigation.

References

[1] C. Mead and M. Mahowald, "A silicon model of early visual processing," *Neural Networks,* vol. 1. pp. 91–97, 1988.

[2] H. Card and W. Moore, "VLSI devices and circuits for neural networks," *Int. J. Neural Systems,* vol. 1, pp. 149–165, 1989.

[3] J. Lazzaro, and C. Mead, "A silicon model of auditory localization," *Neural Computation,* vol. 1, no. 1, pp. 47–57, 1989.

[4] C. Mead, *Analog VLSI and Neural Systems.* Reading, MA: Addison-Wesley, 1989.

[5] D. E. Rumelhart *et al., Parallel Distributed Processing: Explorations in the Microstructure of Cognition,* vol. 1, *Foundations.* Cambridge, MA: MIT. Press, 1986.

[6] A. F. Murray, D. Del Corso, and L. Tarassenko, "Pulse-stream VLSI neural networks mixing analog and digital techniques," *IEEE Trans. Neural Networks,* vol. 2, pp.193–204, Mar. 1991.

[7] M. DeYong, R. Findley, and C. Fields, "Computing with fast modulation: Experiments with biologically-realistic model neurons," in *Proc. Fifth Rocky Mountain Conf. AI,* (Las Cruces, NM), 1990, pp. 111–116.

[8] C. Fields, M. DeYong, R. Findley, (1990). "Computational capabilities of biologically realistic model neurons," in *Proc. Int. Workshop VLSI for Artificial Intelligence and Neural Networks,* (Oxford, England), 1990.

[9] R. Findley, M. DeYong, and C. Fields, "High speed analog computation via VLSI implementable neural networks," in *Proc. Third Microelectronic Education Conf. Exposition,* (San Jose, CA), 1990 pp. 113–123.

[10] C. Miall, "The diversity of neuronal properties," in *The Computing Neuron,* R. Durbin, C. Miall, and G. Hutchinson, Eds. Wokingham, U.K.: Addison-Wesley, 1989, pp. 11–34.

[11] J. Barnden, K. Srinivas, and D. Dharmavaratha, "WTA networks: Time based vs. activation based mechanisms for various selection goals," in *Proc. IEEE Int. Symp. Circuits Syst.,* (New Orleans, LA), 1990, pp. 215–218.

[12] J. H. R. Maunsell and W. T. Newsome, "Visual processing in monkey extrastriate cortex," *Annu. Rev. Neuroscience,* vol. 10, pp. 363–401,1987.

[13] E. A. DeYoe and D. C. VanEssen, "Concurrent processing streams in monkey visual cortex," *Trends in Neuroscience,* vol. 11, no. 5, pp. 219–226, 1988.

[14] A. L. Hodgkin and A. F. Huxley, "Currents carried by sodium and potassium ions through the membrane of the giant squid axon of Loligo," *J. Physiology,* vol. 116, pp. 449–472, 1952.

[15] M. Wilson and J. Bower, "The simulation of large-scale neural net works," in *Methods in Neuronal Modeling,* C. Koch and I. Segev, Eds Cambridge, MA: MIT. Press, 1989, pp. 291–333.

[16] S. W. Kuffler, J. Nicholls, and A. R. Martin, *From Neuron to Brain,* 2nd ed., Cambridge, MA: Sinauer Associates, 1984.

[17] I. Segev, J. W. Fleshman, and R. E. Burke, "Compartmental models of complex neurons," in *Methods in Neuronal Modeling,* C. Koch and I. Segev, Eds. Cambridge, MA: MIT. Press, 1989, pp. 63–96.

[18] D. Junge, *Nerve and Muscle Excitation,* 2nd ed., Cambridge, MA: Sinauer Associates, 1981.

[19] S. Foo, L. Anderson and Y. Takefuji, "Analog components for the VLSI of neural networks," *IEEE Circuits and Devices,* vol. 6, no. 4, pp. 18–26, 1989.

[20] K. Srinivas, "Selection in massively parallel connectionist networks,"Ph.D. dissertation,New Mexico State University, 1991.

[21] J. A. Feldman and D. H. Ballard, "Connectionist models and their properties," *Cognitive Science,* vol. 6, pp. 205–254, 1982.

[22] J. Wang, G. D. Rayson, Z. Lu, and H. Wu, "Coded amperometric electrode arrays for multicomponent analysis," *Anal. Chem,* vol. 62, pp. 1924–1927, 1990.

Neural Network Architectures for Robotic Applications

SUN-YUAN KUNG, FELLOW, IEEE, AND JENQ-NENG HWANG, MEMBER, IEEE

Abstract—**This paper proposes a ring VLSI systolic architecture for implementing artificial neural networks (ANN's) with applications to robotic processing. Key design issues on algorithms, applications, and architectures are examined. A variety of neural networks are considered, including single-layer feedback neural networks, competitive learning networks, and multilayer feed-forward networks. It is demonstrated that the ANN's are suitable to all three levels of robotic processing applications, including task planning, path planning, and path control levels. For these applications, a programmable systolic array is developed, which can exploit the strength of VLSI to provide intensive and pipelined computing. Both the retrieving and learning phases are integrated in the design. The proposed architecture is more versatile than other existing ANN's; therefore, it can accommodate all the useful neural networks for robotic processing.**

I. Introduction

THE POWER of neural networks hinges upon their distinct features of robust processing and adaptive capability in changing and noisy environments. It is estimated that the human brain contains over 100 billion (10^{11}) neurons. Neurons posses tree-like structures (called *dendrites*) specialized to receive incoming signals from other neurons across junctions called *synapses*. From each neuron, there is a single output fiber (called an *axon*) specialized to propagate action potentials so that the activation values of each neuron can be transmitted to many other neurons. Studies of brain neuroanatomy indicate more than 1000 synapses on the input and output of each neuron. In other words, although the neurons' transition time of a few *milliseconds* is about a million-fold times slow than current computer elements, the brain has a thousand-fold greater connectivity than today's supercomputers [16].

It has been postulated that the primary function of neocortical networks in the cerebrum is to form internal representations of classes and subclasses of objects, using perceptron-like algorithms. Moreover, it has been observed that the cerebellum functions as an associative content addressable memory (ACAM) that can be "trained" by the cerebrum. Many examples of biological applications are also available as a useful clue for the development of artificial neural networks (ANN's). In the example of natural vision processing, edge information is extracted in part in the retina via lateral inhibition between retinal neurons. In the cortex, there is a lateral excitation process which computes the brightness of a patch of an image bounded by edges. On the other hand, in primates, depth perception is formed by comparing images from the two eyes. Finding the correct overall assignment (from the several depth assignments existing in each cortical neighborhood) takes numerous trials before the cortical network finds a "solution." Interestingly, this process is analogous to the *relaxation* process used in many numerical algorithms run on current digital computers [6], [15].

The state of the art of the ANN's has embraced a very broad scope of neural processing models, from the earlier experimental research work on digital logic networks [51] in the 1940's, perceptrons [63], Adalines [78], and learning matrix [69] in the 1960's, to the correlation matrix linear models of ACAM networks [5], [7], [35], [79], and the cooperative–competitive neural network models [18], [21] in the 1970's. Extending earlier works by Steinbush [69], Willshaw [79], Amari [5], Cohen [14], Hopfield proposed a very popular iterative computational model for associative retrieval and optimization [25]–[27]. In addition to some significant contributions to the memory and learning models using competitive learning for autonomous feature extraction [66] and delta learning for generalized information storage [50], Rumelhart and his colleagues revived the back-propagation (generalized delta) learning algorithm for the multilayer perceptrons [53], [55], [64], [76], which has been successfully used in many experimental works [11], [20], [43], [44], [67].

There are three important aspects of ANN research can be identified: *algorithms, applications*, and *architectures*.

A. Algorithmic Aspects of ANN's

A basic ANN model consists of a large number of neurons, linked to each other with connection weights (see Fig. 1). Each, say ith, neural processing unit (PU) has an activation value a_i. This value (either discrete or continuous) is propagated through a network of unidirectional connections to other PU's in the network. Associated with each connection, there is a *synaptic weight* denoted as w_{ij} which indicates the effect the jth PU has on the ith PU. From algorithmic point of view, the ANN processing can be divided into two phases: *retrieving* and *learning*.

Retrieving Phase: Suppose that the connectivity pattern and weights of a neural network is known, in response to inputs

Manuscript received December 16, 1987; revised July 5, 1988 and May 1, 1989. This research was supported in part by the National Science Foundation under Grant MIP-87-14689 and by the Innovative Science and Technology Office of the Strategic Defence Initiative Organization, administered by the Office of Naval Research under contract N00014-88-K-0515.

S. Y. Kung is with the Department of Electrical Engineering, Princeton University, Princeton, NJ 08544.

J. N. Hwang was with the Department of Electrical Engineering, Princeton University, Princeton, NJ 08544. He is now with the Department of Electrical Engineering, University of Washington, Seattle, WA 98195.

IEEE Log Number 8929717.

Reprinted from *IEEE Trans. Robotics and Automation*, vol. 5, no. 5, pp. 641–657, Oct. 1989.

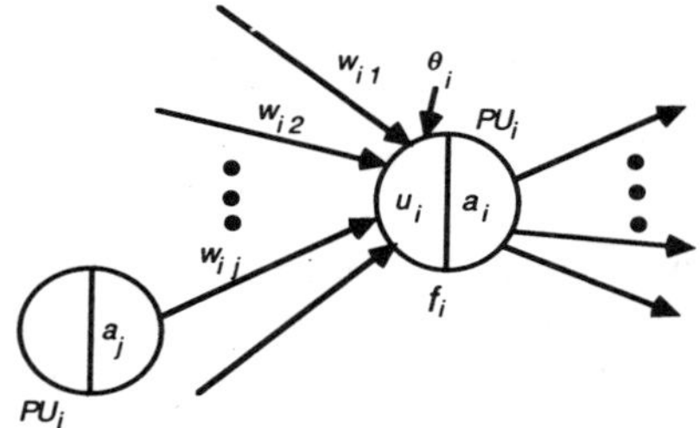

Fig. 1. A basic ANN model with two operations: propagation rule, and nonlinear activation.

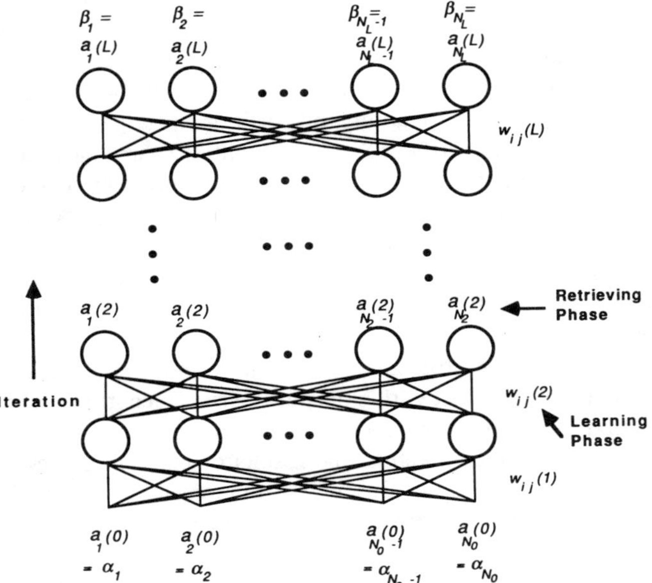

Fig. 2. A generic iterative model (L-iterations) for ANN's, where $w_{ij}(l)$ and N_l may be homogeneous or heterogeneous with respect to l.

(test patterns), the retrieving phase performs the iterative updating of activation values of each neuron based on the *system dynamics* to produce the responding outputs. The system dynamics in the retrieving phase of an ANN model can be written as a generic iterative formulation [41]

$$u_i(l+1) = \sum_{j=1}^{N_l} w_{ij}(l+1)a_j(l) \tag{1}$$

$$a_i(l+1) = f_i(u_i(l+1), \theta_i(l+1)) \tag{2}$$

where $1 \leq i \leq N_{l+1}$ and $0 \leq l \leq L-1$. The iteration index l can represent either *time* or *spatial* iterations.

Two types of inputs can be used to represent the test/training patterns: the stimulus inputs $\{a_i(0)\}$ and the external inputs $\{\theta_i(l)\}$. The initialization activation values are often denoted by α_i, i.e., $\{a_i(0) = \alpha_i\}$. The termination activation values are often denoted by β_i, i.e., $\{a_i(L) = \beta_i\}$.

The system dynamics in (1) and (2) may be graphically represented by an L-level feed-forward neural network (with N_l neural units at lth level) shown in Fig. 2, where one mathematical iteration is corresponding to one level of the network. Each PU, say ith neuron at $(l+1)$th iteration, receives the weighted inputs from other PU's at lth iteration to yield the net input $u_i(l+1)$ according to the *propagation rule* [see (1)]. The net input value $u_i(l+1)$, along with the external input $\theta_i(l+1)$, will determine the new activation value $a_i(l+1)$ by the *nonlinear activation* function f_i [see (2)]. The nonlinear activation functon f_i can be a deterministic function, winner-take-all mechanism [44], or a stochastic decision [1], [60] (for simpler notation, we will denote the position-invariant activation function as f). In some classification applicatons, winner-take-all type of nonlinear mechanism is adopted, which can be easily implemented by lateral inhibitions so that only the neuron receiving largest input is activated.

Learning Phase: Based on the input and/or target training patterns, the learning phase performs the iterative updating of the synaptic weights for all the connections based on the adopted *learning rule*. The weight updating (credit assignment) problem is to find a set of synaptic weights so as to optimize certain predefined measure function E based on a set of training patterns. The learning phase usually involves two steps: In the first step, the input training patterns are processed by the network based on the retrieving phase equations and generate some actual responses. In the second step, the weights are updated according to the responses generated and the chosen learning rules. Often recursive procedures are adopted. A common recursive weight updating formulation (learning rule) for an ANN model can be given [41] by

$$w_{ij}(l) \Leftarrow w_{ij}(l) + \eta \Delta w_{ij}(l). \tag{3}$$

The new weight value can be determined by the current weight value, the updating rate parameter η, and most importantly the increment of weight change. The updating rate η is introduced to regulate the rate of change of each weight at each recursion, it can be a global constant or can be a locally dependent variable $\eta_{ij}(l)$. For a simpler notation, we will use η to represent $\eta_{ij}(l)$ in the following discussions.

The learning recursion can represent either type of the two possible recursions: *pattern* and *sweep*. In the case of pattern recursion, the network updates the synaptic weights after the presentation of each training pattern. On the other hand, in the case of sweep recursion, the network updates the synaptic weights after the presentation of all the training patterns.

B. Applicational Aspects of ANN's

Neural neworks have been successfully applied to many applications which may be grouped into two classes: *optimization* and *associative retrieval/classification*.

Optimization: For the optimization application, the neural networks are used as a state-space search mechanism. The (fixed) synaptic weights are set before the search process is started. There is no explicit learning procedure for setting the weights and the dervation of $\{w_{ij}\}$ usually hinges upon finding a Liapunov energy function for the specific application.[1] For example, if the Hopfield neural network is used for solving an optimization problem, the Liapunov energy function has the

[1] A Liapunov energy function is monotonically decreasing along time evolution, i.e., $(d/dt)\, E \leq 0$.

following form [26], [27]:

$$E = -\frac{1}{2}\sum_i \sum_j w_{ij} a_i a_j - \sum_i \theta_i a_i. \quad (4)$$

In some applications, the Liapunov energy function is directly available (e.g., regularized least squares applications [38]). In others, the Liapunov energy function has to be derived from a given cost function and the constraints in the optimization problem (e.g., traveling salesman problem [27]). Once the synaptic weights are determined from the energy function, then the retrieving phase can be performed by following the system dynamics in (1) and (2). Basically, the iterations execute a gradient descent of the energy function until they converge to a (local) optimal state.

In order to reach the global optimal solution, one can resort to the Boltzmann machine, which adopts the same energy function as the Hopfield network and the simulated annealing technique [24]. Simulated annealing is a search technique that allows the possibility of getting out of the trap of the local optimum by introducing a mechanism of flattening the trap and a possibility based on a stochastic decision of accepting an updating which (temporally) corresponds to a *worse* solution [33]. It is claimed that, with a proper annealing schedule, the state of the Boltzmann machine will gradually move toward the global optimal solution.

Associative Retrieval/Classification: Another very promising application of neural processing is for associative retrieval or associative classification. The associative retrieval is to retrieve the complete pattern, given partial information of the desired pattern (auto-association), or to retrieve a corresponding pattern in subset B, given a pattern in subset A (pattern association). The associative classification is to identify the corresponding category for any test pattern. The retrieving phase uses the same system dynamics equations (1) and (2) as in the optimization applications. For this application class, some learning schemes are often adopted to train the synaptic weights.

Robotic Applications: Most robotic problems currently being attacked can be categorized into either of the three processing levels: task planning (e.g., depth determination and arm–camera coordination), path planning (e.g., robot navigation), and path control (e.g., motor control). Most of these robotic processing can be formulated in terms of optimization or pattern association problems. Therefore, neural networks can naturally be adopted to solve the robotic processing tasks. For example, stereo vision for task planning [40], [70], [71], [80], autonomous robot path planning [31], [74], manipulator position control [46], [75], can be formulated as optimization problems. On the other hand cerebellar model articulation controller (CMAC) [3], [48], [52], sensor/motor control [42], [62], global terrain retrieval [31], and voluntary movement control [32], [58] can be formulated as pattern association tasks. For more detailed discussion of using neural networks for robotic applications, the reader is referred to Section III.

C. Architectural Aspects of ANN's

Neural networks have been proposed to perform computations "in the style of the brain," relying on massive parallelism and massive interconnectivity of their simple processing units ("neurons"). Recently, a discussion on the possible ways to implement neural network models have been conducted in the research community with a major division into "analog" and "digital" cams. In our view, both analog and digital implementations can be useful when used in areas where they are strongest. Analog neural networks must be used in the pre-processing stages for data compression and initial interpretation since the data rates in real-world applications usually exceed the capabilities of digital networks (e.g., vision). On the other hand, preprocessing stages usually require less accuracy thus being suitable for analog networks. In later processing stages, assuming an initial data reduction and interpretation has been performed, the flexibiliy requirements call for the digital implementation where we can take advantage of their capability of partitioning large problems and implementing different networks with the same hardware.

Analog neural network implementations offer great speed but little else since they suffer in terms of accuracy, density, learning capability, flexibility, and do not scale well with the size of problem since scaling introduces additional noise in the device thus limiting the accuracy even more [46]. The main drawback of analog neural networks however is in their inability to solve problems that require a number of neurons greater than the network's physical size since there is no natural way to store intermediate values, reprogram the network, and then combine the results of multiple passes. A related drawback is the difficulty of implementing hierarchies of interconnected networks that cooperate to solve a large problem.

Digital implementations of neural networks, on the other hand, offer a mature and well understood technology, great accuracy, and scale much better than analog implementations. Using digital logic and memory, it is quite easy to partition a large problem so that it can be solved by a smaller (in terms of hardware) implementation. Using the same logic and memory, a digital implementation can realize more than one network and combine the results in hierarchical fashion to solve large problems. The main drawbacks of digital VLSI implementations are their larger silicon area, relatively slower speed, and the great cost of interconnecting processing units. Summarized below are several key considerations for parallel processing and array architecture implementations of neural networks:

- Convergence issues of synchronous (parallel) updating of system dynamics in the retrieving phase.
- The architectural design should ensure that the processing in both the retrieving phase and the learning phase share the same array configuration, storage, and processing hardware. This will not only speed up real-time learning but also avoid the difficulty in the reloading of synaptic weights for retrieval.
- The digital design must identify a proper digital arithmetic technique to efficienctly compute the necessary operations required in both phases.
- The array architecture design should maximize the strength of VLSI in terms of intensive and pipelined computing and yet circumvent its main limitation on communication. It is desirable to find a local interconnectivity of systolic

solution to the implementation of the global interconnectivity of neural networks.

- A digital design must offer a greater flexibility, so a general-purpose programmable array architecture is preferred for implementing a wide variety of neural network algorithms in both retrieving and the learning phases.

This paper is organized as follows: In section II we discuss several neural networks models useful for robotic processing. Section III shows how to apply the neural networks models for all levels of robotic processing. Finally, in Section IV, we propose a universal ring systolic architecture for implementing all the useful neural networks for robotic processing. Implementation considerations of neural processing units are also addressed.

II. Useful Neural Networks for Robotic Processing

Single-layer feedback neural networks have been shown to serve as a state-space search mechanism, which can be useful for pattern association and optimization tasks. Competitive learning networks provide a very efficient unsupervised learning mechanism for a lot of pattern classification applications. Due to the capability of establishing internal representation by the hidden neurons, the feed-forward multilayer perceptrons have demonstrated very powerful retrieving/learning capabilities.

A. Single-Layer Feedback Neural Networks

There are many single-layer neural networks. Some of them are feedback networks, e.g., the Hopfield associative/optimization network [25], [28], the Boltzmann machine [1], and the Rumelhart's memory/learning module [50].

1) Hopfield Associative/Optimization Neural Networks: Hopfield neural networks have found many applications, including pattern recognition [17], vowel classification [8], combinatorial optimization problem [27], [72], linear programming problem [73], computational vision [34], and parameter estimation [61].

A Hopfield associative neural network is a single-layer (time-iterative) feedback network, which consists of N binary-valued neurons linked to each other with symmetric weights $\{w_{ij} = w_{ji}\}$. In the Hopfield model, thresholding elements are added to linear associators to perform iterative feedback auto-association tasks. Moreover, a notion of energy function is adopted to prove that the feedback system exhibits a number of locally stable points (attractors) in the state space, which provides a basic mechanism for signal retrieval and error correction from partial or noisy missing information [25].

Retrieving Phase: The system dynamics in the retrieving phase of a Hopfield associative network is

$$u_i(l+1) = \sum_{j=1}^{N} w_{ij} a_j(l)$$

$$a_i(l+1) = \begin{cases} 1, & \text{if } u_i(l+1) > -\theta_i \\ 0, & \text{if } u_i(l+1) < -\theta_i \\ a_i(l), & \text{if } u_i(l+1) = -\theta_i. \end{cases} \tag{5}$$

The original Hopfield associative network requires each neuron to be updated asynchronously (sequentially) to guarantee the convergence of the system dynamics. The dynamic evolution of the system state can be regarded as an energy minimization that continues until a stable state (local energy minimum) is reached. To demonstrate the convergence, a notion of Liapunov energy function is often instrumental.

In the following, we will show that synchronous (parallel) updating of all the neurons at each iteration is also possible for non-negative definite symmetric weight matrix $\{w_{ij}\}$. To demonstrate the convergence of the parallel updating iterations, it is useful to adopt the following Liapunov energy function $E(l)$ (after the lth updating)

$$E(l) = -\frac{1}{2}\sum_{i=1}^{N}\sum_{j=1}^{N} w_{ij} a_i(l) a_j(l) - \sum_{i=1}^{N} \theta_i a_i(l).$$

If all the neurons are updated in parallel at each time iteration, then the energy change between any two iterations is

$$\begin{aligned} \Delta E(l+1) &= E(l+1) - E(l) \\ &= -\frac{1}{2}\sum_{i=1}^{N}\sum_{j=1}^{N} w_{ij} a_i(l+1) a_j(l+1) - \sum_{i=1}^{N} \theta_i a_i(l+1) \\ &\quad + \frac{1}{2}\sum_{i=1}^{N}\sum_{j=1}^{N} w_{ij} a_i(l) a_j(l) + \sum_{i=1}^{N} \theta_i a_i(l) \\ &\quad -\frac{1}{2}\sum_{i=1}^{N}(a_i(l+1) - a_i(l)) \left[\sum_{j=1}^{N} w_{ij}(a_j(l+1) - a_j(l)\right] \\ &\quad -\frac{1}{2}\sum_{i=1}^{N}(a_i(l+1) - a_i(l)) \left[\sum_{j=1}^{N} w_{ij}(a_j(l+1) - a_j(l)\right] \\ &= -\Delta \boldsymbol{a}^T(l+1)[\boldsymbol{u}(l+1) + \theta] \\ &\quad -\frac{1}{2}\Delta \boldsymbol{a}^T(l+1) \boldsymbol{W} \Delta \boldsymbol{a}(l+1) \\ &= \Delta E_1(l+1) + \Delta E_2(l+1). \end{aligned} \tag{6}$$

Since the nondecreasing thresholding functions are assumed, the signs of $\{\Delta a_i(l+1)\}$ and $\{u_i(l+1) + \theta_i\}$ are the same (see (5)), and $\Delta E_1(l+1) \leq 0$. In order to also guarantee $\Delta E_2(l+1) \leq 0$, the weight matrix should be a non-negative definite matrix, this can be ensured by setting weight matrix to be

$$w_{ij} = \sum_{p=1}^{P} (2\beta_i^{(p)} - 1)(2\beta_j^{(p)} - 1)$$

or

$$\boldsymbol{W} = \sum_{p=1}^{P} [2\boldsymbol{v}^{(p)} - 1][2\boldsymbol{v}^{(p)} - 1]^T \tag{7}$$

where P binary reference patterns, represented by $\{\boldsymbol{v}^{(p)} = [\beta_1^{(p)}, \beta_2^{(p)}, \cdots, \beta_N^{(p)}], p = 1, 2, \cdots, P\}$, are stored in and to be retrieved from the associative network. Note that the $\boldsymbol{W}$ matrix is formed by an outer product without diagonal nullification (as opposed to original Hopfield network [25]), so it is a non-negative definite matrix.

Learning Phase: The weight determination process based on (7) can be interpreted by the recursive weight updating formulation given in (3), i.e.,

$$w_{ij} \Leftarrow w_{ij} + \frac{\eta}{2} \beta_i \beta_j \tag{8}$$

where the recursion index is corresponding to each training pattern used. This leads to the simplest form of Hebbian learning rule [22], and the stored pattern $\{\beta_i\}$ is the desired output (desired auto-associative retrieval information).

Instead of going through the iterative Hebbian weight updating learning procedures, Hopfield used the ensemble average of $\{w_{ij}\}$ over P (training) pattern recursions to determine the fixed connection weights with approximate *zero statistical mean* [25], this leads to the weight determination formulation given in (7).

Hopfield Optimization Neural Networks: In order to imitate the continuous input–output relationship of real neurons and to simulate the integrative time delay due to the capacitance of real neurons [26], Hopfield modified the associative network by adopting the following *sigmoid activation* function $f(x)$:

$$f(x) = \frac{1}{1 + e^{-x/u_0}} \tag{9}$$

and proposed the following dynamic system for the retieving phase [26], [72]:

$$u_i(l+1) = \sum_{j=1}^{N} w_{ij} a_j(l) \tag{10}$$

$$a_i(l+1) = \frac{1}{1 + e^{-u_i'(l+1)/u_0}} \tag{11}$$

where $u_i'(l + 1) = \kappa_1(u_i(l) + \theta_i) + \kappa_2(u_i(l + 1) + \theta_i)$, and κ_1 and κ_2 are proper constants proposed in [72]. Note that when the coefficient u_0 tends to zero, then the activation function of the Hopfield neural network becomes a *step function* as given in (5).

In the continuous state, the convergence with parallel updating can be assured only when the (time step) coefficient (κ_2) is small enough [56]. This will lead us to the small values $\{\Delta \mathbf{a}^T(k + 1)\}$, and the contribution of the energy term $\Delta_k E_2$ in (6) can be neglected, so that $\Delta_k E_1$ dominates the level of the energy change.

2) Boltzmann Machine: If we substitute the sigmoid coefficient u_0 in the Hopfield neural networks (see (11)) by a temperature-control parameter T, then (11) becomes

$$a_i(l+1) = f(u_i(l+1), \theta_i) = \frac{1}{1 + e^{-(u_i(l+1)+\theta_i)/T}} \tag{12}$$

where we let $\kappa_1 = 0$ and $\kappa_2 = 1$ (i.e., no tine delay assumed).

Retrieving Phase: Let us confine the neuron response to be discrete value (0 or 1), and let the deterministic activation operation in the Hopfield neural network be replaced by a stochastic process defined below

$$\Pr\,(a_i(l+1) = 1) = \frac{1}{1 + e^{-(u_i(l+1)+\theta_i)/T}}. \tag{13}$$

More precisely, $a_i(l + 1)$ is set to "1" with probability given in (13), otherwise $a_i(l + 1) = 0$. It was shown in [24] that, the system dynamics given in (13) ensures that "in thermal equilibrium the relative probability of the two global states is determined solely by their energy difference, and follows a Boltzmann distribution." Therefore, this modified version of the Hopfield neural network is called the *Boltzmann machine* [1], [24]. If T follows an *annealing schedule*, i.e., T gradually decreases during the iterations of neural networks, the sigmoid activation operation in (11) becomes an *annealing* operation. With a proper annealing schedule, the states of the Boltzmann machine will gradually move toward the global optimal solution.

Weight Determination: The weight determination procedure, basically the same as that for the Hopfield network, starts with finding the energy function. The synaptic weights can be determined by comparing this energy function with (4). The retrieving phase can be performed by the system dynamics in (13).

Learning Phase: While the retrieving phase of a Boltzmann machine is similar to the Hopfield net, the Boltzmann learning rule has a very different form. It may be regarded as a two-step batch-updating Hebbian rule [1], [4], [65]: In the first step (denoted *phase*$^+$), some outputs of the neural units in the network are clamped to predetermined desired values, and then let the neural net iterates through a proper annealing schedule based on (13) until convergence. In the second step (denoted *phase*$^-$), none of the neural units are clamped and the network again iterates (following the same annealing schedule) until it reaches a new convergent state. If there exists any discrepancy between the convergent states for the *phase*$^+$ and the *phase*$^-$, then the synaptic weights should be adjusted as follows:

$$\Delta w_{ij} = \eta(p_{ij}^+ - p_{ij}^-) \tag{14}$$

where p_{ij}^+ (respectively, p_{ij}^-) represents the correlation of the two neurons i and j in the *phase*$^+$ (respectively, in the *phase*$^-$). They can be determined by the average probability that both neurons are on (i.e., $\beta_i = \beta_j = 1$) over many training patterns. Instead of modifying the weights immediately after the exposure of each training pattern as in the conventional Hebbian learning rule (see (8)), the weight adjusting (i.e., (14)) is performed only after enough training patterns are taken. The goal of this learning rule is to find a set of synaptic weights such that the activation outputs in the *phase*$^-$ match the desired outputs in the *phase*$^+$ as closely as possible.

B. Competitive Learning Networks

Competitive learning is an unsupervised (self-organized) procedure that classify a set of input patterns into a number of disjoint clusters in such a way that the input patterns within each cluster are all similar to one another [23]. Most competitive learning networks are single-layer feed-forward networks using winner-take-all mechanism. It is possible to cascade several individually optimized single-layer competitive learning networks to establish a hierarchical classification network [19], [66].

Retrieving Phase: Without loss of generality, the system dynamics between the inputs $\{\alpha_i\}$ and the outputs $\{\beta_i\}$ in the retrieving phase of a competitive learning network is given

$$u_i = \sum_{j=1}^{N_0} w_{ij}\alpha_j$$

$$\beta_i = \begin{cases} 1, & \text{if } u_i > u_k, \ \forall k \\ 0, & \text{if else} \end{cases} \tag{15}$$

where $1 \le i \le N_1$. The *winner-take-all* nonlinear competition mechanism (e.g., lateral inhibitions) is used in the output layer so that only the neuron receiving largest input is activated.

Learning Phase: The learning phase in the competitive learning network can also be interpreted by the recursive weight updating formulation (see (3))

$$w_{ij} \Leftarrow w_{ij} + \eta\beta_i(\alpha_j - w_{ij}) \tag{16}$$

where one recursion is for one pattern. According to (15), (16) implies that only the weights associated with the winning neuron are updated and all the other weights remain unchanged. This is a special feature of the competitive learning networks.

We hasten to note that the above formulation can only describe the basic feature commonly shared by the competitive learning networks. In actuality, each individual model has almost unexceptionally adopted certain special mechanism.

- Kohonen's self-organized feature map [36], [37] introduced a neighborhood (whose size slowly decreases with each iteration) of a winning neuron in a two-dimensional neuron layer. Weights associated with the winner and the neurons in the neighborhood of the winner are all modified. This has purpose of making the neurons more responsive to the current input pattern.
- The adaptive resonance theory (ART) by Carpenter and Grossberg [12], [21] introduced a *vigilance test* to adaptively create new neuron units for the incoming input patterns which are quite different from the memorized patterns. This test requires additional modifiable feedback weights connecting from output layer to the input layer.
- Rumelhart's competitive learning algorithm [66] introduced a *leaky learning model* to prevent the possibility of totally unlearned neurons. This is done by performing training in (16) over all the weights in the network. However, the weights associated with the winner get much larger η values.
- Neocognitron can be formed as a hierarchical learning system by cascading many single-layer competitive learning networks [19]. The learning for each layer is largely based on single-layer analysis. It progresses stage from the input layer to the output layer.
- Darwin networks were created to explore the brain theory called neural Darwinism. These networks select those neuronal groups (called repertoire) which respond best to the specific input stimuli [62]. Neuron groups in a repertoire respond best to overlapping but similar input patterns due to the randomness of neuronal growth, the response to important unexpected inputs is thus insured [46].

C. Multilayer Perceptrons

Multilayer feed-forward neural networks are *spatially* iterative neural networks, which have several layers of *hidden neuron units* between the input and output neuron layers (see Fig. 1). the weight updating for the hidden layers adopts the mechanism of back-propagated corrective signal from the output layer.

Retrieving Phase: The system dynamics in the retrieving phase of an L-layer perceptron can be described by the following spatially iterative equations, with the iteration index l denoting the *spatial* (layer) iteration:

$$u_i(l+1) = \sum_{j=1}^{N_l} w_{ij}(l+1)a_j(l)$$

$$a_i(l+1) = f(u_i(l+1) + \theta_i(l+1)) = f_i(l+1) \tag{17}$$

where $1 \le i \le N_{l+1}$, $0 \le l \le L-1$, and the nonlinear function f is nondecreasing and differentiable. For simplicity, the external inputs $\{\theta_i(l+1)\}$ are often treated as special modifiable synaptic weights $\{w_{i,0}(l+1)\}$ which have clamped inputs $a_0(l) = 1$.

Learning Phase: The learning phase of an L-layer multilayer perceptron follows an iterative gradient descent approach [54], [55], [64], [76], [77]. Given the pth pair of input/target training patterns, $\{\alpha_i^{(p)}, i = 1, \cdots, N_0\}$, $\{t_j^{(p)}, j = 1, \cdots, N_L\}$, our goal; is to iteratively choose a set of $\{w_{ij}(l), \forall l\}$ for all layers so that the mean squared error between the actual output activations and target activations can be minimized. To be more specific

$$w_{ij}(l) \Leftarrow w_{ij}(l) - \eta\delta_i(l)f_i'(l)a_j(l-1) \tag{18}$$

where $1 \le l \le L$, and $f_i'(l)$ is the derivative of $f_i(l)$ with respect to $u_i(l)$. The back-propagated corrective signal $\delta_i(l)$ can be recursively calculated as shown below. For $L - 1 \ge l \ge 1$

$$\delta_i(l) = \sum_{j=1}^{N_{l+1}} \delta_j(l+1)f_j'(l+1)w_{ji}(l+1). \tag{19}$$

Note that the backward initialization error signal on the top layer $\delta_i(L)$ is defined as $\delta_i(L) = -(t_i - \beta_i)$.

III. Applications of Ann's to the Robotic Processing

This section discusses *how to use the above neural networks for robotic processing*. There are three levels of processing in a hierarchical robotic processing system, task planning, path planning, and path control [68]. In the task planning level the robot will receive the instructions about task plans and manage the information (e.g., target, depth, obstacle locations) about the workspace. The path planning is to produce a sequence of desired path trajectory (e.g., robot positions, arm orientations). The path control is to generate the necessary motor commands (torques and forces) in the joint coordinates to drive the robot or arms to follow the desired trajectory. The computational requirement of the robotic processing is very demanding. At the level of task planning, some primitive image/vision analysis are required to obtain the three-dimensional (3-D) information. At the level of path planning, an optimal path in a workspace has to be selected by using optimization techniques. For the path control level, real-time computing of inverse dynamics and kinematics must be provided.

In this section, neural network techniques for processing the various algorithms in the three levels of robotic processing are discussed. The approach is to formulate each algorithm either as an optimization problem or a pattern association problem. The examples include stereo vision in task planning [49], [70], [71], [80], autonomous robot path planning [31], [74], manipulator position control [58], [75], robotic voluntary movement, and sensor/motor control [32], [42], [62]. Our aim is to demonstrate how these algorithms and some other typical operations in robotic processing can be solved by the neural networks.

A. Task Planing Using Neural Networks

A critical step in the task planning is the determination of the depth information. Traditionally, the sonar sensors have been very popular for this purpose [30]. However, the stereo matching technique is a more reliable but computationally costly alternative. It recovers 3-D depth information from two images taken from two cameras. Usually, the geometry of the imaging system (e.g., the camera postions and their orientations) are known. The main task in the stereo vision then hinges upon the matching between the feature primitives of the two images. This is called the *correspondence problem*, which consists of the detection of the feature primitives (e.g., edges [70]) and the determination of the valid matches of the feature primitives. As discussed next, the latter task may be formulated as a constrained optimization problem [49], [70], [71], [80].

Neural Networks for Constrained Optimization: Consider the following constrained optimization problem [45]:

$$\text{minimize} \quad \phi(\boldsymbol{a})$$
$$\text{subject to the constraints} \quad P_i(\boldsymbol{a})=0, \qquad i=1, 2, \cdots, q \tag{20}$$

where ϕ is a continuous function of n-dimensional vector $\boldsymbol{a}$ (continuous or discrete) and $\{P_i(\boldsymbol{a})\}$ are nonnegative and continuous functions. The above constrained optimization problem may be approximated by an unconstrained optimization problem, which involves minimizing a new energy (or penalty) function [45]

$$\text{minimize} \quad E' = \phi(\boldsymbol{a}) + \sum_{i=1}^{q} c_i P_i(\boldsymbol{a}) \tag{21}$$

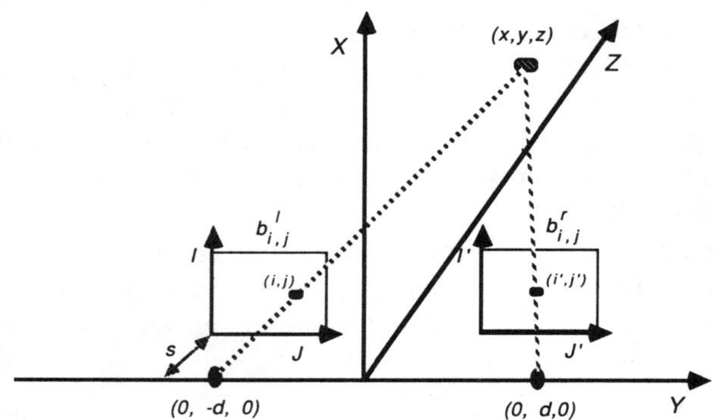

Fig. 3. A standard stereo matching problem involves two camera with focal length s and their focal points lying at (0, $\pm d$, 0).

where $\{c_i\}$ are large positive constants. This naturally leads to the optimization neural network processing and the synaptic weights $\{w_{ij}\}$ and the external inputs $\{\theta_i\}$ can be derived from the (Liapunov) energy function.

Neural Network Formulation for Stereo Matching: As shown in Fig. 3, the stereo matching problem involves two cameras with focal length s and their focal points lying at (0, $\pm d$, 0), respectively. The lines of sight of the cameras are assumed to be parallel to the z-axis. A point (x, y, z) appears at (i, j) pixel of the left image coordinate and at (i', j') pixel of the left image coordinate. Without loss of generality, let us assume that the epipolar scanline is along the horizontal (J-axis) direction, we can have $i = i'$, and the search for candidate match for a point can be limited to the same row in the other image. The value $(j - j')$ is caled the *disparity* value between two matching points. The depth information z can be generated from the disparity value, i.e.,

$$z=\frac{2sd}{(j-j')}.$$

Suppose that there are two $M \times N$ primitives extracted image intensity arrays, $\{b^l_{i,j}\}$ and $\{b^r_{i,j}\}$ as taken by the (left and right) cameras. A 3-D binary matching data array $\{a_{i,j,k}, 1 \le i \le M, 1 \le j \le N, 0 \le k \le D\}$ may be defined to represent the status of matching [49], [80]. When the $a_{i,j,k}$ is 1, this means that the disparity value is k at the point (i, j). In this discussion, the maximal disparity value is limited to D. The goal of the correspondence problem is to minimize the following cost function ϕ:

$$\phi=\phi_1+\phi_2$$
$$=\sum_{i=1}^{M}\sum_{j=1}^{N}\sum_{k=0}^{D}(b^l_{i,j}-b^r_{i,j\oplus k})^2 a_{i,j,k}$$
$$+\lambda\sum_{i=1}^{M}\sum_{j=1}^{N}\sum_{k=0}^{D}\sum_{(i',j')\in\Psi}(a_{i,j,k}-a_{i',j',k})^2 \tag{22}$$

with the constraint

$$P=\sum_{i=1}^{M}\sum_{j=1}^{N}\left(\sum_{k=0}^{D}a_{i,j,k}-1\right)^2=0 \tag{23}$$

where

1) ϕ_1 is a measurement of how two images are matched after alignment in a least squares sense. The symbol $\oplus$ denotes that

$$t_{a \oplus b} = \begin{cases} t_{a+b}, & \text{if } 0 \le a+b \le N \\ 0, & \text{otherwise.} \end{cases}$$

2) ϕ_2 is a measurement of the continuity of the depth value [49], [80], where λ is a proper weighting constant and Ψ denotes the the neighborhood around the pixel (i, j), defined by a square window centered at (i, j).

3) The constraint in (23) is introduced to assure the *uniqueness* property is preserved. That is any point from each image can assume one and only one depth value [9], [49].

Following (21)–(23), the constrained problem can be reformulated as an unconstrained problem i.e.,

$$\min_{a_{i,j,k}} E' = \phi + cP. \tag{24}$$

The Hopfield optimization network or Boltzman machine can be used to solve the above unconstrained problem. Each matching data element $A_{i,j,k}$ can be regarded to be an activation value of one neuron. As discussed in Section I-B, the synaptic weights $w_{i,j,k,l,m,n}$ and the external input $\theta_{i,j,k}$ of the neural network can be derived by equating the energy (penalty) function in (24) to the Liapunov energy function E in (4) [27], [80], i.e.,

$$E = E' = -\frac{1}{2} \sum_{i=1}^{M} \sum_{j=1}^{N} \sum_{k=0}^{D} \sum_{l=1}^{M} \sum_{m=1}^{N} \sum_{n=0}^{D} w_{i,j,k,l,m,n} \cdot a_{i,j,k} a_{l,m,n} - \sum_{i=1}^{M} \sum_{j=1}^{N} \sum_{k=0}^{D} \theta_{i,j,k} a_{i,j,k}.$$

B. Path Planning Using Neural Networks

The problem of path planning for a robot can be stated as follows: Given the locations of a set of obstacles and the initial position and the final target of the robot, to find a continuous path from the initial position to the target while circumventing the obstacles along the way. Two types of planning are often encountered: one is called *global planning*, which derives the optimal path from the overall topographic map. The other is called *local planning*, which relies on information available only to the line of sight of the robot sensors or cameras from particular perspectives. We note that human beings perform very well in the path planning through a two-stage anticipatory planning [31]: First they recall the complete global environment from the local line of sight information via associative retrieval. Then they recursively search for the optimal path from the retrieved global information. In a manner very much similar to that of the human being, artifical neural networks may be adopted to implement the operations in both stages.

1) Global Terrain Retrieval via Pattern Association: The robot navigation area can be divided into 2-D or 3-D grid cells. A binary value may be assigned to each grid cell to indicate the clearness status. Namely, a value "1" implies that the cell is fully occupied, while "-1" implies that the cell is unoccupied. Continuous values between -1 and 1 may also be used to indicate partial occupation of a grid cell or any ambiguity due to vision/sensor processing. The values of the grid cells can be trained and recalled using associative retrieval neural networks. When a robot "perceives" an environment similar to one previously trained, the neural network accepts the (continuous) values of the local grid cells to retrieve a best fitting global terrain features based on the trained data. Possible candidates for this associative retrieval task are the Rumelhart's *memory/learning modules* [50] and multilayer perceptrons [64]. The reasons for adopting such neural networks are that they works well with continuous input values and they exhibit good fault tolerance capabilities in the retrieval.

2) Boltzmann Machine for Optimal Path Finding: Once the information about the global grid cells is retrieved then the neural network is ready to plan an optimal path to reach a remote goal. In order to find the global optimal path, a constrained optimization formulation can be adopted and neural networks applied. The key is to define a cost function ϕ. It is a function of several important variables, e.g., the distance between the current location and the starting position of the robot or the distance between the current location to the goal, etc. Some constraints need to be satisfied, e.g., the path should not be allowed to pass through the grid cells with high clearness values. The robot will stop and start a new path planning process when an unanticipated obstacle is sensed before it reaches the final target position. In order not to be trapped in local optima, simulated annealing techniques such as Boltzmann machine can be adopted. It is important to provide a good (although not optimal) path as an initial state for the optimization process. The recalled global map is very useful to generate such a preliminary path. Based on the map, those unlikely cells, e.g., cells outside the boundary or cells occupied with obstacles, can be ruled out. Then the cells with a minimum traversal distance of the path may be selected to be the preliminary path.

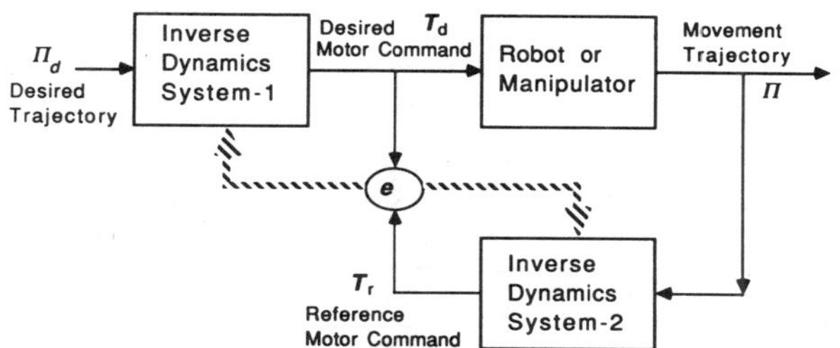

Fig. 4. A simplified schematical diagram for the feed-forward controller.

C. Path Control Using Neural Networks

After the determination of the desired trajectory in the path planning level, the task of path control is to generate the motor control signal (torques and forces) so that the robot may be driven to follow the trajectory.

1) Inverse Dynamics via Multilayer Perceptron: Based on some physiological model [32], [59], the dynamic movement of a robot or a manipulator can be controlled by a feed-forward controller, as shown in Fig. 4. It consists of three major components: two *identical* inverse dynamics systems ($IDS_1 = IDS_2$) and the robot (or manipulator).

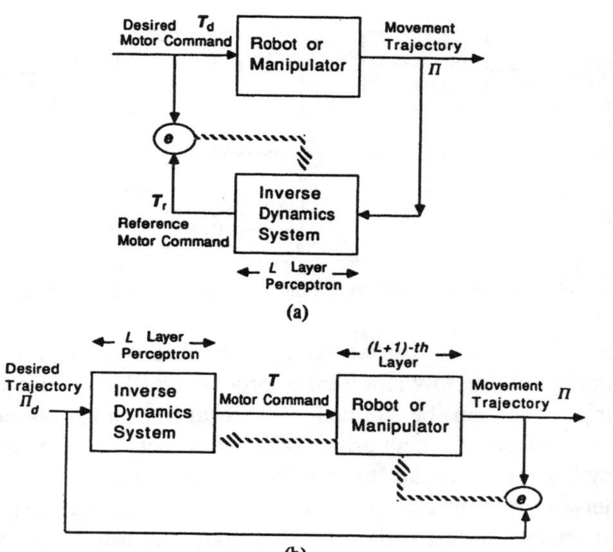

Fig. 5. (a) Generalized learning configuration. (b) Specialized learning configuration. Note that the *IDS* can be modeled by an *L* layer perceptron.

The IDS_1 receives the desired trajectory information Π_d and produces the corresponding desired motor command T_d to drive the robot so that the actual movement of the robot (denoted as Π) may follow as closely as possible the desired trajectory Π_d. The IDS_2, on the other hand, takes the actual movement Π and produces the reference motor command T_r as shown in Fig. 4.

To efficiently train the *IDS* to implement the feed-forward controller as shown in Fig. 4, a two-stages learning procedure can be adopted [32], [58]. The first stage is called *generalized learning* with its configuration shown in Fig. 5(a), the second stage is called *specialized learning* with a different configuration shown in Fig. 5(b). In the generalized learning, a set of desired motor commands, denoted as $\{T_d\}$, are used to drive the robot and the set of resulting trajectories is denoted as $\{\Pi\}$. Then the *IDS* receive $\{\Pi\}$ as input and yield a set of reference motor commands, denoted as $\{T_r\}$. The goal of the generalized learning is to minimize the errors between $\{T_r\}$ and $\{T_d\}$ in the least square sense [59]. After the *IDS* is well trained, if a real input Π' is sufficiently close to one trajectory in the set $\{\Pi\}$, the controller should be able to retrieve a proper motor command $\hat{T}$, making the actual movement $\hat{\Pi}$ closely follows Π.

Due to the lack of knowledge about the operating range of the desired motor commands $\{T_d\}$, an unnecessarily large set of $\{T_d\}$ for the training may have to be used. This difficulty can be overcome by incorporating a specialized learning stage into the controller system (see Fig. 5(b)), in which the *IDS* is trained based on the desired trajectory $\{\Pi_d = [\Pi_1^{(d)}, \cdots, \Pi_m^{(d)}]\}$ and outputs the appropriate motor commands $\{T\}$ to drive the robot. The actual movement trajectory of robot is a function of T, denoted as $\Pi(T) = [\Pi_1, \cdots, \Pi_m]$. When the operating points of the system change or when new training patterns are added, it should be adequate to use specialized learning to fine-tune the system.

A severe weakness in the specialized learning is that in the initial step, the training of *IDS* may be very inefficient due to the lack of knowledge about the dynamic model of the robot. Therefore, it is advantageous to properly combine the generalized learning and the specialized learning.[2] For example, it is possible to first perform the generalized learning until the dynamic model of the robot is approximately learned, then the specialized learning follows to fine-tune the *IDS*.

Multilayer Perceptrons for Generalized Learning: An L-layer neural network can be used to implement the *IDS* in a generalized learning system [59]: An input $T_d = [T_1^{(d)}, \cdots, T_n^{(d)}]$ is selected an applied to the robot to obtain a corresponding Π, and the network is trained to reproduce $T_r = [T_1^{(r)}, T_2^{(r)}, \cdots, T_n^{(r)}]$ at its output from Π (see Fig. 5(a)). The mathematical formulation for the updating of weights $w_{ij}(l)$ using the back propagation learning as given in (18) and (19). Note that the weights training here is based on the least squares errors of $|T_d - T_r|^2$, although the real objective of the robot training should have been minimizing $|\Pi_d - \Pi|^2$.

Multilayer Perceptrons for Specialized Learning: The same L-layer neural network can be used to implement the *IDS* in a specialized learning [59]: Referring to Fig. 5(b), the dynamic model of the robot can be regarded as an additional layer, i.e., the $(L + 1)$th layer. However, the back propagation learning may not be applicable to this last layer, since the layer has no synaptic connections defined in the conventional sense. Instead, a modified back-propagation formula was proposed [58]:

$$\delta_i(L) = \sum_j \delta_j(L+1) \frac{\partial \Pi_j(T)}{\partial T_i}$$

$$\delta_j(L+1) = -(\Pi_j^{(d)} - \Pi_j)$$

where $\Pi_j(T)$ denotes the jth element of the robot movement trajectory. In case the dynamic model of the robot is unknown, the partial derivative can be approximated as

$$\frac{\partial \Pi_j(T)}{\partial T_i} \approx \frac{\Pi_j(T + \Delta T_i I_i) - \Pi_j(T)}{\Delta T_i}$$

where $I_i = [0, 0, \cdots, 1, 0, \cdots, 0]$. For the training of the remaining L layers of the multilayer perceptron, the conventional back propagation learning algorithm (see (18) and (19)) can be adopted.

2) Sensor/Motor Maps Using Competitive Learning: It is important to apply robots to an unforseen environment. For example, various obstacles may enter or leave the working area of the robot and need to be avoided. This can be solved by providing a visual map (sensor/motor topographic map) of the objects in the environment to the robot controller.

Neural architecture with self-organized competitive learning (see (15)) can be used to control adaptive sensory-motor coordination [42]. In the learning phase, visual input signals about the object are processed and combined into a target map through modifiable weights, and producing computed motor signals. The errors between the actual motor signals and the motor signals computed from the visual input are used to

[2] By switching back and forth between the two learning stages, it can also avoid the system to be trapped by local minima [58].

incrementally change the weights to make the later computed motor signals closer to the actual computed signals. After the network is trained, in the retrieving phase the learned sensory–motor correlation is used to recognize and manipulate objects which are similar to those that were experienced in the first stage. This neural architecture is composed of motor map representations interleaved within a sensory topography. This allows any number of topographic sensory inputs to be mapped onto any number of motor outputs.

Darwin networks is a simulated automaton made up of many subnetworks [62]. Inputs to the automaton are from a simulated eye which scans a 2-D input array under control of opponent pair muscles. This eye has a large but low-resolution outer visual field and a smaller, high-resolution inner visual field on its simulated retina. Inputs are also provided by a multiple-jointed arm that can reach and feel objects presented on the input field. Darwin networks is first trained to track objects presented on the input array by coordinated movement of eye muscles. The error signal to learn this task is the distance between the position of the object on the retina and the center or fovea of the retina. Darwin networks is then trained to reach out, touch, and feel around the border of objects using self-organized competitive learning [46].

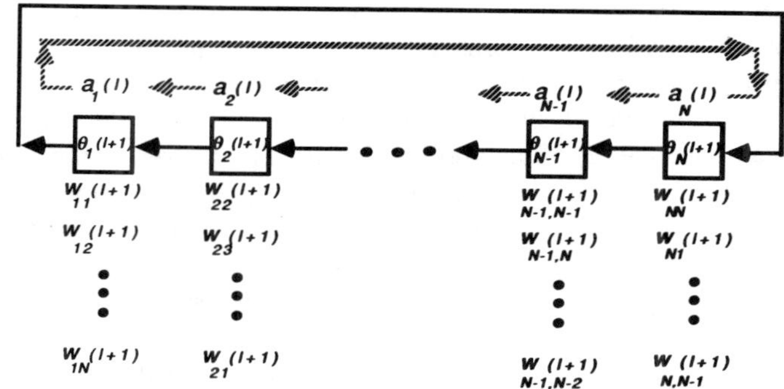

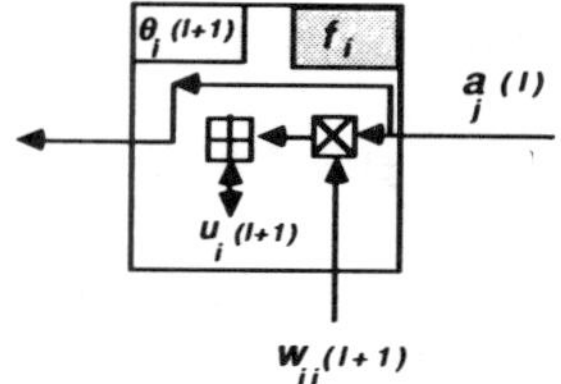

Fig. 6. The ring systolic architecture for consecutive MVM in the retrieving phase at $l + 1$th iteration.

IV. Ring Systolic Architecture for Neural Networks

It is shown that operations in both the retrieving and learning phases of most iterative ANN models can be formulated as *consecutive matrix–vector multiplication, consecutive vector–matrix multiplication*, or *outer-product updating* problems. In terms of the array structure, all these formulations lead to a same universal ring systolic array architecture. In terms of the functional operations, all these formulations calls for a MAC (multiply and accumulation) processor [41].

A. Ring Systolic Design for the Retrieving Phase

Consecutive MVM in the Retrieving Phase: The system dynamics in the retrieving phase of the generic iterative ANN model can be formulated as a consecutive *matrix–vector multiplication* (MVM) problem interleaved with the nonlinear activation function (see (1) and (2)). Without loss of generality and to facilitate the homogeneous architectural design, it is first (which will be relaxed later) assumed that all the iterations in the iterative ANN model have uniform size of N neural units, which is true for single-layer feedback networks, where $\{w_{ij}(l) = w_{ij}\}$, and $\{\theta_i(l) = \theta_i\}$.

$$\boldsymbol{u}(l+1) = \boldsymbol{W}(l+1)\boldsymbol{a}(l)$$
$$\boldsymbol{a}(l+1) = f[\boldsymbol{u}(l+1), \theta(l+1)] \quad (25)$$

where $\boldsymbol{u}(l) = [u_1(l), u_2(l), \cdots, u_N(l)]^T$, $\boldsymbol{a}(l) = [a_1(l), a_2(l), \cdots, a_N(l)]^T$, $\theta(l) = [\theta_1(l), \theta_2(l), \cdots, \theta_N(l)]^T$ and $\boldsymbol{W}(l) = \{w_{ij}(l)\}$. The $f[x]$ operator performs the nonlinear activation function f on each element of the vector $\boldsymbol{x}$.

This uniform size consecutive MVM formulation leads to a ring systolic architecture as shown in Fig. 6 [40], [41]. The pipelining period of this design is 1, which implies 100-percent utilization efficiency [38]. A minor disadvantage of the design is that the (global) spiral communication link is required.

Systolic Processing of Consecutive MVM: At the $(l + 1)$th iteration of the retrieving phase, each processor element (PE), say the ith PE, can be treated as a neuron, and the corresponding incoming synaptic weights $(w_{i1}(l + 1), w_{i2}(l + 1), \cdots, w_{iN}(l + 1))$ are stored in the memory of the ith PE in a circularly shift-up order (by $i - 1$ positions). The operations at the $(l + 1)$th iteration can be described as follows (see Fig. 6):

1) Each neuron activation value $a_i(l)$, created at ith PE, is multiplied with $w_{ii}(l + 1)$. The product is added to the accumulator $u_i(l + 1)$, which has initial value set to be equal to zero (or $\theta_i(l + 1)$ in some networks). After the MAC operation, $a_i(l)$ will move counterclockwise across the ring array and visit each of the other PE's once in N clock units.

2) When $a_i(l)$ arrives at the ith PE, it is multiplied with $w_{ij}(l + 1)$ and the product is added to $u_i(l + 1)$ according to (1).

3) After N clock units, the accumulator $u_i(l + 1)$ collects all the necessary products.

4) One more clock is needed for $a_i(l)$ to returns to the ith PE and the processor is ready for the nonlinear activation operation f_i (see (2)) to create $a_i(l + 1)$ for the next iteration. (Note that N clocks are required to perform the winner-take-all nonlinear mechanism by cycling all the $u_i(l + 1)$ to determine the winner.)

The above procedure can be executed in a fully pipelined fashion. Moreover, it can be recursively executed (with increased l) until L iterations are completed.

B. Ring Systolic Design for the Learning Phase

There are two types of operations required for most weight updating methods: 1) the *outer product updating* (OPU), and 2) the consecutive *vector-matrix multiplication* (VMM).

Both types can be efficiently implemented by the ring systolic ANN based on the same memory storage, processing hardware, and array configuration in the retrieving phase.

1) Operations in the Learning Phase: Given the weight value $w_{ij}(l+1)$ of the previous learning recursion, 1) the new weight values can be calculated based on the OPU; 2) and the back-propagated corrective signal $\delta_i(l)$ of ith iteration can also be computed based on the consecutive VMM operation.

OPU in the Learning Phase: In general, an additive updating formulation of the weights at $(l+1)$th iteration can be summarized by the following OPU equation [65]:

$$w_{ij}(l+1) \Leftarrow w_{ij}(l+1)+\Delta w_{ij}(l+1)$$

$$= w_{ij}(l+1)+g_i(l+1)\cdot h_j(l+1) \text{ or}$$

$$W(l+1) \Leftarrow W(l+1)+g(l+1)h^T(l+1)$$

where $g(l+1) = [g_1(l+1), g_2(l+1), \cdots, g_N(l+1)]^T$ and $\mathbf{h}(l+1) = [h_1(l+1), h_2(l+1), \cdots, h_N(l+1)]^T$. To be more specific

$g_i(l+1)$	$h_j(l+1)$	Learning
β_i	$\eta\beta_j$	Hebbian
$t_i - \beta_i$	$\eta\beta_i$	Delta
β_i	$\eta(\alpha_j - w_{ij})$	Competitive
$\delta_i(l+1)f_i'(l+1)$	$-\eta a_j(l)$	Generalized Delta

Consecutive VMM in the Learning Phase: The back-propagation rule can be formulated as a consecutive VMM operations [41]

$$\delta_i(l)=\sum_{j=1}^{N} g_j(l+1)w_{ji}(l+1)$$

or

$$d^T(l)=g^T(l+1)W(l+1) \tag{26}$$

where $d(l) = [\delta_1(l), \delta_2(l), \cdots, \delta_N(l)]^T$.

2) Systolic Processing in the Learning Phase: The ring systolic array derived for the retrieving phase can be easily adapted to execute in parallel both the OPU and consecutive VMM in the learning phase [41].

Systolic Processing for the OPU: The operations of the OPU in the ring systolic ANN can be briefly described as follows (see Fig. 7):

1) The value of $g_i(l+1)$ is computed and stored in the ith PE. The value $h_j(l+1)$ produced at the jth PE will be cyclically piped (leftward) to all other PE's in the ring systolic ANN during the N clocks.

2) When $h_j(l+1)$ arrives at the ith PE, it is multiplied with the stored value $g_i(l+1)$ to yield $\Delta w_{ij}(l+1)$, which will be added to the old weight $w_{ij}(l+1)$ of the previous recursion to yield the updated $w_{ij}(l+1)$. Note that the old weight data $\{w_{ij}(l+1)\}$ are retrieved in a circularly-shift-up sequence, just like what used in the retrieving phase.

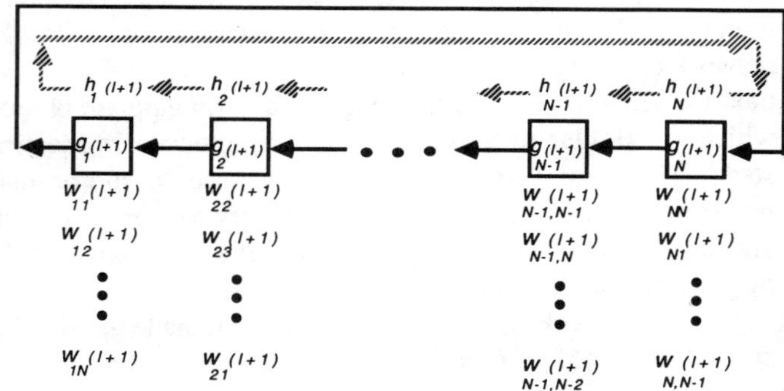

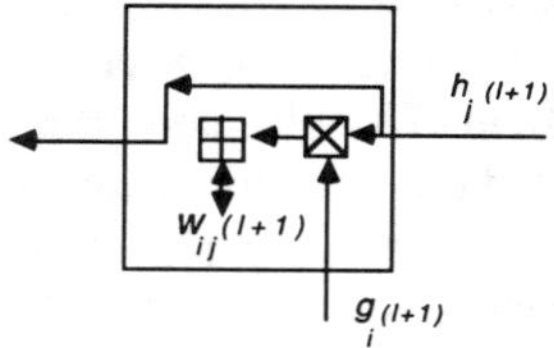

Fig. 7. The ring systolic architecture for OPU in the learning phase.

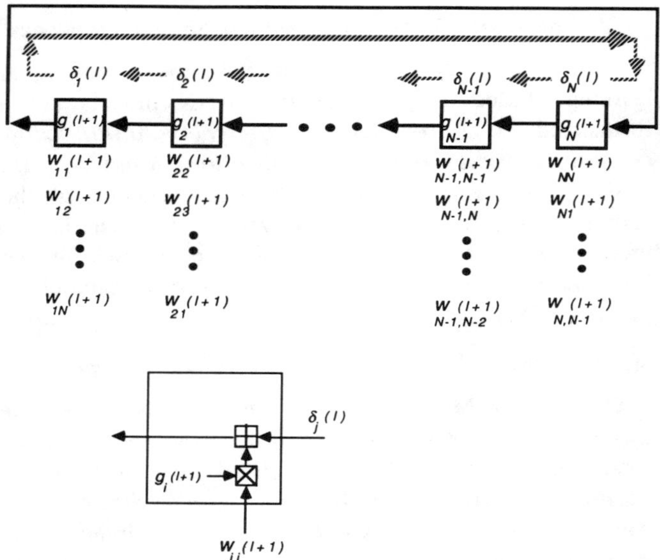

Fig. 8. The ring systolic architecture for consecutive VMM in the learning phase.

3) After N clocks, all the $N \times N$ new weights at $(l+1)$th iteration $\{w_{ij}(l+1)\}$ are generated. The ring systolic ANN is now ready for the weight updating of the next iteration (l).

Systolic Processing for the Consecutive VMM: The operations of the consecutive VMM in the ring systolic ANN can be briefly described as follows (see Fig. 8):

1) The signal $g_j(l+1)$ and the value $w_{ji}(l+1)$ are available in the jth PE at $(l+1)$th iteration. The value $w_{ji}(l+1)$ is then multiplied with $g_j(l+1)$ at jth PE.

2) The product is added to the newly arrived accumulator $\delta_i(l)$. (The parameter $\delta_i(l)$ is initiated at the ith PE with zero initial value and circularly shifted leftward across the ring array.)

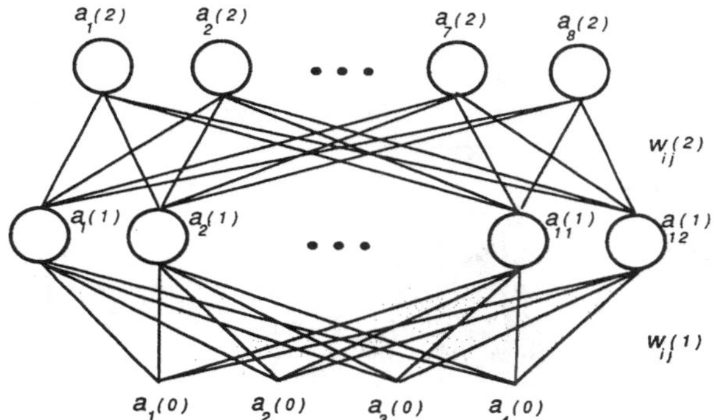

Fig. 9. A two-layer perceptron with configuration (4–12–8).

3) After N such accumulation operations, the accumulator $\delta_i(l)$ will return to ith PE after accumulating all the products

$$\delta_i(l)=\sum_{j=1}^{N} g_j(l+1)w_{ji}(l+1).$$

C. Partitioning Scheme for Large or Nonuniform Networks

Generally speaking, in an overall array architecture system, one seeks to maximize the following performance indicators: effective array configuration, flexibility on problem partitioning, fault-tolerance to improve system reliability, and programmability for adequate software support [38].

In case of large size of neural networks with smaller number of available PE's, it is important to provide hardware and/or software support for an efficient partitioning scheme, which allows large problems to be decomposed into smaller subproblems to be solved on the array. The tasks of several neural networks can be assigned to share the same PE without changing the computation/communication strategy. Based on this approach, this architecture can be easily adapted to nonuniform multilayer perceptron, where each layer has very different number of neurons. By appropriately assigning the tasks of several (equal number of) neurons at the same layer to one specific PE, the computational load of each layer can be uniformly distributed to the N PE's disregarding the number of neurons in each layer. This also guarantees the fully pipelining efficiency.

Without loss of generality, an example is given to illustrate the partitioning scheme for nonuniform networks. We will show that a ring architecture with 4 PE's can be used to implement the retrieving/learning phase of a two-layer perceptron shown in Fig. 9 with configuration (4–12–8, 4 input neurons, 12 hidden neurons, and 8 output neurons). If the number of neurons are not exactly equal to the multiple of 4, it is always possible to artifically pad a suitable number of pseudo (no-operation) neural units (i.e., the weights connected to the units are set to be zero) to match the size.

1) Design for the Retrieving Phase: The consecutive MVM operations can be partitioned based on the *locally parallel globally sequential* (LPGS) scheme [38]:

1) the tasks of the 12 neurons in the first (hidden) layer is uniformly distributed to the 4 PE's, and the associated weights are appropriately arranged as before. Instead of cycling once in the ring array, the four inputs $\{a_i(0)\}$ are cycling three times to generate all the 12 activation values $\{a_i(1)\}$, i.e., $\{a_i(1), i = 1, \cdots, 4\}$, then $\{a_i(1), i = 5, \cdots, 8\}$, and finally $\{a_i(1), i = 9, \cdots, 12\}$ (see Fig. 10(a)).

2) The resulting 12 $\{a_i(1)\}$ will cycle in the ring (4-by-4 sequentially) six times to generate the eight activation values $\{a_i(2)\}$ of the output layer. The first three times to generate the first four activation values $\{a_i(2), i = 1, \cdots, 4\}$, and the second three times to generate the rest of activation values $\{a_i(2), i = 5, \cdots, 8\}$ (see Fig. 10(b)).

2) Design for the Learning Phase:

Systolic Processing of the OPU: The operations of the OPU for the nonuniform multilayer perceptron can again be implemented in the ring systolic ANN. Take for example, the same 4–12–8 two layer perceptron as shown in Fig. 9. Assume that the eight $\{g_i(2) = \delta_i(2)f_i'(2)\}$ and the twelve $\{h_j(2) = \eta a_j(1)\}$ are available at the corresponding PE, so are the twelve $\{g_i(1) = \delta_i(1)f_i'(1)\}$ and the four $\{h_j(1) = \eta a_j(0)\}$. The operations can be briefly described as follows (see Fig. 11):

1) The twelve $\{h_j(2)\}$ will cycle in the ring array in three sequential batches, $\{h_j(2), j = 1, \cdots, 4\}$, $\{h_j(2), j = 5, \cdots, 8\}$, and $\{h_j(2), j = 9, \cdots, 12\}$, to first update all the weights associated with the first four neurons in the output (second) layer $\{w_{ij}(2), i = 1, \cdots, 4, j = 1, \cdots, 12\}$. Then these three sequential batches are again cycled in the ring to update the weights associated with the last four neurons in the output layer $\{w_{ij}(2), i = 5, \cdots, 8, j = 1, \cdots, 12\}$ (see Fig. 11(a)).

2) The same four $\{h_j(1)\}$ will cycle in the ring array three times to update all the weights associated with the twelve neurons in the hidden (first) layer. Each cycling of $\{h_j(1)\}$ will update the weights associated to four neurons in the hidden layer, i.e., $\{w_{ij}(1), i = 1, \cdots, 4, j = 1, \cdots, 4\}$, $\{w_{ij}(1), i = 5, \cdots, 8, j = 1, \cdots, 4\}$, and $\{w_{ij}(1), i = 9, \cdots, 12, j = 1, \cdots, 4\}$ (see Fig. 11(b)).

Systolc Processing of the Consecutive VMM: At the same time when the operations of the OPU are performed, the consecutive VMM can be executed in parallel to compute the $\{\delta_i(1)\}$. The operations can be briefly described as follows

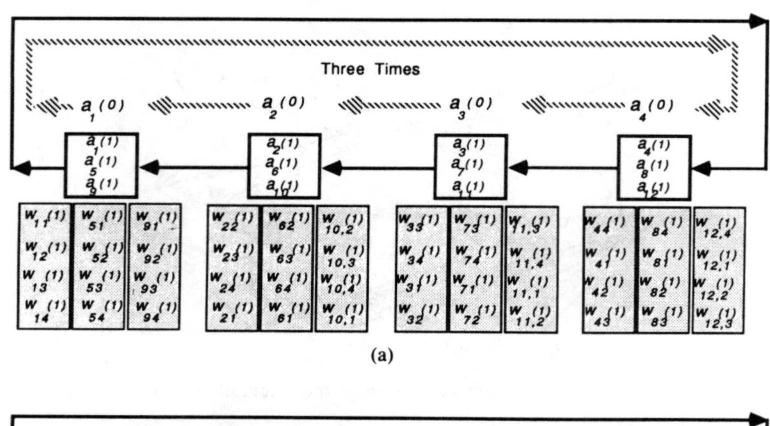

(a)

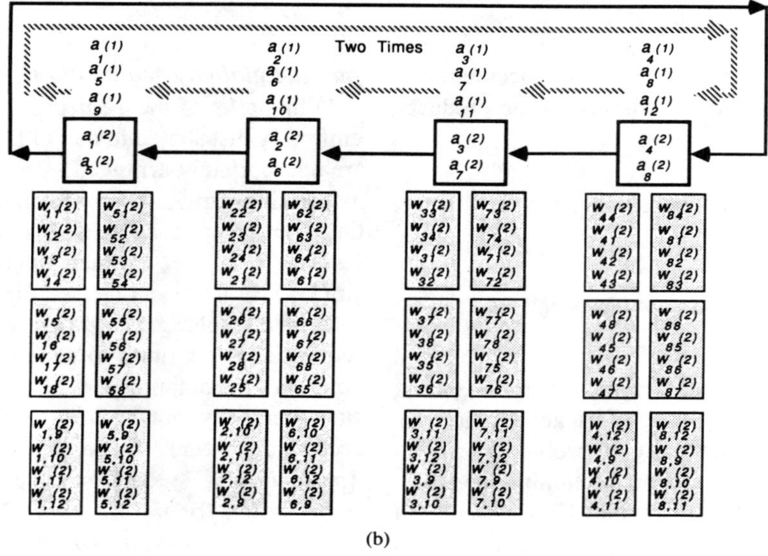

(b)

Fig. 10. The ring architecture with 4 PE's can be used to implement the retrieving phase of this two-layer perceptron. (a) The forward retrieving operations in the first (hidden) layer. (b) The forward retrieving operations in the second (output) layer.

(see Fig. 12):

1) Along with the cycling of each $h_j(2)$ in the ring array for the OPU operation, there are also an accumulator $\delta_j(1)$ cycling in the ring array in three sequential batches, $\{\delta_j(1), j = 1, \cdots, 4\}$, $\{\delta_j(1), j = 5, \cdots, 8\}$, and $\{\delta_j(1), j = 9, \cdots, 12\}$, to accumulate the first half the back-propagated corrective signals, i.e., at the same time as the updating of all the weights associated with the first four neurons in the output layer is done (see Fig. 12(a)), and

$$\delta_i(1) = \sum_{j=1}^{4} g_j(2) w_{ji}(2).$$

2) When the second three-time cycling are performed to update the weights associated with the last four neurons in the output layer, the $\{\delta_j(1)\}$ are again accumulated to form (see Fig. 12(b))

$$\delta_i(1) = \sum_{j=1}^{8} g_j(2) w_{ji}(2).$$

D. Implementation Considerations of Neural Processing Units

Time-Efficient Design Based on a Parallel Array Multiplier: As discussed in Sections IV-A and IV-B, most of the computations involved in the neural processing units are MAC or multiplication (e.g., the calculations of $g_i(l + 1)$, and $h_j(l + 1)$, and $r_{ji}(l + 1)$) operations, so for a *time-efficient* dedicated design, a parallel array multiplier (e.g., Baugh–Wooley multiplier [2], [10]) should be favorably considered.

There is concern about using digital hardware for the nonlinear sigmoid function which is required in many continuous-valued ANN models. Justifed by the simulations, the nonlinear sigmoid function can be well approximated by a piecewise-linear function with 8–16 segments [47]. This again calls for the MAC processor, which greatly simplifies the hardware complexity.

Area-Efficient Design Based on Cordic Processor: For an area-efficient dedicated digital VLSI implementation of the neural processing units, a Cordic processer, which uses about 1/3 silicon area of an array multiplier might be a good

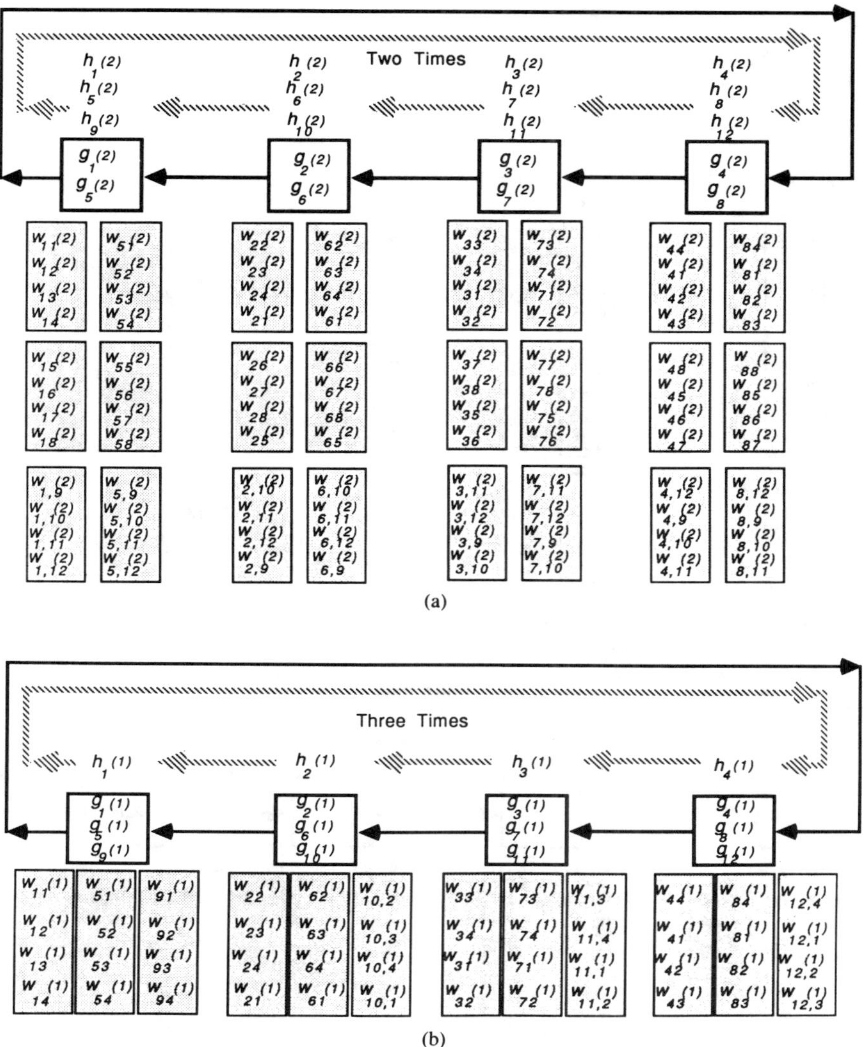

Fig. 11. The ring architecture for the implementation of OPU in a two-layer perceptron with configuration (4–12–8). (a) The OPU operations is the second (output) layer. (b) The OPU operations in the first (hidden) layer.

alternative [2]. A Cordic scheme is an iterative method based on bit-level shift-and-add operations. It is especially suitable for computing a class of rotations [38]. A 2-D vector $v = [x, y]$ can be rotated be an angle α by using a rotation operator $R_\alpha v' = R_\alpha v$.

In a linear mode, the Cordic can be used for MAC (multiple and accumulation) operations, although somewhat slower than the array multiplier. Given x, y, and α, we can get the Cordic output $y + x\alpha$ in the linear mode which is needed for the MAC operations. The key advantage of Cordic is that it may implement the sigmoid activation function by setting the Cordic in the hyperbolic mode, where the inputs are set as $v = [1, 1]$ and $\alpha = u/2$. For a more detailed design, the readers should refer to [2], [28].

V. Conclusion

Due to its robustness and adaptiveness, neural network processing can be very useful to all levels of robotic applications. For real-time processing performance, the neural network architectures for robotic processing will require massive parallel processing. Fortunately, today's VSLI and CAD technologies facilitate practical and cost-effective implementation of large-scale computing networks. In order to fully exploit VLSI's strength, a programmable systolic neural network architecture is proposed in this paper. Thanks to the versatility of the systolic design, most neural network models for robotic applications can be efficiently implemented. One critical issue which requires a closer investigation is the convergence property for both the retrieving and learning phases of the neural networks. For example, the number of hidden units in a multilayer perceptron must be sufficient to provide the discriminating capability required by the given application, but an excessively large number of synaptic weights may lead to costly and unreliable training. Therefore, it s very desirable to have an *a prior* estimate of an optimal number of hidden neurons. For this and to better understand the convergence property, it is essential to develop a theoretical footing based on system theory and numerical analysis

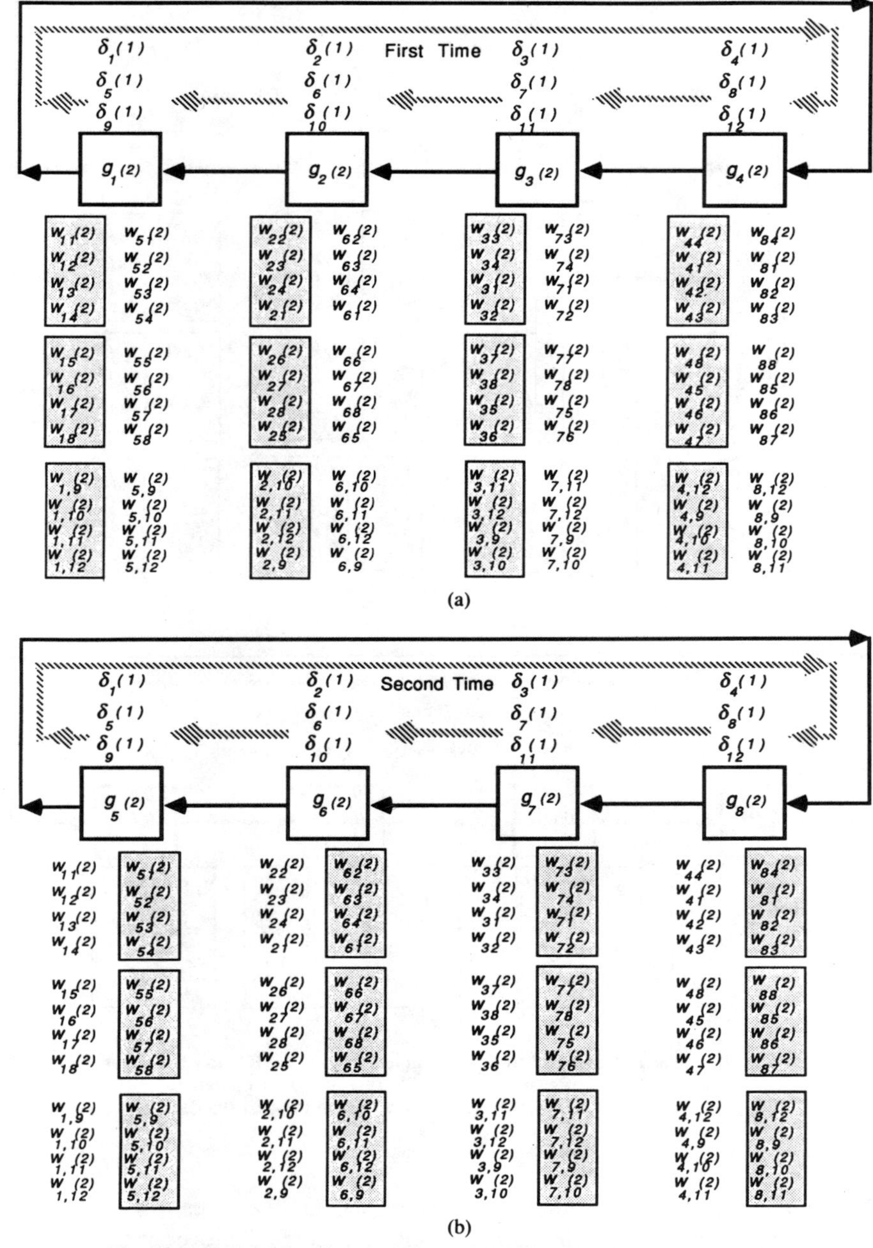

Fig. 12. The ring architecture for the implementation of the consecutive VMM operations of a two-layer perceptron. (a) The accumulation of the first half the back-propagated corrective signals. (b) The accumulation of the second half the back-propagated corrective signals.

[39]. It is also important to compare the neural networks approach with the other conventional methods [44], e.g., simulated annealing [33], hidden Markov model [29], [60], pattern recognition [54], [57], and nonlinear programming [13].

REFERENCES

[1] D. H. Ackley, G. E. Hinton, and T. J. Sejnowski, "A learning algorithm for Boltzmann machines," *Cognitive Sci.*, vol. 9. pp. 147–169, 1985.

[2] H. M. Ahmed, "Alternative arithmetic unit architectures for VLSI digital signal processors," in *VLSI and Modern Signal Processing*. Englewood Cliffs, NJ: Prentice Hall, 1985, ch. 16, pp. 277–306.

[3] J. Albus, "A new approach to manipulator control: The cerebellar model articulation controller (CMAC)," *Trans. ASME, J. Dynamic Syst., Meas., Contr.*, vol. 97, pp. 220–227, Sept. 1975.

[4] J. Alspector and R. B. Allen, "A neuromorphic VLSI learning systems," in P. Losleben, Ed., *Advanced Research on VLSI.* Cambridge, MA: MIT Press, 1987, pp. 313–349.

[5] S.I. Amari, "Characteristics of randomly connected threshold-element network systems," *Proc. IEEE*, vol. 59, pp. 35–47, Jan. 1971.

[6] J. A. Anderson, *Neurocomputing—Paper Collections.* Cambridge, MA: MIT Press, 1988.

[7] ——, "A simple neural network generating an interactive memory," *Math. Biosci.*, vol. 14, pp. 197–220, 1972.

[8] L. E. Atlas, T. Homma, and R. J. Marks II, "A neural network model for vowel classification," in *Proc. IEEE, ICASSP'87* (Dallas, TX, 1987).

[9] D. H. Ballard and C. M. Brown, *Computer Vision.* Englewood Cliffs, NJ: Prentice Hall, 1982.

[10] C. R. Baugh and B. A. Wolley, "A two's complement parallel array multiplication algorithm," *IEEE Trans. Comput.*, vol. C-22, pp. 1045–1047, Dec. 1973.

[11] D. J. Burr, "Experiments with a connectionist text reader," in *Proc. IEEE, 1st Int. Conf. on Neural Networks* (San Diego, CA, 1987), pp. IV 717–IV 724.

[12] G. A. Carpenter and S. Grossberg, "ART2: Self-organization of stable category recognition codes for analog input patterns," in *Proc. IEEE ICNN'87* (San Diego, CA, 1987), pp. II 727–II 736.

[13] L. O. Chua and G. N. Lin, "Nonlinear programming without computation," *IEEE Trans. Circuits Syst.*, vol. CAS-31, pp. 182–188, Feb. 1984.

[14] M. A. Cohen and S. Grossberg, "Absolute stability of global pattern formation and parallel memory storage by competitive neural networks," *IEEE Trans. Syst. Man, Cybern.*, vol. SMC-13, no. 5, pp. 815–825, Sept. 1983.

[15] J. D. Cowan and D. H. Sharp, "Neural nets," Tech. Rep. Math. Dept., Univ. of Chicago, 1987.

[16] R. Eckmiller and C. V. D. Malsburg, *Neural Computers* (NATO ASI Series F, Computer and System Science. New york, NY: Springer-Verlag, 1987.

[17] N. H. Farhat, D. Psaltis, A. Prata, and E. Peak, "Optical implementation of the Hopfield model," *Appl. Opt.*, vol. 24, pp. 1469–1475, May 1985.

[18] K. Fukushima, "Cognitron: A self-organizing multilayered neural network," *Biol. Cybern.*, vol. 20, pp. 121–136, 1975.

[19] K. Fukushima, "Neocognitron: A self-organizing neural network model for a mechanism of pattern recognition unaffected by shift in position," *Biol. Cybern.*, vol. 36, pp. 193–202, Apr. 1980.

[20] R. P. Gorman and T. J. Sejnowski, "Learned classification of sonar targets using a massively parallel network," *IEEE Trans. Acoust., Speech, Signal Process.*, vol. 36, pp. 1135–1140, July 1988.

[21] S. Grossberg, "Adaptive pattern classification and universal recoding: Part 1. Parallel development and coding of neural feature detectors," *Biol. Cybern.*, vol. 23, pp. 121–134, 1976.

[22] D. O. Hebb, *The Organization of Behavior.* New York, NY: Wiley, 1949.

[23] G. E. Hinton, "Connectionist learning procedure," Tech. Rep. CMU-CS-87-115, Carnegie Mellon Univ., Pittsburgh, PA: Sept. 1987.

[24] G. E. Hinton and T. J. Sejnowski, "Learning and relearning in Boltzmann machine," in *Parallel Distributed Processing (PDP): Exploration in the Microstructure of Cognition*, vol. 1, Cambridge, MA: MIT Press, 1986, ch. 7, pp. 282–317.

[25] J. J. Hopfield, "Neural network and physical systems with emergent collective computational abilities," *Proc. Nat. Acad. Sci. USA*, vol. 79, pp. 2554–2558, 1982.

[26] ——, "Neurons with graded response have collective computational properties like those of two-state neurons," *Proc. Nat. Acad. Sci. USA*, vol. 81, pp. 3088–3092, 1984.

[27] J. J. Hopfield and D. W. Tank, "Neural computation of decision in optimization problems," *Biol. Cybern.*, vol. 52, pp. 141–152, 1985.

[28] J. N. Hwang, "Algorithms/applications/architecture of artificial neural nets," Ph.D. dissertation, Dept. of Elect. Eng., Univ. of Southern California, Dec. 1988.

[29] J. N. Hwang, J. A. Vlontzos, and S. Y. Kung, "A neural network architecture for hidden Markov models," to appear in *IEEE Trans. Acoust., Speech, Signal Process.*, Dec. 1989.

[30] C. Jorgensen, W. Hamel, and C. Weisbin, "Autonomous robot navigation," *Byte*, pp. 223–235, Jan. 1986.

[31] C. C. Jorgensen, "Neural network representation of sensor graphs for autonomous robot navigation," in *Proc. IEEE, 1st Int. Conf. on Neural Networks* (San Diego, CA, June 1987), pp. IV 507–IV 515.

[32] M. Kawato, Y. Uno, and M. Isobe, "A hierarchical model for voluntary movement and its application to robotics," in *Proc. IEEE, 1st Int Conf. on Neural Networks* (San Diego, CA, June 1987), pp. IV573–IV582.

[33] S. Kirkpatrick, C. D. Gelatt, Jr., and M. P. Vecci, "Optimization by simulated annealing," *Science*, vol. 220, pp. 671–680, May 1983.

[34] C. Koch, J. Marroquin, and A. Yuille, "Analog neuronal networks in early vision," *Proc. Nat. Acad. Sci.*, vol. 83, pp. 4263–4267, 1986.

[35] T. Kohonen, "Correlation matrix memories," *IEEE Trans. Comput.*, vol. C-21, 1972.

[36] T. Kohonen, *Self-Organization and Associative Memory, Series in Information Science*, vol. 8. New York, NY: Springer-Verlag, 1984.

[37] ——, "Self-organized formation of topologically correct feature map," *Biol. Cybern.*, vol. 43, pp. 59–69, 1982.

[38] S. Y. Kung, *VLSI Array Processors.* Englewood Cliffs, NJ: Prentice-Hall, 1988.

[39] S. Y. Kung and J. N. Hwang, "An algebraic projection analysis for optimal hidden units size and learning rate in back-propagation learning," in *IEEE, Int. Conf. on Neural Networks, ICNN'88* (San Diego, CA, July 1988), vol. 1, pp. 363–370.

[40] ——, "Parallel architectures for artificial neural nets," in *IEEE Int. Conf. on Neural Networks, ICNN'88* (San Diego, CA, July 1988), vol. 2, pp. 165–172.

[41] ——, "A unified systolic architecture for artificial neural networks," *J. Parallel Distributed Comput.*, (Special Issue on Neural Networks), vol. 6, pp. 358–387, Apr. 1989.

[42] M. Kuperstein, "Adaptive visual-motor coordination in multijoint robots using a parallel architecture," in *IEEE Int. Conf. on Automatic Robotics*, pp. 1595–1602, Mar. 1987.

[43] A. Lapedes and R. Farber, "Nonlinear signal processing using neural networks," in *Proc. IEEE, Conf. on Neural Information Processing Systems—Natural and Synthetic* (Denver, CO, Nov. 1987).

[44] R. P. Lippmann, "An introduction to computing with neural nets," *IEEE ASSP Mag.*, vol. 4, pp. 4–22, 1987.

[45] D. G. Luenberger, *Linear and Nonlinear Programming.* Reading, MA: Addison-Wesley, 1984.

[46] J. Lupo, *DARPA: Neural Network Study.* Burke, VA: AFCEA Int. Press, 1988.

[47] W. D. Mao and S. Y. Kung, "Implementation and performance issues of back-propagation neural nets," Tech. Rep. Elec. Eng. Dept., Princeton Univ., Jan. 1989.

[48] D. Marr, "A theory of cerebellar cortex," *J. Physiol. Lond.*, p. 202, 1969.

[49] D. Marr and T. Poggio, "Cooperative computation of stereo disparity," *Science*, vol. 194, pp. 283–287, 1976.

[50] J. L. McClelland and D. E. Rumelhart, "Distributed memory and the representation of general and specific information," *J. Exper. Psychol.: General*, vol. 114, pp. 158–188, 1985.

[51] W. S. McCulloch and W. Pitts, "A logical calculus of the ideas immanent in the nervous activity," *Bull. Math. Biophys.*, vol. 5, no. 115, 1943.

[52] W. T. Miller, "Sensor-based control of robotic manipulator using a general learning algorithm," *IEEE J. Robotics Automat.*, vol. RA-3, pp. 157–165, Apr. 1987.

[53] M. Minsky and S. Papert, *Perceptrons: An Introduction to Computational Geometry.* Cambridge, MA: MIT Press, 1969.

[54] N. J. Nilsson, *Learning Machines.* New York, NY: McGraw-Hill, 1965.

[55] D. Parker, "Learning logic," Tech. Rep. TR-47, Center for Computational Research in Economics and Management Science, MIT, Cambridge, 1985.

[56] B. K. Parsiand B. K. Parsi, "Evaluation of network architectures on test learning tasks," in *Proc. IEEE, 1st Int. Conf. on Neural Networks* (San Diego, CA, June 1987), pp. III785–III790.

[57] T. Pavlidis, *Structural Pattern Recognition.* New York, NY: Springer-Verlag, 1977.

[58] D. Psaltis, A. Sideris, and A. Yamamura, "A hierarchical model for voluntary movement and its application to robotics," in *Proc. IEEE, 1st Int. Conf. on Neural Networks* (San Diego, CA, June 1987), pp. IV551–IV558.

[59] D. Psaltis, K. Wagner, and D. Brady, "Learning in optical neural computers," in *Proc. IEEE, 1st Int. Conf. on Neural Networks* (San Diego, CA, June 1987).

[60] L. R. Rabiner and B H. Juang, "An introduction to hidden Markov models," *IEEE ASSP Mag.*, vol. 3, no. 1, pp. 4–16, Jan. 1986.

[61] R. Rastogi, P. K. Gupta, and R. Kumaresen, "Array signal processing with interconnected neuron-like elements," in *Proc. IEEE ICASSP'87* (Dallas, TX), pp. 54.8.1–54.8.4, 1987.

[62] G. N. Reeke and G. M. Edelman, "Selective neural networks and their implications for recognition automata," *Int. J. Supercomput. Appl.*, pp. 44–69, 1987.

[63] F. Rosenblatt, "The perceptron: A probablistic model for information storage and organization in the brain," *Psych. Rev.*, vol. 65, 1958.

[64] D. E. Rumelhart, G. E. Hinton, and R. J. Williams, "Learning internal representations by error propagation," in *Parallel Distributed Processing (PDP): Exploration in the Microstructure of Cognition*, vol. 1. Cambridge, MA: MIT Press, 1986, ch. 8, pp. 318–362.

[65] D. E. Rumelhart, J. L. McClelland, and the PDP Research Group, *Parallel Distributed Processing (PDP): Exploration in the Microstructure of Cognition*, vol. 1. Cambridge, MA: MIT Press, 1986.

[66] D. E. Rumelhart and D. Zipser, "Feature discovery by competitive

learning, *Cogn. Sci.*, vol. 9, pp. 75–112, 1985.

[67] T. J. Sejnowski, "Parallel networks that learn to pronounce English text," *Complex Syst.*, vol. 1, pp. 145–168. 1987.

[68] K. G. Shin and S. B. Malin, "A structured framework for the control of industrial manipulator," *IEEE Trans. Syst. Man, Cybern.*, vol. SMC-15, pp. 78–94, Jan./Feb. 1985.

[69] K. Steinbush, "The learning matrix, *Kybernetik* (*Biol. Cybern.*), pp. 36–45, 1961.

[70] C. V. Stewart and C. R. Dyer, "Neural network representation of sensor graphs for autonomous robot navigation," in *Proc. IEEE, 1st Int. Conf. on Neural Networks* (San Diego, CA, June 1987), pp. IV215–IV224.

[71] G. Z. Sun, H. H. Chen, and Y. C. Lee, "Neural network representation of sensor graphs for autonomous robot navigation," in *Proc. IEEE, 1st Int. Conf. on Neural Networks* (San Diego, CA, June 1987), pp. IV345–IV356.

[72] M. Takeda and J. W. Goodman, "Neural networks for computation: Number representations and programming complexity," *Appl. Opt.*, vol. 25, pp. 3033–3046, Sept. 1986.

[73] D. W. Tank and J. J. Hopfield, "Simple "neural" optimization networks: An A/D converter, signal decision circuit, and a linear programming circuit," *IEEE Trans. Circuits Syst.*, vol. CAS-33, pp. 533–541, 1986.

[74] C. E. Thorp, "Path relaxation: Path planning for a mobile robot," in *Proc. of the AAAI* (Austin, TX, Aug. 1984), pp. 318–321.

[75] K. Tsutsumi and H. Matsumoto, "Neural network representation of sensor graphs for autonomous robot navigation," in *Proc. IEEE 1st Conf. on Neural Networks* (San Diego, CA, June 1987), pp. IV525–IV534.

[76] P. J. Werbos, "Beyond regrssion: New tools for prediction and analysis in the behavior science," Ph.D. dissertation, Harvard University, Cambridge, MA, 1974.

[77] B. Widrow and R. Winter, "Neural nets for adaptive filtering and adaptive pattern recognition," *IEEE Comput. Mag.*, vol. 21, pp. 25–39, Mar. 1988.

[78] G. Widrow and M. E. Hoff, "Adaptive switching circuit," in *IRE Western Electronic Show and Convention: Convention Record*, pp. 96–104, 1960.

[79] D. J. Willshaw, "Models of distributed associative memory models," Ph.D. dissertation, University of Edinburgh, Scotland, 1971.

[80] Y. T. Zhou and R. Chellappa, "Stereo matching using a neural network," in *Proc. IEEE ICASSP'88* (New York, NY, Apr. 1988), pp. 940–943.

A Fuzzy Inference Engine in Nonlinear Analog Mode and Its Application to a Fuzzy Logic Control

Takeshi Yamakawa
Invited Paper

***Abstract*— This paper is a tutorial to enlighten outsiders or beginners on a utility of a fuzzy system by providing broad scope overview, especially analog mode hardware, with the author's original work. At first, the difference between deterministic words and fuzzy words is explained as well as fuzzy logic. The description of the system by using mathematical equations, linguistic rules and parameter distribution (neural networks) are discussed. The algorithm of fuzzy inference and defuzzification is presented and their hardware implementation is discussed in detail together with an advanced one. The fuzzy logic controller was applied to stabilize a glass with wine and a mouse moving around the plate on the tip of the inverted pendulum.**

I. Introduction

HUMAN beings make tools for their use and also think to control the tools as they desire. A feedback concept is a very important concept to achieve the control of the tools. The first significant work in feedback application was James Watt's flyball governor developed in 1769 for controlling the speed of a steam engine. As modern plants with many inputs and outputs become more and more complex, the description of a modern control system requires a large number of equations. Classical control theory, which deals only with single-input-single-output systems, becomes entirely powerless for multiple-input-multiple-output systems. Since about 1960, modern control theory has been developed to cope with the increased complexity of modern plants and the stringent requirements on accuracy, weight, and cost in military, space, and industrial applications [1]–[3]. This development has been accelerated by a digital computer because it facilitates a solution of simultaneous equations of many unknowns.

Control technologies described above are all based on mathematical equations such as differential equations and relational equations. We, however, occasionally face chemical plants, machines and some other systems to be controlled, the characteristics of which are very difficult to describe with mathematical equations because of their complexity. Even in these cases, human experts may achieve the control by know-hows or control rules which are squeezed out from their long experience and represented by intuitive natural language. For instance, "If the pressure in the chamber *goes high,* then the fuel supply should be *reduced significantly* and the valve should be *a little bit opened.*" Knowledges (know-hows) represented with the intuitive natural language is easily explained and understood by common sense and thus easy to remember. Further more, the number of knowledges is drastically reduced by employing intuitive natural language.

Manuscript received September 10, 1992.
The author is with the Department of Control Engineering and Science, Kyushu Institute of Technology, Iizuka, Fukuoka, 820 Japan.
IEEE Log Number 9207196.

In most cases, the intuitive natural language has its ambiguous boundary of meaning, while numerical values and deterministic linguistic terms used in artificial intelligence are well-defined. This type of natural language is referred to as a *fuzzy linguistic term* and it is characterized by a so called *membership function.* This concept was presented as *fuzzy sets* (strictly speaking, it is fuzzy subsets) by L. A. Zadeh in 1965 [4].

A software or hardware system, which gives a conclusion (output) from a fact (input) and knowledges (control rules), is called an inference engine. If the knowledges include fuzzy linguistic terms, it is referred to as a *fuzzy inference engine* and can be utilized to a *fuzzy logic control.*

In this paper, a fuzzy inference (approximate reasoning) based on fuzzy sets and fuzzy logic and a defuzzification are briefly explained in comparison with other technologies and their hardware implementation in "nonlinear" analog circuits is described. The fuzzy inference engine and defuzzifier are employed to control a liquid-contained inverted pendulum, the dynamics of which is very difficult, if not impossible, to describe with mathematical equations.

II. Fuzzy Sets and Fuzzy Logic

A. Fuzzy Sets [4]–[28]

Linguistic terms and numerical values are classified to three categories in accordance with their meanings which are defined by the characteristic functions.

Deterministic words, e.g., "male" and "female," "dead," and "alive," the personal name "John," have truth values of 0 or 1 corresponding to NO or YES, respectively. In other words, if one is asked, "Are you male?," "Are you alive?," "Is your name John?," then answer can be made with "YES" or "NO." Exact numerical value, e.g., "exactly 26°C," "concentration of 0.1 Mol/1," "38g" etc., are in the same situation. These deterministic words and numerical values have neither flexibilities nor intervals. Those meanings are characterized by a characteristic function as shown in Fig. 1(a). The word "exactly 26°C" means only one single point

Reprinted from *IEEE Trans. on Neural Networks,* vol. 4, no. 3, pp. 496–522, May 1993.

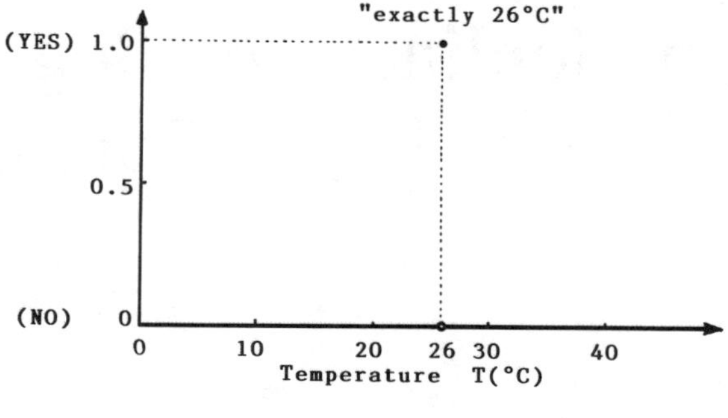

(a)

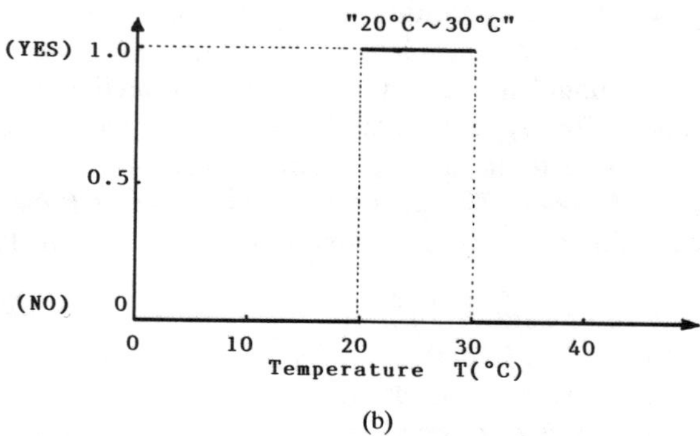

(b)

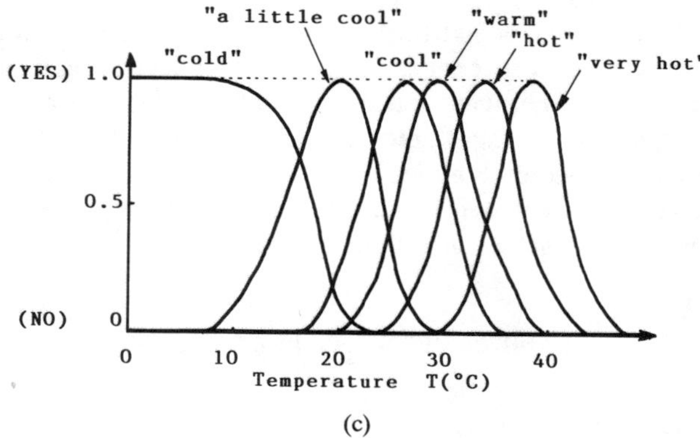

(c)

Fig. 1. (a) A singleton, (b) a crisp interval (crisp sets), and (c) fuzzy sets.

of temperature at 26°C, so that this type of term or value is called a *singleton.*

Even in the scientific analysis, where exact values are preferred, numerical intervals are sometimes used to represent flexible values. For instance, "The comfortable room temperature for human beings is 20°C–30°C." Temperatures 19.9°C, and 30.0001°C are not the case. Thus the characteristic function representing a numerical interval 20°C–30°C" is shown in Fig. 1(b). Truth values for any temperatures are given by YES or NO in this case as well as in a singleton. This interval can be regarded as a set of numerous singletons. Therefore this type of deterministic interval is called *crisp sets.* Crisp sets are also adopted to represent linguistic terms in knowledges in artificial intelligence (AI).

In our daily life, fuzzy natural languages are used for easy and efficient communication. Although these terms are usually intuitive and includes some kinds of uncertainties, they are very easy to select for practical use. For example, "Be careful when you carry *important* documents." "*High* accuracy is a measure of the technology." "Cool it *a little bit more* and a solid will be deposited." The meaning of the fuzzy linguistic term is defined by a characteristic function as shown in Fig. 1(c). This function is specifically called a *membership function,* because it indicates a grade of membership of each element (physical value in the horizontal axis) in a fuzzy linguistic term of interest. For instance, a fuzzy linguistic terms "cold," "a little cool," "cool," "warm," "hot," and "very hot" are indicated in Fig. 1(c). According to a common sense, a grade of membership of 20°C in "a little cool" is undoubtedly 1.0. In other words, you can answer "YES" to the question, "Is a temperature of 20°C included in 'a little cool'?" On the other hand, how about 0°C, 5°C, 40°C, etc.? You may answer "NO" to this question, i.e., grades of membership of 0°C, 5°C, 40°C in "a little cool" are 0. How about 14°C and 23°C? You can not give the answer "NO" nor "YES" to the same question. The grades will be answered to be 0.4 and 0.8 corresponding to 14°C and 23°C, respectively. Of course, strictly speaking, these grades are given by intuition or common sense, so that the shape of membership function changes a little from person to person. In any way, a fuzzy linguistic term can be defined by a membership function, and the membership function exhibits a continuous curve changing from 0 to 1 or vice versa. And this transition region represents a fuzzy boundary of the term. If the fuzzy linguistic term includes a numerical value, e.g., "around 20°C," "much higher than 30%," it is called a *fuzzy number.*

These fuzzy linguistic terms can be regarded as sets of singletons, the grades of which are not only 1 but also ranging from 0 to 1. Therefore, these fuzzy linguistic terms are called *fuzzy sets.* Each singleton is an *element* of the fuzzy sets. A fuzzy linguistic term, elements of which are ordered in the universe of discourse, is called a *fuzzy interval.* An interval on horizontal axis where grades of membership are not zero is called *support.*

Fuzzy sets are defined by *labels* (e.g., high pressure, around 20°C, a little, usually) and *membership functions.* While a modern AI achieves only a symbolic processing with labels, a fuzzy information processing of a future artificial intelligence achieves both of a *symbolic processing* with labels and a *meaning processing* with membership functions.

A membership function can be represented by a piecewise linear function, because important is the continuity between 0 and 1 rather than the curvature of the membership function. Parameters characterizing the shape and the label of the membership function can be also obtained by learning, if needed.

B. Fuzzy Logic [5]–[28]

Fuzzy logic functions which are most popular in engineering field are NOT (Fuzzy Logic Complement), MIN (Minimum; Fuzzy Logic Product, Fuzzy Logic Intersection) and MAX (Maximum; Fuzzy Logic Sum, Fuzzy Logic Union).

Fuzzy Logic Complement of a fuzzy linguistic term "A" is defined by $1 - \mu_A$ and means negation of "A," where μ_A is a membership function of "A." Negation of "hot" is "not hot," and it is easily understood, by Fig. 1(c), that "not hot" does not necessarily mean "cold."

The most popular fuzzy logic functions which implement logical "AND" and logical "OR" are MIN and MAX, respectively, and defined by

$$\begin{aligned} \mathrm{Min}\{\mu_A, \mu_B, \mu_C, \cdots, \mu_Z\} &= \mu_A \wedge \mu_B \wedge \mu_C \wedge \cdots \wedge \mu_Z \\ &= \mu_K \quad (\mu_K \leq \mu_A, \mu_B, \mu_C, \cdots, \mu_Z) \end{aligned} \tag{1}$$

$$\begin{aligned} \mathrm{Max}\{\mu_A, \mu_B, \mu_C, \cdots, \mu_Z\} &= \mu_A \vee \mu_B \vee \mu_C \vee \cdots \vee \mu_Z \\ &= \mu_L \quad (\mu_L \geq \mu_A, \mu_B, \mu_C, \cdots, \mu_Z) \end{aligned} \tag{2}$$

where $\mu_A, \mu_B, \mu_C, \cdots, \mu_Z$ are grades of membership function ranging from 0 to 1.

III. Modeling of the System

In order to control the system of interest, we have to describe the precise behavior of the system to design the controller. In order to use the knowledge, we have to describe the knowledge effectively. The author classifies the description of the system or knowledge to three categories: mathematical equations, linguistic rules and artificial neural networks.

A. Mathematical Equations

The traditional sciences have been based on mathematical equations. Natural scientific phenomena and physical behavior of the artificial system can be modeled with relational equations or differential equations. These equations describe the dynamics or kinetics of the systems or the knowledge about the system in a very simple form. If the relation between the input x and the output $f(x)$ of the system or the relation between the cause x and the result $f(x)$ is obtained as shown in Fig. 2(a) from experiments, $f(x)$ is described as

$$f(x) = \tfrac{1}{30}(x-3)^2. \tag{3}$$

By substitution of the numerical value of x (input value, the fact or the premise) to this equation, the numerical value of $f(x)$ (the output value or the conclusion) is obtained.

A description of the system with this type of equation exhibits significant simplicity. It is, however, very difficult to identify an exact equation from the given relation, especially in the case of many variables. Furthermore, it is also very difficult to reassign this equation, when the relation between x and $f(x)$ is changed. Therefore, this description is not so suitable for complex systems such as nonlinear systems or time-varient systems. As the complexity of the system increases, the possibility to describe the system with mathematical equations diminishes.

B. Linguistic Rules

A relation between x and $f(x)$ can be described with a set of linguistic rules, typical form of which is

$$\text{Rule i:} \quad \text{If } x \text{ is } A_i, \text{ then } f(x) \text{ is } B_i, \qquad (i = 1, 2, \cdots, N) \tag{4}$$

where x and $f(x)$ are independent and dependent variables, respectively, A_i and B_i linguistic constants, and N the number of experimental data. These rules are referred to as *IF-THEN rules* because of their form. An if-clause is referred to as an *antecedent* and a then-clause as a *consequent.*

Linguistic rules can be classified to two categories with respect to linguistic constants A_i and B_i.

1) Linguistic Rules with Well-defined Languages (Crisp IF-THEN Rules): In this category, all the linguistic constants are represented by well-defined languages (crisp information or exact numerical values), e.g., male, 80°C, 60°C–80°C, 98 g, etc., as shown in Fig. 1(a) and (b). Modern AI belongs to this category. Here we have an example shown in Fig. 2(b), which gives the following crisp rules.

$$\begin{array}{ll} \text{Rule 1} & \text{If } x \text{ is } -2, \text{ then } f(x) \text{ is } 25/30 \\ \text{Rule 2} & \text{If } x \text{ is } -1, \text{ then } f(x) \text{ is } 16/30 \\ \text{Rule 3} & \text{If } x \text{ is } 0, \text{ then } f(x) \text{ is } 9/30 \\ \text{Rule 4} & \text{If } x \text{ is } 1, \text{ then } f(x) \text{ is } 4/30 \\ \text{Rule 5} & \text{If } x \text{ is } 2, \text{ then } f(x) \text{ is } 1/30 \\ \text{Rule 6} & \text{If } x \text{ is } 3, \text{ then } f(x) \text{ is } 0 \\ \text{Rule 7} & \text{If } x \text{ is } 4, \text{ then } f(x) \text{ is } 1/30 \\ \text{Rule 8} & \text{If } x \text{ is } 5, \text{ then } f(x) \text{ is } 4/30 \\ \text{Rule 9} & \text{If } x \text{ is } 6, \text{ then } f(x) \text{ is } 9/30 \\ \text{Rule 10} & \text{If } x \text{ is } 7, \text{ then } f(x) \text{ is } 16/30 \\ \text{Rule 11} & \text{If } x \text{ is } 8, \text{ then } f(x) \text{ is } 25/30. \end{array} \tag{5}$$

The advantage of this description is that it is very easy to change the description of the system. For instance, when the system is locally changed as indicated by squares shown in Fig. 2(b), we have only to change the corresponding values of the consequents in the IF-THEN rules (5), (e.g., Rule 7: 1/30 to 1/20, Rule 8: 4/30 to 10/57, Rule 9: 9/30 to 2/5, Rule 10: 16/30 to 5/9) because all the rules are independent of each other. This means that IF-THEN rules are suitable for learning systems, self-organizing systems, adaptive systems, and so on.

On the other hand, there are some disadvantages. When the fact $x = +1$ is given, the conclusion can be obtained to be 4/30 from the data-matching between the fact and the antecedents (if-clause) of rules. This procedure to produce the conclusion is called *inference.* However, the fact $x = +1.5$ cannot give the conclusion from the rules, because there is no antecedent which is exactly matched to the fact $x = +1.5$. This shows that the inference with crisp rules is very weak against a defect of knowledge, a defective fact, a noisy defect or a variation of a fact, and that it needs a large-scale knowledge base for significant performance. Therefore, this type of inference wastes time because of a sequential data-matching between the fact and a large data base.

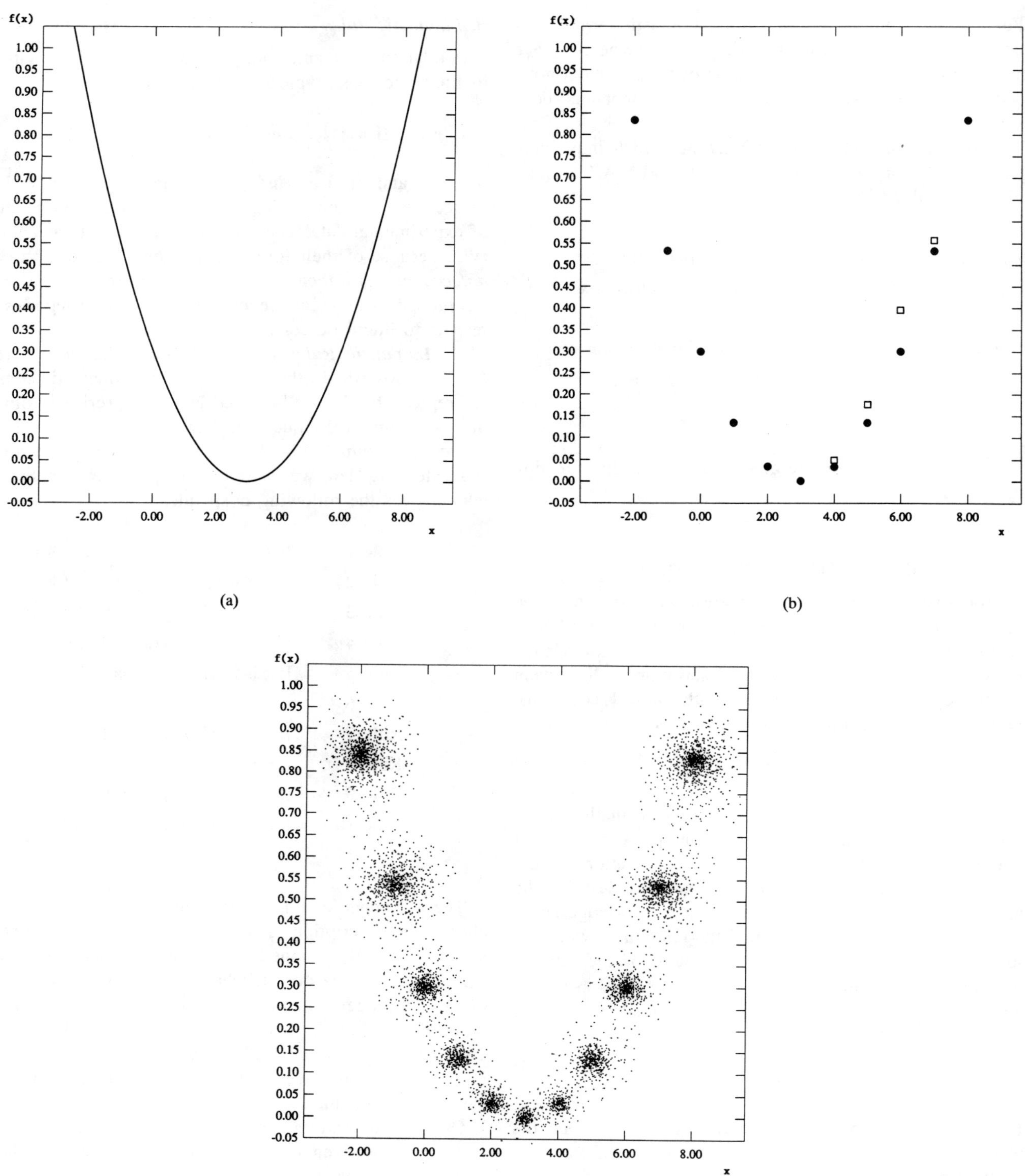

Fig. 2. (a) Example of the system description by a mathematical equation. (b) Example of the system description by a set of linguistic rules with well-defined languages (crisp rules). This description has the advantage of easy reassignment of the rules, when the system is changed as shown by squares. (c) Example of the system description by a set of linguistic rules with ill-defined languages (fuzzy IF-THEN rules, or shortly fuzzy rules). Fuzzy rules give a continuous relationship between x and $f(x)$ through fuzzy inference and defuzzification.

Furthermore, this inference system should not include contradictory rules in a knowledge base. Otherwise, it will produce two contradictory conclusions from one fact.

These disadvantages appear in the modern AI system as well. This type of inference is based only on *symbolic processing* but not on the *processing of the meaning* of linguistic terms.

2) Linguistic Rules with Ill-Defined Languages (Fuzzy IF-THEN Rules): In this category, all the linguistic constraints are represented by ill-defined languages (fuzzy information, uncertain words, approximate numerical values, etc.), e.g., beautiful lady, high temperature, heavy, low power, around 45 kg, etc. The linguistic rules (fuzzy IF-THEN rules or, shortly fuzzy rules) can describe the relation between x and $f(x)$, for example, as shown in Fig. 2(c), and are given in the following form.

$$\begin{array}{lll} \text{Rule 1} & \text{If } x \text{ is } around -2, & \text{then } f(x) \text{ is } around\, 25/30 \\ \text{Rule 2} & \text{If } x \text{ is } around -1, & \text{then } f(x) \text{ is } around\, 16/30 \\ \text{Rule 3} & \text{If } x \text{ is } around\ \ 0, & \text{then } f(x) \text{ is } around\ \ 9/30 \\ \vdots & \vdots & \vdots \\ \text{Rule 11} & \text{If } x \text{ is } around\ \ 8, & \text{then } f(x) \text{ is } around\, 25/30. \end{array} \tag{6}$$

A crisp relation between x and $f(x)$ in Fig. 2(b) is fuzzified to make it continuous as shown in Fig. 2(c). This continuous fuzzified relation (*fuzzy relation*) gives reasonable conclusions for any fact, e.g., $x = -1.5, +3.2, +4.3$, etc., through *fuzzy inference* and *defuzzification,* which are precisely described in Section IV-A. FUZZY INFERENCE (APPROXIMATE REASONING). In other words, fuzzy inference and defuzzification facilitate a reasonable interpolation with much less data. So that the fuzzy inference is called *interpolative inference* and can save the memory hardware. This fuzzy inference with fuzzy rules exhibits the similar function to that of a mathematical equation.

It is much easier to reassign the fuzzy rules rather than the mathematical equation, when the characteristics of the system are changed. In other words, only one or more rules should be added or revised independently of other rules, while the order and the numerous coefficients of the mathematical equation should be recalculated from simultaneous equations.

The distinctive features of this description are as follows.

1. It is suitable for describing a complicated system with a small amount of knowledge.
2. It is easy to select the words to be used in fuzzy rules among the categorized few words.
3. It is easy to remember the knowledge.
4. It is easy for designers to communicate with others by using fuzzy natural languages.

While ordinary inference (modern AI) is based on symbolic processing, fuzzy inference or approximate reasoning is based on both of symbolic processing and meaning processing. Linguistic rules (crisp rules and fuzzy rules) can represent the algorithm of inference explicitly, while an artificial neural network represents it implicitly. So that linguistic rules are called *structured* and an artificial neural network is *unstructured.*

C. Artificial Neural Networks

A system can be described by a distribution of parameters, the typical example of which is a well-known artificial neural network.

An architecture of a neural network is a simple iteration of simple aggregating elements. This aggregating element is a model of a physical neuron in the neural network in the living body. When w_{ij} $(i = 1, 2, \cdots, n)$ is a weight assigned to the signal input p_i to the jth cell and θ_j and q_j are a threshold level and a signal output of the jth cell, respectively, the signal output is characterized by

$$q_j = h\left(\sum_i w_{ij} \cdot p_i - \theta_j\right) \tag{7}$$

where h is a sigmoid function and it is typically described by

$$h(x) = \frac{1}{1 + \exp(-x)}. \tag{8}$$

By employing this neuron model, the relation between x and *f(x)* can be characterized by a neural network as shown in Fig. 3(a), where the parameter distribution such as threshold levels θ_j, weights w_{ij} are obtained, after 10 000 leanings, by the same amount of data (11 data) to IF-THEN rules (5) and (6). The input-output characteristics of this artificial neural net (Fig. 3(a)) is calculated and shown in Fig. 3(b), which shows the ability of interpolation similar to fuzzy inference. It is, however, very difficult to understand, by looking at the parameter distribution, how the neural net behaves. Furthermore, we can not estimate the number of neurons and layers necessary to achieve pattern recognitions or to solve the problems and there is no guarantee of convergence of learning. If the system under modeling changes its behavior, the similar learnings should be applied again to the neural net to reassign the new distribution of weights and thresholds, and sometimes we cannot reach the reasonable goal. In other words, there is less designability in neural networks than in fuzzy systems. This point is one of the reasons why fuzzy systems have been applied to practical use much more than neural nets in Japan.

IV. A Fuzzy Logic Controller

In order to design a controller, the control algorithm should be described by some means. A classical control theory gives differential equations or transfer functions and a modern control theory gives a first-order vector matrix differential equation based on the state-space method. In these approaches, a controller designer has to possess knowledges about mathematics and the system under control.

However, human experts of ripe experience can skillfully control plants, machines, vehicles, etc., even though these systems under control are very complex (time-varient and nonlinear). These human experts mostly utilize know-hows which have been summarized from a long experience including successes and faults, and are represented with IF-THEN rules including fuzzy linguistic terms. They very often succeed to control the systems reasonably with these inexact information and without any calculations such as $\sin \omega t, \cos \omega t, \exp(x)$. This suggests that there is another algorithm which facilitates to control a complicated system by a simple and inexact knowledge base. One candidate is a *fuzzy inference* which produces a conclusion from a *knowledge base* (a set of fuzzy rules) and a *fact.*

A. Fuzzy Inference (Approximate Reasoning) [5]–[28]

The algorithm of the fuzzy inference in case of a deterministic fact is illustrated in the following. Let us consider the control of air conditioner, which is very primitive and

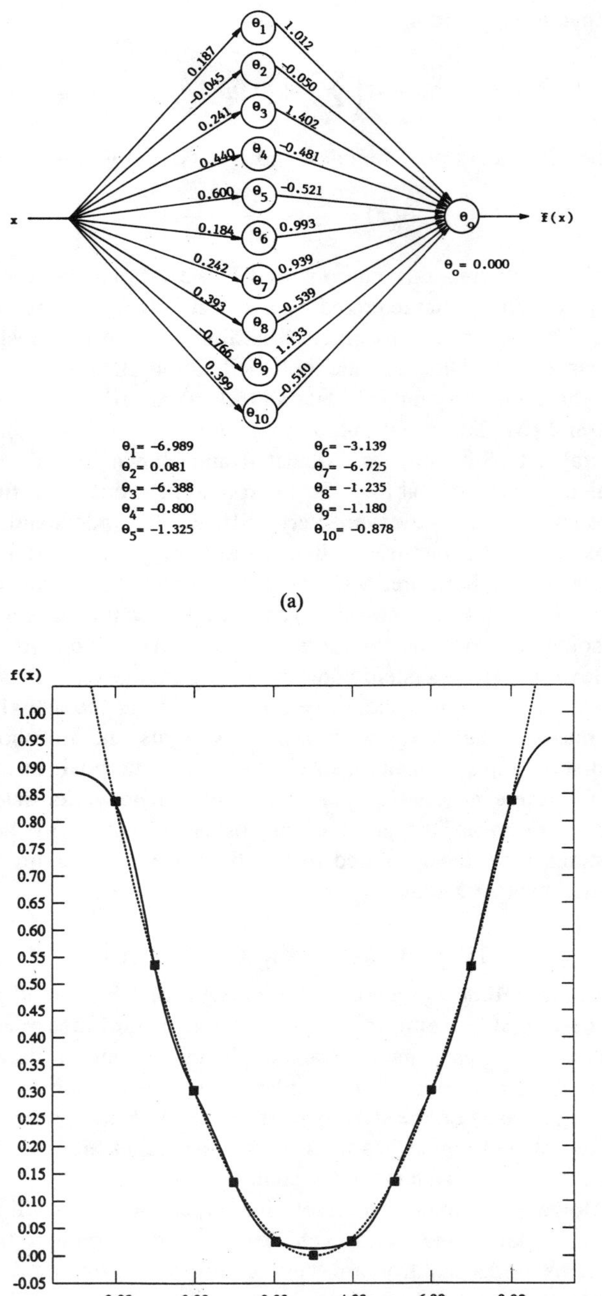

Fig. 3. (a) Example of the system description by a distribution of parameters, e.g., synaptic weights and thresholds in an artificial neural network. However, nobody can estimate, at a glance, how the output $f(x)$ responds to the input x. (b) The input-output characteristics (solid line) of the artificial neural net which was established after 10 000 times of learning by eleven data (squares) same to Fig. 2(b) and (c). The dotted line shows the system under modeling.

possesses no feedback outside. It blows off the air, the temperature and the volume of which can be changed by the dial (but possibly not proportional to the dial) as shown in Fig. 4(a). When the dial is assigned to be positive, warm or hot air is supplied from the air conditioner and when it is assigned to be negative, cool or cold air is supplied. Zero means no air supply. A room temperature T°C is measured with a thermometer as a deterministic value. A human looks

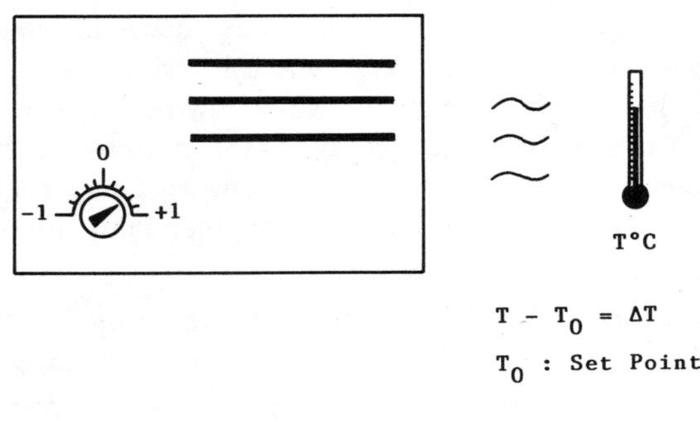

(a)

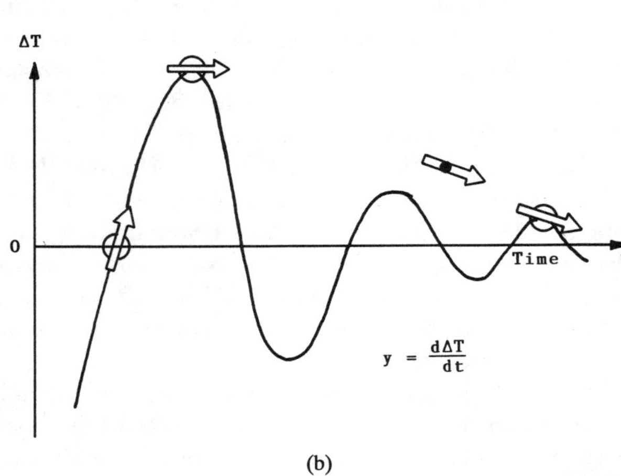

(b)

Fig. 4. (a) A primitive air conditioner and (b) a typical transient response.

at this value and notices the difference ΔT°C between the room temperature and the temperature T_0°C at which the room is desired to be kept. He or she must have a "strategy" to keep the room temperature constant even though the number of persons in the room changes, the outside temperature is changed, location of furnitures is changed and so on. By the strategy, he or she will reasonably changes the dial on the panel of the air conditioner in order to supply the warm or cool air to compensate the temperature change. A typical transient response of the room temperature is shown in Fig. 4(b), where the vertical axis represents the temperature difference ΔT°C, i.e., zero level is a set point of the room temperature.

The strategy may be written by sentences including fuzzy linguistic terms as follows. 1) When the room temperature is approximately equal to the set point and the temperature is rapidly changing higher, i.e., ΔT is *approximately zero* and the temperature change $y = d\Delta T/dt$ is *positively large* (as indicated by the left circle and arrow in Fig. 4(b)), cold air should be blown off rapidly to suppress the increasing temperature, otherwise it will positively deviate from the set point. Thus the dial should be turned to *negative large* (or approximately -1). 2) When the room temperature is high and the temperature does not change, i.e., ΔT is *positively large* and the temperature change $y = d\Delta T/dt$ is *approximately zero* (as indicated by the second circle and arrow in Fig. 4(b)), cold air should be blown off intermediately to decrease the temperature. Thus the dial should be turned to *negative medium* (or

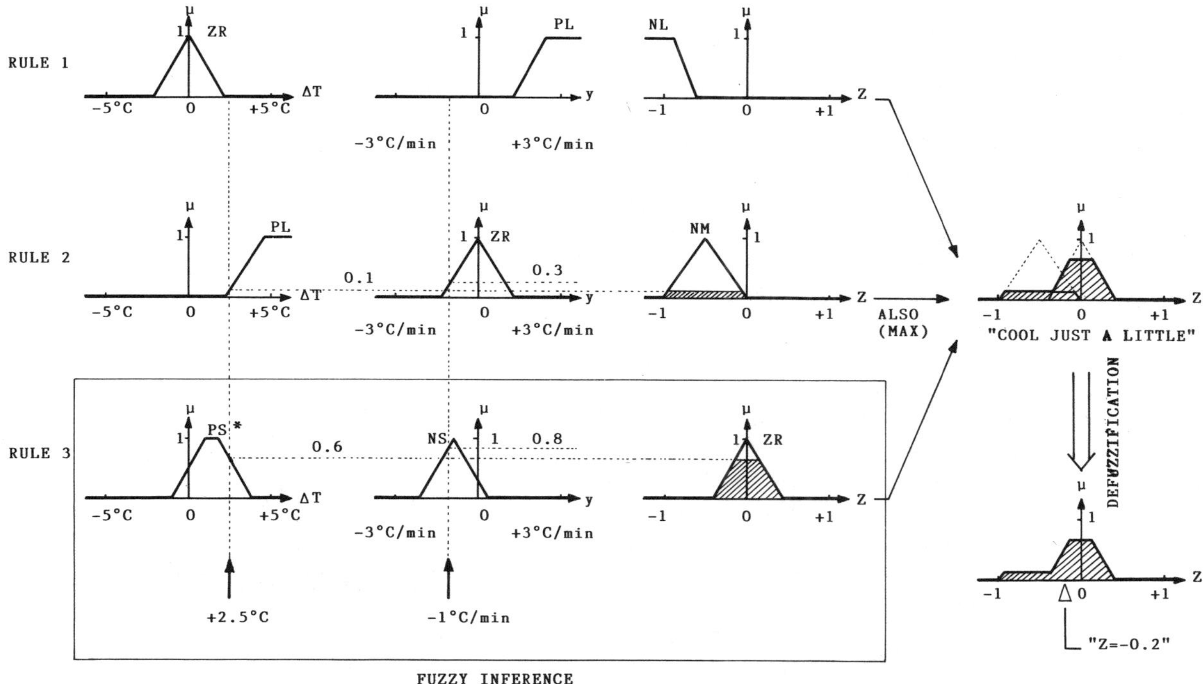

Fig. 5. An aspect of an individual fuzzy inference, aggregation (ALSO) and defuzzification.

approximately -0.5). 3) When the room temperature is a little bit higher than the set point and the temperature is gradually decreasing, i.e., ΔT is *positively small* and the temperature change $y = d\Delta T/dt$ is negatively small (as indicated by the right circle and arrow in Fig. 4(b)), there is no need to supply the cold nor hot air, because the room temperature is asymptotically close to the set point. Thus the dial should be turned to *approximately zero.* Of course, the strategy includes more than three knowledges described above. The other ones are abbreviated here for the simplicity.

The above strategy of temperature control can be rewritten by fuzzy rules as follows, together with a fact (measured values of $\Delta T = +2.5°\mathrm{C}$, $y = -1°\mathrm{C/min}$ indicated by the dot and arrow in Fig. 4(b)), as shown at the bottom of the page, where ΔT and y are variables in an if-clause (antecedent) and "the dial" is a variable in a then-clause (consequent), respectively, in a fuzzy rule, and NL, NM, NS, ZR, PS, and PL are fuzzy constans or fuzzy values and are abbreviations of fuzzy linguistic terms, *negatively large, negatively medium, negatively small* and *approximately zero, positively small* and *positively large,* respectively. When the room temperature is obtained by a thermometer to be 2.5°C higher than the set point and the temperature is decreasing at 1°C/min, what degree do we have to assign the dial to make the room temperature to be close to the set point?

Fig. 5 illustrates the algorithm of fuzzy inference to obtain the dial assignment from the strategy. A fuzzy linguistic term is identified by a label and characterized by a membership function as described in Section II-A. FUZZY SETS, so that a set of rules (9) can be rewritten as Fig. 5 where all the membership functions are defined by piecewise linear function for the simplicity of hardware design. The shapes (slopes, peak points, supports, and shoulders) of the membership functions are assigned by an intuition of a human expert after summarization of his wealth of experience, or assigned by an intuition of a designer of a fuzzy logic controller who has a knowledge about the air conditioner, heat transfer, etc.

Measured deterministic values $\Delta T = +2.5°\mathrm{C}$ and $y = -1°\mathrm{C/min}$ are referred to the antecedent (ΔT is PS* and y is NS) of Rule 3 in order to obtain the compatibility of the fact to the antecedent, where "*" in Fig. 5 represents a trapezoidal membership function, otherwise membership functions are S-

$$\begin{array}{ll}
\text{Rule 1} & \text{If } \Delta T \text{ is } ZR \text{ and } y \text{ is } PL\text{, then the dial should be } NL \\
\text{Rule 2} & \text{If } \Delta T \text{ is } PL \text{ and } y \text{ is } ZR\text{, then the dial should be } NM \\
\text{Rule 3} & \text{If } \Delta T \text{ is } PS \text{ and } y \text{ is } NS\text{, then the dial should be } ZR \\
\vdots & \quad\vdots \qquad\qquad \vdots \qquad\qquad\qquad\qquad\qquad \vdots \\
\text{Fact} & \Delta T \text{ is } +2.5°\mathrm{C} \text{ and } y \text{ is } -1°\mathrm{C/min} \\
\hline
\text{Conclusion} & \qquad\qquad\qquad\qquad\qquad\qquad \text{the dial should be ?}
\end{array} \tag{9}$$

shaped, Z-shaped or triangular. The compatibility is one kind of similarity measure. The compatibility of +2.5°C to PS* and that of −1°C/min to NS are 0.6 and 0.8, respectively. Variables in the antecedent are combined by a conjunction "and," so that the constraint of the antecedent is more severe than the individual variable. So that it is reasonable to evaluate the severe compatibility of the fact, "$\Delta T = +2.5$°C and $y = -1$°C/min," to the antecedent, "ΔT is PS* and y is NS," to be a smaller grade of membership 0.6. This grade is a degree of soft matching between the measured value and the condition of the rule. When a conjunction "or" is used, the bigger compatibility may be adopted as a soft matching degree.

If it is 1, the condition is fully satisfied and thus the then-clause should be entirely adopted. Alternatively, if the degree of soft matching is 0, the then-clause should be rejected. In the fuzzy inference, the degree of soft matching can range from 0 to 1, so that the degree of adopting the then-clause may range from 0 to 1, while it is 0 or 1 in case of unification in AI technology. Since the soft matching degree in this case is 0.6, the then-clause is partially adopted, i.e., the fuzzy constant ZR in the consequent in Rule 3 is weighted by 0.6. There are some weighting methods such as multiplication, minimization, etc. The latter is adopted in this paper. As shown in Fig. 5, the membership function of "dial is ZR" is truncated (minimized) by 0.6 to be an individual conclusion (shaded part) from Rule 3. The procedure to obtain the individual conclusion from each fuzzy rule and the fact is called *fuzzy inference,* and the tool to achieve the fuzzy inference is called a *fuzzy inference engine.* The fuzzy inference is also referred to *approximate reasoning,* because it gives an approximate conclusion represented with a membership function.

There is another rule contributing to produce the final conclusion. By referring measured values to the antecedent of Rule 2, the another individual conclusion (shaded part) is obtained after truncating "dial is NM" by 0.1. Concerning Rule 1, the measured value $y = -1$°C/min exhibits the compatibility 0 to "y is PL," so that the degree of soft matching between the measured value and the condition of Rule 1 is 0. Therefore, we cannot obtain any contribution to the final conclusion from Rule 1. All the rules are implicitly combined by conjunctions "also." Thus the individual conclusions should be aggregated to obtain the final conclusion. The "also" can be interpreted to be a maximization of individual conclusions and the final conclusion is obtained as a trapezoidal membership function with a side lobe as shown in Fig. 5.

B. Defuzzification

A controller employing a fuzzy inference is referred to as a *fuzzy logic controller* (or shortly, *fuzzy controller.)* A fuzzy logic controller delivers a deterministic signal but not a fuzzy signal to the system under control. In case of the air conditioner, we have to assign the dial. The procedure to obtain a deterministic value, on the universe of discourse, from a fuzzy value (membership function) is called *defuzzification* and its tool a *defuzzifier.* The most popular way of defuzzification is a *center-of-gravity method,* or a *centroid method* [22], [58].

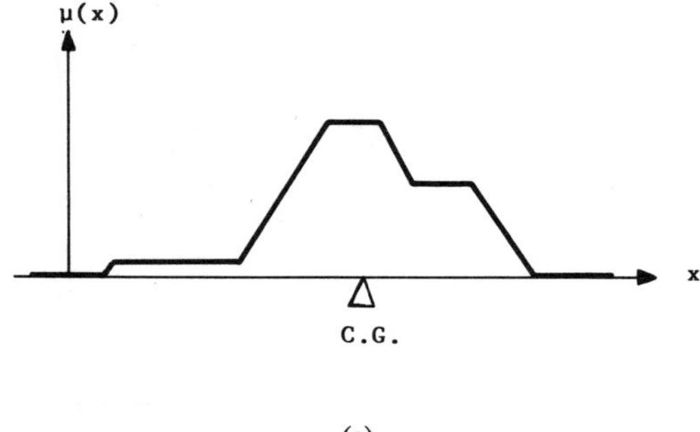

(a)

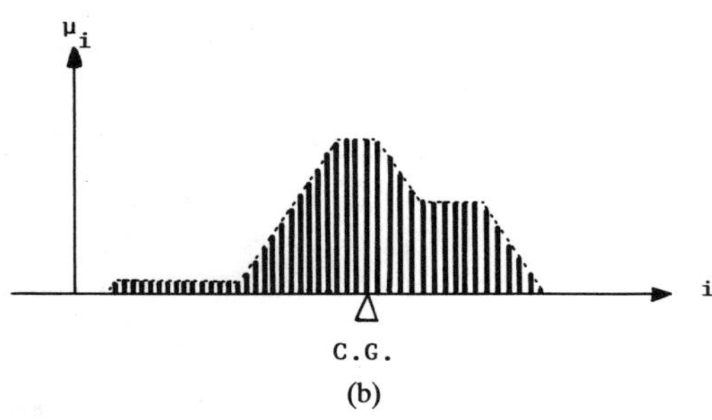

(b)

Fig. 6. Defuzzification (center-of-gravity method) of (a) a continuous membership function and (b) a discrete membership function.

A membership function is very often represented by a sampled data (a set of elements) on the universe of discourse as well as a continuous one as shown in Fig. 6. The center-of-gravity can be calculated from the following equations depending upon whether it is continuous or discrete.

$$\text{C.G.} = \frac{\int x\mu(x)\,dx}{\int \mu(x)\,dx} \quad \text{(for a continuous membership function)} \tag{10}$$

$$\text{C.G.} = \frac{\sum_{i=1}^{n} i \cdot \mu_i}{\sum_{i=1}^{n} \mu_i} \quad \text{(for a discrete membership function)} \tag{11}$$

where n represents the number of elements of the sampled membership function and μ_i the grade of ith element. The center of gravity of the final conclusion in Fig. 5 is obtained to be −0.2. Therefore, the dial should be assigned to be −0.2 in case of $\Delta T = +2.5$°C and $y = -1$°C/min.

The final conclusion is characterized by a relatively complicated membership function and the physical meaning may be represented by a fuzzy linguistic term "cool just a little," because a center of gravity is shifted a little bit to the left by the side lobe.

When the two inputs are changed independently, we can obtain the continuous (sometimes nonlinear) control surface in the 3-D space ($\Delta T, y$, and z) from small amount of fuzzy rules because of the interpolative reasoning.

Consequently we have to develop two kinds of hardware systems. The first one is a *fuzzy inference engine* which achieves an individual fuzzy inference and the second one is a *defuzzifier* which aggregates all the individual conclusions and derives a center of gravity of the membership function of the final conclusion.

V. Hardware Implementation of a Fuzzy Logic Controller in Nonlinear Analog Mode

A fuzzy logic controller, the function of which is characterized by a set of fuzzy IF-THEN rules, accepts and produces deterministic signals. It is constructed by fuzzy inference engines, the number of which is equal to that of rules employed, and a defuzzifier which converts a final conclusion of fuzzy value to a deterministic value.

Fuzzy inference and defuzzification can be accomplished by software or hardware. The software implementation with an ordinary personal computer or a work station takes a long time (for instance, several ms to several hundreds of a ms) to obtain the inference result. On the other hand, the specific hardware implementation accomplishes higher speed processing and exhibits better compactness and lower cost.

The hardware implementation is realized in either digital mode or analog mode. The digital hardware implementation of a fuzzy system originated from M. Togai and H. Watanabe [29] and comes to practical useful chips [30]–[33]. Digital fuzzy chips have distinctive features, such as good programmability, easy design, good compatibility with digital systems, etc. The analog hardware implementation originated from the author's work both in current mode [34] and voltage mode [35], and were followed by many researchers [36]–[42]. Analog fuzzy chips have distinctive features, such as high speed, good compatibility with sensors, etc. Attempts were often made to discuss merits and demerits of digital and analog fuzzy chips. It has not been concluded to choose between them. The adoption should be made according to the circumstances. The interest of this paper is focused on the analog fuzzy hardware system.

Electronic circuits employed in the fuzzy inference engine mainly handle analog signals in this paper. However, these circuits are different from the traditional analog circuits such as an operational amplifier, a multiplier/divider and their derivatives. The circuits employed in the inference engine are *nonlinear analog circuits.*

In this chapter, nonlinear analog circuits, a practical implementation of a fuzzy inference engine, and a defuzzifier are described.

A. Ordinary Analog Circuits and Fuzzy Circuits

There are two ways to implement the algorithm to the hardware system. One is a digital circuit and the other is an analog circuit. Analog circuitry was used to construct an analog computer or some other instrumentations. However, it had less extensibility, so that it is now replaced with digital circuitry.

When we consider the hardware of a fuzzy system, we have to recall the analog system. This is because fuzzy logic handles a continuous grade ranging from 0 to 1, not only 0 or 1. However, an analog circuit peculiar to a fuzzy system is expected to be different from the ordinary analog circuit in the following respects.

1. Since a grade of membership handled in the fuzzy system is 0 through 1 and a resolution of 10% (20 dB) is enough, the designer should not suffer from accuracy including linearity, low thermal drift, low offset, etc.
2. A fuzzy circuit essentially does not need the gain greater than unity, while a high gain and roll off of an amplifier causes unstability (oscillation) and diminishes the extensibility of the system.
3. A fuzzy logic system usually exhibits a narrow signal processing procedure, while an ordinary analog system achieves a deep signal processing, i.e., multistage processing with amplifiers and thus produces a significant cumulative error.
4. A fuzzy circuit should be designed in a simple architecture (with a few devices) to achieve high speed processing, while an ordinary analog circuit is constructed with many devices, some of which are dedicated to increase the linearity, compensate the thermal drift and phase shift, generate the reference voltages, obtain a stable gain, and so on.

The input-output characteristics of intrinsic fuzzy circuits are nonlinear, so that a fuzzy circuitry can be called a *nonlinear analog circuit,* while the traditional operational amplifiers and multipliers are called linear analog circuits.

B. Fuzzy Inference Engine (Rule Chip) [35], [43]–[47]

A fuzzy inference engine described here was designed for PID control, so that three input variables (error, change of error and rate of change of error) and one output were considered. Of course, these input signals are of deterministic values and output signal is a fuzzy value. The architecture of the fuzzy inference engine is shown in Fig. 7, where Rule 3 is programmed (If ΔT is PS* and y is NS, then the dial should be ZR) as an example.

The first stage of the fuzzy inference engine is a *membership function circuit* (MFC), three of which are dedicated to input signals X$'$, Y$'$, and Z$'$. The input-output characteristics of an MFC exhibits the shape of the membership function, which can be assigned externally. Each input terminal has the acceptable range, i.e., minimum and maximum values, which are represented by -1 and $+1$, respectively. In the practical case, -1 and +1 are represented by -5 V and $+5$ V and correspond to -5°C and $+5$°C and also to -3°C/min and $+3$°C/min, respectively. The output voltage of the MFC ranging from 0 V to 5 V corresponds to the grade of membership ranging from 0 to 1. The third membership function is not assigned, because the IF-THEN rules does not include the third variable in the antecedent. The label of the third input variable is referred to as NA (not assigned, don't care). An MFC assigned to be NA produces the grade 1 or the voltage 5 V not to affect other grades at the next MIN stage.

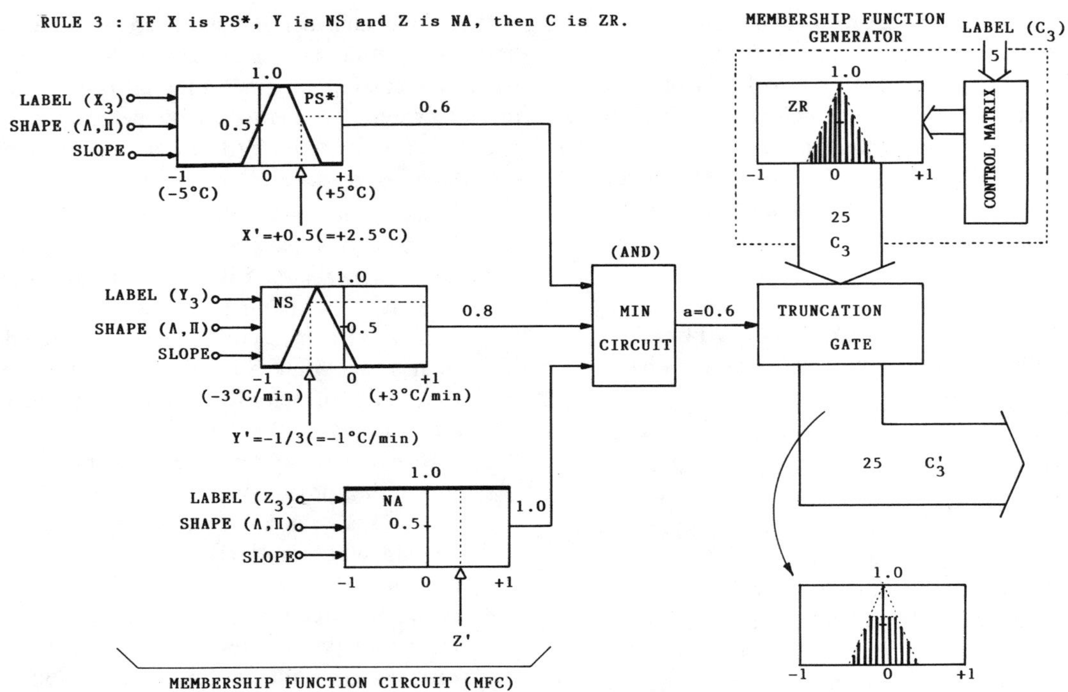

Fig. 7. An architecture of a fuzzy inference engine which accepts three deterministic input values and produces one fuzzy output value.

The second stage is a *MIN circuit* which accepts more than one input signals and produces the lowest values of them at the output terminal. This circuit achieves a conjunction "and" in the antecedent of IF-THEN rules. If variables in the antecedent are connected with "or," the MIN circuit should be replaced with a MAX (maximum) circuit. The output of these circuits represents the degree of soft matching between the antecedent and the inputs and thus feeds to the next stage to weigh the consequent membership function.

The consequent membership function is represented as voltage signal distribution on a data bus. The amplitude of the signal voltage ranges from 0–5 V corresponding to grades 0 to 1, respectively, and the number of signal lines in the data bus is 25 in this paper. We cannot call this data bus as 25 bits but as 25 elements, because each signal voltage is not binary but continuous (analog). This voltage distribution representing the consequent membership function is delivered from a *membership function generator* which has no input ports but a control port. In other words, the voltage distribution is assigned by a 5-bit binary word (LABEL).

A consequent membership function is truncated in a truncation gate by a degree of soft matching and the individual conclusion is obtained at the output data bus C'_3.

Each block is described in detail in the following.

1) MIN Circuit and MAX Circuit: A *MIN circuit* and a *MAX circuit* are the most basic fuzzy logic gates. Fig. 8(a) shows a MIN circuit employing bipolar transistors. PNP transistors and a current source I_{E1} construct a comparator and an NPN transistor and another current source I_{E2} construct a compensator. The potential of the emitter of the PNP transistor is equal to the minimum value among input voltages $X_1, X_2, X_3, \cdots, X_n$ plus 0.7 V (V_{EB} of an active input transistor), because higher input voltages makes corresponding input transistors cut-off. The following compensator (emitter follower) compensates the voltage shift of 0.7 V to produce the output voltage equal to the minimum input voltage. The compensator compensates the thermal drift as well as the voltage shift of emitter junction of the comparator. Fig. 8(b) shows an input-output characteristic and exhibits a nonlinear analog aspect. A transient response of this circuit is shown in Fig. 8(c), which illustrates that the rise and fall times are less than 10 ns. A Max circuit is shown in Fig. 9(a), where PNP transistors and an NPN transistor in a MIN circuit is replaced with NPN transistors and a PNP transistor, respectively. Comparison and compensation are achieved in the similar manner. Fig. 9(b) and (c) show an input-output characteristic and a transient response.

Since all transistors in comparators of a MIN circuit and a MAX circuit are coupled at emitters, the author named these circuits as *emitter coupled fuzzy logic* (ECFL) *gates,* the behavior of which is quite different from that of ordinary emitter coupled logic (ECL) in binary digital logic.

Two or more MIN circuits can be combined at the emitter of input transistors to extend the number of input terminals. In the similar manner, a MAX circuit can be extended.

The distinctive features of emitter coupled fuzzy logic gates are summarized in the following.

1. The output error is insensitive to the transistor size (e.g., base width, collector and emitter areas etc.), a variance in transistor characteristics (e.g., h_{FE}, I_{CBO}, etc.), while the missmatch and the variance give fatal errors in linear analog circuits.

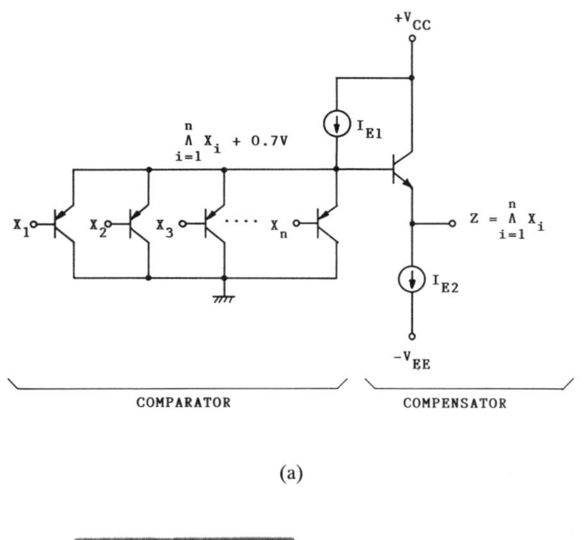

(a)

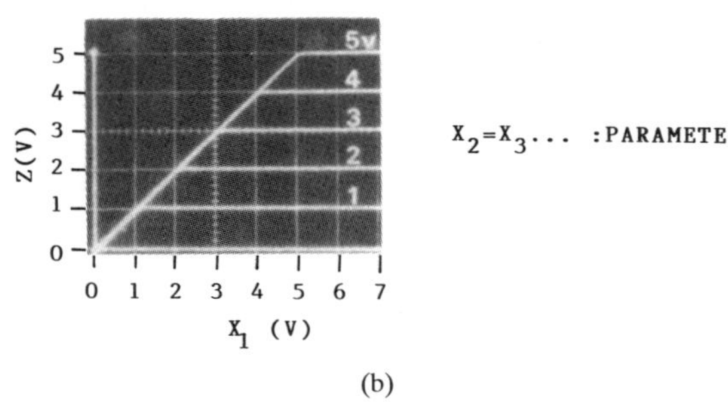

(b)

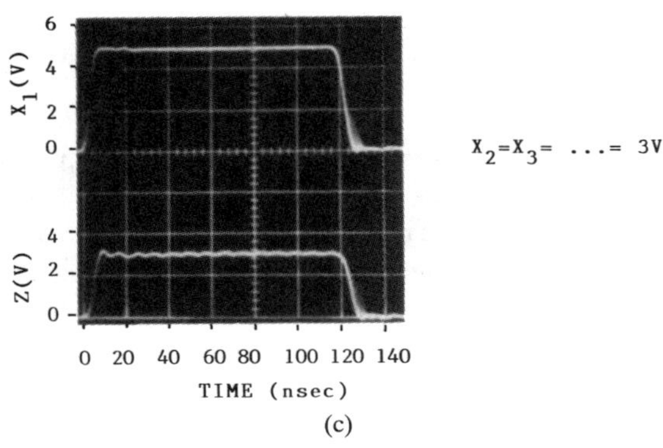

(c)

Fig. 8. An emitter coupled fuzzy logic (ECFL) gate (a MIN circuit). (a) Circuit configuration, (b) input-output characteristics, and (c) transient response.

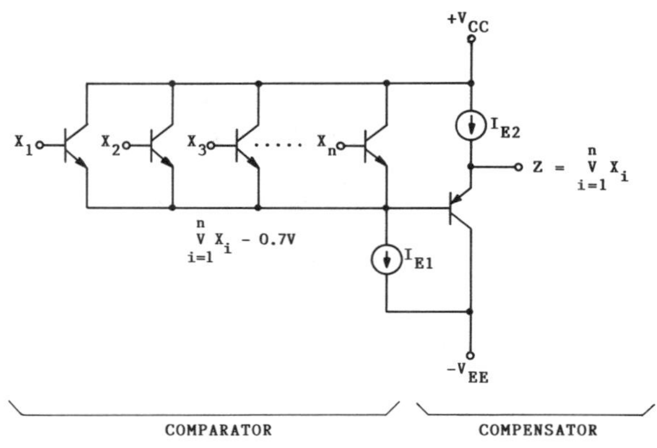

(a)

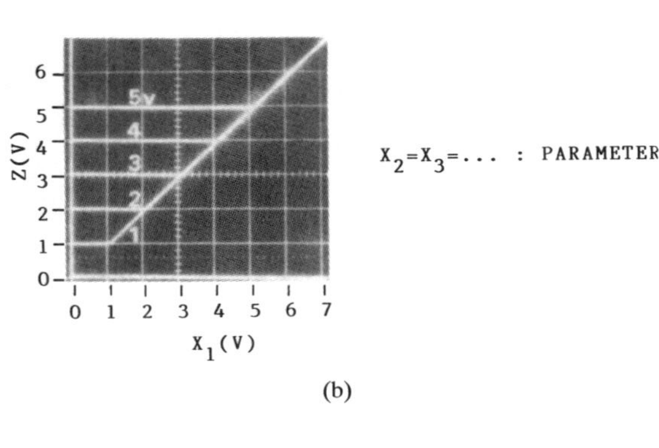

(b)

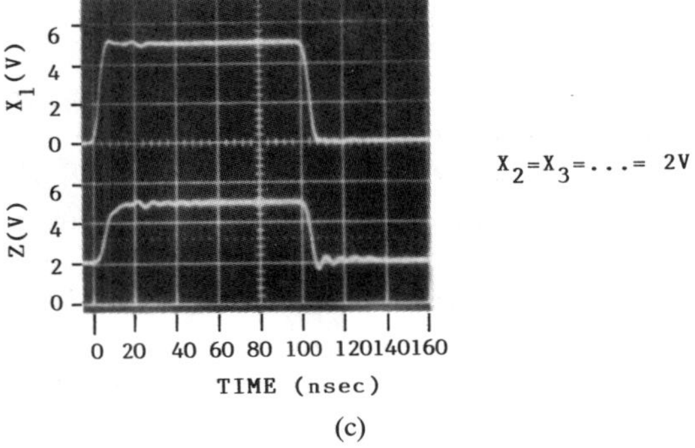

(c)

Fig. 9. An emitter coupled fuzzy logic (ECFL) gate (a MAX circuit). (a) Circuit configuration, (b) input-output characteristics and (c) transient response.

2. Good thermal stability is guaranteed. Experimental results give the output change of 0.2 %F.S. by the temperature change ranging from $-55°C$ to $+125°C$. The output fluctuation is not the problem.
3. The output error is insensitive to the operating currents. The author obtained the experimental results, for the device carefully designed, that the output change caused by the current change of one decade was less than 2 %F.S.
4. These ECFL gates are very robust against the fluctuation of supply voltages, because both a comparator and a compensator are driven by current sources. 5-V fluctuation of supply voltage caused the output fluctuation less than 0.1 %F.S.
5. High speed operation is obtained because no saturation of minority carriers in the base region occurs.
6. The ECFL gates are designed based on emitter followers, thus exhibit high input impedance and low output impedance. It guarantees a large fan-out.
7. The input terminals which are not utilized can be left opened (there is no need to terminate them), because the input transistors of open base are cut-off and does not affect other input transistors. This characteristic is very important in designing a semicustom IC.

2) MFC: Membership functions of variables in the antecedent are implemented by an MFC. In other words, a membership function circuit is one whose input-output charac-

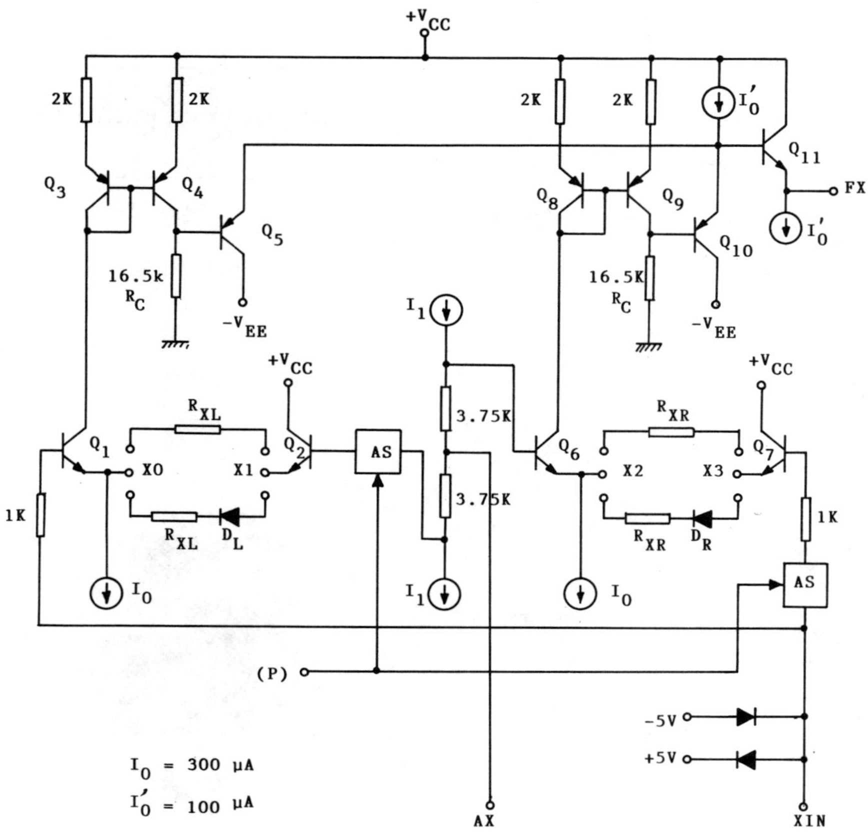

Fig. 10. Circuit configuration of a membership function circuit (MFC).

teristics can be assigned by external signals or devices. Fig. 10 shows the circuit configuration of an MFC which can realize five functions of a Z-function, an S-function, a Π-function (trapezoidal function), a Λ-function (triangular function), and NA (not assigned, don't care). XIN and FX are an input and an output terminals, respectively. AS is an analog switch controlled by an external signal (P).

If (P) is disable, two analog switches turn off and thus two current sources I_0 drive Q_1, Q_3, Q_4 and Q_6, Q_8, Q_9, and base voltages of Q_5 and Q_{10} reach $I_0R_C \approx 5$ V. Q_5, Q_{10}, Q_{11} and two current sources I_0' construct a MIN circuit, so that the output FX is 5 V independent of XIN. This means NA.

If (P) is enabled and X0-X1 are terminated by R_{XL} with X2-X3 opened, the input-output characteristics of this MFC exhibits S-shapes as shown in Fig. 11(a). If (P) is enabled and X2-X3 are terminated by R_{XR} with X0-X1 opened, the input-output characteristics exhibit Z-shapes as shown in Fig. 11(b). If (P) is enabled, and X0-X1 and X2-X3 are terminated by R_{XL} and R_{XR}, respectively, the MFC becomes a Λ-function as shown in Fig. 11(c). If (P) is enabled, and X0-X1 and X2-X3 are terminated by $R_{XL} + D_L$ and $R_{XR} + D_R$, respectively, the MFC becomes a Π-function as shown in Fig. 11(d). The shoulder voltages of S- and Z-functions and the center voltages of Λ- and Π-functions can be assigned by an external voltage AX. Thus labels of the membership functions can be assigned by AX. The positive slope and the negative slope of each function are reciprocally proportional to R_{XL} and R_{XR}, respectively. Therefore, fuzziness of the membership function can be assigned by R_{XL} and/or R_{XR}. Fig. 12(a) and (b) show the input-output characteristics of a triangular membership function for various labels assigned by AX and for various resistors R_{XR} (upper) and R_{XL} (lower), respectively.

3) Membership Function Generator: A membership function of a consequent is sampled to discrete grades, which are represented by voltages ranging 0–5 V and distributed on 25 signal lines (25-element data bus) as shown in Fig. 6(b) or Fig. 7. The shape and the label of the membership function can be changed by reassigning the voltage distribution as shown in Fig. 13. Seven labels (NL, NM, NS, ZR, PS, PM, and PL) and NA (not assigned: all voltages on the data bus are 0 V) are discriminated by a three-bit binary word and Λ-shape, S-shape, Z-shape, and Π-shape are done by a two-bit one, so that these membership functions in the consequent are assigned by a 5-bit binary word (control word). The slope of a membership function shown in Fig. 13 with 25 signal lines (25 elements) guarantees that only four dc voltage sources (1.25 V, 2.5 V, 3.75 V, 5 V) are enough to realize these various types of membership functions.

The desired voltage distribution is obtained by switching each signal line to a desired supply voltage. Fig. 14(a) shows a switch matrix and two emitter follower arrays, and Fig. 14(b)

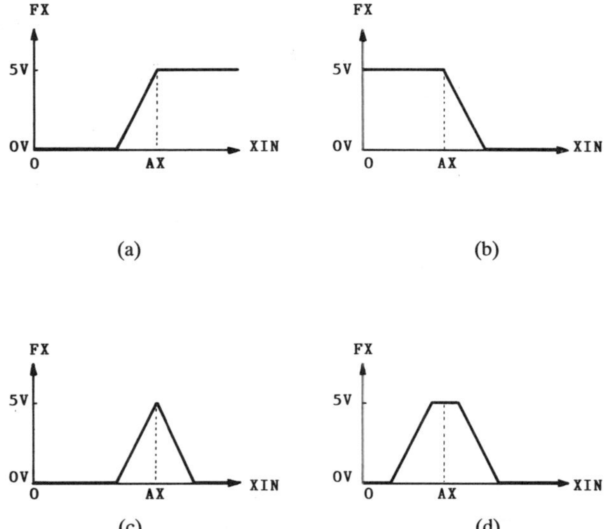

Fig. 11. Four types of membership functions realized by an MFC shown in Fig. 10. (a) S-function, (b) Z-function, (c) Λ-function (triangular function), and (d) Π- function (trapezoidal function).

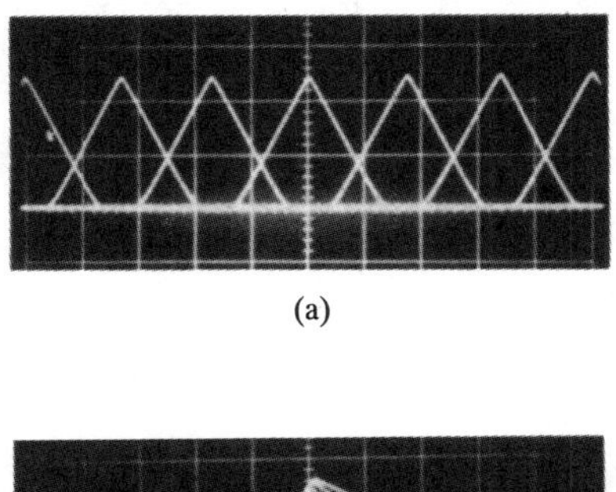

(a)

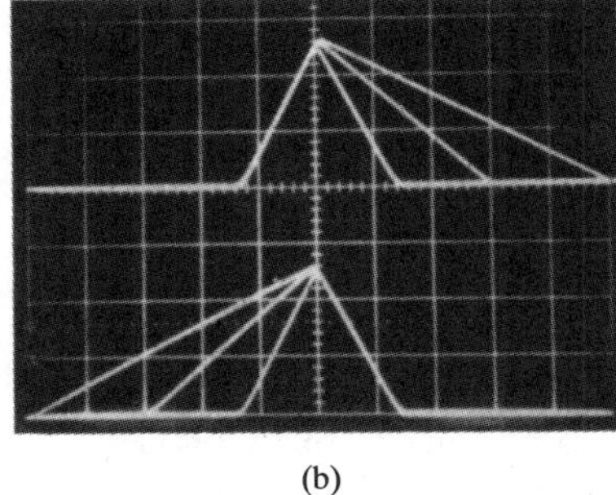

(b)

Fig. 12. Input-output characteristics of a membership function circuit (MFC). (a) Λ-functions for various labels (AX = −5 V, −3.3 V, −1.7 V, 0 V, 1.7 V, 3.3 V, 5 V). (b) Λ-functions for various slopes. (Upper: $R_{XR} = 120\,\Omega, 266\,\Omega, 444\,\Omega, R_{XL} = 120\,\Omega$. Lower: $R_{XL} = 120\,\Omega, 266\,\Omega, 444\,\Omega, R_{XR} = 120\,\Omega$).

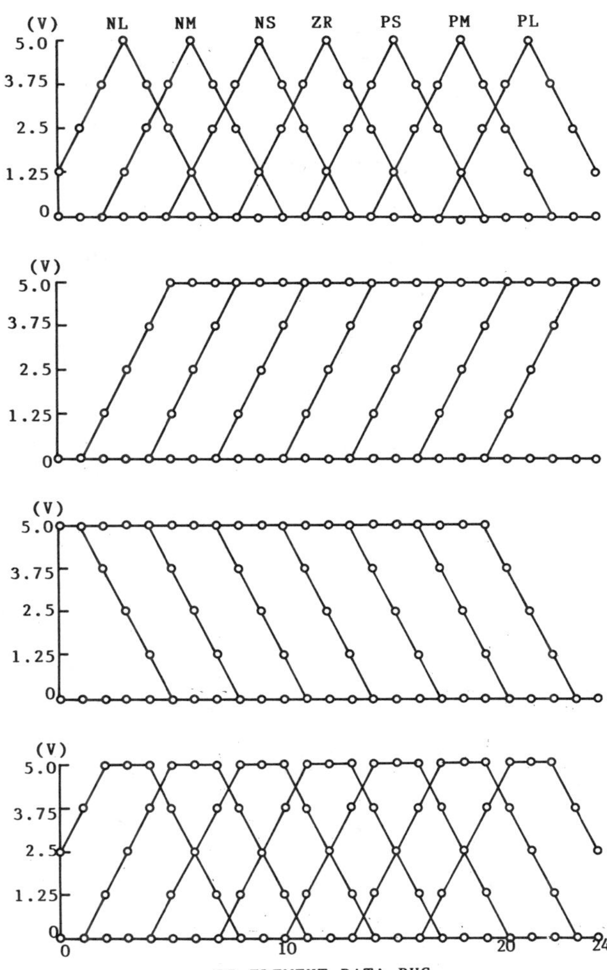

Fig. 13. Voltage distributions on 25-element data bus which represent consequent membership functions such as NL (negatively large), NM (negatively medium), NS (negatively small), ZR (approximately zero), PS (positively small), PM (positively medium), PL (positively large), and NA (all voltages on 25-element data bus are 0 V). NA in a consequent is used for eliminate the rule. Four shapes (triangular, S-shaped, Z-shaped and trapezoidal) are discriminated. Four voltage sources (1.25 V, 2.5 V, 3.75 V, and 5 V) are enough to realize a consequent membership function on 25-element data bus.

shows a control matrix which delivers control signals to analog switches in a switch matrix. In each column of the switch matrix in Fig. 14(a), only one analog switch is turned on by a control signal $G_{i,j}$. In other words, one bit in each binary word C_i is enabled by a 5-bit control word (CS1, CS0, CL2, CL1, CL0) through the 5-input 125-output control matrix. The control matrix can be designed in combinational logic or PLA. Emitter follower arrays are used for impedance transformation and level/temperature compensation. How significant is the number of the element in the consequent membership function? How important is the shape of consequent membership function? They will be discussed in detail in Section V-D-1.

4) Truncation Gate: A truncation gate accepts one fuzzy vector signal (fuzzy word signal) $\boldsymbol{B}$ and one fuzzy scalar signal (ranging from 0 V to 5 V) "a." The former one represents a membership function of consequent and is delivered from a membership function generator. The latter one represents the degree of soft matching between the fact and the antecedent, and is delivered from a membership function circuit. The truncation gate produces a fuzzy vector signal $\boldsymbol{B'}$ truncated by the fuzzy scalar input signal as shown in Fig. 15(a). In other words, the truncation gate accepts one 25-element input data and one analog input and produces one 25-element output data, the voltage distribution of which is truncated by the analog input. This can be constructed by arranging MIN circuits in

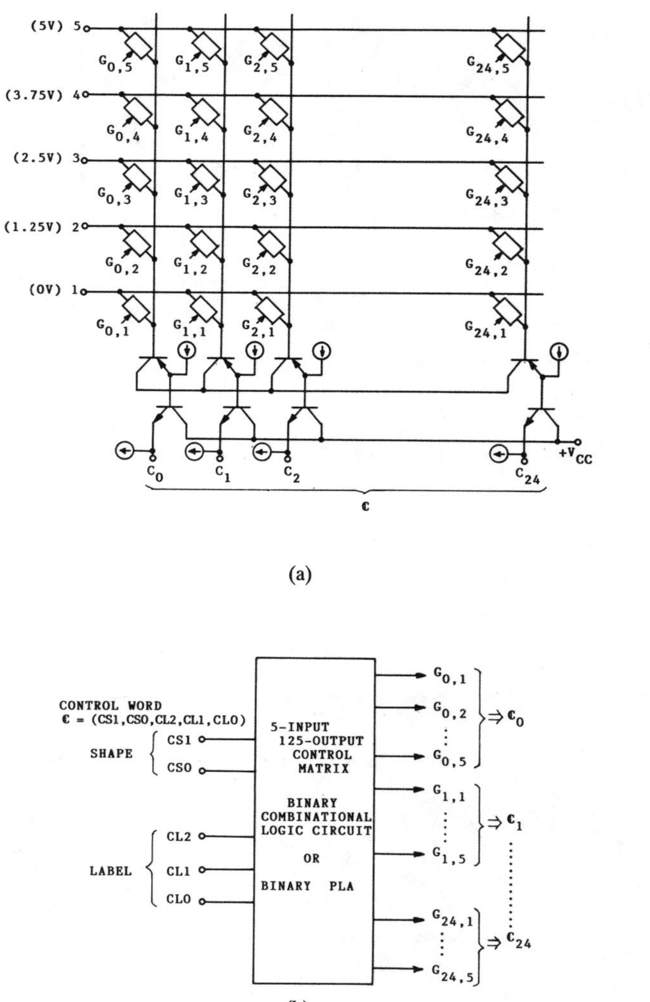

(a)

(b)

Fig. 14. Main blocks of a membership function generator. (a) A switch matrix and an emitter follower array (b) a control matrix.

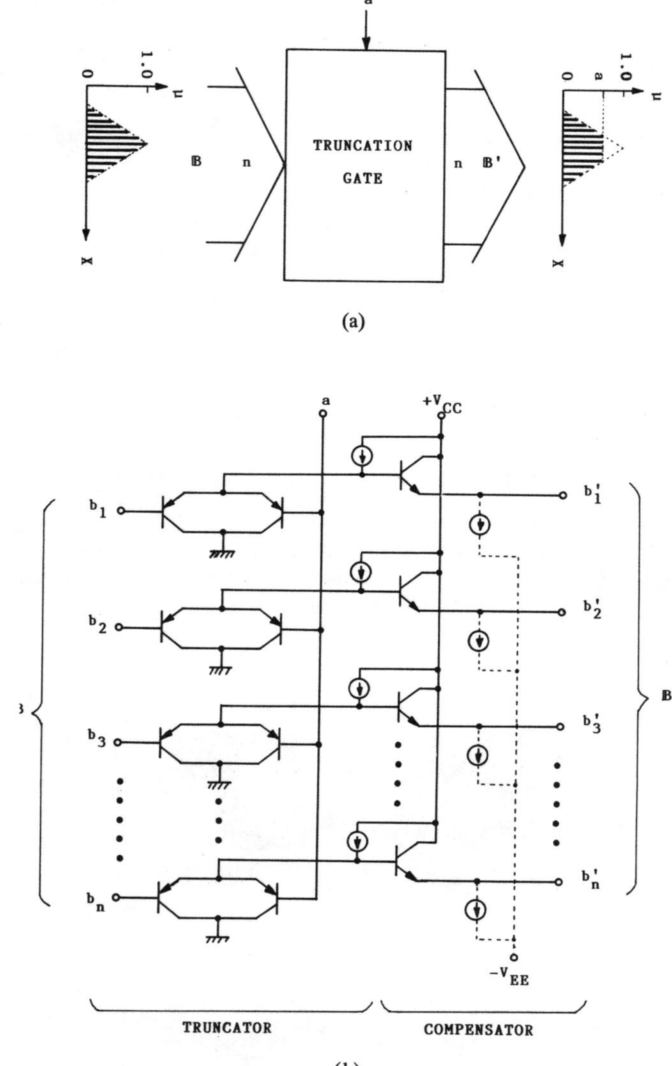

(a)

(b)

Fig. 15. A truncation gate. (a) Its function and (b) its circuit configuration.

an array. The circuit configuration is shown in Fig. 15(b), where one input terminal of each MIN circuit is connected to a common terminal "a." The other input terminal of each MIN circuit is dedicated to the element signal of an input fuzzy word $\boldsymbol{B}$. The output terminal of each MIN circuit produces the element signal of an output fuzzy word $\boldsymbol{B}'$.

5) Open Emitter for Saving a MAX Array: The output ports of fuzzy inference engines should be usually fed to a MAX array to aggregate all the individual conclusions before defuzzification as shown in Fig. 16(a). If all current sources (indicated by a dotted line in Fig. 15(b)) in a comparator of a truncation gate are omitted, all input terminals of a defuzzifier are driven by the flow-in current sources inside the defuzzifier and each input terminal of the defuzzifier is connected to the corresponding terminals of all the fuzzy inference engine outputs as shown in Fig. 16(b), the MAX array can be omitted from this construction. In other words, a fuzzy inference engine with open emitter structure and a defuzzifier involving the current source in the input terminal facilitates *wired MAX* by themselves. Thus the open emitter structure saves a MAX array as well as current sources in compensators.

6) Silicon Implementation: The architecture of the fuzzy inference engines shown in Fig. 17, which is implemented in the monolithic form and called a Rule Chip because one fuzzy inference based on one fuzzy IF-THEN rule (or shortly, fuzzy rule, or control rule), is achieved in this chip.

A LABEL X input for a variable x is a 4-bit word ($\overline{\text{ANAX}}$, AX2, AX1, AX0) to assign the label of the membership function of x. When $\overline{\text{ANAX}}$ is 0, a multiplexor selects the control signal, ANALOG LABEL, to a membership function circuit. Otherwise, the following 3-bit word (AX2, AX1, AX0) assigns one of eight labels as shown in Table I.

The upper and lower limitations of the input signal is assigned by the voltage REF1 and REF2, which can be supplied from a power supply (+10 V) through variable resistors, and detected at two corresponding terminals MONITOR.

REF3 is the terminal where +5 V is usually supplied or a lower voltage is supplied through an external resistor in case of lightening the rule (weight less than 1 or less important than other rules). The weight can be detected at MONITOR corresponding to REF3.

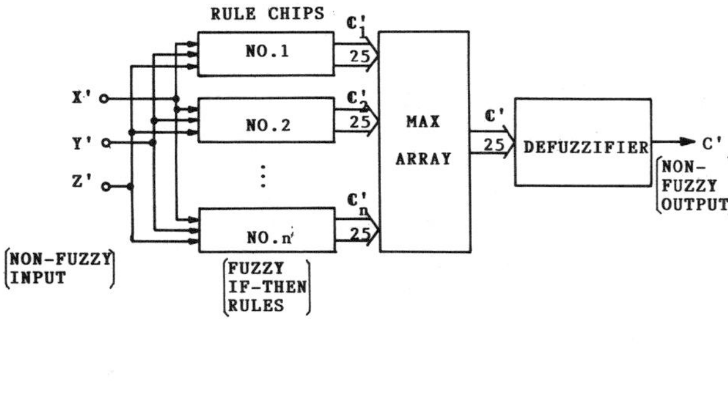

(a)

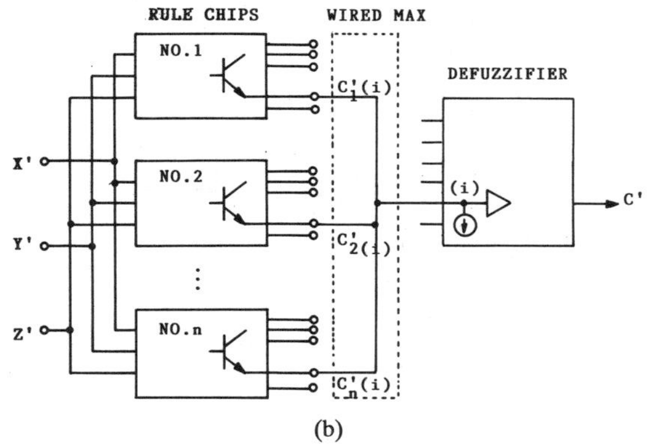

(b)

Fig. 16. (a) Aggregation of individual conclusions (individual inference results) with a MAX array. (b) Aggregation by wired MAX.

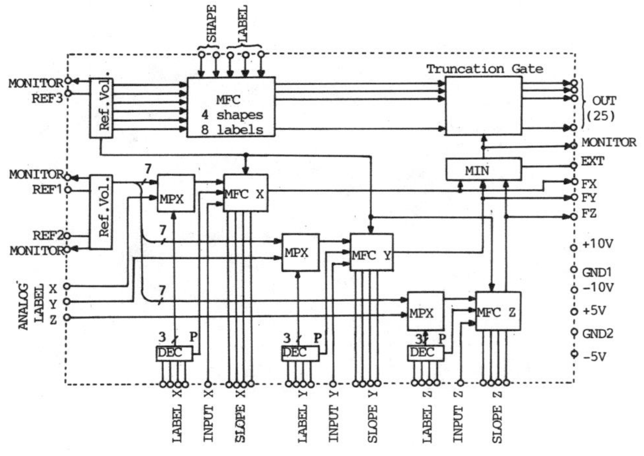

Fig. 17. Architecture of the fuzzy inference engine (rule chip).

A membership function in a consequent can be selected by 5-bit control word as shown in Tables II and III.

When a binary terminal is floating, it is pulled up to a high level equivalent to 1 just like a TTL.

FX, FY, and FZ are the terminals for monitoring the degree of soft matching. The shape of each membership function can be watched by applying a sinusoidal wave to each signal input.

An EXT terminal can be used for combining several rule chips in order to realize more than three variables in the antecedent as described in Section V-B-1. For instance, if three rule chips are connected to each other at terminals EXT, we can implement a rule which has variables in the antecedent up to nine.

TABLE I

	$\overline{\text{ANAX}}$	AX2	AX1	AX0
Analog	0	DON'T CARE		
NL	1	0	0	0
NM	1	0	0	1
NS	1	0	1	0
ZR	1	0	1	1
PS	1	1	0	0
PM	1	1	0	1
PL	1	1	1	0
NA	1	1	1	1

TABLE II

Shape	CS1	CS0
Λ	0	0
S	0	1
Z	1	0
Π	1	1

TABLE III

Label	CL2	CL1	CL0
NL	0	0	0
NM	0	0	1
NS	0	1	0
ZR	0	1	1
PS	1	0	0
PM	1	0	1
PL	1	1	0
NA	1	1	1

The circuit of the fuzzy inference engine is integrated in the monolithic form by a standard BiCMOS process. The design rule of 5 μm, two layers of aluminum and one layer of poli-silicon is adopted. About 600 transistors and about 800 resistors are embedded in a 8 mm × 8 mm silicon chip. Fig. 18(a) shows a microphotograph of the fuzzy inference engine (Rule Chip). This chip is molded in an 84-pin plastic package. The response time of fuzzy inference is 1 μs as shown in Fig. 18(b). This is equivalent to 1 000 000 fuzzy inferences per second (FIPS).

C. Defuzzifier (Defuzzifier Chip) [48]

In order to obtain a deterministic value from a membership function of the final conclusion of fuzzy inference, the defuzzification should be achieved. Fig. 19 shows the defuzzifier circuit. All the individual conclusions are aggregated by wired MAX technology as described in Section V-B-5, that is, the defuzzifier possesses the current-source-input buffer which facilitates a wired MAX with rule chips connected to the input. The output of the buffer is fed to a weighted sum circuit and an ordinary sum circuit, the outputs of which are fed to an analog divider to calculate (15). The weights i in (15) are reciprocally proportional to resistances of a resistor array and thus positive.

(a)

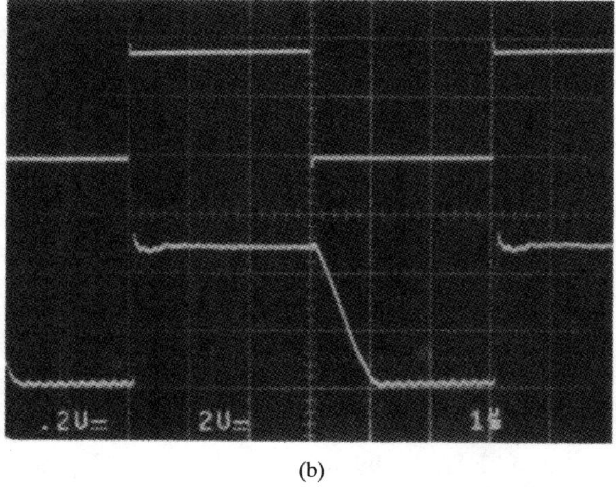

(b)

Fig. 18. (a) A microphotograph of the fuzzy inference engine (rule chip) and (b) its transient response.

Therefore, the calculated value is always positive. In order to obtain the bipolar output signal ranging from −5 V to 5 V, an offset adjustment and an amplification adjustment are achieved by NDC and R_G, respectively. The defuzzifier is implemented on a ceramic base in the hybrid form, and molded in a 44-pin plastic package. The response time of defuzzification is about 5 μs as shown in Fig. 20, which is almost determined by that of an analog divider used in this defuzzifier.

D. Advanced Implementation of a Fuzzy Logic Controller [49],[50]

A direct hardware implementation of the fuzzy inference algorithm shown in Fig. 5, both of inference engines and a defuzzifier, needs too many devices, dissipates high power, and thus the costs are high. So that it is necessary to consider to design another one for each of fuzzy inference and defuzzification.

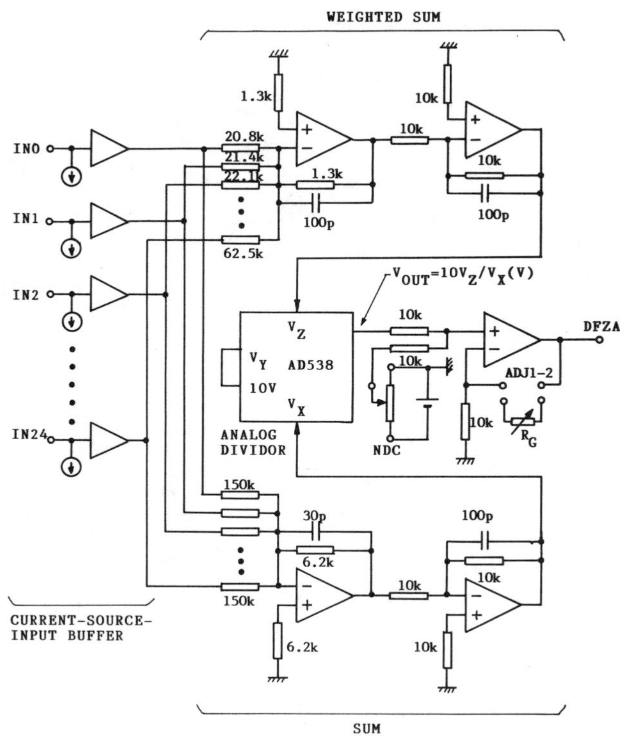

Fig. 19. A circuit configuration of the defuzzifier.

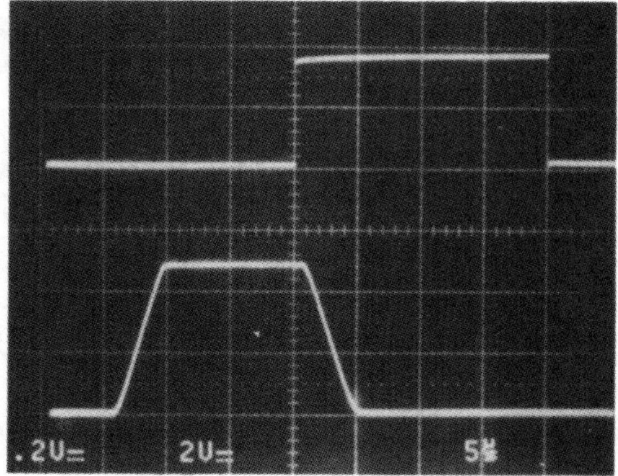

Fig. 20. A step response of the defuzzifier.

1) Fuzzy Inference with a Singleton Consequent: The algorithm of fuzzy inference shown in Fig. 5 is based on the compositional rule of fuzzy inference [27], which is the extension of a classical logic inference to a fuzzy logic and allows us to understand the aspects and the meaning of the fuzzy inference easily. Once the concept is understood, we have to consider other ways equivalent to it, which may be suitable for hardware implementation.

One of the main problems in the fuzzy inference engine is a parallelism in truncation of the consequent. As the number of elements of the consequent membership function increases, the number of devices for switching, buffering, minimizing, and

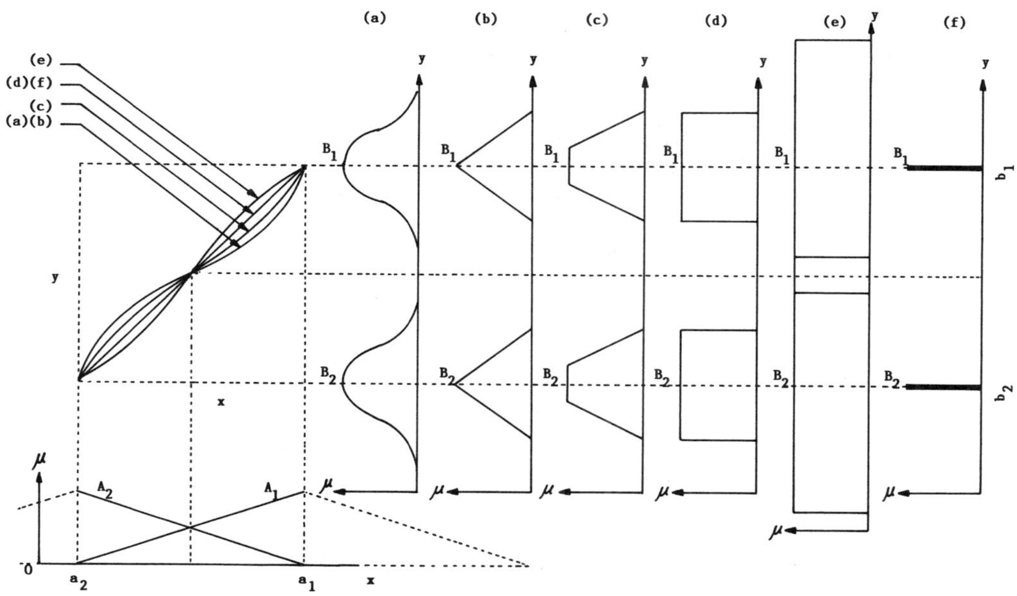

Fig. 21. Comparison of interpolation between some fuzzy inferences.

binary combinational logic circuits increases proportionally. How many elements are needed in the consequent membership function for reasonable inference? Which shape of the membership function in the consequent is the best, though the shape is not essential for the fuzzy inference?

The significant ability of fuzzy inference is the reasonable interpolation between a small number of fuzzy data. Therefore, the attempt is made to compare the abilities of interpolation. Let us consider the case of interpolation between two fuzzy rules as

Rule 1 $\quad$ If x is A_1, then y is B_1.

Rule 2 $\quad$ If x is A_2, then y is B_2.

A_1 and A_2 are assumed to be the fuzzy linguistic terms defined by triangular membership functions as shown in Fig. 21, though the membership functions are meaningless outside the range a_1–a_2. B_1 and B_2 are assigned to be of (a) bell-shape, (b) triangular shape, (c) trapezoidal shape, (d) square shape (crisp interval), (e) overlapped square, and (f) singleton (deterministic value). An inference results of interpolation between these two data is shown in a square formed by dotted lines corresponding to typical values a_1, a_2, b_1, and b_2 in Fig. 21, which is exaggerated to make it easy to understand the differences.

This figure shows that the unoverlapped crisp intervals and the singletons in the consequent facilitate the *linear interpolation,* while all the other cases produce the nonlinear interpolation, the deviation of which depends upon the shapes and overlapping rate. This aspect of singletons is entirely favorable to hardware design and programming. Because we do not need to design the massively parallel processing architecture in a truncation gate and a membership function generator, we do not need to search for the best shape of the membership function in the consequent by computer simulation.

2) Division Reduction by Feedback: In a practical use of fuzzy inference, e.g., a fuzzy logic controller, an output signal of deterministic value, but not fuzzy value, is desired. Thus a defuzzifier is usually attached to the inference engines to construct a fuzzy logic controller.

A significant portion, which determines the operating speed and causes the difficulty of design and the troublesome adjustment, is an analog divider in a defuzzifier. In order to eliminate this analog divider from a defuzzifier, the new architecture should be developed. A center of gravity is calculated by (10) or (11), where a weighted sum of membership grades is divided by a normal sum of them. If the normal sum (denominator) is kept equal to the unity or equal to the constant at any time, no divider is needed. The normal sum is calculated and should be fed back to all the inference engines to reproduce the individual conclusion, the total of which is equal to unity or the constant. It can be realized by a membership function circuit, whose maximum value (or full grade) of membership can be changed by an external signal. The author named this type of membership function circuit as a *grade-controllable membership function circuit (GC-MFC).*

The grade-controllable membership function circuit can be obtained by modifying the ordinary membership function circuit (Fig. 10) as shown in Fig. 22, where two current sources I_0 can be changed by the output current of the V-I converter. The input-output relation of this V-I converter is obtained as

$$I_0 = \frac{1}{R_i} W \tag{12}$$

where W and R_i are a weight input to assign the peak value of the membership function and a resistance shown in Fig. 22, respectively, and the peak value of MFC output is

$$FX_{\max} = R_C \cdot I_0 \tag{13}$$

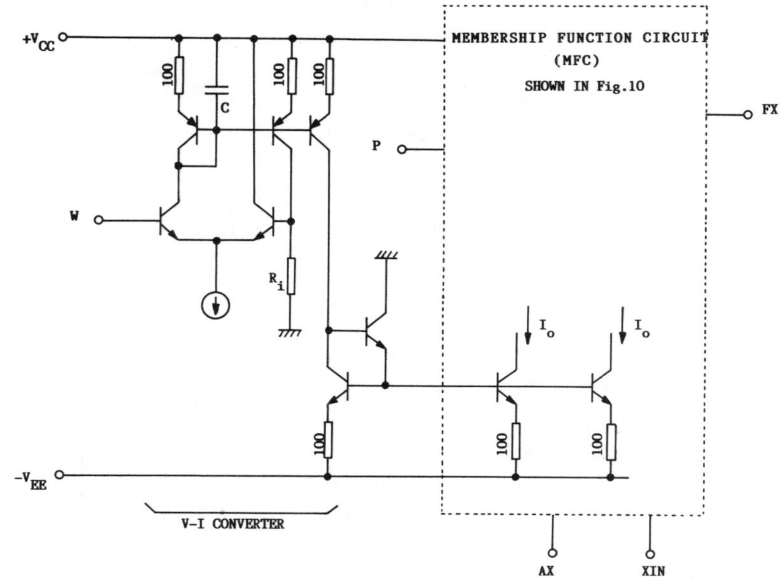

Fig. 22. A grade-controllable membership function circuit (GC-MFC). It is obtained by modifying the ordinary membership function circuit.

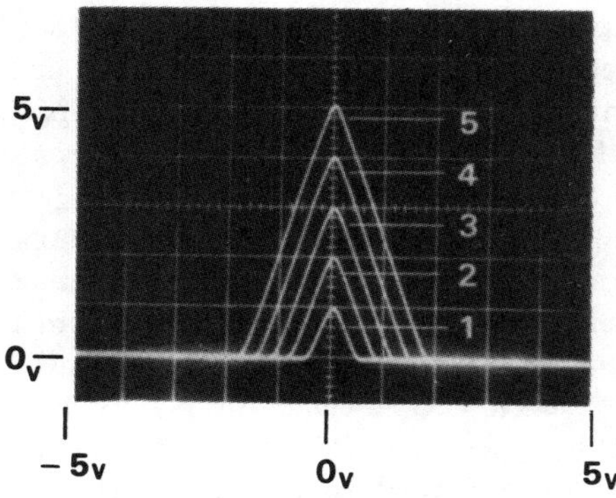

Fig. 23. Input-output characteristics of a grade-controllable membership function circuit. Attached numbers are the control voltages W in volts.

where R_C is an I-V conversion resistance of the grade-controllable membership function circuit (Fig. 22) and is practically shown in Fig. 10.

Thus the peak value of GC-MFC output is obtained as

$$FX_{\max} = \frac{R_C}{R_i} W. \tag{14}$$

Fig. 23 shows the input-output characteristics of this GC-MFC, where $R_C = R_i$ and attached numbers are the control voltages W in volts. This figure illustrates that the input-output characteristics are not multiplied by the control signal, but shifted down and up by it.

A fuzzy logic controller, which achieves fuzzy inferences with fuzzy linguistic terms in IF-clause and singletons in THEN-clause and defuzzifies the final conclusion, is referred to as a *singleton controller.* The example of the rule is "If TEMPERATURE OF CHAMBER grows up *higher* and PRESSURE becomes *a little bit higher,* then FUEL should be *reduced to 3 liters/minute.*"

The GC-MFC's are employed to construct the singleton controller which accomplishes defuzzification without an analog divider as shown in Fig. 24. The ordinary sum of soft matching degrees $a_1 + a_2 + a_3 + \cdots + a_n$ are compared with the constant E_1. When the sum exceeds E_1, all the control inputs W are pulled down by the operational amplifier to make the output of each GC-MFC decrease and thus the ordinary sum decrease. When the ordinary sum becomes below E_1, all the control inputs W are pulled up to increase the ordinary sum. Consequently, the sum $a_1 + a_2 + a_3 + \cdots + a_n$ is equal to the constant E_1 at any time. All of these soft matching degrees $a_1, a_2, a_3, \cdots, a_n$ are weighted by resistors R_j in consequent blocks, respectively, to be summed to produce the weighted sum equal to the center of gravity. The offset E_0 is subtracted from the weighted sum to shift the output level. Precise description on E_0 is presented in the following. Three GC-MFC's in each antecedent block and a resistor R_j in each consequent block are programmed in accordance with the fuzzy rules to be implemented.

The estimation of the resistance R_j depends upon whether the consequent space is unipolar (there exist only the positive values) or bipolar (positive and negative values can be defined). When the output of the singleton controller is fed to an electric heater, the consequent should be unipolar. When it is fed to a dc servo motor, the consequent should be bipolar. A reference diagram shown in Fig. 25 is helpful to estimate the resistance R_j. $R_{\max}$, and $R_{\min}$ should be also estimated so that the resistance ration $R_{\max}/R_{\min}$ should be implemented in the integrated circuit technology with reasonable accuracy. For instance, if E_0 is assigned to be 0 in case of unipolar consequent space in Fig. 25(a), $R_{\max}$ is infinity (open circuit) and R_j should be assigned ranging from $R_{\min}$ to $\infty\,\Omega$ which is impossible to realize. When the consequent space is bipolar and the weighting is achieved by a resistor array, negative weighting cannot be implemented.

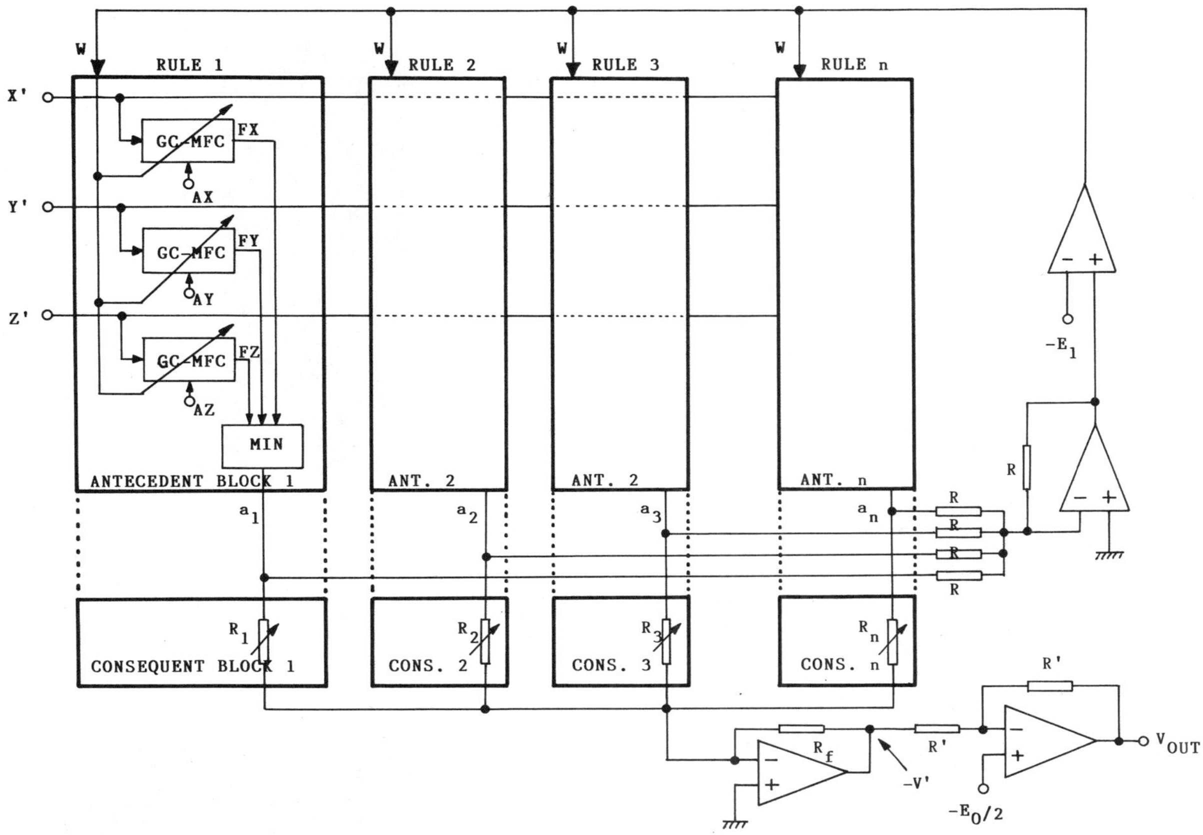

Fig. 24. A singleton controller which achieves fuzzy inferences with singletons in consequents and defuzzify the final conclusion. The defuzzification is accomplished without an analog divider but by employing the grade-controllable membership function circuits and the feedback loops including them. Three GC-MFC's in each antecedent block and a resistor R_j in each consequent block are programmable in accordance with the given fuzzy rules.

Therefore, the voltage space is shifted to the negative in order to calculate the center of gravity in the positive region and shifted back to the positive. The aspect is illustrated in Fig. 25(b), where E_0 is the offset voltage.

As a result, we can get the following equations for the unipolar space of consequent from Fig. 25(a), providing that E_S = 10 V, E_0 = 2 V, $R_f = 10\,\text{k}\Omega$:

$$R_j = \frac{25}{1 + \dfrac{5C_j}{C_{\max}}}\ (\text{k}\Omega) \tag{15}$$

and $R_{\max} = 25\,\text{k}\Omega$ and $R_{\min} = 4.17\,\text{k}\Omega$. Therefore, the resistance ratio is

$$\frac{R_{\max}}{R_{\min}} = 6 \tag{16}$$

and the value is reasonable to realize.

On the other hand, we can get the following equation for the bipolar space of consequent from Fig. 25(b), providing that $E_S = 10\,\text{V}, E_0 = 7\,\text{V}, R_f = 10\,\text{k}\Omega$:

$$R_j = \frac{50}{5\dfrac{C_j}{C_m} + 7}\ (\text{k}\Omega) \tag{17}$$

and $R_{\max} = 25\,\text{k}\Omega$ and $R_{\min} = 4.17\,\text{k}\Omega$.

A very simple example of three fuzzy rules is presented as

$$\begin{array}{ll} \text{Rule 1} & \text{If } x \text{ is } NL, \text{ then } c \text{ is } +5\,\text{V} \\ \text{Rule 2} & \text{If } x \text{ is } PS, \text{ then } c \text{ is } \ \ 0\,\text{V} \\ \text{Rule 3} & \text{If } x \text{ is } PL, \text{ then } c \text{ is } -5\,\text{V}. \end{array} \tag{18}$$

The membership functions in antecedents are assigned as shown in Fig. 26. Assuming that C_m is 5 V, three resistors R_1, R_2, and R_3 in consequent blocks for these three rules are calculated by (17) to be $R_1 = 4.17\,\text{k}\Omega, R_2 = 7.14\,\text{k}\Omega$, and $R_3 = 25\,\text{k}\Omega$, respectively. Fig. 27 shows the input-output characteristic of this singleton controller which verifies good linear interpolation of a singleton controller.

VI. Stabilization of a Glass with Wine

The best way to evaluate a tool is to use it for a typical problem. An inverted pendulum is the most popular example to evaluate the control technology [51]. It is a model of stabilizing a pole on a palm. The author presents a more interesting example to evaluate the control technology. It is stabilization of a glass with wine on a finger as shown in Fig. 28(a). This is very difficult, if not impossible, to control the finger to stabilize the glass, because the glass includes wine which swings back and forth by the external disturbance to change the center of gravity of the glass. How it can be stabilized?

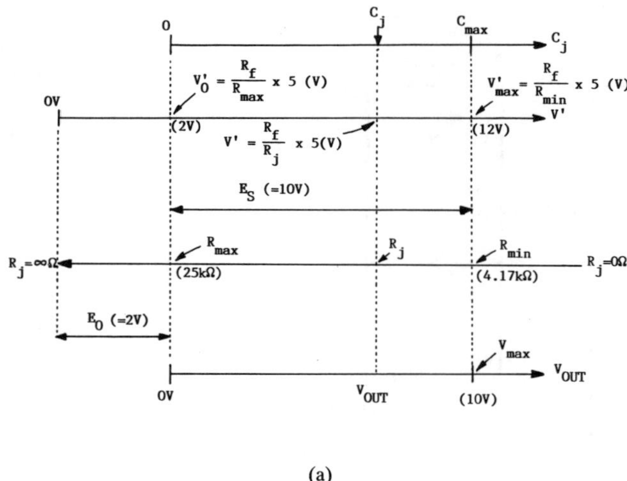

(a)

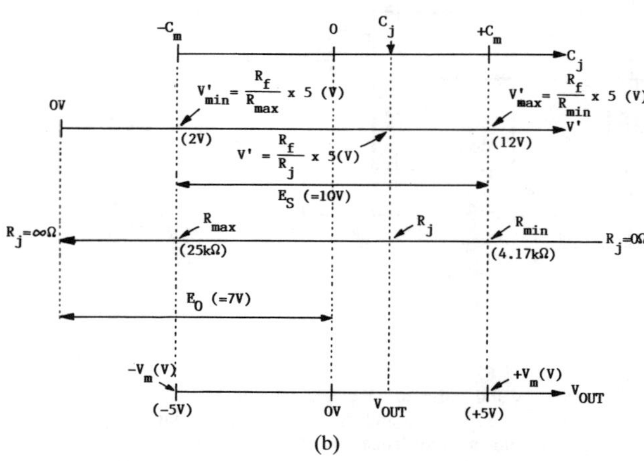

(b)

Fig. 25. A reference diagram of the consequent space which gives the resistance R_j of the jth consequent block from the deterministic value Cj in the consequent of the jth rule. The way of estimating the resistance R_j depends upon whether the consequent is (a) a unipolar space or (b) a bipolar space.

A. Mathematical Approach

An approach to solve this problem, which is very familiar to us, is a mathematical one. At first, we make a simplified model to obtain the mathematical equations which describe the dynamics of this problem. Fig. 28(b) is one example of modeling. The vehicle corresponds to the finger and is driven by the controller not to make the sphere on the pole fall down. The sphere corresponds to the cup, the pole to the support, and the pivot to the tip of the finger. x is a distance from a set point (the place where it is desired to be stabilized). The mathematical equations describing this dynamics are obtained

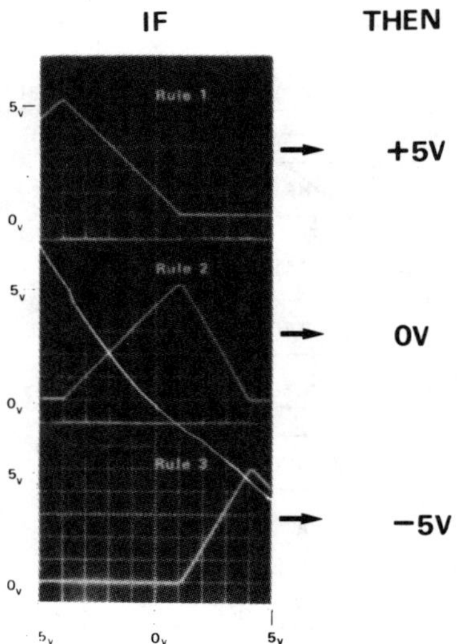

Fig. 26. Membership functions of NL (top), PS (middle), and PL (bottom) which are used in the antecedents of three rules for testing a singleton controller.

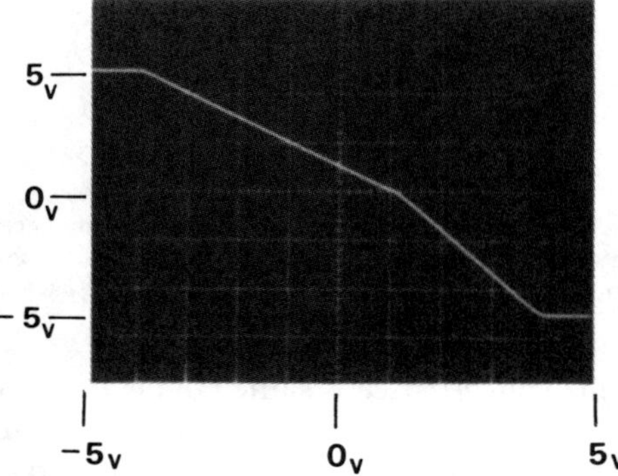

Fig. 27. Input-output characteristic of the singleton controller, the fuzzy rules of which are; Rule 1: If x is NL, then c is +5 V, Rule 2: If x is PS, then c is 0 V, Rule 3: If x is PL, then c is −5 V. This characteristic verifies good linear interpolation of a singleton controller.

as in (19)–(21), shown at the bottom of the page.

$$A = \tfrac{1}{2}\rho\pi(-\tfrac{1}{5}h^5 + rh^4 - \tfrac{5}{3}r^2h^3 + r^3h^2) \tag{22}$$

$$B = \tfrac{1}{2}\rho\pi(\tfrac{1}{2}h^4 - 2rh^3 + 2r^2h^2) \tag{23}$$

$$C = \tfrac{1}{2}\rho\pi(rh^2 - \tfrac{1}{3}h^3) \tag{24}$$

where M and m are the masses of the vehicle and the pole, respectively, $2L$ the length of the pole, r the radius of the sphere, h the depth of wine, ρ the specific gravity, θ_1 and

---

$$B\ddot{x}\cos\theta_2 + B(r+2L)\{(\ddot{\theta}_1 - \ddot{\theta}_2)\cos(\theta_1-\theta_2) - (\dot{\theta}_1^2 + \theta_2^2)\sin(\theta_1-\theta_2)\} = 0 \tag{19}$$

$$\begin{aligned}(C+Lm)\ddot{x}\cos\theta_1 + \{\tfrac{4}{3}mL^2 - 2C(r+2L)\}\ddot{\theta}_1 - Lm + 2C(r+2L)\,\mathrm{g}\sin\theta_1 \\ - B(r+2L)\{\ddot{\theta}_2\cos(\theta_1-\theta_2) + \dot{\theta}_2^2\sin(\theta_1-\theta_2)\} + D\dot{\theta}_1 = 0\end{aligned} \tag{20}$$

$$(2A+M+m)\ddot{x} + (C+Lm)(\ddot{\theta}_1\cos\theta_1 - \dot{\theta}_1^2\sin\theta_1) - B(\ddot{\theta}_2\cos\theta_2 - \dot{\theta}_2^2\sin\theta_2) = u \tag{21}$$

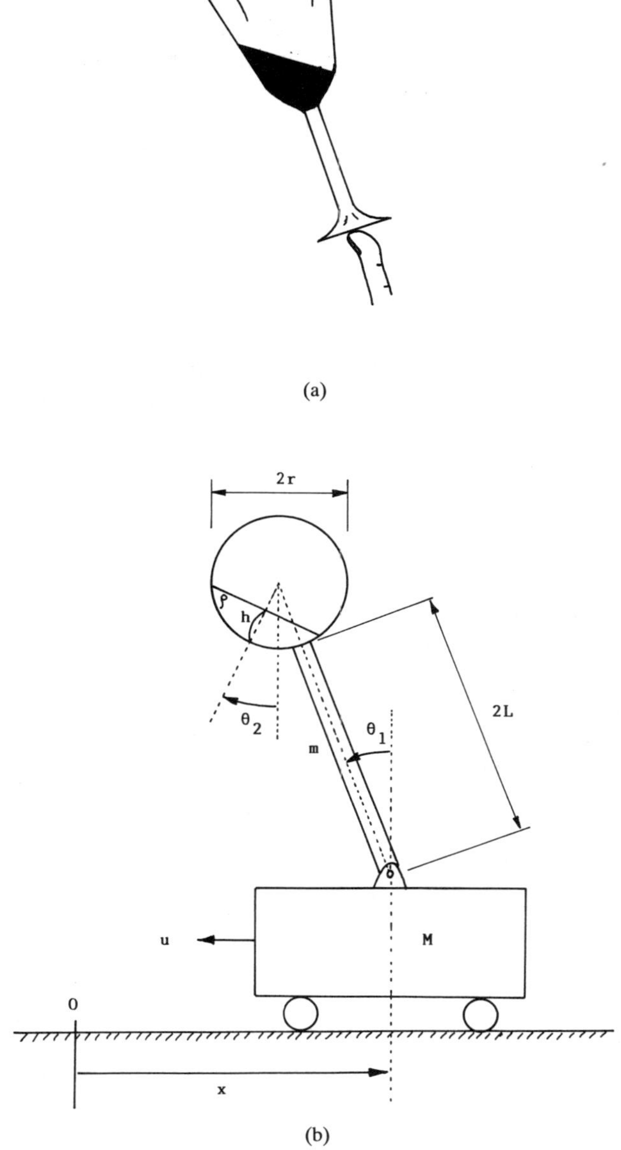

Fig. 28. (a) Stabilization of a glass with wine on a finger tip. (b) A simplified model of the glass stabilization.

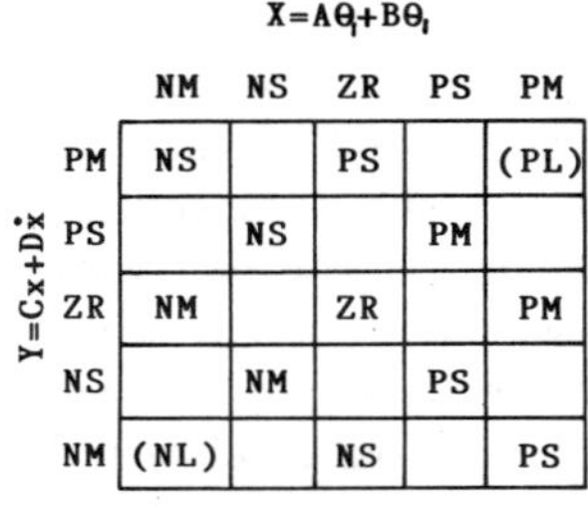

$X = A\theta_1 + B\dot{\theta}_1$ (columns); $Y = Cx + D\dot{x}$ (rows)

	NM	NS	ZR	PS	PM
PM	NS		PS		(PL)
PS		NS		PM	
ZR	NM		ZR		PM
NS		NM		PS	
NM	(NL)		NS		PS

(a)

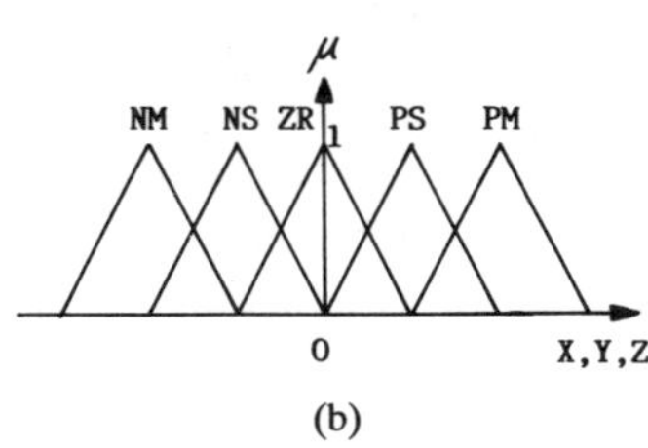

(b)

Fig. 29. (a) A rule map including fuzzy rules for stabilizing a glass with wine. (b) Membership functions which characterize the fuzzy linguistic terms used in the rule map.

θ_2 the angle of the pole, and the wine, respectively. When $\theta_1 = \theta_2$, this system reduces to a problem of the ordinary inverted pendulum.

Although this model is a simplified one, it is very difficult to derive (19)–(24). It needs high level knowledges about mathematics and physics. Even though these equations are derived, the analytical results on the criteria for stability, the dynamic response, the accuracy of position, and so on may not be necessarily reliable because the model shown in Fig. 28(b) is not the exact model. Therefore it is easy to come to a deadlock at this stage.

B. Fuzzy Linguistic Approach

The attempt is made to solve this problem without advanced mathematics but with knowledge like techniques possessed by artists, musicians, sport players, etc. Their knowledge may be represented by the relation between causes and results, or actions and reactions in the form of IF-THEN rules including fuzzy linguistic terms. Let us consider how the glass with wine is stabilized by the fuzzy linguistic approach.

1) Design of a Fuzzy Logic Controller: When we design a fuzzy logic controller, we have to think about the following items.

1. *What are the actuators (motor, heater, speaker, etc.)?* What are the variables in the consequent of fuzzy rules?
2. *What should be measured for controlling the system?* What are the variables in the antecedent of fuzzy rules?
3. *How wide are the ranges of sensors and actuators?* What are the minimum and maximum values of the above variables? All the labels should be assigned within this range.
4. *How many labels are needed to describe the strategy of control?* How precisely the system should be described? If the system should be described precisely, the support of each membership function should be narrow.
5. *What shape of membership functions are suitable for describing the expert's intuition?* Should it be bell-shaped, S-shaped, Z-shaped, triangular, trapezoidal, or of some other shape? The shape should not be necessarily symmetrical.
6. *What is the strategy (a set of fuzzy rules)?* What are the useful fuzzy rules? Which conjunction should be used to combine variables in the antecedent and/or the consequent? Of course, there is no problem, even though the strategy includes contradictory rules or exceptional rules. There is no need to adopt all the combinationally possible rules.

The velocity of the vehicle was adopted as a variable in the consequent of fuzzy rules. Of course the velocity is proportionally related to that of the driving motor.

It is obvious that the pole angle θ_1 and x can be the variable in the antecedent. We have to assign the different rules for positive $\dot{\theta}_1$ and negative $\dot{\theta}_1$ even in case of the same θ_1. So that $\dot{\theta}_1$ is also the variable to be considered in the antecedent. In the similar manner, $\dot{x}$ is also the variable in the antecedent.

The variable gain in the path of each signal flow is very useful to adjust the sensitivity of the fuzzy linguistic term (bandwidth, fuzziness, and label of the word). By extension or compression of the membership function with respect to the universe of discourse (physical quantity such as length, weight, speed, amount of fuel, brightness, etc.), the meaning of the term can be changed.

2) Reduction of Fuzzy IF-THEN Rules: When four variables $\theta_1, \dot{\theta}_1, x, \dot{x}$ are used in the antecedent and 5 labels are adopted for each variable, $5^4 = 625$ rules should be examined. It is not so easy for a designer to do it. To cope with this problem, the number of variables should be reduced. The rule chip does not accept more than three inputs, so that four variables should be reduced to two or three, preferably two, because of the number of possible rules.

The simplest way of variable reduction is a linear combination of two variables to one variable. There are some possibilities.

The first possibility is

$$X = x + 2L\theta_1 \quad \text{(position of the pole tip)} \tag{25}$$

$$Y = \dot{x} + 2L\dot{\theta}_1 \quad \text{(velocity of the pole tip)} \tag{26}$$

where X and Y are the new variables.

Equations (25) and (26) represent the position and the velocity of the tip of the pole, respectively. In order to judge whether the system is stable or not, we need the position and the velocity of the vehicle as well as (25) and (26). Because we cannot judge the stability of the pole with information of the top tip of the pole. We need information on the bottom tip of the pole related to the top tip. Thus (25) and (26) are incomplete.

The second possibility is

$$X = a\theta_1 + b\dot{x} \tag{27}$$

$$Y = c\dot{\theta}_1 + dx \tag{28}$$

where $a, b, c,$ and d are constants. It is impossible to identify the physical meaning of X and Y in (27) and (28), respectively. Since the physical meaning of these new variables are not clear, it is impossible to represent the knowledge (fuzzy rules) by using intuitive fuzzy linguistic terms.

The third possibility is

$$X = A\theta_1 + B\dot{\theta}_1 \tag{29}$$

$$Y = Cx + D\dot{x} \tag{30}$$

where $A, B, C,$ and D are positive constants. When θ_1 is positive and $\dot{\theta}_1$ is also positive, the pole is now falling down to the clockwise and X is big. When θ_1 is positive and $\dot{\theta}_1$ is negative, the pole is now returning to the equilibrium point from the positive angle (we do not feel the *emergency* so much) and X is smaller than the former case. Thus X seems to be a measure of emergency in the angle. In the similar manner, Y seems to be a measure of emergency in the position. Therefore, we can assign the label to these physical measures by intuition. For example, "X is negatively medium (NM) and Y is negatively medium (NM)" means the situation that the pole is falling down to counterclockwise (left hand side) and that the vehicle is now moving to the left from the negative position (i.e., the vehicle is going away from the set point $(x = 0)$)." In this situation, anybody will respond to move the vehicle to the left much faster at the sacrifice of accuracy of position to recover the pole. This strategy gives a fuzzy rule "If X is NM and Y is NM, then $\dot{x}$ should be NL," where negative velocity of the vehicle means the left direction.

In this manner, the fuzzy rules can be assigned in the rule map shown in Fig. 29(a). There is no need to assign all the possible rules. If each membership function is defined as Fig. 29(b), neighboring membership functions penetrate each other. Therefore, a defect of one rule can be compensated (interpolated) by surrounding four rules. Finally, the author obtained eleven rules to stabilize the glass with wine. Two bracketed rules are preferably added against the strong disturbance.

3) Control Systems and Experimental Results: In order to verify the utility of fuzzy inference, it is applied to a controller. The author presents two examples as systems under control. The one is a glass with wine on the plate attached to the inverted pendulum. The other is a mouse moving around on the plate attached to the inverted pendulum.

a) Glass with Wine: The block diagram of this equipment is shown in Fig. 30. A vehicle is driven by a dc servo motor through a flexible steelwire. The position x of the vehicle is detected by a potentiometer connected to the wheel which is driven by the flexible wire. x is differentiated by an analog differentiator and multiplied by a constant D to be fed to the adder. x is also multiplied by a constant C to be fed to the same adder. The output $Y = Cx + D\dot{x}$ is fed to one input of the fuzzy logic controller. On the other hand, the vehicle carries an brushless angle sensor employing hole effect, on the rotating rod of which bottom tip of the pole (inverted pendulum) is fixed. A plate is attached to the tip of this pole and a glass with wine is put on this plate. The angle of the pole θ_1 is detected by the angle sensor to be fed to the differentiator and multiplied by a constant B. θ_1 is also multiplied by a constant A. $A \cdot \theta_1$ and $B \cdot \dot{\theta}_1$ are summed by the adder to be fed to the fuzzy logic controller. A fuzzy logic controller possesses 11 rule chips and 1 defuzzifier chip. A defuzzified inference result $\dot{x}$ is produced from the output of the controller to a servo driver to amplify the electric power which is applied to a dc servo motor to drive the vehicle as it should be.

The glass is successfully stabilized independent of the amount of wine and also independent of the length of support of the glass. This aspect is interesting. In case of mathematical approach, when a glass is changed, or wine is added or reduced, parameters $m, r, L,$ and h should be changed and the stability condition will be changed. However, in case of fuzzy logic control, there is no need to change the rules nor the shapes of membership functions. Fuzzy boundary of a fuzzy linguistic term may absorb the deviation of parameters.

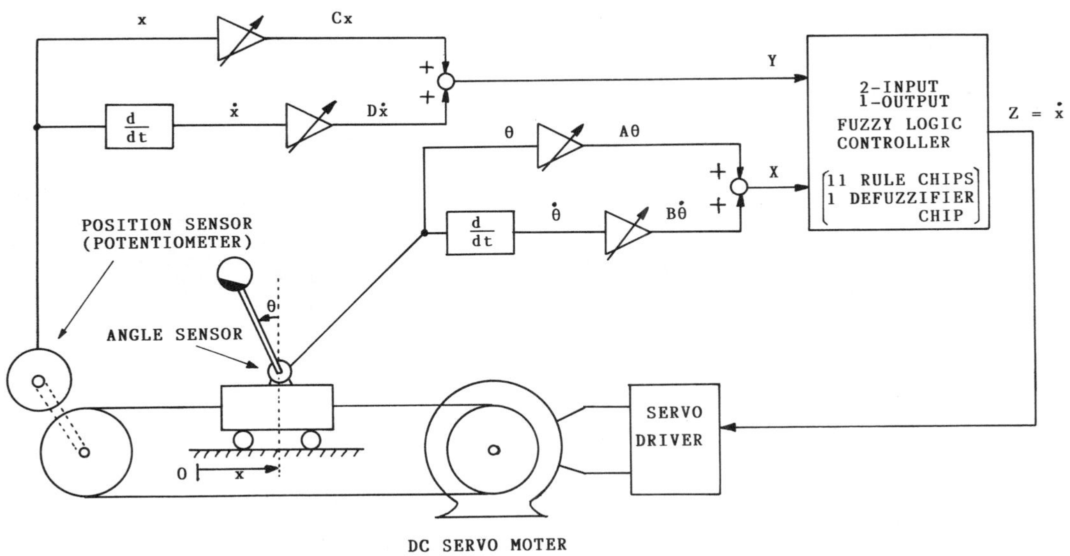

Fig. 30. A block diagram of a wine glass stabilization.

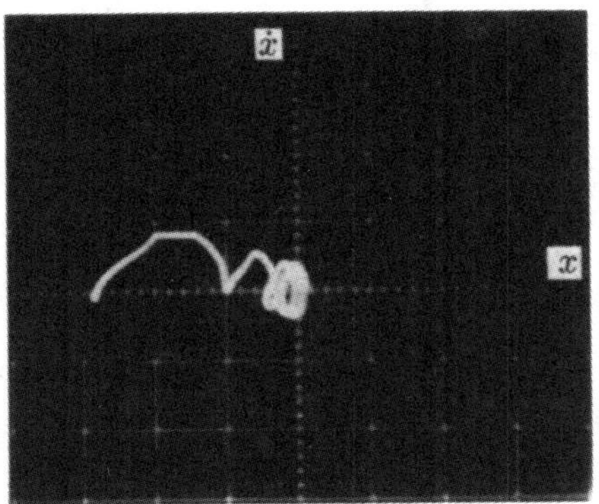

Fig. 31. Phase-plane portrait after applying a negative displacement (left) to the glass.

In order to examine the stability, a disturbance in the position was applied. Fig. 31 shows the phase-plane portrait after applying a negative displacement (left) to the glass. This portrait shows us that the vehicle comes up to the set point, swinging and balancing.

b) Mouse Moving Around: A glass with wine is not living. Therefore the movement or dynamics of this system can be described by mathematics, although it is very difficult. In other words, it may be modeled. On the other hand, let think about a living mouse moving around on the plate attached to the inverted pendulum. We cannot expect the movement nor describe the dynamics, because the center of gravity changes in accordance with the movement of the mouse. Thus we come to a deadlock again by the mathematical approach. However, the fuzzy linguistic approach can stabilize it. Fig. 32 shows the balancing motion. When a mouse moved to the right a little, the vehicle was driven by the fuzzy logic controller to the right to compensate the shift of the center of gravity at the sacrifice of the accuracy of the position.

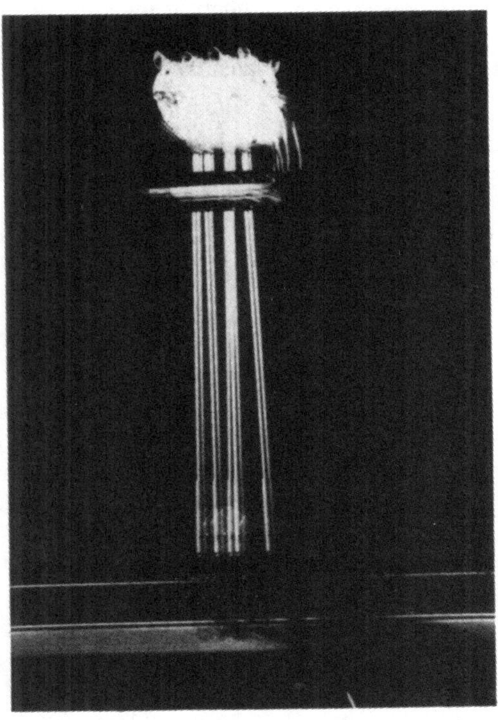

Fig. 32. Stabilization of a moving mouse which changes its center of gravity. The fuzzy logic controller can compensate the movement of the mouse at the sacrifice of the accuracy of the position.

C. Comparison with a Traditional PID Control

It is necessary for a designer to know the differences between a traditional PID control and a fuzzy logic control.

A typical control system (1-input and 1-output) employing a PID controller is shown in Fig. 33. The output state of the system under control is detected by a sensor. It is compared with the input signal (reference input) to derive an error signal

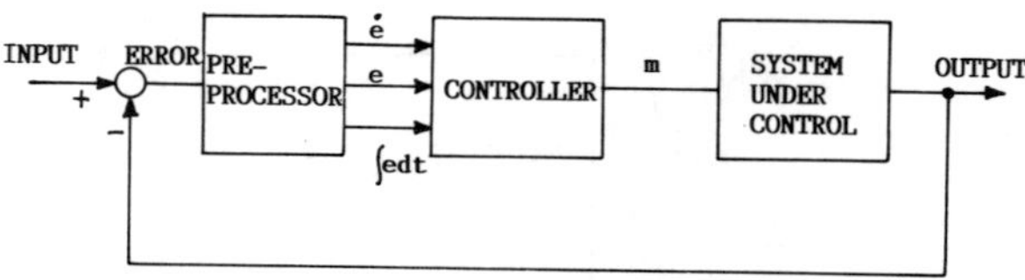

Fig. 33. A typical control system (1-input and 1-output) employing a PID controller.

$e(t)$, where the input signal is externally given and represents a desired state of the system. The error signal is delivered to a controller to produce an appropriate manipulating signal $m(t)$, which change the state of the system under control.

In a control system, there are two main objectives. The first one is to make the state (or output) of the system to be very close or equal, if possible, to the set point (or reference input). In other words, a small steady-state error $e(t)$ or a high steady-state accuracy is desired. The second one is to maintain the transient performance of the system within reasonable limits.

In order to design a controller of high steady-state accuracy and high speed settling, we need a linear combination of three control actions, i.e., proportional control action, integral control action and derivation (PID) control action. It is called a PID control and characterized by the following equation.

$$m(t) = K_P \cdot e(t) + K_I \int e(t)\,dt + K_D \frac{de(t)}{dt}. \tag{31}$$

This controller has three inputs and one output. Let us consider a controller of two inputs and one output to visualize the relationship between input and output. The first and the second terms in (31) is considered and differentiated to

$$\dot{m}(t) = K_P \dot{e}(t) + K_I e(t) \tag{32}$$

which is the linear description of a traditional PI controller.

Three-dimensional control surface in Fig. 34 shows the relation between the change of manipulator output and the linear combination of change of error input and error input, where the constant gains K_P and K_I are $\frac{2}{3}$ and $\frac{1}{3}$, respectively. Input-output characteristics of a fuzzy logic controller can be described by a set of fuzzy rules, antecedents and consequents of which corresponds to inputs and outputs of the controller, respectively. Input signal(s) applied to the controller and the fuzzy rules designed by the system designer produce the output signal(s) through fuzzy inferences and defuzzification(s) as described in Section IV-A and Section IV-B.

Here is presented an example of a fuzzy logic controller which has two inputs and provides one output. These two inputs are assigned to be change of error $\dot{e}(t)$ and error $e(t)$ and one output to the change of manipulating signal (control output) $\dot{m}(t)$ for comparison with the traditional PI controller. A set of fuzzy rules are indicated on the rule map in Table IV and membership functions used in these rules are shown in Fig. 35(a). Fig. 35(b) shows three-dimensional control surface of the *fuzzy PI controller,* the rules of which are given in Table IV.

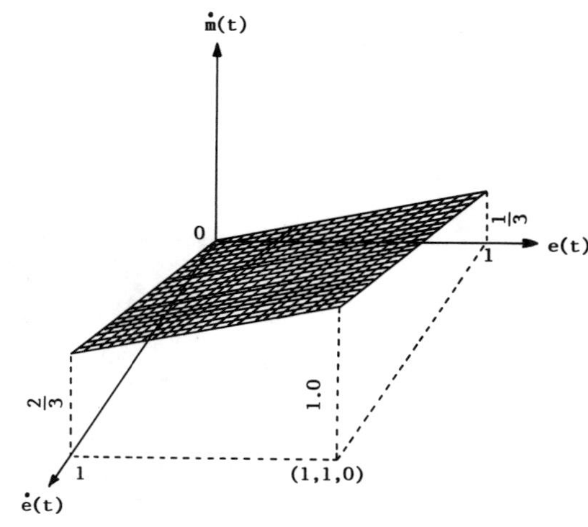

Fig. 34. The relation between the change of manipulator output and the linear combination of change of error input and error input. $K_P = \frac{2}{3}, K_I = \frac{1}{3}$. This can describe only a simple plane.

TABLE IV

$e(t)$	$e(t)$			
$\dot{e}(t)$	ZR	S	M	L
ZR	L	M	S	S
S	M	M	ZR	M
M	ZR	S	[S]	M
L	S	ZR	M	L

ZR: approximately zero, S: small, M: medium, L: large

The solid mesh in Fig. 35(b) is the inference result from fuzzy rules employing the solid membership function for "medium (M)" as shown in Fig. 35(a). When the left slope of the membership function of "medium" is reassigned as a dotted line illustrated in Fig. 35(a), then the circumference of the mesh is changed as illustrated by dotted lines in Fig. 35(b). Let us consider the case of $\dot{e}(t) = 1$ and $e(t) = 0.25$. When the membership function of "medium" is defined as the dotted line in Fig. 35(a), then $(\dot{e}(t), e(t)) = (1, 0.25)$ fires only one rule of 16 rules above. That is "If $\dot{e}(t)$ is L and $e(t)$ is S, then $\dot{m}(t)$ is ZR." So that after the inference followed by defuzzification, $\dot{m}(\dot{e}(t), e(t)) = \dot{m}(1, 0.25) = 0$ is obtained, i.e., (1, 0.25, 0) in Fig. 35(b). On the other hand, when the membership function "medium" is defined as a solid line in Fig. 39(a), then $(\dot{e}(t), e(t)) = 1, 0.25)$ fires two rules: Rule A "If $\dot{e}(t)$ is L and $e(t)$ is S, then $\dot{m}(t)$ is ZR" and Rule B "If $\dot{e}(t)$ is L and $e(t)$ is M, then $\dot{m}(t)$ is M." The degree of soft matching between (1, 0.25) and the antecedents of Rule A and Rule B are 1 and 0.5, respectively. Therefore, the conclusion $\dot{m}(1, 0.25)$ is not exclusively obtained from Rule A, but is affected by Rule B to produce a nonzero value. This means that at the typical point (for example, $e(t) = 0.25$ in "$e(t)$ is small"), grades of membership of other fuzzy linguistic terms should be zero for providing a typical conclusion.

In any way, Fig. 35(b) exhibits a very complicated and curved surface which can never be described by a linear combination of input variables such as (31) or (34). Furthermore, it is very easy to change the control surface (input-output

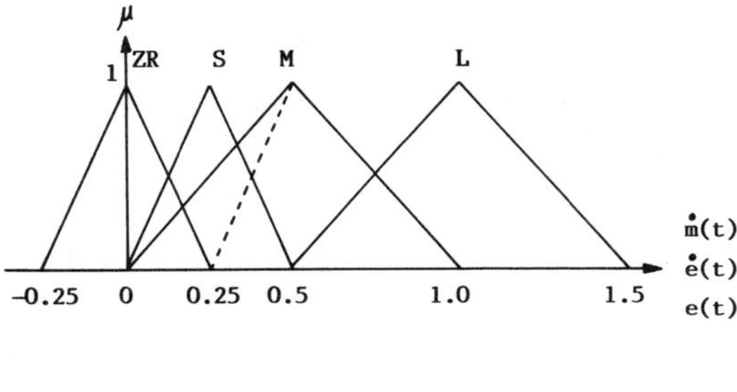

(a)

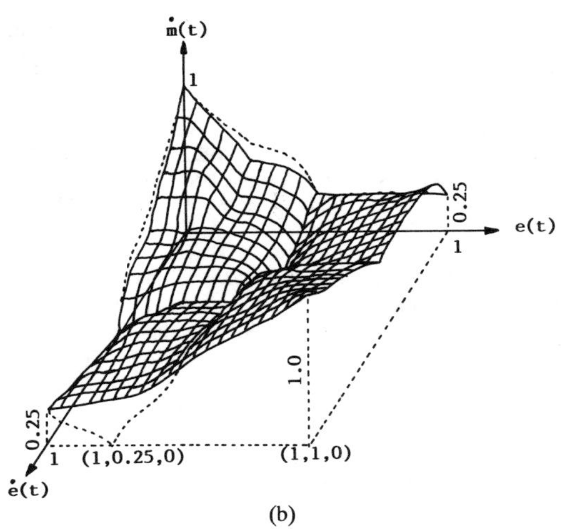

(b)

Fig. 35. (a) Membership functions used for getting the control surface. (b) The control surface of a fuzzy PI controller which achieves fuzzy inference with fuzzy rules given in Table IV. A set of fuzzy rules can describe a complicated control surface. A local change of the surface can be easily accomplished by changing the corresponding rule(s). So that a fuzzy system is suitable for learning systems, self-organizing systems, adaptive systems, etc.

characteristics) of the fuzzy logic controller in accordance with the change of the system under control. If you want to pull up the center of the control surface, you have only to change the label of the consequent in the corresponding rule, i.e., the label S (small) in the bracket in Table IV should be increased to be M (medium). If you want to change exclusively a small part of the control surface, you have to narrow the support of the corresponding membership function and increase the number of labels (fuzzy terms) to achieve the effective assignment of rules. This means the exceptional or irregular points on the control surface can be easily described in a fuzzy logic controller. Thus fuzzy rules are very suitable for describing a sophisticated system by fuzzy linguistic terms (intuitive terms). Of course, the fuzzy inference can be easily modified by the concepts of learning, self-organization, adaptation, etc.

VII. Conclusions

A brief explanation on the fuzzy set and fuzzy logic was made and followed by the comparison between mathematical description, linguistic description, and neural networks. The hardware to implement the fuzzy inference and the defuzzification is also described in detail. Lastly, a fuzzy logic controller was applied to stabilize a glass with wine put on the inverted pendulum and also a mouse moving around the plate on the inverted pendulum.

Distinctive features of a fuzzy logic controller can be summarized as follows.

1. How it should work is described with fuzzy natural languages and the structure of the knowledge is very clear. Therefore, the control strategy, know-hows, knowledge, or data are very easy to represent, very easy to understand, very easy to remember, very easy to debug.
2. Since all the rules are independent of each other, it is easy to update the rule to follow the change of the system under control. Thus a fuzzy system is suitable for a learning system, a self-organizing system, an adaptive system, etc.
3. It can accept the exceptional data (knowledge) and the contradictory data, so that the fuzzy logic controller is suitable for nonlinear and time-variant complicated systems.
4. Compound sensory signals such as outputs from a chemical sensor, an odor sensor and other contaminated cheap sensor can be accepted, as suggested in Section VI-B-2.
5. Since a fuzzy inference is one kind of interpolation with very few data, drastic reduction of data and software/hardware systems can be achieved.

The development of a fuzzy logic hardware system will be changed by another concepts. Several years also, we could find the tendency toward the fusion of fuzzy logic and neural networks [52]–[56]. Fusion of these two has lost much of its novelty now [57]–[60]. Future trends will be a fusion of fuzzy logic and chaos as well as neural networks. A fuzzy system is a modeling of a human brain summarized from the human expert's behavior and chaos is a nonlinear dynamical behavior generated by massive neural networks of the human brain.

Acknowledgment

The author would like to express his sincere thanks to Prof. Edgar Sáchez-Sinencio at Texas A&M University for giving an opportunity to the author to present this tutorial paper.

References

[1] C. L. Phillips and H. Troy Nagle, Ir., *Digital Control System Analysis and Design.* Englewood Cliffs, NJ: Prentice-Hall, 1984.
[2] R. C. Dorf, *Modern Control Systems.* New York: Addison-Wesley, 1980.
[3] K. Ogata, *Modern Control Engineering.* Englewood Cliffs, NJ: Prentice-Hall, 1970.
[4] L. A. Zadeh, "Fuzzy Sets," in *Information and Control.* New York: Academic Press, 1965, Vol. 8, pp. 338–353.
[5] V. Novak, *Fuzzy Sets and Their Applications.* New York: Adam Hilger, 1989.
[6] H.-J. Zimmermann, *Fuzzy Sets, Decision Making, and Expert Systems.* New York: Kluwer Academic, 1987.
[7] A. Kandel, *Fuzzy Expert Systems.* CRC Press, 1992.
[8] A. Kandel, *Fuzzy Mathematical Techniques with Applications.* New York: Addison-Wesley, 1986.
[9] H. J. Zimmermann, *Fuzzy Sets Theory—and Its Applications.* Kluwer Nijhoff, 1985.
[10] G. J. Klir and T. A. Folger, *Fuzzy Sets, Uncertainty, and Information.* Englewood Cliffs, NJ: Prentice Hall, 1988.
[11] D. Dubois and H. Prade, *Fuzzy Sets and Systems: Theory and Applica-*

tions. New York: Academic Press, 1980.
[12] W. J. M. Kickert, *Fuzzy Theories on Decision-Making.* Boston, MA: Kluwer, 1978.
[13] C. V. Negoita and D. Ralescu, *Simulation, Knowledge-Based Computing, and Fuzzy Statistics.* New York: Van Nostrand Reinhold, 1987.
[14] A. Kandel, *Fuzzy Techniques in Pattern Recognition.* New York: Wiley, 1982.
[15] C. V. Negoita, *Fuzzy Systems.* Abacus Press, 1981.
[16] A. Kandel and S. C. Lee, *Fuzzy Switching and Automata.* Crane, Russak & Company, Inc., 1979.
[17] M. Zemankova-Leech and A. Kandel, *Fuzzy Relational Data Bases—A Key to Expert Systems.* Verlag TUV Rheinland, 1984.
[18] J. Kacprzyk, *Multistage Decision-Making Under Fuzziness.* Verlag TUV Rheinland, 1983.
[19] R. de Mori, *Computer Models of Speech Using Fuzzy Algorithm.* New York: Plenum Press, 1983.
[20] M. Smithson, *Fuzzy Set Analysis for Behavioral and Social Sciences.* New York: Springer-Verlag, 1987.
[21] S. Miyamoto, *Fuzzy Sets in Information Retrieval and Cluster Analysis.* New York: Kluwer Academic, 1990.
[22] W. Pedrycz, *Fuzzy Control and Fuzzy Systems.* New York: Wiley, 1989.
[23] J. C. Bezdek, *Pattern Recognition with Fuzzy Objective Function Algorithms.* New York: Plenum Press, 1981.
[24] S. K. Pal and D. K. Dutta Majumdar, *Fuzzy Mathematical Approach to Pattern Recognition.* New York: Wiley, 1986.
[25] A. Kaufmann, *Introduction to the Theory of Fuzzy Subsets - Vol. 1-.* New York: Academic, 1975.
[26] G. E. Lasker, *Applied Systems and Cybernetics: Vol. VI - Fuzzy Sets and Fuzzy Systems, Possibility Theory and Special Topics in Systems Research.* New York: Pergamon Press, 1981.
[27] R. R. Yager, S. Ovchinnikov, R. M. Tong, and H. T. Nguyen, *Fuzzy Sets and Applications: Selected Papers by L. A. Zadeh.* New York: Wiley 1987.
[28] A. Kandel, *Fuzzy Mathematical Techniques with Applications.* New York: Addison-Wesley, 1986.
[29] M. Togai and H Watanabe, "A VLSI implementation of a fuzzy-inference engine: Toward an expert system on a chip," *Information Sciences,* vol. 38, pp. 147–163, 1986.
[30] H. Watanabe, "RISC approach to design of fuzzy processor architecture," in *Proc. IEEE Int. Conf. Fuzzy Systems,* San Diego, CA, Mar. 8–12, 1992, pp. 431–440.
[31] Product Manual, 'AT accelerator board," Togai InfraLogic, Inc., 1989.
[32] M. A. Manzoul and D. Jayabharathi, "Fuzzy controller on FPGA chip," in *Proc. IEEE Int. Conf. Fuzzy Systems,* San Diego, CA, Mar. 8–12, 1992, pp. 1309–1316.
[33] H. Eichfeld, M. Löhner, and M. Müller, "Architecture of a CMOS fuzzy logic controller with optimized memory organization and operator design," in *Proc. IEEE Int. Conf. Fuzzy Systems,* San Diego, CA, Mar. 8–12, 1992, pp. 1317–1323.
[34] T. Yamakawa, U.S. Patents: Fuzzy Logic Circuit, Pat. No. 4,694,418, Sept. 15, 1987, Appl. No. 714809, March 22, 1985; Multi-Functional Fuzzy Logic Circuit, Pat. No. 4,716.540, Dec. 29, 1987, Appl. No. 751447, July 3, 1985, Fuzzy Membership Function Circuit, Pat. No. 4,837,725, June 6 1989, Appl. No. 917952, Oct. 14, 1986. European Patent: Fuzzy Logic Circuit, Pat. No. 0162225, Oct. 26, 1988, Appl. No. 0162225A1, March 22, 1985, Multi-Functional Fuzzy Logic Circuit, Appl. No. 0168004A2, July 5, 1985. Japanese Patent Applications, Tokuganshou No. 59-57121, March 23, 1984, No. 59-57122, Mar. 23, 1984, No. 59-57123, March 23, 1984, No. 59-57124, Mar. 23, 1984, No. 59-57125, Mar. 23, 1984, No. 59-141250, July 6, 1984, No. 59-141251, July 6, 1984, No. 59-141252, July 6, 1984, No. 59-187656, Sept. 6, 1984, No. 59-187657, Sept. 6, 1984, No. 59-263385, Dec. 13, 1984, No. 59-263386, Dec. 13, 1984, No. 60-234640, Oct. 22, 1985, No. 60-234641, Oct. 22, 1985, No. 60-234642, Oct. 22, 1985, No. 60-234643, Oct. 22, 1985, No. 60-234644, Oct. 22, 1985, No. 60-234645, Oct. 22, 1985, No. 60-234646, Oct. 22, 1985, No. 60-234647, Oct. 22, 1985, No. 60-234648, Oct. 22, 1985. T. Yamakawa and T. Miki, "The current mode fuzzy logic integrated circuits fabricated by the standard CMOS process," *IEEE Trans. Computers,* vol. C-35, pp. 161–167, 1986.
[35] T. Yamakawa, U.S. Patents: Fuzzy Logic Computers and Circuits, Pat. No. 4,875,184, Oct. 17, 1989, Appl. No. 116,777, Nov. 5, 1987; Fuzzy Membership Function Circuit, Pat. No. 5,113,366, May 12, 1992, Appl. No. 313,722, Mar. 14, 1989. European Patent: Multi-Functional Fuzzy Logic Circuit, Appl. No. 0168004, July 5, 1985; Australian Patent: MIN/MAX Circuits: Pat. No. 602817, Appl. No. 37139/89, June 28, 1989. Japanese Patent Applications, Tokuganshou, No. 61-268564, Nov. 13, 1986, No. 61-268565, Nov. 13, 1986, No. 61-268566, Nov. 13, 1986, No. 61-268567, Nov. 13, 1986, No. 61-268568, Nov. 13, 1986, No. 61-268569, Nov. 13, 1986, No. 63-49830, March 4, 1988, No. 63-203912, August 18, 1988, No. 63-206008, August 19, 1988, No. 63-206009, August 19, 1988, No. 63-282518, Nov. 10, 1988, No. 63-282519, Nov. 10, 1988. Japanese Patent Applications, Tokuganhei, No. 1-250624, Sept. 28, 1989, No. 2-114039, April 27, 1990, No. 3-322201, Dec. 6, 1991.
[36] T. Yamakawa and H. Kabuo, "A programmable fuzzifier integrated circuit—Synthesis, design, and fabrication," *Information Sciences,* vol. 45, pp. 75–112, 1988.
[37] F. Ueno, T. Inoue, Y. Shirai, and M. Sasaki, "A maximum and minimum circuit with multiple inputs in current mode," *The Transactions of the IEICE,* vol. E-70, no. 40, pp. 392–395, 1987.
[38] M. Sasaki, T. Inoue, Y. Shirai, and F. Ueno, "Fuzzy multiple-input maximum and minimum circuits in current mode and their analysis using bounded-difference equations," *IEEE Trans. Comput.,* vol. C-39, pp. 768–774, 1990.
[39] M. Sasaki, N. Ishikawa, F. Ueno, and T. Inoue, "Current-mode analog fuzzy hardware with voltage input interface and normalization locked loop," *IEICE Trans. Fundamentals,* vol. E75-A, no. 6, pp. 650–654, 1992.
[40] O. Ishizuka, K. Tanno, Z. Tang, and H. Matsumoto, "Design of a fuzzy controller with normalization circuits," in *Proc. IEEE Int. Conf. Fuzzy Systems,* San Diego, CA, Mar. 8–12, 1992, pp. 1303–1308.
[41] M. Morisue and Y. Kogure, "A super conducting fuzzy processor," in *Proc. IEEE Int. Conf. Fuzzy Systems,* San Diego, CA, Mar. 8–12, 1992, pp. 443–450.
[42] J. L. Huertas, S. Sánchez-Solano, A. Barriga, and I. Baturone, "Serial architecture for fuzzy controllers: Hardware inplementation using analog digital VLSI techniques," in *Proc. 2nd Int. Conf. Fuzzy Logic and Neural Networks,* Iizuka, Japan, July 17–22, 1992, pp. 535–538.
[43] T. Yamakawa and K. Sasaki, "Fuzzy memory device," in *Proc. Second IFSA Congress,* Tokyo, July 20–25, 1987, pp. 551–555.
[44] T. Yamakawa, "A simple fuzzy computer hardware system employing MIN & MAX operations—A challenge to 6th generation computer," in *Proc. Second IFSA Congress,* Tokyo, July 20–25, 1987, pp. 827–830.
[45] T. Yamakawa, "High-speed fuzzy controller hardware system: The mega FIPS machine," *Information Sciences,* vol. 45, pp. 113–128, 1988.
[46] T. Yamakawa, "Intrinsic fuzzy electronic circuits for sixth generation computer," in *Fuzzy Computing,* M. M. Gupta and T. Yamakawa, Eds. Amsterdam, The Netherlands: North Holland, 1988, pp. 157–171.
[47] T. Yamakawa, "Fuzzy microprocessors - Rule chip and defuzzifier chip," in *Proc. Int. Workshop on Fuzzy System Applications,* Iizuka, Japan, Aug. 20–24, 1988, pp. 51–52.
[48] T. Yamakawa, Japanese Patent Application, Tokuganshou, No. 63-205007, Aug. 19, 1988.
[49] T. Yamakawa, "An application of a grade-controllable membership function circuit to a singleton-consequent fuzzy logic controller," in *Proc. Third IFSA Congress,* Seattle, WA, Aug. 6–11, 1989, pp. 296–302.
[50] T. Yamakawa, Japanese Patent Application, Tokuganshou, No. 63-206008, Aug. 19, 1988.
[51] T. Yamakawa, "Stabilization of an inverted pendulum by a high-speed fuzzy logic controller hardware system," *Fuzzy Sets and Systems,* vol. 32, pp. 161–180, 1989.
[52] T. Yamakawa, Japanese Patent Applications, Tokuganhei, No. 1-043600, Feb. 23, 1989; No. 1-133690, May 26, 1989.
[53] T. Yamakawa, "A fuzzy neuron and its application to pattern recognition," in *Proc. Third IFSA Congress,* Seattle, WA, Aug. 6–11, 1989, pp. 30–38.
[54] T. Yamakawa, "Pattern recognition hardware system employing a fuzzy neuron," in *Proc. Int. Conf. Fuzzy Logic & Neural Networks,* Iizuka, Japan, July 20–24, 1990, pp. 943–948.
[55] T. Yamakawa, "A fuzzy neuron and its application to a hand-written character recognition system," in *Proc. IEEE Int. Symp. Circuits and Systems,* Singapore, June 11–14, 1991, pp. 1369–1372.
[56] *Proc. Int. Conf. Fuzzy Logic and Neural Networks,* Iizuka, Japan, July 20–24, 1990.
[57] T. Yamakawa, "A design algorithm of membership functions for a fuzzy neuron using example-based learning," in *Proc. IEEE Int. Conf. Fuzzy Systems,* San Diego, CA, Mar. 8–12, 1992, pp. 75–82.
[58] B. Kosko, *Neural Networks and Fuzzy Systems.* Englewood Cliffs, NJ: Prentice Hall, 1992.
[59] *Proc. A Second Int. Conf. Fuzzy Logic and Neural Networks,* Iizuka, Japan, July 17–22, 1992.
[60] T. Yamakawa, E. Uchino, T. Miki, and H. Kusanagi, "A neo fuzzy neuron and its applications to system identification and prediction of the system behavior," in *Proc. Second Int. Conf. Fuzzy Logic & Neural Networks,* Iizuka, Japan, July 17–22, 1992, pp. 477–484.

Author Index

A

Anderson, J. A., 112
Antsaklis, P. J., 81

B

Benson, M. W., 239
Bhat, N. V., 510
Brüwer, M., 476

C

Chan, K. -K., 489
Chang, P.-R., 489
Cotter, N. E., 283
Cowan, J. D., 153
Cruse, H., 476
Cui, X., 451
Cybenko, G., 273

D

De Yong, M. R., 542
Di Massimo, C., 499
Digney, B. L., 460

F

Fields, C., 542
Findley, R. L., 542
Freeman, W. J., 129
Fukushima, K., 218
Funahashi, K., 247
Furukawa, K., 64

G

Gawthrop, P. J., 171
Gelfand, J. J., 466
Girosi, F., 257
Glanz, F. H., 516
Gupta, M. M., 1, 289, 314, 352, 403, 439, 460

H

Handelman, D. A., 466
Hecht-Nielsen, R., 47
Hoffmann, G. W., 239
Hogg, B. W., 483
Hopfield, J. J., 234, 533
Hornik, K., 265
Hunt, K. J., 171, 321
Hwang, J. -N., 554

I

Irwin, G. W., 483

J

Jang, J. S. R., 417
Jin, L., 314

K

Kawato, M., 64
Kohonen, T., 201

Kraft III, L. G., 516
Kung, S. -Y., 554
Kupfermann, I., 53
Kusanagi, H., 427

L

Lane, S. H., 466

M

McAvoy, T., 510
Mead, C., 525
Miki, T., 427
Miller III, W. T., 516
Minderman Jr., P. A., 510
Montague, G. A., 499
Morris, A. J., 499
Mukhopadhyay, S., 90

N

Narendra, K. S., 90, 329
Nguyen, D. H., 308
Nikiforuk, P. N., 314

P

Parthasarathy, K., 329
Passino, K. M., 81
Poggio, T., 257

R

Rao, D. H., 1, 289, 352, 439

S

Sbarbaro, D., 171, 321
Shin, K. G., 451
Stevens, C. F., 101
Stinchcombe, M., 265
Suzuki, R., 64

T

Tank, D. W., 533
Tham, M. T., 499

U

Uchino, E., 427

W

Wang, N. Sun, 510
Wang, S. J., 81
Werbos, P. J., 297
White, H., 265
Widrow, B., 308
Willis, M. J., 499
Wilson, H. R., 153
Wu, Q. H., 483

Y

Yamakawa, T., 427, 571
Yang, W. -H., 489

Z

Zadeh, L. A., 375, 386
Zbikowski, R., 171

Subject Index

A

Acetylcholine-activated channels, 105, 111
ACS, *See* Autonomous control systems (ACS)
Action potential, 6–8
ADALINE (ADAptive LInear NEuron), 4–5, 49, 308–9
Adaptive control, 3, 49
Alternate approach to system analysis, 386–402
  compositional rule of inference, 395–96
  fuzzy algorithms, 388, 396–401
    behavioral, 400–401
    decisional, 401
    definitional, 398
    generational, 398–400
    relational, 400–401
  fuzzy conditional statements, 394–96
  fuzzy relations, 389–90
  fuzzy sets, 388–92
    language and meaning, 391
    operations on, 390–91
  linguistic hedges, 391–92
  linguistic variables, 387–88
    computation of meaning of values of, 392–94
  principle of incompatibility, 386
ANFIS, 417–26
  basics of, 417–18
    adaptive networks, 417–18
  defined, 418–19
  See also Fuzzy controllers
ANN, *See* Artificial neural network (ANN)
Approximate realization of continuous mappings, 247–56
  multilayer neural networks, 248
  theorems, 248–52
    Kolmogorov-Arnold-Sprecher's theorem, 253–54
    proof of, 252–53, 254
Approximation:
  best approximation, 20, 257–64
  by superpositions of sigmoidal functions, 273–82
  and multilayer feedforward networks, 265–72
  and networks, 259–61
Artificial neural networks (ANNs), 2, 4, 169, 171, 179
  algorithmic aspects of, 554–55
    learning phase, 555
    retrieving phase, 554–55
  applicational aspects of, 555–56
    associative retrieval/classification, 556
    optimization, 555–56
    robotic applications, 556
  architectural aspects of, 556–57
  in process engineering, 499–509
    potential application areas, 503–8
    process modeling, 500–503
  and robotic processing, 559–63
Associative learning, 54–55
Associative memory, 23
Autonomous control systems (ACS), 81–89
  characteristics of, 85–86
  control functions, 81–82
  controller, functional architecture of, 83–85
  design methodology, 82
  mathematical models for, 86
Autonomous machines, 1–3
Axon, 8
Axon hillock, 7–8

B

Backpropagation, 18–20, 510–11
  basic backpropagation, 297–301
    approach, 299
    calculating derivatives, 299–301
    code, 300–301
    equations, 300
    simple feedforward networks, 298–99
    supervised learning problem, 297–98
  temporal, 420
  through time, 301–6
    background, 301
    code, 302–3
    equations, 302

example of recurrent network, 301–2
extensions of, 303–6
Behavioral algorithms (fuzzy), 400–401
Best approximation, 20, 257–64
defined, 258–59
and GRBF, 261
Biological neuronal-control, 1–3
conventional design, premises of, 3–4
morphology, 6–9
neural control with learning algorithm, 4–5
Biological processes, neuronal morphology of, 99–100
Brain maps, 202

C

Cell body, 103
Cellular neural networks (CNNs), 179
Cerebellar Model Articulation Controller (CMAC), 5–6, 23, 31–32, 179, 516–22
description of, 517–19
example applications, 519–21
pattern recognition, 520–21
robot control, 519–20
signal processing, 521
properties of, 519
Chaotic system, 22
Chemical process systems:
modeling via neural computation, 510–15
backpropagation, 510–11
biosensor data interpretation, 513–15
dynamic example, 512–13
Classical conditioning, 54–56, 57
CMAC, *See* Cerebellar Model Articulation Controller (CMAC)
CNNs, *See* Computational neural networks (CNNs)
Cognition, 218–33
computer simulation, 225–33
cell responses of each layer, 226
parameters for, 225–26
response to resembling patterns, 231–33
reverse reproduction, 226–28
time-course of synapse organization, 228–31
neural element, 221–22
structure of, 222–25
synapse modification, hypothesis of, 219–21
Cognitive machines, 1–3
Cognitive/psychological computation with neural models, 112–28
associative models, 115–17
biological assumptions, 112–15
categorization, 117–24
example of, 124–26
Competitive learning, 17
early work on, 202–3
Competitive learning networks, and robotic applications, 558–59
Compositional rule of inference, 395–96
Computational neural networks (CNNs), 2, 10–11, 169–244
benefits of, 11
control systems, neural networks for, 171–200
neurons with graded response, 234–38
neurons with hysteresis, 239–44
process, steps in, 10
self-organizing map, 201–17
self-organizing multilayered neural network, 218–33
Conditioned response/stimulus, 54
Connectivity matrix, 24
Containment, fuzzy sets, 376
Continuous mappings, approximate realization of, 247–56
Control functions, autonomous control systems (ACS), 81–82
Control systems:
conventional design, premises of, 3–4
dynamic networks, learning in, 182–84
fuzzy-neural control systems, 373–433
network architectures, 174–79
neural networks for, 171–200, 308–13
representations/identification/control structures, 184–91
static networks, learning in, 179–82
Convexity, fuzzy sets, 381–85
Correspondence problem, 560

D

DACS, *See* Distributed adaptive control systems (DACS)
Decaying-exponential networks, 285
Decisional algorithms (fuzzy), 401
Declarative memory, 58, 61–62
Definitional algorithms (fuzzy), 398
Dendrite, 8
Design methodology, autonomous control systems (ACS), 82
Direct inverse control, 189
Discriminative training, 55
Dishabituation, 54
Distributed adaptive control systems (DACS), 460–65
for quadruped mobile robot, 462–65
simulation/results, 463–65
DNP, *See* Dynamic neural processor (DNP)
DNUs, *See* Dynamic neural units (DNUs)
Dynamical systems:
backpropagation in, 334–37
characterization of, 330
control of, 329–51
adaptive, 344
simulation results, 344–50
identification, 329–51
characterization, 337–38
parallel identification model, 338–39
series-parallel identification model, 339
simulation results, 339–43
input-state-output representation of, 330–31
Dynamic (feedback) neural networks, 21–32
dynamic neural processor (DNP), 27, 28–29
dynamic neural unit (DNU), 24–26, 30
generalized model, 29–30
learning in, 182–84
Dynamic neural processor (DNP), 27, 28–29
architecture/mathematical model of, 440–43
based on neural subpopulations, 440
defined, 435
Dynamic neural units (DNUs), 24–26, 30, 289–94, 352–71, 440–41
function approximation using, 291–93
isolated, 353–60
architectural details, 353–55
computer simulation results, 363–69

function approximation, 355–56
learning/adaptive algorithms, 356–60
terminology, 370–71
mathematical model of, 290–91
multistage, 360–63
nonlinear activation function, 290
proposed model, architectural details of, 289–90

E

Error-based learning algorithms, 17
Excitatory/inhibitory model neurons:
limit cycles, 163–66
localized populations containing, 153–68
model, 154–57
plane phase analysis, 158–63
time coarse graining, 157–58
Excitatory neuron, defined, 8
Exponentiated-function networks, 286–87
Extinction, 55

F

Feedback neural networks, *See* Dynamic (feedback) neural networks
Feedforward neural networks, *See* Static (feedforward) neural networks
Firing, neurons, 7–8
Forward gain, neural masses, 133–38
Fourier networks, 286
Functional approximation, 20–21
Fuzzy algorithms, 388, 396–401
behavioral, 400–401
decisional, 401
definitional, 398
generational, 398–400
relational, 400–401
Fuzzy ARTMAP, 33
Fuzzy conditional statements, 394–96
Fuzzy controllers, 417–26
inverted pendulum system application, 420–23
simulation results, 422–23
simulation settings, 421–22
stage adaptive network, 419–20
trajectory adaptive network, 420
See also ANFIS
Fuzzy inference engine, 571–96
evaluation of technology, example of, 589–95
fuzzy logic controller, 575–79
defuzzification, 578–79
fuzzy interference, 575–78
hardware implementation, 579–89
fuzzy sets/fuzzy logic, 571–73
modeling of system, 573–75
ANNs, 575
linguistic rules, 573–75
mathematical equations, 573
Fuzzy interval, 572
Fuzzy logic, 5, 33–34, 403–16, 571–73
See also Neurons
Fuzzy-neural control systems, 373–433
Fuzzy-neural structure, 32–38
architectures, 34–35
fuzzy logic, 5, 33–34
fuzzy set definition, 34
learning/adaptation in, 36–37
model, 35
unipolar to bipolar transformation, 35–36
Fuzzy sets, 35–36, 375–85, 388–92, 571–73
algebraic operations on, 378–80
complement of, 376
containment, 376
convexity, 381–85
defined, 375–76
fuzzy relations, 389–90
intersections, 377, 378
language and meaning, 391
notation/terminology, 388–89
operations on, 390–91
unions, 376–77, 378

G

Generalized Radial Basis Function (GRBF), and best approximation, 261
Generational algorithms (fuzzy), 398–400
Glial cells (glia), 101
Graded response, neurons with, 234–38

H

Hebbian learning, 17
Heterosynaptic plasticity, internal neural model with, 57–58
Hierarchical neural network model for control/learning of voluntary movement, 64–80
Hopfield associative/optimations neural networks, 557–58
Hysteresis:
with neurons, 30–31, 239–44

I

Identification:
dynamical systems, 329–51
characterization, 337–38
parallel identification model, 338–39
series-parallel identification model, 339
simulation results, 339–43
Imitation learning, 54
Inhibitory neuron, defined, 8
Input-state-output representation of dynamical systems, 330–31
Intelligent control, 90–97
assumptions/theoretical questions, 96
comments/conclusions, 96–97
defined, 90
problems, 92–96
statement of problem, 90–92
Intelligent coordination of multiple systems, 451–59
Intelligent machines, 1–3
Intelligent robotic control, 466–75
declarative/reflexive processing, integration of, 467–68
execution monitor, rule-based, 472

human skill acquisition, features of, 466–67
joint reflexes, 469
neural network training, rule-based supervision of, 468–72
simulation results, 472
swing reflex modulator:
network-based, 470–72
rule-based, 469–70
Internal model control (IMC), 189–91, 322
See also Nonlinear internal model control
Intersections, fuzzy sets, 377, 378
Inverse kinematic transformations:
on-line learning of, 445–46
computer simulation studies, 446
learning scheme, 445–46
Isocline curve, 29
Isolated dynamic neural units (DNUs), 353–60
architectural details, 353–55
computer simulation results, 363–69
function approximation, 355–56
learning/adaptive algorithms, 356–60
terminology, 370–71

K

Kolmogorov-Arnold-Sprecher's theorem, 253–54
Kolmogorov theorem, 20–21

L

Labels, 572
Latency period, defined, 8
Lateral inhibition, defined, 8
Law of effect, 56
Leaky learning model, 559
Learning, 53–63
associative, 54–55
classical conditioning, 54–56, 57
competitive, 17, 202–3
conditioned response/stimulus, 54
defined, 53
in dynamic (feedback) neural networks, 182–84
extinction, 55
Hebbian, 17
nonassociative, 54
operant conditioning, 56–57, 58
process, 53
in static (feedforward) neural networks, 179–82
study of, 53–54
types of, 54
unconditioned response/stimulus, 54
of voluntary movement, 64–80
See also Memory
Learning control system, defined, 4–5
Learning epoch, 16
Learning rules, 16
Learning vector quantization (LVQ) methods, 207–9
Linear analysis:
neural masses, 129–52
AEPs (averaged evoked potential), 131–50
application of, 132
purpose in using, 130–32
Linguistic hedges, 391–92
Linguistic variables, 387–88
computation of meaning of values of, 392–94
Long-term memory, 60

M

Membership functions, defined, 572
Memory:
associative, 23
declarative, 58, 61–62
defined, 53
long-term, 60
memory traces, 60–61
principles of, 59–62
reflexive, 58, 61
stages of, 59–60
See also Learning
Model-based robotic systems, 1
Model reference adaptive control (MRAC), 3
Modified logistic network, 287
Modified sigma-pi and polynomial networks, 286
Multilayer feedforward networks, 295
and approximation theory, 265–72
Multilayer neural networks, 18–20, 248
output tracking control using, 314–20
Multilayer perceptrons, 559
Multilayer static neural networks, 18–20, 248
Multiple-input—single-output (MISO) system, 6
Multiple-loop feedback, and neural masses, 147–49
Multiple-system coordination, 451–59
coordinated control of two 2-link robots, 457–59
knowledge-based coordinator design, 454–56
knowledge representation, 454–55
solution existence/stability analysis, 455–56
NN-based predictor design, 456–57
principle output prediction scheme, 453–54
problems/principles of, 452–53
Multistage dynamic neural networks, 360–63
MVDR beamforming problem:
neural-based canonical nonlinear programming circuit, 491–92
neural-based circuit implementation for, 493–94
neural network approach to, 489–98
illustrated examples, 494–96
problem formulation, 490–91

N

Negative feedback, and neural masses, 141–42
Neo fuzzy neuron, 427–33
behavior prediction of nonlinear dynamical systems by, 428–29
identification of nonlinear dynamical systems by, 428
structure/learning algorithm of, 427–28
Nerve impulse, 6
Neural circuits, 533–41
computation in, 534
model circuits/relation to biology, 534–36
problem solving by, 537–39
symmetry of connections, 536–37

Neural hardware, 523–96
computing with neural circuits, 533–41
fuzzy inference engine, 571–96
neuromorphic electronic systems, 525–32
robotic applications, neural network architectures for, 554–70
VLSI hybrid analog-digital neural processing element, 542–53
Neural masses, 26
compared to neuron, 132–33
defined, 129–30
forward gain, 133–38
linear analysis, 129–52
AEPs (averaged evoked potential), 131–50
application of, 132
purpose in using, 130–32
multiple-loop feedback, characteristics of, 147–49
negative feedback, characteristics of, 141 -42
open-loop response, 138–41
positive feedback, characteristics of, 142–46
root locus display, purpose in using, 146–47
Neural networks, 6
artificial neural networks (ANNs), 2, 4, 169, 171, 179, 449–509, 554–63
best-known, 50
computational neural network (CNN), 2, 10–11, 171–200
for constrained optimization, 560
for control systems, 171–200
defined, 308
dynamic (feedback), 21–32
hardware implementations, 37–38
intelligent control using, 90–97
multilayer, 18–20, 248
MVDR beamforming problem, approach to, 489–98
for nonlinear internal model control, 321–28
path control using, 561–62
path planning using, 561
recurrent (feedback), 23–24, 30
ring systolic architecture for, 563–67
for robotic processing, 557–59
static (feedforward), 11–20, 30
synaptic plasticity, 64–80
task planning using, 560
time-delay neural networks (TDNN), 24, 30
See also Neurocomputing
Neural populations, 26
Neural processors, 27–29
Neural state, defined, 9
Neural subpopulations:
dynamic neural processor (DNP) based on, 440
dynamic neural structures bases, 26–30
Neurocomputing, 47–52
complementary to algorithmic computing, 51–52
defined, 47–48
from concept to hardware, 49–51
how neural networks work, 48–49
learning without programming, 49
See also Neural networks
Neuro-control systems, 435–522
artificial neural networks (ANNs) in process engineering, 499–509
Cerebellar Model Articulation Controller (CMAC), 516–22
chemical process systems, 510–15
distributed adaptive control systems (DACS), 460–65
intelligent coordination of multiple systems, 451–59
intelligent robotic control, 466–75
MVDR beamforming problem, neural network approach to, 489–98
redundant manipulator movement, 476–82
robotic control transformations,
general learning scheme for, 439–50
turbogenerators, neural network regulator for, 483–88
Neuromorphic electronic systems, 525–32
adaptation and learning, 530
adaptive retina, 529–30
computation primitives, 526–27
defined, 527
energy waste, causes of, 526
neural silicon, 530–31
retinal computation, 528–29
scaling laws, 531–32
Neuronal approximations, 245–94
approximate realization of continuous mappings, 247–56
best approximation property, 257–64
by superpositions of a sigmoidal function, 273–82
dynamic neural units (DNUs), 289–96
multilayer feedforward networks, 265–72
Stone-Weierstrass theorem, 283–88
Neuronal morphology of biological processes, 99–168
cognitive/psychological computations with neural models, 112–28
excitatory/inhibitory interactions in localized populations of model neurons, 153–68
neural masses, 129–52
neurons, 101–11
Neuronal morphology for control systems, 295–371
backpropagation through time, 297–307
identification/control of dynamical systems, 329–51
isolated DNUs, 352–71
nonlinear internal model control, 321–28
output tracking control, 314–20
self-learning control systems, 308–13
Neurons, 3, 6–8, 101–11
acetylcholine-activated channels, 105, 111
action potential, 6–8
cell body, 103
defined, 9, 101
firing, 7–8
generalized mathematical model of, 407–8
with graded response, 234–38
with hysteresis, 30–31, 239–44
intrinsic membrane proteins, 107
membrane, 102–3
nerve impulse, 110
propagation of, 106
synapse, 103
synaptic/somatic operations, 403–7
synaptic terminal, 104
Nonlinear functional approximation problem, 21
Nonlinear internal model control:
discrete nonlinear systems, invertibility of, 323
examples, 326–28
learning algorithms, 325–26
network architecture, 323–25
basic elements, 323–24

Gaussian network, 324–25
neural networks for, 321–28
Nonmodel-based robotic systems, 1–2

O

Open-loop response, and neural masses, 138–41
Operant conditioning, 56–57, 58
Operants, defined, 56
Output tracking control, using multilayer neural networks, 314–20

P

Parallel identification model, 338–39
Partial fraction networks, 287
Perceptrons (Minsky/Papert), 5
PN neural processor, 27–28
Positive feedback, and neural masses, 142–46
Post-synaptic potential (PSP), 7
Predictive control, 191
Principle of incompatibility, 386
Process engineering, artificial neural networks (ANNs) in, 499–509
Pseudoconditioning, 54

Q

Quadruped mobile robot, DAC for, 462–65

R

Recurrent (feedback) neural networks, 23–24, 30
Recursive networks, 22
Redundant manipulator movement, 476–82
discussion, 480–82
kinematics, 479–80
Reflexive memory, 58, 61
Refractory period, defined, 9
Relational algorithms (fuzzy), 400–401
Robotic applications:
and competitive learning networks, 558–59
and multilayer perceptrons, 559
neural network architectures for, 554–70
ring systolic architecture, 563–67
and single-layer feedback neural networks, 557–58
Robotic control, 466–75
Robotic coordinate transformations, general learning scheme for, 439–50
Robotic processing:
and artificial neural networks (ANNs), 559–63
neural networks for, 557–59

S

Schwann cell, 101
Self-learning control systems, neural networks for, 308–13
Self-learning fuzzy controllers, *See* Fuzzy controllers
Self-organizing map, 51, 201–17
algorithm, 203–7
brain maps, 202
competitive learning, early work on, 202–3
and LVQ methods, 207–9
role of, 201
semantic map, 211–13
and speech recognition, 209–11
Sensitization, 54
Sensory learning, 54
Series-parallel identification model, 339
Single-layer feedback neural networks, 557–58
Boltzmann machine, 558
Hopfield associative/optimations neural networks, 557–58
Singleton, 572
Soma, defined, 6, 9
Static (feedforward) neural networks, 11–20, 30
extension of, 23–26
learning in, 179–82
multilayer static neural networks, 18–20
synaptic/somatic operations, 11–18
Step functions and perceptron networks, 287–88
Stereo matching, neural network formulation for, 560–61
Stone-Weierstrass theorem, 20, 283–88
advantages of satisfying, 284
applying, 284–85
networks satisfying, 285–88
decaying-exponential networks, 285
exponentiated-function networks, 286–87
Fourier networks, 286
modified logistic network, 287
modified sigma-pi and polynomial networks, 286
partial fraction networks, 287
step functions and perceptron networks, 287–88
terminology, 283–84
Subpopulations, 26–30, 440
Superpositions of sigmoidal functions, approximation by, 273–82
Supervised control, 189
Synaptic delay, 8
Synaptic plasticity, 64–80
control of robotic manipulator, 68–75
heterosynaptic plasticity, internal neural model with, 57–58
System analysis, alternate approach to, 386–402
Systematic desensitization, 57

T

Temporal backpropagation (TBP), 420, 423
Time coarse graining, 157–58
Time-course of synapse organization, 228–31
Time-delay neural networks (TDNN), 24, 30
Trajectory adaptive network, 420
Turbogenerators:
control results, 486
I-O mapping results, 485–86
neural network regulator for, 483–88
parts of, 483–84

U

Unconditioned response/stimulus, 54
Unconventional neural structures, 5, 30–32

Unions, fuzzy sets, 376–77, 378

V

Vigilance tests, 559
VLSI hybrid analog-digital neural processing element, 542–53
  example computation using, 550–51
  fabrication/testing, 548–50
  future applications, 551–52
  structure of, 543–48
    axon hillock model, 547–48
    chemical synapse models, 543–45
    synapse weighting, 545–47
Voluntary movement, of learning/control of, 64–80

W

Weighting, synapse, 545–47
Winner, defined, 17

Editors' Biographies

Madan M. Gupta (Fellow: IEEE and SPIE) received the B. Eng. (Hons.) and the M.Sc. in Electronics-Communications Engineering, from the Birla Engineering College (now the BITS), Pilani, India, in 1961 and 1962, respectively. He received the Ph.D. degree from the University of Warwick, United Kingdom, in 1967 in adaptive control systems. Dr. Gupta is currently Professor of Engineering and the Director of the Intelligent Systems Research Laboratory and the Centre of Excellence on Neuro-Vision Research at the University of Saskatchewan, Canada.

He was elected Fellow of IEEE for his contributions to the theory of fuzzy sets and the adaptive control systems, and the advancement of the diagnosis of cardiovascular disease. He was also elected Fellow of SPIE for his contributions to the field of neuro-vision, neuro-control, and neuro-fuzzy systems.

Dr. Gupta has served the engineering community worldwide in various capacities through societies such as IEEE, IFSA, IFAC, SPIE, NAFIP, UN, CANS-FINS, and ISUMA. He has been elected as a visiting professor and a special advisor, in the areas of high technology, to the European Centre for Peace and Development (ECPD), University for Peace, which was established by the United Nations.

In addition to publishing over 400 research papers, Dr. Gupta has co-authored two books on fuzzy logic with Japanese translation, and has edited fourteen volumes in the field of adaptive control systems, fuzzy logic/computing, neuro-vision, and neuro-control systems.

Dr. Gupta's present research interests are expanded to the areas of neuro-vision, neuro-control and integration of fuzzy-neural systems, neuronal morphology of biological vision systems, intelligent and cognitive robotic systems, cognitive information, new paradigms in information processing, and chaos in neural systems. He is also developing new architectures of computational neural networks (CNNs), and computational fuzzy neural networks (CFNNs) for applications to advanced robotic systems.

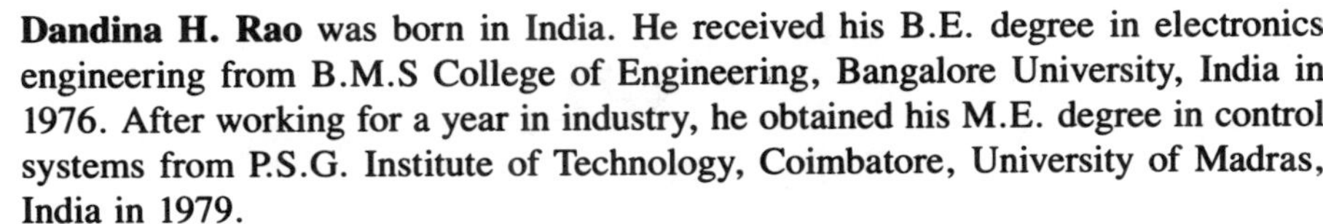

Dandina H. Rao was born in India. He received his B.E. degree in electronics engineering from B.M.S College of Engineering, Bangalore University, India in 1976. After working for a year in industry, he obtained his M.E. degree in control systems from P.S.G. Institute of Technology, Coimbatore, University of Madras, India in 1979.

From April 1979 to October 1982, he worked as a lecturer at M. S. Ramaiah Institute of Technology, Bangalore, India. From November 1982 to September 1988, he was an Assistant Professor at Gogte Institute of Technology, Belgaum, India, where he was also the head of the Department of Electronics and Communication Engineering from 1984–1988. In 1988, he joined the University of Saskatchewan, Canada, where he obtained the Master of Science degree in 1990.

Presently, he is in the process of completing his Ph.D. degree in the field of neuro-control systems. He has coauthored over 40 technical papers and some of his work has appeared in edited books. His research areas of interest are adaptive control, neural networks, and fuzzy logic.